STUDENT SOLUTIONS MANUAL

DANIEL S. MILLER

Niagara County Community College

INTERMEDIATE ALGEBRA FOR COLLEGE STUDENTS

SIXTH EDITION

Robert Blitzer

Miami Dade College

PEARSON

Boston Columbus Indianapolis New York San Francisco Upper Saddle River
Amsterdam Cape Town Dubai London Madrid Milan Munich Paris Montreal Toronto
Delhi Mexico City Sao Paulo Sydney Hong Kong Seoul Singapore Taipei Tokyo

Reproduced by Pearson from electronic files supplied by the author.

Publishing as Pearson, 75 Arlington Street, Boston, MA 02116.

ISBN-13: 978-0-321-76033-3
ISBN-10: 0-321-76033-6

3 17

www.pearsonhighered.com

TABLE OF CONTENTS for STUDENT SOLUTIONS

INTERMEDIATE ALGEBRA FOR COLLEGE STUDENTS 6E

Chapter 1	Algebra, Mathematical Models, and Problem Solving	1
Chapter 2	Functions and Linear Functions	47
Chapter 3	Systems of Linear Equations	89
Chapter 4	Inequalities and Problem Solving	178
Chapter 5	Polynomials, Polynomial Functions, and Factoring	237
Chapter 6	Rational Expressions, Functions, and Equations	300
Chapter 7	Radicals, Radical Functions, and Rational Exponents	417
Chapter 8	Quadratic Equations and Functions	483
Chapter 9	Exponential and Logarithmic Functions	594
Chapter 10	Conic Sections and Systems of Nonlinear Equations	666
Chapter 11	Sequences, Series, and the Binomial Theorem	743

Chapter 1
Algebra, Mathematical Models, and Problem Solving

1.1 Check Points

1. a. 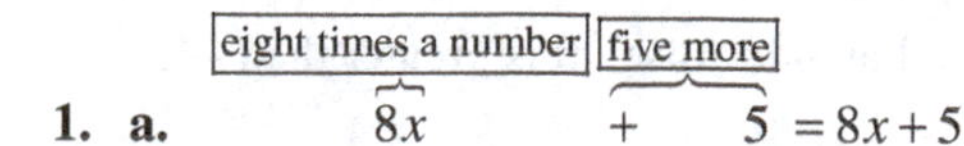

$$\underbrace{8x}_{\text{eight times a number}} + \underbrace{5}_{\text{five more}} = 8x + 5$$

b. $$\underbrace{\frac{x}{7}}_{\text{the quotient of a number and seven}} - \underbrace{2x}_{\text{decreased by twice the number}} = \frac{x}{7} - 2x$$

2. $$\overbrace{23 - 0.12x}^{\text{replace } x \text{ with } 10}$$
$= 23 - 0.12(10)$
$= 23 - 1.2$
$= 21.8$
At age 10, the average neurotic level is 21.8.

3. $$\overbrace{8 + 6(x-3)^2}^{\text{replace } x \text{ with } 13}$$
$= 8 + 6(13-3)^2$
$= 8 + 6(10)^2$
$= 8 + 6(100)$
$= 8 + 600$
$= 608$

4. a. 2010 is 3 years after 2007.
$$S = \overbrace{2.7x^2 + 5.6x + 8}^{\text{replace } x \text{ with } 3}$$
$S = 2.7(3)^2 + 5.6(3) + 8$
$= 2.7(9) + 5.6(3) + 8$
$= 24.3 + 16.8 + 8$
$= 49.1$

b. The model value, 49.1, is the same as the actual data value shown in the figure.

5. a. true; Because the number 13 is an element of the set of integers.

b. true; Because 6 is not an element of $\{7, 8, 9, 10\}$, the statement is true.

6. a. -8 is less than -2; true

b. 7 is greater than -3; true

c. -1 is less than or equal to -4; false

d. 5 is greater than or equal to 5; true

e. 2 is greater than or equal to -14; true

7. a. $\{x \mid -2 \le x < 5\}$

−5 −4 −3 −2 −1 0 1 2 3 4 5

b. $\{x \mid 1 \le x \le 3.5\}$

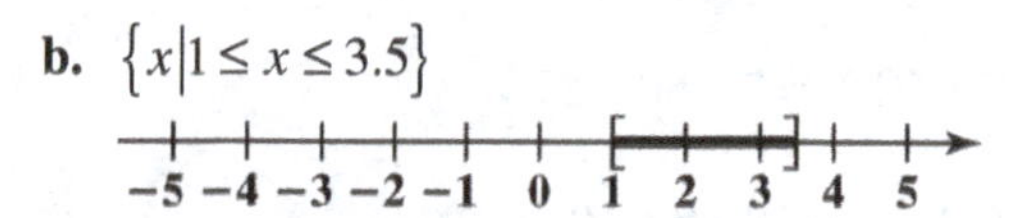

c. $\{x \mid x < -1\}$

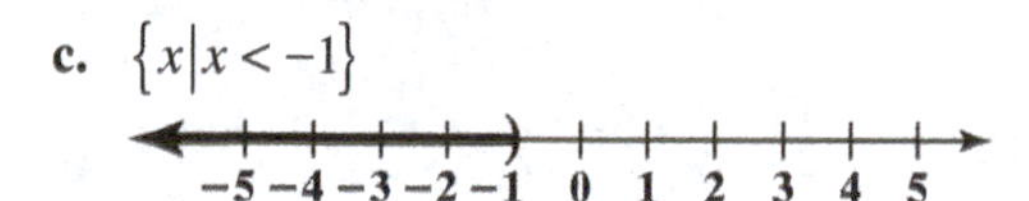

1.1 Concept and Vocabulary Check

1. variable
2. expression
3. *b*th to the *n*th power; base; exponent
4. formula; modeling; models
5. natural
6. whole
7. integers
8. rational
9. irrational
10. rational; irrational
11. left
12. 2; 5; 2; 5
13. greater than
14. less than or equal to

1.1 Exercise Set

1. $x+5$

3. $x-4$

5. $4x$

7. $2x+10$

9. $6-\frac{1}{2}x$

11. $\frac{4}{x}-2$

13. $\frac{3}{5-x}$

15. $7+5(10)=7+50=57$

17. $6(3)-8=18-8=10$

19. $\left(\frac{1}{3}\right)^2+3\left(\frac{1}{3}\right)=\frac{1}{9}+1=1\frac{1}{9}$

21. $7^2-6(7)+3=49-42+3=7+3=10$

23. $4+5(9-7)^3=4+5(2)^3$
$=4+5(8)=4+40=44$

25. $8^2-3(8-2)=64-3(6)$
$=64-18=46$

27. $\{1, 2, 3, 4\}$

29. $\{-7, -6, -5, -4\}$

31. $\{8, 9, 10, \ldots\}$

33. $\{1, 3, 5, 7, 9\}$

35. true; Seven is an integer.

37. true; Seven is a rational number.

39. false; Seven is a rational number.

41. true; Three is not an irrational number.

43. false; $\frac{1}{2}$ is a rational number.

45. true; $\sqrt{2}$ is not a rational number.

47. false; $\sqrt{2}$ is a real number.

49. –6 is less than –2; true

51. 5 is greater than –7; true

53. 0 is less than –4; false. 0 is greater than –4.

55. –4 is less than or equal to 1; true

57. –2 is less than or equal to –6; false. –2 is greater than –6.

59. –2 is less than or equal to –2; true

61. –2 is greater than or equal to –2; true

63. 2 is less than or equal to $-\frac{1}{2}$; false. 2 is greater than $-\frac{1}{2}$.

65. $\{x|1<x\le 6\}$

−2 −1 0 1 2 3 4 5 6 7 8

67. $\{x|-5\le x<2\}$

−5 −4 −3 −2 −1 0 1 2 3 4 5

69. $\{x|-3\le x\le 1\}$

−5 −4 −3 −2 −1 0 1 2 3 4 5

71. $\{x|x>2\}$

−5 −4 −3 −2 −1 0 1 2 3 4 5

73. $\{x|x\ge -3\}$

−5 −4 −3 −2 −1 0 1 2 3 4 5

75. $\{x|x<3\}$

−5 −4 −3 −2 −1 0 1 2 3 4 5

77. $\{x|x<5.5\}$

−2 −1 0 1 2 3 4 5 6 7 8

79. true

81. false; $\{3\}\not\subseteq\{1,2,3,4\}$.

83. true

85. false; The value of $\{x \mid x \text{ is an integer between } -3 \text{ and } 0\} = \{-2,-1\}$, not $\{-3,-2,-1,0\}$.

87. false; Twice the sum of a number and three is represented by $2(x+3)$, not $2x+3$.

89.
$$\begin{aligned} R &= 4.6-0.02x \\ &= 4.6-0.02(20) \\ &= 4.2 \end{aligned}$$
The average resistance to happiness at age 20 is 4.2.

91.
$$\begin{aligned} &[4.6-0.02(30)]-[4.6-0.02(50)] \\ &= 4.0-3.6 \\ &= 0.4 \end{aligned}$$
The difference between the average resistance to happiness at age 30 and at age 50 is 0.4.

93.
$$\begin{aligned} G &= 4.6x^2+5.5x+5 \\ &= 4.6(4)^2+5.5(4)+5 \\ &= 100.6 \\ &\approx 101 \end{aligned}$$
According to the formula, 101 new college programs in green studies were created in 2009. The formula overestimated the actual value by 1 program.

95. $C=\frac{5}{9}(50-32)=\frac{5}{9}(18)=10$

10°C is equivalent to 50°F.

97.
$$\begin{aligned} h &= 4+60t-16t^2 = 4+60(2)-16(2)^2 \\ &= 4+120-16(4) = 4+120-64 \\ &= 124-64 = 60 \end{aligned}$$

Two seconds after it was kicked, the ball's height was 60 feet.

99. – 115. Answers will vary.

117. does not make sense; Explanations will vary. Sample explanation: Many models work for a while and then no longer are valid beyond a certain point.

119. makes sense

121. false; Changes to make the statement true will vary. A sample change is: Every integer is a rational number.

123. true

125. Evaluate the two expressions.
$$2(4+20)=2(24)=48$$
$$2\cdot 4+20=8+20=28$$
Since the bird lover purchases $\frac{1}{7}$ of the birds, the expression has to be a multiple of 7. Since 48 in not a multiple of 7 and 28 is a multiple of 7, we know that the correct expression is $2\cdot 4+20$.

127. $(8+2)\cdot(4-3)=10$ or $8+2\cdot(4-3)=10$

129. –5 and 5 are both a distance of five units from zero on a real number line.

130. $\frac{16+3(2)^4}{12-(10-6)}=\frac{16+3(16)}{12-(4)}=\frac{16+48}{8}=\frac{64}{8}=8$

131.
$$\begin{aligned} &2(3x+5) \\ &= 2(3(4)+5) \\ &= 2(12+5) \\ &= 2(17) \\ &= 34 \end{aligned}$$

$$\begin{aligned} &6x+10 \\ &= 6(4)+10 \\ &= 24+10 \\ &= 34 \end{aligned}$$

1.2 Check Points

1. **a.** $|-6|=6$ because -6 is 6 units from 0.

b. $|4.5|=4.5$ because 4.5 is 4.5 units from 0.

c. $|0|=0$ because 0 is 0 units from 0.

2. **a.** $-10+(-18)=-28$

b. $-0.2+0.9=0.7$

c. $-\frac{3}{5}+\frac{1}{2}=-\frac{6}{10}+\frac{5}{10}=-\frac{1}{10}$

3. **a.** If $x=-8$, then $-x=-(-8)=8$.

b. If $x=\frac{1}{3}$, then $-x=-\frac{1}{3}$.

4. a. $7-10=7+(-10)=-3$

b. $4.3-(-6.2)=4.3+6.2=10.5$

c. $-\frac{4}{5}-\left(-\frac{1}{5}\right)=-\frac{4}{5}+\frac{1}{5}=-\frac{3}{5}$

5. a. $(-5)^2=(-5)(-5)=25$

b. $-5^2=-(5\cdot 5)=-25$

c. $(-4)^3=(-4)(-4)(-4)=-64$

d. $\left(-\frac{3}{5}\right)^4=\left(-\frac{3}{5}\right)\left(-\frac{3}{5}\right)\left(-\frac{3}{5}\right)\left(-\frac{3}{5}\right)=\frac{81}{625}$

6. a. $\frac{32}{-4}=-8$

b. $-\frac{2}{3}\div\left(-\frac{5}{4}\right)=-\frac{2}{3}\cdot\left(-\frac{4}{5}\right)=\frac{8}{15}$

7. $3-5^2+12\div 2(-4)^2$
$=3-25+12\div 2(16)$
$=3-25+6(16)$
$=3-25+96$
$=-22+96$
$=74$

8. $\frac{4+3(-2)^3}{2-(6-9)}$
$=\frac{4+3(-8)}{2-(-3)}$
$=\frac{4-24}{2+3}$
$=\frac{-20}{5}$
$=-4$

9. Commutative Property of Addition: $4x+9=9+4x$
Commutative Property of Multiplication:
$4x+9=x\cdot 4+9$

10. a. $6+(12+x)=(6+12)+x=18+x$

b. $-7(4x)=(-7\cdot 4)x=-28x$

11. $-4(7x+2)=-28x-8$

12. $3x+14x^2+11x+x^2$
$=(14x^2+x^2)+(3x+11x)$
$=(14+1)x^2+(3+11)x$
$=15x^2+14x$

13. $8(2x-5)-4x$
$=16x-40-4x$
$=16x-4x-40$
$=12x-40$

14. $6+4[7-(x-2)]$
$=6+4[7-x+2]$
$=6+4[9-x]$
$=6+36-4x$
$=42-4x$

1.2 Concept and Vocabulary Check

1. negative number
2. 0
3. positive number
4. positive number
5. positive number
6. negative number
7. positive number
8. divide
9. subtract
10. absolute value; 0; a
11. a; $-a$
12. 0; inverse; 0; identity
13. $b+a$
14. $(ab)c$
15. $ab+ac$
16. simplified

1.2 Exercise Set

1. $|-7| = 7$

3. $|4| = 4$

5. $|-7.6| = 7.6$

7. $\left|\frac{\pi}{2}\right| = \frac{\pi}{2}$

9. $|-\sqrt{2}| = \sqrt{2}$

11. $-\left|-\frac{2}{5}\right| = -\frac{2}{5}$

13. $-3+(-8) = -11$

15. $-14+10 = -4$

17. $-6.8+2.3 = -4.5$

19. $\frac{11}{15}+\left(-\frac{3}{5}\right) = \frac{11}{15}+\left(-\frac{9}{15}\right) = \frac{2}{15}$

21. $$-\frac{2}{9}-\frac{3}{4} = -\frac{2}{9}+\left(-\frac{3}{4}\right) = -\frac{8}{36}+\left(-\frac{27}{36}\right) = -\frac{35}{36}$$

23. $-3.7+(-4.5) = -8.2$

25. $0+(-12.4) = -12.4$

27. $12.4+(-12.4) = 0$

29. $x = 11$
$-x = -11$

31. $x = -5$
$-x = 5$

33. $x = 0$
$-x = 0$

35. $3-15 = 3+(-15) = -12$

37. $8-(-10) = 8+10 = 18$

39. $-20-(-5) = -20+5 = -15$

41. $\frac{1}{4}-\frac{1}{2} = \frac{1}{4}+\left(-\frac{1}{2}\right) = \frac{1}{4}+\left(-\frac{2}{4}\right) = -\frac{1}{4}$

43. $-2.3-(-7.8) = -2.3+7.8 = 5.5$

45. $0-\left(-\sqrt{2}\right) = 0+\sqrt{2} = \sqrt{2}$

47. $9(-10) = -90$

49. $(-3)(-11) = 33$

51. $\frac{15}{13}(-1) = -\frac{15}{13}$

53. $-\sqrt{2}\cdot 0 = 0$

55. $(-4)(-2)(-1) = (8)(-1) = -8$

57. $$\begin{aligned} 2(-3)(-1)(-2)(-4) &= (-6)(-1)(-2)(-4) \\ &= (6)(-2)(-4) \\ &= (-12)(-4) \\ &= 48 \end{aligned}$$

59. $(-10)^2 = (-10)(-10) = 100$

61. $-10^2 = -(10)(10) = -100$

63. $(-2)^3 = (-2)(-2)(-2) = -8$

65. $(-1)^4 = (-1)(-1)(-1)(-1) = 1$

67. Since a product with an odd number of negative factors is negative, $(-1)^{33} = -1$.

69. $-\left(-\frac{1}{2}\right)^3 = -\left(-\frac{1}{2}\right)\left(-\frac{1}{2}\right)\left(-\frac{1}{2}\right) = \frac{1}{8}$

71. $\frac{12}{-4} = -3$

73. $\frac{-90}{-2} = 45$

75. $\frac{0}{-4.6} = 0$

77. $-\frac{4.6}{0}$ is undefined.

79. $-\frac{1}{2}\div\left(-\frac{7}{9}\right)=-\frac{1}{2}\cdot\left(-\frac{9}{7}\right)=\frac{9}{14}$

81. $6\div\left(-\frac{2}{5}\right)=\frac{6}{1}\cdot\left(-\frac{5}{2}\right)=-\frac{30}{2}=-15$

83. $4(-5)-6(-3)=-20-(-18)$
$=-20+18=-2$

85. $3(-2)^2-4(-3)^2=3(4)-4(9)$
$=12-36=-24$

87. $8^2-16\div 2^2\cdot 4-3=64-16\div 4\cdot 4-3$
$=64-4\cdot 4-3$
$=64-16-3$
$=48-3$
$=45$

89. $\frac{5\cdot 2-3^2}{\left[3^2-(-2)\right]^2}=\frac{5\cdot 2-9}{[9-(-2)]^2}$
$=\frac{10-9}{(9+2)^2}$
$=\frac{1}{11^2}$
$=\frac{1}{121}$

91. $8-3[-2(2-5)-4(8-6)]$
$=8-3[-2(-3)-4(2)]$
$=8-3[6-8]=8-3[-2]=8+6=14$

93. $\frac{2(-2)-4(-3)}{5-8}=\frac{-4+12}{-3}=\frac{8}{-3}=-\frac{8}{3}$

95. $\frac{(5-6)^2-2|3-7|}{89-3\cdot 5^2}=\frac{(-1)^2-2|-4|}{89-3\cdot 25}$
$=\frac{1-2(4)}{89-75}$
$=\frac{1-8}{14}=\frac{-7}{14}=-\frac{1}{2}$

97. $15-\sqrt{3-(-1)}+12\div 2\cdot 3$
$=15-\sqrt{4}+12\div 2\cdot 3$
$=15-2+12\div 2\cdot 3$
$=15-2+6\cdot 3$
$=15-2+18=13+18=31$

99. $20+1-\sqrt{10^2-(5+1)^2}(-2)$
$=20+1-\sqrt{10^2-6^2}(-2)$
$=20+1-\sqrt{100-36}(-2)$
$=20+1-\sqrt{64}(-2)$
$=20+1-8(-2)=20+1+16=37$

101. Commutative Property of Addition
$4x+10=10+4x$
Commutative Property of Multiplication
$4x+10=x\cdot 4+10$

103. Commutative Property of Addition
$7x-5=-5+7x$
Commutative Property of Multiplication
$7x-5=x\cdot 7-5$

105. $4+(6+x)=(4+6)+x=10+x$

107. $-7(3x)=(-7\cdot 3)x=-21x$

109. $-\frac{1}{3}(-3y)=\left(-\frac{1}{3}\cdot -3\right)y=y$

111. $3(2x+5)=3\cdot 2x+3\cdot 5=6x+15$

113. $-7(2x+3)=-7\cdot 2x+(-7)3$
$=-14x-21$

115. $-(3x-6)=-1\cdot 3x-(-1)6=-3x+6$

117. $7x+5x=(7+5)x=12x$

119. $6x^2-x^2=(6-1)x^2=5x^2$

121. $6x+10x^2+4x+2x^2$
$=6x+4x+10x^2+2x^2$
$=(6+4)x+(10+2)x^2=10x+12x^2$

123. $8(3x-5)-6x$
$=8\cdot 3x-8\cdot 5-6x$
$=24x-40-6x$
$=24x-6x-40$
$=(24-6)x-40=18x-40$

125. $5(3y-2)-(7y+2)$
$=5\cdot3y-5\cdot2-1\cdot7y+(-1)2$
$=15y-10-7y-2$
$=15y-7y-10-2$
$=(15-7)y-12=8y-12$

127. $7-4[3-(4y-5)]$
$=7-4[3-4y+5]$
$=7-12+16y-20$
$=16y-25$

129. $18x^2+4-[6(x^2-2)+5]$
$=18x^2+4-[6x^2-12+5]$
$=18x^2+4-[6x^2-7]$
$=18x^2+4-6x^2+7$
$=18x^2-6x^2+4+7$
$=(18-6)x^2+11=12x^2+11$

131. $x-(x+4)=x-x-4=-4$

133. $6(-5x)=-30x$

135. $5x-2x=3x$

137. $8x-(3x+6)=8x-3x-6=5x-6$

139. $21+(-29)=-8$

141. $21-(-29)=21+29$
$=50$

143. $-3-(-10)=-3+10$
$=7$
The approval rating of France exceeds the approval rating of China by 7.

145. $\frac{-10+(-3)+4}{3}=\frac{-9}{3}$
$=-3$
The average approval rating of China, France, and Israel is -3.

147. $D=-0.2x^2+5(x+12)$
$=-0.2(4)^2+5(4+12)$
$=76.8$
According to the model, college students spent \$76.8 billion in 2010.
The model overestimates the actual value displayed in the graph by \$0.8 billion.

149. **a.** $0.05x+0.12(10,000-x)$
$=0.05x+1200-0.12x$
$=1200-0.07x$

b. $0.05(6000)+0.12(10,000-6000)$
$=0.05(6000)+0.12(4000)$
$=300+480=780$

$1200-0.07(6000)=1200-420$
$=780$
The total interest will be \$780.

151. – 167. Answers will vary.

169. makes sense

171. does not make sense; Explanations will vary. Sample explanation: When there is no number in front of a variable, the coefficient has a value of 1.

173. false; Changes to make the statement true will vary. A sample change is:
$6-2(4+3)=6-2(7)=6-14=-8$

175. false; Changes to make the statement true will vary. A sample change is: $-x-x=-2x$

177. $(8-2)\cdot3-4=14$

179. $\dfrac{9[4-(1+6)]-(3-9)^2}{5+\dfrac{12}{5-\dfrac{6}{2+1}}} = \dfrac{9[4-7]-(-6)^2}{5+\dfrac{12}{5-\dfrac{6}{3}}}$

$= \dfrac{9[-3]-36}{5+\dfrac{12}{5-2}}$

$= \dfrac{-27-36}{5+\dfrac{12}{3}}$

$= \dfrac{-63}{5+4}$

$= \dfrac{-63}{9}$

$= -7$

180. $\dfrac{10}{x} - 4x$

181. $10+2(x-5)^4 = 10+2(7-5)^4$

$= 10+2(2)^4 = 10+2(16)$

$= 10+32 = 42$

182. true; $\dfrac{1}{2}$ is not an irrational number.

183.

x	$y = 4 - x^2$
-3	$y = 4-(-3)^2 = 4-9 = -5$
-2	$y = 4-(-2)^2 = 4-4 = 0$
-1	$y = 4-(-1)^2 = 4-1 = 3$
0	$y = 4-(0)^2 = 4-0 = 4$
1	$y = 4-(1)^2 = 4-1 = 3$
2	$y = 4-(2)^2 = 4-4 = 0$
3	$y = 4-(3)^2 = 4-9 = -5$

184.

x	$y = 1 - x^2$
-3	$y = 1-(-3)^2 = 1-9 = -8$
-2	$y = 1-(-2)^2 = 1-4 = -3$
-1	$y = 1-(-1)^2 = 1-1 = 0$
0	$y = 1-(0)^2 = 1-0 = 1$
1	$y = 1-(1)^2 = 1-1 = 0$
2	$y = 1-(2)^2 = 1-4 = -3$
3	$y = 1-(3)^2 = 1-9 = -8$

185.

x	$y = \lvert x+1 \rvert$
-4	$y = \lvert -4+1 \rvert = \lvert -3 \rvert = 3$
-3	$y = \lvert -3+1 \rvert = \lvert -2 \rvert = 2$
-2	$y = \lvert -2+1 \rvert = \lvert -1 \rvert = 1$
-1	$y = \lvert -1+1 \rvert = \lvert 0 \rvert = 0$
0	$y = \lvert 0+1 \rvert = \lvert 1 \rvert = 1$
1	$y = \lvert 1+1 \rvert = \lvert 2 \rvert = 2$
2	$y = \lvert 2+1 \rvert = \lvert 3 \rvert = 3$

1.3 Check Points

1.

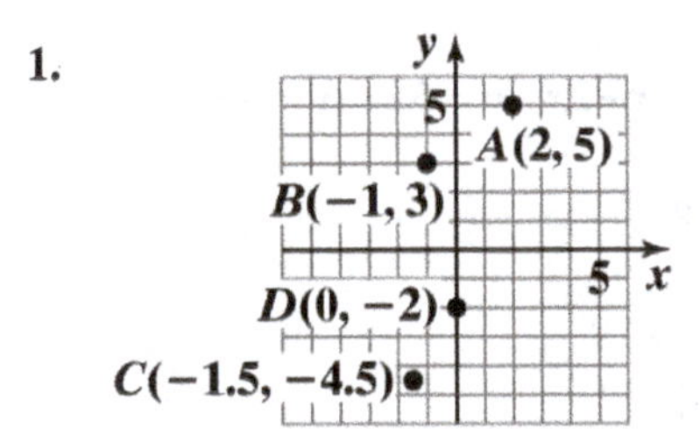

2. Make a table:

x	$y=1-x^2$	(x, y)
−3	$y=1-(-3)^2=-8$	(−3,−8)
−2	$y=1-(-2)^2=-3$	(−2,−3)
−1	$y=1-(-1)^2=0$	(−1,0)
0	$y=1-(0)^2=1$	(0,1)
1	$y=1-(1)^2=0$	(1,0)
2	$y=1-(2)^2=-3$	(2,−3)
3	$y=1-(3)^2=-8$	(3,−8)

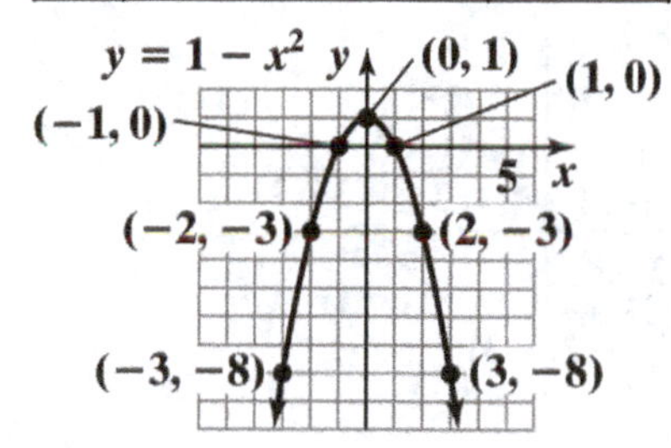

3. Make a table:

x	$y=\lvert x+1\rvert$	(x, y)
−4	$y=\lvert -4+1\rvert=\lvert -3\rvert=3$	(−4,3)
−3	$y=\lvert -3+1\rvert=\lvert -2\rvert=2$	(−3,2)
−2	$y=\lvert -2+1\rvert=\lvert -1\rvert=1$	(−2,1)
−1	$y=\lvert -1+1\rvert=\lvert 0\rvert=0$	(−1,0)
0	$y=\lvert 0+1\rvert=\lvert 1\rvert=1$	(0,1)
1	$y=\lvert 1+1\rvert=\lvert 2\rvert=2$	(1,2)
2	$y=\lvert 2+1\rvert=\lvert 3\rvert=3$	(2,3)

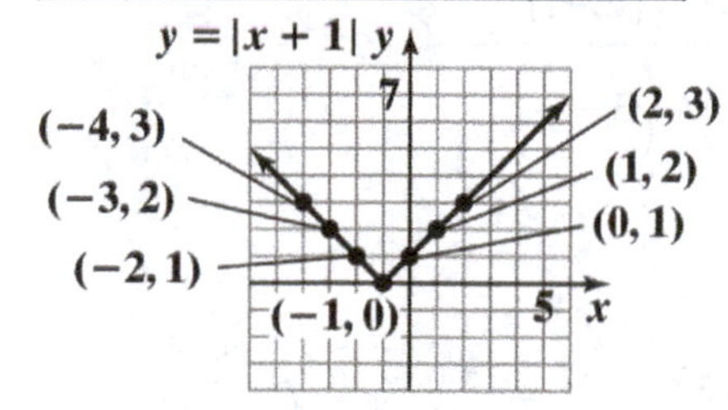

4. a. The drug concentration is increasing from 0 to 3 hours.

b. The drug concentration is decreasing from 3 to 13 hours.

c. The drug's maximum concentration is 0.05 milligram per 100 milliliters, which occurs after 3 hours.

d. None of the drug is left in the body.

5. The minimum x-value is −100, the maximum x-value is 100, and the distance between consecutive tick marks is 50. The minimum y-value is −100, the maximum y-value is 100, and the distance between consecutive tick marks is 10.

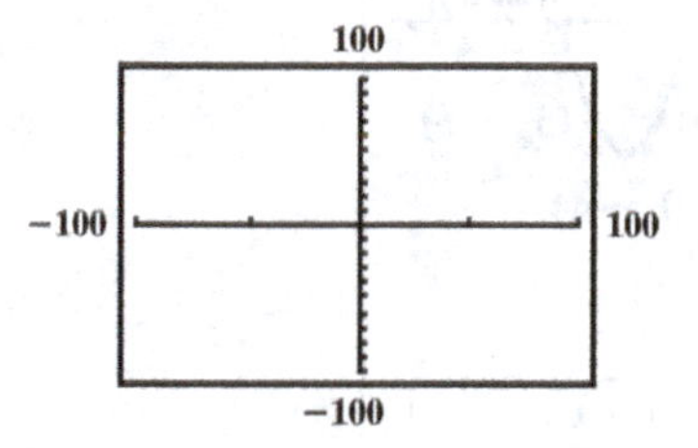

1.3 Concept and Vocabulary Check

1. x-axis
2. y-axis
3. origin
4. quadrants; four
5. x-coordinate; y-coordinate
6. solution; satisfies

1.3 Exercise Set

1. – 9.

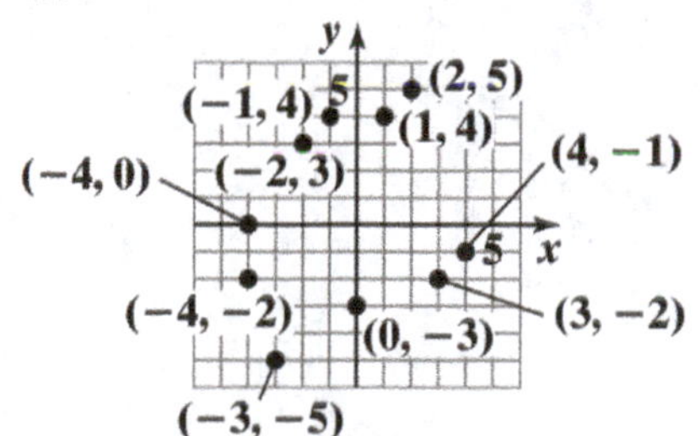

11.

x	(x, y)
−3	(−3,5)
−2	(−2,0)
−1	(−1,−3)
0	(0,−4)
1	(1,−3)
2	(2,0)
3	(3,5)

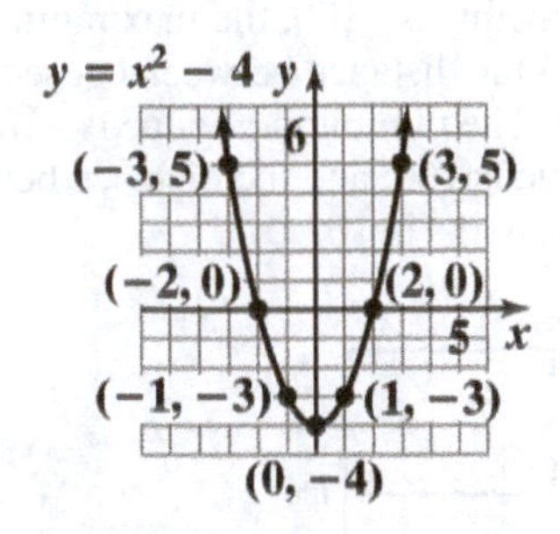

13.

x	(x, y)
−3	(−3,−5)
−2	(−2,−4)
−1	(−1,−3)
0	(0,−2)
1	(1,−1)
2	(2,0)
3	(3,1)

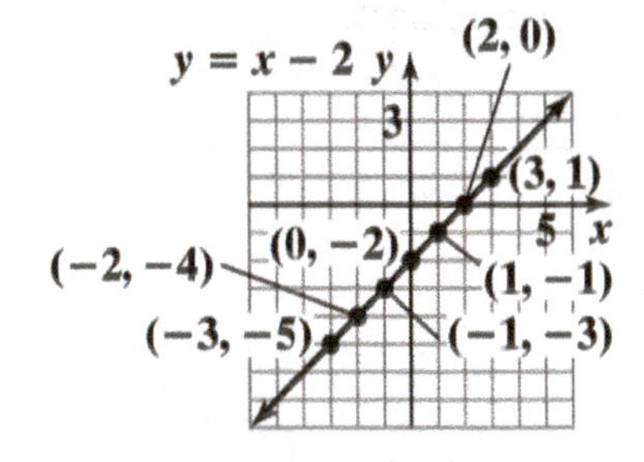

15.

x	(x, y)
−3	(−3,−5)
−2	(−2,−3)
−1	(−1,−1)
0	(0,1)
1	(1,3)
2	(2,5)
3	(3,7)

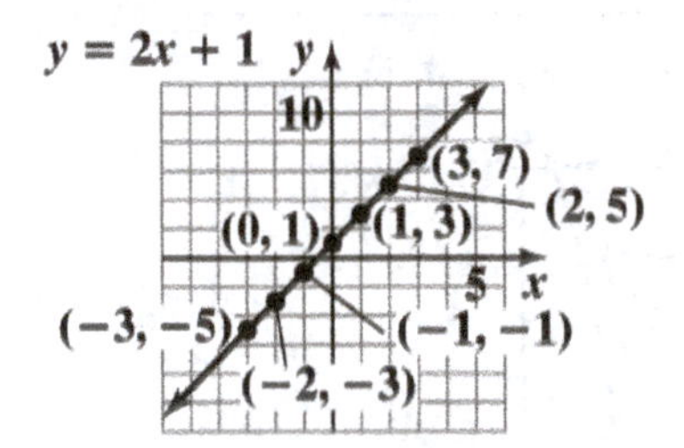

17.

x	(x, y)
−3	$\left(-3, \frac{3}{2}\right)$
−2	(−2,1)
−1	$\left(-1, \frac{1}{2}\right)$
0	(0,0)
1	$\left(1, -\frac{1}{2}\right)$
2	(2,−1)
3	$\left(3, -\frac{3}{2}\right)$

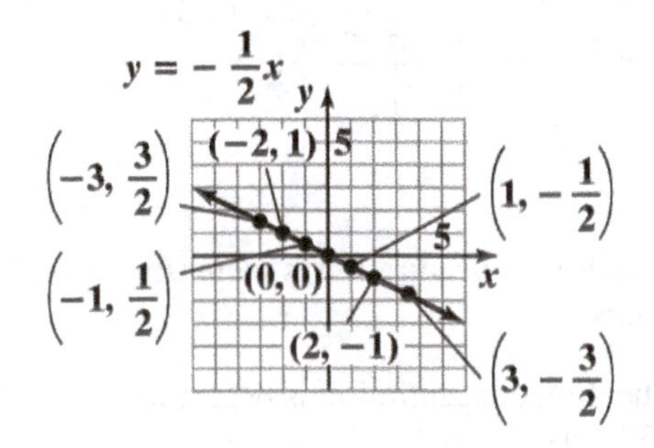

19.

x	(x, y)
–3	(–3,4)
–2	(–2,3)
–1	(–1,2)
0	(0,1)
1	(1,2)
2	(2,3)
3	(3,4)

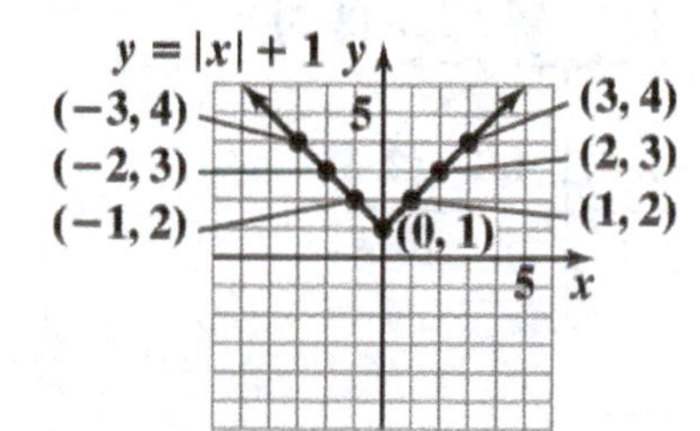

21.

x	(x, y)
–3	(–3,6)
–2	(–2,4)
–1	(–1,2)
0	(0,0)
1	(1,2)
2	(2,4)
3	(3,6)

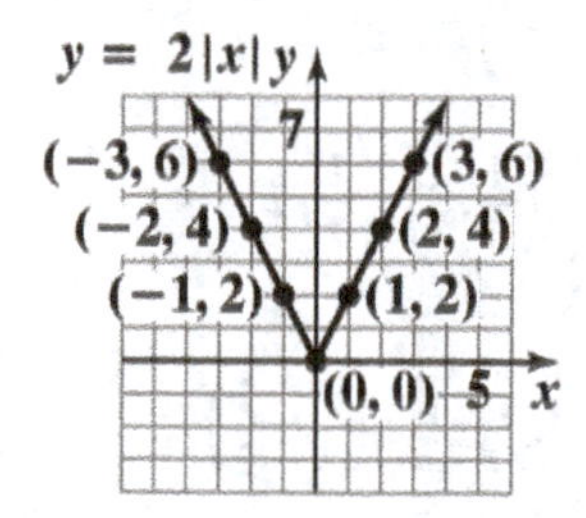

23.

x	(x, y)
–3	(–3,–9)
–2	(–2,–4)
–1	(–1,–1)
0	(0,0)
1	(1,–1)
2	(2,–4)
3	(3,–9)

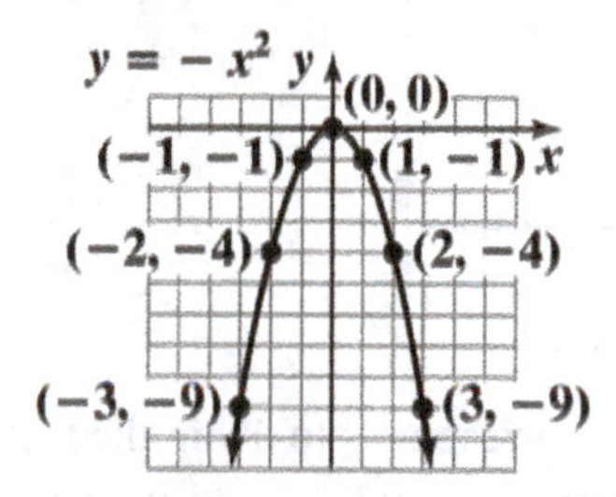

25.

x	(x, y)
–3	(–3,–27)
–2	(–2,–8)
–1	(–1,–1)
0	(0,0)
1	(1,1)
2	(2,8)
3	(3,27)

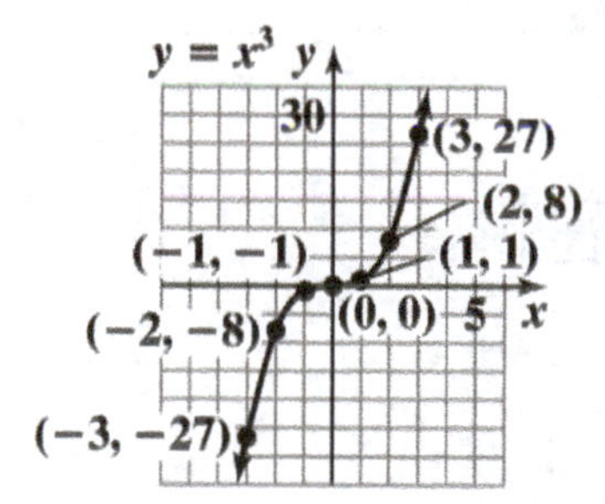

27. [–5, 5, 1] by [–5, 5, 1]
This matches graph c.

29. [–20, 80, 10] by [–30, 70, 10]
This matches graph b.

31. The equation that corresponds to Y_2 in the table is (c), $y_2 = 2 - x$. We can tell because all of the points $(-3,5)$, $(-2,4)$, $(-1,3)$, $(0,2)$, $(1,1)$, $(2,0)$, and $(3,-1)$ are on the line $y = 2 - x$, but all are not on any of the others.

33. No. It passes through the point $(0,2)$.

35. $(2,0)$

37. The graphs of Y_1 and Y_2 intersect at the points $(-2,4)$ and $(1,1)$.

39. $y = 2x + 4$

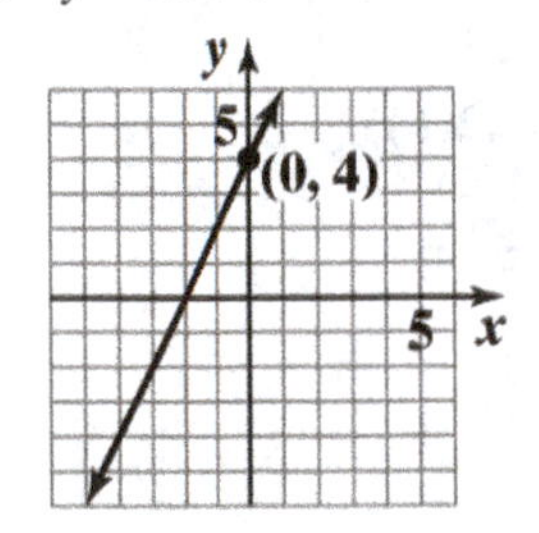

41. $y = 3 - x^2$

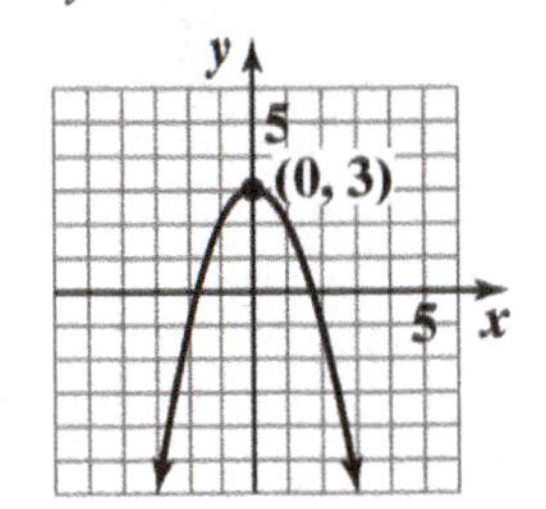

43.

x	(x, y)
–3	$(-3,5)$
–2	$(-2,5)$
–1	$(-1,5)$
0	$(0,5)$
1	$(1,5)$
2	$(2,5)$
3	$(3,5)$

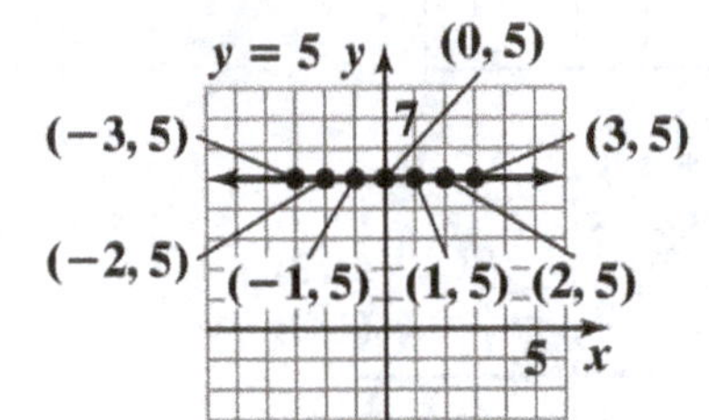

45.

x	(x, y)
–2	$\left(-2, -\frac{1}{2}\right)$
–1	$(-1,-1)$
$-\frac{1}{2}$	$\left(-\frac{1}{2}, -2\right)$
$-\frac{1}{3}$	$\left(-\frac{1}{3}, -3\right)$
$\frac{1}{3}$	$\left(\frac{1}{3}, 3\right)$
$\frac{1}{2}$	$\left(\frac{1}{2}, 2\right)$
1	$(1,1)$
2	$\left(2, \frac{1}{2}\right)$

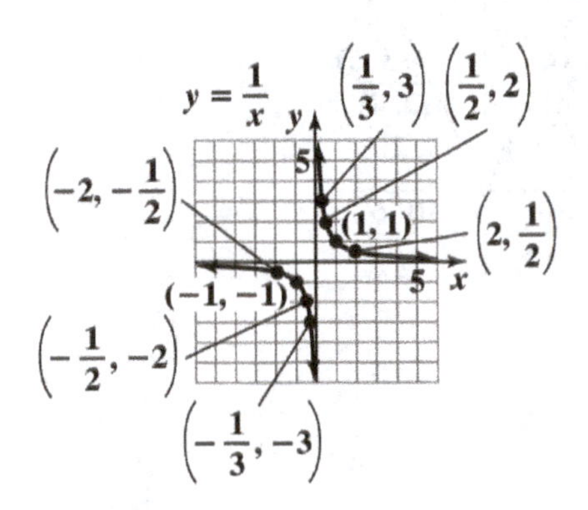

47. The top marginal tax rate in 2010 was 35%.

49. The highest marginal tax rate occurred in 1945 and was about 94%.

51. During the ten-year period from 1950 to 1960, the top marginal tax rate remained constant at about 91%.

53. At age 8, women have the least number of awakenings, averaging about 1 awakening per night.

55. The difference between the number of awakenings for 25-year-old men and women is about 1.9.

57. graph a

59. graph b

61. graph b

63. graph c

65. – 71. Answers will vary.

73. makes sense

75. makes sense

77. false; Changes to make the statement true will vary. A sample change is: If the product of a point's coordinates is positive, the point could be in quadrant I <u>or</u> III.

79. true

81. The four hour day costs \$6 and the five hour day costs \$9. Thus the total cost for the two days is \$15.

83. $|-14.3| = 14.3$

84.
$$\begin{aligned}&[12-(13-17)]-[9-(6-10)]\\&=[12-(-4)]-[9-(-4)]\\&=[12+4]-[9+4]=16-13=3\end{aligned}$$

85.
$$\begin{aligned}6x-5(4x+3)-10&=6x-20x-15-10\\&=(6-20)x-(15+10)\\&=-14x-25\end{aligned}$$

86.
$$\begin{aligned}4x-3&=5x+6\\4(-9)-3&=5(-9)+6\\-36-3&=-45+6\\-39&=-39\end{aligned}$$
The statement is true for $x=-9$.

87.
$$\begin{aligned}&13-3(x+2)\\&=13-3x-6\\&=7-3x\end{aligned}$$

88.
$$\begin{aligned}&10\left(\frac{3x+1}{2}\right)\\&=\frac{10}{1}\cdot\frac{3x+1}{2}\\&=5(3x+1)\\&=15x+5\end{aligned}$$

1.4 Check Points

1.
$$\begin{aligned}4x+5&=29\\4x+5-5&=29-5\\4x&=24\\\frac{4x}{4}&=\frac{24}{4}\\x&=6\end{aligned}$$
The solution set is $\{6\}$.
Check:
$$\begin{aligned}4x+5&=29\\4(6)+5&=29\\24+5&=29\\29&=29\end{aligned}$$

2.
$$\begin{aligned}2x-12+x&=6x-4+5x\\3x-12&=11x-4\\3x-11x&=-4+12\\-8x&=8\\\frac{-8x}{-8}&=\frac{8}{-8}\\x&=-1\end{aligned}$$
The solution set is $\{-1\}$.
Check:
$$\begin{aligned}2x-12+x&=6x-4+5x\\2(-1)-12+(-1)&=6(-1)-4+5(-1)\\-2-12-1&=-6-4-5\\-15&=-15\end{aligned}$$

3. $2(x-3)-17=13-3(x+2)$
$$2x-6-17=13-3x-6$$
$$2x-23=7-3x$$
$$2x+3x=7+23$$
$$5x=30$$
$$\frac{5x}{5}=\frac{30}{5}$$
$$x=6$$
The solution set is $\{6\}$.
Check:
$$2(x-3)-17=13-3(x+2)$$
$$2(6-3)-17=13-3(6+2)$$
$$2(3)-17=13-3(8)$$
$$6-17=13-24$$
$$-11=-11$$

4. $\frac{x+5}{7}+\frac{x-3}{4}=\frac{5}{14}$
$$28\left(\frac{x+5}{7}+\frac{x-3}{4}\right)=28\left(\frac{5}{14}\right)$$
$$\frac{28}{1}\left(\frac{x+5}{7}\right)+\frac{28}{1}\left(\frac{x-3}{4}\right)=\frac{28}{1}\left(\frac{5}{14}\right)$$
$$4(x+5)+7(x-3)=2(5)$$
$$4x+20+7x-21=10$$
$$11x-1=10$$
$$11x=10+1$$
$$11x=11$$
$$\frac{11x}{11}=\frac{11}{11}$$
$$x=1$$
The solution set is $\{1\}$.
Check:
$$\frac{x+5}{7}+\frac{x-3}{4}=\frac{5}{14}$$
$$\frac{1+5}{7}+\frac{1-3}{4}=\frac{5}{14}$$
$$\frac{6}{7}+\frac{-2}{4}=\frac{5}{14}$$
$$\frac{24}{28}+\frac{-14}{28}=\frac{10}{28}$$
$$\frac{10}{28}=\frac{10}{28}$$

5. $4x-7=4(x-1)+3$
$$4x-7=4x-4+3$$
$$4x-7=4x-1$$
$$-7=-1$$
This equation is an inconsistent equation and thus has no solution.
The solution set is $\{\ \}$.

6. $7x+9=9(x+1)-2x$
$$7x+9=9x+9-2x$$
$$7x+9=7x+9$$
$$9=9$$
This equation is an identity and all real numbers are solutions.
The solution set is $\{x|x \text{ is a real number}\}$ or $(-\infty,\infty)$ or $\mathbb{R}$.

7. $T=385x+3129$
$$8904=385x+3129$$
$$8904-3129=385x$$
$$5775=385x$$
$$\frac{5775}{385}=\frac{385x}{385}$$
$$15=x$$
The average cost of tuition and fees at public colleges will reach \$8904 in the school year ending 15 years after 2000, or 2015.

1.4 Concept and Vocabulary Check

1. linear
2. equivalent
3. $b+c$
4. bc
5. apply the distributive property
6. least common denominator; 12
7. inconsistent; $\varnothing$
8. identity; $(-\infty,\infty)$

1.4 Exercise Set

1. $5x+3=18$
$$5x+3-3=18-3$$
$$5x=15$$
$$\frac{5x}{5}=\frac{15}{5}$$
$$x=3$$
The solution set is $\{3\}$.

3.
$$\begin{aligned} 6x-3&=63\\ 6x-3+3&=63+3\\ 6x&=66\\ \frac{6x}{6}&=\frac{66}{6}\\ x&=11 \end{aligned}$$
The solution set is {11}.

5.
$$\begin{aligned} 14-5x&=-41\\ 14-5x-14&=-41-14\\ -5x&=-55\\ \frac{-5x}{-5}&=\frac{-55}{-5}\\ x&=11 \end{aligned}$$
The solution set is {11}.

7.
$$\begin{aligned} 11x-(6x-5)&=40\\ 11x-6x+5&=40\\ 5x+5&=40\\ 5x+5-5&=40-5\\ 5x&=35\\ x&=7 \end{aligned}$$
The solution set is {7}.

9.
$$\begin{aligned} 2x-7&=6+x\\ 2x-x-7&=6+x-x\\ x-7&=6\\ x-7+7&=6+7\\ x&=13 \end{aligned}$$
The solution set is {13}.

11.
$$\begin{aligned} 7x+4&=x+16\\ 7x-x+4&=x-x+16\\ 6x+4&=16\\ 6x+4-4&=16-4\\ 6x&=12\\ \frac{6x}{6}&=\frac{12}{6}\\ x&=2 \end{aligned}$$
The solution set is {2}.

13.
$$\begin{aligned} 8y-3&=11y+9\\ 8y-8y-3&=11y-8y+9\\ -3&=3y+9\\ -3-9&=3y+9-9\\ -12&=3y\\ \frac{-12}{3}&=\frac{3y}{3}\\ -4&=y \end{aligned}$$
The solution set is {–4}.

15.
$$\begin{aligned} 3(x-2)+7&=2(x+5)\\ 3x-6+7&=2x+10\\ 3x-2x-6+7&=2x-2x+10\\ x-6+7&=10\\ x+1&=10\\ x+1-1&=10-1\\ x&=9 \end{aligned}$$
The solution set is {9}.

17.
$$\begin{aligned} 3(x-4)-4(x-3)&=x+3-(x-2)\\ 3x-12-4x+12&=x+3-x+2\\ -x&=5\\ x&=-5 \end{aligned}$$
The solution set is {–5}.

19.
$$\begin{aligned} 16&=3(x-1)-(x-7)\\ 16&=3x-3-x+7\\ 16&=2x+4\\ 16-4&=2x+4-4\\ 12&=2x\\ \frac{12}{2}&=\frac{2x}{2}\\ 6&=x \end{aligned}$$
The solution set is {6}.

21.
$$\begin{aligned} 7(x+1)&=4[x-(3-x)]\\ 7x+7&=4[x-3+x]\\ 7x+7&=4[2x-3]\\ 7x+7&=8x-12\\ 7x-7x+7&=8x-7x-12\\ 7&=x-12\\ 7+12&=x-12+12\\ 19&=x \end{aligned}$$
The solution set is {19}.

23.
$$\begin{aligned} \frac{1}{2}(4z+8)-16&=-\frac{2}{3}(9z-12)\\ 2z+4-16&=-6z+8\\ 2z-12&=-6z+8\\ 8z-12&=8\\ 8z&=20\\ z&=\frac{20}{8}=\frac{5}{2} \end{aligned}$$
The solution set is $\left\{\frac{5}{2}\right\}$.

25. $\frac{x}{3}=\frac{x}{2}-2$

$6\left(\frac{x}{3}\right)=6\left(\frac{x}{2}-2\right)$

$2x=3x-12$

$2x-3x=3x-3x-12$

$-x=-12$

$x=12$

The solution set is $\{12\}$.

27. $20-\frac{x}{3}=\frac{x}{2}$

$6\left(20-\frac{x}{3}\right)=6\left(\frac{x}{2}\right)$

$120-2x=3x$

$120-2x+2x=3x+2x$

$120=5x$

$\frac{120}{5}=\frac{5x}{5}$

$24=x$

The solution set is $\{24\}$.

29. $\frac{3x}{5}=\frac{2x}{3}+1$

$15\left(\frac{3x}{5}\right)=15\left(\frac{2x}{3}+1\right)$

$9x=10x+15$

$9x-10x=10x-10x+15$

$-x=15$

$x=-15$

The solution set is $\{-15\}$.

31. $\frac{3x}{5}-x=\frac{x}{10}-\frac{5}{2}$

$10\left(\frac{3x}{5}-x\right)=10\left(\frac{x}{10}-\frac{5}{2}\right)$

$6x-10x=x-25$

$-4x=x-25$

$-4x-x=x-x-25$

$-5x=-25$

$x=5$

The solution set is $\{5\}$.

33. $\frac{x+3}{6}=\frac{2}{3}+\frac{x-5}{4}$

$12\left(\frac{x+3}{6}\right)=12\left(\frac{2}{3}\right)+12\left(\frac{x-5}{4}\right)$

$2(x+3)=4(2)+3(x-5)$

$2x+6=8+3x-15$

$2x+6=3x-7$

$-x+6=-7$

$-x=-13$

$x=13$

The solution set is $\{13\}$.

35. $\frac{x}{4}=2+\frac{x-3}{3}$

$12\left(\frac{x}{4}\right)=12\left(2+\frac{x-3}{3}\right)$

$3x=24+4(x-3)$

$3x=24+4x-12$

$3x=12+4x$

$3x-4x=12+4x-4x$

$-x=12$

$x=-12$

The solution set is $\{-12\}$.

37. $\frac{x+1}{3}=5-\frac{x+2}{7}$

$21\left(\frac{x+1}{3}\right)=21\left(5-\frac{x+2}{7}\right)$

$7(x+1)=105-3(x+2)$

$7x+7=105-3x-6$

$7x+3x+7=105-3x+3x-6$

$10x+7=99$

$10x=92$

$x=\frac{92}{10}=\frac{46}{5}$

The solution set is $\left\{\frac{46}{5}\right\}$.

39. $5x+9=9(x+1)-4x$

$5x+9=9x+9-4x$

$5x+9=5x+9$

The solution set is $\{x|x \text{ is a real number}\}$ or $(-\infty,\infty)$ or $\mathbb{R}$. The equation is an identity.

41. $3(y+2)=7+3y$

$3y+6=7+3y$

$3y-3y+6=7+3y-3y$

$6=7$

There is no solution. The solution set is $\{\ \}$ or $\varnothing$. The equation is inconsistent.

43. $10x+3=8x+3$
$10x-8x+3=8x-8x+3$
$2x=0$
$x=0$
The solution set is $\{0\}$. The equation is conditional.

45. $\frac{1}{2}(6z+20)-8=2(z-4)$
$3z+10-8=2z-8$
$3z+2=2z-8$
$z+2=-8$
$z=-10$
The solution set is $\{-10\}$. The equation is conditional.

47. $-4x-3(2-2x)=7+2x$
$-4x-6+6x=7+2x$
$2x-6=7+2x$
$-6=7$
There is no solution. The solution set is $\{\ \}$ or $\varnothing$. The equation is inconsistent.

49. $y+3(4y+2)=6(y+1)+5y$
$y+12y+6=6y+6+5y$
$13y+6=11y+6$
$2y+6=6$
$2y=0$
$y=0$
The solution set is $\{0\}$. The equation is conditional.

51. $3(x-4)=3(2-2x)$
$x=2$

53. $-3(x-3)=5(2-x)$
$x=0.5$

55. Solve: $4(x-2)+2=4x-2(2-x)$
$4x-8+2=4x-4+2x$
$4x-6=6x-4$
$-2x-6=-4$
$-2x=2$
$x=-1$
Now, evaluate x^2-x for $x=-1$:
$x^2-x=(-1)^2-(-1)$
$=1-(-1)=1+1=2$

57. Solve for x: $\frac{3(x+3)}{5}=2x+6$
$3(x+3)=5(2x+6)$
$3x+9=10x+30$
$-7x+9=30$
$-7x=21$
$x=-3$

Solve for y: $-2y-10=5y+18$
$-7y-10=18$
$-7y=28$
$y=-4$

Now, evaluate $x^2-(xy-y)$ for $x=-3$ and $y=-4$:
$x^2-(xy-y)=(-3)^2-[-3(-4)-(-4)]$
$=(-3)^2-[12-(-4)]$
$=9-(12+4)=9-16=-7$

59. $\left[(3+6)^2\div 3\right]\cdot 4=-54x$
$\left(9^2\div 3\right)\cdot 4=-54x$
$(81\div 3)\cdot 4=-54x$
$27\cdot 4=-54x$
$108=-54x$
$-2=x$
The solution set is $\{-2\}$.

61. $5-12x=8-7x-\left[6\div 3\left(2+5^3\right)+5x\right]$
$5-12x=8-7x-[6\div 3(2+125)+5x]$
$5-12x=8-7x-[6\div 3\cdot 127+5x]$
$5-12x=8-7x-[2\cdot 127+5x]$
$5-12x=8-7x-[254+5x]$
$5-12x=8-7x-254-5x$
$5-12x=-12x-246$
$5=-246$
The final statement is a contradiction, so the equation has no solution. The solution set is $\varnothing$.

63. $0.7x + 0.4(20) = 0.5(x+20)$

$$0.7x + 8 = 0.5x + 10$$
$$0.2x + 8 = 10$$
$$0.2x = 2$$
$$x = 10$$

The solution set is $\{10\}$.

65. $4x + 13 - \{2x - [4(x-3) - 5]\} = 2(x-6)$

$$4x + 13 - \{2x - [4x - 12 - 5]\} = 2x - 12$$
$$4x + 13 - \{2x - [4x - 17]\} = 2x - 12$$
$$4x + 13 - \{2x - 4x + 17\} = 2x - 12$$
$$4x + 13 - \{-2x + 17\} = 2x - 12$$
$$4x + 13 + 2x - 17 = 2x - 12$$
$$6x - 4 = 2x - 12$$
$$4x - 4 = -12$$
$$4x = -8$$
$$x = -2$$

The solution set is $\{-2\}$.

67. a. Model 1: $T = 1074x + 15{,}145$

$$= 1074(10) + 15{,}145$$
$$= 25{,}885$$

Model 2: $T = 25.5x^2 + 819x + 15{,}527$

$$T = 25.5(10)^2 + 819(10) + 15{,}527$$
$$= 26{,}267$$

Model 1 estimates the cost in 2010 to be $25,885 which means Model 1 underestimates by $388.
Model 2 estimates the cost in 2010 to be $26,267 which means Model 2 underestimates by $6.

b. $T = 1074x + 15{,}145$

$$33{,}403 = 1074x + 15{,}145$$
$$18{,}258 = 1074x$$
$$\frac{18{,}258}{1074} = \frac{1074x}{1074}$$
$$17 = x$$

Tuition and fees will average $33,403 at private four-year colleges in the school year ending 17 years after 2000, or 2017.

69. a. \$52,000

b.
$$\begin{aligned} C &= 1388x + 24{,}963 \\ &= 1388(20) + 24{,}963 \\ &= \$52{,}723 \end{aligned}$$
It describes the estimate from part (a) reasonably well.

c.
$$\begin{aligned} C &= 3x^2 + 1308x + 25{,}268 \\ &= 3(20)^2 + 1308(20) + 25{,}268 \\ &= \$52{,}628 \end{aligned}$$
It describes the estimate from part (a) reasonably well.

71. Model 1:
$$\begin{aligned} C &= 1388x + 24{,}963 \\ &= 1388(0) + 24{,}963 \\ &= \$24{,}963 \end{aligned}$$

Model 2:
$$\begin{aligned} C &= 3x^2 + 1308x + 25{,}268 \\ &= 3(0)^2 + 1308(0) + 25{,}268 \\ &= \$25{,}268 \end{aligned}$$

According to the graph, the cost in 1980 was \$24,900. Thus, Model 1 is the better model. Model 1 overestimates the cost shown in the graph by \$63.

73.
$$\begin{aligned} C &= 1388x + 24{,}963 \\ 77{,}707 &= 1388x + 24{,}963 \\ 52{,}744 &= 1388x \\ \frac{52{,}744}{1388} &= \frac{1388x}{1388} \\ 38 &= x \end{aligned}$$
Model 1 predicts the cost will be \$77,707 38 years after 1980, or 2018.

75. – 85. Answers will vary.

87. $2x + 3(x-4) = 4x - 7$

Let $y_1 = 2x + 3(x-4)$ and let $y_2 = 4x - 7$.

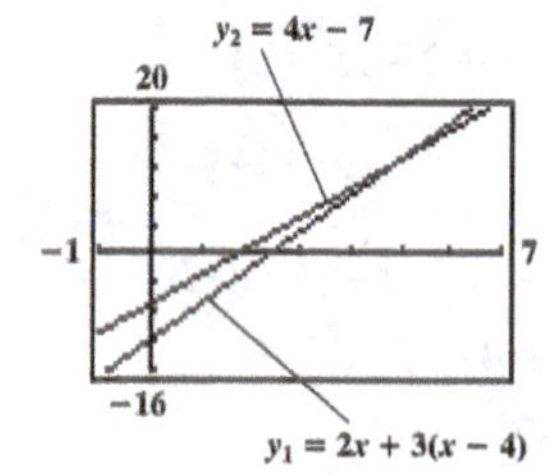

The solution set is $\{5\}$.

89. $\frac{2x-1}{3} - \frac{x-5}{6} = \frac{x-3}{4}$

Let $y_1 = \frac{2x-1}{3} - \frac{x-5}{6}$ and let $y_2 = \frac{x-3}{4}$.

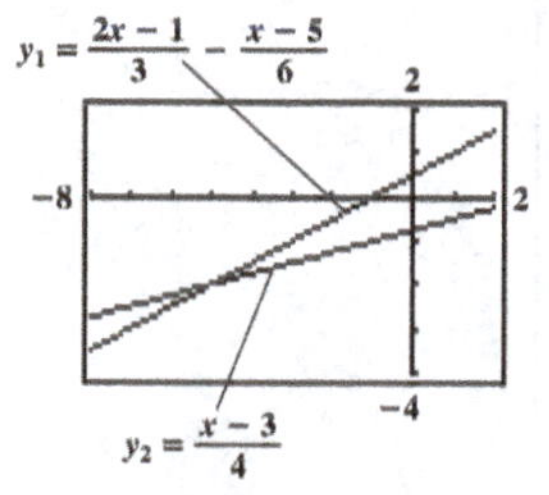

The solution set is $\{-5\}$.

91. makes sense

93. does not make sense; Explanations will vary. Sample explanation: The equation is solved by using the multiplication property.

95. false; Changes to make the statement true will vary. A sample change is: The equations are not equivalent. If the equations were equivalent, they would have the same solution set. 4 cannot be the solution to the first equation, because 4 would make the denominator 0.

97. false; Changes to make the statement true will vary. A sample change is: If a and b are both zero, there are an infinite number of values of x for which the equation is true.

99. Answers will vary.

101.
$$\begin{aligned} \frac{7(-6)+4}{b} + 13 &= -6 \\ \frac{-42+4}{b} + 13 - 13 &= -6 - 13 \\ \frac{-38}{b} &= -19 \\ -38 &= -19b \\ 2 &= b \end{aligned}$$

When $b = 2$, the solution set is $\{-6\}$.

102.
$$-\frac{1}{5} - \left(-\frac{1}{2}\right) = -\frac{1}{5} + \frac{1}{2} = -\frac{1}{5}\cdot\frac{2}{2} + \frac{1}{2}\cdot\frac{5}{5} = -\frac{2}{10} + \frac{5}{10} = \frac{3}{10}$$

103. $4(-3)(-1)(-5) = (-12)(5) = -60$

104.

x	(x,y)
−3	$(-3,5)$
−2	$(-2,0)$
−1	$(-1,-3)$
0	$(0,-4)$
1	$(1,-3)$
2	$(2,0)$
3	$(3,5)$

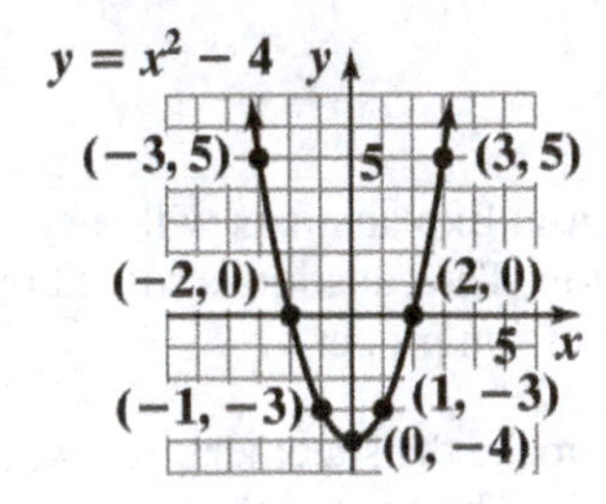

105. a. $3x-4=32$

b. $3x-4=32$

$$3x=36$$
$$x=12$$

The number is 12.

106. $x+44$

107. $20,000-2500x$

Mid-Chapter Check Point – Chapter 1

1. $-5+3(x+5)=-5+3x+15$
$$=3x+10$$

2. $-5+3(x+5)=2(3x-4)$

$$-5+3x+15=6x-8$$
$$3x+10=6x-8$$
$$-3x+10=-8$$
$$-3x=-18$$
$$x=6$$

The solution set is $\{6\}$.

3. $3[7-4(5-2)]=3[7-4(3)]$
$$=3[7-12]$$
$$=3(-5)$$
$$=-15$$

The solution set is $\{-15\}$.

4. $\frac{x-3}{5}-1=\frac{x-5}{4}$

$$20\left(\frac{x-3}{5}-1\right)=20\left(\frac{x-5}{4}\right)$$
$$4(x-3)-20=5(x-5)$$
$$4x-12-20=5x-25$$
$$4x-32=5x-25$$
$$-x-32=-25$$
$$-x=7$$
$$x=-7$$

The solution set is $\{-7\}$.

5. $\frac{-2^4+(-2)^2}{-4-(2-2)}=\frac{-16+4}{-4-0}=\frac{-12}{-4}=3$

6. $7x-[8-3(2x-5)]$
$$=7x-[8-6x+15]$$
$$=7x-[-6x+23]$$
$$=7x+6x-23$$
$$=13x-23$$

7. $3(2x-5)-2(4x+1)=-5(x+3)-2$

$$6x-15-8x-2=-5x-15-2$$
$$-2x-17=-5x-17$$
$$3x-17=-17$$
$$3x=0$$
$$x=0$$

The solution set is $\{0\}$.

8. $3(2x-5)-2(4x+1)-5(x+3)-2$
$$=6x-15-8x-2-5x-15-2$$
$$=(6x-8x-5x)+(-15-2-15-2)$$
$$=-7x-34$$

9. $-4^2\div2+(-3)(-5)=-16\div2+(-3)(-5)$
$$=-8+15$$
$$=7$$

10. $3x+1-(x-5)=2x-4$
$3x+1-x+5=2x-4$
$2x+6=2x-4$
$6=-4$
This is a contradiction, so the equation has no solution. The solution set is $\varnothing$.

11. $\frac{3x}{4}-\frac{x}{3}+1=\frac{4x}{5}-\frac{3}{20}$
$60\left(\frac{3x}{4}-\frac{x}{3}+1\right)=60\left(\frac{4x}{5}-\frac{3}{20}\right)$
$45x-20x+60=48x-9$
$25x+60=48x-9$
$-23x+60=-9$
$-23x=-69$
$x=3$
The solution set is $\{3\}$.

12. $(6-9)(8-12)\div\frac{5^2+4\div 2}{8^2-9^2+8}$
$=(-3)(-4)\div\frac{25+2}{64-81+8}$
$=(-3)(-4)\div\frac{27}{-9}$
$=(-3)(-4)\div(-3)$
$=12\div(-3)$
$=-4$

13. $4x-2(1-x)=3(2x+1)-5$
$4x-2+2x=6x+3-5$
$6x-2=6x-2$
The equation is an identity. The solution set is $\{x|x \text{ is a real number}\}$ or $\mathbb{R}$.

14. $\frac{3\left[4-3(-2)^2\right]}{2^2-2^4}=\frac{3(4-3\cdot 4)}{4-16}$
$=\frac{3(4-12)}{-12}$
$=\frac{3(-8)}{-12}$
$=\frac{-24}{-12}$
$=2$

15. $\{x|-2\le x<0\}$

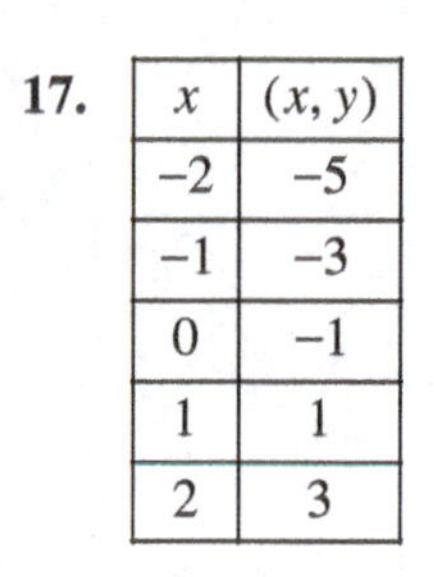

16. $\{x|x\le 0\}$

17.

x	(x, y)
−2	−5
−1	−3
0	−1
1	1
2	3

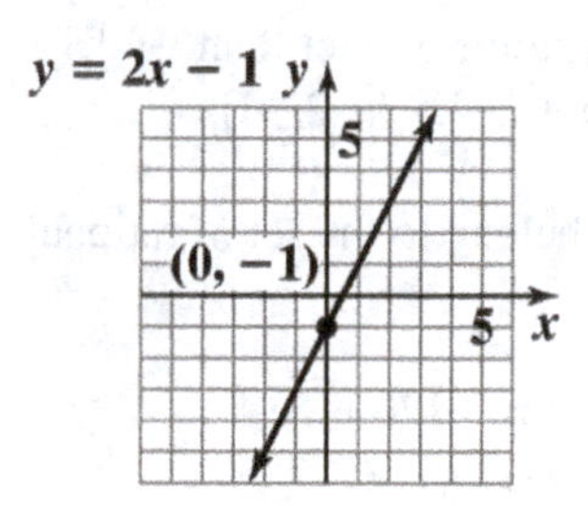

18.

x	(x, y)
−3	−2
−2	−1
−1	0
0	1
1	0
2	−1
3	−2

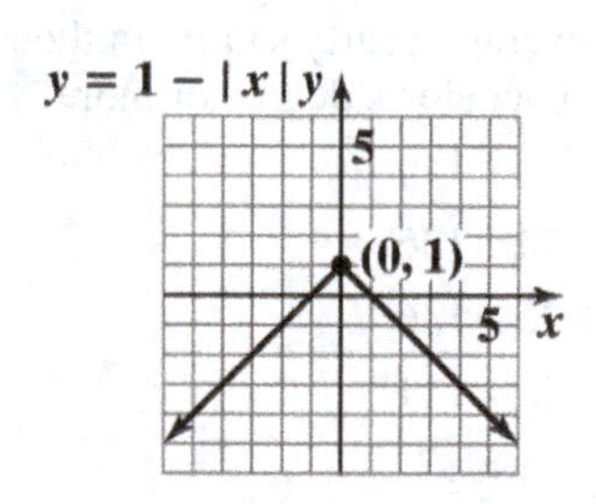

19.

x	(x, y)
–2	6
–1	3
0	2
1	3
2	6

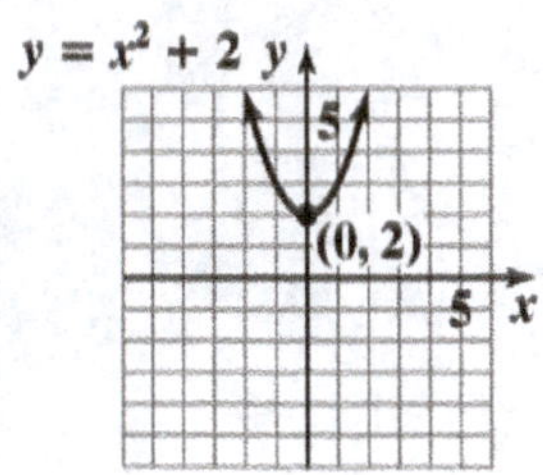

20. true

21. false; $\{x|x$ is a negative greater than $-4\}$ $= \{-3, -2, -1\}$, not $\{-4, -3, -2, -1\}$.

22. false; -17 does belong to the set of rational numbers.

23. true; $-128 \div (2 \cdot 4) > (-128 \div 2) \cdot 4$

$$-128 \div 8 > -64 \cdot 4$$

$$-16 > -256$$

which is true because -16 is to the right of -256 on the number line.

1.5 Check Points

1. Let x = the average yearly salary, in thousands, of women with some college.
Let $x + 4$ = the average yearly salary, in thousands, of women with an associate's degree.
Let $x + 21$ = the average yearly salary, in thousands, of women with a bachelor's degree or more.

$$x + (x+4) + (x+21) = 136$$
$$x + x + 4 + x + 21 = 136$$
$$3x + 25 = 136$$
$$3x = 111$$
$$x = 37$$

$x = 37$, some college: \$37,000
$x + 4 = 41$, associate's degree: \$41,000
$x + 21 = 58$, bachelor's degree: \$58,000

2. Let x = the number of years since 1969.

$$85 - 0.9x = 25$$
$$-0.9x = 25 - 85$$
$$-0.9x = -60$$
$$x = \frac{-60}{-0.9}$$
$$x \approx 67$$

25% of freshmen will respond this way 67 years after 1969, or 2036.

3. Let x = the number of text messages for which the two plans cost the same.

$$15 + 0.08x = 3 + 0.12x$$
$$0.08x - 0.12x = 3 - 15$$
$$-0.04x = -12$$
$$\frac{-0.04x}{-0.04} = \frac{-12}{-0.04}$$
$$x = 300$$

The two plans cost the same for 300 text messages.

4. Let x = the original price of the new computer.

$$x - 0.30x = 840$$
$$0.70x = 840$$
$$\frac{0.70x}{0.70} = \frac{840}{0.70}$$
$$x = 1200$$

The original price of the new computer was \$1200.

5. Let x = the width of the basketball court.
Let $x + 44$ = length of the basketball court.

$$P = 2l + 2w$$
$$288 = 2(x + 44) + 2x$$
$$288 = 2x + 88 + 2x$$
$$288 = 4x + 88$$
$$-4x = -200$$
$$x = 50$$
$$x + 44 = 94$$

The dimensions of the basketball court are 50 feet by 94 feet.

6.
$$2l + 2w = P$$
$$2w = P - 2l$$
$$\frac{2w}{2} = \frac{P - 2l}{2}$$
$$w = \frac{P - 2l}{2}$$

7.
$$V = lwh$$
$$\frac{V}{lw} = \frac{lwh}{lw}$$
$$\frac{V}{lw} = h$$
$$h = \frac{V}{lw}$$

8.
$$\frac{W}{2} - 3H = 53$$
$$\frac{W}{2} = 53 + 3H$$
$$2\left(\frac{W}{2}\right) = 2(53 + 3H)$$
$$W = 106 + 6H$$

9.
$$P = C + MC$$
$$P = C(1+M)$$
$$\frac{P}{1+M} = \frac{C(1+M)}{1+M}$$
$$\frac{P}{1+M} = C$$
$$C = \frac{P}{1+M}$$

1.5 Concept and Vocabulary Check

1. $x + 658.6$
2. $31 + 2.4x$
3. $4 + 0.15x$
4. $x - 0.15x$ or $0.85x$
5. isolated on one side
6. distributive

1.5 Exercise Set

1. Let x = a number.
$$5x - 4 = 26$$
$$5x = 30$$
$$x = 6$$
The number is 6.

3. Let x = a number.
$$x - 0.20x = 20$$
$$0.80x = 20$$
$$x = 25$$
The number is 25.

5. Let x = a number.
$$0.60x + x = 192$$
$$1.6x = 192$$
$$x = 120$$
The number is 120.

7. Let x = a number.
$$0.70x = 224$$
$$x = 320$$
The number is 320.

9. Let x = a number.
Let $x + 26$ = the other number.
$$x + (x + 26) = 64$$
$$x + x + 26 = 64$$
$$2x + 26 = 64$$
$$2x = 38$$
$$x = 19$$
If $x = 19$, then $x + 26 = 45$.
The numbers are 19 and 45.

11.
$$y_1 - y_2 = 2$$
$$(13x - 4) - (5x + 10) = 2$$
$$13x - 4 - 5x - 10 = 2$$
$$8x - 14 = 2$$
$$8x = 16$$
$$x = 2$$

13.
$$y_1 = 8y_2 + 14$$
$$10(2x - 1) = 8(2x + 1) + 14$$
$$20x - 10 = 16x + 8 + 14$$
$$20x - 10 = 16x + 22$$
$$4x = 32$$
$$x = 8$$

15.
$$3y_1 - 5y_2 = y_3 - 22$$
$$3(2x + 6) - 5(x + 8) = (x) - 22$$
$$6x + 18 - 5x - 40 = x - 22$$
$$x - 22 = x - 22$$
$$-22 = -22$$
x is satisfied by all real numbers.

17. Let x = the number of times "sorry" was used.
Let $x+419=$ the number of times "love" was used.
Let $x+32=$ the number of times "thanks" was used.

$$x+(x+419)+(x+32)=1084$$
$$x+x+419+x+32=1084$$
$$3x+451=1084$$
$$3x=633$$
$$x=211$$
$$x+419=630$$
$$x+32=243$$

The word "sorry" was used 211 times, the word "love" was used 630 times, and the word "thanks" was used 243 times.

19. Let x = the measure of the 2nd angle.
Let $2x$ = the measure of the 1st angle.
$x-8$ = the measure of the 3rd angle.

$$x+2x+(x-8)=180$$
$$4x-8=180$$
$$4x=188$$
$$x=47$$

If $x=47$, then $2x=94$ and $x-8=39$. Thus, the measure of the 1st angle is $94°$, the 2nd angle is $47°$, and the 3rd angle is $39°$.

21. Let x = the measure of the first angle.
Let $x+1$ = the measure of the second angle.
Let $x+2$ = the measure of the third angle.

$$x+(x+1)+(x+2)=180$$
$$3x+3=180$$
$$3x=177$$
$$x=59$$

If $x=59$, then $x+1=60$ and $x+2=61$. Thus, the measures of the three angles are $59°$, $60°$, and $61°$.

23. Let x = the number of years since 2000.

$$31+2.4x=67$$
$$2.4x=67-31$$
$$2.4x=36$$
$$x=\frac{36}{2.4}$$
$$x=15$$

67% of American adults will view college education as essential 15 years after 2000, or 2015.

25. Let x = the number of years since 1960.

$$23-0.28x=0$$
$$-0.28x=-23$$
$$\frac{-0.28x}{-0.28}=\frac{-23}{-0.28}$$
$$x\approx 82$$

If this trend continues, corporations will pay zero taxes 82 years after 1960, or 2042.

27. a. Let x = the number of deaths, in thousands, per day.
Let $3x-92$ = the number of births, in thousands, per day.

$$(3x-92)-x=214$$
$$3x-92-x=214$$
$$2x-92=214$$
$$2x=306$$
$$x=153$$
$$3x-92=367$$

births: 367,000
deaths: 153,000

b. $214{,}000\cdot 365=78{,}110{,}000$
≈ 78 million

c. $\dfrac{306\text{ million}}{78\text{ million}}\approx 4$

It will take about 4 years.

29. Let x = the number of months.
The cost for Club A: $25x+40$
The cost for Club B: $30x+15$

$$25x+40=30x+15$$
$$-5x+40=15$$
$$-5x=-25$$
$$x=5$$

The total cost for the clubs will be the same at 5 months. The cost will be $25(5)+40=30(5)+15=\$165$.

31. Let x = the number of bus uses.
Cost without discount pass: $1.25x$
Cost with discount pass: $15+0.75x$

$$1.25x = 15+0.75x$$
$$0.50x = 15$$
$$x = 30$$

The bus must be used 30 times in a month for the costs to be equal.

33. a. Let x = the number of years (after 2008).
College A's enrollment: $13,300+1000x$
College B's enrollment: $26,800-500x$

$$13,300+1000x = 26,800-500x$$
$$13,300+1500x = 26,800$$
$$1500x = 13,500$$
$$x = 9$$

The two colleges will have the same enrollment 9 years after 2008, or 2017.
That year the enrollments will be

$$13,300+1000(9) = 26,800-500(9)$$
$$= 22,300 \text{ students}$$

b. Check points to determine that $y_1 = 13300+1000x$ and $y_2 = 26800-500x$.

35. Let x = the cost of the television set.

$$x-0.20x = 336$$
$$0.80x = 336$$
$$x = 420$$

The television set's price is $420.

37. Let x = the nightly cost.

$$x+0.08x = 162$$
$$1.08x = 162$$
$$x = 150$$

The nightly cost is $150.

39. Let c = the dealer's cost.

$$584 = c+0.25c$$
$$584 = 1.25c$$
$$467.20 = c$$

The dealer's cost is $467.20.

41. Let w = the width of the field.
Let $2w$ = the length of the field.

$$P = 2(\text{length})+2(\text{width})$$
$$300 = 2(2w)+2(w)$$
$$300 = 4w+2w$$
$$300 = 6w$$
$$50 = w$$

If $w = 50$, then $2w = 100$. Thus, the dimensions are 50 yards by 100 yards.

43. Let w = the width of the field.
Let $2w + 6$ = the length of the field.

$$P = 2(\text{length})+2(\text{width})$$
$$228 = 2(2w+6)+2w$$
$$228 = 4w+12+2w$$
$$228 = 6w+12$$
$$216 = 6w$$
$$36 = w$$

If $w = 36$, then $2w+6 = 2(36)+6 = 78$. Thus, the dimensions are 36 feet by 78 feet.

45. Let x = the width of the frame.
Total length: $16+2x$.
Total width: $12+2x$.

$$P = 2(\text{length})+2(\text{width})$$
$$72 = 2(16+2x)+2(12+2x)$$
$$72 = 32+4x+24+4x$$
$$72 = 8x+56$$
$$16 = 8x$$
$$2 = x$$

The width of the frame is 2 inches.

47. Let x = the length of the call.

$$0.43+0.32(x-1)+2.10 = 5.73$$
$$0.43+0.32x-0.32+2.10 = 5.73$$
$$0.32x+2.21 = 5.73$$
$$0.32x = 3.52$$
$$x = 11$$

The person talked for 11 minutes.

49. (from geometry)

$$A = lw$$
$$l = \frac{A}{w}$$

51. (from geometry)

$$A = \frac{1}{2}bh$$
$$2A = bh$$
$$b = \frac{2A}{h}$$

53. (from finance)

$$I = Prt$$
$$P = \frac{I}{rt}$$

55. (from finance)

$$T = D + pm$$
$$T - D = pm$$
$$p = \frac{T - D}{m}$$

57. (from geometry)

$$A = \frac{1}{2}h(a+b)$$
$$2A = h(a+b)$$
$$\frac{2A}{h} = a + b$$
$$a = \frac{2A}{h} - b \text{ or } a = \frac{2A - hb}{h}$$

59. (from geometry)

$$V = \frac{1}{3}\pi r^2 h$$
$$3V = \pi r^2 h$$
$$h = \frac{3V}{\pi r^2}$$

61. (from algebra)

$$y - y_1 = m(x - x_1)$$
$$m = \frac{y - y_1}{x - x_1}$$

63. (from physics)

$$V = \frac{d_1 - d_2}{t}$$
$$Vt = d_1 - d_2$$
$$d_1 = Vt + d_2$$

65. (from algebra)

$$Ax + By = C$$
$$Ax = C - By$$
$$x = \frac{C - By}{A}$$

67. (from physics)

$$s = \frac{1}{2}at^2 + vt$$
$$2s = \cancel{2}\left(\frac{1}{\cancel{2}}at^2\right) + 2vt$$
$$2s = at^2 + 2vt$$
$$2s - at^2 = 2vt$$
$$\frac{2s - at^2}{2t} = \frac{2vt}{2t}$$
$$v = \frac{2s - at^2}{2t}$$

69. (from algebra)

$$L = a + (n-1)d$$
$$L - a = (n-1)d$$
$$\frac{L - a}{d} = n - 1$$
$$n = \frac{L - a}{d} + 1$$

or

$$n = \frac{L - a + d}{d}$$

71. (from geometry)

$$A = 2lw + 2lh + 2wh$$
$$A - 2wh = 2lw + 2lh$$
$$A - 2wh = l(2w + 2h)$$
$$l = \frac{A - 2wh}{2w + 2h}$$

73. (from physics)

$$IR + Ir = E$$
$$I(R + r) = E$$
$$I = \frac{E}{R + r}$$

75. – 79. Answers will vary.

81.

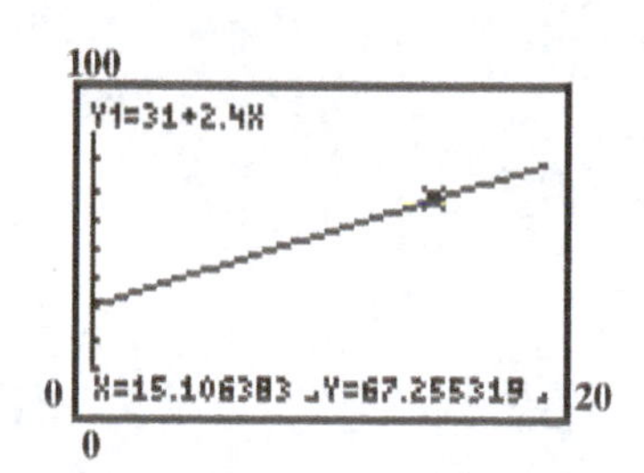

83. does not make sense; Explanations will vary. Sample explanation: The variable may be solved in terms of other variables.

85. makes sense

87. false; Changes to make the statement true will vary.

A sample change is: If $I = prt$, then $t = \frac{I}{pr}$.

89. false; Changes to make the statement true will vary. A sample change is: The solution uses the distributive property.

$$P = C + MC$$
$$P = C(1+M)$$
$$C = \frac{P}{1+M}$$

91. Let x = the original price of the dress. If the reduction in price is 40%, the price paid is 60%.

$$\text{price paid} = 0.60(0.60x)$$
$$72 = 0.60(0.60x)$$
$$72 = 0.36x$$
$$200 = x$$

The original price is $200.

93. Let x = the amount a girl would receive.
$2x$ = the amount Mrs. Ricardo would receive.
$4x$ = the amount a boy would receive.

$$\text{Total Savings} = x + 2x + 4x$$
$$14{,}000 = 7x$$
$$2{,}000 = x$$

Mrs. Ricardo received $4000, the boy received $8000, and the girl received $2000.

95.

$$V = C - \frac{C-S}{L}N$$
$$V = C - \left(\frac{C-S}{L}\right)\frac{N}{1}$$
$$V = C - \frac{CN - SN}{L}$$
$$V = \frac{CL}{L} - \frac{CN - SN}{L}$$
$$V = \frac{CL - CN + SN}{L}$$
$$LV = CL - CN + SN$$
$$LV - SN = CL - CN$$
$$LV - SN = C(L - N)$$
$$C = \frac{LV - SN}{L - N}$$

96. $\{x \mid -4 < x \le 0\}$

−5 −4 −3 −2 −1 0 1 2 3 4 5

97. $\frac{(2+4)^2 + (-1)^5}{12 \div 2 \cdot 3 - 3} = \frac{(6)^2 + (-1)}{6 \cdot 3 - 3} = \frac{36 + (-1)}{18 - 3}$

$= \frac{35}{15} = \frac{7}{3}$

98.

$$\frac{2x}{3} - \frac{8}{3} = x$$
$$3\left(\frac{2x}{3} - \frac{8}{3}\right) = 3(x)$$
$$2x - 8 = 3x$$
$$-8 = x$$

The solution set is $\{-8\}$.

99. **a.** $b^4 \cdot b^3 = (b \cdot b \cdot b \cdot b)(b \cdot b \cdot b) = b^7$

b. $b^5 \cdot b^5 = (b \cdot b \cdot b \cdot b \cdot b)(b \cdot b \cdot b \cdot b \cdot b) = b^{10}$

c. When multiplying exponential expressions with the same base, add the exponents.

100. **a.** $\frac{b^7}{b^3} = \frac{\cancel{b} \cdot \cancel{b} \cdot \cancel{b} \cdot b \cdot b \cdot b \cdot b}{\cancel{b} \cdot \cancel{b} \cdot \cancel{b}} = b^4$

b. $\frac{b^8}{b^2} = \frac{\cancel{b} \cdot \cancel{b} \cdot b \cdot b \cdot b \cdot b \cdot b \cdot b}{\cancel{b} \cdot \cancel{b}} = b^6$

c. When dividing exponential expressions with the same base, subtract the exponents.

101. $\frac{1}{\left(-\frac{1}{2}\right)^3} = \frac{1}{\left(-\frac{2}{1}\right)^{-3}}$

$= \frac{1}{(-2)^{-3}}$

$= (-2)^3$

$= -8$

1.6 Check Points

1. a. $b^6 \cdot b^5 = b^{6+5} = b^{11}$

b. $\left(4x^3y^4\right)\left(10x^2y^6\right) = 4 \cdot 10 \cdot x^3 \cdot x^2 \cdot y^4 \cdot y^6$

$= 40x^{3+2}y^{4+6}$

$= 40x^5y^{10}$

2. a. $\frac{(-3)^6}{(-3)^3} = (-3)^{6-3} = (-3)^3 = -27$

b. $\frac{27x^{14}y^8}{3x^3y^5} = \frac{27}{3}x^{14-3}y^{8-5} = 9x^{11}y^3$

3. a. $7^0 = 1$

b. $(-5)^0 = 1$

c. $-5^0 = -(5^0) = -1$

d. $10x^0 = 10 \cdot 1 = 10$

e. $(10x)^0 = 1$

4. a. $5^{-2} = \frac{1}{5^2} = \frac{1}{25}$

b. $(-3)^{-3} = \frac{1}{(-3)^3} = \frac{1}{-27} = -\frac{1}{27}$

c. $\frac{1}{4^{-2}} = 4^2 = 16$

d. $3x^{-6}y^4 = 3 \cdot \frac{1}{x^6} \cdot y^4 = \frac{3y^4}{x^6}$

5. a. $\frac{7^{-2}}{4^{-3}} = \frac{4^3}{7^2} = \frac{64}{49}$

b. $\frac{1}{5x^{-2}} = \frac{x^2}{5}$

6. a. $\left(x^5\right)^3 = x^{5 \cdot 3} = x^{15}$

b. $\left(y^7\right)^{-2} = y^{(7)(-2)} = y^{-14} = \frac{1}{y^{14}}$

c. $\left(b^{-3}\right)^{-4} = b^{(-3)(-4)} = b^{12}$

7. a. $(2x)^4 = (2)^4(x)^4 = 16x^4$

b. $\left(-3y^2\right)^3 = (-3)^3\left(y^2\right)^3 = -27y^6$

c. $\left(-4x^5y^{-1}\right)^{-2} = (-4)^{-2}\left(x^5\right)^{-2}\left(y^{-1}\right)^{-2}$

$= \frac{1}{(-4)^2} \cdot \frac{1}{\left(x^5\right)^2} \cdot y^2$

$= \frac{y^2}{16x^{10}}$

8. a. $\left(\frac{x^5}{4}\right)^3 = \frac{x^{5 \cdot 3}}{4^3} = \frac{x^{15}}{64}$

b. $\left(\frac{2x^{-3}}{y^2}\right)^4 = \frac{2^4x^{(-3)(4)}}{y^{(2)(4)}} = \frac{16x^{-12}}{y^8} = \frac{16}{x^{12}y^8}$

c. $\left(\frac{x^{-3}}{y^4}\right)^{-5} = \frac{x^{(-3)(-5)}}{y^{(4)(-5)}} = \frac{x^{15}}{y^{-20}} = x^{15}y^{20}$

9. a. $\left(-3x^{-6}y\right)\left(-2x^3y^4\right)^2$

$= \left(-3x^{-6}y\right)(-2)^2\left(x^3\right)^2\left(y^4\right)^2$

$= -3 \cdot x^{-6} \cdot y \cdot 4 \cdot x^6 \cdot y^8$

$= -12 \cdot x^{-6+6} \cdot y^{1+8}$

$= -12x^0y^9$

$= -12y^9$

b. $\left(\frac{10x^3y^5}{5x^6y^{-2}}\right)^2 = \left(2x^{3-6}y^{5+2}\right)^2$

$= \left(2x^{-3}y^7\right)^2 = 4x^{-6}y^{14} = \frac{4y^{14}}{x^6}$

c. $\left(\frac{x^3y^5}{4}\right)^{-3} = \frac{x^{(3)(-3)}y^{(5)(-3)}}{4^{-3}}$

$= \frac{x^{-9}y^{-15}}{4^{-3}} = \frac{4^3}{x^9y^{15}} = \frac{64}{x^9y^{15}}$

1.6 Concept and Vocabulary Check

1. b^{m+n}; add
2. b^{m-n}; subtract
3. 1
4. $\frac{1}{b^n}$
5. false
6. b^n
7. true

1.6 Exercise Set

1. $b^4 \cdot b^7 = b^{4+7} = b^{11}$

3. $x \cdot x^3 = x^{1+3} = x^4$

5. $2^3 \cdot 2^2 = 2^{3+2} = 2^5 = 32$

7. $3x^4 \cdot 2x^2 = 6x^{4+2} = 6x^6$

9. $\left(-2y^{10}\right)\left(-10y^2\right) = 20y^{10+2} = 20y^{12}$

11. $\left(5x^3y^4\right)\left(20x^7y^8\right) = 100x^{3+7}y^{4+8}$
$= 100x^{10}y^{12}$

13. $\left(-3x^4y^0z\right)\left(-7xyz^3\right)$
$= 21x^{(4+1)}y^{0+1}z^{1+3}$
$= 21x^5y^1z^4 = 21x^5yz^4$

15. $\frac{b^{12}}{b^3} = b^{12-3} = b^9$

17. $\frac{15x^9}{3x^4} = 5x^{9-4} = 5x^5$

19. $\frac{x^9y^7}{x^4y^2} = x^{9-4}y^{7-2} = x^5y^5$

21. $\frac{50x^2y^7}{5xy^4} = 10x^{2-1}y^{7-4} = 10xy^3$

23. $\frac{-56a^{12}b^{10}c^8}{7ab^2c^4} = -8a^{12-1}b^{10-2}c^{8-4}$
$= -8a^{11}b^8c^4$

25. $6^0 = 1$

27. $(-4)^0 = 1$

29. $-4^0 = -1$

31. $13y^0 = 13(1) = 13$

33. $(13y)^0 = 1$

35. $3^{-2} = \frac{1}{3^2} = \frac{1}{9}$

37. $(-5)^{-2} = \frac{1}{(-5)^2} = \frac{1}{25}$

39. $-5^{-2} = -\left(5^{-2}\right) = -\frac{1}{5^2} = -\frac{1}{25}$

41. $x^2y^{-3} = \frac{x^2}{y^3}$

43. $8x^{-7}y^3 = \frac{8y^3}{x^7}$

45. $\dfrac{1}{5^{-3}} = 5^3 = 125$

47. $\dfrac{1}{(-3)^{-4}} = (-3)^4 = 81$

49. $\dfrac{x^{-2}}{y^{-5}} = \dfrac{y^5}{x^2}$

51. $\dfrac{a^{-4}b^7}{c^{-3}} = \dfrac{b^7c^3}{a^4}$

53. $\left(x^6\right)^{10} = x^{(6\cdot10)} = x^{60}$

55. $\left(b^4\right)^{-3} = \dfrac{1}{\left(b^4\right)^3} = \dfrac{1}{b^{(4\cdot3)}} = \dfrac{1}{b^{12}}$

57. $\left(7^{-4}\right)^{-5} = 7^{-4\cdot(-5)} = 7^{20}$

59. $(4x)^3 = 4^3x^3 = 64x^3$

61. $\left(-3x^7\right)^2 = (-3)^2x^{7\cdot2} = 9x^{14}$

63. $\left(2xy^2\right)^3 = 8x^{(1\cdot3)}y^{(2\cdot3)} = 8x^3y^6$

65. $\left(-3x^2y^5\right)^2 = (-3)^2x^{(2\cdot2)}y^{(5\cdot2)} = 9x^4y^{10}$

67. $\left(-3x^{-2}\right)^{-3} = (-3)^{-3}\left(x^{-2}\right)^{-3}$

$$= \frac{x^6}{(-3)^3} = \frac{x^6}{-27} = -\frac{x^6}{27}$$

69. $\left(5x^3y^{-4}\right)^{-2} = 5^{-2}\left(x^3\right)^{-2}\left(y^{-4}\right)^{-2}$

$$= 5^{-2}x^{-6}y^8 = \frac{y^8}{25x^6}$$

71. $\left(-2x^{-5}y^4z^2\right)^{-4} = (-2)^{-4}x^{20}y^{-16}z^{-8}$

$$= \frac{x^{20}}{(-2)^4y^{16}z^8}$$

$$= \frac{x^{20}}{16y^{16}z^8}$$

73. $\left(\dfrac{2}{x}\right)^4 = \dfrac{2^4}{x^4} = \dfrac{16}{x^4}$

75. $\left(\dfrac{x^3}{5}\right)^2 = \dfrac{x^{(3\cdot2)}}{5^2} = \dfrac{x^6}{25}$

77. $\left(-\dfrac{3x}{y}\right)^4 = \dfrac{(-3)^4x^4}{y^4} = \dfrac{81x^4}{y^4}$

79. $\left(\dfrac{x^4}{y^2}\right)^6 = \dfrac{x^{(4\cdot6)}}{y^{(2\cdot6)}} = \dfrac{x^{24}}{y^{12}}$

81. $\left(\dfrac{x^3}{y^{-4}}\right)^3 = \dfrac{x^{(3\cdot3)}}{y^{(-4\cdot3)}} = \dfrac{x^9}{y^{-12}} = x^9y^{12}$

83. $\left(\dfrac{a^{-2}}{b^3}\right)^{-4} = \dfrac{a^{(-2\cdot(-4))}}{b^{(3\cdot(-4))}} = \dfrac{a^8}{b^{-12}} = a^8b^{12}$

85. $\dfrac{x^3}{x^9} = x^{3-9} = x^{-6} = \dfrac{1}{x^6}$

87. $\dfrac{20x^3}{-5x^4} = -4x^{3-4} = -4x^{-1} = -\dfrac{4}{x}$

89. $\dfrac{16x^3}{8x^{10}} = 2x^{3-10} = 2x^{-7} = \dfrac{2}{x^7}$

91. $\dfrac{20a^3b^8}{2ab^{13}} = 10a^{3-1}b^{8-13}$

$$= 10a^2b^{-5} = \frac{10a^2}{b^5}$$

93. $x^3 \cdot x^{-12} = x^{3+(-12)} = x^{-9} = \dfrac{1}{x^9}$

95. $\left(2a^5\right)\left(-3a^{-7}\right) = -6a^{5+(-7)}$

$= -6a^{-2} = -\dfrac{6}{a^2}$

97. $\left(-\frac{1}{4}x^{-4}y^5z^{-1}\right)\left(-12x^{-3}y^{-1}z^4\right)$

$= 3x^{-4+(-3)}y^{5+(-1)}z^{-1+4}$

$= 3x^{-7}y^4z^3 = \dfrac{3y^4z^3}{x^7}$

99. $\dfrac{6x^2}{2x^{-8}} = 3x^{2-(-8)} = 3x^{2+8} = 3x^{10}$

101. $\dfrac{x^{-7}}{x^3} = x^{-7-3} = x^{-10} = \dfrac{1}{x^{10}}$

103. $\dfrac{30x^2y^5}{-6x^8y^{-3}} = -5x^{2-8}y^{5-(-3)}$

$= -5x^{-6}y^8 = -\dfrac{5y^8}{x^6}$

105. $\dfrac{-24a^3b^{-5}c^5}{-3a^{-6}b^{-4}c^{-7}} = 8a^{3-(-6)}b^{-5-(-4)}c^{5-(-7)}$

$= 8a^9b^{-1}c^{12} = \dfrac{8a^9c^{12}}{b}$

107. $\left(\dfrac{x^3}{x^{-5}}\right)^2 = \left(x^{3-(-5)}\right)^2 = \left(x^8\right)^2 = x^{16}$

109. $\left(\dfrac{-15a^4b^2}{5a^{10}b^{-3}}\right)^3 = \left(-3a^{4-10}b^{2-(-3)}\right)^3$

$= \left(-3a^{-6}b^{2+3}\right)^3$

$= \left(-3a^{-6}b^5\right)^3$

$= (-3)^3\left(a^{-6}\right)^3\left(b^5\right)^3$

$= -27a^{-18}b^{15}$

$= -\dfrac{27b^{15}}{a^{18}}$

111. $\left(\dfrac{3a^{-5}b^2}{12a^3b^{-4}}\right)^0 = 1$

Recall the Zero Exponent Rule.

113. $\left(\dfrac{x^{-5}y^8}{3}\right)^{-4} = \dfrac{x^{(-5)(-4)}y^{8(-4)}}{3^{-4}}$

$= \dfrac{x^{20}y^{-32}}{3^{-4}} = \dfrac{3^4x^{20}}{y^{32}} = \dfrac{81x^{20}}{y^{32}}$

115. $\left(\dfrac{20a^{-3}b^4c^5}{-2a^{-5}b^{-2}c}\right)^{-2} = \left(10a^{-3-(-5)}b^{4-(-2)}c^{5-1}\right)^{-2}$

$= \dfrac{1}{\left(10a^2b^6c^4\right)^2}$

$= \dfrac{1}{10^2a^{2(2)}b^{6(2)}c^{4(2)}}$

$= \dfrac{1}{100a^4b^{12}c^8}$

117. $\dfrac{9y^4}{x^{-2}} + \left(\dfrac{x^{-1}}{y^2}\right)^{-2} = 9x^2y^4 + \dfrac{x^{(-1)(-2)}}{y^{2(-2)}}$

$= 9x^2y^4 + \dfrac{x^2}{y^{-4}}$

$= 9x^2y^4 + x^2y^4$

$= 10x^2y^4$

119. $\left(\dfrac{3x^4}{y^{-4}}\right)^{-1}\left(\dfrac{2x}{y^2}\right)^3 = \dfrac{3^{-1}x^{4(-1)}}{y^{(-4)(-1)}} \cdot \dfrac{2^3x^{1\cdot3}}{y^{2\cdot3}}$

$= \dfrac{x^{-4}}{3y^4} \cdot \dfrac{8x^3}{y^6}$

$= \dfrac{8x^{-4+3}}{3y^{4+6}}$

$= \dfrac{8x^{-1}}{3y^{10}}$

$= \dfrac{8}{3xy^{10}}$

121. $\left(-4x^3y^{-5}\right)^{-2}\left(2x^{-8}y^{-5}\right)=\dfrac{2x^{-8}y^{-5}}{\left(-4x^3y^{-5}\right)^2}$

$=\dfrac{2x^{-8}y^{-5}}{(-4)^2x^{3\cdot 2}y^{-5\cdot 2}}$

$=\dfrac{2x^{-8}y^{-5}}{16x^6y^{-10}}$

$=\dfrac{y^{-5-(-10)}}{8x^{6-(-8)}}$

$=\dfrac{y^5}{8x^{14}}$

123. $\dfrac{\left(2x^2y^4\right)^{-1}\left(4xy^3\right)^{-3}}{\left(x^2y\right)^{-5}\left(x^3y^2\right)^4}$

$=\dfrac{\left(x^2y\right)^5}{\left(2x^2y^4\right)^1\left(4xy^3\right)^3\left(x^3y^2\right)^4}$

$=\dfrac{x^{2\cdot 5}y^{1\cdot 5}}{\left(2x^2y^4\right)\left(4^3x^{1\cdot 3}y^{3\cdot 3}\right)\left(x^{3\cdot 4}y^{2\cdot 4}\right)}$

$=\dfrac{x^{10}y^5}{\left(2x^2y^4\right)\left(64x^3y^9\right)\left(x^{12}y^8\right)}$

$=\dfrac{x^{10}y^5}{128x^{2+3+12}y^{4+9+8}}$

$=\dfrac{x^{10}y^5}{128x^{17}y^{21}}$

$=\dfrac{1}{128x^{17-10}y^{21-5}}=\dfrac{1}{128x^7y^{16}}$

125. **a.** $A=1000\cdot 2^t=1000\cdot 2^0=1000\cdot 1=1000$
The present aphid population is 1000.

b. $A=1000\cdot 2^t=1000\cdot 2^4=1000\cdot 16=16{,}000$
In four weeks the aphid population will be 16,000.

c. $A=1000\cdot 2^t=1000\cdot 2^{-3}$

$=1000\cdot\dfrac{1}{2^3}=1000\cdot\dfrac{1}{8}=125$

Three weeks ago the aphid population was 125.

127. **a.** $N=\dfrac{25}{1+24\cdot 2^{-t}}=\dfrac{25}{1+24\cdot 2^{-0}}=\dfrac{25}{1+24\cdot 1}=\dfrac{25}{25}=1$
One person started the rumor.

b. $N=\dfrac{25}{1+24\cdot 2^{-t}}=\dfrac{25}{1+24\cdot 2^{-4}}$

$=\dfrac{25}{1+\frac{24}{2^4}}=\dfrac{25}{1+\frac{24}{16}}=\dfrac{25}{1+1.5}=\dfrac{25}{2.5}=10$

After 4 minutes, 10 people in the class had heard the rumor.

129. **a.** At time zero, one person started the rumor. This is represented by the point $(0,1)$.

b. After 4 minutes, 10 people in the class had heard the rumor. This is represented by the point $(4,10)$.

131. Statement d best describes the graph.

133. If $n = 1$,

$d=\dfrac{3\left(2^{n-2}\right)+4}{10}$

$=\dfrac{3\left(2^{1-2}\right)+4}{10}$

$=\dfrac{3\left(2^{-1}\right)+4}{10}$

$=\dfrac{3\left(\frac{1}{2}\right)+4}{10}=\dfrac{1.5+4}{10}=\dfrac{5.5}{10}=0.55$

Mercury is 0.55 astronomical units from the sun.

135. If $n = 5$,

$d=\dfrac{3\left(2^{n-2}\right)+4}{10}=\dfrac{3\left(2^{5-2}\right)+4}{10}=\dfrac{3\left(2^3\right)+4}{10}$

$=\dfrac{3(8)+4}{10}=\dfrac{24+4}{10}=\dfrac{28}{10}=2.8$

Jupiter is 2.8 astronomical units from the Sun. Thus, Jupiter is 1.8 astronomical units farther from the Sun than Earth.

137. – 145. Answers will vary.

147. makes sense

149. does not make sense; Explanations will vary.
Sample explanation: $b^0=1$, so $\dfrac{a^n}{b^0}=\dfrac{a^n}{1}=a^n$.

151. false; Changes to make the statement true will vary. A sample change is: $5^6 \cdot 5^2 = 5^{6+2} = 5^8$

153. false; Changes to make the statement true will vary. A sample change is: $\frac{1}{(-2)^3} = \frac{1}{-8} = -\frac{1}{8}$, but $2^{-3} = \frac{1}{2^3} = \frac{1}{8}$.

155. false; Changes to make the statement true will vary. A sample change is: $2^4 + 2^5 = 16 + 32 = 48$, but $2^9 = 512$.

157. true

159.
$$\begin{aligned}\left(x^{-4n} \cdot x^n\right)^{-3} &= \left(x^{-4n+n}\right)^{-3}\\ &= \left(x^{-3n}\right)^{-3}\\ &= x^{(-3n)(-3)} = x^{9n}\end{aligned}$$

161.
$$\begin{aligned}\left(\frac{x^n y^{3n+1}}{y^n}\right)^{-2} &= \left(x^n y^{(3n+1)-n}\right)^3\\ &= \left(x^n y^{2n+1}\right)^3\\ &= x^{n\cdot 3} y^{(2n+1)\cdot 3}\\ &= x^{3n} y^{6n+3}\end{aligned}$$

162.

x	(x, y)
-3	$(-3,-7)$
-2	$(-2,-5)$
-1	$(-1,0)$
0	$(0,-1)$
1	$(1,1)$
2	$(2,3)$
3	$(3,5)$

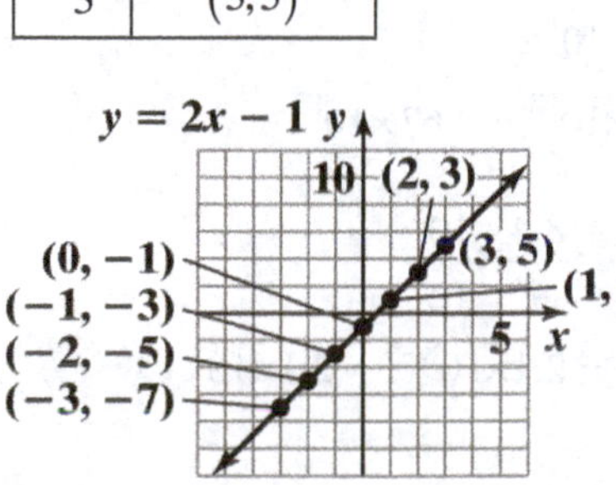

163.
$$\begin{aligned}Ax + By &= C\\ By &= C - Ax\\ y &= \frac{C - Ax}{B}\end{aligned}$$

164. Let w = the width of the playing field.

Let $2w - 5$ = the length of the playing field.

$$\begin{aligned}P &= 2(\text{length}) + 2(\text{width})\\ 230 &= 2(2w-5) + 2w\\ 230 &= 4w - 10 + 2w\\ 230 &= 6w - 10\\ 240 &= 6w\\ 40 &= w\end{aligned}$$

Find the length. $2w - 5 = 2(40) - 5 = 80 - 5 = 75$

The playing field is 40 meters by 75 meters.

165. It moves the decimal point 3 places to the right.

166. It moves the decimal point 2 places to the left.

167. **a.** $10^9 \times 10^{-4} = 10^{9-4} = 10^5 = 100{,}000$

b. $\frac{10^4}{10^{-2}} = 10^4 \times 10^2 = 10^{4+2} = 10^6 = 1{,}000{,}000$

1.7 Check Points

1. **a.** Move the decimal point 7 places to the right.
$-2.6 \times 10^9 = -2{,}600{,}000{,}000$

b. Move the decimal point 6 places to the left.
$3.017 \times 10^{-6} = 0.000003017$

2. **a.** The decimal point must be moved 9 places to the left to get a number whose absolute value is between 1 and 10. Thus the exponent on 10 is 9.
$5{,}210{,}000{,}000 = 5.21 \times 10^9$

b. The decimal point must be moved 8 places to the right to get a number whose absolute value is between 1 and 10. Thus the exponent on 10 is –8.
$-0.00000006893 = -6.893 \times 10^{-8}$

3. 18 million $= 18{,}000{,}000 = 1.8 \times 10^7$

4. a. $(7.1\times10^5)(5\times10^{-7})=(7.1\times5)\times(10^5\times10^{-7})$
$=35.5\times10^{-2}=3.55\times10^{-1}$

b. $\dfrac{1.2\times10^6}{3\times10^{-3}}=\left(\dfrac{1.2}{3}\right)\times\left(\dfrac{10^6}{10^{-3}}\right)$
$=0.4\times10^{6-(-3)}=0.4\times10^9=4\times10^8$

5. $\dfrac{2.75\times10^{12}}{3.06\times10^8}=\left(\dfrac{2.75}{3.06}\right)\times\left(\dfrac{10^{12}}{10^8}\right)$
$\approx0.8987\times10^{12-8}$
$=0.8987\times10^4=8987$
The per capita tax was about $8987 in 2008.

6. $d=rt$
$d=(1.55\times10^3)(20,000)$
$d=(1.55\times10^3)(2\times10^4)$
$d=(1.55\times2)\times(10^3\times10^4)$
$d=3.1\times10^7$
The distance from Venus to Mercury is 3.1×10^7, or 31 million miles.

1.7 Concept and Vocabulary Check

1. a number greater than or equal to 1 and less than 10; integer
2. true
3. false

1.7 Exercise Set

1. $3.8\times10^2=380$

3. $6\times10^{-4}=0.0006$

5. $-7.16\times10^6=-7,160,000$

7. $1.4\times10^0=1.4\times1=1.4$

9. $7.9\times10^{-1}=0.79$

11. $-4.15\times10^{-3}=-0.00415$

13. $-6.00001\times10^{10}=-60,000,100,000$

15. $32,000=3.2\times10^4$

17. $638,000,000,000,000,000=6.38\times10^{17}$

19. $-317=-3.17\times10^2$

21. $-5716=-5.716\times10^3$

23. $0.0027=2.7\times10^{-3}$

25. $-0.00000000504=-5.04\times10^{-9}$

27. $0.007=7\times10^{-3}$

29. $3.14159=3.14159\times10^0$

31. $(3\times10^4)(2.1\times10^3)=(3\times2.1)(10^4\times10^3)$
$=6.3\times10^{4+3}$
$=6.3\times10^7$

33. $(1.6\times10^{15})(4\times10^{-11})=(1.6\times4)(10^{15}\times10^{-11})$
$=6.4\times10^{15+(-11)}$
$=6.4\times10^4$

35. $(6.1\times10^{-8})(2\times10^{-4})=(6.1\times2)(10^{-8}\times10^{-4})$
$=12.2\times10^{-8+(-4)}$
$=12.2\times10^{-12}$
$=1.22\times10^{-11}$

37. $(4.3\times10^8)(6.2\times10^4)$
$=(4.3\times6.2)(10^8\times10^4)$
$=26.66\times10^{8+4}$
$=26.66\times10^{12}$
$=2.666\times10^{13}\approx2.67\times10^{13}$

39. $\dfrac{8.4\times10^8}{4\times10^5}=\dfrac{8.4}{4}\times\dfrac{10^8}{10^5}$
$=2.1\times10^{8-5}=2.1\times10^3$

41. $\frac{3.6\times10^4}{9\times10^{-2}}=\frac{3.6}{9}\times\frac{10^4}{10^{-2}}$
$=0.4\times10^{4-(-2)}$
$=0.4\times10^6=4\times10^5$

43. $\frac{4.8\times10^{-2}}{2.4\times10^6}=\frac{4.8}{2.4}\times\frac{10^{-2}}{10^6}$
$=2\times10^{-2-6}=2\times10^{-8}$

45. $\frac{2.4\times10^{-2}}{4.8\times10^{-6}}=\frac{2.4}{4.8}\times\frac{10^{-2}}{10^{-6}}$
$=0.5\times10^{-2-(-6)}$
$=0.5\times10^4=5\times10^3$

47. $\frac{480,000,000,000}{0.00012}=\frac{4.8\times10^{11}}{1.2\times10^{-4}}$
$=\frac{4.8}{1.2}\times\frac{10^{11}}{10^{-4}}$
$=4\times10^{11-(-4)}$
$=4\times10^{15}$

49. $\frac{0.00072\times0.003}{0.00024}=\frac{\left(7.2\times10^{-4}\right)\left(3\times10^{-3}\right)}{2.4\times10^{-4}}$
$=\frac{7.2\times3}{2.4}\times\frac{10^{-4}\cdot10^{-3}}{10^{-4}}$
$=9\times10^{-3}$

51. $\left(2\times10^{-5}\right)x=1.2\times10^9$
$x=\frac{1.2\times10^9}{2\times10^{-5}}$
$=\frac{1.2}{2}\times\frac{10^9}{10^{-5}}$
$=0.6\times10^{9-(-5)}$
$=0.6\times10^{14}$
$=6\times10^{13}$

53. $\frac{x}{2\times10^8}=-3.1\times10^{-5}$
$x=\left(2\times10^8\right)\left(-3.1\times10^{-5}\right)$
$=[2\cdot(-3.1)]\times\left(10^8\cdot10^{-5}\right)$
$=-6.2\times10^{8+(-5)}=-6.2\times10^3$

55. $x-\left(7.2\times10^{18}\right)=9.1\times10^{18}$
$x=\left(9.1\times10^{18}\right)+\left(7.2\times10^{18}\right)$
$=(9.1+7.2)\times10^{18}$
$=16.3\times10^{18}$
$=1.63\times10^{19}$

57. $\left(-1.2\times10^{-3}\right)x=\left(1.8\times10^{-4}\right)\left(2.4\times10^6\right)$
$x=\frac{\left(1.8\times10^{-4}\right)\left(2.4\times10^6\right)}{-1.2\times10^{-3}}$
$=\frac{1.8\cdot2.4}{-1.2}\times\frac{10^{-4}\cdot10^6}{10^{-3}}$
$=1.8(-2)\times10^{-4+6-(-3)}$
$=-3.6\times10^5$

59. 56.0 billion $=56,000,000,000=5.6\times10^{10}$
Bill Gates is worth $\$5.6\times10^{10}$.

61. $26.5\times10^9-13.5\times10^9=(26.5-13.5)\times10^9$
$=13\times10^9$
$=1.3\times10^{10}$
Christy Walton's worth exceeds Mark Zuckerberg's worth by $\$1.3\times10^{10}$.

63. 20 billion $=20\times10^9=2\times10^{10}$
$\frac{2\times10^{10}}{3\times10^8}=\frac{2}{3}\times\frac{10^{10}}{10^8}$
$\approx0.67\times10^{10-8}$
$=0.67\times10^2$
$=67$
The average American consumes about 67 hotdogs each year.

65. 8 billion $=8\times10^9$
$\frac{8\times10^9}{3.2\times10^7}=\frac{8}{3.2}\times\frac{10^9}{10^7}$
$=2.5\times10^{9-7}$
$=2.5\times10^2=250$
$2.5\times10^2=250$ chickens are raised for food each second in the U.S.

67. a. $\frac{519\times10^9}{48\times10^6}\approx 10.813\times10^3$

$= \$1.0813\times10^4$

$= \$10,813$

b. $\frac{\$10,813}{12}\approx \$901$

69. Medicaid: $\frac{198\times10^9}{53.4\times10^6}\approx 3.708\times10^3$

$= \$3708$

Medicare: $\frac{294\times10^9}{42.3\times10^6}\approx 6.950\times10^3$

$= \$6950$

Medicare provides a greater per person benefit by \$3242.

71. $20,000\left(5.3\times10^{-23}\right)$

$=\left(2\times10^4\right)\left(5.3\times10^{-23}\right)$

$=(2\cdot5.3)\times\left(10^4\cdot10^{-23}\right)$

$=10.6\times10^{4+(-23)}$

$=10.6\times10^{-19}$

$=1.06\times10^{-18}$

The mass of 20,000 oxygen molecules is 1.06×10^{-18} grams.

73. $\frac{365 \text{ days}}{1 \text{ year}}\cdot\frac{24 \text{ hours}}{1 \text{ day}}$

$=8760$ hours/year

$=8.76\times10^3$ hours/year

$\frac{8.76\times10^3 \text{ hours}}{1 \text{ year}}\cdot\frac{60 \text{ minutes}}{1 \text{ hour}}$

$=525.6\times10^3$ minutes/year

$=5.256\times10^5$ minutes/year

$\frac{5.256\times10^5 \text{ minutes}}{1 \text{ year}}\cdot\frac{60 \text{ seconds}}{1 \text{ minute}}$

$=315.36\times10^5$ seconds/year

$=3.1536\times10^7$ seconds/year

There are 3.1536×10^7 seconds in a year.

75. – 79. Answers will vary.

81. does not make sense; Explanations will vary. Sample explanation: That would be less than \$1 per person.

83. makes sense

85. false; Changes to make the statement true will vary. A sample change is: $534.7 = 5.347\times10^2$, not 5.347×10^3.

87. false; Changes to make the statement true will vary. A sample change is:

$\left(7\times10^5\right)+\left(2\times10^{-3}\right)=700,000+0.002$

$=700,000.002$, not $9\times10^2=900$.

89. true

91. $8.2\times10^{-16}+4.3\times10^{-16}$

$=(8.2+4.3)\times10^{-16}$

$=12.5\times10^{-16}=1.25\times10^{-15}$

93. Answers will vary.

94. $9(10x-4)-(5x-10)=90x-36-5x+10$

$=90x-5x-36+10$

$=85x-26$

95. $\frac{4x-1}{10}=\frac{5x+2}{4}-4$

$20\left(\frac{4x-1}{10}\right)=20\left(\frac{5x+2}{4}-4\right)$

$2(4x-1)=5(5x+2)-80$

$8x-2=25x+10-80$

$8x-2=25x-70$

$-2=17x-70$

$68=17x$

$4=x$

96. $\left(8x^4y^{-3}\right)^{-2}=8^{-2}\left(x^4\right)^{-2}\left(y^{-3}\right)^{-2}$

$=8^{-2}x^{-8}y^6=\frac{y^6}{64x^8}$

97. In set 1, each x-coordinate is paired with one and only one y-coordinate.

98. $r^3 - 2r^2 + 5$
$= (-5)^3 - 2(-5)^2 + 5$
$= -125 - 2(25) + 5$
$= -125 - 50 + 5$
$= -170$

99. $5x + 7 = 5(a + h) + 7$
$= 5a + 5h + 7$

Chapter 1 Review

1. $2x - 10$

2. $4 + 6x = 6x + 4$

3. $\frac{9}{x} + \frac{1}{2}x$

4. $x^2 - 7x + 4 = (10)^2 - 7(10) + 4$
$= 100 - 70 + 4$
$= 34$

5. $6 + 2(x - 8)^3 = 6 + 2(11 - 8)^3$
$= 6 + 2(3)^3$
$= 60$

6. $x^4 - (x - y) = (2)^4 - (2 - 1) = 15$

7. $\{1, 2\}$

8. $\{-3, -2, -1, 0, 1\}$

9. false; Zero is not a natural number.

10. true; -2 is a rational number.

11. true; $\frac{1}{3}$ is not an irrational number.

12. Negative five is less than two. True.

13. Negative seven is greater than or equal to negative three. False.

14. Negative seven is less than or equal to negative seven. True.

15. $F = 28 + 6x - 0.6x^2$
$F = 28 + 6(4) - 0.6(4)^2$
$= 42.4$
The model overestimates the actual value by 0.4.

16. $\{x \mid -2 < x \le 3\}$

−5 −4 −3 −2 −1 0 1 2 3 4 5

17. $\{x \mid -1.5 \le x \le 2\}$

−5 −4 −3 −2 −1 0 1 2 3 4 5

18. $\{x \mid x > -1\}$

−5 −4 −3 −2 −1 0 1 2 3 4 5

19. $|-9.7| = 9.7$

20. $|5.003| = 5.003$

21. $|0| = 0$

22. $-2.4 + (-5.2) = -7.6$

23. $-6.8 + 2.4 = -4.4$

24. $-7 - (-20) = -7 + 20 = 13$

25. $(-3)(-20) = 60$

26. $-\frac{3}{5} - \left(-\frac{1}{2}\right) = -\frac{3}{5} + \frac{1}{2}$
$= -\frac{3}{5} \cdot \frac{2}{2} + \frac{1}{2} \cdot \frac{5}{5}$
$= -\frac{6}{10} + \frac{5}{10}$
$= -\frac{1}{10}$

27. $\left(\frac{2}{7}\right)\left(-\frac{3}{10}\right) = -\frac{6}{70} = -\frac{3}{35}$

28. $4(-3)(-2)(-10) = -12(-2)(-10)$
$= -240$

29. $(-2)^4 = 16$

30. $-2^5 = -32$

31. $-\frac{2}{3} \div \frac{8}{5} = -\frac{2}{3} \cdot \frac{5}{8} = -\frac{5}{12}$

32. $\frac{-35}{-5} = 7$

33. $\frac{54.6}{-6} = -9.1$

34.
$$\begin{aligned} x &= -7 \\ -1(x) &= -1(-7) \\ -x &= 7 \end{aligned}$$

35. $-11 - [-17 + (-3)] = -11 - [-20] = 9$

36. $\left(-\frac{1}{2}\right)^3 \cdot 2^4 = -\frac{1}{8} \cdot 16 = -2$

37.
$$\begin{aligned} -3[4-(6-8)] &= -3[4-(-2)] \\ &= -3[6] = -18 \end{aligned}$$

38.
$$\begin{aligned} &8^2 - 36 \div 3^2 \cdot 4 - (-7) \\ &= 64 - 36 \div 9 \cdot 4 + 7 \\ &= 64 - 4 \cdot 4 + 7 = 64 - 16 + 7 \\ &= 48 + 7 = 55 \end{aligned}$$

39. $\frac{(-2)^4 + (-3)^2}{2^2 - (-21)} = \frac{16+9}{4-(-21)} = \frac{25}{25} = 1$

40.
$$\begin{aligned} \frac{(7-9)^3 - (-4)^2}{2 + 2(8) \div 4} &= \frac{(-2)^3 - 16}{2 + 16 \div 4} = \frac{-8-16}{2+4} \\ &= \frac{-24}{6} = -4 \end{aligned}$$

41.
$$\begin{aligned} 4 - (3-8)^2 + 3 \div 6 \cdot 4^2 &= 4 - (-5)^2 + 3 \div 6 \cdot 16 \\ &= 4 - 25 + 3 \div 6 \cdot 16 \\ &= 4 - 25 + \frac{1}{2} \cdot 16 \\ &= 4 - 25 + 8 = -13 \end{aligned}$$

42.
$$\begin{aligned} 5(2x-3) + 7x &= 10x - 15 + 7x \\ &= 17x - 15 \end{aligned}$$

43. $5x + 7x^2 - 4x + 2x^2 = x + 9x^2 = 9x^2 + x$

44.
$$\begin{aligned} 3(4y-5) - (7y+2) &= 12y - 15 - 7y - 2 \\ &= 5y - 17 \end{aligned}$$

45.
$$\begin{aligned} 8 - 2[3 - (5x-1)] &= 8 - 2[3 - 5x + 1] \\ &= 8 - 6 + 10x - 2 = 10x \end{aligned}$$

46.
$$\begin{aligned} 6(2x-3) - 5(3x-2) &= 12x - 18 - 15x + 10 \\ &= -3x - 8 \end{aligned}$$

47. – 49.

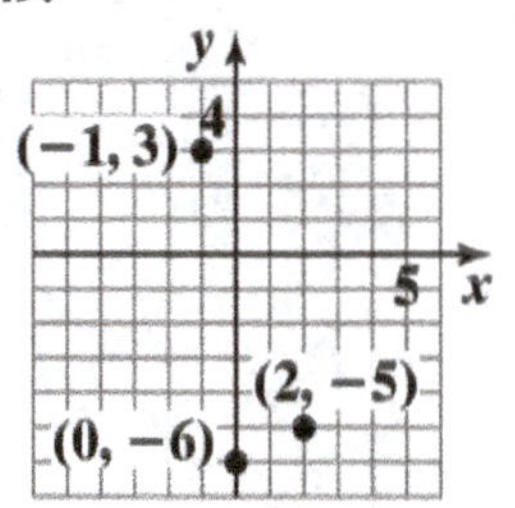

50.

x	(x, y)
−3	(−3, −8)
−2	(−2, −6)
−1	(−1, −4)
0	(0, −2)
1	(1, 0)
2	(2, 2)
3	(3, 4)

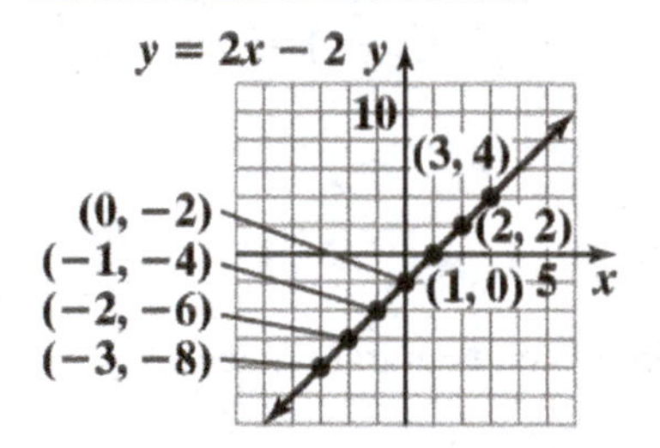

51.

x	(x, y)
−3	$(-3, 6)$
−2	$(-2, 1)$
−1	$(-1, -2)$
0	$(0, -3)$
1	$(1, -2)$
2	$(2, 1)$
3	$(3, 6)$

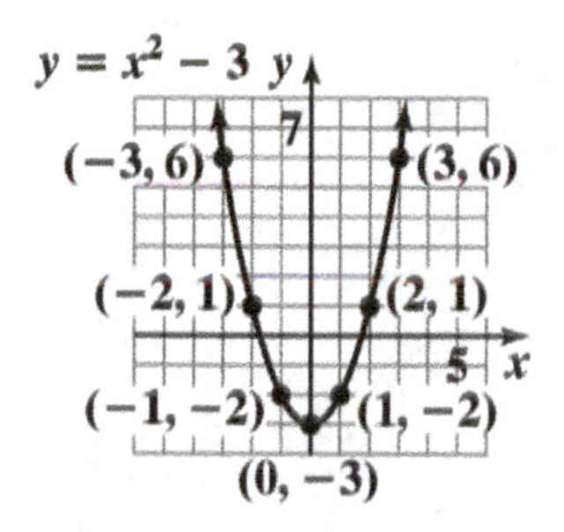

52.

x	(x, y)
−3	$(-3, -3)$
−2	$(-2, -2)$
−1	$(-1, -1)$
0	$(0, 0)$
1	$(1, 1)$
2	$(2, 2)$
3	$(3, 3)$

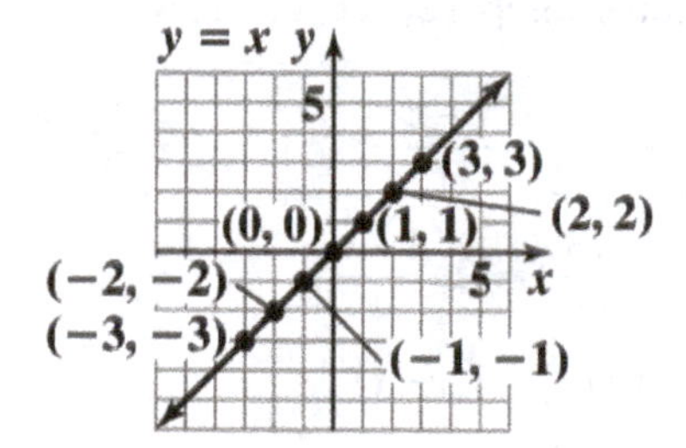

53.

x	(x, y)
−3	$(-3, 1)$
−2	$(-2, 0)$
−1	$(-1, -1)$
0	$(0, -2)$
1	$(1, -1)$
2	$(2, 0)$
3	$(3, 1)$

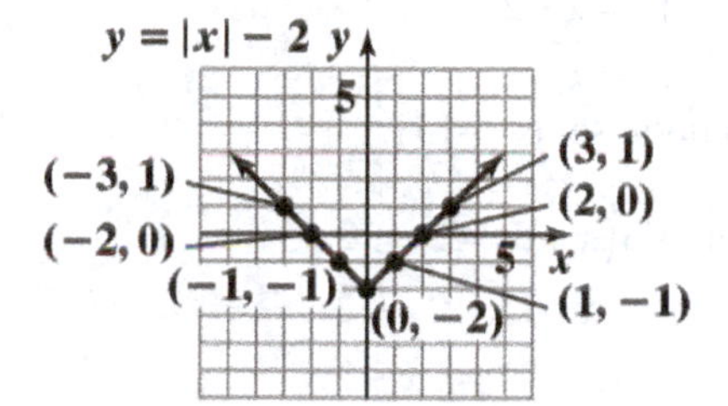

54. The minimum x-value is −20 and the maximum x-value is 40. The distance between tick marks is 10. The minimum y-value is −5 and the maximum y-value is 5. The distance between tick marks is 1.

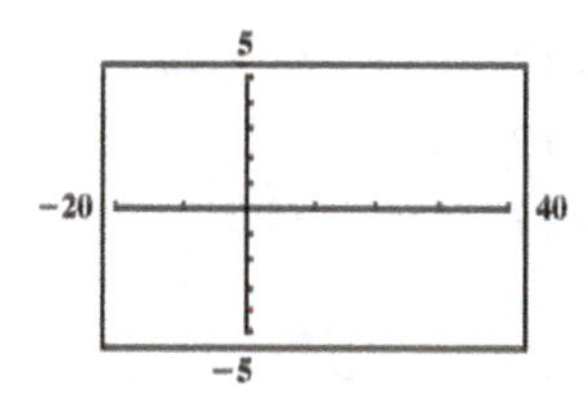

55. 20% of 75-year-old Americans have Alzheimer's.

56. Age 85 represents a 50% prevalence.

57. Answers will vary.

58. Graph c illustrates the description.

59.
$$2x - 5 = 7$$
$$2x = 12$$
$$x = 6$$
The solution set is $\{6\}$.

60.
$$5x + 20 = 3x$$
$$2x + 20 = 0$$
$$2x = -20$$
$$x = -10$$
The solution set is $\{-10\}$.

61. $7(x-4)=x+2$
$$7x-28=x+2$$
$$6x-28=2$$
$$6x=30$$
$$x=5$$
The solution set is $\{5\}$.

62. $1-2(6-x)=3x+2$
$$1-12+2x=3x+2$$
$$-11+2x=3x+2$$
$$-11=x+2$$
$$-13=x$$
The solution set is $\{-13\}$.

63. $2(x-4)+3(x+5)=2x-2$
$$2x-8+3x+15=2x-2$$
$$5x+7=2x-2$$
$$3x+7=-2$$
$$3x=-9$$
$$x=-3$$
The solution set is $\{-3\}$.

64. $2x-4(5x+1)=3x+17$
$$2x-20x-4=3x+17$$
$$-18x-4=3x+17$$
$$-4=21x+17$$
$$-21=21x$$
$$-1=x$$
The solution set is $\{-1\}$.

65. $\frac{2x}{3}=\frac{x}{6}+1$
$$6\left(\frac{2x}{3}\right)=6\left(\frac{x}{6}+1\right)$$
$$4x=x+6$$
$$3x=6$$
$$x=2$$
The solution set is $\{2\}$.

66. $\frac{x}{2}-\frac{1}{10}=\frac{x}{5}+\frac{1}{2}$
$$10\left(\frac{x}{2}-\frac{1}{10}\right)=10\left(\frac{x}{5}+\frac{1}{2}\right)$$
$$5x-1=2x+5$$
$$3x-1=5$$
$$3x=6$$
$$x=2$$
The solution set is $\{2\}$.

67. $\frac{2x}{3}=6-\frac{x}{4}$
$$12\left(\frac{2x}{3}\right)=12\left(6-\frac{x}{4}\right)$$
$$8x=72-3x$$
$$11x=72$$
$$x=\frac{72}{11}$$
The solution set is $\left\{\frac{72}{11}\right\}$.

68. $\frac{x}{4}=2+\frac{x-3}{3}$
$$12\left(\frac{x}{4}\right)=12\left(2+\frac{x-3}{3}\right)$$
$$3x=24+4(x-3)$$
$$3x=24+4x-12$$
$$3x=12+4x$$
$$-x=12$$
$$x=-12$$
The solution set is $\{-12\}$.

69. $\frac{3x+1}{3}-\frac{13}{2}=\frac{1-x}{4}$
$$12\left(\frac{3x+1}{3}-\frac{13}{2}\right)=12\left(\frac{1-x}{4}\right)$$
$$4(3x+1)-6(13)=3(1-x)$$
$$12x+4-78=3-3x$$
$$12x-74=3-3x$$
$$15x-74=3$$
$$15x=77$$
$$x=\frac{77}{15}$$
The solution set is $\left\{\frac{77}{15}\right\}$.

70. $7x+5=5(x+3)+2x$
$7x+5=5x+15+2x$
$7x+5=7x+15$
$5=15$
There is no solution. The solution set is $\varnothing$. The equation is inconsistent.

71. $7x+13=4x-10+3x+23$
$7x+13=7x+13$
The solution set is $(-\infty,\infty)$. The equation is an identity.

72. $7x+13=3x-10+2x+23$
$7x+13=5x-10+23$
$7x+13=5x+13$
$2x+13=13$
$2x=0$
$x=0$
The solution set is $\{0\}$. The equation is conditional.

73. $4(x-3)+5=x+5(x-2)$
$4x-12+5=x+5x-10$
$4x-7=6x-10$
$-2x-7=-10$
$-2x=-3$
$x=\frac{-3}{-2}=\frac{3}{2}$
The solution set is $\left\{\frac{3}{2}\right\}$. The equation is conditional.

74. $(2x-3)2-3(x+1)=(x-2)4-3(x+5)$
$4x-6-3x-3=4x-8-3x-15$
$x-9=x-23$
$-9=-23$
There is no solution. The solution set is $\varnothing$. The equation is inconsistent.

75. **a.** $T=1.4x+20$
$T=1.4(20)+20$
$=48$
According to the model, 48% of households had three or more TVs in 2005. This overestimates the actual value shown in the graph by 2%.

b. $T=1.4x+20$
$62=1.4x+20$
$42=1.4x$
$\frac{42}{1.4}=\frac{1.4x}{1.4}$
$30=x$
According to the model, 62% of households will have three or more TVs 30 years after 1985, or 2015.

76. Let $x=$ the average yearly earnings, in thousands, of marketing majors.
Let $x+19=$ the average yearly earnings, in thousands, of engineering majors.
Let $x+6=$ the average yearly earnings, in thousands, of accounting majors.
$x+(x+19)+(x+6)=196$
$x+x+19+x+6=196$
$3x+25=196$
$3x=171$
$x=57$
$x+19=76$
$x+6=63$
The average yearly earnings for marketing majors, engineering majors, and accounting majors were \$57 thousand, \$76 thousand, and \$63 thousand, respectively.

77. Let $x=$ the measure of the second angle.
$x+10$ = the measure of the first angle.
$2[x+(x+10)]$ = the measure of the 3rd angle.
$x+(x+10)+2[x+(x+10)]=180$
$x+x+10+2x+2x+20=180$
$6x+30=180$
$6x=150$
$x=25$
$x+10=25+10=35$
$2[x+(x+10)]=2[25+35]$
$=2(60)=120$
The angles measure $25°$, $35°$, and $120°$

78. **a.** Let $x =$ the number of years after 2004.

$$575+43x=1177$$
$$43x=602$$
$$x=14$$

The system's income will be \$1177 billion 14 years after 2004, or 2018.

b. 2018 is 14 years after 2004.

$$B=0.07x^2+47.4x+500$$
$$=0.07(14)^2+47.4(14)+500$$
$$\approx 1177$$

The amount paid in benefits for 2018 will be \$1177 billion.

c. In 2018 the \$1177 billion paid in benefits is represented by the point $(2018, 1177)$.

79. Let $x =$ the number of text messages.

Plan A: $C=15+0.05x$

Plan B: $C=5+0.07x$

Set the costs equal to each other.

$$15+0.05x=5+0.07x$$
$$15=5+0.02x$$
$$10=0.02x$$
$$500=x$$

The cost will be the same for 500 text messages.

80. Let $x =$ the original price of the phone.

$$48=x-0.20x$$
$$48=0.80x$$
$$60=x$$

The original price is \$60.

81. Let $x =$ the amount sold to earn \$800 in one week.

$$800=300+0.05x$$
$$500=0.05x$$
$$10{,}000=x$$

Sales must be \$10,000 in one week to earn \$800.

82. Let $w =$ the width of the playing field.

Let $3w - 6 =$ the length of the playing field.

$$P=2(\text{length})+2(\text{width})$$
$$340=2(3w-6)+2w$$
$$340=6w-12+2w$$
$$340=8w-12$$
$$352=8w$$
$$44=w$$

The dimensions are 44 yards by 126 yards.

83. **a.** Let $x =$ the number of years (after 2005).

College A's enrollment: $14{,}100+1500x$

College B's enrollment: $41{,}700-800x$

$$14{,}100+1500x=41{,}700-800x$$

b. Check points to determine that $y_1=14{,}100+1500x$ and $y_2=41{,}700-800x$. Since $y_1=y_2=32{,}100$ when $x=12$, the two colleges will have the same enrollment in the year $2005+12=2017$. That year the enrollments will be 32,100 students.

84.

$$V=\frac{1}{3}Bh$$
$$3V=Bh$$
$$h=\frac{3V}{B}$$

85.

$$y-y_1=m(x-x_1)$$
$$\frac{y-y_1}{m}=x-x_1$$
$$x=\frac{y-y_1}{m}+x_1$$

or

$$x=\frac{y-y_1+mx_1}{m}$$

86.

$$E=I(R+r)$$
$$\frac{E}{I}=R+r$$
$$R=\frac{E}{I}-r \quad \text{or} \quad R=\frac{E-Ir}{I}$$

87.

$$C=\frac{5F-160}{9}$$
$$9C=5F-160$$
$$9C+160=5F$$
$$F=\frac{9C+160}{5} \quad \text{or} \quad F=\frac{9}{5}C+32$$

88.

$$s=vt+gt^2$$
$$s-vt=gt^2$$
$$g=\frac{s-vt}{t^2}$$

89. $T = gr + gvt$
$T = g(r + vt)$
$g = \frac{T}{r + vt}$

90. $(-3x^7)(-5x^6) = 15x^{7+6} = 15x^{13}$

91. $x^2y^{-5} = \frac{x^2}{y^5}$

92. $\frac{3^{-2}x^4}{y^{-7}} = \frac{x^4y^7}{3^2} = \frac{x^4y^7}{9}$

93. $(x^3)^{-6} = x^{3\cdot(-6)} = x^{-18} = \frac{1}{x^{18}}$

94. $(7x^3y)^2 = 7^2x^{3\cdot 2}y^{1\cdot 2} = 49x^6y^2$

95. $\frac{16y^3}{-2y^{10}} = -8y^{3-10} = -8y^{-7} = -\frac{8}{y^7}$

96. $(-3x^4)(4x^{-11}) = -12x^{-7} = -\frac{12}{x^7}$

97. $\frac{12x^7}{4x^{-3}} = 3x^{7-(-3)} = 3x^{10}$

98. $\frac{-10a^5b^6}{20a^{-3}b^{11}} = \frac{-1}{2}a^{5-(-3)}b^{6-11}$
$= \frac{-1}{2}a^8b^{-5} = -\frac{a^8}{2b^5}$

99. $(-3xy^4)(2x^2)^3 = (-3xy^4)(8x^6)$
$= -24x^{1+6}y^4 = -24x^7y^4$

100. $2^{-2} + \frac{1}{2}x^0 = \frac{1}{2^2} + \frac{1}{2}\cdot 1 = \frac{1}{4} + \frac{1}{2} = \frac{3}{4}$

101. $(5x^2y^{-4})^{-3} = \left(\frac{5x^2}{y^4}\right)^{-3} = \left(\frac{y^4}{5x^2}\right)^3 = \frac{y^{12}}{125x^6}$

102. $(3x^4y^{-2})(-2x^5y^{-3}) = \left(\frac{3x^4}{y^2}\right)\left(\frac{-2x^5}{y^3}\right) = -\frac{6x^9}{y^5}$

103. $\left(\frac{3xy^3}{5x^{-3}y^{-4}}\right)^2 = \left(\frac{3x^{1-(-3)}y^{3-(-4)}}{5}\right)^2$
$= \left(\frac{3x^4y^7}{5}\right)^2$
$= \frac{3^2x^{4\cdot 2}y^{7\cdot 2}}{5^2} = \frac{9x^8y^{14}}{25}$

104. $\left(\frac{-20x^{-2}y^3}{10x^5y^{-6}}\right)^{-3} = \left(-2x^{-2-5}y^{3-(-6)}\right)^{-3}$
$= \left(-2x^{-7}y^9\right)^{-3}$
$= (-2)^{-3}x^{(-7)(-3)}y^{9(-3)}$
$= \frac{x^{21}y^{-27}}{(-2)^3}$
$= \frac{x^{21}}{-8y^{27}} = -\frac{x^{21}}{8y^{27}}$

105. $7.16\times 10^6 = 7{,}160{,}000$

106. $1.07\times 10^{-4} = 0.000107$

107. $-41{,}000{,}000{,}000{,}000 = -4.1\times 10^{13}$

108. $0.00809 = 8.09\times 10^{-3}$

109. $(4.2\times 10^{13})(3\times 10^{-6}) = 12.6\times 10^{13+(-6)}$
$= 12.6\times 10^7$
$= 1.26\times 10^8$

110. $\frac{5\times 10^{-6}}{20\times 10^{-8}} = 0.25\times 10^{-6-(-8)}$
$= 0.25\times 10^2 = 2.5\times 10^1$

111. $180(3.2\times 10^4)(5\times 10^6)$
$= (180\times 3.2\times 5)\times(10^4\times 10^6)$
$= 2880\times 10^{10}$
$= 2.880\times 10^3\times 10^{10}$
$= 2.88\times 10^{13}$
The approximate number of red blood cells in the human body of a 180-pound person is 2.88×10^{13}.

Chapter 1 Test

1. $4x-5$

2. $8+2(x-7)^4 = 8+2(10-7)^4$
$= 8+2(3)^4$
$= 8+2(81)$
$= 8+162$
$= 170$

3. $\{-4, -3, -2, -1\}$

4. true; $\frac{1}{4}$ is not a natural number.

5. Negative three is greater than negative one: false

6. $\{x \mid -3 \le x < 2\}$

7. $\{x \mid x \le -1\}$

8. $P = -0.5x^2 + 0.1x + 26.9$
$P = -0.5(2)^2 + 0.1(2) + 26.9$
$= 25.1$
The model estimates that 25.1% of Americans in Group 2 had contact with a police officer. This underestimates the actual number shown in the bar graph by 1.9.

9. $|-17.9| = 17.9$

10. $-10.8 + 3.2 = -7.6$

11. $-\frac{1}{4} - \left(-\frac{1}{2}\right) = -\frac{1}{4} + \frac{1}{2} = -\frac{1}{4} + \frac{2}{4} = \frac{1}{4}$

12. $2(-3)(-1)(-10) = -60$

13. $-\frac{1}{4}\left(-\frac{1}{2}\right) = \frac{1}{8}$

14. $\frac{-27.9}{-9} = 3.1$

15. $24 - 36 \div 4 \cdot 3 = 24 - 9 \cdot 3 = 24 - 27 = -3$

16. $\left(5^2 - 2^4\right) + [9 \div (-3)] = (25-16) + [-3]$
$= (9) + [-3] = 6$

17. $\frac{(8-10)^3 - (-4)^2}{2 + 8(2) \div 4} = \frac{(-2)^3 - 16}{2 + 16 \div 4}$
$= \frac{-8-16}{2+4} = \frac{-24}{6} = -4$

18. $7x - 4(3x+2) - 10 = 7x - 12x - 8 - 10$
$= -5x - 18$

19. $5(2y-6) - (4y-3) = 10y - 30 - 4y + 3$
$= 6y - 27$

20. $9x - [10 - 4(2x-3)]$
$= 9x - [10 - 8x + 12]$
$= 9x - 10 + 8x - 12 = 17x - 22$

21.
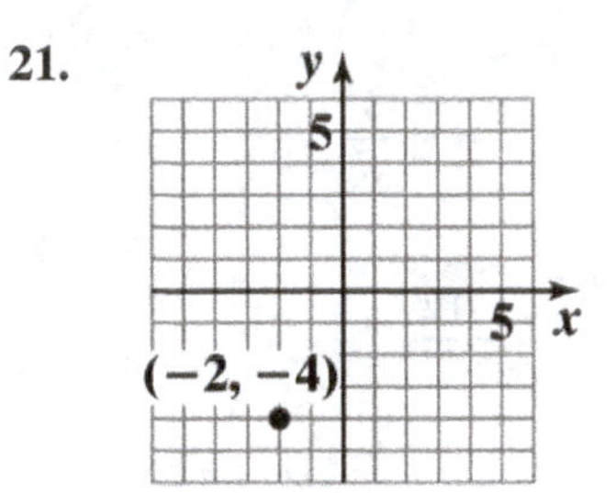

22.

x	(x, y)
−3	(−3,5)
−2	(−2,0)
−1	(−1,−3)
0	(0,−4)
1	(1,−3)
2	(2,0)
3	(3,5)

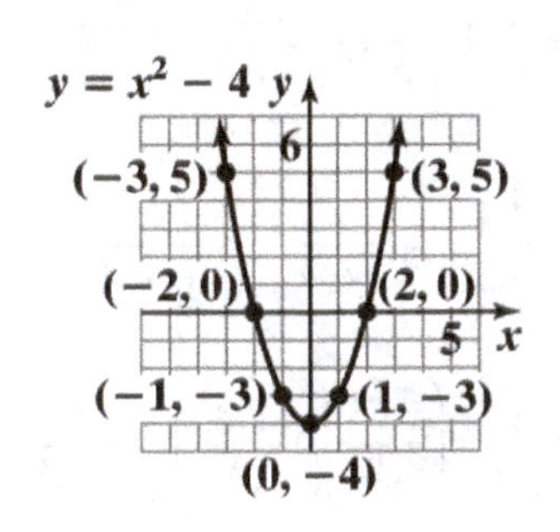

23. $3(2x-4)=9-3(x+1)$
$$6x-12=9-3x-3$$
$$6x-12=6-3x$$
$$9x-12=6$$
$$9x=18$$
$$x=2$$
The solution set is $\{2\}$.

24. $\frac{2x-3}{4}=\frac{x-4}{2}-\frac{x+1}{4}$
$$4\left(\frac{2x-3}{4}\right)=4\left(\frac{x-4}{2}-\frac{x+1}{4}\right)$$
$$2x-3=2(x-4)-(x+1)$$
$$2x-3=2x-8-x-1$$
$$2x-3=x-9$$
$$x-3=-9$$
$$x=-6$$
The solution set is $\{-6\}$.

25. $3(x-4)+x=2(6+2x)$
$$3x-12+x=12+4x$$
$$4x-12=12+4x$$
$$-12=12$$
There is no solution. The solution set is $\{\ \}$ or $\varnothing$.
The equation is inconsistent.

26. Let x = the first number.
Let $2x + 3$ = the second number.
$$x+2x+3=72$$
$$3x+3=72$$
$$3x=69$$
$$x=23$$
Find the second number.
$$2x+3=2(23)+3=46+3=49$$
The first number is 23 and the second number is 49.

27. Let x = the number of years since the car was purchased.
$$\text{Value}=\$13,805-\$1820x$$
$$4705=13,805-1820x$$
$$-9100=-1820x$$
$$5=x$$
The car will have a value of \$4705 in 5 years.

28. Let x = the number of prints.
Photo Shop A: $0.11x+1.60$
Photo Shop B: $0.13x+1.20$
$$0.13x+1.20=0.11x+1.60$$
$$0.02x+1.20=1.60$$
$$0.02x=0.40$$
$$x=20$$
The cost will be the same for 20 prints. That common price is $0.11(20)+1.60=0.13(20)+1.20$
$=\$3.80$

29. Let x = the original selling price.
$$20=x-0.60x$$
$$20=0.40x$$
$$50=x$$
The original price is \$50.

30. Let x = the width of the playing field.
Let $x + 260$ = the length of the playing field.
$$P=2(\text{length})+2(\text{width})$$
$$1000=2(x+260)+2x$$
$$1000=2x+520+2x$$
$$1000=4x+520$$
$$480=4x$$
$$x=120$$
The dimensions of the playing field are 120 yards by 380 yards.

31. $V=\frac{1}{3}lwh$
$$3V=lwh$$
$$h=\frac{3V}{lw}$$

32. $Ax+By=C$
$$By=C-Ax$$
$$y=\frac{C-Ax}{B}$$

33. $(-2x^5)(7x^{-10})=-14x^{5+(-10)}=-14x^{-5}=-\frac{14}{x^5}$

34. $(-8x^{-5}y^{-3})(-5x^2y^{-5})=40x^{-5+2}y^{-3+(-5)}$
$$=40x^{-3}y^{-8}$$
$$=\frac{40}{x^3y^8}$$

35. $\dfrac{-10x^4y^3}{-40x^{-2}y^6} = \dfrac{1}{4}x^{4-(-2)}y^{3-6} = \dfrac{1}{4}x^6y^{-3} = \dfrac{x^6}{4y^3}$

36. $\left(4x^{-5}y^2\right)^{-3} = \left(\dfrac{4y^2}{x^5}\right)^{-3} = \left(\dfrac{x^5}{4y^2}\right)^3 = \dfrac{x^{15}}{64y^6}$

37. $\left(\dfrac{-6x^{-5}y}{2x^3y^{-4}}\right)^{-2} = \left(-3x^{-5-3}y^{1-(-4)}\right)^{-2}$

$= \left(-3x^{-8}y^5\right)^{-2}$

$= (-3)^{-2}x^{(-8)(-2)}y^{5(-2)}$

$= \dfrac{x^{16}y^{-10}}{(-3)^2}$

$= \dfrac{x^{16}}{9y^{10}}$

38. $3.8\times10^{-6} = 0.0000038$

39. $407{,}000{,}000{,}000 = 4.07\times10^{11}$

40. $\dfrac{4\times10^{-3}}{8\times10^{-7}} = 0.5\times10^{-3-(-7)} = 0.5\times10^4 = 5\times10^3$

41. $2\left(6.9\times10^9\right) = 13.8\times10^9 = 1.38\times10^{10}$

The population will be 1.38×10^{10}.

Chapter 2
Functions and Linear Functions

2.1 Check Points

1. The domain is {0, 10, 20, 30, 38}.
   The range is {9.1, 6.7, 10.7, 13.2, 19.6}.

2. **a.** The relation is not a function because an element, 5, in the domain corresponds to two elements in the range.

   b. The relation is a function.

3. **a.** $f(x)=4x+5$
   $f(6)=4(6)+5$
   $f(6)=29$

   b. $g(x)=3x^2-10$
   $g(-5)=3(-5)^2-10$
   $g(-5)=65$

   c. $h(r)=r^2-7r+2$
   $h(-4)=(-4)^2-7(-4)+2$
   $h(-4)=46$

   d. $F(x)=6x+9$
   $F(a+h)=6(a+h)+9$
   $F(a+h)=6a+6h+9$

4. **a.** Every element in the domain corresponds to exactly one element in the range.

   b. The domain is {0, 1, 2, 3, 4}.
   The range is {3, 0, 1, 2}.

   c. $g(1)=0$

   d. $g(3)=2$

   e. $x=0$ and $x=4$.

2.1 Concept and Vocabulary Check

1. relation; domain; range

2. function

3. f, x

4. r, -2

2.1 Exercise Set

1. The relation is a function.
   The domain is {1, 3, 5}.
   The range is {2, 4, 5}.

3. The relation is not a function.
   The domain is {3, 4}.
   The range is {4, 5}.

5. The relation is a function.
   The domain is {-3, -2, -1, 0}.
   The range is {-3, -2, -1, 0}.

7. The relation is not a function.
   The domain is {1}.
   The range is {4, 5, 6}.

9. **a.** $f(0)=0+1=1$

   b. $f(5)=5+1=6$

   c. $f(-8)=-8+1=-7$

   d. $f(2a)=2a+1$

   e. $f(a+2)=(a+2)+1$
   $=a+2+1=a+3$

11. **a.** $g(0)=3(0)-2=0-2=-2$

   b. $g(-5)=3(-5)-2$
   $=-15-2=-17$

   c. $g\left(\frac{2}{3}\right)=3\left(\frac{2}{3}\right)-2=2-2=0$

   d. $g(4b)=3(4b)-2=12b-2$

   e. $g(b+4)=3(b+4)-2$
   $=3b+12-2=3b+10$

13. a. $h(0) = 3(0)^2 + 5 = 3(0) + 5$
$= 0 + 5 = 5$

b. $h(-1) = 3(-1)^2 + 5 = 3(1) + 5$
$= 3 + 5 = 8$

c. $h(4) = 3(4)^2 + 5 = 3(16) + 5$
$= 48 + 5 = 53$

d. $h(-3) = 3(-3)^2 + 5 = 3(9) + 5$
$= 27 + 5 = 32$

e. $h(4b) = 3(4b)^2 + 5 = 3\left(16b^2\right) + 5$
$= 48b^2 + 5$

15. a. $f(0) = 2(0)^2 + 3(0) - 1$
$= 0 + 0 - 1 = -1$

b. $f(3) = 2(3)^2 + 3(3) - 1$
$= 2(9) + 9 - 1$
$= 18 + 9 - 1 = 26$

c. $f(-4) = 2(-4)^2 + 3(-4) - 1$
$= 2(16) - 12 - 1$
$= 32 - 12 - 1 = 19$

d. $f(b) = 2(b)^2 + 3(b) - 1$
$= 2b^2 + 3b - 1$

e. $f(5a) = 2(5a)^2 + 3(5a) - 1$
$= 2\left(25a^2\right) + 15a - 1$
$= 50a^2 + 15a - 1$

17. a. $f(0) = (-0)^3 - (0)^2 - (0) + 7$
$= 7$

b. $f(2) = (-2)^3 - (2)^2 - (2) + 7$
$= -7$

c. $f(-2) = (-(-2))^3 - (-2)^2 - (-2) + 7$
$= 13$

d. $f(1) + f(-1) = \left[(-1)^3 - (1)^2 - (1) + 7\right] + \left[(-(-1))^3 - (-1)^2 - (-1) + 7\right]$
$= 4 + 8$
$= 12$

19. a. $f(0)=\dfrac{2(0)-3}{(0)-4}=\dfrac{0-3}{0-4}$
$=\dfrac{-3}{-4}=\dfrac{3}{4}$

b. $f(3)=\dfrac{2(3)-3}{(3)-4}=\dfrac{6-3}{3-4}$
$=\dfrac{3}{-1}=-3$

c. $f(-4)=\dfrac{2(-4)-3}{(-4)-4}=\dfrac{-8-3}{-8}$
$=\dfrac{-11}{-8}=\dfrac{11}{8}$

d. $f(-5)=\dfrac{2(-5)-3}{(-5)-4}=\dfrac{-10-3}{-9}$
$=\dfrac{-13}{-9}=\dfrac{13}{9}$

e. $f(a+h)=\dfrac{2(a+h)-3}{(a+h)-4}$
$=\dfrac{2a+2h-3}{a+h-4}$

f. Four must be excluded from the domain, because four would make the denominator zero. Division by zero is undefined.

21. a. $f(-2)=6$

b. $f(2)=12$

c. $x=0$

23. a. $h(-2)=2$

b. $h(1)=1$

c. $x=-1$ and $x=1$

25. $g(1)=3(1)-5=3-5=-2$
$f(g(1))=f(-2)=(-2)^2-(-2)+4$
$=4+2+4=10$

27. $\sqrt{3-(-1)}-(-6)^2+6\div-6\cdot 4$
$=\sqrt{3+1}-36+-1\cdot 4$
$=\sqrt{4}-36+-4=2-36-4=-38$

29. $f(-x)-f(x)$
$=(-x)^3+(-x)-5-\left[x^3+x-5\right]$
$=-x^3-x-5-x^3-x+5$
$=-2x^3-2x$

31. a. $f(-2)=3(-2)+5=-6+5=-1$

b. $f(0)=4(0)+7=0+7=7$

c. $f(3)=4(3)+7=12+7=19$

d. $f(-100)+f(100)$
$=3(-100)+5+4(100)+7$
$=-300+5+400+7=112$

33. a. {(Iceland, 9.7), (Finland, 9.6), (New Zealand, 9.6), (Denmark, 9.5)}

b. Yes, the relation is a function because each country in the domain corresponds to exactly one corruption rating in the range.

c. {(9.7, Iceland), (9.6, Finland), (9.6, New Zealand), (9.5, Denmark)}

d. No, the relation is not a function because 9.6 in the domain corresponds to two countries in the range, Finland and New Zealand.

35. – 37. Answers will vary.

39. makes sense

41. makes sense

43. false; Changes to make the statement true will vary. A sample change is: All functions are relations.

45. true

47. true

49. $f(a+h)=3(a+h)+7=3a+3h+7$
$f(a)=3a+7$
$\dfrac{f(a+h)-f(a)}{h}$
$=\dfrac{(3a+3h+7)-(3a+7)}{h}$
$=\dfrac{3a+3h+7-3a-7}{h}=\dfrac{3h}{h}=3$

51. It is given that $f(x+y) = f(x) + f(y)$ and $f(1) = 3$.
To find $f(2)$, rewrite 2 as 1 + 1.
$f(2) = f(1+1) = f(1) + f(1)$
$= 3 + 3 = 6$
Similarly:
$f(3) = f(2+1) = f(2) + f(1)$
$= 6 + 3 = 9$
$f(4) = f(3+1) = f(3) + f(1)$
$= 9 + 3 = 12$
While $f(x+y) = f(x) + f(y)$ is true for this function, it is not true for all functions. It is not true for $f(x) = x^2$, for example.

52. $24 \div 4\left[2-(5-2)\right]^2 - 6$
$= 24 \div 4\left[2-(3)\right]^2 - 6$
$= 24 \div 4(-1)^2 - 6$
$= 24 \div 4(1) - 6$
$= 6(1) - 6 = 6 - 6 = 0$

53. $\left(\dfrac{3x^2y^{-2}}{y^3}\right)^{-2} = \left(\dfrac{3x^2}{y^5}\right)^{-2} = \left(\dfrac{y^5}{3x^2}\right)^2 = \dfrac{y^{10}}{9x^4}$

54. $\dfrac{x}{3} = \dfrac{3x}{5} + 4$
$15\left(\dfrac{x}{3}\right) = 15\left(\dfrac{3x}{5} + 4\right)$
$15\left(\dfrac{x}{3}\right) = 15\left(\dfrac{3x}{5}\right) + 15(4)$
$5x = 3(3x) + 60$
$5x = 9x + 60$
$5x - 9x = 9x - 9x + 60$
$-4x = 60$
$\dfrac{-4x}{-4} = \dfrac{60}{-4}$
$x = -15$
The solution set is $\{-15\}$.

55.

x	$f(x) = 2x$	(x, y)
−2	$f(-2) = 2(-2) = -4$	$(-2,-4)$
−1	$f(-1) = 2(-1) = -2$	$(-1,-2)$
0	$f(0) = 2(0) = 0$	$(0,0)$
1	$f(1) = 2(1) = 2$	$(1,2)$
2	$f(2) = 2(2) = 4$	$(2,4)$

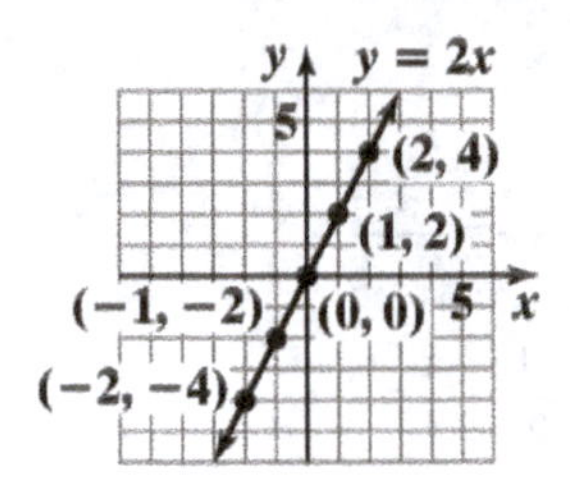

56.

x	$f(x) = 2x+4$	(x, y)
−2	$f(-2) = 2(-2)+4 = 0$	$(-2,0)$
−1	$f(-1) = 2(-1)+4 = 2$	$(-1,2)$
0	$f(0) = 2(0)+4 = 4$	$(0,4)$
1	$f(1) = 2(1)+4 = 6$	$(1,6)$
2	$f(2) = 2(2)+4 = 8$	$(2,8)$

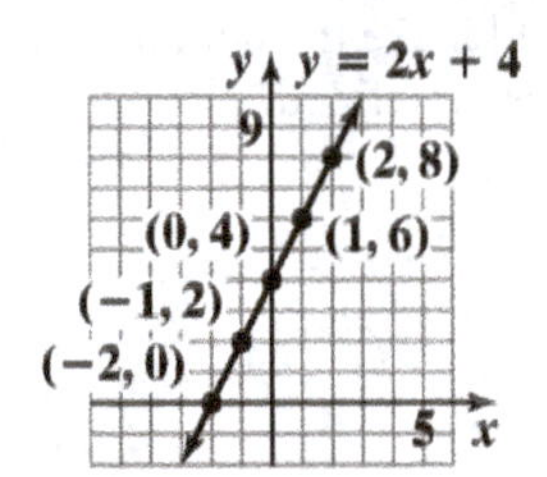

57. **a.** When the x-coordinate is 2, the y-coordinate is 3.

b. When the y-coordinate is 4, the x-coordinates are −3 and 3.

c. $(-\infty, \infty)$

d. $[1, \infty)$

2.2 Check Points

1. $f(x) = 2x$

x	$f(x) = 2x$	(x, y)
-2	$f(-2) = 2(-2) = -4$	$(-2,-4)$
-1	$f(-1) = 2(-1) = -2$	$(-1,-2)$
0	$f(0) = 2(0) = 0$	$(0,0)$
1	$f(1) = 2(1) = 2$	$(1,2)$
2	$f(2) = 2(2) = 4$	$(2,4)$

$g(x) = 2x-3$

x	$g(x) = 2x-3$	(x, y)
-2	$g(-2) = 2(-2)-3 = -7$	$(-2,-7)$
-1	$g(-1) = 2(-1)-3 = -5$	$(-1,-5)$
0	$g(0) = 2(0)-3 = -3$	$(0,-3)$
1	$g(1) = 2(1)-3 = -1$	$(1,-1)$
2	$g(2) = 2(2)-3 = 1$	$(2,1)$

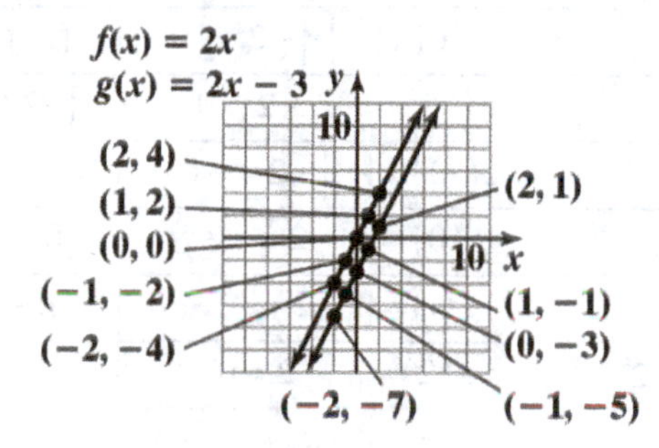

The graph of g is the graph of f shifted down by 3 units.

2. **a.** The graph represents a function. It passes the vertical line test.

   b. The graph represents a function. It passes the vertical line test.

   c. The graph does not represent a function. It fails the vertical line test.

3. **a.** $f(5) = 400$

   b. When x is 9, the function's value is 100. i.e. $f(9) = 100$

   c. The minimum T cell count during the asymptomatic stage is approximately 425.

4. **a.** The domain is $[-2,1]$.
   The range is $[0,3]$.

   b. The domain is $(-2,1]$.
   The range is $[-1,2)$.

   c. The domain is $[-3,0)$.
   The range is $\{-3,-2,-1\}$.

2.2 Concept and Vocabulary Check

1. ordered pairs
2. more than once; function
3. $[1,3)$; domain
4. $[1,\infty)$; range

2.2 Exercise Set

1.

x	$f(x) = x$	(x, y)
-2	$f(-2) = -2$	$(-2,-2)$
-1	$f(-1) = -1$	$(-1,-1)$
0	$f(0) = 0$	$(0,0)$
1	$f(1) = 1$	$(1,1)$
2	$f(2) = 2$	$(2,2)$

x	$g(x) = x+3$	(x, y)
-2	$g(-2) = -2+3 = 1$	$(-2,1)$
-1	$g(-1) = -1+3 = 2$	$(-1,2)$
0	$g(0) = 0+3 = 3$	$(0,3)$
1	$g(1) = 1+3 = 4$	$(1,4)$
2	$g(2) = 2+3 = 5$	$(2,5)$

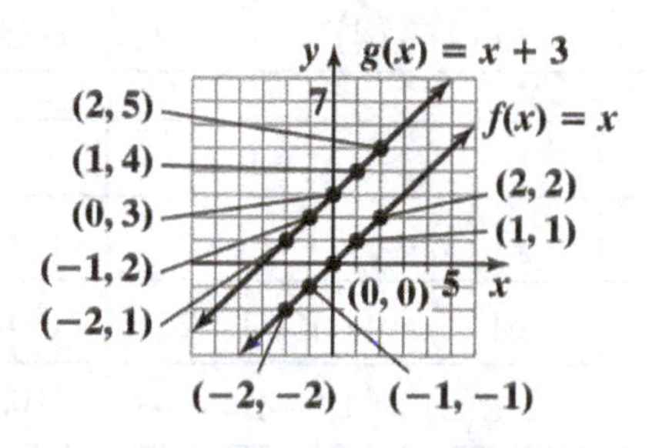

The graph of g is the graph of f shifted up 3 units.

3.

x	$f(x) = -2x$	(x, y)
−2	$f(-2) = -2(-2) = 4$	$(-2, 4)$
−1	$f(-1) = -2(-1) = 2$	$(-1, 2)$
0	$f(0) = -2(0) = 0$	$(0, 0)$
1	$f(1) = -2(1) = -2$	$(1, -2)$
2	$f(2) = -2(2) = -4$	$(2, -4)$

x	$g(x) = -2x - 1$	(x, y)
−2	$g(-2) = -2(-2) - 1 = 3$	$(-2, 3)$
−1	$g(-1) = -2(-1) - 1 = 1$	$(-1, 1)$
0	$g(0) = -2(0) - 1 = -1$	$(0, -1)$
1	$g(1) = -2(1) - 1 = -3$	$(1, -3)$
2	$g(2) = -2(2) - 1 = -5$	$(2, -5)$

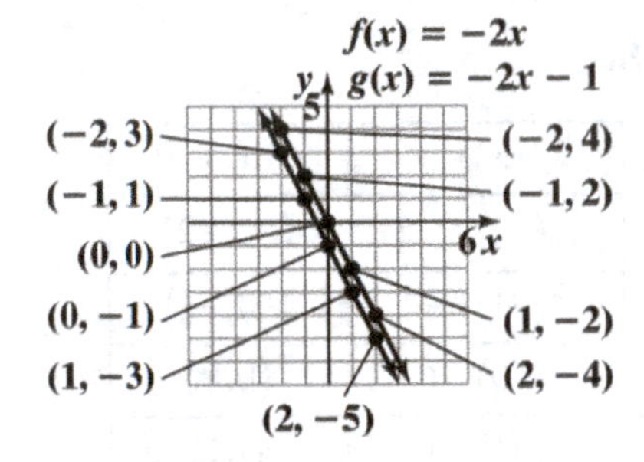

The graph of g is the graph of f shifted down 1 unit.

5.

x	$f(x) = x^2$	(x, y)
-2	$f(-2) = (-2)^2 = 4$	$(-2, 4)$
-1	$f(-1) = (-1)^2 = 1$	$(-1, 1)$
0	$f(0) = (0)^2 = 0$	$(0, 0)$
1	$f(1) = (1)^2 = 1$	$(1, 1)$
2	$f(2) = (2)^2 = 4$	$(2, 4)$

x	$g(x) = x^2 + 1$	(x, y)
-2	$g(-2) = (-2)^2 + 1 = 5$	$(-2, 5)$
-1	$g(-1) = (-1)^2 + 1 = 2$	$(-1, 2)$
0	$g(0) = (0)^2 + 1 = 1$	$(0, 1)$
1	$g(1) = (1)^2 + 1 = 2$	$(1, 2)$
2	$g(2) = (2)^2 + 1 = 5$	$(2, 5)$

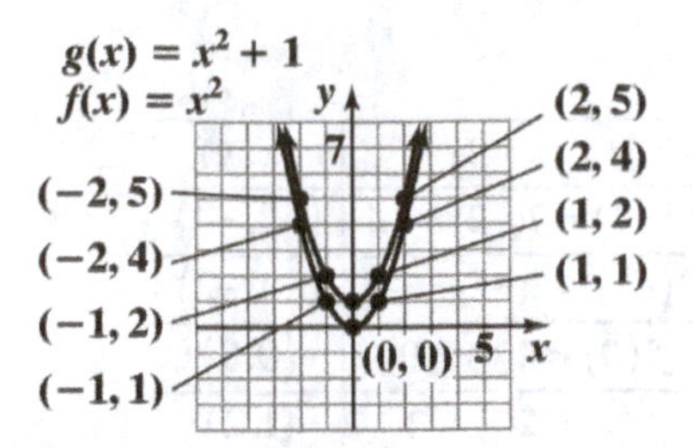

The graph of g is the graph of f shifted up 1 unit.

7.

x	$f(x) = \|x\|$	(x, y)
−2	$f(-2) = \|-2\| = 2$	$(-2, 2)$
−1	$f(-1) = \|-1\| = 1$	$(-1, 1)$
0	$f(0) = \|0\| = 0$	$(0, 0)$
1	$f(1) = \|1\| = 1$	$(1, 1)$
2	$f(2) = \|2\| = 2$	$(2, 2)$

x	$g(x) = \|x\| - 2$	(x, y)
−2	$g(-2) = \|-2\| - 2 = 0$	$(-2, 0)$
−1	$g(-1) = \|-1\| - 2 = -1$	$(-1, -1)$
0	$g(0) = \|0\| - 2 = -2$	$(0, -2)$
1	$g(1) = \|1\| - 2 = -1$	$(1, -1)$
2	$g(2) = \|2\| - 2 = 0$	$(2, 0)$

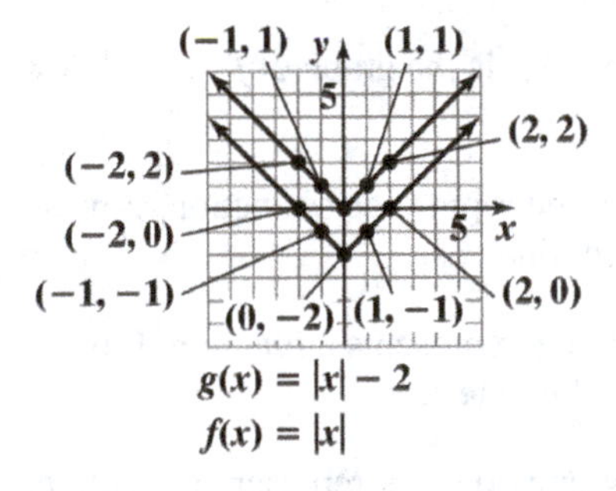

The graph of g is the graph of f shifted down 2 units.

9.

x	$f(x)=x^3$	(x,y)
-2	$f(-2)=(-2)^3=-8$	$(-2,-8)$
-1	$f(-1)=(-1)^3=-1$	$(-1,-1)$
0	$f(0)=(0)^3=0$	$(0,0)$
1	$f(1)=(1)^3=1$	$(1,1)$
2	$f(2)=(2)^3=8$	$(2,8)$

x	$g(x)=x^3+2$	(x,y)
-2	$g(-2)=(-2)^3+2=-6$	$(-2,-6)$
-1	$g(-1)=(-1)^3+2=1$	$(-1,1)$
0	$g(0)=(0)^3+2=2$	$(0,2)$
1	$g(1)=(1)^3+2=3$	$(1,3)$
2	$g(2)=(2)^3+2=10$	$(2,10)$

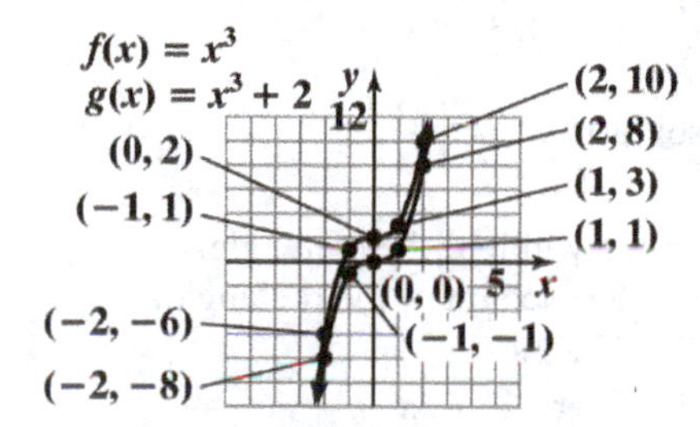

The graph of g is the graph of f shifted up 2 units.

11. The graph represents a function. It passes the vertical line test.

13. The graph does not represent a function. It fails the vertical line test.

15. The graph represents a function. It passes the vertical line test.

17. The graph does not represent a function. It fails the vertical line test.

19. $f(-2)=-4$

21. $f(4)=4$

23. $f(-3)=0$

25. $g(-4)=2$

27. $g(-10)=2$

29. When $x=-2$, $g(x)=1$.

31. The domain is $[0,5)$.
The range is $[-1,5)$.

33. The domain is $[0,\infty)$.
The range is $[1,\infty)$.

35. The domain is $[-2,6]$.
The range is $[-2,6]$.

37. The domain is $(-\infty,\infty)$.
The range is $(-\infty,-2]$.

39. The domain is $\{-5,-2,0,1,3\}$.
The range is $\{2\}$.

41. **a.** The domain is $(-\infty,\infty)$.

b. The range is $[-4,\infty)$.

c. $f(-3)=4$

d. 2 and 6; i.e. $f(2)=f(6)=-2$

e. f crosses the x-axis at $(1,0)$ and $(7,0)$.

f. f crosses the y-axis at $(0,4)$.

g. $f(x)<0$ on the interval $(1,7)$.

h. $f(-8)$ is positive.

43. **a.** $G(30)=-0.01(30)^2+(30)+60=81$
In 2010, the wage gap was 81%. This is represented as $(30,81)$ on the graph.

b. $G(30)$ underestimates the actual data shown by the bar graph by 2%.

45.
$$\begin{aligned} f(20) &= 0.4(20)^2-36(20)+1000 \\ &= 0.4(400)-720+1000 \\ &= 160-720+1000 \\ &= -560+1000=440 \end{aligned}$$
Twenty-year-old drivers have 440 accidents per 50 million miles driven.
This is represented on the graph by point (20,440).

47. The graph reaches its lowest point at $x = 45$.

$$\begin{aligned} f(45) &= 0.4(45)^2 - 36(45) + 1000 \\ &= 0.4(2025) - 1620 + 1000 \\ &= 810 - 1620 + 1000 \\ &= -810 + 1000 \\ &= 190 \end{aligned}$$

Drivers at age 45 have 190 accidents per 50 million miles driven. This is the least number of accidents for any driver between ages 16 and 74.

49. $f(3) = 0.78$

The cost of mailing a first-class letter weighing 3 ounces is $0.78.

51. The cost to mail a letter weighing 1.5 ounces is $0.61.

53. – 55. Answers will vary.

57.

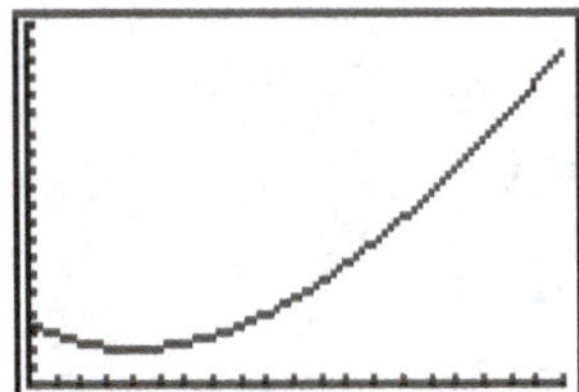

The number of physician's visits per year based on age first decreases and then increases over a person's lifetime.

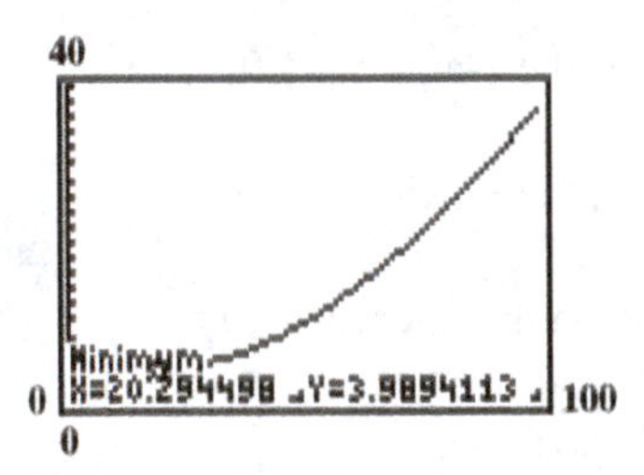

These are the approximate coordinates of the point (20.3, 4.0). The means that the minimum number of physician's visits per year is approximately 4. This occurs around age 20.

59. makes sense

61. does not make sense; Explanations will vary. Sample explanation: The domain is the set of the various ages of the people.

63. true

65. false; Changes to make the statement true will vary. A sample change is: The range of *f* is $[-2, 2)$.

67. false; Changes to make the statement true will vary. A sample change is: $f(0) = 0.6$

69.

$$\begin{aligned} &\sqrt{f(-2.5) - f(1.9)} - [f(-\pi)]^2 + f(-3) \div f(1) \cdot f(\pi) \\ &= \sqrt{2 - (-2)} - [3]^2 + 2 \div (-2)(-4) \\ &= \sqrt{4} - 9 + (-1)(-4) \\ &= 2 - 9 + 4 \\ &= -3 \end{aligned}$$

70. The relation is a function. Every element in the domain corresponds to exactly one element in the range.

71.

$$\begin{aligned} 12 - 2(3x + 1) &= 4x - 5 \\ 12 - 6x - 2 &= 4x - 5 \\ 10 - 6x &= 4x - 5 \\ -6x - 4x &= -5 - 10 \\ -10x &= -15 \\ \frac{-10x}{-10} &= \frac{-15}{-10} \\ x &= \tfrac{3}{2} \end{aligned}$$

The solution set is $\left\{\tfrac{3}{2}\right\}$.

72. Let x = the width of the rectangle.
Let $3x + 8$ = length of the rectangle.

$$\begin{aligned} P &= 2l + 2w \\ 624 &= 2(3x + 8) + 2x \\ 624 &= 6x + 16 + 2x \\ 624 &= 8x + 16 \\ -8x &= -608 \\ x &= 76 \\ 3x + 8 &= 236 \end{aligned}$$

The dimensions of the rectangle are 76 yards by 236 yards.

73. 3 must be excluded from the domain of *f* because it would cause the denominator, $x - 3$, to be equal to zero. Division by 0 is undefined.

74.

$$\begin{aligned} f(4) + g(4) &= \overbrace{(4^2 + 4)}^{f(4)} + \overbrace{(4 - 5)}^{g(4)} \\ &= 20 + (-1) \\ &= 19 \end{aligned}$$

75. $7.4x^2-15x+4046-\left(-3.5x^2+20x+2405\right)=7.4x^2-15x+4046+3.5x^2-20x-2405$

$$=10.9x^2-35x+1641$$

2.3 Check Points

1. **a.** The function contains neither division nor a square root. For every real number, x, the algebraic expression $\frac{1}{2}x+3$ is a real number. Thus, the domain of f is the set of all real numbers.
Domain of f is $(-\infty,\infty)$.

   b. The function $g(x)=\dfrac{7x+4}{x+5}$ contains division. Because division by 0 is undefined, we must exclude from the domain the value of x that causes $x+5$ to be 0. Thus, x cannot equal –5.
Domain of g is $(-\infty,-5)$ or $(-5,\infty)$.

2. **a.** $(f+g)(x)=f(x)+g(x)$
$=\left(3x^2+4x-1\right)+(2x+7)$
$=3x^2+4x-1+2x+7$
$=3x^2+6x+6$

   b. $(f+g)(x)=3x^2+6x+6$
$(f+g)(4)=3(4)^2+6(4)+6$
$=78$

3. **a.** $(f-g)(x)=\dfrac{5}{x}-\dfrac{7}{x-8}$

   b. The domain of f - g is the set of all real numbers that are common to the domain of f and the domain of g. Thus, we must find the domains of f and g.
Note that $f(x)=\dfrac{5}{x}$ is a function involving division. Because division by 0 is undefined, x cannot equal 0.

   The function $g(x)=\dfrac{7}{x-8}$ is also a function involving division. Because division by 0 is undefined, x cannot equal 8.
To be in the domain of f - g, x must be in both the domain of f and the domain of g.

   This means that $x\neq 0$ and $x\neq 8$.
Domain of $f-g=(-\infty,0)$ or $(0,8)$ or $(8,\infty)$.

4. **a.** $(f+g)(5)=f(5)+g(5)=[5^2-2\cdot 5]+[5+3]=23$

   b. $(f-g)(x)=f(x)-g(x)=[x^2-2x]-[x+3]=x^2-3x-3$
$(f-g)(-1)=(-1)^2-3(-1)-3=1$

c. $\left(\frac{f}{g}\right)(x)=\frac{f(x)}{g(x)}=\frac{x^2-2x}{x+3}$

$\left(\frac{f}{g}\right)(7)=\frac{(7)^2-2(7)}{(7)+3}=\frac{35}{10}=\frac{7}{2}$

d. $(fg)(-4)=f(-4)\cdot g(-4)$
$=\left((-4)^2-2(-4)\right)\left((-4)+3\right)$
$=(24)(-1)$
$=-24$

5. a. $(B+D)(x)=B(x)+D(x)$
$=(-2.6x^2+49x+3994)+(-0.6x^2+7x+2412)$
$=-2.6x^2+49x+3994-0.6x^2+7x+2412$
$=-3.2x^2+56x+6406$

b. $(B+D)(x)=-3.2x^2+56x+6406$
$(B+D)(5)=-3.2(3)^2+56(3)+6406$
$=6545.2$

The number of births and deaths in the U.S. in 2003 was 6545.2 thousand.

c. $(B+D)(x)$ overestimates the actual number of births and deaths in 2003 by 7.2 thousand.

2.3 Concept and Vocabulary Check

1. zero
2. negative
3. $f(x)+g(x)$
4. $f(x)-g(x)$
5. $f(x)\cdot g(x)$
6. $\frac{f(x)}{g(x)}$; $g(x)$
7. $(-\infty,\infty)$
8. $(2,\infty)$
9. $(0,3)$; $(3,\infty)$

2.3 Exercise Set

1. Domain of f is $(-\infty,\infty)$.

3. Domain of g is $(-\infty,-4)$ or $(-4,\infty)$.

5. Domain of f is $(-\infty,3)$ or $(3,\infty)$.

7. Domain of g is $(-\infty,5)$ or $(5,\infty)$.

9. Domain of f is $(-\infty,-7)$ or $(-7,9)$ or $(9,\infty)$.

11. $$\begin{aligned}(f+g)(x) &= (3x+1)+(2x-6)\\ &= 3x+1+2x-6\\ &= 5x-5\end{aligned}$$

$$\begin{aligned}(f+g)(5) &= 5(5)-5\\ &= 25-5=20\end{aligned}$$

13. $$\begin{aligned}(f+g)(x) &= (x-5)+\left(3x^2\right)\\ &= x-5+3x^2\\ &= 3x^2+x-5\end{aligned}$$

$$\begin{aligned}(f+g)(5) &= 3(5)^2+5-5\\ &= 3(25)=75\end{aligned}$$

15. $(f+g)(x)$

$$\begin{aligned}&= \left(2x^2-x-3\right)+(x+1)\\ &= 2x^2-x-3+x+1\\ &= 2x^2-2\end{aligned}$$

$$\begin{aligned}(f+g)(5) &= 2(5)^2-2\\ &= 2(25)-2\\ &= 50-2=48\end{aligned}$$

17. $$\begin{aligned}(f+g)(x) &= (5x)+(-2x-3)\\ &= 5x-2x-3\\ &= 3x-3\end{aligned}$$

$$\begin{aligned}(f-g)(x) &= (5x)-(-2x-3)\\ &= 5x+2x+3\\ &= 7x+3\end{aligned}$$

$$\begin{aligned}(fg)(x) &= (5x)(-2x-3)\\ &= -10x^2-15x\end{aligned}$$

$$\left(\frac{f}{g}\right)(x) = \frac{5x}{-2x-3}$$

19. Domain of $f+g=(-\infty,\infty)$.

21. Domain of $f+g=(-\infty,5)$ or $(5,\infty)$.

23. Domain of $f+g=(-\infty,0)$ or $(0,5)$ or $(5,\infty)$.

25. Domain of $f+g=f+g=(-\infty,-3)$ or $(-3,2)$ or $(2,\infty)$.

27. Domain of $f+g=(-\infty,2)$ or $(2,\infty)$.

29. Domain of $f+g=(-\infty,\infty)$.

31.
$$\begin{aligned}(f+g)(x)&=f(x)+g(x)\\&=x^2+4x+2-x\\&=x^2+3x+2\end{aligned}$$
$$(f+g)(3)=(3)^2+3(3)+2=20$$

33. $f(-2)+g(-2)=\left((-2)^2+4(-2)\right)+\left(2-(-2)\right)=-4+4=0$

35.
$$\begin{aligned}(f-g)(x)&=f(x)-g(x)\\&=\left(x^2+4x\right)-\left(2-x\right)\\&=x^2+4x-2+x\\&=x^2+5x-2\end{aligned}$$
$$\begin{aligned}(f-g)(5)&=(5)^2+5(5)-2\\&=25+25-2=48\end{aligned}$$

37. $f(-2)-g(-2)=\left((-2)^2+4(-2)\right)-\left(2-(-2)\right)=-4-4=-8$

39. $(fg)(-2)=f(-2)\cdot g(-2)=\left((-2)^2+4(-2)\right)\cdot\left(2-(-2)\right)=-4(4)=-16$

41. $(fg)(5)=f(5)\cdot g(5)=\left((5)^2+4(5)\right)\cdot\left(2-(5)\right)=45(-3)=-135$

43. $\left(\dfrac{f}{g}\right)(x)=\dfrac{f(x)}{g(x)}=\dfrac{x^2+4x}{2-x}$

$$\left(\frac{f}{g}\right)(1)=\frac{(1)^2+4(1)}{2-(1)}=\frac{1+4}{1}=\frac{5}{1}=5$$

45. $\left(\frac{f}{g}\right)(x)=\frac{f(x)}{g(x)}=\frac{x^2+4x}{2-x}$

$\left(\frac{f}{g}\right)(-1)=\frac{(-1)^2+4(-1)}{2-(-1)}$

$=\frac{1-4}{3}=\frac{-3}{3}=-1$

47. Domain of $f+g=(-\infty,\infty)$.

49. $\left(\frac{f}{g}\right)(x)=\frac{f(x)}{g(x)}=\frac{x^2+4x}{2-x}$

Domain of $\frac{f}{g}=(-\infty,2)$ or $(2,\infty)$.

51. $(f+g)(-3)=f(-3)+g(-3)=4+1=5$

53. $(fg)(2)=f(2)g(2)=(-1)(1)=-1$

55. The domain of $f+g$ is $[-4,3]$.

57. The graph of $f+g$.

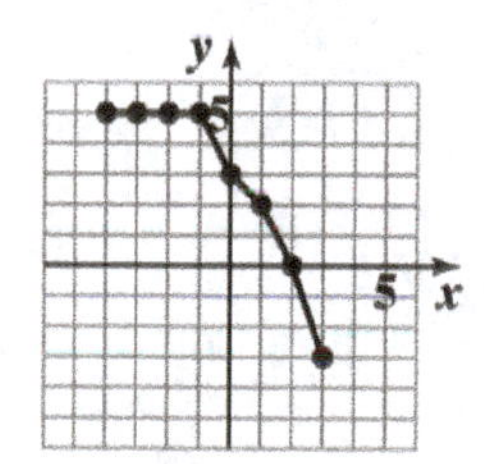

59. $(f+g)(1)-(g-f)(-1)$

$=f(1)+g(1)-[g(-1)-f(-1)]$

$=f(1)+g(1)-g(-1)+f(-1)$

$=-6+-3-(-2)+3$

$=-6+-3+2+3=-4$

61. $(fg)(-2)-\left[\left(\frac{f}{g}\right)(1)\right]^2$

$=f(-2)g(-2)-\left[\frac{f(1)}{g(1)}\right]^2$

$=5\cdot 0-\left[\frac{-6}{-3}\right]^2=0-2^2=0-4=-4$

63. a. $(M+F)(x)=M(x)+F(x)$

$=(1.54x+114.6)+(1.48x+120.6)$

$=3.02x+235.2$

b. $(M+F)(x)=3.02x+235.2$

$(M+F)(20)=3.02(20)+235.2=295.6$

The total U.S. population in 2005 was 295.6 million.

c. The result in part (b) underestimates the actual total by 2.4 million.

65. a. $\left(\frac{M}{F}\right)(x)=\left(\frac{M(x)}{F(x)}\right)=\frac{1.54x+114.6}{1.48x+120.6}$

b. $\left(\frac{M}{F}\right)(x)=\frac{1.54x+114.6}{1.48x+120.6}$

$\left(\frac{M}{F}\right)(15)=\frac{1.54(15)+114.6}{1.48(15)+120.6}\approx 0.964$

In 2000 the ratio of men to women was 0.964.

c. The result in part (b) underestimates the actual ratio of $\frac{138}{143}\approx 0.965$ by about 0.001.

67. – 69. Answers will vary.

71. $y_1=2x+3$ $\quad y_2=2-2x$ $\quad y_3=y_1+y_2$

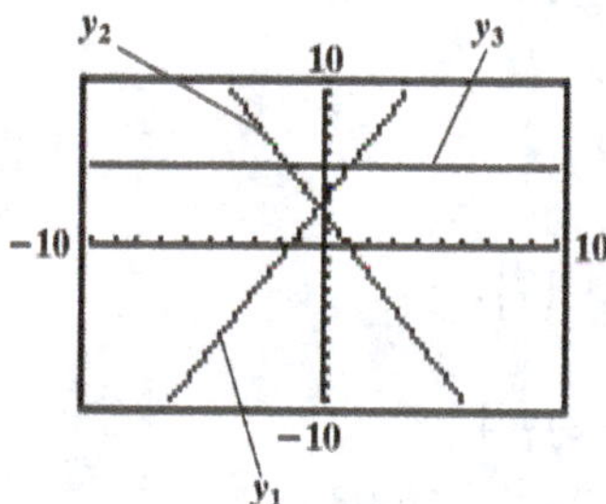

73. $y_1=x$ $\quad y_2=x-4$ $\quad y_3=y_1\cdot y_2$

75. 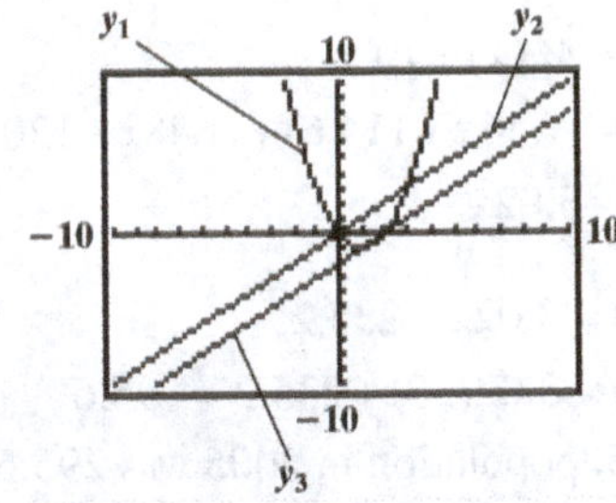

No y-value is displayed because y_3 is undefined at $x = 0$.

77. makes sense

79. makes sense

81. true

83. false; Changes to make the statement true will vary. A sample change is: $f(a)$ or $f(b)$ is 0.

84.
$$R = 3(a+b)$$
$$R = 3a + 3b$$
$$R - 3a = 3b$$
$$b = \frac{R-3a}{3} \text{ or } b = \frac{R}{3} - a$$

85.
$$3(6-x) = 3-2(x-4)$$
$$18-3x = 3-2x+8$$
$$18-3x = 11-2x$$
$$18 = 11+x$$
$$7 = x$$
The solution set is $\{7\}$.

86.
$$f(b+2) = 6(b+2)-4$$
$$= 6b+12-4 = 6b+8$$

87. a.
$$4x-3y = 6$$
$$4x-3(0) = 6$$
$$4x = 6$$
$$x = \frac{3}{2}$$

b.
$$4x-3y = 6$$
$$4(0)-3y = 6$$
$$-3y = 6$$
$$y = -2$$

88. a.

x	$y = 2x+4$	(x, y)
−3	$2(-3)+4 = -2$	$(-3,-2)$
−2	$2(-2)+4 = 0$	$(-2,0)$
−1	$2(-1)+4 = 2$	$(-1,2)$
0	$2(0)+4 = 4$	$(0,4)$
1	$2(1)+4 = 6$	$(1,6)$

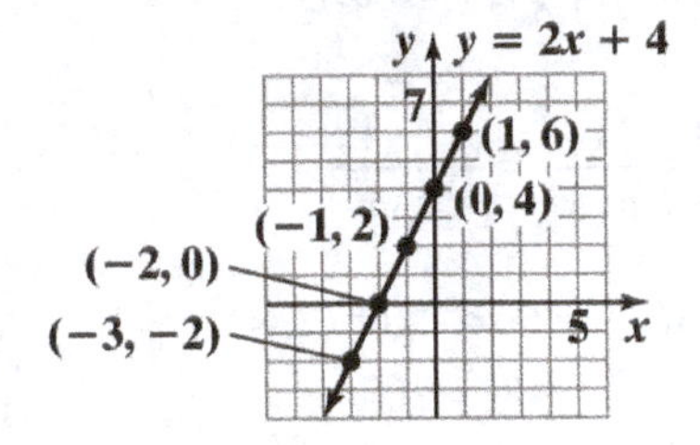

b. The graph crosses the x-axis at the point $(-2,0)$.

c. The graph crosses the y-axis at the point $(0,4)$.

89.
$$5x+3y = -12$$
$$3y = -5x-12$$
$$\frac{3y}{3} = \frac{-5x}{3} - \frac{12}{3}$$
$$y = -\frac{5}{3}x - 4$$

Mid-Chapter Check Point – Chapter 2

1. The relation is not a function.
   The domain is $\{1,2\}$.
   The range is $\{-6,4,6\}$.

2. The relation is a function.
   The domain is $\{0,2,3\}$.
   The range is $\{1,4\}$.

3. The relation is a function.
   The domain is $[-2,2)$.
   The range is $[0,3]$.

4. The relation is not a function.
   The domain is $(-3,4]$.
   The range is $[-1,2]$.

5. The relation is not a function.
The domain is $\{-2,-1,0,1,2\}$.
The range is $\{-2,-1,1,3\}$.

6. The relation is a function.
The domain is $(-\infty,1]$.
The range is $[-1,\infty)$.

7. The graph of f represents the graph of a function because every element in the domain corresponds to exactly one element in the range. It passes the vertical line test.

8. $f(-4)=3$

9. The function $f(x)=4$ when $x=-2$.

10. The function $f(x)=0$ when $x=2$ and $x=-6$.

11. The domain of f is $(-\infty,\infty)$.

12. The range of f is $(-\infty,4]$.

13. The domain is $(-\infty,\infty)$.

14. The domain of g is $(-\infty,-2)$ or $(-2,2)$ or $(2,\infty)$.

15. $f(0)=0^2-3(0)+8=8$
$g(-10)=-2(-10)-5=20-5=15$
$f(0)+g(-10)=8+15=23$

16. $f(-1)=(-1)^2-3(-1)+8=1+3+8=12$
$g(3)=-2(3)-5=-6-5=-11$
$f(-1)-g(3)=12-(-11)=12+11=23$

17. $f(a)=a^2-3a+8$
$g(a+3)=-2(a+3)-5$
$=-2a-6-5=-2a-11$
$f(a)+g(a+3)=a^2-3a+8+-2a-11$
$=a^2-5a-3$

18. $(f+g)(x)=x^2-3x+8+-2x-5$
$=x^2-5x+3$
$(f+g)(-2)=(-2)^2-5(-2)+3$
$=4+10+3=17$

19. $(f-g)(x)=x^2-3x+8-(-2x-5)$
$=x^2-3x+8+2x+5$
$=x^2-x+13$
$(f-g)(5)=(5)^2-5+13$
$=25-5+13=33$

20. $f(-1)=(-1)^2-3(-1)+8$
$=1+3+8=12$
$g(-1)=-2(-1)-5=2-5=-3$
$(fg)(-1)=12(-3)=-36$

21. $\left(\frac{f}{g}\right)(x)=\frac{x^2-3x+8}{-2x-5}$
$\left(\frac{f}{g}\right)(-4)=\frac{(-4)^2-3(-4)+8}{-2(-4)-5}$
$=\frac{16+12+8}{8-5}=\frac{36}{3}=12$

22. The domain of $\frac{f}{g}$ is $\left(-\infty,-\frac{5}{2}\right)$ or $\left(-\frac{5}{2},\infty\right)$.

2.4 Check Points

1. $3x-2y=6$
Find the x–intercept by setting $y=0$.
$3x-2y=6$
$3x-2(0)=6$
$3x=6$
$x=2$
Find the y–intercept by setting $x=0$.
$3x-2y=6$
$3(0)-2y=6$
$-2y=6$
$y=-3$

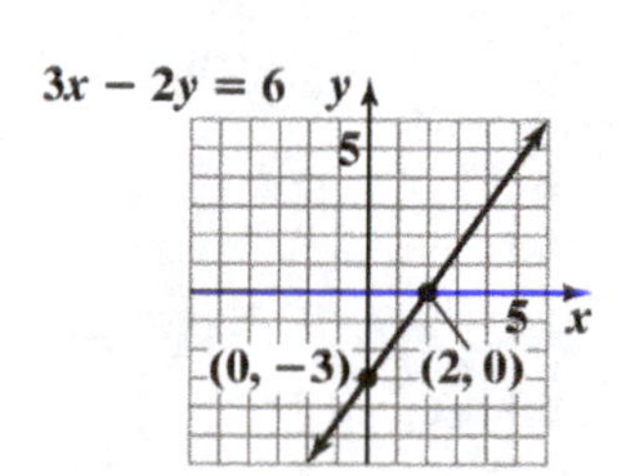

2. a. $m = \dfrac{y_2 - y_1}{x_2 - x_1} = \dfrac{-2-4}{-4-(-3)} = \dfrac{-6}{-1} = 6$

b. $m = \dfrac{y_2 - y_1}{x_2 - x_1} = \dfrac{5-(-2)}{-1-4} = \dfrac{7}{-5} = -\dfrac{7}{5}$

3. First, convert the equation to slope-intercept form by solving the equation for y.

$$8x - 4y = 20$$
$$-4y = -8x + 20$$
$$\frac{-4y}{-4} = \frac{-8x+20}{-4}$$
$$y = 2x - 5$$

In this form, the coefficient of x is the line's slope and the constant term is the y-intercept.
The slope is 2 and the y-intercept is –5.

4. Begin by plotting the y-intercept of –3. Then use the slope of 4 to plot more points.

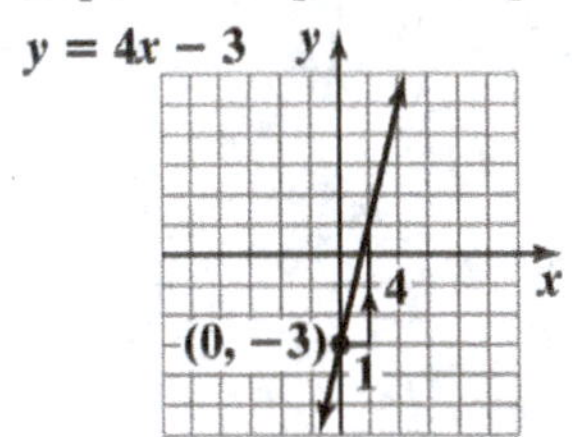

5. Begin by plotting the y-intercept of 0. Then use the slope of $\dfrac{-2}{3}$ to plot more points.

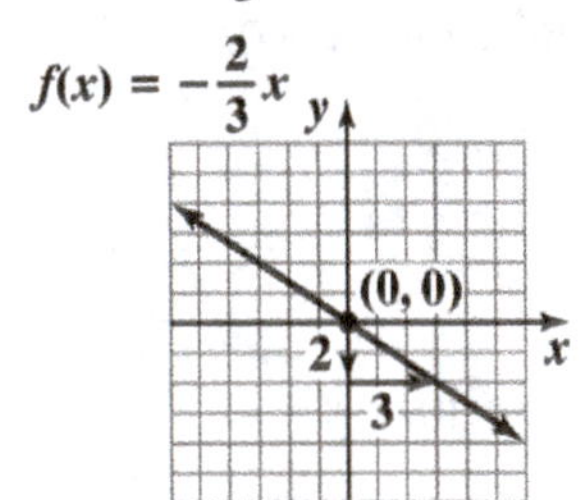

6. $y = 3$ is a horizontal line.

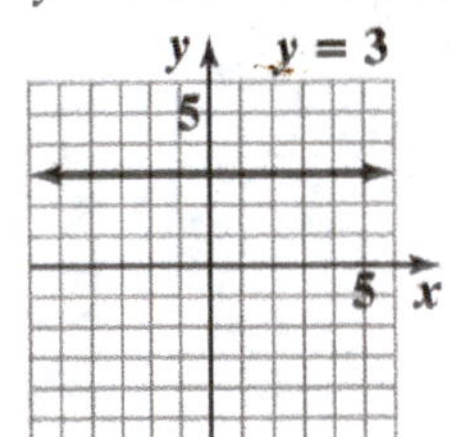

7. $x = -3$ is a vertical line.

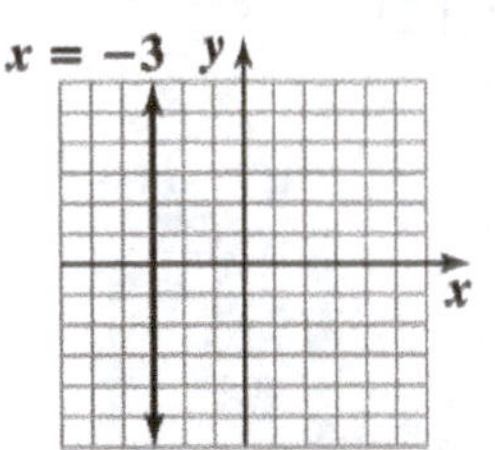

8. $m = \dfrac{y_2 - y_1}{x_2 - x_1} = \dfrac{14.1-15.4}{1970-1965} = \dfrac{-1.3}{5} = -0.26$

From 1965 through 1970, fuel efficiency decreased by 0.26 miles per gallon each year.

9. $m = \dfrac{y_2 - y_1}{x_2 - x_1} = \dfrac{0.05-0.03}{3-1} = \dfrac{0.02}{2} \approx 0.01$

The average rate of change between 1 hour and 3 hours is 0.01. This means that the drug's concentration is increasing at an average rate of 0.01 milligram per 100 milliliters per hour.

10. a. We will use the line segment passing through $(59, 385)$ and $(0, 310)$ to obtain a model. We need values for m, the slope, and b, the y-intercept.

$m = \dfrac{y_2 - y_1}{x_2 - x_1} = \dfrac{385-310}{59-0} = \dfrac{75}{59} \approx 1.27$

The point $(0, 310)$ gives us the y-intercept of 310.

Thus, $C(x) = mx + b$
$C(x) = 1.27x + 310$

b.
$$C(x) = 1.27x + 310$$
$$C(100) = 1.27(100) + 310$$
$$= 437$$

The model predicts the average atmospheric concentration of carbon dioxide will be 437 parts per million in 2050.

2.4 Concept and Vocabulary Check

1. scatterplot; regression
2. standard
3. x-intercept; zero
4. y-intercept; zero
5. $\dfrac{y_2 - y_1}{x_2 - x_1}$
6. positive

7. negative

8. zero

9. undefined

10. $y = mx + b$

11. $(0,3)$; 2; 5

12. horizontal

13. vertical

14. y; x

2.4 Exercise Set

1. $x + y = 4$
   Find the x–intercept by setting $y = 0$.
   $x + y = 4$
   $x + 0 = 4$
   $x = 4$
   Find the y–intercept by setting $x = 0$.
   $x + y = 4$
   $0 + y = 4$
   $y = 4$

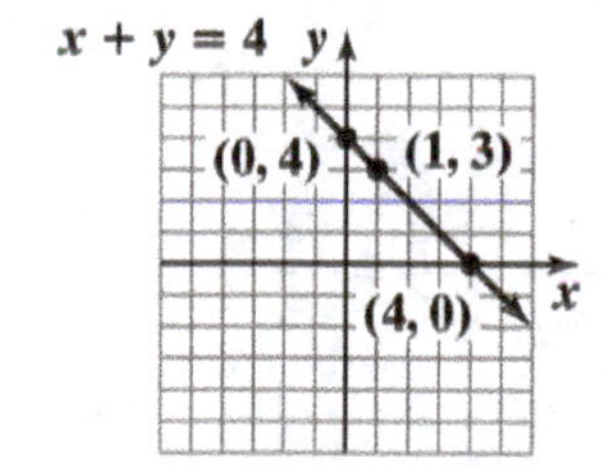

3. $x + 3y = 6$
   Find the x–intercept by setting $y = 0$.
   $x + 3(0) = 6$
   $x = 6$
   Find the y–intercept by setting $x = 0$.
   $(0) + 3y = 6$
   $y = 2$

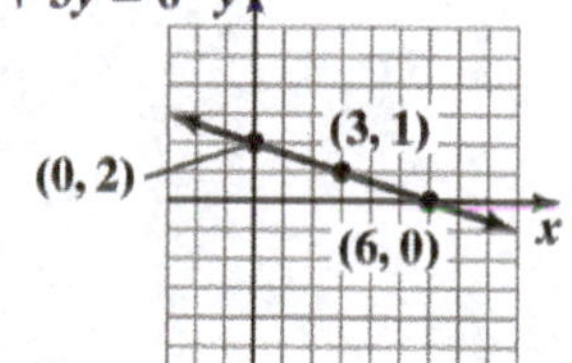

5. $6x - 2y = 12$
   Find the x–intercept by setting $y = 0$.
   $6x - 2(0) = 12$
   $6x = 12$
   $x = 2$
   Find the y–intercept by setting $x = 0$.
   $6(0) - 2y = 12$
   $-2y = 12$
   $y = -6$

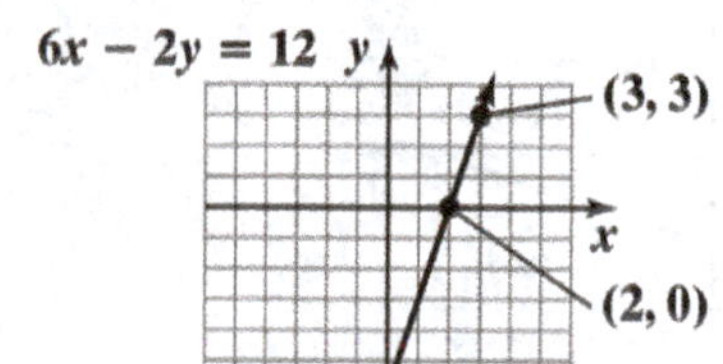

7. $3x - y = 6$
   Find the x–intercept by setting $y = 0$.
   $3x - 0 = 6$
   $3x = 6$
   $x = 2$
   Find the y–intercept by setting $x = 0$.
   $3(0) - y = 6$
   $-y = 6$
   $y = -6$

9. $x - 3y = 9$

Find the x–intercept by setting $y = 0$.

$$x - 3(0) = 9$$
$$x = 9$$

Find the y–intercept by setting $x = 0$.

$$(0) - 3y = 9$$
$$-3y = 9$$
$$y = -3$$

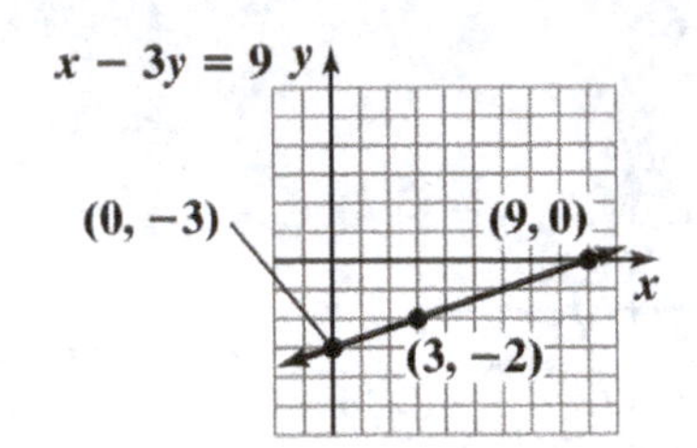

11. $2x = 3y + 6$

Find the x–intercept by setting $y = 0$.

$$2x = 3(0) + 6$$
$$2x = 6$$
$$x = 3$$

Find the y–intercept by setting $x = 0$.

$$2(0) = 3y + 6$$
$$0 = 3y + 6$$
$$-6 = 3y$$
$$-2 = y$$

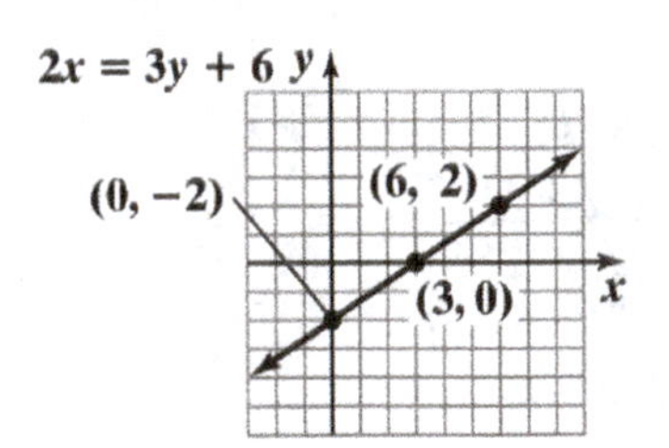

13. $6x - 3y = 15$

Find the x–intercept by setting $y = 0$.

$$6x - 3(0) = 15$$
$$6x = 15$$
$$x = \frac{15}{6} = \frac{5}{2}$$

Find the y–intercept by setting $x = 0$.

$$6(0) - 3y = 15$$
$$-3y = 15$$
$$y = -5$$

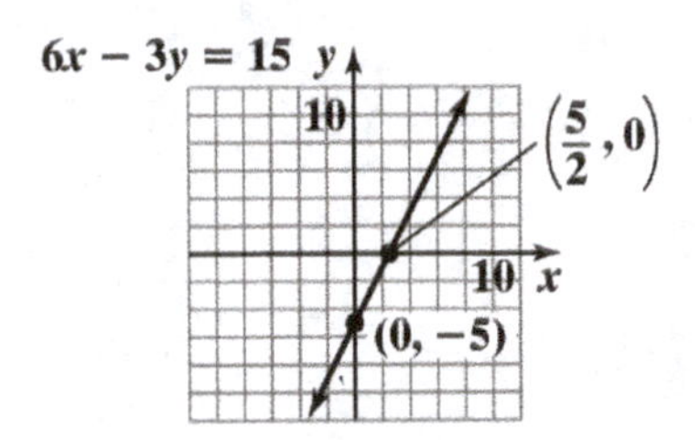

15. $m = \frac{8-4}{3-2} = \frac{4}{1} = 4$

The line rises.

17. $m = \frac{5-4}{2-(-1)} = \frac{1}{2+1} = \frac{1}{3}$

The line rises.

19. $m = \frac{5-5}{-1-2} = \frac{0}{-3} = 0$

The line is horizontal.

21. $m = \frac{-3-1}{-4-(-7)} = \frac{-4}{-4+7} = \frac{-4}{3} = -\frac{4}{3}$

The line falls.

23. $m = \frac{6-(-4)}{-3-(-7)} = \frac{10}{4} = \frac{5}{2}$

The line rises.

25. $m = \frac{\frac{1}{4}-(-2)}{\frac{7}{2}-\frac{7}{2}} = \frac{\frac{1}{4}+2}{0} =$ undefined

undefined slope; The line is vertical.

27. Line 1 goes through $(-3,0)$ and $(0,2)$.

$$m=\frac{2-0}{0-(-3)}=\frac{2}{3}$$

Line 2 goes through $(2,0)$ and $(0,4)$.

$$m=\frac{4-0}{0-2}=\frac{4}{-2}=-2$$

Line 3 goes through $(0,-3)$ and $(2,-4)$.

$$m=\frac{-4-(-3)}{2-0}=\frac{-4+3}{2}=\frac{-1}{2}=-\frac{1}{2}$$

29. $y=2x+1$

$m=2$ $\quad$ $y-\text{intercept}=1$

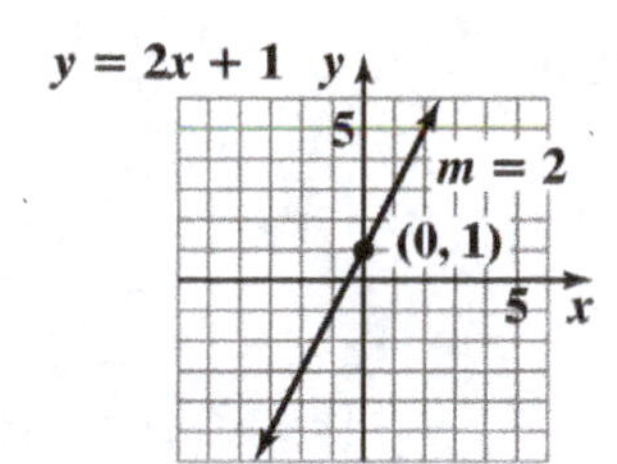

31. $y=-2x+1$

$m=-2$ $\quad$ $y-\text{intercept}=1$

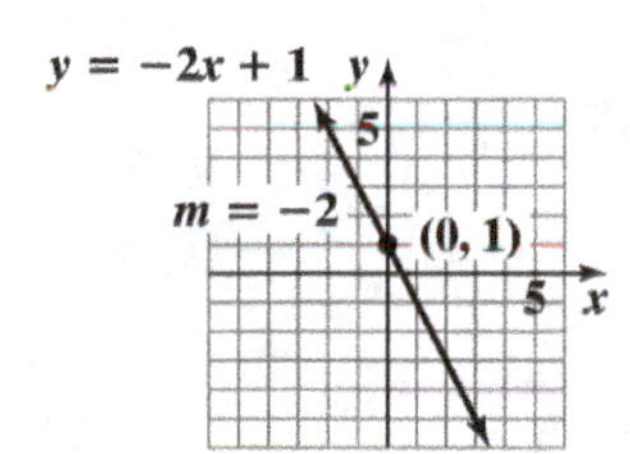

33. $f(x)=\frac{3}{4}x-2$

$m=\frac{3}{4}$ $\quad$ $y-\text{intercept}=-2$

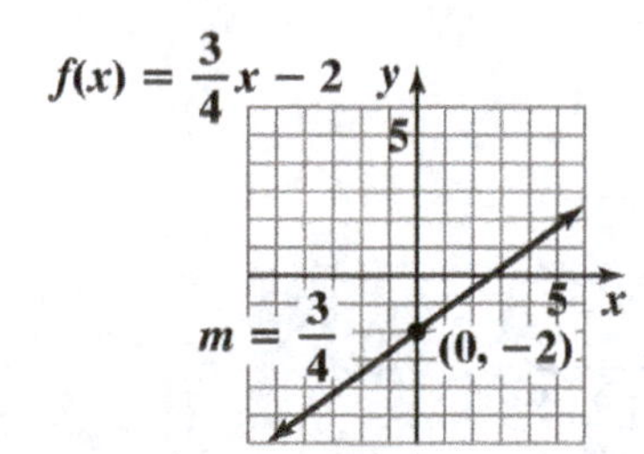

35. $f(x)=-\frac{3}{5}x+7$

$m=-\frac{3}{5}$ $\quad$ $y-\text{intercept}=7$

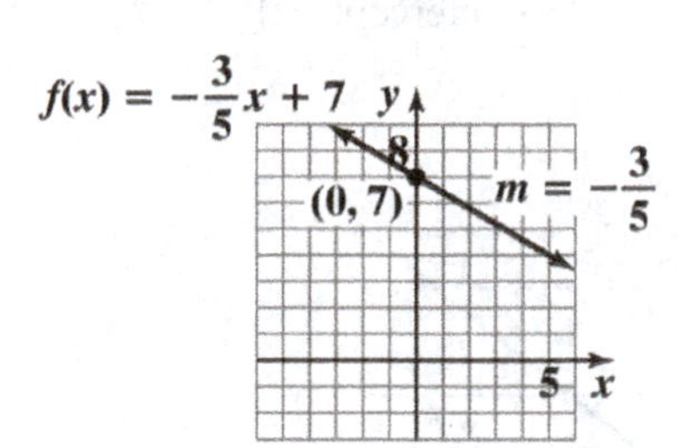

37. $y=-\frac{1}{2}x$

$m=-\frac{1}{2}$ $\quad$ $y-\text{intercept}=0$

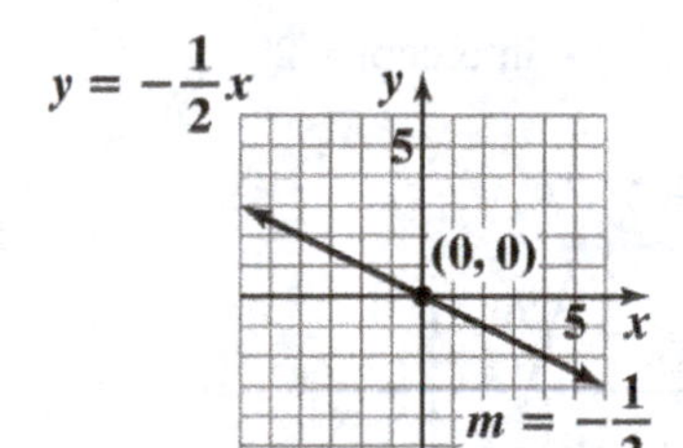

39. $y=-\frac{1}{2}$

$m=0$ $\quad$ $y-\text{intercept}=-\frac{1}{2}$

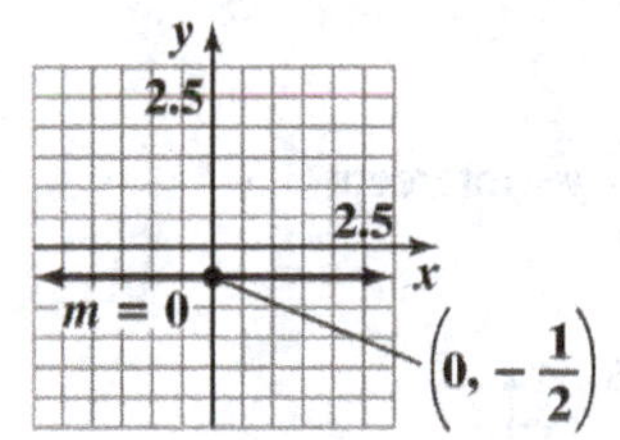

41. a. $2x+y=0$

$y=-2x$

b. $m=-2$ $\quad$ $y-\text{intercept}=0$

c.

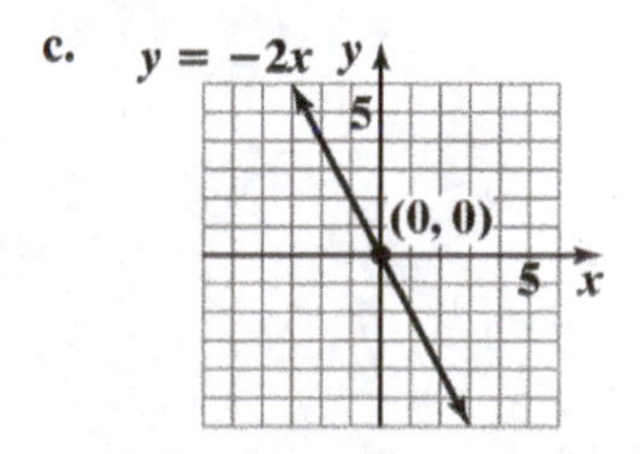

43. a. $5y = 4x$

$$y = \frac{4}{5}x$$

b. $m = \frac{4}{5}$ $\quad y - \text{intercept} = 0$

c.

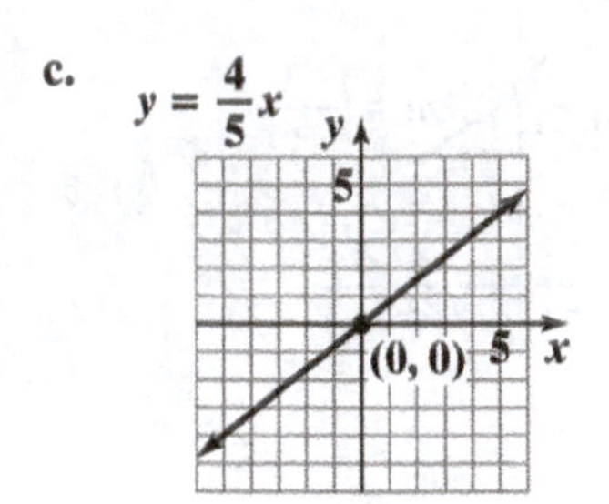

45. a. $3x + y = 2$

$$y = -3x + 2$$

b. $m = -3$ $\quad y - \text{intercept} = 2$

c.

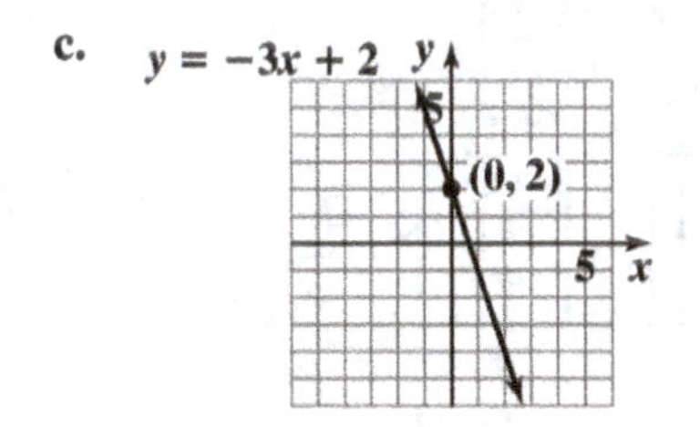

47. a. $5x + 3y = 15$

$$3y = -5x + 15$$

$$y = -\frac{5}{3}x + 5$$

b. $m = -\frac{5}{3}$ $\quad y - \text{intercept} = 5$

c. $y = -\frac{5}{3}x + 5$

49. $y = 3$

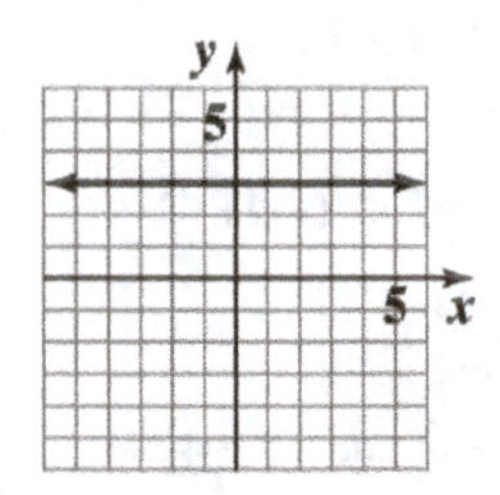

51. $f(x) = -2$

$$y = -2$$

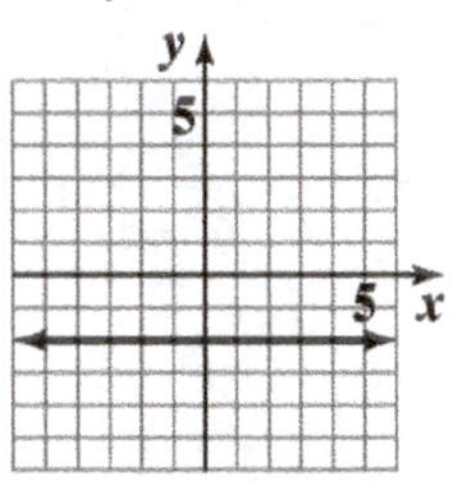

53. $3y = 18$

$$y = 6$$

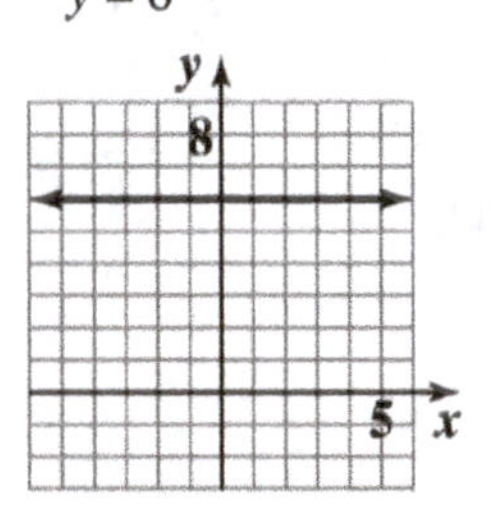

55. $f(x) = 2$

$$y = 2$$

57. $x = 5$

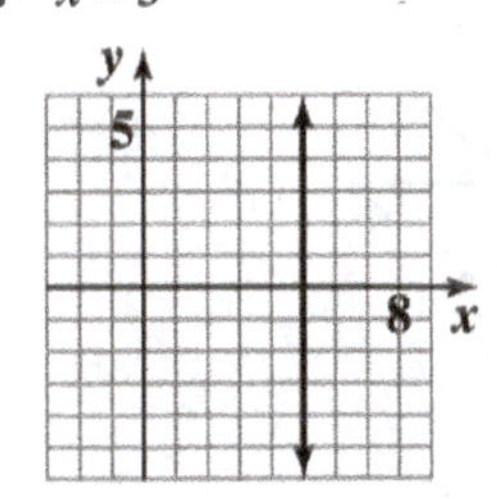

59. $3x = -12$

$x = -4$

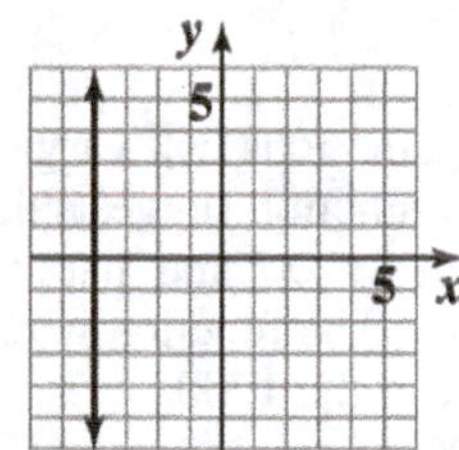

61. $x = 0$

This is the equation of the y–axis.

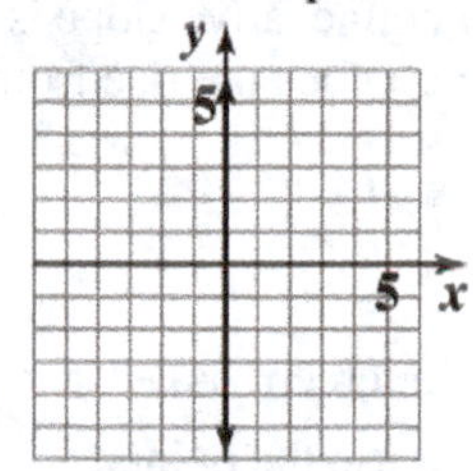

63. $m = \dfrac{0-a}{b-0} = \dfrac{-a}{b} = -\dfrac{a}{b}$

Since a and b are both positive, $-\dfrac{a}{b}$ is negative.

Therefore, the line falls.

65. $m = \dfrac{(b+c)-b}{a-a} = \dfrac{c}{0}$

The slope is undefined.

The line is vertical.

67. $Ax + By = C$

$By = -Ax + C$

$y = -\dfrac{A}{B}x + \dfrac{C}{B}$

The slope is $-\dfrac{A}{B}$ and the $y-$intercept is $\dfrac{C}{B}$.

69. $-3 = \dfrac{4-y}{1-3}$

$-3 = \dfrac{4-y}{-2}$

$6 = 4 - y$

$2 = -y$

$-2 = y$

71. $3x - 4f(x) = 6$

$-4f(x) = -3x + 6$

$f(x) = \dfrac{3}{4}x - \dfrac{3}{2}$

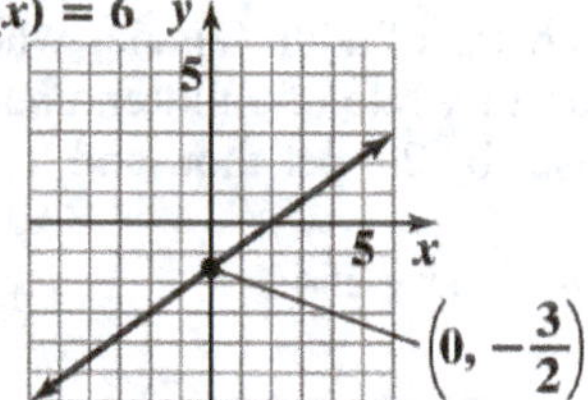

73. Using the slope-intercept form for the equation of a line:

$-1 = -2(3) + b$

$-1 = -6 + b$

$5 = b$

75. The line with slope m_1 is the steepest rising line so its slope is the biggest positive number. Then the line with slope m_3 is next because it is the only other line whose slope is positive. Since the line with slope m_2 is less steep but decreasing, it is next. The slope m_4 is the smallest because it is negative and the line with slope m_4 is steeper than the line with slope m_3, so its slope is more negative.

Decreasing order: m_1, m_3, m_2, m_4

77. The slope is 55.7. This means Smartphone sales are increasing by 55.7 million each year.

79. The slope is –0.52. This means the percentage of U.S. adults who smoke cigarettes is decreasing by 0.52% each year.

81. **a.** 30% of marriages in which the wife is under 18 when she marries end in divorce within the first five years.

b. 50% of marriages in which the wife is under 18 when she marries end in divorce within the first ten years.

c. $m = \dfrac{y_2 - y_1}{x_2 - x_1} = \dfrac{50-30}{10-5} = \dfrac{20}{5} = 4$

There is an average increase of 4% of marriages ending in divorce per year.

83. **a.** The y-intercept is 254. This represents if no women in a country are literate, the mortality rate of children under five is 254 per thousand.

b. $m = \frac{y_2 - y_1}{x_2 - x_1} = \frac{110-254}{60-0} = \frac{-144}{60} = -2.4$

For each 1% of adult females who are literate, the mortality rate of children under five decreases by 2.4 per thousand.

c. $f(x) = -2.4x + 254$

d. $f(50) = -2.4(50) + 254 = 134$

A country where 50% of adult females are literate is predicted to have a mortality rate of children under five of 134 per thousand.

85. $P(x) = 0.725x + 18$

87. – 103. Answers will vary.

105. $y = 2x + 4$

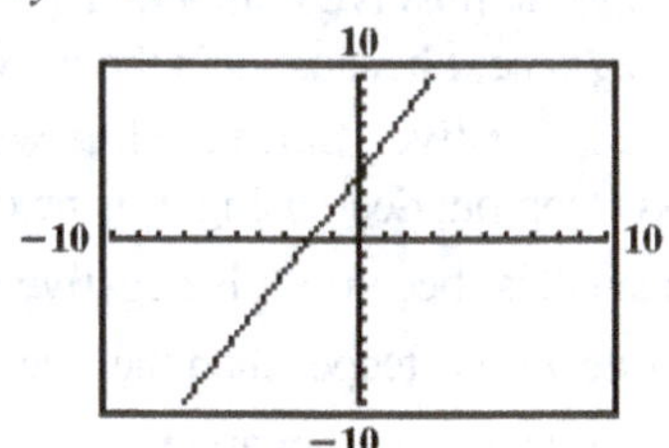

Two points found using [TRACE] are $(0,4)$ and $(2,8)$.

$m = \frac{8-4}{2-0} = \frac{4}{2} = 2$

This is the same as the coefficient of x in the line's equation.

107. $y = -\frac{1}{2}x - 5$

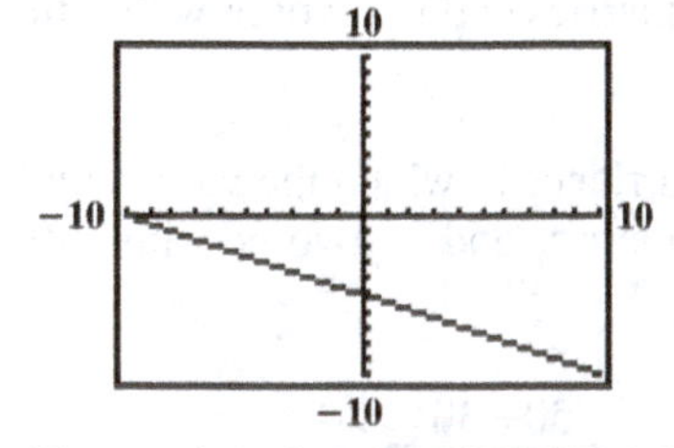

Two points found using [TRACE] are $(0,-5)$ and $(1,-5.5)$. Based on these points, the slope is:

$m = \frac{-5.5-(-5)}{1-0} = \frac{-5.5+5}{1} = \frac{-0.5}{1} = -0.5.$

This is the same as the coefficient of x in the line's equation.

109. does not make sense; Explanations will vary. Sample explanation: Linear functions never change from rising to falling.

111. does not make sense; Explanations will vary. Sample explanation: This function suggests that the average salary in 2000 was $1700, and that there is an annual raise of $49,100. The function would make sense if the x was with the 1700.
i.e. $S(x) = 1700x + 49{,}100$

113. false; Changes to make the statement true will vary. A sample change is: One nonnegative slope is 0. A line with slope equal to zero does not rise from left to right.

115. true

117. We are given that the $x-$intercept is -2 and the $y-$intercept is 4. We can use the points $(-2,0)$ and $(0,4)$ to find the slope.

$m = \frac{4-0}{0-(-2)} = \frac{4}{0+2} = \frac{4}{2} = 2$

Using the slope and one of the intercepts, we can write the line in point-slope form.

$$y - y_1 = m(x - x_1)$$
$$y - 0 = 2(x - (-2))$$
$$y = 2(x+2)$$
$$y = 2x + 4$$
$$-2x + y = 4$$

Find the x– and y–coefficients for the equation of the line with right-hand-side equal to 12. Multiply both sides of $-2x + y = 4$ by 3 to obtain 12 on the right-hand-side.

$$-2x + y = 4$$
$$3(-2x + y) = 3(4)$$
$$-6x + 3y = 12$$

The coefficients are –6 and 3.

119. **a.** $f(x_1 + x_2) = m(x_1 + x_2) + b$
$= mx_1 + mx_2 + b$

b. $f(x_1) + f(x_2)$
$= mx_1 + b + mx_2 + b$
$= mx_1 + mx_2 + 2b$

c. no

120. $\left(\dfrac{4x^2}{y^{-3}}\right)^2=\left(4x^2y^3\right)^2=4^2\left(x^2\right)^2\left(y^3\right)^2$
$=16x^4y^6$

121. $\left(8\times10^{-7}\right)\left(4\times10^3\right)=32\times10^{-4}$
$=\left(3.2\times10^1\right)\times10^{-4}$
$=3.2\times10^{-3}$

122. $5-\left[3(x-4)-6x\right]=5-[3x-12-6x]$
$=5-3x+12+6x$
$=3x+17$

123. $y-5=7(x+4)$
$y-5=7x+28$
$y=7x+33$

124. $y+3=-\dfrac{7}{3}(x-1)$
$y+3=-\dfrac{7}{3}x+\dfrac{7}{3}$
$y+3-3=-\dfrac{7}{3}x+\dfrac{7}{3}-3$
$y=-\dfrac{7}{3}x-\dfrac{2}{3}$

125. a. $x+4y-8=0$
$4y=-x+8$
$\dfrac{4y}{4}=\dfrac{-x+8}{4}$
$y=-\dfrac{1}{4}x+2$

The slope is $-\dfrac{1}{4}$.

b. $-\dfrac{1}{4}\cdot m_2=-1$
$\dfrac{1}{4}\cdot m_2=1$
$m_2=4$

The slope of the second line is 4.

2.5 Check Points

1. Slope $= -2$, passing through $(4, -3)$

   Point-Slope Form

   $$y - y_1 = m(x - x_1)$$
   $$y - (-3) = -2(x - 4)$$
   $$y + 3 = -2(x - 4)$$

   Slope-Intercept Form

   $$y + 3 = -2(x - 4)$$
   $$y + 3 = -2x + 8$$
   $$y = -2x + 5$$
   $$f(x) = -2x + 5$$

2. **a.** Passing through $(6, -3)$ and $(2, 5)$

   First, find the slope.

   $$m = \frac{5-(-3)}{2-6} = \frac{8}{-4} = -2$$

   Then use the slope and one of the points to write the equation in point-slope form.

   $$y - y_1 = m(x - x_1)$$
   $$y - 5 = -2(x - 2)$$

   or

   $$y - y_1 = m(x - x_1)$$
   $$y - (-3) = -2(x - 6)$$
   $$y + 3 = -2(x - 6)$$

   b. Slope-Intercept Form

   $$y - 5 = -2(x - 2)$$
   $$y - 5 = -2x + 4$$
   $$y = -2x + 9$$
   $$f(x) = -2x + 9$$

3. First, find the slope.

   $$m = \frac{79.7 - 74.7}{40 - 10} = \frac{5}{30} \approx 0.17$$

   Then use the slope and one of the points to write the equation in point-slope form.
   Using the point (10, 74.7):

   $$y - y_1 = m(x - x_1)$$
   $$y - 74.7 = 0.17(x - 10)$$
   $$y = 0.17x + 73$$
   $$f(x) = 0.17x + 73$$

   Next, since 2020 is 60 years after 1960, substitute 60 into the function: $f(60) = 0.17(60) + 73 = 83.2$.
   This means that the life expectancy of American women in 2020 is predicted to be 83.2 years.

   Answers vary due to rounding and choice of point. If point (40, 79.7) is chosen, $f(x) = 0.17x + 72.9$ and the life expectancy of American women in 2020 is predicted to be 83.1 years.

4. Since the line is parallel to $y = 3x + 1$, we know it will have slope $m = 3$. We are given that it passes through $(-2, 5)$. We use the slope and point to write the equation in point-slope form.

   $$y - y_1 = m(x - x_1)$$
   $$y - 5 = 3(x - (-2))$$
   $$y - 5 = 3(x + 2)$$

   Solve for y to obtain slope-intercept form.

   $$y - 5 = 3(x + 2)$$
   $$y - 5 = 3x + 6$$
   $$y = 3x + 11$$
   $$f(x) = 3x + 11$$

5. **a.** Solve the given equation for y to obtain slope-intercept form.

   $$x + 3y = 12$$
   $$3y = -x + 12$$
   $$y = -\frac{1}{3}x + 4$$

   Since the slope of the given line is $-\frac{1}{3}$, the slope of any line perpendicular to the given line is 3.

   b. We use the slope of 3 and the point $(-2, -6)$ to write the equation in point-slope form.

   $$y - y_1 = m(x - x_1)$$
   $$y - (-6) = 3(x - (-2))$$
   $$y + 6 = 3(x + 2)$$

   Solve for y to obtain slope-intercept form.

   $$y + 6 = 3(x + 2)$$
   $$y + 6 = 3x + 6$$
   $$y = 3x$$
   $$f(x) = 3x$$

2.5 Concept and Vocabulary Check

1. $y - y_1 = m(x - x_1)$
2. equal/the same
3. -1

4. $-\frac{1}{5}$

5. $\frac{5}{3}$

6. -4; -4

7. $\frac{1}{2}$; -2

2.5 Exercise Set

1. Slope $= 3$, passing through $(2, 5)$
   Point-Slope Form
   $$y - y_1 = m(x - x_1)$$
   $$y - 5 = 3(x - 2)$$
   Slope-Intercept Form
   $$y - 5 = 3(x - 2)$$
   $$y - 5 = 3x - 6$$
   $$y = 3x - 1$$
   $$f(x) = 3x - 1$$

3. Slope $= 5$, passing through $(-2, 6)$
   Point-Slope Form
   $$y - y_1 = m(x - x_1)$$
   $$y - 6 = 5(x - (-2))$$
   $$y - 6 = 5(x + 2)$$
   Slope-Intercept Form
   $$y - 6 = 5(x + 2)$$
   $$y - 6 = 5x + 10$$
   $$y = 5x + 16$$
   $$f(x) = 5x + 16$$

5. Slope $= -4$, passing through $(-3, -2)$
   Point-Slope Form
   $$y - y_1 = m(x - x_1)$$
   $$y - (-2) = -4(x - (-3))$$
   $$y + 2 = -4(x + 3)$$
   Slope-Intercept Form
   $$y + 2 = -4(x + 3)$$
   $$y + 2 = -4x - 12$$
   $$y = -4x - 14$$
   $$f(x) = -4x - 14$$

7. Slope $= -5$, passing through $(-2, 0)$
   Point-Slope Form
   $$y - y_1 = m(x - x_1)$$
   $$y - 0 = -5(x - (-2))$$
   $$y - 0 = -5(x + 2)$$
   Slope-Intercept Form
   $$y - 0 = -5(x + 2)$$
   $$y = -5(x + 2)$$
   $$y = -5x - 10$$
   $$f(x) = -5x - 10$$

9. Slope $= -1$, passing through $\left(-2, -\frac{1}{2}\right)$
   Point-Slope Form
   $$y - y_1 = m(x - x_1)$$
   $$y - \left(-\tfrac{1}{2}\right) = -1(x - (-2))$$
   $$y + \tfrac{1}{2} = -1(x + 2)$$
   Slope-Intercept Form
   $$y + \frac{1}{2} = -1(x + 2)$$
   $$y + \frac{1}{2} = -x - 2$$
   $$y = -x - \frac{5}{2}$$
   $$f(x) = -x - \frac{5}{2}$$

11. Slope $= \frac{1}{4}$, passing through $(0, 0)$
    Point-Slope Form
    $$y - y_1 = m(x - x_1)$$
    $$y - 0 = \tfrac{1}{4}(x - 0)$$
    Slope-Intercept Form
    $$y - 0 = \tfrac{1}{4}(x - 0)$$
    $$y = \tfrac{1}{4}x$$
    $$f(x) = \tfrac{1}{4}x$$

13. Slope $= -\frac{2}{3}$, passing through $(6,-4)$

Point-Slope Form

$$y - y_1 = m(x - x_1)$$
$$y - (-4) = -\tfrac{2}{3}(x-6)$$
$$y + 4 = -\tfrac{2}{3}(x-6)$$

Slope-Intercept Form

$$y + 4 = -\tfrac{2}{3}(x-6)$$
$$y + 4 = -\tfrac{2}{3}x + 4$$
$$y = -\tfrac{2}{3}x$$
$$f(x) = -\tfrac{2}{3}x$$

15. Passing through $(6,3)$ and $(5,2)$

First, find the slope.

$$m = \frac{2-3}{5-6} = \frac{-1}{-1} = 1$$

Then use the slope and one of the points to write the equation in point-slope form.

$$y - y_1 = m(x - x_1)$$
$$y - 3 = 1(x-6)$$

or

$$y - 2 = 1(x-5)$$

Slope-Intercept Form

$$y - 2 = 1(x-5)$$
$$y - 2 = x - 5$$
$$y = x - 3$$
$$f(x) = x - 3$$

17. Passing through $(-2,0)$ and $(0,4)$

First, find the slope.

$$m = \frac{4-0}{0-(-2)} = \frac{4}{2} = 2$$

Then use the slope and one of the points to write the equation in point-slope form.

$$y - y_1 = m(x - x_1)$$
$$y - 4 = 2(x-0)$$

or

$$y - 0 = 2(x-(-2))$$
$$y - 0 = 2(x+2)$$

Slope-Intercept Form

$$y - 0 = 2(x+2)$$
$$y = 2x + 4$$
$$f(x) = 2x + 4$$

19. Passing through $(-6,13)$ and $(-2,5)$

First, find the slope.

$$m = \frac{5-13}{-2-(-6)} = \frac{-8}{-2+6} = \frac{-8}{4} = -2$$

Then use the slope and one of the points to write the equation in point-slope form.

$$y - y_1 = m(x - x_1)$$
$$y - 5 = -2(x-(-2))$$
$$y - 5 = -2(x+2)$$

or

$$y - 13 = -2(x-(-6))$$
$$y - 13 = -2(x+6)$$

Slope-Intercept Form

$$y - 13 = -2(x+6)$$
$$y - 13 = -2x - 12$$
$$y = -2x + 1$$
$$f(x) = -2x + 1$$

21. Passing through $(1,9)$ and $(4,-2)$

First, find the slope.

$$m = \frac{-2-9}{4-1} = \frac{-11}{3} = -\frac{11}{3}$$

Then use the slope and one of the points to write the equation in point-slope form.

$$y - y_1 = m(x - x_1)$$
$$y - (-2) = -\frac{11}{3}(x-4)$$
$$y + 2 = -\frac{11}{3}(x-4)$$

or

$$y - 9 = -\frac{11}{3}(x-1)$$

Slope-Intercept Form

$$y - 9 = -\frac{11}{3}(x-1)$$
$$y - 9 = -\frac{11}{3}x + \frac{11}{3}$$
$$y = -\frac{11}{3}x + \frac{38}{3}$$
$$f(x) = -\frac{11}{3}x + \frac{38}{3}$$

23. Passing through $(-2,-5)$ and $(3,-5)$
First, find the slope.
$$m=\frac{-5-(-5)}{3-(-2)}=\frac{0}{5}=0$$
Then use the slope and one of the points to write the equation in point-slope form.
$$y-y_1=m(x-x_1)$$
$$y-(-5)=0(x-3)$$
$$y+5=0(x-3)$$
or
$$y-(-5)=0(x-(-2))$$
$$y+5=0(x+2)$$
Slope-Intercept Form
$$y+5=0(x+2)$$
$$y+5=0$$
$$y=-5$$
$$f(x)=-5$$

25. Passing through $(7,8)$ with $x-\text{intercept}=3$
If the line has an $x-\text{intercept}=3$, it passes through the point $(3,0)$.
First, find the slope.
$$m=\frac{8-0}{7-3}=\frac{8}{4}=2$$
Then use the slope and one of the points to write the equation in point-slope form.
$$y-y_1=m(x-x_1)$$
$$y-0=2(x-3)$$
$$y-0=2(x-3)$$
or
$$y-8=2(x-7)$$
Slope-Intercept Form
$$y-8=2(x-7)$$
$$y-8=2x-14$$
$$y=2x-6$$
$$f(x)=2x-6$$

27. x-intercept $=2$ and y-intercept $=-1$
If the line has an x-intercept $=2$, it passes through the point $(2,0)$. If the line has a y-intercept $=-1$, it passes through $(0,-1)$.
First, find the slope.
$$m=\frac{-1-0}{0-2}=\frac{-1}{-2}=\frac{1}{2}$$
Then use the slope and one of the points to write the equation in point-slope form.
$$y-y_1=m(x-x_1)$$
$$y-0=\frac{1}{2}(x-2)$$
or
$$y-(-1)=\frac{1}{2}(x-0)$$
$$y+1=\frac{1}{2}(x-0)$$
Slope-Intercept Form
$$y-(-1)=\frac{1}{2}(x-0)$$
$$y+1=\frac{1}{2}x$$
$$y=\frac{1}{2}x-1$$
$$f(x)=\frac{1}{2}x-1$$

29. For $y=5x,\ m=5.$

a. A line parallel to this line would have the same slope, $m=5$.

b. A line perpendicular to it would have slope $m=-\frac{1}{5}$.

31. For $y=-7x,\ m=-7.$

a. A line parallel to this line would have the same slope, $m=-7$.

b. A line perpendicular to it would have slope $m=\frac{1}{7}$.

33. For $y=\frac{1}{2}x+3,\ m=\frac{1}{2}.$

a. A line parallel to this line would have the same slope, $m=\frac{1}{2}$.

b. A line perpendicular to it would have slope $m=-2$.

35. For $y = -\frac{2}{5}x - 1,\ m = -\frac{2}{5}$.

a. A line parallel to this line would have the same slope, $m = -\frac{2}{5}$.

b. A line perpendicular to it would have slope $m = \frac{5}{2}$.

37. To find the slope, we rewrite the equation in slope-intercept form.

$$4x + y = 7$$
$$y = -4x + 7$$

So, $m = -4$.

a. A line parallel to this line would have the same slope, $m = -4$.

b. A line perpendicular to it would have slope $m = \frac{1}{4}$.

39. To find the slope, we rewrite the equation in slope-intercept form.

$$2x + 4y = 8$$
$$4y = -2x + 8$$
$$y = -\frac{1}{2}x + 2$$

So, $m = -\frac{1}{2}$.

a. A line parallel to this line would have the same slope, $m = -\frac{1}{2}$.

b. A line perpendicular to it would have slope $m = 2$.

41. To find the slope, we rewrite the equation in slope-intercept form.

$$2x - 3y = 5$$
$$-3y = -2x + 5$$
$$y = \frac{2}{3}x - \frac{5}{3}$$

So, $m = \frac{2}{3}$.

a. A line parallel to this line would have the same slope, $m = \frac{2}{3}$.

b. A line perpendicular to it would have slope $m = -\frac{3}{2}$.

43. We know that $x = 6$ is a vertical line with undefined slope.

a. A line parallel to it would also be vertical with undefined slope.

b. A line perpendicular to it would be horizontal with slope $m = 0$.

45. Since L is parallel to $y = 2x$, we know it will have slope $m = 2$. We are given that it passes through (4, 2). We use the slope and point to write the equation in point-slope form.

$$y - y_1 = m(x - x_1)$$
$$y - 2 = 2(x - 4)$$

Solve for y to obtain slope-intercept form.

$$y - 2 = 2(x - 4)$$
$$y - 2 = 2x - 8$$
$$y = 2x - 6$$

In function notation, the equation of the line is $f(x) = 2x - 6$.

47. Since L is perpendicular to $y = 2x$, we know it will have slope $m = -\frac{1}{2}$. We are given that it passes through
(2, 4). We use the slope and point to write the equation in point-slope form.

$$y - y_1 = m(x - x_1)$$
$$y - 4 = -\frac{1}{2}(x - 2)$$

Solve for y to obtain slope-intercept form.

$$y - 4 = -\frac{1}{2}(x - 2)$$
$$y - 4 = -\frac{1}{2}x + 1$$
$$y = -\frac{1}{2}x + 5$$

In function notation, the equation of the line is $f(x) = -\frac{1}{2}x + 5$.

49. Since the line is parallel to $y=-4x+3$, we know it will have slope $m=-4$. We are given that it passes through (–8, –10). We use the slope and point to write the equation in point-slope form.

$$y-y_1=m(x-x_1)$$
$$y-(-10)=-4(x-(-8))$$
$$y+10=-4(x+8)$$

Solve for y to obtain slope-intercept form.

$$y+10=-4(x+8)$$
$$y+10=-4x-32$$
$$y=-4x-42$$

In function notation, the equation of the line is $f(x)=-4x-42$.

51. Since the line is perpendicular to $y=\frac{1}{5}x+6$, we know it will have slope $m=-5$. We are given that it passes through (2, –3). We use the slope and point to write the equation in point-slope form.

$$y-y_1=m(x-x_1)$$
$$y-(-3)=-5(x-2)$$
$$y+3=-5(x-2)$$

Solve for y to obtain slope-intercept form.

$$y+3=-5(x-2)$$
$$y+3=-5x+10$$
$$y=-5x+7$$

In function notation, the equation of the line is $f(x)=-5x+7$.

53. To find the slope, we rewrite the equation in slope-intercept form.

$$2x-3y=7$$
$$-3y=-2x+7$$
$$y=\frac{2}{3}x-\frac{7}{3}$$

Since the line is parallel to $y=\frac{2}{3}x-\frac{7}{3}$, we know it will have slope $m=\frac{2}{3}$. We are given that it passes through (–2, 2). We use the slope and point to write the equation in point-slope form.

$$y-y_1=m(x-x_1)$$
$$y-2=\frac{2}{3}(x-(-2))$$
$$y-2=\frac{2}{3}(x+2)$$

Solve for y to obtain slope-intercept form.

$$y-2=\frac{2}{3}(x+2)$$
$$y-2=\frac{2}{3}x+\frac{4}{3}$$
$$y=\frac{2}{3}x+\frac{10}{3}$$

In function notation, the equation of the line is $f(x)=\frac{2}{3}x+\frac{10}{3}$.

55. To find the slope, we rewrite the equation in slope-intercept form.

$$x-2y=3$$
$$-2y=-x+3$$
$$y=\frac{1}{2}x-\frac{3}{2}$$

Since the line is perpendicular to $y=\frac{1}{2}x-\frac{3}{2}$, we know it will have slope $m=-2$. We are given that it passes through (4, –7). We use the slope and point to write the equation in point-slope form.

$$y-y_1=m(x-x_1)$$
$$y-(-7)=-2(x-4)$$
$$y+7=-2(x-4)$$

Solve for y to obtain slope-intercept form.

$$y+7=-2(x-4)$$
$$y+7=-2x+8$$
$$y=-2x+1$$

In function notation, the equation of the line is $f(x)=-2x+1$.

57. Since the line is perpendicular to $x=6$ which is a vertical line, we know the graph of f is a horizontal line with 0 slope. The graph of f passes through $(-1,5)$, so the equation of f is $f(x)=5$.

59. First we need to find the slope of the line with x – intercept of 2 and y – intercept of –4. This line will pass through $(2,0)$ and $(0,-4)$. We use these points to find the slope.

$$m = \frac{-4-0}{0-2} = \frac{-4}{-2} = 2$$

Since the graph of f is perpendicular to this line, it will have slope $m = -\frac{1}{2}$.

Use the point $(-6,4)$ and the slope $-\frac{1}{2}$ to find the equation of the line.

$$\begin{aligned} y - y_1 &= m(x - x_1) \\ y - 4 &= -\frac{1}{2}(x-(-6)) \\ y - 4 &= -\frac{1}{2}(x+6) \\ y - 4 &= -\frac{1}{2}x - 3 \\ y &= -\frac{1}{2}x + 1 \\ f(x) &= -\frac{1}{2}x + 1 \end{aligned}$$

61. First put the equation $3x - 2y = 4$ in slope-intercept form.

$$\begin{aligned} 3x - 2y &= 4 \\ -2y &= -3x + 4 \\ y &= \frac{3}{2}x - 2 \end{aligned}$$

The equation of f will have slope $-\frac{2}{3}$ since it is perpendicular to the line above and the same y – intercept –2.

So the equation of f is $f(x) = -\frac{2}{3}x - 2.$

63. The graph of f is just the graph of g shifted down 2 units. So subtract 2 from the equation of $g(x)$ to obtain the equation of $f(x)$.

$$f(x) = g(x) - 2 = 4x - 3 - 2 = 4x - 5$$

65. To find the slope of the line whose equation is $Ax + By = C$, put this equation in slope-intercept form by solving for y.

$$\begin{aligned} Ax + By &= C \\ By &= -Ax + C \\ y &= -\frac{A}{B}x + \frac{C}{B} \end{aligned}$$

The slope of this line is $m = -\frac{A}{B}$ so the slope of the line that is parallel to it is the same, $-\frac{A}{B}$.

67. a. First, find the slope using $(20, 38.9)$ and $(10, 31.1)$.

$$m = \frac{38.9 - 31.1}{20 - 10} = \frac{7.8}{10} = 0.78$$

Then use the slope and one of the points to write the equation in point-slope form.

$$\begin{aligned} y - y_1 &= m(x - x_1) \\ y - 31.1 &= 0.78(x - 10) \end{aligned}$$

or

$$y - 38.9 = 0.78(x - 20)$$

b.

$$\begin{aligned} y - 31.1 &= 0.78(x - 10) \\ y - 31.1 &= 0.78x - 7.8 \\ y &= 0.78x + 23.3 \\ f(x) &= 0.78x + 23.3 \end{aligned}$$

c. $f(40) = 0.78(40) + 23.3 = 54.5$

The linear function predicts the percentage of never married American females, ages 25 – 29, to be 54.5% in 2020.

69. a/b.

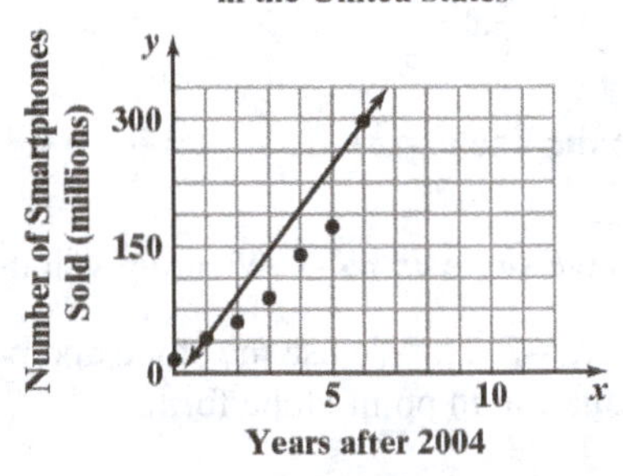

Use the two points $(1, 40.8)$ and $(6, 296.6)$ to find the slope.

$$m = \frac{296.6-40.8}{6-1} = \frac{255.8}{5} = 51.16$$

Then use the slope and one of the points to write the equation in point-slope form.

$$y - y_1 = m(x - x_1)$$
$$y - 40.8 = 51.16(x-1)$$
or
$$y - 296.6 = 51.16(x-6)$$

Solve for y to obtain slope-intercept form.

$$y - 40.8 = 51.16(x-1)$$
$$y - 40.8 = 51.16x - 51.16$$
$$y = 51.16x - 10.36$$
$$f(x) = 51.16x - 10.36$$

c. $f(x) = 51.16x - 10.36$
$$f(11) = 51.16(11) - 10.36$$
$$= 552.4$$

The function predicts that 552.4 million smartphones will be sold in 2015.

71. a. $m = \dfrac{970-582}{2016-2007} = \dfrac{388}{9} \approx 43.1$

The cost of Social Security is projected to increase at a rate of approximately $43.1 billion per year.

b. $m = \dfrac{909-446}{2016-2007} = \dfrac{463}{9} \approx 51.4$

The cost of Medicare is projected to increase at a rate of approximately $51.4 billion per year.

c. No, the slopes are not the same. This means that the cost of Medicare is projected to increase at a faster rate than the cost of Social Security.

73. – 81. Answers will vary.

83. makes sense

85. makes sense

87. true

89. true

91. $By = 8x - 1$
$$y = \frac{8}{B}x - 1$$

Since $\frac{8}{B}$ is the slope, $\frac{8}{B}$ must equal -2.

$$\frac{8}{B} = -2$$
$$8 = -2B$$
$$-4 = B$$

93. Find the slope of the line by using the two points, $(-3,0)$, the x-intercept and $(0,-6)$, the y-intercept.

$$m = \frac{-6-0}{0-(-3)} = \frac{-6}{3} = -2$$

So the equation of the line is $y = -2x - 6$.

Substitute -40 for x:

$$y = -2(-40) - 6 = 80 - 6 = 74$$

This is the y-coordinate of the first ordered pair.

Substitute -200 for y:

$$-200 = -2x - 6$$
$$-194 = -2x$$
$$97 = x$$

This is the x-coordinate of the second ordered pair.

Therefore, the two ordered pairs are $(-40, 74)$ and $(97, -200)$.

95. $f(-2) = 3(-2)^2 - 8(-2) + 5$
$$= 3(4) + 16 + 5$$
$$= 12 + 16 + 5 = 33$$

96. $f(-1) = (-1)^2 - 3(-1) + 4 = 1 + 3 + 4 = 8$
$$g(-1) = 2(-1) - 5 = -2 - 5 = -7$$
$$(fg)(-1) = (f)(-1) \cdot (g)(-1) = 8(-7) = -56$$

97. Let x = the measure of the smallest angle.
$x + 20$ = the measure of the second angle.
$2x$ = the measure of the third angle.

$$x+(x+20)+2x=180$$
$$x+x+20+2x=180$$
$$4x+20=180$$
$$4x=160$$
$$x=40$$

Find the other angles.
$x + 20 = 40 + 20 = 60$
$2x = 2(40) = 80$
The angles are $40°$, $60°$, and $80°$.

98. a.
$$2x-y=-4$$
$$2(-5)-(-6)=-4$$
$$-10+6=-4$$
$$-4=-4, \text{ true}$$
The point satisfies the equation.

b.
$$3x-5y=15$$
$$3(-5)-5(-6)=15$$
$$-15+30=15$$
$$15=15, \text{ true}$$
The point satisfies the equation.

99. The graphs intersect at $(3,-4)$.

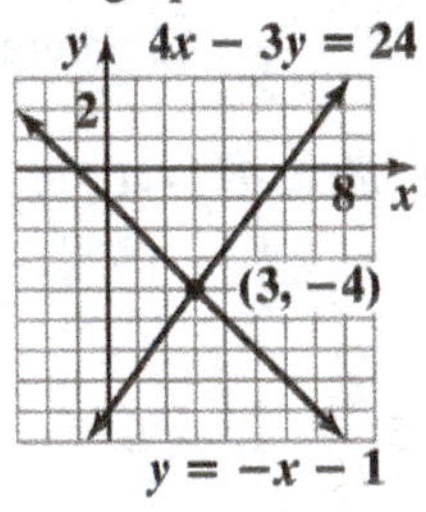

100.
$$7x-2(-2x+4)=3$$
$$7x+4x-8=3$$
$$11x-8=3$$
$$11x=11$$
$$x=1$$
The solution set is $\{1\}$.

Chapter 2 Review

1. The relation is a function.
Domain $\{3, 4, 5\}$
Range $\{10\}$

2. The relation is a function.
Domain $\{1, 2, 3, 4\}$
Range $\{-6, \pi, 12, 100\}$

3. The relation is not a function.
Domain $\{13, 15\}$
Range $\{14, 16, 17\}$

4. a. $f(0)=7(0)-5=0-5=-5$

b. $f(3)=7(3)-5=21-5=16$

c. $f(-10)=7(-10)-5=-75$

d. $f(2a)=7(2a)-5=14a-5$

e. $f(a+2)=7(a+2)-5$
$=7a+14-5=7a+9$

5. a. $g(0)=3(0)^2-5(0)+2=2$

b. $g(5)=3(5)^2-5(5)+2$
$=3(25)-25+2$
$=75-25+2=52$

c. $g(-4)=3(-4)^2-5(-4)+2=70$

d. $g(b)=3(b)^2-5(b)+2$
$=3b^2-5b+2$

e. $g(4a)=3(4a)^2-5(4a)+2$
$=3(16a^2)-20a+2$
$=48a^2-20a+2$

6. g shifts the graph of f down one unit.

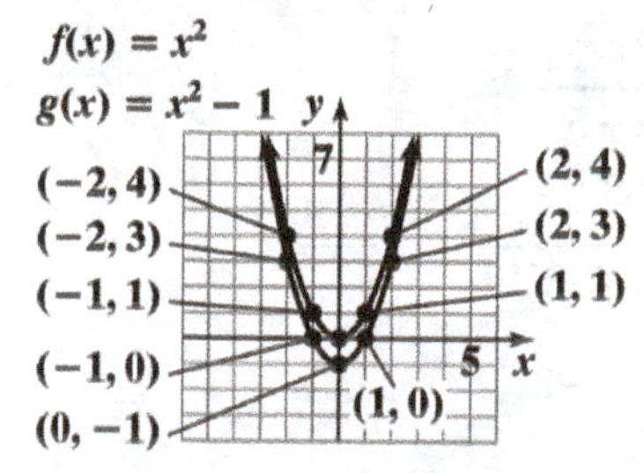

7. g shifts the graph of f up two units.

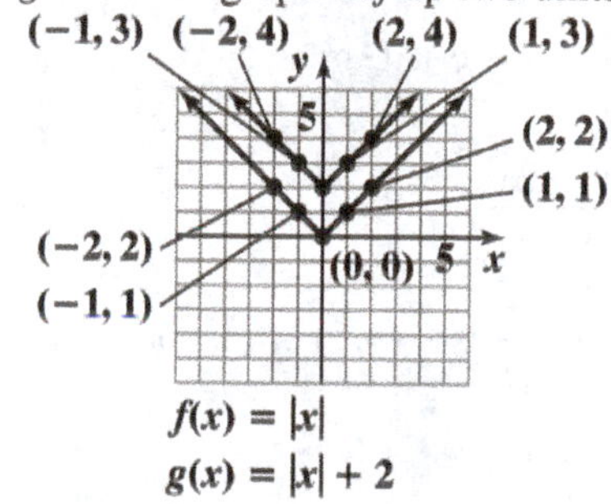

8. The vertical line test shows that this is not the graph of a function.

9. The vertical line test shows that this is the graph of a function.

10. The vertical line test shows that this is the graph of a function.

11. The vertical line test shows that this is not the graph of a function.

12. The vertical line test shows that this is not the graph of a function.

13. The vertical line test shows that this is the graph of a function.

14. $f(-2) = -3$

15. $f(0) = -2$

16. When $x = 3$, $f(x) = -5$.

17. The domain of f is $[-3,5)$.

18. The range of f is $[-5,0]$.

19. a. The eagle's height is a function of its time in flight because every time, t, is associated with at most one height.

b. $f(15) = 0$

At time $t = 15$ seconds, the eagle is at height zero. This means that after 15 seconds, the eagle is on the ground.

c. The eagle's maximum height is 45 meters.

d. For $x = 7$ and 22, $f(x) = 20$. This means that at times 7 seconds and 22 seconds, the eagle is at a height of 20 meters.

e. The eagle began the flight at 45 meters and remained there for approximately 3 seconds. At that time, the eagle descended for 9 seconds. It landed on the ground and stayed there for 5 seconds. The eagle then began to climb back up to a height of 44 meters.

20. The domain of f is $(-\infty,\infty)$.

21. The domain of f is $(-\infty,-8)$ or $(-8,\infty)$.

22. The domain of f is $(-\infty,5)$ or $(5,\infty)$.

23. a. $(f+g)(x) = (4x-5)+(2x+1)$

$= 4x-5+2x+1$

$= 6x-4$

b. $(f+g)(3) = 6(3)-4$

$= 18-4 = 14$

24. a. $(f+g)(x)$

$= \left(5x^2-x+4\right)+(x-3)$

$= 5x^2-x+4+x-3 = 5x^2+1$

b. $(f+g)(3) = 5(3)^2+1 = 5(9)+1$

$= 45+1 = 46$

25. The domain of $f+g$ is $(-\infty,4)$ or $(4,\infty)$.

26. The domain of $f+g$ is $(-\infty,-6)$ or $(-6,-1)$ or $(-1,\infty)$.

27. $f(x)=x^2-2x,\quad g(x)=x-5$

$$\begin{aligned}(f+g)(x)&=(x^2-2x)+(x-5)\\&=x^2-2x+x-5\\&=x^2-x-5\end{aligned}$$

$$\begin{aligned}(f+g)(-2)&=(-2)^2-(-2)-5\\&=4+2-5=1\end{aligned}$$

28. From Exercise 27 we know $(f+g)(x)=x^2-x-5$. We can use this to find $f(3)+g(3)$.

$$\begin{aligned}f(3)+g(3)&=(f+g)(3)\\&=(3)^2-(3)-5\\&=9-3-5=1\end{aligned}$$

29. $f(x)=x^2-2x,\quad g(x)=x-5$

$$\begin{aligned}(f-g)(x)&=(x^2-2x)-(x-5)\\&=x^2-2x-x+5\\&=x^2-3x+5\end{aligned}$$

$$(f-g)(x)=x^2-3x+5$$

$$\begin{aligned}(f-g)(1)&=(1)^2-3(1)+5\\&=1-3+5=3\end{aligned}$$

30. From Exercise 29 we know $(f-g)(x)=x^2-3x+5$. We can use this to find $f(4)-g(4)$.

$$\begin{aligned}f(4)-g(4)&=(f-g)(4)\\&=(4)^2-3(4)+5\\&=16-12+5=9\end{aligned}$$

31. Since $(fg)(-3)=f(-3)\cdot g(-3)$, find $f(-3)$ and $g(-3)$ first.

$$\begin{aligned}f(-3)&=(-3)^2-2(-3)\\&=9+6=15\end{aligned}$$

$$g(-3)=-3-5=-8$$

$$\begin{aligned}(fg)(-3)&=f(-3)\cdot g(-3)\\&=15(-8)=-120\end{aligned}$$

32. $f(x)=x^2-2x,\quad g(x)=x-5$

$$\left(\frac{f}{g}\right)(x)=\frac{x^2-2x}{x-5}$$

$$\begin{aligned}\left(\frac{f}{g}\right)(4)&=\frac{(4)^2-2(4)}{4-5}=\frac{16-8}{-1}\\&=\frac{8}{-1}=-8\end{aligned}$$

33. $(f-g)(x)=x^2-3x+5$

The domain of $f-g$ is $(-\infty,\infty)$.

34. $\left(\frac{f}{g}\right)(x)=\frac{x^2-2x}{x-5}$

The domain of $\frac{f}{g}$ is $(-\infty,5)$ or $(5,\infty)$.

35. $x+2y=4$

Find the x–intercept by setting $y=0$ and the y–intercept by setting $x=0$.

$x+2(0)=4$	$0+2y=4$
$x+0=4$	$2y=4$
$x=4$	$y=2$

Choose another point to use as a check.
Let $x=1$.

$$\begin{aligned}1+2y&=4\\2y&=3\\y&=\frac{3}{2}\end{aligned}$$

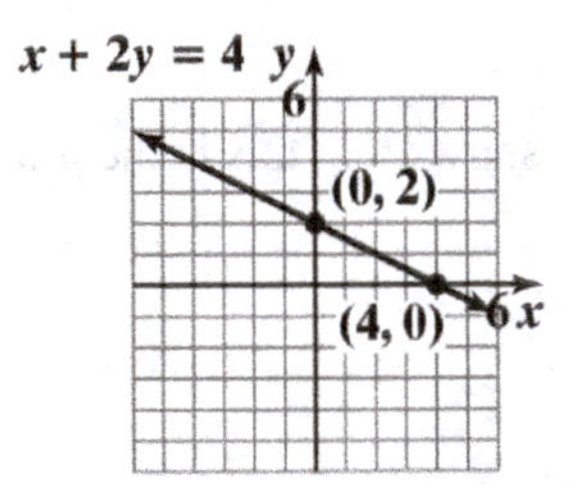

36. $2x-3y=12$

Find the x–intercept by setting $y=0$ and the y–intercept by setting $x=0$.

$$\begin{aligned} 2x-3(0)&=12 \\ 2x+0&=12 \\ 2x&=12 \\ x&=6 \end{aligned} \qquad \begin{aligned} 2(0)-3y&=12 \\ 0-3y&=12 \\ -3y&=12 \\ y&=-4 \end{aligned}$$

Choose another point to use as a check.
Let $x=1$.

$$\begin{aligned} 2(1)-3y&=12 \\ 2-3y&=12 \\ -3y&=10 \\ y&=-\frac{10}{3} \end{aligned}$$

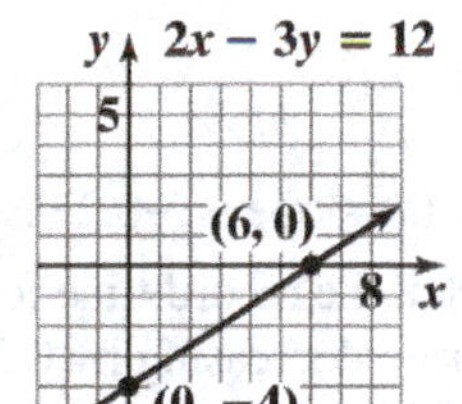

37. $4x=8-2y$

Find the x–intercept by setting $y=0$ and the y–intercept by setting $x=0$.

$$\begin{aligned} 4x&=8-2(0) \\ 4x&=8-0 \\ 4x&=8 \\ x&=2 \end{aligned} \qquad \begin{aligned} 4(0)&=8-2y \\ 0&=8-2y \\ 2y&=8 \\ y&=4 \end{aligned}$$

Choose another point to use as a check. Let $x=1$.

$$\begin{aligned} 4(1)&=8-2y \\ 4&=8-2y \\ -4&=-2y \\ 2&=y \end{aligned}$$

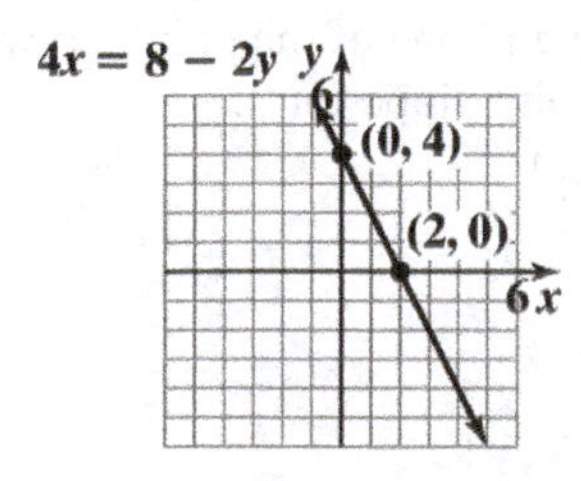

38. $m=\dfrac{2-(-4)}{5-2}=\dfrac{6}{3}=2$

The line through the points rises.

39. $m=\dfrac{3-(-3)}{-2-7}=\dfrac{6}{-9}=-\dfrac{2}{3}$

The line through the points falls.

40. $m=\dfrac{2-(-1)}{3-3}=\dfrac{3}{0}$

m is undefined. The line through the points is vertical.

41. $m=\dfrac{4-4}{-3-(-1)}=\dfrac{0}{-2}=0$

The line through the points is horizontal.

42. $y=2x-1$

$m=2 \qquad y-\text{intercept}=-1$

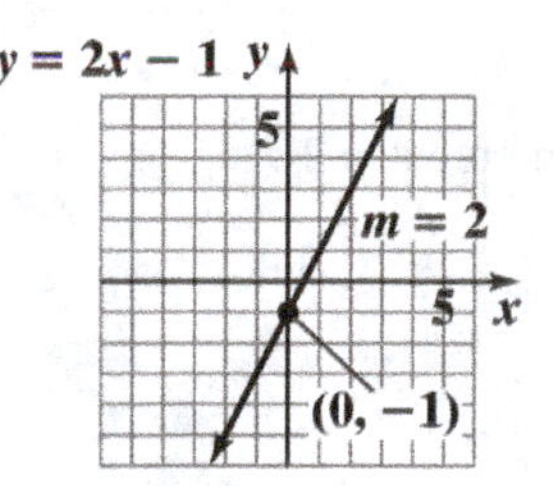

43. $f(x)=-\dfrac{1}{2}x+4$

$m=-\dfrac{1}{2} \qquad y-\text{intercept}=4$

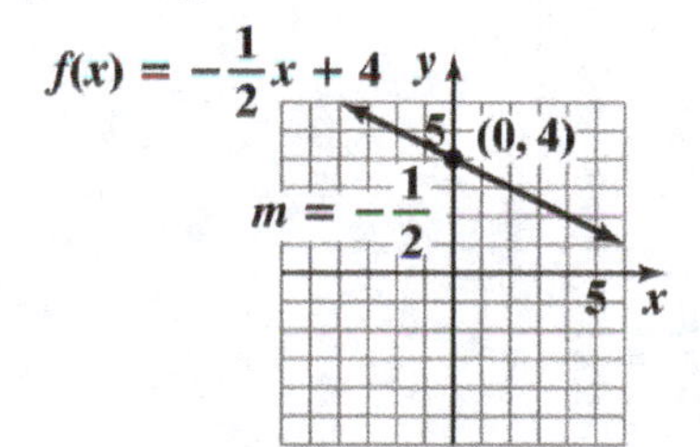

44. $y=\dfrac{2}{3}x$

$m=\dfrac{2}{3} \qquad y-\text{intercept}=0$

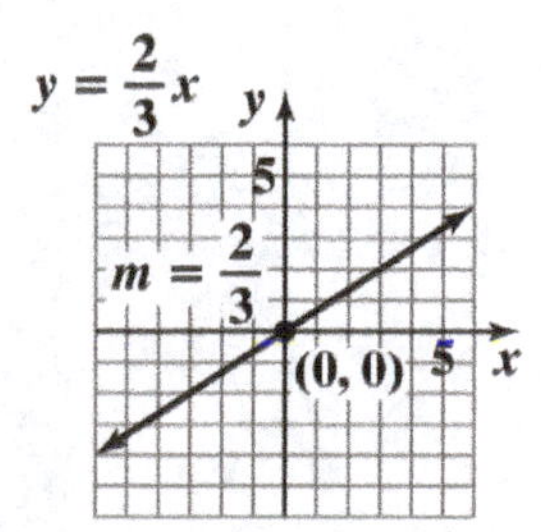

45. To rewrite the equation in slope-intercept form, solve for y.

$$2x + y = 4$$
$$y = -2x + 4$$
$$m = -2 \qquad y\text{-intercept} = 4$$

46.
$$-3y = 5x$$
$$y = -\frac{5}{3}x$$
$$m = -\frac{5}{3} \qquad y\text{-intercept} = 0$$

47.
$$5x + 3y = 6$$
$$3y = -5x + 6$$
$$y = -\frac{5}{3}x + 2$$
$$m = -\frac{5}{3} \qquad y\text{-intercept} = 2$$

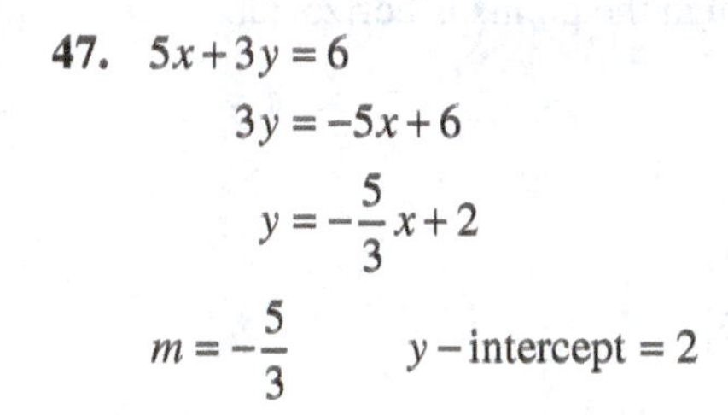

48. $y = 2$

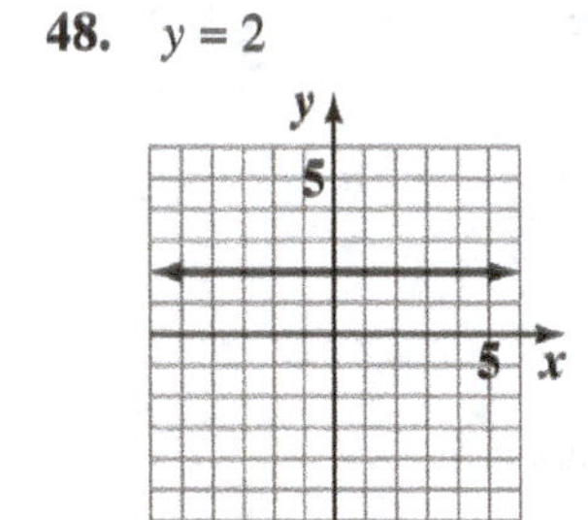

49.
$$7y = -21$$
$$y = -3$$

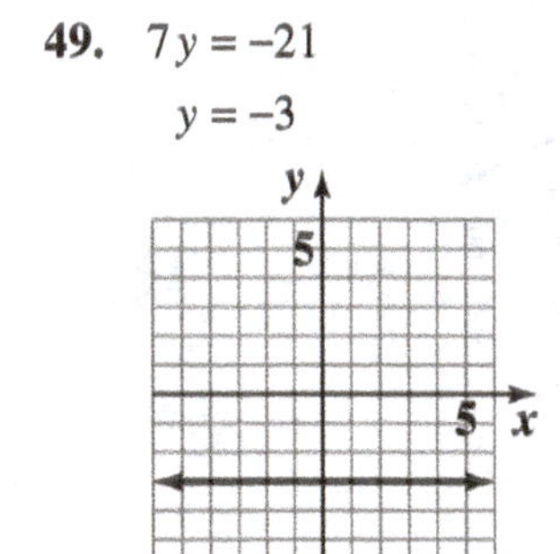

50.
$$f(x) = -4$$
$$y = -4$$

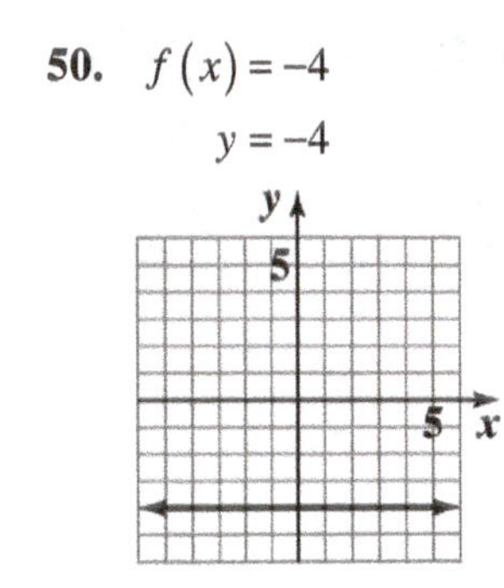

51. $x = 3$

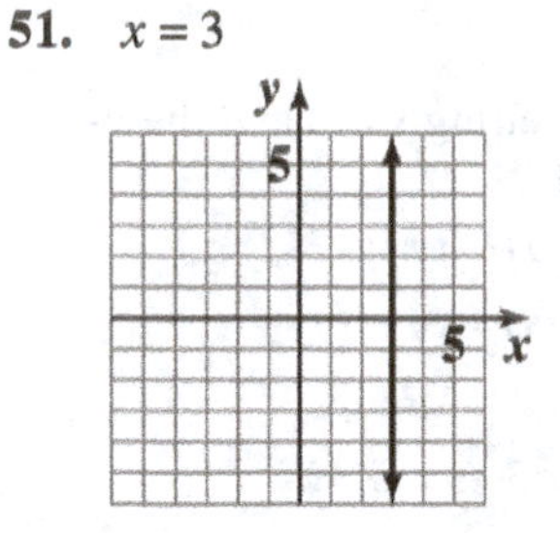

52.
$$2x = -10$$
$$x = -5$$

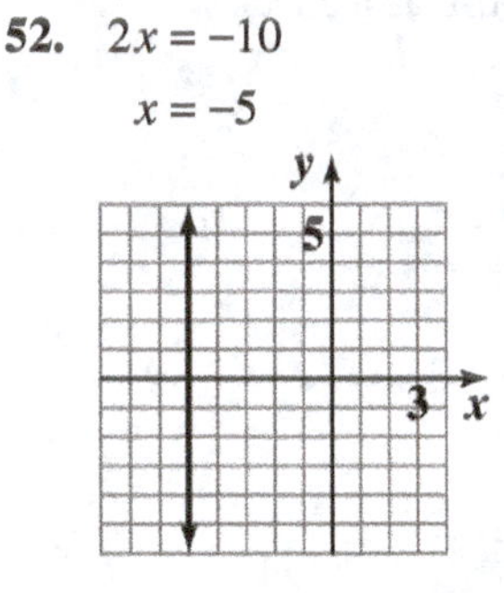

53. In $f(t) = -0.27t + 70.45$, the slope is -0.27. A slope of -0.27 indicates that the record time for the women's 400-meter has been decreasing by 0.27 seconds per year since 1900.

54. **a.** $m = \dfrac{1163 - 617}{1998 - 1994} = \dfrac{546}{4} \approx 137$

There was an average increase of approximately 137 discharges per year.

b. $m = \dfrac{668 - 1273}{2004 - 2001} = \dfrac{-605}{3} \approx -202$

The was an average decrease of approximately 202 discharges per year.

55. **a.** Find the slope of the line by using the two points (0, 32) and (100,212).

$$m = \frac{212 - 32}{100 - 0} = \frac{180}{100} = \frac{9}{5}$$

We use the slope and one of the points to write the equation in point-slope form.

$$y - y_1 = m(x - x_1)$$
$$y - 32 = \frac{9}{5}(x - 0)$$
$$y - 32 = \frac{9}{5}x$$
$$y = \frac{9}{5}x + 32$$
$$F = \frac{9}{5}C + 32$$

b. Let $C = 30$.

$F = \frac{9}{5}(30) + 32 = 54 + 32 = 86$

The Fahrenheit temperature is $86°$ when the Celsius temperature is $30°$.

56. Slope $= -6$, passing through $(-3, 2)$

Point-Slope Form

$y - y_1 = m(x - x_1)$
$y - 2 = -6(x - (-3))$
$y - 2 = -6(x + 3)$

Slope-Intercept Form

$y - 2 = -6(x + 3)$
$y - 2 = -6x - 18$
$y = -6x - 16$
$f(x) = -6x - 16$

57. Passing through $(1, 6)$ and $(-1, 2)$

First, find the slope.

$m = \frac{6 - 2}{1 - (-1)} = \frac{4}{2} = 2$

Then use the slope and one of the points to write the equation in point-slope form.

$y - y_1 = m(x - x_1)$
$y - 6 = 2(x - 1)$

or

$y - y_1 = m(x - x_1)$
$y - 2 = 2(x - (-1))$
$y - 2 = 2(x + 1)$

Slope-Intercept Form

$y - 6 = 2(x - 1)$
$y - 6 = 2x - 2$
$y = 2x + 4$
$f(x) = 2x + 4$

58. Rewrite $3x + y = 9$ in slope-intercept form.

$3x + y = 9$
$y = -3x + 9$

Since the line we are concerned with is parallel to this line, we know it will have slope $m = -3$. We are given that it passes through $(4, -7)$. We use the slope and point to write the equation in point-slope form.

$y - y_1 = m(x - x_1)$
$y - (-7) = -3(x - 4)$
$y + 7 = -3(x - 4)$

Solve for y to obtain slope-intercept form.

$y + 7 = -3(x - 4)$
$y + 7 = -3x + 12$
$y = -3x + 5$

In function notation, the equation of the line is $f(x) = -3x + 5$.

59. The line is perpendicular to $y = \frac{1}{3}x + 4$, so the slope is -3. We are given that it passes through $(-2, 6)$. We use the slope and point to write the equation in point-slope form.

$y - y_1 = m(x - x_1)$
$y - 6 = -3(x - (-2))$
$y - 6 = -3(x + 2)$

Solve for y to obtain slope-intercept form.

$y - 6 = -3(x + 2)$
$y - 6 = -3x - 6$
$y = -3x$

In function notation, the equation of the line is $f(x) = -3x$.

60. **a.** First, find the slope using the points $(3, 66)$ and $(5, 82)$.

$m = \frac{82 - 66}{5 - 3} = \frac{16}{2} = 8$

Then use the slope and one of the points to write the equation in point-slope form.

$y - y_1 = m(x - x_1)$
$y - 66 = 8(x - 3)$

or

$y - 82 = 8(x - 5)$

b. Solve for y to obtain slope-intercept form.

$y - 66 = 8(x - 3)$
$y - 66 = 8x - 24$
$y = 8x + 42$
$f(x) = 8x + 42$

c. $f(x) = 8x + 42$
$f(7) = 8(7) + 42$
$= 98$

The linear function predicts that 98% of libraries will have wireless Internet access in 2012.

Chapter 2 Test

1. The relation is a function.
   Domain {1, 3, 5, 6}
   Range {2, 4, 6}

2. The relation is not a function.
   Domain {2, 4, 6}
   Range {1, 3, 5, 6}

3. $f(a+4)=3(a+4)-2$
   $=3a+12-2=3a+10$

4. $f(-2)=4(-2)^2-3(-2)+6$
   $=4(4)+6+6=16+6+6=28$

5. g shifts the graph of f up 2 units.

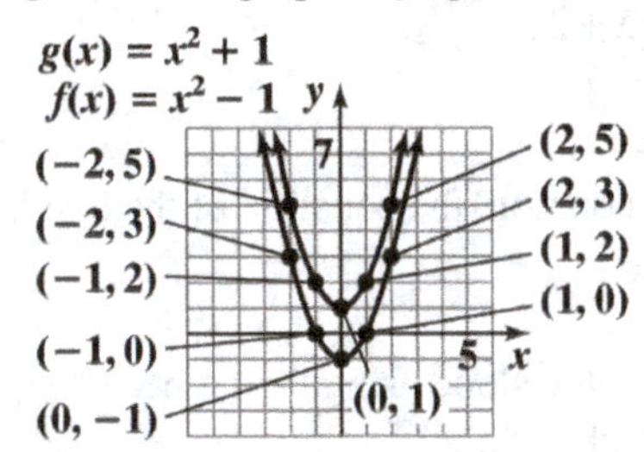

6. The vertical line test shows that this is the graph of a function.

7. The vertical line test shows that this is not the graph of a function.

8. $f(6)=-3$

9. $f(x)=0$ when $x=-2$ and $x=3$.

10. The domain of f is $(-\infty,\infty)$.

11. The range of f is $(-\infty,3]$.

12. The domain of f is $(-\infty,10)$ or $(10,\infty)$.

13. $f(x)=x^2+4x$ and $g(x)=x+2$
    $(f+g)(x)=f(x)+g(x)$
    $=\left(x^2+4x\right)+(x+2)$
    $=x^2+4x+x+2$
    $=x^2+5x+2$
    $(f+g)(3)=(3)^2+5(3)+2$
    $=9+15+2=26$

14. $f(x)=x^2+4x$ and $g(x)=x+2$
    $(f-g)(x)=f(x)-g(x)$
    $=\left(x^2+4x\right)-(x+2)$
    $=x^2+4x-x-2$
    $=x^2+3x-2$
    $(f-g)(-1)=(-1)^2+3(-1)-2$
    $=1-3-2=-4$

15. We know that $(fg)(x)=f(x)\cdot g(x)$. So, to find $(fg)(-5)$, we use $f(-5)$ and $g(-5)$.
    $f(-5)=(-5)^2+4(-5)=25-20=5$
    $g(-5)=-5+2=-3$
    $(fg)(-5)=f(-5)\cdot g(-5)$
    $=5(-3)=-15$

16. $f(x)=x^2+4x$ and $g(x)=x+2$
    $\left(\frac{f}{g}\right)(x)=\frac{x^2+4x}{x+2}$
    $\left(\frac{f}{g}\right)(2)=\frac{(2)^2+4(2)}{2+2}=\frac{4+8}{4}=\frac{12}{4}=3$

17. Domain of $\frac{f}{g}$ is $(-\infty,-2)$ or $(-2,\infty)$.

18. $4x-3y=12$
    Find the x–intercept by setting $y=0$.
    $4x-3(0)=12$
    $4x=12$
    $x=3$
    Find the y–intercept by setting $x=0$.
    $4(0)-3y=12$
    $-3y=12$
    $y=-4$

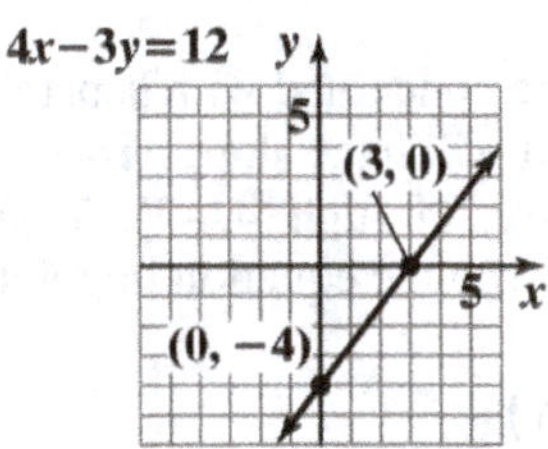

19. $f(x) = -\frac{1}{3}x + 2$

$m = -\frac{1}{3}$ $\quad y - \text{intercept} = 2$

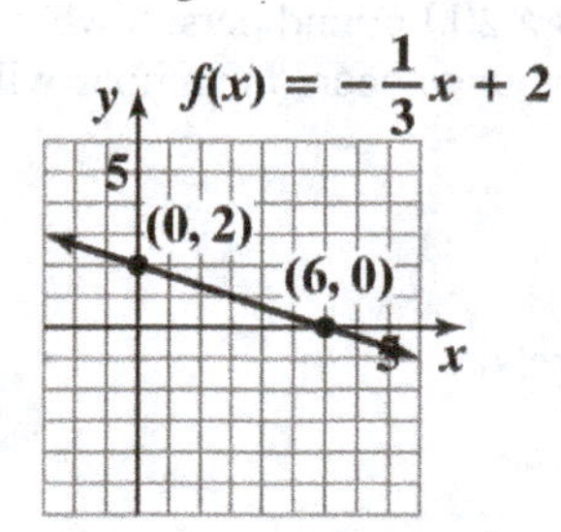

20. $f(x) = 4$

$y = 4$

An equation of the form $y = b$ is a horizontal line.

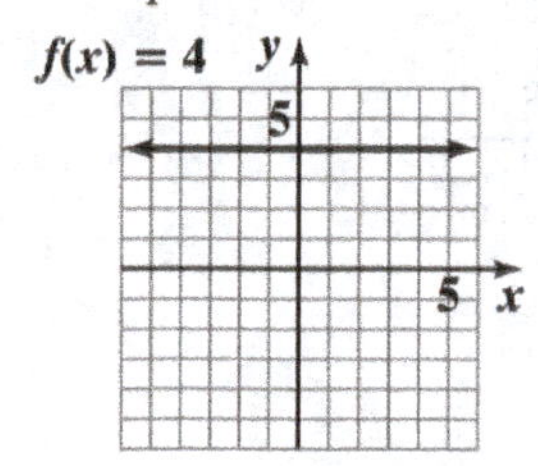

21. $m = \frac{4-2}{1-5} = \frac{2}{-4} = -\frac{1}{2}$

The line through the points falls.

22. $m = \frac{5-(-5)}{4-4} = \frac{10}{0}$

m is undefined

The line through the points is vertical.

23. $V(10) = 3.6(10) + 140$

$= 36 + 140 = 176$

In the year 2005, there were 176 million Super Bowl viewers.

24. The slope is 3.6. This means the number of Super Bowl viewers is increasing at a rate of 3.6 million per year.

25. Passing through $(-1, -3)$ and $(4, 2)$

First, find the slope.

$m = \frac{2-(-3)}{4-(-1)} = \frac{5}{5} = 1$

Then use the slope and one of the points to write the equation in point-slope form.

$y - y_1 = m(x - x_1)$

$y-(-3) = 1(x-(-1))$

$y + 3 = 1(x + 1)$

or

$y - 2 = 1(x - 4)$

$y - 2 = x - 4$

Slope-Intercept Form

$y - 2 = x - 4$

$y = x - 2$

In function notation, the equation of the line is $f(x) = x - 2$.

26. The line is perpendicular to $y = -\frac{1}{2}x - 4$, so the slope is 2. We are given that it passes through $(-2, 3)$. We use the slope and point to write the equation in point-slope form.

$y - y_1 = m(x - x_1)$

$y - 3 = 2(x-(-2))$

$y - 3 = 2(x + 2)$

Solve for y to obtain slope-intercept form.

$y - 3 = 2(x + 2)$

$y - 3 = 2x + 4$

$y = 2x + 7$

In function notation, the equation of the line is $f(x) = 2x + 7$.

27. The line is parallel to $x+2y=5$.
Put this equation in slope-intercept form by solving for y.

$$x+2y=5$$
$$2y=-x+5$$
$$y=-\frac{1}{2}x+\frac{5}{2}$$

Therefore the slopes are the same; $m=-\frac{1}{2}$.

We are given that it passes through $(6,-4)$.

We use the slope and point to write the equation in point-slope form.

$$y-y_1=m(x-x_1)$$
$$y-(-4)=-\frac{1}{2}(x-6)$$
$$y+4=-\frac{1}{2}(x-6)$$

Solve for y to obtain slope-intercept form.

$$y+4=-\frac{1}{2}(x-6)$$
$$y+4=-\frac{1}{2}x+3$$
$$y=-\frac{1}{2}x-1$$

In function notation, the equation of the line is

$$f(x)=-\frac{1}{2}x-1.$$

28. **a.** First, find the slope using the points $(3, 0.053)$ and $(7, 0.121)$.

$$m=\frac{0.121-0.053}{7-3}=\frac{0.068}{4}=0.017$$

Then use the slope and a point to write the equation in point-slope form.

$$y-y_1=m(x-x_1)$$
$$y-0.053=0.017(x-3)$$

or

$$y-0.121=0.017(x-7)$$

b.
$$y-0.053=0.017(x-3)$$
$$y-0.053=0.017x-0.051$$
$$y=0.017x+0.002$$
$$f(x)=0.017x+0.002$$

c.
$$f(x)=0.017x+0.002$$
$$f(8)=0.017(8)+0.002$$
$$=0.138$$

The function predicts that the blood alcohol concentration of a 200-pound person who consumes 8 one-ounce beers in an hour will be 0.138.

Cumulative Review Exercises

1. $\{0, 1, 2, 3\}$

2. False. π is an irrational number.

3.
$$\frac{8-3^2\div 9}{|-5|-[5-(18\div 6)]^2}$$
$$=\frac{8-9\div 9}{5-[5-(3)]^2}=\frac{8-1}{5-[2]^2}$$
$$=\frac{7}{5-4}=\frac{7}{1}=7$$

4.
$$4-(2-9)^0+3^2\div 1+3$$
$$=4-(-7)^0+9\div 1+3=4-1+9\div 1+3$$
$$=4-1+9+3=3+9+3=15$$

5.
$$3-[2(x-2)-5x]$$
$$=3-[2x-4-5x]=3-[-3x-4]$$
$$=3+3x+4=3x+7$$

6.
$$2+3x-4=2(x-3)$$
$$3x-2=2x-6$$
$$x-2=-6$$
$$x=-4$$

The solution set is $\{-4\}$.

7.
$$4x+12-8x=-6(x-2)+2x$$
$$12-4x=-6x+12+2x$$
$$12-4x=-4x+12$$
$$12=12$$
$$0=0$$

The solution set is $\{x|x \text{ is a real number}\}$ or $(-\infty,\infty)$ or $\mathbb{R}$. The equation is an identity.

8. $\frac{x-2}{4}=\frac{2x+6}{3}$
$4(2x+6)=3(x-2)$
$8x+24=3x-6$
$5x+24=-6$
$5x=-30$
$x=-6$
The solution set is $\{-6\}$.

9. Let x = the price before reduction.
$x-0.20x=1800$
$0.80x=1800$
$x=2250$
The price of the computer before the reduction was \$2250.

10. $A=p+prt$
$A-p=prt$
$\frac{A-p}{pr}=t$

11. $\left(3x^4y^{-5}\right)^{-2}=\left(\frac{3x^4}{y^5}\right)^{-2}=\left(\frac{y^5}{3x^4}\right)^2=\frac{y^{10}}{9x^8}$

12. $\left(\frac{3x^2y^{-4}}{x^{-3}y^2}\right)^2=\left(\frac{3x^2x^3}{y^2y^4}\right)^2$
$=\left(\frac{3x^5}{y^6}\right)^2=\frac{9x^{10}}{y^{12}}$

13. $\left(7\times10^{-8}\right)\left(3\times10^2\right)$
$=(7\times3)\left(10^{-8}\times10^2\right)=21\times10^{-6}$
$=(2.1\times10)\times10^{-6}=2.1\left(10\times10^{-6}\right)$
$=2.1\times10^{-5}$

14. The relation is a function.
Domain {1, 2, 3, 4, 6}
Range {5}

15. g shifts the graph of f up three units.

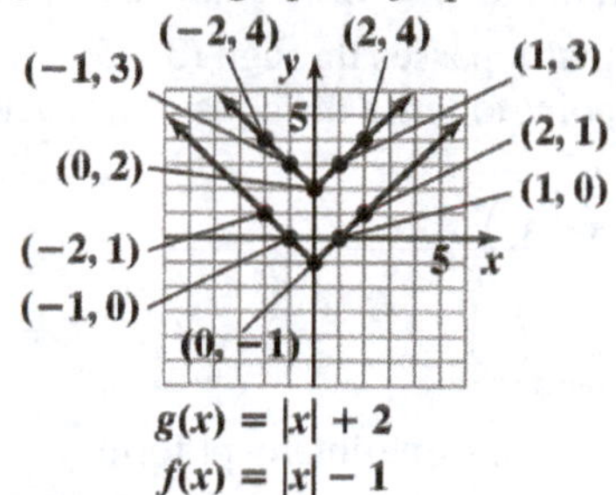

16. The domain of f is $(-\infty,15)$ or $(15,\infty)$.

17. $(f-g)(x)$
$=\left(3x^2-4x+2\right)-\left(x^2-5x-3\right)$
$=3x^2-4x+2-x^2+5x+3$
$=2x^2+x+5$
$(f-g)(-1)=2(-1)^2+(-1)+5$
$=2(1)-1+5=2-1+5=6$

18. $f(x)=-2x+4$
$y=-2x+4$
$m=-2$ $\quad y$-intercept $=4$

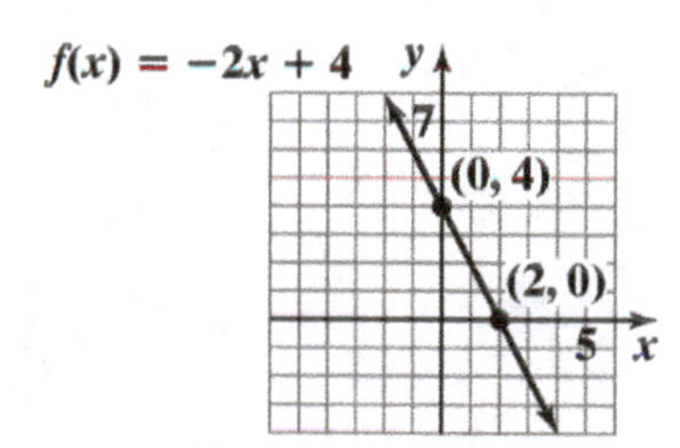

19. $x-2y=6$
Rewrite the equation of the line in slope-intercept form.
$x-2y=6$
$-2y=-x+6$
$y=\frac{1}{2}x-3$
$m=\frac{1}{2}$ $\quad y$-intercept $=-3$

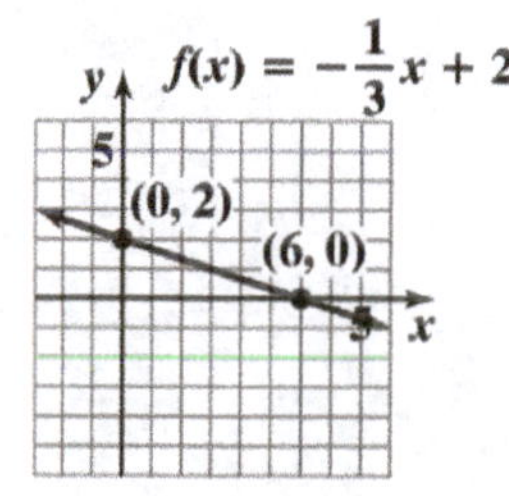

20. The line is parallel to $y = 4x + 7$, so the slope is 4. We are given that it passes through (3, –5). We use the slope and point to write the equation in point-slope form.

$$y - y_1 = m(x - x_1)$$
$$y - (-5) = 4(x - 3)$$
$$y + 5 = 4(x - 3)$$

Solve for y to obtain slope-intercept form.

$$y + 5 = 4(x - 3)$$
$$y + 5 = 4x - 12$$
$$y = 4x - 17$$

In function notation, the equation of the line is $f(x) = 4x - 17$.

Chapter 3
Systems of Linear Equations

3.1 Check Points

1. a.

$$2x+5y=-24 \qquad 3x-5y=14$$
$$2(-7)+5(-2)=-24 \qquad 3(-7)-5(-2)=14$$
$$-24=-24, \text{ true} \qquad -11=14, \text{ false}$$

The pair is not a solution of the system.

b.

$$2x+5y=-24 \qquad 3x-5y=14$$
$$2(-2)+5(-4)=-24 \qquad 3(-2)-5(-4)=14$$
$$-24=-24, \text{ true} \qquad 14=14, \text{ true}$$

The pair is a solution of the system.

2. Graph both equations.

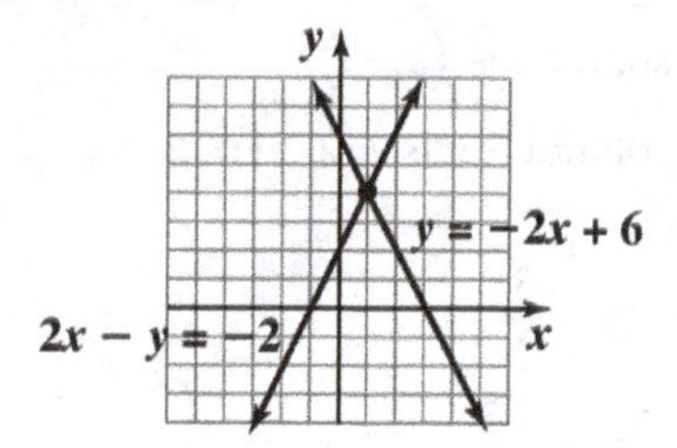

The intersection is $(1,4)$.

The solution set is $\{(1,4)\}$.

3. $y=3x-7$

$5x-2y=8$

Use substitution.

$$5x-2y=8$$
$$5x-2\overbrace{(3x-7)}^{y}=8$$
$$5x-6x+14=8$$
$$-x+14=8$$
$$-x=-6$$
$$x=6$$

Find y.

$$y=3x-7$$
$$y=3(6)-7$$
$$y=11$$

The solution is $(6,11)$.

The solution set is $\{(6,11)\}$.

4. $3x+2y=4$

$2x+y=1$

Solve $2x+y=1$ for y.

$2x+y=1$

$y=-2x+1$

Use substitution.

$3x+2y=4$

$$3x+2\overbrace{(-2x+1)}^{y}=4$$

$3x-4x+2=4$

$-x+2=4$

$-x=2$

$x=-2$

Find y.

$y=-2x+1$

$y=-2(-2)+1$

$y=5$

The solution is $(-2,5)$.

The solution set is $\{(-2,5)\}$.

5. $4x-7y=-16$

$2x+5y=\ 9$

Multiply the second equation by –2.

$$\begin{array}{r} 4x-7y=-16 \\ -4x-10y=-18 \\ \hline -17y=-34 \end{array}$$

$y=2$

Back-substitute to find x.

$2x+5y=9$

$2x+5(2)=9$

$2x+10=9$

$2x=-1$

$x=-\frac{1}{2}$

The solution is $\left(-\frac{1}{2},2\right)$.

The solution set is $\left\{\left(-\frac{1}{2},2\right)\right\}$.

6. $3x=2-4y$

$5y=-1-2x$

Rewrite in the form $Ax+By=C$.

$3x+4y=2$

$2x+5y=-1$

Multiply the first equation by –2. Multiply the second equation by 3.

$$\begin{array}{r} -6x-8y=-4 \\ 6x+15y=-3 \\ \hline 7y=-7 \end{array}$$

$y=-1$

Back-substitute to find x.

$3x=2-4(-1)$

$3x=6$

$x=2$

The solution is $(2,-1)$.

The solution set is $\{(2,-1)\}$.

7. $\frac{3x}{2}-2y=\frac{5}{2}$

$x-\frac{5y}{2}=-\frac{3}{2}$

Rewrite in the form $Ax+By=C$.

$3x-4y=5$

$2x-5y=-3$

Multiply the first equation by –2. Multiply the second equation by 3.

$$\begin{array}{r} -6x+8y=-10 \\ 6x-15y=-9 \\ \hline -7y=-19 \end{array}$$

$y=\frac{19}{7}$

Back-substitute to find x.

$x-\frac{5y}{2}=-\frac{3}{2}$

$x-\frac{5\left(\frac{19}{7}\right)}{2}=-\frac{3}{2}$

$x-\frac{95}{14}=-\frac{3}{2}$

$x=\frac{37}{7}$

The solution is $\left(\frac{37}{7},\frac{19}{7}\right)$.

The solution set is $\left\{\left(\frac{37}{7},\frac{19}{7}\right)\right\}$.

8. $5x-2y=4$
$-10x+4y=7$
Multiply the first equation by 2.
$10x-4y=8$
$\underline{-10x+4y=7}$
$0=15$
Since there are no pairs (x, y) for which 0 will equal 15, the system is inconsistent and has no solution. The solution set is $\varnothing$ or $\{\ \}$.

9. $x=4y-8$
$5x-20y=-40$
Substitute $4y-8$ for x in the second equation.
$5x-20y=-40$
$5(\overbrace{4y-8}^{x})-20y=-40$
$20y-40-20y=-40$
$-40=-40$
Since $-40=-40$ for all values of x and y, the system is dependent. The solution set is $\{(x, y)|x=4y-8\}$ or $\{(x, y)|5x-20y=-40\}$.

3.1 Concept and Vocabulary Check

1. satisfies both equations in the system
2. the intersection point
3. $\{(\frac{1}{3},-2)\}$
4. -2
5. -3
6. $\varnothing$; inconsistent; parallel
7. $\{(x, y)|x=3y+2\}$ or $\{(x, y)|5x-15y=10\}$; dependent; are identical or coincide

3.1 Exercise Set

1. $x-y=12$ $\quad$ $x+y=2$
$7-(-5)=12$ $\quad$ $7+(-5)=2$
$12=12$, true $\quad$ $2=2$, true
The pair is a solution of the system.

3. $3x+4y=2$ $\quad$ $2x+5y=1$
$3(2)+4(-1)=2$ $\quad$ $2(2)+5(-1)=1$
$2=2$, true $\quad$ $-1=1$, false
The pair is not a solution of the system.

5. $y=2x-13$ $\quad$ $4x+9y=-7$
$-3=2(5)-13$ $\quad$ $4(5)+9(-3)=-7$
$-3=10-13$ $\quad$ $20-27=-7$
$-3=-3$ $\quad$ $-7=-7$
The pair is a solution of the system.

7. Graph both equations.

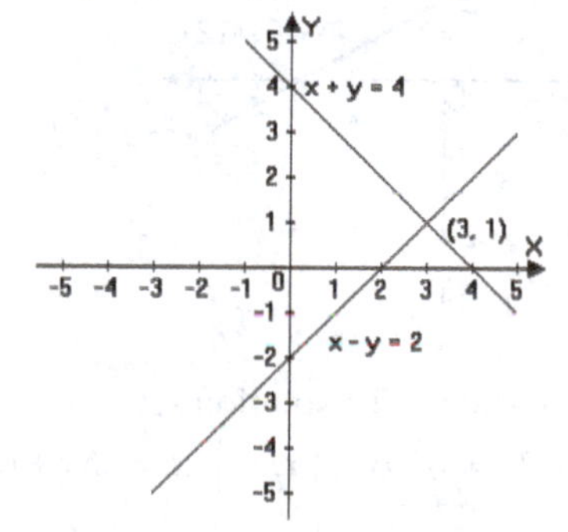

The solution is $(3,1)$.
The solution set is $\{(3,1)\}$.

9. Graph both equations.

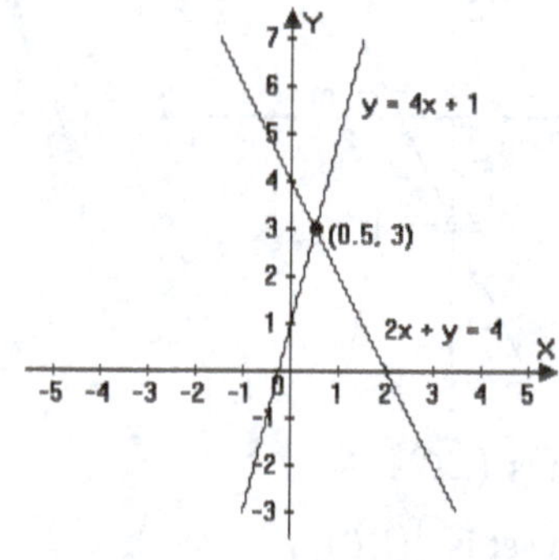

The solution is $\left(\frac{1}{2},3\right)$.

The solution set is $\left\{\left(\frac{1}{2},3\right)\right\}$.

11. Graph both equations.

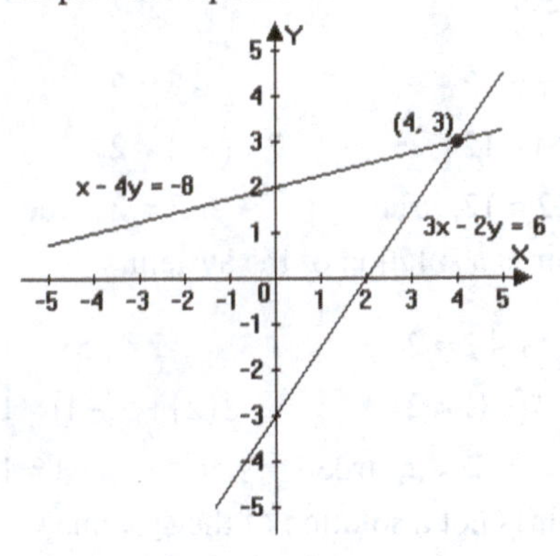

The solution is $(4,3)$.

The solution set is $\{(4,3)\}$.

13. Graph both equations.

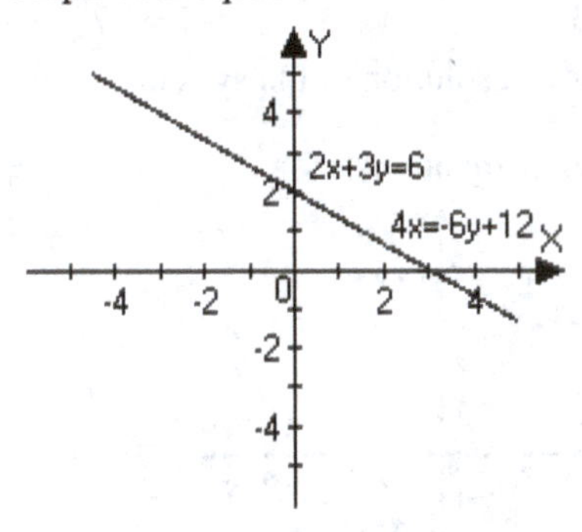

The lines coincide. The solution set is $\{(x, y)|2x+3y=6\}$ or $\{(x, y)|4x=-6y+12\}$.

15. Graph both equations.

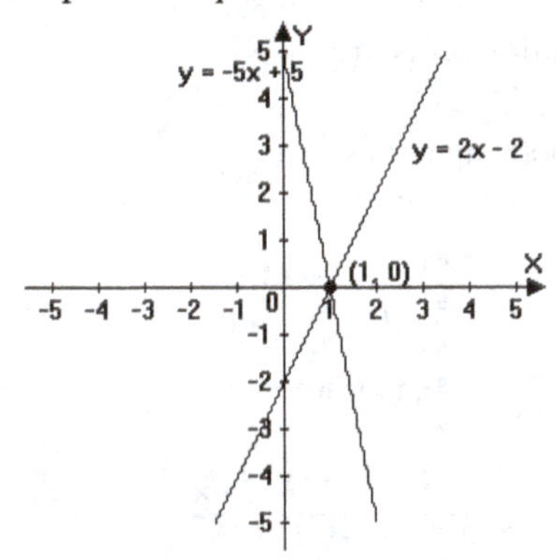

The solution is $(1,0)$.

The solution set is $\{(1,0)\}$.

17. Graph both equations.

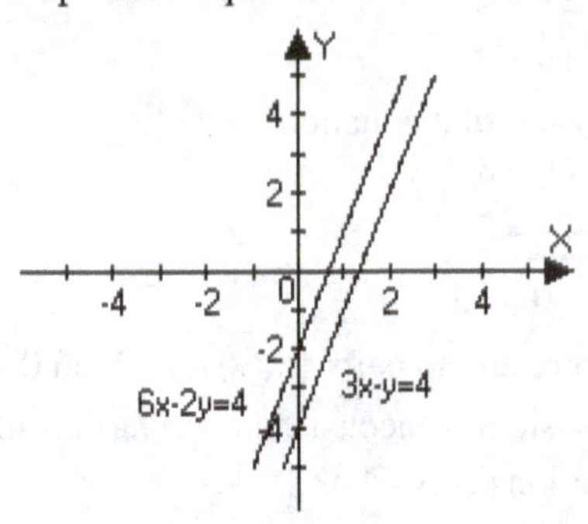

Since the lines do not intersect, there is no solution. The solution set is $\varnothing$ or $\{\ \}$.

19. Graph both equations.

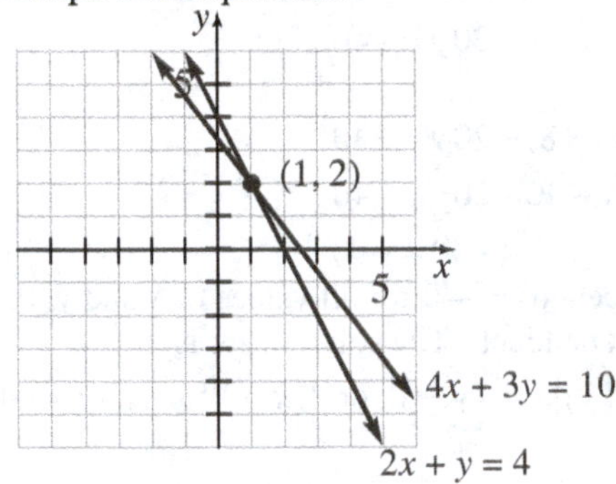

The solution is $(1,2)$.

The solution set is $\{(1,2)\}$.

21. Graph both equations.

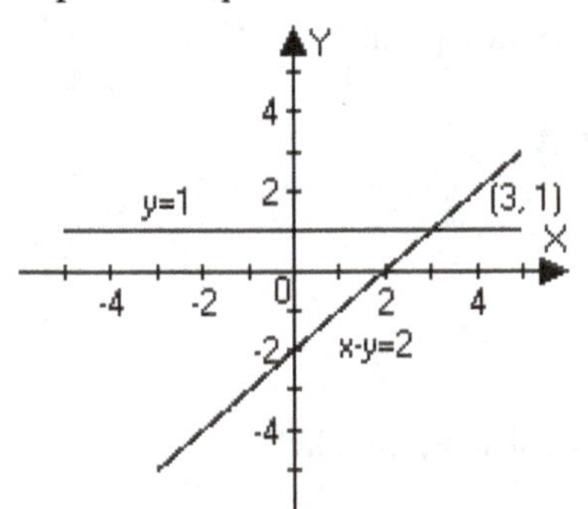

The solution is $(3,1)$.

The solution set is $\{(3,1)\}$.

23. Graph both equations.

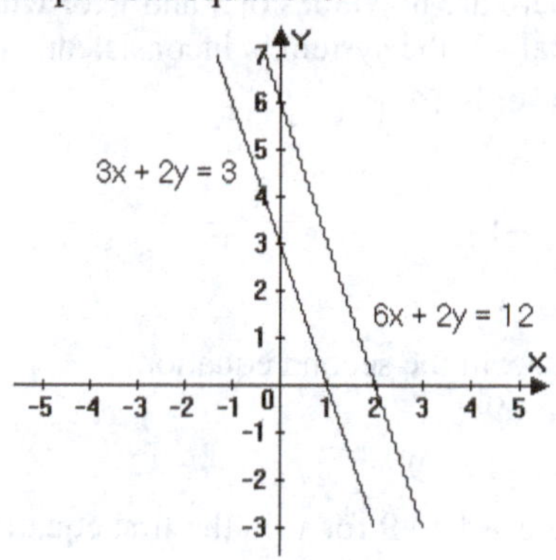

Since the lines do not intersect, there is no solution.
The solution set is $\varnothing$ or $\{\ \}$.

25. $x + y = 6$
$y = 2x$
Substitute $2x$ for y in the first equation.
$x + y = 6$
$x + 2x = 6$
$3x = 6$
$x = 2$
Back-substitute to find y.
$2 + y = 6$
$y = 4$
The solution is $(2,4)$.
The solution set is $\{(2,4)\}$.

27. $2x + 3y = 9$
$x = y + 2$
Substitute $y + 2$ for x in the first equation.
$2x + 3y = 9$
$2(y+2) + 3y = 9$
$2y + 4 + 3y = 9$
$5y + 4 = 9$
$5y = 5$
$y = 1$
Back-substitute to find x.
$x = y + 2$
$x = 1 + 2$
$x = 3$
The solution is $(3,1)$.
The solution set is $\{(3,1)\}$.

29. $y = -3x + 7$
$5x - 2y = 8$
Substitute $-3x + 7$ for y in the second equation.
$5x - 2y = 8$
$5x - 2(-3x+7) = 8$
$5x + 6x - 14 = 8$
$11x - 14 = 8$
$11x = 22$
$x = 2$
Back-substitute to find y.
$y = -3(2) + 7$
$y = -6 + 7$
$y = 1$
The solution is $(2,1)$.
The solution set is $\{(2,1)\}$.

31. $4x + y = 5$
$2x - 3y = 13$
Solve for y in the first equation.
$4x + y = 5$
$y = -4x + 5$
Substitute $-4x + 5$ for y in the second equation.
$2x - 3(-4x+5) = 13$
$2x + 12x - 15 = 13$
$14x - 15 = 13$
$14x = 28$
$x = 2$
Back-substitute to find y.
$4x + y = 5$
$4(2) + y = 5$
$8 + y = 5$
$y = -3$
The solution is $(2,-3)$.
The solution set is $\{(2,-3)\}$.

33. $x-2y=4$
$2x-4y=5$
Solve for x in the first equation.
$x-2y=4$
$x=2y+4$
Substitute $2y+4$ for x in the second equation.
$$2x-4y=5$$
$$2(2y+4)-4y=5$$
$$4y+8-4y=5$$
$$4y+8-4y=5$$
$$8\neq 5$$
The system is inconsistent. There are no values of x and y for which 8 will equal 5. The solution set is $\varnothing$ or $\{\ \}$.

35. $2x+5y=-4$
$3x-y=11$
Solve for y in the second equation.
$3x-y=11$
$-y=-3x+11$
$y=3x-11$
Substitute $3x-11$ for y in the first equation.
$$2x+5y=-4$$
$$2x+5(3x-11)=-4$$
$$2x+15x-55=-4$$
$$17x-55=-4$$
$$17x=51$$
$$x=3$$
Back-substitute to find y.
$$3x-y=11$$
$$3(3)-y=11$$
$$9-y=11$$
$$-y=2$$
$$y=-2$$
The solution is $(3,-2)$.
The solution set is $\{(3,-2)\}$.

37. $2(x-1)-y=-3$
$y=2x+3$
Substitute $2x+3$ for y in the first equation.
$$2(x-1)-y=-3$$
$$2(x-1)-(2x+3)=-3$$
$$2x-2-2x-3=-3$$
$$-5\neq -3$$
Since there are no values of x and y for which -5 will equal -3, the system is inconsistent. The solution set is $\varnothing$ or $\{\ \}$.

39. $\frac{x}{4}-\frac{y}{4}=-1$
$x+4y=-9$
Solve for x in the second equation.
$x+4y=-9$
$x=-4y-9$
Substitute $-4y-9$ for x in the first equation.
$$\frac{x}{4}-\frac{y}{4}=-1$$
$$\frac{-4y-9}{4}-\frac{y}{4}=-1$$
$$4\left(\frac{-4y-9}{4}-\frac{y}{4}\right)=4(-1)$$
$$-4y-9-y=-4$$
$$-5y-9=-4$$
$$-5y=5$$
$$y=-1$$
Back-substitute to find x.
$$x+4y=-9$$
$$x+4(-1)=-9$$
$$x-4=-9$$
$$x=-5$$
The solution is $(-5,-1)$.
The solution set is $\{(-5,-1)\}$.

41. $y=\frac{2}{5}x-2$
$2x-5y=10$
Substitute $\frac{2}{5}x-2$ for y in the second equation.
$$2x-5y=10$$
$$2x-5\left(\frac{2}{5}x-2\right)=10$$
$$2x-2x+10=10$$
$$10=10$$
Since 10 = 10 for all values of x and y, the system is dependent. The solution set is $\left\{(x,y)\,\middle|\,y=\frac{2}{5}x-2\right\}$ or $\{(x,y)\mid 2x-5y=10\}$.

43. Solve by addition.

$$\begin{aligned} x+y&=7\\ x-y&=3\\ \hline 2x&=10\\ x&=5 \end{aligned}$$

Back-substitute to find y.

$$\begin{aligned} x+y&=7\\ 5+y&=7\\ y&=2 \end{aligned}$$

The solution is $(5,2)$.

The solution set is $\{(5,2)\}$.

45. Solve by addition.

$$\begin{aligned} 12x+3y&=15\\ 2x-3y&=13\\ \hline 14x&=28\\ x&=2 \end{aligned}$$

Back-substitute to find y.

$$\begin{aligned} 12(2)+3y&=15\\ 24+3y&=15\\ 3y&=-9\\ y&=-3 \end{aligned}$$

The solution is $(2,-3)$.

The solution set is $\{(2,-3)\}$.

47. $x+3y=2$

$4x+5y=1$

Multiply the first equation by –4.

$$\begin{aligned} -4x-12y&=-8\\ 4x+5y&=1\\ \hline -7y&=-7\\ y&=1 \end{aligned}$$

Back-substitute to find x.

$$\begin{aligned} x+3y&=2\\ x+3(1)&=2\\ x+3&=2\\ x&=-1 \end{aligned}$$

The solution is $(-1,1)$.

The solution set is $\{(-1,1)\}$.

49. $6x-y=-5$

$4x-2y=6$

Multiply the first equation by –2.

$$\begin{aligned} -12x+2y&=10\\ 4x-2y&=6\\ \hline -8x&=16\\ x&=-2 \end{aligned}$$

Back-substitute to find y.

$$\begin{aligned} 6(-2)-y&=-5\\ -12-y&=-5\\ -y&=7\\ y&=-7 \end{aligned}$$

The solution is $(-2,-7)$.

The solution set is $\{(-2,-7)\}$.

51. $3x-5y=11$

$2x-6y=2$

Multiply the first equation by –2 and the second equation by 3.

$$\begin{aligned} -6x+10y&=-22\\ 6x-18y&=6\\ \hline -8y&=-16\\ y&=2 \end{aligned}$$

Back-substitute to find x.

$$\begin{aligned} 2x-6(2)&=2\\ 2x-12&=2\\ 2x&=14\\ x&=7 \end{aligned}$$

The solution is $(7,2)$.

The solution set is $\{(7,2)\}$.

53. $2x-5y=13$

$5x+3y=17$

Multiply the first equation by 3 and the second equation by 5.

$$\begin{aligned} 6x-15y&=39\\ \underline{25x+15y}&\underline{=85}\\ 31x&=124\\ x&=4 \end{aligned}$$

Back-substitute to find *y*.

$$\begin{aligned} 5(4)+3y&=17\\ 20+3y&=17\\ 3y&=-3\\ y&=-1 \end{aligned}$$

The solution is $(4,-1)$.

The solution set is $\{(4,-1)\}$.

55. $2x+6y=8$

$3x+9y=12$

Multiply the first equation by –3 and the second equation by 2.

$$\begin{aligned} -6x-18y&=-24\\ \underline{6x+18y}&\underline{=\ 24}\\ 0&=0 \end{aligned}$$

Since 0 = 0 for all values of *x* and *y*, the system is dependent. The solution set is $\{(x, y)|\ 2x+6y=8\}$ or $\{(x, y)|\ 3x+9y=12\}$.

57. $2x-3y=4$

$4x+5y=3$

Multiply the first equation by –2.

$$\begin{aligned} -4x+6y&=-8\\ \underline{4x+5y}&\underline{=3}\\ 11y&=-5\\ y&=-\frac{5}{11} \end{aligned}$$

Back-substitute to find *x*.

$$\begin{aligned} 2x-3y&=4\\ 2x-3\left(-\frac{5}{11}\right)&=4\\ 2x+\frac{15}{11}&=4\\ 2x&=\frac{29}{11}\\ x&=\frac{29}{22} \end{aligned}$$

The solution is $\left(\frac{29}{22},-\frac{5}{11}\right)$.

The solution set is $\left\{\left(\frac{29}{22},-\frac{5}{11}\right)\right\}$.

59. $3x-7y=1$

$2x-3y=-1$

Multiply the first equation by –2 and the second equation by 3.

$$\begin{aligned} -6x+14y&=-2\\ \underline{6x-\ 9y}&\underline{=-3}\\ 5y&=-5\\ y&=-1 \end{aligned}$$

Back-substitute to find *x*.

$$\begin{aligned} 3x-7(-1)&=1\\ 3x+7&=1\\ 3x&=-6\\ x&=-2 \end{aligned}$$

The solution is $(-2,-1)$.

The solution set is $\{(-2,-1)\}$.

61. $x=y+4$

$3x+7y=-18$

Substitute $y+4$ for x in the second equation.

$$\begin{aligned} 3x+7y&=-18\\ 3(y+4)+7y&=-18\\ 3y+12+7y&=-18\\ 10y+12&=-18\\ 10y&=-30\\ y&=-3 \end{aligned}$$

Back-substitute to find *x*.

$$\begin{aligned} x&=y+4\\ x&=-3+4\\ x&=1 \end{aligned}$$

The solution is $(1,-3)$.

The solution set is $\{(1,-3)\}$.

63. $9x+\frac{4y}{3}=5$

$4x-\frac{y}{3}=5$

Multiply the second equation by 4.

$$\begin{aligned} 9x+\frac{4y}{3} &= 5 \\ 16x-\frac{4y}{3} &= 20 \\ \hline 25x &= 25 \\ x &= 1 \end{aligned}$$

Back-substitute to find y.

$$4x-\frac{y}{3}=5$$
$$4(1)-\frac{y}{3}=5$$
$$4-\frac{y}{3}=5$$
$$-\frac{y}{3}=1$$
$$y=-3$$

The solution is $(1,-3)$.

The solution set is $\{(1,-3)\}$.

65. $\frac{1}{4}x-\frac{1}{9}y=\frac{2}{3}$

$\frac{1}{2}x-\frac{1}{3}y=1$

Multiply the first equation by –2.

$$-2\left(\frac{1}{4}x\right)-(-2)\left(\frac{1}{9}y\right)=-2\left(\frac{2}{3}\right)$$
$$-\frac{1}{2}x+\frac{2}{9}y=-\frac{4}{3}$$

We now have a system of two equations in two variables.

$$-\frac{1}{2}x+\frac{2}{9}y=-\frac{4}{3}$$
$$\frac{1}{2}x-\frac{1}{3}y=\ 1$$

Solve the addition.

$$\begin{aligned} -\frac{1}{2}x+\frac{2}{9}y &= -\frac{4}{3} \\ \frac{1}{2}x-\frac{1}{3}y &= 1 \\ \hline -\tfrac{1}{9}y &= -\tfrac{1}{3} \\ y &= -\tfrac{1}{3}(-9) \\ y &= 3 \end{aligned}$$

Back-substitute to find x.

$$\frac{1}{2}x-\frac{1}{3}y=1$$
$$\frac{1}{2}x-\frac{1}{3}(3)=1$$
$$\frac{1}{2}x-1=1$$
$$\frac{1}{2}x=2$$
$$x=4$$

The solution is $(4,3)$.

The solution set is $\{(4,3)\}$.

67. $x=3y-1$

$2x-6y=-2$

Substitute $3y-1$ for x in the second equation.

$$2x-6y=-2$$
$$2(3y-1)-6y=-2$$
$$6y-2-6y=-2$$
$$-2=-2$$

Since $-2=-2$ for all values of x and y, the system is dependent. The solution set is $\{(x, y)|\ x=3y-1\}$ or $\{(x, y)|\ 2x-6y=-2\}$.

69. $y = 2x + 1$

$y = 2x - 3$

Multiply the first equation by –1.

$$\begin{aligned} -y &= -2x - 1 \\ y &= 2x - 3 \\ \hline 0 &\neq -4 \end{aligned}$$

Since there are no values of x and y for which $0 = -4$, the system is inconsistent. The solution set is $\varnothing$ or $\{\ \}$.

71. $0.4x + 0.3y = 2.3$

$0.2x - 0.5y = 0.5$

Multiply the second equation by –2.

$$\begin{aligned} 0.4x + 0.3y &= 2.3 \\ -0.4x + 1.0y &= -1.0 \\ \hline 1.3y &= 1.3 \\ y &= 1 \end{aligned}$$

Back-substitute to find x.

$$\begin{aligned} 0.2x - 0.5y &= 0.5 \\ 0.2x - 0.5(1) &= 0.5 \\ 0.2x - 0.5 &= 0.5 \\ 0.2x &= 1.0 \\ x &= 5 \end{aligned}$$

The solution is $(5, 1)$.

The solution set is $\{(5, 1)\}$.

73. $5x - 40 = 6y$

$2y = 8 - 3x$

Rewrite the equations in general form.

$5x - 6y = 40$

$3x + 2y = 8$

Multiply the second equation by 3.

$$\begin{aligned} 5x - 6y &= 40 \\ 9x + 6y &= 24 \\ \hline 14x &= 64 \\ x &= \frac{32}{7} \end{aligned}$$

Back-substitute to find y.

$$\begin{aligned} 2y &= 8 - 3x \\ 2y &= 8 - 3\left(\frac{32}{7}\right) \\ 2y &= 8 - \frac{96}{7} \\ 2y &= -\frac{40}{7} \\ y &= -\frac{40}{14} = -\frac{20}{7} \end{aligned}$$

The solution is $\left(\frac{32}{7}, -\frac{20}{7}\right)$.

The solution set is $\left\{\left(\frac{32}{7}, -\frac{20}{7}\right)\right\}$.

75. $3(x + y) = 6$

$3(x - y) = -36$

Divide both equations by 3.

$x + y = 2$

$x - y = -12$

Solve the system by addition.

$$\begin{aligned} x + y &= 2 \\ x - y &= -12 \\ \hline 2x &= -10 \\ x &= -5 \end{aligned}$$

Back-substitute to find y.

$$\begin{aligned} x + y &= 2 \\ -5 + y &= 2 \\ y &= 7 \end{aligned}$$

The solution is $(-5, 7)$.

The solution set is $\{(-5, 7)\}$.

77. $3(x-3)-2y=0$

$2(x-y)=-x-3$

Rewrite the equations in general form.

$3(x-3)-2y=0 \quad 2(x-y)=-x-3$

$3x-9-2y=0 \quad 2x-2y=-x-3$

$3x-2y=9 \quad 3x-2y=-3$

We now have a system of two equations in two variables.

$3x-2y=9$

$3x-2y=-3$

Multiply the second equation by -1.

$3x-2y=9$

$\underline{-3x+2y=3}$

$0 \neq 12$

Since there are no values of x and y for which $0 = 12$, the system is inconsistent. The solution set is $\varnothing$ or $\{\ \}$.

79. $x+2y-3=0$

$12=8y+4x$

Rewrite the equations in general form.

$x+2y-3=0 \quad 12=8y+4x$

$x+2y=3 \quad 8y+4x=12$

$4x+8y=12$

We now have a system of two equations in two variables.

$x+2y=3$

$4x+8y=12$

Multiply the first equation by -4.

$-4x-8y=-12$

$\underline{4x+8y=\ 12}$

$0=0$

Since $0 = 0$ for all values of x and y, the system is dependent. The solution set is

$\{(x, y) \mid x+2y-3=0\}$ or $\{(x, y) \mid 12=8y+4x\}$.

81. $3x+4y=0$

$7x=3y$

After rewriting the second equation in general form, the system becomes

$3x+4y=0$

$7x-3y=0$

Multiply the first equation by 3 and the second equation by 4.

$9x+12y=0$

$\underline{28x-12y=0}$

$37x=0$

$x=0$

Back-substitute to find y.

$7(0)=3y$

$0=3y$

$0=y$

The solution is $(0,0)$.

The solution set is $\{(0,0)\}$.

83. $\frac{x+2}{2}-\frac{y+4}{3}=3$

$\frac{x+y}{5}=\frac{x-y}{2}-\frac{5}{2}$

Start by multiplying each equation by its LCD and simplifying to clear the fractions.

$\frac{x+2}{2}-\frac{y+4}{3}=3$

$\frac{x+y}{5}=\frac{x-y}{2}-\frac{5}{2}$

Start by multiplying each equation by its LCD and simplifying to clear the fractions.

$6\left(\frac{x+2}{2}-\frac{y+4}{3}\right)=6(3)$

$3(x+2)-2(y+4)=18$

$3x+6-2y-8=18$

$3x-2y=20$

$10\left(\frac{x+y}{5}\right)=10\left(\frac{x-y}{2}-\frac{5}{2}\right)$

$2(x+y)=5(x-y)-5(5)$

$2x+2y=5x-5y-25$

$3x-7y=25$

We now need to solve the equivalent system of equations:

$3x-2y=20$

$3x-7y=25$

Subtract the two equations:

$$\begin{aligned} 3x-2y&=20\\ -(3x-7y&=25)\\ \hline 5y&=-5\\ y&=-1 \end{aligned}$$

Back-substitute this value for y and solve for x.

$$\begin{aligned} 3x-2y&=20\\ 3x-2(-1)&=20\\ 3x+2&=20\\ 3x&=18\\ x&=6 \end{aligned}$$

The solution is $(6,-1)$.

The solution set is $\{(6,-1)\}$.

85. $5ax+4y=17$

$ax+7y=22$

Multiply the second equation by -5 and add the equations.

$$\begin{aligned} 5ax+4y&=17\\ -5ax-35y&=-110\\ \hline -31y&=-93\\ y&=3 \end{aligned}$$

Back-substitute into one of the original equations to solve for x.

$$\begin{aligned} ax+7y&=22\\ ax+7(3)&=22\\ ax+21&=22\\ ax&=1\\ x&=\frac{1}{a} \end{aligned}$$

The solution is $\left(\frac{1}{a},3\right)$.

The solution set is $\left\{\left(\frac{1}{a},3\right)\right\}$.

87. $f(-2)=11 \rightarrow -2m+b=11$

$f(3)=-9 \rightarrow 3m+b=-9$

We need to solve the resulting system of equations:

$-2m+b=11$

$3m+b=-9$

Subtract the two equations:

$$\begin{aligned} -2m+b&=11\\ 3m+b&=-9\\ \hline -5m&=20\\ m&=-4 \end{aligned}$$

Back-substitute into one of the original equations to solve for b.

$$\begin{aligned} -2m+b&=11\\ -2(-4)+b&=11\\ 8+b&=11\\ b&=3 \end{aligned}$$

Therefore, $m=-4$ and $b=3$.

89. The solution to a system of linear equations is the point of intersection of the graphs of the equations in the system. If $(6,2)$ is a solution, then we need to find the lines that intersect at that point.

Looking at the graph, we see that the graphs of $x+3y=12$ and $x-y=4$ intersect at the point $(6,2)$. Therefore, the desired system of equations is

$x+3y=12$ or $y=-\frac{1}{3}x+4$

$x-y=4$ $\quad$ $y=x-4$

91. **a.** $-3x+10y=160$

$x+2y=142$

Multiply the second equation by -5 and add the equations.

$$\begin{aligned} -3x+10y&=160\\ -5x-10y&=-710\\ \hline -8x\quad&=-550 \end{aligned}$$

Use the resulting equation to find x.

$$\begin{aligned} \frac{-8x}{-8}&=\frac{-550}{-8}\\ x&=68.75\\ x&\approx 69 \end{aligned}$$

Back-substitute to find y.

$$\begin{aligned} x+2y&=142\\ 68.75+2y&=142\\ 2y&=73.25\\ y&=36.625\\ y&\approx 37 \end{aligned}$$

The percentage of married and never-married adults will be the same about 69 years after 1970, or 2039.

In 2039, about 37% of Americans will belong to each group.

b. The approximate solution in part (a) is described by the point of intersection of the graphs $(69,37)$.

93. a. $y = 0.04x + 5.48$

b. $y = 0.17x + 1.84$

c. Using substitution,

$0.17x + 1.84 = 0.04x + 5.48$

$0.13x = 3.64$

$x = 28$

The costs for Medicare and Medicaid will be the same 28 years after 2000, or 2028.

Back-substitute to find y.

$y = 0.17x + 1.84$

$y = 0.17(28) + 1.84$

$y = 6.6$

In 2028, Medicare and Medicaid will each cost 6.6% of the GDP. After 2028, Medicare will have the greater cost.

95. a. $m = \dfrac{27.3 - 38}{20 - 0} = \dfrac{-10.7}{20} \approx -0.54$

From the point $(0, 38)$ we have that the y-intercept is $b = 38$. Therefore, the equation of the line is $y = -0.54x + 38$.

b. $m = \dfrac{24.2 - 40}{20 - 0} = \dfrac{-15.8}{20} = -0.79$

From the point $(0, 40)$ we have that the y-intercept is $b = 40$. Therefore, the equation of the line is $y = -0.79x + 40$.

c. To find the year when cigarette use is the same, we set the two equations equal to each other and solve for x.

$-0.54x + 38 = -0.79x + 40$

$0.25x = 2$

$x = 8$

Cigarette use was the same for African Americans and Hispanics 8 years after 1985, or 1993.

$y = -0.54x + 38$

$y = -0.54(8) + 38$

$y = 33.68$

At that time about 33.68% of each group used cigarettes.

97. a. $N_d = -5p + 750$

$= -5(120) + 750$

$= -600 + 750$

$= 150$

$N_s = 2.5(120) = 300$

If the price of the televisions is $120, 150 sets can be sold and 300 sets can be supplied.

b. To find the price at which supply and demand are equal, we set the two equations equal to each other and solve for p.

$-5p + 750 = 2.5p$

$750 = 7.5p$

$\dfrac{750}{7.5} = p$

$100 = p$

$N = 2.5(100) = 250$.

Supply and demand will be equal if the price of the televisions is $100. At that price, 250 sets can be supplied and sold.

99. – 107. Answers will vary.

109. makes sense

111. makes sense

113. false; Changes to make the statement true will vary. A sample change is: The addition method *can* be used to eliminate either variable.

115. false; Changes to make the statement true will vary. A sample change is: A system of linear equations can never have exactly two ordered pair solutions. There can be 0, 1, or an infinite number of solutions.

117. Substitute $(2, 1)$ into both equations.

$ax - by = 4$

$a(2) - b(1) = 4$

$2a - b = 4$

$bx + ay = 7$

$b(2) + a(1) = 7$

$2b + a = 7$

The two equations we need to solve now are

$2a - b = 4$

$2b + a = 7$

Solve the second equation for a.

$2b + a = 7$

$a = 7 - 2b$

Substitute $a = 7 - 2b$ into the first equation.

$$2a - b = 4$$
$$2(7 - 2b) - b = 4$$
$$14 - 4b - b = 4$$
$$-5b = -10$$
$$b = 2$$

Back substitute to solve for a.

$$2a - b = 4$$
$$2a - 2 = 4$$
$$2a = 6$$
$$a = 3$$

The solution of the system is $a = 3, b = 2$.

119. Multiply the first equation by $-a_2$ and the second equation by a_1.

$$-a_1a_2x - a_2b_1y = -a_2c_1$$
$$\underline{a_1a_2x + a_1b_2y = \ a_1c_2}$$
$$(a_1b_2 - a_2b_1)y = a_1c_2 - a_2c_1$$
$$y = \frac{a_1c_2 - a_2c_1}{a_1b_2 - a_2b_1}$$

Back-substitute $\frac{a_1c_2 - a_2c_1}{a_1b_2 - a_2b_1}$ for y to find x.

$$-a - b - c = -4$$
$$\underline{9a + 3b + c = 20}$$
$$8a + 2b = 16$$

$$-3x + 3y + 6z = -6$$
$$\underline{2x - 3y + 6z = \ 5}$$
$$-x \quad + 12z = -1$$

$$a_1x = \frac{-a_1b_1c_2 + a_2b_1c_1}{a_1b_2 - a_2b_1} + \frac{a_1b_2c_1 - a_2b_1c_1}{a_1b_2 - a_2b_1}$$
$$a_1x = \frac{-a_1b_1c_2 + \cancel{a_2b_1c_1} + a_1b_2c_1 - \cancel{a_2b_1c_1}}{a_1b_2 - a_2b_1}$$
$$a_1x = \frac{-a_1b_1c_2 + a_1b_2c_1}{a_1b_2 - a_2b_1}$$
$$\frac{\cancel{a_1}x}{\cancel{a_1}} = \frac{\frac{-a_1b_1c_2 + a_1b_2c_1}{a_1b_2 - a_2b_1}}{a_1}$$

$$x = \frac{-a_1b_1c_2 + a_1b_2c_1}{a_1b_2 - a_2b_1} \cdot \frac{1}{a_1}$$
$$x = \frac{(-b_1c_2 + b_2c_1)\cancel{a_1}}{a_1b_2 - a_2b_1} \cdot \frac{1}{\cancel{a_1}}$$
$$x = \frac{-b_1c_2 + b_2c_1}{a_1b_2 - a_2b_1} = \frac{b_2c_1 - b_1c_2}{a_1b_2 - a_2b_1}$$

The solution is $\left(\frac{b_2c_1 - b_1c_2}{a_1b_2 - a_2b_1}, \frac{a_1c_2 - a_2c_1}{a_1b_2 - a_2b_1}\right)$.

The solution set is $\left\{\left(\frac{b_2c_1 - b_1c_2}{a_1b_2 - a_2b_1}, \frac{a_1c_2 - a_2c_1}{a_1b_2 - a_2b_1}\right)\right\}$.

120.

$$6x = 10 + 5(3x - 4)$$
$$6x = 10 + 15x - 20$$
$$6x = 15x - 10$$
$$-9x = -10$$
$$x = \frac{10}{9}$$

The solution is $\frac{10}{9}$.

The solution set is $\left\{\frac{10}{9}\right\}$.

121.

$$\begin{aligned}(4x^2y^4)^2(-2x^5y^0)^3 &= 4^2x^{2\cdot2}y^{4\cdot2}(-2)^3x^{5\cdot3}y^{0\cdot3}\\ &= 4^2x^4y^8(-2)^3x^{15}y^0\\ &= 16(-8)x^4x^{15}y^8y^0\\ &= -128x^{19}y^8\end{aligned}$$

122.

$$f(x) = x^2 - 3x + 7$$
$$f(-1) = (-1)^2 - 3(-1) + 7 = 1 + 3 + 7 = 11$$

123.

$$\begin{aligned}P_1r_1 + P_2r_2 &= x \cdot 0.15 + y \cdot 0.07\\ &= 0.15x + 0.07y\end{aligned}$$

124. $x = (0.30)(50) = 15$ milliliters

125. $\underbrace{80}_{\text{Price per pair}} \cdot \underbrace{x}_{\text{Number of pairs}} = 80x$

3.2 Check Points

1. Let x = the number of calories in hamburger and fries.
   Let y = the number of calories in fettuccine Alfredo.
   $x+2y=4240$
   $2x+y=3980$
   This system can be solved by substitution.
   Solve for x in terms of y.
   $x+2y=4240$
   $x=-2y+4240$
   Substitute this value into the other equation.
   $2x+y=3980$
   $2(\overbrace{-2y+4240}^{x})+y=3980$
   $-4y+8480+y=3980$
   $-3y+8480=3980$
   $-3y=-4500$
   $y=1500$
   Back-substitute to find x.
   $x=-2y+4240$
   $x=-2(1500)+4240$
   $x=1240$
   There are 1240 calories in hamburger and fries and 1500 calories in fettuccine Alfredo.

2. Let x = the amount invested at 9%.
   Let y = the amount invested at 11%.
   $x+y=5000$
   $0.09x+0.11y=487$
   This system can be solved by substitution.
   Solve for y in terms of x.
   $x+y=5000$
   $y=-x+5000$
   Substitute this value into the other equation.
   $0.09x+0.11y=487$
   $0.09x+0.11(\overbrace{-x+5000}^{y})=487$
   $0.09x-0.11x+550=487$
   $-0.02x+550=487$
   $-0.02x=-63$
   $x=3150$
   Back-substitute to find y.
   $y=-x+5000$
   $y=-(3150)+5000$
   $y=1850$
   There was \$3150 invested at 9% and \$1850 invested at 11%.

3. Let x = the number of ounces of 12% acid solution.
Let y = the number of ounces of 20% acid solution.

$$x + y = 160$$
$$0.12x + 0.20y = 0.15(160)$$

This system can be solved by substitution.
Solve for y in terms of x.

$$x + y = 160$$
$$y = -x + 160$$

Substitute this value into the other equation.

$$0.12x + 0.20y = 0.15(160)$$
$$0.12x + 0.20y = 24$$
$$0.12x + 0.20\overbrace{(-x + 160)}^{y} = 24$$
$$-0.08x + 32 = 24$$
$$-0.08x = -8$$
$$x = 100$$

Back-substitute to find y.

$$y = -x + 160$$
$$y = -(100) + 160$$
$$y = 60$$

The chemist should mix 100 ounces of the 12% acid solution and 60 ounces of the 20% acid solution.

4. Let x = the rate of the motorboat in still water.
Let y = the rate of the current.

	Rate	×	Time	=	Distance
Trip with the Current	$x + y$		2		$2(x + y)$
Trip against the Current	$x - y$		3		$3(x - y)$

This gives,

$$2(x + y) = 84$$
$$3(x - y) = 84$$

This system simplifies to:

$$x + y = 42$$
$$x - y = 28$$

This system can be solved by addition.

$$x + y = 42$$
$$\underline{x - y = 28}$$
$$2x \quad = 70$$
$$x = 35$$

Back-substitute to find y.

$$x + y = 42$$
$$35 + y = 42$$
$$y = 7$$

The rate of the motorboat in still water is 35 miles per hour and the rate of the current is 7 miles per hour.

5. **a.** $C(x) = 300,000 + 30x$

b. $R(x) = 80x$

c. $y = 300,000 + 30x$
$y = 80x$
This system can be solved by substitution.
$80x = 300,000 + 30x$
$50x = 300,000$
$x = 6000$
Back-substitute to find y.
$y = 80x$
$y = 80(6000)$
$y = 480,000$
The break-even point is (6000, 480,000). This means the company will break even when it produces and sells 6000 pairs of shoes. At this level, both revenue and costs are $480,000.

6. Profit equals revenue minus cost.
$P(x) = R(x) - C(x)$
$= (80x) - (300,000 + 30x)$
$= 50x - 300,000$

3.2 Concept and Vocabulary Check

1. $1180x + 125y$

2. $0.12x + 0.09y$

3. $0.09x + 0.6y$

4. $x + y$; $x - y$

5. $4(x + y)$

6. revenue; profit

7. break-even point

3.2 Exercise Set

1. Let x = the first number.
Let y = the second number.
$x + y = 7$
$x - y = -1$
Solve by addition:
$x + y = 7$
$x - y = -1$
$2x = 6$
$x = 3$
Back-substitute to find y.
$3 + y = 7$
$y = 4$
The numbers are 3 and 4.

3. Let x = the first number.
Let y = the second number.
$3x - y = 1$
$x + 2y = 12$
Multiply the first equation by 2.
$6x - 2y = 2$
$x + 2y = 12$
$7x = 14$
$x = 2$
Back-substitute to find y.
$x + 2y = 12$
$2 + 2y = 12$
$2y = 10$
$y = 5$
The numbers are 2 and 5.

5. **a.** At the break-even point, $R(x) = C(x)$.
$25500 + 15x = 32x$
$25500 = 17x$
$1500 = x$
$C(x) = 32x$
$C(1500) = 32(1500)$
$= 48000$
Fifteen hundred units must be produced and sold to break even. At this point, there will $48,000 in costs and revenue.

b. $P(x) = R(x) - C(x)$
$= (32x) - (25,500 + 15x)$
$= 32x - 25,500 - 15x$
$= 17x - 25,500$

7. a. At the break-even point, $R(x) = C(x)$.

$$105x + 70,000 = 245x$$
$$70,000 = 140x$$
$$500 = x$$
$$C(x) = 245x$$
$$C(500) = 245(500)$$
$$= 122,500$$

Five hundred units must be produced and sold to break even. At this point, there will $122,500 in costs and revenue.

b. $P(x) = R(x) - C(x)$

$$= (245x) - (105x + 70,000)$$
$$= 245x - 105x - 70,000$$
$$= 140x - 70,000$$

9. Let x = the percentage of parents who would end financial support after a child completes college.
Let y = the percentage of parents who would end financial support after a child completes high school.
Use the given information to set up a system of equations.

$$x + y = 48$$
$$x - y = 34$$

Solve by addition.

$$x + y = 48$$
$$x - y = 34$$
$$2x = 82$$
$$x = 41$$

Back-substitute to find y.

$$x + y = 48$$
$$41 + y = 48$$
$$y = 7$$

41% of parents would end financial support after a child completes college and 7% would end financial support after a child completes high school.

11. Let x = the number of computers sold.
Let y = the number of hard drives sold.

$$x + y = 36$$
$$1180x + 125y = 27,710$$

Multiply the first equation by -125 and add the two equations.

$$-125x - 125y = -4500$$
$$1180x + 125y = 27,710$$
$$1055x = 23,210$$
$$x = 22$$

Back-substitute to solve for y.

$$x + y = 36$$
$$22 + y = 36$$
$$y = 14$$

The store sold 22 computers and 14 hard drives.

13. Let x = the amount invested at 6%.
Let y = the amount invested at 8%.

$$x + y = 7000$$
$$0.06x + 0.08y = 520$$

Solve the first equation for x.

$$x = 7000 - y$$

Substitute this result for x in the second equation.

$$0.06(7000 - y) + 0.08y = 520$$
$$420 - 0.06y + 0.08y = 520$$
$$0.02y = 100$$
$$y = \frac{100}{0.02} = 5000$$

Back-substitute to solve for x.

$$x + y = 7000$$
$$x + 5000 = 7000$$
$$x = 2000$$

$2000 was invested at 6% and $5000 was invested at 8%.

15. Let x = the amount in the first fund.
Let y = the amount in the second fund.

$$0.09x + 0.03y = 900$$
$$0.10x + 0.01y = 860$$

Multiply the second equation by -3 and add the two equations.

$$0.09x + 0.03y = 900$$
$$-0.30x - 0.03y = -2580$$
$$-0.21x = -1680$$
$$x = 8000$$

Back-substitute to solve for y.

$$0.10x + 0.01y = 860$$
$$0.10(8000) + 0.01y = 860$$
$$800 + 0.01y = 860$$
$$0.01y = 60$$
$$y = \frac{60}{0.01} = 6000$$

$8000 was invested in the first fund and $6000 was invested in the second fund.

17. Let x = amount invested with 12% return.
Let y = amount invested with the 5% loss.

$$x + y = 20,000$$
$$0.12x - 0.05y = 1890$$

Multiply the first equation by 0.05 and add the two equations.

$$0.05x + 0.05y = 1000$$
$$\underline{0.12x - 0.05y = 1890}$$
$$0.17x = 2890$$
$$x = 17,000$$

Back-substitute to solve for y.

$$x + y = 20,000$$
$$17,000 + y = 20,000$$
$$y = 3,000$$

$17,000 was invested at 12% return and $3000 was invested at a 5% loss.

19. Let x = gallons of 5% wine.
Let y = gallons of 9% wine.

$$x + y = 200$$
$$0.05x + 0.09y = 0.07(200)$$

or

$$x + y = 200$$
$$0.05x + 0.09y = 14$$

Solve the first equation for x.

$$x + y = 200$$
$$x = 200 - y$$

Substitute this expression for x in the second equation.

$$0.05(200 - y) + 0.09y = 14$$
$$10 - 0.05y + 0.09y = 14$$
$$0.04y = 4$$
$$y = 100$$

Back-substitute and solve for x.

$$x = 200 - y$$
$$= 200 - 100$$
$$= 100$$

The wine company should mix 100 gallons of the 5% California wine with 100 gallons of the 9% French wine.

21. Let x = grams of 18-karat gold.
Let y = grams of 12-karat gold.

$$x + y = 300$$
$$0.75x + 0.5y = 0.58(300)$$

or

$$x + y = 300$$
$$0.75x + 0.5y = 174$$

Solve the first equation for x.

$$x = 300 - y$$

Substitute this result for x into the second equation and solve for y.

$$0.75(300 - y) + 0.5y = 174$$
$$225 - 0.75 + 0.5y = 174$$
$$-0.25y = -51$$
$$y = 204$$

Back-substitute to solve for x.

$$x = 300 - y = 300 - 204 = 96$$

You would need 96 grams of 18-karat gold and 204 grams of 12-karat gold.

23. Let x = pounds of cheaper candy.
Let y = pounds of more expensive candy.

$$x + y = 75$$
$$1.6x + 2.1y = 1.9(75)$$

or

$$x + y = 75$$
$$1.6x + 2.1y = 142.5$$

Multiply the first equation by -1.6 and add the two equations.

$$-1.6x - 1.6y = -120$$
$$\underline{1.6x + 2.1y = 142.5}$$
$$0.5y = 22.5$$
$$y = 45$$

Back-substitute to solve for x.

$$x + 45 = 75$$
$$x = 30$$

The manager should mix 30 pounds of the cheaper candy and 45 pounds of the more expensive candy.

25. Let n = the number of nickels.
Let d = the number of dimes.

$$n + d = 15$$
$$0.05n + 0.1d = 1.10$$

Solve the first equation for n.

$$n = 15 - d$$
$$0.05(15 - d) + 0.1d = 1.10$$
$$0.75 - 0.05d + 0.1d = 1.10$$
$$0.05d = 0.35$$
$$d = 7$$

Back-substitute to solve for n.

$$n + 7 = 15$$
$$n = 8$$

The purse has 8 nickels and 7 dimes.

27. Let x = the speed of the plane in still air.
Let y = the speed of the wind.

	Rate ×	Time =	Distance
Trip with the Wind	$x+y$	5	$5(x+y)$
Trip against the Wind	$x-y$	8	$8(x-y)$

$5(x+y)=800$
$8(x-y)=800$

$5x+5y=800$
$8x-8y=800$

Multiply the first equation by 8 and the second equation by 5.

$40x+40y=6400$
$40x-40y=4000$

$80x=10400$
$x=130$

Back-substitute to find y.

$5x+5y=800$
$5(130)+5y=800$
$650+5y=800$
$5y=150$
$y=30$

The speed of the plane in still air is 130 miles per hour and the speed of the wind is 30 miles per hour.

29. Let x = the crew's rowing rate.
Let y = the rate of the current.

	Rate ×	Time =	Distance
Trip with the Current	$x+y$	2	$2(x+y)$
Trip against the Current	$x-y$	4	$4(x-y)$

$2(x+y)=16$
$4(x-y)=16$

Rewrite the system in $Ax+By=C$ form.

$2x+2y=16$
$4x-4y=16$

Multiply the first equation by –2.

$-4x-4y=-32$
$4x-4y=16$

$-8y=-16$
$y=2$

Back-substitute to find x.

$2x+2(2)=16$
$2x+4=16$
$2x=12$
$x=6$

The crew's rowing rate is 6 kilometers per hour and the rate of the current is 2 kilometers per hour.

31. Let x = the speed in still water.
Let y = the speed of the current.

	Rate	×	Time	=	Distance
Trip with the Current	$x+y$		4		$4(x+y)$
Trip against the Current	$x-y$		6		$6(x-y)$

$$4(x+y)=24$$
$$6(x-y)=\frac{3}{4}(24)$$

Rewrite the system in $Ax+By=C$ form.

$$4x+4y=24$$
$$6x-6y=18$$

Multiply the first equation by –3 and the second equation by 2.

$$-12x-12y=-72$$
$$\underline{12x-12y=36}$$
$$-24y=-36$$
$$y=1.5$$

Back-substitute to find x.

$$4x+4y=24$$
$$4x+4(1.5)=24$$
$$4x+6=24$$
$$4x=18$$
$$x=4.5$$

The speed in still water is 4.5 miles per hour and the speed of the current is 1.5 miles per hour.

33. Let x = the larger score.
Let y = the smaller score.

$$x-y=12$$
$$\frac{x+y}{2}=80$$

Rewrite the second equation in standard form.

$$x-y=12$$
$$x+y=160$$

Add the two equations.

$$x-y=12$$
$$\underline{x+y=160}$$
$$2x \quad =172$$
$$x \quad =86$$

Back-substitute to solve for y.

$$x-y=12$$
$$86-y=12$$
$$-y=-74$$
$$y=74$$

The two test scores are 74 and 86.

35. $x+2y=180$

$(2x-30)+y=180$

Rewrite the second equation in standard form.

$x+2y=180$

$2x+y=210$

Multiply the first equation by -2 and add the equations.

$-2x-4y=-360$

$2x+y=210$

$-3y=-150$

$y=50$

Back-substitute to solve for x.

$x+2y=180$

$x+2(50)=180$

$x+100=180$

$x=80$

The three interior angles measure $80°$, $50°$, and $50°$.

37. Let x = the lot length.
Let y = the lot width.

$2x+2y=220$

$20x+8(2y)=2040$

Rewriting the second equation gives the following equivalent system:

$2x+2y=220$

$20x+16y=2040$

Multiply the first equation by -8 and add the two equations.

$-16x-16y=-1760$

$20x+16y=2040$

$4x=280$

$x=70$

Back-substitute to find y.

$2x+2y=220$

$2(70)+2y=220$

$140+2y=220$

$2y=80$

$y=40$

The lot is 70 feet long and 40 feet wide.

39. Let x = the number of two-seat tables.
Let y = the number of four-seat tables.

$2x+4y=56$

$x+y=17$

Multiply the second equation by -2 and add the two equations.

$2x+4y=56$

$-2x-2y=-34$

$2y=22$

$y=11$

Back-substitute to find x.

$x+y=17$

$x+11=17$

$x=6$

The owners should buy 6 two-seat tables and 11 four-seat tables.

41. At the break-even point, $R(x)=C(x)$.

$50x=10,000+30x$

$20x=10,000$

$20x=10,000$

$x=500$

500 radios must be produced and sold to break even.

43. $R(x)=50x$

$R(200)=50(200)=10,000$

$C(x)=10,000+30x$

$C(200)=10,000+30(200)$

$=10,000+6000=16,000$

$R(200)-C(200)=10,000-16,000$

$=-6000$

This means that if 200 radios are produced and sold the company will lose \$6000.

45. a. $P(x)=R(x)-C(x)$

$=50x-(10,000+30x)$

$=50x-10,000-30x$

$=20x-10,000$

$P(x)=20x-10,000$

b. $P(10,000)=20(10,000)-10,000$

$=200,000-10,000=190,000$

If 10,000 radios are produced and sold the profit will be \$190,000.

47. a. The cost function is:
$C(x) = 18,000 + 20x$

b. The revenue function is:
$R(x) = 80x$

c. At the break-even point, $R(x) = C(x)$.
$80x = 18,000 + 20x$
$60x = 18,000$
$x = 300$

$R(x) = 80x$
$R(300) = 80(300)$
$= 24,000$

When approximately 300 canoes are produced the company will break even with cost and revenue at $24,000.

49. a. The cost function is:
$C(x) = 30,000 + 2500x$

b. The revenue function is:
$R(x) = 3125x$

c. At the break-even point, $R(x) = C(x)$.
$3125x = 30000 + 2500x$
$625x = 30000$
$x = 48$
After 48 sold out performances, the investor will break-even. ($150,000)

51. – 57. Answers will vary.

59. $R(x) = 50x$
$C(x) = 20x + 180$

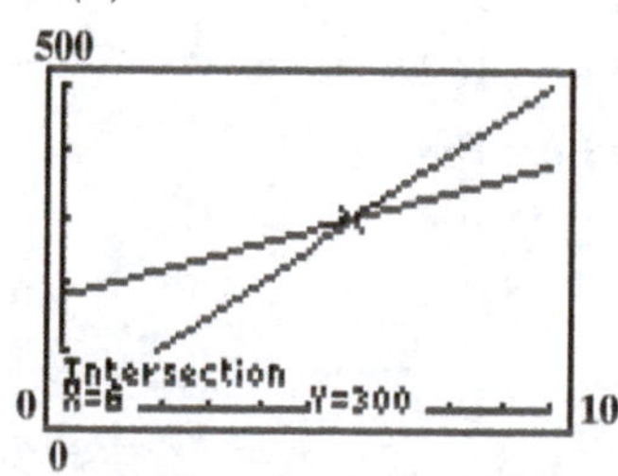

The break-even point is 6 units. When 6 units are produced and sold, cost and revenue are the same at $300.

61. In Exercise 47;
$R(x) = 80x$
$C(x) = 18,000 + 20x$

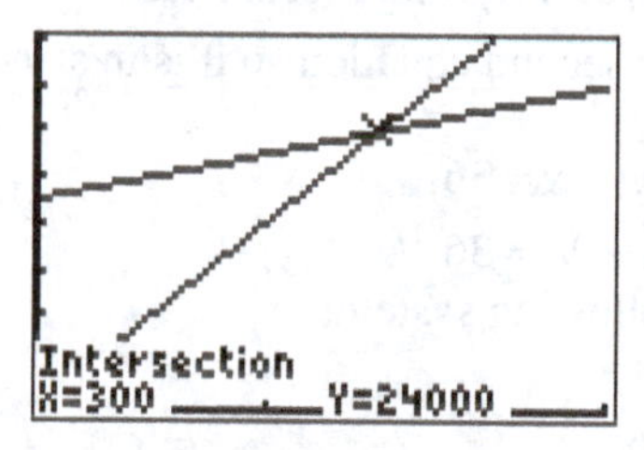

The break-even point is 300 canoes. When 300 canoes are produced and sold, cost and revenue are the same at $24,000.

63. does not make sense; Explanations will vary. Sample explanation: Mixing a 50% acid solution with a 25% acid solution will produce an acid solution that is between 25% and 50%.

65. makes sense

67. Let h = the number of hexagons.
Let s = the number of squares.
We need to solve the following system of equations:
$6h + s = 52$ (band)
$h + 4s = 24$ (pom-pom)
Multiply the first equation by -4 and add the equations.

$$\begin{aligned} -24h - 4s &= -208 \\ h + 4s &= 24 \\ \hline -23h \quad &= -184 \\ h \quad &= 8 \end{aligned}$$

Back-substitute to solve for s.
$6h + s = 52$
$6(8) + s = 52$
$48 + s = 52$
$s = 4$
The students can form 8 hexagons and 4 squares.

69. Let t = the original tens-place digit.
Let u = the original ones-place digit.
$$x+u=14 \quad \text{(digits sum)}$$
$$(10x+u)-(10u+x)=36 \quad \text{(score diff.)}$$
Simplify the second equation so it is in standard form.
$$10x+u-10u-x=36$$
$$9x-9u=36$$
Solve the following system:
$$x+u=14$$
$$9x-9u=36$$
Multiply the first equation by 9 and add the two equations.
$$9x+9u=126$$
$$9x-9u=36$$
$$18x \quad =162$$
$$x \quad =9$$
Back-substitute to solve for u.
$$x+u=14$$
$$9+u=14$$
$$u=5$$
Therefore, your original score was 95.

71. Passing through $(-2,5)$ and $(-6,13)$
First, find the slope.
$$m=\frac{y_2-y_1}{x_2-x_1}=\frac{13-5}{-6-(-2)}=\frac{8}{-4}=-2$$
Use the slope and one of the points to write the equation in point-slope form.
$$y-y_1=m(x-x_1)$$
$$y-5=-2(x-(-2))$$
$$y-5=-2(x+2)$$
or
$$y-y_1=m(x-x_1)$$
$$y-13=-2(x-(-6))$$
$$y-13=-2(x+6)$$
Rewrite the equation in slope-intercept form by solving for y.
$$y-13=-2(x+6)$$
$$y=-2x-12+13$$
$$y=-2x+1$$
In function notation, the equation of the line is $f(x)=-2x+1$.

72. Since the line is parallel to $-x+y=7$, we can use it to obtain the slope. Rewriting the equation in slope-intercept form, we obtain $y=x+7$. The slope is $m=1$. We are given that it passes through (–3, 0). We use the slope and point to write the equation in point-slope form.
$$y-y_1=m(x-x_1)$$
$$y-0=1(x-(-3))$$
$$y-0=1(x+3)$$
Rewrite the equation in slope-intercept form by solving for y.
$$y-0=1(x+3)$$
$$y=x+3$$
In function notation, the equation of the line is $f(x)=x+3$.

73. Since the denominator of a fraction cannot be zero, the domain of g is $(-\infty,3)$ or $(3,\infty)$.

74.
$$2x-y+4z=-8$$
$$2(3)-(2)+4(-3)=-8$$
$$-8=-8, \text{ true}$$
Yes, the ordered triple satisfies the equation.

75.
$$5x-2y-4z=3$$
$$3x+3y+2z=-3$$
Multiply Equation 2 by 2.
$$5x-2y-4z=3$$
$$6x+6y+4z=-6$$
Then add to eliminate z.
$$5x-2y-4z=3$$
$$6x+6y+4z=-6$$
$$11x+4y \quad =-3$$

76.
$$ax^2+bx+c=y$$
$$a(4)^2+b(4)+c=1682$$
$$16a+4b+c=1682$$

3.3 Check Points

1. Test the ordered triple in each equation.

$$x - 2y + 3z = 22$$
$$(-1) - 2(-4) + 3(5) = 22$$
$$22 = 22, \text{ true}$$
$$2x - 3y - z = 5$$
$$2(-1) - 3(-4) - (5) = 5$$
$$5 = 5, \text{ true}$$
$$3x + y - 5z = -32$$
$$3(-1) + (-4) - 5(5) = -32$$
$$-32 = -32, \text{ true}$$

The ordered triple (–1, –4, 5) makes all three equations true, so it is a solution to the system.

2.
$$x + 4y - z = 20$$
$$3x + 2y + z = 8$$
$$2x - 3y + 2z = -16$$

Add the first two equations to eliminate z.

$$x + 4y - z = 20$$
$$\underline{3x + 2y + z = 8}$$
$$4x + 6y = 28$$

Multiply the first equation by 2 and add to the third equation to eliminate z again.

$$2x + 8y - 2z = 40$$
$$\underline{2x - 3y + 2z = -16}$$
$$4x + 5y = 24$$

Solve the system of two equations in two variables.

$$4x + 6y = 28$$
$$4x + 5y = 24$$

Multiply the second equation by –1 and add the equations.

$$4x + 6y = 28$$
$$\underline{-4x - 5y = -24}$$
$$y = 4$$

Back-substitute 4 for y to find x.

$$4x + 6y = 28$$
$$4x + 6(4) = 28$$
$$4x + 24 = 28$$
$$4x = 4$$
$$x = 1$$

Back-substitute into an original equation.

$$3x + 2y + z = 8$$
$$3(1) + 2(4) + z = 8$$
$$11 + z = 8$$
$$z = -3$$

The solution is $(1, 4, -3)$ and the solution set is $\{(1, 4, -3)\}$.

3.
$$2y - z = 7$$
$$x + 2y + z = 17$$
$$2x - 3y + 2z = -1$$

Since the first equation already has only two variables, use the second and third equations to eliminate x.

Multiply the second equation by –2 and add to the third equation.

$$-2x - 4y - 2z = -34$$
$$\underline{2x - 3y + 2z = -1}$$
$$-7y = -35$$
$$\frac{-7y}{-7} = \frac{-35}{-7}$$
$$y = 5$$

Back-substitute 5 for y to find z.

$$2y - z = 7$$
$$2(5) - z = 7$$
$$10 - z = 7$$
$$-z = -3$$
$$z = 3$$

Back-substitute into an original equation to find x.

$$x + 2y + z = 17$$
$$x + 2(5) + (3) = 17$$
$$x + 13 = 17$$
$$x = 4$$

The solution is $(4, 5, 3)$ and the solution set is $\{(4, 5, 3)\}$.

4. Use each ordered pair to write an equation.

$(x, y) = (1, 4)$

$y = ax^2 + bx + c$

$4 = a(1)^2 + b(1) + c$

$4 = a + b + c$

$(x, y) = (2, 1)$

$y = ax^2 + bx + c$

$1 = a(2)^2 + b(2) + c$

$1 = 4a + 2b + c$

$(x, y) = (3, 4)$

$y = ax^2 + bx + c$

$4 = a(3)^2 + b(3) + c$

$4 = 9a + 3b + c$

The system of three equations in three variables is:

$a + b + c = 4$

$4a + 2b + c = 1$

$9a + 3b + c = 4$

Multiplying the first equation by -1 and adding it to the second gives $3a + b = -3$.
Multiplying the first equation by -1 and adding it to the third gives $8a + 2b = 0$.
Solve this system of two equations in two variables.

$3a + b = -3$

$8a + 2b = 0$

Multiply the first equation by -2 and add to the second equation.

$-6a - 2b = 6$

$8a + 2b = 0$

$2a = 6$

$a = 3$

Back-substitute to find b.

$3a + b = -3$

$3(3) + b = -3$

$9 + b = -3$

$b = -12$

Back-substitute into an original equation to find c.

$a + b + c = 4$

$(3) + (-12) + c = 4$

$-9 + c = 4$

$c = 13$

The quadratic function is $y = 3x^2 - 12x + 13$ or $f(x) = 3x^2 - 12x + 13$.

3.3 Concept and Vocabulary Check

1. triple; all

2. -2; -4

3. z; add Equations 1 and 3
4. quadratic
5. curve fitting

3.3 Exercise Set

1. Test the ordered triple in each equation.

$$\begin{aligned} x+y+z&=4 \\ 2-1+3&=4 \\ 4&=4,\ \text{true} \end{aligned} \qquad \begin{aligned} x-2y-z&=1 \\ 2-2(-1)-3&=1 \\ 2+2-3&=1 \\ 1&=1,\ \text{true} \end{aligned} \qquad \begin{aligned} 2x-y-z&=-1 \\ 2(2)-(-1)-3&=-1 \\ 4+1-3&=-1 \\ 2&=-1,\ \text{false} \end{aligned}$$

The ordered triple (2, –1, 3) does not make all three equations true, so it is not a solution.

3. Test the ordered triple in each equation.

$$\begin{aligned} x-2y&=2 \\ 4-2(1)&=2 \\ 4-2&=2 \\ 2&=2,\ \text{true} \end{aligned} \qquad \begin{aligned} 2x+3y&=11 \\ 2(4)+3(1)&=11 \\ 8+3&=11 \\ 11&=11,\ \text{true} \end{aligned} \qquad \begin{aligned} y-4z&=-7 \\ 1-4(2)&=-7 \\ 1-8&=-7 \\ -7&=-7,\ \text{true} \end{aligned}$$

The ordered triple makes all three equations true, so it is a solution.

5. $$\begin{aligned} x+y+2z&=11 \\ x+y+3z&=14 \\ x+2y-z&=5 \end{aligned}$$

Multiply the second equation by –1 and add to the first equation.

$$\begin{aligned} x+y+2z&=11 \\ -x-y-3z&=-14 \\ \hline -z&=-3 \\ z&=3 \end{aligned}$$

Back-substitute 3 for z in the first and third equations.

$$\begin{aligned} x+y+2z&=11 \\ x+y+2(3)&=11 \\ x+y+6&=11 \\ x+y&=5 \end{aligned} \qquad \begin{aligned} x+2y-z&=5 \\ x+2y-3&=5 \\ x+2y&=8 \end{aligned}$$

We now have two equations in two variables.

$$\begin{aligned} x+y&=5 \\ x+2y&=8 \end{aligned}$$

Multiply the first equation by –1 and solve by addition.

$$\begin{aligned} -x-y&=-5 \\ x+2y&=8 \\ \hline y&=3 \end{aligned}$$

Back-substitute 3 for y into one of the equations in two variables.

$$\begin{aligned} x+y&=5 \\ x+3&=5 \\ x&=2 \end{aligned}$$

The solution is $(2,3,3)$ and the solution set is $\{(2,3,3)\}$.

7. $4x - y + 2z = 11$

$x + 2y - z = -1$

$2x + 2y - 3z = -1$

Multiply the second equation by –4 and add to the first equation.

$$\begin{array}{r} 4x - y + 2z = 11 \\ \underline{-4x - 8y + 4z = 4} \\ -9y + 6z = 15 \end{array}$$

Multiply the second equation by –2 and add it to the third equation.

$$\begin{array}{r} -2x - 4y + 2z = 2 \\ \underline{2x + 2y - 3z = -1} \\ -2y - z = 1 \end{array}$$

We now have two equations in two variables.

$-9y + 6z = 15$

$-2y - z = 1$

Multiply the second equation by 6 and solve by addition.

$$\begin{array}{r} -9y + 6z = 15 \\ \underline{-12y - 6z = 6} \\ -21y = 21 \\ y = -1 \end{array}$$

Back-substitute –1 for y in one of the equations in two variables.

$$\begin{aligned} -2y - z &= 1 \\ -2(-1) - z &= 1 \\ 2 - z &= 1 \\ -z &= -1 \\ z &= 1 \end{aligned}$$

Back-substitute –1 for y and 1 for z in one of the original equations in three variables.

$$\begin{aligned} x + 2y - z &= -1 \\ x + 2(-1) - 1 &= -1 \\ x - 2 - 1 &= -1 \\ x - 3 &= -1 \\ x &= 2 \end{aligned}$$

The solution is $(2, -1, 1)$ and the solution set is $\{(2, -1, 1)\}$.

9. $3x + 2y - 3z = -2$

$2x - 5y + 2z = -2$

$4x - 3y + 4z = 10$

Multiply the second equation by –2 and add to the third equation.

$$\begin{array}{r} -4x + 10y - 4z = 4 \\ \underline{4x - 3y + 4z = 10} \\ 7y = 14 \\ y = 2 \end{array}$$

Back-substitute 2 for y in the first and third equations to obtain two equations in two unknowns.

$$\begin{aligned} 3x + 2y - 3z &= -2 \\ 3x + 2(2) - 3z &= -2 \\ 3x + 4 - 3z &= -2 \\ 3x - 3z &= -6 \end{aligned}$$

$$\begin{aligned} 4x - 3y + 4z &= 10 \\ 4x - 3(2) + 4z &= 10 \\ 4x - 6 + 4z &= 10 \\ 4x + 4z &= 16 \end{aligned}$$

The system of two equations in two variables becomes:

$3x - 3z = -6$

$4x + 4z = 16$

Multiply the first equation by –4 and the second equation by 3.

$$\begin{array}{r} -12x + 12z = 24 \\ \underline{12x + 12z = 48} \\ 24z = 72 \\ z = 3 \end{array}$$

Back-substitute 3 for z to find x.

$$\begin{aligned} 3x - 3z &= -6 \\ 3x - 3(3) &= -6 \\ 3x - 9 &= -6 \\ 3x &= 3 \\ x &= 1 \end{aligned}$$

The solution is $(1, 2, 3)$ and the solution set is $\{(1, 2, 3)\}$.

11. $2x-4y+3z=17$

$x+2y-z=0$

$4x-y-z=6$

Multiply the second equation by –1 and add it to the third equation.

$-x-2y+z=0$

$4x-y-z=6$

$3x-3y=6$

Multiply the second equation by 3 and add it to the first equation.

$2x-4y+3z=17$

$3x+6y-3z=0$

$5x+2y=17$

The system in two variables becomes:

$3x-3y=6$

$5x+2y=17$

Multiply the first equation by 2 and the second equation by 3 and solve by addition.

$6x-6y=12$

$15x+6y=51$

$21x=63$

$x=3$

Back-substitute 3 for x in one of the equations in two variables.

$3x-3y=6$

$3(3)-3y=6$

$9-3y=6$

$-3y=-3$

$y=1$

Back-substitute 3 for x and 1 for y in one of the original equations in three variables.

$x+2y-z=0$

$3+2(1)-z=0$

$3+2-z=0$

$5-z=0$

$5=z$

The solution is $(3,1,5)$ and the solution set is $\{(3,1,5)\}$.

13. $2x+y=2$

$x+y-z=4$

$3x+2y+z=0$

Add the second and third equations together to obtain an equation in two variables.

$x+y-z=4$

$3x+2y+z=0$

$4x+3y=4$

Use this equation and the first equation in the original system to write two equations in two variables.

$2x+y=2$

$4x+3y=4$

Multiply the first equation by –2 and solve by addition.

$-4x-2y=-4$

$4x+3y=4$

$y=0$

Back-substitute 0 for y in one of the equations in two unknowns.

$2x+y=2$

$2x+0=2$

$2x=2$

$x=1$

Back-substitute 1 for x and 0 for y in one of the equations in three unknowns.

$x+y-z=4$

$1+0-z=4$

$1-z=4$

$-z=3$

$z=-3$

The solution is $(1,0,-3)$ and the solution set is $\{(1,0,-3)\}$.

15. $x+y=-4$

$y-z=1$

$2x+y+3z=-21$

Multiply the first equation by –1 and add to the second equation.

$-x-y=4$

$y-z=1$

$-x-z=5$

Multiply the second equation by –1 and add to the third equation.

$-y+z=-1$

$2x+y+3z=-21$

$2x+4z=-22$

The system of two equations in two variables becomes:

$$-x - z = 5$$
$$2x + 4z = -22$$

Multiply the first equation by 2 and add to the second equation.

$$\begin{aligned} -2x - 2z &= 10 \\ \underline{2x + 4z} &\underline{= -22} \\ 2z &= -12 \\ z &= -6 \end{aligned}$$

Back-substitute –6 for z in one of the equations in two variables.

$$\begin{aligned} -x - z &= 5 \\ -x - (-6) &= 5 \\ -x + 6 &= 5 \\ -x &= -1 \\ x &= 1 \end{aligned}$$

Back-substitute 1 for x in the first equation of the original system.

$$\begin{aligned} x + y &= -4 \\ 1 + y &= -4 \\ y &= -5 \end{aligned}$$

The solution is $(1, -5, -6)$ and the solution set is $\{(1, -5, -6)\}$.

17. $2x + y + 2z = 1$
$3x - y + z = 2$
$x - 2y - z = 0$

Add the first and second equations to eliminate y.

$$\begin{aligned} 2x + y + 2z &= 1 \\ \underline{3x - y + z} &\underline{= 2} \\ 5x \quad + 3z &= 3 \end{aligned}$$

Multiply the second equation by –2 and add to the third equation.

$$\begin{aligned} -6x + 2y - 2z &= -4 \\ \underline{x - 2y - z} &\underline{= 0} \\ -5x \quad - 3z &= -4 \end{aligned}$$

We obtain two equations in two variables.

$$5x + 3z = 3$$
$$-5x - 3z = -4$$

Adding the two equations, we obtain:

$$\begin{aligned} 5x + 3z &= 3 \\ \underline{-5x - 3z} &\underline{= -4} \\ 0 &= -1 \end{aligned}$$

The system is inconsistent. There are no values of x, y, and z for which $0 = -1$. The solution set is $\varnothing$ or $\{\ \}$.

19. $5x - 2y - 5z = 1$
$10x - 4y - 10z = 2$
$15x - 6y - 15z = 3$

Multiply the first equation by –2 and add to the second equation.

$$\begin{aligned} -10x + 4y + 10z &= -2 \\ \underline{10x - 4y - 10z} &\underline{= 2} \\ 0 &= 0 \end{aligned}$$

The system is dependent and has infinitely many solutions.

21. $3(2x + y) + 5z = -1$
$2(x - 3y + 4z) = -9$
$4(1 + x) = -3(z - 3y)$

Rewrite each equation and obtain the system of three equations in three variables.

$$6x + 3y + 5z = -1$$
$$2x - 6y + 8z = -9$$
$$4x - 9y + 3z = -4$$

Multiply the second equation by –3 and add to the first equation.

$$\begin{aligned} 6x + 3y + 5z &= -1 \\ \underline{-6x + 18y - 24z} &\underline{= 27} \\ 21y - 19z &= 26 \end{aligned}$$

Multiply the second equation by –2 and add to the third equation.

$$\begin{aligned} -4x + 12y - 16z &= 18 \\ \underline{4x - 9y + 3z} &\underline{= -4} \\ 3y - 13z &= 14 \end{aligned}$$

The system of two variables in two equations is:

$$21y - 19z = 26$$
$$3y - 13z = 14$$

Multiply the second equation by –7 and add to the third equation.

$$\begin{aligned} 21y - 19z &= 26 \\ \underline{-21y + 91z} &\underline{= -98} \\ 72z &= -72 \\ z &= -1 \end{aligned}$$

Back-substitute –1 for z in one of the equations in two variables to find y.

$$\begin{aligned} 3y - 13z &= 14 \\ 3y - 13(-1) &= 14 \\ 3y + 13 &= 14 \\ 3y &= 1 \\ y &= \frac{1}{3} \end{aligned}$$

Back-substitute –1 for z and $\frac{1}{3}$ for y in one of the original equations in three variables.

$$6x+3y+5z=-1$$
$$6x+1-5=-1$$
$$6x-4=-1$$
$$6x=3$$
$$x=\frac{1}{2}$$

The solution is $\left(\frac{1}{2},\frac{1}{3},-1\right)$ and the solution set is $\left\{\left(\frac{1}{2},\frac{1}{3},-1\right)\right\}$.

23. Use each ordered pair to write an equation.

$(x, y) = (-1, 6)$

$$y=ax^2+bx+c$$
$$6=a(-1)^2+b(-1)+c$$
$$6=a-b+c$$

$(x, y) = (1, 4)$

$$y=ax^2+bx+c$$
$$4=a(1)^2+b(1)+c$$
$$4=a+b+c$$

$(x, y) = (2, 9)$

$$y=ax^2+bx+c$$
$$9=a(2)^2+b(2)+c$$
$$9=a(4)+2b+c$$
$$9=4a+2b+c$$

The system of three equations in three variables is:

$$a-b+c=6$$
$$a+b+c=4$$
$$4a+2b+c=9$$

Add the first and second equations.

$$a-b+c=6$$
$$a+b+c=4$$
$$2a+2c=10$$

Multiply the first equation by 2 and add to the third equation.

$$2a-2b+2c=12$$
$$4a+2b+c=9$$
$$6a+3c=21$$

The system of two equations in two variables becomes:

$$2a+2c=10$$
$$6a+3c=21$$

Multiply the first equation by –3 and add to the second equation.

$$-6a-6c=-30$$
$$6a+3c=21$$
$$-3c=-9$$
$$c=3$$

Back-substitute 3 for c in one of the equations in two variables.

$$2a+2c=10$$
$$2a+2(3)=10$$
$$2a+6=10$$
$$2a=4$$
$$a=2$$

Back-substitute 3 for c and 2 for a in one of the equations in three variables.

$$a+b+c=4$$
$$2+b+3=4$$
$$b+5=4$$
$$b=-1$$

The quadratic function is $y=2x^2-x+3$.

25. Use each ordered pair to write an equation.

$(x, y) = (-1, -4)$

$$y=ax^2+bx+c$$
$$-4=a(-1)^2+b(-1)+c$$
$$-4=a-b+c$$

$(x, y) = (1, -2)$

$$y=ax^2+bx+c$$
$$-2=a(1)^2+b(1)+c$$
$$-2=a+b+c$$

$(x, y) = (2, 5)$

$$y=ax^2+bx+c$$
$$5=a(2)^2+b(2)+c$$
$$5=a(4)+2b+c$$
$$5=4a+2b+c$$

The system of three equations in three variables is:

$$a-b+c=-4$$
$$a+b+c=-2$$
$$4a+2b+c=5$$

Multiply the second equation by –1 and add to the first equation.

$$\begin{aligned} a-b+c&=-4\\ -a-b-c&=2\\ \hline -2b&=-2\\ b&=1 \end{aligned}$$

Back-substitute 4 for b in first and third equations to obtain two equations in two variables.

$$\begin{aligned} a-b+c&=-4 \\ a-1+c&=-4 \\ a+c&=-3 \end{aligned} \qquad \begin{aligned} 4a+2b+c&=5\\ 4a+2(1)+c&=5\\ 4a+2+c&=5\\ 4a+c&=3 \end{aligned}$$

The system of two equations in two variables becomes:

$$\begin{aligned} a+c&=-3\\ 4a+c&=3 \end{aligned}$$

Multiply the first equation by –1 and add to the second equation.

$$\begin{aligned} -a-c&=3\\ 4a+c&=3\\ \hline 3a&=6\\ a&=2 \end{aligned}$$

Back-substitute 2 for a and 1for b in one of the equations in three variables.

$$\begin{aligned} a-b+c&=-4\\ 2-1+c&=-4\\ 1+c&=-4\\ c&=-5 \end{aligned}$$

The quadratic function is $y=2x^2+x-5$.

27. Let x = the first number.
Let y = the second number.
Let z = the third number.

$$\begin{aligned} x+y+z&=16\\ 2x+3y+4z&=46\\ 5x-y&=31 \end{aligned}$$

Multiply the first equation by –4 and add to the second equation.

$$\begin{aligned} -4x-4y-4z&=-64\\ 2x+3y+4z&=46\\ \hline -2x-y&=-18 \end{aligned}$$

The system of two equations in two variables becomes:

$$\begin{aligned} 5x-y&=31\\ -2x-y&=-18 \end{aligned}$$

Multiply the first equation by –1 and add to the second equation.

$$\begin{aligned} -5x+y&=-31\\ -2x-y&=-18\\ \hline -7x&=-49\\ x&=7 \end{aligned}$$

Back-substitute 7 for x in one of the equations in two variables.

$$\begin{aligned} 5x-y&=31\\ 5(7)-y&=31\\ 35-y&=31\\ -y&=-4\\ y&=4 \end{aligned}$$

Back-substitute 7 for x and 4 for y in one of the equations in two variables.

$$\begin{aligned} x+y+z&=16\\ 7+4+z&=16\\ 11+z&=16\\ z&=5 \end{aligned}$$

The numbers are 7, 4 and 5.

29. Simplify each equation.

$$\begin{aligned} \frac{x+2}{6}-\frac{y+4}{3}+\frac{z}{2}&=0\\ 6\left(\frac{x+2}{6}-\frac{y+4}{3}+\frac{z}{2}\right)&=6(0)\\ (x+2)-2(y+4)+3z&=0\\ x+2-2y-8+3z&=0\\ x-2y+3z&=6 \end{aligned}$$

$$\begin{aligned} \frac{x+1}{2}+\frac{y-1}{2}-\frac{z}{4}&=\frac{9}{2}\\ 4\left(\frac{x+1}{2}+\frac{y-1}{2}-\frac{z}{4}\right)&=4\left(\frac{9}{2}\right)\\ 2(x+1)+2(y-1)-z&=18\\ 2x+2+2y-2-z&=18\\ 2x+2y-z&=18 \end{aligned}$$

$$\begin{aligned} \frac{x-5}{4}+\frac{y+1}{3}+\frac{z-2}{2}&=\frac{19}{4}\\ 12\left(\frac{x-5}{4}+\frac{y+1}{3}+\frac{z-2}{2}\right)&=12\left(\frac{19}{4}\right)\\ 3(x-5)+4(y+1)+6(z-2)&=57\\ 3x-15+4y+4+6z-12&=57\\ 3x+4y+6z&=80 \end{aligned}$$

Now solve the equivalent system.

$$x-2y+3z=6$$
$$2x+2y-z=18$$
$$3x+4y+6z=80$$

Add the first two equations together.

$$x-2y+3z=6$$
$$\underline{2x+2y-z=18}$$
$$3x+2z=24$$

Multiply the second equation by -2 and add it to the third equation.

$$-4x-4y+2z=-36$$
$$\underline{3x+4y+6z=80}$$
$$-x+8z=44$$

Using the two reduced equations, we solve the system.

$$3x+2z=24$$
$$-x+8z=44$$

Multiply the second equation by 3 and add the equations.

$$3x+2z=24$$
$$\underline{-3x+24z=132}$$
$$26z=156$$
$$z=6$$

Back-substitute to find x.

$$-x+8(6)=44$$
$$-x+48=44$$
$$-x=-4$$
$$x=4$$

Back substitute to find y.

$$x-2y+3z=6$$
$$4-2y+3(6)=6$$
$$-2y=-16$$
$$y=8$$

The solution is $(4,8,6)$ and the solution set is $\{(4,8,6)\}$.

31. Selected points may vary, but the equation will be the same.

$$y=ax^2+bx+c$$

Use the points $(2,-2)$, $(4,1)$, and $(6,-2)$ to get the system

$$4a+2b+c=-2$$
$$16a+4b+c=1$$
$$36a+6b+c=-2$$

Multiply the first equation by -1 and add to the second equation.

$$-4a-2b-c=2$$
$$\underline{16a+4b+c=1}$$
$$12a+2b=3$$

Multiply the first equation by -1 and add to the third equation.

$$-4a-2b-c=2$$
$$\underline{36a+6b+c=-2}$$
$$32a+4b=0$$

Using the two reduced equations, we get the system

$$12a+2b=3$$
$$32a+4b=0$$

Multiply the first equation by -2 and add to the second equation.

$$-24a-4b=-6$$
$$\underline{32a+4b=0}$$
$$8a=-6$$
$$a=-\frac{3}{4}$$

Back-substitute to solve for b.

$$12a+2b=3$$
$$12\left(-\frac{3}{4}\right)+2b=3$$
$$-9+2b=3$$
$$2b=12$$
$$b=6$$

Back-substitute to solve for c.

$$4a+2b+c=-2$$
$$4\left(-\frac{3}{4}\right)+2(6)+c=-2$$
$$-3+12+c=-2$$
$$c=-11$$

The equation is $y=-\frac{3}{4}x^2+6x-11$.

33. $ax - by - 2cz = 21$
$ax + by + cz = 0$
$2ax - by + cz = 14$
Add the first two equations.
$$\begin{aligned} ax - by - 2cz &= 21 \\ ax + by + cz &= 0 \\ \hline 2ax - cz &= 21 \end{aligned}$$
Multiply the first equation by -1 and add to the third equation.
$$\begin{aligned} -ax + by + 2cz &= -21 \\ 2ax - by + cz &= 14 \\ \hline ax + 3cz &= -7 \end{aligned}$$
Use the two reduced equations to get the following system:
$$2ax - cz = 21$$
$$ax + 3cz = -7$$
Multiply the second equation by -2 and add the equations.
$$\begin{aligned} 2ax - cz &= 21 \\ -2ax - 6cz &= 14 \\ \hline -7cz &= 35 \\ z &= -\frac{5}{c} \end{aligned}$$
Back-substitute to solve for x.
$$\begin{aligned} ax + 3cz &= -7 \\ ax + 3c\left(-\frac{5}{c}\right) &= -7 \\ ax - 15 &= -7 \\ ax &= 8 \\ x &= \frac{8}{a} \end{aligned}$$
Back-substitute to solve for y.
$$\begin{aligned} ax + by + cz &= 0 \\ a\left(\frac{8}{a}\right) + by + c\left(-\frac{5}{c}\right) &= 0 \\ 8 + by - 5 &= 0 \\ by &= -3 \\ y &= -\frac{3}{b} \end{aligned}$$
The solution is $\left(\frac{8}{a}, -\frac{3}{b}, -\frac{5}{c}\right)$ and the solution set is $\left\{\left(\frac{8}{a}, -\frac{3}{b}, -\frac{5}{c}\right)\right\}$.

35. a. $(0,5)$, $(50,31)$, $(100,15)$

b. Substituting each ordered pair gives:
$$\begin{aligned} ax^2 + bx + c &= y \\ a(0)^2 + b(0) + c &= 5 \\ a(50)^2 + b(50) + c &= 31 \\ a(100)^2 + b(100) + c &= 15 \end{aligned}$$
Simplifying gives the following system:
$$\begin{aligned} 0a + 0b + c &= 5 \\ 2500a + 50b + c &= 31 \\ 10,000a + 100b + c &= 15 \end{aligned}$$

37. a. Using the three ordered pairs, $(1,224)$, $(3,176)$, and $(4,104)$, we get the following system:
$$\begin{aligned} a + b + c &= 224 \\ 9a + 3b + c &= 176 \\ 16a + 4b + c &= 104 \end{aligned}$$
Multiply the first equation by -1 and add to the second equation.
$$\begin{aligned} -a - b - c &= -224 \\ 9a + 3b + c &= 176 \\ \hline 8a + 2b &= -48 \end{aligned}$$
Multiply the first equation by -1 and add to the third equation.
$$\begin{aligned} -a - b - c &= -224 \\ 16a + 4b + c &= 104 \\ \hline 15a + 3b &= -120 \end{aligned}$$
Using the two reduced equations, we get the following system:
$$8a + 2b = -48$$
$$15a + 3b = -120$$
Multiply the first equation by -3 and multiply the second equation by 2, then add to the equations.
$$\begin{aligned} -24a - 6b &= 144 \\ 30a + 6b &= -240 \\ \hline 6a &= -96 \\ a &= -16 \end{aligned}$$
Back-substitute to solve for b.
$$\begin{aligned} 8a + 2b &= -48 \\ 8(-16) + 2b &= -48 \\ -128 + 2b &= -48 \\ 2b &= 80 \\ b &= 40 \end{aligned}$$

Back-substitute to solve for c.

$$a+b+c=224$$
$$-16+40+c=224$$
$$c=200$$

The function is $y=-16x^2+40x+200$.

b. When $x=5$, we get

$$y=-16(5)^2+40(5)+200$$
$$=-16(25)+200+200$$
$$=-400+400$$
$$=0$$

After 5 seconds, the ball hits the ground.

39. Let x = annual spending in 2010 per person on housing.
Let y = annual spending in 2010 per person on vehicles/gas.
Let z = annual spending in 2010 per person on health care.

$$x+y+z=13,840$$
$$x-y=3864$$
$$x-z=695$$

Solve the second equation for y.

$$x-y=3864$$
$$-y=-x+3864$$
$$y=x-3864$$

Solve the third equation for z.

$$x-z=695$$
$$-z=-x+695$$
$$z=x-695$$

Substitute the expressions for x and z into the first equation and solve for y.

$$x+y+z=13,840$$
$$x+\overbrace{(x-3864)}^{y}+\overbrace{(x-695)}^{z}=13,840$$
$$x+x-3864+x-695=13,840$$
$$3x-4559=13,840$$
$$3x=18,399$$
$$x=6133$$

Back-substitute to solve for y and z.

$$y=x-3864$$
$$=6133-3864$$
$$=2269$$

$$z=x-695$$
$$=6133-695$$
$$=5438$$

The annual spending in 2010 per person on housing is \$6133, on vehicles/gas is \$2269, and on health care is \$5438.

41. Let x = the amount invested at 8%.
Let y = the amount invested at 10%.
Let z = the amount invested at 12%.

$$x+y+z=6700$$
$$0.08x+0.10y+0.12z=716$$
$$z-x-y=300$$

Rewrite the system in $Ax+By+Cz=D$ form.

$$x+y+z=6700$$
$$0.08x+0.10y+0.12z=716$$
$$-x-y+z=300$$

Add the first and third equations to find z.

$$x+y+z=6700$$
$$\underline{-x-y+z=300}$$
$$2z=7000$$
$$z=3500$$

Back-substitute 3500 for z to obtain two equations in two variables.

$$x+y+z=6700$$
$$x+y+3500=6700$$
$$x+y=3200$$

$$0.08x+0.10y+0.12(3500)=716$$
$$0.08x+0.10y+420=716$$
$$0.08x+0.10y=296$$

The system of two equations in two variables becomes:

$$x+y=3200$$
$$0.08x+0.10y=296$$

Multiply the second equation by -10 and add it to the first equation.

$$x+y=3200$$
$$\underline{-0.8x+-y=-2960}$$
$$0.2x=240$$
$$x=1200$$

Back-substitute 1200 for x in one of the equations in two variables.

$$x+y=3200$$
$$1200+y=3200$$
$$y=2000$$

\$1200 was invested at 8%, \$2000 was invested at 10%, and \$3500 was invested at 12%.

43. Let x = the number of \$8 tickets.
Let y = the number of \$10 tickets.
Let z = the number of \$12 tickets.

$$\begin{aligned} x + y + z &= 400 \\ 8x + 10y + 12z &= 3700 \\ x + y &= 7z \end{aligned}$$

Rewrite the system in $Ax + By + Cz = D$ form.

$$\begin{aligned} x + y + z &= 400 \\ 8x + 10y + 12z &= 3700 \\ x + y - 7z &= 0 \end{aligned}$$

Multiply the first equation by –1 and add to the third equation.

$$\begin{aligned} -x - y - z &= -400 \\ x + y - 7z &= 0 \\ \hline -8z &= -400 \\ z &= 50 \end{aligned}$$

Back-substitute 50 for z in two of the original equations to obtain two of equations in two variables.

$$\begin{aligned} x + y + z &= 400 \\ x + y + 50 &= 400 \\ x + y &= 350 \end{aligned}$$

$$\begin{aligned} 8x + 10y + 12z &= 3700 \\ 8x + 10y + 12(50) &= 3700 \\ 8x + 10y + 600 &= 3700 \\ 8x + 10y &= 3100 \end{aligned}$$

The system of two equations in two variables becomes:

$$\begin{aligned} x + y &= 350 \\ 8x + 10y &= 3100 \end{aligned}$$

Multiply the first equation by –8 and add to the second equation.

$$\begin{aligned} -8x - 8y &= -2800 \\ 8x + 10y &= 3100 \\ \hline 2y &= 300 \\ y &= 150 \end{aligned}$$

Back-substitute 50 for z and 150 for y in one of the original equations in three variables.

$$\begin{aligned} x + y + z &= 400 \\ x + 150 + 50 &= 400 \\ x + 200 &= 400 \\ x &= 200 \end{aligned}$$

There were 200 \$8 tickets, 150 \$10 tickets, and 50 \$12 tickets sold.

45. Let A = the number of servings of A.
Let B = the number of servings of B.
Let C = the number of servings of C.

$$\begin{aligned} 40A + 200B + 400C &= 660 \\ 5A + 2B + 4C &= 25 \\ 30A + 10B + 300C &= 425 \end{aligned}$$

Multiply the second equation by –8 and add to the first equation to obtain an equation in two variables.

$$\begin{aligned} 40A + 200B + 400C &= 660 \\ -40A - 16B - 32C &= -200 \\ \hline 184B + 368C &= 460 \end{aligned}$$

Multiply the second equation by –6 and add to the third equation to obtain an equation in two variables.

$$\begin{aligned} -30A - 12B - 24C &= -150 \\ 30A + 10B + 300C &= 425 \\ \hline -2B + 276C &= 275 \end{aligned}$$

The system of two equations in two variables becomes:

$$\begin{aligned} 184B + 368C &= 460 \\ -2B + 276C &= 275 \end{aligned}$$

Multiply the second equation by 92 and eliminate B.

$$\begin{aligned} 184B + 368C &= 460 \\ -184B + 25392C &= 25300 \\ \hline 25760C &= 25760 \\ C &= 1 \end{aligned}$$

Back-substitute 1 for C in one of the equations in two variables.

$$\begin{aligned} -2B + 276C &= 275 \\ -2B + 276(1) &= 275 \\ -2B + 276 &= 275 \\ -2B &= -1 \\ B &= \frac{1}{2} \end{aligned}$$

Back-substitute 1 for C and $\frac{1}{2}$ for B in one of the original equations in three variables.

$$\begin{aligned} 5A + 2B + 4C &= 25 \\ 5A + 2\left(\frac{1}{2}\right) + 4(1) &= 25 \\ 5A + 1 + 4 &= 25 \\ 5A + 5 &= 25 \\ 5A &= 20 \\ A &= 4 \end{aligned}$$

To meet the requirements, 4 ounces of Food A, $\frac{1}{2}$ ounce of Food B, and 1 ounce of Food C should be used.

47. – 53. Answers will vary.

55. does not make sense; Explanations will vary. Sample explanation: The third variable could possibly have the same variable as one of the other two.

57. makes sense

59. false; Changes to make the statement true will vary. A sample change is: The given ordered triple is one solution to the equation, but there are an infinite number of other ordered triples which satisfy the equation.

61. true

63. $x + y + z = 180$

$(2x+5) + y = 180$

$(2x-5) + z = 180$

Rewrite the system in standard form as

$x + y + z = 180$

$2x + y = 175$

$2x + z = 185$

Multiply the first equation by -1 and add to the second equation to obtain an equation with two variables.

$$\begin{array}{r} -x - y - z = -180 \\ \underline{2x + y \qquad = 175} \\ x - z = -5 \end{array}$$

Combine this equation with the third equation to make a system of two equations.

$$\begin{array}{r} x - z = -5 \\ \underline{2x + z = 185} \\ 3x = 180 \\ x = 60 \end{array}$$

Back-substitute to find z.

$x - z = -5$

$60 - z = -5$

$-z = -65$

$z = 65$

Back-substitute to find y.

$x + y + z = 180$

$60 + y + 65 = 180$

$y = 55$

The angles measure $55°$, $60°$, and $65°$.

65. Let x = height of the table.
Let y = length of the wood blocks.
Let z = width of the wood blocks.

From the problem, we have the following two equations.

$x + y - z = 32$

$x - y + z = 28$

Add the two equations.

$$\begin{array}{r} x + y - z = 32 \\ \underline{x - y + z = 28} \\ 2x \qquad = 60 \\ x = 30 \end{array}$$

The height of the table is 30 centimeters.

66. $f(x) = -\frac{3}{4}x + 3$

Use the slope and the y–intercept to graph the line.

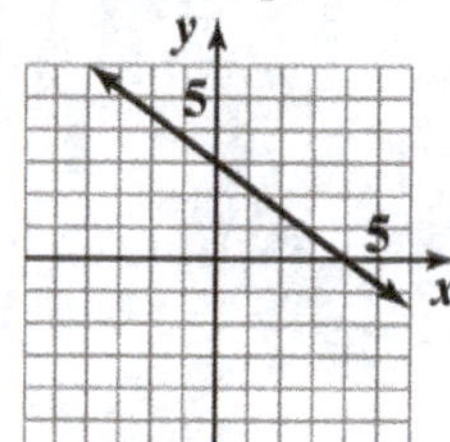

67. $-2x + y = 6$

Rewrite the equation in slope-intercept form.

$-2x + y = 6$

$y = 2x + 6$

Use the slope and the y–intercept to graph the line.

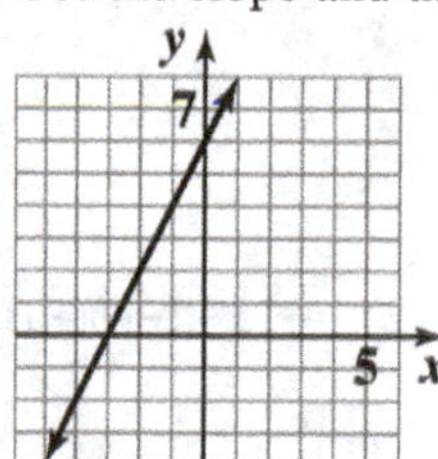

68. $f(x) = -5$

This line is the horizontal line, $y = -5$.

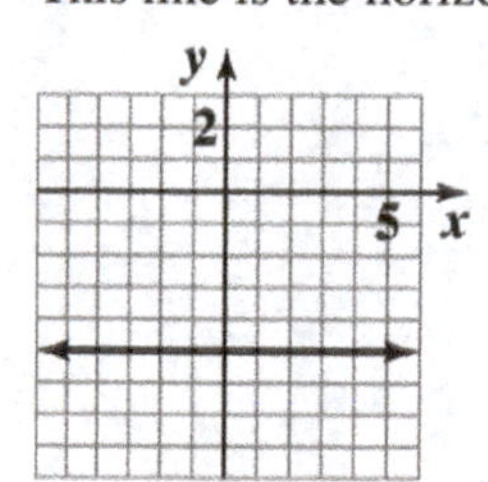

69. $x+2y=-1$
$y=1$

The value for y is given and thus the value of x can be found by back-substitution.

$$x+2y=-1$$
$$x+2(1)=-1$$
$$x+2=-1$$
$$x=-3$$

The solution is $(-3,\ 1)$ and the solution set is $\{(-3,\ 1)\}$.

70. $x+y+2z=19$
$y+2z=13$
$z=5$

The value for y is given and the other variables be found by back-substitution.

$$y+2z=13$$
$$y+2(5)=13$$
$$y+10=13$$
$$y=3$$

$$x+y+2z=19$$
$$x+(3)+2(5)=19$$
$$x+13=19$$
$$x=6$$

The solution is $(6,3,5)$ and the solution set is $\{(6,3,5)\}$.

71. $\begin{bmatrix} 1 & 2 & -1 \\ 4 & -3 & -15 \end{bmatrix}$

$$\begin{bmatrix} 1 & 2 & -1 \\ 4+(-4)(1) & -3+(-4)(2) & -15+(-4)(-1) \end{bmatrix} = \begin{bmatrix} 1 & 2 & -1 \\ 0 & -11 & -11 \end{bmatrix}$$

Mid-Chapter Check Point – Chapter 3

1. $x=3y-7$
$4x+3y=2$

Since the first equation is solved for x already, we will use substitution.

Let $x=3y-7$ in the second equation and solve for y.

$$4(3y-7)+3y=2$$
$$12y-28+3y=2$$
$$15y=30$$
$$y=2$$

Substitute this value for y in the first equation.

$$x=3(2)-7=6-7=-1$$

The solution is $(-1,2)$ and the solution set is $\{(-1,2)\}$.

2. $3x+4y=-5$
$2x-3y=8$

Multiply the first equation by 3 and the second equation by 4, then add the equations.

$$9x+12y=-15$$
$$\underline{8x-12y=32}$$
$$17x=17$$
$$x=1$$

Back-substitute to solve for y.

$$3x+4y=-5$$
$$3(1)+4y=-5$$
$$3+4y=-5$$
$$4y=-8$$
$$y=-2$$

The solution is $(1,-2)$ and the solution set is $\{(1,-2)\}$.

3. $\frac{2x}{3}+\frac{y}{5}=6$

$\frac{x}{6}-\frac{y}{2}=-4$

Multiply the first equation by 15 and the second equation by 6 to eliminate the fractions.

$15\left(\frac{2x}{3}+\frac{y}{5}\right)=15(6)$

$10x+3y=90$

$6\left(\frac{x}{6}-\frac{y}{2}\right)=6(-4)$

$x-3y=-24$

We now need to solve the equivalent system.

$10x+3y=90$

$x-3y=-24$

Add the two equations to eliminate y.

$10x+3y=90$

$x-3y=-24$

$11x=66$

$x=6$

Back-substitute to solve for y.

$x-3y=-24$

$6-3y=-24$

$-3y=-30$

$y=10$

The solution is $(6,10)$ and the solution set is $\{(6,10)\}$.

4. $y=4x-5$

$8x-2y=10$

Since the first equation is already solved for y, we will use substitution.

Let $y=4x-5$ in the second equation and solve for x.

$8x-2(4x-5)=10$

$8x-8x+10=10$

$10=10$

This statement is an identity. The system is dependent so there are an infinite number of solutions. The solution set is $\{(x,y)\mid y=4x-5\}$ or $\{(x,y)\mid 8x-2y=10\}$.

5. $2x+5y=3$

$3x-2y=1$

Multiply the first equation by 3 and the second equation by -2, then add the equations.

$6x+15y=9$

$-6x+4y=-2$

$19y=7$

$y=\frac{7}{19}$

Back-substitute to solve for x.

$2x+5y=3$

$2x+5\left(\frac{7}{19}\right)=3$

$2x+\frac{35}{19}=3$

$2x=\frac{22}{19}$

$x=\frac{11}{19}$

The solution is $\left(\frac{11}{19},\frac{7}{19}\right)$ and the solution set is $\left\{\left(\frac{11}{19},\frac{7}{19}\right)\right\}$.

6. $\frac{x}{12}-y=\frac{1}{4}$

$4x-48y=16$

Solve the first equation for y.

$\frac{x}{12}-y=\frac{1}{4}$

$-y=-\frac{x}{12}+\frac{1}{4}$

$y=\frac{x}{12}-\frac{1}{4}$

Let $y=\frac{x}{12}-\frac{1}{4}$ in the second equation and solve for x.

$4x-48\left(\frac{x}{12}-\frac{1}{4}\right)=16$

$4x-4x+12=16$

$12=16$

This statement is a contradiction. The system is inconsistent so there is no solution. The solution is $\{\ \}$ or $\varnothing$.

7. $2x-y+2z=-8$

$x+2y-3z=9$

$3x-y-4z=3$

Multiply the first equation by 2 and add to the second equation.

$$\begin{array}{r} 4x-2y+4z=-16 \\ \underline{x+2y-3z=9} \\ 5x+z=-7 \end{array}$$

Multiply the first equation by -1 and add to the third equation.

$$\begin{array}{r} -2x+y-2z=8 \\ \underline{3x-y-4z=3} \\ x-6z=11 \end{array}$$

Use the two reduced equations to get the following system:

$5x+z=-7$

$x-6z=11$

Multiply the first equation by 6 and add to the second equation.

$$\begin{array}{r} 30x+6z=-42 \\ \underline{x-6z=11} \\ 31x=-31 \\ x=-1 \end{array}$$

Back-substitute to solve for z.

$$\begin{aligned} 5x+z&=-7 \\ 5(-1)+z&=-7 \\ -5+z&=-7 \\ z&=-2 \end{aligned}$$

Back-substitute to solve for y.

$$\begin{aligned} 2x-y+2z&=-8 \\ 2(-1)-y+2(-2)&=-8 \\ -2-y-4&=-8 \\ -y&=-2 \\ y&=2 \end{aligned}$$

The solution is $(-1,2,-2)$ and the solution set is $\{(-1,2,-2)\}$.

8. $x\quad -3z=-5$

$2x-y+2z=16$

$7x-3y-5z=19$

Multiply the second equation by -3 and add to the third equation.

$$\begin{array}{r} -6x+3y-6z=-48 \\ \underline{7x-3y-5z=19} \\ x-11z=-29 \end{array}$$

Use this reduced equation and the original first equation to obtain the following system:

$x-3z=-5$

$x-11z=-29$

Multiply the second equation by -1 and add to the first equation.

$$\begin{array}{r} x-3z=-5 \\ \underline{-x+11z=29} \\ 8z=24 \\ z=3 \end{array}$$

Back-substitute to solve for x.

$$\begin{aligned} x-3z&=-5 \\ x-3(3)&=-5 \\ x-9&=-5 \\ x&=4 \end{aligned}$$

Back-substitute to solve for y.

$$\begin{aligned} 2x-y+2z&=16 \\ 2(4)-y+2(3)&=16 \\ 8-y+6&=16 \\ -y&=2 \\ y&=-2 \end{aligned}$$

The solution is $(4,-2,3)$ and the solution set is $\{(4,-2,3)\}$.

9. Graph the two lines by using the intercepts.

$2x-y=4$

x-intercept:
$$\begin{aligned} 2x-y&=4 \\ 2x-0&=4 \\ 2x&=4 \\ x&=2 \end{aligned}$$

y-intercept:
$$\begin{aligned} 2x-y&=4 \\ 2(0)-y&=4 \\ -y&=4 \\ y&=-4 \end{aligned}$$

$x + y = 5$

x-intercept: $x + y = 5$

$x + 0 = 5$

$x = 5$

y-intercept: $x + y = 5$

$0 + y = 5$

$y = 5$

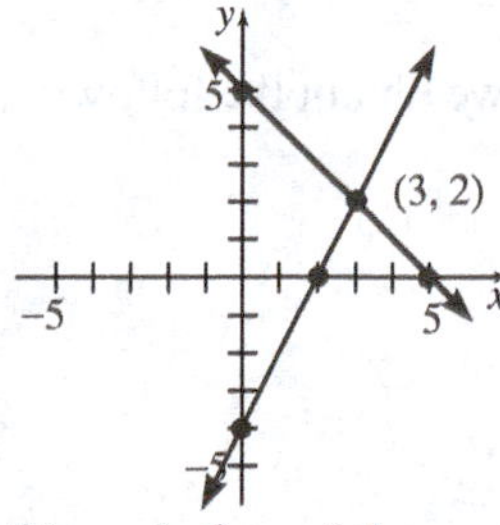

The solution of the system is the intersection point of the graphs. Therefore, the solution is $(3, 2)$ and the solution set is $\{(3, 2)\}$.

10. Graph the two lines by using the slope and y-intercept.

$y = x + 3$

y-intercept: $b = 3$

slope: $m = 1 = \frac{1}{1}$

Plot the points $(0, 3)$ and $(0+1, 3+1) = (1, 4)$

$y = -\frac{1}{2}x$

y-intercept: $b = 0$

slope: $m = -\frac{1}{2} = \frac{-1}{2}$

Plot the points $(0, 0)$ and $(0+2, 0-1) = (2, -1)$.

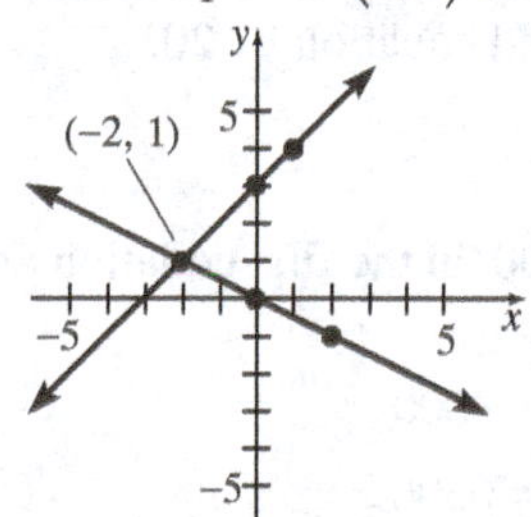

The solution of the system is the intersection point of the graphs. Therefore, the solution is $(-2, 1)$ and the solution set is $\{(-2, 1)\}$.

11. a. $C(x) = 400{,}000 + 20x$

b. $R(x) = 100x$

c. $P(x) = R(x) - C(x)$

$= 100x - (400{,}000 + 20x)$

$= 80x - 400{,}000$

d. The break-even point is the point where cost and revenue are the same. We need to solve the system

$y = 400{,}000 + 20x$

$y = 100x$

Let $y = 400{,}000 + 20x$ in the second equation and solve for x.

$400{,}000 + 20x = 100x$

$400{,}000 = 80x$

$5000 = x$

Back-substitute to solve for y.

$y = 100x$

$= 100(5000)$

$= 500{,}000$

Thus, the break-even point is $(5000,\ 500{,}000)$.

The company will break even when it produces and sells 5000 PDAs. At this level, the cost and revenue will both be \$500,000.

12. Let x = the number of roses.

Let y = the number of carnations.

$x + y = 20$

$3x + 1.5y = 39$

Solve the first equation for x.

$x + y = 20$

$x = 20 - y$

Substitute this expression for x in the second equation and solve for y.

$3(20 - y) + 1.5y = 39$

$60 - 3y + 1.5y = 39$

$-1.5y = -21$

$y = 14$

Back-substitute to solve for x.

$x = 20 - y = 20 - 14 = 6$

There are 6 roses and 14 carnations in the bouquet.

13. Let x = the amount invested at 5%.
Let y = the amount invested at 6%.

$$x + y = 15,000$$
$$0.05x + 0.06y = 837$$

Solve the first equation for x.

$$x + y = 15,000$$
$$x = 15,000 - y$$

Substitute this expression for x in the second equation and solve for y.

$$0.05(15,000 - y) + 0.06y = 837$$
$$750 - 0.05y + 0.06y = 837$$
$$0.01y = 87$$
$$y = 8700$$

Back-substitute to solve for x.

$$x = 15,000 - y$$
$$= 15,000 - 8700$$
$$= 6300$$

You invested \$6300 at 5% and \$8700 at 6%.

14. Let x = gallons of 13% nitrogen.
Let y = gallons of 18% nitrogen.

$$x + y = 50$$
$$0.13x + 0.18y = 0.16(50)$$

or

$$x + y = 50$$
$$0.13x + 0.18y = 8$$

Solve the first equation for x.

$$x + y = 50$$
$$x = 50 - y$$

Substitute this expression for x in the second equation and solve for y.

$$0.13(50 - y) + 0.18y = 8$$
$$6.5 - 0.13y + 0.18y = 8$$
$$0.05y = 1.5$$
$$y = 30$$

Back-substitute to solve for x.

$$x = 50 - y = 50 - 30 = 20$$

The manager should mix 20 gallons of the 13% nitrogen with 30 gallons of the 18% nitrogen.

15. Let w = the rate of the water (current).
Let r = your average rowing rate.
For this problem we will make use of the distance traveled formula: $d = r \cdot t$
In addition, remember that when you go *with* the current you add the rate of the current to your rowing rate. If you go *against* the current, you subtract the rate of the current from your rowing rate.

With this in mind, we obtain the following system:

$$9 = (r + w)(2)$$
$$9 = (r - w)(6)$$

or

$$2r + 2w = 9$$
$$6r - 6w = 9$$

Multiply the first equation by 3 and add the two equations.

$$6r + 6w = 27$$
$$\underline{6r - 6w = 9}$$
$$12r = 36$$
$$r = 3$$

Back-substitute to solve for w.

$$2r + 2w = 9$$
$$2(3) + 2w = 9$$
$$6 + 2w = 9$$
$$2w = 3$$
$$w = 1.5$$

Your rowing rate in still water is 3 miles per hour; the rate of the current is 1.5 miles per hour.

16. Let x = the amount invested at 2%.
Let y = the amount invested at 5%.

$$x + y = 8000$$
$$0.05y = 0.02x + 85$$

Multiply the second equation by 20.

$$x + y = 8000$$
$$y = 0.4x + 1700$$

Let $y = 0.4x + 1700$ in the first equation and solve for x.

$$x + (0.4x + 1700) = 8000$$
$$1.4x = 6300$$
$$x = 4500$$

Back-substitute to solve for y.

$$x + y = 8000$$
$$4500 + y = 8000$$
$$y = 3500$$

You invested \$4500 at 2% and \$3500 at 5%.

17. Using the points $(-1,0)$, $(1,4)$, and $(2,3)$ in the equation $y = ax^2 + bx + c$, we get the following system of equations:

$$a - b + c = 0$$
$$a + b + c = 4$$
$$4a + 2b + c = 3$$

Add the first two equations.

$$a - b + c = 0$$
$$\underline{a + b + c = 4}$$
$$2a + 2c = 4$$

Multiply the first equation by 2 and add to the third equation.

$$2a - 2b + 2c = 0$$
$$\underline{4a + 2b + c = 3}$$
$$6a + 3c = 3$$

Using the two reduced equations, we get the following system of equations:

$$2a + 2c = 4$$
$$6a + 3c = 3$$

Multiply the first equation by -3 and add to the second equation.

$$-6a - 6c = -12$$
$$\underline{6a + 3c = 3}$$
$$-3c = -9$$
$$c = 3$$

Back-substitute to solve for a.

$$2a + 2c = 4$$
$$2a + 2(3) = 4$$
$$2a + 6 = 4$$
$$2a = -2$$
$$a = -1$$

Back-substitute to solve for b.

$$a + b + c = 4$$
$$-1 + b + 3 = 4$$
$$b = 2$$

The equation is $y = -x^2 + 2x + 3$.

18. Let n = the number of nickels.
Let d = the number of dimes.
Let q = the number of quarters.
From the problem statement, we have

$$n + d + q = 26$$
$$0.05n + 0.10d + 0.25q = 4.00$$
$$q = n + d - 2$$

If we multiply the second equation by 20 and rearrange the third equation, we get the following equivalent system:

$$n + d + q = 26$$
$$n + 2d + 5q = 80$$
$$n + d - q = 2$$

Multiply the first equation by -1 and add to the second equation.

$$-n - d - q = -26$$
$$\underline{n + 2d + 5q = 80}$$
$$d + 4q = 54$$

Multiply the third equation by -1 and add to the first equation.

$$n + d + q = 26$$
$$\underline{-n - d + q = -2}$$
$$2q = 24$$
$$q = 12$$

Back-substitute to solve for d.

$$d + 4q = 54$$
$$d + 4(12) = 54$$
$$d + 48 = 54$$
$$d = 6$$

Back-substitute to solve for n.

$$n + d + q = 26$$
$$n + 6 + 12 = 26$$
$$n = 8$$

The collection contains 8 nickels, 6 dimes, and 12 quarters.

3.4 Check Points

1. a. $\left[\begin{array}{ccc|c} 4 & 12 & -20 & 8 \\ 1 & 6 & -3 & 7 \\ -3 & -2 & 1 & -9 \end{array}\right] R_1 \leftrightarrow R_2 = \left[\begin{array}{ccc|c} 1 & 6 & -3 & 7 \\ 4 & 12 & -20 & 8 \\ -3 & -2 & 1 & -9 \end{array}\right]$

b. $\left[\begin{array}{ccc|c} 4 & 12 & -20 & 8 \\ 1 & 6 & -3 & 7 \\ -3 & -2 & 1 & -9 \end{array}\right] \frac{1}{4}R_1 = \left[\begin{array}{ccc|c} 1 & 3 & -5 & 2 \\ 1 & 6 & -3 & 7 \\ -3 & -2 & 1 & -9 \end{array}\right]$

c. $\left[\begin{array}{ccc|c} 4 & 12 & -20 & 8 \\ 1 & 6 & -3 & 7 \\ -3 & -2 & 1 & -9 \end{array}\right] 3R_2 + R_3 = \left[\begin{array}{ccc|c} 4 & 12 & -20 & 8 \\ 1 & 6 & -3 & 7 \\ 0 & 16 & -8 & 12 \end{array}\right]$

2. $2x - y = -4$

$x + 3y = 5$

Write the augmented matrix for the system.

$\left[\begin{array}{cc|c} 2 & -1 & -4 \\ 1 & 3 & 5 \end{array}\right]$

We want a 1 in the upper left position. One way to do this is to interchange row 1 and row 2.

$\left[\begin{array}{cc|c} 2 & -1 & -4 \\ 1 & 3 & 5 \end{array}\right] R_1 \leftrightarrow R_2 = \left[\begin{array}{cc|c} 1 & 3 & 5 \\ 2 & -1 & -4 \end{array}\right]$

Now we want a 0 below the 1 in the first column.

$\left[\begin{array}{cc|c} 1 & 3 & 5 \\ 2 & -1 & -4 \end{array}\right] -2R_1 + R_2 = \left[\begin{array}{cc|c} 1 & 3 & 5 \\ 0 & -7 & -14 \end{array}\right]$

Next we want a 1 in the second row, second column.

$\left[\begin{array}{cc|c} 1 & 3 & 5 \\ 0 & -7 & -14 \end{array}\right] \frac{1}{-7}R_2 = \left[\begin{array}{cc|c} 1 & 3 & 5 \\ 0 & 1 & 2 \end{array}\right]$

The resulting system is:

$x + 3y = 5$

$y = 2$

Back-substitute 2 for y in the first equation.

$x + 3y = 5$

$x + 3(2) = 5$

$x + 6 = 5$

$x = -1$

$(-1, 2)$ satisfies both equations.

The solution set is $\{(-1, 2)\}$.

3. $2x + y + 2z = 18$
$x - y + 2z = 9$
$x + 2y - z = 6$

Write the augmented matrix for the system.

$$\left[\begin{array}{ccc|c} 2 & 1 & 2 & 18 \\ 1 & -1 & 2 & 9 \\ 1 & 2 & -1 & 6 \end{array}\right]$$

We want a 1 in the upper left position. One way to do this is to interchange row 1 and row 2.

$$\left[\begin{array}{ccc|c} 2 & 1 & 2 & 18 \\ 1 & -1 & 2 & 9 \\ 1 & 2 & -1 & 6 \end{array}\right] \quad R_1 \leftrightarrow R_2 = \left[\begin{array}{ccc|c} 1 & -1 & 2 & 9 \\ 2 & 1 & 2 & 18 \\ 1 & 2 & -1 & 6 \end{array}\right]$$

Now we want zeros below the 1 in the first column.

$$\left[\begin{array}{ccc|c} 1 & -1 & 2 & 9 \\ 2 & 1 & 2 & 18 \\ 1 & 2 & -1 & 6 \end{array}\right] \quad -2R_1 + R_2 = \left[\begin{array}{ccc|c} 1 & -1 & 2 & 9 \\ 0 & 3 & -2 & 0 \\ 1 & 2 & -1 & 6 \end{array}\right]$$

$$\left[\begin{array}{ccc|c} 1 & -1 & 2 & 9 \\ 0 & 3 & -2 & 0 \\ 1 & 2 & -1 & 6 \end{array}\right] \quad -R_1 + R_3 = \left[\begin{array}{ccc|c} 1 & -1 & 2 & 9 \\ 0 & 3 & -2 & 0 \\ 0 & 3 & -3 & -3 \end{array}\right]$$

Next we want a 1 in the second row, second column.

$$\left[\begin{array}{ccc|c} 1 & -1 & 2 & 9 \\ 0 & 3 & -2 & 0 \\ 0 & 3 & -3 & -3 \end{array}\right] \quad \frac{1}{3}R_2 = \left[\begin{array}{ccc|c} 1 & -1 & 2 & 9 \\ 0 & 1 & -\frac{2}{3} & 0 \\ 0 & 3 & -3 & -3 \end{array}\right]$$

Now we want a zero below the 1 in the second row, second column.

$$\left[\begin{array}{ccc|c} 1 & -1 & 2 & 9 \\ 0 & 1 & -\frac{2}{3} & 0 \\ 0 & 3 & -3 & -3 \end{array}\right] \quad -3R_2 + R_3 = \left[\begin{array}{ccc|c} 1 & -1 & 2 & 9 \\ 0 & 1 & -\frac{2}{3} & 0 \\ 0 & 0 & -1 & -3 \end{array}\right]$$

Next we want a 1 in the third row, third column.

$$\left[\begin{array}{ccc|c} 1 & -1 & 2 & 9 \\ 0 & 1 & -\frac{2}{3} & 0 \\ 0 & 0 & -1 & -3 \end{array}\right] \quad -R_3 = \left[\begin{array}{ccc|c} 1 & -1 & 2 & 9 \\ 0 & 1 & -\frac{2}{3} & 0 \\ 0 & 0 & 1 & 3 \end{array}\right]$$

The resulting system is:

$x - y + 2z = 9$

$y - \frac{2}{3}z = 0$

$z = 3$

Back-substitute 3 for z in the second equation.

$y - \frac{2}{3}(3) = 0$

$y - 2 = 0$

$y = 2$

Back-substitute 2 for y and 3 for z in the first equation.

$$x - y + 2z = 9$$
$$x - (2) + 2(3) = 9$$
$$x - 2 + 6 = 9$$
$$x + 4 = 9$$
$$x = 5$$

$(5, 2, 3)$ satisfies both equations.

The solution set is $\{(5, 2, 3)\}$.

3.4 Concept and Vocabulary Check

1. matrix; elements

2. $\left[\begin{array}{cc|c} 3 & -2 & -6 \\ 4 & 5 & -8 \end{array}\right]$

3. $\left[\begin{array}{ccc|c} 2 & 1 & 4 & -4 \\ 3 & 0 & 1 & 1 \\ 4 & 3 & 1 & 8 \end{array}\right]$

4. 2; first; $\frac{1}{2}$

5. −3; 2; 3; second; −2; third

6. false

7. true

3.4 Exercise Set

1. $\left[\begin{array}{cc|c} 2 & 2 & 5 \\ 1 & -\frac{3}{2} & 5 \end{array}\right]$ $R_1 \leftrightarrow R_2 = \left[\begin{array}{cc|c} 1 & -\frac{3}{2} & 5 \\ 2 & 2 & 5 \end{array}\right]$

3. $\left[\begin{array}{cc|c} -6 & 8 & -12 \\ 3 & 5 & -2 \end{array}\right]$ $-\frac{1}{6}R_1 = \left[\begin{array}{cc|c} 1 & -\frac{4}{3} & 2 \\ 3 & 5 & -2 \end{array}\right]$

5. $\left[\begin{array}{cc|c} 1 & -3 & 5 \\ 2 & 6 & 4 \end{array}\right]$ $-2R_1 + R_2 = \left[\begin{array}{cc|c} 1 & -3 & 5 \\ 0 & 12 & -6 \end{array}\right]$

7. $\left[\begin{array}{cc|c} 1 & -\frac{3}{2} & \frac{7}{2} \\ 3 & 4 & 2 \end{array}\right]$ $-3R_1 + R_2 = \left[\begin{array}{cc|c} 1 & -\frac{3}{2} & \frac{7}{2} \\ 0 & \frac{17}{2} & -\frac{17}{2} \end{array}\right]$

9. $\left[\begin{array}{ccc|c} 2 & -6 & 4 & 10 \\ 1 & 5 & -5 & 0 \\ 3 & 0 & 4 & 7 \end{array}\right] \frac{1}{2}R_1 = \left[\begin{array}{ccc|c} 1 & -3 & 2 & 5 \\ 1 & 5 & -5 & 0 \\ 3 & 0 & 4 & 7 \end{array}\right]$

11. $\left[\begin{array}{ccc|c} 1 & -3 & 2 & 0 \\ 3 & 1 & -1 & 7 \\ 2 & -2 & 1 & 3 \end{array}\right] -3R_1 + R_2 = \left[\begin{array}{ccc|c} 1 & -3 & 2 & 0 \\ 0 & 10 & -7 & 7 \\ 2 & -2 & 1 & 3 \end{array}\right]$

13. $\left[\begin{array}{ccc|c} 1 & 1 & -1 & 6 \\ 2 & -1 & 1 & -3 \\ 3 & -1 & -1 & 4 \end{array}\right] \begin{array}{l} \\ -2R_1 + R_2 = \\ -3R_1 + R_3 \end{array} \left[\begin{array}{ccc|c} 1 & 1 & -1 & 6 \\ 0 & -3 & 3 & -15 \\ 0 & -4 & 2 & -14 \end{array}\right]$

15. $\left[\begin{array}{cc|c} 1 & 1 & 6 \\ 1 & -1 & 2 \end{array}\right] -R_1 + R_2 = \left[\begin{array}{cc|c} 1 & 1 & 6 \\ 0 & -2 & -4 \end{array}\right] -\frac{1}{2}R_2$

$= \left[\begin{array}{cc|c} 1 & 1 & 6 \\ 0 & 1 & 2 \end{array}\right]$

The resulting system is:

$x + y = 6$

$y = 2$

Back-substitute 2 for y in the first equation.

$x + 2 = 6$

$x = 4$

The solution set is $\{(4, 2)\}$.

17. $\left[\begin{array}{cc|c} 2 & 1 & 3 \\ 1 & -3 & 12 \end{array}\right] R_1 \leftrightarrow R_2$

$= \left[\begin{array}{cc|c} 1 & -3 & 12 \\ 2 & 1 & 3 \end{array}\right] -2R_1 + R_2$

$= \left[\begin{array}{cc|c} 1 & -3 & 12 \\ 0 & 7 & -21 \end{array}\right] \frac{1}{7}R_2$

$= \left[\begin{array}{cc|c} 1 & -3 & 12 \\ 0 & 1 & -3 \end{array}\right]$

The system is:

$x - 3y = 12$

$y = -3$

Back-substitute –3 for y in the first equation.

$x - 3y = 12$

$x - 3(-3) = 12$

$x + 9 = 12$

$x = 3$

The solution set is $\{(3, -3)\}$.

19. $\left[\begin{array}{cc|c} 5 & 7 & -25 \\ 11 & 6 & -8 \end{array}\right] \quad \frac{1}{5}R_1$

$= \left[\begin{array}{cc|c} 1 & \frac{7}{5} & -5 \\ 11 & 6 & -8 \end{array}\right] \quad -11R_1 + R_2$

$= \left[\begin{array}{cc|c} 1 & \frac{7}{5} & -5 \\ 0 & -\frac{47}{5} & 47 \end{array}\right] \quad -\frac{5}{47}R_2$

$= \left[\begin{array}{cc|c} 1 & \frac{7}{5} & -5 \\ 0 & 1 & -5 \end{array}\right]$

The resulting system is:

$x + \frac{7}{5}y = -5$

$y = -5$

Back-substitute –5 for y in the first equation.

$x + \frac{7}{5}y = -5$

$x + \frac{7}{5}(-5) = -5$

$x - 7 = -5$

$x = 2$

The solution set is $\{(2, -5)\}$.

21. $\left[\begin{array}{cc|c} 4 & -2 & 5 \\ -2 & 1 & 6 \end{array}\right] \quad \frac{1}{4}R_1$

$= \left[\begin{array}{cc|c} 1 & -\frac{1}{2} & \frac{5}{4} \\ -2 & 1 & 6 \end{array}\right] \quad 2R_1 + R_2$

$= \left[\begin{array}{cc|c} 1 & -\frac{1}{2} & \frac{5}{4} \\ 0 & 0 & \frac{17}{2} \end{array}\right]$

The resulting system is:

$x - \frac{1}{2}y = \frac{5}{4}$

$0x + 0y = \frac{17}{2}$

This is a contradiction. The system is inconsistent. There is no solution.

23. $\left[\begin{array}{cc|c} 1 & -2 & 1 \\ -2 & 4 & -2 \end{array}\right] \quad 2R_1 + R_2$

$= \left[\begin{array}{cc|c} 1 & -2 & 1 \\ 0 & 0 & 0 \end{array}\right]$

The resulting system is:

$x - 2y = 1$

$0x + 0y = 0$

The system is dependent. There are infinitely many solutions.

25. $\left[\begin{array}{ccc|c} 1 & 1 & -1 & -2 \\ 2 & -1 & 1 & 5 \\ -1 & 2 & 2 & 1 \end{array}\right] \quad -2R_1 + R_2$

$= \left[\begin{array}{ccc|c} 1 & 1 & -1 & -2 \\ 0 & -3 & 3 & 9 \\ -1 & 2 & 2 & 1 \end{array}\right] \quad R_1 + R_3$

$= \left[\begin{array}{ccc|c} 1 & 1 & -1 & -2 \\ 0 & -3 & 3 & 9 \\ 0 & 3 & 1 & -1 \end{array}\right] \quad R_2 + R_3$

$= \left[\begin{array}{ccc|c} 1 & 1 & -1 & -2 \\ 0 & -3 & 3 & 9 \\ 0 & 0 & 4 & 8 \end{array}\right] \quad \begin{array}{l} -\frac{1}{3}R_2 \\ \frac{1}{4}R_3 \end{array}$

$= \left[\begin{array}{ccc|c} 1 & 1 & -1 & -2 \\ 0 & 1 & -1 & -3 \\ 0 & 0 & 1 & 2 \end{array}\right]$

The resulting system is:

$x + y - z = -2$

$y - z = -3$

$z = 2$

Back-substitute 2 for z to find y.

$y - z = -3$

$y - 2 = -3$

$y = -1$

Back-substitute 2 for z and –1 for y to find x.

$x + y - z = -2$

$x - 1 - 2 = -2$

$x - 3 = -2$

$x = 1$

The solution set is $\{(1, -1, 2)\}$.

27. $\left[\begin{array}{ccc|c} 1 & 3 & 0 & 0 \\ 1 & 1 & 1 & 1 \\ 3 & -1 & -1 & 11 \end{array}\right] -R_1+R_2$

$=\left[\begin{array}{ccc|c} 1 & 3 & 0 & 0 \\ 0 & -2 & 1 & 1 \\ 3 & -1 & -1 & 11 \end{array}\right] -3R_1+R_3$

$=\left[\begin{array}{ccc|c} 1 & 3 & 0 & 0 \\ 0 & -2 & 1 & 1 \\ 0 & -10 & -1 & 11 \end{array}\right] -\frac{1}{2}R_2$

$=\left[\begin{array}{ccc|c} 1 & 3 & 0 & 0 \\ 0 & 1 & -\frac{1}{2} & -\frac{1}{2} \\ 0 & -10 & -1 & 11 \end{array}\right] -\frac{1}{10}R_3$

$=\left[\begin{array}{ccc|c} 1 & 3 & 0 & 0 \\ 0 & 1 & -\frac{1}{2} & -\frac{1}{2} \\ 0 & 1 & \frac{1}{10} & -\frac{11}{10} \end{array}\right] -R_2+R_3$

$=\left[\begin{array}{ccc|c} 1 & 3 & 0 & 0 \\ 0 & 1 & -\frac{1}{2} & -\frac{1}{2} \\ 0 & 0 & \frac{3}{5} & -\frac{3}{5} \end{array}\right] \frac{5}{3}R_3$

$=\left[\begin{array}{ccc|c} 1 & 3 & 0 & 0 \\ 0 & 1 & -\frac{1}{2} & -\frac{1}{2} \\ 0 & 0 & 1 & -1 \end{array}\right]$

The resulting system is:

$$x+3y = 0$$
$$y-\frac{1}{2}z=-\frac{1}{2}$$
$$z=-1$$

Back-substitute –1 for z and solve for y.

$$y-\frac{1}{2}z=-\frac{1}{2}$$
$$y-\frac{1}{2}(-1)=-\frac{1}{2}$$
$$y+\frac{1}{2}=-\frac{1}{2}$$
$$y=-1$$

Back-substitute –1 for y to find x.

$$x+3y=0$$
$$x+3(-1)=0$$
$$x-3=0$$
$$x=3$$

The solution set is $\{(3,-1,-1)\}$.

29. $\left[\begin{array}{ccc|c} 2 & 2 & 7 & -1 \\ 2 & 1 & 2 & 2 \\ 4 & 6 & 1 & 15 \end{array}\right] \frac{1}{2}R_1$

$=\left[\begin{array}{ccc|c} 1 & 1 & \frac{7}{2} & -\frac{1}{2} \\ 2 & 1 & 2 & 2 \\ 4 & 6 & 1 & 15 \end{array}\right] -2R_1+R_2$

$=\left[\begin{array}{ccc|c} 1 & 1 & \frac{7}{2} & -\frac{1}{2} \\ 0 & -1 & -5 & 3 \\ 4 & 6 & 1 & 15 \end{array}\right] -R_2$

$=\left[\begin{array}{ccc|c} 1 & 1 & \frac{7}{2} & -\frac{1}{2} \\ 0 & 1 & 5 & -3 \\ 4 & 6 & 1 & 15 \end{array}\right] -4R_1+R_3$

$=\left[\begin{array}{ccc|c} 1 & 1 & \frac{7}{2} & -\frac{1}{2} \\ 0 & 1 & 5 & -3 \\ 0 & 2 & -13 & 17 \end{array}\right] -2R_2+R_3$

$=\left[\begin{array}{ccc|c} 1 & 1 & \frac{7}{2} & -\frac{1}{2} \\ 0 & 1 & 5 & -3 \\ 0 & 0 & -23 & 23 \end{array}\right] -\frac{1}{23}R_3$

$=\left[\begin{array}{ccc|c} 1 & 1 & \frac{7}{2} & -\frac{1}{2} \\ 0 & 1 & 5 & -3 \\ 0 & 0 & 1 & -1 \end{array}\right]$

The resulting system is:

$$x+y+\frac{7}{2}z=-\frac{1}{2}$$
$$y+5z=-3$$
$$z=-1$$

Back-substitute –1 for z to find y.

$$y+5z=-3$$
$$y+5(-1)=-3$$
$$y-5=-3$$
$$y=2$$

Back-substitute –1 for z and 2 for y to find x.

$$x+y+\frac{7}{2}z=-\frac{1}{2}$$
$$x+2+\frac{7}{2}(-1)=-\frac{1}{2}$$
$$x+2-\frac{7}{2}=-\frac{1}{2}$$
$$x-\frac{3}{2}=-\frac{1}{2}$$
$$x=1$$

The solution set is $\{(1,2,-1)\}$.

31. $\left[\begin{array}{ccc|c} 1 & 1 & 1 & 6 \\ 1 & 0 & -1 & -2 \\ 0 & 1 & 3 & 11 \end{array}\right] R_2 \leftrightarrow R_3$

$=\left[\begin{array}{ccc|c} 1 & 1 & 1 & 6 \\ 0 & 1 & 3 & 11 \\ 1 & 0 & -1 & -2 \end{array}\right] -R_1+R_3$

$=\left[\begin{array}{ccc|c} 1 & 1 & 1 & 6 \\ 0 & 1 & 3 & 11 \\ 0 & -1 & -2 & -8 \end{array}\right] R_2+R_3$

$=\left[\begin{array}{ccc|c} 1 & 1 & 1 & 6 \\ 0 & 1 & 3 & 11 \\ 0 & 0 & 1 & 3 \end{array}\right]$

The resulting system is:

$$x+y+z=6$$
$$y+3z=11$$
$$z=3$$

Back-substitute 3 for z to find y.

$$y+3z=11$$
$$y+3(3)=11$$
$$y+9=11$$
$$y=2$$

Back-substitute 3 for z and 2 for y to find x.

$$x+y+z=6$$
$$x+2+3=6$$
$$x+5=6$$
$$x=1$$

The solution set is $\{(1,2,3)\}$.

33. $\left[\begin{array}{ccc|c} 1 & -1 & 3 & 4 \\ 2 & -2 & 6 & 7 \\ 3 & -1 & 5 & 14 \end{array}\right]$ $-2R_1+R_2$ and $-3R_1+R_3$

$=\left[\begin{array}{ccc|c} 1 & -1 & 3 & 4 \\ 0 & 0 & 0 & -1 \\ 0 & 2 & -4 & 2 \end{array}\right]$

The resulting system is:

$$x-y+3z=4$$
$$0x+0y+0z=-1$$
$$2y-4z=2$$

The second row is a contradiction, since $0x+0y+0z$ cannot equal -1. We conclude that the system is inconsistent and there is no solution.

35. $\left[\begin{array}{ccc|c} 1 & -2 & 1 & 4 \\ 5 & -10 & 5 & 20 \\ -2 & 4 & -2 & -8 \end{array}\right] \frac{1}{5}R_2$

$=\left[\begin{array}{ccc|c} 1 & -2 & 1 & 4 \\ 1 & -2 & 1 & 4 \\ -2 & 4 & -2 & -8 \end{array}\right]$

R_1 and R_2 are the same. The system is dependent and there are infinitely many solutions.

37. $\left[\begin{array}{ccc|c} 1 & 1 & 0 & 1 \\ 0 & 1 & 2 & -2 \\ 2 & 0 & -1 & 0 \end{array}\right] -2R_1+R_3$

$=\left[\begin{array}{ccc|c} 1 & 1 & 0 & 1 \\ 0 & 1 & 2 & -2 \\ 0 & -2 & -1 & -2 \end{array}\right] 2R_2+R_3$

$=\left[\begin{array}{ccc|c} 1 & 1 & 0 & 1 \\ 0 & 1 & 2 & -2 \\ 0 & 0 & 3 & -6 \end{array}\right] \frac{1}{3}R_3$

$=\left[\begin{array}{ccc|c} 1 & 1 & 0 & 1 \\ 0 & 1 & 2 & -2 \\ 0 & 0 & 1 & -2 \end{array}\right]$

The resulting system is:

$$x+y=1$$
$$y+2z=-2$$
$$z=-2$$

Back-substitute –2 for z to find y.

$$y+2z=-2$$
$$y+2(-2)=-2$$
$$y-4=-2$$
$$y=2$$

Back-substitute 2 for y to find x.

$$x+y=1$$
$$x+2=1$$
$$x=-1$$

The solution set is $\{(-1,2,-2)\}$.

39. The system is

$$w-x+y+z=3$$
$$x-2y-z=0$$
$$y+6z=17$$
$$z=3$$

Back-substitute $z=3$ to solve for y.

$$y+6(3)=17$$
$$y+18=17$$
$$y=-1$$

Back-substitute $z=3$ and $y=-1$ to solve for x.

$$x-2(-1)-(3)=0$$
$$x+2-3=0$$
$$x=1$$

Back-substitute $x=1$, $y=-1$ and $z=3$ to solve for w.

$$w-(1)+(-1)+(3)=3$$
$$w-1-1+3=3$$
$$w=2$$

The solution set is $\{(2,1,-1,3)\}$.

41. $$\left[\begin{array}{cccc|c} 1 & -1 & 1 & 1 & 3 \\ 0 & 1 & -2 & -1 & 0 \\ 2 & 0 & 3 & 4 & 11 \\ 5 & 1 & 2 & 4 & 6 \end{array}\right]\begin{array}{l} \\ \\ -2R_1+R_3 \\ -5R_1+R_4 \end{array}$$

$$=\left[\begin{array}{cccc|c} 1 & -1 & 1 & 1 & 3 \\ 0 & 1 & -2 & -1 & 0 \\ 0 & 2 & 1 & 2 & 5 \\ 0 & 6 & -3 & -1 & -9 \end{array}\right]$$

43. $$\left[\begin{array}{cccc|c} 1 & 1 & 1 & 1 & 4 \\ 2 & 1 & -2 & -1 & 0 \\ 1 & -2 & -1 & -2 & -2 \\ 3 & 2 & 1 & 3 & 4 \end{array}\right]\begin{array}{l} \\ -2R_1+R_2 \\ -1R_1+R_3 \\ -3R_1+R_4 \end{array}$$

$$=\left[\begin{array}{cccc|c} 1 & 1 & 1 & 1 & 4 \\ 0 & -1 & -4 & -3 & -8 \\ 0 & -3 & -2 & -3 & -6 \\ 0 & -1 & -2 & 0 & -8 \end{array}\right]\begin{array}{l} \\ -1R_2 \\ \\ \\ \end{array}$$

$$=\left[\begin{array}{cccc|c} 1 & 1 & 1 & 1 & 4 \\ 0 & 1 & 4 & 3 & 8 \\ 0 & -3 & -2 & -3 & -6 \\ 0 & -1 & -2 & 0 & -8 \end{array}\right]\begin{array}{l} \\ \\ 3R_2+R_3 \\ R_2+R_4 \end{array}$$

$$=\left[\begin{array}{cccc|c} 1 & 1 & 1 & 1 & 4 \\ 0 & 1 & 4 & 3 & 8 \\ 0 & 0 & 10 & 6 & 18 \\ 0 & 0 & 2 & 3 & 0 \end{array}\right]\begin{array}{l} \\ \\ \frac{1}{2}R_4 \\ R_3 \end{array}$$

$$=\left[\begin{array}{cccc|c} 1 & 1 & 1 & 1 & 4 \\ 0 & 1 & 4 & 3 & 8 \\ 0 & 0 & 1 & \frac{3}{2} & 0 \\ 0 & 0 & 10 & 6 & 18 \end{array}\right]\begin{array}{l} \\ \\ \\ -10R_3+R_4 \end{array}$$

$$=\left[\begin{array}{cccc|c} 1 & 1 & 1 & 1 & 4 \\ 0 & 1 & 4 & 3 & 8 \\ 0 & 0 & 1 & \frac{3}{2} & 0 \\ 0 & 0 & 0 & -9 & 18 \end{array}\right]\begin{array}{l} \\ \\ \\ -\frac{1}{9}R_4 \end{array}$$

$$=\left[\begin{array}{cccc|c} 1 & 1 & 1 & 1 & 4 \\ 0 & 1 & 4 & 3 & 8 \\ 0 & 0 & 1 & \frac{3}{2} & 0 \\ 0 & 0 & 0 & 1 & -2 \end{array}\right]$$

The resulting system is

$$w+x+y+z=4$$
$$x+4y+3z=8$$
$$y+\frac{3}{2}z=0$$
$$z=-2$$

Back-substitute $z=-2$ to solve for y.

$$y+\frac{3}{2}(-2)=0$$
$$y-3=0$$
$$y=3$$

Back-substitute $y=3$ and $z=-2$ to solve for x.

$$x+4(3)+3(-2)=8$$
$$x+12-6=8$$
$$x=2$$

Back-substitute $x=2$, $y=3$, and $z=-2$ to solve for w.

$$w+(2)+(3)+(-2)=4$$
$$w+5-2=4$$
$$w=1$$

The solution set is $\{(1,2,3,-2)\}$.

45. a. Use each ordered pair to write an equation as follows:

$$(t,s(t))=(1,40)$$
$$s(t)=at^2+bt+c$$
$$40=a(1)^2+b(1)+c$$
$$40=a+b+c$$

$$(t,s(t))=(2,48)$$
$$s(t)=at^2+bt+c$$
$$48=a(2)^2+b(2)+c$$
$$48=4a+2b+c$$

$$(t,s(t))=(3,24)$$
$$s(t)=at^2+bt+c$$
$$24=a(3)^2+b(3)+c$$
$$24=9a+3b+c$$

The system of three equations in three variables is:

$$a+b+c=40$$
$$4a+2b+c=48$$
$$9a+3b+c=24$$

$$\left[\begin{array}{ccc|c}1&1&1&40\\4&2&1&48\\9&3&1&24\end{array}\right]\begin{array}{l}\\-4R_1+R_2\\-9R_1+R_3\end{array}$$

$$=\left[\begin{array}{ccc|c}1&1&1&40\\0&-2&-3&-112\\0&-6&-8&-336\end{array}\right]\begin{array}{l}\\-\frac{1}{2}R_2\\ \\ \end{array}$$

$$=\left[\begin{array}{ccc|c}1&1&1&40\\0&1&\frac{3}{2}&56\\0&-6&-8&-336\end{array}\right]\begin{array}{l}\\ \\R_3+6R_2\end{array}$$

$$=\left[\begin{array}{ccc|c}1&1&1&40\\0&1&\frac{3}{2}&56\\0&0&1&0\end{array}\right]$$

The resulting system is:

$$a+b+\ \ c=40$$
$$b+\frac{3}{2}c=56$$
$$c=0$$

Back-substitute to find b.

$$b+\frac{3}{2}c=56$$
$$b+\frac{3}{2}(0)=56$$
$$b=56$$

Back-substitute to find a.

$$a+b+c=40$$
$$a+(56)+(0)=40$$
$$a+56=40$$
$$a=-16$$

The quadratic function is $s(t)=-16t^2+56t$.

b. $s(t)=-16t^2+56t$

$s(3.5)=-16(3.5)^2+56(3.5)=0$

The ball hits the ground 3.5 seconds after it is thrown. This is represented by the point (3.5, 0).

47. From the problem statement, we have the following equations:

$$y=x+z+22$$
$$2x=y+7$$
$$x+y+z=100$$

Writing the equations in standard form gives

$$x-y+z=-22$$
$$2x-y\quad=7$$
$$x+y+z=100$$

The corresponding augmented matrix is

$$\left[\begin{array}{ccc|c}1&-1&1&-22\\2&-1&0&7\\1&1&1&100\end{array}\right]\begin{array}{l}\\-2R_1+R_2\\-1R_1+R_3\end{array}$$

$$=\left[\begin{array}{ccc|c}1&-1&1&-22\\0&1&-2&51\\0&2&0&122\end{array}\right]\begin{array}{l}\\ \\-2R_2+R_3\end{array}$$

$$=\left[\begin{array}{ccc|c}1&-1&1&-22\\0&1&-2&51\\0&0&4&20\end{array}\right]\begin{array}{l}\\ \\ \frac{1}{4}R_4\end{array}$$

$$=\left[\begin{array}{ccc|c}1&-1&1&-22\\0&1&-2&51\\0&0&1&5\end{array}\right]$$

The resulting system of equations is

$$\begin{aligned} x - y + z &= -22 \\ y - 2z &= 51 \\ z &= 5 \end{aligned}$$

Back-substitute $z = 5$ to solve for y.

$$\begin{aligned} y - 2(5) &= 51 \\ y - 10 &= 51 \\ y &= 61 \end{aligned}$$

Back-substitute $y = 61$ and $z = 5$ to solve for x.

$$\begin{aligned} x - 61 + 5 &= -22 \\ x &= 34 \end{aligned}$$

Therefore, 34% of the single women polled responded "Yes", 61% responded "No", and 5% responded "Not Sure".

49. – 57. Answers will vary.

59. makes sense

61. does not make sense; Explanations will vary. Sample explanation: If zeroes appear to the left of the vertical bar, but 6 appears on the right, the system has no solution.

63. false; Changes to make the statement true will vary. A sample change is: The augmented matrix for this system is

$$\left[\begin{array}{ccc|c} 1 & -3 & 0 & 5 \\ 0 & 1 & -2 & 7 \\ 2 & 0 & 1 & 4 \end{array}\right].$$

65. false; Changes to make the statement true will vary. A sample change is: It is row j that will change.

67.

$$\begin{aligned} f(x) &= -3x + 10 \\ f(2a-1) &= -3(2a-1) + 10 \\ &= -6a + 3 + 10 \\ &= -6a + 13 \end{aligned}$$

68. $f(x)=3x$ and $g(x)=2x-3$

$$(fg)(x)=f(x)\cdot g(x)=3x(2x-3)$$
$$=6x^2-9x$$
$$(fg)(-1)=6(-1)^2-9(-1)$$
$$=6(1)+9=6+9=15$$

69. $\dfrac{-4x^8y^{-12}}{12x^{-3}y^{24}}=\dfrac{-4x^8x^3}{12y^{24}y^{12}}=\dfrac{-x^{11}}{3y^{36}}=-\dfrac{x^{11}}{3y^{36}}$

70. $2(-5)-(-3)(4)$
$$=-10-(-12)$$
$$=-10+12$$
$$=2$$

71. $\dfrac{2(-5)-1(-4)}{5(-5)-6(-4)}$
$$=\frac{-10+4}{-25+24}$$
$$=\frac{-6}{-1}$$
$$=6$$

72. $2(-30-(-3))-3(6-9)+(-1)(1-15)$
$$=2(-30+3)-3(-3)+(-1)(-14)$$
$$=2(-27)-3(-3)+(-1)(-14)$$
$$=-54+9+14$$
$$=-31$$

3.5 Check Points

1. a. $\begin{vmatrix}10 & 9\\ 6 & 5\end{vmatrix}=10(5)-6(9)=50-54=-4$

b. $\begin{vmatrix}4 & 3\\ -5 & -8\end{vmatrix}=4(-8)-(-5)(3)=-32+15=-17$

2. $D=\begin{vmatrix}5 & 4\\3 & -6\end{vmatrix}=5(-6)-3(4)=-30-12=-42$

$D_x=\begin{vmatrix}12 & 4\\24 & -6\end{vmatrix}=12(-6)-24(4)=-72-96=-168$

$D_y=\begin{vmatrix}5 & 12\\3 & 24\end{vmatrix}=5(24)-3(12)=120-36=84$

$x=\dfrac{D_x}{D}=\dfrac{-168}{-42}=4$

$y=\dfrac{D_y}{D}=\dfrac{84}{-42}=-2$

The solution set is $\{(4,-2)\}$.

3. $\begin{vmatrix}2 & 1 & 7\\-5 & 6 & 0\\-4 & 3 & 1\end{vmatrix}=2\begin{vmatrix}6 & 0\\3 & 1\end{vmatrix}-(-5)\begin{vmatrix}1 & 7\\3 & 1\end{vmatrix}-4\begin{vmatrix}1 & 7\\6 & 0\end{vmatrix}$

$$\begin{aligned}&=2(6(1)-3(0))+5(1(1)-3(7))-4(1(0)-6(7))\\&=2(6)+5(-20)-4(-42)\\&=12-100+168\\&=80\end{aligned}$$

4. $3x-2y+z=16$
$2x+3y-z=-9$
$x+4y+3z=2$

$D=\begin{vmatrix}3 & -2 & 1\\2 & 3 & -1\\1 & 4 & 3\end{vmatrix}=3\begin{vmatrix}3 & -1\\4 & 3\end{vmatrix}-2\begin{vmatrix}-2 & 1\\4 & 3\end{vmatrix}+1\begin{vmatrix}-2 & 1\\3 & -1\end{vmatrix}=58$

$D_x=\begin{vmatrix}16 & -2 & 1\\-9 & 3 & -1\\2 & 4 & 3\end{vmatrix}=16\begin{vmatrix}3 & -1\\4 & 3\end{vmatrix}-(-9)\begin{vmatrix}-2 & 1\\4 & 3\end{vmatrix}+2\begin{vmatrix}-2 & 1\\3 & -1\end{vmatrix}=116$

$D_y=\begin{vmatrix}3 & 16 & 1\\2 & -9 & -1\\1 & 2 & 3\end{vmatrix}=3\begin{vmatrix}-9 & -1\\2 & 3\end{vmatrix}-2\begin{vmatrix}16 & 1\\2 & 3\end{vmatrix}+1\begin{vmatrix}16 & 1\\-9 & -1\end{vmatrix}=-174$

$D_z=\begin{vmatrix}3 & -2 & 16\\2 & 3 & -9\\1 & 4 & 2\end{vmatrix}=3\begin{vmatrix}3 & -9\\4 & 2\end{vmatrix}-2\begin{vmatrix}-2 & 16\\4 & 2\end{vmatrix}+1\begin{vmatrix}-2 & 16\\3 & -9\end{vmatrix}=232$

$x=\dfrac{D_x}{D}=\dfrac{116}{58}=2$ $\quad y=\dfrac{D_y}{D}=\dfrac{-174}{58}=-3$ $\quad z=\dfrac{D_z}{D}=\dfrac{232}{58}=4$

The solution set is $\{(2,-3,4)\}$.

3.5 Concept and Vocabulary Check

1. $5\cdot3-2\cdot4=15-8=7$

2. $x=\dfrac{\begin{vmatrix}8 & 1\\ -2 & -1\end{vmatrix}}{\begin{vmatrix}1 & 1\\ 1 & -1\end{vmatrix}}$ and $y=\dfrac{\begin{vmatrix}1 & 8\\ 1 & -2\end{vmatrix}}{\begin{vmatrix}1 & 1\\ 1 & -1\end{vmatrix}}$

3. $3\begin{vmatrix}3 & 1\\ 1 & 1\end{vmatrix}-4\begin{vmatrix}2 & 1\\ 1 & 1\end{vmatrix}+5\begin{vmatrix}2 & 1\\ 3 & 1\end{vmatrix}$

4. $y=\dfrac{\begin{vmatrix}3 & -8 & 4\\ 2 & 11 & -2\\ 1 & 4 & -2\end{vmatrix}}{\begin{vmatrix}3 & 1 & 4\\ 2 & 3 & -2\\ 1 & -3 & -2\end{vmatrix}}$

3.5 Exercise Set

1. $\begin{vmatrix}5 & 7\\ 2 & 3\end{vmatrix}=5(3)-2(7)=15-14=1$

3. $\begin{vmatrix}-4 & 1\\ 5 & 6\end{vmatrix}=-4(6)-5(1)=-24-5=-29$

5. $\begin{vmatrix}-7 & 14\\ 2 & -4\end{vmatrix}=-7(-4)-2(14)=28-28=0$

7. $\begin{vmatrix}-5 & -1\\ -2 & -7\end{vmatrix}=-5(-7)-(-2)(-1)=35-2=33$

9. $\begin{vmatrix}\frac{1}{2} & \frac{1}{2}\\ \frac{1}{8} & -\frac{3}{4}\end{vmatrix}=\frac{1}{2}\left(-\frac{3}{4}\right)-\frac{1}{8}\left(\frac{1}{2}\right)=-\frac{3}{8}-\frac{1}{16}=-\frac{6}{16}-\frac{1}{16}=-\frac{7}{16}$

11. $D=\begin{vmatrix}1 & 1\\ 1 & -1\end{vmatrix}=1(-1)-1(1)=-1-1=-2$

$D_x=\begin{vmatrix}7 & 1\\ 3 & -1\end{vmatrix}=7(-1)-3(1)=-7-3=-10$

$D_y=\begin{vmatrix}1 & 7\\ 1 & 3\end{vmatrix}=1(3)-1(7)=3-7=-4$

$x=\frac{D_x}{D}=\frac{-10}{-2}=5$; $y=\frac{D_y}{D}=\frac{-4}{-2}=2$

The solution set is $\{(5,2)\}$.

13. $D=\begin{vmatrix}12 & 3\\ 2 & -3\end{vmatrix}=12(-3)-2(3)=-36-6=-42$

$D_x=\begin{vmatrix}15 & 3\\ 13 & -3\end{vmatrix}=15(-3)-13(3)=-45-39=-84$

$D_y=\begin{vmatrix}12 & 15\\ 2 & 13\end{vmatrix}=12(13)-2(15)=156-30=126$

$x=\dfrac{D_x}{D}=\dfrac{-84}{-42}=2;\quad y=\dfrac{D_y}{D}=\dfrac{126}{-42}=-3$

The solution set is $\{(2,-3)\}$.

15. $D=\begin{vmatrix}4 & -5\\ 2 & 3\end{vmatrix}=4(3)-2(-5)=12+10=22$

$D_x=\begin{vmatrix}17 & -5\\ 3 & 3\end{vmatrix}=17(3)-3(-5)=51+15=66$

$D_y=\begin{vmatrix}4 & 17\\ 2 & 3\end{vmatrix}=4(3)-2(17)=12-34=-22$

$x=\dfrac{D_x}{D}=\dfrac{66}{22}=3;\quad y=\dfrac{D_y}{D}=\dfrac{-22}{22}=-1$

The solution set is $\{(3,-1)\}$.

17. $D=\begin{vmatrix}1 & -3\\ 3 & -4\end{vmatrix}=1(-4)-3(-3)=-4+9=5$

$D_x=\begin{vmatrix}4 & -3\\ 12 & -4\end{vmatrix}=4(-4)-12(-3)=-16+36=20$

$D_y=\begin{vmatrix}1 & 4\\ 3 & 12\end{vmatrix}=1(12)-3(4)=12-12=0$

$x=\dfrac{D_x}{D}=\dfrac{20}{5}=4;\quad y=\dfrac{D_y}{D}=\dfrac{0}{5}=0$

The solution set is $\{(4,0)\}$.

19. $D=\begin{vmatrix}3 & -4\\ 2 & 2\end{vmatrix}=3(2)-2(-4)=6+8=14$

$D_x=\begin{vmatrix}4 & -4\\ 12 & 2\end{vmatrix}=4(2)-12(-4)=8+48=56$

$D_y=\begin{vmatrix}3 & 4\\ 2 & 12\end{vmatrix}=3(12)-2(4)=36-8=28$

$x=\dfrac{D_x}{D}=\dfrac{56}{14}=4;\quad y=\dfrac{D_y}{D}=\dfrac{28}{14}=2$

The solution set is $\{(4,2)\}$.

21. First, rewrite the system in standard form.

$2x-3y=2$

$5x+4y=51$

$$D=\begin{vmatrix}2 & -3\\5 & 4\end{vmatrix}=2(4)-5(-3)=8+15=23$$

$$D_x=\begin{vmatrix}2 & -3\\51 & 4\end{vmatrix}=2(4)-51(-3)=8+153=161$$

$$D_y=\begin{vmatrix}2 & 2\\5 & 51\end{vmatrix}=2(51)-5(2)=102-10=92$$

$$x=\frac{D_x}{D}=\frac{161}{23}=7;\quad y=\frac{D_y}{D}=\frac{92}{23}=4$$

The solution set is $\{(7,4)\}$.

23. First, rewrite the system in standard form.

$3x+3y=2$.

$2x+2y=3$

$$D=\begin{vmatrix}3 & 3\\2 & 2\end{vmatrix}=3(2)-2(3)=6-6=0$$

$$D_x=\begin{vmatrix}2 & 3\\3 & 2\end{vmatrix}=2(2)-3(3)=4-9=-5$$

$$D_y=\begin{vmatrix}3 & 2\\2 & 3\end{vmatrix}=3(3)-2(2)=9-4=5$$

Because $D=0$ and at least one of the determinants for the numerators is not zero, the system is inconsistent and the solution set is $\varnothing$. Using matrices, we would get:

$$\left[\begin{array}{cc|c}3 & 3 & 2\\2 & 2 & 3\end{array}\right]\ \frac{1}{3}R_1$$

$$=\left[\begin{array}{cc|c}1 & 1 & 2/3\\2 & 2 & 3\end{array}\right]\ -2R_1+R_2$$

$$=\left[\begin{array}{cc|c}1 & 1 & 2/3\\0 & 0 & 5/3\end{array}\right]$$

This is a contradiction. There are no values for x and y for which $0 = 5/3$. The solution set is $\varnothing$ and the system is inconsistent.

25. First, rewrite the system in standard form.

$3x+4y=16$

$6x+8y=32$

$$D=\begin{vmatrix}3 & 4\\6 & 8\end{vmatrix}=3(8)-6(4)=24-24=0$$

$$D_x=\begin{vmatrix}16 & 4\\32 & 8\end{vmatrix}=16(8)-32(4)=128-128=0$$

$$D_y=\begin{vmatrix}3 & 16\\6 & 32\end{vmatrix}=3(32)-6(16)=96-96=0$$

Since $D=0$ and all determinants in the numerators are 0, the equations in the system are dependent and there are infinitely many solutions.

27. $\begin{vmatrix} 3 & 0 & 0 \\ 2 & 1 & -5 \\ 2 & 5 & -1 \end{vmatrix} = 3\begin{vmatrix} 1 & -5 \\ 5 & -1 \end{vmatrix} - 2\begin{vmatrix} 0 & 0 \\ 5 & -1 \end{vmatrix} + 2\begin{vmatrix} 0 & 0 \\ 1 & -5 \end{vmatrix}$

$= 3(1(-1)-5(-5)) - 2(0(-1)-5(0)) + 2(0(-5)-1(0)) = 3(24)-2(0)+2(0) = 72$

29. $\begin{vmatrix} 3 & 1 & 0 \\ -3 & 4 & 0 \\ -1 & 3 & -5 \end{vmatrix} = 3\begin{vmatrix} 4 & 0 \\ 3 & -5 \end{vmatrix} - (-3)\begin{vmatrix} 1 & 0 \\ 3 & -5 \end{vmatrix} + (-1)\begin{vmatrix} 1 & 0 \\ 4 & 0 \end{vmatrix}$

$= 3(4(-5)-3(0)) + 3(1(-5)-3(0)) - 1(1(0)-4(0)) = 3(-20)+3(-5)-1(0) = -75$

31. $\begin{vmatrix} 1 & 1 & 1 \\ 2 & 2 & 2 \\ -3 & 4 & -5 \end{vmatrix} = 1\begin{vmatrix} 2 & 2 \\ 4 & -5 \end{vmatrix} - 2\begin{vmatrix} 1 & 1 \\ 4 & -5 \end{vmatrix} + (-3)\begin{vmatrix} 1 & 1 \\ 2 & 2 \end{vmatrix}$

$= 1(2(-5)-4(2)) - 2(1(-5)-4(1)) - 3(1(2)-2(1)) = 1(-18)-2(-9)-3(0) = 0$

33.
$$\begin{aligned} x + y + z &= 0 \\ 2x - y + z &= -1 \\ -x + 3y - z &= -8 \end{aligned}$$

$D = \begin{vmatrix} 1 & 1 & 1 \\ 2 & -1 & 1 \\ -1 & 3 & -1 \end{vmatrix} = 1\begin{vmatrix} -1 & 1 \\ 3 & -1 \end{vmatrix} - 2\begin{vmatrix} 1 & 1 \\ 3 & -1 \end{vmatrix} - 1\begin{vmatrix} 1 & 1 \\ -1 & 1 \end{vmatrix} = 4$

$D_x = \begin{vmatrix} 0 & 1 & 1 \\ -1 & -1 & 1 \\ -8 & 3 & -1 \end{vmatrix} = 0\begin{vmatrix} -1 & 1 \\ 3 & -1 \end{vmatrix} - (-1)\begin{vmatrix} 1 & 1 \\ 3 & -1 \end{vmatrix} - 8\begin{vmatrix} 1 & 1 \\ -1 & 1 \end{vmatrix} = -20$

$D_y = \begin{vmatrix} 1 & 0 & 1 \\ 2 & -1 & 1 \\ -1 & -8 & -1 \end{vmatrix} = 1\begin{vmatrix} -1 & 1 \\ -8 & -1 \end{vmatrix} - 2\begin{vmatrix} 0 & 1 \\ -8 & -1 \end{vmatrix} - 1\begin{vmatrix} 0 & 1 \\ -1 & 1 \end{vmatrix} = -8$

$D_z = \begin{vmatrix} 1 & 1 & 0 \\ 2 & -1 & -1 \\ -1 & 3 & -8 \end{vmatrix} = 1\begin{vmatrix} -1 & -1 \\ 3 & -8 \end{vmatrix} - 2\begin{vmatrix} 1 & 0 \\ 3 & -8 \end{vmatrix} - 1\begin{vmatrix} 1 & 0 \\ -1 & -1 \end{vmatrix} = 28$

$x = \frac{D_x}{D} = \frac{-20}{4} = -5 \qquad y = \frac{D_y}{D} = \frac{-8}{4} = -2 \qquad z = \frac{D_z}{D} = \frac{28}{4} = 7$

The solution set is $\{(-5, -2, 7)\}$.

35. $4x-5y-6z=-1$
$x-2y-5z=-12$
$2x-y\quad=7$

$$D=\begin{vmatrix}4&-5&-6\\1&-2&-5\\2&-1&0\end{vmatrix}=4\begin{vmatrix}-2&-5\\-1&0\end{vmatrix}-1\begin{vmatrix}-5&-6\\-1&0\end{vmatrix}+2\begin{vmatrix}-5&-6\\-2&-5\end{vmatrix}=12$$

$$D_x=\begin{vmatrix}-1&-5&-6\\-12&-2&-5\\7&-1&0\end{vmatrix}=-1\begin{vmatrix}-2&-5\\-1&0\end{vmatrix}-(-12)\begin{vmatrix}-5&-6\\-1&0\end{vmatrix}+7\begin{vmatrix}-5&-6\\-2&-5\end{vmatrix}=24$$

$$D_y=\begin{vmatrix}4&-1&-6\\1&-12&-5\\2&7&0\end{vmatrix}=4\begin{vmatrix}-12&-5\\7&0\end{vmatrix}-1\begin{vmatrix}-1&-6\\7&0\end{vmatrix}+2\begin{vmatrix}-1&-6\\-12&-5\end{vmatrix}=-36$$

$$D_z=\begin{vmatrix}4&-5&-1\\1&-2&-12\\2&-1&7\end{vmatrix}=4\begin{vmatrix}-2&-12\\-1&7\end{vmatrix}-1\begin{vmatrix}-5&-1\\-1&7\end{vmatrix}+2\begin{vmatrix}-5&-1\\-2&-12\end{vmatrix}=48$$

$x=\frac{D_x}{D}=\frac{24}{12}=2$ $\quad y=\frac{D_y}{D}=\frac{-36}{12}=-3$ $\quad z=\frac{D_z}{D}=\frac{48}{12}=4$

The solution set is $\{(2,-3,4)\}$.

37. $x+y+z=4$
$x-2y+z=7$
$x+3y+2z=4$

$$D=\begin{vmatrix}1&1&1\\1&-2&1\\1&3&2\end{vmatrix}=1\begin{vmatrix}-2&1\\3&2\end{vmatrix}-1\begin{vmatrix}1&1\\3&2\end{vmatrix}+1\begin{vmatrix}1&1\\-2&1\end{vmatrix}=-3$$

$$D_x=\begin{vmatrix}4&1&1\\7&-2&1\\4&3&2\end{vmatrix}=4\begin{vmatrix}-2&1\\3&2\end{vmatrix}-7\begin{vmatrix}1&1\\3&2\end{vmatrix}+4\begin{vmatrix}1&1\\-2&1\end{vmatrix}=-9$$

$$D_y=\begin{vmatrix}1&4&1\\1&7&1\\1&4&2\end{vmatrix}=1\begin{vmatrix}7&1\\4&2\end{vmatrix}-1\begin{vmatrix}4&1\\4&2\end{vmatrix}+1\begin{vmatrix}4&1\\7&1\end{vmatrix}=3$$

$$D_z=\begin{vmatrix}1&1&4\\1&-2&7\\1&3&4\end{vmatrix}=1\begin{vmatrix}-2&7\\3&4\end{vmatrix}-1\begin{vmatrix}1&4\\3&4\end{vmatrix}+1\begin{vmatrix}1&4\\-2&7\end{vmatrix}=-6$$

$x=\frac{D_x}{D}=\frac{-9}{-3}=3$ $\quad y=\frac{D_y}{D}=\frac{3}{-3}=-1$ $\quad z=\frac{D_z}{D}=\frac{-6}{-3}=2$

The solution set is $\{(3,-1,2)\}$.

39. $x \quad +2z = 4$
$2y - z = 5$
$2x + 3y \quad = 13$

$$D = \begin{vmatrix} 1 & 0 & 2 \\ 0 & 2 & -1 \\ 2 & 3 & 0 \end{vmatrix} = 1\begin{vmatrix} 2 & -1 \\ 3 & 0 \end{vmatrix} - 0\begin{vmatrix} 0 & 2 \\ 3 & 0 \end{vmatrix} + 2\begin{vmatrix} 0 & 2 \\ 2 & -1 \end{vmatrix} = -5$$

$$D_x = \begin{vmatrix} 4 & 0 & 2 \\ 5 & 2 & -1 \\ 13 & 3 & 0 \end{vmatrix} = 4\begin{vmatrix} 2 & -1 \\ 3 & 0 \end{vmatrix} - 5\begin{vmatrix} 0 & 2 \\ 3 & 0 \end{vmatrix} + 13\begin{vmatrix} 0 & 2 \\ 2 & -1 \end{vmatrix} = -10$$

$$D_y = \begin{vmatrix} 1 & 4 & 2 \\ 0 & 5 & -1 \\ 2 & 13 & 0 \end{vmatrix} = 1\begin{vmatrix} 5 & -1 \\ 13 & 0 \end{vmatrix} - 0\begin{vmatrix} 4 & 2 \\ 13 & 0 \end{vmatrix} + 2\begin{vmatrix} 4 & 2 \\ 5 & -1 \end{vmatrix} = -15$$

$$D_z = \begin{vmatrix} 1 & 0 & 4 \\ 0 & 2 & 5 \\ 2 & 3 & 13 \end{vmatrix} = 1\begin{vmatrix} 2 & 5 \\ 3 & 13 \end{vmatrix} - 0\begin{vmatrix} 0 & 4 \\ 3 & 13 \end{vmatrix} + 2\begin{vmatrix} 0 & 4 \\ 2 & 5 \end{vmatrix} = -5$$

$$x = \frac{D_x}{D} = \frac{-10}{-5} = 2 \qquad y = \frac{D_y}{D} = \frac{-15}{-5} = 3 \qquad z = \frac{D_z}{D} = \frac{-5}{-5} = 1$$

The solution set is $\{(2,3,1)\}$.

41. $$\begin{vmatrix} \begin{vmatrix} 3 & 1 \\ -2 & 3 \end{vmatrix} & \begin{vmatrix} 7 & 0 \\ 1 & 5 \end{vmatrix} \\ \begin{vmatrix} 3 & 0 \\ 0 & 7 \end{vmatrix} & \begin{vmatrix} 9 & -6 \\ 3 & 5 \end{vmatrix} \end{vmatrix} = \begin{vmatrix} 3(3)-(-2)(1) & 7(5)-1(0) \\ 3(7)-0(0) & 9(5)-3(-6) \end{vmatrix} = \begin{vmatrix} 11 & 35 \\ 21 & 63 \end{vmatrix} = -42$$

43. From $D = \begin{vmatrix} 2 & -4 \\ 3 & 5 \end{vmatrix}$ we obtain the coefficients of the variables in our equations:

$2x - 4y = c_1$
$3x + 5y = c_2$

From $D_x = \begin{vmatrix} 8 & -4 \\ -10 & 5 \end{vmatrix}$ we obtain the constant coefficients: 8 and -10

$2x - 4y = 8$
$3x + 5y = -10$

45. $$\begin{vmatrix} -2 & x \\ 4 & 6 \end{vmatrix} = 32$$

$-2(6) - 4(x) = 32$
$-12 - 4x = 32$
$-4x = 44$
$x = -11$

The solution set is $\{-11\}$.

47.
$$\begin{vmatrix} 1 & x & -2 \\ 3 & 1 & 1 \\ 0 & -2 & 2 \end{vmatrix} = -8$$

$$1\begin{vmatrix} 1 & 1 \\ -2 & 2 \end{vmatrix} - 3\begin{vmatrix} x & -2 \\ -2 & 2 \end{vmatrix} + 0\begin{vmatrix} x & -2 \\ 1 & 1 \end{vmatrix} = -8$$
$$1[1(2)-(-2)(1)]-3[2x-(-2)(-2)] = -8$$
$$(2+2)-3(2x-4) = -8$$
$$-6x+16 = -8$$
$$-6x = -24$$
$$x = 4$$

The solution set is $\{4\}$.

49. $$\text{Area} = \pm\frac{1}{2}\begin{vmatrix} 3 & -5 & 1 \\ 2 & 6 & 1 \\ -3 & 5 & 1 \end{vmatrix} = \pm\frac{1}{2}\left[3\begin{vmatrix} 6 & 1 \\ 5 & 1 \end{vmatrix} - 2\begin{vmatrix} -5 & 1 \\ 5 & 1 \end{vmatrix} - 3\begin{vmatrix} -5 & 1 \\ 6 & 1 \end{vmatrix}\right]$$
$$= \pm\frac{1}{2}[3(6(1)-5(1))-2(-5(1)-5(1))-3(-5(1)-6(1))]$$
$$= \pm\frac{1}{2}[3(6-5)-2(-5-5)-3(-5-6)]$$
$$= \pm\frac{1}{2}[3(1)-2(-10)-3(-11)]$$
$$\pm\frac{1}{2}[56]$$
$$= \pm 28$$

The area is 28 square units.

51. $$\begin{vmatrix} 3 & -1 & 1 \\ 0 & -3 & 1 \\ 12 & 5 & 1 \end{vmatrix} = 3\begin{vmatrix} -3 & 1 \\ 5 & 1 \end{vmatrix} - 0\begin{vmatrix} -1 & 1 \\ 5 & 1 \end{vmatrix} + 12\begin{vmatrix} -1 & 1 \\ -3 & 1 \end{vmatrix}$$
$$= 3(-3(1)-5(1))+12(-1(1)-(-3)1)$$
$$= 3(-3-5)+12(-1+3)$$
$$= -24+24$$
$$= 0$$

Because the determinant is equal to zero, the points are collinear.

53. $$\begin{vmatrix} x & y & 1 \\ 3 & -5 & 1 \\ -2 & 6 & 1 \end{vmatrix} = x\begin{vmatrix} -5 & 1 \\ 6 & 1 \end{vmatrix} - 3\begin{vmatrix} y & 1 \\ 6 & 1 \end{vmatrix} - 2\begin{vmatrix} y & 1 \\ -5 & 1 \end{vmatrix}$$
$$= x(-5(1)-6(1))-3(y(1)-6(1))-2(y(1)-(-5)1)$$
$$= x(-5-6)-3(y-6)-2(y+5)$$
$$= x(-11)-3y+18-2y-10$$
$$= -11x-5y+8$$

To find the equation of the line, set the determinant equal to zero.

$-11x-5y+8=0$

Solve for y to obtain slope-intercept form.

$$-11x-5y+8=0$$
$$-5y=11x-8$$
$$y=-\frac{11}{5}x+\frac{8}{5}$$

55. – 63. Answers will vary.

65. does not make sense; Explanations will vary. Sample explanation: Determinants must have the same number of rows and columns.

67. does not make sense; Explanations will vary. Sample explanation: When using Cramer's rule, the number of determinants we set up is one more than the number of variables.

69. true

71. false; Changes to make the statement true will vary. A sample change is: Despite determinants being different (i.e., all entries are not identical), they can have the same value. This means that the numerators of the x and y when using Cramer's rule can have the same value, without being the same determinant. As a result, x and y can have the same value.

73. $\begin{vmatrix} a_1 & b_1 \\ a_2 & b_2 \end{vmatrix} = a_1b_2 - a_2b_1$

Switch the columns and re-evaluate.

$\begin{vmatrix} b_1 & a_1 \\ b_2 & a_2 \end{vmatrix} = a_2b_1 - a_1b_2 = -1(a_1b_2 - a_2b_1)$

The value is multiplied by -1.

75. We are given two points, (x_1, y_1) and (x_2, y_2). To find the equation of a line using two points, we first find the slope, and then use the slope and one of the points to write the equation of the line in point-slope form. Here, the slope is: $m=\frac{y_2-y_1}{x_2-x_1}$. Using point slope form, we obtain: $y-y_1=\frac{y_2-y_1}{x_2-x_1}(x-x_1)$. To determine that this is equivalent to what is obtained when the determinant is set equal to zero, we multiply as follows:

$$y-y_1=\frac{y_2-y_1}{x_2-x_1}(x-x_1)$$
$$(x_2-x_1)(y-y_1)=(y_2-y_1)(x-x_1)$$
$$x_2y-x_2y_1-x_1y+x_1y_1=xy_2-x_1y_2-xy_1+x_1y_1$$

Now, evaluate the determinant to see if they are equivalent.

$$\begin{vmatrix} x & y & 1 \\ x_1 & y_1 & 1 \\ x_2 & y_2 & 1 \end{vmatrix} = 0$$

$$x\begin{vmatrix} y_1 & 1 \\ y_2 & 1 \end{vmatrix} - x_1\begin{vmatrix} y & 1 \\ y_2 & 1 \end{vmatrix} + x_2\begin{vmatrix} y & 1 \\ y_1 & 1 \end{vmatrix} = 0$$

$$x\left(y_1(1) - y_2(1)\right) - x_1\left(y(1) - y_2(1)\right) + x_2\left(y(1) - y_1(1)\right) = 0$$

$$x\left(y_1 - y_2\right) - x_1\left(y - y_2\right) + x_2\left(y - y_1\right) = 0$$

$$xy_1 - xy_2 - x_1y + x_1y_2 + x_2y - x_2y_1 = 0$$

$$x_2y - x_2y_1 - x_1y = xy_2 - x_1y_2 - xy_1$$

When we compare, we see that the equations not the same. Using the Addition Property of Equality, we add x_1y_1 to both sides of the equation and see that the expressions are the equivalent. This shows that the equation of a line through (x_1, y_1) and (x_2, y_2) is given by the determinant $\begin{vmatrix} x & y & 1 \\ x_1 & y_1 & 1 \\ x_2 & y_2 & 1 \end{vmatrix} = 0.$

76. $6x - 4 = 2 + 6(x - 1)$

$6x - 4 = 2 + 6x - 6$

$6x - 4 = 6x - 4$

$0 = 0$

Since $0 = 0$ for all x, the solution set is $(-\infty, \infty)$ or $\mathbb{R}$.

77. $-2x + 3y = 7$

$3y = 2x + 7$

$y = \frac{2x+7}{3}$ or $y = \frac{2}{3}x + \frac{7}{3}$

78. $\frac{4x+1}{3} = \frac{x-3}{6} + \frac{x+5}{6}$

$6\left(\frac{4x+1}{3}\right) = 6\left(\frac{x-3}{6}\right) + 6\left(\frac{x+5}{6}\right)$

$2(4x+1) = x - 3 + x + 5$

$8x + 2 = 2x + 2$

$6x + 2 = 2$

$6x = 0$

$x = \frac{0}{6}$

$x = 0$

The solution set is $\{0\}$.

79. $$\frac{x+3}{4}=\frac{x-2}{3}+\frac{1}{4}$$
$$12\left(\frac{x+3}{4}\right)=12\left(\frac{x-2}{3}\right)+12\left(\frac{1}{4}\right)$$
$$3(x+3)=4(x-2)+3$$
$$3x+9=4x-8+3$$
$$3x+9=4x-5$$
$$-x=-14$$
$$x=14$$
The solution set is $\{14\}$.

80. $$-2x-4=x+5$$
$$-2x-x=4+5$$
$$-3x=9$$
$$\frac{-3x}{-3}=\frac{9}{-3}$$
$$x=-3$$
The solution set is $\{-3\}$.

81. $$2(x+4)>2x+3$$
$$2x+8>2x+3$$
$$2x-2x+8>2x-2x+3$$
$$8>3$$
The resulting statement is always true.
Thus, the solution set is $(-\infty,\infty)$ or $\mathbb{R}$.

Chapter 3 Review Exercises

1. $$2x-5y=-2$$
$$2(4)-5(2)=-2$$
$$8-10=-2$$
$$-2=-2,\ \text{true}$$
$$3x+4y=4$$
$$3(4)+4(2)=4$$
$$12+8=4$$
$$20=4,\ \text{false}$$
The pair is not a solution of the system.

2. $$-x+2y=11$$
$$-(-5)+2(3)=11$$
$$5+6=11$$
$$11=11,\ \text{true}$$
$$y=-\frac{x}{3}+\frac{4}{3}$$
$$3=-\frac{(-5)}{3}+\frac{4}{3}$$
$$3=\frac{5}{3}+\frac{4}{3}$$
$$3=\frac{9}{3}$$
$$3=3,\ \text{true}$$
The pair is a solution of the system.

3. $$x+y=5$$
$$y=-x+5$$
$$m=-1$$
$$y-\text{intercept}=5$$
$$3x-y=3$$
$$-y=-3x+3$$
$$y=3x-3$$
$$m=3;\quad y-\text{intercept}=-3$$

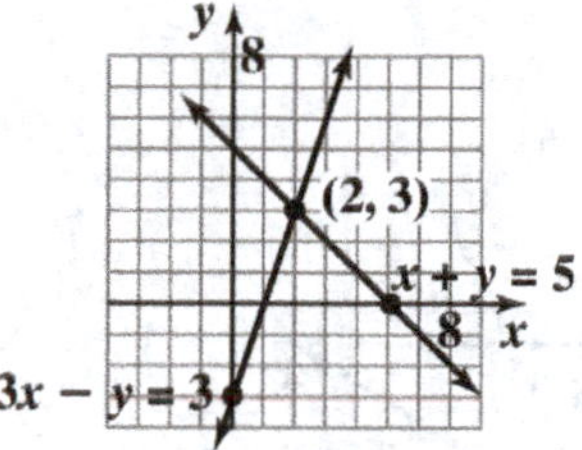

The solution set is $\{(2,3)\}$.

4. $$3x-2y=6$$
$$-2y=-3x+6$$
$$y=\frac{3}{2}x-3$$
$$m=\frac{3}{2}$$
$$y-\text{intercept}=-3$$

$$6x-4y=12$$
$$-4y=-6x+12$$
$$y=\frac{6}{4}x-3$$
$$y=\frac{3}{2}x-3$$
$$m=\frac{3}{2}$$
$$y-\text{intercept}=-3$$

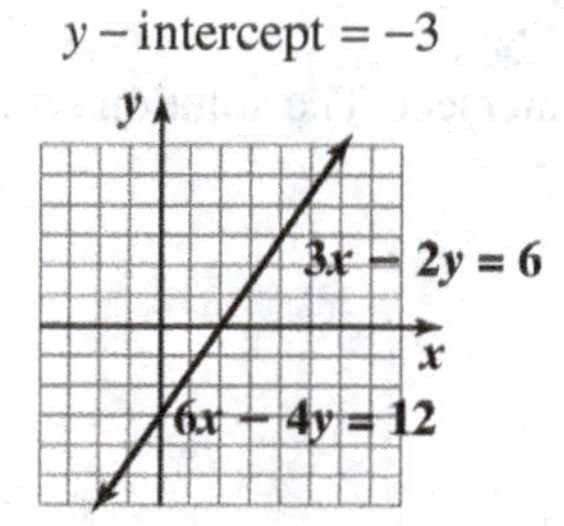

Since the lines coincide, the solution set is $\{(x,y)|3x-2y=6\}$ or $\{(x,y)|6x-4y=12\}$.

5. $y = \frac{3}{5}x - 3$

$m = \frac{3}{5}$

$y - \text{intercept} = -3$

$2x - y = -4$
$-y = -2x - 4$
$y = 2x + 4$
$m = 2$
$y - \text{intercept} = 4$

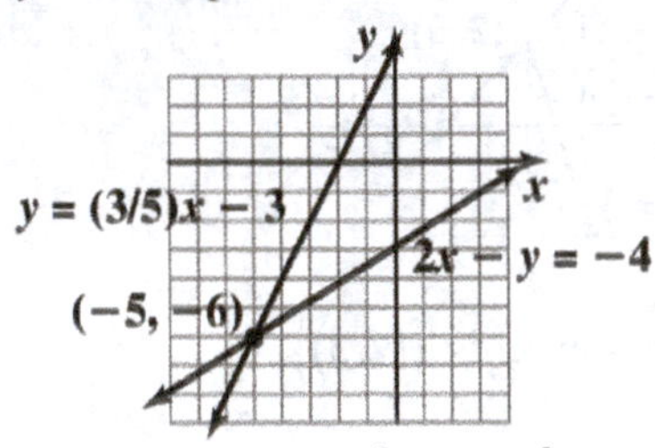

The solution set is $\{(-5,-6)\}$.

6. $y = -x + 4$
$m = -1$
$y - \text{intercept} = 4$

$3x + 3y = -6$
$3y = -3x - 6$
$y = -x - 2$
$m = -1$
$y - \text{intercept} = -2$

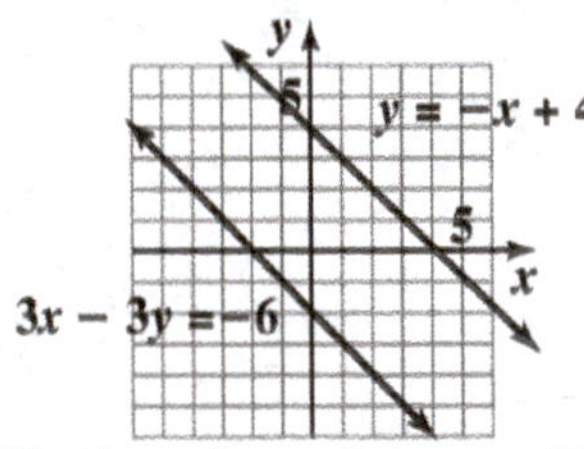

The lines do not intersect. The solution set is $\varnothing$ or $\{\ \}$.

7. $2x - y = 2$
$x + 2y = 11$
Solve the second equation for x.
$x + 2y = 11$
$x = 11 - 2y$
Substitute for x in the first equation solve for y.
$2(11 - 2y) - y = 2$
$22 - 4y - y = 2$
$22 - 5y = 2$
$-5y = -20$
$y = 4$
Back-substitute 4 for y to find x.
$2x - 4 = 2$
$2x = 6$
$x = 3$
The solution set is $\{(3,4)\}$.

8. $y = -2x + 3$
$3x + 2y = -17$
Substitute for y in the second equation and solve for x.
$3x + 2y = -17$
$3x + 2(-2x + 3) = -17$
$3x - 4x + 6 = -17$
$-x + 6 = -17$
$-x = -23$
$x = 23$
Back-substitute 23 for x to find y.
$y = -2(23) + 3$
$y = -46 + 3$
$y = -43$
The solution set is $\{(23,-43)\}$.

9. $3x+2y=-8$
$2x+5y=\ 2$
Multiply the first equation by –2 and the second equation by 3 and solve by addition.
$$\begin{aligned}-6x-\ 4y&=16\\ 6x+15y&=\ 6\\ \hline 11y&=22\\ y&=2\end{aligned}$$
Back-substitute 2 for y in one of the original equations.
$$\begin{aligned}3x+2(2)&=-8\\ 3x+4&=-8\\ 3x&=-12\\ x&=-4\end{aligned}$$
The solution set is $\{(-4,2)\}$.

10. $5x-2y=14$
$3x+4y=11$
Multiply the first equation by 2 and add to the second equation.
$$\begin{aligned}10x-4y&=28\\ 3x+4y&=11\\ \hline 13x&=39\\ x&=3\end{aligned}$$
Back-substitute 3 for x in one of the original equations.
$$\begin{aligned}5x-2y&=14\\ 5(3)-2y&=14\\ 15-2y&=14\\ -2y&=-1\\ y&=\frac{1}{2}\end{aligned}$$
The solution set is $\left\{\left(3,\frac{1}{2}\right)\right\}$.

11. $y=4-x$
$3x+3y=12$
Substitute for x in the first equation solve for y.
$$\begin{aligned}3x+3y&=12\\ 3x+3(4-x)&=12\\ 3x+12-3x&=12\\ 12&=12\end{aligned}$$
The system has infinitely many solutions. The solution set is $\{(x,y)|y=4-x\}$ or $\{(x,y)|3x+3y=12\}$.

12. $\frac{x}{8}+\frac{3y}{4}=\frac{19}{8}$
$-\frac{x}{2}+\frac{3y}{4}=\frac{1}{2}$
To clear fractions, multiply the first equation by 8 and the second equation by 4.
$$\begin{aligned}8\left(\frac{x}{8}\right)+8\left(\frac{3y}{4}\right)&=8\left(\frac{19}{8}\right)\\ x+2(3y)&=19\\ x+6y&=19\end{aligned}$$
$$\begin{aligned}4\left(-\frac{x}{2}\right)+4\left(\frac{3y}{4}\right)&=4\left(\frac{1}{2}\right)\\ 2(-x)+3y&=2(1)\\ -2x+3y&=2\end{aligned}$$
The system becomes
$x+6y=19$
$-2x+3y=\ 2$
Multiply the first equation by 2 and add to the second equation.
$$\begin{aligned}2x+12y&=38\\ -2x+\ 3y&=\ 2\\ \hline 15y&=40\\ y&=\frac{40}{15}\\ y&=\frac{8}{3}\end{aligned}$$
Back-substitute $\frac{8}{3}$ for y in one of the equations to find x.
$$\begin{aligned}x+6\left(\frac{8}{3}\right)&=19\\ x+2(8)&=19\\ x+16&=19\\ x&=3\end{aligned}$$
The solution set is $\left\{\left(3,\frac{8}{3}\right)\right\}$.

13. $x-2y+3=0$

$2x-4y+7=0$

Rewrite the system is standard form.

$x-2y=-3$

$2x-4y=-7$

Multiply the first equation by –2 and add to the second equation.

$$\begin{aligned} -2x+4y &= 6 \\ 2x-4y &= -7 \\ \hline 0 &= -1 \end{aligned}$$

There are no values for x and y that will make $0=-1$, so the solution set is $\varnothing$ or $\{\ \}$.

14. Let t = the cost of the televisions.
Let s = the cost of the stereos.

$3t+4s=2530$

$4t+3s=2510$

Multiply the first equation by –4 and the second by 3 and solve by addition.

$$3t+4s=2530$$
$$-4(3t)+(-4)(4s)=-4(2530)$$
$$-12t-16s=-10120$$
$$4t+3s=2510$$
$$3(4t)+3(3s)=3(2510)$$
$$12t+9s=7530$$

The system of two equations in two variables:

$$\begin{aligned} -12t-16s &= -10120 \\ 12t+\ 9s &= \ \ 7530 \\ \hline -7s &= -2590 \\ s &= 370 \end{aligned}$$

Back-substitute 370 for s in one of the original equations to find t.

$$\begin{aligned} 3t+4s &= 2530 \\ 3t+4(370) &= 2530 \\ 3t+1480 &= 2530 \\ 3t &= 1050 \\ t &= 350 \end{aligned}$$

The televisions cost \$350 each and the stereos costs \$370 each.

15. Let x = the amount invested at 4%.
Let y = the amount invested at 7%.

$x+\quad y=9000$

$0.04x+0.07y=555$

Multiply the first equation by –0.04 and add.

$$\begin{aligned} -0.04x-0.04y &= -360 \\ 0.04x+0.07y &= 555 \\ \hline 0.03y &= 195 \\ y &= 6500 \end{aligned}$$

Back-substitute 6500 for y in one of the original equations to find x.

$$\begin{aligned} x+y &= 9000 \\ x+6500 &= 9000 \\ x &= 2500 \end{aligned}$$

There was \$2500 invested at 4% and \$6500 invested at 7%.

16. Let x = the amount of the 34% solution.
Let y = the amount of the 4% solution.

$x+\quad y=100$

$0.34x+0.04y=0.07(100)$

Simplified, the system becomes

$x+\quad y=100$

$0.34x+0.04y=\ \ 7$

Multiply the first equation by –0.34 and add to the second equation.

$$\begin{aligned} -0.34x-0.34y &= -34 \\ 0.34x+0.04y &= \ \ 7 \\ \hline -0.30 &= -27 \\ y &= 90 \end{aligned}$$

Back-substitute 90 for y to find x.

$$\begin{aligned} x+y &= 100 \\ x+90 &= 100 \\ x &= 10 \end{aligned}$$

10 ml of the 34% solution and 90 ml of the 4% solution must be used.

17. Let r = the speed of the plane in still air.
Let w = the speed of the wind.

	Rate	×	Time	=	Distance
Trip with the Wind	$r+w$		3		$3(r+w)$
Trip against the Wind	$r-w$		4		$4(r-w)$

$3(r+w)=2160$

$4(r-w)=2160$

Simplified, the system becomes

$3r+3w=2160$

$4r-4w=2160$

Multiply the first equation by 4, the second equation by 3, and solve by addition.

$$\begin{aligned} 12r+12w&=8640 \\ \underline{12r-12w}&\underline{=6480} \\ 24r&=15120 \\ r&=630 \end{aligned}$$

Back-substitute 630 for r to find w.

$$\begin{aligned} 3r+3w&=2160 \\ 3(630)+3w&=2160 \\ 1890+3w&=2160 \\ 3w&=270 \\ w&=90 \end{aligned}$$

The speed of the plane in still air is 630 miles per hour and the speed of the wind is 90 miles per hour.

18. Let l = the length of the table.
Let w = the width of the table.

$2l+2w=34$

$4l-3w=33$

Multiply the first equation by –2 and solve by addition.

$$\begin{aligned} -4l-4w&=-68 \\ \underline{4l-3w}&\underline{=\ \ 33} \\ -7w&=-35 \\ w&=5 \end{aligned}$$

Back-substitute 5 for w to find l.

$$\begin{aligned} 2l+2w&=34 \\ 2l+2(5)&=34 \\ 2l+10&=34 \\ 2l&=24 \\ l&=12 \end{aligned}$$

The dimensions of the table are 12 feet by 5 feet.

19. $C(x) = 22500 + 40x$

$C(400) = 22500 + 40(400)$

$= 22500 + 16000 = 38500$

$R(x) = 85x$

$R(400) = 85(400) = 34000$

$P(x) = R(x) - C(x)$

$P(400) = R(400) - C(400)$

$= 34000 - 38500 = -4500$

There is a $4500 loss when 400 calculators are sold.

20. $R(x) = C(x)$

$85x = 22500 + 40x$

$45x = 22500$

$x = 500$

$R(x) = 85x$

$R(500) = 85(500) = 42500$

The break-even point is $(500,\ 42,500)$. This means that when 500 calculators are produced and sold, the cost is the same as the revenue at $42,500. At this point, there is no net loss or gain.

21. $P(x) = R(x) - C(x)$

$= 85x - (22,500 + 40x)$

$= 85x - 22,500 - 40x = 45x - 22,500$

22. **a.** $C(x) = 60,000 + 200x$

b. $R(x) = 450x$

c. $R(x) = C(x)$

$450x = 60,000 + 200x$

$250x = 60,000$

$x = 240$

$R(x) = 450x$

$R(240) = 450(240) = 108,000$

The break-even point is $(240,\ 108,000)$. This means that when 240 desks are produced and sold, the cost is the same as the revenue at $108,000. At this point, there is no net loss or gain.

23. $x + y + z = 0$

$-3 + (-2) + 5 = 0$

$-5 + 5 = 0$

$0 = 0$, true

$2x - 3y + z = 5$

$2(-3) - 3(-2) + 5 = 5$

$-6 + 6 + 5 = 5$

$5 = 5$, true

$4x + 2y + 4z = 3$

$4(-3) + 2(-2) + 4(5) = 3$

$-12 - 4 + 20 = 3$

$4 = 3$, false

The ordered triple (–3, –2, 5) does not satisfy all three equations, so it is not a solution.

24. $2x - y + z = 1$

$3x - 3y + 4z = 5$

$4x - 2y + 3z = 4$

Multiply the first equation by –2 and add to the third.

$-4x + 2y - 2z = -2$

$\underline{4x - 2y + 3z = 4}$

$z = 2$

Back-substitute 2 for z in two of the original equations to obtain a system of two equations in two variables.

$2x - y + z = 1$ $3x - 3y + 4z = 5$

$2x - y + 2 = 1$ $3x - 3y + 4(2) = 5$

$2x - y = -1$ $3x - 3y + 8 = 5$

$3x - 3y = -3$

The system of two equations in two variables becomes:

$2x - y = -1$

$3x - 3y = -3$

Multiply the first equation by –3 and solve by addition.

$$\begin{aligned} -6x+3y &= 3 \\ 3x-3y &= -3 \\ \hline -3x &= 0 \\ x &= 0 \end{aligned}$$

Back-substitute 0 for x to find y.

$$\begin{aligned} 2x-y &= -1 \\ 2(0)-y &= -1 \\ -y &= -1 \\ y &= 1 \end{aligned}$$

The solution set is $\{(0,1,2)\}$.

25. $x+2y-z=5$
$2x-y+3z=0$
$2y+z=1$

Multiply the first equation by –2 and add to the second equation.

$$\begin{aligned} -2x-4y+2z &= -10 \\ 2x-y+3z &= 0 \\ \hline -5y+5z &= -10 \end{aligned}$$

We now have two equations in two variables.

$$\begin{aligned} 2y+z &= 1 \\ -5y+5z &= -10 \end{aligned}$$

Multiply the first equation by –5 and solve by addition.

$$\begin{aligned} -10y-5z &= -5 \\ -5y+5z &= -10 \\ \hline -15y &= -15 \\ y &= 1 \end{aligned}$$

Back-substitute 1 for y to find z.

$$\begin{aligned} 2(1)+z &= 1 \\ 2+z &= 1 \\ z &= -1 \end{aligned}$$

Back-substitute 1 for y and –1 for z to find x.

$$\begin{aligned} x+2y-z &= 5 \\ x+2(1)-(-1) &= 5 \\ x+2+1 &= 5 \\ x+3 &= 5 \\ x &= 2 \end{aligned}$$

The solution set is $\{(2,1,-1)\}$.

26. $3x-4y+4z=7$
$x-y-2z=2$
$2x-3y+6z=5$

Multiply the second equation by –3 and add to the third equation.

$$\begin{aligned} -3x+3y+6z &= -6 \\ 2x-3y+6z &= 5 \\ \hline -x \qquad +12z &= -1 \end{aligned}$$

Multiply the second equation by –4 and add to the first equation.

$$\begin{aligned} 3x-4y+4z &= 7 \\ -4x+4y+8z &= -8 \\ \hline -x \qquad +12z &= -1 \end{aligned}$$

The system of two equations in two variables becomes:

$$\begin{aligned} -x+12z &= -1 \\ -x+12z &= -1 \end{aligned}$$

The two equations in two variables are identical. The system is dependent. There are an infinite number of solutions to the system.

27. Use each ordered pair to write an equation as follows:

$(x, y) = (1, 4)$

$$\begin{aligned} y &= ax^2+bx+c \\ 4 &= a(1)^2+b(1)+c \\ 4 &= a+b+c \end{aligned}$$

$(x, y) = (3, 20)$

$$\begin{aligned} y &= ax^2+bx+c \\ 20 &= a(3)^2+b(3)+c \\ 20 &= a(9)+3b+c \\ 20 &= 9a+3b+c \end{aligned}$$

$(x, y) = (-2, 25)$

$$\begin{aligned} y &= ax^2+bx+c \\ 25 &= a(-2)^2+b(-2)+c \\ 25 &= a(4)-2b+c \\ 25 &= 4a-2b+c \end{aligned}$$

The system of three equations in three variables is:

$$a + b + c = 4$$
$$9a + 3b + c = 20$$
$$4a - 2b + c = 25$$

Multiply the first equation by –1 and add to the second equation.

$$-a - b - c = -4$$
$$\underline{9a + 3b + c = 20}$$
$$8a + 2b = 16$$

Multiply the first equation by –1 and add to the third equation.

$$-a - b - c = -4$$
$$\underline{4a - 2b + c = 25}$$
$$3a - 3b = 21$$

The system of two equations in two variables becomes:

$$8a + 2b = 16$$
$$3a - 3b = 21$$

Multiply the first equation by 3, the second equation by 2 and solve by addition.

$$24a + 6b = 48$$
$$\underline{6a - 6b = 42}$$
$$30a = 90$$
$$a = 3$$

Back-substitute 3 for a to find b.

$$3(3) - 3b = 21$$
$$9 - 3b = 21$$
$$-3b = 12$$
$$b = -4$$

Back-substitute 3 for a and –4 for b to find c.

$$a + b + c = 4$$
$$3 + (-4) + c = 4$$
$$-1 + c = 4$$
$$c = 5$$

The quadratic function is $y = 3x^2 - 4x + 5$.

28. Let x = number of deaths, in millions, in the 20th century from war.
Let y = number of deaths, in millions, in the 20th century from famine.
Let z = number of deaths, in millions, in the 20th century from tobacco.

$$x + y + z = 306$$
$$x - y = 13$$
$$x - z = 53$$

Solve the second equation for y.

$$x - y = 13$$
$$-y = -x + 13$$
$$y = x - 13$$

Solve the third equation for z.

$$x - z = 53$$
$$-z = -x + 53$$
$$z = x - 53$$

Substitute the expressions for x and z into the first equation and solve for y.

$$x \quad + \quad y \quad + \quad z \quad = 306$$
$$x + \overbrace{(x-13)}^{y} + \overbrace{(x-53)}^{z} = 306$$
$$x + x - 13 + x - 53 = 306$$
$$3x - 66 = 306$$
$$3x = 372$$
$$x = 124$$

Back-substitute to solve for y and z.

$$y = x - 13$$
$$= 124 - 13$$
$$= 111$$
$$z = x - 53$$
$$= 124 - 53$$
$$= 71$$

The number of deaths in the 20th century from war is 124 million, from famine is 111 million, and from tobacco is 71 million.

29. $\left[\begin{array}{cc|c} 1 & -8 & 3 \\ 0 & 7 & -14 \end{array}\right] \; \frac{1}{7}R_2$

$= \left[\begin{array}{cc|c} 1 & -8 & 3 \\ 0 & 1 & -2 \end{array}\right]$

30. $\left[\begin{array}{cc|c} 1 & -3 & 1 \\ 2 & 1 & -5 \end{array}\right] \; -2R_1 + R_2$

$= \left[\begin{array}{cc|c} 1 & -3 & 1 \\ 0 & 7 & -7 \end{array}\right]$

31. $\left[\begin{array}{ccc|c} 2 & -2 & 1 & -1 \\ 1 & 2 & -1 & 2 \\ 6 & 4 & 3 & 5 \end{array}\right] \quad \frac{1}{2}R_1$

$= \left[\begin{array}{ccc|c} 1 & -1 & \frac{1}{2} & -\frac{1}{2} \\ 1 & 2 & -1 & 2 \\ 6 & 4 & 3 & 5 \end{array}\right]$

32. $\left[\begin{array}{ccc|c} 1 & 2 & 2 & 2 \\ 0 & 1 & -1 & 2 \\ 0 & 5 & 4 & 1 \end{array}\right] \quad -5R_2 + R_3$

$= \left[\begin{array}{ccc|c} 1 & 2 & 2 & 2 \\ 0 & 1 & -1 & 2 \\ 0 & 0 & 9 & -9 \end{array}\right]$

33. $\left[\begin{array}{cc|c} 1 & 4 & 7 \\ 3 & 5 & 0 \end{array}\right] \quad -3R_1 + R_2$

$= \left[\begin{array}{cc|c} 1 & 4 & 7 \\ 0 & -7 & -21 \end{array}\right] \quad -\frac{1}{7}R_2$

$= \left[\begin{array}{cc|c} 1 & 4 & 7 \\ 0 & 1 & 3 \end{array}\right]$

The resulting system is

$$\begin{aligned} x + 4y &= 7 \\ y &= 3 \end{aligned}$$

Back-substitute 3 for y to find x.

$$\begin{aligned} x + 4y &= 7 \\ x + 4(3) &= 7 \\ x + 12 &= 7 \\ x &= -5 \end{aligned}$$

The solution set is $\{(-5, 3)\}$.

34. $\left[\begin{array}{cc|c} 2 & -3 & 8 \\ -6 & 9 & 4 \end{array}\right] \quad 3R_1 + R_2$

$= \left[\begin{array}{cc|c} 2 & -3 & 8 \\ 0 & 0 & 28 \end{array}\right]$

This is a contradiction. R_2 states that $0x + 0y = 28$
$0 = 28$

There are no values of x and y for which $0 = 28$. The system is inconsistent and the solution set is $\varnothing$ or $\{\ \}$.

35. $\left[\begin{array}{ccc|c} 1 & 2 & 3 & -5 \\ 2 & 1 & 1 & 1 \\ 1 & 1 & -1 & 8 \end{array}\right] \quad -2R_1 + R_2$

$= \left[\begin{array}{ccc|c} 1 & 2 & 3 & -5 \\ 0 & -3 & -5 & 11 \\ 1 & 1 & -1 & 8 \end{array}\right] \quad -R_1 + R_3$

$= \left[\begin{array}{ccc|c} 1 & 2 & 3 & -5 \\ 0 & -3 & -5 & 11 \\ 0 & -1 & -4 & 13 \end{array}\right] \quad -\frac{1}{3}R_2$

$= \left[\begin{array}{ccc|c} 1 & 2 & 3 & -5 \\ 0 & 1 & \frac{5}{3} & -\frac{11}{3} \\ 0 & -1 & -4 & 13 \end{array}\right] \quad R_2 + R_3$

$= \left[\begin{array}{ccc|c} 1 & 2 & 3 & -5 \\ 0 & 1 & \frac{5}{3} & -\frac{11}{3} \\ 0 & 0 & -\frac{7}{3} & \frac{28}{3} \end{array}\right] \quad -\frac{3}{7}R_3$

$= \left[\begin{array}{ccc|c} 1 & 2 & 3 & -5 \\ 0 & 1 & \frac{5}{3} & -\frac{11}{3} \\ 0 & 0 & 1 & -4 \end{array}\right]$

The resulting system is

$$\begin{aligned} x + 2y + 3z &= -5 \\ y + \frac{5}{3}z &= -\frac{11}{3} \\ z &= -4 \end{aligned}$$

Back-substitute –4 for z to find y.

$$\begin{aligned} y + \frac{5}{3}z &= -\frac{11}{3} \\ y + \frac{5}{3}(-4) &= -\frac{11}{3} \\ y - \frac{20}{3} &= -\frac{11}{3} \\ y &= \frac{9}{3} \\ y &= 3 \end{aligned}$$

Back-substitute 3 for y and –4 for z to find x.

$$\begin{aligned} x+2y+3z &= -5 \\ x+2(3)+3(-4) &= -5 \\ x+6-12 &= -5 \\ x-6 &= -5 \\ x &= 1 \end{aligned}$$

The solution set is $\{(1,3,-4)\}$.

36. $\left[\begin{array}{rrr|r} 1 & -2 & 1 & 0 \\ 0 & 1 & -3 & -1 \\ 0 & 2 & 5 & -2 \end{array}\right] \; -2R_2+R_3$

$=\left[\begin{array}{rrr|r} 1 & -2 & 1 & 0 \\ 0 & 1 & -3 & -1 \\ 0 & 0 & 11 & 0 \end{array}\right] \frac{1}{11}R_3$

$=\left[\begin{array}{rrr|r} 1 & -2 & 1 & 0 \\ 0 & 1 & -3 & -1 \\ 0 & 0 & 1 & 0 \end{array}\right]$

The resulting system is

$$\begin{aligned} x-2y+\ z &= \ 0 \\ y-3z &= -1 \\ z &= \ 0 \end{aligned}$$

Back-substitute 0 for z to find y.

$$\begin{aligned} y-3z &= -1 \\ y-\cancel{3(0)} &= -1 \\ y &= -1 \end{aligned}$$

Back-substitute -1 for y and 0 for z to find x.

$$\begin{aligned} x-2y+z &= 0 \\ x-2(-1)+0 &= 0 \\ x+2 &= 0 \\ x &= -2 \end{aligned}$$

The solution set is $\{(-2,-1,0)\}$.

37. $\begin{vmatrix} 3 & 2 \\ -1 & 5 \end{vmatrix} = 3(5)-(-1)2 = 15+2 = 17$

38. $\begin{vmatrix} -2 & -3 \\ -4 & -8 \end{vmatrix} = -2(-8)-(-4)(-3)$

$= 16-12 = 4$

39. $\begin{vmatrix} 2 & 4 & -3 \\ 1 & -1 & 5 \\ -2 & 4 & 0 \end{vmatrix} = 2\begin{vmatrix} -1 & 5 \\ 4 & 0 \end{vmatrix} - 1\begin{vmatrix} 4 & -3 \\ 4 & 0 \end{vmatrix} - 2\begin{vmatrix} 4 & -3 \\ -1 & 5 \end{vmatrix}$

$= 2\left(\cancel{-1(0)}-4(5)\right)-1\left(\cancel{4(0)}-4(-3)\right)$

$\qquad -2\left(4(5)-(-1)(-3)\right)$

$= 2(-20)-1(12)-2(20-3)$

$= -40-12-2(17) = -40-12-34 = -86$

40. $\begin{vmatrix} 4 & 7 & 0 \\ -5 & 6 & 0 \\ 3 & 2 & -4 \end{vmatrix} = 4\begin{vmatrix} 6 & 0 \\ 2 & -4 \end{vmatrix} - (-5)\begin{vmatrix} 7 & 0 \\ 2 & -4 \end{vmatrix} + 3\begin{vmatrix} 7 & 0 \\ 6 & 0 \end{vmatrix}$

$= 4\left(6(-4)-\cancel{2(0)}\right)+5\left(7(-4)-\cancel{2(0)}\right)$

$\qquad +3\left(\cancel{7(0)}-\cancel{6(0)}\right)$

$= 4(-24)+5(-28)+\cancel{3(0)} = -96-140 = -236$

41. $D = \begin{vmatrix} 1 & -2 \\ 3 & 2 \end{vmatrix} = 1(2)-3(-2) = 2+6 = 8$

$D_x = \begin{vmatrix} 8 & -2 \\ -1 & 2 \end{vmatrix} = 8(2)-(-1)(-2) = 16-2 = 14$

$D_y = \begin{vmatrix} 1 & 8 \\ 3 & -1 \end{vmatrix} = 1(-1)-3(8) = -1-24 = -25$

$x = \frac{D_x}{D} = \frac{14}{8} = \frac{7}{4};\ y = \frac{D_y}{D} = \frac{-25}{8} = -\frac{25}{8}$

The solution set is $\left\{\left(\frac{7}{4}, -\frac{25}{8}\right)\right\}$.

42. $D=\begin{vmatrix}7 & 2\\ 2 & 1\end{vmatrix}=7(1)-2(2)=7-4=3$

$D_x=\begin{vmatrix}0 & 2\\ -3 & 1\end{vmatrix}=0(1)-(-3)(2)=6$;

$D_y=\begin{vmatrix}7 & 0\\ 2 & -3\end{vmatrix}=7(-3)-2(0)=-21$

$x=\frac{D_x}{D}=\frac{6}{3}=2$; $y=\frac{D_y}{D}=\frac{-21}{3}=-7$

The solution set is $\{(2,-7)\}$.

43.
$$D=\begin{vmatrix}1 & 2 & 2\\ 2 & 4 & 7\\ -2 & -5 & -2\end{vmatrix}$$
$$=1\begin{vmatrix}4 & 7\\ -5 & -2\end{vmatrix}-2\begin{vmatrix}2 & 2\\ -5 & -2\end{vmatrix}-2\begin{vmatrix}2 & 2\\ 4 & 7\end{vmatrix}$$
$$=1(4(-2)-(-5)7)-2(2(-2)-(-5)2)$$
$$-2(2(7)-4(2))$$
$$=1(-8+35)-2(-4+10)-2(14-8)$$
$$=1(27)-2(6)-2(6)$$
$$=27-12-12$$
$$=3$$

$$D_x=\begin{vmatrix}5 & 2 & 2\\ 19 & 4 & 7\\ 8 & -5 & -2\end{vmatrix}$$
$$=5\begin{vmatrix}4 & 7\\ -5 & -2\end{vmatrix}-19\begin{vmatrix}2 & 2\\ -5 & -2\end{vmatrix}+8\begin{vmatrix}2 & 2\\ 4 & 7\end{vmatrix}$$
$$=5(4(-2)-(-5)7)-19(2(-2)-(-5)2)$$
$$+8(2(7)-4(2))$$
$$=5(-8+35)-19(-4+10)+8(14-8)$$
$$=5(27)-19(6)+8(6)$$
$$=135-114+48$$
$$=69$$

$$D_y=\begin{vmatrix}1 & 5 & 2\\ 2 & 19 & 7\\ -2 & 8 & -2\end{vmatrix}$$
$$=1\begin{vmatrix}19 & 7\\ 8 & -2\end{vmatrix}-2\begin{vmatrix}5 & 2\\ 8 & -2\end{vmatrix}-2\begin{vmatrix}5 & 2\\ 19 & 7\end{vmatrix}$$
$$=1(19(-2)-(8)7)-2(5(-2)-8(2))$$
$$-2(5(7)-19(2))$$
$$=1(-38-56)-2(-10-16)-2(35-38)$$
$$=1(-94)-2(-26)-2(-3)$$
$$=-94+52+6$$
$$=-36$$

$$D_z=\begin{vmatrix}1 & 2 & 5\\ 2 & 4 & 19\\ -2 & -5 & 8\end{vmatrix}$$
$$=1\begin{vmatrix}4 & 19\\ -5 & 8\end{vmatrix}-2\begin{vmatrix}2 & 5\\ -5 & 8\end{vmatrix}-2\begin{vmatrix}2 & 5\\ 4 & 19\end{vmatrix}$$
$$=1(4(8)-(-5)19)-2(2(8)-(-5)5)$$
$$-2(2(19)-4(5))$$
$$=1(32+95)-2(16+25)-2(38-20)$$
$$=1(127)-2(41)-2(18)$$
$$=127-82-36$$
$$=9$$

$x=\frac{D_x}{D}=\frac{69}{3}=23$; $y=\frac{D_y}{D}=\frac{-36}{3}=-12$;

$z=\frac{D_z}{D}=\frac{9}{3}=3$

The solution set is $\{(23,-12,3)\}$.

44. Rewrite the system in standard form.

$$\begin{aligned} 2x + y \quad &= -4 \\ y - 2z &= 0 \\ 3x \quad - 2z &= -11 \end{aligned}$$

$$D = \begin{vmatrix} 2 & 1 & 0 \\ 0 & 1 & -2 \\ 3 & 0 & -2 \end{vmatrix}$$

$$= 2\begin{vmatrix} 1 & -2 \\ 0 & -2 \end{vmatrix} - 0\begin{vmatrix} 1 & 0 \\ 0 & -2 \end{vmatrix} + 3\begin{vmatrix} 1 & 0 \\ 1 & -2 \end{vmatrix}$$

$$= 2\left(1(-2) - 0(-2)\right) + 3\left(1(-2) - 1(0)\right)$$

$$= 2(-2) + 3(-2)$$

$$= -4 - 6$$

$$= -10$$

$$D_x = \begin{vmatrix} -4 & 1 & 0 \\ 0 & 1 & -2 \\ -11 & 0 & -2 \end{vmatrix}$$

$$= -4\begin{vmatrix} 1 & -2 \\ 0 & -2 \end{vmatrix} - 0\begin{vmatrix} 1 & 0 \\ 0 & -2 \end{vmatrix} - 11\begin{vmatrix} 1 & 0 \\ 1 & -2 \end{vmatrix}$$

$$= -4\left(1(-2) - 0(-2)\right) - 11\left(1(-2) - 1(0)\right)$$

$$= -4(-2) - 11(-2)$$

$$= 8 + 22$$

$$= 30$$

$$D_y = \begin{vmatrix} 2 & -4 & 0 \\ 0 & 0 & -2 \\ 3 & -11 & -2 \end{vmatrix}$$

$$= 2\begin{vmatrix} 0 & -2 \\ -11 & -2 \end{vmatrix} - 0\begin{vmatrix} -4 & 0 \\ -11 & -2 \end{vmatrix} + 3\begin{vmatrix} -4 & 0 \\ 0 & -2 \end{vmatrix}$$

$$= 2\left(0(-2) - (-11)(-2)\right) + 3\left(-4(-2) - 0(0)\right)$$

$$= 2(-22) + 3(8)$$

$$= -44 + 24$$

$$= -20$$

$$D_z = \begin{vmatrix} 2 & 1 & -4 \\ 0 & 1 & 0 \\ 3 & 0 & -11 \end{vmatrix}$$

$$= 2\begin{vmatrix} 1 & 0 \\ 0 & -11 \end{vmatrix} - 0\begin{vmatrix} 1 & -4 \\ 0 & -11 \end{vmatrix} + 3\begin{vmatrix} 1 & -4 \\ 1 & 0 \end{vmatrix}$$

$$= 2\left(1(-11) - 0(0)\right) + 3\left(1(0) - 1(-4)\right)$$

$$= 2(-11) + 3(4)$$

$$= -22 + 12$$

$$= -10$$

$$x = \frac{D_x}{D} = \frac{30}{-10} = -3\ ;\ y = \frac{D_y}{D} = \frac{-20}{-10} = 2\ ;\ z = \frac{D_z}{D} = \frac{-10}{-10} = 1$$

The solution set is $\{(-3, 2, 1)\}$.

45. Use each ordered pair to write an equation as follows:

$(x, y) = (20, 400)$	$(x, y) = (40, 150)$	$(x, y) = (60, 400)$
$y = ax^2 + bx + c$	$y = ax^2 + bx + c$	$y = ax^2 + bx + c$
$400 = a(20)^2 + b(20) + c$	$150 = a(40)^2 + b(40) + c$	$400 = a(60)^2 + b(60) + c$
$400 = a(400) + 20b + c$	$150 = a(1600) + 40b + c$	$400 = a(3600) + 60b + c$
$400 = 400a + 20b + c$	$150 = 1600a + 40b + c$	$400 = 3600a + 60b + c$

The system of three equations in three variables is:

$$400a + 20b + c = 400$$
$$1600a + 40b + c = 150$$
$$3600a + 60b + c = 400$$

$$D = \begin{vmatrix} 400 & 20 & 1 \\ 1600 & 40 & 1 \\ 3600 & 60 & 1 \end{vmatrix} = 400\begin{vmatrix} 40 & 1 \\ 60 & 1 \end{vmatrix} - 1600\begin{vmatrix} 20 & 1 \\ 60 & 1 \end{vmatrix} + 3600\begin{vmatrix} 20 & 1 \\ 40 & 1 \end{vmatrix}$$

$$= 400\left(40(1) - 60(1)\right) - 1600\left(20(1) - 60(1)\right) + 3600\left(20(1) - 40(1)\right)$$

$$= 400(40 - 60) - 1600(20 - 60) + 3600(20 - 40)$$

$$= 400(-20) - 1600(-40) + 3600(-20) = -8000 + 64000 - 72000 = -16000$$

$$D_a = \begin{vmatrix} 400 & 20 & 1 \\ 150 & 40 & 1 \\ 400 & 60 & 1 \end{vmatrix} = 400\begin{vmatrix} 40 & 1 \\ 60 & 1 \end{vmatrix} - 150\begin{vmatrix} 20 & 1 \\ 60 & 1 \end{vmatrix} + 400\begin{vmatrix} 20 & 1 \\ 40 & 1 \end{vmatrix}$$

$$= 400\left(40(1) - 60(1)\right) - 150\left(20(1) - 60(1)\right) + 400\left(20(1) - 40(1)\right)$$

$$= 400(40 - 60) - 150(20 - 60) + 400(20 - 40)$$

$$= 400(-20) - 150(-40) + 400(-20) = -8000 + 6000 - 8000 = -10000$$

$$D_b = \begin{vmatrix} 400 & 400 & 1 \\ 1600 & 150 & 1 \\ 3600 & 400 & 1 \end{vmatrix} = 400\begin{vmatrix} 150 & 1 \\ 400 & 1 \end{vmatrix} - 1600\begin{vmatrix} 400 & 1 \\ 400 & 1 \end{vmatrix} + 3600\begin{vmatrix} 400 & 1 \\ 150 & 1 \end{vmatrix}$$

$$= 400(150(1) - 400(1)) - 1600(400(1) - 400(1)) + 3600(400(1) - 150(1))$$

$$= 400(150 - 400) - 1600(400 - 400) + 3600(400 - 150)$$

$$= 400(-250) - 1600(0) + 3600(250)$$

$$= 400(-250) + 3600(250) = -100,000 + 900,000 = 800,000$$

$$D_c = \begin{vmatrix} 400 & 20 & 400 \\ 1600 & 40 & 150 \\ 3600 & 60 & 400 \end{vmatrix} = 400\begin{vmatrix} 40 & 150 \\ 60 & 400 \end{vmatrix} - 1600\begin{vmatrix} 20 & 400 \\ 60 & 400 \end{vmatrix} + 3600\begin{vmatrix} 20 & 400 \\ 40 & 150 \end{vmatrix}$$

$$= 400(40(400) - 60(150)) - 1600(20(400) - 60(400)) + 3600(20(150) - 40(400))$$

$$= 400(16000 - 9000) - 1600(8000 - 24000) + 3600(3000 - 16000)$$

$$= 400(7000) - 1600(-16000) + 3600(-13000)$$

$$= 2800000 + 25600000 - 46800000 = -18400000$$

$$a = \frac{D_a}{D} = \frac{-10000}{-16000} = \frac{5}{8}; \quad b = \frac{D_b}{D} = \frac{800000}{-16000} = -50; \quad c = \frac{D_c}{D} = \frac{-18400000}{-16000} = 1150$$

The quadratic function is $f(x) = \frac{5}{8}x^2 - 50x + 1150.$

To predict the average number of automobile accidents in which 30-year-olds and 50-year-olds are involved, find $f(30)$ and $f(50)$.

$$f(30) = \frac{5}{8}(30)^2 - 50(30) + 1150 \qquad f(50) = \frac{5}{8}(50)^2 - 50(50) + 1150$$

$$= \frac{5}{8}(900) - 1500 + 1150 \qquad = \frac{5}{8}(2500) - 2500 + 1150$$

$$= 562.5 - 1500 + 1150 \qquad = 1562.5 - 2500 + 1150$$

$$= 212.5 \qquad = 212.5$$

Both 30-year-olds and 50-year-olds are involved in approximately 212.5 accidents per day in the United States.

Chapter 3 Test

1. $x+y=6$ $\qquad$ $4x-y=4$

$y=-x+6$ $\qquad$ $-y=-4x+4$

$m=-1$ $\qquad$ $y=4x-4$

$y-\text{intercept}=6$ $\qquad$ $m=4$

$y-\text{intercept}=-4$

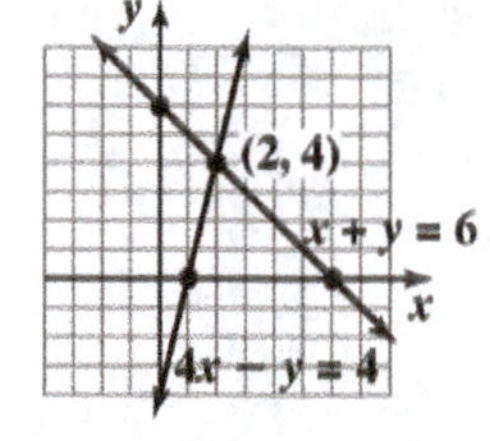

The solution set is $\{(2,4)\}$.

2. $5x+4y=10$
$3x+5y=-7$

Multiply the first equation by –5, and the second equation by 4, and solve by addition.

$$-25x-20y=-50$$
$$\underline{12x+20y=-28}$$
$$-13x=-78$$
$$x=6$$

Back-substitute 6 for x to find y.

$$5x+4y=10$$
$$5(6)+4y=10$$
$$30+4y=10$$
$$4y=-20$$
$$y=-5$$

The solution set is $\{(6,-5)\}$.

3. $x=y+4$
$3x+7y=-18$

Substitute $y + 4$ for x to find y.

$$3x+7y=-18$$
$$3(y+4)+7y=-18$$
$$3y+12+7y=-18$$
$$10y+12=-18$$
$$10y=-30$$
$$y=-3$$

Back-substitute –3 for y to find x.

$$x=y+4$$
$$x=-3+4$$
$$x=1$$

The solution set is $\{(1,-3)\}$.

4. $4x=2y+6$
$y=2x-3$

Substitute $2x - 3$ for y to find x.

$$4x=2y+6$$
$$4x=2(2x-3)+6$$
$$4x=4x-6+6$$
$$0=-6+6$$
$$0=0$$

The system is dependent. There are infinitely many solutions. The solution set is $\{(x,y)|4x=2y+6\}$ or $\{(x,y)\,|\,y=2x-3\}$.

5. Let x = the number of one-bedroom condominiums.
Let y = the number of two-bedroom condominiums.

$$x+y=50$$
$$120x+150y=7050$$

Solve the first equation for x.

$$x+y=50$$
$$x=50-y$$

Substitute $x=50-y$ into the second equation and solve for y.

$$120(50-y)+150y=7050$$
$$6000-120y+150y=7050$$
$$30y=1050$$
$$y=35$$

Back-substitute $y=35$ to solve for x.

$$x+y=50$$
$$x+35=50$$
$$x=15$$

15 one-bedroom and 35 two-bedroom condominiums were sold.

6. Let x = the amount invested at 6%.
Let y = the amount invested at 7%.

$$\begin{aligned} x + \quad y &= 9000 \\ 0.06x + 0.07y &= 610 \end{aligned}$$

Multiply the first equation by –0.06 and add to the second equation.

$$\begin{aligned} -0.06x - 0.06y &= -540 \\ 0.06x + 0.07y &= 610 \\ \hline 0.01y &= 70 \\ y &= 7000 \end{aligned}$$

Back-substitute 7000 for y to find x.

$$\begin{aligned} x + y &= 9000 \\ x + 7000 &= 9000 \\ x &= 2000 \end{aligned}$$

There is \$2000 invested at 6% and \$7000 invested at 7%.

7. Let x = the number of ounces of 6% solution.
Let y = the number of ounces of 9% solution.

$$\begin{aligned} x + y &= 36 \\ 0.06x + 0.09y &= 0.08(36) \end{aligned}$$

Rewrite the system in standard form.

$$\begin{aligned} x + \quad y &= 36 \\ 0.06x + 0.09y &= 2.88 \end{aligned}$$

Multiply the first equation by –0.06 and add to the second equation.

$$\begin{aligned} -0.06x - 0.06y &= -2.16 \\ 0.06x + 0.09y &= 2.88 \\ \hline 0.03y &= 0.72 \\ y &= 24 \end{aligned}$$

Back-substitute 24 for y to find x.

$$\begin{aligned} x + y &= 36 \\ x + 24 &= 36 \\ x &= 12 \end{aligned}$$

12 ounces of 6% peroxide solution and 24 ounces of 9% peroxide solution must be used.

8. Let r = the speed of the paddleboat in still water.
Let c = the speed of the current.

	Rate ×	Time =	Distance
Trip with the Current	$r+c$	3	$3(r+c)$
Trip against the Current	$r-c$	4	$4(r-c)$

$$3(r+c)=48$$
$$4(r-c)=48$$

Simplified, the system becomes

$$3r+3c=48$$
$$4r-4c=48$$

Multiply the first equation by 4, the second equation by 3, and solve by addition.

$$\begin{aligned} 12r+12c&=192 \\ \underline{12r-12c}&\underline{=144} \\ 24r&=336 \\ r&=14 \end{aligned}$$

Back-substitute 14 for r to find c.

$$\begin{aligned} 3r+3c&=48 \\ 3(14)+3c&=48 \\ 42+3c&=48 \\ 3c&=6 \\ c&=2 \end{aligned}$$

The speed of the paddleboat in still water is 14 miles per hour and the speed of the current is 2 miles per hour.

9. Let x = the number of computers produced.
$C(x)=360,000+850x$

10. Let x = the number of computers sold.
$R(x)=1150x$

11.

$$\begin{aligned} R(x)&=C(x) \\ 1150x&=360,000+850x \\ 300x&=360,000 \\ x&=1200 \end{aligned}$$

$$R(x)=1150x$$

$$\begin{aligned} R(1200)&=1150(1200) \\ &=1,380,000 \end{aligned}$$

The break-even point is
(1200, 1,380,000). When 1200 computers are produced and sold, the revenue will equal the cost at $1,380,000.

12.

$$\begin{aligned} P(x)&=R(x)-C(x) \\ &=125x-(40x+350,000) \\ &=125x-40x-350,000 \\ &=85x-350,000 \end{aligned}$$

13. $x + y + z = 6$
$3x + 4y - 7z = 1$
$2x - y + 3z = 5$

Multiply the first equation by 7 and add to the second equation.

$$\begin{aligned} 7x + 7y + 7z &= 42 \\ 3x + 4y - 7z &= 1 \\ \hline 10x + 11y &= 43 \end{aligned}$$

Multiply the first equation by –3 and add to the third equation.

$$\begin{aligned} -3x - 3y - 3z &= -18 \\ 2x - y + 3z &= 5 \\ \hline -x - 4y &= -13 \end{aligned}$$

The system of two equations in two variables.

$$\begin{aligned} 10x + 11y &= 43 \\ -x - 4y &= -13 \end{aligned}$$

Multiply the second equation by 10 and solve by addition.

$$\begin{aligned} 10x + 11y &= 43 \\ -10x - 40y &= -130 \\ \hline -29y &= -87 \\ y &= 3 \end{aligned}$$

Back-substitute 3 for y to find x.

$$\begin{aligned} -x - 4y &= -13 \\ -x - 4(3) &= -13 \\ -x - 12 &= -13 \\ -x &= -1 \\ x &= 1 \end{aligned}$$

Back-substitute 1 for x and 3 for y to find z.

$$\begin{aligned} x + y + z &= 6 \\ 1 + 3 + z &= 6 \\ 4 + z &= 6 \\ z &= 2 \end{aligned}$$

The solution set is $\{(1, 3, 2)\}$.

14. $\left[\begin{array}{ccc|c} 1 & 0 & -4 & 5 \\ 6 & -1 & 2 & 10 \\ 2 & -1 & 4 & -3 \end{array}\right] \quad -6R_1 + R_2$

$= \left[\begin{array}{ccc|c} 1 & 0 & -4 & 5 \\ 0 & -1 & 26 & -20 \\ 2 & -1 & 4 & -3 \end{array}\right]$

15. $\left[\begin{array}{cc|c} 2 & 1 & 6 \\ 3 & -2 & 16 \end{array}\right] \quad \frac{1}{2}R_1$

$=\left[\begin{array}{cc|c} 1 & 1/2 & 3 \\ 3 & -2 & 16 \end{array}\right] \quad -3R_1+R_2$

$=\left[\begin{array}{cc|c} 1 & 1/2 & 3 \\ 0 & -7/2 & 7 \end{array}\right] \quad -\frac{2}{7}R_2$

$=\left[\begin{array}{cc|c} 1 & 1/2 & 3 \\ 0 & 1 & -2 \end{array}\right]$

The resulting system is

$$x+\frac{1}{2}y=3$$
$$y=-2$$

Back-substitute –2 for y to find x:

$$x+\frac{1}{2}y=3$$
$$x+\frac{1}{2}(-2)=3$$
$$x-1=3$$
$$x=4$$

The solution set is $\{(4,-2)\}$.

16. $\left[\begin{array}{ccc|c} 1 & -4 & 4 & -1 \\ 2 & -1 & 5 & 6 \\ -1 & 3 & -1 & 5 \end{array}\right] \quad -2R_1+R_2$

$=\left[\begin{array}{ccc|c} 1 & -4 & 4 & -1 \\ 0 & 7 & -3 & 8 \\ -1 & 3 & -1 & 5 \end{array}\right] \quad \frac{1}{7}R_2$

$=\left[\begin{array}{ccc|c} 1 & -4 & 4 & -1 \\ 0 & 1 & -3/7 & 8/7 \\ -1 & 3 & -1 & 5 \end{array}\right] \quad R_1+R_3$

$=\left[\begin{array}{ccc|c} 1 & -4 & 4 & -1 \\ 0 & 1 & -3/7 & 8/7 \\ 0 & -1 & 3 & 4 \end{array}\right] \quad R_2+R_3$

$=\left[\begin{array}{ccc|c} 1 & -4 & 4 & -1 \\ 0 & 1 & -3/7 & 8/7 \\ 0 & 0 & 18/7 & 36/7 \end{array}\right] \quad \frac{7}{18}R_3$

$=\left[\begin{array}{ccc|c} 1 & -4 & 4 & -1 \\ 0 & 1 & -3/7 & 8/7 \\ 0 & 0 & 1 & 2 \end{array}\right]$

The resulting system is

$$\begin{aligned} x-4y+\ 4z &= -1 \\ y-\frac{3}{7}z &= \frac{8}{7} \\ z &= 2 \end{aligned}$$

Back-substitute 2 for z to find y.

$$\begin{aligned} y-\frac{3}{7}(2) &= \frac{8}{7} \\ y-\frac{6}{7} &= \frac{8}{7} \\ y &= \frac{14}{7} \\ y &= 2 \end{aligned}$$

Back-substitute 2 for y and 2 for z to find x.

$$\begin{aligned} x-4y+4z &= -1 \\ x-4(2)+4(2) &= -1 \\ x-8+8 &= -1 \\ x &= -1 \end{aligned}$$

The solution set is $\{(-1,2,2)\}$.

17. $\begin{vmatrix} -1 & -3 \\ 7 & 4 \end{vmatrix} = -1(4)-7(-3) = -4+21 = 17$

18. $\begin{vmatrix} 3 & 4 & 0 \\ -1 & 0 & -3 \\ 4 & 2 & 5 \end{vmatrix} = 3\begin{vmatrix} 0 & -3 \\ 2 & 5 \end{vmatrix} - (-1)\begin{vmatrix} 4 & 0 \\ 2 & 5 \end{vmatrix} + 4\begin{vmatrix} 4 & 0 \\ 0 & -3 \end{vmatrix}$

$$\begin{aligned} &= 3\left(\cancel{0(5)}-2(-3)\right)-(-1)\left(4(5)-\cancel{2(0)}\right)+4\left(4(-3)-\cancel{0(0)}\right) \\ &= 3(6)+1(20)+4(-12) \\ &= -10 \end{aligned}$$

19. $D = \begin{vmatrix} 4 & -3 \\ 3 & -1 \end{vmatrix} = 4(-1)-3(-3) = -4+9 = 5$

$D_x = \begin{vmatrix} 14 & -3 \\ 3 & -1 \end{vmatrix} = 14(-1)-3(-3) = -14+9 = -5$

$D_y = \begin{vmatrix} 4 & 14 \\ 3 & 3 \end{vmatrix} = 4(3)-3(14) = 12-42 = -30$

$x = \frac{D_x}{D} = \frac{-5}{5} = -1 \qquad y = \frac{D_y}{D} = \frac{-30}{5} = -6$

The solution set is $\{(-1,-6)\}$.

20. $D = \begin{vmatrix} 2 & 3 & 1 \\ 3 & 3 & -1 \\ 1 & -2 & -3 \end{vmatrix}$

$$= 2\begin{vmatrix} 3 & -1 \\ -2 & -3 \end{vmatrix} - 3\begin{vmatrix} 3 & 1 \\ -2 & -3 \end{vmatrix} + 1\begin{vmatrix} 3 & 1 \\ 3 & -1 \end{vmatrix}$$
$$= 2\big(3(-3)-(-2)(-1)\big) - 3\big(3(-3)-(-2)1\big) + 1\big(3(-1)-3(1)\big)$$
$$= 2(-11) - 3(-7) + 1(-6)$$
$$= -7$$

$D_x = \begin{vmatrix} 2 & 3 & 1 \\ 0 & 3 & -1 \\ 1 & -2 & -3 \end{vmatrix}$

$$= 2\begin{vmatrix} 3 & -1 \\ -2 & -3 \end{vmatrix} - \cancel{0\begin{vmatrix} 3 & 1 \\ -2 & -3 \end{vmatrix}} + 1\begin{vmatrix} 3 & 1 \\ 3 & -1 \end{vmatrix}$$
$$= 2\big(3(-3)-(-2)(-1)\big) + 1\big(3(-1)-3(1)\big)$$
$$= 2(-11) + 1(-6)$$
$$= -28$$

$D_y = \begin{vmatrix} 2 & 2 & 1 \\ 3 & 0 & -1 \\ 1 & 1 & -3 \end{vmatrix}$

$$= 2\begin{vmatrix} 0 & -1 \\ 1 & -3 \end{vmatrix} - 3\begin{vmatrix} 2 & 1 \\ 1 & -3 \end{vmatrix} + 1\begin{vmatrix} 2 & 1 \\ 0 & -1 \end{vmatrix}$$
$$= 2\big(\cancel{0(-3)} - 1(-1)\big) - 3\big(2(-3)-1(1)\big) + 1\big(2(-1) - \cancel{0(1)}\big)$$
$$= 2(1) - 3(-7) + 1(-2)$$
$$= 21$$

$D_z = \begin{vmatrix} 2 & 3 & 2 \\ 3 & 3 & 0 \\ 1 & -2 & 1 \end{vmatrix}$

$$= 2\begin{vmatrix} 3 & 0 \\ -2 & 1 \end{vmatrix} - 3\begin{vmatrix} 3 & 2 \\ -2 & 1 \end{vmatrix} + 1\begin{vmatrix} 3 & 2 \\ 3 & 0 \end{vmatrix}$$
$$= 2\big(3(1) - \cancel{(-2)(0)}\big) - 3\big(3(1)-(-2)2\big) + 1\big(\cancel{3(0)} - 3(2)\big)$$
$$= 2(3) - 3(7) + 1(-6)$$
$$= -21$$

$x = \dfrac{D_x}{D} = \dfrac{-28}{-7} = 4 \qquad y = \dfrac{D_y}{D} = \dfrac{21}{-7} = -3 \qquad z = \dfrac{D_z}{D} = \dfrac{-21}{-7} = 3$

The solution set is $\{(4, -3, 3)\}$.

Cumulative Review Exercises (Chapters 1 – 3)

1. $\frac{6(8-10)^3+(-2)}{(-5)^2(-2)}=\frac{6(-2)^3-2}{(25)(-2)}$
$=\frac{6(-8)-2}{-50}$
$=\frac{-50}{-50}$
$=1$

2. $7x-[5-2(4x-1)]=7x-[5-8x+2]$
$=7x-[7-8x]$
$=7x-7+8x$
$=15x-7$

3. $5-2(3-x)=2(2x+5)+1$
$5-6+2x=4x+10+1$
$-1+2x=4x+11$
$-1-2x=11$
$-2x=12$
$x=-6$
The solution set is $\{-6\}$.

4. $\frac{3x}{5}+4=\frac{x}{3}$
$15\left(\frac{3x}{5}+4\right)=15\left(\frac{x}{3}\right)$
$9x+60=5x$
$4x=-60$
$x=-15$
The solution set is $\{-15\}$.

5. $3x-4=2(3x+2)-3x$
$3x-4=6x+4-3x$
$3x-4=3x+4$
$-4=4$
This is a contradiction. There is no value of x for which –4 equals 4. The solution set is $\varnothing$ or $\{\ \}$.

6. Let x = the amount of the sales.
$200+0.05x=0.15x$
$200=0.10x$
$2000=x$

For sales of $2000, the earnings will be the same under either pay arrangement.

7. $\frac{-5x^6y^{-10}}{20x^{-2}y^{20}}=\frac{-x^6x^2}{4y^{20}y^{10}}=-\frac{x^8}{4y^{30}}$

8. $f(x)=-4x+5$
$f(a+2)=-4(a+2)+5$
$=-4a-8+5$
$=-4a-3$

9. To find the domain, set the denominator equal to zero.
$x+3=0$
$x=-3$
The domain is $\{x|x \text{ is a real number and } x\neq-3\}$.

10. $(f-g)(x)=f(x)-g(x)$
$=2x^2-5x+2-(x^2-2x+3)$
$=2x^2-5x+2-x^2+2x-3$
$=x^2-3x-1$

$(f-g)(3)=(3)^2-3(3)-1$
$=9-9-1$
$=-1$

11. $f(x)=-\frac{2}{3}x+2$
$y=-\frac{2}{3}x+2$
y – intercept = 2
$m=-\frac{2}{3}$

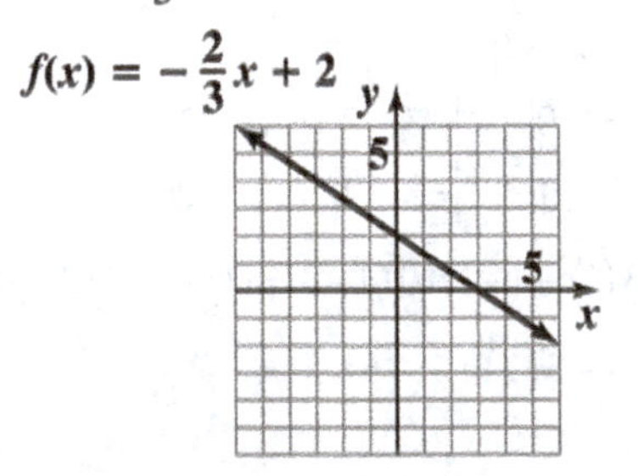

12. $2x - y = 6$
$$-y = -2x + 6$$
$$y = 2x - 6$$
y – intercept = –6
m = 2

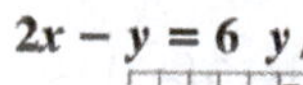

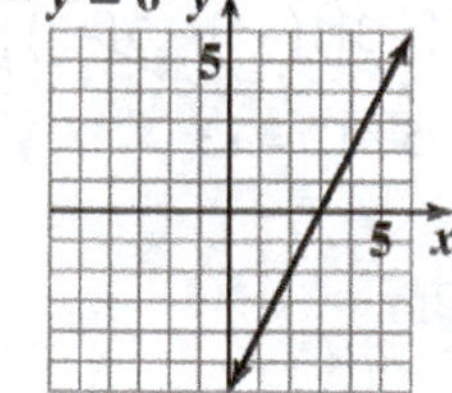

13. First, find the slope.
$$m = \frac{4-(-2)}{2-4} = \frac{4+2}{-2} = \frac{6}{-2} = -3$$

Use the slope and one of the points to write the line in point-slope form.
$$y - y_1 = m(x - x_1)$$
$$y - 4 = -3(x - 2)$$

Solve for y to obtain slope-intercept form.
$$y - 4 = -3(x - 2)$$
$$y - 4 = -3x + 6$$
$$y = -3x + 10$$

14. Parallel lines have the same slope. Find the slope of the given line.
$$3x + y = 6$$
$$y = -3x + 6$$

We obtain, $m = -3$.

Use the slope and the point to put the line in point-slope form.
$$y - y_1 = m(x - x_1)$$
$$y - 0 = -3(x - (-1))$$
$$y - 0 = -3(x + 1)$$

Solve for y to obtain slope-intercept form.
$$y - 0 = -3(x + 1)$$
$$y = -3x - 3$$

15. $3x + 12y = 25$
$2x - 6y = 12$

Multiply the second equation by 2 and solve by addition.
$$\begin{array}{r} 3x + 12y = 25 \\ \underline{4x - 12y = 24} \\ 7x = 49 \\ x = 7 \end{array}$$

Back-substitute 7 for x to find y.
$$2x - 6y = 12$$
$$2(7) - 6y = 12$$
$$14 - 6y = 12$$
$$-6y = -2$$
$$y = \frac{-2}{-6}$$
$$y = \frac{1}{3}$$

The solution set is $\left\{\left(7, \frac{1}{3}\right)\right\}$.

16. $x + 3y - z = 5$
$-x + 2y + 3z = 13$
$2x - 5y - z = -8$

Add the first and second equations to eliminate x.
$$\begin{array}{r} x + 3y - z = 5 \\ \underline{-x + 2y + 3z = 13} \\ 5y + 2z = 18 \end{array}$$

Multiply the first equation by –2 and add to the third equation.
$$\begin{array}{r} -2x - 6y + 2z = -10 \\ \underline{2x - 5y - z = -8} \\ -11y + z = -18 \end{array}$$

The system of two equations in two variables becomes:
$$5y + 2z = 18$$
$$-11y + z = -18$$

Multiply the second equation by –2 and solve by addition.

$$\begin{aligned} 5y+2z&=18\\ 22y-2z&=36\\ \hline 27y&=54\\ y&=2 \end{aligned}$$

Back-substitute 2 for y to find z.

$$\begin{aligned} 5y+2z&=18\\ 5(2)+2z&=18\\ 10+2z&=18\\ 2z&=8\\ z&=4 \end{aligned}$$

Back-substitute 2 for y and 4 for z in one of the equations in three variables.

$$\begin{aligned} x+3y-z&=5\\ x+3(2)-4&=5\\ x+6-4&=5\\ x+2&=5\\ x&=3 \end{aligned}$$

The solution set is $\{(3,2,4)\}$.

17. Let x = the number of pads.
Let y = the number of pens.

$$\begin{aligned} 2x+19y&=5.40\\ 7x+4y&=6.40 \end{aligned}$$

Multiply the first equation by 7 and the second equation by –2 and solve by addition.

$$\begin{aligned} 14x+133y&=37.80\\ -14x-8y&=-12.80\\ \hline 125y&=25.00\\ y&=0.20 \end{aligned}$$

Back-substitute .20 for y to find x.

$$\begin{aligned} 7x+4y&=6.40\\ 7x+4(.20)&=6.40\\ 7x+.80&=6.40\\ 7x&=5.60\\ x&=0.80 \end{aligned}$$

Pads cost \$0.80 each and pens cost \$0.20 each.

18. $\begin{vmatrix} 0 & 1 & -2\\ -7 & 0 & -4\\ 3 & 0 & 5 \end{vmatrix}$

$$= 0\begin{vmatrix} 0 & -4\\ 0 & 5\end{vmatrix} - (-7)\begin{vmatrix} 1 & -2\\ 0 & 5\end{vmatrix} + 3\begin{vmatrix} 1 & -2\\ 0 & -4\end{vmatrix}$$

$$= 7\left(1(5)-0(-2)\right)+3\left(1(-4)-0(-2)\right)$$

$$= 7(5)+3(-4)=35-12=23$$

19. $\left[\begin{array}{ccc|c} 2 & 3 & -1 & -1\\ 1 & 2 & 3 & 2\\ 3 & 5 & -2 & -3 \end{array}\right]$ $R_1 \leftrightarrow R_2$

$$= \left[\begin{array}{ccc|c} 1 & 2 & 3 & 2\\ 2 & 3 & -1 & -1\\ 3 & 5 & -2 & -3 \end{array}\right] \begin{array}{l} \\ R_2-2R_1\\ R_3-3R_1 \end{array}$$

$$= \left[\begin{array}{ccc|c} 1 & 2 & 3 & 2\\ 0 & -1 & -7 & -5\\ 0 & -1 & -11 & -9 \end{array}\right] \begin{array}{l} \\ -R_2\\ \ \end{array}$$

$$= \left[\begin{array}{ccc|c} 1 & 2 & 3 & 2\\ 0 & 1 & 7 & 5\\ 0 & -1 & -11 & -9 \end{array}\right] \begin{array}{l} \\ \\ R_3+R_2+ \end{array}$$

$$= \left[\begin{array}{ccc|c} 1 & 2 & 3 & 2\\ 0 & 1 & 7 & 5\\ 0 & 0 & -4 & -4 \end{array}\right] \begin{array}{l} \\ \\ -\tfrac{1}{4}R_3 \end{array}$$

$$= \left[\begin{array}{ccc|c} 1 & 2 & 3 & 2\\ 0 & 1 & 7 & 5\\ 0 & 0 & 1 & 1 \end{array}\right]$$

The resulting system is

$$\begin{aligned} x+2y+3z&=2\\ y+7z&=5\\ z&=1 \end{aligned}$$

Back-substitute 1 for z to find y.

$$\begin{aligned} y+7z&=5\\ y+7(1)&=5\\ y+7&=5\\ y&=-2 \end{aligned}$$

Back-substitute –2 for y and 1 for z to find x.

$$\begin{aligned} x+2y+3z&=2 \\ x+2(-2)+3(1)&=2 \\ x-4+3&=2 \\ x-1&=2 \\ x&=3 \end{aligned}$$

The solution set is $\{(3,-2,1)\}$.

20. $D=\begin{vmatrix} 3 & 4 \\ -2 & 1 \end{vmatrix}=3(1)-(-2)4$

$=3+8$

$=11$

$D_x=\begin{vmatrix} -1 & 4 \\ 8 & 1 \end{vmatrix}=-1(1)-8(4)$

$=-1-32$

$=-33$

$D_y=\begin{vmatrix} 3 & -1 \\ -2 & 8 \end{vmatrix}=3(8)-(-2)(-1)$

$=24-2$

$=22$

$$x=\frac{D_x}{D}=\frac{-33}{11}=-3$$

$$y=\frac{D_y}{D}=\frac{22}{11}=2$$

The solution set is $\{(-3,2)\}$.

Chapter 4
Inequalities and Problem Solving

4.1 Check Points

1. $4x-3>-23$

$4x>-20$

$x>-5$

$(-5,\infty)$

−8 −7 −6 −5 −4 −3 −2 −1 0 1 2

2. $3x+1>7x-15$

$-4x>-16$

$\frac{-4x}{-4}<\frac{-16}{-4}$

$x<4$

$(-\infty,4)$

−3 −2 −1 0 1 2 3 4 5 6

3. $\frac{x-4}{2}\geq\frac{x-2}{3}+\frac{5}{6}$

$6\left(\frac{x-4}{2}\right)\geq 6\left(\frac{x-2}{3}+\frac{5}{6}\right)$

$3(x-4)\geq 2(x-2)+5$

$3x-12\geq 2x-4+5$

$3x-12\geq 2x+1$

$x\geq 13$

$[13,\infty)$

6 7 8 9 10 11 12 13 14 15 16

4. a. $3(x+1)>3x+2$

$3x+3>3x+2$

$3>2$

This expression is always true.

$\mathbb{R}$ or $(-\infty,\infty)$

b. $x+1\leq x-1$

$1\leq -1$

This expression is always false.

$\varnothing$

5. Let $x=$ number of miles driven in a week.

$$\underbrace{260}_{\text{Cost for Basic Rental}} < \underbrace{80+0.25x}_{\text{Cost for Continental}}$$

$260<80+0.25x$

$180<0.25x$

$\frac{180}{0.25}<\frac{0.25x}{0.25}$

$720<x$

$x>720$

Driving more than 720 miles per week makes Basic Rental a better deal.

4.1 Concept and Vocabulary Check

1. $<b+c$

2. $<bc$

3. $>bc$

4. adding 4; dividing; -3; direction; $>$; $<$

5. $\varnothing$

6. $(-\infty,\infty)$

7. $x\geq 7$

8. $x\leq 7$

9. $x\leq 7$

10. $x\geq 7$

4.1 Exercise Set

1. $5x+11<26$

$5x<15$

$x<3$

The solution set is $(-\infty,3)$.

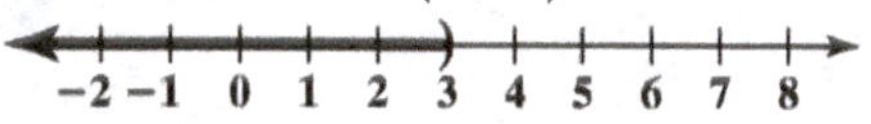

3. $3x-8\geq 13$

$3x\geq 21$

$x\geq 7$

The solution set is $[7,\infty)$.

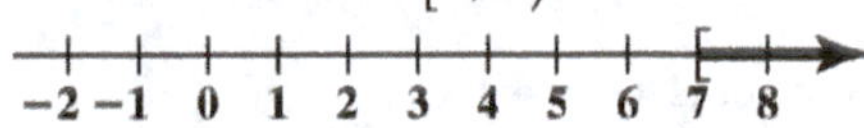

5. $-9x\geq 36$

$x\leq -4$

The solution set is $(-\infty,-4]$.

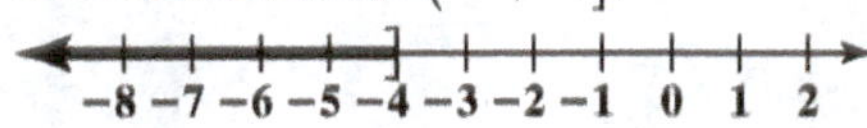

7. $8x-11\leq 3x-13$

$5x-11\leq -13$

$5x\leq -2$

$x\leq -\frac{2}{5}$

The solution set is $\left(-\infty,-\frac{2}{5}\right]$.

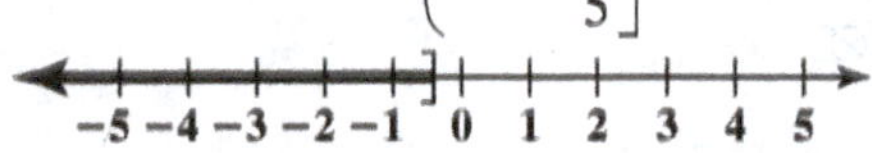

9. $4(x+1)+2\geq 3x+6$

$4x+4+2\geq 3x+6$

$4x+6\geq 3x+6$

$x+6\geq 6$

$x\geq 0$

The solution set is $[0,\infty)$.

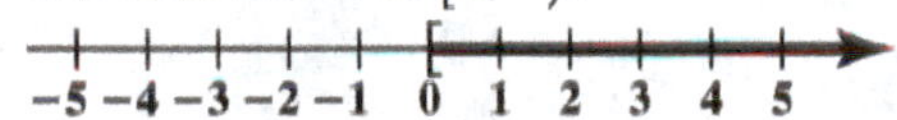

11. $2x-11<-3(x+2)$

$2x-11<-3x-6$

$5x-11<-6$

$5x<5$

$x<1$

The solution set is $(-\infty,1)$.

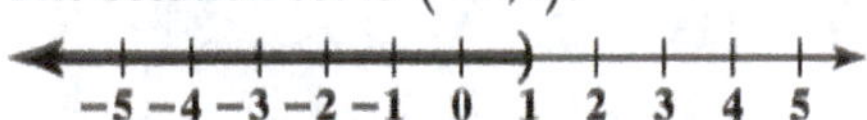

13. $1-(x+3)\geq 4-2x$

$1-x-3\geq 4-2x$

$-x-2\geq 4-2x$

$x-2\geq 4$

$x\geq 6$

The solution set is $[6,\infty)$.

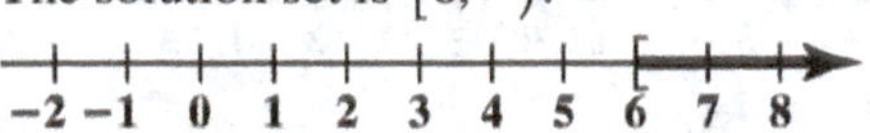

15. $\frac{x}{4}-\frac{1}{2}\leq\frac{x}{2}+1$

$4\left(\frac{x}{4}\right)-4\left(\frac{1}{2}\right)\leq 4\left(\frac{x}{2}\right)+4(1)$

$x-2\leq 2x+4$

$-x-2\leq 4$

$-x\leq 6$

$x\geq -6$

The solution set is $[-6,\infty)$.

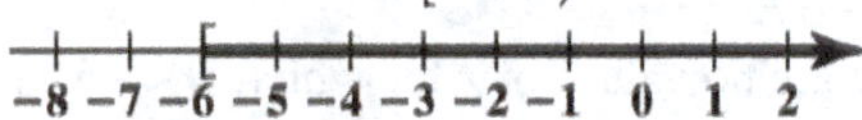

17. $1-\frac{x}{2}>4$

$2(1)-2\left(\frac{x}{2}\right)>2(4)$

$2-x>8$

$-x>6$

$x<-6$

The solution set is $(-\infty,-6)$.

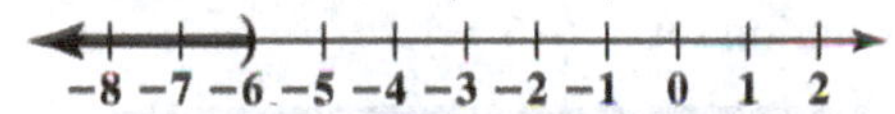

19. $\frac{x-4}{6}\geq\frac{x-2}{9}+\frac{5}{18}$

$18\left(\frac{x-4}{6}\right)\geq 18\left(\frac{x-2}{9}+\frac{5}{18}\right)$

$3(x-4)\geq 2(x-2)+5$

$3x-12\geq 2x-4+5$

$3x-12\geq 2x+1$

$x\geq 13$

The solution set is $[13,\infty)$.

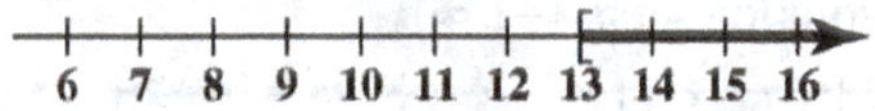

21. $4(3x-2)-3x<3(1+3x)-7$

$12x-8-3x<3+9x-7$

$9x-8<9x-4$

$-8<-4$

Regardless of the value of x, $-8<-4$ is always true. Therefore, the solution set is $(-\infty, \infty)$.

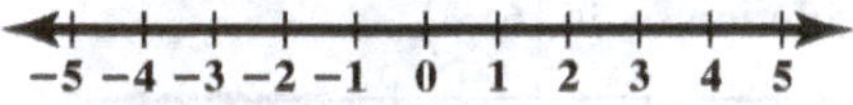

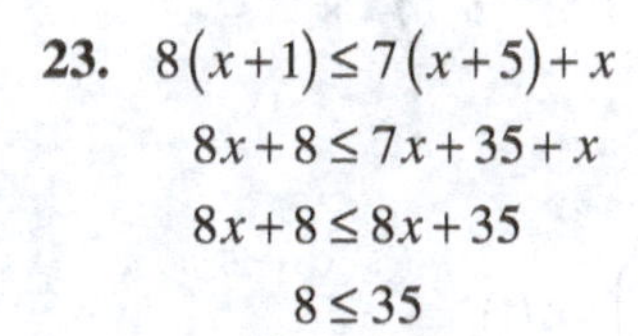

23. $8(x+1)\le 7(x+5)+x$

$8x+8\le 7x+35+x$

$8x+8\le 8x+35$

$8\le 35$

Regardless of the value of x, $8\le 35$ is always true. Therefore, the solution set is $(-\infty, \infty)$.

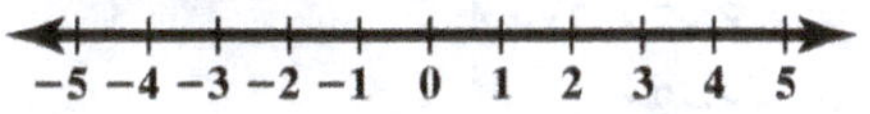

25. $3x<3(x-2)$

$3x<3x-6$

$0<-6$

There are no values of x for which $0<-6$. Therefore, the solution set is $\varnothing$.

27. $7(x+4)-13<12+13(3+x)$

$7x+28-13<12+39+13x$

$7x+15<13x+51$

$-6x+15<51$

$-6x<36$

$\frac{-6x}{-6}>\frac{36}{-6}$

$x>-6$

The solution set is $(-6, \infty)$.

29. $6-\frac{2}{3}(3x-12)\le\frac{2}{5}(10x+50)$

$6-2x+8\le 4x+20$

$-2x+14\le 4x+20$

$-6x\le 6$

$\frac{-6x}{-6}\ge\frac{6}{-6}$

$x\ge -1$

The solution set is $[-1, \infty)$.

31. $3[3(x+5)+8x+7]+5[3(x-6)-2(3x-5)]<2(4x+3)$

$$3[3x+15+8x+7]+5[3x-18-6x+10]<8x+6$$
$$3[11x+22]+5[-3x-8]<8x+6$$
$$33x+66-15x-40<8x+6$$
$$18x+26<8x+6$$
$$10x+26<6$$
$$10x<-20$$
$$x<-2$$

The solution set is $(-\infty,-2)$.

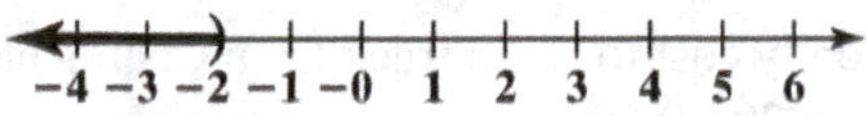

33. $f(x)>g(x)$

$$3x+2>5x-8$$
$$-2x+2>-8$$
$$-2x>-10$$
$$\frac{-2x}{-2}<\frac{-10}{-2}$$
$$x<5$$

The solution set is $(-\infty,5)$.

35. $f(x)\le g(x)$

$$\frac{1}{4}(8-12x)\le\frac{2}{5}(10x+15)$$
$$2-3x\le 4x+6$$
$$2-7x\le 6$$
$$-7x\le 4$$
$$\frac{-7x}{-7}\ge\frac{4}{-7}$$
$$x\ge-\frac{4}{7}$$

The solution set is $\left[-\frac{4}{7},\infty\right)$.

37. $1-(x+3)+2x\ge 4$

$$1-x-3+2x\ge 4$$
$$-2+x\ge 4$$
$$x\ge 6$$

The solution set is $[6,\infty)$.

39. $2(x+3)>6-\{4[x-(3x-4)-x]+4\}$

$2x+6>6-\{4[x-3x+4-x]+4\}$

$2x+6>6-\{4[-3x+4]+4\}$

$2x+6>6-\{-12x+16+4\}$

$2x+6>6-\{-12x+20\}$

$2x+6>6+12x-20$

$2x+6>12x-14$

$6>10x-14$

$20>10x$

$2>x$

$x<2$

The solution set is $(-\infty, 2)$.

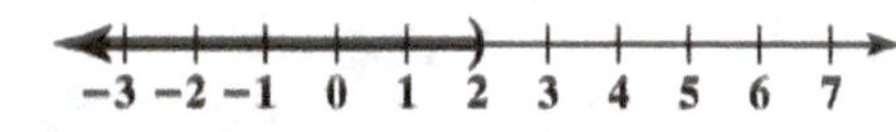

41. $ax+b>c, a<0$

$ax+b-b>c-b$

$ax>c-b$

$\frac{ax}{a}<\frac{c-b}{a}, a<0$

$x<\frac{c-b}{a}$

43. $(-\infty,-3]$

45. $(-1.4,\infty)$

47. $(0,4)$

49. passion ≤ intimacy or intimacy ≥ passion

51. passion < commitment or commitment > passion

53. 9, after 3 years

55. $3.1x+25.8>63$

$3.1x>37.2$

$x>12$

Since x is the number of years after 1994, we calculate 1994+12=2006. 63% of voters will use electronic systems after 2006.

57. $W<M$

$-0.19t+57<-0.15t+50$

$-0.04t+57<50$

$-0.04t<-7$

$t>175$

Interval notation: $(175,\infty)$

The women's winning time will be less than the men's winning time after 1900 + 175 = 2075.

59. The daily rental cost to rent a truck from Basic Rental is $C_B=50+0.20x$.

The daily cost to rent a truck from Continental is $C_C=20+0.50x$.

$C_B<C_C$

$50+0.20x<20+0.50x$

$50-0.30x<20$

$-0.30x<-30$

$x>100$

Basic Rental is a better deal when the truck is driven more than 100 miles in a day.

61. The tax bill assessed under the first tax bill is $T_1=1800+0.03x$.

The tax bill assessed under the second tax bill is $T_2=200+0.08x$.

$T_1<T_2$

$1800+0.03x<200+0.08x$

$1800-0.05x<200$

$-0.05x<-1600$

$x>32000$

The first tax bill is a better deal when the assessed value of the home is greater than $32,000.

63. The cost is $C=10,000+0.40x$.

The revenue is $R=2x$.

$C<R$

$10,000+0.40x<2x$

$10,000-1.6x<0$

$-1.6x<-10,000$

$x>6250$

The company will make a profit gain when more than 6,250 tapes are produced and sold each week.

65. Let x = the number of bags of cement that can be lifted safely in the elevator.

$$W_{\text{operator}} + W_{\text{cement}} \le 3000$$
$$200 + 70x \le 3000$$
$$70x \le 2800$$
$$x \le 40$$

At most, 40 bags of cement can be lifted safely on the elevator per trip.

67. – 73. Answers will vary.

75. $-2(x+4) > 6x+16$

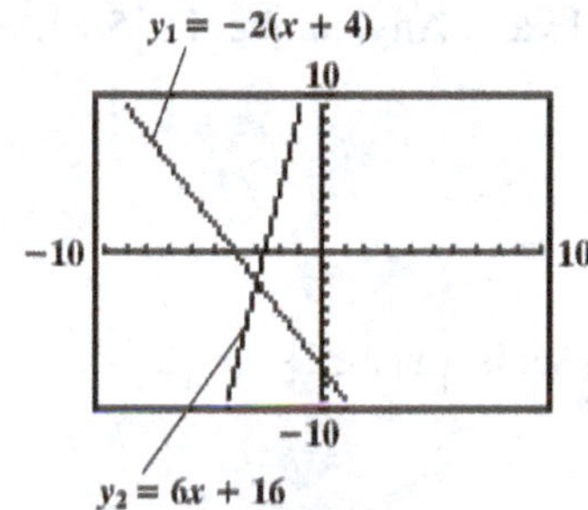

Moving from left to right on the graphing calculator screen, we see that the graph of $-2(x+4)$ is above the graph of $6x+16$ from $-\infty$ to -3. The solution set is $(-\infty, -3)$.

77. Graph $y = 12x - 10$ and $y = 2(x-4) + 10x$.

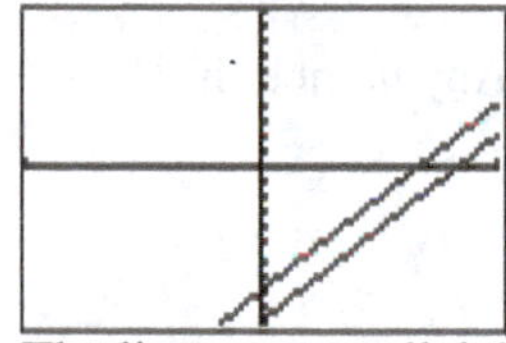

The lines are parallel. They do not intersect. There is no solution.
When solving the inequality algebraically, we arrive at a *false* statement:

$$12x - 10 > 2(x-4) + 10x$$
$$12x - 10 > 2x - 8 + 10x$$
$$12x - 10 > 12x - 8$$
$$-10 > -8 \text{ false}$$

There are no x-values that solve the inequality. The solution set is $\varnothing$.

79. a. Plan A: $4 + 0.10x$
Plan B: $2 + 0.15x$

b. Window: [0,50,1] by [0,10,1]

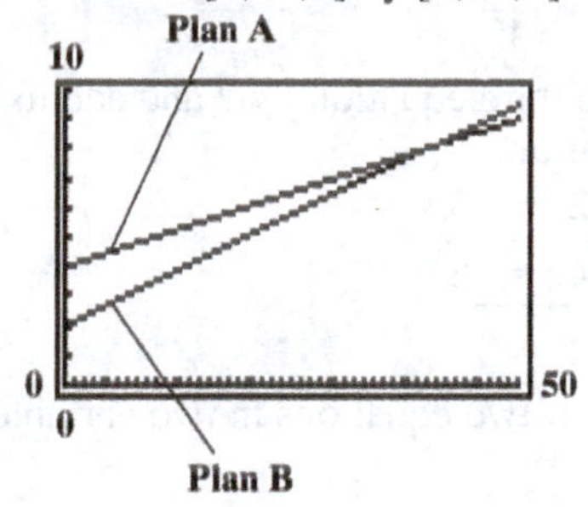

c. Plan A is better than Plan B for more than 40 checks per month.

d.
$$A < B$$
$$4 + 0.10x < 2 + 0.15x$$
$$4 < 2 + 0.05x$$
$$0.05x + 2 > 4$$
$$0.05x > 2$$
$$x > 40$$

81. makes sense

83. makes sense

85. false; Changes to make the statement true will vary. A sample change is: 4 is not the smallest real number in the solution set of $2x > 6$. For example, 3.1 is a smaller real number in the solution set.

87. true

89. Since $x > y$, then $y - x < 0$. When multiplying both sides of the inequality by $(y-x)$, remember to *flip* the inequality.

$$2 > 1$$
$$2(y-x) < 1(y-x)$$
$$2y - 2x < y - x$$
$$y - 2x < -x$$
$$y < x$$

90. $f(-4) = (-4)^2 - 2(-4) + 5 = 16 + 8 + 5 = 29$

91. Add the first and third equations to eliminate y.

$$\begin{array}{r} 2x - y - z = -3 \\ -x + y + 2z = 4 \\ \hline x + z = 1 \end{array}$$

Multiply the third equation by 2 and add to the second equation.

$$\begin{array}{r} 3x - 2y - 2z = -5 \\ -2x + 2y + 4z = 8 \\ \hline x + 2z = 3 \end{array}$$

The system of two equations in two variables becomes:

$x + z = 1$

$x + 2z = 3$

Multiply the second equation by –1 and solve for z.

$$\begin{array}{r} x + z = 1 \\ -x - 2z = -3 \\ \hline -z = -2 \\ z = 2 \end{array}$$

Back-substitute 2 for z to find x.

$x + z = 1$

$x + 2 = 1$

$x = -1$

Back-substitute 2 for z and –1 and x in one of the original equations in three variables to find y.

$2x - y - z = -3$

$2(-1) - y - 2 = -3$

$-2 - y - 2 = -3$

$-y - 4 = -3$

$-y = 1$

$y = -1$

The solution is $(-1, -1, 2)$.

The solution set is $\{(-1, -1, 2)\}$.

92. $$\left(\frac{2x^4y^{-2}}{4xy^3}\right)^3 = \frac{2^3x^{12}y^{-6}}{4^3x^3y^9} = \frac{8}{64}x^9y^{-15} = \frac{x^9}{8y^{15}}$$

93. a. $\{3, 4\}$

b. $\{1, 2, 3, 4, 5, 6, 7\}$

94. a. $x - 3 < 5$

$x < 8$

The solution set is $\{x | x < 8\}$ or $(-\infty, 8)$.

b. $2x + 4 < 14$

$2x < 10$

$x < 5$

The solution set is $\{x | x < 5\}$ or $(-\infty, 5)$.

c. Answers will vary. Any number less than 5.

d. Answers will vary. Any number in $[5, 8)$.

95. a. $2x - 6 \ge -4$

$2x \ge 2$

$x \ge 1$

The solution set is $\{x | x \ge 1\}$ or $[1, \infty)$.

b. $5x + 2 \ge 17$

$5x \ge 15$

$x \ge 3$

The solution set is $\{x | x \ge 3\}$ or $[3, \infty)$.

c. Answers will vary. Any number greater than or equal to 3.

d. Answers will vary. Any number in $[1, 3)$.

4.2 Check Points

1. Elements 3 and 7 are common to both sets.

$\{3, 4, 5, 6, 7\} \cap \{3, 7, 8, 9\} = \{3, 7\}$

2. Solve and graph each inequality, and graph the intersection.

$x + 2 < 5$ and $2x - 4 < -2$

$x < 3$ $2x < 2$

$x < 1$

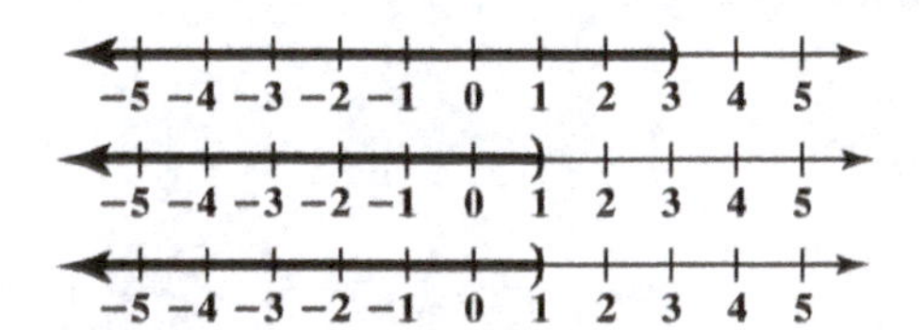

The solution set is $(-\infty, 1)$.

3. Solve and graph each inequality, and graph the intersection.

$4x-5>7$ and $5x-2<3$

$4x>12$ $\qquad 5x<5$

$x>3$ $\qquad x<1$

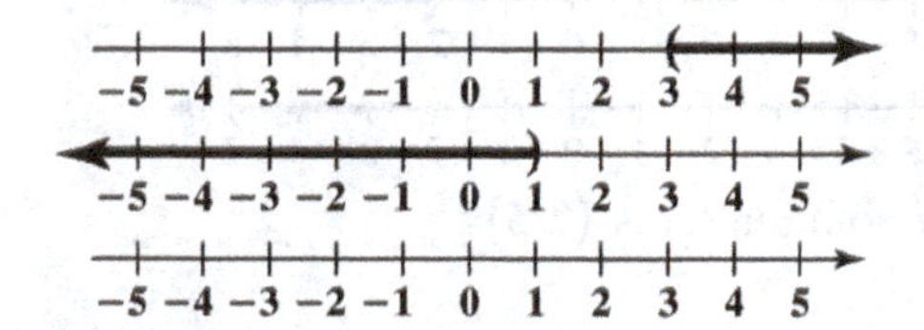

Since the two sets do not intersect, the solution set is $\varnothing$.

4.

$$1\le 2x+3<11$$
$$1-3\le 2x+3-3<11-3$$
$$-2\le 2x<8$$
$$\frac{-2}{2}\le\frac{2x}{2}<\frac{8}{2}$$
$$-1\le x<4$$

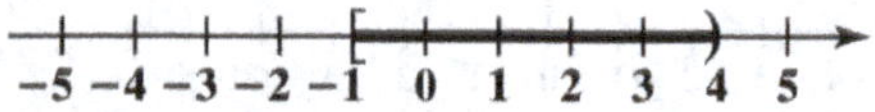

The solution set is $[-1,4)$.

5. $\{3,4,5,6,7\}\cup\{3,7,8,9\}=\{3,4,5,6,7,8,9\}$

6. Solve and graph each inequality, and graph the intersection.

$3x-5\le -2$ or $10-2x<4$

$3x\le 3$ $\qquad -2x<-6$

$x\le 1$ $\qquad \dfrac{-2x}{-2}>\dfrac{-6}{-2}$

$x>3$

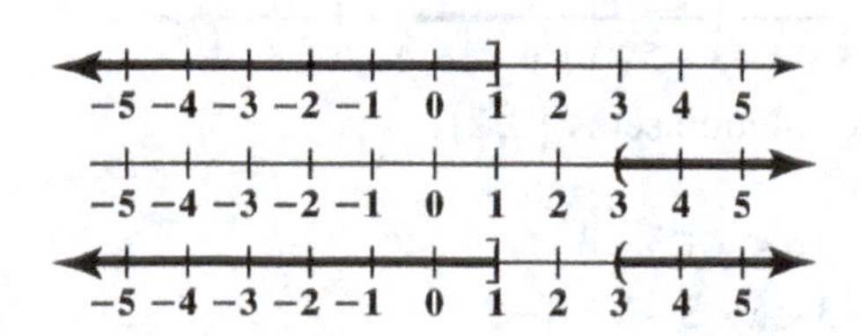

The solution set is $(-\infty,1]\cup(3,\infty)$.

7. Solve and graph each inequality, and graph the intersection.

$2x+5\ge 3$ or $2x+3<3$

$2x\ge -2$ $\qquad 2x<0$

$x\ge -1$ $\qquad x<0$

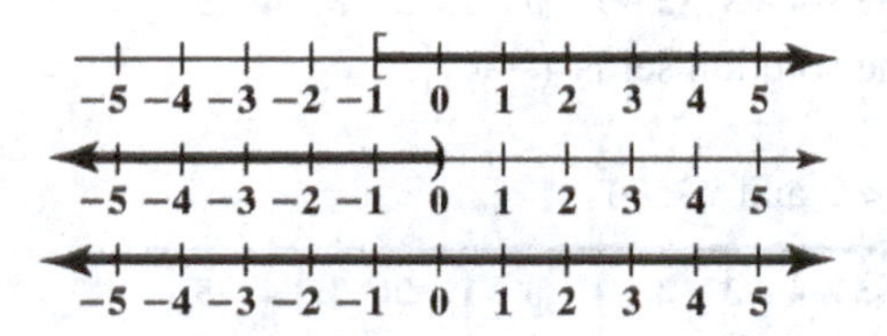

The solution set is $(-\infty,\infty)$.

4.2 Concept and Vocabulary Check

1. intersection; $A\cap B$
2. union; $A\cup B$
3. $(-\infty,9)$
4. $(-\infty,12)$
5. middle

4.2 Exercise Set

1. $\{1,2,3,4\}\cap\{2,4,5\}=\{2,4\}$

3. $\{1,3,5,7\}\cap\{2,4,6,8,10\}=\{\ \}$ or $\varnothing$

5. $\{a,b,c,d\}\cap\varnothing=\varnothing$

7. $x>3$ and $x>6$

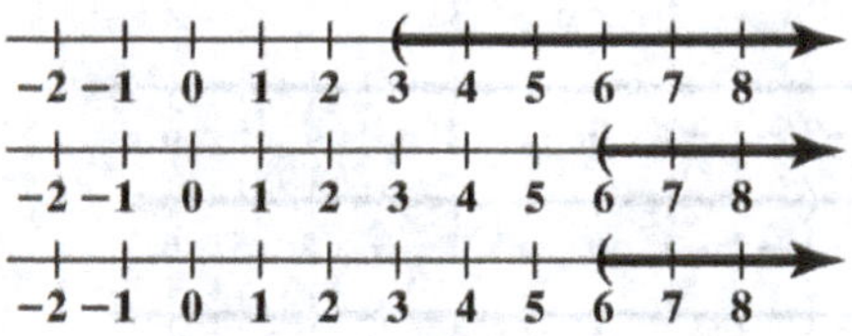

The solution set is $(6,\infty)$.

9. $x \le 5$ and $x \le 1$

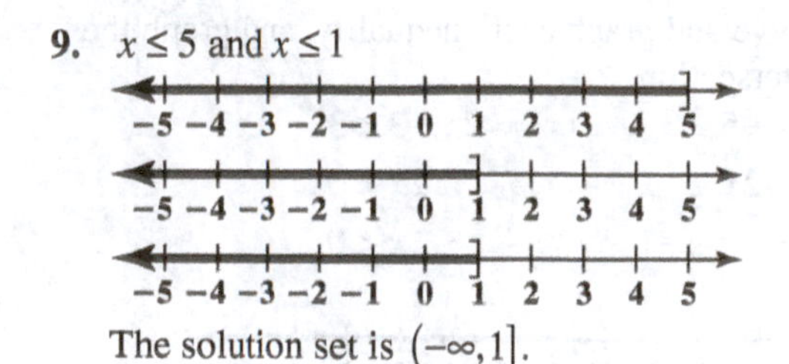

The solution set is $(-\infty, 1]$.

11. $x < 2$ and $x \ge -1$

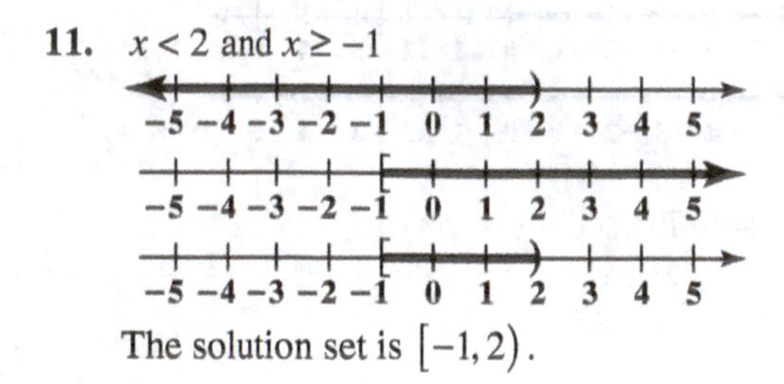

The solution set is $[-1, 2)$.

13. $x > 2$ and $x < -1$

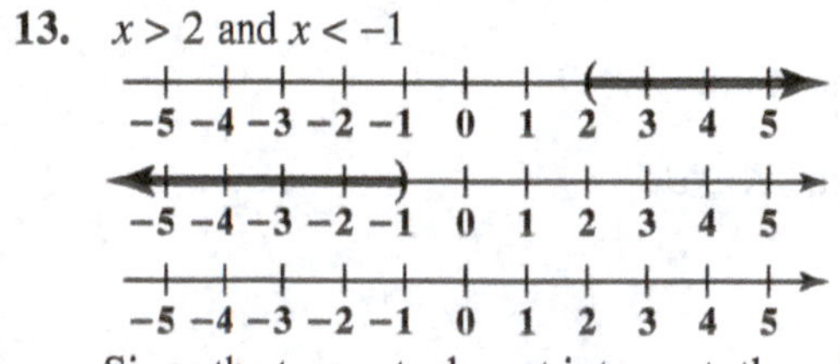

Since the two sets do not intersect, the solution set is $\varnothing$.

15. $5x < -20$ and $3x > -18$

$x < -4$ $\quad$ $x > -6$

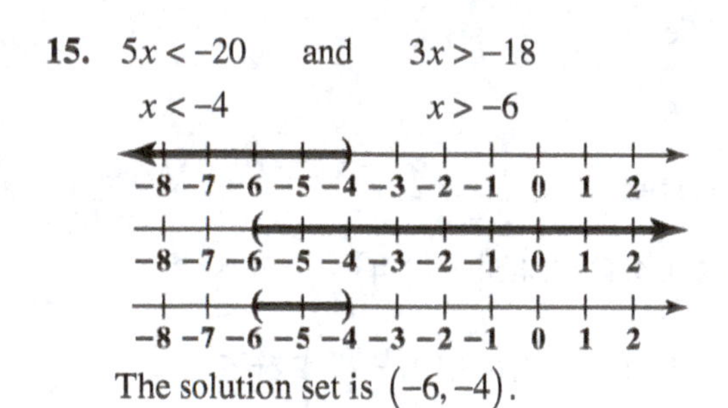

The solution set is $(-6, -4)$.

17. $x - 4 \le 2$ and $3x + 1 > -8$

$x \le 6$ $\quad$ $3x > -9$

$x > -3$

The solution set is $(-3, 6]$.

19. $2x > 5x - 15$ and $7x > 2x + 10$

$-3x > -15$ $\quad$ $5x > 10$

$x < 5$ $\quad$ $x > 2$

The solution set is $(2, 5)$.

21. $4(1 - x) < -6$ and $\frac{x-7}{5} \le -2$

$4 - 4x < -6$

$-4x < -10$ $\quad$ $5\left(\frac{x-7}{5}\right) \le 5(-2)$

$x > \frac{5}{2}$ $\quad$ $x - 7 \le -10$

$x \le -3$

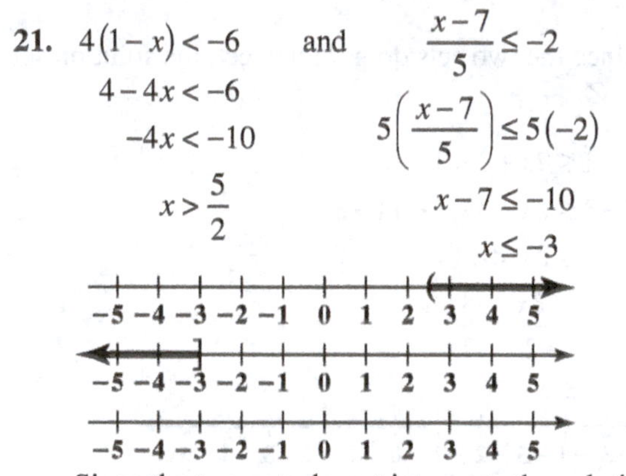

Since the two sets do not intersect, the solution set is $\varnothing$.

23. $x - 1 \le 7x - 1$ and $4x - 7 < 3 - x$

$-1 \le 6x - 1$ $\quad$ $5x - 7 < 3$

$0 \le 6x$ $\quad$ $5x < 10$

$0 \le x$ $\quad$ $x < 2$

$x \ge 0$

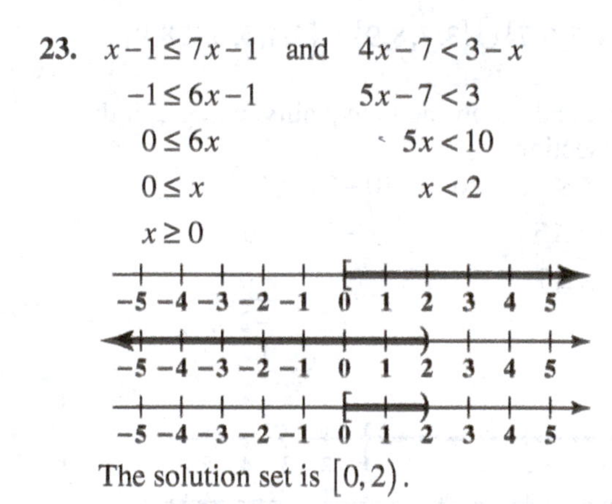

The solution set is $[0, 2)$.

25. $6 < x + 3 < 8$

$6 - 3 < x + 3 - 3 < 8 - 3$

$3 < x < 5$

The solution set is $(3, 5)$.

27.
$$-3 \le x-2 < 1$$
$$-3+2 \le x-2+2 < 1+2$$
$$-1 \le x < 3$$

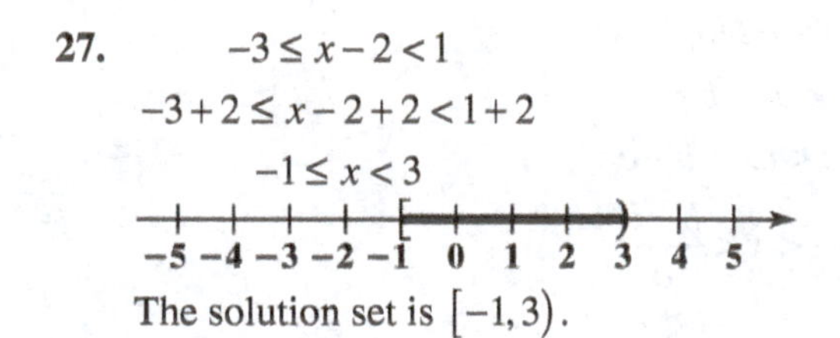

The solution set is $[-1,3)$.

29.
$$-11 < 2x-1 \le -5$$
$$-11+1 < 2x-1+1 \le -5+1$$
$$-10 < 2x \le -4$$
$$-5 < x \le -2$$

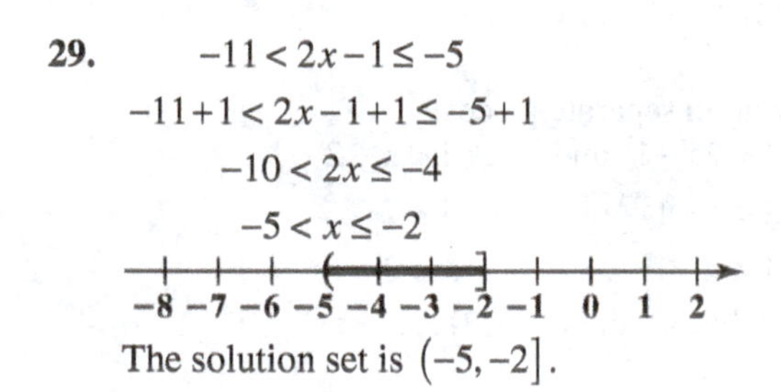

The solution set is $(-5,-2]$.

31.
$$-3 \le \frac{2x}{3} - 5 < -1$$
$$-3+5 \le \frac{2x}{3} - 5 + 5 < -1+5$$
$$2 \le \frac{2x}{3} < 4$$
$$3(2) \le 3\left(\frac{2x}{3}\right) < 3(4)$$
$$6 \le 2x < 12$$
$$3 \le x < 6$$

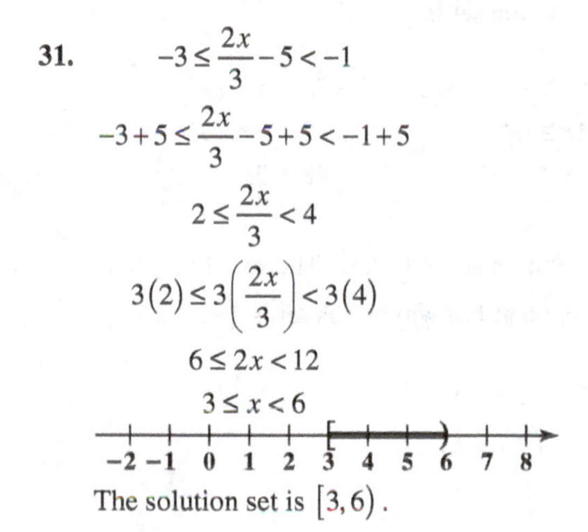

The solution set is $[3,6)$.

33. $\{1,2,3,4\} \cup \{2,4,5\} = \{1,2,3,4,5\}$

35. $\{1,3,5,7\} \cup \{2,4,6,8,10\}$
$= \{1,2,3,4,5,6,7,8,10\}$

37. $\{a,e,i,o,u\} \cup \varnothing = \{a,e,i,o,u\}$

39. $x > 3$ or $x > 6$

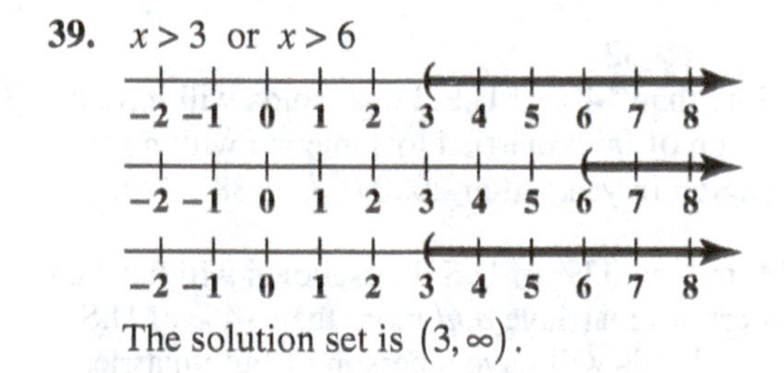

The solution set is $(3,\infty)$.

41. $x \le 5$ or $x \le 1$

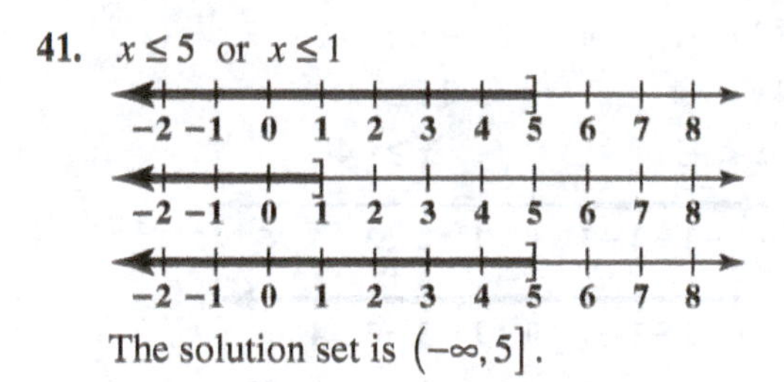

The solution set is $(-\infty,5]$.

43. $x < 2$ or $x \ge -1$

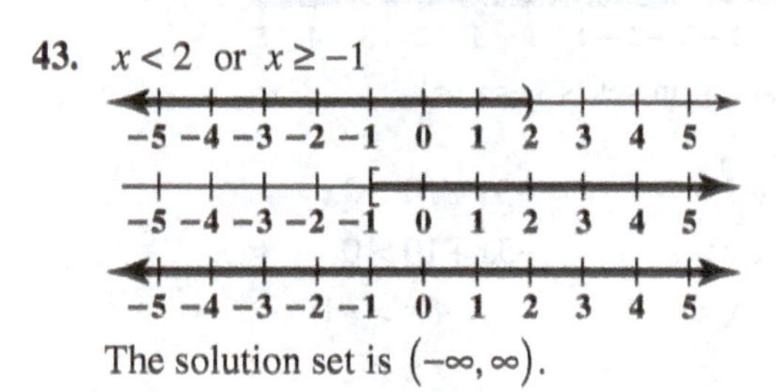

The solution set is $(-\infty,\infty)$.

45. $x \ge 2$ or $x < -1$

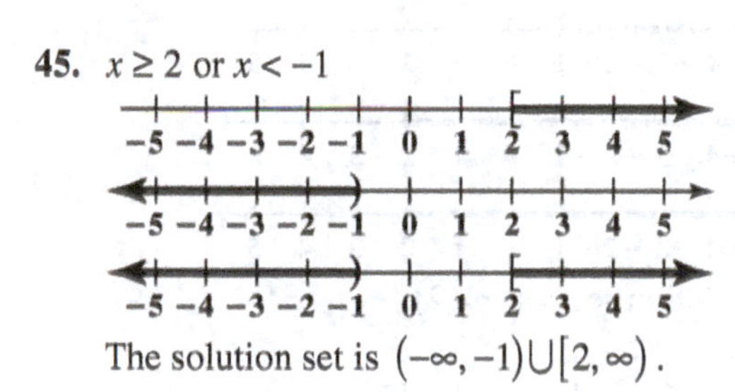

The solution set is $(-\infty,-1) \cup [2,\infty)$.

47. $3x > 12$ or $2x < -6$
$x > 4$ $\quad$ $x < -3$

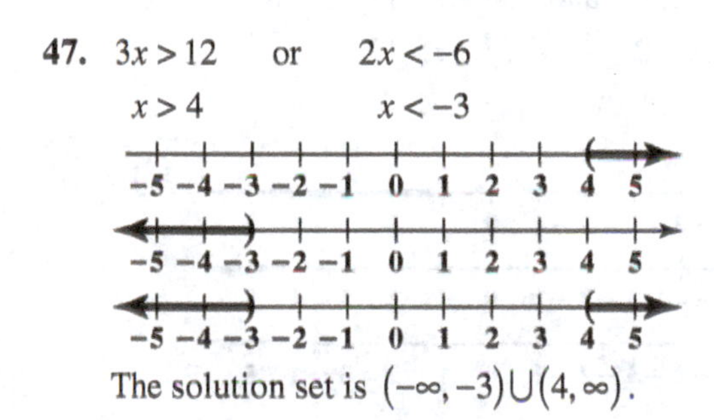

The solution set is $(-\infty,-3) \cup (4,\infty)$.

49. $3x+2 \le 5$ or $5x-7 \ge 8$
$3x \le 3$ $\quad$ $5x \ge 15$
$x \le 1$ $\quad$ $x \ge 3$

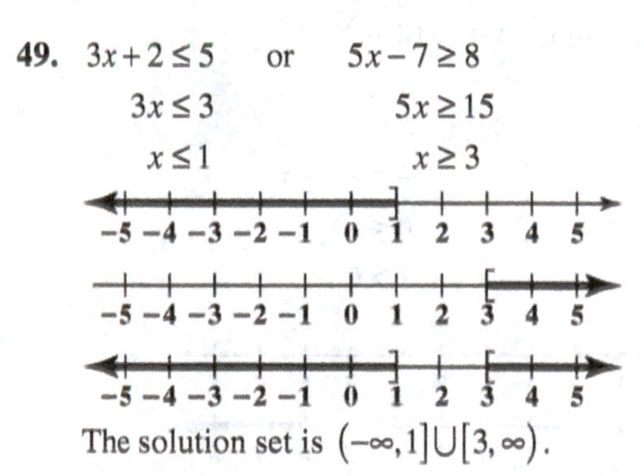

The solution set is $(-\infty,1] \cup [3,\infty)$.

51. $4x+3<-1$ or $2x-3\geq -11$

$4x<-4$ $\quad 2x\geq -8$

$x<-1$ $\quad x\geq -4$

The solution set is $(-\infty,\infty)$.

53. $-2x+5>7$ or $-3x+10>2x$

$-2x>2$ $\quad -5x+10>0$

$x<-1$ $\quad -5x>-10$

$x<2$

The solution set is $(-\infty,2)$.

55. $2x+3\geq 5$ and $3x-1>11$

$2x\geq 2$ $\quad 3x>12$

$x\geq 1$ $\quad x>4$

The solution set is $(4,\infty)$.

57. $3x-1<-1$ or $4-x<-2$

$3x<0$ $\quad 4<-2+x$

$x<0$ $\quad 6<x$

$x>6$

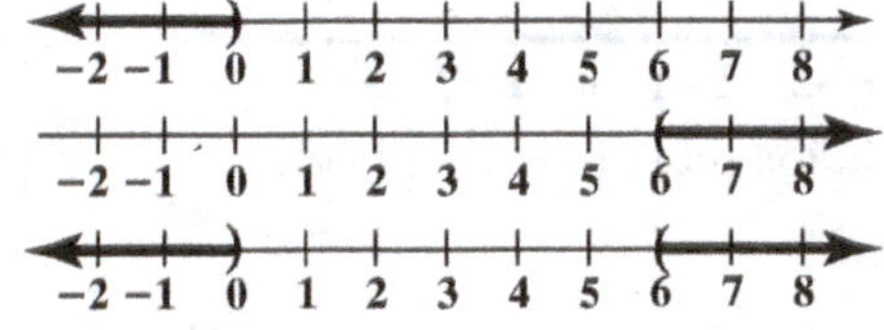

The solution set is $(-\infty,0)\cup(6,\infty)$.

59. $a>0,\ b>0,\ c>0$

$-c<ax-b<c$

$b-c<ax<b+c$

$\frac{b-c}{a}<x<\frac{b+c}{a}$

61. $[-1,3]$

63. Solving in separate pieces:

$x-2<2x-1$ and $2x-1<x+2$

$-2<x-1$ $\quad x-1<2$

$-1<x$ $\quad x<3$

The solution set is $(-1,3)$.

65. The solution set is $[-1,2)$.

67. $5-4x\geq 1$ and $3-7x<31$

$-4x\geq -4$ $\quad 3<7x+31$

$x\leq 1$ $\quad -28<7x$

$-4<x$

The solution set is $(-4,1]$. The set of negative integers that fall within this set is $\{-3,-2,-1\}$.

69. a. $I=\frac{1}{4}x+26$

$\frac{1}{4}x+26>33$

$\frac{1}{4}x>7$

$x>28$

More than 33% of U.S. households will have an interfaith marriage in years after 2016 (i.e. $1988+28$).

b. $N=\frac{1}{4}x+6$

$\frac{1}{4}x+6>14$

$\frac{1}{4}x>8$

$x>32$

More than 14% of U.S. households will have a person of faith married to someone with no religion in years after 2020 (i.e. $1988+32$).

c. More than 33% of U.S. households will have an interfaith marriage *and* more than 14% of U.S. households will have a person of faith married to someone with no religion in years after 2020.

d. More than 33% of U.S. households will have an interfaith marriage *or* more than 14% of U.S. households will have a person of faith married to someone with no religion in years after 2016.

71. $28 \le 20 + 0.40(x - 60) \le 40$

$$28 \le 20 + 0.40x - 24 \le 40$$
$$28 \le 0.40x - 4 \le 40$$
$$32 \le 0.40x \le 44$$
$$80 \le x \le 110$$

Between 80 and 110 minutes, inclusive.

73. Let x = the score on the fifth exam.

$$80 \le \frac{70 + 75 + 87 + 92 + x}{5} < 90$$
$$80 \le \frac{324 + x}{5} < 90$$
$$5(80) \le 5\left(\frac{324 + x}{5}\right) < 5(90)$$
$$400 \le 324 + x < 450$$
$$400 - 324 \le 324 - 324 + x < 450 - 324$$
$$76 \le x < 126$$

A grade between 76 and 125 is needed on the fifth exam.
Because the inequality states the score must be less than 126, we say 125 is the highest possible score. In interval notation, we can use parentheses to exclude the maximum value. The range of scores can be expressed as $[76, 126)$.
If the highest grade is 100, the grade would need to be between 76 and 100, inclusive.

75. Let x = the number of times the bridge is crossed per three month period.
The cost with the 3-month pass is $C_3 = 7.50 + 0.50x$.
The cost with the 6-month pass is $C_6 = 30$.
Because we need to buy two 3-month passes per 6-month pass, we multiply the cost with the 3-month pass by 2.

$$2(7.50 + 0.50x) < 30$$
$$15 + x < 30$$
$$x < 15$$

We also must consider the cost without purchasing a pass. We need this cost to be less than the cost with a 3-month pass.

$$3x > 7.50 + 0.50x$$
$$2.50x > 7.50$$
$$x > 3$$

The 3-month pass is the best deal when making more than 3 but less than 15 crossings per 3-month period.

77. – 81. Answers will vary.

83. $1 < x + 3 < 9$
Using the intersection feature, find the range of the x-values of the points lying between the two constant functions.

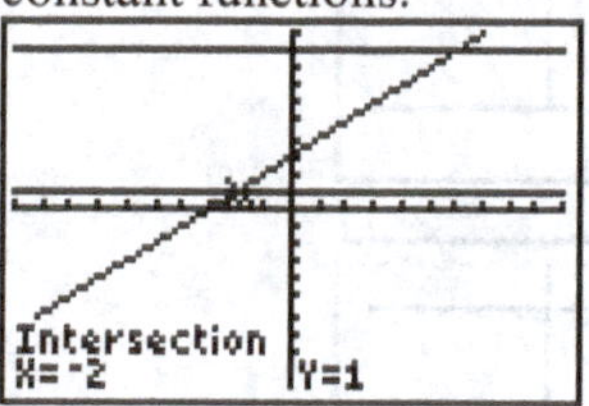

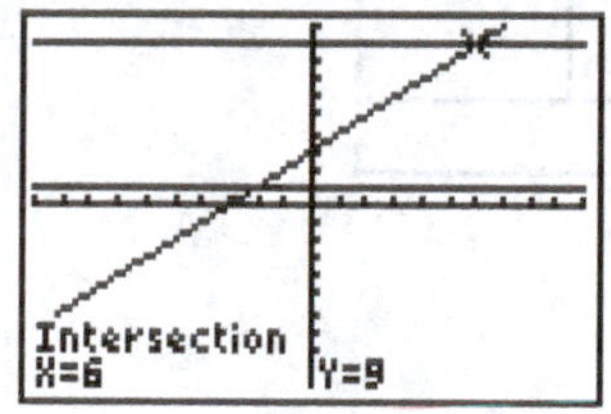

The solution set is $(-2, 6)$.

85. $1 \le 4x - 7 \le 3$
Using the intersection feature, find the range of the x-values of the points lying between the two constant functions.

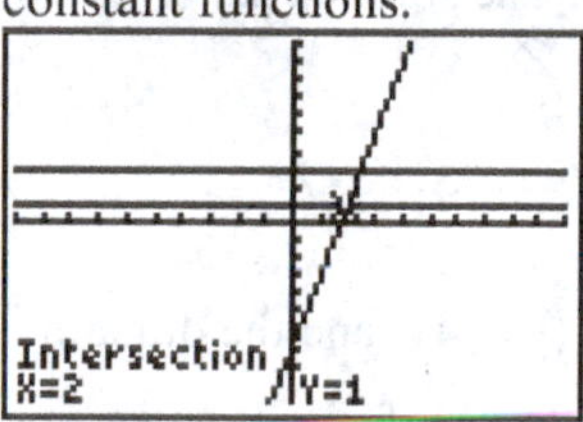

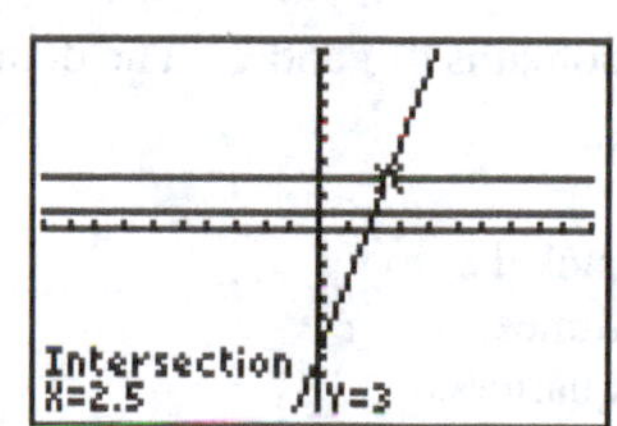

The solution set is $\left[2, \frac{5}{2}\right]$.

87. a.

X	Y1
-4	-1
-2	1
0	3
2	5
4	7
6	9
8	11

Y1=X+3

b.

X	Y1
-8	-2
-6	-1
-4	0
-2	1
0	2
2	3
4	4

Y1=(X+4)/2

c.

X	Y1
1.5	-1
1.75	0
2	1
2.25	2
2.5	3
2.75	4
3	5

Y1=4X−7

d.

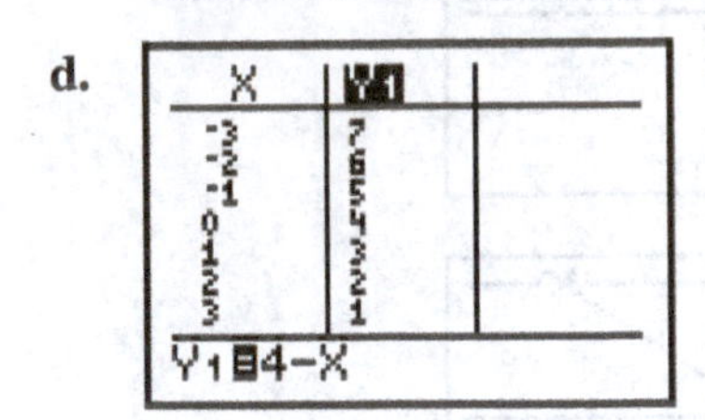

89. makes sense

91. makes sense

93. false; Changes to make the statement true will vary. A sample change is: $(-\infty, 3) \cup (-\infty, -2) = (-\infty, 3)$.

95. false; Changes to make the statement true will vary. A sample change is: The solution set of $x < a$ and $x > a$ is $\varnothing$.

97. The domain of $f = (-\infty, 4]$.

99. The domain of $f = (-\infty, 4]$, and the domain of $g = [-1, \infty)$. The domain of $f + g$ is the intersection of the domains of f and g. The domain of $f + g = [-1, 4]$.

101. Let n = number of nickels.
Let d = number of dimes.
Let q = number of quarters.

$$\begin{aligned}
320 &\le 5n + 10d + 25q \le 545 \\
320 &\le 5n + 10(2n-3) + 25(2n+2) \le 545 \\
320 &\le 5n + 20n - 30 + 50n + 50 \le 545 \\
320 &\le 75n + 20 \le 545 \\
30 &\le 75n \le 525 \\
4 &\le n \le 7
\end{aligned}$$

The least possible number of nickels was 4 and the greatest possible number of nickels was 7.

102.
$$\begin{aligned}
(g-f)(x) &= g(x) - f(x) \\
&= (2x-5) - (x^2 - 3x + 4) \\
&= 2x - 5 - x^2 + 3x - 4 \\
&= -x^2 + 5x - 9 \\
(g-f)(-1) &= -(-1)^2 + 5(-1) - 9 \\
&= -1 - 5 - 9 = -15
\end{aligned}$$

103.
$$\begin{aligned}
4x - 2y &= 8 \\
-2y &= -4x + 8 \\
y &= 2x - 4
\end{aligned}$$

The slope of this line is 2. The slope of the line perpendicular to this line is $-\frac{1}{2}$.

$$\begin{aligned}
y - 2 &= -\frac{1}{2}(x-4) \\
y - 2 &= -\frac{1}{2}x + 2 \\
y &= -\frac{1}{2}x + 4 \\
f(x) &= -\frac{1}{2}x + 4
\end{aligned}$$

104.
$$\begin{aligned}
4 - [2(x-4) - 5] &= 4 - [2x - 8 - 5] \\
&= 4 - [2x - 13] \\
&= 4 - 2x + 13 \\
&= 17 - 2x
\end{aligned}$$

105.
$$\begin{aligned}
1 - 4x &= 3 & \text{or} \quad 1 - 4x &= -3 \\
-4x &= 2 & -4x &= -4 \\
x &= -\tfrac{1}{2} & x &= 1
\end{aligned}$$

106.
$$\begin{aligned}
3x - 1 &= -(x+5) & \text{or} \quad 3x - 1 &= x + 5 \\
3x - 1 &= -x - 5 & 2x &= 6 \\
4x &= -4 & x &= 3 \\
x &= -1 & &
\end{aligned}$$

107. a. $|2x+3| \ge 5$

$|2(-5)+3| \ge 5$

$|-10+3| \ge 5$

$|-7| \ge 5$

$7 \ge 5$, true

-5 satisfies the inequality.

b. $|2x+3| \ge 5$

$|2(0)+3| \ge 5$

$|0+3| \ge 5$

$|3| \ge 5$

$3 \ge 5$, false

0 does not satisfy the inequality.

4.3 Check Points

1. $|2x-1| = 5$

Rewrite without absolute value bars.

$|u| = c$ means $u = c$ or $u = -c$.

$2x-1=5$ or $2x-1=-5$

$2x=6$ $\quad$ $2x=-4$

$x=3$ $\quad$ $x=-2$

The solution set is $\{-2, 3\}$.

2. First, isolate $|1-3x|$.

$2|1-3x|-28=0$

$2|1-3x|=28$

$|1-3x|=14$

Rewrite without absolute value bars.

$|u| = c$ means $u = c$ or $u = -c$.

$1-3x=14$ or $1-3x=-14$

$-3x=13$ $\quad$ $-3x=-15$

$x=-\frac{13}{3}$ $\quad$ $x=5$

The solution set is $\left\{-\frac{13}{3}, 5\right\}$.

3. $|2x-7| = |x+3|$

$2x-7=x+3$ or $2x-7=-(x+3)$

$x=10$ $\quad$ $2x-7=-x-3$

$3x=4$

$x=\frac{4}{3}$

The solution set is $\left\{\frac{4}{3}, 10\right\}$.

4. $|x-2| < 5$

Rewrite without absolute value bars.

$|u| < c$ means $-c < u < c$.

$-5 < x-2 < 5$

$-5+2 < x-2+2 < 5+2$

$-3 < x < 7$

The solution set is $(-3, 7)$.

−3 −2 −1 0 1 2 3 4 5 6 7

5. First, isolate the absolute value expression on one side of the inequality.

$-3|5x-2|+20 \ge -19$

$-3|5x-2| \ge -39$

$\frac{-3|5x-2|}{-3} \le \frac{-39}{-3}$

$|5x-2| \le 13$

Rewrite without absolute value bars.

$|u| \le c$ means $-c \le u \le c$.

$-13 \le 5x-2 \le 13$

$-13+2 \le 5x-2+2 \le 13+2$

$-11 \le 5x \le 15$

$\frac{-11}{5} \le \frac{5x}{5} \le \frac{15}{5}$

$-\frac{11}{5} \le x \le 3$

The solution set is $\left[\frac{-11}{5}, 3\right]$.

−5 −4 −3 −2 −1 0 1 2 3 4 5

6. Rewrite without absolute value bars.
$|u| \ge c$ means $u \le -c$ or $u \ge c$.
$|2x-5| \ge 3$
$2x-5 \ge 3$ or $2x-5 \le -3$
$2x \ge 8$ $\quad 2x \le 2$
$x \ge 4$ $\quad x \le 1$
The solution set is $(-\infty,1] \cup [4,\infty)$.

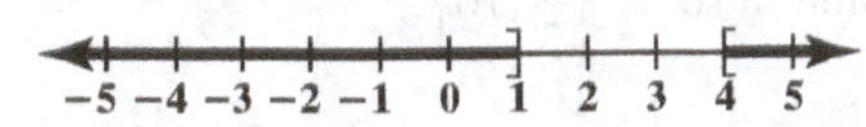

7. $|x-41| \le 3.2$
Rewrite without absolute value bars.
$|u| \le c$ means $-c \le u \le c$.
$-3.2 \le x-41 \le 3.2$
$-3.2+41 \le x-41+41 \le 3.2+41$
$37.8 \le x \le 44.2$
The solution set is $[37.8, 44.2]$.
The percentage of U.S. adults in the population who dread going to the dentist is between a low of 37.8% and a high of 44.2%.

4.3 Concept and Vocabulary Check

1. c; $-c$
2. v; $-v$
3. $-c$; c
4. $-c$; c
5. $<$
6. $<$
7. C
8. E
9. A
10. B
11. D
12. F

4.3 Exercise Set

1. $|x| = 8$
$x = 8$ or $x = -8$
The solution set is $\{-8, 8\}$.

3. $|x-2| = 7$
$x-2 = 7$ or $x-2 = -7$
$x = 9$ $\quad x = -5$
The solution set is $\{-5, 9\}$.

5. $|2x-1| = 7$
$2x-1 = 7$ or $2x-1 = -7$
$2x = 8$ $\quad 2x = -6$
$x = 4$ $\quad x = -3$
The solution set is $\{-3, 4\}$.

7. $\left|\frac{4x-2}{3}\right| = 2$
$\frac{4x-2}{3} = 2$ or $\frac{4x-2}{3} = -2$
$4x-2 = 3(2)$ $\quad 4x-2 = 3(-2)$
$4x-2 = 6$ $\quad 4x-2 = -6$
$4x = 8$ $\quad 4x = -4$
$x = 2$ $\quad x = -1$

The solution set is $\{-1, 2\}$.

9. $|x| = -8$
The solution set is $\varnothing$. There are no values of x for which the absolute value of x is a negative number.

11. $|x+3| = 0$
Since the absolute value of the expression equals zero, we set the expression equal to zero and solve.
$x+3 = 0$
$x = -3$
The solution set is $\{-3\}$.

13. $2|y+6| = 10$
$|y+6| = 5$
$y+6 = 5$ or $y+6 = -5$
$y = -1$ $\quad y = -11$
The solution set is $\{-11, -1\}$.

15. $3|2x-1|=21$

$|2x-1|=7$

$2x-1=7$ or $2x-1=-7$

$2x=8$ $\quad$ $2x=-6$

$x=4$ $\quad$ $x=-3$

The solution set is $\{-3,4\}$.

17. $|6y-2|+4=32$

$|6y-2|=28$

$6y-2=28$ or $6y-2=-28$

$6y=30$ $\quad$ $6y=-26$

$y=5$ $\quad$ $y=-\frac{26}{6}$

$y=-\frac{13}{3}$

The solution set is $\left\{-\frac{13}{3},5\right\}$.

19. $7|5x|+2=16$

$7|5x|=14$

$|5x|=2$

$5x=2$ or $5x=-2$

$x=\frac{2}{5}$ $\quad$ $x=-\frac{2}{5}$

The solution set is $\left\{-\frac{2}{5},\frac{2}{5}\right\}$.

21. $|x+1|+5=3$

$|x+1|=-2$

The solution set is $\varnothing$. By definition, absolute values are always zero or positive.

23. $|4y+1|+10=4$

$|4y+1|=-6$

The solution set is $\varnothing$. By definition, absolute values are always zero or positive.

25. $|2x-1|+3=3$

$|2x-1|=0$

Since the absolute value of the expression equals zero, we set the expression equal to zero and solve.

$2x-1=0$

$2x=1$

$x=\frac{1}{2}$

The solution set is $\left\{\frac{1}{2}\right\}$.

27. $|5x-8|=|3x+2|$

$5x-8=3x+2$ or $5x-8=-3x-2$

$2x-8=2$ $\quad$ $8x-8=-2$

$2x=10$ $\quad$ $8x=6$

$x=5$ $\quad$ $x=\frac{6}{8}=\frac{3}{4}$

The solution set is $\left\{\frac{3}{4},5\right\}$.

29. $|2x-4|=|x-1|$

$2x-4=x-1$ or $2x-4=-x+1$

$x-4=-1$ $\quad$ $3x-4=1$

$x=3$ $\quad$ $3x=5$

$x=\frac{5}{3}$

The solution set is $\left\{\frac{5}{3},3\right\}$.

31. $|2x-5|=|2x+5|$

$2x-5=2x+5$ or $2x-5=-2x-5$

$-5=5$, false $\quad$ $4x-5=-5$

$4x=0$

$x=0$

The solution set is $\{0\}$.

33. $|x-3|=|5-x|$

$x-3=5-x$ or $x-3=-(5-x)$

$2x-3=5$ $\quad$ $x-3=-5+x$

$2x=8$ $\quad$ $-3=-5$, false

$x=4$

The solution set is $\{4\}$.

35. $|2y-6|=|10-2y|$

$2y-6=10-2y$ or $2y-6=-10+2y$

$4y-6=10$ $\quad -6=-10$, false

$4y=16$

$y=4$

The solution set is $\{4\}$.

37. $\left|\frac{2x}{3}-2\right|=\left|\frac{x}{3}+3\right|$

$\frac{2x}{3}-2=\frac{x}{3}+3$ or

$3\left(\frac{2x}{3}\right)-3(2)=3\left(\frac{x}{3}\right)+3(3)$

$2x-6=x+9$

$x-6=9$

$x=15$

$\frac{2x}{3}-2=-\left(\frac{x}{3}+3\right)$

$\frac{2x}{3}-2=-\frac{x}{3}-3$

$3\left(\frac{2x}{3}\right)-3(2)=3\left(-\frac{x}{3}\right)-3(3)$

$2x-6=-x-9$

$3x-6=-9$

$3x=-3$

$x=-1$

The solution set is $\{-1,15\}$.

39. $|x|<3$

$-3<x<3$

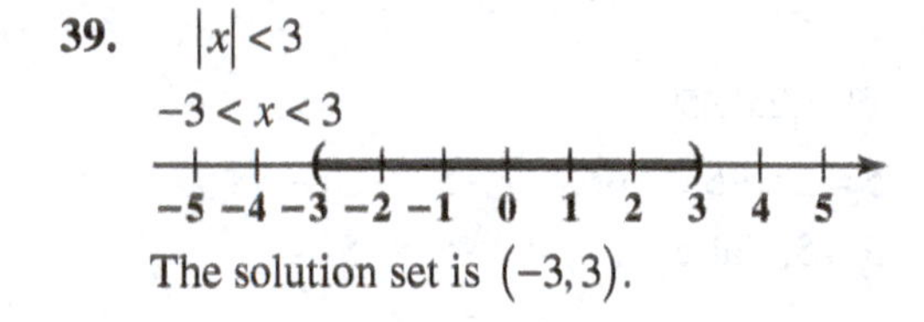

The solution set is $(-3,3)$.

41. $|x-2|<1$

$-1<x-2<1$

$-1+2<x-2+2<1+2$

$1<x<3$

The solution set is $(1,3)$.

43. $|x+2|\le 1$

$-1\le x+2\le 1$

$-1-2\le x+2-2\le 1-2$

$-3\le x\le -1$

The solution set is $[-3,-1]$.

45. $|2x-6|<8$

$-8<2x-6<8$

$-8+6<2x-6+6<8+6$

$-2<2x<14$

$-1<x<7$

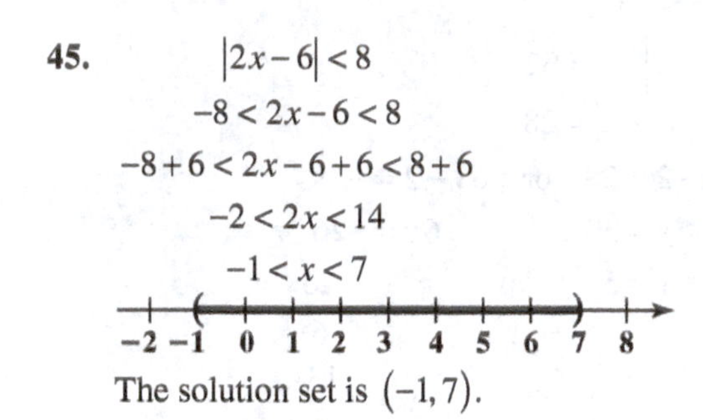

The solution set is $(-1,7)$.

47. $|x|>3$

$x<-3$ or $x>3$

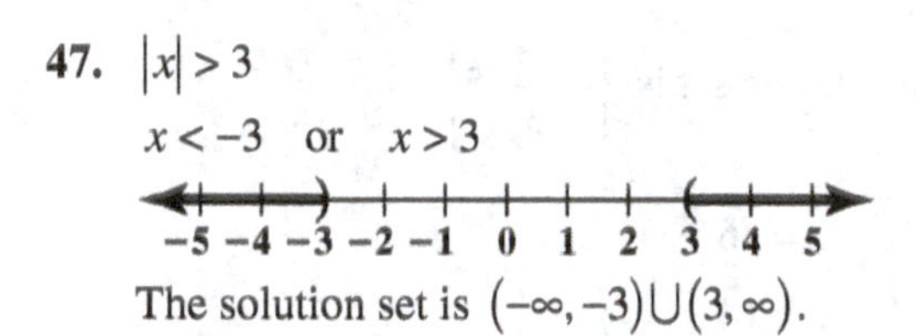

The solution set is $(-\infty,-3)\cup(3,\infty)$.

49. $|x+3|>1$

$x+3<-1$ or $x+3>1$

$x<-4$ $\quad x>-2$

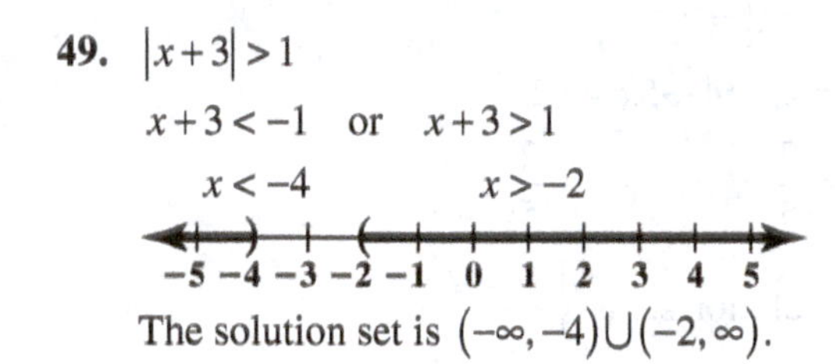

The solution set is $(-\infty,-4)\cup(-2,\infty)$.

51. $|x-4|\ge 2$

$x-4\le -2$ or $x-4\ge 2$

$x\le 2$ $\quad x\ge 6$

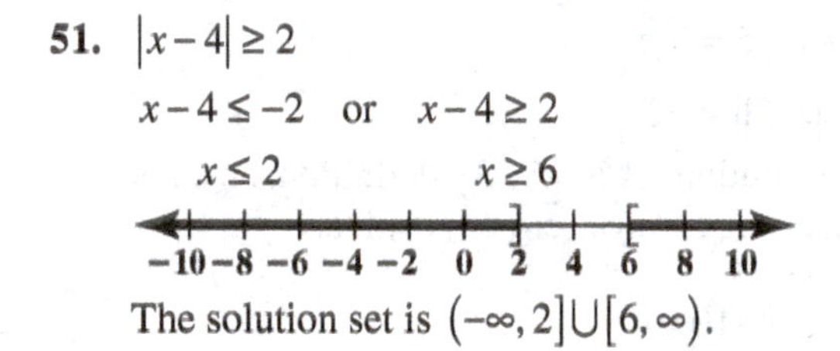

The solution set is $(-\infty,2]\cup[6,\infty)$.

53. $|3x-8|>7$

$3x-8<-7$ or $3x-8>7$

$3x<1$ $\quad 3x>15$

$x<\frac{1}{3}$ $\quad x>5$

The solution set is $\left(-\infty,\frac{1}{3}\right)\cup(5,\infty)$.

55.
$$|2(x-1)+4| \le 8$$
$$|2x-2+4| \le 8$$
$$|2x+2| \le 8$$
$$-8 \le 2x+2 \le 8$$
$$-8-2 \le 2x+2-2 \le 8-2$$
$$-10 \le 2x \le 6$$
$$-5 \le x \le 3$$

The solution set is $[-5,3]$.

57. $\left|\frac{2y+6}{3}\right| < 2$

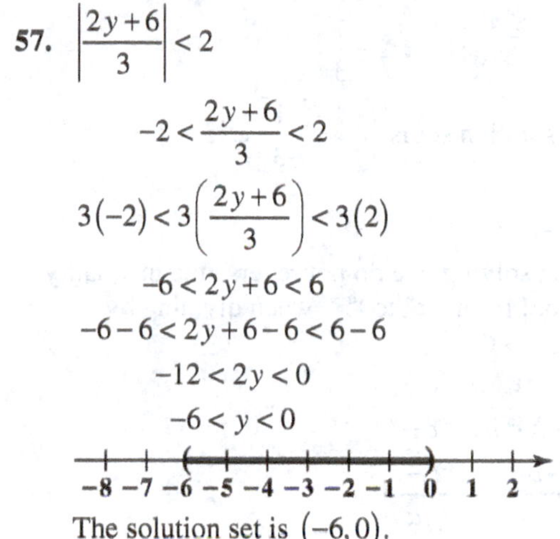

$$-2 < \frac{2y+6}{3} < 2$$
$$3(-2) < 3\left(\frac{2y+6}{3}\right) < 3(2)$$
$$-6 < 2y+6 < 6$$
$$-6-6 < 2y+6-6 < 6-6$$
$$-12 < 2y < 0$$
$$-6 < y < 0$$

The solution set is $(-6,0)$.

59. $\left|\frac{2x+2}{4}\right| \ge 2$

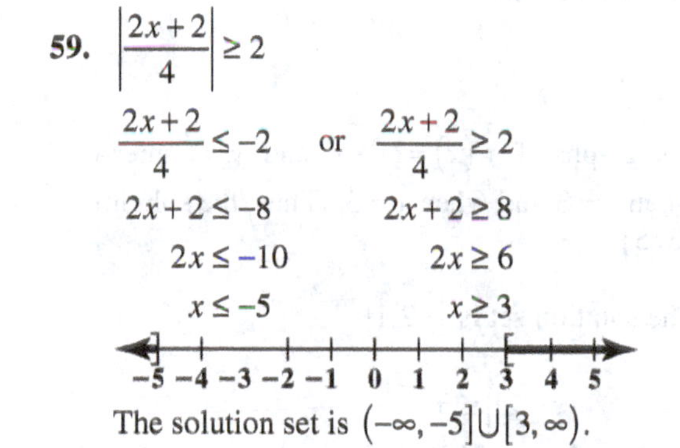

$\frac{2x+2}{4} \le -2$ or $\frac{2x+2}{4} \ge 2$

$2x+2 \le -8$ $\quad$ $2x+2 \ge 8$

$2x \le -10$ $\quad$ $2x \ge 6$

$x \le -5$ $\quad$ $x \ge 3$

The solution set is $(-\infty,-5] \cup [3,\infty)$.

61. $\left|3-\frac{2x}{3}\right| > 5$

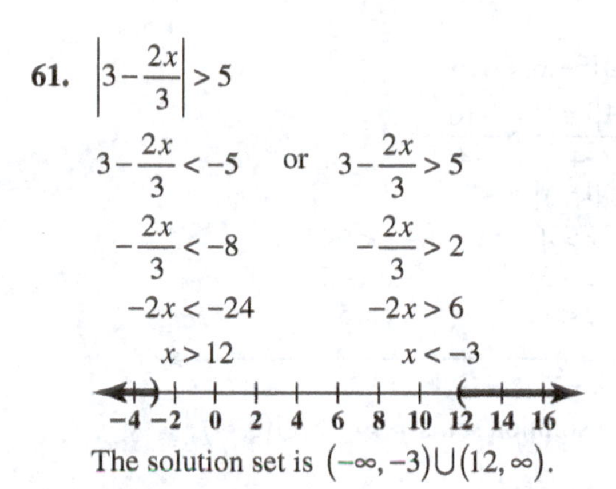

$3-\frac{2x}{3} < -5$ or $3-\frac{2x}{3} > 5$

$-\frac{2x}{3} < -8$ $\quad$ $-\frac{2x}{3} > 2$

$-2x < -24$ $\quad$ $-2x > 6$

$x > 12$ $\quad$ $x < -3$

The solution set is $(-\infty,-3) \cup (12,\infty)$.

63. $|x-2| < -1$

The solution set is $\varnothing$. Since all absolute values are zero or positive, there are no values of x that will make the absolute value of the expression less than -1.

65. $|x+6| > -10$

Since all absolute values are zero or positive, we know that when simplified, the left hand side will be a positive number. We also know that any positive number is greater than any negative number. This means that regardless of the value of x, the left hand side will be greater than the right hand side of the inequality. The solution set is $(-\infty,\infty)$.

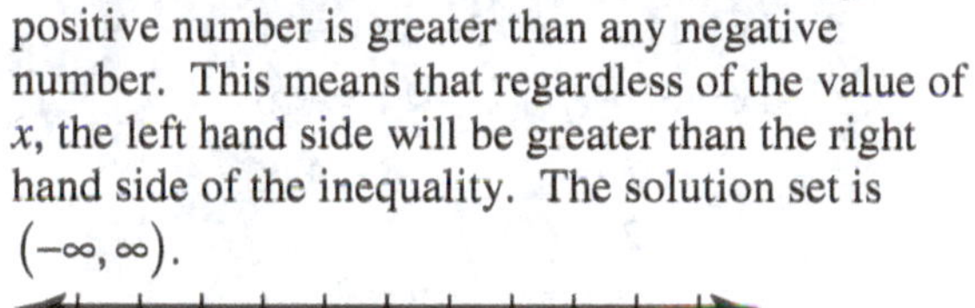

67.

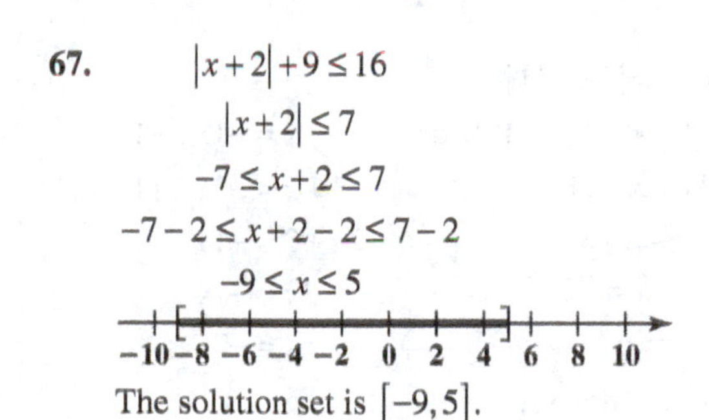

$$|x+2|+9 \le 16$$
$$|x+2| \le 7$$
$$-7 \le x+2 \le 7$$
$$-7-2 \le x+2-2 \le 7-2$$
$$-9 \le x \le 5$$

The solution set is $[-9,5]$.

69.
$$2|2x-3|+10 > 12$$
$$2|2x-3| > 2$$
$$|2x-3| > 1$$

$2x-3 < -1$ or $2x-3 > 1$

$2x < 2$ $\quad$ $2x > 4$

$x < 1$ $\quad$ $x > 2$

The solution set is $(-\infty,1) \cup (2,\infty)$.

71. $-4|1-x| < -16$

$\frac{-4|1-x|}{-4} > \frac{-16}{-4}$

$|1-x| > 4$

$1-x > 4$ or $1-x < -4$

$-x > 3 \qquad -x < -5$

$x < -3 \qquad x > 5$

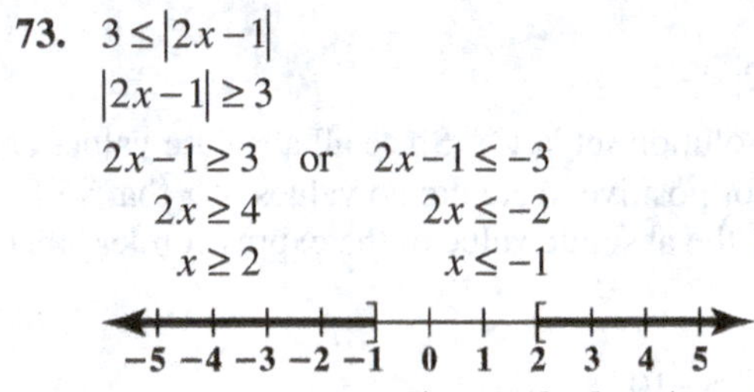

The solution set is $(-\infty,-3)\cup(5,\infty)$.

73. $3 \le |2x-1|$

$|2x-1| \ge 3$

$2x-1 \ge 3$ or $2x-1 \le -3$

$2x \ge 4 \qquad 2x \le -2$

$x \ge 2 \qquad x \le -1$

-5 -4 -3 -2 -1 0 1 2 3 4 5

The solution set is $(-\infty,-1]\cup[2,\infty)$.

75. $|5-4x| = 11$

$5-4x = 11$ or $5-4x = -11$

$-4x = 6 \qquad -4x = -16$

$x = -\frac{3}{2} \qquad y = 4$

The solution set is $\left\{-\frac{3}{2}, 4\right\}$.

77. $|3-x| = |3x+11|$

$3-x = 3x+11$ or $3-x = -(3x+11)$

$-4x+3 = 11 \qquad 3-x = -3x-11$

$-4x = 8 \qquad 2x+3 = -11$

$x = -2 \qquad 2x = -14$

$x = -7$

The solution set is $\{-7,-2\}$.

79. $|-1+3(x+1)| \le 5$

$-5 \le -1+3x+3 \le 5$

$-5 \le 3x+2 \le 5$

$-7 \le 3x \le 3$

$-\frac{7}{3} \le x \le 1$

The solution set is $\left[-\frac{7}{3}, 1\right]$.

81. $|2x-3|+1 > 6$

$|2x-3| > 5$

$2x-3 > 5$ or $2x-3 < -5$

$2x > 8 \qquad 2x < -2$

$x > 4 \qquad x < -1$

The solution set is $(-\infty,-1)\cup(4,\infty)$.

83. Let x be the number.

$|4-3x| \ge 5$

$3x-4 \ge 5$ or $3x-4 \le -5$

$3x \ge 9 \qquad 3x \le -1$

$x \ge 3 \qquad x \le -\frac{1}{3}$

The solution set is $\left(-\infty,-\frac{1}{3}\right]\cup[3,\infty)$.

85. $|ax+b| < c$

When solving, we do not reverse the inequality symbol from "<" to ">" when dividing by a since $a > 0$.

$-c < ax+b < c$

$-c-b < ax < c-b$

$\frac{-c-b}{a} < x < \frac{c-b}{a}$

The solution set is $\left(\frac{-c-b}{a}, \frac{c-b}{a}\right)$.

87. $|4-x| = 1$

The graphs of $f(x) = |4-x|$ and $y = 1$ intersect when $x = 3$ and when $x = 5$. Thus, the solution set is $\{3, 5\}$.

89. The solution set is $[-2,1]$

91. $|x-21| \le 3$

$-3 \le x-21 \le 3$

$-3+21 \le x-21+21 \le 3+21$

$18 \le x \le 24$

Interval notation: [18, 24]

The percentage of job interviewers in the population turned off by the job applicant being arrogant is between a low of 18% and a high of 24%. The margin of error is ±3%.

93.
$$|T-57| \le 7$$
$$-7 \le T-57 \le 7$$
$$-7+57 \le T-57+57 \le 7+57$$
$$50 \le T \le 64$$

The monthly average temperature for San Francisco, California ranges from $50°F$ to $64°F$, inclusive.

95.
$$|x-8.6| \le 0.01$$
$$-0.01 \le x-8.6 \le 0.01$$
$$-0.01+8.6 \le x-8.6+8.6 \le 0.01+8.6$$
$$8.59 \le x \le 8.61$$

The length of the machine part must be between 8.59 and 8.61 centimeters, inclusive.

97. $\left|\dfrac{h-50}{5}\right| \ge 1.645$

$$\frac{h-50}{5} \le -1.645 \quad \text{or} \quad \frac{h-50}{5} \ge 1.645$$
$$h-50 \le 5(-1.645) \qquad h-50 \ge 5(1.645)$$
$$h-50 \le -8.225 \qquad h-50 \ge 8.225$$
$$h \le 41.775 \qquad h \ge 58.225$$

The coin would be considered unfair if the tosses resulted in 41 or less heads, or 59 or more heads.

99. – 103. Answers will vary.

105. $|x+1| = 5$

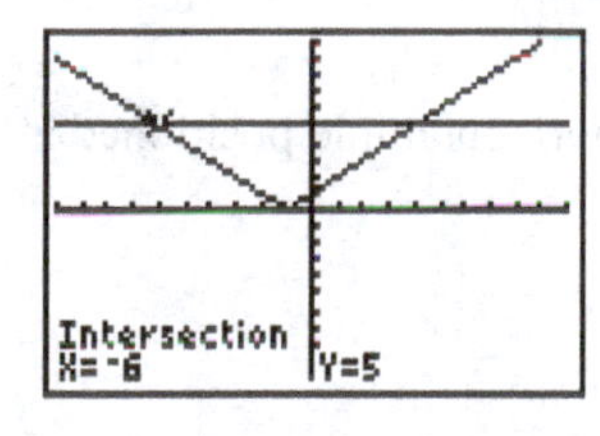

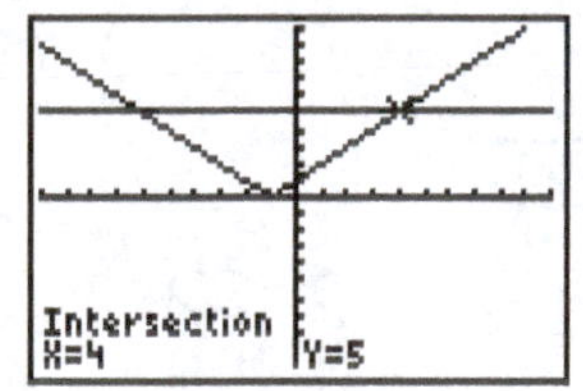

The solutions are –6 and 4 and the solution set is $\{-6, 4\}$.

107. $|2x-3| = |9-4x|$

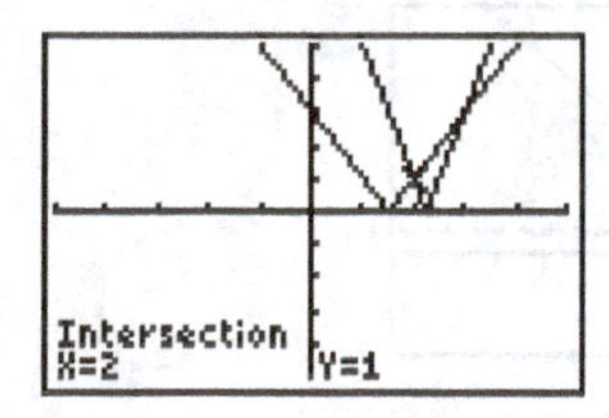

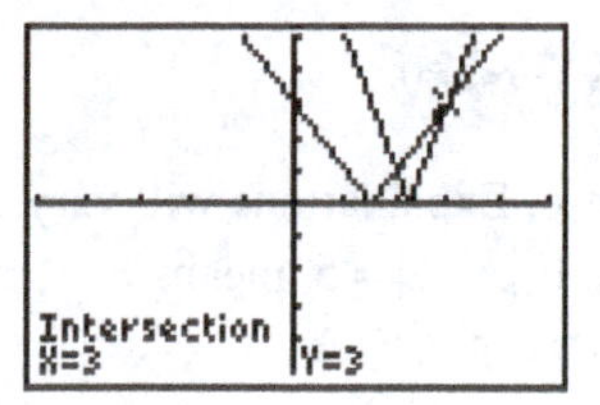

The solutions are 2 and 3 and the solution set is $\{2, 3\}$.

109. $\left|\dfrac{2x-1}{3}\right| < \dfrac{5}{3}$

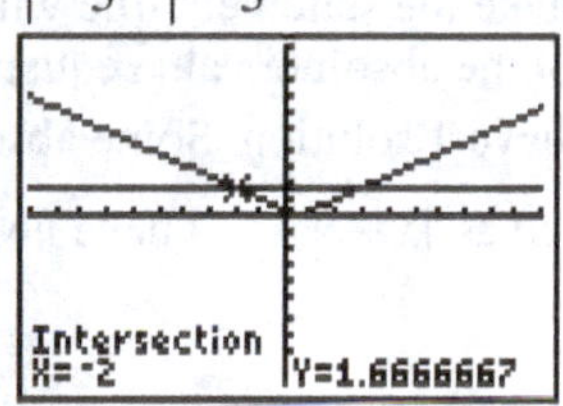

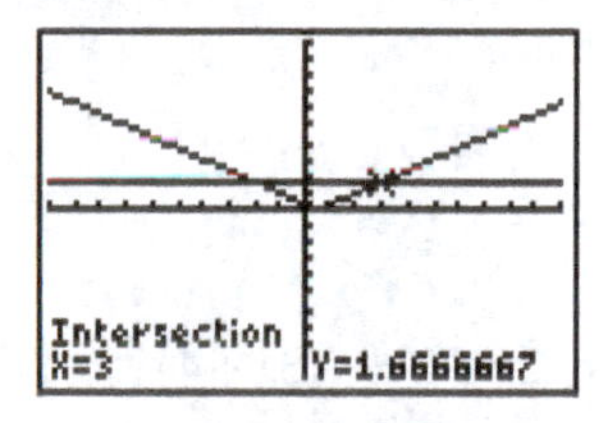

The solution set is $(-2, 3)$.

111. $|2x-1| > 7$

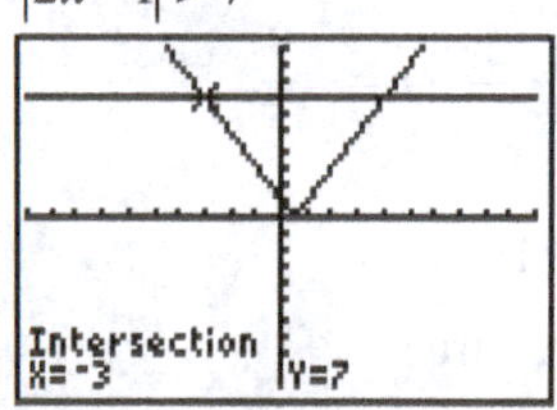

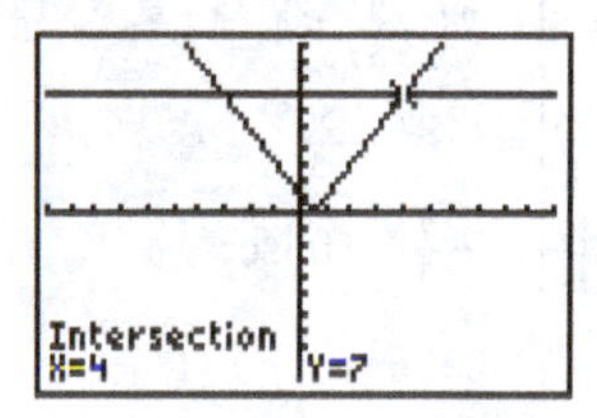

The solution set is $(-\infty, -3) \cup (4, \infty)$.

113. $|x+4|>-1$

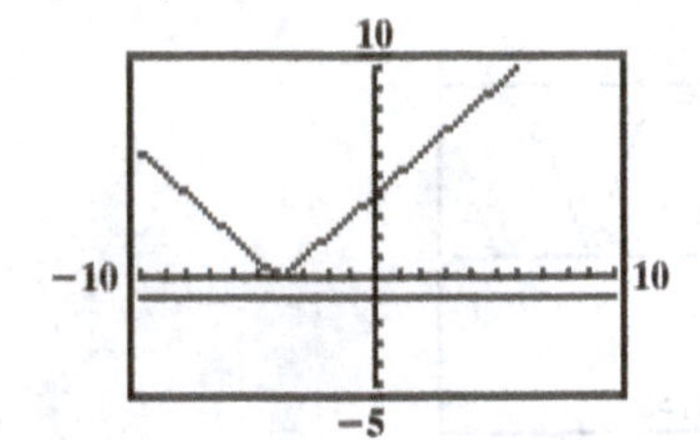

The solution set is $(-\infty, \infty)$.

115. does not make sense; Explanations will vary. Sample explanation: $|x-2|=5$ means $x-2=-5$ or $x-2=5$.

117. does not make sense; Explanations will vary. Sample explanation: By adding 9 to both sides, we see that the absolute value expression must be less than 5. This is possible so the inequality does have a solution.

119. false; Changes to make the statement true will vary. A sample change is: Some absolute value equations, such as $|x-3|=0$ have 1 solution. Some absolute value equations, such as $|x+5|=-2$ have no solution.

121. true

123. **a.** $|x-4|<3$

b. $|x-4|\geq 3$

125. $|2x+5|=3x+4$

$$\begin{array}{lll} 2x+5=3x+4 & \text{or} & 2x+5=-(3x+4) \\ -x+5=4 & & 2x+5=-3x-4 \\ -x=-1 & & 5x+5=-4 \\ x=1 & & 5x=-9 \\ & & x=-\frac{9}{5} \end{array}$$

Check:

$x=1$: $|2(1)+5|=3(1)+4$

$|7|=7$, true

$x=-\frac{9}{5}$: $\left|2\left(-\frac{9}{5}\right)+5\right|=3\left(-\frac{9}{5}\right)+4$

$\left|-\frac{18}{5}+\frac{25}{5}\right|=-\frac{27}{5}+\frac{20}{5}$

$\left|\frac{7}{5}\right|=-\frac{7}{5}$, false

Thus, 1 checks and $-\frac{9}{5}$ does not check, so the solution set is $\{1\}$.

126. $3x-5y=15$

$$-5y=-3x+15$$

$$y=\frac{3}{5}x-3$$

The y-intercept is –3 and the slope is $\frac{3}{5}$.

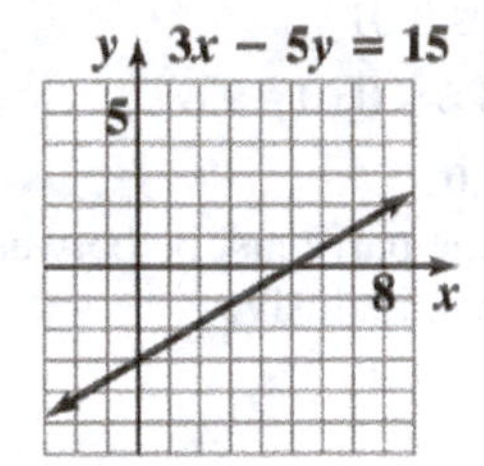

127. $f(x)=-\frac{2}{3}x$

The y-intercept is 0 and the slope is $-\frac{2}{3}$.

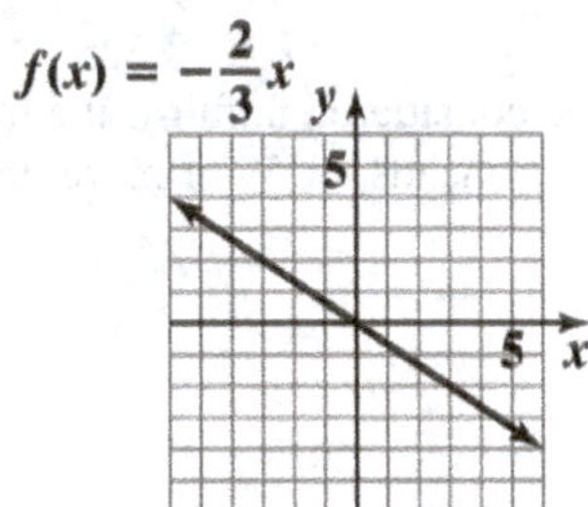

128. $f(x)=-2$ is the horizontal line positioned at $y=-2$.

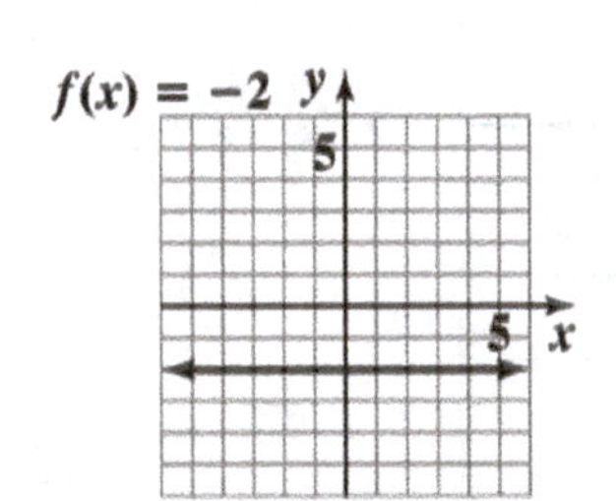

Mid-Chapter Check Point – Chapter 4

1. $4-3x \ge 12-x$

$4 \ge 12+2x$

$-8 \ge 2x$

$-4 \ge x$

$x \le -4$

The solution set is $(-\infty,-4]$.

2. $5 \le 2x-1 < 9$

$6 \le 2x < 10$

$3 \le x < 5$

The solution set is $[3,5)$.

3. $|4x-7|=5$

$4x-7=5$ or $4x-7=-5$

$4x=12$ $\quad$ $4x=2$

$x=3$ $\quad$ $x=\frac{1}{2}$

The solution set is $\left\{\frac{1}{2},3\right\}$.

4. $-10-3(2x+1) > 8x+1$

$-10-6x-3 > 8x+1$

$-6x-13 > 8x+1$

$-13 > 14x+1$

$-14 > 14x$

$-1 > x$

$x < -1$

The solution set is $(-\infty,-1)$.

5. $2x+7<-11$ or $-3x-2<13$

$2x<-18$ $\quad$ $-3x<15$

$x<-9$ $\quad$ $x>-5$

The solution set is $(-\infty,-9)\cup(-5,\infty)$.

6. $|3x-2| \le 4$

$-4 \le 3x-2 \le 4$

$-2 \le 3x \le 6$

$-\frac{2}{3} \le x \le 2$

The solution set is $\left[-\frac{2}{3},2\right]$.

7. $|x+5|=|5x-8|$

$x+5=5x-8$ or $x+5=-(5x-8)$

$-4x+5=-8$ $\quad$ $x+5=-5x+8$

$-4x=-13$ $\quad$ $6x+5=8$

$x=\frac{13}{4}$ $\quad$ $6x=3$

$x=\frac{1}{2}$

The solution set is $\left\{\frac{1}{2},\frac{13}{4}\right\}$.

8. $5-2x \ge 9$ and $5x+3>-17$

$5 \ge 2x+9$ $\quad$ $5x>-20$

$-4 \ge 2x$ $\quad$ $x>-4$

$-2 \ge x$

$x \le -2$

The solution set is $(-4,-2]$.

9. $3x-2>-8$ or $2x+1<9$

$3x>-6$ $\quad$ $2x<8$

$x>-2$ $\quad$ $x<4$

The union of these sets is the entire number line.
The solution set is $(-\infty,\infty)$.

10. $\frac{x}{2}+3 \le \frac{x}{3}+\frac{5}{2}$

$6\left(\frac{x}{2}+3\right) \le 6\left(\frac{x}{3}+\frac{5}{2}\right)$

$3x+18 \le 2x+15$

$x+18 \le 15$

$x \le -3$

The solution set is $(-\infty,-3]$.

11. $\frac{2}{3}(6x-9)+4>5x+1$

$4x-6+4>5x+1$

$4x-2>5x+1$

$-2>x+1$

$-3>x$

$x<-3$

The solution set is $(-\infty,-3)$.

12. $|5x+3|>2$

$5x+3>2$ or $5x+3<-2$

$5x>-1$ $\quad$ $5x<-5$

$x>-\frac{1}{5}$ $\quad$ $x<-1$

The solution set is $(-\infty,-1)\cup\left(-\frac{1}{5},\infty\right)$.

13. $7-\left|\frac{x}{2}+2\right|\le 4$

$-\left|\frac{x}{2}+2\right|\le -3$

$\left|\frac{x}{2}+2\right|\ge 3$

$|x+4|\ge 6$

$x+4\ge 6$ or $x+4\le -6$

$x\ge 2$ $\quad$ $x\le -10$

The solution set is $(-\infty,-10]\cup[2,\infty)$.

14. $5(x-2)-3(x+4)\ge 2x-20$

$5x-10-3x-12\ge 2x-20$

$2x-22\ge 2x-20$

$-22\ge -20$, false

There are no solutions to the inequality. The solution set is $\varnothing$.

15. $\frac{x+3}{4}<\frac{1}{3}$

$3x+9<4$

$3x<-5$

$x<-\frac{5}{3}$

The solution set is $\left(-\infty,-\frac{5}{3}\right)$.

16. $5x+1\ge 4x-2$ and $2x-3>5$

$x+1\ge -2$ $\quad$ $2x>8$

$x\ge -3$ $\quad$ $x>4$

The solution set is $(4,\infty)$.

17. $3-|2x-5|=-6$

$-|2x-5|=-9$

$|2x-5|=9$

$2x-5=9$ or $2x-5=-9$

$2x=14$ $\quad$ $2x=-4$

$x=7$ $\quad$ $x=-2$

The solution set is $\{-2,7\}$.

18. $3+|2x-5|=-6$

$|2x-5|=-9$

Since absolute values cannot be negative, there are no solutions. The solution set is $\varnothing$.

19. Let x = number of miles.

$24+0.20x\le 40$

$0.20x\le 16$

$x\le 80$

No more than 80 miles per day.

20. Let x = grade on the fifth exam.

$80\le\frac{95+79+91+86+x}{5}<90$

$80\le\frac{351+x}{5}<90$

$400\le x+351<450$

$49\le x<99$

$[49,99)$

21. Let x = amount invested.

$x(0.075)\ge 9000$

$x\ge 120{,}000$

The retiree should invest at least $120,000.

22. Let x = the number of discs produced.
The cost function: $C(x) = 0.18x + 60,000$
The cost function: $R(x) = 0.30x$
The profit function:
$$P(x) = R(x) - C(x)$$
$$= 0.30x - (0.18x + 60,000)$$
$$= 0.12x - 60,000$$
We need
$$P(x) \geq 30,000$$
$$0.12x - 60,000 \geq 30,000$$
$$0.12x \geq 90,000$$
$$\frac{0.12x}{0.12} \geq \frac{90,000}{0.12}$$
$$x \geq 750,000$$
The company should produce and sell at least 750,000 compact discs each month.

4.4 Check Points

1. $4x - 2y \geq 8$
First, graph the equation $4x - 2y = 8$ with a solid line.
Set $y = 0$ to find the x-intercept.
$$4x - 2y = 8$$
$$4x - 2(0) = 8$$
$$4x = 8$$
$$x = 2$$
Set $x = 0$ to find the y-intercept.
$$4x - 2y = 8$$
$$4(0) - 2y = 8$$
$$-2y = 8$$
$$y = -4$$
Next, use the origin as a test point.
$$4x - 2y \geq 8$$
$$4(0) - 2(0) \geq 8$$
$$0 \geq 8, \text{ false}$$
Since the statement is false, shade the half-plane that does not contain the test point.

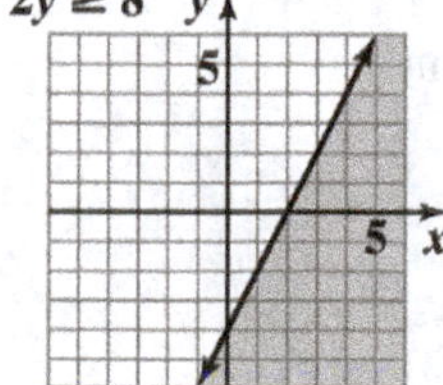

2. $y > -\frac{3}{4}x$

First, graph the equation $y = -\frac{3}{4}x$ with a dashed line.

x	$y = -\frac{3}{4}x$	(x, y)
–4	$-\frac{3}{4}(-4) =$	(–4, 3)
0	$-\frac{3}{4}(0) =$	(0, 0)
4	$-\frac{3}{4}(4) =$	(4, –3)

Next, use a test point such as (4, 0).
$$y > -\frac{3}{4}x$$
$$4 > -\frac{3}{4}(0)$$
$$4 > 0, \text{ true}$$
Since the statement is true, shade the half-plane that contains the test point.

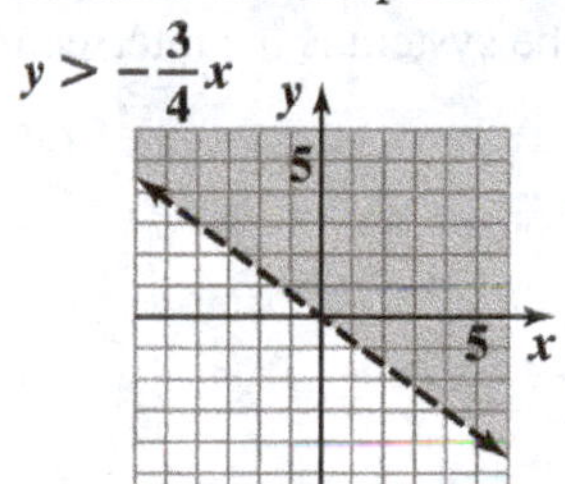

3. a. $y > 1$
Graph the line $y = 1$ with a dashed line.
Since the inequality is of the form $y > a$, shade the half-plane above the line.

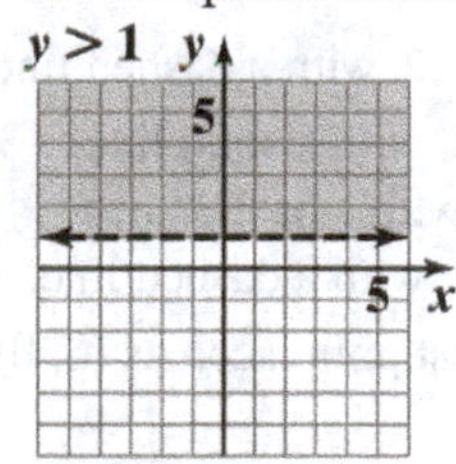

b. $x \leq -2$
Graph the line $x = -2$ with a solid line.
Since the inequality is of the form $x \leq a$, shade the half-plane to the left of the line.

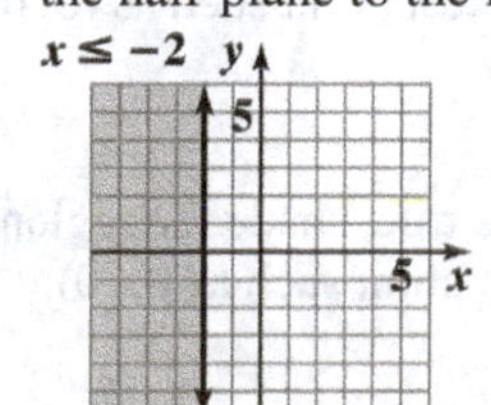

4. Point $B=(60,20)$. Check this point in each of the three inequalities for grasslands.

$T \geq 35$ $\quad$ $5T-7P \geq 70$ $\quad$ $3T-35P \leq -140$

$60 \geq 35$, true $\quad$ $5(60)-7(20) \geq 70$ $\quad$ $3(60)-35(20) \leq -140$

$160 \geq 70$, true $\quad$ $-520 \leq -140$, true

Since the three inequalities for grasslands are true, the point $B=(60,20)$ does describe where grasslands occur.

5. $x-3y<6$

$2x+3y \geq -6$

Graph the line $x-3y=6$ with a dashed line. Graph the line $2x+3y=-6$ with a solid line.

For $x-3y<6$ use a test point such as (0, 0).

$x-3y<6$

$0-3(0)<6$

$0<6$, true

Since the statement is true, shade the half-plane that contains the test point.

For $2x+3y \geq -6$ use a test point such as (0, 0).

$2x+3y \geq -6$

$2(0)+3(0) \geq -6$

$0 \geq -6$, true

Since the statement is true, shade the half-plane that contains the test point.

The solution set of the system is the intersection (the overlap) of the two half-planes.

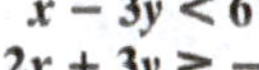

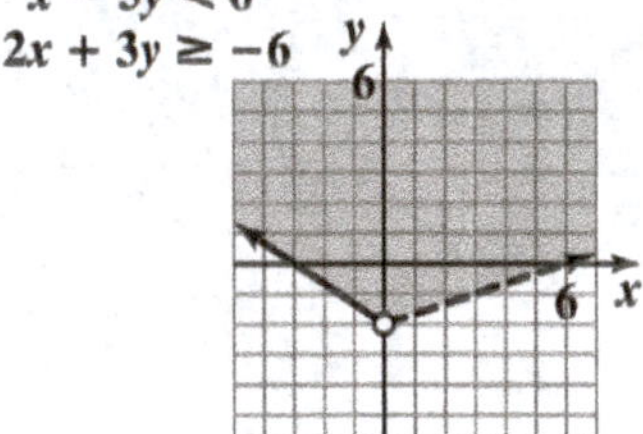

6. $x+y<2$

$-2 \leq x<1$

$y>-3$

Graph the line $x+y=2$ with a dashed line.

Graph the line $x=-2$ with a solid line.

Graph the line $x=1$ with a dashed line.

Graph the line $y=-3$ with a dashed line.

For $x+y<2$ use a test point such as (0, 0).

$x+y<2$

$0+0<2$

$0<2$, true

Since the statement is true, shade the half-plane below the line, as it contains the test point.

For $-2 \leq x<1$ use a test point such as (0, 0).

$-2 \leq x<1$

$-2 \leq 0<1$, true

Since the statement is true, shade the region between the lines.

For $y>-3$ use a test point such as (0, 0).

$y>-3$

$0>-3$, true

Since the statement is true, shade the half-plane above the line, as it contains the test point.
The solution set of the system is the intersection of the shaded regions.

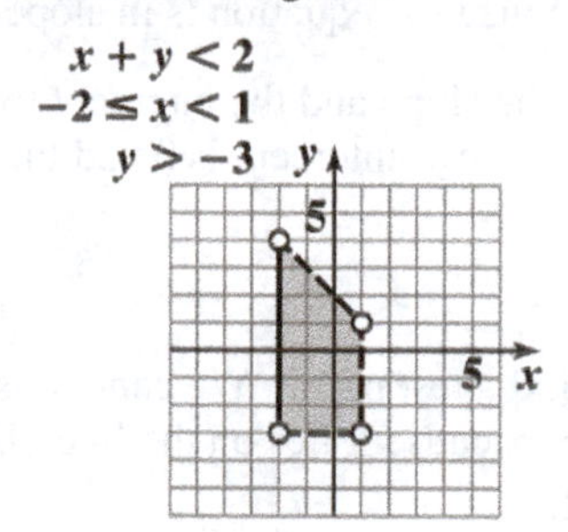

4.4 Concept and Vocabulary Check

1. solution; x; y; $5 > 1$
2. graph
3. half-plane
4. false
5. true
6. false
7. $x - y < 1$; $2x + 3y \ge 12$
8. false

4.4 Exercise Set

1. $x + y \ge 3$
First, graph the equation $x + y = 3$. Rewrite the equation in slope-intercept form by solving for y.
$x + y = 3$
$y = -x + 3$
y-intercept = 3
slope = −1
Next, use the origin as a test point.
$x + y \ge 3$
$0 + 0 \ge 3$
$0 \ge 3$
This is a false statement. This means that the point $(0,0)$ will not fall in the shaded half-plane.

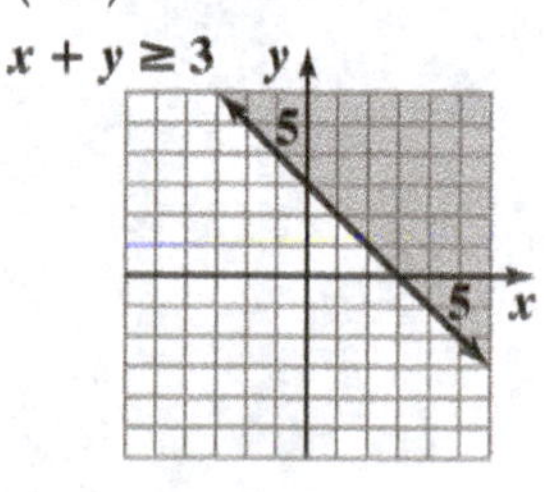

3. $x - y < 5$
First, graph the equation $x - y = 5$. Rewrite the equation in slope-intercept form by solving for y.
$x - y = 5$
$-y = -x + 5$
$y = x - 5$
y-intercept = −5
slope = 1
Next, use the origin as a test point.
$x - y < 5$
$0 - 0 < 5$
$0 < 5$
This is a true statement. This means that the point $(0,0)$ will fall in the shaded half-plane.

$x - y < 5$

5. $x + 2y > 4$
First, graph the equation $x + 2y = 4$. Rewrite the equation in slope-intercept form by solving for y.
$x + 2y = 4$
$2y = -x + 4$
$y = -\frac{1}{2}x + 2$
y-intercept = 2
slope = $-\frac{1}{2}$
Next, use the origin as a test point.
$0 + 2(0) > 4$
$0 + 0 > 4$
$0 > 4$
This is a false statement. This means that the point $(0,0)$ will not fall in the shaded half-plane.

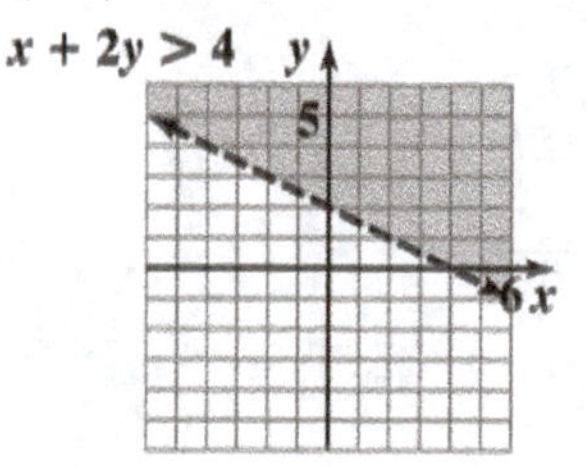

7. $3x - y \le 6$

First, graph the equation $3x - y = 6$. Rewrite the equation in slope-intercept form by solving for y.

$3x - y = 6$

$-y = -3x + 6$

$y = 3x - 6$

y-intercept $= -6$ slope $= 3$

Next, use the origin as a test point.

$3(0) - 0 \le 6$

$0 - 0 \le 6$

$0 \le 6$

This is a true statement. This means that the point $(0,0)$ will fall in the shaded half-plane.

9. $\frac{x}{2} + \frac{y}{3} < 1$

First, graph the equation $\frac{x}{2} + \frac{y}{3} = 1$. Rewrite the equation in slope-intercept form by solving for y.

$\frac{x}{2} + \frac{y}{3} = 1$

$6\left(\frac{x}{2}\right) + 6\left(\frac{y}{3}\right) = 6(1)$

$3x + 2y = 6$

$2y = -3x + 6$

$y = -\frac{3}{2}x + 3$

y-intercept $= 3$ slope $= -\frac{3}{2}$

Next, use the origin as a test point.

$\frac{0}{2} + \frac{0}{3} < 1$

$0 + 0 < 1$

$0 < 1$

This is a true statement. This means that the point $(0,0)$ will fall in the shaded half-plane.

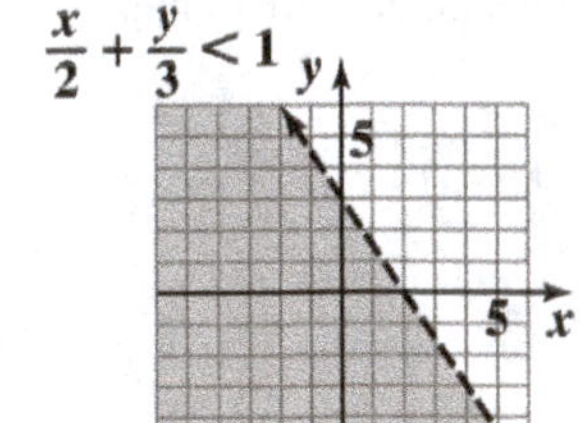

11. $y > \frac{1}{3}x$

Replacing the inequality symbol with an equal sign, we have $y = \frac{1}{3}x$. Since the equation is in slope-intercept form, use the slope and the intercept to graph the equation. The y–intercept is 0 and the slope is $\frac{1}{3}$.

Next, we need to find a test point. We cannot use the origin this time, because it lies on the line. Use $(1,1)$ as a test point.

$1 > \frac{1}{3}(1)$

$1 > \frac{1}{3}$

This is a true statement, so we know the point $(1,1)$ lies in the shaded half-plane.

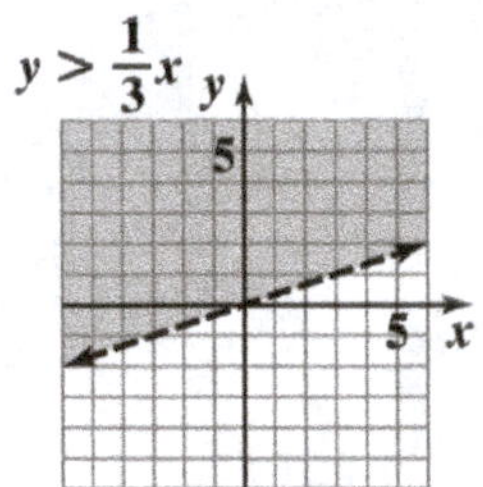

13. $y \le 3x + 2$

First, graph the equation $y = 3x + 2$. Since the equation is in slope-intercept form, use the slope and the intercept to graph the equation. The y–intercept is 2 and the slope is 3.

Next, use the origin as a test point.

$0 \le 3(0) + 2$

$0 \le 2$

This is a true statement. This means that the point $(0,0)$ will fall in the shaded half-plane.

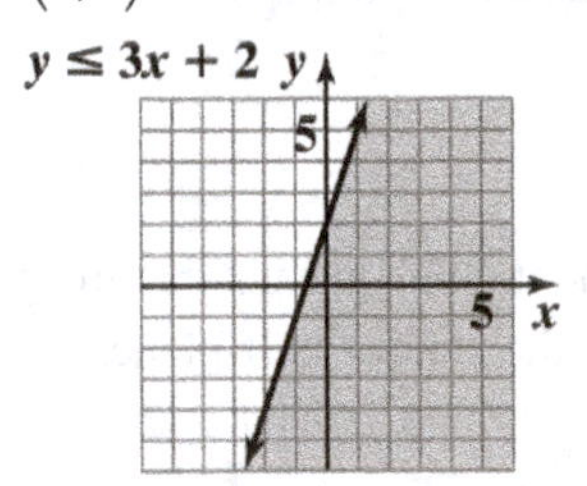

15. $y < -\frac{1}{4}x$

Replacing the inequality symbol with an equal sign, we have $y = -\frac{1}{4}x$. Since the equation is in slope-intercept form, use the slope and the intercept to graph the equation. The y–intercept is 0 and the slope is $-\frac{1}{4}$.

Next, we need to find a test point. We cannot use the origin this time, because it lies on the line. Use $(1,1)$ as a test point.

$1 < -\frac{1}{4}(1)$

$1 < -\frac{1}{4}$

This is a false statement, so we know the point $(1,1)$ does not lie in the shaded half-plane.

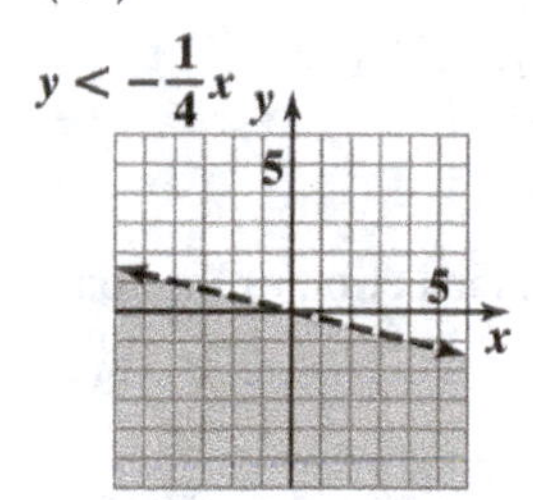

17. $x \le 2$

Replacing the inequality symbol with an equal sign, we have $x = 2$. We know that equations of the form $x = a$ are vertical lines with x-intercept $= a$.

Next, use the origin as a test point.

$x \le 2$

$0 \le 2$

This is a true statement, so we know the point $(0,0)$ lies in the shaded half-plane.

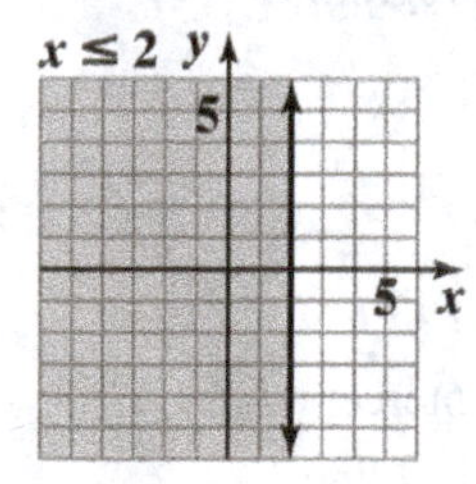

19. $y > -4$

Replacing the inequality symbol with an equal sign, we have $y = -4$. We know that equations of the form $y = b$ are horizontal lines with y-intercept $= b$.

Next, use the origin as a test point.

$y > -4$

$0 > -4$

This is a true statement, so we know the point $(0,0)$ lies in the shaded half- plane.

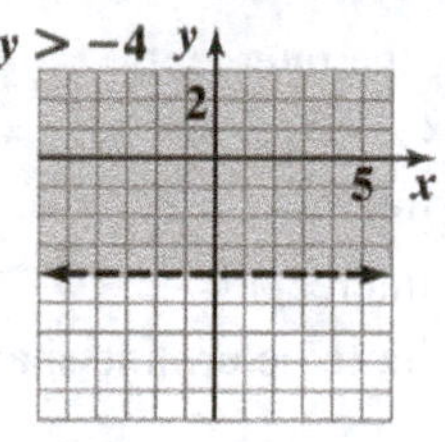

21. $y \ge 0$

Replacing the inequality symbol with an equal sign, we have $y = 0$. We know that equations of the form $y = b$ are horizontal lines with y-intercept $= b$. In this case, we have $y = 0$, the equation of the x-axis.

Next, we need to find a test point. We cannot use the origin, because it lies on the line. Use $(1,1)$ as a test point.

$y \ge 0$

$1 \ge 0$

This is a true statement, so we know the point $(1,1)$ lies in the shaded half-plane.

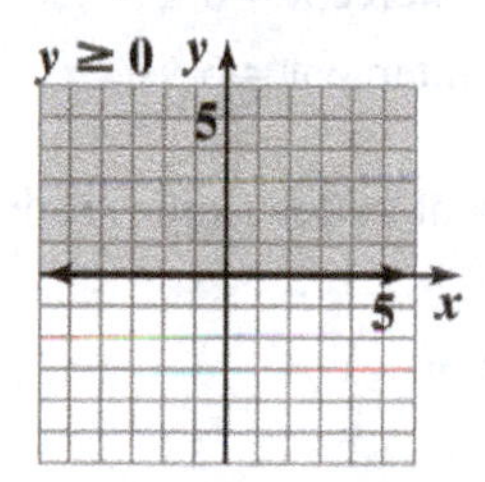

23. $3x + 6y \le 6$

$2x + y \le 8$

Graph the equations using the intercepts.

$3x + 6y = 6$	$2x + y = 8$
$x-$intercept $= 2$	$x-$intercept $= 4$
$y-$intercept $= 1$	$y-$intercept $= 8$

Use the origin as a test point to determine shading.

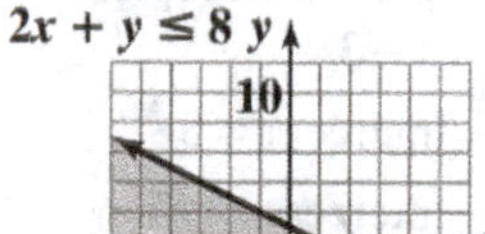
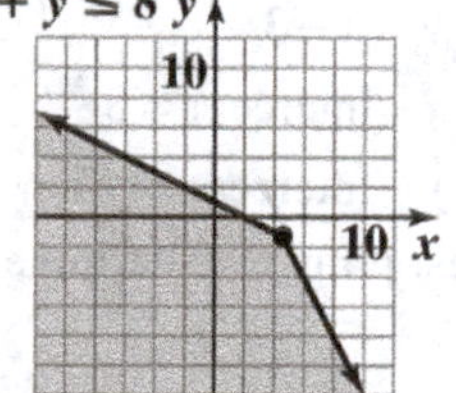

The solution set is the intersection of the shaded half-planes.

25. $2x-5y\le 10$

$3x-2y>6$

Graph the equations using the intercepts.

$2x-5y=10$	$3x-2y=6$
$x-\text{intercept}=5$	$x-\text{intercept}=2$
$y-\text{intercept}=-2$	$y-\text{intercept}=-3$

Use the origin as a test point to determine shading.

$2x-5y\le 10$
$3x-2y>6$

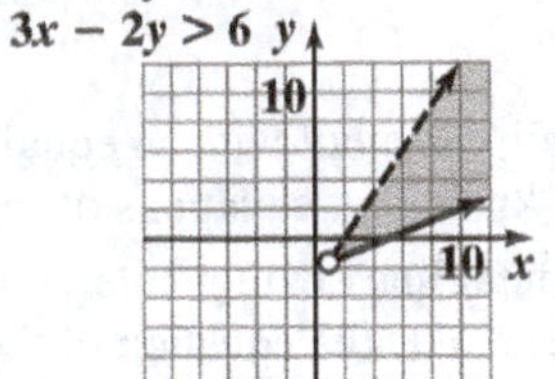

The solution set is the intersection of the shaded half-planes.

27. $y>2x-3$

$y<-x+6$

Graph the equations using the intercepts.

$y=2x-3$	$y=-x+6$
$x-\text{intercept}=\frac{3}{2}$	$x-\text{intercept}=6$
$y-\text{intercept}=-3$	$y-\text{intercept}=6$

Use the origin as a test point to determine shading.

$y>2x-3$
$y<-x+6$

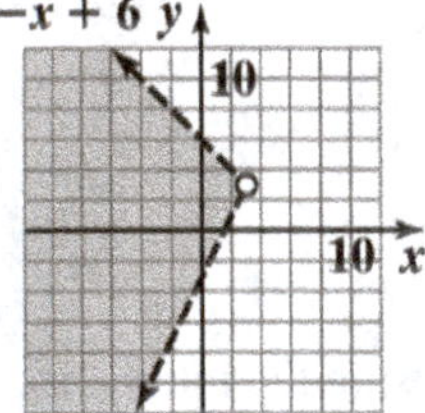

The solution set is the intersection of the shaded half-planes.

29. $x+2y\le 4$

$y\ge x-3$

Graph the equations using the intercepts.

$x+2y=4$	$y=x-3$
$x-\text{intercept}=4$	$x-\text{intercept}=3$
$y-\text{intercept}=2$	$y-\text{intercept}=-3$

Use the origin as a test point to determine shading.

$x+2y\le 4$
$y\ge x-3$

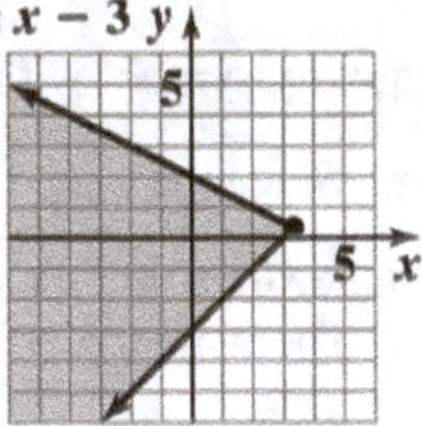

The solution set is the intersection of the shaded half-planes.

31. $x\le 2$

$y\ge -1$

Graph the vertical line, $x=2$, and the horizontal line, $y=-1$. Use the origin as a test point to determine shading.

$x\le 2$
$y\ge -1$

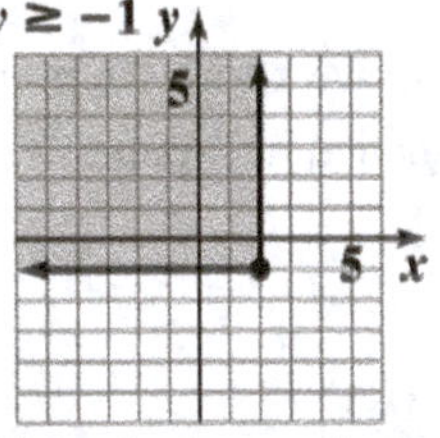

The solution set is the intersection of the shaded half-planes.

33. $-2\le x<5$

Since x lies between -2 and 5, graph the two vertical lines, $x=-2$ and $x=5$. Since x lies between -2 and 5, shade between the two vertical lines.

$-2\le x<5$

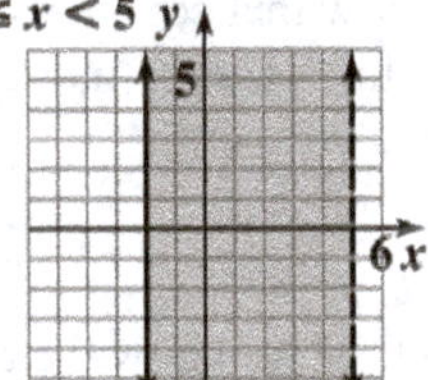

The solution is the shaded region.

35. $x-y\le 1$

$x\ge 2$

Graph the equations.

$x-y=1$	$x=2$
$x-\text{intercept}=1$	$x-\text{intercept}=2$
$y-\text{intercept}=-1$	vertical line

Use the origin as a test point to determine shading.

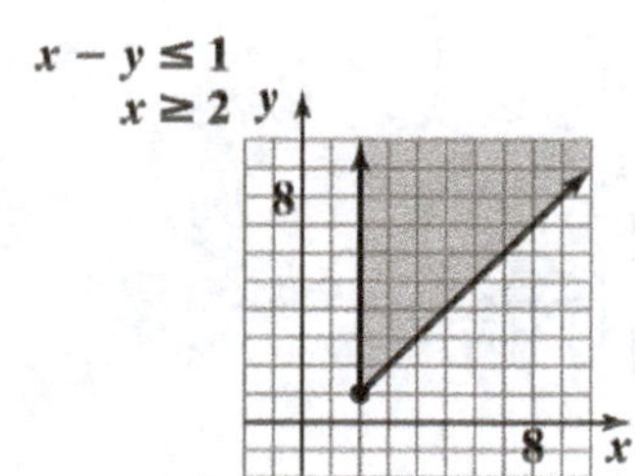

The solution set is the intersection of the shaded half-planes.

37. $x + y > 4$

$x + y < -1$

Graph the equations using the intercepts.

$x + y = 4$	$x + y = -1$
$x - \text{intercept} = 4$	$x - \text{intercept} = -1$
$y - \text{intercept} = 4$	$y - \text{intercept} = -1$

Use the origin as a test point to determine shading.

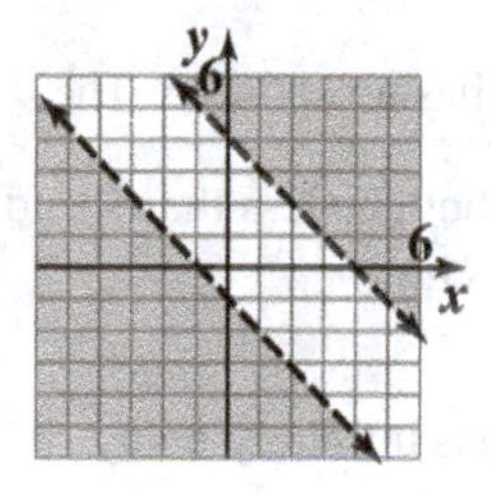

The solution set is the intersection of the shaded half-planes. Since the shaded half-planes do not intersect, there is no solution. The solution set is $\varnothing$ or $\{\ \}$.

39. $x + y > 4$

$x + y > -1$

Graph the equations using the intercepts.

$x + y = 4$	$x + y = -1$
$x - \text{intercept} = 4$	$x - \text{intercept} = -1$
$y - \text{intercept} = 4$	$y - \text{intercept} = -1$

Use the origin as a test point to determine shading.

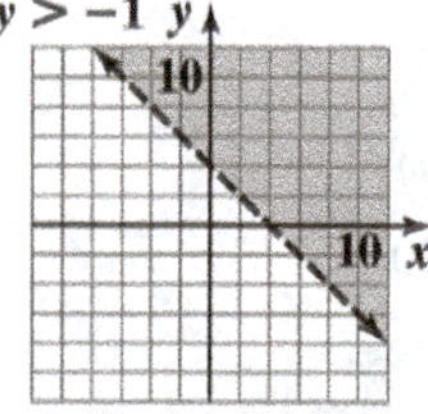

The solution set is the intersection of the shaded half-planes.

41. $x - y \le 2$

$x \ge -2$

$y \le 3$

Graph the equations using the intercepts.

$x - y = 2$	$y = 3$
$x - \text{intercept} = 2$	$y - \text{intercept} = 3$
$y - \text{intercept} = -2$	horizontal line

$x = -2$

$x - \text{intercept} = -2$

vertical line

Use the origin as a test point to determine shading.

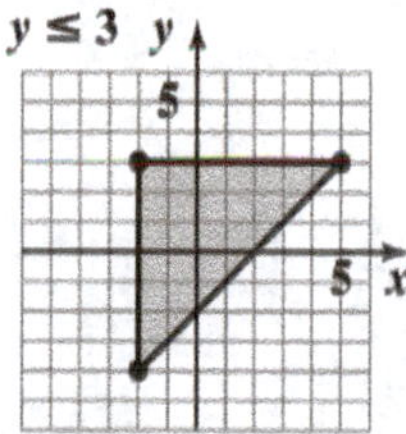

The solution set is the intersection of the shaded half-planes.

43. $x \ge 0$

$y \ge 0$

$2x + 5y \le 10$

$3x + 4y \le 12$

Since both x and y are greater than 0, we are concerned only with the first quadrant. Graph the other equations using the intercepts.

$2x + 5y = 10$	$3x + 4y = 12$
$x - \text{intercept} = 5$	$x - \text{intercept} = 4$
$y - \text{intercept} = 2$	$y - \text{intercept} = 3$

Use the origin as a test point to determine shading.

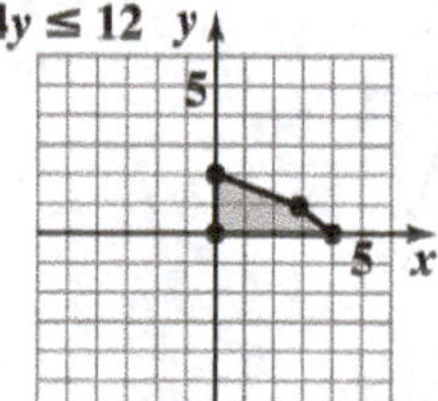

The solution set is the intersection of the shaded half-planes.

45. $3x + y \le 6$

$2x - y \le -1$

$x \ge -2$

$y \le 4$

Graph the equations using the intercepts.

$z = 25x + 55y$

$= 25(30) + 55(10)$

$= 750 + 550 = 1300$

$z = 25x + 55y$

$= 25(70) + 55(10)$

$= 1750 + 550 = 2300$

Use the origin as a test point to determine shading.

$3x + y \le 6$
$2x - y \le -1$
$x \ge -2$
$y \le 4$

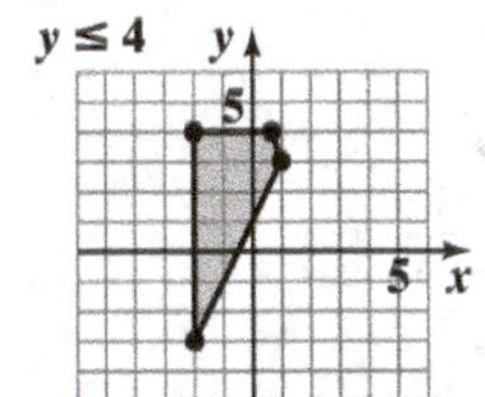

The solution set is the intersection of the shaded half-planes. Because all inequalities are greater than or equal to or less than or equal to, the boundaries of the shaded half-planes are also included in the solution set.

47. $y \ge -2x + 4$

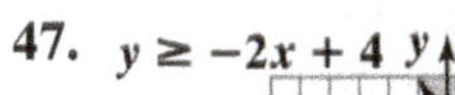

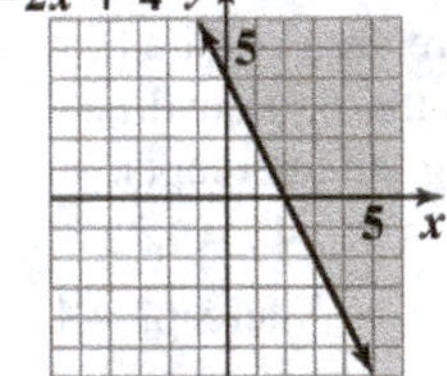

49. $x + y \le 4$
$3x + y \le 6$

51. $-2 \le x \le 2$
$-3 \le y \le 3$

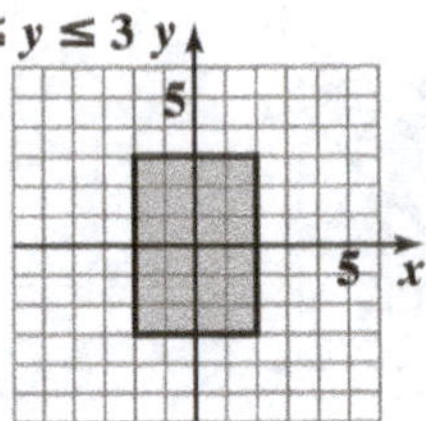

53. Find the union of solutions of $y > \frac{3}{2}x - 2$ and $y < 4$.

$y > \frac{3}{2}x - 2$ or $y < 4$

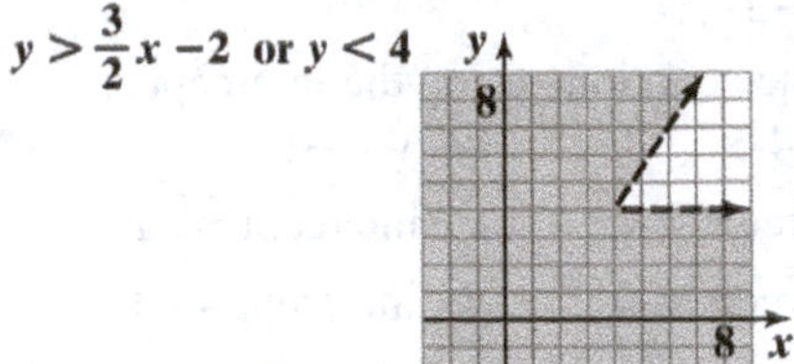

55. The system $\begin{array}{l} 3x + 3y < 9 \\ 3x + 3y > 9 \end{array}$ has no solution. The number $3x + 3y$ cannot both be less than 9 and greater than 9 at the same time.

57. The system $\begin{array}{l} 3x + y \le 9 \\ 3x + y \ge 9 \end{array}$ has infinitely many solutions. The solutions are all points on the line $3x + y = 9$.

59. **a.** The coordinates of point A are $(20, 150)$. This means that a 20 year-old person with a heart rate of 150 beats per minute falls within the target zone.

b. $10 \le a \le 70$

$10 \le 20 \le 70$, true

$H \ge 0.7(220 - a)$

$150 \ge 0.7(220 - 20)$

$150 \ge 140$, true

$H \le 0.8(220 - a)$

$150 \le 0.8(220 - 20)$

$150 \le 160$, true

Since point *A* makes all three inequalities true, it is a solution of the system.

61. $10 \le a \le 70$

$H \ge 0.6(220 - a)$

$H \le 0.7(220 - a)$

63. a.

$$\begin{aligned} y &\ge 0 \\ x + y &\ge 5 \\ x &\ge 1 \\ 200x + 100y &\le 700 \end{aligned}$$

b. $y \ge 0$

$x + y \ge 5$

$x \ge 1$

$200x + 100y \le 700$

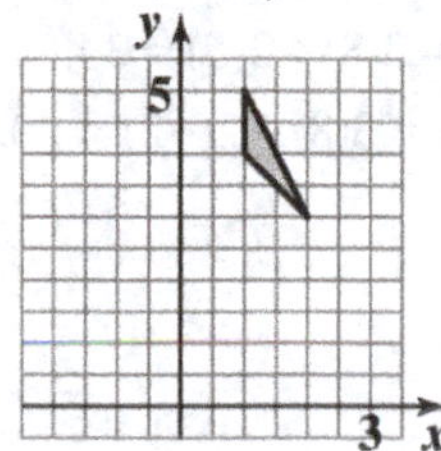

c. 2 nights

65. – 73. Answers will vary.

75. $y \le 4x + 4$

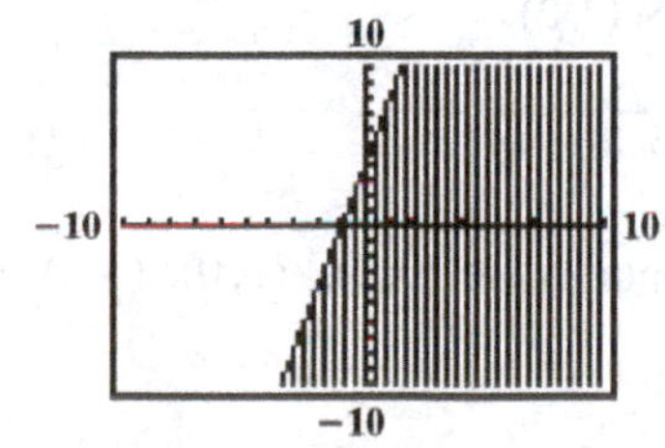

77. $2x + y \le 6$

$y \le -2x + 6$

79. Answers will vary.

81. Answers will vary. For example, verify Exercise 23.

$3x + 6y \le 6$

$2x + y \le 8$

First solve both inequalities for y.

$3x + 6y \le 6$ $\qquad$ $2x + y \le 8$

$6y \le -3x + 6$ $\qquad$ $y \le -2x + 8$

$y \le -\frac{1}{2}x + 1$

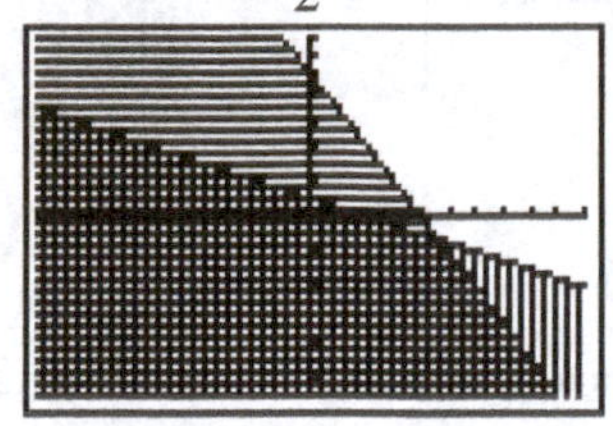

83. makes sense

85. makes sense

87. false; Changes to make the statement true will vary. A sample change is: The graph of $y \ge -x + 1$ has a solid line that falls from left to right.

89. true

91. The shading is above a dashed line of $y = -1$ and to the right of a solid line of $x = -2$. Thus, the system of inequalities is $x \ge -2$

$y > -1.$

93. The slope of the linear equation is

$$m = \frac{y_2 - y_1}{x_2 - x_1} = \frac{6 - (-8)}{4 - (-3)} = \frac{14}{7} = 2.$$

Use the point slope form to find the of the linear equation.

$y - y_1 = m(x - x_1)$

$y - 6 = 2(x - 4)$

or

$y - 6 = 2x - 8$

$y = 2x - 2$

The point (1, 1) is above the line so the inequality is $y \ge 2x - 2$.

95. $y \ge nx + b$

$y \le mx + b$

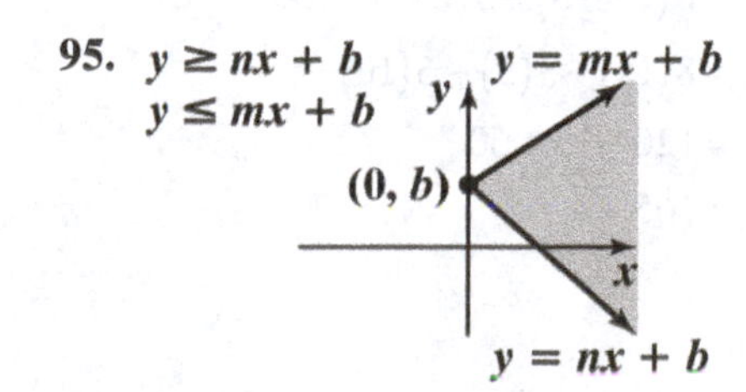

96. $3x - y = 8$
$x - 5y = -2$

$$\left[\begin{array}{cc|c} 3 & -1 & 8 \\ 1 & -5 & -2 \end{array}\right] R_1 \leftrightarrow R_2$$
$$= \left[\begin{array}{cc|c} 1 & -5 & -2 \\ 3 & -1 & 8 \end{array}\right] -3R_1 + R_2$$
$$= \left[\begin{array}{cc|c} 1 & -5 & -2 \\ 0 & 14 & 14 \end{array}\right] \frac{1}{14}R_2 = \left[\begin{array}{cc|c} 1 & -5 & -2 \\ 0 & 1 & 1 \end{array}\right]$$

$x - 5y = -2$
$y = 1$

Since we know $y = 1$, we can use back-substitution to find x.

$x - 5(1) = -2$
$x - 5 = -2$
$x = 3$

The solution is $(3, 1)$.

97. $y = 3x - 2$
$y = -2x + 8$

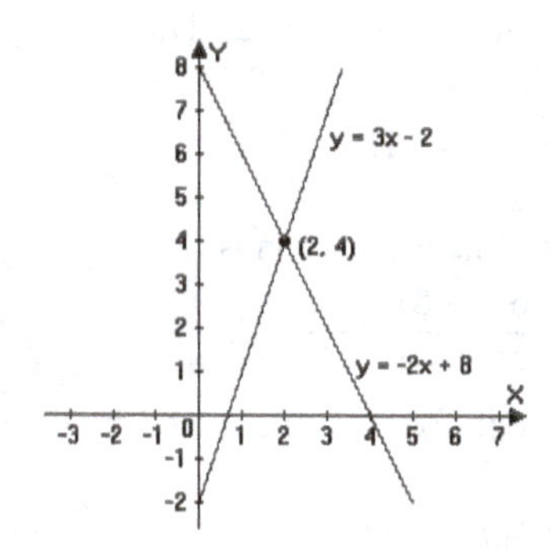

The solution is $(2, 4)$.

98. $$\begin{vmatrix} 8 & 2 & -1 \\ 3 & 0 & 5 \\ 6 & -3 & 4 \end{vmatrix} = 8\begin{vmatrix} 0 & 5 \\ -3 & 4 \end{vmatrix} - 3\begin{vmatrix} 2 & -1 \\ -3 & 4 \end{vmatrix} + 6\begin{vmatrix} 2 & -1 \\ 0 & 5 \end{vmatrix}$$
$= 8(0+15) - 3(8-3) + 6(10-0)$
$= 8(15) - 3(5) + 6(10)$
$= 120 - 15 + 60$
$= 165$

99. a. $x + y \geq 6$
$x \leq 8$
$y \leq 5$

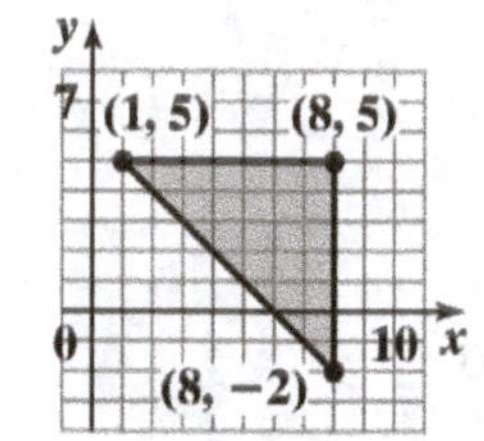

b. The corner points are $(1, 5)$, $(8, 5)$, and $(8, -2)$.

c. At $(1, 5)$, $3x + 2y = 3(1) + 2(5) = 13$.
At $(8, 5)$, $3x + 2y = 3(8) + 2(5) = 34$.
At $(8, -2)$, $3x + 2y = 3(8) + 2(-2) = 20$.

100. a. $x \geq 0$
$y \geq 0$
$3x - 2y \leq 6$
$y \leq -x + 7$

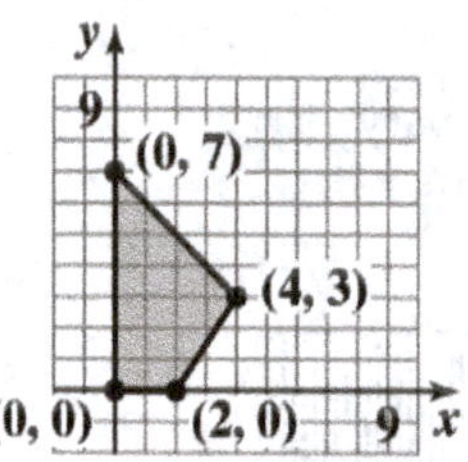

b. The corner points are $(0, 0)$, $(2, 0)$, $(4, 3)$, and $(0, 7)$.

c. At $(0, 0)$, $2x + 5y = 2(0) + 5(0) = 0$.
At $(2, 0)$, $2x + 5y = 2(2) + 5(0) = 4$.
At $(4, 3)$, $2x + 5y = 2(4) + 5(3) = 23$.
At $(0, 7)$, $2x + 5y = 2(0) + 5(7) = 35$.

101. $20x + 10y \leq 80{,}000$

4.5 Check Points

1. The total profit is 25 times the number of bookshelves, x, plus 55 times the number of desks, y. The objective function is $z = 25x + 55y$.

2. Not more than a total of 80 bookshelves and desks can be manufactured per day. This is represented by the inequality $x + y \leq 80$.

3. Objective function: $z = 25x + 55y$

   Constraints: $x + y \leq 80$

   $30 \leq x \leq 80$

   $10 \leq y \leq 30$

4. Graph the constraints and find the corners, or vertices, of the region of intersection.

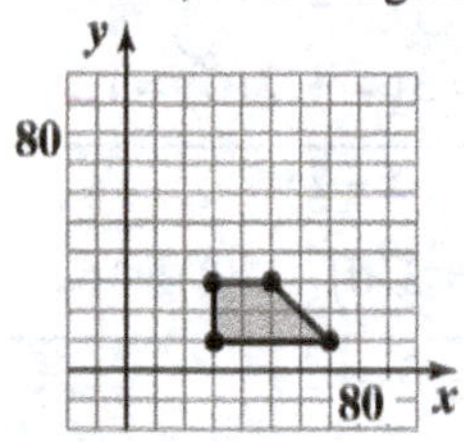

Find the value of the objective function at each corner of the graphed region.

Corner (x, y)	Objective Function $z = 25x + 55y$
$(30, 10)$	$z = 25(30) + 55(10)$ $= 750 + 550$ $= 1300$
$(30, 30)$	$z = 25(30) + 55(30)$ $= 750 + 1650$ $= 2400$
$(50, 30)$	$z = 25(50) + 55(30)$ $= 1250 + 1650$ $= 2900$ (Maximum)
$(70, 10)$	$z = 25(70) + 55(10)$ $= 1750 + 550$ $= 2300$

The maximum value of z is 2900 and it occurs at the point (50, 30).
In order to maximize profit, 50 bookshelves and 30 desks must be produced each day for a profit of $2900.

5. Objective Function: $z = 3x + 5y$

   Constraints: $x \geq 0, \quad y \geq 0$

   $x + y \geq 1$

   $x + y \leq 6$

   Graph the region that represents the intersection of the constraints:

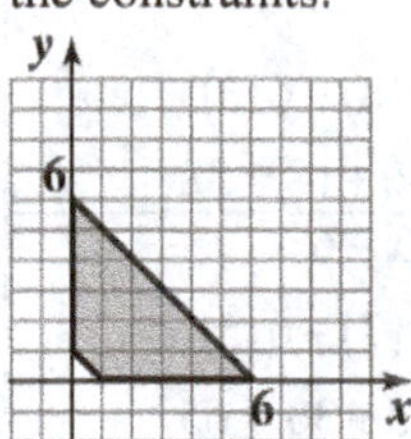

Find the value of the objective function at each corner of the graphed region.

Corner (x, y)	Objective Function $z = 3x + 5y$
$(0,1)$	$z = 3(0) + 5(1) = 5$
$(1, 0)$	$z = 3(1) + 5(0) = 3$
$(0, 6)$	$z = 3(0) + 5(6) = 30$ (Maximum)
$(6, 0)$	$z = 3(6) + 5(0) = 18$

4.5 Concept and Vocabulary Check

1. linear programming

2. objective

3. constraints; corner

4.5 Exercise Set

1.

Corner (x, y)	Objective Function $z = 5x+6y$
(1, 2)	$z = 5x+6y$ $= 5(1)+6(2)$ $= 5+12 = 17$
(8, 3)	$z = 5x+6y$ $= 5(8)+6(3)$ $= 40+18 = 58$
(7, 5)	$z = 5x+6y$ $= 5(7)+6(5)$ $= 35+30 = 65$
(2, 10)	$z = 5x+6y$ $= 5(2)+6(10)$ $= 10+60$ $= 70$

The maximum value is 70 and the minimum is 17.

3.

Corner (x, y)	Objective Function $z = 40x+50y$
(0, 0)	$z = 40x+50y$ $= 40(0)+50(0)$ $= 0+0 = 0$
(8, 0)	$z = 40x+50y$ $= 40(8)+50(0) = 320$
(4, 9)	$z = 40x+50y$ $= 40(4)+50(9)$ $= 160+450 = 610$
(0, 8)	$z = 40x+50y$ $= 40(0)+50(8) = 400$

The maximum value is 610 and the minimum value is 0.

5. Objective Function: $z = 3x+2y$

Constraints: $x \geq 0,\ y \geq 0$

$2x+y \leq 8$

$x+y \geq 4$

a.

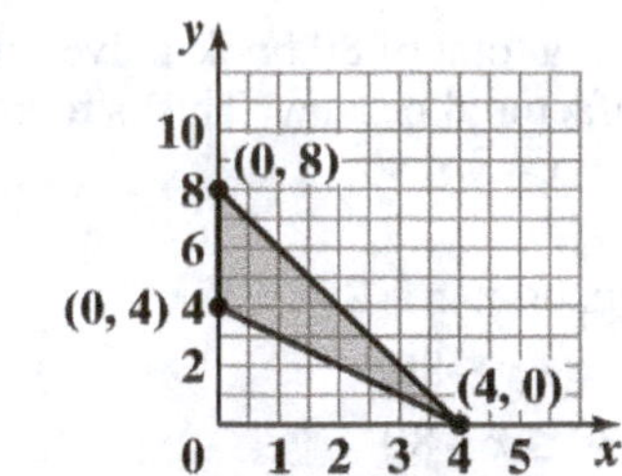

b.

Corner (x, y)	Objective Function $z = 3x+2y$
(4, 0)	$z = 3x+2y$ $= 3(4)+2(0) = 12$
(0, 8)	$z = 3x+2y$ $= 3(0)+2(8) = 16$
(0, 4)	$z = 3x+2y$ $= 3(0)+2(4) = 8$

c. The maximum value is 16. It occurs at the point (0, 8).

7. Objective Function: $z = 4x+y$

Constraints: $x \geq 0,\ y \geq 0$

$2x+3y \leq 12$

$x+y \geq 3$

a.

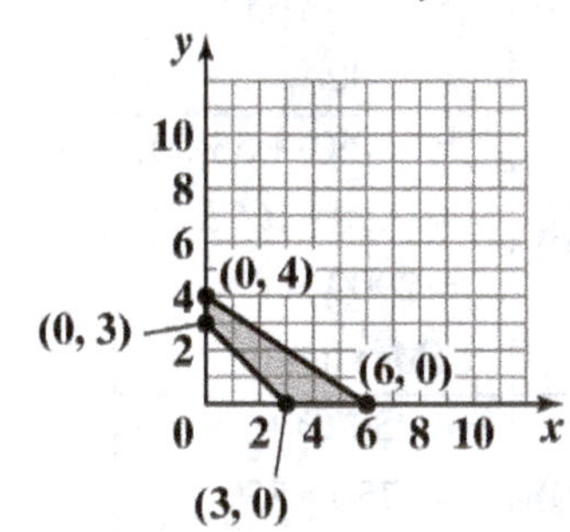

b.

Corner (x, y)	Objective Function $z = 4x + y$
(3, 0)	$z = 4x + y$ $= 4(3) + 0 = 12$
(6, 0)	$z = 4x + y$ $= 4(6) + 0 = 24$
(0, 3)	$z = 4x + y$ $= 4(0) + 3 = 3$
(0, 4)	$z = 4x + y$ $= 4(0) + 4 = 4$

c. The maximum value is 24. It occurs at the point (6, 0).

9. Objective Function: $z = 3x - 2y$

Constraints: $1 \le x \le 5$

$y \ge 2$

$x - y \ge -3$

a.

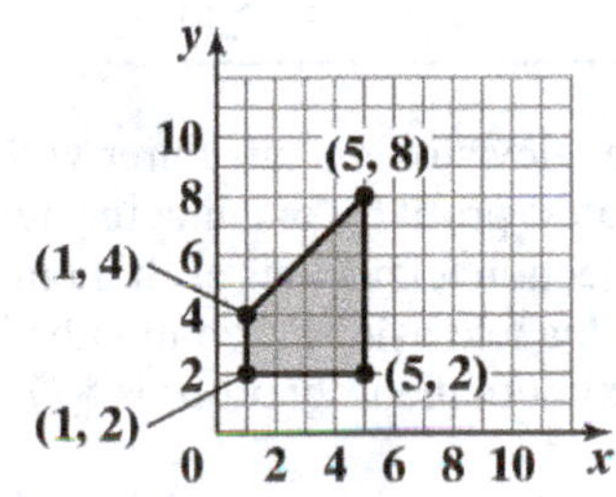

b.

Corner (x, y)	Objective Function $z = 3x - 2y$
(1, 2)	$z = 3x - 2y$ $= 3(1) - 2(2) = -1$
(5, 2)	$z = 3x - 2y$ $= 3(5) - 2(2) = 11$
(5, 8)	$z = 3x - 2y$ $= 3(5) - 2(8) = -1$
(1, 4)	$z = 3x - 2y$ $= 3(1) - 2(4) = -5$

c. The maximum value is 11. It occurs at the point (5, 2).

11. Objective Function: $z = 4x + 2y$

Constraints: $x \ge 0,\ y \ge 0$

$2x + 3y \le 12$

$3x + 2y \le 12$

$x + y \ge 2$

a.

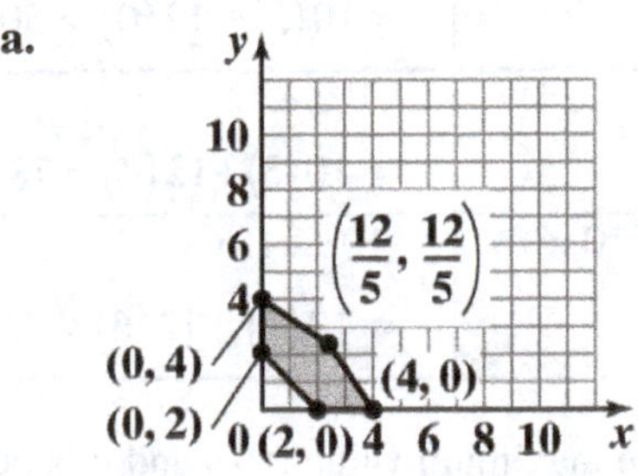

b.

Corner (x, y)	Objective Function $z = 4x + 2y$
(2, 0)	$z = 4x + 2y$ $= 4(2) + 2(0) = 8$
(4, 0)	$z = 4x + 2y$ $= 4(4) + 2(0) = 16$
(2.4,2.4)	$z = 4x + 2y$ $= 4(2.4) + 2(2.4)$ $= 9.6 + 4.8 = 14.4$
(0, 4)	$z = 4x + 2y$ $= 4(0) + 2(4) = 8$
(0, 2)	$z = 4x + 2y$ $= 4(0) + 2(2) = 4$

c. The maximum value is 16. It occurs at the point (4, 0).

13. Objective Function: $z = 10x + 12y$

Constraints: $x \ge 0,\ y \ge 0$

$2x + y \le 10$

$2x + 3y \le 18$

a.

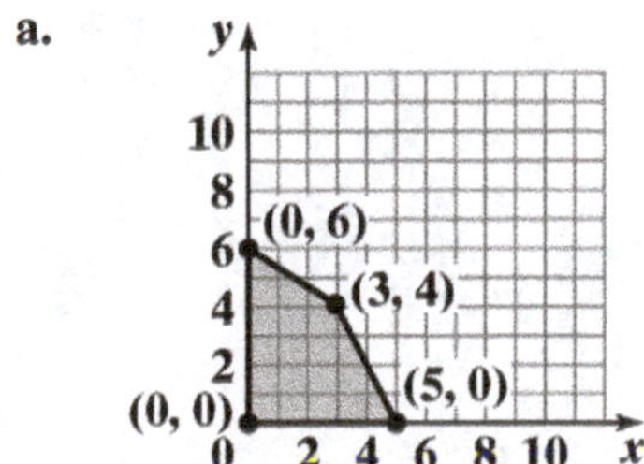

b.

Corner (x, y)	Objective Function $z = 10x + 12y$
(0, 0)	$10x + 12y$ $= 10(0) + 12(0) = 0$
(5, 0)	$10x + 12y$ $= 10(5) + 12(0) = 50$
(3, 4)	$10x + 12y$ $= 10(3) + 12(4) = 78$
(0, 6)	$10x + 12y$ $= 10(0) + 12(6) = 72$

c. The maximum value is 78 and it occurs at the point (3, 4).

15. a. The objective function is $z = 125x + 200y$.

b. Since we can make at most 450 rear-projection televisions, we have $x \le 450$. Since we can make at most 200 plasma televisions, we have $y \le 200$.

Since we can spend at most $360,000 per month, we have $600x + 900y \le 360{,}000$.

c.

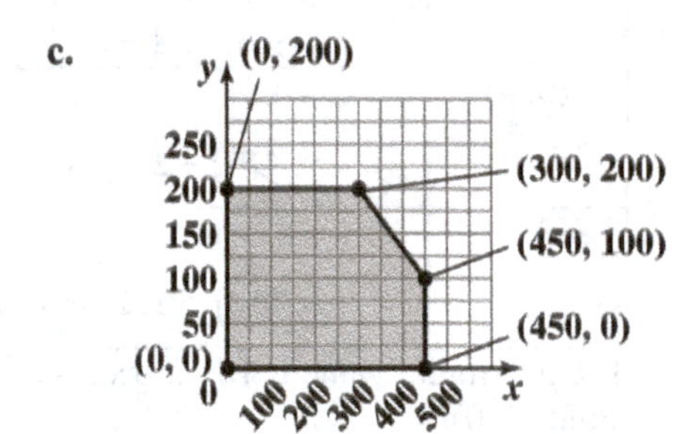

d.

Corner (x, y)	Objective Function $z = 125x + 200y$
(0, 0)	$125x + 200y$ $= 125(0) + 200(0) = 0$
(0, 200)	$125x + 200y$ $= 125(0) + 200(200)$ $= 40{,}000$
(300, 200)	$125x + 200y$ $= 125(300) + 200(200)$ $= 37{,}500 + 40{,}000$ $= 77{,}500$
(450, 100)	$125x + 200y$ $= 125(450) + 200(100)$ $= 56{,}250 + 20{,}000$ $= 76{,}250$
(450, 0)	$125x + 200y$ $= 125(450) + 200(0)$ $= 56{,}250$

e. The television manufacturer will make the greatest profit by manufacturing <u>300</u> rear-projection televisions each month and <u>200</u> plasma televisions each month. The maximum monthly profit is <u>$77,500</u>.

17. Let x = the number of model A bicycles produced.
Let y = the number of model B bicycles produced.
The objective function is $z = 25x + 15y$.

The assembling constraint is $5x + 4y \le 200$.

The painting constraint is $2x + 3y \le 108$.

We also know that x and y must either be zero or a positive number. We cannot make a negative number of bicycles.

Next, graph the constraints.

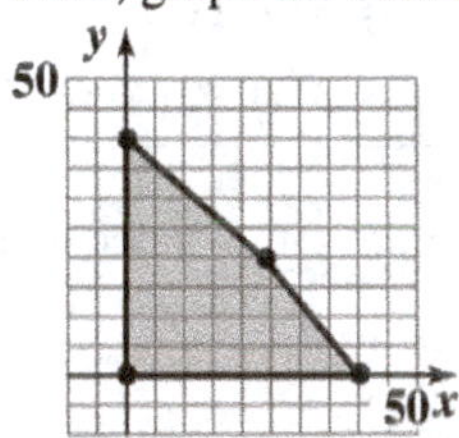

Using the graph, find the value of the objective function at each of the corner points.

Corner (x, y)	Objective Function $z = 25x + 15y$
(0, 0)	$25x + 15y$ $= 25(0) + 15(0)$ $= 0$
(40, 0)	$25x + 15y$ $= 25(40) + 15(0)$ $= 1000$
(24, 20)	$25x + 15y$ $= 25(24) + 15(20)$ $= 600 + 300$ $= 900$
(0, 36)	$25x + 15y$ $= 25(0) + 15(36)$ $= 540$

The maximum of 1000 occurs at the point $(40, 0)$. This means that the company should produce 40 of model A and none of model B each week for a profit of $1000.

19. Let x = the number of cartons of food.
Let y = the number of cartons of clothing.
The objective function is $z = 12x + 5y$.
The weight constraint is $50x + 20y \leq 19,000$.
The volume constraint is $20x + 10y \leq 8000$.
We also know that x and y must either be zero or a positive number. We cannot have a negative number of cartons of food or clothing.

Next, graph the constraints.

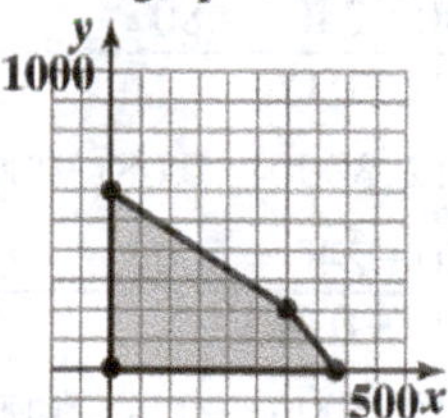

Using the graph, find the value of the objective function at each of the corner points.

Corner (x, y)	Objective Function $z = 12x + 5y$
(0, 0)	$12x + 5y$ $= 12(0) + 5(0) = 0$
(380, 0)	$12x + 5y$ $= 12(380) + 5(0)$ $= 4560$
(300, 200)	$12x + 5y$ $= 12(300) + 5(200)$ $= 3600 + 1000 = 4600$
(0, 600)	$12x + 5y$ $= 12(0) + 5(600)$ $= 3000$

The maximum of 4600 occurs at the point (300, 200). This means that to maximize the number of people who are helped, 300 boxes of food and 200 boxes of clothing should be sent.

21. Let x = the number of parents.
Let y = the number of students.
The objective function is $z = 2x + y$.
The seating constraint is $x + y \leq 150$.
The two parents per student constraint is $2y \geq x$.
We also know that x and y must either be zero or a positive number. We cannot have a negative number of parents or students.

Next, graph the constraints.

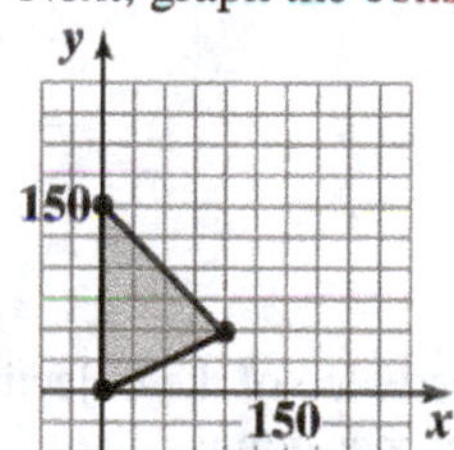

Using the graph, find the value of the objective function at each of the corner points.

Corner (x, y)	Objective Function $z = 2x + y$
(0, 0)	$2x+y$ $= 2(0)+0=0$
(100, 50)	$2x+y$ $= 2(100)+50$ $= 200+50 = 250$
(0, 150)	$2x+y$ $= 2(0)+150 = 150$

The maximum of 250 occurs at the point (100, 50). This means that to maximize the amount of money raised, 100 parents and 50 students should attend.

23. Let x = the number of Boeing 727s.
Let y = the number of Falcon 20s.
The objective function is $z = x + y$.

The hourly operating cost constraint is $1400x + 500y \le 35,000$.

The total payload constraint is $42000x + 6000y \ge 672,000$.

The 727 constraint is $x \le 20$.
We also know that x and y must either be zero or a positive number. We cannot have a negative number of aircraft.

Next, graph the constraints.

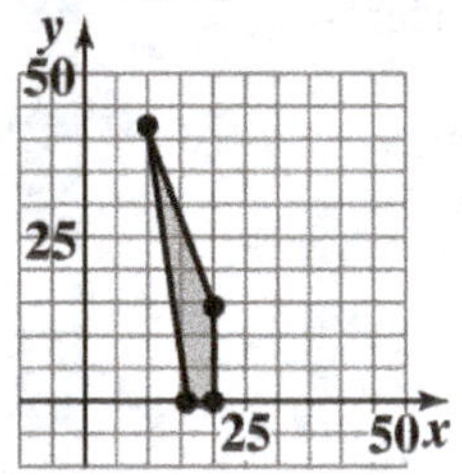

Using the graph, find the value of the objective function at each of the corner points.

Corner (x, y)	Objective Function $z = x + y$
(16, 0)	$z = x + y$ $= 16+0 = 16$
(20, 0)	$z = x + y$ $= 20+0 = 20$
(20, 14)	$z = x + y$ $= 20+14 = 34$
(10, 42)	$z = x + y$ $= 10+42 = 52$

The maximum of 52 occurs at the point (10, 42). This means that to maximize the number of aircraft, 10 Boeing 727s and 42 Falcon 20s should be purchased.

25. – 27. Answers will vary.

29. does not make sense; Explanations will vary. Sample explanation: Solving a linear programming problem does not require graphing the objective function.

31. makes sense

33. Let x = the amount invested in stocks.
Let y = the amount invested in bonds.
The objective function is $z = 0.12x + 0.08y$.

The total money constraint is $x + y \le 10000$.

The minimum bond investment constraint is $y \ge 3000$.

The minimum stock investment constraint is $x \ge 2000$.

The stock versus bond constraint is $y \ge x$.

We also know that x and y must either be zero or a positive number. We cannot invest a negative amount of money.

Next, graph the constraints.

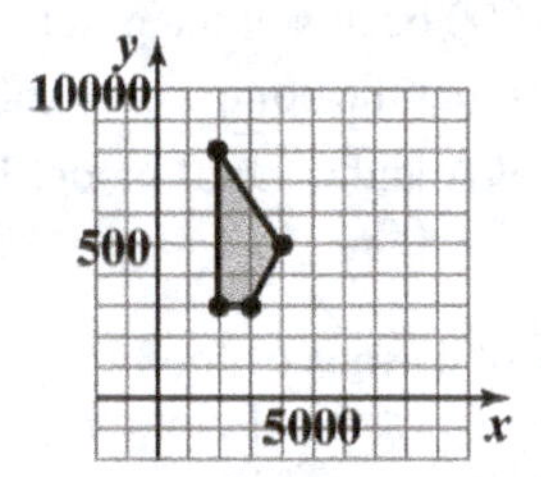

Using the graph, find the value of the objective function at each of the corner points.

Corner (x, y)	Objective Function $z = 0.12x + 0.08y$
(2000,3000)	$0.12x+0.08y$ $= 0.12(2000)+0.08(3000)$ $= 240+240 = 480$
(3000,3000)	$0.12x+0.08y$ $= 0.12(3000)+0.08(3000)$ $= 360+240 = 600$
(5000,5000)	$0.12x+0.08y$ $= 0.12(5000)+0.08(5000)$ $= 600+400 = 1000$
(2000,8000)	$0.12x+0.08y$ $= 0.12(2000)+0.08(8000)$ $= 240+640 = 880$

The maximum of 1000 occurs at the point (5000, 5000). This means that to maximize the return on the investment, \$5000 should be invested in stocks and \$5000 should be invested in bonds.

35. $\left(2x^4y^3\right)\left(3xy^4\right)^3 = 2x^4y^33^3x^3y^{12}$
$= 2\cdot 3^3x^4x^3y^3y^{12}$
$= 2\cdot 27x^7y^{15}$
$= 54x^7y^{15}$

36.
$$3P = \frac{2L-W}{4}$$
$$12P = 2L-W$$
$$12P+W = 2L$$
$$\frac{12P+W}{2} = L$$
$$\frac{12P}{2}+\frac{W}{2} = L$$
$$L = 6P+\frac{W}{2}$$
or
$$L = \frac{12P+W}{2}$$

37. $f(-1) = (-1)^3+2(-1)^2-5(-1)+4$
$= -1+2(1)+5+4$
$= -1+2+5+4 = 10$

38. $(-9x^3+7x^2-5x+3)+(13x^3+2x^2-8x-6)$
$= -9x^3+7x^2-5x+3+13x^3+2x^2-8x-6$
$= 4x^3+9x^2-13x-3$

39. $(7x^3-8x^2+9x-6)-(2x^3-6x^2-3x+9)$
$= 7x^3-8x^2+9x-6-2x^3+6x^2+3x-9$
$= 5x^3-2x^2+12x-15$

40. a. Function g rises to the left and falls to the right.

b. Function f falls to the left and rises to the right.

Chapter 4 Review

1. $-6x+3\le 15$
$-6x\le 12$
$\frac{-6x}{-6}\ge\frac{12}{-6}$
$x\ge -2$

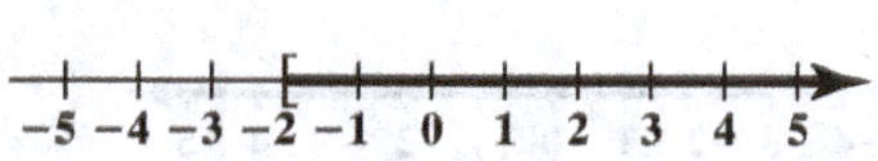

The solution set is $\{x|x\ge -2\}$ or $[2,\infty)$.

2. $6x-9\ge -4x-3$
$10x-9\ge -3$
$10x\ge 6$
$x\ge\frac{6}{10}$
$x\ge\frac{3}{5}$

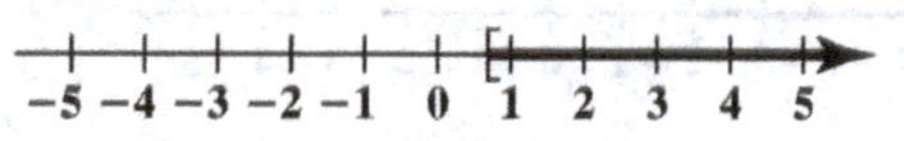

The solution set is $\left\{x\middle|x\ge\frac{3}{5}\right\}$ or $\left[\frac{3}{5},\infty\right)$.

3. $\frac{x}{3}-\frac{3}{4}-1>\frac{x}{2}$
$12\left(\frac{x}{3}\right)-12\left(\frac{3}{4}\right)-12(1)>12\left(\frac{x}{2}\right)$
$4x-3(3)-12>6x$
$4x-9-12>6x$
$4x-21>6x$
$-2x-21>0$
$-2x>21$
$x<-\frac{21}{2}$

The solution set is $\left\{x\middle|x<-\frac{21}{2}\right\}$ or $\left(-\infty,-\frac{21}{2}\right)$.

4. $6x+5>-2(x-3)-25$

$6x+5>-2x+6-25$

$6x+5>-2x-19$

$8x+5>-19$

$8x>-24$

$x>-3$

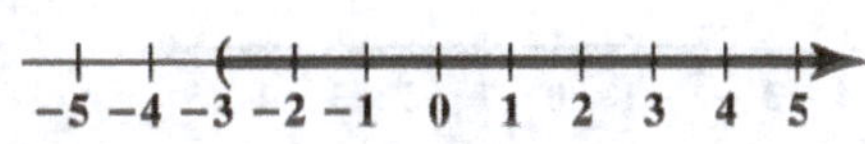

The solution set is $\{x|x>-3\}$ or $(-3,\infty)$.

5. $3(2x-1)-2(x-4)\geq 7+2(3+4x)$

$6x-3-2x+8\geq 7+6+8x$

$4x+5\geq 13+8x$

$-4x+5\geq 13$

$-4x\geq 8$

$x\leq -2$

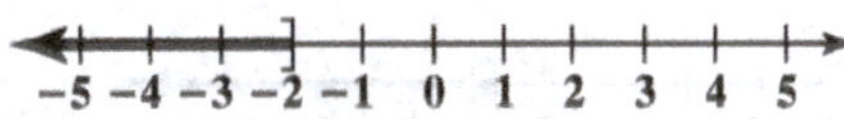

The solution set is $\{x|x\leq -2\}$ or $(-\infty,-2]$.

6. $2x+7\leq 5x-6-3x$

$2x+7\leq 2x-6$

$7\leq -6$

This is a contradiction. Seven is not less than or equal to –6. There are no values of x for which $7\leq -6$. The solution set is $\varnothing$ or $\{\ \}$.

7. Let x = the number of checks written per month. The cost using the first method is $c_1=11+0.06x$. The cost using the second method is $c_2=4+0.20x$. The first method is a better deal if it costs less than the second method.

$c_1<c_2$

$11+0.06x<4+0.20x$

$11-0.14x<4$

$-0.14x<-7$

$\frac{-0.14x}{-0.14}>\frac{-7}{-0.14}$

$x>50$

The first method is a better deal when more than 50 checks per month are written.

8. Let x = the amount of sales per month in dollars. The salesperson's commission is $c=500+0.20x$. We are looking for the amount of sales, x, the salesman must make to receive more than \$3200 in income.

$c>3200$

$500+0.20x>3200$

$0.20x>2700$

$x>13500$

The salesman must sell more than \$13,500 to receive a total income that exceeds \$3200 per month.

9. $A\cap B=\{a,c\}$

10. $A\cap C=\{a\}$

11. $A\cup B=\{a,b,c,d,e\}$

12. $A\cup C=\{a,b,c,d,f,g\}$

13. $x\leq 3$ and $x<6$

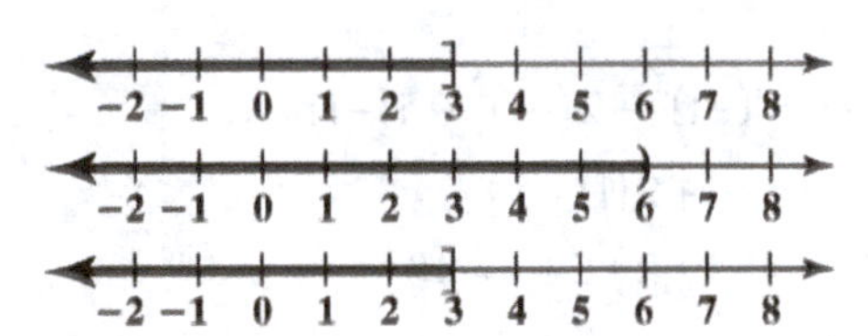

The solution set is $\{x|x\leq 3\}$ or $(-\infty,3]$.

14. $x\leq 3$ or $x<6$

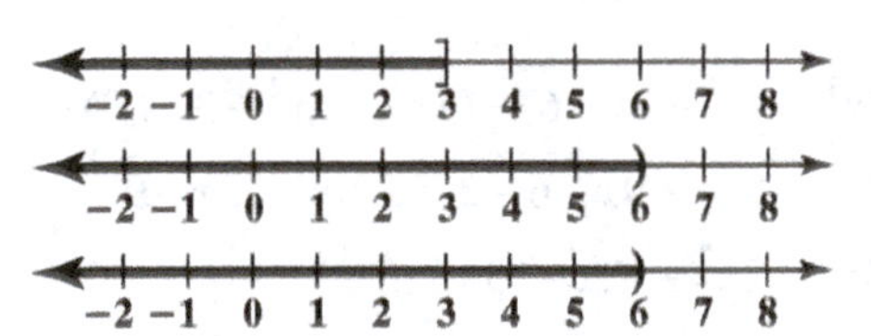

The solution set is $\{x|x<6\}$ or $(-\infty,6)$.

15. $-2x<-12$ and $x-3<5$

$\frac{-2x}{-2}>\frac{-12}{-2}$ $\quad x<8$

$x>6$

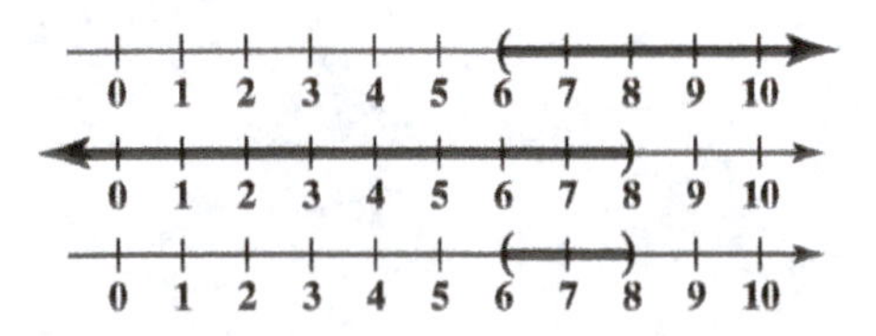

The solution set is $\{x|6<x<8\}$ or $(6,8)$.

16. $5x+3\le 18$ and $2x-7\le -5$

$5x\le 15$ $\quad 2x\le 2$

$x\le 3$ $\quad x\le 1$

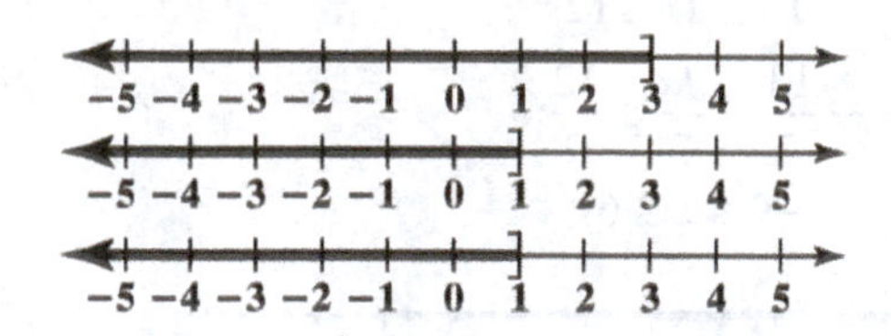

The solution set is $\{x|x\le 1\}$ or $(-\infty,1]$.

17. $2x-5>-1$ and $3x<3$

$2x>4$ $\quad x<1$

$x>2$

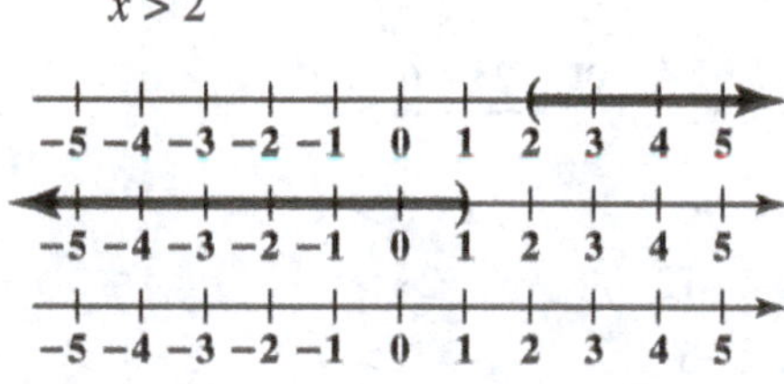

Since the two sets do not intersect, the solution set is $\varnothing$ or $\{\ \}$.

18. $2x-5>-1$ or $3x<3$

$2x>4$ $\quad x<1$

$x>2$

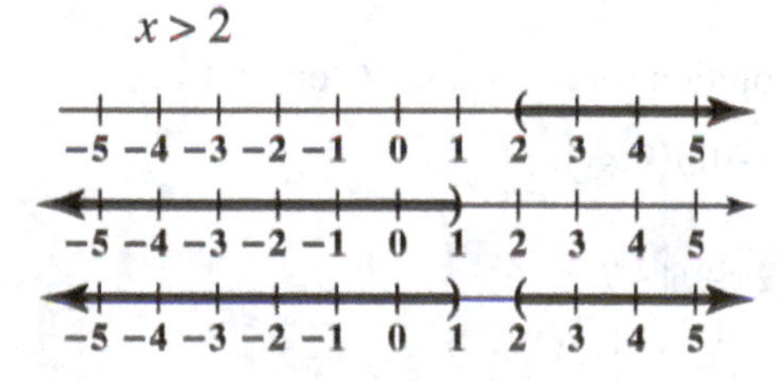

The solution set is $\{x|x<1 \text{ or } x>2\}$ or $(-\infty,1)\cup(2,\infty)$.

19. $x+1\le -3$ or $-4x+3<-5$

$x\le -4$ $\quad -4x<-8$

$x>2$

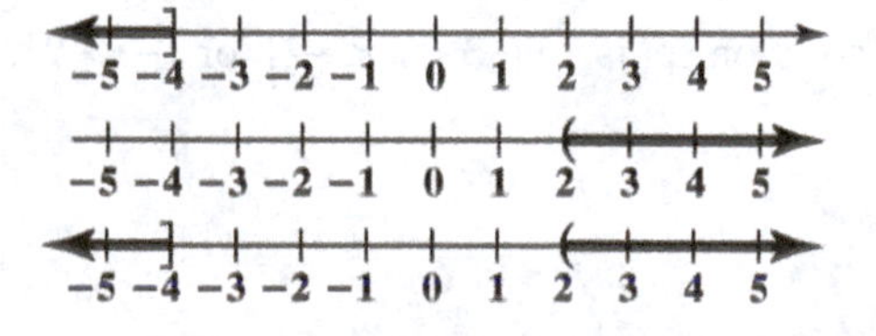

The solution set is $\{x|x\le -4 \text{ or } x>2\}$ or $(-\infty,-4]\cup(2,\infty)$.

20. $5x-2\le -22$ or $-3x-2>4$

$5x\le -20$ $\quad -3x>6$

$x\le -4$ $\quad x<-2$

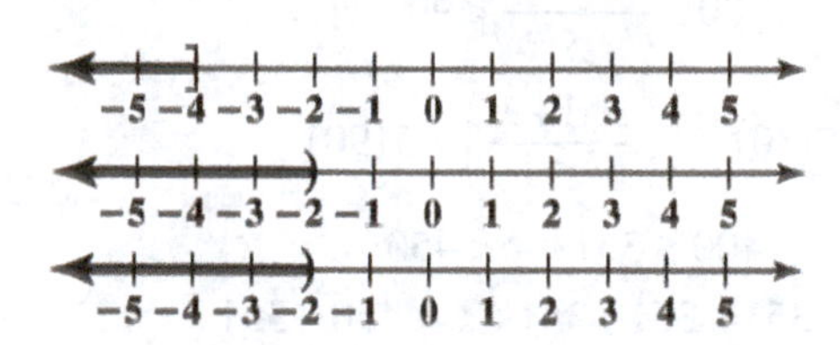

The solution set is $\{x|x<-2\}$ or $(-\infty,-2)$.

21. $5x+4\ge -11$ or $1-4x\ge 9$

$5x\ge -15$ $\quad -4x\ge 8$

$x\ge -3$ $\quad x\le -2$

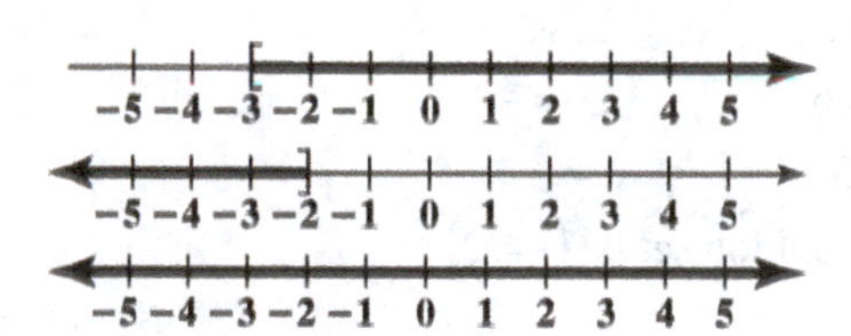

The solution set is $\mathbb{R}$, $(-\infty,\infty)$ or $\{x|x \text{ is a real number}\}$.

22.

$$-3<x+2\le 4$$
$$-3-2<x+2-2\le 4-2$$
$$-5<x\le 2$$

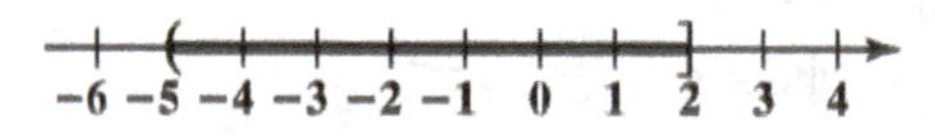

The solution set is $\{x|-5<x\le 2\}$ or $(-5,2]$.

23.

$$-1\le 4x+2\le 6$$
$$-1-2\le 4x+2-2\le 6-2$$
$$-3\le 4x\le 4$$
$$-\frac{3}{4}\le \frac{4x}{4}\le \frac{4}{4}$$
$$-\frac{3}{4}\le x\le 1$$

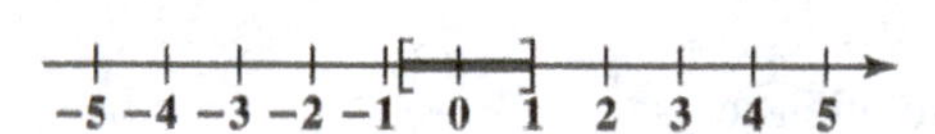

The solution set is $\left\{x\middle|-\frac{3}{4}\le x\le 1\right\}$ or $\left[-\frac{3}{4},1\right]$.

24. Let x = the grade on the fifth exam.

$$80 \le \frac{95+79+91+86+x}{5} < 90$$
$$80 \le \frac{351+x}{5} < 90$$
$$5(80) \le 5\left(\frac{351+x}{5}\right) < 5(90)$$
$$400 \le 351+x < 450$$
$$400-351 \le 351-351+x < 450-351$$
$$49 \le x < 99$$

You need to score at least 49% and less than 99% on the exam to receive a B. In interval notation, the range is $[49\%, 99\%)$.

25. $|2x+1| = 7$

$2x+1=7$ or $2x+1=-7$

$2x=6$ $\quad$ $2x=-8$

$x=3$ $\quad$ $x=-4$

The solution set is $\{-4, 3\}$.

26. $|3x+2| = -5$

There are no values of x for which the absolute value of $3x + 2$ is a negative number. By definition, absolute values are always positive. The solution set is $\varnothing$ or $\{\ \}$.

27. $2|x-3|-7=10$

$$2|x-3| = 17$$
$$|x-3| = 8.5$$

$x-3=8.5$ or $x-3=-8.5$

$x=11.5$ $\quad$ $x=-5.5$

The solution set is $\{-5.5, 11.5\}$.

28. $|4x-3| = |7x+9|$

$4x-3=7x+9$ or $4x-3=-7x-9$

$-3x-3=9$ $\quad$ $11x-3=-9$

$-3x=12$ $\quad$ $11x=-6$

$x=-4$ $\quad$ $x=-\frac{6}{11}$

The solution set is $\left\{-4, -\frac{6}{11}\right\}$.

29.

$$|2x+3| \le 15$$
$$-15 \le 2x+3 \le 15$$
$$-15-3 \le 2x+3-3 \le 15-3$$
$$-18 \le 2x \le 12$$
$$-\frac{18}{2} \le \frac{2x}{2} \le \frac{12}{2}$$
$$-9 \le x \le 6$$

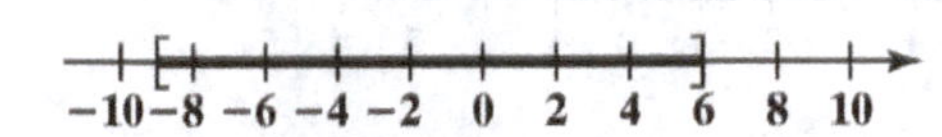

The solution set is $\{x \mid -9 \le x \le 6\}$ or $[-9, 6]$.

30. $\left|\frac{2x+6}{3}\right| > 2$

$\frac{2x+6}{3} < -2$ or $\frac{2x+6}{3} > 2$

$2x+6 < -6$ $\quad$ $2x+6 > 6$

$2x < -12$ $\quad$ $2x > 0$

$x < -6$ $\quad$ $x > 0$

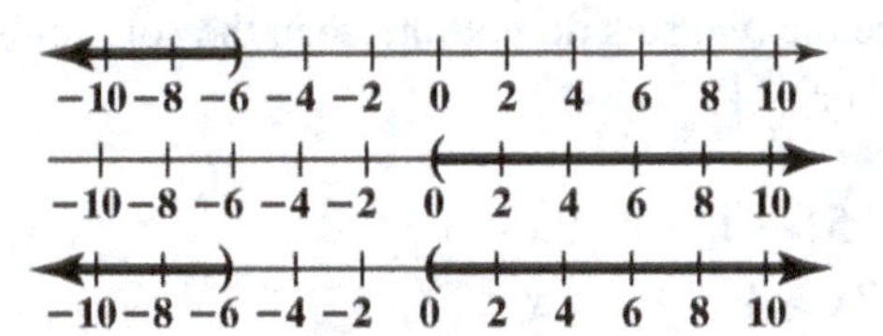

The solution set is $\{x \mid x < -6 \text{ or } x > 0\}$ or $(-\infty, -6) \cup (0, \infty)$.

31.

$$|2x+5|-7 < -6$$
$$|2x+5| < 1$$
$$-1 < 2x+5 < 1$$
$$-1-5 < 2x+5-5 < 1-5$$
$$-6 < 2x < -4$$
$$-3 < x < -2$$

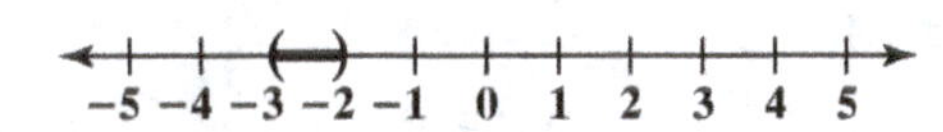

The solution set is $\{x \mid -3 < x < -2\}$ or $(-3, -2)$.

32. $-4|x+2|+5 \le -7$

$-4|x+2| \le -12$

$\frac{-4|x+2|}{-4} \ge \frac{-12}{-4}$

$|x+2| \ge 3$

$x+2 \ge 3$ or $x+2 \le -3$

$x \ge 1$ $\quad x \le -5$

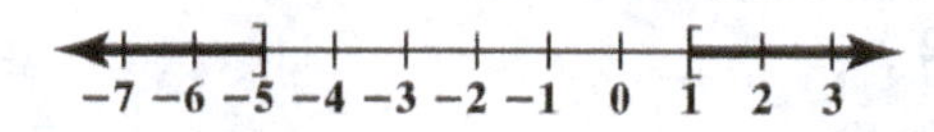

The solution set is $\{x | x \le -5 \text{ or } x \ge 1\}$ or $(-\infty, -5] \cup [1, \infty)$.

33. $|2x-3|+4 \le -10$

$|2x-3| \le -14$

There are no values of x for which the absolute value of $2x-3$ is a negative number. By definition, absolute values are always positive. The solution set is $\varnothing$ or $\{ \ \}$.

34. $|h-6.5| \le 1$

$-1 \le h-6.5 \le 1$

$5.5 \le h \le 7.5$

Approximately 90% of the population sleeps between 5.5 hours and 7.5 hours daily, inclusive.

35. $3x-4y > 12$

First, find the intercepts to the equation $3x-4y=12$.

Find the x–intercept by setting $y = 0$.

$3x-4y=12$

$3x-4(0)=12$

$3x=12$

$x=4$

Find the y–intercept by setting $x = 0$.

$3x-4y=12$

$3(0)-4y=12$

$-4y=12$

$y=-3$

Next, use the origin as a test point.

$3x-4y>12$

$3(0)-4(0)>12$

$0>12$

This is a false statement. This means that the point, $(0,0)$, will not fall in the shaded half-plane.

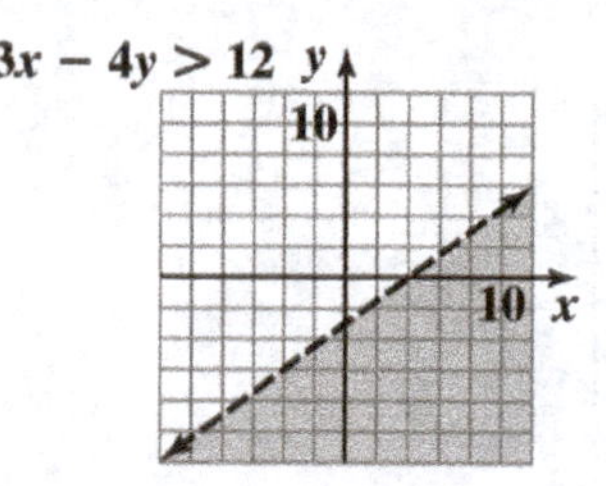

36. $x-3y \le 6$

First, find the intercepts to the equation $x-3y=6$. Find the x–intercept by setting $y=0$, find the y–intercept by setting $x = 0$.

$x-3y=6 \qquad x-3y=6$

$x-3(0)=6 \qquad 0-3y=6$

$x=6 \qquad -3y=6$

$y=-2$

Next, use the origin as a test point.

$0-3(0) \le 6$

$0 \le 6$

This is a true statement. This means that the point, $(0,0)$, will fall in the shaded half-plane.

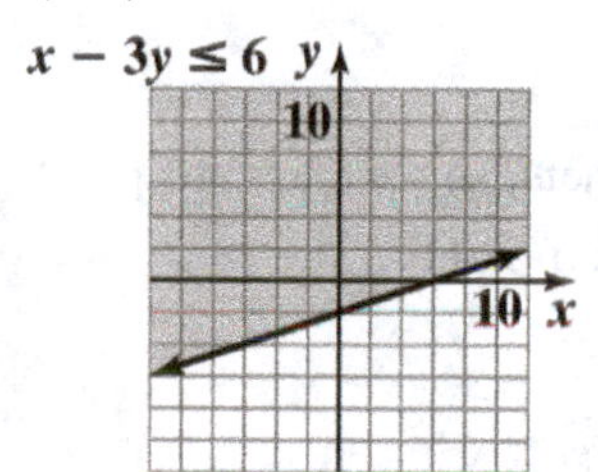

37. $y \le -\frac{1}{2}x+2$

Replacing the inequality symbol with an equal sign, we have $y=-\frac{1}{2}x+2$. Since the equation is in slope-intercept form, use the slope and the intercept to graph the equation. The y–intercept is 2 and the slope is $-\frac{1}{2}$.

Next, use the origin as a test point.

$y \le -\frac{1}{2}x+2$

$0 \le -\frac{1}{2}(0)+2$

$0 \le 2$

This is a true statement. This means that the point $(0,0)$ will fall in the shaded half-plane.

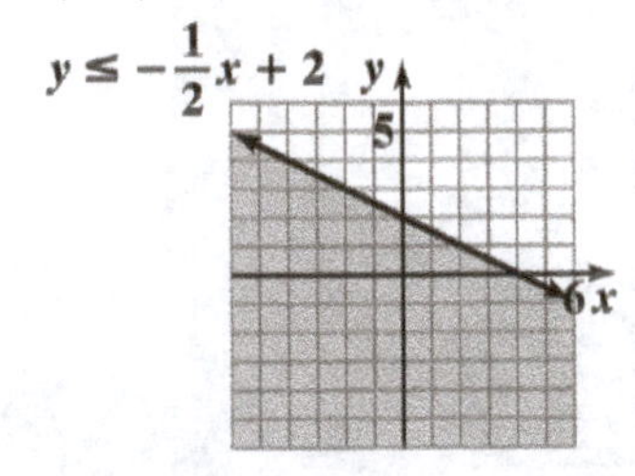

38. $y > \frac{3}{5}x$

Replacing the inequality symbol with an equal sign, we have $y = \frac{3}{5}x$. Since the equation is in slope-intercept form, use the slope and the intercept to graph the equation. The y–intercept is 0 and the slope is $\frac{3}{5}$.

Next, we need to find a test point. We cannot use the origin this time, because it lies on the line. Use $(1,1)$ as a test point.

$1 > \frac{3}{5}(1)$

$1 > \frac{3}{5}$

This is a true statement, so we know the point $(1,1)$ lies in the shaded half-plane.

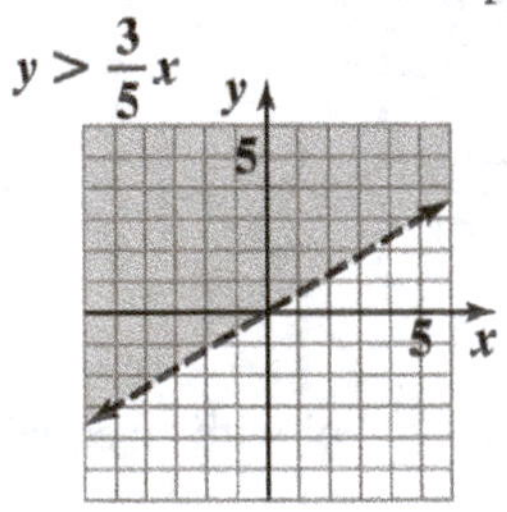

39. $x \le 2$

Replacing the inequality symbol with an equal sign, we have $x = 2$. We know that equations of the form $x = a$ are vertical lines with x–intercept = a. Next, use the origin as a test point.

$x \le 2$

$0 \le 2$

This is a true statement, so we know the point $(0,0)$ lies in the shaded half-plane.

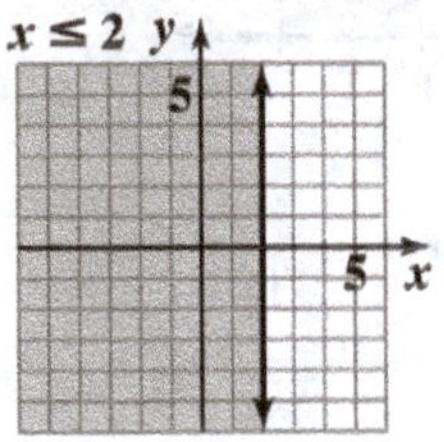

40. $y > -3$

Replacing the inequality symbol with an equal sign, we have $y = -3$. We know that equations of the form $y = b$ are horizontal lines with y–intercept = b. Next, use the origin as a test point.

$y > -3$

$0 > -3$

This is a true statement, so we know the point $(0,0)$ lies in the shaded half-plane.

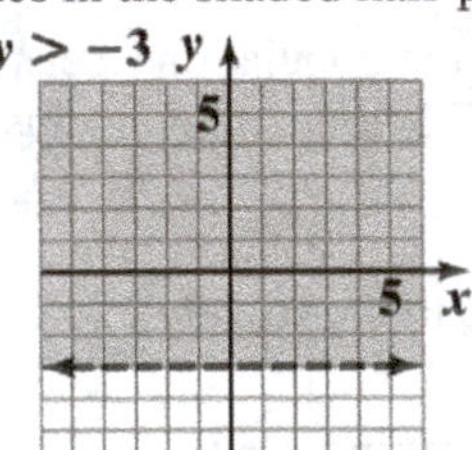

41. $2x - y \le 4$

$x + y \ge 5$

First consider $2x - y \le 4$. If we solve for y in $2x - y = 4$, we can graph the line using the slope and the y-intercept.

$2x - y = 4$

$-y = -2x + 4$

$y = 2x - 4$

y-intercept = -4

slope = 2

Now, use the origin as a test point.

$2(0) - 0 \le 4$

$0 \le 4$

This is a true statement. This means that the point $(0,0)$ will fall in the shaded half-plane.

Next consider $x + y \geq 5$. If we solve for y in $x + y = 5$, we can graph using the slope and the y-intercept.

$x + y = 5$

$y = -x + 5$

y-intercept = 5

slope = −1

Now, use the origin as a test point.

$0 + 0 \geq 5$

$0 \geq 5$

This is a false statement. This means that the point $(0,0)$ will not fall in the shaded half-plane.

Next, graph each of the inequalities. The solution to the system is the intersection of the shaded half-planes.

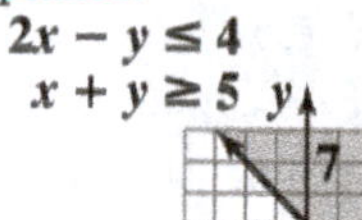

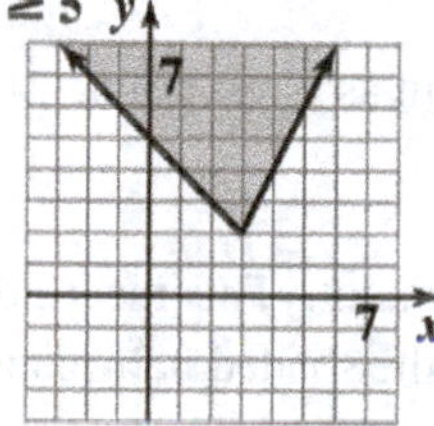

42. $y < -x + 4$

$y > x - 4$

First consider $y < -x + 4$. Change the inequality symbol to an equal sign. The line $y = -x + 4$ is in slope-intercept form and can be graphed using the slope and the y–intercept.

y–intercept = 4

slope = −1

Now, use the origin as a test point.

$0 < -0 + 4$

$0 < 4$

This is a true statement. This means that the point $(0,0)$ will fall in the shaded half-plane.

Next consider $y > x - 4$. Change the inequality symbol to an equal sign. The line $y = x - 4$ is in slope-intercept form and can be graphed using the slope and the y–intercept.

y–intercept = −4

slope = 1

Now, use the origin as a test point.

$0 > 0 - 4$

$0 > -4$

This is a true statement. This means that the point $(0,0)$ will fall in the shaded half-plane.

Next, graph each of the inequalities. The solution to the system is the intersection of the shaded half-planes.

$y < -x + 4$

$y > x - 4$

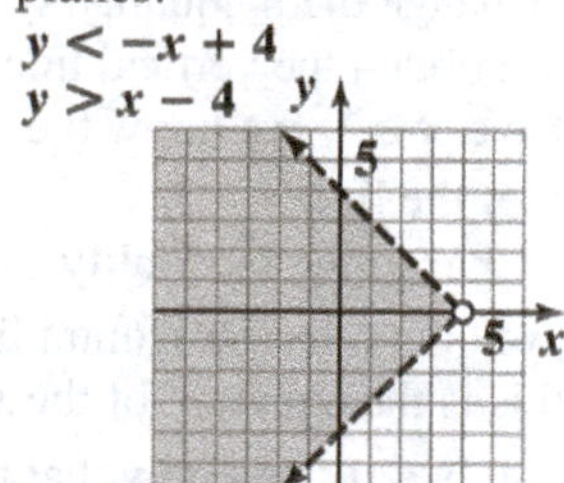

43. $-3 \leq x < 5$

Rewrite the three part inequality as two separate inequalities. We have $-3 \leq x$ and $x < 5$. We replace the inequality symbols with equal signs and obtain $-3 = x$ and $x = 5$. Equations of the form $x = a$ are vertical lines with x–intercept = a.

We know the shading in the graph will be between $x = -3$ and $x = 5$ because in the original inequality we see that x lies between −3 and 5.

$-3 \leq x < 5$

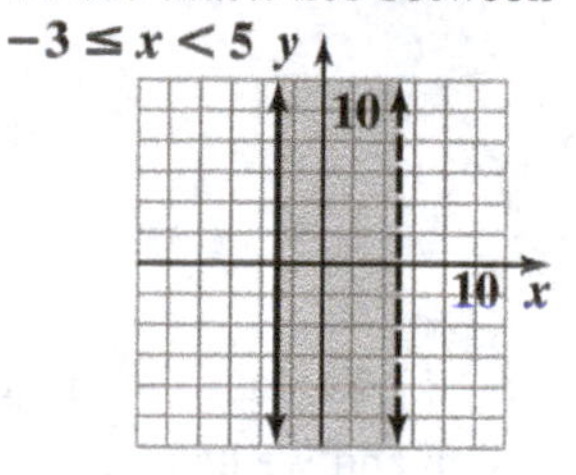

44. $-2 < y \leq 6$

Rewrite the three part inequality as two separate inequalities. We have $-2 < y$ and $y \leq 6$. We replace the inequality symbols with equal signs and obtain $-2 = y$ and $y = 6$. Equations of the form $y = b$ are vertical lines with y–intercept = b.

We know the shading in the graph will be between $y = -2$ and $y = 6$ because in the original inequality we see that y lies between −2 and 6.

$-2 < y \leq 6$

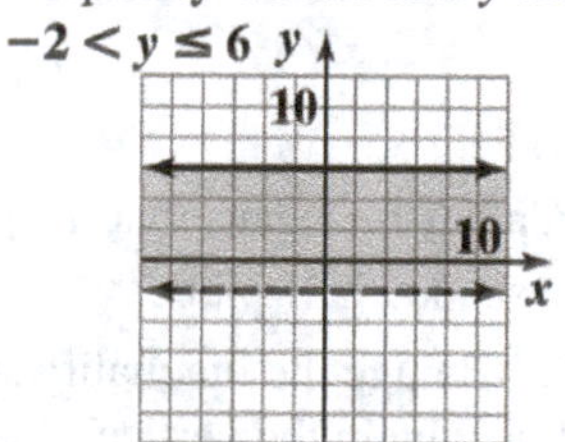

45. $x \geq 3$

$y \leq 0$

First consider $x \geq 3$. Change the inequality symbol to an equal sign and we obtain the vertical line $x = 3$. Because we have $x \geq 3$, we know the shading is to the right of the line $x = 3$.

Next consider $y \leq 0$. Change the inequality symbol to an equal sign and we obtain the horizontal line $y = 0$. (Recall that this is the equation of the x–axis.) Because we have $y \leq 0$, we know that the shading will be below the x–axis.

Next, graph each of the inequalities. The solution to the system is the intersection of the shaded half-planes.

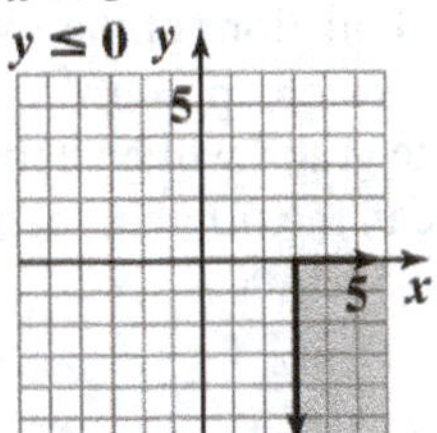

46. $2x - y > -4$

$x \geq 0$

First consider $2x - y > -4$. Replace the inequality symbol with an equal sign and we have $2x - y = -4$. Solve for y to obtain slope-intercept form.

$2x - y = -4$

$-y = -2x - 4$

$y = 2x + 4$

y–intercept = 4

slope = 2

Now, use the origin as a test point.

$2x - y > -4$

$2(0) - 0 > -4$

$0 > -4$

This is a true statement. This means that the point $(0, 0)$ will fall in the shaded half-plane.

Next consider $x \geq 0$. Change the inequality symbol to an equal sign and we obtain the vertical line $x = 0$. (Recall that this is the equation of the y–axis.) Because we have $x \geq 0$, we know that the shading will be to the right of the y–axis.

Next, graph each of the inequalities. The solution to the system is the intersection of the shaded half-planes.

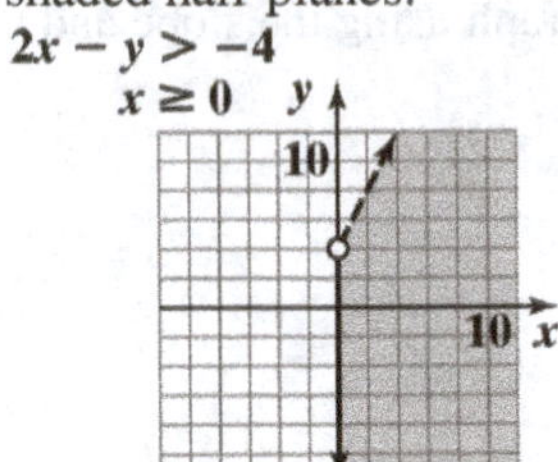

47. $x + y \leq 6$

$y \geq 2x - 3$

First consider $x + y \leq 6$. Replace the inequality symbol with an equal sign and we have $x + y = 6$. Solve for y to obtain slope-intercept form.

$x + y = 6$

$y = -x + 6$

y-intercept = 6

slope = –1

Now, use the origin as a test point.

$0 + 0 \leq 6$

$0 \leq 6$

This is a true statement. This means that the point $(0, 0)$ will fall in the shaded half-plane.

Next consider $y \geq 2x - 3$. Replace the inequality symbol with an equal sign and we have $y = 2x - 3$. The equation is in slope-intercept form, so we can use the slope and the y-intercept to graph the line.

y-intercept = –3

slope = 2

Now, use the origin as a test point.

$y \geq 2x - 3$

$0 \geq 2(0) - 3$

$0 \geq -3$

This is a true statement. This means that the point $(0, 0)$ will fall in the shaded half-plane.

Next, graph each of the inequalities. The solution to the system is the intersection of the shaded half-planes.

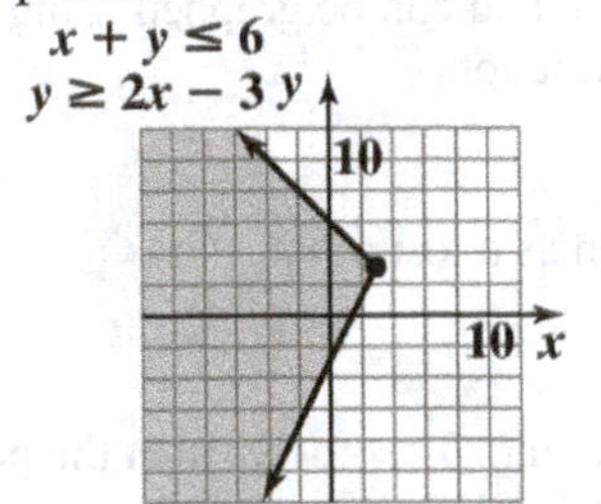

48. $3x+2y \ge 4$
$x-y \le 3$
$x \ge 0,\ y \ge 0$

First consider $3x+2y \ge 4$. Replace the inequality symbol with an equal sign and we have $3x+2y=4$. Solve for y to obtain slope-intercept form.

$$3x+2y=4$$
$$2y=-3x+4$$
$$y=-\frac{3}{2}x+2$$

y–intercept = 2 slope = $-\frac{3}{2}$

Now, use the origin as a test point.

$$3x+2y \ge 4$$
$$3(0)+2(0) \ge 4$$
$$0 \ge 4$$

This is a false statement. This means that the point $(0,0)$ will not fall in the shaded half-plane.

Now consider $x-y \le 3$. Replace the inequality symbol with an equal sign and we have $x-y=3$. Solve for y to obtain slope-intercept form.

$$x-y=3$$
$$-y=-x+3$$
$$y=x-3$$

y–intercept = –3 slope = 1

Now, use the origin as a test point.

$$x-y \le 3$$
$$0-0 \le 3$$
$$0 \le 3$$

This is a true statement. This means that the point $(0,0)$ will fall in the shaded half-plane.

Now consider the inequalities $x \ge 0$ and $y \ge 0$. The inequalities mean that both x and y will be positive. This means that we only need to consider quadrant I.

Next, graph each of the inequalities. The solution to the system is the intersection of the shaded half-planes.

$3x + 2y \ge 4$
$x - y \le 3$
$x \ge 0, y \ge 0$

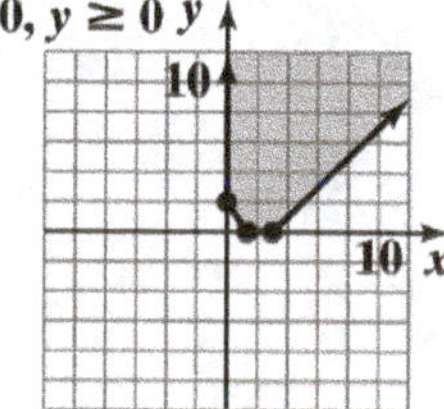

49. $2x-y>2$
$2x-y<-2$

First consider $2x-y>2$. Replace the inequality symbol with an equal sign and we have $2x-y=2$. Solve for y to obtain slope-intercept form.

$$2x-y=2$$
$$-y=-2x+2$$
$$y=2x-2$$

y–intercept = –2 slope = 2

Now, use the origin as a test point.

$$2x-y>2$$
$$2(0)-0>2$$
$$0>2$$

This is a false statement. This means that the point $(0,0)$ will not fall in the shaded half-plane.

Now consider $2x-y<-2$. Replace the inequality symbol with an equal sign and we have $2x-y=-2$. Solve for y to obtain slope-intercept form.

$$2x-y=-2$$
$$-y=-2x-2$$
$$y=2x+2$$

y–intercept = 2 slope = 2

Now, use the origin as a test point.

$$2x-y<-2$$
$$2(0)-0<-2$$
$$0<-2$$

This is a false statement. This means that the point $(0,0)$ will not fall in the shaded half-plane.

Next, graph each of the inequalities. The solution to the system is the intersection of the shaded half-planes.

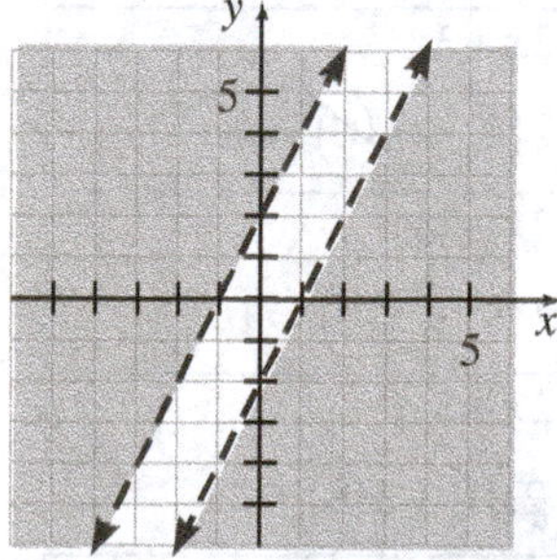

The graphs of the inequalities do not intersect, so there is no solution. The solution set is $\varnothing$ or $\{\ \}$.

50.

Corner (x, y)	Objective Function $z=2x+3y$
$(1, 0)$	$z=2x+3y$ $=2(1)+3(0)=2$
$(4, 0)$	$z=2x+3y$ $=2(4)+3(0)=8$
$(2, 2)$	$z=2x+3y$ $=2(2)+3(2)$ $=4+6=10$
$\left(\frac{1}{2},\frac{1}{2}\right)$	$z=2x+3y$ $=2\left(\frac{1}{2}\right)+3\left(\frac{1}{2}\right)$ $=\frac{2}{2}+\frac{3}{2}=\frac{5}{2}$

The maximum value is 10 and the minimum is 2.

51. Objective Function: $z=2x+3y$

Constraints: $x\ge 0,\quad y\ge 0$
$x+y\le 8$
$3x+2y\ge 6$

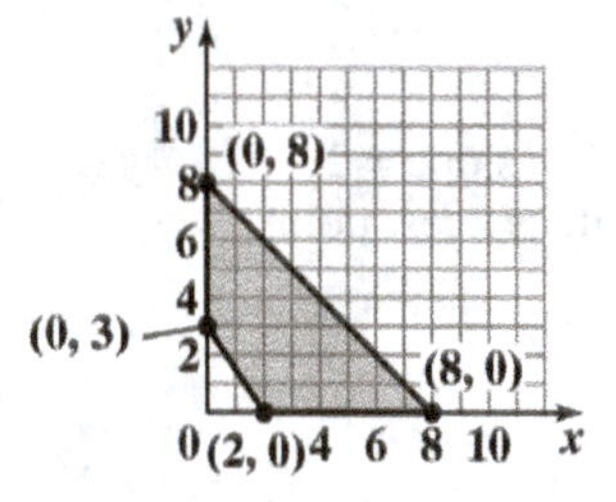

Using the graph, find the value of the objective function at each of the corner points.

Corner (x, y)	Objective Function $z=2x+3y$
$(2, 0)$	$z=2x+3y$ $=2(2)+3(0)=4$
$(8, 0)$	$z=2x+3y$ $=2(8)+3(0)=16$
$(0, 8)$	$z=2x+3y$ $=2(0)+3(8)=24$
$(0, 3)$	$z=2x+3y$ $=2(0)+3(3)=9$

The maximum of 24 occurs at the point $(0, 8)$.

52. Objective Function: $z=x+4y$

Constraints: $0\le x\le 5$
$0\le y\le 7$
$x+y\ge 3$

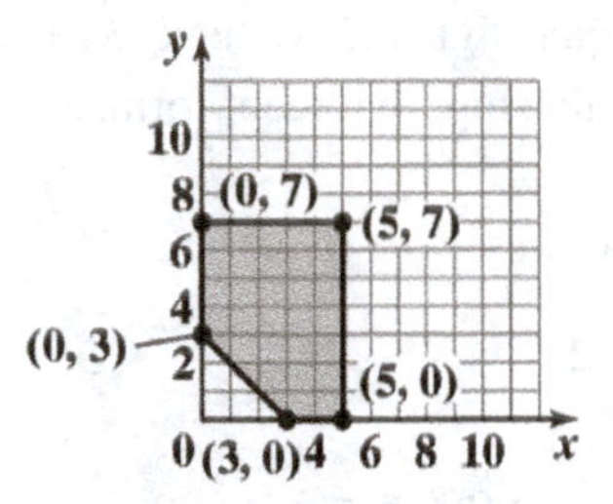

Using the graph, find the value of the objective function at each of the corner points.

Corner (x, y)	Objective Function $z=x+4y$
$(3, 0)$	$z=x+4y=3+4(0)=3$
$(5, 0)$	$z=x+4y=5+4(0)=5$
$(5, 7)$	$z=x+4y$ $=5+4(7)=5+28=33$
$(0, 7)$	$z=x+4y=0+4(7)=28$
$(0, 3)$	$z=x+4y=0+4(3)=12$

The maximum of 33 occurs at the point $(5, 7)$.

53. Objective Function: $z=5x+6y$

Constraints: $x\ge 0,\quad y\ge 0$
$y\le x$
$2x+y\le 12$
$2x+3y\ge 6$

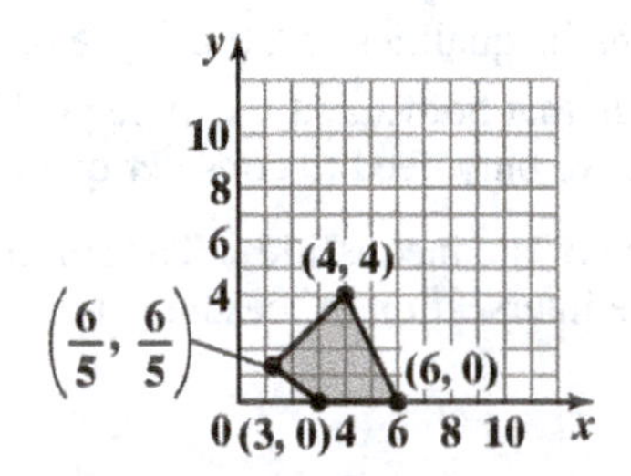

Using the graph, find the value of the objective function at each of the corner points.

Corner (x, y)	Objective Function $z = 5x + 6y$
(3, 0)	$z = 5x + 6y$ $= 5(3) + 6(0) = 15$
(6, 0)	$z = 5x + 6y$ $= 5(6) + 6(0) = 30$
(4, 4)	$z = 5x + 6y$ $= 5(4) + 6(4)$ $= 20 + 24 = 44$
(1.2, 1.2)	$z = 5x + 6y$ $= 5(1.2) + 6(1.2)$ $= 6 + 7.2 = 13.2$

The maximum of 44 occurs at the point (4, 4).

54. **a.** The objective function is $z = 500x + 350y$.

b. The paper constraint is $x + y \leq 200$.
The minimum writing paper constraint is $x \geq 10$.
The minimum newsprint constraint is $y \geq 80$.

c.

d.

Corner (x, y)	Objective Function $z = 500x + 350y$
(10, 80)	$z = 500x + 350y$ $= 500(10) + 350(80)$ $= 5000 + 28,000$ $= 33,000$
(120, 80)	$z = 500x + 350y$ $= 500(120) + 350(80)$ $= 60,000 + 28,000$ $= 88,000$
(10, 190)	$z = 500x + 350y$ $= 500(10) + 350(190)$ $= 5000 + 66,500$ $= 71,500$

e. The company will make the greatest profit by producing 120 units of writing paper and 80 units of newsprint each day. The maximum daily profit is \$88,000.

55. Let x = the number of model A produced.
Let y = the number of model B produced.
The objective function is $z = 25x + 40y$.

The cutting department labor constraint is $0.9x + 1.8y \leq 864$.

The assembly department labor constraint is $0.8x + 1.2y \leq 672$.

We also know that x and y are either zero or a positive number. We cannot have a negative number of units produced.
Next, graph the constraints.

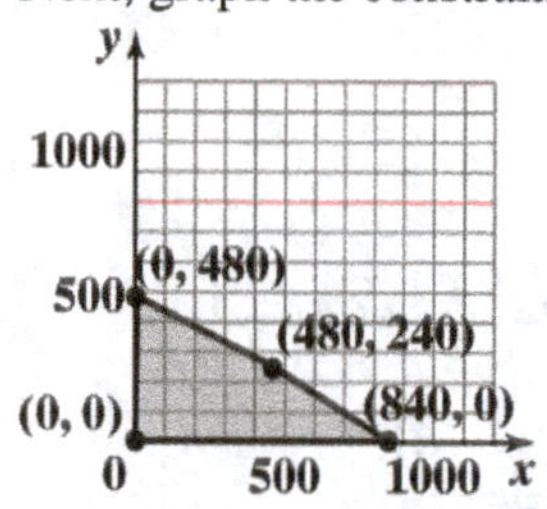

Using the graph, find the value of the objective function at each of the corner points.

Corner (x, y)	Objective Function $z = 25x + 40y$
(0, 0)	$25x + 40y$ $= 25(0) + 40(0)$ $= 0$
(840, 0)	$25x + 40y$ $= 25(840) + 40(0)$ $= 21,000$
(480, 240)	$25x + 40y$ $= 25(480) + 40(240)$ $= 12,000 + 9600$ $= 21,600$
(0, 480)	$25x + 40y$ $= 25(0) + 40(480)$ $= 19,200$

The maximum of 21,600 occurs at the point (480, 240). This means that to maximize the profit, 480 of model A and 240 of model B should be manufactured monthly. This would result in a profit of \$21,600.

Chapter 4 Test

1. $3(x+4) \geq 5x-12$

$3x+12 \geq 5x-12$

$-2x+12 \geq -12$

$-2x \geq -24$

$\frac{-2x}{-2} \leq \frac{-24}{-2}$

$x \leq 12$

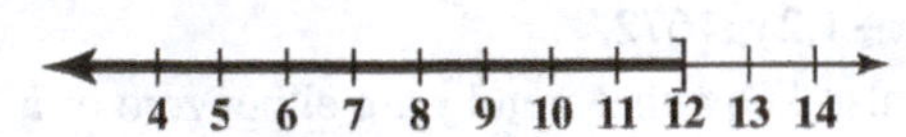

The solution set is $(-\infty, 12]$.

2. $\frac{x}{6}+\frac{1}{8} \leq \frac{x}{2}-\frac{3}{4}$

$24\left(\frac{x}{6}\right)+24\left(\frac{1}{8}\right) \leq 24\left(\frac{x}{2}\right)-24\left(\frac{3}{4}\right)$

$4x+3 \leq 12x-6(3)$

$4x+3 \leq 12x-18$

$-8x+3 \leq -18$

$-8x \leq -21$

$\frac{-8x}{-8} \geq \frac{-21}{-8}$

$x \geq \frac{21}{8}$

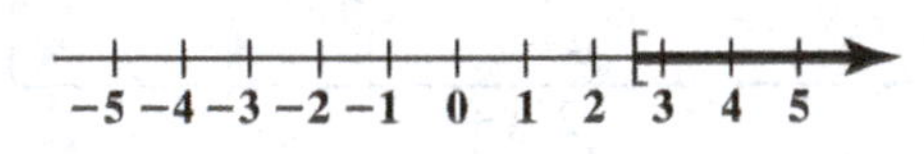

The solution set is $\left[\frac{21}{8}, \infty\right)$.

3. Let x = the number of minutes.
The monthly cost using Plan A is $C_A = 25$.
The monthly cost using Plan B is $C_B = 13+0.06x$.
For Plan A to be better deal, it must cost less than Plan B.

$C_A < C_B$

$25 < 13+0.06x$

$12 < 0.06x$

$200 < x$

$x > 200$

Plan A is a better deal when more than 200 minutes of calls are made per month.

4. $\{2,4,6,8,10\} \cap \{4,6,12,14\} = \{4,6\}$

5. $\{2,4,6,8,10\} \cup \{4,6,12,14\}$
$= \{2,4,6,8,10,12,14\}$

6. $2x+4 < 2$ and $x-3 > -5$

$2x < -2$ $\quad$ $x > -2$

$x < -1$

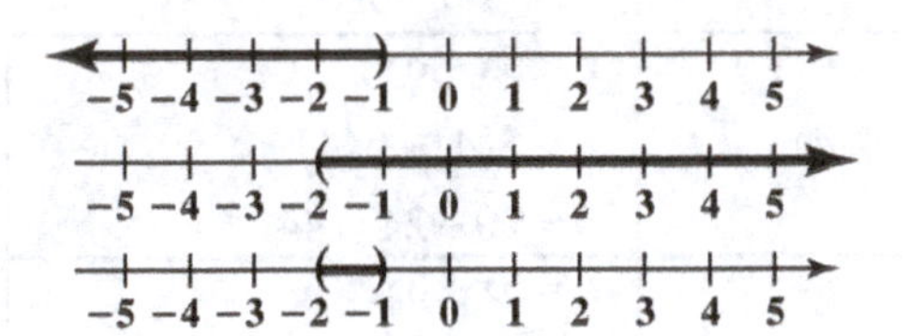

The solution set is $(-2, -1)$.

7. $x+6 \geq 4$ and $2x+3 \geq -2$

$x \geq -2$ $\quad$ $2x \geq -5$

$x \geq -\frac{5}{2}$

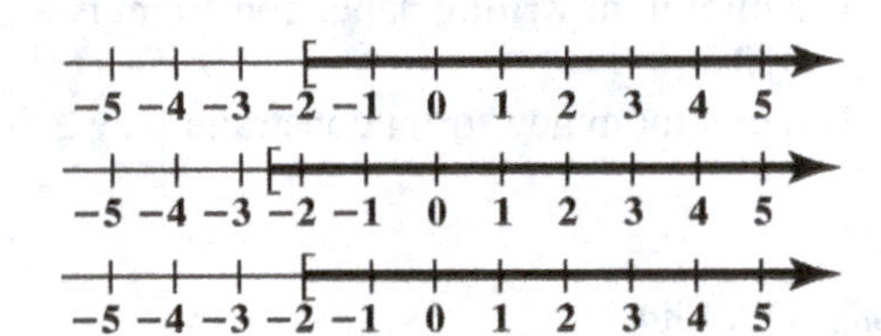

The solution set is $[-2, \infty)$.

8. $2x-3 < 5$ or $3x-6 \leq 4$

$2x < 8$ $\quad$ $3x \leq 10$

$x < 4$ $\quad$ $x \leq \frac{10}{3}$

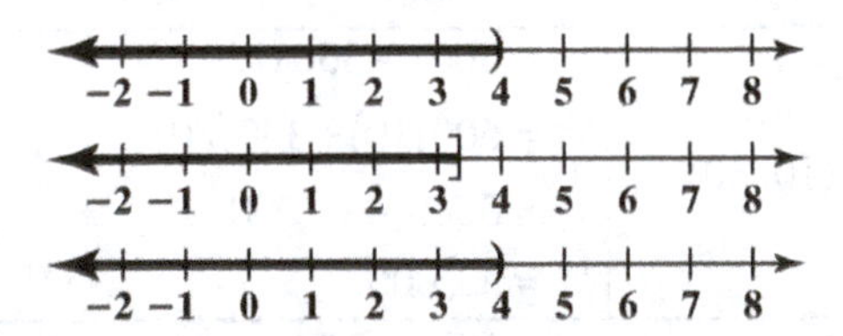

The solution set is $(-\infty, 4)$.

9. $x+3 \le -1$ or $-4x+3 < -5$

$x \le -4$ $\quad -4x < -8$

$x > 2$

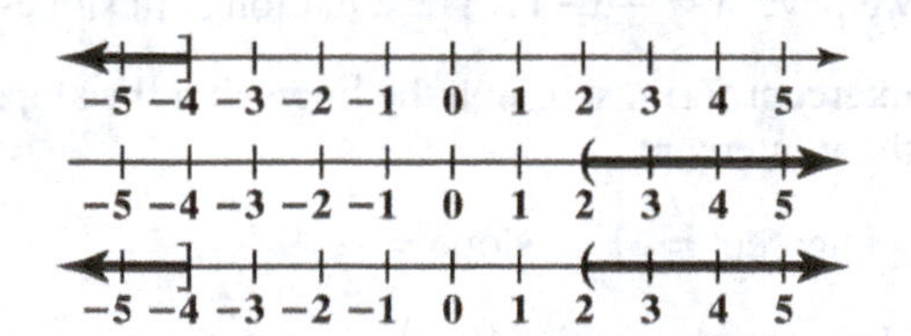

The solution set is $(-\infty,-4] \cup (2,\infty)$.

10. $-3 \le \dfrac{2x+5}{3} < 6$

$3(-3) \le 3\left(\dfrac{2x+5}{3}\right) < 3(6)$

$-9 \le 2x+5 < 18$

$-9-5 \le 2x+5-5 < 18-5$

$-14 \le 2x < 13$

$-7 \le x < \dfrac{13}{2}$

The solution set is $\left[-7, \dfrac{13}{2}\right)$.

11. $|5x+3| = 7$

$5x+3 = 7$ or $5x+3 = -7$

$5x = 4$ $\quad 5x = -10$

$x = \dfrac{4}{5}$ $\quad x = -2$

The solutions are -2 and $\dfrac{4}{5}$ and the solution set is $\left\{-2, \dfrac{4}{5}\right\}$.

12. $|6x+1| = |4x+15|$

$6x+1 = 4x+15$

$2x+1 = 15$

$2x = 14$

$x = 7$

or

$6x+1 = -(4x+15)$

$6x+1 = -4x-15$

$10x+1 = -15$

$10x = -16$

$x = -\dfrac{16}{10} = -\dfrac{8}{5}$

The solutions are $-\dfrac{8}{5}$ and 7 and the solution set is $\left\{-\dfrac{8}{5}, 7\right\}$.

13. $|2x-1| < 7$

$-7 < 2x-1 < 7$

$-7+1 < 2x-1+1 < 7+1$

$-6 < 2x < 8$

$-3 < x < 4$

The solution set is $(-3, 4)$.

14. $|2x-3| \ge 5$

$2x-3 \le -5$ or $2x-3 \ge 5$

$2x \le -2$ $\quad 2x \ge 8$

$x \le -1$ $\quad x \ge 4$

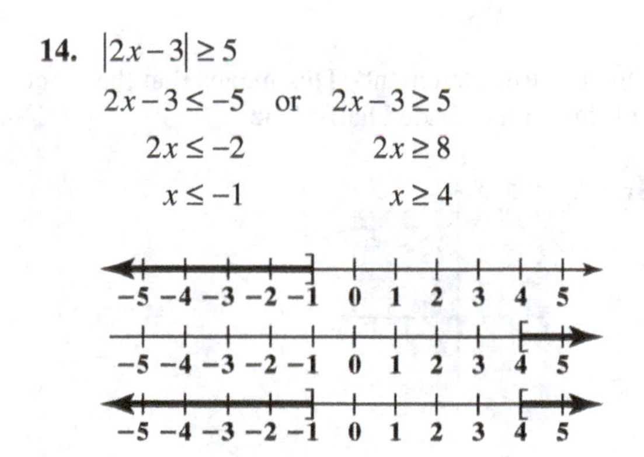

The solution set is $(-\infty,-1] \cup [4,\infty)$.

15. $|b-98.6|>8$

$b-98.6>-8$ or $b-98.6>8$

$b>90.6$ $\quad$ $b>106.6$

Interval notation: $(-\infty, 90.6)\cup(106.6, \infty)$

Hypothermia occurs when the body temperature is below $90.6°F$ and hyperthermia occurs when the body temperature is over $106.6°F$.

16. $3x-2y<6$

First, find the intercepts to the equation $3x-2y=6$.

Find the x–intercept by setting $y=0$.

$$3x-2y=6$$
$$3x-2(0)=6$$
$$3x=6$$
$$x=2$$

Find the y–intercept by setting $x=0$.

$$3x-2y=6$$
$$3(0)-2y=6$$
$$-2y=6$$
$$y=-3$$

Next, use the origin as a test point.

$$3x-2y<6$$
$$3(0)-2(0)<6$$
$$0<6$$

This is a true statement. This means that the point will fall in the shaded half-plane.

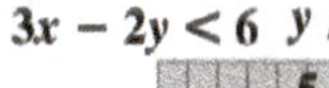

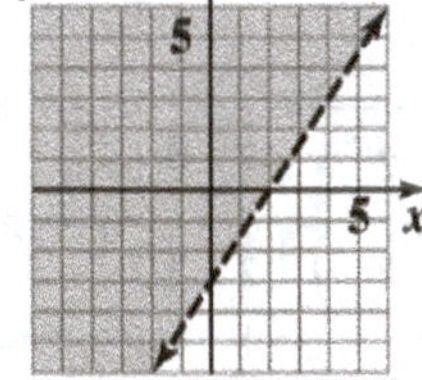

17. $y\geq\frac{1}{2}x-1$

Replacing the inequality symbol with an equal sign, we have $y=\frac{1}{2}x-1$. The equation is in slope-intercept form, so graph the line using the slope and the y-intercept.

y-intercept $=-1$ $\quad$ slope $=\frac{1}{2}$

Now, use the origin, $(0,0)$, as a test point.

$$y\geq\frac{1}{2}x-1$$
$$0\geq\frac{1}{2}(0)-1$$
$$0\geq-1$$

This is a true statement. This means that the point will fall in the shaded half-plane.

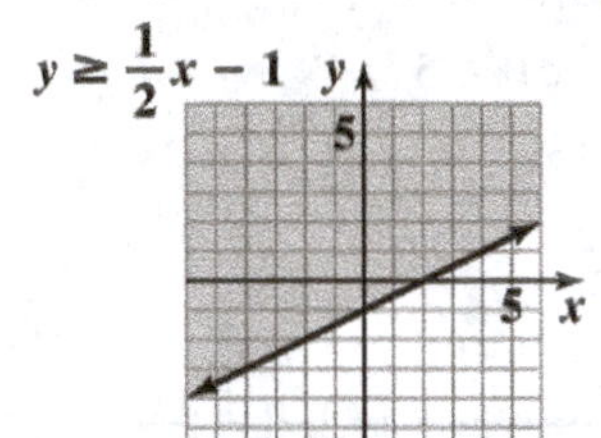

18. $y\leq-1$

Replacing the inequality symbol with an equal sign, we have $y=-1$. Equations of the form $y=b$ are horizontal lines with y–intercept $=b$, so this is a horizontal line at $y=-1$.

Next, use the origin as a test point.

$$y\leq-1$$
$$0\leq-1$$

This is a false statement, so we know the point $(0,0)$ does not lie in the shaded half-plane.

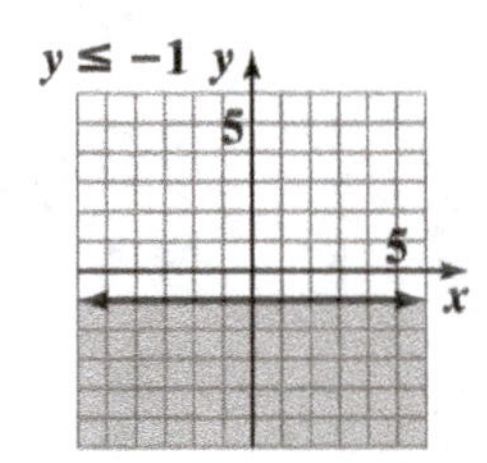

19. $x+y\ge 2$

$x-y\ge 4$

First consider $x+y\ge 2$. If we solve for y in $x+y=2$, we can graph the line using the slope and the y–intercept.

$x+y=2$

$y=-x+2$

y-intercept $=2$ slope $=-1$

Now, use the origin as a test point.

$x+y\ge 2$

$0+0\ge 2$

$0\ge 2$

This is a false statement. This means that the point will not fall in the shaded half-plane.

Next consider $x-y\ge 4$. If we solve for y in $x-y=4$, we can graph using the slope and the y-intercept.

$x-y=4$

$-y=-x+4$

$y=x-4$

y-intercept $=-4$ slope $=1$

Now, use the origin as a test point.

$x-y\ge 4$

$0-0\ge 4$

$0\ge 4$

This is a false statement. This means that the point will not fall in the shaded half-plane.

Next, graph each of the inequalities. The solution to the system is the intersection of the shaded half-planes.

$x+y\ge 2$

$x-y\ge 4$

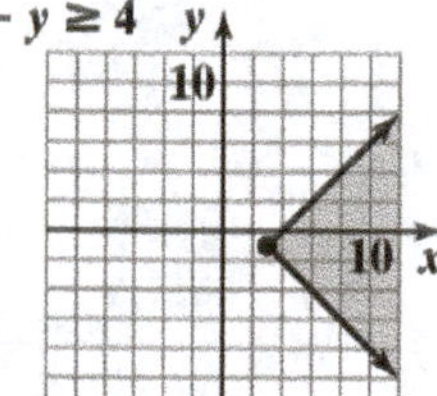

20. $3x+y\le 9$

$2x+3y\ge 6$

$x\ge 0,\ \ y\ge 0$

First consider $3x+y\le 9$. If we solve for y in $3x+y=9$, we can graph the line using the slope and the y–intercept.

$3x+y=9$

$y=-3x+9$

y-intercept $=9$ slope $=-3$

Now, use the origin as a test point.

$3x+y\le 9$

$3(0)+0\le 9$

$0\le 9$

This is a true statement. This means that the point will fall in the shaded half-plane.

Next consider $2x+3y\ge 6$. If we solve for y in $2x+3y=6$, we can graph using the slope and the y-intercept.

$2x+3y=6$

$3y=-2x+6$

$y=-\dfrac{2}{3}x+2$

y-intercept $=2$ slope $=-\dfrac{2}{3}$

Now, use the origin as a test point.

$2x+3y\ge 6$

$2(0)+3(0)\ge 6$

$0\ge 6$

This is a false statement. This means that the point will not fall in the shaded half-plane.

Next consider the inequalities $x\ge 0$ and $y\ge 0$. When x and y are both positive, we are only concerned with the first quadrant of the coordinate system.

Graph each of the inequalities. The solution to the system is the intersection of the shaded half-planes.

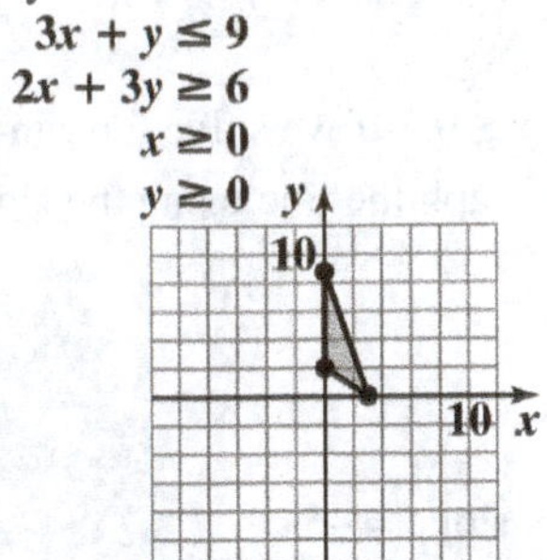

21. $-2 < x \le 4$

Rewrite the three part inequality as two separate inequalities. We have $-2 < x$ and $x \le 4$. We replace the inequality symbols with equal signs and obtain $-2 = x$ and $x = 4$. Equations of the form $x = a$ are vertical lines with x–intercept = a.

We know the shading will be between $x = -2$ and $x = 4$ because in the original inequality we see that x lies between –2 and 4.

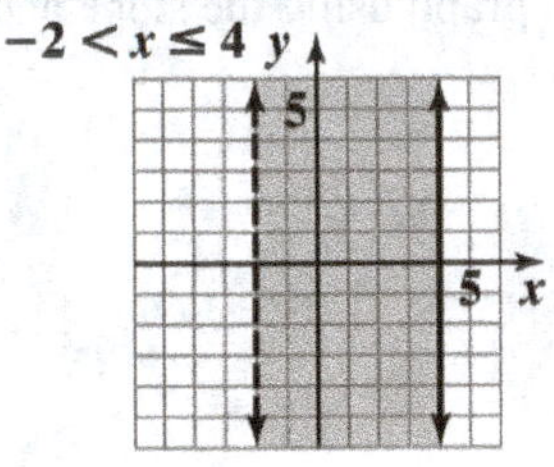

22. Objective Function: $z = 3x + 5y$

Constraints: $x \ge 0,\ y \ge 0$

$x + y \le 6$

$x \ge 2$

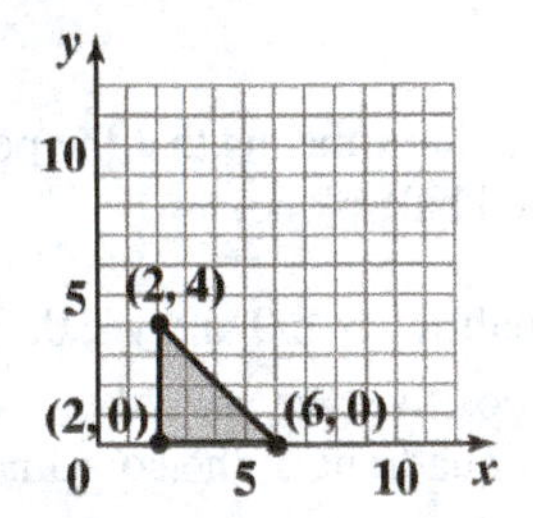

Using the graph, find the value of the objective function at each of the corner points.

Corner (x, y)	Objective Function $z = 3x + 5y$
(2, 0)	$z = 3x + 5y$ $= 3(2) + 5(0) = 6$
(6, 0)	$z = 3x + 5y$ $= 3(6) + 5(0) = 18$
(2, 4)	$z = 3x + 5y$ $= 3(2) + 5(4)$ $= 6 + 20 = 26$

The maximum of 26 occurs at the point (2, 4).

23. Let x = the number of regular jet skis produced. Let y = the number of deluxe jet skis produced. The objective function is $z = 200x + 250y$.

The regular jet ski demand constraint is $x \ge 50$.
The deluxe jet ski demand constraint is $y \ge 75$.

The quality constraint is $x + y \le 150$.

We also know that x and y are either zero or a positive number. We cannot have a negative number of units produced.
Next, graph the constraints.

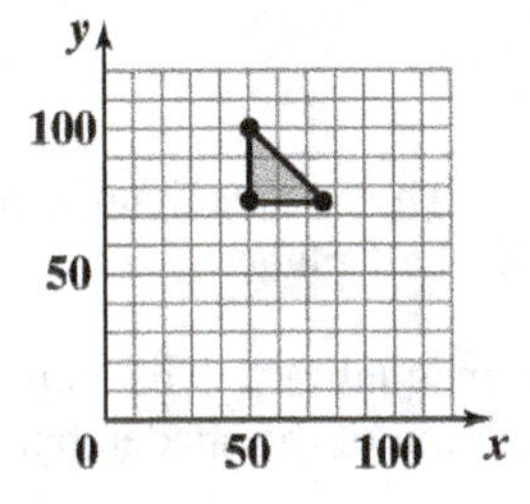

Using the graph, find the value of the objective function at each of the corner points.

Corner (x, y)	Objective Function $z = 200x + 250y$
(50, 75)	$z = 200x + 250y$ $= 200(50) + 250(75)$ $= 10,000 + 18,750$ $= 28,750$
(75, 75)	$z = 200x + 250y$ $= 200(75) + 250(75)$ $= 15,000 + 18,750$ $= 33,750$
(50, 100)	$z = 200x + 250y$ $= 200(50) + 250(100)$ $= 10,000 + 25,000$ $= 35,000$

The maximum of 35,000 occurs at the point (50, 100). This means that to maximize the profit, 50 regular jet skis and 100 deluxe jet skis should be manufactured weekly. This would result in a profit of $35,000.

Cumulative Review Exercises (Chapters 1 – 4)

1.
$$\begin{aligned}5(x+1)+2 &= x-3(2x+1)\\ 5x+5+2 &= x-6x-3\\ 5x+7 &= -5x-3\\ 10x+7 &= -3\\ 10x &= -10\\ x &= -1\end{aligned}$$

The solution set is $\{-1\}$.

2.
$$\begin{aligned}\frac{2(x+6)}{3} &= 1+\frac{4x-7}{3}\\ 3\left(\frac{2(x+6)}{3}\right) &= 3(1)+3\left(\frac{4x-7}{3}\right)\\ 2(x+6) &= 3+4x-7\\ 2x+12 &= 4x-4\\ -2x+12 &= -4\\ -2x &= -16\\ x &= 8\end{aligned}$$

The solution set is $\{8\}$.

3.
$$\frac{-10x^2y^4}{15x^7y^{-3}} = \frac{-10}{15}x^{2-7}y^{4-(-3)} = -\frac{2}{3}x^{-5}y^7 = -\frac{2y^7}{3x^5}$$

4.
$$\begin{aligned}f(x) &= x^2-3x+4\\ f(-3) &= (-3)^2-3(-3)+4\\ &= 9+9+4 = 22\\ f(2a) &= (2a)^2-3(2a)+4\\ &= 4a^2-6a+4\end{aligned}$$

5.
$$f(x) = 3x^2-4x+1$$
$$g(x) = x^2-5x-1$$

$$\begin{aligned}(f-g)(x) &= f(x)-g(x)\\ &= (3x^2-4x+1)-(x^2-5x-1)\\ &= 3x^2-4x+1-x^2+5x+1\\ &= 2x^2+x+2\end{aligned}$$

$$\begin{aligned}(f-g)(2) &= 2(2)^2+2+2\\ &= 2(4)+2+2\\ &= 8+2+2 = 12\end{aligned}$$

6. Since the line we are concerned with is perpendicular to the line, $y = 2x-3$, we know the slopes are negative reciprocals. The slope of the line will be the negative reciprocal of 2 which is $-\frac{1}{2}$. Using the slope and the point, $(2,3)$, write the equation of the line in point-slope form.

$$y-y_1 = m(x-x_1)$$
$$y-3 = -\frac{1}{2}(x-2)$$

Solve for y to write the equation in function notation.

$$\begin{aligned}y-3 &= -\frac{1}{2}(x-2)\\ y-3 &= -\frac{1}{2}x+1\\ y &= -\frac{1}{2}x+4\\ f(x) &= -\frac{1}{2}x+4\end{aligned}$$

7. $f(x) = 2x + 1$

$y = 2x + 1$

Find the x–intercept by setting $y = 0$, and the y–intercept by setting $x = 0$.

$$\begin{aligned} y &= 2x + 1 \\ 0 &= 2x + 1 \\ -1 &= 2x \\ -\frac{1}{2} &= x \end{aligned} \qquad \begin{aligned} y &= 2x + 1 \\ y &= 2(0) + 1 \\ y &= 1 \end{aligned}$$

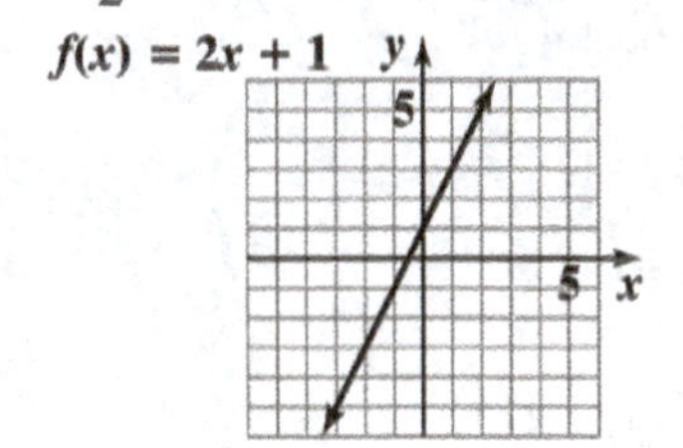

8. $y > 2x$

Consider the line $y = 2x$. Since the line is in slope-intercept form, we know that the slope is 2 and the y–intercept is 0. Use this information to graph the line.

Since the origin, (0, 0), lies on the line, we cannot use it as a test point. Instead, use the point (1, 1).

$y > 2x$

$1 > 2(1)$

$1 > 2$

This is a false statement. This means that the point (1, 1) does not lie in the shaded region.

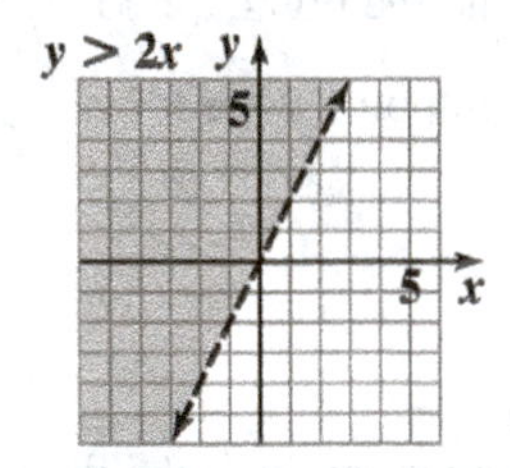

9. $2x - y \geq 6$

Graph the equation using the intercepts.

$2x - y = 6$

$x - \text{intercept} = 3$

$y - \text{intercept} = -6$

Use the origin as a test point to determine shading.

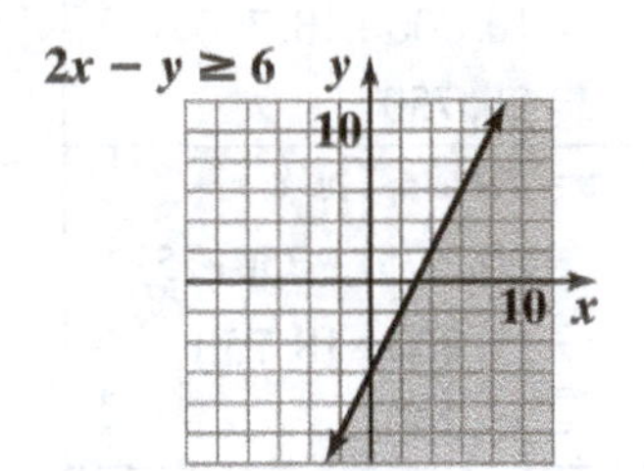

10. $f(x) = -1$

$y = -1$

Equations of the form $y = b$ are horizontal lines with y–intercept = b. This is the horizontal line at $y = -1$.

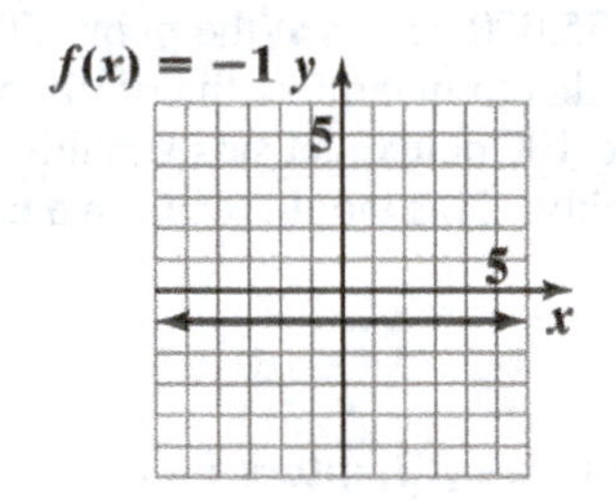

11.

$$\begin{aligned} 3x - y + z &= -15 \\ x + 2y - z &= 1 \\ 2x + 3y - 2z &= 0 \end{aligned}$$

Add the first two equations to eliminate z.

$$\begin{aligned} 3x - y + z &= -15 \\ x + 2y - z &= 1 \\ \hline 4x + y &= -14 \end{aligned}$$

Multiply the first equation by 2 and add to the third equation.

$$\begin{aligned} 6x - 2y + 2z &= -30 \\ 2x + 3y - 2z &= 0 \\ \hline 8x + y &= -30 \end{aligned}$$

The system of two equations in two variables becomes as follows.

$4x + y = -14$

$8x + y = -30$

Multiply the first equation by –1 and add to the second equation.

$$\begin{aligned} -4x - y &= 14 \\ 8x + y &= -30 \\ \hline 4x &= -16 \\ x &= -4 \end{aligned}$$

Back-substitute –4 for x to find y.

$$\begin{aligned} 4(-4) + y &= -14 \\ -16 + y &= -14 \\ y &= 2 \end{aligned}$$

Back-substitute 2 for y and –4 for x to find z.

$$\begin{aligned} 3x - y + z &= -15 \\ 3(-4) - 2 + z &= -15 \\ -12 - 2 + z &= -15 \\ -14 + z &= -15 \\ z &= -1 \end{aligned}$$

The solution is $(-4, 2, -1)$ and the solution set is $\{(-4, 2, -1)\}$.

12. $2x - y = -4$

$x + 3y = 5$

$$\left[\begin{array}{cc|c} 2 & -1 & -4 \\ 1 & 3 & 5 \end{array}\right] \quad R_1 \leftrightarrow R_2$$

$$= \left[\begin{array}{cc|c} 1 & 3 & 5 \\ 2 & -1 & -4 \end{array}\right] \quad -2R_1 + R_2$$

$$= \left[\begin{array}{cc|c} 1 & 3 & 5 \\ 0 & -7 & -14 \end{array}\right] \quad -\frac{1}{7}R_2$$

$$= \left[\begin{array}{cc|c} 1 & 3 & 5 \\ 0 & 1 & 2 \end{array}\right]$$

The resulting system is:

$x + 3y = 5$

$y = 2.$

Back-substitute 2 for y to find x.

$$\begin{aligned} x + 3(2) &= 5 \\ x + 6 &= 5 \\ x &= -1 \end{aligned}$$

The solution is $(-1, 2)$ and the solution set is $\{(-1, 2)\}$.

13. $$\begin{vmatrix} 4 & 3 \\ -1 & -5 \end{vmatrix} = 4(-5) - (-1)3 = -20 + 3 = -17$$

14. Let x = the number of rooms with a kitchen.
Let y = the number of rooms without a kitchen.

$x + y = 60$

$90x + 80y = 5260$

Solve the first equation for y.

$x + y = 60$

$y = 60 - x$

Substitute $60 - x$ for y to find x.

$$\begin{aligned} 90x + 80y &= 5260 \\ 90x + 80(60 - x) &= 5260 \\ 90x + 4800 - 80x &= 5260 \\ 10x + 4800 &= 5260 \\ 10x &= 460 \\ x &= 46 \end{aligned}$$

Back-substitute 46 for x to find y.

$y = 60 - x = 60 - 46 = 14$

There are 46 rooms with kitchens and 14 rooms without kitchens.

15. Using the vertical line test, we see that graphs a. and b. are functions.

16.
$$\begin{aligned} \frac{x}{4} - \frac{3}{4} - 1 &\le \frac{x}{2} \\ 4\left(\frac{x}{4}\right) - 4\left(\frac{3}{4}\right) - 4(1) &\le 4\left(\frac{x}{2}\right) \\ x - 3 - 4 &\le 2x \\ x - 7 &\le 2x \\ x &\le 2x + 7 \\ -x &\le 7 \\ x &\ge -7 \end{aligned}$$

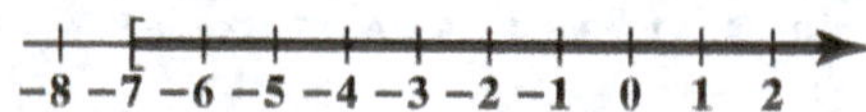

The solution set is $[-7, \infty)$.

17. $2x + 5 \le 11$ and $-3x > 18$

$2x \le 6$ $\qquad x < -6$

$x \le 3$

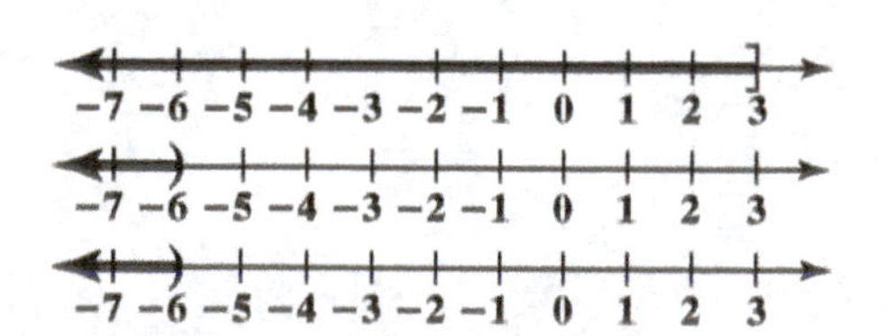

The solution set is $(-\infty, -6)$.

18. $x-4\ge 1$ or $-3x+1\ge -5-x$
$x\ge 5$ $\quad -2x+1\ge -5$
$-2x\ge -6$
$x\le 3$

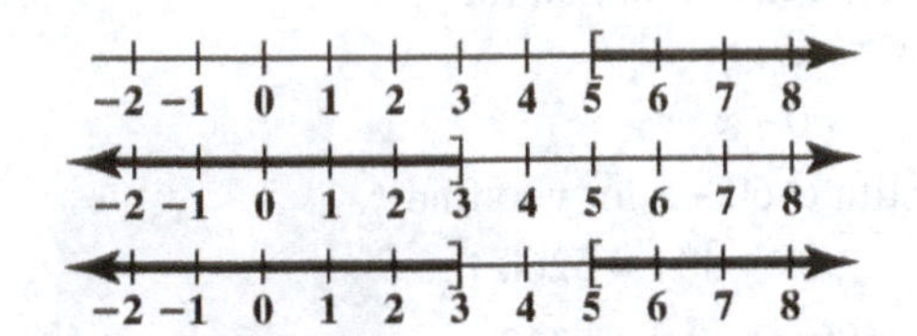

The solution set is $(-\infty,3]\cup[5,\infty)$.

19. $|2x+3|\le 17$
$-17\le 2x+3\le 17$
$-20\le 2x\le 14$
$-10\le x\le 7$

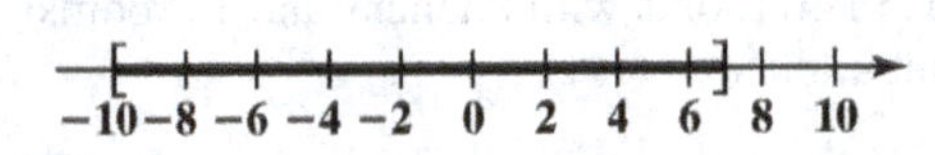

The solution set is $[-10,7]$.

20. $|3x-8|>7$
$3x-8<-7$ or $3x-8>7$
$3x<1$ $\quad 3x>15$
$x<\frac{1}{3}$ $\quad x>5$

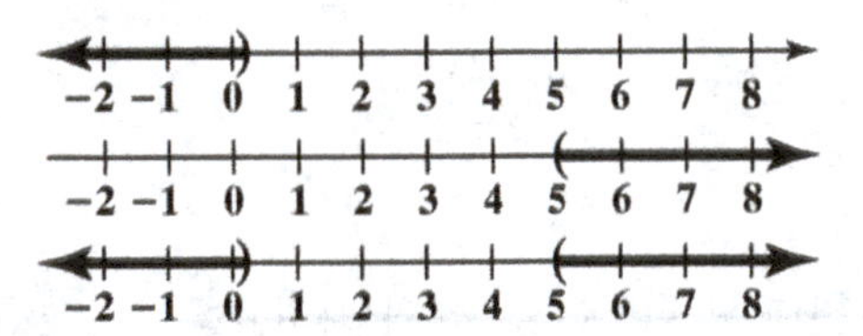

The solution set is $\left(-\infty,\frac{1}{3}\right)\cup(5,\infty)$.

Chapter 5
Polynomials, Polynomial Functions, and Factoring

5.1 Check Points

1. The degree of the polynomial is 9, the leading term is $8x^4y^5$, and the leading coefficient is 8.

Term	Coefficient	Degree
$8x^4y^5$	8	$4+5=9$
$-7x^3y^2$	-7	$3+2=5$
$-x^2y$	-1	$2+1=3$
$-5x$	-5	1
11	11	0

2. $f(x)=4x^3-3x^2-5x+6$
$$f(2)=4(2)^3-3(2)^2-5(2)+6$$
$$=32-12-10+6$$
$$=16$$

3. The leading coefficient is positive and the degree is even. Therefore the graph rises to the left and to the right.

4. This model would not be appropriate over long time periods. Because the leading coefficient is negative, the graph falls to the right. At some point the ratio would be negative, which is not possible.

5. The graph does not show the end behavior of the function. The graph should fall to the left.

6. $(-7x^3+4x^2+3)+(4x^3+6x^2-13)$
$$=-7x^3+4x^2+3+4x^3+6x^2-13$$
$$=\underbrace{-7x^3+4x^3}_{-3x^3}+\underbrace{4x^2+6x^2}_{+10x^2}+\underbrace{3-13}_{-10}$$
$$=-3x^3+10x^2-10$$

7. $(7xy^3-5xy^2-3y)+(2xy^3+8xy^2-12y-9)$
$$\begin{array}{l} 7xy^3-5xy^2-3y \\ \underline{2xy^3+8xy^2-12y-9} \\ 9xy^3+3xy^2-15y-9 \end{array}$$

8. $(14x^3-5x^2+x-9)-(4x^3-3x^2-7x+1)$
$$=(14x^3-5x^2+x-9)+(-4x^3+3x^2+7x-1)$$
$$=14x^3-5x^2+x-9-4x^3+3x^2+7x-1$$
$$=14x^3-4x^3-5x^2+3x^2+x+7x-9-1$$
$$=10x^3-2x^2+8x-10$$

9. $(6x^2y^5-2xy^3-8)-(-7x^2y^5-4xy^3+2)$
$$=6x^2y^5-2xy^3-8+7x^2y^5+4xy^3-2$$
$$=6x^2y^5+7x^2y^5-2xy^3+4xy^3-8-2$$
$$=13x^2y^5+2xy^3-10$$

5.1 Concept and Vocabulary Check

1. whole
2. standard
3. monomial
4. binomial
5. trinomial
6. n
7. $n+m$
8. greatest; leading; leading
9. true
10. false
11. end; leading
12. falls; rises
13. rises; falls
14. rises; rises
15. falls; falls
16. true
17. true

18. like

19. $-3x^3$

20. $-3x^3y$

21. $2x^5$

22. $10x^5y^2$

23. $12xy^2 - 12y^2$

5.1 Exercise Set

1. The coefficient of $-x^4$ is –1 and the degree is 4.
The coefficient of x^2 is 1 and the degree is 2.
The degree of the polynomial is 4.
The leading term is $-x^4$ and the leading coefficient is –1.

3. The coefficient of $5x^3$ is 5 and the degree is 3.
The coefficient of $7x^2$ is 7 and the degree is 2.
The coefficient of $-x$ is –1 and the degree is 1.
The coefficient of 9 is 9 and the degree is 0.
The degree of the polynomial is 3.
The leading term is $5x^3$ and the leading coefficient is 5.

5. The coefficient of $3x^2$ is 3 and the degree is 2.
The coefficient of $-7x^4$ is –7 and the degree is 4.
The coefficient of $-x$ is –1 and the degree is 1.
The coefficient of 6 is 6 and the degree is 0.
The degree of the polynomial is 4.
The leading term is $-7x^4$ and the leading coefficient is –7.

7. The coefficient of x^3y^2 is 1 and the degree is 5.
The coefficient of $-5x^2y^7$ is –5 and the degree is 9.
The coefficient of $6y^2$ is 6 and the degree is 2.
The coefficient of –3 is –3 and the degree is 0.
The degree of the polynomial is 9.
The leading term is $-5x^2y^7$ and the leading coefficient is –5.

9. The coefficient of x^5 is 1 and the degree is 5.
The coefficient of $3x^2y^4$ is 3 and the degree is 6.
The coefficient of $7xy$ is 7 and the degree is 2.
The coefficient of $9x$ is 9 and the degree is 1.
The coefficient of –2 is –2 and the degree is 0.
The degree of the polynomial is 6.
The leading term is $3x^2y^4$ and the leading coefficient is 3.

11.
$$\begin{aligned} f(x) &= x^2 - 5x + 6 \\ f(3) &= (3)^2 - 5(3) + 6 \\ &= 9 - 15 + 6 = 0 \end{aligned}$$

13.
$$\begin{aligned} f(x) &= x^2 - 5x + 6 \\ f(-1) &= (-1)^2 - 5(-1) + 6 \\ &= 1 + 5 + 6 = 12 \end{aligned}$$

15.
$$\begin{aligned} g(x) &= 2x^3 - x^2 + 4x - 1 \\ g(3) &= 2(3)^3 - (3)^2 + 4(3) - 1 \\ &= 2(27) - 9 + 12 - 1 \\ &= 54 - 9 + 12 - 1 = 56 \end{aligned}$$

17.
$$\begin{aligned} g(x) &= 2x^3 - x^2 + 4x - 1 \\ g(-2) &= 2(-2)^3 - (-2)^2 + 4(-2) - 1 \\ &= 2(-8) - 4 - 8 - 1 \\ &= -16 - 4 - 8 - 1 \\ &= -29 \end{aligned}$$

19.
$$\begin{aligned} g(x) &= 2x^3 - x^2 + 4x - 1 \\ g(0) &= 2(0)^3 - (0)^2 + 4(0) - 1 \\ &= 2(0) - 0 - 0 - 1 \\ &= 0 - 0 - 0 - 1 \\ &= -1 \end{aligned}$$

21. polynomial function

23. not a polynomial function

25. Since the degree of the polynomial is 4, an even number, and the leading coefficient is –1, the graph will fall to the left and to the right. The graph of the polynomial is graph (b).

27. Since the degree of the polynomial is 2, an even number, and the leading coefficient is 1, the graph will rise to the left and to the right. The graph of the polynomial is graph (a).

29. $\left(-6x^3+5x^2-8x+9\right)+\left(17x^3+2x^2-4x-13\right)=-6x^3+5x^2-8x+9+17x^3+2x^2-4x-13$
$=-6x^3+17x^3+5x^2+2x^2-8x-4x+9-13$
$=11x^3+7x^2-12x-4$

31. $\left(\frac{2}{5}x^4+\frac{2}{3}x^3+\frac{5}{8}x^2+7\right)+\left(-\frac{4}{5}x^4+\frac{1}{3}x^3-\frac{1}{4}x^2-7\right)=\frac{2}{5}x^4+\frac{2}{3}x^3+\frac{5}{8}x^2+7-\frac{4}{5}x^4+\frac{1}{3}x^3-\frac{1}{4}x^2-7$
$=\frac{2}{5}x^4-\frac{4}{5}x^4+\frac{2}{3}x^3+\frac{1}{3}x^3+\frac{5}{8}x^2-\frac{1}{4}x^2+7-7$
$=-\frac{2}{5}x^4+\frac{3}{3}x^3+\left(\frac{5}{8}-\frac{2}{8}\right)x^2$
$=-\frac{2}{5}x^4+\frac{3}{3}x^3+\frac{3}{8}x^2$
$=-\frac{2}{5}x^4+x^3+\frac{3}{8}x^2$

33. $\left(7x^2y-5xy\right)+\left(2x^2y-xy\right)=7x^2y-5xy+2x^2y-xy$
$=7x^2y+2x^2y-5xy-xy$
$=9x^2y-6xy$

35. $\left(5x^2y+9xy+12\right)+\left(-3x^2y+6xy+3\right)=5x^2y+9xy+12-3x^2y+6xy+3$
$=5x^2y-3x^2y+9xy+6xy+12+3$
$=2x^2y+15xy+15$

37. $\left(9x^4y^2-6x^2y^2+3xy\right)+\left(-18x^4y^2-5x^2y-xy\right)=9x^4y^2-6x^2y^2+3xy-18x^4y^2-5x^2y-xy$
$=9x^4y^2-18x^4y^2-6x^2y^2-5x^2y+3xy-xy$
$=-9x^4y^2-6x^2y^2-5x^2y+2xy$

39. $\left(x^{2n}+5x^n-8\right)+\left(4x^{2n}-7x^n+2\right)=x^{2n}+5x^n-8+4x^{2n}-7x^n+2$
$=x^{2n}+4x^{2n}+5x^n-7x^n-8+2$
$=5x^{2n}-2x^n-6$

41. $\left(17x^3-5x^2+4x-3\right)-\left(5x^3-9x^2-8x+11\right)=17x^3-5x^2+4x-3-5x^3+9x^2+8x-11$
$=17x^3-5x^3-5x^2+9x^2+4x+8x-3-11$
$=12x^3+4x^2+12x-14$

43. $\left(13y^5+9y^4-5y^2+3y+6\right)-\left(-9y^5-7y^3+8y^2+11\right)$
$=13y^5+9y^4-5y^2+3y+6+9y^5+7y^3-8y^2-11$
$=13y^5+9y^5+9y^4+7y^3-5y^2-8y^2+3y+6-11$
$=22y^5+9y^4+7y^3-13y^2+3y-5$

45. $\left(x^3+7xy-5y^2\right)-\left(6x^3-xy+4y^2\right)=x^3+7xy-5y^2-6x^3+xy-4y^2$
$=x^3-6x^3+7xy+xy-5y^2-4y^2$
$=-5x^3+8xy-9y^2$

47. $\left(3x^4y^2+5x^3y-3y\right)-\left(2x^4y^2-3x^3y-4y+6x\right)=3x^4y^2+5x^3y-3y-2x^4y^2+3x^3y+4y-6x$
$=3x^4y^2-2x^4y^2+5x^3y+3x^3y-3y+4y-6x$
$=x^4y^2+8x^3y+y-6x$

49. $\left(7y^{2n}+y^n-4\right)-\left(6y^{2n}-y^n-1\right)=7y^{2n}+y^n-4-6y^{2n}+y^n+1$
$=7y^{2n}-6y^{2n}+y^n+y^n-4+1$
$=y^{2n}+2y^n-3$

51. $\left(3a^2b^4-5ab^2+7ab\right)-\left(-5a^2b^4-8ab^2-ab\right)=3a^2b^4-5ab^2+7ab+5a^2b^4+8ab^2+ab$
$=3a^2b^4+5a^2b^4-5ab^2+8ab^2+7ab+ab$
$=8a^2b^4+3ab^2+8ab$

53. $\left(x^3+2x^2y-y^3\right)-\left(-4x^3-x^2y+xy^2+3y^3\right)=x^3+2x^2y-y^3+4x^3+x^2y-xy^2-3y^3$
$=x^3+4x^3+2x^2y+x^2y-xy^2-y^3-3y^3$
$=5x^3+3x^2y-xy^2-4y^3$

55. $\left(6x^4-5x^3+2x\right)+\left[\left(4x^3+3x^2-1\right)-\left(x^4-2x^2+7x-3\right)\right]$
$=6x^4-5x^3+2x+\left[4x^3+3x^2-1-x^4+2x^2-7x+3\right]$
$=6x^4-5x^3+2x+\left[-x^4+4x^3+3x^2+2x^2-7x-1+3\right]$
$=6x^4-5x^3+2x+\left[-x^4+4x^3+5x^2-7x+2\right]$
$=6x^4-5x^3+2x-x^4+4x^3+5x^2-7x+2$
$=6x^4-x^4-5x^3+4x^3+5x^2+2x-7x+2$
$=5x^4-x^3+5x^2-5x+2$

57. $\left[\left(-6x^2y^2-x^2-1\right)+\left(5x^2y^2+2x^2-1\right)\right]-\left(9x^2y^2-3x^2-5\right)$
$=\left[-6x^2y^2-x^2-1+5x^2y^2+2x^2-1\right]-9x^2y^2+3x^2+5$
$=\left[-6x^2y^2+5x^2y^2-x^2+2x^2-1-1\right]-9x^2y^2+3x^2+5$
$=\left[-x^2y^2+x^2-2\right]-9x^2y^2+3x^2+5$
$=-x^2y^2+x^2-2-9x^2y^2+3x^2+5$
$=-x^2y^2-9x^2y^2+x^2+3x^2-2+5$
$=-10x^2y^2+4x^2+3$

For Exercises 59-63, $f(x) = -3x^3 - 2x^2 - x + 4$; $g(x) = x^3 - x^2 - 5x - 4$; $h(x) = -2x^3 + 5x^2 - 4x + 1$.

59. $$\begin{aligned}(f-g)(x) &= \left(-3x^3 - 2x^2 - x + 4\right) - \left(x^3 - x^2 - 5x - 4\right)\\ &= -3x^3 - 2x^2 - x + 4 - x^3 + x^2 + 5x + 4\\ &= -3x^3 - x^3 - 2x^2 + x^2 - x + 5x + 4 + 4\\ &= -4x^3 - x^2 + 4x + 8\end{aligned}$$

$$\begin{aligned}(f-g)(-1) &= -4(-1)^3 - (-1)^2 + 4(-1) + 8\\ &= 4 - 1 - 4 + 8\\ &= 7\end{aligned}$$

61. $$\begin{aligned}(f+g-h)(x) &= \left(-3x^3 - 2x^2 - x + 4\right) + \left(x^3 - x^2 - 5x - 4\right) - \left(-2x^3 + 5x^2 - 4x + 1\right)\\ &= -3x^3 - 2x^2 - x + 4 + x^3 - x^2 - 5x - 4 + 2x^3 - 5x^2 + 4x - 1\\ &= -3x^3 + x^3 + 2x^3 - 2x^2 - x^2 - 5x^2 - x - 5x + 4x + 4 - 4 - 1\\ &= -8x^2 - 2x - 1\end{aligned}$$

$$(f+g-h)(-2) = -8(-2)^2 - 2(-2) - 1 = -8(4) + 4 - 1 = -32 + 3 = -29$$

63. $$\begin{aligned}2f(x) - 3g(x) &= 2\left(-3x^3 - 2x^2 - x + 4\right) - 3\left(x^3 - x^2 - 5x - 4\right)\\ &= -6x^3 - 4x^2 - 2x + 8 - 3x^3 + 3x^2 + 15x + 12\\ &= -6x^3 - 3x^3 - 4x^2 + 3x^2 - 2x + 15x + 8 + 12\\ &= -9x^3 - x^2 + 13x + 20\end{aligned}$$

65. a. $$\begin{aligned}f(x) &= 0.76x^3 - 30x^2 - 882x + 37{,}807\\ f(40) &= 0.76(40)^3 - 30(40)^2 - 882(40) + 37{,}807\\ &= 3167\end{aligned}$$

The world tiger population in 2010 (40 years after 1970) was about 3167.
This is represented by the point $(40, 3167)$.

b. This underestimates the actual data shown in the bar graph by 33.

67. The leading coefficient is positive, thus the graph rises to the right.
If conservation efforts fail, the model will not be useful. The model indicates an increasing world tiger population that will actually decrease without conservation efforts.

69. The leading coefficient is –0.75 and the degree is 4. This means that the graph will fall to the right. Over time, the number of viral particles in the body will go to zero.

71. In the polynomial, $f(x) = -x^4 + 21x^2 + 100$, the leading coefficient is –1 and the degree is 4. Applying the Leading Coefficient Test, we know that even-degree polynomials with negative leading coefficient will fall to the left and to the right. Since the graph falls to the right, we know that the elk population will die out over time.

73 – 83. Answers will vary.

85. The graph falls to the left and r, so the polynomial must be of even degree with a negative leading coefficient. Functions will vary. An example is $-x^4 - 2x^3$.

87. The graph falls to the left and rises to the right, so the polynomial must be of odd degree with a positive leading coefficient. Functions will vary. An example is $x^3 + 3x^2$.

89. $f(x) = -2x^3 + 6x^2 + 3x - 1$

Since we have an odd-degree polynomial with a negative coefficient, the Leading Coefficient Test predicts that the graph will rise to the left and fall to the right.

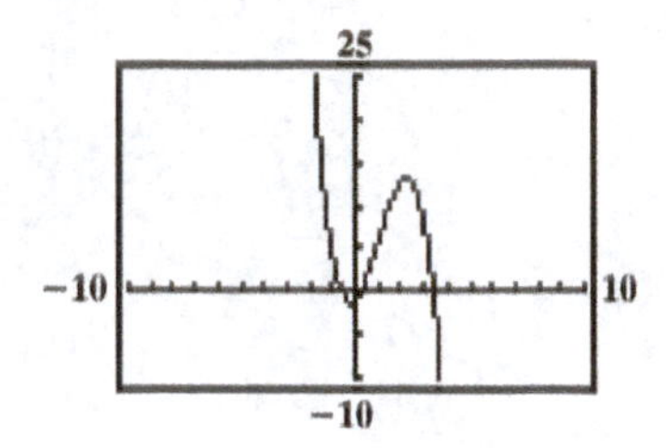

91. $f(x) = -x^5 + 5x^4 - 6x^3 + 2x + 20$

Since we have an odd-degree polynomial with a negative coefficient, the Leading Coefficient Test predicts that the graph will rise to the left and fall to the right.

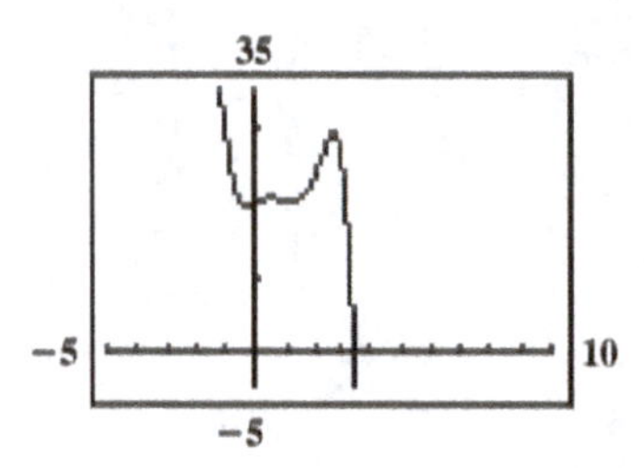

93. $f(x) = -x^4 + 2x^3 - 6x$

$g(x) = -x^4$

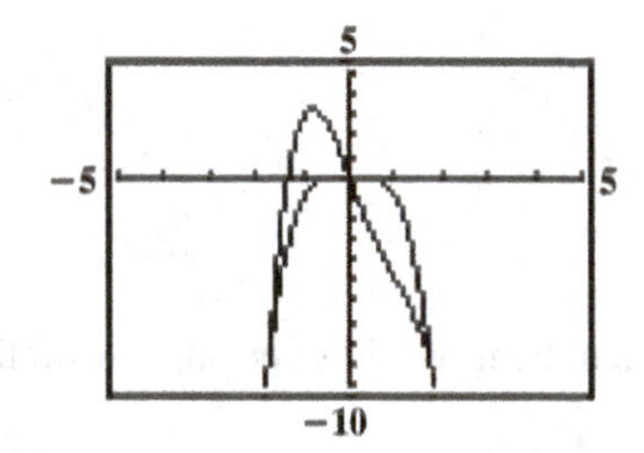

95. does not make sense; Explanations will vary. Sample explanation: The leading coefficient is the coefficient of the term of highest degree, which may or may not be the first term.

97. makes sense

99. false; Changes to make the statement true will vary. A sample change is: If two polynomials of degree 2 are added, the result is not necessarily a polynomial of degree 2. If the leading coefficients of the two polynomials are opposites, then the sum of the two polynomials will be of lower degree.

101. true

103. $\left(y^{3n} - 7y^{2n} + 3\right) - \left(-3y^{3n} - 2y^{2n} - 1\right) + \left(6y^{3n} - y^{2n} + 1\right)$

$= y^{3n} - 7y^{2n} + 3 + 3y^{3n} + 2y^{2n} + 1 + 6y^{3n} - y^{2n} + 1$

$= 10y^{3n} - 6y^{2n} + 5$

105. $9(x-1) = 1 + 3(x-2)$

$9x - 9 = 1 + 3x - 6$

$9x - 9 = 3x - 5$

$6x - 9 = -5$

$6x = 4$

$x = \frac{4}{6} = \frac{2}{3}$

The solution set is $\left\{\frac{2}{3}\right\}$.

106. $2x - 3y < -6$

Replacing the inequality symbol with an equal sign, we have $2x - 3y = -6$. Solve for y to obtain slope-intercept form.

$2x - 3y = -6$

$-3y = -2x - 6$

$y = \frac{2}{3}x + 2$

The slope is $\frac{2}{3}$ and the y–intercept is 2. Next, use the origin as a test point.

$2(0) - 3(0) < -6$

$0 < -6$

This is a false statement. This means that the point, $(0, 0)$, will not fall in the shaded half-plane.

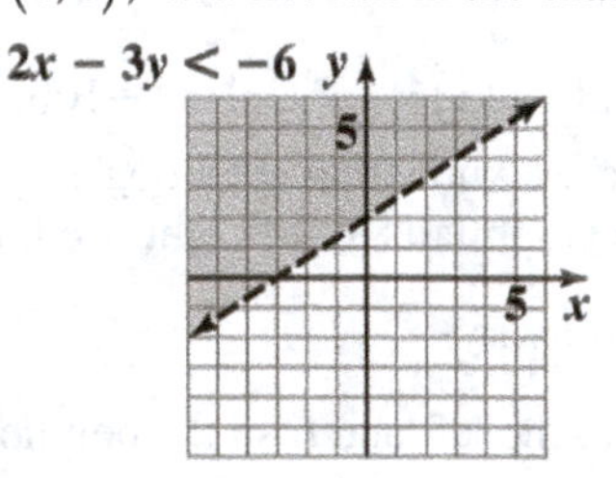

107. Since the line is parallel to $3x - y = 9$, it will have the same slope. To find the slope, put the equation in slope intercept form.

$$3x - y = 9$$
$$-y = -3x + 9$$
$$y = 3x - 9$$

The slope is $m = 3$.

Use the slope and the point, $(-2, 5)$, to write the equation of the line in point-slope form.

$$y - y_1 = m(x - x_1)$$
$$y - 5 = 3(x - (-2))$$
$$y - 5 = 3(x + 2)$$

Solve for y to obtain slope-intercept form.

$$y - 5 = 3x + 6$$
$$y = 3x + 11$$

108. $(2x^3y^2)(5x^4y^7) = 2 \cdot 5 \cdot x^3x^4 \cdot y^2y^7 = 10x^7y^9$

109. $2x^4(8x^4 + 3x) = 2x^4 \cdot 8x^4 + 2x^4 \cdot 3x = 16x^8 + 6x^5$

110. $3x(x^2 + 4x + 5) + 7(x^2 + 4x + 5) = 3x^3 + 12x^2 + 15x + 7x^2 + 28x + 35$
$= 3x^3 + 19x^2 + 43x + 35$

5.2 Check Points

1. a. $(6x^5y^7)(-3x^2y^4) = 6(-3)x^5 \cdot x^2 \cdot y^7 \cdot y^4$
$= -18x^{5+2}y^{7+4}$
$= -18x^7y^{11}$

b. $(10x^4y^3z^6)(3x^6y^3z^2) = 10 \cdot 3 \cdot x^4 \cdot x^6 \cdot y^3 \cdot y^3 \cdot z^6 \cdot z^2$
$= 30x^{4+6}y^{3+3}z^{6+2}$
$= 30x^{10}y^6z^8$

2. a. $6x^4(2x^5 - 3x^2 + 4) = 6x^4 \cdot 2x^5 - 6x^4 \cdot 3x^2 + 6x^4 \cdot 4$
$= 12x^9 - 18x^6 + 24x^4$

b. $2x^4y^3(5xy^6 - 4x^3y^4 - 5) = 2x^4y^3 \cdot 5xy^6 - 2x^4y^3 \cdot 4x^3y^4 - 2x^4y^3 \cdot 5$
$= 10x^5y^9 - 8x^7y^7 - 10x^4y^3$

3. $(3x + 2)(2x^2 - 2x + 1) = 3x(2x^2 - 2x + 1) + 2(2x^2 - 2x + 1)$
$= 6x^3 - 6x^2 + 3x + 4x^2 - 4x + 2$
$= 6x^3 - 2x^2 - x + 2$

4. $(4xy^2+2y)(3xy^4-2xy^2+y) = 4xy^2(3xy^4-2xy^2+y)+2y(3xy^4-2xy^2+y)$
$= 12x^2y^6-8x^2y^4+4xy^3+6xy^5-4xy^3+2y^2$
$= 12x^2y^6-8x^2y^4+6xy^5+2y^2$

5. a. $(x+5)(x+3) = \overbrace{x\cdot x}^{\text{F}}+\overbrace{x\cdot 3}^{\text{O}}+\overbrace{5\cdot x}^{\text{I}}+\overbrace{5\cdot 3}^{\text{L}}$
$= x^2+3x+5x+15$
$= x^2+8x+15$

b. $(7x+4y)(2x-y) = \overbrace{7x\cdot 2x}^{\text{F}}+\overbrace{7x(-y)}^{\text{O}}+\overbrace{4y\cdot 2x}^{\text{I}}+\overbrace{4y(-y)}^{\text{L}}$
$= 14x^2-7xy+8xy-4y^2$
$= 14x^2+xy-4y^2$

c. $(4x^3-5)(x^3-3x) = 4x^3\cdot x^3+4x^3(-3x)-5\cdot x^3-5(-3x)$
$= 4x^6-12x^4-5x^3+15x$

6. a. Use the special-product formula $(A+B)^2 = A^2+2AB+B^2$.
$(A+B)^2 = A^2+2AB+B^2$
$(x+8)^2 = x^2+2\cdot x\cdot 8+8^2$
$= x^2+16x+64$

b. Use the special-product formula $(A+B)^2 = A^2+2AB+B^2$.
$(A+B)^2 = A^2+2AB+B^2$
$(4x+5y)^2 = (4x)^2+2\cdot 4x\cdot 5y+(5y)^2$
$= 16x^2+40xy+25y^2$

7. a. Use the special-product formula $(A-B)^2 = A^2-2AB+B^2$.
$(A-B)^2 = A^2-2AB+B^2$
$(x-5)^2 = x^2-2\cdot x\cdot 5+5^2$
$= x^2-10x+25$

b. Use the special-product formula $(A-B)^2 = A^2-2AB+B^2$.
$(A-B)^2 = A^2-2AB+B^2$
$(2x-6y^4)^2 = (2x)^2-2\cdot 2x\cdot 6y^4+(6y^4)^2$
$= 4x^2-24xy^4+36y^8$

8. a. Use the special-product formula $(A+B)(A-B) = A^2-B^2$.
$(A+B)(A-B) = A^2-B^2$
$(x+3)(x-3) = x^2-3^2$
$= x^2-9$

b. Use the special-product formula $(A+B)(A-B)=A^2-B^2$.

$$(A+B)(A-B)=A^2-B^2$$
$$(5x+7y)(5x-7y)=(5x)^2-(7y)^2$$
$$=25x^2-49y^2$$

c. Use the special-product formula $(A+B)(A-B)=A^2-B^2$.

$$(A+B)(A-B)=A^2-B^2$$
$$(5ab^2-4a)(5ab^2+4a)=(5ab^2)^2-(4a)^2$$
$$=25a^2b^4-16a^2$$

9. a. Group the first two terms and use the special-product formula $(A+B)(A-B)=A^2-B^2$.

$$(3x+2+5y)(3x+2-5y)=[(3x+2)+5y][(3x+2)-5y]$$
$$=(3x+2)^2-(5y)^2$$
$$=(3x)^2+2\cdot 3x\cdot 2+2^2-(5y)^2$$
$$=9x^2+12x+4-25y^2$$

b. Group the first two terms and use the special-product formula $(A+B)^2=A^2+2AB+B^2$.

$$(A+B)^2=A^2+2AB+B^2$$
$$(2x+y+3)^2=[(2x+y)+3]^2$$
$$=(2x+y)^2+2\cdot(2x+y)\cdot 3+3^2$$
$$=4x^2+4xy+y^2+12x+6y+9$$

10. a. $(fg)(x)=f(x)\cdot g(x)$

$$=(x-3)(x-7)$$
$$=x^2-7x-3x+21$$
$$=x^2-10x+21$$

b. $(fg)(x)=x^2-10x+21$

$$(fg)(2)=(2)^2-10(2)+21=5$$

11. a. $f(x)=x^2-5x+4$

$$f(a+3)=(a+3)^2-5(a+3)+4$$
$$=a^2+6a+9-5a-15+4$$
$$=a^2+a-2$$

b. $f(x)=x^2-5x+4$

$$f(a+h)-f(a)=\left((a+h)^2-5(a+h)+4\right)-\left(a^2-5a+4\right)$$
$$=\left(a^2+2ah+h^2-5a-5h+4\right)-\left(a^2-5a+4\right)$$
$$=a^2+2ah+h^2-5a-5h+4-a^2+5a-4$$
$$=2ah+h^2-5h$$

5.2 Concept and Vocabulary Check

1. add

2. distributive; $4x^5 - 8x^2 + 6$; $7x^3$

3. $5x$; 3; like

4. $3x^2$; $5x$; $21x$; 35

5. $A^2 + 2AB + B^2$; squared; product of the terms; squared

6. $A^2 - 2AB + B^2$; minus; product of the terms; plus

7. $A^2 - B^2$; minus

8. x; $a + h$

5.2 Exercise Set

1. $$\begin{aligned}(3x^2)(5x^4) &= 3(5)x^2 \cdot x^4\\ &= 15x^{2+4}\\ &= 15x^6\end{aligned}$$

3. $$\begin{aligned}(3x^2y^4)(5xy^7) &= 3(5)x^2 \cdot x \cdot y^4 \cdot y^7\\ &= 15x^{2+1}y^{4+7}\\ &= 15x^3y^{11}\end{aligned}$$

5. $$\begin{aligned}&(-3xy^2z^5)(2xy^7z^4)\\ &= -3(2)x \cdot x \cdot y^2 \cdot y^7 \cdot z^5 \cdot z^4\\ &= -6x^{1+1}y^{2+7}z^{5+4}\\ &= -6x^2y^9z^9\end{aligned}$$

7. $$\begin{aligned}&(-8x^{2n}y^{n-5})\left(-\frac{1}{4}x^ny^3\right)\\ &= (-8)\left(-\frac{1}{4}\right)x^{2n} \cdot x^n \cdot y^{n-5} \cdot y^3\\ &= 2x^{2n+n}y^{n-5+3}\\ &= 2x^{3n}y^{n-2}\end{aligned}$$

9. $$\begin{aligned}4x^2(3x+2) &= 4x^2 \cdot 3x + 4x^2 \cdot 2\\ &= 12x^3 + 8x^2\end{aligned}$$

11. $$\begin{aligned}2y(y^2 - 5y) &= 2y \cdot y^2 - 2y \cdot 5y\\ &= 2y^3 - 10y^2\end{aligned}$$

13. $$\begin{aligned}&5x^3(2x^5 - 4x^2 + 9)\\ &= 5x^3 \cdot 2x^5 - 5x^3 \cdot 4x^2 + 5x^3 \cdot 9\\ &= 10x^8 - 20x^5 + 45x^3\end{aligned}$$

15. $$\begin{aligned}4xy(7x + 3y) &= 4xy \cdot 7x + 4xy \cdot 3y\\ &= 28x^2y + 12xy^2\end{aligned}$$

17. $$\begin{aligned}&3ab^2(6a^2b^3 + 5ab)\\ &= 3ab^2 \cdot 6a^2b^3 + 3ab^2 \cdot 5ab\\ &= 18a^3b^5 + 15a^2b^3\end{aligned}$$

19. $$\begin{aligned}&-4x^2y(3x^4y^2 - 7xy^3 + 6)\\ &= -4x^2y \cdot 3x^4y^2 + 4x^2y \cdot 7xy^3 - 4x^2y \cdot 6\\ &= -12x^6y^3 + 28x^3y^4 - 24x^2y\end{aligned}$$

21. $$\begin{aligned}&-4x^n\left(3x^{2n} - 5x^n + \frac{1}{2}x\right)\\ &= -4x^n \cdot 3x^{2n} + 4x^n \cdot 5x^n - 4x^n \cdot \frac{1}{2}x\\ &= -12x^{3n} + 20x^{2n} - 2x^{n+1}\end{aligned}$$

23. $$\begin{array}{r}x^2 + 2x + 5\\ \underline{x - 3}\\ x^3 + 2x^2 + 5x\\ \underline{-3x^2 - 6x - 15}\\ x^3 - x^2 - x - 15\end{array}$$

25. $$\begin{array}{r}x^2 + x + 1\\ \underline{x - 1}\\ x^3 + x^2 + x\\ \underline{-x^2 - x - 1}\\ x^3 \qquad\quad - 1\end{array}$$

27. $$\begin{array}{r}a^2 + ab + b^2\\ \underline{a - b}\\ a^3 + a^2b + ab^2\\ \underline{-a^2b - ab^2 - b^3}\\ a^3 \qquad\qquad - b^3\end{array}$$

29.
$$\begin{array}{r} x^2+2x-1 \\ \underline{x^2+3x-4} \\ x^4+2x^3-x^2 \\ 3x^3+6x^2-3x \\ \underline{-4x^2-8x+4} \\ x^4+5x^3+x^2-11x+4 \end{array}$$

31.
$$\begin{array}{r} x^2-3xy+y^2 \\ \underline{x-y} \\ x^3-3x^2y+xy^2 \\ \underline{-x^2y+3xy^2-y^3} \\ x^3-4x^2y+4xy^2-y^3 \end{array}$$

33.
$$\begin{array}{r} x^2y^2-2xy+4 \\ \underline{xy+2} \\ x^3y^3-2x^2y^2+4xy \\ \underline{2x^2y^2-4xy+8} \\ x^3y^3 \qquad\qquad +8 \end{array}$$

35. $(x+4)(x+7)=x^2+7x+4x+28$
$=x^2+11x+28$

37. $(y+5)(y-6)=y^2-6y+5y-30$
$=y^2-y-30$

39. $(5x+3)(2x+1)=10x^2+5x+6x+3$
$=10x^2+11x+3$

41. $(3y-4)(2y-1)=6y^2-3y-8y+4$
$=6y^2-11y+4$

43. $(3x-2)(5x-4)=15x^2-12x-10x+8$
$=15x^2-22x+8$

45. $(x-3y)(2x+7y)$
$=2x^2+7xy-6xy-21y^2$
$=2x^2+xy-21y^2$

47. $(7xy+1)(2xy-3)$
$=14x^2y^2-21xy+2xy-3$
$=14x^2y^2-19xy-3$

49. $(x-4)(x^2-5)=x^3-5x-4x^2+20$
$=x^3-4x^2-5x+20$

51. $(8x^3+3)(x^2-5)$
$=8x^3\cdot x^2-8x^3\cdot 5+3\cdot x^2-3\cdot 5$
$=8x^5-40x^3+3x^2-15$

53. $(3x^n-y^n)(x^n+2y^n)$
$=3x^n\cdot x^n+3x^n\cdot 2y^n-y^n\cdot x^n-y^n\cdot 2y^n$
$=3x^{2n}+6x^ny^n-x^ny^n-2y^{2n}$
$=3x^{2n}+5x^ny^n-2y^{2n}$

55. $(x+3)^2=x^2+2(3x)+9$
$=x^2+6x+9$

57. $(y-5)^2=y^2+2(-5y)+25$
$=y^2-10y+25$

59. $(2x+y)^2=4x^2+2(2xy)+y^2$
$=4x^2+4xy+y^2$

61. $(5x-3y)^2=25x^2+2(-15xy)+9y^2$
$=25x^2-30xy+9y^2$

63. $(2x^2+3y)^2=4x^4+2(6x^2y)+9y^2$
$=4x^4+12x^2y+9y^2$

65. $(4xy^2-xy)^2$
$=16x^2y^4+2(-4x^2y^3)+x^2y^2$
$=16x^2y^4-8x^2y^3+x^2y^2$

67. $(a^n+4b^n)^2$
$=(a^n)^2+2(a^n)(4b^n)+(4b^n)^2$
$=a^{2n}+8a^nb^n+16b^{2n}$

69. $(x+4)(x-4)=(x)^2-(4)^2$
$=x^2-16$

71. $(5x+3)(5x-3)=(5x)^2-(3)^2$
$=25x^2-9$

73. $(4x+7y)(4x-7y)=(4x)^2-(7y)^2$
$=16x^2-49y^2$

75. $(y^3+2)(y^3-2)=(y^3)^2-(2)^2$
$=y^6-4$

77. $(1-y^5)(1+y^5)=(1)^2-(y^5)^2$
$=1-y^{10}$

79. $(7xy^2-10y)(7xy^2+10y)$
$=(7xy^2)^2-(10y)^2$
$=49x^2y^4-100y^2$

81. $(5a^n-7)(5a^n+7)=(5a^n)^2-(7)^2$
$=25a^{2n}-49$

83. $[(2x+3)+4y][(2x+3)-4y]$
$=(2x+3)^2-16y^2$
$=4x^2+2(6x)+9-16y^2$
$=4x^2+12x+9-16y^2$

85. $(x+y+3)(x+y-3)$
$=((x+y)+3)((x+y)-3)$
$=(x+y)^2-9$
$=x^2+2xy+y^2-9$

87. $(5x+7y-2)(5x+7y+2)$
$=((5x+7y)-2)((5x+7y)+2)$
$=(5x+7y)^2-4$
$=25x^2+2(35xy)+49y^2-4$
$=25x^2+70xy+49y^2-4$

89. $[5y+(2x+3)][5y-(2x+3)]$
$=25y^2-(2x+3)^2$
$=25y^2-(4x^2+2(6x)+9)$
$=25y^2-(4x^2+12x+9)$
$=25y^2-4x^2-12x-9$

91.
$$\begin{array}{l} x+y+1 \\ \underline{x+y+1} \\ x^2+xy\quad +x \\ \qquad xy \qquad +y^2+y \\ \underline{\qquad\qquad x \qquad +y\ +1} \\ x^2+2xy+2x+y^2+2y+1 \end{array}$$
or
$x^2+2xy+y^2+2x+2y+1$

93. $(x+1)(x-1)(x^2+1)=(x^2-1)(x^2+1)$
$=(x^2)^2-(1)^2$
$=x^4-1$

95. a. $(fg)(x)=f(x)\cdot g(x)$
$=(x-2)(x+6)$
$=x^2+6x-2x-12$
$=x^2+4x-12$

b. $(fg)(-1)=(-1)^2+4(-1)-12$
$=1-4-12=-15$

c. $(fg)(0)=(0)^2+4(0)-12$
$=0+0-12=-12$

97. a. $(fg)(x)$
$=f(x)\cdot g(x)$
$=(x-3)(x^2+3x+9)$
$=x(x^2+3x+9)-3(x^2+3x+9)$
$=x^3+3x^2+9x-3x^2-9x-27$
$=x^3-27$

b. $(fg)(-2)=(-2)^3-27$
$=-8-27=-35$

c. $(fg)(0)=(0)^3-27=-27$

99. a. $f(a+2)=(a+2)^2-3(a+2)+7$
$=a^2+4a+4-3a-6+7=a^2+a+5$

b. $f(a+h)-f(a)=(a+h)^2-3(a+h)+7-(a^2-3a+7)$
$=\cancel{a^2}+2ah+h^2-\cancel{3a}-3h+\cancel{7}-\cancel{a^2}+\cancel{3a}-\cancel{7}$
$=2ah+h^2-3h$

101. a. $f(a+2)=3(a+2)^2+2(a+2)-1$
$=3(a^2+4a+4)+2a+4-1$
$=3a^2+12a+12+2a+4-1$
$=3a^2+14a+15$

b. $f(a+h)-f(a)=3(a+h)^2+2(a+h)-1-(3a^2+2a-1)$
$=3(a^2+2ah+h^2)+2a+2h-1-3a^2-2a+1$
$=\cancel{3a^2}+6ah+3h^2+\cancel{2a}+2h-\cancel{1}-\cancel{3a^2}-\cancel{2a}+\cancel{1}$
$=6ah+3h^2+2h$

103. $(3x+4y)^2-(3x-4y)^2=\left[(3x)^2+2(3x)(4y)+(4y)^2\right]-\left[(3x)^2-2(3x)(4y)+(4y)^2\right]$
$=(9x^2+24xy+16y^2)-(9x^2-24xy+16y^2)$
$=9x^2+24xy+16y^2-9x^2+24xy-16y^2$
$=48xy$

105. $(5x-7)(3x-2)-(4x-5)(6x-1)$
$=\left[15x^2-10x-21x+14\right]-\left[24x^2-4x-30x+5\right]$
$=(15x^2-31x+14)-(24x^2-34x+5)$
$=15x^2-31x+14-24x^2+34x-5$
$=-9x^2+3x+9$

107. $(2x+5)(2x-5)(4x^2+25)$
$=\left[(2x)^2-5^2\right](4x^2+25)$
$=(4x^2-25)(4x^2+25)$
$=(4x^2)^2-(25)^2$
$=16x^4-625$

109. $(x-1)^3 = (x-1)(x-1)^2$
$= (x-1)(x^2-2x+1)$
$= x(x^2-2x+1)-1(x^2-2x+1)$
$= x^3-2x^2+x-x^2+2x-1$
$= x^3-3x^2+3x-1$

111. $\frac{(2x-7)^5}{(2x-7)^3} = (2x-7)^{5-3}$
$= (2x-7)^2$
$= (2x)^2-2(2x)(7)+(7)^2$
$= 4x^2-28x+49$

113. **a.** $x^2+6x+4x+24 = x^2+10x+24$

b. $(x+6)(x+4) = x^2+4x+6x+24$
$= x^2+10x+24$

115. **a.** $(x+9)(x+3) = x^2+3x+9x+27$
$= x^2+12x+27$

b. $(x+5)(x+1) = x^2+x+5x+5$
$= x^2+6x+5$

c. $(x^2+12x+27)-(x^2+6x+5)$
$= x^2+12x+27-x^2-6x-5$
$= 6x+22$

117. **a.** $(8-2x)(10-2x) = 80-16x-20x+4x^2$
$= 80-36x+4x^2$
$= 4x^2-36x+80$

b. $x(80-36x+4x^2)$
$= 80x-36x^2+4x^3$
$= 4x^3-36x^2+80x$

119. **a.** $V(x) = x\cdot(x+10)(30-2x)$
$= x\cdot(-2x^2+10x+300)$
$= -2x^3+10x^2+300x$

b. Since the degree of the polynomial is 3, an odd number, and the leading coefficient is negative, the graph will rise to the left and fall to the right.

c. no; Since the graph falls to the right, the volume will eventually be negative, which is not possible.

d. $V(x) = -2x^3 + 10x^2 + 300x$

$V(10) = -2(10)^3 + 10(10)^2 + 300(10) = 2000$

Carry-on luggage with a depth of 10 inches has a volume of 2000 cubic inches.

e. The point $(10, 2000)$ represents carry-on luggage with a depth of 10 inches and a volume of 2000 cubic inches.

f. $\{x | 0 < x < 15\}$ or $(0,15)$, although answer may vary as long as they are within this interval. Values for x outside this interval would cause at least one dimension to be negative. Since these dimensions are for luggage, they must each be greater than zero. Values for x inside this interval will ensure that all dimensions are positive. But it can be argued that very small positive dimensions are not reasonable for luggage. Thus, it can be claimed that a reasonable interval should be smaller than the one listed here.

121. – 129. Answers will vary.

131.

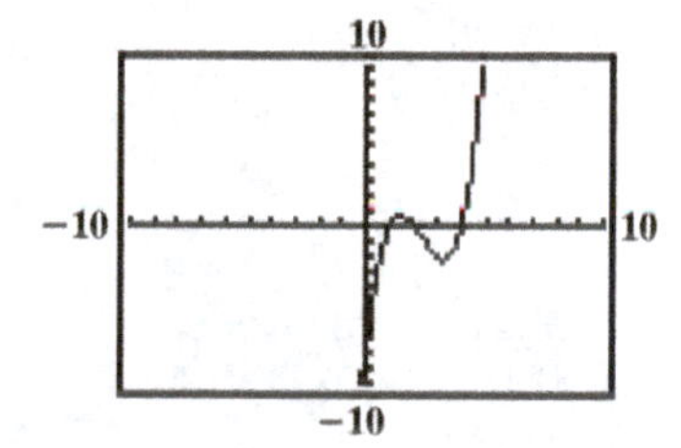

The functions y_1 and y_2 are the same.

Verify by multiplying.

$$\begin{array}{r} x^2 - 3x + 2 \\ x - 4 \\ \hline x^3 - 3x^2 + 2x \\ -4x^2 + 12x - 8 \\ \hline x^3 - 7x^2 + 14x - 8 \end{array}$$

133.

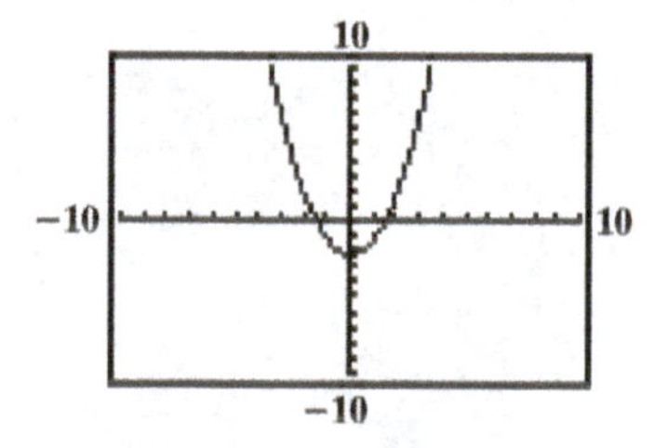

The functions y_1 and y_2 are the same.

Verify by multiplying.
$(x + 1.5)(x - 1.5) = x^2 - 2.25$

135. makes sense

137. makes sense, although this preference may vary

139. false; Changes to make the statement true will vary. A sample change is: $f(a+h) \neq f(a) + f(h)$.

141. false; Changes to make the statement true will vary. A sample change is: $(x+1)^2 = x^2 + 2 \cdot x \cdot 1 + 1^2$
$= x^2 + 2x + 1$

143.
$$\begin{aligned} \text{Area} &= x \cdot 3 + x(x-1) \\ &= 3x + x^2 - x \\ &= x^2 + 2x \end{aligned}$$

145.
$$\begin{aligned} V_{\text{total}} &= V_1 + V_2 \\ &= (x+5)(x+1)(x+2) + (x+5)(x)(x-1) \\ &= (x+5)\left[(x+1)(x+2) + (x)(x-1)\right] \\ &= (x+5)\left[x^2 + 3x + 2 + x^2 - x\right] \\ &= (x+5)\left[2x^2 + 2x + 2\right] \\ &= 2x^3 + 10x^2 + 2x^2 + 10x + 2x + 10 \\ &= 2x^3 + 12x^2 + 12x + 10 \end{aligned}$$

147. Let x be an odd integer. The next consecutive odd integer is $x+2$.

Then $x(x+2) = (x+2)^2 - 22$
$$\begin{aligned} x^2 + 2x &= x^2 + 4x + 4 - 22 \\ x^2 + 2x &= x^2 + 4x - 18 \\ 2x &= 4x - 18 \\ -2x &= -18 \\ x &= 9 \end{aligned}$$
Thus, the next consecutive odd integer is $x + 2 = 9 + 2 = 11$.
The integers are 9 and 11.

148. $|3x+4| \geq 10$
$$\begin{aligned} 3x + 4 &\leq -10 & &\text{or} & 3x + 4 &\geq 10 \\ 3x &\leq -14 & & & 3x &\geq 6 \\ x &\leq -\frac{14}{3} & & & x &\geq 2 \end{aligned}$$

The solution set is $\left(-\infty, -\frac{14}{3}\right] \cup [2, \infty)$.

149. $2-6x\le 20$
$-6x\le 18$
$x\ge -3$
The solution set is $[-3,\infty)$.

150. $8,034,000,000 = 8.034\times 10^9$

151. **a.** $3x^3\cdot\boxed{?}=9x^5$
$\boxed{?}=\dfrac{9x^5}{3x^3}$
$\boxed{?}=3x^2$
Thus, $3x^3\cdot\boxed{3x^2}=9x^5$.

b. $2x^3y^2\cdot\boxed{?}=12x^5y^4$
$\boxed{?}=\dfrac{12x^5y^4}{2x^3y^2}$
$\boxed{?}=6x^2y^2$
Thus, $2x^3y^2\cdot\boxed{6x^2y^2}=12x^5y^4$.

152. $(x-5)(x^2+3)=x^3-5x^2+3x-15$

153. $(x+4)(3x-2y)=3x^2-2xy+12x-8y$

5.3 Check Points

1. $20x^2+30x=\overbrace{10x}^{\text{GCF}}\cdot 2x+\overbrace{10x}^{\text{GCF}}\cdot 3$
$=10x(2x+3)$

2. **a.** $9x^4+21x^2=\overbrace{3x^2}^{\text{GCF}}\cdot 3x^2+\overbrace{3x^2}^{\text{GCF}}\cdot 7$
$=3x^2(3x^2+7)$

b. $15x^3y^2-25x^4y^3=\overbrace{5x^3y^2}^{\text{GCF}}\cdot 3-\overbrace{5x^3y^2}^{\text{GCF}}\cdot 5xy$
$=5x^3y^2(3-5xy)$

c. $16x^4y^5-8x^3y^4+4x^2y^3=\overbrace{4x^2y^3}^{\text{GCF}}\cdot 4x^2y^2-\overbrace{4x^2y^3}^{\text{GCF}}\cdot 2xy+\overbrace{4x^2y^3}^{\text{GCF}}\cdot 1$
$=4x^2y^3(4x^2y^2-2xy+1)$

3. $-2x^3+10x^2-6x=-2x\cdot x^2-2x\cdot(-5x)-2x\cdot 3$
$=-2x(x^2-5x+3)$

4. a. $3\overbrace{(x-4)}^{\text{GCF}}+7a\overbrace{(x-4)}^{\text{GCF}}=\overbrace{(x-4)}^{\text{GCF}}(3+7a)$
$=(x-4)(3+7a)$

b. $7x\overbrace{(a+b)}^{\text{GCF}}-\overbrace{(a+b)}^{\text{GCF}}=\overbrace{(a+b)}^{\text{GCF}}(7x-1)$
$=(a+b)(7x-1)$

5. $x^3-4x^2+5x-20=\overbrace{x^3-4x^2}^{\text{common factor is } x^2}+\overbrace{5x-20}^{\text{common factor is 5}}$
$=x^2(x-4)+5(x-4)$
$=(x-4)(x^2+5)$

6. $4x^2+20x-3xy-15y=\overbrace{4x^2+20x}^{\text{common factor is } 4x}+\overbrace{-3xy-15y}^{\text{common factor is } -3y}$
$=4x(x+5)-3y(x+5)$
$=(x+5)(4x-3y)$

5.3 Concept and Vocabulary Check

1. factoring
2. greatest common factor; smallest/least
3. false
4. $-2x$
5. false

5.3 Exercise Set

1. $10x^2+4x=2x\cdot5x+2x\cdot2$
$=2x(5x+2)$

3. $y^2-4y=y\cdot y-y\cdot4=y(y-4)$

5. $x^3+5x^2=x^2\cdot x+x^2\cdot5=x^2(x+5)$

7. $12x^4-8x^2=4x^2\cdot3x^2-4x^2\cdot2$
$=4x^2(3x^2-2)$

9. $32x^4 + 2x^3 + 8x^2$
$= 2x^2 \cdot 16x^2 + 2x^2 \cdot x + 2x^2 \cdot 4$
$= 2x^2\left(16x^2 + x + 4\right)$

11. $4x^2y^3 + 6xy = 2xy \cdot 2xy^2 + 2xy \cdot 3$
$= 2xy\left(2xy^2 + 3\right)$

13. $30x^2y^3 - 10xy^2$
$= 10xy^2 \cdot 3xy - 10xy^2 \cdot 1$
$= 10xy^2(3xy - 1)$

15. $12xy - 6xz + 4xw$
$= 2x \cdot 6y - 2x \cdot 3z + 2x \cdot 2w$
$= 2x(6y - 3z + 2w)$

17. $15x^3y^6 - 9x^4y^4 + 12x^2y^5$
$= 3x^2y^4 \cdot 5xy^2 - 3x^2y^4 \cdot 3x^2 + 3x^2y^4 \cdot 4y$
$= 3x^2y^4\left(5xy^2 - 3x^2 + 4y\right)$

19. $25x^3y^6z^2 - 15x^4y^4z^4 + 25x^2y^5z^3$
$= 5x^2y^4z^2 \cdot 5xy^2 - 5x^2y^4z^2 \cdot 3x^2z^2$
$+ 5x^2y^4z^2 \cdot 5yz$
$= 5x^2y^4z^2\left(5xy^2 - 3x^2z^2 + 5yz\right)$

21. $15x^{2n} - 25x^n = 5x^n \cdot 3x^n - 5x^n \cdot 5$
$= 5x^n\left(3x^n - 5\right)$

23. $-4x + 12 = -4 \cdot x + (-4)(-3)$
$= -4(x - 3)$

25. $-8x - 48 = -8 \cdot x + (-8)6$
$= -8(x + 6)$

27. $-2x^2 + 6x - 14$
$= -2 \cdot x^2 + (-2)(-3x) + (-2)(7)$
$= -2\left(x^2 + (-3x) + 7\right)$
$= -2\left(x^2 - 3x + 7\right)$

29. $-5y^2 + 40x = -5 \cdot y^2 + (-5)(-8x)$
$= -5\left(y^2 + (-8x)\right)$
$= -5\left(y^2 - 8x\right)$

31. $-4x^3 + 32x^2 - 20x$
$= -4x \cdot x^2 + (-4x)(-8x) + (-4x)(5)$
$= -4x\left(x^2 + (-8x) + 5\right)$
$= -4x\left(x^2 - 8x + 5\right)$

33. $-x^2 - 7x + 5$
$= -1 \cdot x^2 + (-1)(7x) + (-1)(-5)$
$= -1\left(x^2 + 7x + (-5)\right)$
$= -1\left(x^2 + 7x - 5\right)$

35. $4(x + 3) + a(x + 3)$
$= (x + 3)(4 + a)$

37. $x(y - 6) - 7(y - 6)$
$= (y - 6)(x - 7)$

39. $3x(x + y) - (x + y)$
$= 3x(x + y) - 1(x + y)$
$= (x + y)(3x - 1)$

41. $4x^2(3x - 1) + 3x - 1$
$= 4x^2(3x - 1) + 1(3x - 1)$
$= (3x - 1)\left(4x^2 + 1\right)$

43. $(x + 2)(x + 3) + (x - 1)(x + 3)$
$= (x + 3)(x + 2 + x - 1)$
$= (x + 3)(2x + 1)$

45. $x^2 + 3x + 5x + 15$
$= x(x + 3) + 5(x + 3)$
$= (x + 3)(x + 5)$

47. $x^2 + 7x - 4x - 28$
$= x(x + 7) - 4(x + 7)$
$= (x + 7)(x - 4)$

49. $x^3 - 3x^2 + 4x - 12$
$= x^2(x-3) + 4(x-3)$
$= (x-3)(x^2+4)$

51. $xy - 6x + 2y - 12$
$= x(y-6) + 2(y-6)$
$= (y-6)(x+2)$

53. $xy + x - 7y - 7$
$= x(y+1) - 7(y+1)$
$= (y+1)(x-7)$

55. $10x^2 - 12xy + 35xy - 42y^2$
$= 2x(5x-6y) + 7y(5x-6y)$
$= (5x-6y)(2x+7y)$

57. $4x^3 - x^2 - 12x + 3$
$= x^2(4x-1) - 3(4x-1)$
$= (4x-1)(x^2-3)$

59. $x^2 - ax - bx + ab$
$= x(x-a) - b(x-a)$
$= (x-a)(x-b)$

61. $x^3 - 12 - 3x^2 + 4x$
$= x^3 - 3x^2 + 4x - 12$
$= x^2(x-3) + 4(x-3)$
$= (x-3)(x^2+4)$

63. $ay - by + bx - ax$
$= y(a-b) + x(b-a)$
$= y(a-b) + x(-1)(a-b)$
$= y(a-b) - x(a-b)$
$= (a-b)(y-x)$

65. $ay^2 + 2by^2 - 3ax - 6bx$
$= y^2 \cdot a + y^2 \cdot 2b + (-3x) \cdot a + (-3x) \cdot 2b$
$= y^2(a+2b) - 3x(a+2b)$
$= (a+2b)(y^2-3x)$

67. $x^n y^n + 3x^n + y^n + 3$
$= x^n \cdot y^n + x^n \cdot 3 + 1 \cdot y^n + 1 \cdot 3$
$= x^n(y^n+3) + 1(y^n+3)$
$= (y^n+3)(x^n+1)$

69. $ab - c - ac + b$
$= ab + b - ac - c$
$= a \cdot b + 1 \cdot b + (-c) \cdot a + (-c) \cdot 1$
$= b(a+1) + (-c)(a+1)$
$= (a+1)(b-c)$

71. $x^3 - 5 + 4x^3y - 20y$
$= 1 \cdot x^3 - 1 \cdot 5 + 4y \cdot x^3 - 4y \cdot 5$
$= 1(x^3-5) + 4y(x^3-5)$
$= (x^3-5)(1+4y)$

73. $2y^7(3x-1)^5 - 7y^6(3x-1)^4$
$= y^6(3x-1)^4 \cdot 2y(3x-1) - y^6(3x-1)^4 \cdot 7$
$= y^6(3x-1)^4(2y(3x-1)-7)$
$= y^6(3x-1)^4(6xy-2y-7)$

75. $ax^2 + 5ax - 2a + bx^2 + 5bx - 2b$
$= a \cdot x^2 + a \cdot 5x - a \cdot 2 + b \cdot x^2 + b \cdot 5x - b \cdot 2$
$= a(x^2+5x-2) + b(x^2+5x-2)$
$= (x^2+5x-2)(a+b)$

77. $ax + ay + az - bx - by - bz + cx + cy + cz$
$= a \cdot x + a \cdot y + a \cdot z - b \cdot x - b \cdot y - b \cdot z + c \cdot x + c \cdot y + c \cdot z$
$= a(x+y+z) - b(x+y+z) + c(x+y+z)$
$= (x+y+z)(a-b+c)$

79. $f(t) = -16t^2 + 40t$

a. $f(2) = -16(2)^2 + 40(2)$
$= -16(4) + 80$
$= -64 + 80$
$= 16$

After 2 seconds, the ball will be at a height of 16 feet.

b. $f(2.5) = -16(2.5)^2 + 40(2.5)$
$= -16(6.25) + 100$
$= -100 + 100 = 0$

After 2.5 seconds, the ball will hit the ground.

c. $-16t^2 + 40t = -4t \cdot 4t - 4t(-10)$
$= -8t(2t - 5)$

d. $f(t) = -8t(2t - 5)$
$f(2) = -8(2)(2(2) - 5)$
$= -16(4 - 5)$
$= -16(-1) = 16$

$f(2.5) = -8(2.5)(2(2.5) - 5)$
$= -20(5 - 5)$
$= -20(0) = 0$

These are the same answers as parts (a) and (b). This shows that the factorization is equivalent to the original polynomial and that the factorization is correct.

81. a. $(x - 0.4x) - 0.4(x - 0.4x)$
$= (x - 0.4x)(1 - 0.4)$
$= (0.6x)(0.6)$
$= 0.36x$

b. No, the computer is not selling at 20% of its original price. The computer is selling at 36% of its original price.

83. $A = P + Pr + (P + Pr)r$
$= (P + Pr)1 + (P + Pr)r$
$= (P + Pr)(1 + r)$
$= (P \cdot 1 + P \cdot r)(1 + r)$
$= P(1 + r)(1 + r)$
$= P(1 + r)^2$

85. $A = \pi r^2 + 2rl$
$= r \cdot \pi r + r \cdot 2l$
$= r(\pi r + 2l)$

87. – 91. Answers will vary.

93. $x^2 - 4x = x(x - 4)$

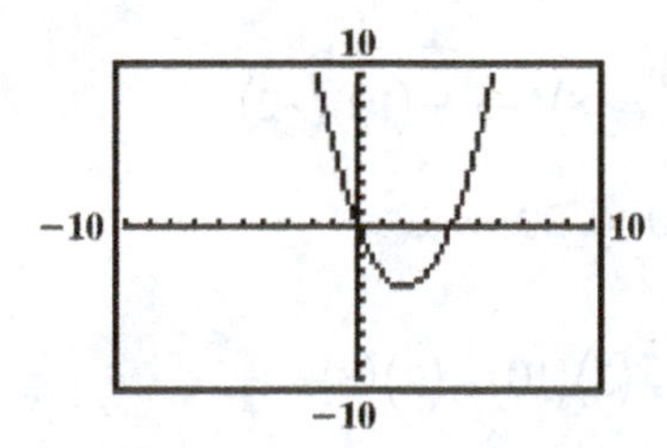

The graphs coincide. The polynomial is factored correctly.

95. $x^2 + 2x + x + 2 = x(x + 2) + 1$

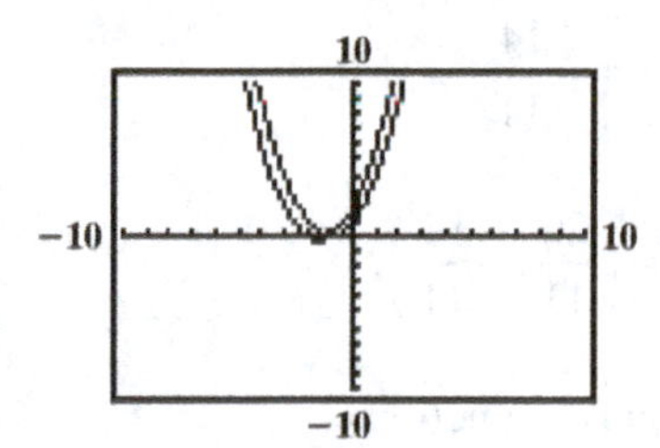

The graphs do not coincide. Factor the polynomial correctly.
$x^2 + 2x + x + 2 = x(x + 2) + 1(x + 2)$
$= (x + 2)(x + 1)$

97. does not make sense; Explanations will vary. Sample explanation: You can check your factoring by multiplication.

99. makes sense

101. false; Changes to make the statement true will vary. A sample change is: It is necessary to write the 1 when the expression is written is factored form.

103. true

105. $x^{4n} + x^{2n} + x^{3n} = x^{2n}\left(x^{2n} + 1 + x^n\right)$

107. $8y^{2n+4} + 16y^{2n+3} - 12y^{2n}$
$= 4y^{2n}\left(2y^4 + 4y^3 - 3\right)$

109. Answers will vary; an example is $6x^2 - 4x + 9x - 6$.

110. $D = \begin{vmatrix} 3 & -2 \\ 2 & -5 \end{vmatrix} = (3)(-5)-(2)(-2)$
$= -15+4 = -11$

$D_x = \begin{vmatrix} 8 & -2 \\ 10 & -5 \end{vmatrix} = (8)(-5)-(10)(-2)$
$= -40+20 = -20$

$D_y = \begin{vmatrix} 3 & 8 \\ 2 & 10 \end{vmatrix} = (3)(10)-(2)(8)$
$= 30-16 = 14$

$x = \frac{D_x}{D} = \frac{-20}{-11} = \frac{20}{11}$

$y = \frac{D_y}{D} = \frac{14}{-11} = -\frac{14}{11}$

The solution is $\left(\frac{20}{11}, -\frac{14}{11}\right)$.

111. a. The relation is a function.

b. The relation is not a function.

112. Let w = the width of the rectangle.
Let $2w + 2$ = the length of the rectangle.
$P = 2l + 2w$
$22 = 2(2w+2)+2w$
$22 = 4w+4+2w$
$22 = 6w+4$
$18 = 6w$
$3 = w$
Find the length.
$2w+2 = 2(3)+2 = 6+2 = 8$
The length is 8 feet and the width is 3 feet.

113. $(x+3)(x+\boxed{?}) = x^2+7x+12$

This will be true if $3\cdot\boxed{?} = 12$. Thus, $\boxed{?} = 4$.

114. $(x-\boxed{?})(x-12) = x^2-14x+24$

This will be true if $-\boxed{?}\cdot(-12) = 24$. Thus, $\boxed{?} = 2$.

115. $(x+3y)(x-\boxed{?}y) = x^2-4xy-21y^2$

This will be true if $3y\cdot(-\boxed{?}y) = -21y^2$ which means $-3\boxed{?} = -21$ Thus, $\boxed{?} = 7$.

5.4 Check Points

1. $x^2+6x+8 = (x+4)(x+2)$
factor pair: $4(2) = 8$
sum: $4+2 = 6$

2. $x^2-9x+20 = (x-5)(x-4)$
factor pair: $-5(-4) = 20$
sum: $-5+(-4) = -9$

3. $y^2+19y-66 = (y+22)(y-3)$
factor pair: $22(-3) = -66$
sum: $22+(-3) = 19$

4. $x^2-5xy+6y^2 = (x-3y)(x-2y)$
factor pair: $-3(-2) = 6$
sum: $-3+(-2) = -5$

5. First, factor out the GCF.
$3x^3-15x^2-42x = 3x(x^2-5x-14)$
Next, factor the trinomial.
$3x^3-15x^2-42x = 3x(x^2-5x-14)$
$= 3x(x-7)(x+2)$

6. Let $u = x^3$.
$x^6-7x^3+10 = (x^3)^2-7x^3+10$
$= u^2-7u+10$
$= (u-5)(u-2)$
Now substitute x^3 for u.
$= (u-5)(u-2)$
$= (x^3-5)(x^3-2)$

7. $3x^2-20x+28 = (3x-14)(x-2)$

8. $6x^6+19x^5-7x^4 = x^4(6x^2+19x-7)$
$= x^4(3x-1)(2x+7)$

9. $2x^2-7xy+3y^2 = (2x-y)(x-3y)$

10. Let $u = y^2$.

$$3y^4 + 10y^2 - 8 = 3(y^2)^2 + 10y^2 - 8$$
$$= 3u^2 + 10u - 8$$
$$= (3u-2)(u+4)$$

Now substitute y^2 for u.

$$= (3u-2)(u+4)$$
$$= (3y^2-2)(y^2+4)$$

11. $8x^2 - 22x + 5 = \overset{a}{8}x^2 \overset{b}{-22}x + \overset{c}{5}$

Multiply a and c.

$ac = 8(5) = 40$

Find the factors of ac whose sum is b.

Product: $-20(-2) = 40$

Sum: $-20 + (-2) = -22$

Rewrite the middle term using the factors found in the previous step.

$$8x^2 - 22x + 5 = 8x^2 - 20x - 2x + 5$$
$$= \overbrace{8x^2 - 20x}^{\text{Find GCF}} \overbrace{-2x + 5}^{\text{Find GCF}}$$
$$= 4x(2x-5) - 1(2x-5)$$
$$= (2x-5)(4x-1)$$

5.4 Concept and Vocabulary Check

1. completely
2. greatest common factor
3. +5
4. –4
5. +16
6. $-3y$
7. $2x$; $2x$; -18
8. -11
9. $2x+9$
10. $-6y$

5.4 Exercise Set

1. $x^2 + 5x + 6 = (x+3)(x+2)$

3. $x^2 + 8x + 12 = (x+6)(x+2)$

5. $x^2 + 9x + 20 = (x+5)(x+4)$

7. $y^2 + 10y + 16 = (y+8)(y+2)$

9. $x^2 - 8x + 15 = (x-5)(x-3)$

11. $y^2 - 12y + 20 = (y-10)(y-2)$

13. $a^2 + 5a - 14 = (a+7)(a-2)$

15. $x^2 + x - 30 = (x+6)(x-5)$

17. $x^2 - 3x - 28 = (x-7)(x+4)$

19. $y^2 - 5y - 36 = (y-9)(y+4)$

21. $x^2 - x + 7$

The trinomial is not factorable. There are no factors of 7 that add up to –1. The polynomial is prime.

23. $x^2 - 9xy + 14y^2 = (x-7y)(x-2y)$

25. $x^2 - xy - 30y^2 = (x-6y)(x+5y)$

27. $x^2 + xy + y^2$

The trinomial is not factorable. There are no factors of 1 that add up to 1. The polynomial is prime.

29. $a^2 - 18ab + 80b^2 = (a-10b)(a-8b)$

31. $3x^2 + 3x - 18 = 3(x^2 + x - 6)$
$$= 3(x+3)(x-2)$$

33. $2x^3 - 14x^2 + 24x = 2x(x^2 - 7x + 12)$
$$= 2x(x-4)(x-3)$$

35.
$$3y^3 - 15y^2 + 18y$$
$$= 3y(y^2 - 5y + 6)$$
$$= 3y(y-3)(y-2)$$

37. $2x^4 - 26x^3 - 96x^2$
$= 2x^2\left(x^2 - 13x - 48\right)$
$= 2x^2\left(x - 16\right)\left(x + 3\right)$

39. Let $u = x^3$
$x^6 - x^3 - 6 = \left(x^3\right)^2 - x^3 - 6$
$= u^2 - u - 6$
$= (u - 3)(u + 2)$
Substitute x^3 for u.
$= \left(x^3 - 3\right)\left(x^3 + 2\right)$

41. Let $u = x^2$
$x^4 - 5x^2 - 6 = \left(x^2\right)^2 - 5x^2 - 6$
$= u^2 - 5u - 6$
$= (u - 6)(u + 1)$
Substitute x^2 for u.
$= (x^2 - 6)(x^2 + 1)$

43. Let $u = x + 1$
$(x+1)^2 + 6(x+1) + 5$
$= u^2 + 6u + 5$
$= (u + 5)(u + 1)$

Substitute $x + 1$ for u.
$= \left((x+1)+5\right)\left((x+1)+1\right)$
$= (x+1+5)(x+1+1)$
$= (x+6)(x+2)$

45. $3x^2 + 8x + 5 = (3x + 5)(x + 1)$

47. $5x^2 + 56x + 11 = (5x + 1)(x + 11)$

49. $3y^2 + 22y - 16 = (3y - 2)(y + 8)$

51. $4y^2 + 9y + 2 = (y + 2)(4y + 1)$

53. $10x^2 + 19x + 6 = (5x + 2)(2x + 3)$

55. $8x^2 - 18x + 9 = (4x - 3)(2x - 3)$

57. $6y^2 - 23y + 15 = (6y - 5)(y - 3)$

59. $6y^2 + 14y + 3$
The trinomial is not factorable. The polynomial is prime.

61. $3x^2 + 4xy + y^2 = (3x + y)(x + y)$

63. $6x^2 - 7xy - 5y^2 = (2x + y)(3x - 5y)$

65. $15x^2 - 31xy + 10y^2 = (3x - 5y)(5x - 2y)$

67. $3a^2 - ab - 14b^2 = (3a - 7b)(a + 2b)$

69. $15x^3 - 25x^2 + 10x = 5x\left(3x^2 - 5x + 2\right)$
$= 5x(3x - 2)(x - 1)$

71. $24x^4 + 10x^3 - 4x^2 = 2x^2\left(12x^2 + 5x - 2\right)$
$= 2x^2(3x + 2)(4x - 1)$

73. $15y^5 - 2y^4 - y^3 = y^3\left(15y^2 - 2y - 1\right)$
$= y^3(3y - 1)(5y + 1)$

75. $24x^2 + 3xy - 27y^2 = 3\left(8x^2 + xy - 9y^2\right)$
$= 3(8x + 9y)(x - y)$

77. $6a^2b - 2ab - 60b = 2b\left(3a^2 - a - 30\right)$
$= 2b(3a - 10)(a + 3)$

79. $12x^2y - 34xy^2 + 14y^3 = 2y\left(6x^2 - 17xy + 7y^2\right)$
$= 2y(3x - 7y)(2x - y)$

81. $13x^3y^3 + 39x^3y^2 - 52x^3y = 13x^3y\left(y^2 + 3y - 4\right)$
$= 13x^3y(y + 4)(y - 1)$

83. Let $u = x^2$
$2x^4 - x^2 - 3$
$= 2\left(x^2\right)^2 - x^2 - 3$
$= 2u^2 - u - 3$
$= (2u - 3)(u + 1)$
Substitute x^2 for u.
$= \left(2x^2 - 3\right)\left(x^2 + 1\right)$

85. Let $u = x^3$

$$2x^6 + 11x^3 + 15 = 2\left(x^3\right)^2 + 11x^3 + 15$$
$$= 2u^2 + 11u + 15$$
$$= (2u+5)(u+3)$$

Substitute x^3 for u.

$$= \left(2x^3+5\right)\left(x^3+3\right)$$

87. Let $u = y^5$

$$2y^{10} + 7y^5 + 3 = 2\left(y^5\right)^2 + 7y^5 + 3$$
$$= 2u^2 + 7u + 3$$
$$= (2u+1)(u+3)$$

Substitute y^5 for u.

$$= \left(2y^5+1\right)\left(y^5+3\right)$$

89. Let $u = x+1$

$$5(x+1)^2 + 12(x+1) + 7 = 5u^2 + 12u + 7$$
$$= (5u+7)(u+1)$$

Substitute $x+1$ for u.

$$= (5(x+1)+7)((x+1)+1)$$
$$= (5x+5+7)(x+1+1)$$
$$= (5x+12)(x+2)$$

91. Let $u = x-3$

$$2(x-3)^2 - 5(x-3) - 7 = 2u^2 - 5u - 7$$
$$= (2u-7)(u+1)$$

Substitute $x-3$ for u.

$$= (2(x-3)-7)((x-3)+1)$$
$$= (2x-6-7)(x-3+1)$$
$$= (2x-13)(x-2)$$

93. $x^2 - 0.5x + 0.06$

Since $(0.3)\cdot(0.2) = 0.06$ and $0.3 + 0.2 = 0.5$, we get

$$x^2 - 0.5x + 0.06 = (x-0.3)(x-0.2)$$

95. $x^2 - \frac{3}{49} + \frac{2}{7}x = x^2 + \frac{2}{7}x - \frac{3}{49}$

Since $\left(\frac{3}{7}\right)\left(-\frac{1}{7}\right) = -\frac{3}{49}$ and $\frac{3}{7} + \left(-\frac{1}{7}\right) = \frac{2}{7}$, we get

$$x^2 - \frac{3}{49} + \frac{2}{7}x = \left(x+\frac{3}{7}\right)\left(x-\frac{1}{7}\right)$$

97. $acx^2 - bcx + adx - bd$

$$= cx(ax-b) + d(ax-b)$$
$$= (ax-b)(cx+d)$$

99. $-4x^5y^2 + 7x^4y^3 - 3x^3y^4$

$$= \left(-x^3y^2\right)\cdot 4x^2 - \left(-x^3y^2\right)\cdot 7xy + \left(-x^3y^2\right)\cdot 3y^2$$
$$= -x^3y^2\left(4x^2 - 7xy + 3y^2\right)$$
$$= -x^3y^2(4x-3y)(x-y)$$

101. $(fg)(x) = 3x^2 - 22x + 39$

$$= (3x-13)(x-3)$$

$f(x) = 3x - 13$ and $g(x) = x - 3$, or vice versa.

103. $x^2 + x^2 + x + x + x + x + x + x + x + 1 + 1 + 1$

$$= 2x^2 + 7x + 3$$

$$2x^2 + 7x + 3 = (2x+1)(x+3)$$

The dimensions of the rectangle are $2x+1$ by $x+3$.

105. **a.** $f(1) = -16(1)^2 + 16(1) + 32$

$$= -16(1) + 16 + 32$$
$$= -16 + 16 + 32$$
$$= 32$$

After 1 second, the diver will be 32 feet above the water.

b. $f(2) = -16(2)^2 + 16(2) + 32$

$$= -16(4) + 32 + 32$$
$$= -64 + 32 + 32 = 0$$

After 2 seconds, the diver will hit the water.

c. $-16t^2 + 16t + 32$
$= -16\left(t^2 - t - 2\right)$
$= -16(t-2)(t+1)$

d. $f(t) = -16(t-2)(t+1)$
$f(1) = -16(1-2)(1+1)$
$= -16(-1)(2) = 32$
$f(2) = -16(2-2)(2+1)$
$= -16(0)(3) = 0$

107. a. $x^2 + x + x + x + 1 + 1$
$= x^2 + 3x + 2$

b. $(x+2)(x+1)$

c. Answers will vary.

109. – 115. Answers will vary.

117. $x^2 + 7x + 12 = (x+4)(x+3)$

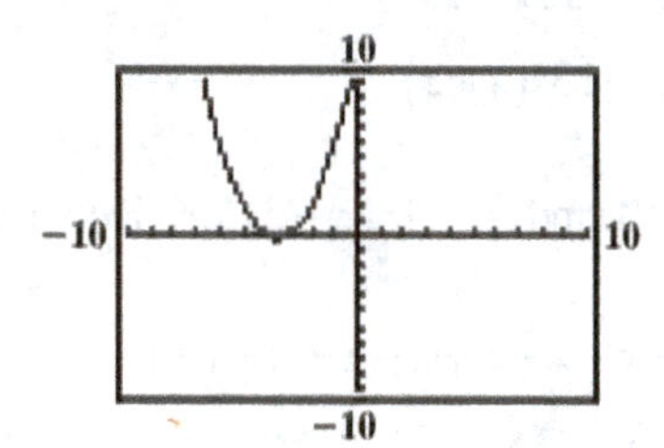

The graphs coincide. The polynomial is factored correctly.

119. $6x^3 + 5x^2 - 4x = x(3x+4)(2x-1)$

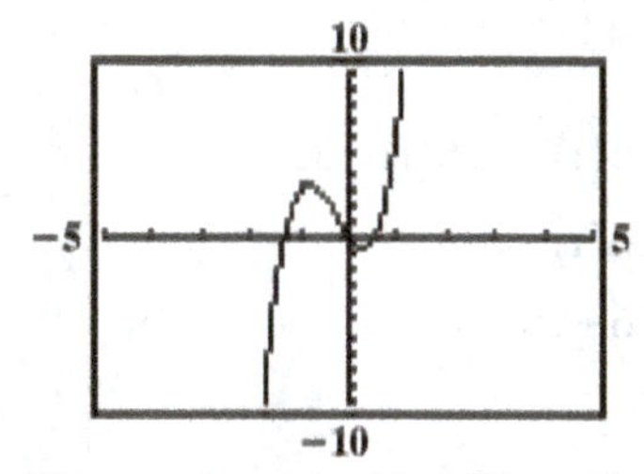

The graphs coincide. The polynomial is factored correctly.

121. Answers will vary.

123. makes sense

125. makes sense

127. true

129. false; Changes to make the statement true will vary. A sample change is: The correct factorization is $12x^2 - 19xy + 5y^2 = (4x-5y)(3x-y)$.

131. Use the factors of 3 and 5 to find b. The factors of 3 are 1 and 3 and the factors of 5 are 1 and 5. Multiply the combinations of the factors as follows.
$(3x+5)(x+1) = 3x^2 + 3x + 5x + 5$
$= 3x^2 + 8x + 5$
or
$(3x+1)(x+5) = 3x^2 + 15x + x + 5$
$= 3x^2 + 16x + 5$

In addition to 8 and 16, –8 and –16 can also be obtained.
$(3x-5)(x-1) = 3x^2 - 3x - 5x + 5$
$= 3x^2 - 8x + 5$
or
$(3x-1)(x-5) = 3x^2 - 15x - x + 5$
$= 3x^2 - 16x + 5$

The integers are –16, –8, 8 and 16.

133. Let $u = x^n$
$4x^{2n} - 9x^n + 5m = 4\left(x^n\right)^2 - 9x^n + 5$
$= 4u^2 - 9u + 5$
$= (4u-5)(u-1)$
Substitute x^n for u.
$= \left(4x^n - 5\right)\left(x^n - 1\right)$

135. Let $u = b^n$.
$b^{2n+2} + 3b^{n+2} - 10b^2 = b^2\left(b^{2n} + 3b^n - 10\right)$
$= b^2\left(\left(b^n\right)^2 + 3b^n - 10\right)$
$= b^2\left(u^2 + 3u - 10\right)$
$= b^2(u-2)(u+5)$
Substitute b^n for u.
$= b^2\left(b^n - 2\right)\left(b^n + 5\right)$

137. $2d^{n+2} - 5d^{n+1} + 3d^n = d^n\left(2d^2 - 5d + 3\right)$
$= d^n(2d-3)(d-1)$

138. $-2x \le 6$ and $-2x+3<-7$

$x \ge -3$ $\quad -2x<-10$

$x>5$

The solution set is $(5,\infty)$.

139. Multiply the second equation by –2 and add to the first equation.

$$\begin{aligned} 2x-\ y-2z&=-1\\ -2x+4y+2z&=-2\\ \hline 3y&=-3\\ y&=-1 \end{aligned}$$

Add the second and third equations to eliminate z.

$$\begin{aligned} x-2y-\ z&=\ 1\\ x+\ y+\ z&=\ 4\\ \hline 2x-\ y\quad &=\ 5 \end{aligned}$$

Now, back-substitute –1 for y to find x.

$$\begin{aligned} 2x-y&=5\\ 2x-(-1)&=5\\ 2x+1&=5\\ 2x&=4\\ x&=2 \end{aligned}$$

Now back-substitute 2 for x and –1 for y to find z.

$$\begin{aligned} x+y+z&=4\\ 2+(-1)+z&=4\\ 1+z&=4\\ z&=3 \end{aligned}$$

The solution is $(2,-1,3)$.

140. $4x^3+8x^2-5x-10=4x^2(x+2)-5(x+2)$

$=(x+2)(4x^2-5)$

141. $x^2+14x+49=(x+7)(x+7)$

$=(x+7)^2$

142. $x^2-8x+16=(x-4)(x-4)$

$=(x-4)^2$

143. $x^2-25=(x+5)(x-5)$

Mid-Chapter Check Point – Chapter 5

1. $\left(-8x^3+6x^2-x+5\right)-\left(-7x^3+2x^2-7x-12\right)$
$=-8x^3+6x^2-x+5+7x^3-2x^2+7x+12$
$=-8x^3+7x^3+6x^2-2x^2-x+7x+5+12$
$=-x^3+4x^2+6x+17$

2. $\left(6x^2yz^4\right)\left(-\frac{1}{3}x^5y^2z\right)$
$=(6)\left(-\frac{1}{3}\right)x^2\cdot x^5\cdot y\cdot y^2\cdot z^4\cdot z$
$=-2x^7y^3z^5$

3. $5x^2y\left(6x^3y^2-7xy-\frac{2}{5}\right)$
$=5x^2y\cdot 6x^3y^2-5x^2y\cdot 7xy-5x^2y\cdot\frac{2}{5}$
$=30x^5y^3-35x^3y^2-2x^2y$

4. $(3x-5)\left(x^2+3x-8\right)$
$=3x\left(x^2+3x-8\right)-5\left(x^2+3x-8\right)$
$=3x^3+9x^2-24x-5x^2-15x+40$
$=3x^3+4x^2-39x+40$

5. $\left(x^2-2x+1\right)\left(2x^2+3x-4\right)=x^2\left(2x^2+3x-4\right)-2x\left(2x^2+3x-4\right)+1\left(2x^2+3x-4\right)$
$=2x^4+3x^3-4x^2-4x^3-6x^2+8x+2x^2+3x-4$
$=2x^4-x^3-8x^2+11x-4$

6. $\left(x^2-2x+1\right)-\left(2x^2+3x-4\right)=x^2-2x+1-2x^2-3x+4=-x^2-5x+5$

7. $\left(6x^3y-11x^2y-4y\right)+\left(-11x^3y+5x^2y-y-6\right)-\left(-x^3y+2y-1\right)$
$=6x^3y-11x^2y-4y-11x^3y+5x^2y-y-6+x^3y-2y+1$
$=6x^3y-11x^3y+x^3y-11x^2y+5x^2y-4y-y-2y-6+1$
$=-4x^3y-6x^2y-7y-5$

8. $(2x+5)(4x-1)=2x\cdot 4x-2x\cdot 1+5\cdot 4x-5\cdot 1$
$=8x^2-2x+20x-5$
$=8x^2+18x-5$

9. $(2xy-3)(5xy+2)$
$= 2xy\cdot 5xy + 2xy\cdot 2 - 3\cdot 5xy - 3\cdot 2$
$= 10x^2y^2 + 4xy - 15xy - 6$
$= 10x^2y^2 - 11xy - 6$

10. $(3x-2y)(3x+2y) = (3x)^2 - (2y)^2$
$= 9x^2 - 4y^2$

11. $(3xy+1)(2x^2-3y)$
$= 3xy\cdot 2x^2 - 3xy\cdot 3y + 1\cdot 2x^2 - 1\cdot 3y$
$= 6x^3y - 9xy^2 + 2x^2 - 3y$

12. $(7x^3y+5x)(7x^3y-5x) = (7x^3y)^2 - (5x)^2$
$= 49x^6y^2 - 25x^2$

13. $3(x+h)^2 - 2(x+h) + 5 - (3x^2 - 2x + 5)$
$= 3(x^2 + 2xh + h^2) - 2x - 2h + 5 - 3x^2 + 2x - 5$
$= 3x^2 + 6xh + 3h^2 - 2x - 2h + 5 - 3x^2 + 2x - 5$
$= 3x^2 - 3x^2 - 2x + 2x + 6xh + 3h^2 - 2h + 5 - 5$
$= 6xh + 3h^2 - 2h$

14. $(x^2-3)^2 = (x^2)^2 - 2(x^2)(3) + 3^2$
$= x^4 - 6x^2 + 9$

15. $(x^2-3)(x^3+5x+2)$
$= x^2(x^3+5x+2) - 3(x^3+5x+2)$
$= x^5 + 5x^3 + 2x^2 - 3x^3 - 15x - 6$
$= x^5 + 2x^3 + 2x^2 - 15x - 6$

16. $(2x+5y)^2 = (2x)^2 + 2(2x)(5y) + (5y)^2$
$= 4x^2 + 20xy + 25y^2$

17. $(x+6+3y)(x+6-3y)$
$= ((x+6)+3y)((x+6)-3y)$
$= (x+6)^2 - (3y)^2$
$= x^2 + 12x + 36 - 9y^2$

18. $(x+y+5)^2$
$= ((x+y)+5)^2$
$= (x+y)^2 + 2(x+y)(5) + 5^2$
$= x^2 + 2xy + y^2 + 10x + 10y + 25$

19. $x^2 - 5x - 24$
We need two factors of -24 whose sum is -5. Since the product is negative, the factors will have opposite signs. Since the sum is negative, the factor with the larger absolute value will be negative.

$(-8)(3) = -24$ and $(-8)+(3) = -5$

$x^2 - 5x - 24 = (x-8)(x+3)$

20. $15xy + 5x + 6y + 2 = 5x(3y+1) + 2(3y+1)$
$= (3y+1)(5x+2)$

21. $5x^2 + 8x - 4$

$a\cdot c = 5\cdot(-4) = -20$
We need two factors of -20 whose sum is 8. Since the product is negative, the factors will have opposite signs. Since the sum is positive, the factor with the larger absolute value will be positive.

$(10)(-2) = -20$ and $(10)+(-2) = 8$

$5x^2 + 8x - 4 = 5x^2 + 10x - 2x - 4$
$= 5x(x+2) - 2(x+2)$
$= (x+2)(5x-2)$

22. $35x^2 + 10x - 50 = 5(7x^2 + 2x - 10)$
The reduced polynomial (in parentheses) cannot be factored further over the set of integers.

23. $9x^2 - 9x - 18 = 9(x^2 - x - 2)$
$= 9(x-2)(x+1)$

24. $10x^3y^2 - 20x^2y^2 + 35x^2y$
$= 5x^2y(2xy - 4y + 7)$

25. $18x^2+21x+5$

$a\cdot c=18\cdot 5=90$

We need two factors of 90 whose sum is 21. Since both the product and sum are positive, the two factors will be positive.

$(15)(6)=90$ and $(15)+(6)=21$

$$\begin{aligned}18x^2+21x+5&=18x^2+15x+6x+5\\&=3x(6x+5)+1(6x+5)\\&=(6x+5)(3x+1)\end{aligned}$$

26. $12x^2-9xy-16x+12y$

$$\begin{aligned}&=3x(4x-3y)-4(4x-3y)\\&=(4x-3y)(3x-4)\end{aligned}$$

27. $9x^2-15x+4=(3x-1)(3x-4)$

$a\cdot c=9\cdot 4=36$

We need two factors of 36 whose sum is -15. Since the product is positive and the sum is negative, the two factors will be negative.

$(-12)(-3)=36$ and $(-12)+(-3)=-15$

$$\begin{aligned}9x^2-15x+4&=9x^2-3x-12x+4\\&=3x(3x-1)-4(3x-1)\\&=(3x-1)(3x-4)\end{aligned}$$

28. $3x^6+11x^3+10$

Let $t=x^3$. Then $t^2=\left(x^3\right)^2=x^6$.

$$\begin{aligned}3t^2+11t+10&=(3t+5)(t+2)\\&=\left(3x^3+5\right)\left(x^3+2\right)\end{aligned}$$

29. $$\begin{aligned}25x^3+25x^2-14x&=x\left(25x^2+25x-14\right)\\&=x(5x-2)(5x+7)\end{aligned}$$

30. $$\begin{aligned}2x^4-6x-x^3y+3y&=2x\left(x^3-3\right)-y\left(x^3-3\right)\\&=\left(x^3-3\right)(2x-y)\end{aligned}$$

5.5 Check Points

1. a. $$\begin{aligned}16x^2-25&=(4x)^2-5^2\\&=(4x+5)(4x-5)\end{aligned}$$

b. $$\begin{aligned}100y^6-9x^4&=(10y^3)^2-(3x^2)^2\\&=(10y^3+3x^2)(10y^3-3x^2)\end{aligned}$$

2. $$\begin{aligned}6y-6x^2y^7&=6y(1-x^2y^6)\\&=6y(1+xy^3)(1-xy^3)\end{aligned}$$

3. $$\begin{aligned}16x^4-81&=(4x^2+9)(4x^2-9)\\&=(4x^2+9)(2x+3)(2x-3)\end{aligned}$$

4. $$\begin{aligned}x^3+7x^2-4x-28&=(x^3+7x^2)+(-4x-28)\\&=x^2(x+7)-4(x+7)\\&=(x+7)(x^2-4)\\&=(x+7)(x+2)(x-2)\end{aligned}$$

5. a. $$\begin{aligned}x^2+6x+9&=x^2+2\cdot x\cdot 3+3^2\\&=(x+3)^2\end{aligned}$$

b. $$\begin{aligned}16x^2+40xy+25y^2&=(4x)^2+2\cdot 4x\cdot 5y+(5y)^2\\&=(4x+5y)^2\end{aligned}$$

c. $$\begin{aligned}4y^4-20y^2+25&=(2y^2)^2-2\cdot 2y^2\cdot 5+(5)^2\\&=(2y^2-5)^2\end{aligned}$$

6. $$\begin{aligned}x^2+10x+25-y^2&=(x^2+10x+25)-y^2\\&=(x+5)^2-y^2\\&=(x+5+y)(x+5-y)\end{aligned}$$

7. $$\begin{aligned}a^2-b^2+4b-4&=a^2-(b^2-4b+4)\\&=a^2-(b-2)^2\\&=(a+b-2)[a-(b-2)]\\&=(a+b-2)(a-b+2)\end{aligned}$$

8. a. $x^3+27=x^3+3^3$
$=(x+3)(x^2-x\cdot 3+3^2)$
$=(x+3)(x^2-3x+9)$

b. $x^6+1000y^3$
$=(x^2)^3+(10y)^3$
$=(x^2+10y)\left((x^2)^2-x^2\cdot 10y+(10y)^2\right)$
$=(x^2+10y)(x^4-10x^2y+100y^2)$

9. a. $x^3-8=x^3-2^3$
$=(x-2)(x^2+x\cdot 2+2^2)$
$=(x-2)(x^2+2x+4)$

b. $1-27x^3y^3=1^3-(3xy)^3$
$=(1-3xy)\left(1^2+1\cdot 3xy+(3xy)^2\right)$
$=(1-3xy)(1+3xy+9x^2y^2)$

5.5 Concept and Vocabulary Check

1. $(A+B)(A-B)$
2. $(A+B)^2$
3. $(A-B)^2$
4. $(A+B)(A^2-AB+B^2)$
5. $(A+B)(A^2+AB+B^2)$
6. $4x$; $4x$
7. $(b+3)$; $(b+3)$
8. -7
9. $4x$
10. $+3$; $-3x$
11. -10 ; $+100$
12. false
13. true
14. false
15. true
16. false

5.5 Exercise Set

1. $x^2-4=x^2-2^2$
$=(x+2)(x-2)$

3. $9x^2-25=(3x)^2-5^2$
$=(3x+5)(3x-5)$

5. $9-25y^2=3^2-(5y)^2$
$=(3+5y)(3-5y)$

7. $36x^2-49y^2=(6x)^2-(7y)^2$
$=(6x+7y)(6x-7y)$

9. $x^2y^2-1=(xy)^2-1^2$
$=(xy+1)(xy-1)$

11. $9x^4-25y^6=(3x^2)^2-(5y^3)^2$
$=(3x^2+5y^3)(3x^2-5y^3)$

13. $x^{14}-y^4=(x^7)^2-(y^2)^2$
$=(x^7+y^2)(x^7-y^2)$

15. $(x-3)^2-y^2$
$=\left((x-3)+y\right)\left((x-3)-y\right)$
$=(x-3+y)(x-3-y)$

17. $a^2-(b-2)^2$
$=\left(a+(b-2)\right)\left(a-(b-2)\right)$
$=(a+b-2)(a-b+2)$

19. $x^{2n}-25=(x^n)^2-5^2$
$=(x^n+5)(x^n-5)$

21. $1-a^{2n}=1^2-(a^n)^2$
$=(1+a^n)(1-a^n)$

23. $2x^3-8x=2x(x^2-4)$
$=2x(x^2-2^2)$
$=2x(x+2)(x-2)$

25. $50-2y^2=2(25-y^2)$
$=2(5^2-y^2)$
$=2(5+y)(5-y)$

27. $8x^2-8y^2=8(x^2-y^2)$
$=8(x+y)(x-y)$

29. $2x^3y-18xy=2xy(x^2-9)$
$=2xy(x^2-3^2)$
$=2xy(x+3)(x-3)$

31. $a^3b^2-49ac^2=a(a^2b^2-49c^2)$
$=a((ab)^2-(7c)^2)$
$=a(ab+7c)(ab-7c)$

33. $5y-5x^2y^7=5y(1-x^2y^6)$
$=5y\left(1^2-(xy^3)^2\right)$
$=5y(1+xy^3)(1-xy^3)$

35. $8x^2+8y^2=8(x^2+y^2)$

37. x^2+25y^2
prime

39. $x^4-16=(x^2)^2-4^2$
$=(x^2+4)(x^2-4)$
$=(x^2+4)(x^2-2^2)$
$=(x^2+4)(x+2)(x-2)$

41. $81x^4-1=(9x^2)^2-1^2$
$=(9x^2+1)(9x^2-1)$
$=(9x^2+1)((3x)^2-1^2)$
$=(9x^2+1)(3x+1)(3x-1)$

43. $2x^5-2xy^4$
$=2x(x^4-y^4)$
$=2x\left((x^2)^2-(y^2)^2\right)$
$=2x(x^2+y^2)(x^2-y^2)$
$=2x(x^2+y^2)(x+y)(x-y)$

45. $x^3+3x^2-4x-12$
$=(x^3+3x^2)+(-4x-12)$
$=x^2(x+3)+(-4)(x+3)$
$=(x+3)(x^2-4)$
$=(x+3)(x^2-2^2)$
$=(x+3)(x+2)(x-2)$

47. x^3-7x^2-x+7
$=(x^3-7x^2)+(-x+7)$
$=x^2(x-7)+(-1)(x-7)$
$=(x-7)(x^2-1)$
$=(x-7)(x^2-1^2)$
$=(x-7)(x+1)(x-1)$

49. $x^2+4x+4=x^2+2\cdot x\cdot 2+2^2$
$=(x+2)^2$

51. $x^2-10x+25=x^2-2\cdot x\cdot 5+5^2$
$=(x-5)^2$

53. $x^4-4x^2+4=(x^2)^2-2\cdot x^2\cdot 2+2^2$
$=(x^2-2)^2$

55. $9y^2+6y+1=(3y)^2+2\cdot y\cdot 3+1^2$
$=(3y+1)^2$

57. $64y^2-16y+1=(8y)^2-2\cdot y\cdot 8+1^2$
$=(8y-1)^2$

59. $x^2-12xy+36y^2$
$=x^2-2\cdot x\cdot 6y+(6y)^2$
$=(x-6y)^2$

61. $x^2-8xy+64y^2$
prime
Because the first and third terms of the polynomial are perfect squares, check to see if the trinomial is a perfect square. The middle term would have to be $-16xy$ for the polynomial to be a perfect square. Since this is not the case and the polynomial cannot be factored in any other way, we conclude that the polynomial is prime.

63. $9x^2+48xy+64y^2$
$=(3x)^2+2\cdot 3x\cdot 8y+(8y)^2$
$=(3x+8y)^2$

65. $x^2-6x+9-y^2$
$=(x^2-6x+9)-y^2$
$=(x^2-2\cdot x\cdot 3+3^2)-y^2$
$=(x-3)^2-y^2$
$=((x-3)+y)((x-3)-y)$
$=(x-3+y)(x-3-y)$

67. $x^2+20x+100-x^4$
$=(x^2+20x+100)-x^4$
$=(x^2+2\cdot x\cdot 10+10^2)-x^4$
$=(x+10)^2-(x^2)^2$
$=((x+10)+x^2)((x+10)-x^2)$
$=(x+10+x^2)(x+10-x^2)$

69. $9x^2-30x+25-36y^2$
$=(9x^2-30x+25)-36y^2$
$=((3x)^2-2\cdot 3x\cdot 5+5^2)-36y^2$
$=(3x-5)^2-(6y)^2$
$=((3x-5)+6y)((3x-5)-6y)$
$=(3x-5+6y)(3x-5-6y)$

71. x^4-x^2-2x-1
$=x^4-(x^2+2x+1)$
$=x^4-(x^2+2\cdot x\cdot 1+1^2)$
$=(x^2)^2-(x+1)^2$
$=(x^2+(x+1))(x^2-(x+1))$
$=(x^2+x+1)(x^2-x-1)$

73. $z^2-x^2+4xy-4y^2$
$=z^2-(x^2-4xy+4y^2)$
$=z^2-(x^2-2\cdot x\cdot 2y+(2y)^2)$
$=z^2-(x-2y)^2$
$=(z+(x-2y))(z-(x-2y))$
$=(z+x-2y)(z-x+2y)$

75. $x^3+64=x^3+4^3$
$=(x+4)(x^2-x\cdot 4+4^2)$
$=(x+4)(x^2-4x+16)$

77. $x^3-27=x^3-3^3$
$=(x-3)(x^2+3\cdot x+3^2)$
$=(x-3)(x^2+3x+9)$

79. $8y^3+1$
$=(2y)^3+1^3$
$=(2y+1)((2y)^2-2y\cdot 1+1^2)$
$=(2y+1)(4y^2-2y+1)$

81. $125x^3 - 8$
$= (5x)^3 - 2^3$
$= (5x-2)\left(25x^2 + 5x \cdot 2 + 2^2\right)$
$= (5x-2)\left(25x^2 + 10x + 4\right)$

83. $x^3y^3 + 27$
$= (xy)^3 + 3^3$
$= (xy+3)\left((xy)^2 - xy \cdot 3 + 3^2\right)$
$= (xy+3)\left(x^2y^2 - 3xy + 9\right)$

85. $64x - x^4$
$= x\left(64 - x^3\right)$
$= x\left(4^3 - x^3\right)$
$= x(4-x)\left(4^2 + 4 \cdot x + x^2\right)$
$= x(4-x)\left(16 + 4x + x^2\right)$

87. $x^6 + 27y^3$
$= \left(x^2\right)^3 + (3y)^3$
$= \left(x^2 + 3y\right)\left(\left(x^2\right)^2 - x^2 \cdot 3y + (3y)^2\right)$
$= \left(x^2 + 3y\right)\left(x^4 - 3x^2y + 9y^2\right)$

89. $125x^6 - 64y^6$
$= \left(5x^2\right)^3 - \left(4y^2\right)^3$
$= \left(5x^2 - 4y^2\right)\left(\left(5x^2\right)^2 + 5x^2 4y^2 + \left(4y^2\right)^2\right)$
$= \left(5x^2 - 4y^2\right)\left(25x^4 + 20x^2y^2 + 16y^4\right)$

91. $x^9 + 1$
$= \left(x^3\right)^3 + 1^3$
$= \left(x^3 + 1\right)\left(\left(x^3\right)^2 - x^3 \cdot 1 + 1^2\right)$
$= \left(x^3 + 1\right)\left(x^6 - x^3 + 1\right)$
$= \left(x^3 + 1^3\right)\left(x^6 - x^3 + 1\right)$
$= (x+1)\left(x^2 - x \cdot 1 + 1^2\right)\left(x^6 - x^3 + 1\right)$
$= (x+1)\left(x^2 - x + 1\right)\left(x^6 - x^3 + 1\right)$

93. $(x-y)^3 - y^3$
$= \left((x-y) - y\right)\left((x-y)^2 + y(x-y) + y^2\right)$
$= (x-2y)\left(x^2 - 2xy + y^2 + xy - y^2 + y^2\right)$
$= (x-2y)\left(x^2 - xy + y^2\right)$

95. $0.04x^2 + 0.12x + 0.09$
$= (0.2x + 0.3)^2$ or $\frac{1}{100}(2x+3)^2$

97. $8x^4 - \frac{x}{8} = x\left(8x^3 - \frac{1}{8}\right)$
$= x\left(2x - \frac{1}{2}\right)\left((2x)^2 + 2x \cdot \frac{1}{2} + \left(\frac{1}{2}\right)^2\right)$
$= x\left(2x - \frac{1}{2}\right)\left(4x^2 + x + \frac{1}{4}\right)$

99. $x^6 - 9x^3 + 8$
$= \left(x^3 - 1\right)\left(x^3 - 8\right)$
$= (x-1)\left(x^2 + x + 1\right)(x-2)(x^2 + 2x + 4)$

101. $x^8 - 15x^4 - 16$
$= \left(x^4 + 1\right)\left(x^4 - 16\right)$
$= \left(x^4 + 1\right)\left(x^2 + 4\right)\left(x^2 - 4\right)$
$= \left(x^4 + 1\right)\left(x^2 + 4\right)(x+2)(x-2)$

103. $x^5 - x^3 - 8x^2 + 8$
$= x^3(x^2 - 1) - 8(x^2 - 1)$
$= \left(x^2 - 1\right)\left(x^3 - 8\right)$
$= (x+1)(x-1)(x-2)\left(x^2 + 2x + 4\right)$

105. a. $(A+B)^2$

b. A^2; AB; AB; B^2

c. $A^2 + AB + AB + B^2 = A^2 + 2AB + B^2$

d. $A^2 + 2AB + B^2 = (A+B)^2$
This is the perfect square trinomial factoring technique.

107. $(5x)^2 - 3^2 = (5x+3)(5x-3)$

109. $(7x)^2 - 4 \cdot 3^2 = (7x)^2 - 4 \cdot 9$
$= (7x)^2 - 36$
$= (7x+6)(7x-6)$

111. $V_{\text{shaded}} = V_{\text{outside}} - V_{\text{inside}}$
$= a \cdot a \cdot 3a - b \cdot b \cdot 3a$
$= 3a^3 - 3ab^2$
$= 3a\left(a^2 - b^2\right)$
$= 3a(a+b)(a-b)$

113. – 115. Answers will vary.

117. $x^2 + 4x + 4 = (x+4)^2$

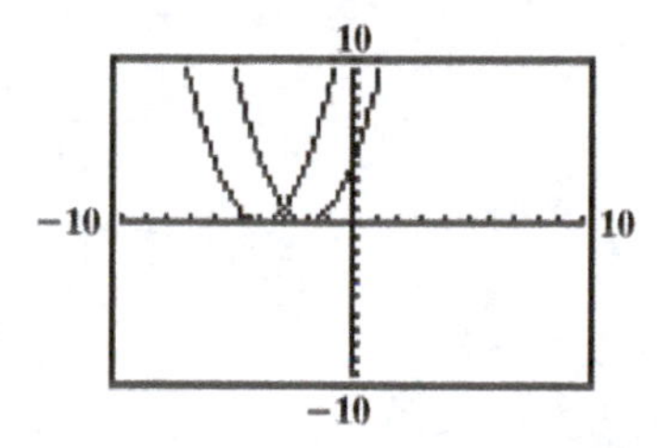

The graphs do not coincide. The polynomial is not factored correctly.

$x^2 + 4x + 4 = (x+2)^2$

119. $25 - \left(x^2 + 4x + 4\right) = (x+7)(x-3)$

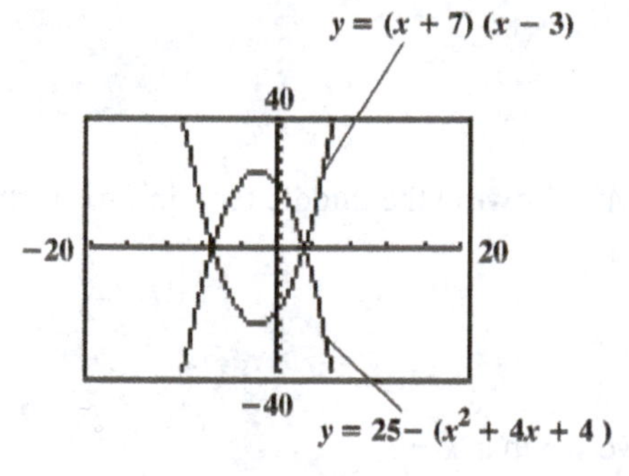

The graphs do not coincide. The polynomial is not factored correctly.

$25 - \left(x^2 + 4x + 4\right) = (5 + (x+2))(5 - (x+2))$
$= (5 + x + 2)(5 - x - 2)$
$= (x+7)(3-x)$

121. $(x-3)^2 + 8(x-3) + 16 = (x-1)^2$

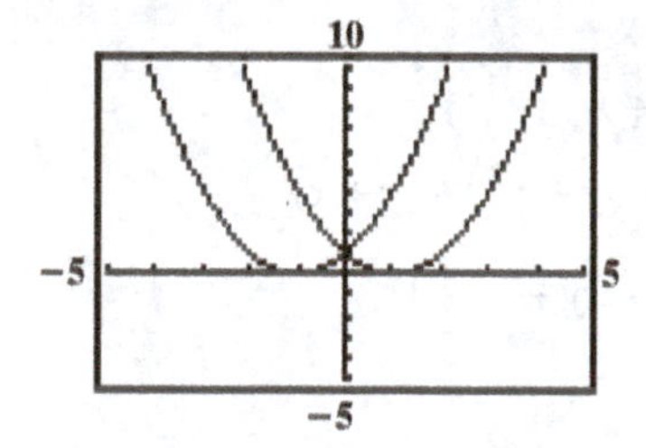

The graphs do not coincide. The polynomial is not factored correctly.

$(x-3)^2 + 8(x-3) + 16 = ((x-3)+4)^2$
$= (x+1)^2$

123. $(x+1)^3 + 1 = (x+1)\left(x^2 + x + 1\right)$

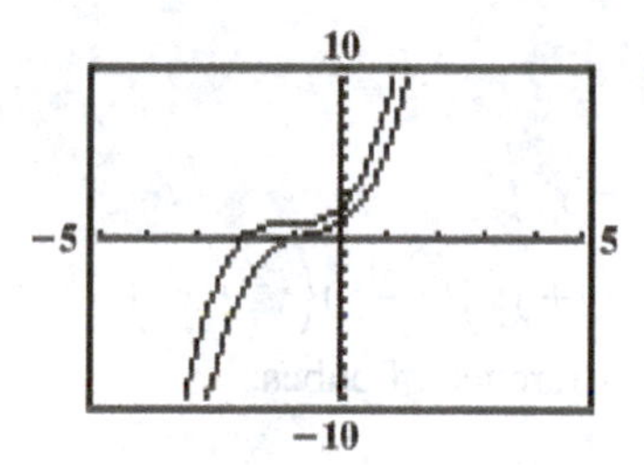

The graphs do not coincide. The factorization is not correct.

$(x+1)^3 + 1 = ((x+1)+1)\left((x+1)^2 - (x+1) + 1^2\right)$
$= (x+1+1)\left(\left(x^2 + 2x + 1\right) - x - 1 + 1\right)$
$= (x+2)\left(x^2 + 2x + 1 - x - 1 + 1\right)$
$= (x+2)\left(x^2 + x + 1\right)$

125. makes sense

127. does not make sense; Explanations will vary. Sample explanation: To factor completely you must factor out the common factor of 4.

129. false; Changes to make the statement true will vary. A sample change is: $9x^2 + 30x + 25 = (3x+5)^2$

131. false; Changes to make the statement true will vary. A sample change is: $x^3 - 64 = (x-4)(x^2 + 4x + 16)$

133.
$$y^3 + x + x^3 + y$$
$$= y^3 + x^3 + y + x$$
$$= (y+x)\left(y^2 - xy + x^2\right) + (y+x)$$
$$= (y+x)\left((y^2 - xy + x^2 + 1\right)$$

135.
$$x^{3n} + y^{12n}$$
$$= \left(x^n\right)^3 + \left(y^{4n}\right)^3$$
$$= \left(x^n + y^{4n}\right)\left(\left(x^n\right)^2 - x^n y^{4n} + \left(y^{4n}\right)^2\right)$$
$$= \left(x^n + y^{4n}\right)\left(x^{2n} - x^n y^{4n} + y^{8n}\right)$$

137. Factoring as a difference of squares:
$$x^6 - y^6$$
$$= \left(x^3\right)^2 - \left(y^3\right)^2$$
$$= \left(x^3 + y^3\right)\left(x^3 - y^3\right)$$
$$= (x+y)\left(x^2 - xy + y^2\right)(x-y)\left(x^2 + xy + y^2\right)$$
Factoring as a difference of cubes:
$$x^6 - y^6$$
$$= \left(x^2\right)^3 - \left(y^2\right)^3$$
$$= \left(x^2 - y^2\right)\left(x^4 + x^2 y^2 + y^4\right)$$
$$= (x+y)(x-y)\left(x^4 + x^2 y^2 + y^4\right)$$
Set the factorizations equal.
$$(x+y)(x-y)\left(x^4 + x^2 y^2 + y^4\right) = (x+y)\left(x^2 - xy + y^2\right)(x-y)\left(x^2 + xy + y^2\right)$$
$$x^4 + x^2 y^2 + y^4 = \left(x^2 - xy + y^2\right)\left(x^2 + xy + y^2\right)$$

139. In a perfect square trinomial, the middle term is $2AB$ and the first term is A^2. Rewrite the middle term in $2AB$ form and the first term in A^2 form.
$$64x^2 - 16x + k$$
$$= (8x)^2 - 2 \cdot 8x \cdot 1 + k$$
This means that B must be 1. The last term is $B^2 = (1)^2 = 1$. From this, we see that $k = 1$.

140. $2x+2\geq 12$ and $\dfrac{2x-1}{3}\leq 7$

$$2x\geq 10 \qquad 2x-1\leq 21$$
$$x\geq 5 \qquad 2x\leq 22$$
$$x\leq 11$$

The solution set is $[5,11]$.

141. $3x-2y=-8$
$x+6y=\ \ 4$

Write the augmented matrix and solve by Gaussian elimination.

$$\left[\begin{array}{cc|c}3 & -2 & -8\\ 1 & 6 & 4\end{array}\right]=\left[\begin{array}{cc|c}1 & 6 & 4\\ 3 & -2 & -8\end{array}\right]$$
$$=\left[\begin{array}{cc|c}1 & 6 & 4\\ 0 & -20 & -20\end{array}\right]$$
$$=\left[\begin{array}{cc|c}1 & 6 & 4\\ 0 & 1 & 1\end{array}\right]$$

$x+6y=4$
$y=1$

Back-substitute 1 for y, to find x.

$x+6(1)=4$
$x+6=4$
$x=-2$

The solution is $(-2,1)$.

142. $3x^2+21x-xy-7y=(3x^2+21x)+(-xy-7y)$
$=3x(x+7)-y(x+7)$
$=(x+7)(3x-y)$

143. $2x^3+8x^2+8x=2x(x^2+4x+4)$
$=2x(x+2)^2$

144. $5x^3-40x^2y+35y^2=5x(x^2-8xy+7y^2)$
$=5x(x-y)(x-7y)$

145. $9b^2x+9b^2y-16x-16y=(9b^2x+9b^2y)+(-16x-16y)$
$=9b^2(x+y)-16(x+y)$
$=(x+y)(9b^2-16)$
$=(x+y)(3b+4)(3b-4)$

5.6 Check Points

1. $3x^3-30x^2+75x=3x(x^2-10x+25)$
$$=3x(x-5)^2$$

2. $3x^2y-12xy-36y=3y(x^2-4x-12)$
$$=3y(x+2)(x-6)$$

3. $16a^2x-25y-25x+16a^2y=16a^2x+16a^2y-25y-25x$
$$=(16a^2x+16a^2y)+(-25y-25x)$$
$$=16a^2(x+y)-25(y+x)$$
$$=16a^2(x+y)-25(x+y)$$
$$=(x+y)(16a^2-25)$$
$$=(x+y)(4a+5)(4a-5)$$

4. $x^2-36a^2+20x+100=x^2+20x+100-36a^2$
$$=(x^2+20x+100)-36a^2$$
$$=(x+10)^2-36a^2$$
$$=(x+10+6a)(x+10-6a)$$

5. $x^{10}+512x=x(x^9+512)$
$$=x\left((x^3)^3+8^3\right)$$
$$=x(x^3+8)(x^6-8x^3+64)$$
$$=x\overbrace{(x+2)(x^2-2x+4)}^{x^3+8}(x^6-8x^3+64)$$
$$=x(x+2)(x^2-2x+4)(x^6-8x^3+64)$$

5.6 Concept and Vocabulary Check

1. b
2. e
3. h
4. c
5. d
6. f
7. a
8. g

5.6 Exercise Set

1. $x^3-16x=x(x^2-16)$
$=x(x+4)(x-4)$

3. $3x^2+18x+27=3(x^2+6x+9)$
$=3(x+3)^2$

5. $81x^3-3=3(27x^3-1)$
$=3(3x-1)(9x^2+3x+1)$

7. $x^2y-16y+32-2x^2$
$=(x^2y-16y)+(-2x^2+32)$
$=y(x^2-16)+(-2)(x^2-16)$
$=(x^2-16)(y+(-2))$
$=(x+4)(x-4)(y-2)$

9. $4a^2b-2ab-30b$
$=2b(2a^2-a-15)$
$=2b(2a+5)(a-3)$

11. $ay^2-4a-4y^2+16$
$=a(y^2-4)+(-4)(y^2-4)$
$=(y^2-4)(a+(-4))$
$=(y+2)(y-2)(a-4)$

13. $11x^5-11xy^2=11x(x^4-y^2)$
$=11x(x^2+y)(x^2-y)$

15. $4x^5-64x$
$=4x(x^4-16)$
$=4x(x^2+4)(x^2-4)$
$=4x(x^2+4)(x+2)(x-2)$

17. $x^3-4x^2-9x+36$
$=x^2(x-4)+(-9)(x-4)$
$=(x-4)(x^2+(-9))$
$=(x-4)(x^2-9)$
$=(x-4)(x+3)(x-3)$

19. $2x^5+54x^2=2x^2(x^3+27)$
$=2x^2(x+3)(x^2-3x+9)$

21. $3x^4y-48y^5$
$=3y(x^4-16y^4)$
$=3y(x^2+4y^2)(x^2-4y^2)$
$=3y(x^2+4y^2)(x+2y)(x-2y)$

23. $12x^3+36x^2y+27xy^2$
$=3x(4x^2+12xy+9y^2)$
$=3x(2x+3y)^2$

25. $x^2-12x+36-49y^2$
$=(x^2-12x+36)-49y^2$
$=(x-6)^2-(7y)^2$
$=((x-6)+7y)((x-6)-7y)$
$=(x-6+7y)(x-6-7y)$

27. $4x^2+25y^2$
Prime
The sum of two squares with no common factor other than 1 is a prime polynomial.

29. $12x^3y-12xy^3=12xy(x^2-y^2)$
$=12xy(x+y)(x-y)$

31. $6bx^2+6by^2=6b(x^2+y^2)$

33. $x^4-xy^3+x^3y-y^4$
$=x(x^3-y^3)+y(x^3-y^3)$
$=(x^3-y^3)(x+y)$
$=(x-y)(x^2+xy+y^2)(x+y)$

35. $x^2-4a^2+12x+36$
$=x^2+12x+36-4a^2$
$=\left(x^2+12x+36\right)-4a^2$
$=(x+6)^2-4a^2$
$=((x+6)+2a)((x+6)-2a)$
$=(x+6+2a)(x+6-2a)$

37. $5x^3+x^6-14$
$=x^6+5x^3-14$
$=\left(x^3+7\right)\left(x^3-2\right)$

39. $4x-14+2x^3-7x^2$
$=2(2x-7)+x^2(2x-7)$
$=(2x-7)\left(2+x^2\right)$

41. $54x^3-16y^3$
$=2\left(27x^3-8y^3\right)$
$=2(3x-2y)\left(9x^2+6xy+4y^2\right)$

43. $x^2+10x-y^2+25$
$=x^2+10x+25-y^2$
$=(x+5)^2-y^2$
$=((x+5)+y)((x+5)-y)$
$=(x+5+y)(x+5-y)$

45. x^8-y^8
$=\left(x^4+y^4\right)\left(x^4-y^4\right)$
$=\left(x^4+y^4\right)\left(x^2+y^2\right)\left(x^2-y^2\right)$
$=\left(x^4+y^4\right)\left(x^2+y^2\right)(x+y)(x-y)$

47. $x^3y-16xy^3$
$=xy\left(x^2-16y^2\right)$
$=xy(x+4y)(x-4y)$

49. $x+8x^4$
$=x\left(1+8x^3\right)$
$=x(1+2x)\left(1-2x+4x^2\right)$

51. $16y^2-4y-2$
$=2\left(8y^2-2y-1\right)$
$=2(4y+1)(2y-1)$

53. $14y^3+7y^2-10y$
$=y\left(14y^2+7y-10\right)$

55. $27x^2+36xy+12y^2$
$=3\left(9x^2+12xy+4y^2\right)$
$=3(3x+2y)^2$

57. $12x^3+3xy^2=3x\left(4x^2+y^2\right)$

59. $x^6y^6-x^3y^3$
$=x^3y^3\left(x^3y^3-1\right)$
$=x^3y^3(xy-1)\left(x^2y^2+xy+1\right)$

61. $(x+5)(x-3)+(x+5)(x-7)$
$=(x+5)((x-3)+(x-7))$
$=(x+5)(2x-10)$
$=(x+5)\cdot 2(x-5)$
$=2(x+5)(x-5)$

63. $a^2(x-y)+4(y-x)$
$=a^2(x-y)+4(-1)(x-y)$
$=(x-y)\left(a^2-4\right)$
$=(x-y)(a+2)(a-2)$

65. $(c+d)^4-1$
$=\left((c+d)^2+1\right)\left((c+d)^2-1\right)$
$=\left((c+d)^2+1\right)((c+d)-1)((c+d)+1)$
$=\left((c+d)^2+1\right)(c+d-1)(c+d+1)$ or
$=\left(c^2+2cd+d^2+1\right)(c+d-1)(c+d+1)$

67. $p^3 - pq^2 + p^2q - q^3$
$= p^3 + p^2q - pq^2 - q^3$
$= p^2(p+q) - q^2(p+q)$
$= (p+q)(p^2 - q^2)$
$= (p+q)(p+q)(p-q)$
$= (p+q)^2(p-q)$

69. $x^4 - 5x^2y^2 + 4y^4$
$= (x^2 - 4y^2)(x^2 - y^2)$
$= (x-2y)(x+2y)(x+y)(x-y)$

71. $(x+y)^2 + 6(x+y) + 9$
$= ((x+y)+3)((x+y)+3)$
$= (x+y+3)^2$

73. $(x-y)^4 - 4(x-y)^2$
$= (x-y)^2((x-y)^2 - 4)$
$= (x-y)^2((x-y)+2)((x-y)-2)$
$= (x-y)^2(x-y+2)(x-y-2)$

75. $2x^2 - 7xy^2 + 3y^4 = (2x - y^2)(x - 3y^2)$

77. $x^3 - y^3 - x + y$
$= (x-y)(x^2 + xy + y^2) - (x-y)$
$= (x-y)(x^2 + xy + y^2 - 1)$

79. $x^6y^3 + x^3 - 8x^3y^3 - 8$
$= x^3(x^3y^3 + 1) - 8(x^3y^3 + 1)$
$= (x^3y^3 + 1)(x^3 - 8)$
$= (xy+1)(x^2y^2 - xy + 1)(x-2)(x^2 + 2x + 4)$

81. **a.** $x(x+y) - y(x+y)$

b. $x(x+y) - y(x+y)$
$= (x+y)(x-y)$

83. **a.** $xy + xy + xy + 3x(x) = 3xy + 3x^2$

b. $3xy + 3x^2 = 3x(y+x)$

85. **a.** $4x(2x) - 2(\pi x^2) = 8x^2 - 2\pi x^2$

b. $8x^2 - 2\pi x^2 = 2x^2(4 - \pi)$

87. Answers will vary.

89. $4x^2 - 12x + 9 = (4x-3)^2$

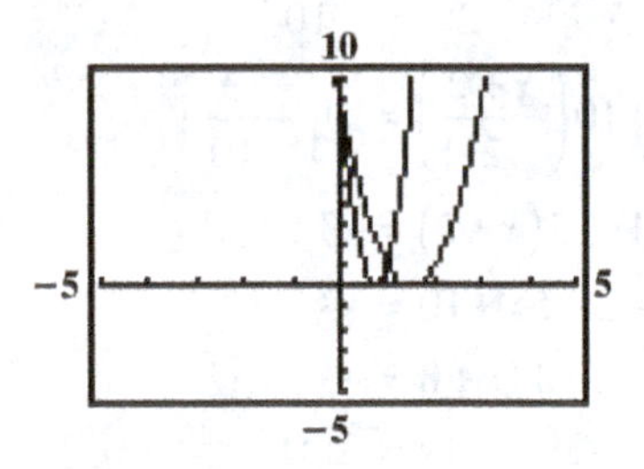

The graphs do not coincide. The factorization is not correct.

$4x^2 - 12x + 9 = (2x-3)^2$

91. $x^4 - 16 = (x^2+4)(x+2)(x-2)$

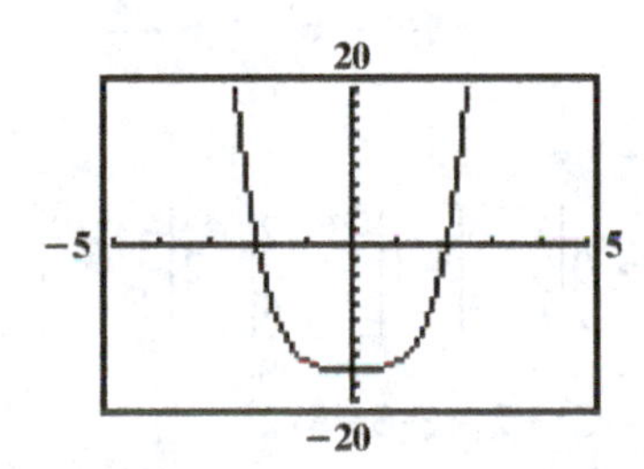

The graphs coincide so the factorization is correct.

93. Answers will vary.

95. makes sense

97. does not make sense; Explanations will vary. Sample explanation: The three dimensions do not necessarily correspond to the three factors stated. Other dimensions can give the same product.

99. true

101. false; Changes to make the statement true will vary. A sample change is: $x^3 - 64 = (x-4)(x^2 + 4x + 16)$

103. $3x^{n+2} - 13x^{n+1} + 4x^n = x^n(3x^2 - 13x + 4)$
$= x^n(3x-1)(x-4)$

105. $x^4+4=x^4+4x^2+4-4x^2$
$$=\left(x^2+2\right)^2-4x^2$$
$$=\left((x^2+2)+2x\right)\left((x^2+2)-2x\right)$$
$$=\left(x^2+2x+2\right)\left(x^2-2x+2\right)$$

107.
$$\frac{3x-1}{5}+\frac{x+2}{2}=-\frac{3}{10}$$
$$10\left(\frac{3x-1}{5}\right)+10\left(\frac{x+2}{2}\right)=10\left(-\frac{3}{10}\right)$$
$$2(3x-1)+5(x+2)=-3$$
$$6x-2+5x+10=-3$$
$$11x+8=-3$$
$$11x=-11$$
$$x=-1$$
The solution set is $\{-1\}$.

108. $\left(4x^3y^{-1}\right)^2\left(2x^{-3}y\right)^{-1}=4^2x^6y^{-2}2^{-1}x^3y^{-1}$
$$=4^2 2^{-1}x^6x^3y^{-2}y^{-1}$$
$$=16\cdot\frac{1}{2}x^9y^{-3}=\frac{8x^9}{y^3}$$

109. $\begin{vmatrix}0&-3&2\\1&5&3\\-2&1&4\end{vmatrix}=0\begin{vmatrix}5&3\\1&4\end{vmatrix}-1\begin{vmatrix}-3&2\\1&4\end{vmatrix}-2\begin{vmatrix}-3&2\\5&3\end{vmatrix}$
$$=-1(-3\cdot4-1\cdot2)-2(-3\cdot3-5\cdot2)$$
$$=-1(-12-2)-2(-9-10)$$
$$=-1(-14)-2(-19)=14+38=52$$

110. If $x=4$, then $x-4$ is 0. Thus, the product is 0.

111. If $t=6$, then $t-6$ is 0. Thus, the product is 0.

112.
$$x^2+(x+7)^2=(x+8)^2$$
$$x^2+(x+7)^2-(x+8)^2=0$$
$$x^2+x^2+14x+49-(x^2+16x+64)=0$$
$$x^2+x^2+14x+49-x^2-16x-64=0$$
$$x^2-2x-15=0$$
$$(x-5)(x+3)=0$$

5.7 Check Points

1.
$$2x^2-9x=5$$
$$2x^2-9x-5=0$$
$$(2x+1)(x-5)=0$$
Apply the zero product principle.
$2x+1=0$ or $x-5=0$
$x=-\frac{1}{2}$ $\quad x=5$

The solution set is $\left\{-\frac{1}{2},5\right\}$.

2. a.
$$3x^2=2x$$
$$3x^2-2x=0$$
$$x(3x-2)=0$$
Apply the zero product principle.
$x=0$ or $3x-2=0$
$x=\frac{2}{3}$

The solution set is $\left\{0,\frac{2}{3}\right\}$.

b.
$$x^2+7=10x-18$$
$$x^2-10x+25=0$$
$$(x-5)(x-5)=0$$
Apply the zero product principle.
$x-5=0$
$x=5$
The solution set is $\{5\}$.

c.
$$(x-2)(x+3)=6$$
$$x^2+x-6=6$$
$$x^2+x-12=0$$
$$(x+4)(x-3)=0$$
Apply the zero product principle.
$x+4=0$ or $x-3=0$
$x=-4$ $\quad x=3$

The solution set is $\{-4,3\}$.

3. $2x^3+3x^2=8x+12$

$2x^3+3x^2-8x-12=0$

$(2x^3+3x^2)+(-8x-12)=0$

$x^2(2x+3)-4(2x+3)=0$

$(2x+3)(x^2-4)=0$

$(2x+3)(x+2)(x-2)=0$

Apply the zero product principle.

$2x+3=0$ or $x+2=0$ or $x-2=0$

$x=-\frac{3}{2}$ $\quad x=-2$ $\quad x=2$

The solution set is $\left\{-2,-\frac{3}{2},2\right\}$.

4. $s(t)=-16t^2+32t+384$

$336=-16t^2+32t+384$

$0=-16t^2+32t+48$

$\frac{0}{-16}=\frac{-16t^2+32t+48}{-16}$

$0=t^2-2t-3$

$0=(t+1)(t-3)$

$t+1=0$ or $t-3=0$

$t=-1$ $\quad t=3$

Reject –1 because time cannot be negative.
The ball's height will be 336 feet after 3 seconds.
This is represented by the point $(3, 336)$.

5. Let $x=$ the width of the path.

$(12+2x)(16+2x)=320$

$192+24x+32x+4x^2=320$

$4x^2+56x+192=320$

$4x^2+56x-128=0$

$4(x^2+14x-32)=0$

$4(x+16)(x-2)=0$

$x+16=0$ or $x-2=0$

$x=-16$ $\quad x=2$

Reject –16 because width cannot be negative. The width of the path is 2 feet.

6. Let $x=$ the distance from the base of the tree to the stake.
Let $x+2=$ the length of the wire.
Use the Pythagorean Theorem.

$\text{leg}^2+\text{leg}^2=\text{hypotenuse}^2$

$x^2+(x+1)^2=(x+2)^2$

$x^2+x^2+2x+1=x^2+4x+4$

$2x^2+2x+1=x^2+4x+4$

$x^2-2x-3=0$

$(x+1)(x-3)=0$

$x+1=0$ or $x-3=0$

$x=-1$ $\quad x=3$

The length of the wire is given by $x+2=3+2=5$.
The length of the wire is 5 feet.

5.7 Concept and Vocabulary Check

1. quadratic
2. $A=0$ or $B=0$
3. x-intercepts
4. subtracting $20x$
5. subtracting $8x$; adding 12
6. polynomial; 0; descending; highest/greatest
7. right; hypotenuse; legs
8. right; legs; the square of the length of the hypotenuse

5.7 Exercise Set

1. $x^2+x-12=0$

$(x+4)(x-3)=0$

Apply the zero product principle.

$x+4=0$ $\quad x-3=0$

$x=-4$ $\quad x=3$

The solution set is $\{-4,3\}$.

3. $x^2+6x=7$

$x^2+6x-7=0$

$(x+7)(x-1)=0$

Apply the zero product principle.

$x+7=0 \qquad x-1=0$

$x=-7 \qquad x=1$

The solution set is $\{-7,1\}$.

5. $3x^2+10x-8=0$

$(3x-2)(x+4)=0$

Apply the zero product principle.

$3x-2=0 \qquad x+4=0$

$3x=2 \qquad x=-4$

$x=\frac{2}{3}$

The solution set is $\left\{-4,\frac{2}{3}\right\}$.

7. $5x^2=8x-3$

$5x^2-8x+3=0$

$(5x-3)(x-1)=0$

Apply the zero product principle.

$5x-3=0 \qquad x-1=0$

$5x=3 \qquad x=1$

$x=\frac{3}{5}$

The solution set is $\left\{\frac{3}{5},1\right\}$.

9. $3x^2=2-5x$

$3x^2+5x-2=0$

$(3x-1)(x+2)=0$

Apply the zero product principle.

$3x-1=0 \qquad x+2=0$

$3x=1 \qquad x=-2$

$x=\frac{1}{3}$

The solution set is $\left\{-2,\frac{1}{3}\right\}$.

11. $x^2=8x$

$x^2-8x=0$

$x(x-8)=0$

Apply the zero product principle.

$x=0 \qquad x-8=0$

$x=8$

The solution set is $\{0,8\}$.

13. $3x^2=5x$

$3x^2-5x=0$

$x(3x-5)=0$

Apply the zero product principle.

$x=0 \qquad 3x-5=0$

$3x=5$

$x=\frac{5}{3}$

The solution set is $\left\{0,\frac{5}{3}\right\}$.

15. $x^2+4x+4=0$

$(x+2)^2=0$

Apply the zero product principle.

$x+2=0$

$x=-2$

The solution set is $\{-2\}$.

17. $x^2=14x-49$

$x^2-14x+49=0$

$(x-7)^2=0$

Apply the zero product principle.

$x-7=0$

$x=7$

The solution is 7 and the solution set is $\{7\}$.

19. $9x^2 = 30x - 25$

$9x^2 - 30x + 25 = 0$

$(3x-5)^2 = 0$

Apply the zero product principle.

$3x - 5 = 0$

$3x = 5$

$x = \frac{5}{3}$

The solution set is $\left\{\frac{5}{3}\right\}$.

21. $x^2 - 25 = 0$

$(x+5)(x-5) = 0$

Apply the zero product principle.

$x + 5 = 0 \qquad x - 5 = 0$

$x = -5 \qquad x = 5$

The solution set is $\{-5, 5\}$.

23. $9x^2 = 100$

$9x^2 - 100 = 0$

$(3x+10)(3x-10) = 0$

Apply the zero product principle.

$3x + 10 = 0 \qquad 3x - 10 = 0$

$3x = -10 \qquad 3x = 10$

$x = -\frac{10}{3} \qquad x = \frac{10}{3}$

The solution set is $\left\{-\frac{10}{3}, \frac{10}{3}\right\}$.

25. $x(x-3) = 18$

$x^2 - 3x = 18$

$x^2 - 3x - 18 = 0$

$(x-6)(x+3) = 0$

Apply the zero product principle.

$x - 6 = 0 \qquad x + 3 = 0$

$x = 6 \qquad x = -3$

The solution set is $\{-3, 6\}$.

27. $(x-3)(x+8) = -30$

$x^2 + 8x - 3x - 24 = -30$

$x^2 + 5x - 24 = -30$

$x^2 + 5x + 6 = 0$

$(x+3)(x+2) = 0$

Apply the zero product principle.

$x + 3 = 0 \qquad x + 2 = 0$

$x = -3 \qquad x = -2$

The solution set is $\{-3, -2\}$.

29. $x(x+8) = 16(x-1)$

$x^2 + 8x = 16x - 16$

$x^2 - 8x + 16 = 0$

$(x-4)^2 = 0$

Apply the zero product principle.

$x - 4 = 0$

$x = 4$

The solution set is $\{4\}$.

31. $(x+1)^2 - 5(x+2) = 3x + 7$

$(x^2 + 2x + 1) - 5x - 10 = 3x + 7$

$x^2 + 2x + 1 - 5x - 10 = 3x + 7$

$x^2 - 3x - 9 = 3x + 7$

$x^2 - 6x - 16 = 0$

$(x-8)(x+2) = 0$

Apply the zero product principle.

$x - 8 = 0 \qquad x + 2 = 0$

$x = 8 \qquad x = -2$

The solution set is $\{-2, 8\}$.

33. $x(8x+1) = 3x^2 - 2x + 2$

$8x^2 + x = 3x^2 - 2x + 2$

$5x^2 + 3x - 2 = 0$

$(5x-2)(x+1) = 0$

Apply the zero product principle.

$5x - 2 = 0 \qquad x + 1 = 0$

$5x = 2 \qquad x = -1$

$x = \frac{2}{5}$

The solution set is $\left\{-1, \frac{2}{5}\right\}$.

35.
$$\frac{x^2}{18}+\frac{x}{2}+1=0$$
$$18\left(\frac{x^2}{18}\right)+18\left(\frac{x}{2}\right)+18(1)=18(0)$$
$$x^2+9x+18=0$$
$$(x+3)(x+6)=0$$
Apply the zero product principle.
$x+3=0 \quad x+6=0$
$x=-3 \quad x=-6$
The solution set is $\{-6,-3\}$.

37. $x^3+4x^2-25x-100=0$
$$x^2(x+4)-25(x+4)=0$$
$$(x+4)(x^2-25)=0$$
$$(x+4)(x+5)(x-5)=0$$
Apply the zero product principle.
$x+4=0 \quad x+5=0 \quad x-5=0$
$x=-4 \quad x=-5 \quad x=5$
The solution set is $\{-5,-4,5\}$.

39.
$$x^3-x^2=25x-25$$
$$x^3-x^2-25x+25=0$$
$$x^2(x-1)-25(x-1)=0$$
$$(x-1)(x^2-25)=0$$
$$(x-1)(x+5)(x-5)=0$$
Apply the zero product principle.
$x-1=0 \quad x+5=0 \quad x-5=0$
$x=1 \quad x=-5 \quad x=5$
The solution set is $\{-5,1,5\}$.

41.
$$3x^4-48x^2=0$$
$$3x^2(x^2-16)=0$$
$$3x^2(x+4)(x-4)=0$$
Apply the zero product principle.
$3x^2=0 \quad x+4=0 \quad x-4=0$
$x=0 \quad x=-4 \quad x=4$
The solution set is $\{-4,0,4\}$.

43. $x^4-4x^3+4x^2=0$
$$x^2(x^2-4x+4)=0$$
$$x^2(x-2)^2=0$$
Apply the zero product principle.
$x=0 \quad x-2=0$
$x=2$
The solution set is $\{0,2\}$.

45. $2x^3+16x^2+30x=0$
$$2x(x^2+8x+15)=0$$
$$2x(x+5)(x+3)=0$$
Apply the zero product principle.
$2x=0 \quad x+5=0 \quad x+3=0$
$x=0 \quad x=-5 \quad x=-3$
The solution set is $\{-5,-3,0\}$.

47. $x^2-6x+8=0$
$$(x-4)(x-2)=0$$
Apply the zero product principle.
$x-4=0 \quad x-2=0$
$x=4 \quad x=2$
The x-intercepts are 2 and 4. This corresponds to graph d.

49. $x^2+6x+8=0$
$$(x+4)(x+2)=0$$
Apply the zero product principle.
$x+4=0 \quad x+2=0$
$x=-4 \quad x=-2$
The x-intercepts are –4 and –2. This corresponds to graph c.

51. $x(x+1)^3-42(x+1)^2=0$
$$(x+1)^2(x(x+1)-42)=0$$
$$(x+1)^2(x^2+x-42)=0$$
$$(x+1)^2(x+7)(x-6)=0$$
Apply the zero product principle.
$x+1=0 \quad \text{or} \quad x+7=0 \quad \text{or} \quad x-6=0$
$x=-1 \quad x=-7 \quad x=6$
The solution set is $\{-7,-1,6\}$.

53. $-4x[x(3x-2)-8](25x^2-40x+16)=0$

$-4x\left[3x^2-2x-8\right](5x-4)(5x-4)=0$

$-4x(3x+4)(x-2)(5x-4)^2=0$

Apply the zero product principle.

$-4x=0 \quad 3x+4=0 \quad x-2=0 \quad 5x-4=0$

$x=0 \quad 3x=-4 \quad x=2 \quad 5x=4$

$x=-\frac{4}{3} \quad x=2 \quad x=\frac{4}{5}$

The solution set is $\left\{-\frac{4}{3}, 0, \frac{4}{5}, 2\right\}$.

55. $f(c)=c^2-4c-27$

$5=c^2-4c-27$

$0=c^2-4c-32$

$0=(c-8)(c+4)$

Apply the zero product principle.

$c-8=0 \quad c+4=0$

$c=8 \quad c=-4$

The solutions are -4 and 8.

57. $f(c)=2c^3+c^2-8c+2$

$6=2c^3+c^2-8c+2$

$0=2c^3+c^2-8c-4$

$0=c^2(2c+1)-4(2c+1)$

$0=(2c+1)(c^2-4)$

$0=(2c+1)(c-2)(c+2)$

Apply the zero product principle.

$2c+1=0 \quad c-2=0 \quad c+2=0$

$2c=-1 \quad c=2 \quad c=-2$

$c=-\frac{1}{2}$

The solutions are $-2, -\frac{1}{2}$, and 2.

59. Let x be the number.

$(x-1)(x+4)=24$

$x^2+3x-4=24$

$x^2+3x-28=0$

$(x+7)(x-4)=0$

Apply the zero product principle.

$x+7=0 \quad x-4=0$

$x=-7 \quad x=4$

61. Let x be the number.

$3x-5=(x-1)^2$

$3x-5=(x-1)(x-1)$

$3x-5=x^2-2x+1$

$0=x^2-5x+6$

$0=(x-2)(x-3)$

Apply the zero product principle.

$x-2=0$ or $x-3=0$

$x=2 \quad x=3$

63. $f(x)=\frac{3}{x^2+4x-45}$

The numbers that make the denominator 0 must be excluded from the domain. So, set the denominator equal to 0 and solve.

$x^2+4x-45=0$

$(x+9)(x-5)=0$

$x+9=0 \quad x-5=0$

$x=-9 \quad x=5$

So, -9 and 5 must be excluded from the domain.

65. $-16t^2+8t+8=0$

$-8\left(2t^2-t-1\right)=0$

$-8(2t+1)(t-1)=0$

Apply the zero product principle.

$2t+1=0 \quad t-1=0$

$2t+1=0 \quad t=1$

$t=-\frac{1}{2}$

The solution set is $\left(-\frac{1}{2}, 1\right)$. Disregard $-\frac{1}{2}$ because we can't have a negative time measurement. The gymnast will reach the ground at $t=1$ second. The tick marks are then 0.25, 0.5, 0.75, and 1.

67. $f(x)=\dfrac{x^2-x}{2}$

$21=\dfrac{x^2-x}{2}$

$42=x^2-x$

$0=x^2-x-42$

$0=(x+6)(x-7)$

$x+6=0$ or $x-7=0$

$x=-6$ $\quad x=7$

7 players entered the tournament.

69. In Exercise 67, we found that if 21 games had been played, then 7 players had entered the tournament. This corresponds to the point $(7,21)$ on the graph.

71. Let w = the width.
Let $w+3$ = the length.

Area = lw

$54=(w+3)w$

$54=w^2+3w$

$0=w^2+3w-54$

$0=(w+9)(w-6)$

Apply the zero product principle.

$w+9=0$ $\quad w-6=0$

$w=-9$ $\quad w=6$

Disregard –9 because we can't have a negative length measurement. The width is 6 feet and the length is $6+3=9$ feet.

73. Let x = the length of the side of the original square. Let $x+3$ = the length of the side of the new, larger square.

$(x+3)^2=64$

$x^2+6x+9=64$

$x^2+6x-55=0$

$(x+11)(x-5)=0$

Apply the zero product principle.

$x+11=0$ $\quad x-5=0$

$x=-11$ $\quad x=5$

The solution set is $\{-11,5\}$. Disregard -11 because we can't have a negative length measurement. This means that x, the length of the side of the original square, is 5 inches.

75. Let x = the width of the path.

$(20+2x)(10+2x)=600$

$200+40x+20x+4x^2=600$

$200+60x+4x^2=600$

$4x^2+60x+200=600$

$4x^2+60x-400=0$

$4(x^2+15x-100)=0$

$4(x+20)(x-5)=0$

Apply the zero product principle.

$4(x+20)=0$ $\quad x-5=0$

$x+20=0$ $\quad x=5$

$x=-20$

Disregard –20 because the width cannot be negative. The width of the path is 5 meters.

77. a. $(2x+12)(2x+10)-10(12)$

$=4x^2+20x+24x+\cancel{120}-\cancel{120}$

$=4x^2+44x$

b. $4x^2+44x=168$

$4x^2+44x-168=0$

$4(x^2+11x-42)=0$

$4(x+14)(x-3)=0$

Apply the zero product principle.

$4(x+14)=0$ $\quad x-3=0$

$x+14=0$ $\quad x=3$

$x=-14$

Disregard –14 because the width cannot be negative. The width of the border is 3 feet.

79. Volume = lwh

$$200 = x \cdot x \cdot 2$$
$$200 = 2x^2$$
$$0 = 2x^2 - 200$$
$$0 = 2(x^2 - 100)$$
$$0 = 2(x+10)(x-10)$$

Apply the zero product principle.

$$2(x+10) = 0 \qquad x - 10 = 0$$
$$x + 10 = 0 \qquad x = 10$$
$$x = -10$$

Disregard -10 because the width cannot be negative. The length and width of the open box are both 10 inches.

81.

$$\text{leg}^2 + \text{leg}^2 = \text{hypotenuse}^2$$
$$(2x+2)^2 + x^2 = 13^2$$
$$4x^2 + 8x + 4 + x^2 = 169$$
$$5x^2 + 8x + 4 = 169$$
$$5x^2 + 8x - 165 = 0$$
$$(5x+33)(x-5) = 0$$

Apply the zero product principle.

$$5x + 33 = 0 \qquad x - 5 = 0$$
$$5x = -33 \qquad x = 5$$
$$x = -\frac{33}{5}$$

The width of the closet is 5 feet. The length is $2x + 2 = 2(5) + 2 = 10 + 2 = 12$ feet.

83. Let x = the length of the wire.

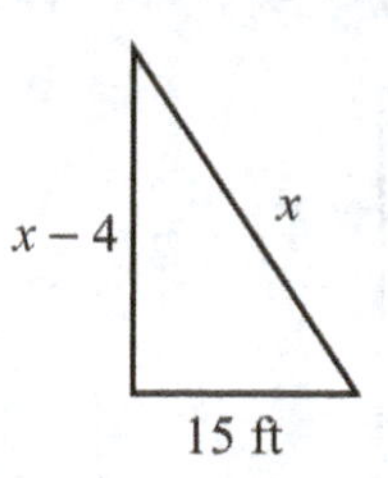

$$\text{leg}^2 + \text{leg}^2 = \text{hypotenuse}^2$$
$$(x-4)^2 + 15^2 = x^2$$
$$x^2 - 8x + 16 + 225 = x^2$$
$$-8x + 241 = 0$$
$$-8x = -241$$
$$-8x = -241$$
$$x = \frac{241}{8} = 30\frac{1}{8}$$

The length of the wire is $30\frac{1}{8}$ feet.

85. – 95. Answers will vary.

97. $y = x^3 + 3x^2 - x - 3$

$$x + 3 = 0 \qquad x + 1 = 0 \qquad x - 1 = 0$$
$$x = -3 \qquad x = -1 \qquad x = 1$$

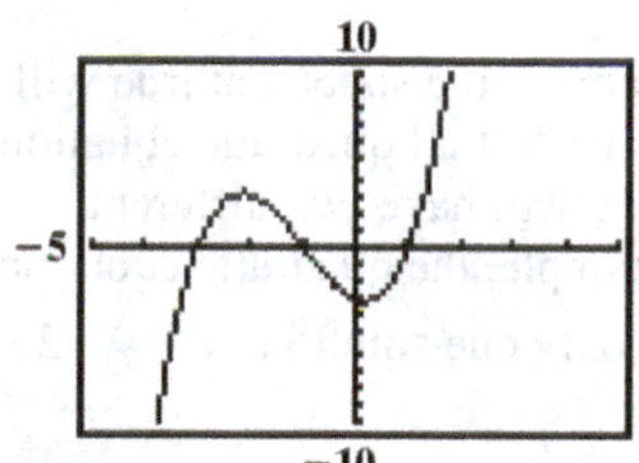

The x-intercepts are -3, -1, and 1. Use the intercepts to solve the equation.

$$x^3 + 3x^2 - x - 3 = 0$$
$$x^2(x+3) - 1(x+3) = 0$$
$$(x+3)(x^2 - 1) = 0$$
$$(x+3)(x+1)(x-1) = 0$$

Apply the zero product principle.
The solution set is $\{-3, -1, 1\}$.

99. $y = -x^4 + 4x^3 - 4x^2$

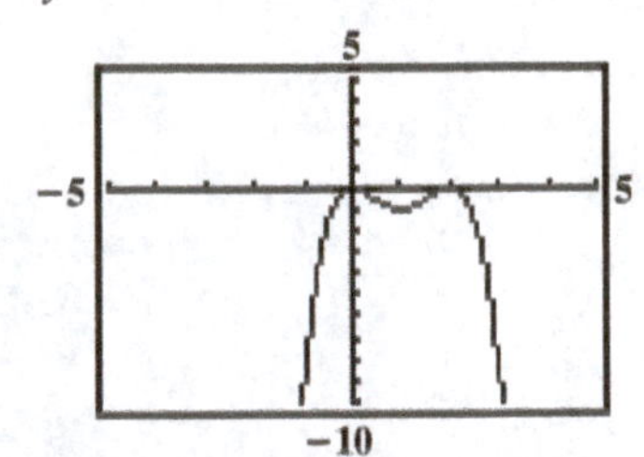

The x-intercepts are 0 and 2. Use the intercepts to solve the equation.

$$-x^4 + 4x^3 - 4x^2 = 0$$
$$-x^2\left(x^2 - 4x + 4\right) = 0$$
$$-x^2(x-2)^2 = 0$$

Apply the zero product principle.

$$-x^2 = 0 \qquad (x-2)^2 = 0$$
$$x^2 = 0 \qquad x - 2 = 0$$
$$x = 0 \qquad x = 2$$

The solutions are 0 and 2 and the solution set is $\{0, 2\}$.

101. makes sense

103. makes sense

105. false; Changes to make the statement true will vary. A sample change is: Not all quadratic equations solved by factoring will have two different solutions. For example, the quadratic equation $(x+2)^2 = 0$ has only one solution of $x = -2$.

107. true

109. Answers will vary; An example is $x^2 - 4x - 21 = 0$

111. $|3x - 2| = 8$

$$3x - 2 = -8 \quad \text{or} \quad 3x - 2 = 8$$
$$3x = -6 \qquad 3x = 10$$
$$x = -2 \qquad x = \frac{10}{3}$$

The solutions are -2 and $\frac{10}{3}$ and the solution set is $\left\{-2, \frac{10}{3}\right\}$.

112. $3(5-7)^2 + \sqrt{16} + 12 \div (-3)$

$$= 3(-2)^2 + 4 + (-4)$$
$$= 3(4) + 4 + (-4)$$
$$= 12 + 4 + (-4)$$
$$= 12$$

113. Let x = the amount invested at 5%.
Let y = the amount invested at 8%.

$$x + \quad y = 3000$$
$$0.05x + 0.08y = 189$$

Multiply the first equation by –0.05 and add to the second equation.

$$-0.05x - 0.05y = -150$$
$$\underline{0.05x + 0.08y = \ \ 189}$$
$$0.03y = \ \ 39$$
$$y = 1300$$

Back-substitute 1300 for y in one of the original equations.

$$x + y = 3000$$
$$x + 1300 = 3000$$
$$x = 1700$$

\$1700 was invested at 5% and \$1300 was invested at 8%.

114. $f(x) = \dfrac{120x}{100 - x}$

$$f(20) = \frac{120(20)}{100 - 20}$$
$$= \frac{2400}{80}$$
$$= 30$$

115. $f(x) = \dfrac{4}{x-2}$

The domain is $(-\infty, 2) \cup (2, \infty)$.

116. $\dfrac{x^2 - 7x - 18}{2x^2 + 3x - 2} = \dfrac{(x-9)(x+2)}{(2x-1)(x+2)}$

$$= \frac{(x-9)\cancel{(x+2)}}{(2x-1)\cancel{(x+2)}}$$
$$= \frac{x-9}{2x-1}$$

Chapter 5 Review

1. The coefficient of $-5x^3$ is –5 and the degree is 3.
The coefficient of $7x^2$ is 7 and the degree is 2.
The coefficient of $-x$ is –1 and the degree is 1.
The coefficient of 2 is 2 and the degree is 0.
The degree of the polynomial is 3.
The leading term is $-5x^3$ and the leading coefficient is –5.

2. The coefficient of $8x^4y^2$ is 8 and the degree is 6.
The coefficient of $-7xy^6$ is –7 and the degree is 7.
The coefficient of $-x^3y$ is –1 and the degree is 4.
The degree of the polynomial is 7.
The leading term is $-7xy^6$ and the leading coefficient is –7.

3. $f(x) = x^3 - 4x^2 + 3x - 1$
$f(-2) = (-2)^3 - 4(-2)^2 + 3(-2) - 1$
$= -8 - 4(4) - 6 - 1$
$= -8 - 16 - 6 - 1 = -31$

4. **a.** $f(x) = 10x^3 - 200x^2 + 1230x - 2016$
$f(10) = 10(10)^3 - 200(10)^2 + 1230(10) - 2016$
$= 284$
There were 284,000 record daily high temperatures in the United States in the 2000s.

b. Using the model in part (a) underestimates the actual number shown in the graph by 6000.

5. Since the degree of the polynomial is 3, an odd number, and the leading coefficient is –1, the graph will rise to the left and fall to the right. The graph of the polynomial is graph (c).

6. Since the degree of the polynomial is 6, an even number, and the leading coefficient is 1, the graph will rise to the left and rise to the right. The graph of the polynomial is graph (b).

7. Since the degree of the polynomial is 5, an odd number, and the leading coefficient is 1, the graph will fall to the left and rise to the right. The graph of the polynomial is graph (a).

8. Since the degree of the polynomial is 4, an even number, and the leading coefficient is –1, the graph will fall to the left and fall to the right. The graph of the polynomial is graph (d).

9. **a.** The graph of function *f* rises to the right.

b. No, the model will not be useful. The model indicates increasing deforestation despite a declining rate in which the forest is being cut down.

c. The graph of function *g* falls to the right.

d. No, the model will not be useful. The model indicates the amount of forest cleared, in square kilometers, will eventually be negative, which is not possible.

10. $\left(-8x^3 + 5x^2 - 7x + 4\right) + \left(9x^3 - 11x^2 + 6x - 13\right)$
$= -8x^3 + 5x^2 - 7x + 4 + 9x^3 - 11x^2 + 6x - 13$
$= -8x^3 + 9x^3 + 5x^2 - 11x^2 - 7x + 6x + 4 - 13$
$= x^3 - 6x^2 - x - 9$

11. $\left(7x^3y - 13x^2y - 6y\right) + \left(5x^3y + 11x^2y - 8y - 17\right)$
$= 7x^3y - 13x^2y - 6y + 5x^3y + 11x^2y - 8y - 17$
$= 7x^3y + 5x^3y - 13x^2y + 11x^2y - 6y - 8y - 17$
$= 12x^3y - 2x^2y - 14y - 17$

12. $\left(7x^3 - 6x^2 + 5x - 11\right) - \left(-8x^3 + 4x^2 - 6x - 3\right)$
$= 7x^3 - 6x^2 + 5x - 11 + 8x^3 - 4x^2 + 6x + 3$
$= 7x^3 + 8x^3 - 6x^2 - 4x^2 + 5x + 6x - 11 + 3$
$= 15x^3 - 10x^2 + 11x - 8$

13. $\left(4x^3y^2 - 7x^3y - 4\right) - \left(6x^3y^2 - 3x^3y + 4\right)$
$= 4x^3y^2 - 7x^3y - 4 - 6x^3y^2 + 3x^3y - 4$
$= 4x^3y^2 - 6x^3y^2 - 7x^3y + 3x^3y - 4 - 4$
$= -2x^3y^2 - 4x^3y - 8$

14. $\left(x^3 + 4x^2y - y^3\right) - \left(-2x^3 - x^2y + xy^2 + 7y^3\right)$
$= x^3 + 4x^2y - y^3 + 2x^3 + x^2y - xy^2 - 7y^3$
$= x^3 + 2x^3 + 4x^2y + x^2y - xy^2 - y^3 - 7y^3$
$= 3x^3 + 5x^2y - xy^2 - 8y^3$

15. $\left(4x^2yz^5\right)\left(-3x^4yz^2\right) = 4x^2yz^5(-3)x^4yz^2$
$= 4(-3)x^2x^4yyz^5z^2$
$= -12x^6y^2z^7$

16. $6x^3\left(\frac{1}{3}x^5 - 4x^2 - 2\right)$
$= 6x^3\left(\frac{1}{3}x^5\right) - 6x^3\left(4x^2\right) - 6x^3(2)$
$= 6\left(\frac{1}{3}\right)x^3x^5 - 6(4)x^3x^2 - 6(2)x^3$
$= 2x^8 - 24x^5 - 12x^3$

17. $7xy^2\left(3x^4y^2 - 5xy - 1\right)$
$= 7xy^2\left(3x^4y^2\right) - 7xy^2(5xy) - 7xy^2(1)$
$= 7(3)xx^4y^2y^2 - 7(5)xxy^2y - 7xy^2$
$= 21x^5y^4 - 35x^2y^3 - 7xy^2$

18.
$$\begin{array}{r} 3x^2 + 7x - 4 \\ \underline{2x + 5} \\ 6x^3 + 14x^2 - 8x \\ \underline{15x^2 + 35x - 20} \\ 6x^3 + 29x^2 + 27x - 20 \end{array}$$

19.
$$\begin{array}{l} x^2 + x - 1 \\ \underline{x^2 + 3x + 2} \\ x^4 + x^3 - x^2 \\ \quad 3x^3 + 3x^2 - 3x \\ \underline{\qquad 2x^2 + 2x - 2} \\ x^4 + 4x^3 + 4x^2 - x - 2 \end{array}$$

20. $(4x-1)(3x-5) = 12x^2 - 20x - 3x + 5$
$= 12x^2 - 23x + 5$

21. $(3xy-2)(5xy+4)$
$= 15x^2y^2 + 12xy - 10xy - 8$
$= 15x^2y^2 + 2xy - 8$

22. Two methods can be used to multiply the binomials.
Using the FOIL Method:
$(3x+7y)^2 = (3x+7y)(3x+7y)$
$= 9x^2 + 21xy + 21xy + 49y^2$
$= 9x^2 + 42xy + 49y^2$
Recognizing a Perfect Square Trinomial:
$(3x+7y)^2 = (3x)^2 + 2(3x)(7y) + (7y)^2$
$= 9x^2 + 2 \cdot 3x \cdot 7y + 49y^2$
$= 9x^2 + 42xy + 49y^2$

23. Two methods can be used to multiply the binomials.
Using the FOIL Method:
$\left(x^2 - 5y\right)^2$
$= \left(x^2 - 5y\right)\left(x^2 - 5y\right)$
$= x^4 - 5x^2y - 5x^2y + 25y^2$
$= x^4 - 10x^2y + 25y^2$
Recognizing a Perfect Square Trinomial:
$\left(x^2 - 5y\right)^2$
$= \left(x^2\right)^2 + 2\left(x^2\right)(-5y) + (-5y)^2$
$= x^4 + 2 \cdot x^2(-5y) + 25y^2$
$= x^4 - 10x^2y + 25y^2$

24. Two methods can be used to multiply the binomials.
Using the FOIL Method:
$(2x+7y)(2x-7y)$
$= 4x^2 - 14xy + 14xy - 49y^2$
$= 4x^2 - 49y^2$
Recognizing the Difference of Two Squares:
$(2x+7y)(2x-7y) = (2x)^2 - (7y)^2$
$= 4x^2 - 49y^2$

25. Two methods can be used to multiply the binomials.
Using the FOIL Method:
$\left(3xy^2 - 4x\right)\left(3xy^2 + 4x\right)$
$= 9x^2y^4 + 12x^2y^2 - 12x^2y^2 - 16x^2$
$= 9x^2y^4 - 16x^2$
Recognizing the Difference of Two Squares:
$\left(3xy^2 - 4x\right)\left(3xy^2 + 4x\right)$
$= \left(3xy^2\right)^2 - (4x)^2$
$= 9x^2y^4 - 16x^2$

26. Two methods can be used to multiply the binomials.
Using the FOIL Method:
$[(x+3)+5y][(x+3)-5y]$
$= (x+3)^2 - 5(x+3)y + 5(x+3)y - 25y^2$
$= (x+3)^2 - 25y^2$
$= x^2 + 6x + 9 - 25y^2$
Recognizing the Difference of Two Squares:
$[(x+3)+5y][(x+3)-5y]$
$= (x+3)^2 - (5y)^2 = x^2 + 6x + 9 - 25y^2$

27.
$$\begin{array}{l} x+y+4 \\ \underline{x+y+4} \\ x^2+xy+4x \\ \quad xy \qquad +y^2+4y \\ \underline{\qquad\quad 4x \qquad\quad +4y+16} \\ x^2+2xy+8x+y^2+8y+16 \end{array}$$
or
$$x^2+2xy+y^2+8x+8y+16$$

28. $f(x)=x-3$ and $g(x)=2x+5$

$$\begin{aligned}(fg)(x)&=f(x)\cdot g(x)\\ &=(x-3)(2x+5)\\ &=2x^2+5x-6x-15\\ &=2x^2-x-15\end{aligned}$$

$$\begin{aligned}(fg)(-4)&=2(-4)^2-(-4)-15\\ &=2(16)+4-15\\ &=32+4-15\\ &=21\end{aligned}$$

29. $f(x)=x^2-7x+2$

a. $f(a-1)$
$$\begin{aligned}&=(a-1)^2-7(a-1)+2\\ &=a^2-2a+1-7a+7+2\\ &=a^2-9a+10\end{aligned}$$

b. $f(a+h)-f(a)$
$$\begin{aligned}&=(a+h)^2-7(a+h)\\ &\quad +2-(a^2-7a+2)\\ &=\cancel{a^2}+2ah+h^2\cancel{-7a}-7h\\ &\quad \cancel{+2}\cancel{-a^2}\cancel{+7a}\cancel{-2}\\ &=2ah+h^2-7h\end{aligned}$$

30. $16x^3+24x^2=8x^2(2x+3)$

31. $2x-36x^2=2x(1-18x)$

32. $21x^2y^2-14xy^2+7xy=7xy(3xy-2y+1)$

33. $18x^3y^2-27x^2y=9x^2y(2xy-3)$

34. $-12x^2+8x-48=-4(3x^2-2x+12)$

35.
$$\begin{aligned}-x^2-11x+14&=-1(x^2+11x-14)\\ &=-(x^2+11x-14)\end{aligned}$$

36.
$$\begin{aligned}x^3-x^2-2x+2&=x^2(x-1)-2(x-1)\\ &=(x-1)(x^2-2)\end{aligned}$$

37.
$$\begin{aligned}xy-3x-5y+15&=x(y-3)-5(y-3)\\ &=(y-3)(x-5)\end{aligned}$$

38. $5ax-15ay+2bx-6by$
$$\begin{aligned}&=5a(x-3y)+2b(x-3y)\\ &=(x-3y)(5a+2b)\end{aligned}$$

39. $x^2+8x+15=(x+5)(x+3)$

40. $x^2+16x-80=(x+20)(x-4)$

41. $x^2+16xy-17y^2=(x+17y)(x-y)$

42.
$$\begin{aligned}3x^3-36x^2+33x&=3x(x^2-12x+11)\\ &=3x(x-11)(x-1)\end{aligned}$$

43. $3x^2+22x+7=(3x+1)(x+7)$

44. $6x^2-13x+6=(2x-3)(3x-2)$

45. $5x^2-6xy-8y^2=(5x+4y)(x-2y)$

46.
$$\begin{aligned}6x^3+5x^2-4x&=x(6x^2+5x-4)\\ &=x(2x-1)(3x+4)\end{aligned}$$

47. $2x^2+11x+15=(2x+5)(x+3)$

48. Let $u=x^3$
$$\begin{aligned}x^6+x^3-30&=(x^3)^2+x^3-30\\ &=u^2+u-30\\ &=(u+6)(u-5)\\ &=(x^3+6)(x^3-5)\end{aligned}$$

49. Let $u = x^2$

$$\begin{aligned} x^4 - 10x^2 - 39 &= \left(x^2\right)^2 - 10x^2 - 39 \\ &= u^2 - 10u - 39 \\ &= (u-13)(u+3) \\ &= \left(x^2-13\right)\left(x^2+3\right) \end{aligned}$$

50. Let $u = x+5$

$$\begin{aligned} (x+5)^2 + 10(x+5) + 24 &= u^2 + 10u + 24 \\ &= (u+6)(u+4) \\ &= ((x+5)+6)((x+5)+4) \\ &= (x+11)(x+9) \end{aligned}$$

51. Let $u = x^3$

$$\begin{aligned} 5x^6 + 17x^3 + 6 &= 5\left(x^3\right)^2 + 17x^3 + 6 \\ &= 5u^2 + 17u + 6 \\ &= (5u+2)(u+3) \\ &= \left(5x^3+2\right)\left(x^3+3\right) \end{aligned}$$

52. $4x^2 - 25 = (2x+5)(2x-5)$

53. $1 - 81x^2y^2 = (1+9xy)(1-9xy)$

54. $x^8 - y^6 = \left(x^4+y^3\right)\left(x^4-y^3\right)$

55.
$$\begin{aligned} (x-1)^2 - y^2 &= ((x-1)+y)((x-1)-y) \\ &= (x-1+y)(x-1-y) \end{aligned}$$

56.
$$\begin{aligned} x^2 + 16x + 64 &= x^2 + 2\cdot x\cdot 8 + 8^2 \\ &= (x+8)^2 \end{aligned}$$

57.
$$\begin{aligned} 9x^2 - 6x + 1 &= (3x)^2 - 2\cdot 3x\cdot 1 + 1^2 \\ &= (3x-1)^2 \end{aligned}$$

58.
$$\begin{aligned} 25x^2 + 20xy + 4y^2 &= (5x)^2 + 2\cdot 5x\cdot 2y + (2y)^2 \\ &= (5x+2y)^2 \end{aligned}$$

59. $49x^2 + 7x + 1$ is prime.

60.
$$\begin{aligned} 25x^2 - 40xy + 16y^2 &= (5x)^2 - 2\cdot 5x\cdot 4y + (4y)^2 \\ &= (5x-4y)^2 \end{aligned}$$

61.
$$\begin{aligned} x^2 + 18x + 81 - y^2 &= \left(x^2+18x+81\right) - y^2 \\ &= (x+9)^2 - y^2 \\ &= ((x+9)+y)((x+9)-y) \\ &= (x+9+y)(x+9-y) \end{aligned}$$

62.
$$\begin{aligned} &z^2 - 25x^2 + 10x - 1 \\ &= z^2 - \left(25x^2 - 10x + 1\right) \\ &= z^2 - \left((5x)^2 - 2\cdot 5x\cdot 1 + 1^2\right) \\ &= z^2 - (5x-1)^2 \\ &= (z+(5x-1))(z-(5x-1)) \\ &= (z+5x-1)(z-5x+1) \end{aligned}$$

63.
$$\begin{aligned} 64x^3 + 27 &= (4x)^3 + (3)^3 \\ &= (4x+3)\left((4x)^2 - 4x\cdot 3 + 3^2\right) \\ &= (4x+3)\left(16x^2 - 12x + 9\right) \end{aligned}$$

64.
$$\begin{aligned} 125x^3 - 8 &= (5x)^3 - (2)^3 \\ &= (5x-2)\left((5x)^2 + 5x\cdot 2 + 2^2\right) \\ &= (5x-2)\left(25x^2 + 10x + 4\right) \end{aligned}$$

65.
$$\begin{aligned} x^3y^3 + 1 &= (xy)^3 + 1^3 \\ &= (xy+1)\left((xy)^2 - xy\cdot 1 + 1^2\right) \\ &= (xy+1)\left(x^2y^2 - xy + 1\right) \end{aligned}$$

66. $15x^2 + 3x = 3x(5x+1)$

67.
$$\begin{aligned} 12x^4 - 3x^2 &= 3x^2\left(4x^2-1\right) \\ &= 3x^2\left((2x)^2 - 1^2\right) \\ &= 3x^2(2x+1)(2x-1) \end{aligned}$$

68.
$$\begin{aligned} &20x^4 - 24x^3 + 28x^2 - 12x \\ &= 4x\left(5x^3 - 6x^2 + 7x - 3\right) \end{aligned}$$

69.
$$\begin{aligned} x^3 - 15x^2 + 26x &= x\left(x^2 - 15x + 26\right) \\ &= x(x-2)(x-13) \end{aligned}$$

70. $-2y^4+24y^3-54y^2=-2y^2\left(y^2-12y+27\right)$
$=-2y^2(y-9)(y-3)$

71. $9x^2-30x+25=(3x)^2-2\cdot 3x\cdot 5+5^2$
$=(3x-5)^2$

72. $5x^2-45=5\left(x^2-9\right)$
$=5(x+3)(x-3)$

73. $2x^3-x^2-18x+9=x^2(2x-1)-9(2x-1)$
$=(2x-1)\left(x^2-9\right)$
$=(2x-1)(x+3)(x-3)$

74. $6x^2-23xy+7y^2=(3x-y)(2x-7y)$

75. $2y^3+12y^2+18y=2y\left(y^2+6y+9\right)$
$=2y\left(y^2+2\cdot y\cdot 3+3^2\right)$
$=2y(y+3)^2$

76. $x^2+6x+9-4a^2=\left(x^2+6x+9\right)-4a^2$
$=\left(x^2+2\cdot x\cdot 3+3^2\right)-(2a)^2$
$=(x+3)^2-(2a)^2$
$=((x+3)+2a)((x+3)-2a)$
$=(x+3+2a)(x+3-2a)$

77. $8x^3-27=(2x)^3-3^3$
$=(2x-3)\left((2x)^2+2x\cdot 3+3^2\right)$
$=(2x-3)\left(4x^2+6x+9\right)$

78. $x^5-x=x\left(x^4-1\right)$
$=x\left(\left(x^2\right)^2-1^2\right)$
$=x\left(x^2+1\right)\left(x^2-1\right)$
$=x\left(x^2+1\right)\left(x^2-1^2\right)$
$=x\left(x^2+1\right)(x+1)(x-1)$

79. Let $u=x^2$
$x^4-6x^2+9=\left(x^2\right)^2-6x^2+9$
$=u^2-6u+9$
$=u^2-2\cdot u\cdot 3+3^2$
$=(u-3)^2$
$=\left(x^2-3\right)^2$

80. x^2+xy+y^2 is prime.

81. $4a^3+32=4\left(a^3+8\right)$
$=4\left(a^3+2^3\right)$
$=4(a+2)\left(a^2-a\cdot 2+2^2\right)$
$=4(a+2)\left(a^2-2a+4\right)$

82. $x^4-81=\left(x^2\right)^2-9^2$
$=\left(x^2+9\right)\left(x^2-9\right)$
$=\left(x^2+9\right)\left(x^2-3^2\right)$
$=\left(x^2+9\right)(x+3)(x-3)$

83. $ax+3bx-ay-3by=x(a+3b)-y(a+3b)$
$=(a+3b)(x-y)$

84. $27x^3-125y^3=(3x)^3-(5y)^3$
$=(3x-5y)\left((3x)^2+3x\cdot 5y+(5y)^2\right)$
$=(3x-5y)\left(9x^2+15xy+25y^2\right)$

85. $10x^3y+22x^2y-24xy=2xy\left(5x^2+11x-12\right)$
$=2xy(5x-4)(x+3)$

86. Let $u=x^3$
$6x^6+13x^3-5=6\left(x^3\right)^2+13x^3-5$
$=6u^2+13u-5$
$=(2u+5)(3u-1)$
$=\left(2x^3+5\right)\left(3x^3-1\right)$

87. $2x+10+x^2y+5xy = 2(x+5)+xy(x+5)$
$= (x+5)(2+xy)$

88. $y^3+2y^2-25y-50 = y^2(y+2)-25(y+2)$
$= (y+2)(y^2-25)$
$= (y+2)(y+5)(y-5)$

89. Let $u = a^4$

$a^8-1 = \left(a^4\right)^2-1$
$= u^2-1$
$= (u+1)(u-1)$
$= \left(a^4+1\right)\left(a^4-1\right)$
$= \left(a^4+1\right)(a^2+1)(a^2-1)$
$= \left(a^4+1\right)(a^2+1)(a+1)(a-1)$

90. $9(x-4)+y^2(4-x) = 9(x-4)+(-1)y^2(x-4)$
$= (x-4)\left(9-y^2\right)$
$= (x-4)(3-y)(3+y)$

91. a. $2xy+2y^2$

b. $2xy+2y^2 = 2y(x+y)$

92. a. x^2-4y^2

b. $x^2-4y^2 = x^2-(2y)^2$
$= (x+2y)(x-2y)$

93. $x^2+6x+5=0$
$(x+5)(x+1)=0$

Apply the zero product principle.

$x+5=0 \qquad x+1=0$
$x=-5 \qquad x=-1$

The solution set is $\{-5,-1\}$.

94. $3x^2 = 22x-7$
$3x^2-22x+7=0$
$(3x-1)(x-7)=0$

Apply the zero product principle.

$3x-1=0 \qquad x-7=0$
$3x=1 \qquad x=7$
$x=\frac{1}{3}$

The solution set is $\left\{\frac{1}{3},7\right\}$.

95. $(x+3)(x-2)=50$
$x^2-2x+3x-6=50$
$x^2+x-6=50$
$x^2+x-56=0$
$(x+8)(x-7)=0$

$x+8=0 \qquad x-7=0$
$x=-8 \qquad x=7$

The solution set is $\{-8,7\}$.

96. $3x^2 = 12x$
$3x^2-12x=0$
$3x(x-4)=0$

$3x=0 \qquad x-4=0$
$x=0 \qquad x=4$

The solution set is $\{0,4\}$.

97. $x^3+5x^2 = 9x+45$
$x^3+5x^2-9x-45=0$
$\left(x^3+5x^2\right)+(-9x-45)=0$
$x^2(x+5)+(-9)(x+5)=0$
$(x+5)\left(x^2-9\right)=0$
$(x+5)(x+3)(x-3)=0$

$x+5=0 \qquad x+3=0 \qquad x-3=0$
$x=-5 \qquad x=-3 \qquad x=3$

The solution set is $\{-5,-3,3\}$.

98. $-16t^2+128t+144=0$

$-16\left(t^2-8t-9\right)=0$

$-16(t-9)(t+1)=0$

$-16(t-9)=0 \qquad t+1=0$

$t-9=0 \qquad t=-1$

$t=9$

Disregard –1 because time cannot be negative. The rocket will hit the water at $t=9$ seconds.

99. **a.** $d(x)=\dfrac{x^2}{20}+x$

$40=\dfrac{x^2}{20}+x$

$800=x^2+20x$

$0=x^2+20x-800$

$0=(x+40)(x-20)$

$x+40=0 \quad \text{or} \quad x-20=0$

$x=-40 \qquad x=20$

The car was traveling 20 miles per hour.

b. A car traveling 20 miles per hour takes 40 feet to stop. This is represented on the graph as the point $(20,40)$.

c. As a car's speed increases, its stopping distance gets longer at increasingly greater rates.

100. Let w = the width of the sign.
Let $w+3$ = the length of the sign.
Area = lw

$54=(w+3)w$

$54=w^2+3w$

$0=w^2+3w-54$

$0=(w+9)(w-6)$

$w+9=0 \qquad w-6=0$

$w=-9 \qquad w=6$

Disregard –9 because length cannot be negative. The width is 6 feet and the length is $6+3=9$ feet.

101. Let x = the width of the border.

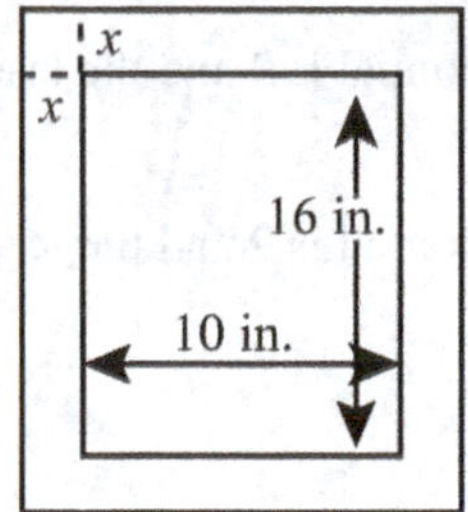

Area = lw

$280=(2x+16)(2x+10)$

$280=4x^2+20x+32x+160$

$0=4x^2+52x-120$

$0=4\left(x^2+13x-30\right)$

$0=4(x+15)(x-2)$

$4(x+15)=0 \qquad x-2=0$

$x+15=0 \qquad x=2$

$x=-15$

Disregard –15 because length cannot be negative. The width of the frame is 2 inches.

102. $\text{leg}^2+\text{leg}^2=\text{hypotenuse}^2$

$x^2+(2x+20)^2=(2x+30)^2$

$x^2+4x^2+80x+400=4x^2+120x+900$

$x^2-40x-500=0$

$(x-50)(x+10)=0$

$x-50=0 \qquad x+10=0$

$x=50 \qquad x=-10$

Disregard –10 because length cannot be negative. We have $x=50$ yards. Find the lengths of the other sides.

$2x+20 \qquad 2x+30$

$=2(50)+20 \qquad =2(50)+30$

$=100+20 \qquad =100+30$

$=120 \qquad =130$

The three sides are 50 yards, 120 yards, and 130 yards.

Chapter 5 Test

1. The degree of the polynomial is 3 and the leading coefficient is –6.

2. The degree of the polynomial is 9 and the leading coefficient is 7.

3. $f(x) = 3x^3 + 5x^2 - x + 6$
$$\begin{aligned} f(0) &= 3(0)^3 + 5(0)^2 - 0 + 6 \\ &= 3(0) + 5(0) + 6 \\ &= 0 + 0 + 6 \\ &= 6 \end{aligned}$$
$$\begin{aligned} f(-2) &= 3(-2)^3 + 5(-2)^2 - (-2) + 6 \\ &= 3(-8) + 5(4) + 2 + 6 \\ &= -24 + 20 + 2 + 6 \\ &= 4 \end{aligned}$$

4. Since the degree of the polynomial is 2, an even number, and the leading coefficient is negative, the graph will fall to the left and fall to the right.

5. Since the degree of the polynomial is 3, an odd number, and the leading coefficient is positive, the graph will fall to the left and rise to the right.

6. $$\begin{aligned} &(4x^3y - 19x^2y - 7y) + (3x^3y + x^2y + 6y - 9) \\ &= 4x^3y - 19x^2y - 7y + 3x^3y + x^2y + 6y - 9 \\ &= 4x^3y + 3x^3y - 19x^2y + x^2y - 7y + 6y - 9 \\ &= 7x^3y - 18x^2y - y - 9 \end{aligned}$$

7. $$\begin{aligned} &(6x^2 - 7x - 9) - (-5x^2 + 6x - 3) \\ &= 6x^2 - 7x - 9 + 5x^2 - 6x + 3 \\ &= 6x^2 + 5x^2 - 7x - 6x - 9 + 3 \\ &= 11x^2 - 13x - 6 \end{aligned}$$

8. $$\begin{aligned} (-7x^3y)(-5x^4y^2) &= -7(-5)x^3x^4yy^2 \\ &= 35x^7y^3 \end{aligned}$$

9. $$\begin{array}{r} x^2 - 3xy - y^2 \\ x - y \\ \hline x^3 - 3x^2y - xy^2 \\ -\ x^2y + 3xy^2 + y^3 \\ \hline x^3 - 4x^2y + 2xy^2 + y^3 \end{array}$$

10. $$\begin{aligned} (7x - 9y)(3x + y) &= 21x^2 + 7xy - 27xy - 9y^2 \\ &= 21x^2 - 20xy - 9y^2 \end{aligned}$$

11. $$\begin{aligned} (2x - 5y)(2x + 5y) &= 4x^2 + \cancel{10xy} - \cancel{10xy} - 25y^2 \\ &= 4x^2 - 25y^2 \end{aligned}$$

12. $$\begin{aligned} (4y - 7)^2 &= (4y)^2 + 2 \cdot 4y \cdot (-7) + (-7)^2 \\ &= 16y^2 - 56y + 49 \end{aligned}$$

13. $$\begin{aligned} &[(x+2) + 3y][(x+2) - 3y] \\ &= (x+2)^2 - 3(x+2)y + 3(x+2)y - 9y^2 \\ &= (x+2)^2 - 9y^2 \\ &= x^2 + 4x + 4 - 9y^2 \end{aligned}$$

14. $f(x) = x + 2$ and $g(x) = 3x - 5$
$$\begin{aligned} (fg)(x) &= f(x) \cdot g(x) \\ &= (x+2)(3x-5) \\ &= 3x^2 - 5x + 6x - 10 \\ &= 3x^2 + x - 10 \end{aligned}$$
$$\begin{aligned} (fg)(-5) &= 3(-5)^2 + (-5) - 10 \\ &= 3(25) - 5 - 10 \\ &= 75 - 5 - 10 \\ &= 60 \end{aligned}$$

15. $f(x) = x^2 - 5x + 3$
$$\begin{aligned} &f(a+h) - f(a) \\ &= (a+h)^2 - 5(a+h) + 3 - (a^2 - 5a + 3) \\ &= a^2 + 2ah + h^2 - 5a - 5h + 3 - a^2 + 5a - 3 \\ &= 2ah + h^2 - 5h \end{aligned}$$

16. $14x^3 - 15x^2 = x^2(14x - 15)$

17. $$\begin{aligned} 81y^2 - 25 &= (9y)^2 - 5^2 \\ &= (9y+5)(9y-5) \end{aligned}$$

18. $x^3+3x^2-25x-75$
$=x^2(x+3)-25(x+3)$
$=(x+3)(x^2-25)$
$=(x+3)(x^2-5^2)$
$=(x+3)(x+5)(x-5)$

19. $25x^2-30x+9=(5x)^2-2\cdot 5x\cdot 3+3^2$
$=(5x-3)^2$

20. $x^2+10x+25-9y^2=(x^2+10x+25)-9y^2$
$=(x+5)^2-(3y)^2$
$=((x+5)+3y)((x+5)-3y)$
$=(x+5+3y)(x+5-3y)$

21. x^4+1 is prime.

22. $y^2-16y-36=(y-18)(y+2)$

23. $14x^2+41x+15=(2x+5)(7x+3)$

24. $5x^3-5=5(x^3-1)$
$=5(x^3-1^3)$
$=5(x-1)(x^2+x\cdot 1+1^2)$
$=5(x-1)(x^2+x+1)$

25. $12x^2-3y^2=3(4x^2-y^2)$
$=3((2x)^2-y^2)$
$=3(2x+y)(2x-y)$

26. $12x^2-34x+10=2(6x^2-17x+5)$
$=2(3x-1)(2x-5)$

27. $3x^4-3=3(x^4-1)$
$=3\left((x^2)^2-1^2\right)$
$=3(x^2+1)(x^2-1)$
$=3(x^2+1)(x^2-1^2)$
$=3(x^2+1)(x+1)(x-1)$

28. $x^8-y^8=(x^4)^2-(y^4)^2$
$=(x^4+y^4)(x^4-y^4)$
$=(x^4+y^4)\left((x^2)^2-(y^2)^2\right)$
$=(x^4+y^4)(x^2+y^2)(x^2-y^2)$
$=(x^4+y^4)(x^2+y^2)(x+y)(x-y)$

29. $12x^2y^4+8x^3y^2-36x^2y$
$=4x^2y(3y^3+2xy-9)$

30. Let $u=x^3$
$x^6-12x^3-28=(x^3)^2-12x^3-28$
$=u^2-12u-28$
$=(u-14)(u+2)$ Substitute x^3 for u.
$=(x^3-14)(x^3+2)$

31. Let $u=x^2$
$x^4-2x^2-24=(x^2)^2-2x^2-24$
$=u^2-2u-24$
$=(u-6)(u+4)$ Substitute x^2 for u.
$=(x^2-6)(x^2+4)$

32. $12x^2y-27xy+6y=3y(4x^2-9x+2)$
$=3y(x-2)(4x-1)$

33. $y^4-3y^3+2y^2-6y=y(y^3-3y^2+2y-6)$
$=y(y^2(y-3)+2(y-3))$
$=y(y-3)(y^2+2)$

34. $3x^2 = 5x + 2$

$3x^2 - 5x - 2 = 0$

$(3x+1)(x-2) = 0$

Apply the zero product principle.

$3x + 1 = 0 \qquad x - 2 = 0$

$3x = -1 \qquad x = 2$

$x = -\frac{1}{3}$

The solution set is $\left\{-\frac{1}{3}, 2\right\}$.

35. $(5x+4)(x-1) = 2$

$5x^2 - 5x + 4x - 4 = 2$

$5x^2 - x - 4 = 2$

$5x^2 - x - 6 = 0$

$(5x-6)(x+1) = 0$

Apply the zero product principle.

$5x - 6 = 0 \qquad x + 1 = 0$

$5x = 6 \qquad x = -1$

$x = \frac{6}{5}$

The solution set is $\left\{-1, \frac{6}{5}\right\}$.

36. $15x^2 - 5x = 0$

$5x(3x-1) = 0$

Apply the zero product principle.

$5x = 0 \qquad 3x - 1 = 0$

$x = 0 \qquad 3x = 1$

$x = \frac{1}{3}$

The solution set is $\left\{0, \frac{1}{3}\right\}$.

37. $x^3 - 4x^2 - x + 4 = 0$

$x^2(x-4) - 1(x-4) = 0$

$(x-4)(x^2-1) = 0$

$(x-4)(x+1)(x-1) = 0$

Apply the zero product principle.

$x - 4 = 0 \qquad x + 1 = 0 \qquad x - 1 = 0$

$x = 4 \qquad x = -1 \qquad x = 1$

The solution set is $\{-1, 1, 4\}$.

38. $-16t^2 + 48t + 448 = 0$

$-16(t^2 - 3t - 28) = 0$

$-16(t-7)(t+4) = 0$

Apply the zero product principle.

$-16(t-7) = 0 \qquad t + 4 = 0$

$t - 7 = 0 \qquad t = -4$

$t = 7$

Disregard –4 because we can't have a negative time measurement. The baseball will hit the water at $t = 7$ seconds.

39. Let l be the length of the room.
Let $w = 2l - 7$ be the width of the room.

Area $= lw$

$15 = l(2l - 7)$

$15 = 2l^2 - 7l$

$0 = 2l^2 - 7l - 15$

$0 = (2l+3)(l-5)$

Apply the zero product principle.

$2l + 3 = 0 \qquad l - 5 = 0$

$2l = -3 \qquad l = 5$

$l = -\frac{3}{2}$

Disregard $-\frac{3}{2}$ because we can't have a negative length measurement. The length is 5 yards and the width is $2l - 7 = 2(5) - 7 = 10 - 7 = 3$ yards.

40. $\text{leg}^2 + \text{leg}^2 = \text{hypotenuse}^2$

$x^2 + 12^2 = (2x-3)^2$

$x^2 + 144 = 4x^2 - 12x + 9$

$0 = 3x^2 - 12x - 135$

$0 = 3(x^2 - 4x - 45)$

$0 = 3(x-9)(x+5)$

Apply the zero product principle.

$3(x-9) = 0 \qquad x + 5 = 0$

$x - 9 = 0 \qquad x = -5$

$x = 9$

Disregard –5 because we can't have a negative length measurement. The lengths of the sides are 9 and 12 units and the length of the hypotenuse is $2x - 3 = 2(9) - 3 = 18 - 3 = 15$ units .

Cumulative Review Exercises (Chapters 1 – 5)

1. $8(x+2)-3(2-x)=4(2x+6)-2$

$8x+16-6+3x=8x+24-2$

$11x+10=8x+22$

$3x=12$

$x=4$

The solution set is $\{4\}$.

2. $2x+4y=-6$

$x=2y-5$

Rewrite the system in standard form.

$2x+4y=-6$

$x-2y=-5$

Multiply the second equation by 2 and add to the first equation.

$2x+4y=-6$

$2x-4y=-10$

$4x=-16$

$x=-4$

Back-substitute –4 for x in one of the original equations.

$2x+4y=-6$

$2(-4)+4y=-6$

$-8+4y=-6$

$4y=2$

$y=\frac{1}{2}$

The solution is $\left(-4,\frac{1}{2}\right)$ and the solution set is $\left\{\left(-4,\frac{1}{2}\right)\right\}$.

3. $2x-y+3z=0$

$2y+z=1$

$x+2y-z=5$

Multiply the third equation by –2 and add to the first equation.

$2x-y+3z=0$

$-2x-4y+2z=-10$

$-5y+5z=-10$

We now have 2 equations in 2 variables.

$2y+z=1$

$-5y+5z=-10$

Multiply the first equation by –5 and add to the second equation.

$-10y-5z=-5$

$-5y+5z=-10$

$-15y=-15$

$y=1$

Back-substitute 1 for y in one of the equations in 2 variables.

$2y+z=1$

$2(1)+z=1$

$2+z=1$

$z=-1$

Back-substitute 1 for y and –1 for z in one of the original equations in 3 variables.

$2x-y+3z=0$

$2x-1+3(-1)=0$

$2x-1-3=0$

$2x-4=0$

$2x=4$

$x=2$

The solution is $(2,1,-1)$ and the solution set is $\{(2,1,-1)\}$.

4. $2x+4<10$ and $3x-1>5$

$2x<6$ $\quad$ $3x>6$

$x<3$ $\quad$ $x>2$

The solution set is $(2,3)$.

5. $|2x-5|\geq 9$

$2x-5\leq -9$ or $2x-5\geq 9$

$2x\leq -4$ $\quad$ $2x\geq 14$

$x\leq -2$ $\quad$ $x\geq 7$

The solution set is $(-\infty,-2]\cup[7,\infty)$.

6. $2x^2=7x-5$

$2x^2-7x+5=0$

$(2x-5)(x-1)=0$

Apply the zero product principle.

$2x-5=0$ $\quad$ $x-1=0$

$2x=5$ $\quad$ $x=1$

$x=\frac{5}{2}$

The solution set is $\left\{1,\frac{5}{2}\right\}$.

7. $2x^3 + 6x^2 = 20x$

$2x^3 + 6x^2 - 20x = 0$

$2x\left(x^2 + 3x - 10\right) = 0$

$2x(x+5)(x-2) = 0$

Apply the zero product principle.

$2x = 0 \quad x+5 = 0 \quad x-2 = 0$

$x = 0 \quad x = -5 \quad x = 2$

The solution set is $\{-5, 0, 2\}$.

8. $x = \dfrac{ax+b}{c}$

$cx = ax + b$

$cx - ax = b$

$x(c-a) = b$

$x = \dfrac{b}{c-a}$

9. First, find the slope.

$m = \dfrac{5-(-3)}{2-(-2)} = \dfrac{5+3}{2+2} = \dfrac{8}{4} = 2$

Write the equation in point-slope form.

$y - y_1 = m(x - x_1)$

$y - 5 = 2(x-2)$

$y - 5 = 2x - 4$

$y = 2x + 1$

Write the equation of the line using function notation.

$f(x) = 2x + 1$

10. Let x = the number of votes for the loser.
Let y = the number of votes for the winner.

$x + y = 2800$

$x + 160 = y$

Rewrite the system in standard form.

$x + y = 2800$

$x - y = -160$

Add the equations to eliminate y.

$x + y = 2800$

$x - y = -160$

$2x = 2640$

$x = 1320$

Back-substitute 1320 for x to find y.

$x + 160 = y$

$1320 + 160 = y$

$1480 = y$

The solution set is $\{(1320, 1480)\}$. The loser received 1320 votes and the winner received 1480 votes.

11. $f(x) = -\dfrac{1}{3}x + 1$

First, find the intercepts. To find the y–intercept, set $x = 0$.

$y = -\dfrac{1}{3}(0) + 1$

$y = 0 + 1$

$y = 1$

To find the x–intercept, set $y = 0$.

$0 = -\dfrac{1}{3}x + 1$

$-3(-1) = -3\left(-\dfrac{1}{3}x\right)$

$3 = x$

Graph the line using the intercepts.

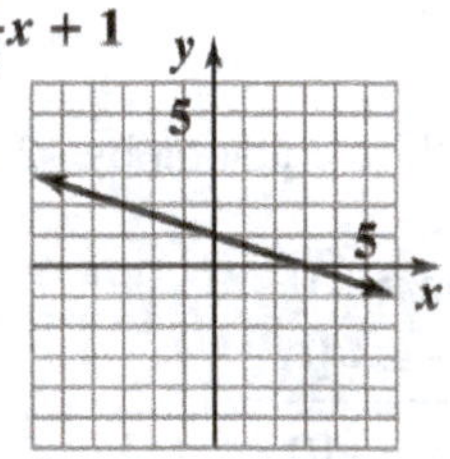

$f(x) = -\frac{1}{3}x + 1$

12. $4x-5y<20$

First, find the intercepts. To find the y–intercept, set $x=0$.

$$4(0)-5y=20$$
$$0-5y=20$$
$$-5y=20$$
$$y=-4$$

To find the x–intercept, set $y=0$.

$$4x-5(0)=20$$
$$4x-0=20$$
$$4x=20$$
$$x=5$$

Graph the inequality using the intercepts.

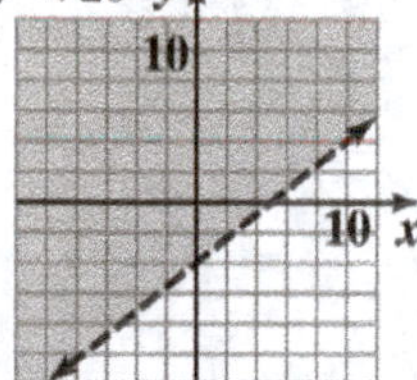

13. $y\le -1$

We know that $y=-1$ is a horizontal line and the y–intercept is -1. This is sufficient information to graph the inequality.

y ≤ −1

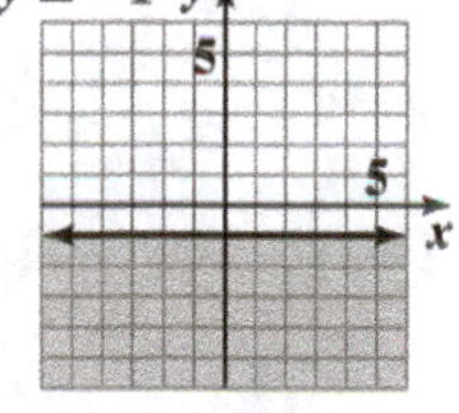

14. $\dfrac{-8x^3y^6}{16x^9y^{-4}}=\dfrac{-8}{16}x^{3-9}y^{6-(-4)}$

$$=\frac{-8}{16}x^{3-9}y^{6+4}$$
$$=\frac{-1}{2}x^{-6}y^{10}$$
$$=-\frac{y^{10}}{2x^6}$$

15. $0.0000706=7.06\times10^{-5}$

16. $\left(3x^2-y\right)^2=\left(3x^2\right)^2-2\cdot3x^2\cdot y+y^2$

$$=9x^4-6x^2y+y^2$$

17. $\left(3x^2-y\right)\left(3x^2+y\right)$

$$=\left(3x^2\right)^2+\cancel{3x^2y}-\cancel{3x^2y}-y^2$$
$$=9x^4-y^2$$

18. $x^3-3x^2-9x+27$

$$=x^2(x-3)-9(x-3)$$
$$=(x-3)\left(x^2-9\right)$$
$$=(x-3)\left(x^2-3^2\right)$$
$$=(x-3)(x+3)(x-3)$$
$$=(x-3)^2(x+3)$$

19. x^6-x^2

$$=x^2\left(x^4-1\right)$$
$$=x^2\left(\left(x^2\right)^2-1^2\right)$$
$$=x^2\left(x^2+1\right)\left(x^2-1\right)$$
$$=x^2\left(x^2+1\right)\left(x^2-1^2\right)$$
$$=x^2\left(x^2+1\right)(x+1)(x-1)$$

20. $14x^3y^2-28x^4y^2$

$$=14x^3y^2(1-2x)$$

Chapter 6
Rational Expressions, Functions, and Equations

6.1 Check Points

1. a.
$$f(x)=\frac{120x}{100-x}$$
$$f(40)=\frac{120(40)}{100-40}=\frac{4800}{60}=80$$
The cost to remove 40% of the lake's pollutants is $80 thousand.
This is represented by the point $(40, 80)$.

b.
$$f(x)=\frac{120x}{100-x}$$
$$f(60)=\frac{120(60)}{100-60}=\frac{7200}{40}=180$$
The cost to remove 60% of the lake's pollutants is $180 thousand.
This is represented by the point $(60, 180)$.

2. The domain is all real numbers except those which make the denominator zero. Set the denominator equal to zero and solve.
$$2x^2+5x-3=0$$
$$(2x-1)(x+3)=0$$
$$2x-1=0 \quad \text{or} \quad x+3=0$$
$$x=\tfrac{1}{2} \qquad\qquad x=-3$$
Domain of $f = \left(-\infty,-3\right)\cup\left(-3,\tfrac{1}{2}\right)\cup\left(\tfrac{1}{2},\infty\right)$.

3. $$\frac{x^2+7x+10}{x+2}=\frac{(x+5)(x+2)}{1(x+2)}=\frac{(x+5)\cancel{(x+2)}}{1\cancel{(x+2)}}=x+5$$

4. a. $$\frac{x^2-2x-15}{3x^2+8x-3}=\frac{(x-5)(x+3)}{(3x-1)(x+3)}=\frac{(x-5)\cancel{(x+3)}}{(3x-1)\cancel{(x+3)}}=\frac{x-5}{3x-1}$$

b. $$\frac{3x^2+9xy-12y^2}{9x^3-9xy^2}=\frac{3(x^2+3xy-4y^2)}{9x(x^2-y^2)}=\frac{3(x+4y)(x-y)}{9x(x+y)(x-y)}=\frac{x+4y}{3x(x+y)}$$

5. $$\frac{x+4}{x-7}\cdot\frac{x^2-4x-21}{x^2-16}=\frac{x+4}{x-7}\cdot\frac{(x+3)(x-7)}{(x+4)(x-4)}=\frac{\cancel{x+4}}{\cancel{x-7}}\cdot\frac{(x+3)\cancel{(x-7)}}{\cancel{(x+4)}(x-4)}=\frac{x+3}{x-4}$$

6. $\frac{4x+8}{6x-3x^2}\cdot\frac{3x^2-4x-4}{9x^2-4}=\frac{4(x+2)}{3x(2-x)}\cdot\frac{(3x+2)(x-2)}{(3x+2)(3x-2)}=\frac{4(x+2)}{3x(2-x)}\cdot\frac{(3x+2)\overset{-1}{(x-2)}}{(3x+2)(3x-2)}=\frac{-4(x+2)}{3x(3x-2)}$

7. a. $(9x^2-49)\div\frac{3x-7}{9}=\frac{9x^2-49}{1}\div\frac{3x-7}{9}$

$$=\frac{9x^2-49}{1}\cdot\frac{9}{3x-7}$$

$$=\frac{(3x+7)(3x-7)}{1}\cdot\frac{9}{1(3x-7)}$$

$$=\frac{(3x+7)(3x-7)}{1}\cdot\frac{9}{1(3x-7)}$$

$$=9(3x+7)$$

b. $\frac{x^2-x-12}{5x}\div\frac{x^2-10x+24}{x^2-6x}=\frac{x^2-x-12}{5x}\cdot\frac{x^2-6x}{x^2-10x+24}$

$$=\frac{(x+3)(x-4)}{5x}\cdot\frac{x(x-6)}{(x-6)(x-4)}$$

$$=\frac{(x+3)(x-4)}{5x}\cdot\frac{x(x-6)}{(x-6)(x-4)}$$

$$=\frac{x+3}{5}$$

6.1 Concept and Vocabulary Check

1. polynomial; polynomial

2. zero

3. asymptote

4. asymptote

5. factoring; common factors

6. $x+5$

7. false

8. -1

9. numerators; denominators

10. multiplicative inverse/reciprocal; $\frac{S}{R}=\frac{PS}{QR}$

11. $\frac{x^2}{70}$

12. $\frac{10}{7}$

6.1 Exercise Set

1. $f(x)=\frac{x^2-9}{x+3}$

$f(-2)=\frac{(-2)^2-9}{-2+3}=\frac{4-9}{1}=\frac{-5}{1}=-5$

$f(0)=\frac{0^2-9}{0+3}=-3$

$f(5)=\frac{5^2-9}{5+3}=\frac{25-9}{8}=\frac{16}{8}=2$

3. $f(x)=\frac{x^2-2x-3}{4-x}$

$f(-1)=\frac{(-1)^2-2(-1)-3}{4-(-1)}$

$=\frac{1+2-3}{4+1}=\frac{0}{5}=0$

$f(4)$ does not exist because division by zero is undefined.

$f(6)=\frac{6^2-2(6)-3}{4-6}=\frac{36-12-3}{-2}=-\frac{21}{2}$

5. $g(t)=\frac{2t^3-5}{t^2+1}$

$g(-1)=\frac{2(-1)^3-5}{(-1)^2+1}=\frac{2(-1)-5}{1+1}$

$=\frac{-2-5}{2}=-\frac{7}{2}$

$g(0)=\frac{2(0)^3-5}{0^2+1}=\frac{-5}{1}=-5$

$g(2)=\frac{2(2)^3-5}{(2)^2+1}=\frac{2(8)-5}{4+1}=\frac{16-5}{5}=\frac{11}{5}$

7. The domain is all real numbers except those which make the denominator zero. Set the denominator equal to zero and solve.

$x-5=0$

$x=5$

Domain of $f=(-\infty,5)\cup(5,\infty)$.

9. The domain is all real numbers except those which make the denominator equal to zero. Set the denominator equal to zero and solve.

$x-1=0 \qquad x+3=0$

$x=1 \qquad x=-3$

Domain of $f=(-\infty,-3)\cup(-3,1)\cup(1,\infty)$.

11. The domain is all real numbers except those which make the denominator equal to zero. Set the denominator equal to zero and solve.

$x+5=0$

$x=-5$

Domain of $f=(-\infty,-5)\cup(-5,\infty)$.

13. The domain is all real numbers except those which make the denominator equal to zero. Set the denominator equal to zero and solve.

$x^2-8x+15=0$

$(x-5)(x-3)=0$

$x-5=0 \qquad x-3=0$

$x=5 \qquad x=3$

Domain of $f=(-\infty,3)\cup(3,5)\cup(5,\infty)$.

15. The domain is all real numbers except those which make the denominator equal to zero. Set the denominator equal to zero and solve.

$3x^2-2x-8=0$

$(3x+4)(x-2)=0$

$3x+4=0 \qquad x-2=0$

$3x=-4 \qquad x=2$

$x=-\frac{4}{3}$

Domain of $f=\left(-\infty,-\frac{4}{3}\right)\cup\left(-\frac{4}{3},2\right)\cup(2,\infty)$.

17. $f(4)=4$

19. Domain of $f=(-\infty,-2)\cup(-2,2)\cup(2,\infty)$.

Range of $f=(-\infty,0]\cup(3,\infty)$

21. As x decreases, the value of the function approaches 3. The equation of the horizontal asymptote is $y=3$.

23. There is no point on the graph with an x–coordinate of –2.

25. The graph is not continuous. Furthermore, it neither rises nor falls without bound to the left or the right.

27. $\dfrac{x^2-4}{x-2}=\dfrac{(x+2)\cancel{(x-2)}}{1\cancel{(x-2)}}$

$=x+2$

29. $\dfrac{x+2}{x^2-x-6}=\dfrac{1\cancel{(x+2)}}{(x-3)\cancel{(x+2)}}$

$=\dfrac{1}{x-3}$

31. $\dfrac{4x+20}{x^2+5x}=\dfrac{4\cancel{(x+5)}}{x\cancel{(x+5)}}$

$=\dfrac{4}{x}$

33. $\dfrac{4y-20}{y^2-25}=\dfrac{4\cancel{(y-5)}}{(y+5)\cancel{(y-5)}}$

$=\dfrac{4}{y+5}$

35. $\dfrac{3x-5}{25-9x^2}=\dfrac{1\overset{-1}{\cancel{(3x-5)}}}{(5+3x)\cancel{(5-3x)}}$

$=\dfrac{-1}{5+3x}$ or $-\dfrac{1}{5+3x}$

Or, by the commutative property of addition, we have $\dfrac{-1}{3x+5}$ or $-\dfrac{1}{3x+5}$.

37. $\dfrac{y^2-49}{y^2-14y+49}=\dfrac{(y+7)\cancel{(y-7)}}{(y-7)\cancel{(y-7)}}$

$=\dfrac{y+7}{y-7}$

39. $\dfrac{x^2+7x-18}{x^2-3x+2}=\dfrac{(x+9)\cancel{(x-2)}}{\cancel{(x-2)}(x-1)}$

$=\dfrac{x+9}{x-1}$

41. $\dfrac{3x+7}{3x+10}$

The rational expression cannot be simplified.

43. $\dfrac{x^2-x-12}{16-x^2}=\dfrac{\overset{-1}{\cancel{(x-4)}}(x+3)}{\cancel{(4-x)}(4+x)}$

$=-\dfrac{x+3}{4+x}$ or $-\dfrac{x+3}{x+4}$

45. $\dfrac{x^2+3xy-10y^2}{3x^2-7xy+2y^2}=\dfrac{(x+5y)\cancel{(x-2y)}}{(3x-y)\cancel{(x-2y)}}$

$=\dfrac{x+5y}{3x-y}$

47. $\dfrac{x^3-8}{x^2-4}=\dfrac{\cancel{(x-2)}(x^2+2x+4)}{(x+2)\cancel{(x-2)}}$

$=\dfrac{x^2+2x+4}{x+2}$

49. $\dfrac{x^3+4x^2-3x-12}{x+4}=\dfrac{x^2(x+4)-3(x+4)}{x+4}$

$=\dfrac{\cancel{(x+4)}(x^2-3)}{1\cancel{(x+4)}}$

$=x^2-3$

51. $\dfrac{x-3}{x+7}\cdot\dfrac{3x+21}{2x-6}=\dfrac{1\cancel{(x-3)}}{1\cancel{(x+7)}}\cdot\dfrac{3\cancel{(x+7)}}{2\cancel{(x-3)}}=\dfrac{3}{2}$

53. $\dfrac{x^2-49}{x^2-4x-21}\cdot\dfrac{x+3}{x}$

$=\dfrac{(x+7)\cancel{(x-7)}}{\cancel{(x-7)}\cancel{(x+3)}}\cdot\dfrac{1\cancel{(x+3)}}{x}$

$=\dfrac{x+7}{x}$

55. $\dfrac{x^2-9}{x^2-x-6}\cdot\dfrac{x^2+5x+6}{x^2+x-6}$

$=\dfrac{(x+3)\cancel{(x-3)}}{\cancel{(x-3)}\cancel{(x+2)}}\cdot\dfrac{\cancel{(x+3)}\cancel{(x+2)}}{\cancel{(x+3)}(x-2)}$

$=\dfrac{x+3}{x-2}$

57. $\frac{x^2+4x+4}{x^2+8x+16}\cdot\frac{(x+4)^3}{(x+2)^3}$

$=\frac{(x+2)^2}{(x+4)^2}\cdot\frac{(x+4)^3}{(x+2)^3}$

$=\frac{\cancel{(x+2)^2}}{\cancel{(x+4)^2}}\cdot\frac{\overset{x+4}{\cancel{(x+4)^3}}}{\underset{x+2}{\cancel{(x+2)^3}}}$

$=\frac{x+4}{x+2}$

59. $\frac{8y+2}{y^2-9}\cdot\frac{3-y}{4y^2+y}$

$=\frac{2\cancel{(4y+1)}}{(y+3)\cancel{(y-3)}}\cdot\frac{1\overset{-1}{\cancel{(3-y)}}}{y\cancel{(4y+1)}}$

$=\frac{-2}{y(y+3)}$ or $-\frac{2}{y(y+3)}$

61. $\frac{y^3-8}{y^2-4}\cdot\frac{y+2}{2y}$

$=\frac{\cancel{(y-2)}(y^2+2y+4)}{\cancel{(y+2)}\cancel{(y-2)}}\cdot\frac{1\cancel{(y+2)}}{2y}$

$=\frac{y^2+2y+4}{2y}$

63. $(x-3)\cdot\frac{x^2+x+1}{x^2-5x+6}$

$=\frac{1\cancel{(x-3)}}{1}\cdot\frac{x^2+x+1}{\cancel{(x-3)}(x-2)}$

$=\frac{x^2+x+1}{x-2}$

65. $\frac{x^2+xy}{x^2-y^2}\cdot\frac{4x-4y}{x}$

$=\frac{x(x+y)}{(x+y)(x-y)}\cdot\frac{4(x-y)}{x}$

$=\frac{\cancel{x}\cancel{(x+y)}}{\cancel{(x+y)}\cancel{(x-y)}}\cdot\frac{4\cancel{(x-y)}}{\cancel{x}}=4$

67. $\dfrac{x^2+2xy+y^2}{x^2-2xy+y^2}\cdot\dfrac{4x-4y}{3x+3y}$

$=\dfrac{(x+y)^2}{(x-y)^2}\cdot\dfrac{4(x-y)}{3(x+y)}$ (cancelled: $(x+y)^2$ to $x+y$, $(x-y)^2$ to $x-y$)

$=\dfrac{4(x+y)}{3(x-y)}$

69. $\dfrac{4a^2+2ab+b^2}{2a+b}\cdot\dfrac{4a^2-b^2}{8a^3-b^3}$

$=\dfrac{4a^2+2ab+b^2}{2a+b}\cdot\dfrac{(2a-b)(2a+b)}{(2a-b)(4a^2+2ab+b^2)}$

$=1$

71. $\dfrac{10z^2+13z-3}{3z^2-8z+5}\cdot\dfrac{2z^2-3z-2z+3}{25z^2-10z+1}\cdot\dfrac{15z^2-28z+5}{4z^2-9}$

$=\dfrac{(5z-1)(2z+3)}{(3z-5)(z-1)}\cdot\dfrac{z(2z-3)-1(2z-3)}{(5z-1)(5z-1)}\cdot\dfrac{(5z-1)(3z-5)}{(2z-3)(2z+3)}$

$=\dfrac{(5z-1)(2z+3)}{(3z-5)(z-1)}\cdot\dfrac{(z-1)(2z-3)}{(5z-1)(5z-1)}\cdot\dfrac{(5z-1)(3z-5)}{(2z-3)(2z+3)}$

$=1$

73. $\dfrac{x+5}{7}\div\dfrac{4x+20}{9}$

$=\dfrac{x+5}{7}\cdot\dfrac{9}{4x+20}$

$=\dfrac{1(x+5)}{7}\cdot\dfrac{9}{4(x+5)}=\dfrac{9}{28}$

75. $\dfrac{4}{y-6}\div\dfrac{40}{7y-42}$

$=\dfrac{4}{y-6}\cdot\dfrac{7y-42}{40}$

$=\dfrac{4}{1(y-6)}\cdot\dfrac{7(y-6)}{4\cdot 10}=\dfrac{7}{10}$

77. $\dfrac{x^2-2x}{15} \div \dfrac{x-2}{5}$

$= \dfrac{x^2-2x}{15} \cdot \dfrac{5}{x-2}$

$= \dfrac{x\cancel{(x-2)}}{3\cdot\cancel{5}} \cdot \dfrac{\cancel{5}}{1\cancel{(x-2)}} = \dfrac{x}{3}$ or $\dfrac{1}{3}x$

79. $\dfrac{y^2-25}{2y-2} \div \dfrac{y^2+10y+25}{y^2+4y-5}$

$= \dfrac{y^2-25}{2y-2} \cdot \dfrac{y^2+4y-5}{y^2+10y+25}$

$= \dfrac{\cancel{(y+5)}(y-5)}{2\cancel{(y-1)}} \cdot \dfrac{\cancel{(y+5)}\cancel{(y-1)}}{\cancel{(y+5)^2}}$

$= \dfrac{y-5}{2}$

81. $\left(x^2-16\right) \div \dfrac{x^2+3x-4}{x^2+4}$

$= \dfrac{x^2-16}{1} \div \dfrac{x^2+3x-4}{x^2+4}$

$= \dfrac{x^2-16}{1} \cdot \dfrac{x^2+4}{x^2+3x-4}$

$= \dfrac{\cancel{(x+4)}(x-4)}{1} \cdot \dfrac{x^2+4}{\cancel{(x+4)}(x-1)}$

$= \dfrac{(x-4)\left(x^2+4\right)}{x-1}$

83. $\dfrac{y^2-4y-21}{y^2-10y+25} \div \dfrac{y^2+2y-3}{y^2-6y+5}$

$= \dfrac{y^2-4y-21}{y^2-10y+25} \cdot \dfrac{y^2-6y+5}{y^2+2y-3}$

$= \dfrac{(y-7)\cancel{(y+3)}}{\cancel{(y-5)^2}_{\,y-5}} \cdot \dfrac{\cancel{(y-5)}\cancel{(y-1)}}{\cancel{(y+3)}\cancel{(y-1)}}$

$= \dfrac{y-7}{y-5}$

85. $\dfrac{8x^3-1}{4x^2+2x+1} \div \dfrac{x-1}{(x-1)^2}$

$= \dfrac{8x^3-1}{4x^2+2x+1} \cdot \dfrac{(x-1)^2}{x-1}$

$= \dfrac{(2x-1)\cancel{\left(4x^2+2x+1\right)}}{1\cancel{\left(4x^2+2x+1\right)}} \cdot \dfrac{\overset{x-1}{\cancel{(x-1)^2}}}{1\cancel{(x-1)}}$

$= (2x-1)(x-1)$

87. $\dfrac{x^2-4y^2}{x^2+3xy+2y^2} \div \dfrac{x^2-4xy+4y^2}{x+y}$

$= \dfrac{x^2-4y^2}{x^2+3xy+2y^2} \cdot \dfrac{x+y}{x^2-4xy+4y^2}$

$= \dfrac{\cancel{(x+2y)}\cancel{(x-2y)}}{\cancel{(x+2y)}\cancel{(x+y)}} \cdot \dfrac{1\cancel{(x+y)}}{\cancel{(x-2y)^2}_{\,x-2y}}$

$= \dfrac{1}{x-2y}$

89. $\dfrac{x^4-y^8}{x^2+y^4} \div \dfrac{x^2-y^4}{3x^2}$

$= \dfrac{x^4-y^8}{x^2+y^4} \cdot \dfrac{3x^2}{x^2-y^4}$

$= \dfrac{\cancel{\left(x^2+y^4\right)}\left(x^2-y^4\right)}{1\cancel{\left(x^2+y^4\right)}} \cdot \dfrac{3x^2}{x^2-y^4}$

$= \dfrac{x^2-y^4}{1} \cdot \dfrac{3x^2}{x^2-y^4}$

$= \dfrac{\cancel{x^2-y^4}}{1} \cdot \dfrac{3x^2}{\cancel{x^2-y^4}}$

$= 3x^2$

91. $\dfrac{x^3-4x^2+x-4}{2x^3-8x^2+x-4}\cdot\dfrac{2x^3+2x^2+x+1}{x^4-x^3+x^2-x}$

$=\dfrac{x^2(x-4)+1(x-4)}{2x^2(x-4)+1(x-4)}\cdot\dfrac{2x^2(x+1)+1(x+1)}{x^3(x-1)+x(x-1)}$

$=\dfrac{(x-4)(x^2+1)}{(x-4)(2x^2+1)}\cdot\dfrac{(x+1)(2x^2+1)}{(x-1)(x^3+x)}$

$=\dfrac{\cancel{(x-4)}(x^2+1)}{\cancel{(x-4)}\cancel{(2x^2+1)}}\cdot\dfrac{(x+1)\cancel{(2x^2+1)}}{(x-1)(x^3+x)}$

$=\dfrac{(x^2+1)(x+1)}{(x-1)x(x^2+1)}=\dfrac{\cancel{(x^2+1)}(x+1)}{(x-1)x\cancel{(x^2+1)}}$

$=\dfrac{x+1}{x(x-1)}$

93. $\dfrac{ax-ay+3x-3y}{x^3+y^3}\div\dfrac{ab+3b+ac+3c}{xy-x^2-y^2}$

$=\dfrac{ax-ay+3x-3y}{x^3+y^3}\cdot\dfrac{xy-x^2-y^2}{ab+3b+ac+3c}$

$=\dfrac{a(x-y)+3(x-y)}{(x+y)(x^2-xy+y^2)}\cdot\dfrac{(-1)(x^2-xy+y^2)}{b(a+3)+c(a+3)}$

$=\dfrac{(-1)(x-y)(a+3)(x^2-xy+y^2)}{(x+y)(x^2-xy+y^2)(a+3)(b+c)}$

$=\dfrac{(-1)(x-y)\cancel{(a+3)}\cancel{(x^2-xy+y^2)}}{(x+y)\cancel{(x^2-xy+y^2)}\cancel{(a+3)}(b+c)}$

$=\dfrac{-(x-y)}{(x+y)(b+c)}$

95. $\dfrac{a^2b+b}{3a^2-4a-20}\cdot\dfrac{a^2+5a}{2a^2+11a+5}\div\dfrac{ab^2}{6a^2-17a-10}=\dfrac{a^2b+b}{3a^2-4a-20}\cdot\dfrac{a^2+5a}{2a^2+11a+5}\cdot\dfrac{6a^2-17a-10}{ab^2}$

$=\dfrac{b(a^2+1)}{(3a-10)(a+2)}\cdot\dfrac{a(a+5)}{(2a+1)(a+5)}\cdot\dfrac{(3a-10)(2a+1)}{ab^2}$

$=\dfrac{\cancel{b}(a^2+1)}{\cancel{(3a-10)}(a+2)}\cdot\dfrac{\cancel{a}\cancel{(a+5)}}{\cancel{(2a+1)}\cancel{(a+5)}}\cdot\dfrac{\cancel{(3a-10)}\cancel{(2a+1)}}{\cancel{a}\,\cancel{b^2}_{\,b}}$

$=\dfrac{a^2+1}{b(a+2)}$

97. $\dfrac{a-b}{4c}\div\left(\dfrac{b-a}{c}\div\dfrac{a-b}{c^2}\right)$

$=\dfrac{a-b}{4c}\div\left(\dfrac{b-a}{c}\cdot\dfrac{c^2}{a-b}\right)$

$=\dfrac{a-b}{4c}\div\left(\dfrac{-\cancel{(a-b)}}{\cancel{c}}\cdot\dfrac{\cancel{c}\cdot c}{\cancel{(a-b)}}\right)$

$=\dfrac{a-b}{4c}\div\dfrac{-c}{1}$

$=\dfrac{a-b}{4c}\cdot\dfrac{-1}{c}$

$=-\dfrac{a-b}{4c^2}$ or $\dfrac{b-a}{4c^2}$

99. $\dfrac{f(a+h)-f(a)}{h}=\dfrac{\left[7(a+h)-4\right]-\left[7a-4\right]}{h}$

$=\dfrac{\cancel{7a}+7h-\cancel{4}-\cancel{7a}+\cancel{4}}{h}$

$=\dfrac{7\cancel{h}}{\cancel{h}}=7$

101. $\dfrac{f(a+h)-f(a)}{h}$

$=\dfrac{\left[(a+h)^2-5(a+h)+3\right]-\left[a^2-5a+3\right]}{h}$

$=\dfrac{\cancel{a^2}+2ah+h^2\cancel{-5a}-5h\cancel{+3}\cancel{-a^2}\cancel{+5a}\cancel{-3}}{h}$

$=\dfrac{2ah+h^2-5h}{h}$

$=\dfrac{\cancel{h}(2a+h-5)}{\cancel{h}}=2a+h-5$

103. $f(x)=\dfrac{(x+2)^2}{1-2x}$ and $g(x)=\dfrac{x+2}{2x-1}$

$\left(\dfrac{f}{g}\right)(x)=\dfrac{f(x)}{g(x)}=\dfrac{\dfrac{(x+2)^2}{1-2x}}{\dfrac{x+2}{2x-1}}$

$=\dfrac{(x+2)^2}{1-2x}\cdot\dfrac{2x-1}{x+2}$

$=\dfrac{(x+2)^2(2x-1)}{(-1)(2x-1)(x+2)}$

$=\dfrac{(x+2)^{\cancel{2}}\,\cancel{(2x-1)}}{(-1)\,\cancel{(2x-1)}\,\cancel{(x+2)}}$

$=-(x+2)$

To determine the domain, we need to exclude any values for x that make either $f(x)$ or $g(x)$ have division by 0. In addition, we need to exclude all values for x such that $g(x)=0$.

$1-2x=0 \qquad 2x-1=0 \qquad x+2=0$

$-2x=-1 \qquad 2x=1 \qquad x=-2$

$x=\dfrac{1}{2} \qquad x=\dfrac{1}{2}$

We need to exclude the values $g(x)$: and $x=-2$.

Domain: $(-\infty,-2)\cup\left(-2,\dfrac{1}{2}\right)\cup\left(\dfrac{1}{2},\infty\right)$.

105. $f(60)=\dfrac{130(60)}{100-60}=\dfrac{7800}{40}=195$. This corresponds to the point $(60,195)$. The cost to inoculate 60% of the population against a particular strain of flu is \$195,000,000.

107. The value 100 must be excluded from the domain. We cannot inoculate 100% of the population.

109. The minimum occurs after 6 minutes.

$f(x)=\dfrac{6.5x^2-20.4x+234}{x^2+36}$

$f(6)=\dfrac{6.5(6)^2-20.4(6)+234}{(6)^2+36}=4.8$

The minimum pH level is about 4.8.

111. The normal level, shown at time 0, is 6.5. Over time, the pH level rises back to normal.

113. $P(10) = \frac{100(10-1)}{10} = \frac{100(9)}{10}$
$= \frac{900}{10} = 90.$

This is represented by the point $(10, 90)$ and refers to an incidence ratio of 10. From the chart, we know that smokers between the ages of 55 and 64 are 10 times more likely than nonsmokers to die from lung cancer. Also, 90% of the deaths from lung cancer in this group are smoking-related.

115. The horizontal asymptote of the graph is $y = 100$. This means that as incidence ratio increases the percentage of smoking-related deaths increases towards 100%, although the percentage will never actually reach 100%.

117. – 127. Answers will vary.

129. $\frac{x^2+x}{3x} \cdot \frac{6x}{x+1} = 2x$

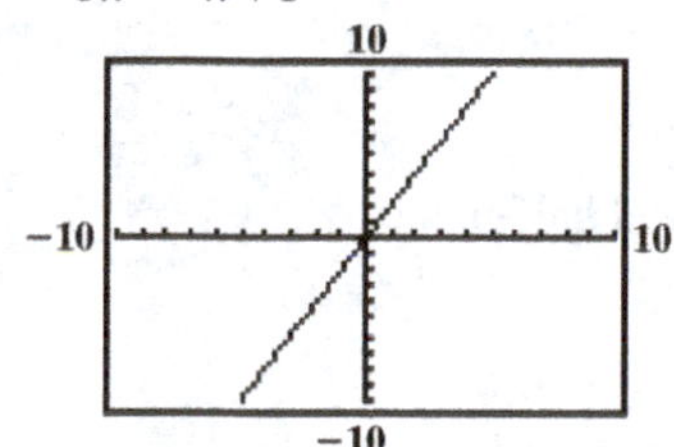

The graphs coincide. The multiplication is correct.

131. $\frac{x^2-9}{x+4} \div \frac{x-3}{x+4} = x-3$

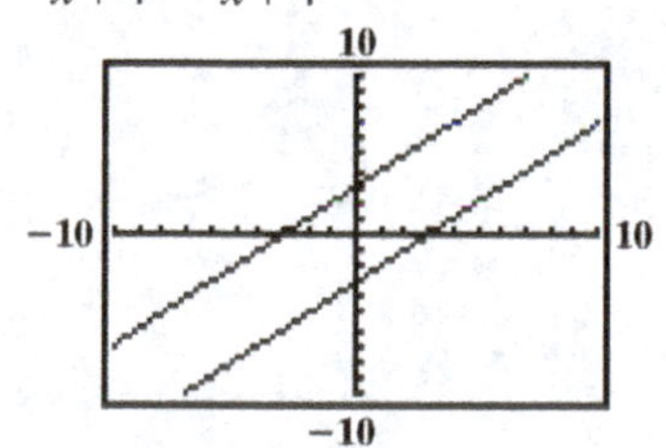

The graphs do not coincide. The division is incorrect.

$\frac{x^2-9}{x+4} \div \frac{x-3}{x+4}$

$= \frac{x^2-9}{x+4} \cdot \frac{x+4}{x-3}$

$= \frac{(x+3)\cancel{(x-3)}}{\cancel{x+4}} \cdot \frac{\cancel{x+4}}{\cancel{x-3}}$

$= x+3$

133. Answers will vary.

135. makes sense

137. does not make sense; Explanations will vary. Sample explanation: You only exclude the values that make the denominator equal to zero, not the numerator.

139. false; Changes to make the statement true will vary. A sample change is:

$\frac{x^2-25}{x-5} = \frac{(x-5)(x+5)}{x-5} = x+5$

141. false; Changes to make the statement true will vary. A sample change is: Because $x(x-3)+5(x-3) = (x-3)(x+5)$, the domain of $f(x)$ is $(-\infty,-5)\cup(-5,3)\cup(3,\infty)$.

143. Table of values:

x	$y = f(x) = \frac{x^2-x-2}{x-2}$	(x, y)
−2	$y = \frac{(-2)^2-(-2)-2}{-2-2} = -1$	$(-2,-1)$
0	$y = \frac{0^2-0-2}{0-2} = 1$	$(0,1)$
2	$y = \frac{2^2-2-2}{2-2} = \frac{0}{0}$ indeterminate	hole in graph at $(2,3)$
4	$y = \frac{4^2-4-2}{4-2} = 5$	$(4,5)$

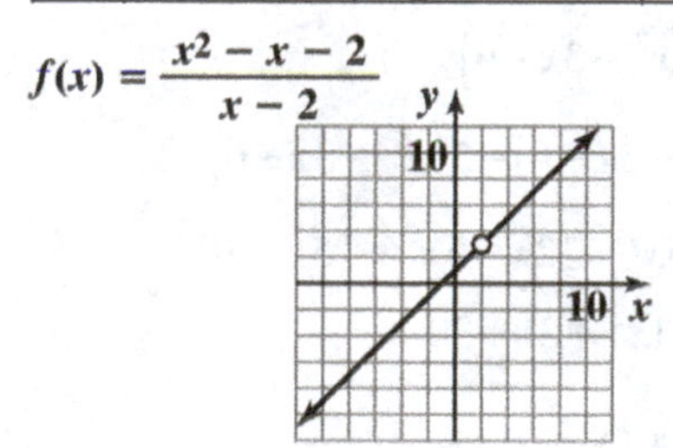

145. $\dfrac{y^{2n}-1}{y^{2n}+3y^n+2} \div \dfrac{y^{2n}+y^n-12}{y^{2n}-y^n-6} = \dfrac{y^{2n}-1}{y^{2n}+3y^n+2} \cdot \dfrac{y^{2n}-y^n-6}{y^{2n}+y^n-12}$

$= \dfrac{(y^n+1)(y^n-1)}{(y^n+2)(y^n+1)} \cdot \dfrac{(y^n-3)(y^n+2)}{(y^n+4)(y^n-3)} = \dfrac{y^n-1}{y^n+4}$

147. $4x-5y \geq 20$

First, find the intercepts to the equation $4x-5y=20$.

Find the x–intercept by setting $y=0$.

$4x-5(0)=20$

$4x=20$

$x=5$

Find the y–intercept by setting $x=0$.

$4(0)-5y=20$

$-5y=20$

$y=-4$

Next, use the origin, $(0,0)$, as a test point.

$4(0)-5(0) \geq 20$

$0 \geq 20$

This is a false statement. This means that the point, $(0,0)$, will not fall in the shaded half-plane.

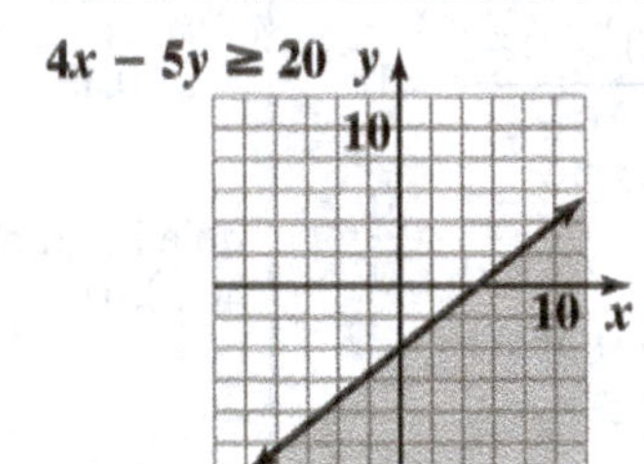

148. $(2x-5)(x^2-3x-6)$

$=2x(x^2-3x-6)-5(x^2-3x-6)$

$=2x^3-6x^2-12x-5x^2+15x+30$

$=2x^3-11x^2+3x+30$

149. $\left(\dfrac{ab^{-3}c^{-4}}{4a^5b^{10}c^{-3}}\right)^{-2}$

$=\left(4^{-1}a^{1-5}b^{-3-10}c^{-4-(-3)}\right)^{-2}$

$=\left(4^{-1}a^{-4}b^{-13}c^{-1}\right)^{-2}$

$=4^{-1(-2)}a^{-4(-2)}b^{-13(-2)}c^{-1(-2)}$

$=4^2a^8b^{26}c^2=16a^8b^{26}c^2$

150. $\frac{7}{10}-\frac{3}{10}=\frac{4}{10}=\frac{2}{5}$

151. $\frac{1}{2}+\frac{2}{3}=\frac{3}{6}+\frac{4}{6}=\frac{7}{6}$

152. $\frac{7}{15}-\frac{3}{10}=\frac{14}{30}-\frac{9}{30}=\frac{5}{30}=\frac{1}{6}$

6.2 Check Points

1.
$$\begin{aligned}\frac{x^2-5x-15}{x^2+5x+6}+\frac{2x+5}{x^2+5x+6}&=\frac{x^2-5x-15+2x+5}{x^2+5x+6}\\&=\frac{x^2-3x-10}{x^2+5x+6}\\&=\frac{(x-5)(x+2)}{(x+3)(x+2)}\\&=\frac{x-5}{x+3}\end{aligned}$$

2.
$$\begin{aligned}\frac{5x-y}{x^2-y^2}-\frac{4x-2y}{x^2-y^2}&=\frac{5x-y-(4x-2y)}{x^2-y^2}\\&=\frac{5x-y-4x+2y}{x^2-y^2}\\&=\frac{x+y}{x^2-y^2}\\&=\frac{x+y}{(x+y)(x-y)}\\&=\frac{1}{x-y}\end{aligned}$$

3. Factor each denominator:

$6x^2=3\cdot 2x^2$

$9x=3\cdot 3x$

List the factors of the first denominator:

$3, 2, x^2$

Add any unlisted factors from the second denominator:

$3, 3, 2, x^2$

The LCD is the product of all the factors in the final list:

$3\cdot 3\cdot 2\cdot x^2=18x^2$

4. Factor each denominator:

$5x^2+15x=5x(x+3)$

$x^2+6x+9=(x+3)(x+3)$

List the factors of the first denominator:

$5, x, (x+3)$

Add any unlisted factors from the second denominator:

$5, x, (x+3), (x+3)$

The LCD is the product of all the factors in the final list:

$5\cdot x\cdot(x+3)\cdot(x+3)=5x(x+3)(x+3)=5x(x+3)^2$

5. The LCD is $18x^2$. Rewrite each rational expression with LCD as the denominator.

$$\begin{aligned}\frac{7}{6x^2}+\frac{2}{9x}&=\frac{7}{6x^2}\cdot\frac{3}{3}+\frac{2}{9x}\cdot\frac{2x}{2x}\\&=\frac{21}{18x^2}+\frac{4x}{18x^2}\\&=\frac{4x+21}{18x^2}\end{aligned}$$

6. The LCD is $(x-4)(x+4)$. Rewrite each rational expression with LCD as the denominator.

$$\begin{aligned}\frac{x}{x-4}+\frac{x-2}{x+4}&=\frac{x(x+4)}{(x-4)(x+4)}+\frac{(x-2)(x-4)}{(x+4)(x-4)}\\&=\frac{x^2+4x}{(x-4)(x+4)}+\frac{x^2-6x+8}{(x-4)(x+4)}\\&=\frac{2x^2-2x+8}{(x-4)(x+4)}\\&=\frac{2(x^2-x+4)}{(x-4)(x+4)}\end{aligned}$$

7. The LCD is $(x-3)(x+1)(x-2)$. Rewrite each rational expression with LCD as the denominator.

$$\begin{aligned}\frac{2x-3}{x^2-5x+6}-\frac{x+4}{x^2-2x-3}&=\frac{2x-3}{(x-3)(x-2)}-\frac{x+4}{(x-3)(x+1)}\\&=\frac{(2x-3)(x+1)}{(x-3)(x-2)(x+1)}-\frac{(x+4)(x-2)}{(x-3)(x+1)(x-2)}\\&=\frac{2x^2-x-3}{(x-3)(x-2)(x+1)}-\frac{x^2+2x-8}{(x-3)(x+1)(x-2)}\\&=\frac{2x^2-x-3-(x^2+2x-8)}{(x-3)(x+1)(x-2)}\\&=\frac{2x^2-x-3-x^2-2x+8}{(x-3)(x+1)(x-2)}\\&=\frac{x^2-3x+5}{(x-3)(x+1)(x-2)}\end{aligned}$$

8. The LCD is $(y+2)(y-2)$. Rewrite each rational expression with LCD as the denominator.

$$\begin{aligned}\frac{y-1}{y-2}+\frac{y-6}{y^2-4}-\frac{y+1}{y+2}&=\frac{y-1}{y-2}+\frac{y-6}{(y-2)(y+2)}-\frac{y+1}{y+2}\\&=\frac{(y-1)(y+2)}{(y-2)(y+2)}+\frac{y-6}{(y+2)(y-2)}-\frac{(y+1)(y-2)}{(y+2)(y-2)}\\&=\frac{y^2+y-2}{(y-2)(y+2)}+\frac{y-6}{(y+2)(y-2)}-\frac{y^2-y-2}{(y+2)(y-2)}\\&=\frac{y^2+y-2+y-6-(y^2-y-2)}{(y+2)(y-2)}\\&=\frac{y^2+y-2+y-6-y^2+y+2}{(y+2)(y-2)}\\&=\frac{3y-6}{(y+2)(y-2)}\\&=\frac{3(y-2)}{(y+2)(y-2)}\\&=\frac{3}{y+2}\end{aligned}$$

9. Since the denominators are opposites, multiply the numerator and denominator of the second rational expression by -1.

$$\begin{aligned}\frac{4x-7y}{x-3y}+\frac{x-2y}{3y-x}&=\frac{4x-7y}{x-3y}+\frac{(-1)}{(-1)}\cdot\frac{x-2y}{3y-x}\\&=\frac{4x-7y}{x-3y}+\frac{2y-x}{x-3y}\\&=\frac{4x-7y+2y-x}{x-3y}\\&=\frac{3x-5y}{x-3y}\end{aligned}$$

6.2 Concept and Vocabulary Check

1. $\frac{P+Q}{R}$; numerators; common denominator

2. $\frac{P-Q}{R}$; numerators; common denominator

3. $\frac{x-5+y}{3}$

4. $\frac{-1}{-1}$

5. factor denominators

6. $x+3$ and $x-2$; $x+3$ and $x+1$; $(x+3)(x-2)(x+1)$

7. $2x$

8. $3y+4$

9. -1

6.2 Exercise Set

1. $\frac{2}{9x}+\frac{4}{9x}=\frac{2+4}{9x}=\frac{6}{9x}=\frac{2\cdot\cancel{3}}{3\cdot\cancel{3}x}=\frac{2}{3x}$

3. $\frac{x}{x-5}+\frac{9x+3}{x-5}=\frac{x+9x+3}{x-5}=\frac{10x+3}{x-5}$

5. $\frac{x^2-2x}{x^2+3x}+\frac{x^2+x}{x^2+3x}=\frac{x^2-2x+x^2+x}{x^2+3x}$
$=\frac{2x^2-x}{x^2+3x}$
$=\frac{\cancel{x}(2x-1)}{\cancel{x}(x+3)}$
$=\frac{2x-1}{x+3}$

7. $\frac{y^2}{y^2-9}+\frac{9-6y}{y^2-9}=\frac{y^2+9-6y}{y^2-9}$
$=\frac{\cancel{(y-3)^2}^{\,y-3}}{(y+3)\cancel{(y-3)}}$
$=\frac{y-3}{y+3}$

9. $\frac{3x}{4x-3}-\frac{2x-1}{4x-3}=\frac{3x-(2x-1)}{4x-3}$
$=\frac{3x-2x+1}{4x-3}$
$=\frac{x+1}{4x-3}$

11. $\frac{x^2-2}{x^2+6x-7}-\frac{19-4x}{x^2+6x-7}$
$=\frac{x^2-2-(19-4x)}{x^2+6x-7}=\frac{x^2-2-19+4x}{x^2+6x-7}$
$=\frac{x^2+4x-21}{x^2+6x-7}=\frac{\cancel{(x+7)}(x-3)}{\cancel{(x+7)}(x-1)}$
$=\frac{x-3}{x-1}$

13. $\frac{20y^2+5y+1}{6y^2+y-2}-\frac{8y^2-12y-5}{6y^2+y-2}$
$=\frac{20y^2+5y+1-(8y^2-12y-5)}{6y^2+y-2}$
$=\frac{20y^2+5y+1-8y^2+12y+5}{6y^2+y-2}$
$=\frac{12y^2+17y+6}{6y^2+y-2}$
$=\frac{(4y+3)\cancel{(3y+2)}}{\cancel{(3y+2)}(2y-1)}$
$=\frac{4y+3}{2y-1}$

15. $\frac{2x^3-3y^3}{x^2-y^2}-\frac{x^3-2y^3}{x^2-y^2}$
$=\frac{2x^3-3y^3-(x^3-2y^3)}{x^2-y^2}$
$=\frac{2x^3-3y^3-x^3+2y^3}{x^2-y^2}$
$=\frac{x^3-y^3}{x^2-y^2}$
$=\frac{\cancel{(x-y)}(x^2+xy+y^2)}{(x+y)\cancel{(x-y)}}$
$=\frac{x^2+xy+y^2}{x+y}$

17. $25x^2=5^2\cdot x^2$
$35x=5\cdot 7x$
$\text{LCD}=5^2\cdot 7x^2=175x^2$

19. $x-5 = x-5$

$x^2-25=(x+5)(x-5)$

$\text{LCD}=(x+5)(x-5)$

21. $y^2-100 = (y+10)(y-10)$

$y(y-10)=y(y-10)$

$\text{LCD}=y(y+10)(y-10)$

23. $x^2-16=(x+4)(x-4)$

$x^2-8x+16=(x-4)^2$

$\text{LCD}=(x+4)(x-4)^2$

25. $y^2-5y-6=(y-6)(y+1)$

$y^2-4y-5=(y-5)(y+1)$

$\text{LCD}=(y-6)(y-5)(y+1)$

27. $2y^2+7y+6=(2y+3)(y+2)$

$y^2-4=(y+2)(y-2)$

$2y^2-3y-2=(y-2)(2y+1)$

$\text{LCD}=(2y+3)(y+2)(y-2)(2y+1)$

29. $\frac{3}{5x^2}+\frac{10}{x}$

The LCD is $5x^2$

$=\frac{3}{5x^2}+\frac{10\cdot 5x}{x\cdot 5x}$

$=\frac{3}{5x^2}+\frac{50x}{5x^2}$

$=\frac{3+50x}{5x^2}$

31. $\frac{4}{x-2}+\frac{3}{x+1}$

The LCD is $(x-2)(x+1)$.

$=\frac{4}{(x-2)}\cdot\frac{(x+1)}{(x+1)}+\frac{3}{(x+1)}\cdot\frac{(x-2)}{(x-2)}$

$=\frac{4(x+1)+3(x-2)}{(x-2)(x+1)}$

$=\frac{4x+4+3x-6}{(x-2)(x+1)}$

$=\frac{7x-2}{(x-2)(x+1)}$

33. $\frac{3x}{x^2+x-2}+\frac{2}{x^2-4x+3}=\frac{3x}{(x+2)(x-1)}+\frac{2}{(x-1)(x-3)}$

The LCD is $(x+2)(x-1)(x-3)$.

$$=\frac{3x(x-3)}{(x+2)(x-1)(x-3)}+\frac{2(x+2)}{(x+2)(x-1)(x-3)}=\frac{3x(x-3)+2(x+2)}{(x+2)(x-1)(x-3)}$$

$$=\frac{3x^2-9x+2x+4}{(x+2)(x-1)(x-3)}=\frac{3x^2-7x+4}{(x+2)(x-1)(x-3)}=\frac{(3x-4)\cancel{(x-1)}}{(x+2)\cancel{(x-1)}(x-3)}=\frac{3x-4}{(x+2)(x-3)}$$

35. $\frac{x-6}{x+5}+\frac{x+5}{x-6}$

Since the denominators have no common factors, the LCD is $(x+5)(x-6)$.

$$=\frac{(x-6)(x-6)}{(x+5)(x-6)}+\frac{(x+5)(x+5)}{(x+5)(x-6)}$$

$$=\frac{(x-6)(x-6)+(x+5)(x+5)}{(x+5)(x-6)}$$

$$=\frac{x^2-12x+36+x^2+10x+25}{(x+5)(x-6)}$$

$$=\frac{2x^2-2x+61}{(x+5)(x-6)}$$

37. $\frac{3x}{x^2-25}-\frac{4}{x+5}$

$$=\frac{3x}{(x+5)(x-5)}-\frac{4}{x+5}$$

The LCD is $(x+5)(x-5)$.

$$=\frac{3x}{(x+5)(x-5)}-\frac{4(x-5)}{(x+5)(x-5)}$$

$$=\frac{3x-4(x-5)}{(x+5)(x-5)}=\frac{3x-4x+20}{(x+5)(x-5)}$$

$$=\frac{-x+20}{(x+5)(x-5)}$$

39. $\frac{3y+7}{y^2-5y+6}-\frac{3}{y-3}$

$=\frac{3y+7}{(y-3)(y-2)}-\frac{3}{y-3}$

The LCD is $(y-3)(y-2)$.

$=\frac{3y+7}{(y-3)(y-2)}-\frac{3(y-2)}{(y-3)(y-2)}$

$=\frac{3y+7-3(y-2)}{(y-3)(y-2)}$

$=\frac{3y+7-3y+6}{(y-3)(y-2)}$

$=\frac{13}{(y-3)(y-2)}$

41. $\frac{x^2-6}{x^2+9x+18}-\frac{x-4}{x+6}$

$=\frac{x^2-6}{(x+3)(x+6)}-\frac{x-4}{x+6}$

The LCD is $(x+3)(x+6)$.

$=\frac{x^2-6}{(x+3)(x+6)}-\frac{(x-4)(x+3)}{(x+3)(x+6)}$

$=\frac{x^2-6-(x-4)(x+3)}{(x+3)(x+6)}$

$=\frac{x^2-6-(x^2-x-12)}{(x+3)(x+6)}$

$=\frac{x^2-6-x^2+x+12}{(x+3)(x+6)}$

$=\frac{x+6}{(x+3)(x+6)}$

$=\frac{1\cancel{(x+6)}}{(x+3)\cancel{(x+6)}}$

$=\frac{1}{x+3}$

43. $\frac{4x+1}{x^2+7x+12}+\frac{2x+3}{x^2+5x+4}=\frac{4x+1}{(x+3)(x+4)}+\frac{2x+3}{(x+4)(x+1)}$

The LCD is $(x+3)(x+4)(x+1)$.

$$=\frac{(4x+1)(x+1)}{(x+3)(x+4)(x+1)}+\frac{(2x+3)(x+3)}{(x+3)(x+4)(x+1)}=\frac{(4x+1)(x+1)+(2x+3)(x+3)}{(x+3)(x+4)(x+1)}$$

$$=\frac{4x^2+5x+1+2x^2+9x+9}{(x+3)(x+4)(x+1)}=\frac{6x^2+14x+10}{(x+3)(x+4)(x+1)}=\frac{2(3x^2+7x+5)}{(x+3)(x+4)(x+1)}$$

45. $\frac{x+4}{x^2-x-2}-\frac{2x+3}{x^2+2x-8}=\frac{x+4}{(x-2)(x+1)}-\frac{2x+3}{(x+4)(x-2)}$

The LCD is $(x-2)(x+1)(x+4)$.

$$=\frac{(x+4)(x+4)}{(x-2)(x+1)(x+4)}-\frac{(2x+3)(x+1)}{(x+4)(x+1)(x-2)}=\frac{(x+4)(x+4)-(2x+3)(x+1)}{(x-2)(x+1)(x+4)}$$

$$=\frac{x^2+8x+16-(2x^2+5x+3)}{(x-2)(x+1)(x+4)}=\frac{x^2+8x+16-2x^2-5x-3}{(x-2)(x+1)(x+4)}=\frac{-x^2+3x+13}{(x-2)(x+1)(x+4)}$$

$$=\frac{-(x^2-3x-13)}{(x-2)(x+1)(x+4)}=-\frac{x^2-3x-13}{(x-2)(x+1)(x+4)}$$

47. $4+\frac{1}{x-3}=\frac{4}{1}+\frac{1}{x-3}$

The LCD is $(x-3)$.

$$\frac{4}{1}+\frac{1}{x-3}=\frac{4(x-3)}{x-3}+\frac{1}{x-3}$$

$$=\frac{4(x-3)+1}{x-3}$$

$$=\frac{4x-12+1}{x-3}$$

$$=\frac{4x-11}{x-3}$$

49. $\frac{y-7}{y^2-16}+\frac{7-y}{16-y^2}$

The LCD is y^2-16.

$$=\frac{y-7}{y^2-16}+\frac{(-1)}{(-1)}\cdot\frac{7-y}{16-y^2}$$

$$=\frac{y-7}{y^2-16}+\frac{-7+y}{-16+y^2}$$

$$=\frac{y-7}{y^2-16}+\frac{y-7}{y^2-16}$$

$$=\frac{y-7+y-7}{y^2-16}=\frac{2y-14}{(y+4)(y-4)}$$

51. $\frac{x+7}{3x+6}+\frac{x}{4-x^2}=\frac{x+7}{3(x+2)}+\frac{x}{(2+x)(2-x)}=\frac{x+7}{3(x+2)}+\frac{x}{(x+2)(2-x)}$

The LCD is $3(x+2)(2-x)$.

$$=\frac{(x+7)(2-x)}{3(x+2)(2-x)}+\frac{3x}{3(x+2)(2-x)}=\frac{(x+7)(2-x)+3x}{3(x+2)(2-x)}=\frac{2x-x^2+14-7x+3x}{3(x+2)(2-x)}$$

$$=\frac{-x^2-2x+14}{3(x+2)(2-x)}=\frac{-1}{-1}\cdot\frac{-x^2-2x+14}{3(x+2)(2-x)}=\frac{x^2+2x-14}{3(x+2)(x-2)}$$

53. $\frac{2x}{x-4}+\frac{64}{x^2-16}-\frac{2x}{x+4}=\frac{2x}{x-4}+\frac{64}{(x+4)(x-4)}-\frac{2x}{x+4}$

The LCD is $(x+4)(x-4)$.

$$=\frac{2x(x+4)}{(x+4)(x-4)}+\frac{64}{(x+4)(x-4)}-\frac{2x(x-4)}{(x+4)(x-4)}=\frac{2x(x+4)+64-2x(x-4)}{(x+4)(x-4)}$$

$$=\frac{2x^2+8x+64-2x^2+8x}{(x+4)(x-4)}=\frac{16x+64}{(x+4)(x-4)}=\frac{16\cancel{(x+4)}}{\cancel{(x+4)}(x-4)}=\frac{16}{x-4}$$

55. $\frac{5x}{x^2-y^2}-\frac{7}{y-x}=\frac{5x}{(x+y)(x-y)}-\frac{(-1)7}{(-1)(y-x)}$

$$=\frac{5x}{(x+y)(x-y)}-\frac{-7}{(-y+x)}=\frac{5x}{(x+y)(x-y)}-\frac{-7}{(x-y)}$$

The LCD is $(x+y)(x-y)$.

$$=\frac{5x}{(x+y)(x-y)}-\frac{-7(x+y)}{(x+y)(x-y)}=\frac{5x-(-7)(x+y)}{(x+y)(x-y)}=\frac{5x+7(x+y)}{(x+y)(x-y)}$$

$$=\frac{5x+7x+7y}{(x+y)(x-y)}=\frac{12x+7y}{(x+y)(x-y)}=\frac{12x+7y}{x^2-y^2}$$

57. $\frac{3}{5x+6}-\frac{4}{x-2}+\frac{x^2-x}{5x^2-4x-12}=\frac{3}{5x+6}-\frac{4}{x-2}+\frac{x^2-x}{(5x+6)(x-2)}$

The LCD is $(5x+6)(x-2)$.

$$=\frac{3(x-2)}{(5x+6)(x-2)}-\frac{4(5x+6)}{(5x+6)(x-2)}+\frac{x^2-x}{(5x+6)(x-2)}=\frac{3(x-2)-4(5x+6)+x^2-x}{(5x+6)(x-2)}$$

$$=\frac{3x-6-20x-24+x^2-x}{(5x+6)(x-2)}=\frac{x^2-18x-30}{(5x+6)(x-2)}$$

59. $\dfrac{3x-y}{x^2-9xy+20y^2}+\dfrac{2y}{x^2-25y^2}=\dfrac{3x-y}{(x-5y)(x-4y)}+\dfrac{2y}{(x+5y)(x-5y)}$

The LCD is $(x+5y)(x-5y)(x-4y)$.

$$=\frac{(3x-y)(x+5y)}{(x+5y)(x-5y)(x-4y)}+\frac{2y(x-4y)}{(x+5y)(x-5y)(x-4y)}=\frac{(3x-y)(x+5y)+2y(x-4y)}{(x+5y)(x-5y)(x-4y)}$$

$$=\frac{3x^2+14xy-5y^2+2xy-8y^2}{(x+5y)(x-5y)(x-4y)}=\frac{3x^2+16xy-13y^2}{(x+5y)(x-5y)(x-4y)}$$

61. $\dfrac{3x}{x^2-4}+\dfrac{5x}{x^2+x-2}-\dfrac{3}{x^2-4x+4}=\dfrac{3x}{(x+2)(x-2)}+\dfrac{5x}{(x+2)(x-1)}-\dfrac{3}{(x-2)^2}$

The LCD is $(x+2)(x-2)^2(x-1)$.

$$=\frac{3x(x-2)(x-1)}{(x+2)(x-2)^2(x-1)}+\frac{5x(x-2)^2}{(x+2)(x-2)^2(x-1)}-\frac{3(x+2)(x-1)}{(x+2)(x-2)^2(x-1)}$$

$$=\frac{3x(x-2)(x-1)+5x(x-2)^2-3(x+2)(x-1)}{(x+2)(x-2)^2(x-1)}$$

$$=\frac{3x\left(x^2-3x+2\right)+5x\left(x^2-4x+4\right)-3\left(x^2+x-2\right)}{(x+2)(x-2)^2(x-1)}$$

$$=\frac{3x^3-9x^2+6x+5x^3-20x^2+20x-3x^2-3x+6}{(x+2)(x-2)^2(x-1)}=\frac{8x^3-32x^2+23x+6}{(x+2)(x-2)^2(x-1)}$$

63. $\dfrac{6a+5b}{6a^2+5ab-4b^2}-\dfrac{a+2b}{9a^2-16b^2}=\dfrac{6a+5b}{(3a+4b)(2a-b)}-\dfrac{a+2b}{(3a+4b)(3a-4b)}$

The LCD is $(3a+4b)(2a-b)(3a-4b)$.

$$=\frac{(6a+5b)(3a-4b)}{(3a+4b)(2a-b)(3a-4b)}-\frac{(a+2b)(2a-b)}{(3a+4b)(2a-b)(3a-4b)}$$

$$=\frac{(6a+5b)(3a-4b)-(a+2b)(2a-b)}{(3a+4b)(2a-b)(3a-4b)}=\frac{18a^2-9ab-20b^2-\left(2a^2+3ab-2b^2\right)}{(3a+4b)(2a-b)(3a-4b)}$$

$$=\frac{18a^2-9ab-20b^2-2a^2-3ab+2b^2}{(3a+4b)(2a-b)(3a-4b)}=\frac{16a^2-12ab-18b^2}{(3a+4b)(2a-b)(3a-4b)}$$

$$=\frac{2\left(8a^2-6ab-9b^2\right)}{(3a+4b)(2a-b)(3a-4b)}=\frac{2(4a+3b)(2a-3b)}{(3a+4b)(2a-b)(3a-4b)}$$

65. $\dfrac{1}{m^2+m-2}-\dfrac{3}{2m^2+3m-2}+\dfrac{2}{2m^2-3m+1}$

$=\dfrac{1}{(m+2)(m-1)}-\dfrac{3}{(2m-1)(m+2)}+\dfrac{2}{(2m-1)(m-1)}$

The LCD is $(m+2)(m-1)(2m-1)$.

$=\dfrac{1(2m-1)}{(m+2)(m-1)(2m-1)}-\dfrac{3(m-1)}{(2m-1)(m+2)}+\dfrac{2(m+2)}{(2m-1)(m-1)}$

$=\dfrac{2m-1-3m+3+2m+4}{(m+2)(m-1)(2m-1)}$

$=\dfrac{m+6}{(m+2)(m-1)(2m-1)}$

67. $\left(\dfrac{2x+3}{x+1}\cdot\dfrac{x^2+4x-5}{2x^2+x-3}\right)-\dfrac{2}{x+2}=\left(\dfrac{\cancel{(2x+3)}}{x+1}\cdot\dfrac{(x+5)\cancel{(x-1)}}{\cancel{(2x+3)}\cancel{(x-1)}}\right)-\dfrac{2}{x+2}=\dfrac{x+5}{x+1}-\dfrac{2}{x+2}$

$=\dfrac{(x+5)(x+2)}{(x+1)(x+2)}-\dfrac{2(x+1)}{(x+1)(x+2)}=\dfrac{(x+5)(x+2)-2(x+1)}{(x+1)(x+2)}=\dfrac{x^2+2x+5x+10-2x-2}{(x+1)(x+2)}=\dfrac{x^2+5x+8}{(x+1)(x+2)}$

69. $\left(2-\dfrac{6}{x+1}\right)\left(1+\dfrac{3}{x-2}\right)=\left(\dfrac{2(x+1)}{(x+1)}-\dfrac{6}{(x+1)}\right)\left(\dfrac{(x-2)}{(x-2)}+\dfrac{3}{(x-2)}\right)$

$=\left(\dfrac{2x+2-6}{x+1}\right)\left(\dfrac{x-2+3}{x-2}\right)=\left(\dfrac{2x-4}{x+1}\right)\left(\dfrac{x+1}{x-2}\right)=\dfrac{2\cancel{(x-2)}\cancel{(x+1)}}{\cancel{(x+1)}\cancel{(x-2)}}=2$

71. $\left(\dfrac{1}{x+h}-\dfrac{1}{x}\right)\div h=\left(\dfrac{x}{x(x+h)}-\dfrac{(x+h)}{x(x+h)}\right)\div h=\left(\dfrac{x-x-h}{x(x+h)}\right)\div\dfrac{h}{1}=\dfrac{-h}{x(x+h)}\cdot\dfrac{1}{h}=-\dfrac{1}{x(x+h)}$

73. $\left(\dfrac{1}{a^3-b^3}\cdot\dfrac{ac+ad-bc-bd}{1}\right)-\dfrac{c-d}{a^2+ab+b^2}=\left(\dfrac{1}{(a-b)(a^2+ab+b^2)}\cdot\dfrac{a(c+d)-b(c+d)}{1}\right)-\dfrac{c-d}{a^2+ab+b^2}$

$=\left(\dfrac{1}{\cancel{(a-b)}(a^2+ab+b^2)}\cdot\dfrac{(c+d)\cancel{(a-b)}}{1}\right)-\dfrac{c-d}{a^2+ab+b^2}=\dfrac{c+d}{a^2+ab+b^2}-\dfrac{c-d}{a^2+bd+b^2}$

$=\dfrac{c+d-c+d}{a^2+ab+b^2}=\dfrac{2d}{a^2+ab+b^2}$

75. $f(x)=\dfrac{2x-3}{x+5}$ and $g(x)=\dfrac{x^2-4x-19}{x^2+8x+15}$

$$(f-g)(x)=f(x)-g(x)=\frac{2x-3}{x+5}-\frac{x^2-4x-19}{x^2+8x+15}=\frac{2x-3}{x+5}-\frac{x^2-4x-19}{(x+5)(x+3)}$$
$$=\frac{(2x-3)(x+3)}{(x+5)(x+3)}-\frac{x^2-4x-19}{(x+5)(x+3)}=\frac{2x^2+6x-3x-9-x^2+4x+19}{(x+5)(x+3)}$$
$$=\frac{x^2+7x+10}{(x+5)(x+3)}=\frac{\cancel{(x+5)}(x+2)}{\cancel{(x+5)}(x+3)}=\frac{x+2}{x+3}$$

To find the domain of $(f-g)(x)$ we need to find the intersection of the domains for the individual functions. The domain of f is $(-\infty,-5)\cup(-5,\infty)$. The domain of g is $(-\infty,-5)\cup(-5,-3)\cup(-3,\infty)$. Therefore, the domain of $(f-g)(x)$ is $(-\infty,-5)\cup(-5,-3)\cup(-3,\infty)$.

77. $$T(0)=\frac{470}{0+70}+\frac{250}{0+65}$$
$$=\frac{470}{70}+\frac{250}{65}$$
$$=6.7+3.8=10.5\approx 11$$

This corresponds to the point $(0,11)$ on the graph. If you drive zero miles per hour over the speed limit, total driving time is approximately 11 hours.

79. $$T(x)=\frac{470}{x+70}+\frac{250}{x+65}$$
$$=\frac{470(x+65)}{(x+70)(x+65)}+\frac{250(x+70)}{(x+70)(x+65)}$$
$$=\frac{470(x+65)+250(x+70)}{(x+70)(x+65)}$$
$$=\frac{470x+30,550+250x+17,500}{(x+70)(x+65)}$$
$$=\frac{720x+48,050}{(x+70)(x+65)}$$
$$T(0)=\frac{\cancel{720(0)}+48050}{(0+70)(0+65)}=\frac{48,050}{(70)(65)}$$
$$\approx 11$$

81. Answers will vary. In order to make the trip in 9 hours, you need to drive approximately 12 miles per hour over the speed limit.

83. **a.** $$f(x)=\frac{27,725(x-14)}{x^2+9}-5x$$
$$f(20)=\frac{27,725(20-14)}{(20)^2+9}-5(20)$$
$$\approx 307$$

This is represented by the point $(20,307)$.

b. $f(x)=\dfrac{27,725(x-14)}{x^2+9}-5x$

$=\dfrac{27,725(x-14)}{x^2+9}-\dfrac{5x(x^2+9)}{x^2+9}$

$=\dfrac{27,725x-388,150}{x^2+9}-\dfrac{5x^3+45x}{x^2+9}$

$=\dfrac{27,725x-388,150-(5x^3+45x)}{x^2+9}$

$=\dfrac{27,725x-388,150-5x^3-45x}{x^2+9}$

$=\dfrac{-5x^3+27,680x-388,150}{x^2+9}$

c. The greatest number occurs at about 25 years.

$f(x)=\dfrac{-5x^3+27,680x-388,150}{x^2+9}$

$f(25)=\dfrac{-5(25)^3+27,680(25)-388,150}{(25)^2+9}$

≈ 356

The number of arrests, per 100,000 drivers, is about 356 at age 25.

85. $P=2\left(\dfrac{x}{x+7}\right)+2\left(\dfrac{x}{x+8}\right)=\dfrac{2x}{x+7}+\dfrac{2x}{x+8}=\dfrac{2x(x+8)}{(x+7)(x+8)}+\dfrac{2x(x+7)}{(x+7)(x+8)}$

$=\dfrac{2x(x+8)+2x(x+7)}{(x+7)(x+8)}=\dfrac{2x^2+16x+2x^2+14x}{(x+7)(x+8)}=\dfrac{4x^2+30x}{(x+7)(x+8)}=\dfrac{2x(2x+15)}{(x+7)(x+8)}$

87. – 89. Answers will vary.

91. Answers will vary; $\dfrac{b+a}{ab}$

93. makes sense

95. does not make sense; Explanations will vary. Sample explanation: Since the denominators are opposites, the fastest way to add is to multiply the numerator and denominator of the second rational expression by -1.

97. false; Changes to make the statement true will vary. A sample change is: $\dfrac{2}{x+3}+\dfrac{3}{x+4}=\dfrac{2(x+4)}{(x+3)(x+4)}+\dfrac{3(x+3)}{(x+4)(x+3)}$

$=\dfrac{2x+8}{(x+3)(x+4)}+\dfrac{3x+9}{(x+4)(x+3)}$

$=\dfrac{5x+17}{(x+3)(x+4)}$

99. false; Changes to make the statement true will vary. A sample change is: $6+\dfrac{1}{x}=\dfrac{6x}{x}+\dfrac{1}{x}=\dfrac{6x+1}{x}$

101. $\frac{1}{x^n-1}-\frac{1}{x^n+1}-\frac{1}{x^{2n}-1}$

$=\frac{x^n+1}{x^{2n}-1}-\frac{x^n-1}{x^{2n}-1}-\frac{1}{x^{2n}-1}$

$=\frac{x^n+1-x^n+1-1}{x^{2n}-1}$

$=\frac{1}{x^{2n}-1}$

103. $(x-y)^{-1}+(x-y)^{-2}$

$=\frac{1}{(x-y)}+\frac{1}{(x-y)^2}$

$=\frac{(x-y)}{(x-y)(x-y)}+\frac{1}{(x-y)^2}$

$=\frac{x-y+1}{(x-y)^2}$

104. $\left(\frac{3x^2y^{-2}}{y^3}\right)^{-2}=\left(\frac{3x^2}{y^2y^3}\right)^{-2}=\left(\frac{3x^2}{y^5}\right)^{-2}=\left(\frac{y^5}{3x^2}\right)^2=\frac{y^{10}}{9x^4}$

105.

$|3x-1|\le 14$

$-14\le 3x-1\le 14$

$-14+1\le 3x-1+1\le 14+1$

$-13\le 3x\le 15$

$-\frac{13}{3}\le x\le 5$

The solution set is $\left[-\frac{13}{3},5\right]$.

106. $50x^3-18x=2x(25x^2-9)=2x(5x+3)(5x-3)$

107. $x^2y^2\left(\frac{1}{x}+\frac{y}{x^2}\right)=\frac{x^2y^2}{x}+\frac{y(x^2y^2)}{x^2}$

$=xy^2+y^3$

108. $x(x+h)\left(\frac{1}{x+h}-\frac{1}{x}\right)=\frac{x(x+h)}{x+h}-\frac{x(x+h)}{x}$

$=x-(x+h)$

$=x-x-h$

$=-h$

109. $\frac{x^2-1}{x^2} \div \frac{x^2-4x+3}{x^2} = \frac{x^2-1}{x^2} \cdot \frac{x^2}{x^2-4x+3}$

$= \frac{x^2-1}{x^2-4x+3}$

$= \frac{(x+1)(x-1)}{(x-3)(x-1)}$

$= \frac{x+1}{x-3}$

6.3 Check Points

1. The LCD is y^2.

$$\frac{\frac{x}{y}-1}{\frac{x^2}{y^2}-1} = \frac{y^2}{y^2} \cdot \frac{\frac{x}{y}-1}{\frac{x^2}{y^2}-1}$$

$$= \frac{\frac{x \cdot y^2}{y} - 1 \cdot y^2}{\frac{x^2 \cdot y^2}{y^2} - 1 \cdot y^2}$$

$$= \frac{xy - y^2}{x^2 - y^2}$$

$$= \frac{y(x-y)}{(x+y)(x-y)}$$

$$= \frac{y}{x+y}$$

2. The LCD is $x(x+7)$.

$$\frac{\frac{1}{x+7}-\frac{1}{x}}{7} = \frac{x(x+7)}{x(x+7)} \cdot \frac{\frac{1}{x+7}-\frac{1}{x}}{7}$$

$$= \frac{\frac{x(x+7)}{x+7} - \frac{x(x+7)}{x}}{7x(x+7)}$$

$$= \frac{x-(x+7)}{7x(x+7)}$$

$$= \frac{x-x-7}{7x(x+7)}$$

$$= \frac{-7}{7x(x+7)}$$

$$= \frac{-1}{x(x+7)}$$

$$= -\frac{1}{x(x+7)}$$

3. The LCD of the numerator is $(x+1)(x-1)$.
The LCD of the denominator is $(x+1)(x-1)$.

$$\frac{\frac{x+1}{x-1}-\frac{x-1}{x+1}}{\frac{x-1}{x+1}+\frac{x+1}{x-1}} = \frac{\frac{(x+1)(x+1)}{(x-1)(x+1)}-\frac{(x-1)(x-1)}{(x+1)(x-1)}}{\frac{(x-1)(x-1)}{(x+1)(x-1)}+\frac{(x+1)(x+1)}{(x-1)(x+1)}}$$

$$= \frac{\frac{x^2+2x+1}{(x-1)(x+1)}-\frac{x^2-2x+1}{(x+1)(x-1)}}{\frac{x^2-2x+1}{(x+1)(x-1)}+\frac{x^2+2x+1}{(x-1)(x+1)}}$$

$$= \frac{\frac{x^2+2x+1-(x^2-2x+1)}{(x+1)(x-1)}}{\frac{x^2-2x+1+x^2+2x+1}{(x+1)(x-1)}}$$

$$= \frac{\frac{x^2+2x+1-x^2+2x-1}{(x+1)(x-1)}}{\frac{x^2-2x+1+x^2+2x+1}{(x+1)(x-1)}}$$

$$= \frac{\frac{4x}{(x+1)(x-1)}}{\frac{2x^2+2}{(x+1)(x-1)}}$$

$$= \frac{4x}{(x+1)(x-1)} \cdot \frac{(x+1)(x-1)}{2x^2+2}$$

$$= \frac{4x}{2(x^2+1)}$$

$$= \frac{2x}{x^2+1}$$

4. Rewrite the expression without negative exponents. Then multiply the numerator and denominator by the LCD of x^2.

$$\frac{1-4x^{-2}}{1-7x^{-1}+10x^{-2}}=\frac{1-\frac{4}{x^2}}{1-\frac{7}{x}+\frac{10}{x^2}}$$

$$=\frac{x^2}{x^2}\cdot\frac{1-\frac{4}{x^2}}{1-\frac{7}{x}+\frac{10}{x^2}}$$

$$=\frac{1\cdot x^2-\frac{4\cdot x^2}{x^2}}{1\cdot x^2-\frac{7\cdot x^2}{x}+\frac{10\cdot x^2}{x^2}}$$

$$=\frac{x^2-4}{x^2-7x+10}$$

$$=\frac{(x+2)(x-2)}{(x-5)(x-2)}$$

$$=\frac{x+2}{x-5}$$

6.3 Concept and Vocabulary Check

1. complex; complex

2. $\frac{7x+5}{5x+x^2}$

3. $\frac{1}{x+3}$; $\frac{1}{x}$; x; $x+3$; -3; $\frac{1}{x(x+3)}$

6.3 Exercise Set

1. $$\frac{4+\frac{2}{x}}{1-\frac{3}{x}}=\frac{x}{x}\cdot\frac{4+\frac{2}{x}}{1-\frac{3}{x}}=\frac{x\cdot4+\cancel{x}\cdot\frac{2}{\cancel{x}}}{x\cdot1-\cancel{x}\cdot\frac{3}{\cancel{x}}}=\frac{4x+2}{x-3}$$

3. $$\frac{\frac{3}{x}+\frac{x}{3}}{\frac{x}{3}-\frac{3}{x}}=\frac{3x}{3x}\cdot\frac{\frac{3}{x}+\frac{x}{3}}{\frac{x}{3}-\frac{3}{x}}=\frac{3\cancel{x}\cdot\frac{3}{\cancel{x}}+\cancel{3}x\cdot\frac{x}{\cancel{3}}}{\cancel{3}x\cdot\frac{x}{\cancel{3}}-3\cancel{x}\cdot\frac{3}{\cancel{x}}}=\frac{3\cdot3+x\cdot x}{x\cdot x-3\cdot3}=\frac{9+x^2}{x^2-9}=\frac{x^2+9}{(x+3)(x-3)}$$

5. $$\frac{\frac{1}{x}+\frac{1}{y}}{\frac{1}{x}-\frac{1}{y}}=\frac{xy}{xy}\cdot\frac{\frac{1}{x}+\frac{1}{y}}{\frac{1}{x}-\frac{1}{y}}=\frac{\cancel{x}y\cdot\frac{1}{\cancel{x}}+x\cancel{y}\cdot\frac{1}{\cancel{y}}}{\cancel{x}y\cdot\frac{1}{\cancel{x}}-x\cancel{y}\cdot\frac{1}{\cancel{y}}}=\frac{y+x}{y-x}$$

7. $$\frac{8x^{-2}-2x^{-1}}{10x^{-1}-6x^{-2}}=\frac{\frac{8}{x^2}-\frac{2}{x}}{\frac{10}{x}-\frac{6}{x^2}}=\frac{x^2}{x^2}\cdot\frac{\frac{8}{x^2}-\frac{2}{x}}{\frac{10}{x}-\frac{6}{x^2}}=\frac{\cancel{x^2}\cdot\frac{8}{\cancel{x^2}}-x^{\cancel{2}}\cdot\frac{2}{\cancel{x}}}{x^{\cancel{2}}\cdot\frac{10}{\cancel{x}}-\cancel{x^2}\cdot\frac{6}{\cancel{x^2}}}=\frac{8-x\cdot 2}{x\cdot 10-6}$$

$$=\frac{8-2x}{10x-6}=\frac{\cancel{2}(4-x)}{\cancel{2}(5x-3)}=\frac{4-x}{5x-3}$$

9. $$\frac{\frac{1}{x-2}}{1-\frac{1}{x-2}}=\frac{x-2}{x-2}\cdot\frac{\frac{1}{x-2}}{1-\frac{1}{x-2}}=\frac{\cancel{(x-2)}\cdot\frac{1}{\cancel{x-2}}}{(x-2)\cdot 1-\cancel{(x-2)}\cdot\frac{1}{\cancel{x-2}}}=\frac{1}{x-2-1}=\frac{1}{x-3}$$

11. $$\frac{\frac{1}{x+5}-\frac{1}{x}}{5}=\frac{x(x+5)}{x(x+5)}\cdot\frac{\frac{1}{x+5}-\frac{1}{x}}{5}=\frac{x\cancel{(x+5)}\cdot\frac{1}{\cancel{x+5}}-\cancel{x}(x+5)\cdot\frac{1}{\cancel{x}}}{5x(x+5)}=\frac{x-(x+5)}{5x(x+5)}$$

$$=\frac{\cancel{x-x}-5}{5x(x+5)}=-\frac{\cancel{5}}{\cancel{5}x(x+5)}=-\frac{1}{x(x+5)}$$

13. $$\frac{\frac{4}{x+4}}{\frac{1}{x+4}-\frac{1}{x}}=\frac{x(x+4)}{x(x+4)}\cdot\frac{\frac{4}{x+4}}{\frac{1}{x+4}-\frac{1}{x}}=\frac{x\cancel{(x+4)}\cdot\frac{4}{\cancel{x+4}}}{x\cancel{(x+4)}\cdot\frac{1}{\cancel{x+4}}-\cancel{x}(x+4)\cdot\frac{1}{\cancel{x}}}$$

$$=\frac{x\cdot 4}{x-(x+4)}=\frac{4x}{\cancel{x-x}-4}=-\frac{\cancel{4}x}{\cancel{4}}=-x$$

15. $$\frac{\frac{1}{x-1}+1}{\frac{1}{x+1}-1}=\frac{(x+1)(x-1)}{(x+1)(x-1)}\cdot\frac{\frac{1}{x-1}+1}{\frac{1}{x+1}-1}=\frac{(x+1)\cancel{(x-1)}\cdot\frac{1}{\cancel{x-1}}+(x+1)(x-1)\cdot 1}{\cancel{(x+1)}(x-1)\cdot\frac{1}{\cancel{x+1}}-(x+1)(x-1)\cdot 1}=\frac{(x+1)+\left(x^2-1\right)}{(x-1)-\left(x^2-1\right)}$$

$$=\frac{x+1+x^2-1}{x-1-x^2+1}=\frac{x^2+x}{x-x^2}=\frac{\cancel{x}(x+1)}{\cancel{x}(1-x)}=\frac{x+1}{-1(x-1)}=-\frac{x+1}{x-1}$$

17. $$\frac{x^{-1}+y^{-1}}{(x+y)^{-1}}=\frac{\frac{1}{x}+\frac{1}{y}}{\frac{1}{(x+y)}}=\frac{xy(x+y)}{xy(x+y)}\cdot\frac{\frac{1}{x}+\frac{1}{y}}{\frac{1}{(x+y)}}=\frac{\cancel{x}y(x+y)\cdot\frac{1}{\cancel{x}}+x\cancel{y}(x+y)\cdot\frac{1}{\cancel{y}}}{xy\cancel{(x+y)}\cdot\frac{1}{\cancel{(x+y)}}}$$

$$=\frac{y(x+y)+x(x+y)}{xy}=\frac{(x+y)(y+x)}{xy}=\frac{(x+y)(x+y)}{xy}$$

19. $$\frac{\frac{x+2}{x-2}-\frac{x-2}{x+2}}{\frac{x-2}{x+2}+\frac{x+2}{x-2}}=\frac{(x+2)(x-2)}{(x+2)(x-2)}\cdot\frac{\frac{x+2}{x-2}-\frac{x-2}{x+2}}{\frac{x-2}{x+2}+\frac{x+2}{x-2}}$$

$$=\frac{(x+2)\cancel{(x-2)}\cdot\frac{x+2}{\cancel{x-2}}-\cancel{(x+2)}(x-2)\cdot\frac{x-2}{\cancel{x+2}}}{\cancel{(x+2)}(x-2)\cdot\frac{x-2}{\cancel{x+2}}+(x+2)\cancel{(x-2)}\cdot\frac{x+2}{\cancel{x-2}}}$$

$$=\frac{(x+2)(x+2)-(x-2)(x-2)}{(x-2)(x-2)+(x+2)(x+2)}=\frac{x^2+4x+4-\left(x^2-4x+4\right)}{x^2-\cancel{4x}+4+x^2+\cancel{4x}+4}$$

$$=\frac{\cancel{x^2}+4x+\cancel{4}-\cancel{x^2}+4x-\cancel{4}}{2x^2+8}=\frac{\overset{4}{\cancel{8}}\,x}{\cancel{2}\left(x^2+4\right)}=\frac{4x}{x^2+4}$$

21. $$\frac{\frac{2}{x^3y}+\frac{5}{xy^4}}{\frac{5}{x^3y}-\frac{3}{xy}}=\frac{x^3y^4}{x^3y^4}\cdot\frac{\frac{2}{x^3y}+\frac{5}{xy^4}}{\frac{5}{x^3y}-\frac{3}{xy}}=\frac{\cancel{x^3}\,\overset{y^3}{\cancel{y^4}}\cdot\frac{2}{\cancel{x^3}\,\cancel{y}}+\overset{x^2}{\cancel{x^3}}\,\cancel{y^4}\cdot\frac{5}{\cancel{x}\,\cancel{y^4}}}{\cancel{x^3}\,\underset{y^3}{\cancel{y^4}}\cdot\frac{5}{\cancel{x^3}\,\cancel{y}}-\underset{x^2}{\cancel{x^3}}\,\underset{y^3}{\cancel{y^4}}\cdot\frac{3}{\cancel{x}\,\cancel{y}}}$$

$$=\frac{y^3\cdot2+x^2\cdot5}{y^3\cdot5-x^2y^3\cdot3}=\frac{2y^3+5x^2}{5y^3-3x^2y^3}=\frac{2y^3+5x^2}{y^3\left(5-3x^2\right)}$$

23. $$\frac{\frac{3}{x+2}-\frac{3}{x-2}}{\frac{5}{x^2-4}}=\frac{\frac{3}{x+2}-\frac{3}{x-2}}{\frac{5}{(x+2)(x-2)}}=\frac{(x+2)(x-2)}{(x+2)(x-2)}\cdot\frac{\frac{3}{x+2}-\frac{3}{x-2}}{\frac{5}{(x+2)(x-2)}}$$

$$=\frac{\cancel{(x+2)}(x-2)\cdot\frac{3}{\cancel{x+2}}-(x+2)\cancel{(x-2)}\cdot\frac{3}{\cancel{x-2}}}{\cancel{(x+2)(x-2)}\cdot\frac{5}{\cancel{(x+2)(x-2)}}}$$

$$=\frac{(x-2)\cdot3-(x+2)\cdot3}{5}=\frac{3x-6-(3x+6)}{5}=\frac{3x-6-3x-6}{5}=-\frac{12}{5}$$

25. $$\frac{3a^{-1}+3b^{-1}}{4a^{-2}-9b^{-2}}=\frac{\frac{3}{a}+\frac{3}{b}}{\frac{4}{a^2}-\frac{9}{b^2}}=\frac{a^2b^2}{a^2b^2}\cdot\frac{\frac{3}{a}+\frac{3}{b}}{\frac{4}{a^2}-\frac{9}{b^2}}=\frac{a^{\cancel{2}}b^2\cdot\frac{3}{\cancel{a}}+a^2b^{\cancel{2}}\cdot\frac{3}{\cancel{b}}}{\cancel{a^2}b^2\cdot\frac{4}{\cancel{a^2}}-a^2\cancel{b^2}\cdot\frac{9}{\cancel{b^2}}}$$

$$=\frac{ab^2\cdot3+a^2b\cdot3}{b^2\cdot4-a^2\cdot9}=\frac{3ab^2+3a^2b}{4b^2-9a^2}=\frac{3ab(b+a)}{(2b+3a)(2b-3a)}$$

27. $$\frac{\frac{4x}{x^2-4}-\frac{5}{x-2}}{\frac{2}{x-2}+\frac{3}{x+2}}=\frac{\frac{4x}{(x+2)(x-2)}-\frac{5}{x-2}}{\frac{2}{x-2}+\frac{3}{x+2}}=\frac{(x+2)(x-2)}{(x+2)(x-2)}\cdot\frac{\frac{4x}{(x+2)(x-2)}-\frac{5}{x-2}}{\frac{2}{x-2}+\frac{3}{x+2}}$$

$$=\frac{(x+2)(x-2)\cdot\frac{4x}{(x+2)(x-2)}-(x+2)(x-2)\cdot\frac{5}{x-2}}{(x+2)(x-2)\cdot\frac{2}{x-2}+(x+2)(x-2)\cdot\frac{3}{x+2}}$$

$$=\frac{4x-(x+2)\cdot 5}{(x+2)\cdot 2+(x-2)\cdot 3}=\frac{4x-5x-10}{2x+4+3x-6}=\frac{-x-10}{5x-2}=-\frac{x+10}{5x-2}$$

29. $$\frac{\frac{2y}{y^2+4y+3}}{\frac{1}{y+3}+\frac{2}{y+1}}=\frac{\frac{2y}{(y+3)(y+1)}}{\frac{1}{y+3}+\frac{2}{y+1}}$$

$$=\frac{(y+3)(y+1)}{(y+3)(y+1)}\cdot\frac{\frac{2y}{(y+3)(y+1)}}{\frac{1}{y+3}+\frac{2}{y+1}}$$

$$=\frac{(y+3)(y+1)\cdot\frac{2y}{(y+3)(y+1)}}{(y+3)(y+1)\cdot\frac{1}{y+3}+(y+3)(y+1)\cdot\frac{2}{y+1}}$$

$$=\frac{2y}{(y+1)+(y+3)\cdot 2}=\frac{2y}{y+1+2y+6}=\frac{2y}{3y+7}$$

31. $$\frac{\frac{2}{a^2}-\frac{1}{ab}-\frac{1}{b^2}}{\frac{1}{a^2}-\frac{3}{ab}+\frac{2}{b^2}}=\frac{a^2b^2}{a^2b^2}\cdot\frac{\frac{2}{a^2}-\frac{1}{ab}-\frac{1}{b^2}}{\frac{1}{a^2}-\frac{3}{ab}+\frac{2}{b^2}}=\frac{a^2b^2\cdot\frac{2}{a^2}-a^2b^2\cdot\frac{1}{ab}-a^2b^2\cdot\frac{1}{b^2}}{a^2b^2\cdot\frac{1}{a^2}-a^2b^2\cdot\frac{3}{ab}+a^2b^2\cdot\frac{2}{b^2}}$$

$$=\frac{b^2\cdot 2-ab-a^2}{b^2-ab\cdot 3+a^2\cdot 2}=\frac{2b^2-ab-a^2}{b^2-3ab+2a^2}=\frac{(2b+a)(b-a)}{(b-2a)(b-a)}=\frac{2b+a}{b-2a}$$

33. $$\frac{\frac{2x}{x^2-25}+\frac{1}{3x-15}}{\frac{5}{x-5}+\frac{3}{4x-20}}=\frac{\frac{2x}{(x+5)(x-5)}+\frac{1}{3(x-5)}}{\frac{5}{x-5}+\frac{3}{4(x-5)}}=\frac{12(x+5)(x-5)}{12(x+5)(x-5)}\cdot\frac{\frac{2x}{(x+5)(x-5)}+\frac{1}{3(x-5)}}{\frac{5}{x-5}+\frac{3}{4(x-5)}}$$

$$=\frac{12\cancel{(x+5)}\cancel{(x-5)}\cdot\frac{2x}{\cancel{(x+5)}\cancel{(x-5)}}+\overset{4}{\cancel{12}}(x+5)\cancel{(x-5)}\cdot\frac{1}{\cancel{3}\cancel{(x-5)}}}{12(x+5)\cancel{(x-5)}\cdot\frac{5}{\cancel{x-5}}+\overset{3}{\cancel{12}}(x+5)\cancel{(x-5)}\cdot\frac{3}{\cancel{4}\cancel{(x-5)}}}$$

$$=\frac{12\cdot 2x+4(x+5)}{12(x+5)\cdot 5+3(x+5)\cdot 3}=\frac{24x+4x+20}{60(x+5)+9(x+5)}$$

$$=\frac{28x+20}{(x+5)(60+9)}=\frac{4(7x+5)}{69(x+5)}$$

35. $$\frac{\frac{3}{x+2y}-\frac{2y}{x^2+2xy}}{\frac{3y}{x^2+2xy}+\frac{5}{x}}=\frac{\frac{3}{x+2y}-\frac{2y}{x(x+2y)}}{\frac{3y}{x(x+2y)}+\frac{5}{x}}=\frac{x(x+2y)}{x(x+2y)}\cdot\frac{\frac{3}{x+2y}-\frac{2y}{x(x+2y)}}{\frac{3y}{x(x+2y)}+\frac{5}{x}}$$

$$=\frac{x\cancel{(x+2y)}\cdot\frac{3}{\cancel{x+2y}}-\cancel{x(x+2y)}\cdot\frac{2y}{\cancel{x(x+2y)}}}{\cancel{x(x+2y)}\cdot\frac{3y}{\cancel{x(x+2y)}}+\cancel{x}(x+2y)\cdot\frac{5}{\cancel{x}}}$$

$$=\frac{x\cdot 3-2y}{3y+(x+2y)\cdot 5}=\frac{3x-2y}{3y+5x+10y}=\frac{3x-2y}{5x+13y}$$

37. $$\frac{\frac{2}{m^2-3m+2}+\frac{2}{m^2-m-2}}{\frac{2}{m^2-1}+\frac{2}{m^2+4m+3}}=\frac{\frac{2}{(m-2)(m-1)}+\frac{2}{(m-2)(m+1)}}{\frac{2}{(m-1)(m+1)}+\frac{2}{(m+3)(m+1)}}$$

$$=\frac{\frac{2(m+1)}{(m-2)(m-1)(m+1)}+\frac{2(m-1)}{(m-2)(m-1)(m+1)}}{\frac{2(m+3)}{(m+3)(m-1)(m+1)}+\frac{2(m-1)}{(m+3)(m-1)(m+1)}}$$

$$=\frac{\frac{2m+2+2m-2}{(m-2)(m-1)(m+1)}}{\frac{2m+6+2m-2}{(m+3)(m-1)(m+1)}}=\frac{\frac{4m}{(m-2)(m-1)(m+1)}}{\frac{4m+4}{(m+3)(m-1)(m+1)}}$$

$$=\frac{\cancel{4}m}{(m-2)\cancel{(m-1)}\cancel{(m+1)}}\cdot\frac{(m+3)\cancel{(m-1)}\cancel{(m+1)}}{\cancel{4}(m+1)}$$

$$=\frac{m(m+3)}{(m-2)(m+1)}$$

39. $$\frac{\frac{2}{a^2+2a-8}+\frac{1}{a^2+5a+4}}{\frac{1}{a^2-5a+6}+\frac{2}{a^2-a-2}}=\frac{\frac{2}{(a+4)(a-2)}+\frac{1}{(a+4)(a+1)}}{\frac{1}{(a-3)(a-2)}+\frac{2}{(a+1)(a-2)}}$$

$$=\frac{\frac{2(a+1)}{(a+4)(a+1)(a-2)}+\frac{1(a-2)}{(a+4)(a+1)(a-2)}}{\frac{1(a+1)}{(a+1)(a-3)(a-2)}+\frac{2(a-3)}{(a+1)(a-3)(a-2)}}$$

$$=\frac{\frac{2a+2+a-2}{(a+4)(a+1)(a-2)}}{\frac{a+1+2a-6}{(a+1)(a-3)(a-2)}}=\frac{3a}{(a+4)\cancel{(a+1)}\cancel{(a-2)}}\cdot\frac{\cancel{(a+1)}(a-3)\cancel{(a-2)}}{(3a-5)}$$

$$=\frac{3a(a-3)}{(a+4)(3a-5)}$$

41. $$\frac{\frac{x-1}{x^2-4}}{1+\frac{1}{x-2}}-\frac{1}{x-2}=\frac{\frac{x-1}{(x-2)(x+2)}}{\frac{x-2}{x-2}+\frac{1}{x-2}}-\frac{1}{x-2}=\frac{\frac{x-1}{(x-2)(x+2)}}{\frac{x-1}{x-2}}-\frac{1}{x-2}$$

$$=\frac{\cancel{(x-1)}}{\cancel{(x-2)}(x+2)}\cdot\frac{\cancel{(x-2)}}{\cancel{(x-1)}}-\frac{1}{x-2}$$

$$=\frac{1}{x+2}-\frac{1}{x-2}=\frac{x-2}{(x+2)(x-2)}-\frac{x+2}{(x+2)(x-2)}$$

$$=\frac{x-2-x-2}{(x+2)(x-2)}=\frac{-4}{(x+2)(x-2)}$$

43. $$\frac{3}{1-\frac{3}{3+x}}-\frac{3}{\frac{3}{3-x}-1}=\frac{3}{\frac{3+x}{3+x}-\frac{3}{3+x}}-\frac{3}{\frac{3}{3-x}-\frac{3-x}{3-x}}$$

$$=\frac{3}{\frac{3+x-3}{3+x}}-\frac{3}{\frac{3-3+x}{3-x}}=\frac{3}{\frac{x}{3+x}}-\frac{3}{\frac{x}{3-x}}$$

$$=3\left(\frac{3+x}{x}\right)-3\left(\frac{3-x}{x}\right)=\frac{9+3x}{x}-\frac{9-3x}{x}$$

$$=\frac{9+3x-9+3x}{x}=\frac{6x}{x}=6$$

45. $$\frac{x}{1-\frac{1}{1+\frac{1}{x}}}=\frac{x}{1-\frac{x}{x}\cdot\frac{1}{1+\frac{1}{x}}}=\frac{x}{1-\frac{x\cdot 1}{x\cdot 1+\cancel{x}\cdot\frac{1}{\cancel{x}}}}=\frac{x}{1-\frac{x}{x+1}}=\frac{x+1}{x+1}\cdot\frac{x}{1-\frac{x}{x+1}}$$

$$=\frac{(x+1)x}{(x+1)1-(x+1)\frac{x}{x+1}}=\frac{x(x+1)}{(x+1)-\cancel{(x+1)}\frac{x}{\cancel{x+1}}}=\frac{x(x+1)}{x+1-x}=\frac{x(x+1)}{1}=x(x+1)$$

47. $f(x)=\dfrac{1+x}{1-x}$

$$f\left(\frac{1}{x+3}\right)=\frac{1+\frac{1}{x+3}}{1-\frac{1}{x+3}}=\frac{x+3}{x+3}\cdot\frac{1+\frac{1}{x+3}}{1-\frac{1}{x+3}}=\frac{(x+3)+1}{(x+3)-1}=\frac{x+3+1}{x+3-1}=\frac{x+4}{x+2}$$

49. $f(x)=\dfrac{3}{x}$

$$\frac{f(a+h)-f(a)}{h}=\frac{\frac{3}{a+h}-\frac{3}{a}}{h}=\frac{\frac{3a}{a(a+h)}-\frac{3(a+h)}{a(a+h)}}{h}$$

$$=\frac{\frac{3a-3a-3h}{a(a+h)}}{h}=\frac{3a-3a-3h}{a(a+h)}\cdot\frac{1}{h}$$

$$=\frac{-3\cancel{h}}{a(a+h)}\cdot\frac{1}{\cancel{h}}$$

$$=-\frac{3}{a(a+h)}$$

51. a. $A=\dfrac{Pi}{1-\frac{1}{(1+i)^n}}=\dfrac{(1+i)^n}{(1+i)^n}\cdot\dfrac{Pi}{1-\frac{1}{(1+i)^n}}=\dfrac{Pi(1+i)^n}{(1+i)^n\cdot 1-\cancel{(1+i)^n}\cdot\frac{1}{\cancel{(1+i)^n}}}=\dfrac{Pi(1+i)^n}{(1+i)^n-1}$

b. $A=\dfrac{Pi(1+i)^n}{(1+i)^n-1}=\dfrac{(20000)(0.01)(1+0.01)^{48}}{(1+0.01)^{48}-1}=\dfrac{(20000)(0.01)(1.01)^{48}}{(1.01)^{48}-1}$

$$=\frac{(20000)(0.01)(1.612)}{(1.612)-1}=\frac{322.4}{0.612}=526.80$$

You will pay approximately \$527 each month.

53. $R=\dfrac{1}{\frac{1}{R_1}+\frac{1}{R_2}+\frac{1}{R_3}}=\dfrac{R_1R_2R_3}{R_1R_2R_3}\cdot\dfrac{1}{\frac{1}{R_1}+\frac{1}{R_2}+\frac{1}{R_3}}$

$$=\frac{R_1R_2R_3\cdot 1}{\cancel{R_1}R_2R_3\cdot\frac{1}{\cancel{R_1}}+R_1\cancel{R_2}R_3\cdot\frac{1}{\cancel{R_2}}+R_1R_2\cancel{R_3}\cdot\frac{1}{\cancel{R_3}}}=\frac{R_1R_2R_3}{R_2R_3+R_1R_3+R_1R_2}$$

$$R=\frac{R_1R_2R_3}{R_2R_3+R_1R_3+R_1R_2}=\frac{4\cdot 8\cdot 12}{8\cdot 12+4\cdot 12+4\cdot 8}=\frac{384}{96+48+32}=\frac{384}{176}\approx 2.18$$

The combined resistance is approximately 2.18 ohms.

55. – 57. Answers will vary.

59. $\dfrac{x-\dfrac{1}{2x+1}}{1-\dfrac{x}{2x+1}}=2x-1$

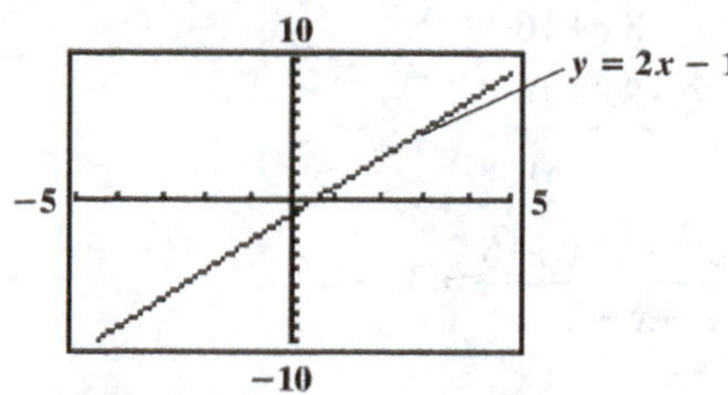

The graphs coincide. The complex fraction is simplified correctly.

61. $\dfrac{\dfrac{1}{x}+\dfrac{1}{3}}{\dfrac{1}{3x}}=x+\dfrac{1}{3}$

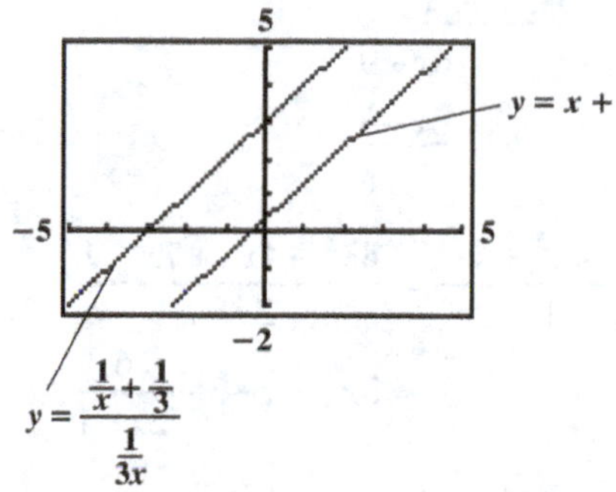

The graphs do not coincide. Simplify the complex fraction.

$$\dfrac{\dfrac{1}{x}+\dfrac{1}{3}}{\dfrac{1}{3x}}=\dfrac{3x}{3x}\cdot\dfrac{\dfrac{1}{x}+\dfrac{1}{3}}{\dfrac{1}{3x}}$$

$$=\dfrac{3\cancel{x}\cdot\dfrac{1}{\cancel{x}}+\cancel{3}x\cdot\dfrac{1}{\cancel{3}}}{\cancel{3x}\cdot\dfrac{1}{\cancel{3x}}}$$

$$=\dfrac{3+x}{1}=x+3$$

63. does not make sense; Explanations will vary. Sample explanation: You must also multiply the denominator by xy.

65. does not make sense; Explanations will vary. Sample explanation: The 3 and the 1 must also be multiplied by the LCD.

67. $\dfrac{\dfrac{x+h}{x+h+1}-\dfrac{x}{x+1}}{h}$

$$=\dfrac{\dfrac{(x+h)(x+1)}{(x+h+1)(x+1)}-\dfrac{x(x+h+1)}{(x+h+1)(x+1)}}{h}$$

$$=\dfrac{\dfrac{(x+h)(x+1)-x(x+h+1)}{(x+h+1)(x+1)}}{h}$$

$$=\dfrac{\dfrac{\cancel{x^2}+\cancel{x}+\cancel{hx}+h-\cancel{x^2}-\cancel{hx}-\cancel{x}}{(x+h+1)(x+1)}}{h}$$

$$=\dfrac{\dfrac{h}{(x+h+1)(x+1)}}{h}$$

$$=\dfrac{\dfrac{h}{(x+h+1)(x+1)}}{\dfrac{h}{1}}$$

$$=\dfrac{\cancel{h}}{(x+h+1)(x+1)}\cdot\dfrac{1}{\cancel{h}}$$

$$=\dfrac{1}{(x+h+1)(x+1)}$$

69. $f(x)=\dfrac{1}{x+1};\ f(a)=\dfrac{1}{a+1}$

$$f(f(a))=\dfrac{1}{\dfrac{1}{a+1}+1}$$

$$=\dfrac{1}{\dfrac{1}{a+1}+\dfrac{a+1}{a+1}}$$

$$=\dfrac{1}{\dfrac{a+2}{a+1}}=\dfrac{a+1}{a+2}$$

Therefore, $f(f(a))=\dfrac{a+1}{a+2}$.

71.
$$x^2+27=12x$$
$$x^2-12x+27=0$$
$$(x-9)(x-3)=0$$

Apply the zero product principle.

$x-9=0 \qquad x-3=0$

$x=9 \qquad\quad x=3$

The solution set is $\{3,9\}$.

72. $\left(4x^2-y\right)^2=\left(4x^2\right)^2+2\cdot 4x^2\left(-y\right)+\left(-y\right)^2$
$=16x^4-8x^2y+y^2$

73.
$$\begin{aligned}-4&<3x-7<8\\-4+7&<3x-7+7<8+7\\3&<3x<15\\1&<x<5\end{aligned}$$
The solution set is $\{x|1<x<5\}$ or $(1,5)$.

74. $\dfrac{8x^4y^5}{4x^3y^2}=\dfrac{8}{4}x^{4-3}y^{5-2}=2xy^3$

75.
$$\begin{array}{r}35\\21\overline{)737}\\\underline{63}\\107\\\underline{105}\\2\end{array}$$
Writing as quotient+$\dfrac{\text{remainder}}{\text{divisor}}$ gives $35+\dfrac{2}{21}$.

76. $6x^2+3x-(6x^2-4x)=6x^2+3x-6x^2+4x=7x$

6.4 Check Points

1. $\dfrac{16x^3-32x^2+2x+4}{4x}$
$=\dfrac{16x^3}{4x}-\dfrac{32x^2}{4x}+\dfrac{2x}{4x}+\dfrac{4}{4x}$
$=4x^2-8x+\dfrac{1}{2}+\dfrac{1}{x}$

2. $\dfrac{15x^4y^5-5x^3y^4+10x^2y^2}{5x^2y^3}$
$=\dfrac{15x^4y^5}{5x^2y^3}-\dfrac{5x^3y^4}{5x^2y^3}+\dfrac{10x^2y^2}{5x^2y^3}$
$=3x^2y^2-xy+\dfrac{2}{y}$

3.
$$\begin{array}{r}3x-8\\x-2\overline{)3x^2-14x+16}\\\underline{3x^2-6x}\\-8x+16\\\underline{-8x+16}\\0\end{array}$$
Thus, $\dfrac{3x^2-14x+16}{x-2}=3x-8$

4. Write the dividend in descending powers of x.
$-9+7x-4x^2+4x^3=4x^3-4x^2+7x-9$
$$\begin{array}{r}2x^2-\ x+3\\2x-1\overline{)4x^3-4x^2+7x-9}\\\underline{4x^3-2x^2}\\-2x^2+7x\\\underline{-2x^2+x}\\6x-9\\\underline{6x-3}\\-6\end{array}$$
$\dfrac{-9+7x-4x^2+4x^3}{2x-1}=\dfrac{4x^3-4x^2+7x-9}{2x-1}$
$=2x^2-x+3-\dfrac{6}{2x-1}$

5. Rewrite the dividend with the missing power of x and divide.
$$\begin{array}{r}2x^2+7x+14\\x^2-2x\overline{)2x^4+3x^3+0x^2-7x-10}\\\underline{2x^4-4x^3}\\7x^3-0x^2\\\underline{7x^3-14x^2}\\14x^2-7x\\\underline{14x^2-28x}\\21x-10\end{array}$$
Thus,
$\dfrac{2x^4+3x^3-7x-10}{x^2-2x}=2x^2+7x+14+\dfrac{21x-10}{x^2-2x}$

6.4 Concept and Vocabulary Check

1. $16x^2-32x^2+2x+4$; $4x$

2. $2x^3+0x^2+6x-4$

3. $6x^3$; $3x$; $2x^2$; $7x^2$

4. $2x^2$; $5x-2$; $10x^3-4x^2$; $10x^3+6x^2$

5. $6x^2+10x$; $6x^2+8x$; $18x$; -4; $18x-4$

6. 9; $3x-5$; 9; $3x-5+\frac{9}{2x+1}$

7. divisor; quotient; remainder; dividend

6.4 Exercise Set

1. $\frac{25x^7-15x^5+10x^3}{5x^3}$
$=\frac{25x^7}{5x^3}-\frac{15x^5}{5x^3}+\frac{10x^3}{5x^3}$
$=5x^4-3x^2+2$

3. $\frac{18x^3+6x^2-9x-6}{3x}$
$=\frac{18x^3}{3x}+\frac{6x^2}{3x}-\frac{9x}{3x}-\frac{6}{3x}$
$=6x^2+2x-3-\frac{2}{x}$

5. $\frac{28x^3-7x^2-16x}{4x^2}=\frac{28x^3}{4x^2}-\frac{7x^2}{4x^2}-\frac{16x}{4x^2}$
$=7x-\frac{7}{4}-\frac{4}{x}$

7. $\frac{25x^8-50x^7+3x^6-40x^5}{-5x^5}$
$=\frac{25x^8}{-5x^5}-\frac{50x^7}{-5x^5}+\frac{3x^6}{-5x^5}-\frac{40x^5}{-5x^5}$
$=-5x^3+10x^2-\frac{3}{5}x+8$

9. $\frac{18a^3b^2-9a^2b-27ab^2}{9ab}$
$=\frac{18a^3b^2}{9ab}-\frac{9a^2b}{9ab}-\frac{27ab^2}{9ab}$
$=2a^2b-a-3b$

11. $\frac{36x^4y^3-18x^3y^2-12x^2y}{6x^3y^3}$
$=\frac{36x^4y^3}{6x^3y^3}-\frac{18x^3y^2}{6x^3y^3}-\frac{12x^2y}{6x^3y^3}$
$=6x-\frac{3}{y}-\frac{2}{xy^2}$

13.
$$\begin{array}{r} x+3 \\ x+5\overline{)x^2+8x+15} \\ \underline{x^2+5x} \\ 3x+15 \\ \underline{3x+15} \\ 0 \end{array}$$
$\frac{x^2+8x+15}{x+5}=x+3$

15.
$$\begin{array}{r} x^2+x-2 \\ x-3\overline{)x^3-2x^2-5x+6} \\ \underline{x^3-3x^2} \\ x^2-5x \\ \underline{x^2-3x} \\ -2x+6 \\ \underline{-2x+6} \\ 0 \end{array}$$
$\frac{x^3-2x^2-5x+6}{x-3}=x^2+x-2$

17.
$$\begin{array}{r} x-2 \\ x-5\overline{)x^2-7x+12} \\ \underline{x^2-5x} \\ -2x+12 \\ \underline{-2x+10} \\ 2 \end{array}$$
$\frac{x^2-7x+12}{x-5}=x-2+\frac{2}{x-5}$

19.
$$\begin{array}{r} x+5 \\ 2x+3\overline{)2x^2+13x+5} \\ \underline{2x^2+3x} \\ 10x+5 \\ \underline{10x+15} \\ -10 \end{array}$$

$$\frac{2x^2+13x+5}{2x+3}=x+5-\frac{10}{2x+3}$$

21.
$$\begin{array}{r} x^2+2x+3 \\ x+1\overline{)x^3+3x^2+5x+4} \\ \underline{x^3+x^2} \\ 2x^2+5x \\ \underline{2x^2+2x} \\ 3x+4 \\ \underline{3x+3} \\ 1 \end{array}$$

$$\frac{x^3+3x^2+5x+4}{x+1}=x^2+2x+3+\frac{1}{x+1}$$

23.
$$\begin{array}{r} 2y^2+3y-1 \\ 2y+3\overline{)4y^3+12y^2+7y-3} \\ \underline{4y^3+6y^2} \\ 6y^2+7y \\ \underline{6y^2+9y} \\ -2y-3 \\ \underline{-2y-3} \\ 0 \end{array}$$

$$\frac{4y^3+12y^2+7y-3}{2y+3}=2y^2+3y-1$$

25.
$$\begin{array}{r} 3x^2-3x+1 \\ 3x+2\overline{)9x^3-3x^2-3x+4} \\ \underline{9x^3+6x^2} \\ -9x^2-3x \\ \underline{-9x^2-6x} \\ 3x+4 \\ \underline{3x+2} \\ 2 \end{array}$$

$$\frac{9x^3-3x^2-3x+4}{3x+2}=3x^2-3x+1+\frac{2}{3x+2}$$

27. $(4x^3-6x-11)\div(2x-4)$

Rewrite the dividend with the missing power of x and divide.

$$\begin{array}{r} 2x^2+4x+5 \\ 2x-4\overline{)4x^3+0x^2-6x-11} \\ \underline{4x^3-8x^2} \\ 8x^2-6x \\ \underline{8x^2-16x} \\ 10x-11 \\ \underline{10x-20} \\ 9 \end{array}$$

$$\frac{4x^3-6x-11}{2x-4}=2x^2+4x+5+\frac{9}{2x-4}$$

29. $(4y^3-5y)\div(2y-1)$

Rewrite the dividend with the missing powers of y and divide. Notice: There is no constant term in the polynomial; a 0 is added for the purposes of long division.

$$\begin{array}{r} 2y^2+y-2 \\ 2y-1\overline{)4y^3+0y^2-5y+0} \\ \underline{4y^3-2y^2} \\ 2y^2-5y \\ \underline{2y^2-y} \\ -4y+0 \\ \underline{-4y+2} \\ -2 \end{array}$$

$$\frac{4y^3-5y}{2y-1}=2y^2+y-2-\frac{2}{2y-1}$$

31. $\left(4y^4-17y^2+14y-3\right)\div(2y-3)$

Rewrite the dividend with the missing power of y and divide.

$$
\begin{array}{r}
2y^3+3y^2-4y+1 \\
2y-3\overline{\big)4y^4+0y^3-17y^2+14y-3} \\
\underline{4y^3-6y^3} \\
6y^3-17y^2 \\
\underline{6y^3-\ 9y^2} \\
-8y^2+14y \\
\underline{-8y^2+12y} \\
2y-3 \\
\underline{2y-3} \\
0
\end{array}
$$

$$\frac{4y^4-17y^2+14y-3}{2y-3}=2y^3+3y^2-4y+1$$

33. $\left(4x^4+3x^3+4x^2+9x-6\right)\div\left(x^2+3\right)$

Rewrite the divisor with the missing power of x and divide.

$$
\begin{array}{r}
4x^2+3x-8 \\
x^2+0x+3\overline{\big)4x^4+3x^3+4x^2+9x-6} \\
\underline{4x^4+0x^3+12x^2} \\
3x^3-8x^2+9x \\
\underline{3x^3+0x^2+9x} \\
-8x^2+0x-6 \\
\underline{-8x^2+0x-24} \\
18
\end{array}
$$

$$\frac{4x^4+3x^3+4x^2+9x-6}{x^2+3}$$
$$=4x^2+3x-8+\frac{18}{x^2+3}$$

35. $\left(15x^4+3x^3+4x^2+4\right)\div\left(3x^2-1\right)$

Rewrite the dividend and the divisor with the missing powers of x and divide.

$$
\begin{array}{r}
5x^2+\ x+3 \\
3x^2+0x-1\overline{\big)15x^4+3x^3+4x^2+0x+4} \\
\underline{15x^4+0x^3-5x^2} \\
3x^3+9x^2+0x \\
\underline{3x^3+0x^2-\ x} \\
9x^2+\ x+4 \\
\underline{9x^2+0x-3} \\
x+7
\end{array}
$$

$$\frac{15x^4+3x^3+4x^2+4}{3x^2-1}=5x^2+x+3+\frac{7+x}{3x^2-1}$$
$$\text{or } 5x^2+x+3+\frac{x+7}{3x^2-1}$$

37. $f(x)=8x^3-38x^2+49x-10$

$g(x)=4x-1$

$$\left(\frac{f}{g}\right)(x)=\frac{f(x)}{g(x)}=\frac{8x^3-38x^2+49x-10}{4x-1}$$

$$
\begin{array}{r}
2x^2-9x+10 \\
4x-1\overline{\big)8x^3-38x^2+49x-10} \\
\underline{8x^3-2x^2} \\
-36x^2+49x \\
\underline{-36x^2+\ 9x} \\
40x-10 \\
\underline{40x-10} \\
0
\end{array}
$$

$$\left(\frac{f}{g}\right)(x)=2x^2-9x+10$$

39. $f(x) = 2x^4 - 7x^3 + 7x^2 - 9x + 10$

$g(x) = 2x - 5$

$$\left(\frac{f}{g}\right)(x) = \frac{f(x)}{g(x)} = \frac{2x^4 - 7x^3 + 7x^2 - 9x + 10}{2x - 5}$$

$$\begin{array}{r}
x^3 - x^2 + x - 2 \\
2x - 5\overline{\big)\,2x^4 - 7x^3 + 7x^2 - 9x + 10} \\
\underline{2x^4 - 5x^3} \qquad\qquad\qquad\quad \\
-2x^3 + 7x^2 \qquad\qquad \\
\underline{-2x^3 + 5x^2} \qquad\qquad \\
2x^2 - 9x \qquad \\
\underline{2x^2 - 5x} \qquad \\
-4x + 10 \\
\underline{-4x + 10} \\
0
\end{array}$$

$$\left(\frac{f}{g}\right)(x) = \frac{f(x)}{g(x)} = x^3 - x^2 + x - 2$$

41.

$$\begin{array}{r}
x^3 - x^2y + xy^2 - y^3 \\
x + y\overline{\big)\,x^4 + 0x^3 + 0x^2 + 0x + y^4} \\
\underline{x^4 + x^3y} \qquad\qquad\qquad\quad \\
-x^3y + 0x^2 \qquad\qquad \\
\underline{-x^3y - x^2y^2} \qquad\qquad \\
x^2y^2 + 0x \qquad \\
\underline{x^2y^2 + xy^3} \qquad \\
-xy^3 + y^4 \\
\underline{-xy^3 - y^4} \\
2y^4
\end{array}$$

$$\frac{x^4 + y^4}{x + y} = x^3 - x^2y + xy^2 - y^3 + \frac{2y^4}{x + y}$$

43.

$$\begin{array}{r}
3x^2 + 2x - 1 \\
x^2 + x + 2\overline{\big)\,3x^4 + 5x^3 + 7x^2 + 3x - 2} \\
\underline{3x^4 + 3x^3 + 6x^2} \qquad\qquad \\
2x^3 + x^2 + 3x \qquad \\
\underline{2x^3 + 2x^2 + 4x} \qquad \\
-x^2 - x - 2 \\
\underline{-x^2 - x - 2} \\
0
\end{array}$$

$$\frac{3x^4 + 5x^3 + 7x^2 + 3x - 2}{x^2 + x + 2} = 3x^2 + 2x - 1$$

45.

$$\begin{array}{r}
4x - 7 \\
x^2 + x + 1\overline{\big)\,4x^3 - 3x^2 + x + 1} \\
\underline{4x^3 + 4x^2 + 4x} \qquad \\
-7x^2 - 3x + 1 \\
\underline{-7x^2 - 7x - 7} \\
4x + 8
\end{array}$$

$$\frac{4x^3 - 3x^2 + x + 1}{x^2 + x + 1} = 4x - 7 + \frac{4x + 8}{x^2 + x + 1}$$

47.

$$\begin{array}{r}
x^3 + x^2 - x - 3 \\
x^2 - x + 2\overline{\big)\,x^5 + 0x^4 + 0x^3 + 0x^2 + 0x - 1} \\
\underline{x^5 - x^4 + 2x^3} \qquad\qquad\qquad\quad \\
x^4 - 2x^3 + 0x^2 \qquad\qquad \\
\underline{x^4 - x^3 + 2x^2} \qquad\qquad \\
-x^3 - 2x^2 + 0x \qquad \\
\underline{-x^3 + x^2 - 2x} \qquad \\
-3x^2 + 2x - 1 \\
\underline{-3x^2 + 3x - 6} \\
-x + 5
\end{array}$$

$$\frac{x^5 - 1}{x^2 - x + 2} = x^3 + x^2 - x - 3 + \frac{-x + 5}{x^2 - x + 2}$$

49.
$$\begin{array}{r} 4x^2+5xy-y^2 \\ x-3y\overline{)4x^3-7x^2y-16xy^2+3y^3} \\ \underline{4x^3-12x^2y} \qquad\qquad\qquad \\ 5x^2y-16xy^2 \qquad\quad \\ \underline{5x^2y-15xy^2} \qquad\quad \\ -xy^2+3y^3 \\ \underline{-xy^2+3y^3} \\ 0 \end{array}$$

$$\frac{4x^3-7x^2y-16xy^2+3y^3}{x-3y}=4x^2+5xy-y^2$$

51. $f(x)=3x^3+4x^2-x-4$

$g(x)=-5x^3+22x^2-28x-12$

$h(x)=4x+1$

$$\left(\frac{f-g}{h}\right)(x)=\frac{f(x)-g(x)}{h(x)}$$

$$=\frac{(3x^3+4x^2-x-4)-(-5x^3+22x^2-28x-12)}{4x+1}$$

$$=\frac{3x^3+4x^2-x-4+5x^3-22x^2+28x+12}{4x+1}$$

$$=\frac{8x^3-18x^2+27x+8}{4x+1}$$

$$\begin{array}{r} 2x^2-5x+8 \\ 4x+1\overline{)8x^3-18x^2+27x+8} \\ \underline{8x^3+2x^2} \qquad\qquad\quad \\ -20x^2+27x \qquad \\ \underline{-20x^2-5x} \qquad \\ 32x+8 \\ \underline{32x+8} \\ 0 \end{array}$$

Therefore, $\left(\frac{f-g}{h}\right)(x)=2x^2-5x+8$.

Domain: $\left(-\infty,-\frac{1}{4}\right)\cup\left(-\frac{1}{4},\infty\right)$.

53. $ax+2x+4=3a^3+10a^2+6a$

$x(a+2)+4=3a^3+10a^2+6a$

$x(a+2)=3a^3+10a^2+6a-4$

$$x=\frac{3a^3+10a^2+6a-4}{a+2}$$

$$\begin{array}{r} 3a^2+4a-2 \\ a+2\overline{)3a^3+10a^2+6a-4} \\ \underline{3a^3+6a^2} \qquad\qquad \\ 4a^2+6a \qquad \\ \underline{4a^2+8a} \qquad \\ -2a-4 \\ \underline{-2a-4} \\ 0 \end{array}$$

Therefore, $x=3a^2+4a-2$.

55. $f(30)=\dfrac{80(30)-8000}{30-110}=70$

With a tax rate percentage of 30%, the government tax revenue will be \$70 ten billions or \$700 billion. This is shown on the graph as the point $(30, 70)$.

57. $(80x-8000)\div(x-110)$

$$\begin{array}{r} 80 \\ x-110\overline{)80x-8000} \\ \underline{80x-8800} \\ 800 \end{array}$$

$$\frac{80x-8000}{x-110}=80+\frac{800}{x-110}$$

$$f(x)=80+\frac{800}{x-110}$$

$$f(30)=80+\frac{800}{30-110}=70$$

This is the same answer obtained in Exercise 55.

59. - 63. Answers will vary.

65. $(6x^2+16x+8)\div(3x+2)=2x+4$

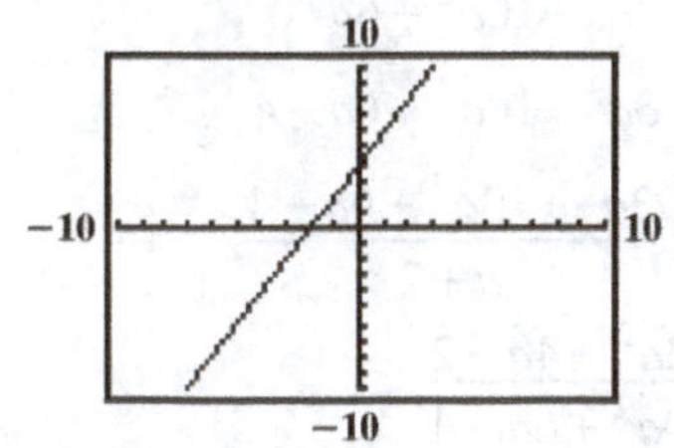

The graphs coincide. The division is correct.

67. $(3x^4+4x^3-32x^2-5x-20)\div(x+4)$

$=3x^3-8x^2+5$

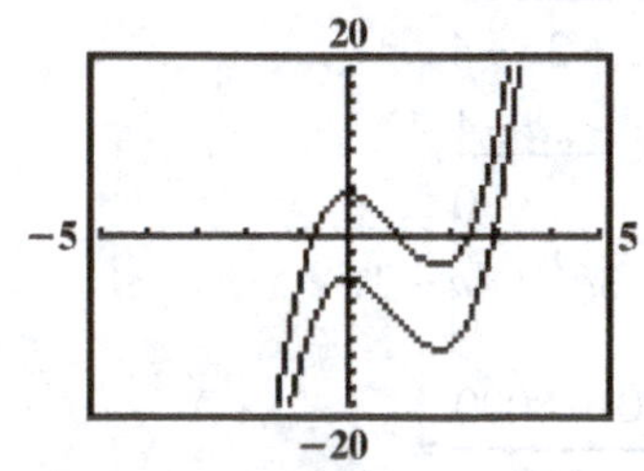

The graphs do not coincide, so the division has not been performed correctly.

$$\begin{array}{r} 3x^3-8x^2-5 \\ x+4\overline{)3x^4+4x^3-32x^2-5x-20} \\ \underline{3x^4+12x^3} \qquad\qquad\qquad \\ -8x^3-32x^2 \qquad\quad \\ \underline{-8x^3-32x^2} \qquad\quad \\ 0-5x-20 \\ \underline{-5x-20} \\ 0 \end{array}$$

$(3x^4+4x^3-32x^2-5x-20)\div(x+4)$

$=3x^3-8x^2-5$

69. makes sense

71. makes sense

73. false; Changes to make the statement true will vary. A sample change is: The dividend does not always have the divisor as a factor.

75. false; Changes to make the statement true will vary. A sample change is: The long division process should be continued until the degree of the remainder is less than the degree of the divisor.

77.
$$\begin{array}{r} x^{2n}+\ x^n+\ 3 \\ x^n-5\overline{)x^{3n}-4x^{2n}-2x^n-12} \\ \underline{x^{3n}-5x^{2n}} \qquad\qquad\quad \\ x^{2n}-2x^n \qquad\quad \\ \underline{x^{2n}-5x^n} \qquad\quad \\ 3x^n-12 \\ \underline{3x^n-15} \\ 3 \end{array}$$

$$\frac{x^{3n}-4x^{2n}-2x^n-12}{x^n-5}=x^{2n}+x^n+3+\frac{3}{x^n-5}$$

79. Let k = the unknown polynomial.

$$\frac{2x^2-7x+9}{k}=2x-3+\frac{3}{k}$$

$$\frac{2x^2-7x+9}{k}-\frac{3}{k}=2x-3$$

$$\frac{2x^2-7x+9-3}{k}=2x-3$$

$$\frac{2x^2-7x+6}{k}=2x-3$$

$$2x^2-7x+6=k(2x-3)$$

$$\frac{2x^2-7x+6}{2x-3}=k$$

Solve for k using long division.

$$\begin{array}{r} x-2 \\ 2x-3\overline{)2x^2-7x+6} \\ \underline{2x^2-3x} \qquad \\ -4x+6 \\ \underline{-4x+6} \\ 0 \end{array}$$

The unknown polynomial is $x-2$.

81. $|2x-3|>4$

$2x-3<-4$ or $2x-3>4$

$x<-\frac{1}{2}$ $\qquad 2x>7$

$x>\frac{7}{2}$

The solution set is $\left(-\infty,-\frac{1}{2}\right)\cup\left(\frac{7}{2},\infty\right)$.

82. $40,610,000=4.061\times10^7$

83. $2x-4[x-3(2x+1)]$
$=2x-4[x-6x-3]$
$=2x-4[-5x-3]$
$=2x+20x+12$
$=22x+12$

84. a. $\dfrac{5x^3+6x+8}{x+2}=5x^2-10x+26-\dfrac{44}{x+2}$

b.

$$\begin{array}{r|rrrr} -2 & 5 & 0 & 6 & 8 \\ & & -10 & 20 & -52 \\ \hline & 5 & \boxed{-10} & \boxed{26} & \boxed{-44} \end{array}$$

The numbers are the coefficients of the quotient and the remainder.

85. a. $\dfrac{3x^3-4x^2+2x-1}{x+1}=3x^2-7x+9-\dfrac{10}{x+1}$

b.

$$\begin{array}{r|rrrr} -1 & 3 & -4 & 2 & -1 \\ & & -3 & 7 & -9 \\ \hline & 3 & \boxed{-7} & \boxed{9} & \boxed{-10} \end{array}$$

The numbers are the coefficients of the quotient and the remainder.

86.

$$\begin{array}{r} 2x^2+3x-2 \\ x-3\overline{)2x^3-3x^2-11x+6} \\ \underline{2x^3-6x^2} \qquad\qquad \\ 3x^2-11x \quad\; \\ \underline{3x^2-9x} \quad\;\; \\ -2x+6 \\ \underline{-2x+6} \\ 0 \end{array}$$

Thus, $2x^3-3x^2-11x+6=(x-3)(2x^2+3x-2)$
Now factor the quadratic expression.
$2x^3-3x^2-11x+6=(x-3)(2x^2+3x-2)$
$=(x-3)(2x-1)(x+2)$

Mid-Chapter Check Point – Chapter 6

1. $\dfrac{x^2-x-6}{x^2+3x-18}=\dfrac{\cancel{(x-3)}(x+2)}{(x+6)\cancel{(x-3)}}=\dfrac{x+2}{x+6}$

2. $$\frac{2x^2-8x-11}{x^2+3x-4}+\frac{x^2+14x-13}{x^2+3x-4}=\frac{2x^2-8x-11+x^2+14x-13}{x^2+3x-4}$$
$$=\frac{2x^2+x^2-8x+14x-11-13}{x^2+3x-4}$$
$$=\frac{3x^2+6x-24}{x^2+3x-4}$$
$$=\frac{3\left(x^2+2x-8\right)}{x^2+3x-4}$$
$$=\frac{3\cancel{(x+4)}(x-2)}{\cancel{(x+4)}(x-1)}$$
$$=\frac{3(x-2)}{x-1}$$

3. $$\frac{x^3-27}{4x^2-4x}\cdot\frac{4x}{x-3}=\frac{\cancel{(x-3)}\left(x^2+3x+9\right)}{\cancel{4x}(x-1)}\cdot\frac{\cancel{4x}}{\cancel{(x-3)}}=\frac{x^2+3x+9}{x-1}$$

4. $$5+\frac{7}{x-2}=\frac{5(x-2)}{x-2}+\frac{7}{x-2}=\frac{5x-10+7}{x-2}=\frac{5x-3}{x-2}$$

5. $$\frac{x-\frac{4}{x+6}}{\frac{1}{x+6}+x}=\frac{\frac{x(x+6)}{x+6}-\frac{4}{x+6}}{\frac{1}{x+6}+\frac{x(x+6)}{x+6}}=\frac{\frac{x^2+6x-4}{x+6}}{\frac{x^2+6x+1}{x+6}}=\frac{x^2+6x-4}{\cancel{(x+6)}}\cdot\frac{\cancel{(x+6)}}{x^2+6x+1}=\frac{x^2+6x-4}{x^2+6x+1}$$

6. $$x-4\overline{\big)2x^4-13x^3+17x^2+18x-24}$$ quotient: $2x^3-5x^2-3x+6$

$$\begin{array}{l}
\underline{2x^4-8x^3} \\
\quad -5x^3+17x^2 \\
\quad \underline{-5x^3+20x^2} \\
\qquad -3x^2+18x \\
\qquad \underline{-3x^2+12x} \\
\qquad\qquad 6x-24 \\
\qquad\qquad \underline{6x-24} \\
\qquad\qquad\quad 0
\end{array}$$

$$\left(2x^4-13x^3+17x^2+18x-24\right)\div(x-4)=2x^3-5x^2-3x+6$$

7. $$\frac{x^3y-y^3x}{x^2y-xy^2}=\frac{xy\left(x^2-y^2\right)}{xy(x-y)}=\frac{\cancel{xy}\,\cancel{(x-y)}(x+y)}{\cancel{xy}\,\cancel{(x-y)}}=x+y$$

8. $$\frac{28x^8y^3-14x^6y^2+3x^2y^2}{7x^2y}=\frac{28x^8y^3}{7x^2y}-\frac{14x^6y^2}{7x^2y}+\frac{3x^2y^2}{7x^2y}=4x^6y^2-2x^4y+\frac{3}{7}y$$

9. $$\frac{2x-1}{x+6}-\frac{x+3}{x-2}=\frac{(2x-1)(x-2)}{(x+6)(x-2)}-\frac{(x+6)(x+3)}{(x+6)(x-2)}=\frac{2x^2-4x-x+2}{(x+6)(x-2)}-\frac{x^2+3x+6x+18}{(x+6)(x-2)}$$
$$=\frac{2x^2-5x+2}{(x+6)(x-2)}-\frac{x^2+9x+18}{(x+6)(x-2)}=\frac{2x^2-5x+2-x^2-9x-18}{(x+6)(x-2)}$$
$$=\frac{2x^2-x^2-5x-9x+2-18}{(x+6)(x-2)}$$
$$=\frac{x^2-14x-16}{(x+6)(x-2)}$$

10. $$\frac{3}{x-2}-\frac{2}{x+2}-\frac{x}{x^2-4}=\frac{3}{x-2}-\frac{2}{x+2}-\frac{x}{(x-2)(x+2)}$$
$$=\frac{3(x+2)}{(x-2)(x+2)}-\frac{2(x-2)}{(x-2)(x+2)}-\frac{x}{(x-2)(x+2)}$$
$$=\frac{3x+6}{(x-2)(x+2)}-\frac{2x-4}{(x-2)(x+2)}-\frac{x}{(x-2)(x+2)}$$
$$=\frac{3x+6-2x+4-x}{(x-2)(x+2)}=\frac{3x-2x-x+6+4}{(x-2)(x+2)}$$
$$=\frac{10}{(x-2)(x+2)}$$

11. $$\frac{3x^2-7x-6}{3x^2-13x-10}\div\frac{2x^2-x-1}{4x^2-18x-10}=\frac{3x^2-7x-6}{3x^2-13x-10}\cdot\frac{4x^2-18x-10}{2x^2-x-1}$$
$$=\frac{\cancel{(3x+2)}(x-3)}{\cancel{(3x+2)}(x-5)}\cdot\frac{2\left(2x^2-9x-5\right)}{(2x+1)(x-1)}$$
$$=\frac{2(x-3)\cancel{(2x+1)}\cancel{(x-5)}}{\cancel{(x-5)}\cancel{(2x+1)}(x-1)}$$
$$=\frac{2(x-3)}{x-1}$$

12. $$\frac{3}{7-x}+\frac{x-2}{x-7}=\frac{-3}{-(7-x)}+\frac{x-2}{x-7}=\frac{-3}{x-7}+\frac{x-2}{x-7}=\frac{-3+x-2}{x-7}=\frac{x-5}{x-7}$$

13.
$$\begin{array}{r|l}
 & \quad 2x^2-x-3 \\
3x^2-1 & 6x^4-3x^3-11x^2+2x+4 \\
 & \underline{6x^4 \qquad\quad -2x^2} \\
 & \qquad -3x^3-9x^2 \\
 & \qquad \underline{-3x^3 \qquad\quad +x} \\
 & \qquad\qquad -9x^2\ +x \\
 & \qquad\qquad \underline{-9x^2 \qquad +3} \\
 & \qquad\qquad\qquad x+1
\end{array}$$

$$\left(6x^4-3x^3-11x^2+2x+4\right)\div\left(3x^2-1\right)=2x^2-x-3+\frac{x+1}{3x^2-1}$$

14. $\dfrac{5+\dfrac{2}{x}}{3-\dfrac{1}{x}}=\dfrac{\dfrac{5x}{x}+\dfrac{2}{x}}{\dfrac{3x}{x}-\dfrac{1}{x}}=\dfrac{\dfrac{5x+2}{x}}{\dfrac{3x-1}{x}}=\dfrac{5x+2}{\cancel{x}}\cdot\dfrac{\cancel{x}}{3x-1}=\dfrac{5x+2}{3x-1}$

15.
$$\begin{aligned}\frac{x}{x^2-7x+6}-\frac{x}{x^2-2x-24}&=\frac{x}{(x-6)(x-1)}-\frac{x}{(x-6)(x+4)}\\&=\frac{x(x+4)}{(x-6)(x-1)(x+4)}-\frac{x(x-1)}{(x-6)(x-1)(x+4)}\\&=\frac{x(x+4)-x(x-1)}{(x-6)(x-1)(x+4)}=\frac{x^2+4x-x^2+x}{(x-6)(x-1)(x+4)}\\&=\frac{x^2-x^2+4x+x}{(x-6)(x-1)(x+4)}=\frac{5x}{(x-6)(x-1)(x+4)}\end{aligned}$$

16. $\dfrac{\dfrac{3}{x+1}+\dfrac{4}{x}}{\dfrac{4}{x}}=\dfrac{x(x+1)}{x(x+1)}\cdot\dfrac{\dfrac{3}{x+1}+\dfrac{4}{x}}{\dfrac{4}{x}}=\dfrac{3x+4(x+1)}{4(x+1)}=\dfrac{3x+4x+4}{4(x+1)}=\dfrac{7x+4}{4(x+1)}$

17.
$$\begin{aligned}\frac{x^2-x-6}{x+1}\div\left(\frac{x^2-9}{x^2-1}\cdot\frac{x-1}{x+3}\right)&=\frac{(x-3)(x+2)}{(x+1)}\div\left(\frac{(x-3)\cancel{(x+3)}\cancel{(x-1)}}{\cancel{(x-1)}(x+1)\cancel{(x+3)}}\right)\\&=\frac{(x-3)(x+2)}{(x+1)}\div\left(\frac{x-3}{x+1}\right)\\&=\frac{\cancel{(x-3)}(x+2)}{\cancel{(x+1)}}\cdot\frac{\cancel{(x+1)}}{\cancel{(x-3)}}\\&=x+2\end{aligned}$$

18.
$$\begin{array}{r l}
 & 16x^2-8x+4\\
4x+2\,\big) & 64x^3+0x^2+0x+4\\
 & \underline{64x^3+32x^2}\\
 & -32x^2+0x\\
 & \underline{-32x-16x}\\
 & 16x+4\\
 & \underline{16x+8}\\
 & -4
\end{array}$$

$$\begin{aligned}(64x^3+4)\div(4x+2)&=16x^2-8x+4-\frac{4}{4x+2}\\&=16x^2-8x+4-\frac{2(2)}{2(2x+1)}\\&=16x^2-8x+4-\frac{2}{2x+1}\end{aligned}$$

19. $\frac{x+1}{x^2+x-2}-\frac{1}{x^2-3x+2}+\frac{2x}{x^2-4}$

$$=\frac{x+1}{(x+2)(x-1)}-\frac{1}{(x-2)(x-1)}+\frac{2x}{(x-2)(x+2)}$$

$$=\frac{(x+1)(x-2)}{(x-2)(x+2)(x-1)}-\frac{1(x+2)}{(x-2)(x+2)(x-1)}+\frac{2x(x-1)}{(x-2)(x+2)(x-1)}$$

$$=\frac{x^2-2x+x-2}{(x-2)(x+2)(x-1)}-\frac{x+2}{(x-2)(x+2)(x-1)}+\frac{2x^2-2x}{(x-2)(x+2)(x-1)}$$

$$=\frac{x^2-x-2-x-2+2x^2-2x}{(x-2)(x+2)(x-1)}$$

$$=\frac{x^2+2x^2-x-x-2x-2-2}{(x-2)(x+2)(x-1)}$$

$$=\frac{3x^2-4x-4}{(x-2)(x+2)(x-1)}$$

$$=\frac{(3x+2)\cancel{(x-2)}}{\cancel{(x-2)}(x+2)(x-1)}$$

$$=\frac{3x+2}{(x+2)(x-1)}$$

20. $f(x)=\frac{5x-10}{x^2+5x-14}$

To find the domain, we first set the denominator equal to 0 and solve the resulting equation.

$x^2+5x-14=0$

$(x+7)(x-2)=0$

$x+7=0 \quad \text{or} \quad x-2=0$

$x=-7 \qquad\qquad x=2$

Since the values -7 and 2 make the denominator 0, we must exclude these values from the domain. Therefore, we have Domain: $(-\infty,-7)\cup(-7,2)\cup(2,\infty)$

$f(x)=\frac{5x-10}{x^2+5x-14}=\frac{5\cancel{(x-2)}}{(x+7)\cancel{(x-2)}}=\frac{5}{x+7}$ where $x\neq-7.x\neq 2$.

6.5 Check Points

1. $(x^3 - 7x - 6) \div (x + 2) = x^2 - 2x - 3$

$$\begin{array}{r|rrrr} -2 & 1 & 0 & -7 & -6 \\ & & -2 & 4 & 6 \\ \hline & 1 & -2 & -3 & 0 \end{array}$$

2. $f(-4) = -105$

$$\begin{array}{r|rrrr} -4 & 3 & 4 & -5 & 3 \\ & & -12 & 32 & -108 \\ \hline & 3 & -8 & 27 & -105 \end{array}$$

3. Synthetic division shows -1 is a solution.

$$\begin{array}{r|rrrr} -1 & 15 & 14 & -3 & -2 \\ & & -15 & 1 & 2 \\ \hline & 15 & -1 & -2 & 0 \end{array}$$

Next, continue factoring to find all solution.

$$15x^3 + 14x^2 - 3x - 2 = 0$$
$$(x+1)(15x^2 - x - 2) = 0$$
$$(x+1)(5x-2)(3x+1) = 0$$
$$x + 1 = 0 \quad \text{or} \quad 5x - 2 = 0 \quad \text{or} \quad 3x + 1 = 0$$
$$x = -1 \qquad 5x = 2 \qquad 3x = -1$$
$$x = \frac{2}{5} \qquad x = -\frac{1}{3}$$

The solution set is $\left\{-1, -\frac{1}{3}, \frac{2}{5}\right\}$.

6.5 Concept and Vocabulary Check

1. 4; 1; 5; −7; 1

2. −5; 4; 0; −8; −2

3. true

4. $f(c)$

6.5 Exercise Set

1. $(2x^2 + x - 10) \div (x - 2)$

$$\begin{array}{r|rrr} 2 & 2 & 1 & -10 \\ & & 4 & 10 \\ \hline & 2 & 5 & 0 \end{array}$$

$$(2x^2 + x - 10) \div (x - 2) = 2x + 5$$

3. $(3x^2 + 7x - 20) \div (x + 5)$

$$\begin{array}{r|rrr} -5 & 3 & 7 & -20 \\ & & -15 & 40 \\ \hline & 3 & -8 & 20 \end{array}$$

$$(3x^2 + 7x - 20) \div (x + 5) = 3x - 8 + \frac{20}{x+5}$$

5. $(4x^3 - 3x^2 + 3x - 1) \div (x - 1)$

$$\begin{array}{r|rrrr} 1 & 4 & -3 & 3 & -1 \\ & & 4 & 1 & 4 \\ \hline & 4 & 1 & 4 & 3 \end{array}$$

$$(4x^3 - 3x^2 + 3x - 1) \div (x - 1)$$
$$= 4x^2 + x + 4 + \frac{3}{x-1}$$

7. $(6x^5 - 2x^3 + 4x^2 - 3x + 1) \div (x - 2)$

$$\begin{array}{r|rrrrrr} 2 & 6 & 0 & -2 & 4 & -3 & 1 \\ & & 12 & 24 & 44 & 96 & 186 \\ \hline & 6 & 12 & 22 & 48 & 93 & 187 \end{array}$$

$$(6x^5 - 2x^3 + 4x^3 - 3x + 1) \div (x - 2)$$
$$= 6x^4 + 12x^3 + 22x^2 + 48x + 93 + \frac{187}{x-2}$$

9. $\left(x^2-5x-5x^3+x^4\right)\div(5+x)$

Rewrite the polynomials in descending order.

$\left(x^4-5x^3+x^2-5x\right)\div(x+5)$

$$\begin{array}{r|rrrrr} -5 & 1 & -5 & 1 & -5 & 0 \\ & & -5 & 50 & -255 & 1300 \\ \hline & 1 & -10 & 51 & -260 & 1300 \end{array}$$

$\left(x^2-5x-5x^3+x^4\right)\div(5+x)$

$= x^3-10x^2+51x-260+\dfrac{1300}{5+x}$

11. $\left(3x^3+2x^2-4x+1\right)\div\left(x-\dfrac{1}{3}\right)$

$$\begin{array}{r|rrrr} \frac{1}{3} & 3 & 2 & -4 & 1 \\ & & 1 & 1 & -1 \\ \hline & 3 & 3 & -3 & 0 \end{array}$$

$\left(3x^3+2x^2-4x-1\right)\div\left(x-\dfrac{1}{3}\right)=3x^2+3x-3$

13. $\dfrac{x^5+x^3-2}{x-1}$

$$\begin{array}{r|rrrrrr} 1 & 1 & 0 & 1 & 0 & 0 & -2 \\ & & 1 & 1 & 2 & 2 & 2 \\ \hline & 1 & 1 & 2 & 2 & 2 & 0 \end{array}$$

$\dfrac{x^5+x^3-2}{x-1}=x^4+x^3+2x^2+2x+2$

15. $\dfrac{x^4-256}{x-4}$

$$\begin{array}{r|rrrrr} 4 & 1 & 0 & 0 & 0 & -256 \\ & & 4 & 16 & 64 & 256 \\ \hline & 1 & 4 & 16 & 64 & 0 \end{array}$$

$\dfrac{x^4-256}{x-4}=x^3+4x^2+16x+64$

17. $\dfrac{2x^5-3x^4+x^3-x^2+2x-1}{x+2}$

$$\begin{array}{r|rrrrrr} -2 & 2 & -3 & 1 & -1 & 2 & -1 \\ & & -4 & 14 & -30 & 62 & -128 \\ \hline & 2 & -7 & 15 & -31 & 64 & -129 \end{array}$$

$\dfrac{2x^5-3x^4+x^3-x^2+2x-1}{x+2}$

$=2x^4-7x^3+15x^2-31x+64-\dfrac{129}{x+2}$

19. $f(x)=2x^3-11x^2+7x-5$

$$\begin{array}{r|rrrr} 4 & 2 & -11 & 7 & -5 \\ & & 8 & -12 & -20 \\ \hline & 2 & -3 & -5 & -25 \end{array}$$

$f(4)=-25$

21. $f(x)=3x^3-7x^2-2x+5$

$$\begin{array}{r|rrrr} -3 & 3 & -7 & -2 & 5 \\ & & -9 & 48 & -138 \\ \hline & 3 & -16 & 46 & -133 \end{array}$$

$f(-3)=-133$

23. $f(x)=x^4+5x^3+5x^2-5x-6$

$$\begin{array}{r|rrrrr} 3 & 1 & 5 & 5 & -5 & -6 \\ & & 3 & 24 & 87 & 246 \\ \hline & 1 & 8 & 29 & 82 & 240 \end{array}$$

$f(3)=240$

25. $f(x)=2x^4-5x^3-x^2+3x+2$

$$\begin{array}{r|rrrrr} -\frac{1}{2} & 2 & -5 & -1 & 3 & 2 \\ & & -1 & 3 & -1 & -1 \\ \hline & 2 & -6 & 2 & 2 & 1 \end{array}$$

$f\left(-\frac{1}{2}\right)=1$

27. $x^3 - 4x^2 + x + 6 = 0$

$$\begin{array}{r|rrrr} -1 & 1 & -4 & 1 & 6 \\ & & -1 & 5 & -6 \\ \hline & 1 & -5 & 6 & 0 \end{array}$$

The remainder is zero and –1 is a solution to the equation.

$x^3 - 4x^2 + x + 6 = (x+1)(x^2 - 5x + 6)$

To solve the equation, set it equal to zero and factor.

$$x^3 - 4x^2 + x + 6 = 0$$
$$(x+1)(x^2 - 5x + 6) = 0$$
$$(x+1)(x-3)(x-2) = 0$$

Apply the zero product principle.

$$x+1=0 \quad x-3=0 \quad x-2=0$$
$$x=-1 \quad x=3 \quad x=2$$

The solution set is $\{-1, 2, 3\}$.

29. $2x^3 - 5x^2 + x + 2 = 0$

$$\begin{array}{r|rrrr} 2 & 2 & -5 & 1 & 2 \\ & & 4 & -2 & -2 \\ \hline & 2 & -1 & -1 & 0 \end{array}$$

The remainder is zero and 2 is a solution to the equation.

$$2x^3 - 5x^2 + x + 2 = (x-2)(2x^2 - x - 1)$$

To solve the equation, we set it equal to zero and factor.

$$2x^3 - 5x^2 + x + 2 = 0$$
$$(x-2)(2x^2 - x - 1) = 0$$
$$(x-2)(2x+1)(x-1) = 0$$

Apply the zero product principle.

$$x-2=0 \quad 2x+1=0 \quad x-1=0$$
$$x=2 \quad 2x=-1 \quad x=1$$
$$x=-\frac{1}{2}$$

The solution set is $\left\{-\frac{1}{2}, 1, 2\right\}$.

31. $6x^3 + 25x^2 - 24x + 5 = 0$

$$\begin{array}{r|rrrr} -5 & 6 & 25 & -24 & 5 \\ & & -30 & 25 & -5 \\ \hline & 6 & -5 & 1 & 0 \end{array}$$

The remainder is zero and 3 is a solution to the equation.

$$6x^3 + 25x^2 - 24x + 5$$
$$= (x+5)(6x^2 - 5x + 1)$$

To solve the equation, we set it equal to zero and factor.

$$6x^3 + 25x^2 - 24x + 5 = 0$$
$$(x+5)(6x^2 - 5x + 1) = 0$$
$$(x+5)(3x-1)(2x-1) = 0$$

Apply the zero product principle.

$$x+5=0 \quad 3x-1=0 \quad 2x-1=0$$
$$x=-5 \quad 3x=1 \quad 2x=1$$
$$x=\frac{1}{3} \quad x=\frac{1}{2}$$

The solution set is $\left\{-5, \frac{1}{3}, \frac{1}{2}\right\}$.

33. The graph indicates that 2 is a solution to the equation.

$$\begin{array}{r|rrrr} 2 & 1 & 2 & -5 & -6 \\ & & 2 & 8 & 6 \\ \hline & 1 & 4 & 3 & 0 \end{array}$$

The remainder is 0, so 2 is a solution.

$$x^3 + 2x^2 - 5x - 6 = 0$$
$$(x-2)(x^2 + 4x + 3) = 0$$
$$(x-2)(x+3)(x+1) = 0$$

The solution set is $\{-3, -1, 2\}$.

35. The table indicates that 1 is a solution to the equation.

$$\begin{array}{r|rrrr} 1 & 6 & -11 & 6 & -1 \\ & & 6 & -5 & 1 \\ \hline & 6 & -5 & 1 & 0 \end{array}$$

The remainder is 0, so 1 is a solution.

$$6x^3-11x^2+6x-1=0$$
$$(x-1)(6x^2-5x+1)=0$$
$$(x-1)(3x-1)(2x-1)=0$$

The solution set is $\left\{\frac{1}{3},\frac{1}{2},1\right\}$.

37. $\left(22x-24+7x^3-29x^2+4x^4\right)(x+4)^{-1}$

$$=\frac{22x-24+7x^3-29x^2+4x^4}{x+4}$$

$$\begin{array}{r|rrrrr} -4 & 4 & 7 & -29 & 22 & -24 \\ & & -16 & 36 & -28 & 24 \\ \hline & 4 & -9 & 7 & -6 & 0 \end{array}$$

Therefore,

$$\left(22x-24+7x^3-29x^2+4x^4\right)(x+4)^{-1}$$
$$=4x^3-9x^2+7x-6$$

39. $A=l\cdot w$ so

$$l=\frac{A}{w}=\frac{0.5x^3-0.3x^2+0.22x+0.06}{x+0.2}$$

$$\begin{array}{r|rrrr} -0.2 & 0.5 & -0.3 & 0.22 & 0.06 \\ & & -0.1 & 0.08 & -0.06 \\ \hline & 0.5 & -0.4 & 0.3 & 0 \end{array}$$

Therefore, the length of the rectangle is $0.5x^2-0.4x+0.3$ units.

41. a. $14x^3-17x^2-16x-177=0$

$$\begin{array}{r|rrrr} 3 & 14 & -17 & -16 & -177 \\ & & 42 & 75 & 177 \\ \hline & 14 & 25 & 59 & 0 \end{array}$$

The remainder is 0 so 3 is a solution.

$$14x^3-17x^2-16x-177$$
$$=(x-3)\left(14x^2+25x+59\right)$$

b. $f(x)=14x^3-17x^2-16x+34$

We need to find x when $f(x)=211$.

$$f(x)=14x^3-17x^2-16x+34$$
$$211=14x^3-17x^2-16x+34$$
$$0=14x^3-17x^2-16x-177$$

This is the equation obtained in part **a.** One solution is 3. It can be used to find other solutions (if they exist).

$$14x^3-17x^2-16x-177=0$$
$$(x-3)\left(14x^2+25x+59\right)=0$$

The polynomial $14x^2+25x+59$ cannot be factored, so the only solution is $x=3$. The female moth's abdominal width is 3 millimeters.

43. – 45. Answers will vary.

47. Exercise 27:

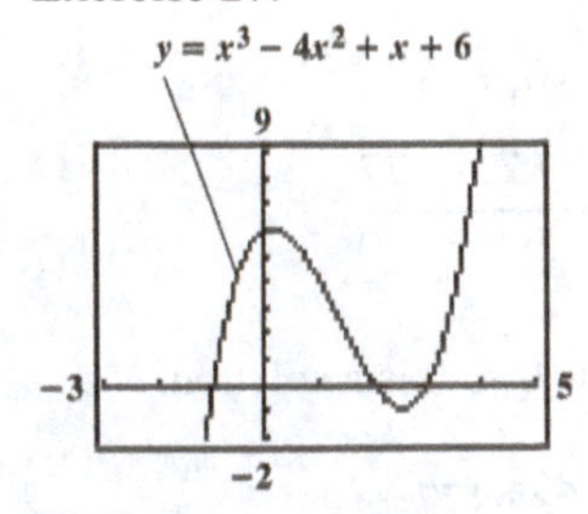

Exercise 28:

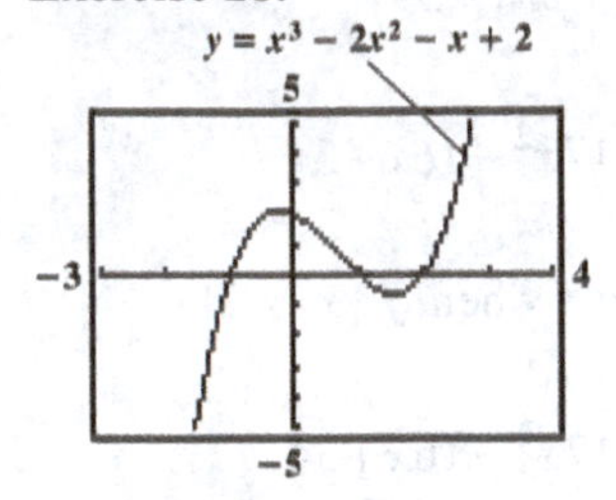

Exercise 29:

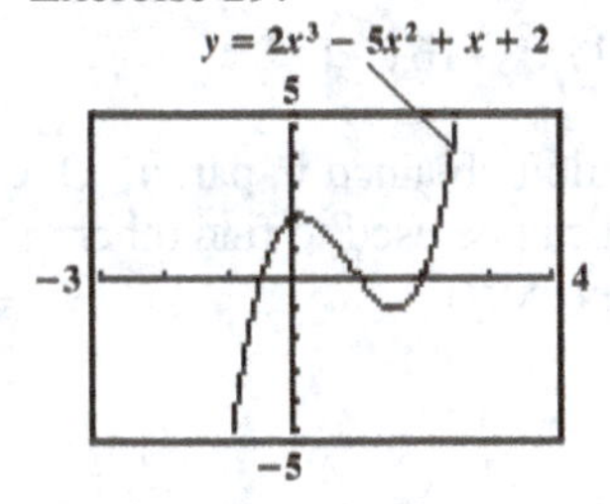

Exercise 30:

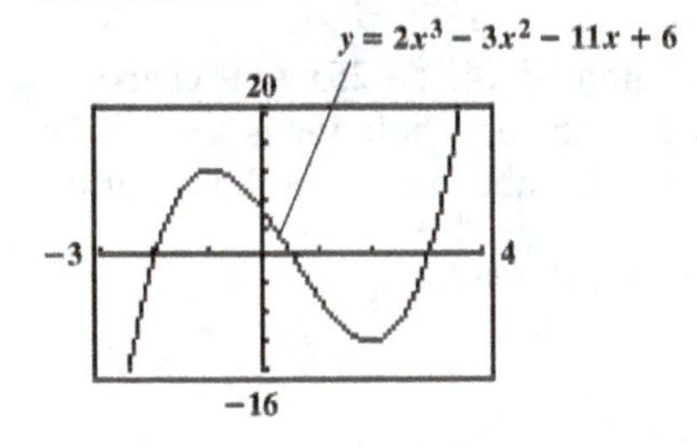

Exercise 31:

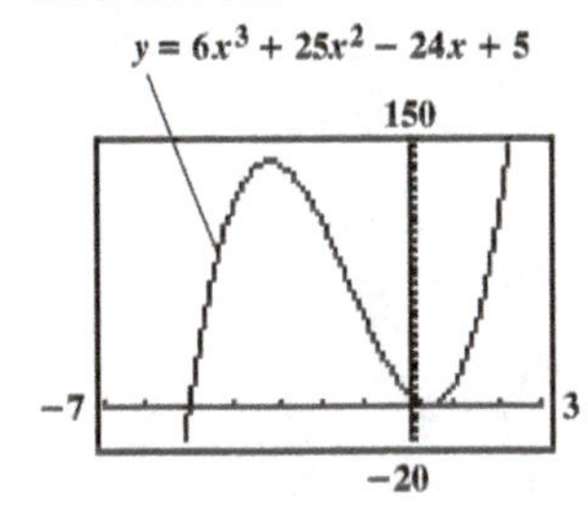

Exercise 32:

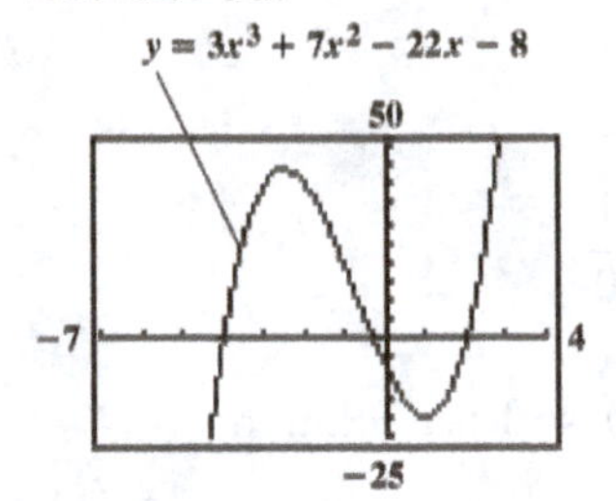

49. makes sense

51. makes sense

53. $x^4 - 4x^3 - 9x^2 + 16x + 20 = 0$

$$\begin{array}{r|rrrrr} 5 & 1 & -4 & -9 & 16 & 20 \\ & & 5 & 5 & -20 & -20 \\ \hline & 1 & 1 & -4 & -4 & 0 \end{array}$$

The remainder is zero and 5 is a solution to the equation.

$$x^4 - 4x^3 - 9x^2 + 16x + 20$$
$$= (x-5)(x^3 + x^2 - 4x - 4)$$

To solve the equation, we set it equal to zero and factor.

$$(x-5)(x^3 + x^2 - 4x - 4) = 0$$
$$(x-5)(x^2(x+1) - 4(x+1)) = 0$$
$$(x-5)(x+1)(x^2 - 4) = 0$$
$$(x-5)(x+1)(x+2)(x-2) = 0$$

Apply the zero product principle.

$x - 5 = 0 \qquad x + 1 = 0$
$x = 5 \qquad x = -1$

$x + 2 = 0 \qquad x - 2 = 0$
$x = -2 \qquad x = 2$

The solutions are −2, −1, 2 and 5 and the solution set is $\{-2, -1, 2, 5\}$.

54. $4x+3-13x-7<2(3-4x)$
$$-9x-4<6-8x$$
$$-x-4<6$$
$$-x<10$$
$$x>-10$$

The solution set is $(-10,\infty)$.

55. $2x(x+3)+6(x-3)=-28$
$$2x^2+6x+6x-18=-28$$
$$2x^2+12x-18=-28$$
$$2x^2+12x+10=0$$
$$2(x^2+6x+5)=0$$
$$2(x+5)(x+1)=0$$

Apply the zero product principle.

$2(x+5)=0 \qquad x+1=0$

$x+5=0 \qquad x=-1$

$x=-5$

The solution set is $\{-5,-1\}$.

56. $7x-6y=17$

$3x+\ \ y=18$

$$D=\begin{vmatrix}7 & -6\\ 3 & 1\end{vmatrix}=7(1)-3(-6)$$
$$=7+18=25$$

$$D_x=\begin{vmatrix}17 & -6\\ 18 & 1\end{vmatrix}=17(1)-18(-6)$$
$$=17+108=125$$

$$D_y=\begin{vmatrix}7 & 17\\ 3 & 18\end{vmatrix}=7(18)-3(17)$$
$$=126-51=75$$

$$x=\frac{D_x}{D}=\frac{125}{25}=5;\quad y=\frac{D_y}{D}=\frac{75}{25}=3$$

The solution is $(5,3)$ and the solution set is $\{(5,3)\}$.

57. LCD: $2\cdot 3\cdot x=6x$

58. LCD: $3\cdot x=3x$

59. LCD: $(x+3)(x-3)$

6.6 Check Points

1. The restriction is $x\neq 0$.
$$\frac{x+6}{2x}+\frac{x+24}{5x}=2$$
$$10x\cdot\left(\frac{x+6}{2x}+\frac{x+24}{5x}\right)=10x\cdot 2$$
$$\frac{10x(x+6)}{2x}+\frac{10x(x+24)}{5x}=10x\cdot 2$$
$$5(x+6)+2(x+24)=20x$$
$$5x+30+2x+48=20x$$
$$7x+78=20x$$
$$78=13x$$
$$x=6$$

The proposed result makes the original equation true.

The solution set is $\{6\}$.

2. The restrictions are $x\neq -1$ and $x\neq -6$.
$$\frac{x-3}{x+1}=\frac{x-2}{x+6}$$
$$(x+1)(x+6)\left(\frac{x-3}{x+1}\right)=(x+1)(x+6)\left(\frac{x-2}{x+6}\right)$$
$$(x+6)(x-3)=(x+1)(x-2)$$
$$x^2+3x-18=x^2-x-2$$
$$4x=16$$
$$x=4$$

The proposed result makes the original equation true.

The solution set is $\{4\}$.

3. The restriction is $x\neq -1$.
$$\frac{8x}{x+1}=4-\frac{8}{x+1}$$
$$(x+1)\frac{8x}{x+1}=(x+1)\left(4-\frac{8}{x+1}\right)$$
$$8x=4(x+1)-8$$
$$8x=4x+4-8$$
$$8x=4x-4$$
$$4x=-4$$
$$x=-1$$

The proposed result is restricted because $x\neq -1$.

The solution set is $\{\ \}$.

4. The restriction is $x \neq 0$.

$$\frac{x}{2}+\frac{12}{x}=5$$
$$2x\left(\frac{x}{2}+\frac{12}{x}\right)=2x\cdot 5$$
$$x^2+24=10x$$
$$x^2-10x+24=0$$
$$(x-6)(x-4)=0$$
$$x-6=0 \quad \text{or} \quad x-4=0$$
$$x=6 \qquad\qquad x=4$$

The solution set is $\{4, 6\}$.

5. The restrictions are $x \neq 3$ and $x \neq 4$.

$$\frac{3}{x-3}+\frac{5}{x-4}=\frac{x^2-20}{x^2-7x+12}$$
$$\frac{3}{x-3}+\frac{5}{x-4}=\frac{x^2-20}{(x-3)(x-4)}$$
$$(x-3)(x-4)\left(\frac{3}{x-3}+\frac{5}{x-4}\right)=(x-3)(x-4)\frac{x^2-20}{(x-3)(x-4)}$$
$$3(x-4)+5(x-3)=x^2-20$$
$$3x-12+5x-15=x^2-20$$
$$8x-27=x^2-20$$
$$0=x^2-8x+7$$
$$0=(x-7)(x-1)$$
$$x-7=0 \quad \text{or} \quad x-1=0$$
$$x=7 \qquad\qquad x=1$$

The solution set is $\{1, 7\}$.

6.
$$f(x)=\frac{120x}{100-x}$$
$$120=\frac{120x}{100-x}$$
$$(100-x)120=(100-x)\frac{120x}{100-x}$$
$$12{,}000-120x=120x$$
$$12{,}000=240x$$
$$x=50$$

If voters commit \$120 thousand, 50% of the lake's pollutants can be removed.

6.6 Concept and Vocabulary Check

1. LCD

2. 0

3. $2x$

4. $(x+5)(x+1)$

5. $x \neq 2$; $x \neq -4$

6. $5(x+3)+3(x+4)=12x+9$

7. true

6.6 Exercise Set

1. $\frac{1}{x}+2=\frac{3}{x}$

So that the denominator will not equal zero, x cannot be zero. To eliminate fractions, multiply by the LCD, x.

$$x\left(\frac{1}{x}+2\right)=x\left(\frac{3}{x}\right)$$
$$x\cdot\frac{1}{x}+x\cdot 2=3$$
$$1+2x=3$$
$$2x=2$$
$$x=1$$

The solution set is $\{1\}$.

3. $\frac{5}{x}+\frac{1}{3}=\frac{6}{x}$

So that the denominator will not equal zero, x cannot equal 0. To eliminate fractions, multiply by the LCD, $3x$.

$$3x\left(\frac{5}{x}+\frac{1}{3}\right)=3x\left(\frac{6}{x}\right)$$
$$3x\cdot\frac{5}{x}+3x\cdot\frac{1}{3}=3\cdot 6$$
$$3\cdot 5+x\cdot 1=18$$
$$15+x=18$$
$$x=3$$

The solution set is $\{3\}$.

5. $\dfrac{x-2}{2x}+1=\dfrac{x+1}{x}$

So that the denominator will not equal zero, x cannot equal 0. To eliminate fractions, multiply by the LCD, $2x$.

$$2x\left(\frac{x-2}{2x}+1\right)=2x\left(\frac{x+1}{x}\right)$$
$$2x\cdot\frac{x-2}{2x}+2x\cdot 1=2(x+1)$$
$$x-2+2x=2x+2$$
$$-2+3x=2x+2$$
$$-2+x=2$$
$$x=4$$

The solution set is $\{4\}$.

7. $\dfrac{3}{x+1}=\dfrac{5}{x-1}$

So that the denominator will not equal zero, x cannot equal 1 or -1. To eliminate fractions, multiply by the LCD, $(x+1)(x-1)$.

$$(x+1)(x-1)\frac{3}{x+1}=(x+1)(x-1)\frac{5}{x-1}$$
$$(x-1)\cdot 3=(x+1)\cdot 5$$
$$3x-3=5x+5$$
$$-3=2x+5$$
$$-8=2x$$
$$-4=x$$

The solution set is $\{-4\}$.

9. $\dfrac{x-6}{x+5}=\dfrac{x-3}{x+1}$

So that the denominator will not equal zero, x cannot equal -5 or -1. To eliminate fractions, multiply by the LCD, $(x+5)(x+1)$.

$$(x+5)(x+1)\frac{x-6}{x+5}=(x+5)(x+1)\frac{x-3}{x+1}$$
$$(x+1)(x-6)=(x+5)(x-3)$$
$$\cancel{x^2}-5x-6=\cancel{x^2}+2x-15$$
$$-5x-6=2x-15$$
$$-7x-6=-15$$
$$-7x=-9$$
$$x=\frac{9}{7}$$

The solution set is $\left\{\frac{9}{7}\right\}$.

11. $\frac{x+6}{x+3}=\frac{3}{x+3}+2$

So that the denominator will not equal zero, x cannot equal –3. To eliminate fractions, multiply by the LCD, $x+3$.

$$\begin{aligned}(x+3)\left(\frac{x+6}{x+3}\right)&=(x+3)\left(\frac{3}{x+3}+2\right)\\ x+6&=3+2x+6\\ x+6&=2x+9\\ 6&=x+9\\ -3&=x\end{aligned}$$

Since –3 would result in a zero denominator, we disregard it and conclude that the solution set is $\varnothing$.

13. $1-\frac{4}{x+7}=\frac{5}{x+7}$

So that the denominator will not equal zero, x cannot equal –7. To eliminate fractions, multiply by the LCD, $x+7$.

$$\begin{aligned}(x+7)\left(1-\frac{4}{x+7}\right)&=(x+7)\left(\frac{5}{x+7}\right)\\ (x+7)\cdot 1-(x+7)\cdot\frac{4}{x+7}&=5\\ x+7-4&=5\\ x+3&=5\\ x&=2\end{aligned}$$

The solution set is $\{2\}$.

15. $\frac{4x}{x+2}+\frac{2}{x-1}=4$

So that the denominator will not equal zero, x cannot equal –2 or 1. To eliminate fractions, multiply by the LCD, $(x+2)(x-1)$.

$$\begin{aligned}(x+2)(x-1)\cdot\left(\frac{4x}{x+2}+\frac{2}{x-1}\right)&=(x+2)(x-1)\cdot 4\\ (x-1)4x+(x+2)2&=\left(x^2+x-2\right)4\\ 4x^2-4x+2x+4&=4x^2+4x-8\\ \cancel{4x^2}-2x+4&=\cancel{4x^2}+4x-8\\ -2x+4&=4x-8\\ -6x+4&=-8\\ -6x&=-12\\ x&=2\end{aligned}$$

The solution set is $\{2\}$.

17. $$\frac{8}{x^2-9}+\frac{4}{x+3}=\frac{2}{x-3}$$
$$\frac{8}{(x+3)(x-3)}+\frac{4}{x+3}=\frac{2}{x-3}$$

So that the denominator will not equal zero, x cannot equal -3 or 3. To eliminate fractions, multiply by the LCD, $(x+3)(x-3)$.

$$(x+3)(x-3)\cdot\left(\frac{8}{(x+3)(x-3)}+\frac{4}{x+3}\right)=(x+3)(x-3)\cdot\frac{2}{x-3}$$
$$8+4(x-3)=2(x+3)$$
$$8+4x-12=2x+6$$
$$4x-4=2x+6$$
$$2x-4=6$$
$$2x=10$$
$$x=5$$

The solution set is $\{5\}$.

19. $x+\frac{7}{x}=-8$

So that the denominator will not equal zero, x cannot equal 0. To eliminate fractions, multiply by the LCD, x.

$$x\left(x+\frac{7}{x}\right)=x(-8)$$
$$x^2+7=-8x$$
$$x^2+8x+7=0$$
$$(x+7)(x+1)=0$$

Apply the zero product principle.

$x+7=0$ $\quad$ $x+1=0$
$x=-7$ $\quad$ $x=-1$

The solution set is $\{-7,-1\}$.

21. $\frac{6}{x}-\frac{x}{3}=1$

So that the denominator will not equal zero, x cannot equal 0. To eliminate fractions, multiply by the LCD, $3x$.

$$3x\left(\frac{6}{x}-\frac{x}{3}\right)=3x(1)$$

$$3x\left(\frac{6}{x}\right)-3x\left(\frac{x}{3}\right)=3x$$

$$3(6)-x(x)=3x$$
$$18-x^2=3x$$
$$0=x^2+3x-18$$
$$0=(x+6)(x-3)$$

Apply the zero product principle.

$x+6=0$ $\quad\quad$ $x-3=0$

$x=-6$ $\quad\quad$ $x=3$

The solution set is $\{-6,3\}$.

23. $\frac{x+6}{3x-12}=\frac{5}{x-4}+\frac{2}{3}$

$$\frac{x+6}{3(x-4)}=\frac{5}{x-4}+\frac{2}{3}$$

So that the denominator will not equal zero, x cannot equal 4. To eliminate fractions, multiply by the LCD, $3(x-4)$.

$$3(x-4)\left(\frac{x+6}{3(x-4)}\right)=3(x-4)\left(\frac{5}{x-4}+\frac{2}{3}\right)$$
$$x+6=3(x-4)\cdot\frac{5}{x-4}+3(x-4)\cdot\frac{2}{3}$$
$$x+6=3\cdot 5+2(x-4)$$
$$x+6=15+2x-8$$
$$x+6=7+2x$$
$$6=7+x$$
$$-1=x$$

The solution set is $\{-1\}$.

25. $\frac{1}{x-1}+\frac{1}{x+1}=\frac{2}{x^2-1}$

$$\frac{1}{x-1}+\frac{1}{x+1}=\frac{2}{(x+1)(x-1)}$$

So that the denominator will not equal zero, x cannot equal -1 or 1. To eliminate fractions, multiply by the LCD, $(x+1)(x-1)$.

$$(x+1)(x-1)\left(\frac{1}{x-1}+\frac{1}{x+1}\right)=(x+1)(x-1)\cdot\frac{2}{(x+1)(x-1)}$$

$$(x+1)+(x-1)=2$$

$$x+1+x-1=2$$

$$x+x=2$$

$$2x=2$$

$$x=1$$

Since 1 would result in a zero denominator, we disregard it and conclude that the solution set is $\varnothing$.

27. $\frac{5}{x+4}+\frac{3}{x+3}=\frac{12x+19}{x^2+7x+12}$

$$\frac{5}{x+4}+\frac{3}{x+3}=\frac{12x+19}{(x+4)(x+3)}$$

So that the denominator will not equal zero, x cannot equal -4 or -3. To eliminate fractions, multiply by the LCD, $(x+4)(x+3)$.

$$(x+4)(x+3)\left(\frac{5}{x+4}+\frac{3}{x+3}\right)=(x+4)(x+3)\left(\frac{12x+19}{(x+4)(x+3)}\right)$$

$$(x+4)(x+3)\cdot\left(\frac{5}{x+4}\right)+(x+4)(x+3)\cdot\left(\frac{3}{x+3}\right)=12x+19$$

$$(x+3)\cdot 5+(x+4)\cdot 3=12x+19$$

$$5x+15+3x+12=12x+19$$

$$8x+27=12x+19$$

$$-4x+27=19$$

$$-4x=-8$$

$$x=2$$

The solution set is $\{2\}$.

29. $\frac{4x}{x+3}-\frac{12}{x-3}=\frac{4x^2+36}{x^2-9}$

$\frac{4x}{x+3}-\frac{12}{x-3}=\frac{4x^2+36}{(x+3)(x-3)}$

So that the denominator will not equal zero, x cannot equal –3 or 3. To eliminate fractions, multiply by the LCD, $(x+3)(x-3)$.

$$(x+3)(x-3)\left(\frac{4x}{x+3}-\frac{12}{x-3}\right)=(x+3)(x-3)\left(\frac{4x^2+36}{(x+3)(x-3)}\right)$$

$$(x+3)(x-3)\left(\frac{4x}{x+3}\right)-(x+3)(x-3)\left(\frac{12}{x-3}\right)=4x^2+36$$

$$(x-3)(4x)-(x+3)(12)=4x^2+36$$

$$4x^2-12x-(12x+36)=4x^2+36$$

$$\cancel{4x^2}-12x-12x-36=\cancel{4x^2}+36$$

$$-24x-36=36$$

$$-24x=72$$

$$x=-3$$

Since –3 would result in a zero denominator, we disregard it and conclude that the solution set is $\varnothing$.

31. $\frac{4}{x^2+3x-10}+\frac{1}{x^2+9x+20}=\frac{2}{x^2+2x-8}$

$\frac{4}{(x+5)(x-2)}+\frac{1}{(x+5)(x+4)}=\frac{2}{(x+4)(x-2)}$

So that the denominator will not equal zero, x cannot equal –5, –4, or 2. To eliminate fractions, multiply by the LCD, $(x+5)(x+4)(x-2)$.

$$(x+5)(x+4)(x-2)\left(\frac{4}{(x+5)(x-2)}+\frac{1}{(x+5)(x+4)}\right)=(x+5)(x+4)(x-2)\left(\frac{2}{(x+4)(x-2)}\right)$$

$$(x+4)\cdot 4+(x-2)\cdot 1=(x+5)\cdot 2$$

$$4x+16+x-2=2x+10$$

$$5x+14=2x+10$$

$$3x+14=10$$

$$3x=-4$$

$$x=-\frac{4}{3}$$

The solution set is $\left\{-\frac{4}{3}\right\}$.

33. $\dfrac{3y}{y^2+5y+6}+\dfrac{2}{y^2+y-2}=\dfrac{5y}{y^2+2y-3}$

$$\frac{3y}{(y+2)(y+3)}+\frac{2}{(y+2)(y-1)}=\frac{5y}{(y+3)(y-1)}$$

So that the denominator will not equal zero, x cannot equal $-3, -2,$ or 1. To eliminate fractions, multiply by the LCD $(y+2)(y+3)(y-1)$.

$$(y+2)(y+3)(y-1)\left(\frac{3y}{(y+2)(y+3)}+\frac{2}{(y+2)(y-1)}\right)+=(y+2)(y+3)(y-1)\frac{5y}{(y+3)(y-1)}$$

$$3y(y-1)+2(y+3)=5y(y+2)$$
$$3y^2-3y+2y+6=5y^2+10y$$
$$0=2y^2+11y-6$$
$$0=(2y-1)(y+6)$$
$$y=\frac{1}{2} \text{ or } y=-6$$

The solution set is $\left\{-6,\dfrac{1}{2}\right\}$.

35. $g(x)=\dfrac{x}{2}+\dfrac{20}{x}$; Since $g(a)=7$, then

$$7=\frac{a}{2}+\frac{20}{a}$$
$$(2a)7=(2a)\left(\frac{a}{2}+\frac{20}{a}\right)$$
$$14a=a^2+40$$
$$a^2-14a+40=0$$
$$(a-10)(a-4)=0$$
$$a=10 \text{ or } a=4$$

37. $g(x)=\dfrac{5}{x+2}+\dfrac{25}{x^2+4x+4}$;

$=\dfrac{5}{x+2}+\dfrac{25}{(x+2)^2}$

Since $g(a)=20$ then

$$20=\frac{5}{a+2}+\frac{25}{(a+2)^2}$$

$$20(a+2)^2=(a+2)^2\left(\frac{5}{a+2}+\frac{25}{(a+2)^2}\right)$$

$$20(a+2)^2=5(a+2)+25$$

$$20\left(a^2+4a+4\right)=5a+10+25$$

$$20a^2+80a+80=5a+35$$

$$20a^2+75a+45=0$$

$$5\left(4a^2+15a+9\right)=0$$

$$5(4a+3)(a+3)=0$$

$a=-\dfrac{3}{4}$ or $a=-3$

39. $$\frac{x+2}{x^2-x}-\frac{6}{x^2-1}=\frac{x+2}{x(x-1)}-\frac{6}{(x-1)(x+1)}=\frac{(x+1)}{(x+1)}\frac{x+2}{x(x-1)}-\frac{x}{x}\cdot\frac{6}{(x-1)(x+1)}$$

$$=\frac{(x+1)(x+2)}{x(x-1)(x+1)}-\frac{6x}{x(x-1)(x+1)}=\frac{(x+1)(x+2)-6x}{x(x-1)(x+1)}$$

$$=\frac{x^2+3x+2-6x}{x(x-1)(x+1)}=\frac{x^2-3x+2}{x(x-1)(x+1)}=\frac{(x-2)(x-1)}{x(x-1)(x+1)}$$

$$=\frac{x-2}{x(x+1)}$$

41. In Exercise 39, the left side of this equation was simplified;

$$\frac{x+2}{x^2-x}-\frac{6}{x^2-1}=0$$

$$\frac{x-2}{x(x+1)}=0$$

$$x(x+1)\frac{x-2}{x(x+1)}=0\cdot x(x+1)$$

$$x-2=0$$

$$x=2$$

The solution set is $\{2\}$.

43.
$$\frac{1}{x^3-8}+\frac{3}{(x-2)(x^2+2x+4)}=\frac{2}{x^2+2x+4}$$
$$\frac{1}{x^3-8}+\frac{3}{(x-2)(x^2+2x+4)}-\frac{2}{x^2+2x+4}=0$$
$$\frac{1}{(x-2)(x^2+2x+4)}+\frac{3}{(x-2)(x^2+2x+4)}-\frac{2}{x^2+2x+4}=0$$
$$\frac{4}{(x-2)(x^2+2x+4)}-\frac{(x-2)}{(x-2)}\frac{2}{x^2+2x+4}=0$$
$$\frac{4-2(x-2)}{(x-2)\left(x^2+2x+4\right)}=0$$
$$\frac{4-2x+4}{(x-2)\left(x^2+2x+4\right)}=0$$
$$\frac{-2x+8}{(x-2)(x^2+2x+4)}=0$$
$$(x-2)\left(x^2+2x+4\right)\frac{-2x+8}{(x-2)\left(x^2+2x+4\right)}=0(x-2)\left(x^2+2x+4\right)$$
$$-2x+8=0$$
$$-2x=-8$$
$$x=4$$

The solution set is $\{4\}$.

45. $\frac{1}{x^3-8}+\frac{3}{(x-2)(x^2+2x+4)}-\frac{2}{x^2+2x+4}$

$$=\frac{1}{(x-2)(x^2+2x+4)}+\frac{3}{(x-2)(x^2+2x+4)}-\frac{2}{x^2+2x+4}$$
$$=\frac{4}{(x-2)(x^2+2x+4)}-\frac{(x-2)}{(x-2)}\frac{2}{x^2+2x+4}=\frac{4-2(x-2)}{(x-2)\left(x^2+2x+4\right)}$$
$$=\frac{4-2x+4}{(x-2)\left(x^2+2x+4\right)}=\frac{-2x+8}{(x-2)\left(x^2+2x+4\right)}$$
$$=\frac{-2(x-4)}{(x-2)\left(x^2+2x+4\right)}$$

47. First find $f(a)$ and $g(a)$;

$$f(a)=\frac{a+2}{a+3};\quad g(a)=\frac{a+1}{a^2+2a-3}=\frac{a+1}{(a+3)(a-1)}$$

Then

$$f(a)=g(a)+1$$

$$\frac{a+2}{a+3}=\frac{a+1}{(a+3)(a-1)}+1$$

$$\frac{a+2}{a+3}-\frac{a+1}{(a+3)(a-1)}-1=0$$

$$(a+3)(a-1)\left(\frac{a+2}{a+3}-\frac{a+1}{(a+3)(a-1)}-1\right)=0(a+3)(a-1)$$

$$(a-1)(a+2)-(a+1)-(a+3)(a-1)=0$$

$$(a^2+a-2)-a-1-(a^2+2a-3)=0$$

$$a^2-3-a^2-2a+3=0$$

$$-2a=0$$

$$a=0$$

49. First find $(f+g)(a)$ and $h(a)$;

$$(f+g)(a)=f(a)+g(a)=\frac{5}{a-4}+\frac{3}{a-3}=\frac{(a-3)}{(a-3)}\cdot\frac{5}{a-4}+\frac{(a-4)}{(a-4)}\cdot\frac{3}{a-3}$$

$$=\frac{5(a-3)+3(a-4)}{(a-3)(a-4)}=\frac{5a-15+3a-12}{(a-3)(a-4)}=\frac{8a-27}{(a-3)(a-4)}$$

$$h(a)=\frac{a^2-20}{a^2-7a+12}=\frac{a^2-20}{(a-4)(a-3)}$$

$$\frac{8a-27}{(a-3)(a-4)}=\frac{a^2-20}{(a-4)(a-3)}$$

$$(a-3)(a-4)\frac{8a-27}{(a-3)(a-4)}=(a-3)(a-4)\frac{a^2-20}{(a-4)(a-3)}$$

$$8a-27=a^2-20$$

$$a^2-8a+7=0$$

$$(a-1)(a-7)=0$$

$$a=1 \text{ or } a=7$$

51.
$$f(x)=\frac{250x}{100-x}$$
$$375=\frac{250x}{100-x}$$
$$(100-x)(375)=(100-x)\left(\frac{250x}{100-x}\right)$$
$$37500-375x=250x$$
$$37500=625x$$
$$60=x$$

If the government commits \$375 million for the project, 60% of pollutants can be removed.

53. $f(x)=\frac{5x+30}{x}$
$$8=\frac{5x+30}{x}$$
$$x(8)=x\left(\frac{5x+30}{x}\right)$$
$$8x=5x+30$$
$$3x=30$$
$$x=10$$
The students will remember an average of 8 words after 10 days. This is shown on the graph as the point (10, 8).

55. The horizontal asymptote is $y = 5$. The means that on average, the students will remember only about 5 words over an extended period of time.

57. $f(x)=\frac{0.9x-0.4}{0.9x+0.1}$
$$0.95=\frac{0.9x-0.4}{0.9x+0.1}$$
$$\frac{0.95}{1}=\frac{0.9x-0.4}{0.9x+0.1}$$

$$0.95(0.9x+0.1)=0.9x-0.4$$
$$0.855x+0.095=0.9x-0.4$$
$$0.095=0.045x-0.4$$
$$0.495=0.045x$$
$$11=x$$
It will take 11 learning trials for 0.95 or 95% of the responses to be correct. This is shown on the graph as the point (11, 0.95).

59. As the number of learning trials increases, the proportion of correct responses increases. Initially the proportion of correct responses increases rapidly, but slows down over time.

61.
$$C(x)=\frac{x+0.1(500)}{x+500}$$
$$0.28=\frac{x+0.1(500)}{x+500}$$
$$0.28=\frac{x+50}{x+500}$$
$$0.28(x+500)=(x+500)\frac{x+50}{x+500}$$
$$0.28x+140=x+50$$
$$0.72x=90$$
$$x=125$$
125 liters of pure peroxide should be added to produce a new product that is 28% peroxide.

63. – 69. Answers will vary.

71. $\frac{50}{x}=2x$

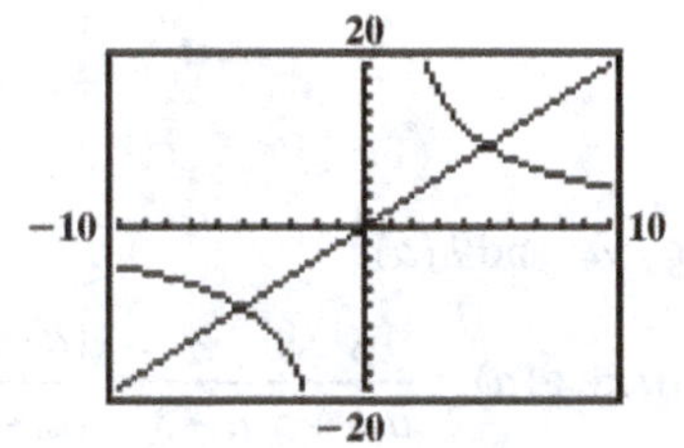

Check the solutions by substituting –5 and 5 in the original equation.

$x=5$	$x=-5$
$\frac{50}{5}=2(5)$	$\frac{50}{-5}=2(-5)$
$10=10$	$-10=-10$

The solutions check. The solution set is $\{-5,5\}$.

73. $\frac{2}{x}=x+1$

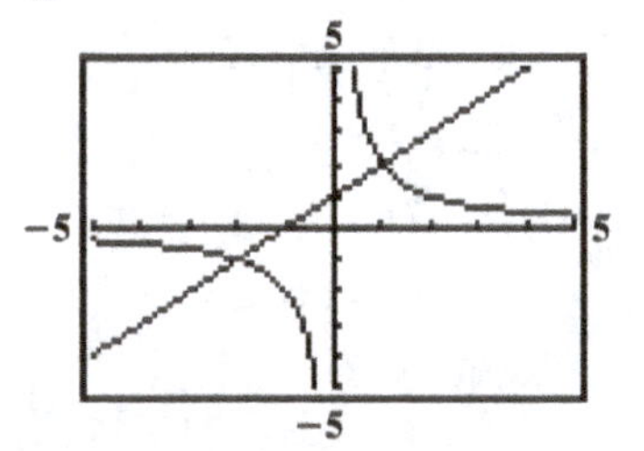

Check the solutions by substituting –2 and 1 in the original equation.

$$x = -2 \qquad x = 1$$
$$\frac{2}{-2} = -2+1 \qquad \frac{2}{1} = 1+1$$
$$-1 = -1 \qquad 2 = 2$$

The solutions check. The solution set is $\{-2,1\}$.

75. does not make sense; Explanations will vary. Sample explanation: Some rational equations have no solution.

77. makes sense

79. false; Changes to make the statement true will vary. A sample change is: Not all of the solutions of the resulting equation are solutions of the rational equation. Solutions which will result in a denominator of zero must be discarded.

81. false; Changes to make the statement true will vary. A sample change is: Zero does not satisfy the equation. Zero results in a denominator of zero.

83.
$$\frac{\left(\frac{1}{x+1}-\frac{x}{x-1}\right)}{\left(\frac{x}{x+1}-\frac{1}{x-1}\right)} = -1$$
$$\frac{(x+1)(x-1)\left(\frac{1}{x+1}-\frac{x}{x-1}\right)}{(x+1)(x-1)\left(\frac{x}{x+1}-\frac{1}{x-1}\right)} = -1$$
$$\frac{x-1-x(x+1)}{x(x-1)-(x+1)} = -1$$
$$\frac{x-1-x^2-x}{x^2-x-x-1} = -1$$
$$\frac{-x^2-1}{x^2-2x-1} = -1$$
$$-x^2-1 = (-1)(x^2-2x-1)$$
$$-x^2-1 = -x^2+2x+1$$
$$-2 = 2x$$
$$-1 = x$$

$x=-1$ is not a possible solution because -1 makes the denominator of the original equation 0. So, there is no solution or $\varnothing$.

85. Answers will vary. One example is $\frac{1}{x}+3=\frac{1}{x}+5$.

87. $x+2y \ge 2$
$x-y \ge -4$

First consider $x+2y \ge 2$. If we solve for y in $x+2y=2$, we can graph using the slope and the y-intercept.

$$x+2y = 2$$
$$2y = -x+2$$
$$y = -\frac{1}{2}x+1$$

y-intercept = 1; slope = $-\frac{1}{2}$

Now, use the origin, $(0,0)$, as a test point.

$$x+2y \ge 2$$
$$0+2(0) \ge 2$$
$$0 \ge 2$$

This is a false statement. This means that the point $(0,0)$ will not fall in the shaded half-plane.

Next consider $x-y \ge -4$. If we solve for y in $x-y=-4$, we can graph using the slope and the y-intercept.

$$x-y = -4$$
$$-y = -x-4$$
$$y = x+4 \qquad y\text{-intercept} = 4;\ \text{slope} = 1$$

Now, use the origin, $(0,0)$, as a test point.

$$x-y \ge -4$$
$$0-0 \ge -4$$
$$0 \ge -4$$

This is a true statement. This means that the point $(0,0)$ will fall in the shaded half-plane.

Next, graph each of the inequalities. The solution to the system is the intersection of the shaded half-planes.

$x + 2y \ge 2$
$x - y \ge -4$

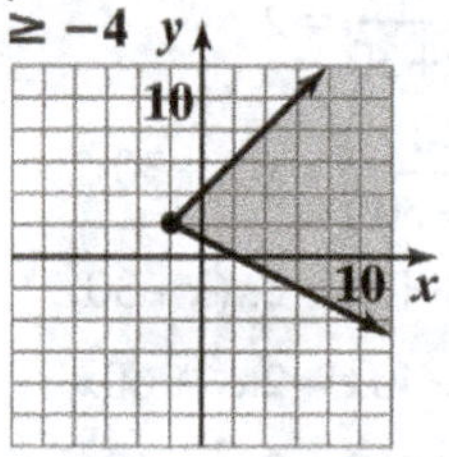

88.
$$\frac{x-4}{2}-\frac{1}{5}=\frac{7x+1}{20}$$
$$20\left(\frac{x-4}{2}-\frac{1}{5}\right)=20\left(\frac{7x+1}{20}\right)$$
$$20\left(\frac{x-4}{2}\right)-20\left(\frac{1}{5}\right)=7x+1$$
$$10(x-4)-4(1)=7x+1$$
$$10x-40-4=7x+1$$
$$10x-44=7x+1$$
$$3x-44=1$$
$$3x=45$$
$$x=15$$

The solution set is $\{15\}$.

89.
$$C=\frac{5F-160}{9}$$
$$9C=(9)\frac{5F-160}{9}$$
$$9C=5F-160$$
$$5F=9C+160$$
$$\frac{5F}{5}=\frac{9C+160}{5}$$
$$F=\frac{9C+160}{5}$$

90.
$$qf+pf=pq$$
$$pf-pq=-qf$$
$$p(f-q)=-qf$$
$$p=\frac{-qf}{f-q}$$
$$p=\frac{qf}{q-f}$$

91.
$$\frac{40}{x}+\frac{40}{x+30}=2$$
$$x(x+30)\left(\frac{40}{x}+\frac{40}{x+30}\right)=x(x+30)2$$
$$40(x+30)+40x=2x(x+30)$$
$$40x+1200+40x=2x^2+60x$$
$$80x+1200=2x^2+60x$$
$$0=2x^2-20x-1200$$
$$0=x^2-10x-600$$
$$0=(x+20)(x-30)$$

$$x+20=0 \quad \text{or} \quad x-30=0$$
$$x=-20 \qquad\qquad x=30$$

The solution set is $\{-20, 30\}$.

92.
$$\frac{980}{450+x}=\frac{820}{450-x}$$
$$(450+x)(450-x)\frac{980}{450+x}=(450+x)(450-x)\frac{820}{450-x}$$
$$(450-x)980=(450+x)820$$
$$441,000-980x=369,000+820x$$
$$72,000=1800x$$
$$x=40$$

The average rate of the wind is 40 miles per hour.

6.7 Check Points

1.
$$a=\frac{b}{x+2}$$
$$(x+2)(a)=(x+2)\left(\frac{b}{x+2}\right)$$
$$ax+2a=b$$
$$ax=b-2a$$
$$\frac{ax}{a}=\frac{b-2a}{a}$$
$$x=\frac{b-2a}{a}$$

2.
$$\frac{1}{x}+\frac{1}{y}=\frac{1}{z}$$
$$xyz\left(\frac{1}{x}+\frac{1}{y}\right)=xyz\left(\frac{1}{z}\right)$$
$$xyz\left(\frac{1}{x}\right)+xyz\left(\frac{1}{y}\right)=xy$$
$$yz+xz=xy$$
$$xz-xy=-yz$$
$$x(z-y)=-yz$$
$$x=\frac{-yz}{z-y}$$
$$x=\frac{yz}{y-z}$$

3. a. $C(x) = 500,000 + 400x$

b. $\overline{C}(x) = \dfrac{500,000 + 400x}{x}$

c.
$$\overline{C}(x) = \frac{500,000 + 400x}{x}$$
$$450 = \frac{500,000 + 400x}{x}$$
$$x \cdot 450 = x \cdot \frac{500,000 + 400x}{x}$$
$$450x = 500,000 + 400x$$
$$50x = 500,000$$
$$x = 10,000$$

10,000 wheelchairs must be produced each month for the company to reach an average cost of $450 per chair.

4. Let $x =$ the walking rate.
Let $4x =$ the cycling rate.

	Distance	Rate	Time $= \dfrac{\text{Distance}}{\text{Rate}}$
Cycling	40	$4x$	$\dfrac{40}{4x}$
Walking	5	x	$\dfrac{5}{x}$

$$\overbrace{\frac{40}{4x}}^{\text{time cycling}} + \overbrace{\frac{5}{x}}^{\text{time walking}} = \overbrace{5}^{\text{total time}}$$
$$4x\left(\frac{40}{4x} + \frac{5}{x}\right) = 4x \cdot 5$$
$$\frac{4x \cdot 40}{4x} + \frac{4x \cdot 5}{x} = 4x \cdot 5$$
$$40 + 20 = 20x$$
$$60 = 20x$$
$$x = 3$$
$$4x = 12$$

The cyclist was riding at a rate of 12 miles per hour.

5. Let $x =$ months it takes if both machines work together.

	Fractional part of job completed in 1 month	Time working together	Fractional part of job completed in x months
Slower machine	$\dfrac{1}{18}$	x	$\dfrac{x}{18}$
Faster machine	$\dfrac{1}{9}$	x	$\dfrac{x}{9}$

$$\overbrace{\frac{x}{18}}^{\substack{\text{Fractional part}\\\text{of job completed}\\\text{by slower machine}}} + \overbrace{\frac{x}{9}}^{\substack{\text{Fractional part}\\\text{of job completed}\\\text{by fasterer machine}}} = \overbrace{1}^{\substack{\text{one}\\\text{whole}\\\text{job}}}$$

$$18\left(\frac{x}{18}+\frac{x}{9}\right)=18\cdot 1$$

$$\frac{18\cdot x}{18}+\frac{18\cdot x}{9}=18$$

$$x+2x=18$$

$$3x=18$$

$$x=6$$

The two machines can complete the job together in 6 months.

6. Let x = number of hours it takes the experienced carpenter working alone.
Let $3x$ = number of hours it takes the apprentice working alone.

	Fractional part of job completed in 1 hour	Time working together	Fractional part of job completed in 6 hours
Experienced carpenter	$\frac{1}{x}$	6	$\frac{6}{x}$
Apprentice	$\frac{1}{3x}$	6	$\frac{6}{3x}$

$$\overbrace{\frac{6}{x}}^{\substack{\text{Fractional part}\\\text{of job completed}\\\text{by the experienced}\\\text{carpenter}}} + \overbrace{\frac{6}{3x}}^{\substack{\text{Fractional part}\\\text{of job completed}\\\text{by the apprentice}}} = \overbrace{1}^{\substack{\text{one}\\\text{whole}\\\text{job}}}$$

$$3x\left(\frac{6}{x}+\frac{6}{3x}\right)=3x\cdot 1$$

$$\frac{3x\cdot 6}{x}+\frac{3x\cdot 6}{3x}=3x$$

$$18+6=3x$$

$$3x=24$$

$$x=8$$

The experienced carpenter can panel a room in 8 hours and the apprentice in 24 hours, working alone.

6.7 Concept and Vocabulary Check

1. xyz
2. fixed; variable
3. number of units produced
4. distance traveled; rate of travel
5. 1
6. $\frac{x}{19}$

6.7 Exercise Set

1.
$$\frac{V_1}{V_2}=\frac{P_2}{P_1}$$
$$P_1V_2\left(\frac{V_1}{V_2}\right)=P_1V_2\left(\frac{P_2}{P_1}\right)$$
$$P_1(V_1)=V_2(P_2)$$
$$P_1V_1=P_2V_2$$
$$\frac{P_1V_1}{V_1}=\frac{P_2V_2}{V_1}$$
$$P_1=\frac{P_2V_2}{V_1}$$

3.
$$\frac{1}{p}+\frac{1}{q}=\frac{1}{f}$$
$$fpq\left(\frac{1}{p}+\frac{1}{q}\right)=fpq\left(\frac{1}{f}\right)$$
$$fpq\left(\frac{1}{p}\right)+fpq\left(\frac{1}{q}\right)=pq$$
$$fq+fp=pq$$
$$f(q+p)=pq$$
$$\frac{f(q+p)}{q+p}=\frac{pq}{q+p}$$
$$f=\frac{pq}{q+p}$$

5.
$$P=\frac{A}{1+r}$$
$$(1+r)(P)=(1+r)\left(\frac{A}{1+r}\right)$$
$$P+Pr=A$$
$$Pr=A-P$$
$$\frac{Pr}{P}=\frac{A-P}{P}$$
$$r=\frac{A-P}{P}$$

7.
$$F=\frac{Gm_1m_2}{d^2}$$
$$d^2(F)=d^2\left(\frac{Gm_1m_2}{d^2}\right)$$
$$d^2F=Gm_1m_2$$
$$\frac{d^2F}{Gm_2}=\frac{Gm_1m_2}{Gm_2}$$
$$m_1=\frac{d^2F}{Gm_2}$$

9.
$$z=\frac{x-\overline{x}}{s}$$
$$s(z)=s\left(\frac{x-\overline{x}}{s}\right)$$
$$zs=x-\overline{x}$$
$$x=\overline{x}+zs$$

11.
$$I=\frac{E}{R+r}$$
$$(R+r)(I)=(R+r)\left(\frac{E}{R+r}\right)$$
$$IR+Ir=E$$
$$IR=E-Ir$$
$$\frac{IR}{I}=\frac{E-Ir}{I}$$
$$R=\frac{E-Ir}{I}$$

13.
$$f=\frac{f_1f_2}{f_1+f_2}$$
$$(f_1+f_2)(f)=(f_1+f_2)\left(\frac{f_1f_2}{f_1+f_2}\right)$$
$$ff_1+ff_2=f_1f_2$$
$$ff_2=f_1f_2-ff_1$$
$$ff_2=f_1(f_2-f)$$
$$\frac{ff_2}{f_2-f}=\frac{f_1(f_2-f)}{f_2-f}$$
$$f_1=\frac{ff_2}{f_2-f}$$

15. Approximately 50,000 wheelchairs must be produced each month for the average cost to be $410 per chair.

17. The horizontal asymptote is $y = 400$. This means that despite the number of wheelchairs produced, the cost will approach but never reach $400.

19. a. $C(x) = 100,000 + 100x$

b. $\overline{C}(x) = \dfrac{100,000 + 100x}{x}$

c.

$$300 = \frac{100,000 + 100x}{x}$$
$$x(300) = x\left(\frac{100,000 + 100x}{x}\right)$$
$$300x = 100,000 + 100x$$
$$200x = 100,000$$
$$x = 500$$

500 mountain bikes must be produced each month for the average cost to be $300.

21. The running rate is 5 miles per hour.

23. The time increases as the running rate is close to zero miles per hour.

25.

	d	R	$t = \frac{d}{r}$
Car	300	X	$\frac{300}{x}$
Bus	180	$x-20$	$\frac{180}{x-20}$

$$\frac{300}{x} = \frac{180}{x-20}$$
$$x(x-20)\left(\frac{300}{x}\right) = x(x-20)\left(\frac{180}{x-20}\right)$$
$$(x-20)(300) = 180x$$
$$300x - 6000 = 180x$$
$$120x = 6000$$
$$x = 50$$

The average rate for the car is 50 miles per hour and the average rate for the bus is $x - 20 = 50 - 20 = 30$ miles per hour.

27.

	d	r	$t=\frac{d}{r}$
To Campus	5	$x+9$	$\frac{5}{x+9}$
From Campus	5	x	$\frac{5}{x}$

$$\frac{5}{x+9}+\frac{5}{x}=\frac{7}{6}$$

$$6x(x+9)\left(\frac{5}{x+9}+\frac{5}{x}\right)=6x(x+9)\left(\frac{7}{6}\right)$$

$$6x(x+9)\left(\frac{5}{x+9}\right)+6x(x+9)\left(\frac{5}{x}\right)=x(x+9)(7)$$

$$6x(5)+6(x+9)(5)=7x(x+9)$$

$$30x+30(x+9)=7x^2+63x$$

$$30x+30x+270=7x^2+63x$$

$$60x+270=7x^2+63x$$

$$0=7x^2+3x-270$$

$$0=(7x+45)(x-6)$$

Apply the zero product principle.

$$7x+45=0 \qquad x-6=0$$

$$7x=-45 \qquad x=6$$

$$x=-\tfrac{45}{7}$$

The solution set is $\left\{-\frac{45}{7},6\right\}$. We disregard $-\frac{45}{7}$ because we can't have a negative time measurement. The average rate on the trip home is 6 miles per hour.

29. The time with the current is $\frac{20}{7+x}$. The time against the current is $\frac{8}{7-x}$.

$$\frac{20}{7+x}=\frac{8}{7-x}$$

$$(7+x)(7-x)\left(\frac{20}{7+x}\right)=(7+x)(7-x)\left(\frac{8}{7-x}\right)$$

$$(7-x)(20)=(7+x)(8)$$

$$140-20x=56+8x$$

$$84=28x$$

$$3=x$$

The rate of the current is 3 miles per hour.

31.

	D	r	$t = \frac{d}{r}$
With Stream	2400	$x + 100$	$\frac{2400}{x+100}$
Against Stream	1600	$x - 100$	$\frac{1600}{x-100}$

$$\frac{2400}{x+100} = \frac{1600}{x-100}$$
$$(x+100)(x-100)\left(\frac{2400}{x+100}\right) = (x+100)(x-100)\left(\frac{1600}{x-100}\right)$$
$$(x-100)2400 = 1600(x+100)$$
$$2400x - 240000 = 1600x + 160000$$
$$800x = 400000$$
$$x = 500$$

The airplane's average rate in calm air is 500 miles per hour.

33. Think of the speed of the sidewalk as a "current." Walking with or against the movement of the sidewalk is the same as paddling with or against a current.

	d	r	$t = \frac{d}{r}$
With Sidewalk	100	$x + 1.8$	$\frac{100}{x+1.8}$
Against Sidewalk	40	$x - 1.8$	$\frac{40}{x-1.8}$

$$\frac{100}{x+1.8} = \frac{40}{x-1.8}$$
$$(x+1.8)(x-1.8)\left(\frac{100}{x+1.8}\right) = (x+1.8)(x-1.8)\left(\frac{40}{x-1.8}\right)$$
$$(x-1.8)(100) = (x+1.8)(40)$$
$$100x - 180 = 40x + 72$$
$$60x = 252$$
$$x = 4.2$$

The walking speed on a nonmoving sidewalk is 4.2 feet per second.

35.

	d	r	$t=\frac{d}{r}$
Fast Runner	x	8	$\frac{x}{8}$
Slow Runner	x	6	$\frac{x}{6}$

$$\frac{x}{6}-\frac{x}{8}=\frac{1}{2}$$
$$24\left(\frac{x}{6}-\frac{x}{8}\right)=24\left(\frac{1}{2}\right)$$
$$24\left(\frac{x}{6}\right)-24\left(\frac{x}{8}\right)=12$$
$$4x-3x=12$$
$$x=12$$

Each person ran 12 miles.

37.

	Part Done in 1 Minute	Time Working Together	Part Done in x Minutes
You	$\frac{1}{45}$	x	$\frac{x}{45}$
Your Sister	$\frac{1}{30}$	x	$\frac{x}{30}$

$$\frac{x}{45}+\frac{x}{30}=1$$
$$90\left(\frac{x}{45}+\frac{x}{30}\right)=90(1)$$
$$90\left(\frac{x}{45}\right)+90\left(\frac{x}{30}\right)=90$$
$$2x+3x=90$$
$$5x=90$$
$$x=18$$

If they work together, it will take 18 minutes to wash the car. There is enough time to finish the job before the parents return home.

39.

	Part Done in 1 hr.	Time Working Together	Part Done in x Hours
First Pipe	$\frac{1}{6}$	x	$\frac{x}{6}$
Second Pipe	$\frac{1}{12}$	x	$\frac{x}{12}$

$$\frac{x}{6}+\frac{x}{12}=1$$
$$12\left(\frac{x}{6}+\frac{x}{12}\right)=12(1)$$
$$12\left(\frac{x}{6}\right)+12\left(\frac{x}{12}\right)=12$$
$$2x+x=12$$
$$3x=12$$
$$x=4$$

If both pipes are used, the pool will be filled in 4 hours.

41.

	Part Done in 1 hr.	Time Working Together	Part Done in 3 Hours
You	$\frac{1}{x}$	3	$\frac{3}{x}$
Your Cousin	$\frac{1}{4}$	3	$\frac{3}{4}$

$$\frac{3}{x}+\frac{3}{4}=1$$
$$4x\left(\frac{3}{x}\right)+4x\left(\frac{3}{4}\right)=4x(1)$$
$$4(3)+x(3)=4x$$
$$12+3x=4x$$
$$12=x$$

Working alone, it would take you 12 hours to finish the job.

43.

	Part Done in 1 Hour	Time Working Together	Part Done in x Hours
Crew 1	$\frac{1}{20}$	x	$\frac{x}{20}$
Crew 2	$\frac{1}{30}$	x	$\frac{x}{30}$
Crew 3	$\frac{1}{60}$	x	$\frac{x}{60}$

$$\frac{x}{20}+\frac{x}{30}+\frac{x}{60}=1$$

$$60\left(\frac{x}{20}+\frac{x}{30}+\frac{x}{60}\right)=60(1)$$

$$60\left(\frac{x}{20}\right)+60\left(\frac{x}{30}\right)+60\left(\frac{x}{60}\right)=60$$

$$3x+2x+x=60$$

$$6x=60$$

$$x=10$$

If the three crews work together, it will take 10 hours to dispense the food and water.

45.

	Part done in 1 hour	Time together	Part done in 6 hrs.
Old	$\frac{1}{x+5}$	6	$\frac{6}{x+5}$
New	$\frac{1}{x}$	6	$\frac{6}{x}$

$$\frac{6}{x+5}+\frac{6}{x}=1$$

$$x(x+5)\left(\frac{6}{x+5}+\frac{6}{x}\right)=x(x+5)(1)$$

$$x(x+5)\left(\frac{6}{x+5}\right)+x(x+5)\left(\frac{6}{x}\right)=x(x+5)$$

$$x(6)+(x+5)(6)=x^2+5x$$

$$6x+6x+30=x^2+5x$$

$$12x+30=x^2+5x$$

$$0=x^2-7x-30$$

$$0=(x-10)(x+3)$$

Apply the zero product principle.

$x-10=0$ $\qquad x+3=0$

$x=10$ $\qquad x=-3$

We disregard –3 because we can't have a negative time measurement. Working alone, it would take the new copying machine 10 hours to make all the copies.

47.

	Part Done in 1 Minute	Time Working Together	Part Done in x Minutes
Faucet	$\frac{1}{5}$	x	$\frac{x}{5}$
Drain	$\frac{1}{10}$	x	$\frac{x}{10}$

$$\frac{x}{5}-\frac{x}{10}=1$$
$$10\left(\frac{x}{5}-\frac{x}{10}\right)=10(1)$$
$$10\left(\frac{x}{5}\right)-10\left(\frac{x}{10}\right)=10$$
$$2x-x=10$$
$$x=10$$

It will take 10 minutes for the sink to fill if the drain is left open.

49. Let x be the number.

$$\frac{4x}{x+5}=\frac{3}{2}$$
$$2(x+5)\frac{4x}{x+5}=2(x+5)\frac{3}{2}$$
$$8x=3(x+5)$$
$$8x=3x+15$$
$$5x=15$$
$$x=3$$

51. Let x be the number.

$$2x+2\cdot\frac{1}{x}=\frac{20}{3}$$
$$(3x)2x+(3x)2\cdot\frac{1}{x}=(3x)\frac{20}{3}$$
$$6x^2+6=20x$$
$$6x^2-20x+6=0$$
$$2(3x^2-10x+3)=0$$
$$2(3x-1)(x-3)=0$$

Using the zero product principle,

$x=\frac{1}{3}$ and $x=3$.

53. Let x be the number of consecutive hits. You already have 35 hits in 140 times at bat.

$$\frac{x+35}{x+140}=0.30$$

$$(x+140)\frac{x+35}{x+140}=0.30(x+140)$$

$$x+35=0.30x+42$$

$$0.70x=7$$

$$x=10$$

You must have 10 consecutive hits to increase your batting average to 0.30.

55.

	Part Done in 1 Hour	Time Working Together	Part Done in x Hours
First Pipe	$\frac{1}{a}$	x	$\frac{x}{a}$
Second Pipe	$\frac{1}{b}$	x	$\frac{x}{b}$

$$\frac{x}{a}+\frac{x}{b}=1$$

$$ab\left(\frac{x}{a}+\frac{x}{b}\right)=ab(1)$$

$$ab\left(\frac{x}{a}\right)+ab\left(\frac{x}{b}\right)=ab$$

$$bx+ax=ab$$

$$(a+b)x=ab$$

$$x=\frac{ab}{a+b}$$

If both pipes are used, the pool will be filled in $\frac{ab}{a+b}$ hours.

57. – 63. Answers will vary.

65. Exercise 45.

$$\frac{6}{x+5}+\frac{6}{x}=1$$

10

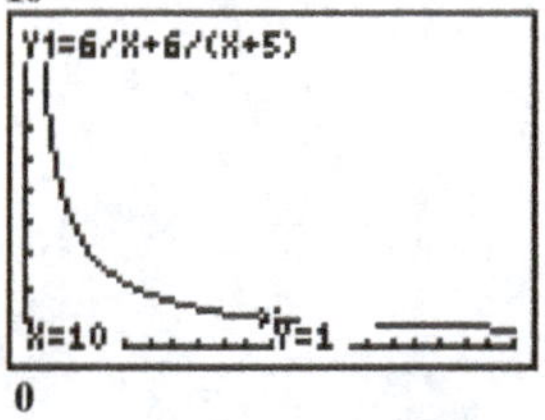

Y1=6/X+6/(X+5)

0 X=10 Y=1 20

0

Exercise 46.

$$\frac{12}{x+10}+\frac{12}{x}=1$$

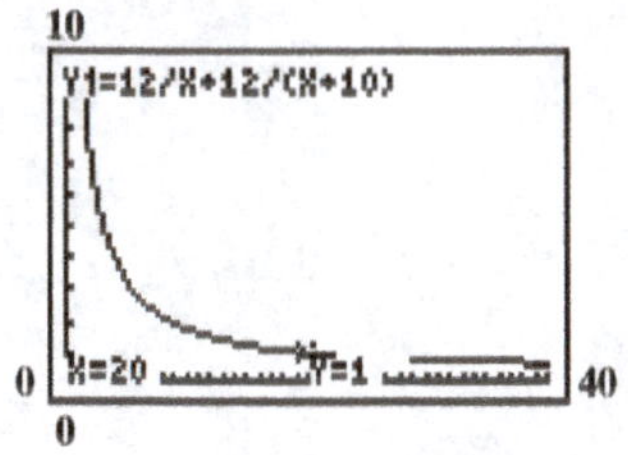

10

Y1=12/X+12/(X+10)

0 X=20 Y=1 40

0

67. makes sense

69. makes sense

71. false; Changes to make the statement true will vary. A sample change is: As production level increases, the average cost for a company to produce each unit of its product decreases.

73. true

75.

$$\frac{1}{s}=f+\frac{1-f}{p}$$

$$ps\left(\frac{1}{s}\right)=ps\left(f+\frac{1-f}{p}\right)$$

$$p=fps+s(1-f)$$

$$p=fps+s-sf$$

$$p-s=fps-sf$$

$$p-s=f(ps-s)$$

$$f=\frac{p-s}{ps-s} \text{ or } \frac{p-s}{s(p-1)}$$

77.

	Part Done in 1 Hour	Time Working Together	Part Done in 12 Hours
Mrs. L & Mr. T	$\frac{1}{x}+\frac{1}{x+4}$	2	$\frac{2}{x}+\frac{2}{x+4}$
Mr. T Alone	$\frac{1}{x+4}$	7	$\frac{7}{x+4}$

$$\frac{2}{x}+\frac{2}{x+4}+\frac{7}{x+4}=1$$
$$x(x+4)\left(\frac{2}{x}+\frac{2}{x+4}+\frac{7}{x+4}\right)=x(x+4)(1)$$
$$x(x+4)\left(\frac{2}{x}\right)+x(x+4)\left(\frac{2}{x+4}\right)+x(x+4)\left(\frac{7}{x+4}\right)=x(x+4)$$
$$(x+4)(2)+x(2)+x(7)=x^2+4x$$
$$2x+8+2x+7x=x^2+4x$$
$$11x+8=x^2+4x$$
$$0=x^2-7x-8$$
$$0=(x-8)(x+1)$$
$$x-8=0 \qquad x+1=0$$
$$x=8 \qquad \cancel{x=-1}$$

Working alone, it would take Mrs. Lovett 8 hours to prepare the pies.

78. $x^2+4x+4-9y^2$

$=\left(x^2+4x+4\right)-9y^2$

$=(x+2)^2-(3y)^2$

$=((x+2)+3y)((x+2)-3y)$

$=(x+2+3y)(x+2-3y)$

79. $2x+5y=-5$

$x+2y=-1$

$\left[\begin{array}{cc|c} 2 & 5 & -5 \\ 1 & 2 & -1 \end{array}\right] \quad R_1 \leftrightarrow R_2$

$=\left[\begin{array}{cc|c} 1 & 2 & -1 \\ 2 & 5 & -5 \end{array}\right] \quad -2R_1+R_2$

$=\left[\begin{array}{cc|c} 1 & 2 & -1 \\ 0 & 1 & -3 \end{array}\right]$

The resulting system is:

$x+2y=-1$

$y=-3.$

Since we know $y=-3$, we can use back-substitution to find x.

$$x+2y=-1$$
$$x+2(-3)=-1$$
$$x-6=-1$$
$$x=5$$

The solution set is $\{(5,-3)\}$.

80.
$$\begin{aligned} x+y+z&=4 \\ 2x+5y\quad&=1 \\ x-y-2z&=0 \end{aligned}$$

Multiply the first equation by 2 and add to the third equation.

$$\begin{aligned} 2x+2y+2z&=8 \\ x-y-2z&=0 \\ \hline 3x+y\quad&=8 \end{aligned}$$

We can use this equation and the second equation in the original system to form a system of two equations in two variables.

$$\begin{aligned} 2x+5y&=1 \\ 3x+y&=8 \end{aligned}$$

Multiply the second equation by –5 and solve by addition.

$$\begin{aligned} 2x+5y&=1 \\ -15x-5y&=-40 \\ \hline -13x\quad&=-39 \\ x&=3 \end{aligned}$$

Back-substitute 3 for x to find y.

$$\begin{aligned} 2x+5y&=1 \\ 2(3)+5y&=1 \\ 6+5y&=1 \end{aligned}$$

$$\begin{aligned} 5y&=-5 \\ y&=-1 \end{aligned}$$

Back-substitute –1 for y and 3 for x to find z.

$$\begin{aligned} x+y+z&=4 \\ 3+(-1)+z&=4 \\ 2+z&=4 \\ z&=2 \end{aligned}$$

The solution is $(3,-1,2)$ and the solution set is $\{(3,-1,2)\}$.

81. **a.** Substitute to find k.

$$\begin{aligned} y&=kx^2 \\ y&=kx^2 \\ 64&=k\cdot 2^2 \\ 64&=4k \\ k&=16 \end{aligned}$$

b.
$$\begin{aligned} y&=kx^2 \\ y&=16x^2 \end{aligned}$$

c.
$$\begin{aligned} y&=16x^2 \\ y&=16\cdot 5^2 \\ y&=400 \end{aligned}$$

82. **a.** Substitute to find k.

$$\begin{aligned} y&=\frac{k}{x} \\ 12&=\frac{k}{8} \\ k&=12\cdot 8 \\ k&=96 \end{aligned}$$

b.
$$\begin{aligned} y&=\frac{k}{x} \\ y&=\frac{96}{x} \end{aligned}$$

c.
$$\begin{aligned} y&=\frac{96}{x} \\ y&=\frac{96}{3} \\ y&=32 \end{aligned}$$

83.
$$\begin{aligned} S&=\frac{kA}{P} \\ 12,000&=\frac{k\cdot 60,000}{40} \\ 12,000&=\frac{k\cdot 60,000}{40} \\ 12,000&=1500k \\ k&=8 \end{aligned}$$

6.8 Check Points

1. Since W varies directly with t, we have $W = kt$.
Use the given values to find k.

$W = kt$

$30 = k \cdot 5$

$\frac{30}{5} = \frac{k \cdot 5}{5}$

$6 = k$

The equation becomes $W = 6t$.
Find W when $t = 11$.

$W = 6t$

$W = 6 \cdot 11$

$W = 66$

An 11 minute shower will use 66 gallons of water.

2. Beginning with $y = kx^2$, we will use s for the stopping distance and v for the speed of the car.
Use the given values to find k.

$s = kv^2$

$200 = k \cdot 60^2$

$200 = k \cdot 3600$

$\frac{200}{3600} = k$

$k = \frac{1}{18}$

The equation becomes $s = kv^2$

$s = \frac{1}{18}v^2$

$s = \frac{v^2}{18}$

Find s when $v = 100$.

$s = \frac{v^2}{18}$

$s = \frac{100^2}{18}$

$s = \frac{100^2}{18}$

$s \approx 556$

About 556 feet are required to stop a car traveling 100 miles per hour.

3. Beginning with $y = \frac{k}{x}$, we will use l for the length of the string and f for the frequency.
Use the given values to find k.

$f = \frac{k}{l}$

$640 = \frac{k}{8}$

$8 \cdot 640 = 8 \cdot \frac{k}{8}$

$5120 = k$

The equation becomes $f = \frac{k}{l}$

$f = \frac{5120}{l}$

Find f when $l = 10$.

$f = \frac{5120}{l}$

$f = \frac{5120}{10}$

$f = 512$

A string length of 10 inches will vibrate at 512 cycles per second.

4. Let m = the number of minutes needed to solve an exercise set.
Let p = the number of people working on the problems.
Let x = the number of problems in the exercise set.

Use $m = \frac{kx}{p}$ to find k.

$m = \frac{kx}{p}$

$32 = \frac{k16}{4}$

$32 = 4k$

$k = 8$

Thus, $m = \frac{8x}{p}$.

Find m when $p = 8$ and $x = 24$.

$m = \frac{8 \cdot 24}{8}$

$m = 24$

It will take 24 minutes for 8 people to solve 24 problems.

5. Find k: $V = khr^2$

$$120\pi = k \cdot 10 \cdot 6^2$$
$$120\pi = k \cdot 360$$
$$\frac{120\pi}{360} = \frac{k \cdot 360}{360}$$
$$\frac{\pi}{3} = k$$

Thus, $V = \frac{\pi}{3}hr^2 = \frac{\pi hr^2}{3}$.

$$V = \frac{\pi hr^2}{3}$$
$$V = \frac{\pi \cdot 2 \cdot 12^2}{3} = 96\pi$$

The volume of a cone having a radius of 12 feet and a height of 2 feet is 96π cubic feet.

6.8 Concept and Vocabulary Check

1. $y = kx$; constant of variation
2. $y = kx^n$
3. $y = \frac{k}{x}$
4. $y = \frac{kx}{z}$
5. $y = kxz$
6. directly; inversely
7. jointly; inversely

6.8 Exercise Set

1. Since y varies directly with x, we have $y = kx$. Use the given values to find k.

$$y = kx$$
$$65 = k \cdot 5$$
$$\frac{65}{5} = \frac{k \cdot 5}{5}$$
$$13 = k$$

The equation becomes $y = 13x$.
When $x = 12$, $y = 13x = 13 \cdot 12 = 156$.

3. Since y varies inversely with x, we have $y = \frac{k}{x}$. Use the given values to find k.

$$y = \frac{k}{x}$$
$$12 = \frac{k}{5}$$
$$5 \cdot 12 = 5 \cdot \frac{k}{5}$$
$$60 = k$$

The equation becomes $y = \frac{60}{x}$.

When $x = 2$, $y = \frac{60}{2} = 30$.

5. Since y varies inversely as x and inversely as the square of z, we have $y = \frac{kx}{z^2}$. Use the given values to find k.

$$y = \frac{kx}{z^2}$$
$$20 = \frac{k(50)}{5^2}$$
$$20 = \frac{k(50)}{25}$$
$$20 = 2k$$
$$10 = k$$

The equation becomes $y = \frac{10x}{z^2}$.

When $x = 3$ and $z = 6$,

$$y = \frac{10x}{z^2} = \frac{10(3)}{6^2} = \frac{10(3)}{36} = \frac{30}{36} = \frac{5}{6}.$$

7. Since y varies jointly as x and y, we have $y = kxz$. Use the given values to find k.

$$y = kxz$$
$$25 = k(2)(5)$$
$$25 = k(10)$$
$$\frac{25}{10} = \frac{k(10)}{10}$$
$$\frac{5}{2} = k$$

The equation becomes $y = \frac{5}{2}xz$.

When $x = 8$ and $z = 12$,

$$y = \frac{5}{2}(8)(12) = \frac{5}{\cancel{2}}(\overset{4}{\cancel{8}})(12) = 240.$$

9. Since y varies jointly as a and b and inversely as the square root of c, we have $y = \frac{kab}{\sqrt{c}}$.

Use the given values to find k.

$$y = \frac{kab}{\sqrt{c}}$$
$$12 = \frac{k(3)(2)}{\sqrt{25}}$$
$$12 = \frac{k(6)}{5}$$
$$12(5) = \frac{k(6)}{5}(5)$$
$$60 = 6k$$
$$\frac{60}{6} = \frac{6k}{6}$$
$$10 = k$$

The equation becomes $y = \frac{10ab}{\sqrt{c}}$.

When $a = 5$, $b = 3$, $c = 9$,

$$y = \frac{10ab}{\sqrt{c}} = \frac{10(5)(3)}{\sqrt{9}} = \frac{150}{3} = 50.$$

11. $x = kyz$;

Solving for y:

$$x = kyz$$
$$\frac{x}{kz} = \frac{kyz}{yz}.$$
$$y = \frac{x}{kz}$$

13. $x = \frac{kz^3}{y}$;

Solving for y:

$$x = \frac{kz^3}{y}$$
$$xy = y \cdot \frac{kz^3}{y}$$
$$xy = kz^3$$
$$\frac{xy}{x} = \frac{kz^3}{x}$$
$$y = \frac{kz^3}{x}$$

15. $x = \frac{kyz}{\sqrt{w}}$;

Solving for y:

$$x = \frac{kyz}{\sqrt{w}}$$
$$x\left(\sqrt{w}\right) = \left(\sqrt{w}\right)\frac{kyz}{\sqrt{w}}$$
$$x\sqrt{w} = kyz$$
$$\frac{x\sqrt{w}}{kz} = \frac{kyz}{kz}$$
$$y = \frac{x\sqrt{w}}{kz}$$

17. $x = kz(y + w)$;

Solving for y:

$$x = kz(y + w)$$
$$x = kzy + kzw$$
$$x - kzw = kzy$$
$$\frac{x - kzw}{kz} = \frac{kzy}{kz}$$
$$y = \frac{x - kzw}{kz}$$

19. $x = \frac{kz}{y - w}$;

Solving for y:

$$x = \frac{kz}{y - w}$$
$$(y - w)x = (y - w)\frac{kz}{y - w}$$
$$xy - wx = kz$$
$$xy = kz + wx$$
$$\frac{xy}{x} = \frac{kz + wx}{x}$$
$$y = \frac{kz + wx}{x}$$

21. Since T varies directly as B, we have $T = kB$.

Use the given values to find k.

$$T = kB$$
$$3.6 = k(4)$$
$$\frac{3.6}{4} = \frac{k(4)}{4}$$
$$0.9 = k$$

The equation becomes $T = 0.9B$.

When $B = 6$, $T = 0.9(6) = 5.4$. The tail length is 5.4 feet.

23. Since B varies directly as D, we have $B = kD$.
Use the given values to find k.

$$B = kD$$
$$8.4 = k(12)$$
$$\frac{8.4}{12} = \frac{k(12)}{12}$$
$$k = \frac{8.4}{12} = 0.7$$

The equation becomes $B = 0.7D$.
When $B = 56$,

$$56 = 0.7D$$
$$\frac{56}{0.7} = \frac{0.7D}{0.7}$$
$$D = \frac{56}{0.7} = 80$$

It was dropped from 80 inches.

25. Since a man's weight varies directly as the cube of his height, we have $w = kh^3$.
Use the given values to find k.

$$w = kh^3$$
$$170 = k(70)^3$$
$$170 = k(343,000)$$
$$\frac{170}{343,000} = \frac{k(343,000)}{343,000}$$
$$0.000496 = k$$

The equation becomes $w = 0.000496h^3$.
When $h = 107$,

$$w = 0.000496(107)^3$$
$$= 0.000496(1,225,043) \approx 607.$$

Robert Wadlow's weight was approximately 607 pounds.

27. Since the banking angle varies inversely as the turning radius, we have $B = \frac{k}{r}$.
Use the given values to find k.

$$B = \frac{k}{r}$$
$$28 = \frac{k}{4}$$
$$28(4) = 28\left(\frac{k}{4}\right)$$
$$112 = k$$

The equation becomes $B = \frac{112}{r}$.

When $r = 3.5$, $B = \frac{112}{r} = \frac{112}{3.5} = 32$.

The banking angle is 32° when the turning radius is 3.5 feet.

29. a. Use $L = \frac{k}{R}$ to find k.

$$L = \frac{k}{R}$$
$$30 = \frac{k}{63}$$
$$63 \cdot 30 = 63 \cdot \frac{k}{63}$$
$$1890 = k$$

Thus, $L = \frac{1890}{R}$.

b. This is an approximate model.

c.
$$L = \frac{1890}{R}$$
$$L = \frac{1890}{27} = 70$$

The average life span of an elephant is 70 years.

31. a. A mammal with a life span of 20 years will have a heart rate of about 90 beats per minute.

b.
$$L = \frac{1890}{R}$$
$$20 = \frac{1890}{R}$$
$$20R = 1890$$
$$R = \frac{1890}{20}$$
$$R \approx 95$$

The model estimates a mammal with a life span of 20 years will have a heart rate of about 95 beats per minute.

c. The data for horses is represented on the graph by the point $(63, 30)$.

33. Since intensity varies inversely as the square of the distance, we have $I = \frac{k}{d^2}$.

Use the given values to find k.

$$I = \frac{k}{d^2}.$$
$$62.5 = \frac{k}{3^2}$$
$$62.5 = \frac{k}{9}$$
$$9(62.5) = 9\left(\frac{k}{9}\right)$$
$$562.5 = k$$

The equation becomes $I = \frac{562.5}{d^2}$.

When $d = 2.5$, $I = \frac{562.5}{2.5^2} = \frac{562.5}{6.25} = 90$

The intensity is 90 milliroentgens per hour.

35. Since index varies directly as weight and inversely as the square of one's height, we have $I = \frac{kw}{h^2}$.

Use the given values to find k.

$$I = \frac{kw}{h^2}$$
$$35.15 = \frac{k(180)}{60^2}$$
$$35.15 = \frac{k(180)}{3600}$$
$$(3600)35.15 = \frac{k(180)}{3600}$$
$$126540 = k(180)$$
$$k = \frac{126540}{180} = 703$$

The equation becomes $I = \frac{703w}{h^2}$.

When $w = 170$ and $h = 70$, $I = \frac{703(170)}{(70)^2} \approx 24.4$.

This person has a BMI of 24.4 and is not overweight.

37. Since heat loss varies jointly as the area and temperature difference, we have $L = kAD$.

Use the given values to find k.

$$L = kAD$$
$$1200 = k(3 \cdot 6)(20)$$
$$1200 = 360k$$
$$\frac{1200}{360} = \frac{360k}{360}$$
$$k = \tfrac{10}{3}$$

The equation becomes $L = \frac{10}{3}AD$

When $A = 6 \cdot 9 = 54$, $D = 10$,

$L = \frac{10}{3}(9 \cdot 6)(10) = 1800$.

The heat loss is 1800 Btu.

39. Since intensity varies inversely as the square of the distance from the sound source, we have $I = \frac{k}{d^2}$.

If you move to a seat twice as far, then $d = 2d$ .So we have $I = \frac{k}{(2d)^2} = \frac{k}{4d^2} = \frac{1}{4} \cdot \frac{k}{d^2}$. The intensity will be multiplied by a factor of $\frac{1}{4}$. So the sound intensity is $\frac{1}{4}$ of what it was originally.

41. a. Since the average number of phone calls varies jointly as the product of the populations and inversely as the square of the distance, we have $C = \frac{kP_1P_2}{d^2}$.

b. Use the given values to find k.

$$C = \frac{kP_1P_2}{d^2}$$
$$326,000 = \frac{k(777,000)(3,695,000)}{(420)^2}$$
$$326,000 = \frac{k\left(2.87 \times 10^{12}\right)}{176,400}$$
$$326,000 = 16,269,841.27k$$
$$0.02 \approx k$$

The equation becomes $C = \frac{0.02P_1P_2}{d^2}$.

c. $C = \dfrac{0.02(650,000)(490,000)}{(400)^2}$

$\approx 39,813$

The average number of calls is approximately 39,813 daily phone calls.

43. a.

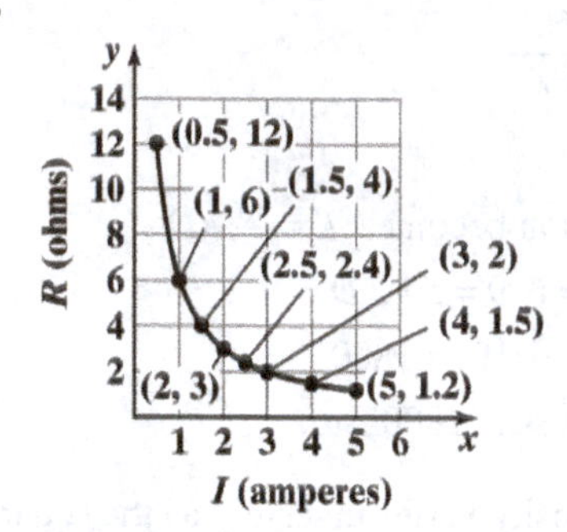

b. Current varies inversely as resistance. Answers will vary.

c. Since the current varies inversely as resistance we have $R = \dfrac{k}{I}$. Use one of the given ordered pairs to find k.

$$12 = \frac{k}{0.5}$$

$$12(0.5) = \frac{k}{0.5}(0.5)$$

$$k = 6$$

The equation becomes $R = \dfrac{6}{I}$.

45. – 47. Answers will vary.

49. z varies directly as the square root of x and inversely as the square root of y.

51. Answers will vary.

53. does not make sense; Explanations will vary. Sample explanation: This would cause division by 0. Division by zero is undefined.

55. makes sense

57. Since wind pressure varies directly as the square of the wind velocity, we have $P = kv^2$. If the wind speed doubles then the value of v has been multiplied by two. In the formula, $P = k(2v)^2 = k(4v^2) = 4kv^2$. Then the wind pressure will be multiplied by a factor of 4. So if the wind speed doubles, the wind pressure is 4 times more destructive.

59. Since the brightness of a source point varies inversely as the square of its distance from an observer, we have $B = \dfrac{k}{d^2}$. We can now see things that are only $\dfrac{1}{50}$ as bright.

$$B = \frac{1}{50} \cdot \frac{k}{d^2} = \frac{k}{50d^2}$$

$$= \frac{k}{(7.07)^2 d^2} = \frac{k}{(7.07d)^2}$$

The distance that can be seen is about 7.07 times farther with the space telescope.

60. $\begin{vmatrix} -1 & 2 \\ 3 & -4 \end{vmatrix} = -1(-4) - 3(2)$

$= 4 - 6 = -2$

61. $x^2y - 9y - 3x^2 + 27$

$= y(x^2 - 9) - 3(x^2 - 9)$

$= (x^2 - 9)(y - 3)$

$= (x + 3)(x - 3)(y - 3)$

62. $7xy + x^2y^2 - 5x^3 - 7$

The degree of the polynomial is 4.

63. $f(x) = \sqrt{3x + 12}$

$f(-1) = \sqrt{3(-1) + 12}$

$= \sqrt{9}$

$= 3$

64. $f(x) = \sqrt{3x + 12}$

$f(-1) = \sqrt{3(8) + 12}$

$= \sqrt{36}$

$= 6$

65. Domain: $[-4, \infty)$

Range: $[0, \infty)$

Chapter 6 Review

1. $f(x)=\frac{x^2+2x-3}{x^2-4}$

a. $f(4)=\frac{(4)^2+2(4)-3}{(4)^2-4}=\frac{16+8-3}{16-4}=\frac{21}{12}=\frac{7}{4}$

b. $f(0)=\frac{0^2+2(0)-3}{0^2-4}=\frac{0+0-3}{0-4}=\frac{-3}{-4}=\frac{3}{4}$

c. $f(-2)=\frac{(-2)^2+2(-2)-3}{(-2)^2-4}=\frac{4-4-3}{4-4}=\frac{-3}{0}$

Division by zero is undefined. $f(-2)$ does not exist.

d. $f(-3)=\frac{(-3)^2+2(-3)-3}{(-3)^2-4}=\frac{9-6-3}{9-4}=\frac{0}{5}=0$

2. The domain is all real numbers except those that make the denominator zero. To find these values, set the denominator equal to zero and solve.

$(x-3)(x+4)=0$

Apply the zero product principle.

$x-3=0 \qquad x+4=0$

$x=3 \qquad x=-4$

The domain of f is $(-\infty,-4)\cup(-4,3)\cup(3,\infty)$.

3. The domain is all real numbers except those that make the denominator zero. To find these values, set the denominator equal to zero and solve.

$x^2+x-2=0$

$(x+2)(x-1)=0$

Apply the zero product principle.

$x+2=0 \qquad x-1=0$

$x=-2 \qquad x=1$

The domain of f is $(-\infty,-2)\cup(-2,1)\cup(1,\infty)$.

4. $\frac{5x^3-35x}{15x^2}=\frac{\cancel{5x}(x^2-7)}{3x\cdot\cancel{5x}}=\frac{x^2-7}{3x}$

5. $\dfrac{x^2+6x-7}{x^2-49}=\dfrac{(x+7)(x-1)}{(x+7)(x-7)}=\dfrac{\cancel{(x+7)}(x-1)}{\cancel{(x+7)}(x-7)}=\dfrac{x-1}{x-7}$

6. $\dfrac{6x^2+7x+2}{2x^2-9x-5}=\dfrac{(3x+2)(2x+1)}{(2x+1)(x-5)}=\dfrac{(3x+2)\cancel{(2x+1)}}{\cancel{(2x+1)}(x-5)}=\dfrac{3x+2}{x-5}$

7. $\dfrac{x^2+4}{x^2-4}$; cannot be simplified

8. $\dfrac{x^3-8}{x^2-4}=\dfrac{\cancel{(x-2)}(x^2+2x+4)}{(x+2)\cancel{(x-2)}}=\dfrac{x^2+2x+4}{x+2}$

9. $\dfrac{5x^2-5}{3x+12}\cdot\dfrac{x+4}{x-1}=\dfrac{5(x^2-1)}{3\cancel{(x+4)}}\cdot\dfrac{\cancel{x+4}}{x-1}=\dfrac{5(x^2-1)}{3(x-1)}=\dfrac{5(x+1)\cancel{(x-1)}}{3\cancel{(x-1)}}=\dfrac{5(x+1)}{3}$

10. $\dfrac{2x+5}{4x^2+8x-5}\cdot\dfrac{4x^2-4x+1}{x+1}=\dfrac{\cancel{2x+5}}{\cancel{(2x+5)}\cancel{(2x-1)}}\cdot\dfrac{\cancel{(2x-1)}(2x-1)}{x+1}=\dfrac{2x-1}{x+1}$

11. $\dfrac{x^2-9x+14}{x^3+2x^2}\cdot\dfrac{x^2-4}{x^2-4x+4}=\dfrac{(x-7)\cancel{(x-2)}}{x^2\cancel{(x+2)}}\cdot\dfrac{\cancel{(x+2)}\cancel{(x-2)}}{\cancel{(x-2)}\cancel{(x-2)}}=\dfrac{x-7}{x^2}$

12. $\dfrac{1}{x^2+8x+15}\div\dfrac{3}{x+5}=\dfrac{1}{x^2+8x+15}\cdot\dfrac{x+5}{3}=\dfrac{1}{\cancel{(x+5)}(x+3)}\cdot\dfrac{\cancel{x+5}}{3}=\dfrac{1}{3(x+3)}$

13. $\dfrac{x^2+16x+64}{2x^2-128}\div\dfrac{x^2+10x+16}{x^2-6x-16}=\dfrac{x^2+16x+64}{2x^2-128}\cdot\dfrac{x^2-6x-16}{x^2+10x+16}$

$$=\frac{\cancel{(x+8)}(x+8)}{2(x^2-64)}\cdot\frac{(x-8)\cancel{(x+2)}}{\cancel{(x+8)}\cancel{(x+2)}}$$

$$=\frac{(x+8)(x-8)}{2(x^2-64)}$$

$$=\frac{\cancel{(x+8)}\cancel{(x-8)}}{2\cancel{(x+8)}\cancel{(x-8)}}$$

$$=\frac{1}{2}$$

14. $\frac{y^2-16}{y^3-64} \div \frac{y^2-3y-18}{y^2+5y+6} = \frac{y^2-16}{y^3-64} \cdot \frac{y^2+5y+6}{y^2-3y-18}$

$= \frac{(y+4)\cancel{(y-4)}}{\cancel{(y-4)}\left(y^2+4y+16\right)} \cdot \frac{\cancel{(y+3)}(y+2)}{(y-6)\cancel{(y+3)}}$

$= \frac{(y+4)(y+2)}{(y-6)\left(y^2+4y+16\right)}$

15. $\frac{x^2-4x+4-y^2}{2x^2-11x+15} \cdot \frac{x^4y}{x-2+y} \div \frac{x^3y-2x^2y-x^2y^2}{3x-9} = \frac{(x-2)(x-2)-y^2}{(2x-5)(x-3)} \cdot \frac{x^4y}{x-2+y} \div \frac{x^2y(x-2-y)}{3(x-3)}$

$= \frac{(x-2)^2-y^2}{(2x-5)\cancel{(x-3)}} \cdot \frac{x^4y}{(x-2+y)} \cdot \frac{3\cancel{(x-3)}}{x^2y(x-2-y)}$

$= \frac{\cancel{((x-2)+y)}\cancel{((x-2)-y)}}{2x-5} \cdot \frac{x^2}{\cancel{(x-2+y)}} \cdot \frac{3}{\cancel{(x-2-y)}}$

$= \frac{3x^2}{2x-5}$

16. **a.** 50 deer were introduced into the habitat.

b. After 10 years, the population is 150 deer.

c. The equation of the horizontal asymptote is $y = 225$. This means that the deer population will increase over time to 225, but will never reach it.

17. $\frac{4x+1}{3x-1} + \frac{8x-5}{3x-1} = \frac{4x+1+8x-5}{3x-1} = \frac{12x-4}{3x-1} = \frac{4(3x-1)}{3x-1} = 4$

18. $\frac{2x-7}{x^2-9} - \frac{x-4}{x^2-9} = \frac{2x-7-(x-4)}{x^2-9} = \frac{2x-7-x+4}{x^2-9} = \frac{\cancel{x-3}}{(x+3)\cancel{(x-3)}} = \frac{1}{x+3}$

19. $\frac{4x^2-11x+4}{x-3} - \frac{x^2-4x+10}{x-3} = \frac{4x^2-11x+4-\left(x^2-4x+10\right)}{x-3}$

$= \frac{4x^2-11x+4-x^2+4x-10}{x-3}$

$= \frac{3x^2-7x-6}{x-3}$

$= \frac{\cancel{(x-3)}(3x+2)}{\cancel{x-3}}$

$= 3x+2$

20. $9x^3 = 3^2x^2$

$12x = 2^2 \cdot 3x$

$\text{LCD} = 2^2 \cdot 3^2x^3 = 4 \cdot 9x^3 = 36x^3$

21. $x^2+2x-35=(x+7)(x-5)$

$x^2+9x+14=(x+7)(x+2)$

$\text{LCD}=(x+7)(x-5)(x+2)$

22. $\frac{1}{x}+\frac{2}{x-5}=\frac{1(x-5)}{x(x-5)}+\frac{2x}{x(x-5)}=\frac{1(x-5)+2x}{x(x-5)}=\frac{x-5+2x}{x(x-5)}=\frac{3x-5}{x(x-5)}$

23.
$$\begin{aligned}\frac{2}{x^2-5x+6}+\frac{3}{x^2-x-6}&=\frac{2}{(x-3)(x-2)}+\frac{3}{(x-3)(x+2)}\\&=\frac{2(x+2)}{(x-3)(x-2)(x+2)}+\frac{3(x-2)}{(x-3)(x-2)(x+2)}\\&=\frac{2(x+2)+3(x-2)}{(x-3)(x-2)(x+2)}\\&=\frac{2x+4+3x-6}{(x-3)(x-2)(x+2)}\\&=\frac{5x-2}{(x-3)(x-2)(x+2)}\end{aligned}$$

24.
$$\begin{aligned}\frac{x-3}{x^2-8x+15}+\frac{x+2}{x^2-x-6}&=\frac{x-3}{(x-3)(x-5)}+\frac{\cancel{x+2}}{(x-3)(\cancel{x+2})}=\frac{x-3}{(x-3)(x-5)}+\frac{1}{x-3}\\&=\frac{x-3}{(x-3)(x-5)}+\frac{1(x-5)}{(x-3)(x-5)}=\frac{x-3+1(x-5)}{(x-3)(x-5)}\\&=\frac{x-3+x-5}{(x-3)(x-5)}=\frac{2x-8}{(x-3)(x-5)}\text{ or }\frac{2(x-4)}{(x-3)(x-5)}\end{aligned}$$

25.
$$\begin{aligned}\frac{3x^2}{9x^2-16}-\frac{x}{3x+4}&=\frac{3x^2}{(3x+4)(3x-4)}-\frac{x}{3x+4}=\frac{3x^2}{(3x+4)(3x-4)}-\frac{x(3x-4)}{(3x+4)(3x-4)}\\&=\frac{3x^2-x(3x-4)}{(3x+4)(3x-4)}=\frac{3x^2-3x^2+4x}{(3x+4)(3x-4)}=\frac{4x}{(3x+4)(3x-4)}\end{aligned}$$

26.
$$\begin{aligned}\frac{y}{y^2+5y+6}-\frac{2}{y^2+3y+2}&=\frac{y}{(y+3)(y+2)}-\frac{2}{(y+2)(y+1)}\\&=\frac{y(y+1)}{(y+3)(y+2)(y+1)}-\frac{2(y+3)}{(y+3)(y+2)(y+1)}\\&=\frac{y(y+1)-2(y+3)}{(y+3)(y+2)(y+1)}=\frac{y^2+y-2y-6}{(y+3)(y+2)(y+1)}=\frac{y^2-y-6}{(y+3)(y+2)(y+1)}\\&=\frac{(y-3)\cancel{(y+2)}}{(y+3)\cancel{(y+2)}(y+1)}=\frac{y-3}{(y+3)(y+1)}\end{aligned}$$

27. $\frac{x}{x+3}+\frac{x}{x-3}-\frac{9}{x^2-9}=\frac{x}{x+3}+\frac{x}{x-3}-\frac{9}{(x+3)(x-3)}$

$=\frac{x(x-3)}{(x+3)(x-3)}+\frac{x(x+3)}{(x-3)(x+3)}-\frac{9}{(x+3)(x-3)}$

$=\frac{x(x-3)+x(x+3)-9}{(x+3)(x-3)}=\frac{x^2-3x+x^2+3x-9}{(x+3)(x-3)}=\frac{2x^2-9}{(x+3)(x-3)}$

28. $\frac{3x^2}{x-y}+\frac{3y^2}{y-x}=\frac{3x^2}{x-y}+\frac{-1(3y^2)}{-1(y-x)}=\frac{3x^2}{x-y}+\frac{-3y^2}{x-y}=\frac{3x^2-3y^2}{x-y}=\frac{3(x^2-y^2)}{x-y}$

$=\frac{3(x+y)\cancel{(x-y)}}{\cancel{x-y}}=3(x+y)$ or $3x+3y$

29. $\frac{\frac{3}{x}-3}{\frac{8}{x}-8}=\frac{x}{x}\cdot\frac{\frac{3}{x}-3}{\frac{8}{x}-8}=\frac{x\cdot\frac{3}{x}-x\cdot 3}{x\cdot\frac{8}{x}-x\cdot 8}=\frac{3-3x}{8-8x}=\frac{3\cancel{(1-x)}}{8\cancel{(1-x)}}=\frac{3}{8}$

30. $\frac{\frac{5}{x}+1}{1-\frac{25}{x^2}}=\frac{x^2}{x^2}\cdot\frac{\frac{5}{x}+1}{1-\frac{25}{x^2}}=\frac{x^2\cdot\frac{5}{x}+x^2\cdot 1}{x^2\cdot 1-x^2\cdot\frac{25}{x^2}}=\frac{5x+x^2}{x^2-25}=\frac{x\cancel{(5+x)}}{\cancel{(x+5)}(x-5)}=\frac{x}{x-5}$

31. $\frac{3-\frac{1}{x+3}}{3+\frac{1}{x+3}}=\frac{x+3}{x+3}\cdot\frac{3-\frac{1}{x+3}}{3+\frac{1}{x+3}}=\frac{(x+3)\cdot 3-(x+3)\cdot\frac{1}{x+3}}{(x+3)\cdot 3+(x+3)\cdot\frac{1}{x+3}}=\frac{3x+9-1}{3x+9+1}=\frac{3x+8}{3x+10}$

32. $\frac{\frac{4}{x+3}}{\frac{2}{x-2}-\frac{1}{x^2+x-6}}=\frac{\frac{4}{x+3}}{\frac{2}{x-2}-\frac{1}{(x+3)(x-2)}}=\frac{(x+3)(x-2)}{(x+3)(x-2)}\cdot\frac{\frac{4}{x+3}}{\frac{2}{x-2}-\frac{1}{(x+3)(x-2)}}$

$=\frac{(x-2)4}{(x+3)2-1}=\frac{4x-8}{2x+6-1}=\frac{4x-8}{2x+5}=\frac{4(x-2)}{2x+5}$

33. $$\frac{\frac{2}{x^2-x-6}+\frac{1}{x^2-4x+3}}{\frac{3}{x^2+x-2}-\frac{2}{x^2+5x+6}}=\frac{\frac{2}{(x-3)(x+2)}+\frac{1}{(x-3)(x-1)}}{\frac{3}{(x+2)(x-1)}-\frac{2}{(x+2)(x+3)}}$$

$$=\frac{\frac{2(x-1)}{(x-3)(x+2)(x-1)}+\frac{1(x+2)}{(x-3)(x+2)(x-1)}}{\frac{3(x+3)}{(x+2)(x-1)(x+3)}-\frac{2(x-1)}{(x+2)(x-1)(x+3)}}=\frac{\frac{2(x-1)+(x+2)}{(x-3)(x+2)(x-1)}}{\frac{3(x+3)-2(x-1)}{(x+2)(x-1)(x+3)}}=\frac{\frac{2x-2+x+2}{(x-3)(x+2)(x-1)}}{\frac{3x+9-2x+2}{(x+2)(x-1)(x+3)}}$$

$$=\frac{\frac{3x}{(x-3)(x+2)(x-1)}}{\frac{x+11}{(x+2)(x-1)(x+3)}}=\frac{3x}{(x-3)\cancel{(x+2)}\cancel{(x-1)}}\cdot\frac{\cancel{(x+2)}\cancel{(x-1)}(x+3)}{x+11}=\frac{3x(x+3)}{(x-3)(x+11)}=\frac{3x^2+9x}{x^2+8x-33}$$

34. $$\frac{x^{-2}+x^{-1}}{x^{-2}-x^{-1}}=\frac{\frac{1}{x^2}+\frac{1}{x}}{\frac{1}{x^2}-\frac{1}{x}}=\frac{x^2}{x^2}\cdot\frac{\frac{1}{x^2}+\frac{1}{x}}{\frac{1}{x^2}-\frac{1}{x}}=\frac{x^2\cdot\frac{1}{x^2}+x^2\cdot\frac{1}{x}}{x^2\cdot\frac{1}{x^2}-x^2\cdot\frac{1}{x}}=\frac{1+x}{1-x}$$

35. $$\frac{15x^3-30x^2+10x-2}{5x^2}=\frac{15x^3}{5x^2}-\frac{30x^2}{5x^2}+\frac{10x}{5x^2}-\frac{2}{5x^2}=3x-6+\frac{2}{x}-\frac{2}{5x^2}$$

36. $$\frac{36x^4y^3+12x^2y^3-60x^2y^2}{6xy^2}=\frac{36x^4y^3}{6xy^2}+\frac{12x^2y^3}{6xy^2}-\frac{60x^2y^2}{6xy^2}=6x^3y+2xy-10x$$

37.
$$\begin{array}{r}
3x-7 \\
2x+3\overline{)6x^2-5x+5} \\
\underline{6x^2+9x} \\
-14x+5 \\
\underline{-14x-21} \\
26
\end{array}$$

$$\frac{6x^2-5x+5}{2x+3}=3x-7+\frac{26}{2x+3}$$

38.
$$\begin{array}{r}
2x^2-4x+1 \\
5x-3\overline{)10x^3-26x^2+17x-13} \\
\underline{10x^3-6x^2} \\
-20x^2+17x \\
\underline{-20x^2+12x} \\
5x-13 \\
\underline{5x-3} \\
-10
\end{array}$$

$$\frac{10x^3-26x^2+17x-13}{5x-3}=2x^2-4x+1-\frac{10}{5x-3}$$

39.

$$\begin{array}{r}
x^5+5x^4+8x^3+16x^2+33x+63 \\
x-2\overline{)x^6+3x^5-2x^4+0x^3+x^2-3x+2} \\
\underline{x^6-2x^5} \\
5x^5-2x^4 \\
\underline{5x^5-10x^4} \\
8x^4+0x^3 \\
\underline{8x^4-16x^3} \\
16x^3+x^2 \\
\underline{16x^3-32x^2} \\
33x^2-3x \\
\underline{33x^2-66x} \\
63x+2 \\
\underline{63x-126} \\
128
\end{array}$$

$$\frac{x^6+3x^5-2x^4+x^2-3x+2}{x-2}=x^5+5x^4+8x^3+16x^2+33x+63+\frac{128}{x-2}$$

40.

$$\begin{array}{r}
2x^2+3x-1 \\
2x^2+0x+1\overline{)4x^4+6x^3+0x^2+3x-1} \\
\underline{4x^4+0x^3+2x^2} \\
6x^3-2x^2+3x \\
\underline{6x^3+0x^2+3x} \\
-2x^2+0x-1 \\
\underline{-2x^2+0x-1} \\
0
\end{array}$$

$$\frac{4x^4+6x^3+3x-1}{2x^2+1}=2x^2+3x-1$$

41. $\left(4x^3-3x^2-2x+1\right)\div(x+1)$

$$\begin{array}{r|rrrr}
-1 & 4 & -3 & -2 & 1 \\
 & & -4 & 7 & -5 \\
\hline
 & 4 & -7 & 5 & -4
\end{array}$$

$\left(4x^3-3x^2-2x+1\right)\div(x+1)$

$=4x^2-7x+5-\dfrac{4}{x+1}$

42. $\left(3x^4-2x^2-10x-20\right)\div(x-2)$

$$\begin{array}{r|rrrrr} 2 & 3 & 0 & -2 & -10 & -20 \\ & & 6 & 12 & 20 & 20 \\ \hline & 3 & 6 & 10 & 10 & 0 \end{array}$$

$$\left(3x^4-2x^2-10x-20\right)\div(x-2)$$
$$=3x^3+6x^2+10x+10$$

43. $\left(x^4+16\right)\div(x+4)$

$$\begin{array}{r|rrrrr} -4 & 1 & 0 & 0 & 0 & 16 \\ & & -4 & 16 & -64 & 256 \\ \hline & 1 & -4 & 16 & -64 & 272 \end{array}$$

$$\left(x^4+16\right)\div(x+4)$$
$$=x^3-4x^2+16x-64+\frac{272}{x+4}$$

44. $f(x)=2x^3-5x^2+4x-1$

Divide $f(x)$ by $x-2$.

$$\begin{array}{r|rrrr} 2 & 2 & -5 & 4 & -1 \\ & & 4 & -2 & 4 \\ \hline & 2 & -1 & 2 & 3 \end{array}$$

$$f(2)=3$$

45. $f(x)=3x^4+7x^3+8x^2+2x+4$

$$\begin{array}{r|rrrrr} -\frac{1}{3} & 3 & 7 & 8 & 2 & 4 \\ & & -1 & -2 & -2 & 0 \\ \hline & 3 & 6 & 6 & 0 & 4 \end{array}$$

$$f\left(-\frac{1}{3}\right)=4$$

46. To show that –2 is a solution to the equation, show that when the polynomial is divided by $x+2$ the remainder is zero.

$$\begin{array}{r|rrrr} -2 & 2 & -1 & -8 & 4 \\ & & -4 & 10 & -4 \\ \hline & 2 & -5 & 2 & 0 \end{array}$$

Since the remainder is zero, -2 is a solution to the equation.

47. $x^4-x^3-7x^2+x+6=0$

$$\begin{array}{r|rrrrr} 4 & 1 & -1 & -7 & 1 & 6 \\ & & 4 & 12 & 20 & 84 \\ \hline & 1 & 3 & 5 & 21 & 90 \end{array}$$

Since the remainder is not zero, 4 is not a solution to the equation.

48. To show that $\frac{1}{2}$ is a solution to the equation, show that when the polynomial is divided by $x-\frac{1}{2}$ the remainder is zero.

$$\begin{array}{r|rrrr} \frac{1}{2} & 6 & 1 & -4 & 1 \\ & & 3 & 2 & -1 \\ \hline & 6 & 4 & -2 & 0 \end{array}$$

$$6x^3+x^2-4x+1=\left(x-\frac{1}{2}\right)\left(6x^2+4x-2\right)$$

To solve the equation, we set it equal to zero and factor.

$$\left(x-\frac{1}{2}\right)\left(6x^2+4x-2\right)=0$$
$$\left(x-\frac{1}{2}\right)\left(2\left(3x^2+2x-1\right)\right)=0$$
$$\left(x-\frac{1}{2}\right)\left(2(3x-1)(x+1)\right)=0$$
$$2\left(x-\frac{1}{2}\right)(3x-1)(x+1)=0$$

$$x=-1,\frac{1}{3},\frac{1}{2}\text{ or }\left\{-1,\frac{1}{3},\frac{1}{2}\right\}$$

49. $\frac{3}{x}+\frac{1}{3}=\frac{5}{x}$

So that denominators will not equal zero, x cannot equal zero. To eliminate fractions, multiply by the LCD, $3x$.

$$\frac{3}{x}+\frac{1}{3}=\frac{5}{x}$$
$$3x\left(\frac{3}{x}+\frac{1}{3}\right)=3x\left(\frac{5}{x}\right)$$
$$3x\left(\frac{3}{x}\right)+3x\left(\frac{1}{3}\right)=3(5)$$
$$3(3)+x(1)=15$$
$$9+x=15$$
$$x=6$$

The solution set is $\{6\}$.

50. $\frac{5}{3x+4}=\frac{3}{2x-8}$

To find the restrictions on x, set the denominators equal to zero and solve.

$$3x+4=0 \qquad 2x-8=0$$
$$3x=-4 \qquad 2x=8$$
$$x=-\frac{4}{3} \qquad x=4$$

To eliminate fractions, multiply by the LCD, $(3x+4)(2x-8)$.

$$\frac{5}{3x+4}=\frac{3}{2x-8}$$
$$(3x+4)(2x-8)\left(\frac{5}{3x+4}\right)=(3x+4)(2x-8)\left(\frac{3}{2x-8}\right)$$
$$(2x-8)(5)=(3x+4)(3)$$
$$10x-40=9x+12$$
$$x-40=12$$
$$x=52$$

The solution set is $\{52\}$.

51. $\frac{1}{x-5}-\frac{3}{x+5}=\frac{6}{x^2-25}$

$$\frac{1}{x-5}-\frac{3}{x+5}=\frac{6}{(x+5)(x-5)}$$

So that denominators will not equal zero, x cannot equal 5 or –5. To eliminate fractions, multiply by the LCD, $(x+5)(x-5)$.

$$\frac{1}{x-5}-\frac{3}{x+5}=\frac{6}{(x+5)(x-5)}$$

$$(x+5)(x-5)\left(\frac{1}{x-5}-\frac{3}{x+5}\right)=(x+5)(x-5)\left(\frac{6}{(x+5)(x-5)}\right)$$

$$(x+5)(x-5)\left(\frac{1}{x-5}\right)-(x+5)(x-5)\left(\frac{3}{x+5}\right)=6$$

$$(x+5)(1)-(x-5)(3)=6$$

$$x+5-3x+15=6$$

$$-2x+20=6$$

$$-2x=-14$$

$$x=7$$

The solution set is $\{7\}$.

52. $\frac{x+5}{x+1}-\frac{x}{x+2}=\frac{4x+1}{x^2+3x+2}$

$\frac{x+5}{x+1}-\frac{x}{x+2}=\frac{4x+1}{(x+2)(x+1)}$

So that denominators will not equal zero, x cannot equal –1 or –2. To eliminate fractions, multiply by the LCD, $(x+2)(x+1)$.

$$\frac{x+5}{x+1}-\frac{x}{x+2}=\frac{4x+1}{(x+2)(x+1)}$$

$$(x+2)(x+1)\left(\frac{x+5}{x+1}-\frac{x}{x+2}\right)=(x+2)(x+1)\left(\frac{4x+1}{(x+2)(x+1)}\right)$$

$$(x+2)(x+1)\left(\frac{x+5}{x+1}\right)-(x+2)(x+1)\left(\frac{x}{x+2}\right)=4x+1$$

$$(x+2)(x+5)-(x+1)(x)=4x+1$$

$$x^2+7x+10-\left(x^2+x\right)=4x+1$$

$$x^2+7x+10-x^2-x=4x+1$$

$$6x+10=4x+1$$

$$2x+10=1$$

$$2x=-9$$

$$x=-\frac{9}{2}$$

The solution set is $\left\{-\frac{9}{2}\right\}$.

53. $\frac{2}{3}-\frac{5}{3x}=\frac{1}{x^2}$

So that denominators will not equal zero, x cannot equal zero. To eliminate fractions, multiply by the LCD, $3x^2$.

$$3x^2\left(\frac{2}{3}-\frac{5}{3x}\right)=3x^2\left(\frac{1}{x^2}\right)$$
$$3x^2\left(\frac{2}{3}\right)-3x^2\left(\frac{5}{3x}\right)=3(1)$$
$$x^2(2)-x(5)=3$$
$$2x^2-5x=3$$
$$2x^2-5x-3=0$$
$$(2x+1)(x-3)=0$$

Apply the zero product principle.

$$2x+1=0 \qquad x-3=0$$
$$2x=-1 \qquad x=3$$
$$x=-\frac{1}{2}$$

The solution set is $\left\{-\frac{1}{2},3\right\}$.

54. $\frac{2}{x-1}=\frac{1}{4}+\frac{7}{x+2}$

So that denominators will not equal zero, x cannot equal 1 or –2. To eliminate fractions, multiply by the LCD, $4(x-1)(x+2)$.

$$4(x-1)(x+2)\left(\frac{2}{x-1}\right)=4(x-1)(x+2)\left(\frac{1}{4}+\frac{7}{x+2}\right)$$
$$4(x+2)(2)=4(x-1)(x+2)\left(\frac{1}{4}\right)+4(x-1)(x+2)\left(\frac{7}{x+2}\right)$$
$$8x+16=(x-1)(x+2)+4(x-1)(7)$$
$$8x+16=x^2+x-2+28x-28$$
$$8x+16=x^2+29x-30$$
$$0=x^2+21x-46$$
$$0=(x+23)(x-2)$$

Apply the zero product principle.

$$x+23=0 \qquad x-2=0$$
$$x=-23 \qquad x=2$$

The solution set is $\{-23,2\}$.

55. $\frac{2x+7}{x+5}-\frac{x-8}{x-4}=\frac{x+18}{x^2+x-20}$

$\frac{2x+7}{x+5}-\frac{x-8}{x-4}=\frac{x+18}{(x+5)(x-4)}$

So that denominators will not equal zero, x cannot equal –5 or 4. To eliminate fractions, multiply by the LCD, $(x+5)(x-4)$.

$$\frac{2x+7}{x+5}-\frac{x-8}{x-4}=\frac{x+18}{(x+5)(x-4)}$$

$$(x+5)(x-4)\left(\frac{2x+7}{x+5}-\frac{x-8}{x-4}\right)=(x+5)(x-4)\left(\frac{x+18}{(x+5)(x-4)}\right)$$

$$(x+5)(x-4)\left(\frac{2x+7}{x+5}\right)-(x+5)(x-4)\left(\frac{x-8}{x-4}\right)=x+18$$

$$(x-4)(2x+7)-(x+5)(x-8)=x+18$$

$$2x^2-x-28-\left(x^2-3x-40\right)=x+18$$

$$2x^2-x-28-x^2+3x+40=x+18$$

$$x^2+2x+12=x+18$$

$$x^2+x-6=0$$

$$(x+3)(x-2)=0$$

Apply the zero product principle.

$x+3=0$ $\qquad$ $x-2=0$

$x=-3$ $\qquad$ $x=2$

The solution set is $\{-3,2\}$.

56.

$$f(x)=\frac{4x}{100-x}$$

$$16=\frac{4x}{100-x}$$

$$(100-x)(16)=(100-x)\left(\frac{4x}{100-x}\right)$$

$$1600-16x=4x$$

$$1600=20x$$

$$80=x$$

80% of the pollutants can be removed for $16 million.

57.
$$P = \frac{R-C}{n}$$
$$n(P) = n\left(\frac{R-C}{n}\right)$$
$$nP = R - C$$
$$nP + C = R$$
$$C = R - nP$$

58.
$$\frac{P_1V_1}{T_1} = \frac{P_2V_2}{T_2}$$
$$T_1T_2\left(\frac{P_1V_1}{T_1}\right) = T_1T_2\left(\frac{P_2V_2}{T_2}\right)$$
$$P_1T_2V_1 = P_2T_1V_2$$
$$\frac{P_1T_2V_1}{P_2V_2} = \frac{P_2T_1V_2}{P_2V_2}$$
$$T_1 = \frac{P_1T_2V_1}{P_2V_2}$$

59. $T = \frac{A-P}{Pr}$
$$Pr(T) = Pr\left(\frac{A-P}{Pr}\right)$$
$$PrT = A - P$$
$$PrT + P = A$$
$$P(rT+1) = A$$
$$\frac{P(rT+1)}{rT+1} = \frac{A}{rT+1}$$
$$P = \frac{A}{rT+1}$$

60. $\frac{1}{R} = \frac{1}{R_1} + \frac{1}{R_2}$
$$RR_1R_2\left(\frac{1}{R}\right) = RR_1R_2\left(\frac{1}{R_1} + \frac{1}{R_2}\right)$$
$$R_1R_2 = RR_1R_2\left(\frac{1}{R_1}\right) + RR_1R_2\left(\frac{1}{R_2}\right)$$
$$R_1R_2 = RR_2 + RR_1$$
$$R_1R_2 = R(R_2 + R_1)$$
$$\frac{R_1R_2}{R_2 + R_1} = \frac{R(R_2 + R_1)}{R_2 + R_1}$$
$$R = \frac{R_1R_2}{R_2 + R_1}$$

61. $I = \frac{nE}{R+nr}$

$$\begin{aligned}
(R+nr)(I) &= (R+nr)\left(\frac{nE}{R+nr}\right) \\
IR + Inr &= nE \\
IR &= nE - Inr \\
IR &= n(E - Ir) \\
\frac{IR}{E-Ir} &= \frac{n(E-Ir)}{E-Ir} \\
n &= \frac{IR}{E-Ir}
\end{aligned}$$

62. **a.** $C(x) = 50,000 + 25x$

b. $\overline{C}(x) = \frac{50,000+25x}{x}$

c.
$$\begin{aligned}
35 &= \frac{50,000+25x}{x} \\
x(35) &= x\left(\frac{50,000+25x}{x}\right) \\
35x &= 50,000 + 25x \\
10x &= 50,000 \\
x &= 5000
\end{aligned}$$

5000 graphing calculators must be produced each month to have an average cost of $35.

63.

	d	r	$t = \frac{d}{r}$
Riding	60	$3x$	$\frac{60}{3x}$
Walking	8	x	$\frac{8}{x}$

$$\begin{aligned}
\frac{60}{3x} + \frac{8}{x} &= 7 \\
3x\left(\frac{60}{3x} + \frac{8}{x}\right) &= 3x(7) \\
3x\left(\frac{60}{3x}\right) + 3x\left(\frac{8}{x}\right) &= 21x \\
60 + 3(8) &= 21x \\
60 + 24 &= 21x \\
84 &= 21x \\
4 &= x
\end{aligned}$$

The cyclist was riding at a rate of $3x = 3(4) = 12$ miles per hour.

64.

	d	r	$t=\dfrac{d}{r}$
Down Stream	12	$x+3$	$\dfrac{12}{x+3}$
Up Stream	12	$x-3$	$\dfrac{12}{x-3}$

$$\frac{12}{x+3}+\frac{12}{x-3}=3$$
$$(x+3)(x-3)\left(\frac{12}{x+3}+\frac{12}{x-3}\right)=(x+3)(x-3)(3)$$
$$(x+3)(x-3)\left(\frac{12}{x+3}\right)+(x+3)(x-3)\left(\frac{12}{x-3}\right)=3\left(x^2-9\right)$$
$$(x-3)(12)+(x+3)(12)=3x^2-27$$
$$12x-36+12x+36=3x^2-27$$
$$24x=3x^2-27$$
$$0=3x^2-24x-27$$
$$0=3\left(x^2-8x-9\right)$$
$$0=3(x-9)(x+1)$$
$$3(x-9)=0 \qquad x+1=0$$
$$x-9=0 \qquad x=-1$$
$$x=9$$

The solutions are –1 and 9. We disregard –1 because we cannot have a negative rate. The boat's rate in still water is 9 miles per hour.

65.

	Part Done in 1 Hour	Time Working Together	Part Done in x Hours
First Person	$\dfrac{1}{3}$	x	$\dfrac{x}{3}$
Second Person	$\dfrac{1}{6}$	x	$\dfrac{x}{6}$

$$\frac{x}{3}+\frac{x}{6}=1$$
$$6\left(\frac{x}{3}+\frac{x}{6}\right)=6(1)$$
$$6\left(\frac{x}{3}\right)+6\left(\frac{x}{6}\right)=6$$
$$2x+x=6$$
$$3x=6$$
$$x=2$$

If they work together, it will take 2 hours to clean the house. There is not enough time to finish the job before the TV program starts.

66.

	Part Done in 1 Hour	Time Working Together	Part Done in 20 Hours
Fast Crew	$\frac{1}{x-9}$	20	$\frac{20}{x-9}$
Slow Crew	$\frac{1}{x}$	20	$\frac{20}{x}$

$$\frac{20}{x-9}+\frac{20}{x}=1$$
$$x(x-9)\left(\frac{20}{x-9}+\frac{20}{x}\right)=x(x-9)(1)$$
$$x(x-9)\left(\frac{20}{x-9}\right)+x(x-9)\left(\frac{20}{x}\right)=x^2-9x$$
$$x(20)+(x-9)(20)=x^2-9x$$
$$20x+20x-180=x^2-9x$$
$$40x-180=x^2-9x$$
$$0=x^2-49x+180$$
$$0=(x-45)(x-4)$$

Apply the zero product principle.

$$x-45=0 \qquad x-4=0$$
$$x=45 \qquad x=4$$

The solutions are 45 and 4. We disregard 4, because the fast crew's rate would be $4-9=-5$. No crew can do the job in a negative number of hours. It would take the slow crew 45 hours and the fast crew $45-9=36$ hours to complete the job working alone.

67.

	Part Done in 1 Minute	Time Working Together	Part Done in x Minutes
Faucet	$\frac{1}{60}$	x	$\frac{x}{60}$
Drain	$\frac{1}{80}$	x	$\frac{x}{80}$

$$\frac{x}{60}-\frac{x}{80}=1$$
$$240\left(\frac{x}{60}-\frac{x}{80}\right)=240(1)$$
$$240\left(\frac{x}{60}\right)-240\left(\frac{x}{80}\right)=240$$
$$4x-3x=240$$
$$x=240$$

It will take 240 minutes or 4 hours to fill the pond.

68. Since the profit varies directly as the number of products sold, we have $p = kn$. Use the given values to find k.

$$p = kn.$$
$$1175 = k(25)$$
$$\frac{1175}{25} = \frac{k(25)}{25}$$
$$47 = k$$

The equation becomes $p = 47n$.

When $n = 105$ products, $p = 47(105) = 4935$.

If 105 products are sold, the company's profit is $4935.

69. Since distance varies directly as the square of the time, we have $d = kt^2$.

Use the given values to find k.

$$d = kt^2$$
$$144 = k(3)^2$$
$$144 = k(9)$$
$$\frac{144}{9} = \frac{k(9)}{9}$$
$$16 = k$$

The equation becomes $d = 16t^2$. When $t = 10$, $d = 16(10)^2 = 16(100) = 1600$.

A skydiver will fall 1600 feet in 10 seconds.

70. Since the pitch of a musical tone varies inversely as its wavelength, we have $p = \frac{k}{w}$.

Use the given values to find k.

$$p = \frac{k}{w}$$
$$660 = \frac{k}{1.6}$$
$$660(1.6) = 1.6\left(\frac{k}{1.6}\right)$$
$$1056 = k$$

The equation becomes $p = \frac{1056}{w}$.

When $w = 2.4$, $p = \frac{1056}{2.4} = 440$.

The tone's pitch is 440 vibrations per second.

71. Since loudness varies inversely as the square of the distance, we have $l = \frac{k}{d^2}$.

Use the given values to find k.

$$l = \frac{k}{d^2}$$
$$28 = \frac{k}{8^2}$$
$$28 = \frac{k}{64}$$
$$64(28) = 64\left(\frac{k}{64}\right)$$
$$1792 = k$$

The equation becomes $l = \frac{1792}{d^2}$.

When $d = 4$, $l = \frac{1792}{(4)^2} = \frac{1792}{16} = 112$.

At a distance of 4 feet, the loudness of the stereo is 112 decibels.

72. Since time varies directly as the number of computers and inversely as the number of workers, we have $t = \frac{kn}{w}$.

Use the given values to find k.

$$t = \frac{kn}{w}$$
$$10 = \frac{k(30)}{6}$$
$$10 = 5k$$
$$\frac{10}{5} = \frac{5k}{5}$$
$$2 = k$$

The equation becomes $t = \frac{2n}{w}$.

When $n = 40$ and $w = 5$, $t = \frac{2(40)}{5} = \frac{80}{5} = 16$.

It will take 16 hours for 5 workers to assemble 40 computers.

73. Since the volume varies jointly as height and the area of the base, we have $v = kha$.

Use the given values to find k.

$$175 = k(15)(35)$$
$$175 = k(525)$$
$$\frac{175}{525} = \frac{k(525)}{525}$$
$$\frac{1}{3} = k$$

The equation becomes $v = \frac{1}{3}ha$. When $h = 20$ feet and $a = 120$ square feet, $v = \frac{1}{3}(20)(120) = 800$.

If the height is 20 feet and the area is 120 square feet, the volume will be 800 cubic feet.

Chapter 6 Test

1. The domain is all real numbers except those that make the denominator zero. To find these values, set the denominator equal to zero and solve.

$$x^2 - 7x + 10 = 0$$
$$(x-5)(x-2) = 0$$

Apply the zero product principle.

$$x - 5 = 0 \qquad x - 2 = 0$$
$$x = 5 \qquad x = 2$$

The domain of f is $(-\infty, 2) \cup (2, 5) \cup (5, \infty)$.

$$f(x) = \frac{x^2 - 2x}{x^2 - 7x + 10} = \frac{x\cancel{(x-2)}}{(x-5)\cancel{(x-2)}} = \frac{x}{x-5}$$

2. $$\frac{x^2}{x^2-16} \cdot \frac{x^2+7x+12}{x^2+3x} = \frac{x^2}{(x+4)(x-4)} \cdot \frac{(x+4)(x+3)}{x(x+3)} = \frac{x^{\cancel{2}}}{\cancel{(x+4)}(x-4)} \cdot \frac{\cancel{(x+4)}\cancel{(x+3)}}{\cancel{x}\cancel{(x+3)}} = \frac{x}{x-4}$$

3. $$\frac{x^3+27}{x^2-1} \div \frac{x^2-3x+9}{x^2-2x+1} = \frac{x^3+27}{x^2-1} \cdot \frac{x^2-2x+1}{x^2-3x+9} = \frac{(x+3)\cancel{(x^2-3x+9)}}{(x+1)\cancel{(x-1)}} \cdot \frac{(x-1)^{\cancel{2}}}{\cancel{x^2-3x+9}} = \frac{(x+3)(x-1)}{x+1}$$

4. $$\frac{x^2+3x-10}{x^2+4x+3} \cdot \frac{x^2+x-6}{x^2+10x+25} \cdot \frac{x+1}{x-2} = \frac{\cancel{(x+5)}\cancel{(x-2)}}{\cancel{(x+3)}\cancel{(x+1)}} \cdot \frac{\cancel{(x+3)}(x-2)}{\cancel{(x+5)}(x+5)} \cdot \frac{\cancel{x+1}}{\cancel{x-2}} = \frac{x-2}{x+5}$$

5. $$\frac{x^2-6x-16}{x^3+3x^2+2x}\cdot\left(x^2-3x-4\right)\div\frac{x^2-7x+12}{3x}=\frac{(x-8)(x+2)}{x(x^2+3x+2)}\cdot(x-4)(x+1)\div\frac{(x-4)(x-3)}{3x}$$
$$=\frac{(x-8)\cancel{(x+2)}}{\cancel{x}\cancel{(x+1)}\cancel{(x+2)}}\cdot\cancel{(x-4)}\cancel{(x+1)}\cdot\frac{3\cancel{x}}{\cancel{(x-4)}(x-3)}$$
$$=\frac{3(x-8)}{x-3}=\frac{3x-24}{x-3}$$

6. $$\frac{x^2-5x-2}{6x^2-11x-35}-\frac{x^2-7x+5}{6x^2-11x-35}=\frac{x^2-5x-2-\left(x^2-7x+5\right)}{6x^2-11x-35}$$
$$=\frac{x^2-5x-2-x^2+7x-5}{6x^2-11x-35}$$
$$=\frac{2x-7}{6x^2-11x-35}$$
$$=\frac{\cancel{2x-7}}{\left(\cancel{2x-7}\right)(3x+5)}=\frac{1}{3x+5}$$

7. $$\frac{x}{x+3}+\frac{5}{x-3}=\frac{x(x-3)}{(x+3)(x-3)}+\frac{5(x+3)}{(x+3)(x-3)}=\frac{x(x-3)+5(x+3)}{(x+3)(x-3)}=\frac{x^2-3x+5x+15}{(x+3)(x-3)}=\frac{x^2+2x+15}{(x+3)(x-3)}$$

8. $$\frac{2}{x^2-4x+3}+\frac{3x}{x^2+x-2}=\frac{2}{(x-3)(x-1)}+\frac{3x}{(x-1)(x+2)}$$
$$=\frac{2(x+2)}{(x-3)(x-1)(x+2)}+\frac{3x(x-3)}{(x-3)(x-1)(x+2)}$$
$$=\frac{2(x+2)+3x(x-3)}{(x-3)(x-1)(x+2)}=\frac{2x+4+3x^2-9x}{(x-3)(x-1)(x+2)}$$
$$=\frac{3x^2-7x+4}{(x-3)(x-1)(x+2)}=\frac{(3x-4)\cancel{(x-1)}}{(x-3)\cancel{(x-1)}(x+2)}=\frac{3x-4}{(x+2)(x-3)}$$

9. $$\frac{5x}{x^2-4}-\frac{2}{x^2+x-2}=\frac{5x}{(x+2)(x-2)}-\frac{2}{(x+2)(x-1)}$$
$$=\frac{5x(x-1)}{(x+2)(x-2)(x-1)}-\frac{2(x-2)}{(x+2)(x-2)(x-1)}$$
$$=\frac{5x(x-1)-2(x-2)}{(x+2)(x-2)(x-1)}=\frac{5x^2-5x-(2x-4)}{(x+2)(x-2)(x-1)}$$
$$=\frac{5x^2-5x-2x+4}{(x+2)(x-2)(x-1)}=\frac{5x^2-7x+4}{(x+2)(x-2)(x-1)}$$

10. $$\frac{x-4}{x-5}-\frac{3}{x+5}-\frac{10}{x^2-25}=\frac{(x-4)(x+5)}{(x+5)(x-5)}-\frac{3(x-5)}{(x+5)(x-5)}-\frac{10}{(x+5)(x-5)}$$
$$=\frac{(x-4)(x+5)-3(x-5)-10}{(x+5)(x-5)}=\frac{x^2+x-20-3x+15-10}{(x+5)(x-5)}$$
$$=\frac{x^2-2x-15}{(x+5)(x-5)}=\frac{\cancel{(x-5)}(x+3)}{(x+5)\cancel{(x-5)}}=\frac{x+3}{x+5}$$

11. $$\frac{1}{10-x}+\frac{x-1}{x-10}=\frac{-1(1)}{-1(10-x)}+\frac{x-1}{x-10}=\frac{-1}{x-10}+\frac{x-1}{x-10}=\frac{-1+x-1}{x-10}=\frac{x-2}{x-10}$$

12. $$\frac{\frac{x}{4}-\frac{1}{x}}{1+\frac{x+4}{x}}=\frac{4x}{4x}\cdot\frac{\frac{x}{4}-\frac{1}{x}}{1+\frac{x+4}{x}}=\frac{\cancel{4}x\cdot\frac{x}{\cancel{4}}-4\cancel{x}\cdot\frac{1}{\cancel{x}}}{4x\cdot1+4\cancel{x}\cdot\frac{x+4}{\cancel{x}}}=\frac{x^2-4}{4x+4(x+4)}=\frac{(x+2)(x-2)}{4x+4x+16}$$
$$=\frac{(x+2)(x-2)}{8x+16}=\frac{\cancel{(x+2)}(x-2)}{8\cancel{(x+2)}}=\frac{x-2}{8}$$

13. $$\frac{\frac{1}{x}-\frac{3}{x+2}}{\frac{2}{x^2+2x}}=\frac{\frac{1}{x}-\frac{3}{x+2}}{\frac{2}{x(x+2)}}=\frac{x(x+2)}{x(x+2)}\cdot\frac{\frac{1}{x}-\frac{3}{x+2}}{\frac{2}{x(x+2)}}=\frac{x(x+2)\cdot\frac{1}{x}-x(x+2)\cdot\frac{3}{x+2}}{x(x+2)\cdot\frac{2}{x(x+2)}}$$
$$=\frac{x+2-3x}{2}=\frac{-2x+2}{2}=\frac{-\cancel{2}(x-1)}{\cancel{2}}=-(x-1)=1-x$$

14. $$\frac{12x^4y^3+16x^2y^3-10x^2y^2}{4x^2y}=\frac{12x^4y^3}{4x^2y}+\frac{16x^2y^3}{4x^2y}-\frac{10x^2y^2}{4x^2y}=3x^2y^2+4y^2-\frac{5y}{2}$$

15.
$$\begin{array}{r}
3x^2-3x+1\\
3x+2\overline{)9x^3-3x^2-3x+4}\\
\underline{9x^3+6x^2}\qquad\qquad\\
-9x^2-3x\qquad\\
\underline{-9x^2-6x}\qquad\\
3x+4\\
\underline{3x+2}\\
2
\end{array}$$

$$\frac{9x^3-3x^2-3x+4}{3x+2}=3x^2-3x+1+\frac{2}{3x+2}$$

16.
$$\begin{array}{r} 3x^2+2x+3 \\ x^2+0x-1\overline{)3x^4+2x^3+0x^2-8x+6} \\ \underline{3x^4+0x^3-3x^2} \qquad\qquad \\ 2x^3+3x^2-8x \qquad \\ \underline{2x^3+0x^2-2x} \qquad \\ 3x^2-6x+6 \\ \underline{3x^2+0x-3} \\ -6x+9 \end{array}$$

$$\frac{3x^4+2x^3-8x+6}{x^2-1}=3x^2+2x+3+\frac{-6x+9}{x^2-1}=3x^2+2x+3+\frac{9-6x}{x^2-1}$$

17. $\left(3x^4+11x^3-20x^2+7x+35\right)\div(x+5)$

$$\begin{array}{r|rrrrr} -5 & 3 & 11 & -20 & 7 & 35 \\ & & -15 & 20 & 0 & -35 \\ \hline & 3 & -4 & 0 & 7 & 0 \end{array}$$

$\left(3x^4+11x^3-20x^2+7x+35\right)\div(x+5)=3x^3-4x^2+7$

18. Divide $f(x)$ by $x-(-2)=x+2$.

$$\begin{array}{r|rrrrr} -2 & 1 & -2 & -11 & 5 & 34 \\ & & -2 & 8 & 6 & -22 \\ \hline & 1 & -4 & -3 & 11 & 12 \end{array}$$

$f(-2)=12$

19. $2x^3-3x^2-11x+6$

$$\begin{array}{r|rrrr} -2 & 2 & -3 & -11 & 6 \\ & & -4 & 14 & -6 \\ \hline & 2 & -7 & 3 & 0 \end{array}$$

Since the remainder is 0, -2 is a solution.

20. $\dfrac{x}{x+4}=\dfrac{11}{x^2-16}+2$

$\dfrac{x}{x+4}=\dfrac{11}{(x-4)(x+4)}+\dfrac{2(x-4)(x+4)}{(x-4)(x+4)}$

$\dfrac{x}{x+4}=\dfrac{11+2(x-4)(x+4)}{(x-4)(x+4)}$

So that denominators will not equal zero, x cannot equal 4 and -4. To eliminate fractions, multiply by the LCD, $(x-4)(x+4)$.

$$(x-4)(x+4)\frac{x}{x+4}=(x-4)(x+4)\frac{11+2(x-4)(x+4)}{(x-4)(x+4)}$$
$$x(x-4)=11+2(x-4)(x+4)$$
$$x^2-4x=11+2(x^2-16)$$
$$x^2-4x=11+2x^2-32$$
$$0=x^2+4x-21$$
$$0=(x+7)(x-3)$$
$$x=-7 \text{ or } x=3$$

The solution set is $\{-7, 3\}$.

21. $\dfrac{x+1}{x^2+2x-3}-\dfrac{1}{x+3}=\dfrac{1}{x-1}$

$\dfrac{x+1}{(x+3)(x-1)}-\dfrac{1}{x+3}=\dfrac{1}{x-1}$

So that denominators will not equal zero, x cannot equal 1 or –3. To eliminate fractions, multiply by the LCD, $(x+3)(x-1)$.

$$(x+3)(x-1)\left(\frac{x+1}{(x+3)(x-1)}-\frac{1}{x+3}\right)=(x+3)(x-1)\left(\frac{1}{x-1}\right)$$
$$(x+3)(x-1)\left(\frac{x+1}{(x+3)(x-1)}\right)-(x+3)(x-1)\left(\frac{1}{x+3}\right)=(x+3)(1)$$
$$x+1-(x-1)=x+3$$
$$x+1-x+1=x+3$$
$$2=x+3$$
$$-1=x$$

The solution set is $\{-1\}$.

22.

$$f(t)=\frac{250(3t+5)}{t+25}$$
$$125=\frac{250(3t+5)}{t+25}$$
$$(t+25)(125)=(t+25)\left(\frac{250(3t+5)}{t+25}\right)$$
$$125t+3125=250(3t+5)$$
$$125t+3125=750t+1250$$
$$-625t+3125=1250$$
$$-625t=-1875$$
$$t=3$$

It will take 3 years for the elk population to reach 125.

23.

$$R=\frac{as}{a+s}$$
$$(a+s)R=(a+s)\left(\frac{as}{a+s}\right)$$
$$aR+Rs=as$$
$$aR=as-Rs$$
$$aR-as=-Rs$$
$$a(R-s)=-Rs$$
$$\frac{a(R-s)}{R-s}=-\frac{Rs}{R-s}$$
$$a=-\frac{Rs}{R-s}\text{ or }\frac{Rs}{s-R}$$

24. **a.** $C(x)=300,000+10x$

b. $\overline{C}(x)=\frac{300,000+10x}{x}$

c.

$$25=\frac{300,000+10x}{x}$$
$$x(25)=x\left(\frac{300,000+10x}{x}\right)$$
$$25x=300,000+10x$$
$$15x=300,000$$
$$x=20,000$$

20,000 players must be produced for the average cost to be $25.

25.

	Part Done in 1 Hour	Time Working Together	Part Done in x Hours
Fill Pipe	$\frac{1}{3}$	x	$\frac{x}{3}$
Drain Pipe	$\frac{1}{4}$	x	$\frac{x}{4}$

$$\frac{x}{3}-\frac{x}{4}=1$$
$$12\left(\frac{x}{3}-\frac{x}{4}\right)=12(1)$$
$$12\left(\frac{x}{3}\right)-12\left(\frac{x}{4}\right)=12$$
$$4x-3x=12$$
$$x=12$$

It will take 12 hours to fill the pool.

26.

	d	r	$t=\frac{d}{r}$
down stream	3	$20+x$	$\frac{3}{20+x}$
up stream	2	$20-x$	$\frac{2}{20-x}$

$$\frac{3}{20+x}=\frac{2}{20-x}$$
$$(20+x)(20-x)\cdot\frac{3}{20+x}=(20+x)(20-x)\cdot\frac{2}{20-x}$$
$$(20-x)3=(20+x)2$$
$$60-3x=40+2x$$
$$20=5x$$
$$x=4$$

The current's rate is 4 miles per hour.

27. Since intensity varies inversely as the square of the distance, we have $I = \frac{k}{d^2}$.

Use the given values to find k.

$$I = \frac{k}{d^2}$$
$$20 = \frac{k}{15^2}$$
$$20 = \frac{k}{225}$$
$$225(20) = 225\left(\frac{k}{225}\right)$$
$$4500 = k$$

The equation becomes $I = \frac{4500}{d^2}$. When $d = 10$, $I = \frac{4500}{10^2} = \frac{4500}{100} = 45$.

At a distance of 10 feet, the light's intensity is 45 foot-candles.

Cumulative Review Exercises (Chapters 1-6)

1. $2x+5 \le 11$ and $-3x > 18$

$$2x \le 6 \qquad \frac{-3x}{-3} < \frac{18}{-3}$$
$$x \le 3 \qquad x < -6$$

−7 −6 −5 −4 −3 −2 −1 0 1 2 3

−7 −6 −5 −4 −3 −2 −1 0 1 2 3

−7 −6 −5 −4 −3 −2 −1 0 1 2 3

The solution set is $(-\infty, -6)$.

2.
$$2x^2 = 7x + 4$$
$$2x^2 - 7x - 4 = 0$$
$$(2x+1)(x-4) = 0$$

Apply the zero product principle.

$$2x+1=0 \qquad x-4=0$$
$$2x=-1 \qquad x=4$$
$$x=-\frac{1}{2}$$

The solution set is $\left\{-\frac{1}{2}, 4\right\}$.

3. $$\begin{aligned} 4x+3y+3z &= 4 \\ 3x \qquad +2z &= 2 \\ 2x-5y \qquad &= -4 \end{aligned}$$

Multiply the first equation by 5 and the third equation by 3.

$$\begin{aligned} 20x+15y+15z &= 20 \\ \underline{6x-15y \qquad} &\underline{= -12} \\ 26x \qquad +15z &= 8 \end{aligned}$$

We now have a system of two equations in two variables.

$$\begin{aligned} 3x+\ 2z &= 2 \\ 26x+15z &= 8 \end{aligned}$$

Multiply the first equation by –15 and the second equation by 2.

$$\begin{aligned} -45x-30z &= -30 \\ \underline{52x+30z} &\underline{= 16} \\ 7x \qquad &= -14 \\ x \qquad &= -2 \end{aligned}$$

Back-substitute –2 for x to find z.

$$\begin{aligned} 3x+2z &= 2 \\ 3(-2)+2z &= 2 \\ -6+2z &= 2 \\ 2z &= 8 \\ z &= 4 \end{aligned}$$

Back-substitute–2 for x to find y.

$$\begin{aligned} 2x-5y &= -4 \\ 2(-2)-5y &= -4 \\ -4-5y &= -4 \\ -5y &= 0 \\ y &= 0 \end{aligned}$$

The solution is $(-2, 0, 4)$ and the solution set is $\{(-2, 0, 4)\}$.

4. $$\begin{gathered} |3x-4| \le 10 \\ -10 \le 3x-4 \le 10 \\ -10+4 \le 3x-4+4 \le 10+4 \\ -6 \le 3x \le 14 \\ \frac{-6}{3} \le \frac{3x}{3} \le \frac{14}{3} \\ -2 \le x \le \frac{14}{3} \end{gathered}$$

The solution set is $\left[-2, \frac{14}{3}\right]$.

5. $$\frac{x}{x-8}+\frac{6}{x-2}=\frac{x^2}{x^2-10x+16}$$
$$\frac{x}{x-8}+\frac{6}{x-2}=\frac{x^2}{(x-8)(x-2)}$$

So that denominators will not equal zero, x cannot equal 2 or 8. To eliminate fractions, multiply by the LCD, $(x-8)(x-2)$.

$$(x-8)(x-2)\left(\frac{x}{x-8}+\frac{6}{x-2}\right)=(x-8)(x-2)\left(\frac{x^2}{(x-8)(x-2)}\right)$$
$$(x-8)(x-2)\left(\frac{x}{x-8}\right)+(x-8)(x-2)\left(\frac{6}{x-2}\right)=x^2$$
$$(x-2)(x)+(x-8)(6)=x^2$$
$$\cancel{x^2}-2x+6x-48=\cancel{x^2}$$
$$4x-48=0$$
$$4x=48$$
$$x=12$$

The solution set is $\{12\}$.

6. $$I=\frac{2R}{w+2s}$$
$$(w+2s)(I)=(w+2s)\left(\frac{2R}{w+2s}\right)$$
$$Iw+2Is=2R$$
$$2Is=2R-Iw$$
$$\frac{2Is}{2I}=\frac{2R-Iw}{2I}$$
$$s=\frac{2R-Iw}{2I}$$

7. $2x - y = 4$
$x + y = 5$

$2x - y = 4$
$-y = -2x + 4$
$y = 2x - 4$
$m = 2$
$y\text{-intercept} = -4$

$x + y = 5$
$y = -x + 5$
$m = -1$
$y\text{-intercept} = 5$

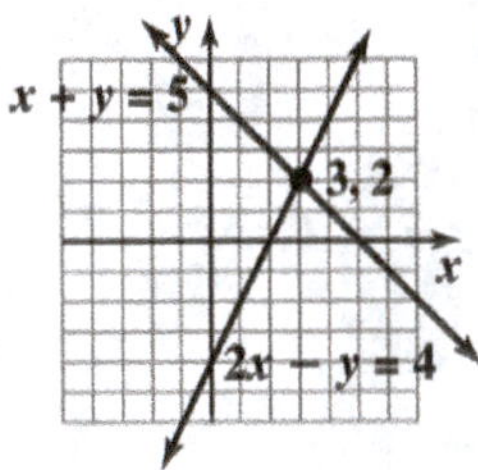

The solution set is $\{(3, 2)\}$.

8. Slope $= -3$, passing through $(1, -5)$
Point-Slope Form

$$y - y_1 = m(x - x_1)$$
$$y - (-5) = -3(x - 1)$$
$$y + 5 = -3(x - 1)$$

Slope-Intercept Form

$$y + 5 = -3(x - 1)$$
$$y + 5 = -3x + 3$$
$$y = -3x - 2$$

In function notation, the equation of the line is $f(x) = -3x - 2$.

9. $y = |x| + 2$

x	(x, y)
-3	$(-3, 5)$
-2	$(-2, 4)$
-1	$(-1, 3)$
0	$(0, 2)$
1	$(1, 3)$
2	$(2, 4)$
3	$(3, 5)$

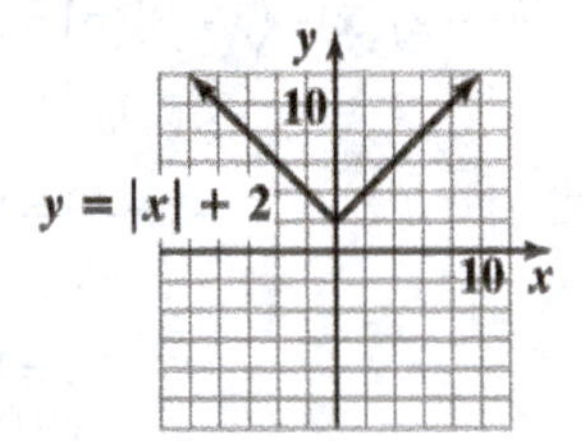

10. $y \geq 2x - 1$
$x \geq 1$

First consider $y \geq 2x - 1$. Replace the inequality symbol with an equal sign and we have $y = 2x - 1$. Since the equation is in slope-intercept form, we know that the y-intercept is -1 and the slope is 2.

Now, use the origin, $(0, 0)$, as a test point.

$$y \geq 2x - 1$$
$$0 \geq 2(0) - 1$$
$$0 \geq -1$$

This is a true statement. This means that the point $(0, 0)$ will fall in the shaded half-plane.

Next consider $x \geq 1$. Replace the inequality symbol with an equal sign and we have $x = 1$. We know that equations of the form $x = b$ are vertical lines through the point $(b, 0)$. Since the inequality is greater than or equal to, we know that the shading will extend from $x = 1$ toward ∞.

Next, graph each of the inequalities. The solution to the system is the intersection of the shaded half-planes.

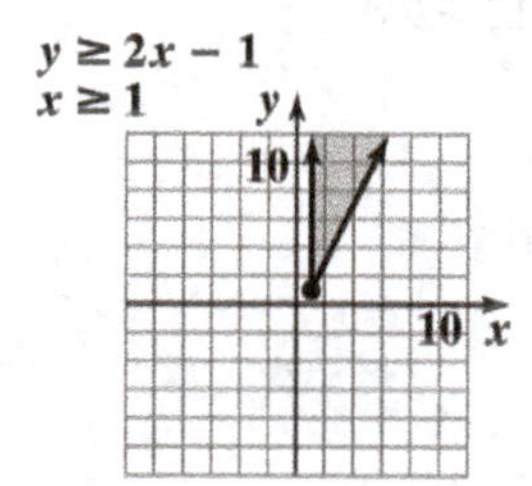

11. $2x - y < 4$

First graph the line, $2x - y = 4$. Solve for y to obtain slope-intercept form.

$$2x - y = 4$$
$$-y = -2x + 4$$
$$y = 2x - 4$$

slope = 2

y–intercept = –4

Now, use the origin, $(0,0)$, as a test point.

$$2x - y < 4$$
$$2(0) - 0 < 4$$
$$0 - 0 < 4$$
$$0 < 4$$

This is a true statement. This means that the point $(0,0)$ will fall in the shaded half-plane.

Next, graph the inequality.

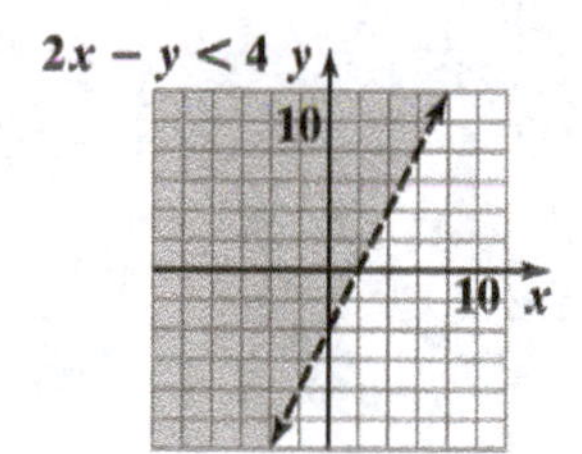

12. $[(x+2)+3y][(x+2)-3y]$

$$= [(x+2)^2 - (3y)^2]$$
$$= [(x^2 + 4x + 4) - 9y^2]$$
$$= x^2 + 4x + 4 - 9y^2$$

13. $\dfrac{2x^2 + x - 1}{2x^2 - 9x + 4} \div \dfrac{6x + 15}{3x^2 - 12x}$

$$= \frac{2x^2 + x - 1}{2x^2 - 9x + 4} \cdot \frac{3x^2 - 12x}{6x + 15}$$
$$= \frac{(2x-1)(x+1)}{(2x-1)(x-4)} \cdot \frac{3x(x-4)}{3(2x+5)}$$
$$= \frac{x(x+1)}{2x+5}$$

14. $\dfrac{3x}{x^2 - 9x + 20} - \dfrac{5}{2x - 8}$

$$= \frac{3x}{(x-4)(x-5)} - \frac{5}{2(x-4)}$$

The LCD is $2(x-4)(x-5)$.

$$\frac{3x}{(x-4)(x-5)} - \frac{5}{2(x-4)}$$
$$= \frac{2 \cdot 3x}{2(x-4)(x-5)} - \frac{5(x-5)}{2(x-4)(x-5)}$$
$$= \frac{6x - 5(x-5)}{2(x-4)(x-5)}$$
$$= \frac{6x - 5x + 25}{2(x-4)(x-5)}$$
$$= \frac{x + 25}{2(x-4)(x-5)}$$

15.

$$\begin{array}{r} 3x+4 \\ x+2 \overline{)3x^2 + 10x + 10} \\ \underline{3x^2 + 6x} \\ 4x + 10 \\ \underline{4x + 8} \\ 2 \end{array}$$

$$\frac{3x^2 + 10x + 10}{x+2} = 3x + 4 + \frac{2}{x+2}$$

16. $xy - 6x + 2y - 12$

$$= x(y-6) + 2(y-6)$$
$$= (y-6)(x+2)$$

17. $24x^3y+16x^2y-30xy$
$=2xy\left(12x^2+8x-15\right)$
$=2xy(2x+3)(6x-5)$

18. $s(t)=-16t^2+48t+64$

$$0=-16t^2+48t+64$$
$$0=-16\left(t^2-3t-4\right)$$
$$0=-16(t-4)(t+1)$$

Apply the zero product principle.

$-16(t-4)=0 \qquad t+1=0$
$t-4=0 \qquad t=-1$
$t=4$

The solutions are –1 and 4. Disregard –1 because we can't have a negative time measurement. The ball will hit the ground in 4 seconds.

19. Let x = the cost for the basic cable service.

Let y = the cost for a movie channel.

$x+y=35$
$x+2y=45$

Solve the first equation for x.

$x+y=35$
$x=35-y$

Substitute $35-y$ for x in the second equation to find y.

$$x+2y=45$$
$$(35-y)+2y=45$$
$$35-y+2y=45$$
$$35+y=45$$
$$y=10$$

Back-substitute 10 for y to find x.

$x=35-10$
$x=25$

The cost of basic cable is \$25 and the cost for each movie channel is \$10.

20.

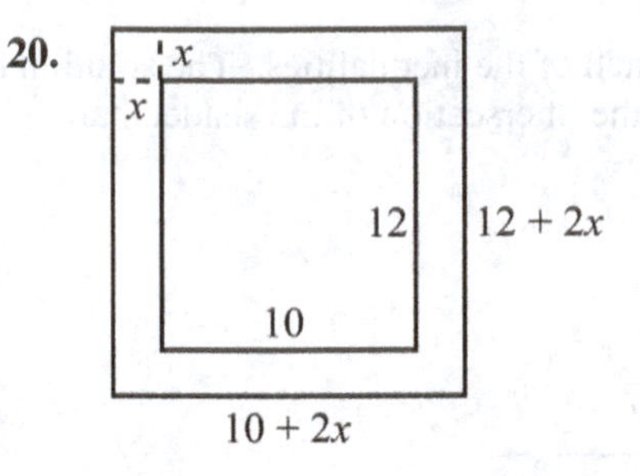

$$A=lw$$
$$168=(12+2x)(10+2x)$$
$$168=120+44x+4x^2$$
$$0=-48+44x+4x^2$$
$$0=4x^2+44x-48$$
$$0=4\left(x^2+11x-12\right)$$
$$0=4(x+12)(x-1)$$

Apply the zero product principle.

$4(x+12)=0 \qquad x-1=0$
$x+12=0 \qquad x=1$
$x=-12$

The solutions are –12 and 1. Disregard –12 because a length measurement cannot be negative. The width of the rock border is 1 foot.

Chapter 7
Radicals, Radical Functions, and Rational Exponents

7.1 Check Points

1. a. $\sqrt{64} = 8$ because $8^2 = 64$

   b. $-\sqrt{49} = -7$ because $(-7)^2 = 49$

   c. $\sqrt{\frac{16}{25}} = \frac{4}{5}$ because $\left(\frac{4}{5}\right)^2 = \frac{16}{25}$

   d. $\sqrt{0.0081} = 0.09$ because $0.09^2 = 0.0081$

   e. $\sqrt{9+16} = \sqrt{25} = 5$

   f. $\sqrt{9} + \sqrt{16} = 3 + 4 = 7$

2. a. $f(x) = \sqrt{12x - 20}$
   $f(3) = \sqrt{12(3) - 20}$
   $= \sqrt{36 - 20}$
   $= \sqrt{16}$
   $= 4$

   b. $g(x) = -\sqrt{9 - 3x}$
   $g(-5) = -\sqrt{9 - 3(-5)}$
   $= -\sqrt{9 + 15}$
   $= -\sqrt{24}$
   ≈ -4.90

3. $f(x) = \sqrt{9x - 27}$
   $9x - 27 \geq 0$
   $9x \geq 27$
   $x \geq 3$
   Domain of *f* is $\{x | x \geq 3\}$ or $[3, \infty)$.

4. $M(x) = 0.7\sqrt{x} + 12.5$
   Because 2014 is 18 years after 1996, substitute 18 for *x*.
   $M(18) = 0.7\sqrt{18} + 12.5 \approx 15.5$
   The model indicates that there will be about 15.5 non-program minutes in an hour of prime-time cable in 2014.

5. a. $\sqrt{(-7)^2} = \sqrt{49} = 7$

   b. $\sqrt{(x+8)^2} = |x+8|$

   c. $\sqrt{49x^{10}} = |7x^5|$

   d. $\sqrt{x^2 - 6x + 9} = \sqrt{(x-3)^2} = |x-3|$

6. a. $f(x) = \sqrt[3]{x-6}$
   $f(33) = \sqrt[3]{33-6}$
   $= \sqrt[3]{27}$
   $= 3$

   b. $g(x) = \sqrt[3]{2x+2}$
   $g(-5) = \sqrt[3]{2(-5)+2}$
   $= \sqrt[3]{-8}$
   $= -2$

7. $\sqrt[3]{-27x^3} = \sqrt[3]{(-3x)^3} = -3x$

8. a. $\sqrt[4]{16} = 2$

   b. $-\sqrt[4]{16} = -2$

   c. $\sqrt[4]{-16}$ is not a real number.

   d. $\sqrt[5]{-1} = -1$

9. a. $\sqrt[4]{(x+6)^4} = |x+6|$

   b. $\sqrt[5]{(3x-2)^5} = 3x - 2$

   c. $\sqrt[6]{(-8)^6} = 8$

7.1 Concept and Vocabulary Check

1. principal
2. 8^2
3. $[0, \infty)$
4. $5x - 20 \geq 0$
5. $|a|$
6. 10^3
7. $(-5)^3$
8. a
9. $(-\infty, \infty)$
10. nth; index
11. $|a|$; a
12. true
13. false
14. true
15. false

7.1 Exercise Set

1. $\sqrt{36} = 6$ because $6^2 = 36$

3. $-\sqrt{36} = -6$ because $(-6)^2 = 36$

5. $\sqrt{-36}$
   not a real number

7. $\sqrt{\frac{1}{25}} = \frac{1}{5}$ because $\left(\frac{1}{5}\right)^2 = \frac{1}{25}$

9. $-\sqrt{\frac{9}{16}} = -\frac{3}{4}$ because $\left(\frac{3}{4}\right)^2 = \frac{9}{16}$

11. $\sqrt{0.81} = 0.9$ because $(0.9)^2 = 0.81$

13. $-\sqrt{0.04} = -0.2$ because $(0.2)^2 = 0.04$

15. $\sqrt{25-16} = \sqrt{9} = 3$

17. $\sqrt{25} - \sqrt{16} = 5 - 4 = 1$

19. $\sqrt{16-25} = \sqrt{-9}$
    not a real number

21. $f(x) = \sqrt{x-2}$
    $f(18) = \sqrt{18-2} = \sqrt{16} = 4$
    $f(3) = \sqrt{3-2} = \sqrt{1} = 1$
    $f(2) = \sqrt{2-2} = \sqrt{0} = 0$
    $f(-2) = \sqrt{-2-2} = \sqrt{-4}$
    not a real number

23. $g(x) = -\sqrt{2x+3}$
    $g(11) = -\sqrt{2(11)+3}$
    $= -\sqrt{22+3}$
    $= -\sqrt{25} = -5$
    $g(1) = -\sqrt{2(1)+3}$
    $= -\sqrt{2+3}$
    $= -\sqrt{5} \approx -2.24$
    $g(-1) = -\sqrt{2(-1)+3}$
    $= -\sqrt{-2+3}$
    $= -\sqrt{1} = -1$
    $g(-2) = -\sqrt{2(-2)+3}$
    $= -\sqrt{-4+3} = -\sqrt{-1}$
    not a real number

25. $h(x) = \sqrt{(x-1)^2}$
    $h(5) = \sqrt{(5-1)^2} = \sqrt{(4)^2} = |4| = 4$
    $h(3) = \sqrt{(3-1)^2} = \sqrt{(2)^2} = |2| = 2$
    $h(0) = \sqrt{(0-1)^2} = \sqrt{(-1)^2} = |-1| = 1$
    $h(-5) = \sqrt{(-5-1)^2} = \sqrt{(-6)^2}$
    $= |-6| = 6$

27. To find the domain, set the radicand greater than or equal to zero and solve.
$x-3\ge 0$
$x\ge 3$
The domain of f is $[3,\infty)$. This corresponds to graph (c).

29. To find the domain, set the radicand greater than or equal to zero and solve.
$3x+15\ge 0$
$3x\ge -15$
$x\ge -5$
The domain of f is $[-5,\infty)$. This corresponds to graph (d).

31. To find the domain, set the radicand greater than or equal to zero and solve.
$6-2x\ge 0$
$-2x\ge -6|$
$x\le 3$
The domain of f is $\sqrt{\frac{2x}{3}}\cdot\sqrt{\frac{3}{2}}=\sqrt{\frac{2x}{3}\cdot\frac{3}{2}}$
$=\sqrt{\frac{12x}{13}\cdot\frac{13}{12}}$
$=\sqrt{x}$

This corresponds to graph (e).

33. $\sqrt{5^2}=|5|=5$

35. $\sqrt{(-4)^2}=|-4|=4$

37. $\sqrt{(x-1)^2}=|x-1|$

39. $\sqrt{36x^4}=\sqrt{(6x^2)^2}=|6x^2|=6x^2$

41. $-\sqrt{100x^6}=-\sqrt{(10x^3)^2}$
$=-|10x^3|=-10|x^3|$

43. $\sqrt{x^2+12x+36}=\sqrt{(x+6)^2}=|x+6|$

45. $-\sqrt{x^2-8x+16}=-\sqrt{(x-4)^2}$
$=-|x-4|$

47. $\sqrt[3]{27}=3$ because $3^3=27$

49. $\sqrt[3]{-27}=-3$ because $(-3)^3=-27$

51. $\sqrt[3]{\frac{1}{125}}=\frac{1}{5}$ because
$\sqrt[7]{7x^2y}\cdot\sqrt[7]{11x^3y^2}=\sqrt[7]{7x^2y\cdot 11x^3y^2}$
$=\sqrt[7]{7\cdot 11x^2x^3yy^2}$
$=\sqrt[7]{77x^5y^3}$

53. $\sqrt[3]{\frac{-27}{1000}}=-\frac{3}{10}$ because $\left(-\frac{3}{10}\right)^3=\frac{-27}{1000}$

55. $f(x)=\sqrt[3]{x-1}$
$f(28)=\sqrt[3]{28-1}=\sqrt[3]{27}=3$
$f(9)=\sqrt[3]{9-1}=\sqrt[3]{8}=2$
$f(0)=\sqrt[3]{0-1}=\sqrt[3]{-1}=-1$
$f(-63)=\sqrt[3]{-63-1}=\sqrt[3]{-64}=-4$

57. $g(x)=-\sqrt[3]{8x-8}$
$g(2)=-\sqrt[3]{8(2)-8}=-\sqrt[3]{16-8}$
$=-\sqrt[3]{8}=-2$
$g(1)=-\sqrt[3]{8(1)-8}=-\sqrt[3]{8-8}$
$=-\sqrt[3]{0}=-0=0$
$g(0)=-\sqrt[3]{8(0)-8}=-\sqrt[3]{-8}$
$=-(-2)=2$

59. $\sqrt[4]{1}=1$ because $1^4=1$

61. $\sqrt[4]{16}=2$ because $2^4=16$

63. $-\sqrt[4]{16}=-2$ because $2^4=16$

65. $\sqrt[4]{-16}$
not a real number

67. $\sqrt{75x}=\sqrt{25\cdot 3x}$
$=\sqrt{25}\cdot\sqrt{3x}=5\sqrt{3x}$

69. $\sqrt[6]{-1}$
not a real number

71. $-\sqrt[4]{256} = -4$ because $4^4 = 256$

73. $\sqrt[6]{64} = 2$ because $2^6 = 64$

75. $-\sqrt[5]{32} = -2$ because $2^5 = 32$

77. $\sqrt[3]{x^3} = x$

79. $\sqrt[4]{y^4} = |y|$

81. $\sqrt[3]{-8x^3} = -2x$

83. $\sqrt[3]{(-5)^3} = -5$

85. $\sqrt[4]{(-5)^4} = |-5| = 5$

87. $\sqrt[4]{(x+3)^4} = |x+3|$

89. $\sqrt[5]{-32(x-1)^5} = -2(x-1)$

91.

x	$f(x) = \sqrt{x}+3$
0	$f(0) = \sqrt{0}+3 = 0+3 = 3$
1	$f(1) = \sqrt{1}+3 = 1+3 = 4$
4	$f(4) = \sqrt{4}+3 = 2+3 = 5$
9	$f(9) = \sqrt{9}+3 = 3+3 = 6$

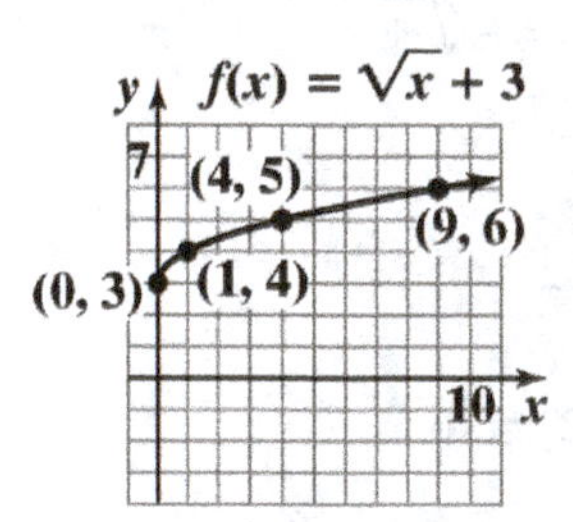

Domain: $[0, \infty)$
Range: $[3, \infty)$

93.

x	$f(x) = \sqrt{x-3}$
3	$f(3) = \sqrt{3-3} = \sqrt{0} = 0$
4	$f(4) = \sqrt{4-3} = \sqrt{1} = 1$
7	$f(7) = \sqrt{7-3} = \sqrt{4} = 2$
12	$f(12) = \sqrt{12-3} = \sqrt{9} = 3$

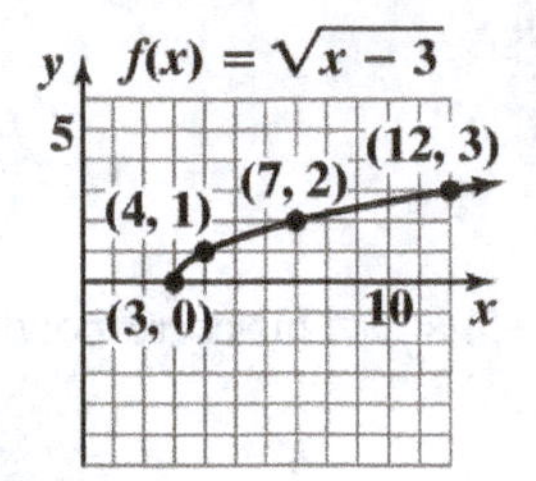

Domain: $[3, \infty)$
Range: $[0, \infty)$

95. The domain of the cube root function is all real numbers, so we only need to worry about the square root in the denominator. We need the radicand of the square root to be ≥ 0, but we also cannot divide by 0. Therefore, we have

$$30 - 2x > 0$$
$$-2x > -30$$
$$x < 15$$

The domain of f is $(-\infty, 15)$.

97. From the numerator, we need $x - 1 \geq 0$. From the denominator, we need $3 - x > 0$. We need to solve the two inequalities. The domain of the function is the overlap of the two solution sets.

$$x - 1 \geq 0 \quad \text{and} \quad 3 - x > 0$$
$$x \geq 1 \qquad\qquad -x > -3$$
$$x < 3$$

We need $x \geq 1$ and $x < 3$. Therefore, the domain of f is $[1, 3)$.

99. $\sqrt[3]{\sqrt[4]{16} + \sqrt{625}} = \sqrt[3]{2+25} = \sqrt[3]{27} = 3$

101. a. $f(x) = 2.9\sqrt{x} + 20.1$

$f(48) = 2.9\sqrt{48} + 20.1 \approx 40.2$

The model estimates the median height of boys who are 48 months to be 40.2 inches. This underestimates the actual median height by 0.6 inches.

b. Find $f(0)$ and $f(10)$.

$f(x) = 2.9\sqrt{x} + 20.1$

$f(0) = 2.9\sqrt{0} + 20.1 = 20.1$

$f(10) = 2.9\sqrt{10} + 20.1 \approx 29.3$

Average rate of change is

$m = \dfrac{f(10) - f(0)}{10 - 0} = \dfrac{29.3 - 20.1}{10} \approx 0.9$ inches per month.

c. Find $f(50)$ and $f(60)$.

$f(x) = 2.9\sqrt{x} + 20.1$

$f(50) = 2.9\sqrt{50} + 20.1 \approx 40.6$

$f(60) = 2.9\sqrt{60} + 20.1 \approx 42.6$

Average rate of change is

$m = \dfrac{f(60) - f(50)}{60 - 50} = \dfrac{42.6 - 40.6}{10} \approx 0.2$ inches per month.

This is a much smaller rate of change.
This is shown on the graph because the graph is not as steep between 50 and 60 as it is between 0 and 10.

103. $f(245) = \sqrt{20(245)} = \sqrt{4900} = 70$

The officer should not believe the motorist. The model predicts that the motorist's speed was 70 miles per hour. This is well above the 50 miles per hour speed limit.

105. – 113. Answers will vary.

115. $y_1 = \sqrt{x}$ $\quad y_2 = \sqrt{x+4}$

$y_3 = \sqrt{x-3}$

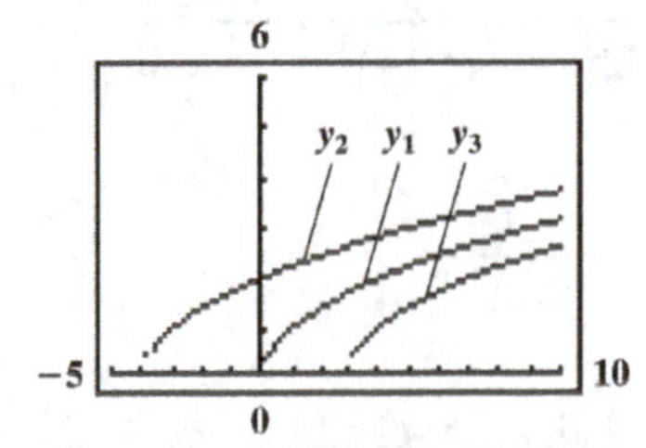

The graphs have the same shape, but differ in their orientation along the x–axis. The graphs are shifted left or right from $x = 0$.

117. $f(x) = \sqrt{x}$ $\quad g(x) = -\sqrt{x}$

$h(x) = \sqrt{-x}$ $\quad k(x) = -\sqrt{-x}$

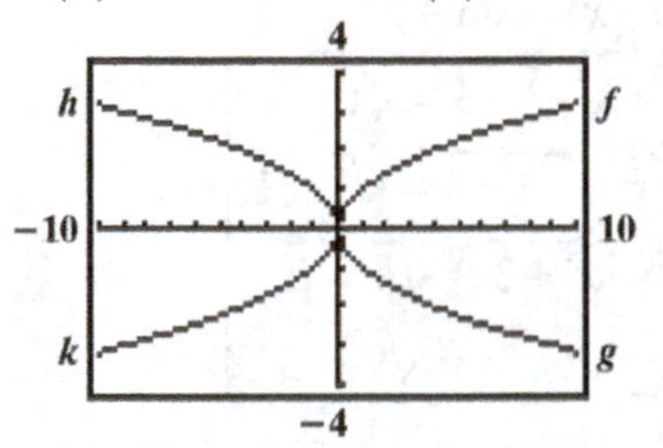

Function	Domain	Range
$f(x) = \sqrt{x}$	$[0, \infty)$	$[0, \infty)$
$g(x) = -\sqrt{x}$	$[0, \infty)$	$(-\infty, 0]$
$h(x) = \sqrt{-x}$	$(-\infty, 0]$	$[0, \infty)$
$k(x) = -\sqrt{-x}$	$(-\infty, 0]$	$(-\infty, 0]$

119. does not make sense; Explanations will vary. Sample explanation: Because the negative is raised to an even power *first,* this expression will simplify to positive 8.

121. make sense

123. false; Changes to make the statement true will vary. A sample change is: Because the expression is a cube root, the radicand is not required to be greater than zero.

125. false; Changes to make the statement true will vary.

A sample change is:
$$\sqrt{x^6} = |x^3|$$
$$\sqrt{(-2)^6} = |(-2)^3|$$
$$\sqrt{64} = |-8|$$
$$8 = 8$$

127. Answers will vary. One example is $f(x) = \sqrt{5-x}$.

129.
$$\sqrt{(2x+3)^{10}} = \sqrt{\left((2x+3)^5\right)^2}$$
$$= \left|(2x+3)^5\right|$$

131. $h(x)=\sqrt{x+3}$

x	$h(x)=\sqrt{x+3}$
−3	$h(-3)=\sqrt{-3+3}=\sqrt{0}=0$
−2	$h(-2)=\sqrt{-2+3}=\sqrt{1}=1$
1	$h(1)=\sqrt{1+3}=\sqrt{4}=2$
6	$h(6)=\sqrt{6+3}=\sqrt{9}=3$

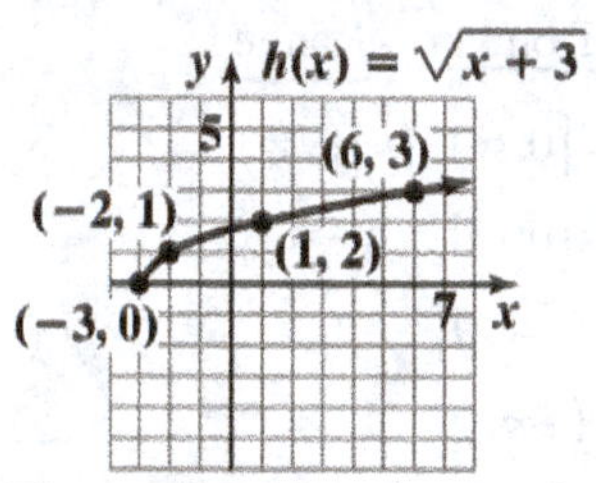

The graph of h is the graph of f shifted three units to the left.

132. $3x-2[x-3(x+5)]=3x-2[x-3x-15]$
$=3x-2[-2x-15]$
$=3x+4x+30$
$=7x+30$

133. $\left(-3x^{-4}y^{3}\right)^{-2}=(-3)^{-2}\left(x^{-4}\right)^{-2}\left(y^{3}\right)^{-2}$
$=\dfrac{1}{(-3)^2}x^8y^{-6}$
$=\dfrac{x^8}{(-3)^2y^6}=\dfrac{x^8}{9y^6}$

134. $|3x-4|>11$
$3x-4<-11$ or $3x-4>11$
$3x<-7$ $\quad$ $3x>15$
$x<-\dfrac{7}{3}$ $\quad$ $x>5$

The solution set is $\left(-\infty,-\dfrac{7}{3}\right)\cup(5,\infty)$.

135. $(2^3x^5)(2^4x^{-6})=2^{3+4}x^{5+(-6)}=2^7x^{-1}=\dfrac{2^7}{x}=\dfrac{128}{x}$

136. $\dfrac{32x^2}{16x^5}=2x^{2-5}=2x^{-3}=\dfrac{2}{x^3}$

137. $(x^{-2}y^3)^4=x^{-2\cdot4}y^{3\cdot4}=x^{-8}y^{12}=\dfrac{y^{12}}{x^8}$

7.2 Check Points

1. a. $25^{1/2}=\sqrt{25}=5$

b. $(-8)^{1/3}=\sqrt[3]{-8}=-2$

c. $\left(5xy^2\right)^{1/4}=\sqrt[4]{5xy^2}$

2. a. $\sqrt[4]{5xy}=(5xy)^{1/4}$

b. $\sqrt[5]{\dfrac{a^3b}{2}}=\left(\dfrac{a^3b}{2}\right)^{1/5}$

3. a. $8^{4/3}=\left(\sqrt[3]{8}\right)^4=2^4=16$

b. $25^{3/2}=\left(\sqrt{25}\right)^3=5^3=125$

c. $-81^{3/4}=-\left(\sqrt[4]{81}\right)^3=-3^3=-27$

4. a. $\sqrt[3]{6^4}=6^{4/3}$

b. $\left(\sqrt[5]{2xy}\right)^7=(2xy)^{7/5}$

5. a. $100^{-1/2}=\dfrac{1}{100^{1/2}}=\dfrac{1}{\sqrt{100}}=\dfrac{1}{10}$

b. $8^{-1/3}=\dfrac{1}{8^{1/3}}=\dfrac{1}{\sqrt[3]{8}}=\dfrac{1}{2}$

c. $32^{-3/5}=\dfrac{1}{32^{3/5}}=\dfrac{1}{\left(\sqrt[5]{32}\right)^3}=\dfrac{1}{2^3}=\dfrac{1}{8}$

d. $(3xy)^{-5/9}=\dfrac{1}{(3xy)^{5/9}}$

6. a. $7^{\frac{1}{2}}\cdot7^{\frac{1}{3}}=7^{\frac{1}{2}+\frac{1}{3}}=7^{\frac{5}{6}}$

b. $\dfrac{50x^{\frac{1}{3}}}{10x^{\frac{4}{3}}}=5\cdot x^{\frac{1}{3}-\frac{4}{3}}=5x^{-1}=\dfrac{5}{x}$

c. $\left(9.1^{\frac{2}{5}}\right)^{\frac{3}{4}} = 9.1^{\frac{2}{5}\cdot\frac{3}{4}} = 9.1^{\frac{3}{10}}$

d. $\left(x^{-\frac{3}{5}}y^{\frac{1}{4}}\right)^{\frac{1}{3}} = x^{-\frac{3}{5}\cdot\frac{1}{3}}y^{\frac{1}{4}\cdot\frac{1}{3}} = x^{-\frac{1}{5}}y^{\frac{1}{12}} = \dfrac{y^{\frac{1}{12}}}{x^{\frac{1}{5}}}$

7. a. $\sqrt[6]{x^3} = x^{\frac{3}{6}} = x^{\frac{1}{2}} = \sqrt{x}$

b. $\sqrt[3]{8a^{12}} = \left(8a^{12}\right)^{\frac{1}{3}} = 8^{\frac{1}{3}}a^{12\cdot\frac{1}{3}} = 2a^4$

c. $\sqrt[8]{x^4y^2} = \left(x^4y^2\right)^{\frac{1}{8}} = \left(x^4\right)^{\frac{1}{8}}\left(y^2\right)^{\frac{1}{8}}$

$= x^{\frac{2}{4}}y^{\frac{1}{4}} = \left(x^2y\right)^{\frac{1}{4}} = \sqrt[4]{x^2y}$

d. $\dfrac{\sqrt{x}}{\sqrt[3]{x}} = \dfrac{x^{\frac{1}{2}}}{x^{\frac{1}{3}}} = x^{\frac{1}{2}-\frac{1}{3}} = x^{\frac{3}{6}-\frac{2}{6}} = x^{\frac{1}{6}} = \sqrt[6]{x}$

e. $\sqrt{\sqrt[3]{x}} = \left(x^{\frac{1}{3}}\right)^{\frac{1}{2}} = x^{\frac{1}{3}\cdot\frac{1}{2}} = x^{\frac{1}{6}} = \sqrt[6]{x}$

7.2 Concept and Vocabulary Check

1. 36; 6
2. 8; 2
3. $\sqrt[n]{a}$
4. 16; 2; 8
5. $(\sqrt[n]{a})^m$ or $\sqrt[n]{a^m}$
6. $\dfrac{5}{3}$
7. $16^{-\frac{3}{2}} = \dfrac{1}{16^{\frac{3}{2}}} = \dfrac{1}{\left(\sqrt{16}\right)^3} = \dfrac{1}{4^3} = \dfrac{1}{64}$

7.2 Exercise Set

1. $49^{1/2} = \sqrt{49} = 7$

3. $(-27)^{1/3} = \sqrt[3]{-27} = -3$

5. $-16^{1/4} = -\sqrt[4]{16} = -2$

7. $(xy)^{1/3} = \sqrt[3]{xy}$

9. $\left(2xy^3\right)^{1/5} = \sqrt[5]{2xy^3}$

11. $81^{3/2} = \left(\sqrt{81}\right)^3 = 9^3 = 729$

13. $125^{2/3} = \left(\sqrt[3]{125}\right)^2 = 5^2 = 25$

15. $(-32)^{3/5} = \left(\sqrt[5]{-32}\right)^3 = (-2)^3 = -8$

17. $27^{2/3} + 16^{3/4} = \left(\sqrt[3]{27}\right)^2 + \left(\sqrt[4]{16}\right)^3$

$= 3^2 + 2^3$

$= 9 + 8 = 17$

19. $(xy)^{4/7} = \left(\sqrt[7]{xy}\right)^4$ or $\sqrt[7]{(xy)^4}$

21. $\sqrt{7} = 7^{1/2}$

23. $\sqrt[3]{5} = 5^{1/3}$

25. $\sqrt[5]{11x} = (11x)^{1/5}$

27. $\sqrt[5]{y^{17}} = \sqrt[5]{y^{15}\cdot y^2} = \sqrt[5]{y^{15}}\cdot\sqrt[5]{y^2}$

$= y^3\sqrt[5]{y^2}$

29. $\sqrt[5]{x^3} = x^{3/5}$

31. $\sqrt[5]{x^2y} = \left(x^2y\right)^{1/5}$

33. $\left(\sqrt{19xy}\right)^3 = (19xy)^{3/2}$

35. $\left(\sqrt[6]{7xy^2}\right)^5 = \left(7xy^2\right)^{5/6}$

37. $2x\sqrt[3]{y^2} = 2xy^{2/3}$

39. $49^{-1/2} = \dfrac{1}{49^{1/2}} = \dfrac{1}{\sqrt{49}} = \dfrac{1}{7}$

41. $\sqrt[5]{64x^6y^{17}} = \sqrt[5]{32 \cdot 2x^5xy^{15}y^2}$
$= \sqrt[5]{32x^5y^{15}} \cdot \sqrt[5]{2xy^2}$
$= 2xy^3\sqrt[5]{2xy^2}$

43. $16^{-3/4} = \dfrac{1}{16^{3/4}} = \dfrac{1}{\left(\sqrt[4]{16}\right)^3} = \dfrac{1}{2^3} = \dfrac{1}{8}$

45. $8^{-2/3} = \dfrac{1}{8^{2/3}} = \dfrac{1}{\left(\sqrt[3]{8}\right)^2} = \dfrac{1}{2^2} = \dfrac{1}{4}$

47. $\left(\dfrac{8}{27}\right)^{-1/3} = \left(\dfrac{27}{8}\right)^{1/3} = \sqrt[3]{\dfrac{27}{8}} = \dfrac{3}{2}$

49. $(-64)^{-2/3} = \dfrac{1}{(-64)^{2/3}} = \dfrac{1}{\left(\sqrt[3]{-64}\right)^2}$
$= \dfrac{1}{(-4)^2} = \dfrac{1}{16}$

51. $(2xy)^{-7/10} = \dfrac{1}{(2xy)^{7/10}}$
$= \dfrac{1}{\sqrt[10]{(2xy)^7}}$ or $\dfrac{1}{\left(\sqrt[10]{2xy}\right)^7}$

53. $5xz^{-1/3} = \dfrac{5xz^{-1/3}}{1} = \dfrac{5x}{z^{1/3}}$

55. $3^{3/4} \cdot 3^{1/4} = 3^{(3/4)+(1/4)}$
$= 3^{4/4} = 3^1 = 3$

57. $\dfrac{16^{3/4}}{16^{1/4}} = 16^{(3/4)-(1/4)} = 16^{2/4}$
$= 16^{1/2} = \sqrt{16} = 4$

59. $x^{1/2} \cdot x^{1/3} = x^{(1/2)+(1/3)}$
$= x^{(3/6)+(2/6)} = x^{5/6}$

61. $\dfrac{x^{4/5}}{x^{1/5}} = x^{(4/5)-(1/5)} = x^{3/5}$

63. $\dfrac{x^{1/3}}{x^{3/4}} = x^{(1/3)-(3/4)} = x^{(4/12)-(9/12)}$
$= x^{-5/12} = \dfrac{1}{x^{5/12}}$

65. $\left(5^{\frac{2}{3}}\right)^3 = 5^{\frac{2}{3}\cdot 3} = 5^2 = 25$

67. $\left(y^{-2/3}\right)^{1/4} = y^{(-2/3)\cdot(1/4)} = y^{-2/12}$
$= y^{-1/6} = \dfrac{1}{y^{1/6}}$

69. $\left(2x^{1/5}\right)^5 = 2^5x^{(1/5)\cdot 5} = 32x^1 = 32x$

71. $\left(25x^4y^6\right)^{1/2} = 25^{1/2}\left(x^4\right)^{1/2}\left(y^6\right)^{1/2}$
$= \sqrt{25}x^{4(1/2)}y^{6(1/2)}$
$= 5x^2y^3$

73. $\left(x^{1/2}y^{-3/5}\right)^{1/2} = \left(\dfrac{x^{1/2}y^{-3/5}}{1}\right)^{1/2}$
$= \left(\dfrac{x^{1/2}}{y^{3/5}}\right)^{1/2}$
$= \dfrac{x^{(1/2)\cdot(1/2)}}{y^{(3/5)\cdot(1/2)}} = \dfrac{x^{1/4}}{y^{3/10}}$

75. $\dfrac{3^{1/2} \cdot 3^{3/4}}{3^{1/4}} = 3^{(1/2)+(3/4)-(1/4)}$
$= 3^{(2/4)+(3/4)-(1/4)}$
$= 3^{4/4} = 3^1 = 3$

77. $\dfrac{\left(3y^{1/4}\right)^3}{y^{1/12}} = \dfrac{3^3y^{(1/4)\cdot 3}}{y^{1/12}} = \dfrac{27y^{3/4}}{y^{1/12}}$
$= 27y^{(3/4)-(1/12)}$
$= 27y^{(9/12)-(1/12)}$
$= 27y^{8/12} = 27y^{2/3}$

79. $\sqrt[8]{x^2} = x^{2/8} = x^{1/4} = \sqrt[4]{x}$

81. $\sqrt[3]{8a^6} = 8^{1/3}a^{6/3} = 2a^2$

83. $\sqrt[5]{x^{10}y^{15}} = x^{10/5}y^{15/5} = x^2y^3$

85. $\left(\sqrt[3]{xy}\right)^{18} = (xy)^{18/3} = (xy)^6 = x^6y^6$

87. $\sqrt[10]{(3y)^2} = (3y)^{2/10} = (3y)^{1/5} = \sqrt[5]{3y}$

89. $\left(\sqrt[6]{2a}\right)^4 = (2a)^{4/6} = (2a)^{2/3}$
$= \left(4a^2\right)^{1/3} = \sqrt[3]{4a^2}$

91. $\sqrt[9]{x^6y^3} = x^{6/9}y^{3/9}$
$= x^{2/3}y^{1/3} = \sqrt[3]{x^2y}$

93. $\sqrt{2}\cdot\sqrt[3]{2} = 2^{1/2}\cdot 2^{1/3} = 2^{(1/2)+(1/3)}$
$= 2^{(3/6)+(2/6)} = 2^{5/6}$
$= \sqrt[6]{2^5}$ or $\sqrt[6]{32}$

95. $\sqrt[5]{x^2}\cdot\sqrt{x} = \left(x^2\right)^{1/5}\cdot x^{1/2}$
$= x^{2/5}\cdot x^{1/2} = x^{(2/5)+(1/2)}$
$= x^{(4/10)+(5/10)} = x^{9/10}$
$= \sqrt[10]{x^9}$

97. $\sqrt[4]{a^2b}\cdot\sqrt[3]{ab} = \left(a^2b\right)^{1/4}\cdot(ab)^{1/3}$
$= a^{1/2}b^{1/4}\cdot a^{1/3}b^{1/3}$
$= a^{(1/2)+(1/3)}b^{(1/4)+(1/3)}$
$= a^{(6/12)+(4/12)}b^{(3/12)+(4/12)}$
$= a^{10/12}b^{7/12}$
$= \sqrt[12]{a^{10}b^7}$

99. $\frac{\sqrt[4]{x}}{\sqrt[5]{x}} = \frac{x^{1/4}}{x^{1/5}} = x^{(1/4)-(1/5)}$
$= x^{(5/20)-(4/20)}$
$= x^{1/20} = \sqrt[20]{x}$

101. $\frac{\sqrt[3]{y^2}}{\sqrt[6]{y}} = \frac{y^{2/3}}{y^{1/6}} = y^{(2/3)-(1/6)}$
$= y^{(4/6)-(1/6)} = y^{3/6}$
$= y^{1/2} = \sqrt{y}$

103. $\sqrt[4]{\sqrt{x}} = \sqrt[4]{x^{1/2}} = \left(x^{1/2}\right)^{1/4}$
$= x^{(1/2)\cdot(1/4)} = x^{1/8}$
$= \sqrt[8]{x}$

105. $\sqrt{\sqrt{x^2y}} = \sqrt{\left(x^2y\right)^{1/2}} = \left(\left(x^2y\right)^{1/2}\right)^{1/2}$
$= \left(x^2y\right)^{(1/2)\cdot(1/2)} = \left(x^2y\right)^{1/4}$
$= \sqrt[4]{x^2y}$

107. $\sqrt[4]{\sqrt[3]{2x}} = \sqrt[4]{(2x)^{1/3}} = \left((2x)^{1/3}\right)^{1/4}$
$= (2x)^{(1/3)\cdot(1/4)} = (2x)^{1/12}$
$= \sqrt[12]{2x}$

109. $\left(\sqrt[4]{x^3y^5}\right)^{12} = \left(\left(x^3y^5\right)^{1/4}\right)^{12}$
$= \left(x^3y^5\right)^{(1/4)\cdot 12}$
$= \left(x^3y^5\right)^{12/4} = \left(x^3y^5\right)^3$
$= x^{3\cdot 3}y^{5\cdot 3} = x^9y^{15}$

111. $\frac{\sqrt[4]{a^5b^5}}{\sqrt{ab}} = \frac{a^{5/4}b^{5/4}}{a^{1/2}b^{1/2}}$
$= a^{(5/4)-(1/2)}b^{(5/4)-(1/2)}$
$= a^{(5/4)-(2/4)}b^{(5/4)-(2/4)}$
$= a^{3/4}b^{3/4} = \left(a^3b^3\right)^{1/4}$
$= \sqrt[4]{a^3b^3}$

113. $x^{1/3}\left(x^{1/3} - x^{2/3}\right) = x^{1/3}\cdot x^{1/3} - x^{1/3}\cdot x^{2/3}$
$= x^{(1/3)+(1/3)} - x^{(1/3)+(2/3)}$
$= x^{2/3} - x^{3/3}$
$= x^{2/3} - x$

115. $\left(x^{1/2}-3\right)\left(x^{1/2}+5\right)$
$= x^{1/2}\cdot x^{1/2}+x^{1/2}\cdot 5-3\cdot x^{1/2}-3\cdot 5$
$= x^{(1/2)+(1/2)}+5x^{1/2}-3x^{1/2}-15$
$= x^{2/2}+2x^{1/2}-15$
$= x+2x^{1/2}-15$

117. $6x^{1/2}+2x^{3/2}=3\cdot 2x^{1/2}+2x^{(1/2)+(2/2)}$
$=3\cdot 2x^{1/2}+2x^{1/2}\cdot x^{2/2}$
$=3\cdot 2x^{1/2}+2x^{1/2}\cdot x$
$=2x^{1/2}(3+x)$

119. $15x^{1/3}-60x=15x^{1/3}-60x^{3/3}$
$=15x^{1/3}-60x^{(1/3)+(2/3)}$
$=15x^{1/3}-60x^{1/3}x^{2/3}$
$=15x^{1/3}\cdot 1-15x^{1/3}\cdot 4x^{2/3}$
$=15x^{1/3}\left(1-4x^{2/3}\right)$

121. $\left(49x^{-2}y^{4}\right)^{-1/2}\left(xy^{1/2}\right)$
$=(49)^{-1/2}\left(x^{-2}\right)^{-1/2}\left(y^{4}\right)^{-1/2}\left(xy^{1/2}\right)$
$=\dfrac{1}{49^{1/2}}x^{(-2)(-1/2)}y^{(4)(-1/2)}\left(xy^{1/2}\right)$
$=\dfrac{1}{7}x^{1}y^{-2}\cdot xy^{1/2}=\dfrac{1}{7}x^{1+1}y^{-2+(1/2)}$
$=\dfrac{1}{7}x^{2}y^{-3/2}=\dfrac{x^{2}}{7y^{3/2}}$

123. $\left(\dfrac{x^{-5/4}y^{1/3}}{x^{-3/4}}\right)^{-6}=\left(x^{(-5/4)-(-3/4)}y^{1/3}\right)^{-6}$
$=\left(x^{-2/4}y^{1/3}\right)^{-6}=x^{(-2/4)(-6)}y^{(1/3)(-6)}$
$=x^{3}y^{-2}=\dfrac{x^{3}}{y^{2}}$

125. $f(8)=29(8)^{1/3}$
$=29\sqrt[3]{8}$
$=29(2)=58$

There are 58 plant species on an 8 square mile island.

127. $f(x) = 70x^{3/4}$

$f(80) = 70(80)^{3/4} \approx 1872$

A person who weighs 80 kilograms needs about 1872 calories per day to maintain life.

129. **a.** $C = 35.74 + 0.6215t - 35.74v^{4/25} + 0.4275tv^{4/25}$

b. $C = 35.74 + 0.6215(25) - 35.74(30)^{4/25} + 0.4275(25)(30)^{4/25} \approx 8°\text{F}$

131. $C = 35.74 + 0.6215t - 35.74v^{4/25} + 0.4275t \cdot v^{4/25}$

a. For $t = 0$, we get

$C(v) = 35.74 - 35.74v^{4/25}$

b. $C(25) = 35.74 - 35.74(25)^{4/25} \approx -24$

When the air temperature is $0°\text{F}$ and the wind speed is 25 miles per hour, the windchill temperature is $-24°\text{F}$.

c. The solution to part (b) is represented by the point $(25, -24)$ on the graph.

133. $L + 1.25\sqrt{S} - 9.8\sqrt[3]{D} \le 16.296$

a. $L + 1.25S^{1/2} - 9.8D^{1/3} \le 16.296$

b.

$$
\begin{aligned}
L + 1.25S^{1/2} - 9.8D^{1/3} &\le 16.296 \\
20.85 + 1.25(276.4)^{1/2} - 9.8(18.55)^{1/3} &\le 16.296 \\
20.85 + 1.25\sqrt{276.4} - 9.8\sqrt[3]{18.55} &\le 16.296 \\
20.85 + 1.25(16.625) - 9.8(2.647) &\le 16.296 \\
20.85 + 20.781 - 25.941 &\le 16.296 \\
15.69 &\le 16.296
\end{aligned}
$$

The yacht is eligible to enter the America's Cup.

135. – 143. Answers will vary.

145.

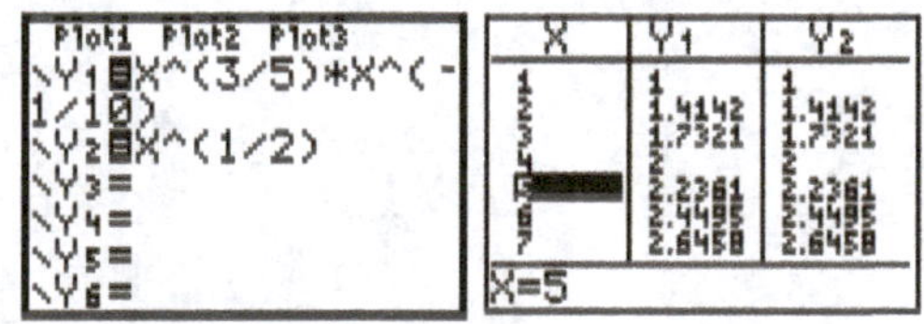

The simplification is correct.

147.

```
Plot1 Plot2 Plot3
\Y1■X^(1/4)/(X^(
1/2)*X^(-3/4))
\Y2■1/X^(1/2)
\Y3=
\Y4=
\Y5=
\Y6=
```

X	Y1	Y2
1	1	1
2	1.4142	.70711
3	1.7321	.57735
4	2	.5
5	2.2361	.44721
6	2.4495	.40825
7	2.6457	.37796

X=1

The simplification is not correct.

$$\frac{x^{1/4}}{x^{1/2}\cdot x^{-3/4}}=x^{(1/4)-[(1/2)+(-3/4)]}$$
$$=x^{(1/4)-(2/4)+(3/4)}$$
$$=x^{2/4}$$
$$=x^{1/2}$$

```
Plot1 Plot2 Plot3
\Y1■X^(1/4)/(X^(
1/2)*X^(-3/4))
\Y2■X^(1/2)
\Y3=
\Y4=
\Y5=
\Y6=
```

X	Y1	Y2
1	1	1
2	1.4142	1.4142
3	1.7321	1.7321
4	2	2
5	2.2361	2.2361
6	2.4495	2.4495
7	2.6458	2.6458

X=7

The new simplification is correct.

149. does not make sense; Explanations will vary. Sample explanation: It is often easier to find the *n*th root before raising the expression to the *m*th power.

151. does not make sense; Explanations will vary. Sample explanation: In the top line, $5\cdot 5^{1/2}$ was incorrectly simplified to $25^{1/2}$. The order of operations does not allow the multiplication to occur before the exponent is evaluated.

153. false; Changes to make the statement true will vary. A sample change is: $(a+b)^{1/n}\neq a^{1/n}+b^{1/n}$. Do not confuse $(a+b)^{1/n}$ with $(ab)^{1/n}$. We can simplify $(ab)^{1/n}$ as $a^{1/n}b^{1/n}$.

155. true

157.
$$\frac{8^{-4/3}+2^{-2}}{16^{-3/4}+2^{-1}}=\frac{\frac{1}{8^{4/3}}+\frac{1}{2^2}}{\frac{1}{16^{3/4}}+\frac{1}{2^1}}=\frac{\frac{1}{\left(\sqrt[3]{8}\right)^4}+\frac{1}{4}}{\frac{1}{\left(\sqrt[4]{16}\right)^3}+\frac{1}{2}}$$

$$=\frac{\frac{1}{(2)^4}+\frac{1}{4}}{\frac{1}{(2)^3}+\frac{1}{2}}=\frac{\frac{1}{16}+\frac{1}{4}}{\frac{1}{8}+\frac{1}{2}}=\frac{\frac{1}{16}+\frac{4}{16}}{\frac{1}{8}+\frac{4}{8}}$$

$$=\frac{\frac{5}{16}}{\frac{5}{8}}=\frac{5}{16}\div\frac{5}{8}=\frac{1}{2}\frac{\cancel{5}}{\cancel{16}}\cdot\frac{1}{1}\frac{\cancel{8}}{\cancel{5}}=\frac{1}{2}$$

The boy is allowed to eat half of the cake. The professor will eat half of what's left over or $\frac{1}{2}\cdot\frac{1}{2}=\frac{1}{4}$ of the cake.

159. First simplify.

$$f(x) = (x-3)^{1/2}(x+4)^{-1/2}$$
$$= \frac{(x-3)^{1/2}(x+4)^{-1/2}}{1}$$
$$= \frac{(x-3)^{1/2}}{(x+4)^{1/2}} = \frac{\sqrt{x-3}}{\sqrt{x+4}}$$

To find the domain, we need to consider the radicals and the denominator of the fraction. Since radicands have to be greater than or equal to zero, we can write inequalities and solve. But since the denominator cannot be zero, that radicand must be strictly greater than zero.

$x-3 \geq 0$ and $x+4 > 0$

$x \geq 3$ $\qquad$ $x > -4$

The domain of the function is the intersection of the sets, $[3, \infty)$.

160. $m = \dfrac{y_2 - y_1}{x_2 - x_1} = \dfrac{3-1}{4-5} = \dfrac{2}{-1} = -2$

$y - 3 = -2(x-4)$

Solve for y to write the equation in slope–intercept form.

$y - 3 = -2x + 8$

$y = -2x + 11$

or

$f(x) = -2x + 11$

161. $y \leq -\dfrac{3}{2}x + 3$

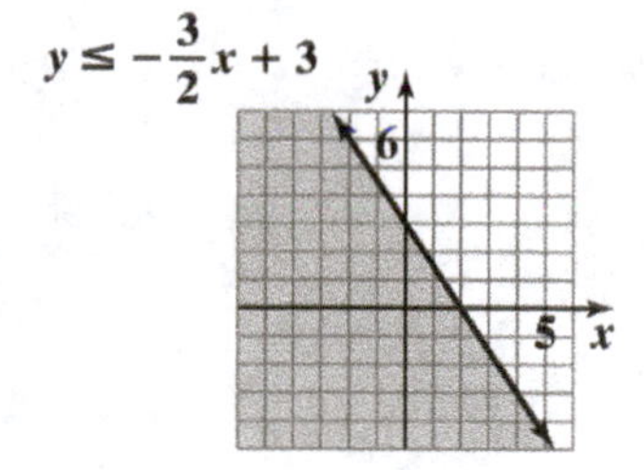

162. $D = \begin{vmatrix} 5 & -3 \\ 7 & 1 \end{vmatrix} = 5(1) - 7(-3) = 5 + 21 = 26$

$D_x = \begin{vmatrix} 3 & -3 \\ 25 & 1 \end{vmatrix} = 3(1) - 25(-3) = 3 + 75 = 78$

$D_y = \begin{vmatrix} 5 & 3 \\ 7 & 25 \end{vmatrix} = 5(25) - 7(3) = 125 - 21 = 104$

$x = \dfrac{D_x}{D} = \dfrac{78}{26} = 3 \qquad y = \dfrac{D_y}{D} = \dfrac{104}{26} = 4$

The solution set is $\{(3, 4)\}$.

163. a. $\sqrt{16}\cdot\sqrt{4}=4\cdot 2=8$

b. $\sqrt{16\cdot 4}=\sqrt{64}=8$

c. $\sqrt{16}\cdot\sqrt{4}=\sqrt{16\cdot 4}$

164. a. $\sqrt{300}\approx 17.32$

b. $10\sqrt{3}\approx 17.32$

c. $\sqrt{300}=10\sqrt{3}$

165. a. $\sqrt[3]{x^{21}}=x^{21/3}=x^7$

b. $\sqrt[6]{y^{24}}=y^{24/6}=y^4$

7.3 Check Points

1. a. $\sqrt{5}\cdot\sqrt{11}=\sqrt{5\cdot 11}=\sqrt{55}$

b. $\sqrt{x+4}\cdot\sqrt{x-4}=\sqrt{(x+4)(x-4)}=\sqrt{x^2-16}$

c. $\sqrt[3]{6}\cdot\sqrt[3]{10}=\sqrt[3]{6\cdot 10}=\sqrt[3]{60}$

d. $\sqrt[7]{2x}\cdot\sqrt[7]{6x^3}=\sqrt[7]{12x^4}$

2. a. $\sqrt{80}=\sqrt{16\cdot 5}=\sqrt{16}\cdot\sqrt{5}=4\sqrt{5}$

b. $\sqrt[3]{40}=\sqrt[3]{8\cdot 5}=\sqrt[3]{8}\cdot\sqrt[3]{5}=2\sqrt[3]{5}$

c. $\sqrt[4]{32}=\sqrt[4]{16\cdot 2}=\sqrt[4]{16}\cdot\sqrt[4]{2}=2\sqrt[4]{2}$

d. $\sqrt{200x^2y}=\sqrt{100x^2}\cdot\sqrt{2y}=10|x|\sqrt{2y}$

3.
$$\begin{aligned} f(x)&=\sqrt{3x^2-12x+12}\\ &=\sqrt{3(x^2-4x+4)}\\ &=\sqrt{3(x-2)^2}\\ &=\sqrt{3}\cdot\sqrt{(x-2)^2}\\ &=\sqrt{3}\cdot|x-2| \end{aligned}$$

4. $\sqrt{x^9y^{11}z^3}=\sqrt{x^8y^{10}z^2}\cdot\sqrt{xyz}=x^4y^5z\sqrt{xyz}$

5. $\sqrt[3]{40x^{10}y^{14}}=\sqrt[3]{8\cdot 5\cdot x^9\cdot x\cdot y^{12}\cdot y^2}=\sqrt[3]{8x^9y^{12}}\sqrt[3]{5xy^2}=2x^3y^4\sqrt[3]{5xy^2}$

6. $\sqrt[5]{32x^{12}y^2z^8} = \sqrt[5]{32 \cdot x^{10} \cdot x^2 \cdot y^2 \cdot z^5 \cdot z^3} = \sqrt[5]{32x^{10}z^5}\sqrt[5]{x^2y^2z^3} = 2x^2z\sqrt[5]{x^2y^2z^3}$

7. a. $\sqrt{6} \cdot \sqrt{2} = \sqrt{12} = \sqrt{4 \cdot 3} = \sqrt{4} \cdot \sqrt{3} = 2\sqrt{3}$

b. $10\sqrt[3]{16} \cdot 5\sqrt[3]{2} = 50\sqrt[3]{16 \cdot 2} = 50\sqrt[3]{32} = 50\sqrt[3]{8 \cdot 4} = 50\sqrt[3]{8} \cdot \sqrt[3]{4} = 50 \cdot 2 \cdot \sqrt[3]{4} = 100\sqrt[3]{4}$

c. $\sqrt[4]{4x^2y} \cdot \sqrt[4]{8x^6y^3} = \sqrt[4]{32x^8y^4} = \sqrt[4]{16x^8y^4} \cdot \sqrt[4]{2} = 2x^2y\sqrt[4]{2}$

7.3 Concept and Vocabulary Check

1. $\sqrt[n]{ab}$

2. 77

3. 8; 5; $2\sqrt[3]{5}$

4. $\sqrt{5}\,|x+1|$

5. x^5; x^4; x^5

7.3 Exercise Set

1. $\sqrt{3} \cdot \sqrt{5} = \sqrt{3 \cdot 5} = \sqrt{15}$

3. $\sqrt[3]{2} \cdot \sqrt[3]{9} = \sqrt[3]{2 \cdot 9} = \sqrt[3]{18}$

5. $\sqrt[4]{11} \cdot \sqrt[4]{3} = \sqrt[4]{11 \cdot 3} = \sqrt[4]{33}$

7. $\sqrt{3x} \cdot \sqrt{11y} = \sqrt{3x \cdot 11y} = \sqrt{33xy}$

9. $\sqrt[5]{6x^3} \cdot \sqrt[5]{4x} = \sqrt[5]{6x^3 \cdot 4x} = \sqrt[5]{24x^4}$

11. $\sqrt{x+3} \cdot \sqrt{x-3} = \sqrt{(x+3)(x-3)} = \sqrt{x^2-9}$

13. $\sqrt[6]{x-4} \cdot \sqrt[6]{(x-4)^4} = \sqrt[6]{(x-4)(x-4)^4}$

$= \sqrt[6]{(x-4)^5}$

15. $\sqrt{\frac{2x}{3}} \cdot \sqrt{\frac{3}{2}} = \sqrt{\frac{2x}{3} \cdot \frac{3}{2}} = \sqrt{\frac{\cancel{2}x}{\cancel{3}} \cdot \frac{\cancel{3}}{\cancel{2}}} = \sqrt{x}$

17. $\sqrt[4]{\frac{x}{7}} \cdot \sqrt[4]{\frac{3}{y}} = \sqrt[4]{\frac{x}{7} \cdot \frac{3}{y}} = \sqrt[4]{\frac{3x}{7y}}$

19. $\sqrt[7]{7x^2y}\cdot\sqrt[7]{11x^3y^2} = \sqrt[7]{7x^2y\cdot 11x^3y^2}$
$= \sqrt[7]{7\cdot 11x^2x^3yy^2}$
$= \sqrt[7]{77x^5y^3}$

21. $\sqrt{50} = \sqrt{25\cdot 2} = \sqrt{25}\cdot\sqrt{2} = 5\sqrt{2}$

23. $\sqrt{45} = \sqrt{9\cdot 5} = \sqrt{9}\cdot\sqrt{5} = 3\sqrt{5}$

25. $\sqrt{75x} = \sqrt{25\cdot 3x} = \sqrt{25}\cdot\sqrt{3x} = 5\sqrt{3x}$

27. $\sqrt[3]{16} = \sqrt[3]{8\cdot 2} = \sqrt[3]{8}\cdot\sqrt[3]{2} = 2\sqrt[3]{2}$

29. $\sqrt[3]{27x^3} = \sqrt[3]{27\cdot x^3} = \sqrt[3]{27}\cdot\sqrt[3]{x^3} = 3x$

31. $\sqrt[3]{-16x^2y^3} = \sqrt[3]{-8\cdot 2x^2y^3}$
$= \sqrt[3]{-8y^3}\cdot\sqrt[3]{2x^2}$
$= -2y\sqrt[3]{2x^2}$

33. $f(x) = \sqrt{36(x+2)^2} = 6|x+2|$

35. $f(x) = \sqrt[3]{32(x+2)^3}$
$= \sqrt[3]{8\cdot 4(x+2)^3}$
$= \sqrt[3]{8(x+2)^3}\cdot\sqrt[3]{4}$
$= 2(x+2)\sqrt[3]{4}$

37. $f(x) = \sqrt{3x^2-6x+3}$
$= \sqrt{3(x^2-2x+1)}$
$= \sqrt{3(x-1)^2}$
$= |x-1|\sqrt{3}$

39. $\sqrt{x^7} = \sqrt{x^6\cdot x} = \sqrt{x^6}\cdot\sqrt{x} = x^3\sqrt{x}$

41. $\sqrt{x^8y^9} = \sqrt{x^8y^8y} = \sqrt{x^8y^8}\sqrt{y}$
$= x^4y^4\sqrt{y}$

43. $\sqrt{48x^3} = \sqrt{16\cdot 3x^2x} = \sqrt{16x^2}\cdot\sqrt{3x}$
$= 4x\sqrt{3x}$

45. $\sqrt[3]{y^8} = \sqrt[3]{y^6\cdot y^2} = \sqrt[3]{y^6}\cdot\sqrt[3]{y^2}$
$= y^2\sqrt[3]{y^2}$

47. $\sqrt[3]{x^{14}y^3z} = \sqrt[3]{x^{12}x^2y^3z} = \sqrt[3]{x^{12}y^3}\cdot\sqrt[3]{x^2z}$
$= x^4y\sqrt[3]{x^2z}$

49. $\sqrt[3]{81x^8y^6} = \sqrt[3]{27\cdot 3x^6x^2y^6}$
$= \sqrt[3]{27x^6y^6}\cdot\sqrt[3]{3x^2}$
$= 3x^2y^2\sqrt[3]{3x^2}$

51. $\sqrt[3]{(x+y)^5} = \sqrt[3]{(x+y)^3\cdot(x+y)^2}$
$= \sqrt[3]{(x+y)^3}\cdot\sqrt[3]{(x+y)^2}$
$= (x+y)\sqrt[3]{(x+y)^2}$

53. $\sqrt[5]{y^{17}} = \sqrt[5]{y^{15}\cdot y^2} = \sqrt[5]{y^{15}}\cdot\sqrt[5]{y^2}$
$= y^3\sqrt[5]{y^2}$

55. $\sqrt[5]{64x^6y^{17}} = \sqrt[5]{32\cdot 2x^5xy^{15}y^2}$
$= \sqrt[5]{32x^5y^{15}}\cdot\sqrt[5]{2xy^2}$
$= 2xy^3\sqrt[5]{2xy^2}$

57. $\sqrt[4]{80x^{10}} = \sqrt[4]{16\cdot 5x^8x^2}$
$= \sqrt[4]{16x^8}\cdot\sqrt[4]{5x^2}$
$= 2x^2\sqrt[4]{5x^2}$

59. $\sqrt[4]{(x-3)^{10}} = \sqrt[4]{(x-3)^8(x-3)^2}$
$= \sqrt[4]{(x-3)^8}\cdot\sqrt[4]{(x-3)^2}$
$= (x-3)^2\sqrt[4]{(x-3)^2}$
or
$= (x-3)^2\sqrt{x-3}$

61. $\sqrt{12}\cdot\sqrt{2} = \sqrt{12\cdot 2} = \sqrt{24}$
$= \sqrt{4\cdot 6} = \sqrt{4}\cdot\sqrt{6}$
$= 2\sqrt{6}$

63. $\sqrt{5x}\cdot\sqrt{10y} = \sqrt{5x\cdot 10y} = \sqrt{50xy}$
$= \sqrt{25\cdot 2xy} = 5\sqrt{2xy}$

65. $\sqrt{12x}\cdot\sqrt{3x}=\sqrt{12x\cdot 3x}$
$=\sqrt{36x^2}=6x$

67. $\sqrt{50xy}\cdot\sqrt{4xy^2}=\sqrt{50xy\cdot 4xy^2}$
$=\sqrt{200x^2y^3}$
$=\sqrt{100\cdot 2x^2y^2y}$
$=\sqrt{100x^2y^2}\cdot\sqrt{2y}$
$=10xy\sqrt{2y}$

69. $2\sqrt{5}\cdot 3\sqrt{40}=2\cdot 3\sqrt{5\cdot 40}=6\sqrt{200}$
$=6\sqrt{100\cdot 2}=6\sqrt{100}\cdot\sqrt{2}$
$=6\cdot 10\sqrt{2}=60\sqrt{2}$

71. $\sqrt[3]{12}\cdot\sqrt[3]{4}=\sqrt[3]{12\cdot 4}=\sqrt[3]{48}=\sqrt[3]{8\cdot 6}$
$=\sqrt[3]{8}\cdot\sqrt[3]{6}=2\sqrt[3]{6}$

73. $\sqrt{5x^3}\cdot\sqrt{8x^2}=\sqrt{5x^3\cdot 8x^2}=\sqrt{40x^5}$
$=\sqrt{4\cdot 10x^4x}$
$=\sqrt{4x^4}\cdot\sqrt{10x}$
$=2x^2\sqrt{10x}$

75. $\sqrt[3]{25x^4y^2}\cdot\sqrt[3]{5xy^{12}}$
$=\sqrt[3]{25x^4y^2\cdot 5xy^{12}}$
$=\sqrt[3]{125x^5y^{14}}$
$=\sqrt[3]{125x^3x^2y^{12}y^2}$
$=\sqrt[3]{125x^3y^{12}}\cdot\sqrt[3]{x^2y^2}$
$=5xy^4\sqrt[3]{x^2y^2}$

77. $\sqrt[4]{8x^2y^3z^6}\cdot\sqrt[4]{2x^4yz}$
$=\sqrt[4]{8x^2y^3z^6\cdot 2x^4yz}$
$=\sqrt[4]{16x^6y^4z^7}$
$=\sqrt[4]{16x^4x^2y^4z^4z^3}$
$=\sqrt[4]{16x^4y^4z^4}\cdot\sqrt[4]{x^2z^3}$
$=2xyz\sqrt[4]{x^2z^3}$

79. $\sqrt[5]{8x^4y^6z^2}\cdot\sqrt[5]{8xy^7z^4}$
$=\sqrt[5]{8x^4y^6z^2\cdot 8xy^7z^4}$
$=\sqrt[5]{64x^5y^{13}z^6}$
$=\sqrt[5]{32\cdot 2x^5y^{10}z^5z}$
$=\sqrt[5]{32x^5y^{10}z^5}\cdot\sqrt[5]{2y^3z}$
$=2xy^2z\sqrt[5]{2y^3z}$

81. $\sqrt[3]{x-y}\cdot\sqrt[3]{(x-y)^7}$
$=\sqrt[3]{(x-y)\cdot(x-y)^7}$
$=\sqrt[3]{(x-y)^8}$
$=\sqrt[3]{(x-y)^6(x-y)^2}$
$=\sqrt[3]{(x-y)^6}\cdot\sqrt[3]{(x-y)^2}$
$=(x-y)^2\sqrt[3]{(x-y)^2}$

83. $-2x^2y\left(\sqrt[3]{54x^3y^7z^2}\right)$
$=-2x^2y\sqrt[3]{27\cdot 2x^3y^6yz^2}$
$=-2x^2y\sqrt[3]{27x^3y^6}\cdot\sqrt[3]{2yz^2}$
$=-2x^2y\cdot 3xy^2\cdot\sqrt[3]{2yz^2}$
$=-6x^3y^3\sqrt[3]{2yz^2}$

85. $-3y\left(\sqrt[5]{64x^3y^6}\right)=-3y\sqrt[5]{32\cdot 2x^3y^5y}$
$=-3y\sqrt[5]{32y^5}\cdot\sqrt[5]{2x^3y}$
$=-3y\cdot 2y\sqrt[5]{2x^3y}$
$=-6y^2\sqrt[5]{2x^3y}$

87. $\left(-2xy^2\sqrt{3x}\right)\left(xy\sqrt{6x}\right)=-2x^2y^3\sqrt{3x\cdot 6x}$
$=-2x^2y^3\sqrt{18x^2}$
$=-2x^2y^3\sqrt{9x^2\cdot 2}$
$=-2x^2y^3\sqrt{9x^2}\cdot\sqrt{2}$
$=-2x^2y^3(3x)\sqrt{2}$
$=-6x^3y^3\sqrt{2}$

89. $\left(2x^2y\sqrt[4]{8xy}\right)\left(-3xy^2\sqrt[4]{2x^2y^3}\right)$
$= -6x^3y^3\sqrt[4]{8xy \cdot 2x^2y^3}$
$= -6x^3y^3\sqrt[4]{16x^3y^4}$
$= -6x^3y^3\sqrt[4]{16y^4 \cdot x^3}$
$= -6x^3y^3(2y)\sqrt[4]{x^3}$
$= -12x^3y^4\sqrt[4]{x^3}$

91. $d(x) = \sqrt{\dfrac{3x}{2}}$
$d(72) = \sqrt{\dfrac{3(72)}{2}}$
$= \sqrt{3(36)}$
$= \sqrt{3} \cdot \sqrt{36}$
$= 6\sqrt{3} \approx 10.4$ miles
A passenger on the pool deck can see roughly 10.4 miles.

93. $W(x) = 4\sqrt{2x}$
$W(6) = 4\sqrt{2(6)} = 4\sqrt{12}$
$= 4\sqrt{4 \cdot 3} = 4\sqrt{4} \cdot \sqrt{3}$
$= 4 \cdot 2\sqrt{3}$
$= 8\sqrt{3} \approx 14$ feet per second
A dinosaur with a leg length of 6 feet has a walking speed of about 14 feet per second.

95. a. $C(32) = \dfrac{7.644}{\sqrt[4]{32}} = \dfrac{7.644}{\sqrt[4]{16 \cdot 2}}$
$= \dfrac{7.644}{2\sqrt[4]{2}} = \dfrac{3.822}{\sqrt[4]{2}}$

The cardiac index of a 32-year-old is $\dfrac{3.822}{\sqrt[4]{2}}$.

b. $\dfrac{3.822}{\sqrt[4]{2}} = \dfrac{3.822}{1.189} = \dfrac{3.822}{1.189} \approx 3.21$
The cardiac index of a 32-year-old is 3.21 liters per minute per square meter. This is shown on the graph as the point (32, 3.21).

97. – 101. Answers will vary.

103. $\sqrt{x^4} = x^2$

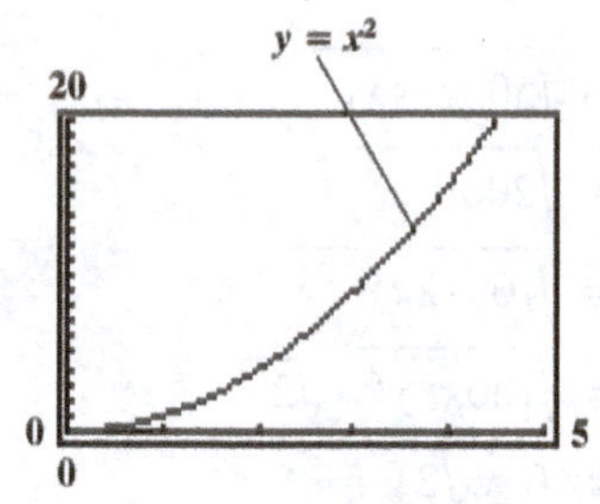

The graphs coincide, so the simplification is correct.

105. $\sqrt{3x^2 - 6x + 3} = (x-1)\sqrt{3}$

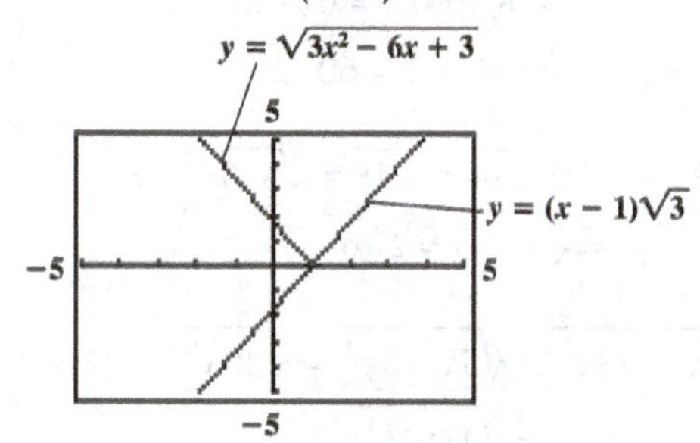

The graphs do not coincide. Correct the simplification.
$\sqrt{3x^2 - 6x + 3} = \sqrt{3\left(x^2 - 2x + 1\right)}$
$= \sqrt{3(x-1)^2} = |x-1|\sqrt{3}$

107. makes sense

109. makes sense

111. false; Changes to make the statement true will vary. A sample change is:
$2\sqrt{5} \cdot 6\sqrt{5} = 12\sqrt{5 \cdot 5} = 12 \cdot 5 = 60$

113. false; Changes to make the statement true will vary. A sample change is: $\sqrt{12} = \sqrt{4 \cdot 3} = 2\sqrt{3}$

115. If a number is tripled, its square root is multiplied by $\sqrt{3}$. For example, consider the number 4 and its triple, 12.
$\sqrt{4} = 2$ $\qquad$ $\sqrt{12} = \sqrt{4 \cdot 3} = 2\sqrt{3}$
Thus, if a number is tripled, its square root is multiplied by $\sqrt{3}$.

117. $(fg)(x) = f(x) \cdot g(x)$

$$2x = \sqrt[3]{2x} \cdot g(x)$$
$$\frac{2x}{\sqrt[3]{2x}} = g(x)$$
$$\frac{(2x)^1}{(2x)^{\frac{1}{3}}} = g(x)$$
$$(2x)^{1-\frac{1}{3}} = g(x)$$
$$(2x)^{\frac{2}{3}} = g(x)$$
$$\sqrt[3]{(2x)^2} = g(x)$$
$$g(x) = \sqrt[3]{4x^2}$$

119. $2x - 1 \le 21$ and $2x + 2 \ge 12$

$2x \le 22$ $\quad$ $2x \ge 10$

$x \le 11$ $\quad$ $x \ge 5$

The intersection of these intervals is $5 \le x \le 11$.
Interval notation: $[5, 11]$

120. Multiply the first equation by –3, the second equation by 2 and solve by addition.

$$-15x - 6y = -6$$
$$8x + 6y = -8$$
$$-7x = -14$$
$$x = 2$$

Back-substitute 2 for x to find y.

$$5x + 2y = 2$$
$$5(2) + 2y = 2$$
$$10 + 2y = 2$$
$$2y = -8$$
$$y = -4$$

The solution is $(2, -4)$.
The solution set is $\{(2, -4)\}$.

121. $64x^3 - 27 = (4x - 3)(16x^2 + 12x + 9)$

122. **a.** $21x + 10x = 31x$

b. $21\sqrt{2} + 10\sqrt{2} = 31\sqrt{2}$

123. **a.** $4x - 12x = -8x$

b. $4\sqrt[3]{2} - 12\sqrt[3]{2} = -8\sqrt[3]{2}$

124. $\dfrac{\sqrt[4]{7y^5}}{\sqrt[4]{x^{12}}} = \dfrac{\sqrt[4]{7y^5}}{x^3} = \dfrac{\sqrt[4]{y^4} \cdot \sqrt[4]{7y}}{x^3} = \dfrac{y\sqrt[4]{7y}}{x^3}$

7.4 Check Points

1. **a.** $8\sqrt{13} + 2\sqrt{13} = 10\sqrt{13}$

b. $9\sqrt[3]{7} - 6x\sqrt[3]{7} + 12\sqrt[3]{7} = (9 - 6x + 12)\sqrt[3]{7}$
$= (21 - 6x)\sqrt[3]{7}$

c. $7\sqrt[4]{3x} - 2\sqrt[4]{3x} + 2\sqrt[3]{3x} = (7 - 2)\sqrt[4]{3x} + 2\sqrt[3]{3x}$
$= 5\sqrt[4]{3x} + 2\sqrt[3]{3x}$

2. **a.** $3\sqrt{20} + 5\sqrt{45} = 3\sqrt{4 \cdot 5} + 5\sqrt{9 \cdot 5}$
$= 3 \cdot 2\sqrt{5} + 5 \cdot 3\sqrt{5}$
$= 6\sqrt{5} + 15\sqrt{5}$
$= 21\sqrt{5}$

b. $3\sqrt{12x} - 6\sqrt{27x} = 3\sqrt{4 \cdot 3x} - 6\sqrt{9 \cdot 3x}$
$= 3 \cdot 2\sqrt{3x} - 6 \cdot 3\sqrt{3x}$
$= 6\sqrt{3x} - 18\sqrt{3x}$
$= -12\sqrt{3x}$

c. $8\sqrt{5} - 6\sqrt{2}$ cannot be simplified.

3. **a.** $3\sqrt[3]{24} - 5\sqrt[3]{81} = 3\sqrt[3]{8 \cdot 3} - 5\sqrt[3]{27 \cdot 3}$
$= 3 \cdot 2\sqrt[3]{3} - 5 \cdot 3\sqrt[3]{3}$
$= 6\sqrt[3]{3} - 15\sqrt[3]{3}$
$= -9\sqrt[3]{3}$

b. $5\sqrt[3]{x^2y} + \sqrt[3]{27x^5y^4} = 5\sqrt[3]{x^2y} + \sqrt[3]{27x^3y^3x^2y}$
$= 5\sqrt[3]{x^2y} + \sqrt[3]{27x^3y^3}\sqrt[3]{x^2y}$
$= 5\sqrt[3]{x^2y} + 3xy\sqrt[3]{x^2y}$
$= (5 + 3xy)\sqrt[3]{x^2y}$

4. a. $\sqrt[3]{\dfrac{24}{125}}=\dfrac{\sqrt[3]{24}}{\sqrt[3]{125}}=\dfrac{\sqrt[3]{8\cdot 3}}{\sqrt[3]{125}}=\dfrac{\sqrt[3]{8}\cdot\sqrt[3]{3}}{\sqrt[3]{125}}=\dfrac{2\sqrt[3]{3}}{5}$

b. $\sqrt{\dfrac{9x^3}{y^{10}}}=\dfrac{\sqrt{9x^3}}{\sqrt{y^{10}}}=\dfrac{\sqrt{9x^2\cdot x}}{\sqrt{y^{10}}}=\dfrac{\sqrt{9x^2}\sqrt{x}}{\sqrt{y^{10}}}=\dfrac{3x\sqrt{x}}{y^5}$

c. $\sqrt[3]{\dfrac{8y^7}{x^{12}}}=\dfrac{\sqrt[3]{8y^7}}{\sqrt[3]{x^{12}}}=\dfrac{\sqrt[3]{8y^6\cdot y}}{\sqrt[3]{x^{12}}}=\dfrac{2y^2\sqrt[3]{y}}{x^4}$

5. a. $\dfrac{\sqrt{40x^5}}{\sqrt{2x}}=\sqrt{\dfrac{40x^5}{2x}}=\sqrt{20x^4}=\sqrt{4x^4\cdot 5}=\sqrt{4x^4}\sqrt{5}=2x^2\sqrt{5}$

b. $\dfrac{\sqrt{50xy}}{2\sqrt{2}}=\dfrac{1}{2}\cdot\sqrt{\dfrac{50xy}{2}}=\dfrac{1}{2}\cdot\sqrt{25xy}=\dfrac{1}{2}\cdot 5\sqrt{xy}=\dfrac{5\sqrt{xy}}{2}$

c. $\dfrac{\sqrt[3]{48x^7y}}{\sqrt[3]{6xy^{-2}}}=\sqrt[3]{\dfrac{48x^7y}{6xy^{-2}}}=\sqrt[3]{\dfrac{48}{6}\cdot x^{7-1}\cdot y^{1-(-2)}}=\sqrt[3]{8x^6\cdot y^3}=2x^2y$

7.4 Concept and Vocabulary Check

1. 5; 8; $13\sqrt{3}$

2. 9; 4; 3; 2; $\sqrt{3}$

3. 27; 8; 3; 2; $5\sqrt[3]{2}$

4. $\dfrac{\sqrt[n]{a}}{\sqrt[n]{b}}$

5. 8; 27; $\dfrac{2}{3}$

6. $72x^3$; $2x$; $36x^2$; $6x$

7.4 Exercise Set

1. $8\sqrt{5}+3\sqrt{5}=(8+3)\sqrt{5}=11\sqrt{5}$

3. $9\sqrt[3]{6}-2\sqrt[3]{6}=(9-2)\sqrt[3]{6}=7\sqrt[3]{6}$

5. $4\sqrt[5]{2}+3\sqrt[5]{2}-5\sqrt[5]{2}=(4+3-5)\sqrt[5]{2}$
$=2\sqrt[5]{2}$

7. $3\sqrt{13}-2\sqrt{5}-2\sqrt{13}+4\sqrt{5}$
$=3\sqrt{13}-2\sqrt{13}-2\sqrt{5}+4\sqrt{5}$
$=(3-2)\sqrt{13}+(-2+4)\sqrt{5}$
$=\sqrt{13}+2\sqrt{5}$

9. $3\sqrt{5}-\sqrt[3]{x}+4\sqrt{5}+3\sqrt[3]{x}$
$=3\sqrt{5}+4\sqrt{5}-\sqrt[3]{x}+3\sqrt[3]{x}$
$=(3+4)\sqrt{5}+(-1+3)\sqrt[3]{x}$
$=7\sqrt{5}+2\sqrt[3]{x}$

11. $\sqrt{3}+\sqrt{27}=\sqrt{3}+\sqrt{9\cdot 3}=\sqrt{3}+3\sqrt{3}$
$=(1+3)\sqrt{3}=4\sqrt{3}$

13. $7\sqrt{12}+\sqrt{75}=7\sqrt{4\cdot 3}+\sqrt{25\cdot 3}$
$=7\cdot 2\sqrt{3}+5\sqrt{3}$
$=14\sqrt{3}+5\sqrt{3}$
$=(14+5)\sqrt{3}=19\sqrt{3}$

15. $\left(\sqrt{2}-\sqrt{7}\right)\left(\sqrt{3}-\sqrt{5}\right)$
$=\sqrt{2}\sqrt{3}-\sqrt{2}\sqrt{5}-\sqrt{7}\sqrt{3}+\sqrt{7}\sqrt{5}$
$=\sqrt{6}-\sqrt{10}-\sqrt{21}+\sqrt{35}$

17. $5\sqrt[3]{16}+\sqrt[3]{54}=5\sqrt[3]{8\cdot 2}+\sqrt[3]{27\cdot 2}$
$=5\cdot 2\sqrt[3]{2}+3\sqrt[3]{2}$
$=10\sqrt[3]{2}+3\sqrt[3]{2}$
$=(10+3)\sqrt[3]{2}=13\sqrt[3]{2}$

19. $3\sqrt{45x^3}+\sqrt{5x}=3\sqrt{9\cdot 5x^2x}+\sqrt{5x}$
$=3\cdot 3x\sqrt{5x}+\sqrt{5x}$
$=9x\sqrt{5x}+\sqrt{5x}$
$=(9x+1)\sqrt{5x}$

21. $\frac{3}{\sqrt[4]{x}}=\frac{3}{\sqrt[4]{x}}\cdot\frac{\sqrt[4]{x^3}}{\sqrt[4]{x^3}}=\frac{3\sqrt[4]{x^3}}{\sqrt[4]{xx^3}}$

23. $\sqrt[3]{54x^4}-\sqrt[3]{16x}=\sqrt[3]{27\cdot 2x^3x}-\sqrt[3]{8\cdot 2x}$
$=3x\sqrt[3]{2x}-2\sqrt[3]{2x}$
$=(3x-2)\sqrt[3]{2x}$

25. $\sqrt{9x-18}+\sqrt{x-2}$
$=\sqrt{9(x-2)}+\sqrt{x-2}$
$=3\sqrt{x-2}+\sqrt{x-2}$
$=(3+1)\sqrt{x-2}$
$=4\sqrt{x-2}$

27. $2\sqrt[3]{x^4y^2}+3x\sqrt[3]{xy^2}$
$=2\sqrt[3]{x^3xy^2}+3x\sqrt[3]{xy^2}$
$=2x\sqrt[3]{xy^2}+3x\sqrt[3]{xy^2}$
$=(2x+3x)\sqrt[3]{xy^2}$
$=5x\sqrt[3]{xy^2}$

29. $\sqrt{\frac{11}{4}}=\frac{\sqrt{11}}{\sqrt{4}}=\frac{\sqrt{11}}{2}$

31. $\sqrt[3]{\frac{19}{27}}=\frac{\sqrt[3]{19}}{\sqrt[3]{27}}=\frac{\sqrt[3]{19}}{3}$

33. $\sqrt{\frac{x^2}{36y^8}}=\frac{\sqrt{x^2}}{\sqrt{36y^8}}=\frac{x}{6y^4}$

35. $\sqrt{\frac{8x^3}{25y^6}}=\frac{\sqrt{8x^3}}{\sqrt{25y^6}}=\frac{\sqrt{4\cdot 2x^2x}}{5y^3}$
$=\frac{2x\sqrt{2x}}{5y^3}$

37. $\sqrt[3]{\frac{x^4}{8y^3}}=\frac{\sqrt[3]{x^4}}{\sqrt[3]{8y^3}}=\frac{\sqrt[3]{x^3x}}{2y}=\frac{x\sqrt[3]{x}}{2y}$

39. $\sqrt[3]{\frac{50x^8}{27y^{12}}}=\frac{\sqrt[3]{50x^8}}{\sqrt[3]{27y^{12}}}=\frac{\sqrt[3]{50x^6x^2}}{3y^4}$
$=\frac{x^2\sqrt[3]{50x^2}}{3y^4}$

41. $\sqrt[4]{\frac{9y^6}{x^8}}=\frac{\sqrt[4]{9y^6}}{\sqrt[4]{x^8}}=\frac{\sqrt[4]{9y^4y^2}}{x^2}=\frac{y\sqrt[4]{9y^2}}{x^2}$

43. $\sqrt[5]{\dfrac{64x^{13}}{y^{20}}} = \dfrac{\sqrt[5]{64x^{13}}}{\sqrt[5]{y^{20}}} = \dfrac{\sqrt[5]{32 \cdot 2x^{10}x^{3}}}{y^4}$

$= \dfrac{2x^2\sqrt[5]{2x^3}}{y^4}$

45. $\dfrac{\sqrt{40}}{\sqrt{5}} = \sqrt{\dfrac{40}{5}} = \sqrt{8} = \sqrt{4 \cdot 2} = 2\sqrt{2}$

47. $\dfrac{\sqrt[3]{48}}{\sqrt[3]{6}} = \sqrt[3]{\dfrac{48}{6}} = \sqrt[3]{8} = 2$

49. $\dfrac{\sqrt{54x^3}}{\sqrt{6x}} = \sqrt{\dfrac{54x^3}{6x}} = \sqrt{9x^2} = 3x$

51. $\dfrac{\sqrt{x^5y^3}}{\sqrt{xy}} = \sqrt{\dfrac{x^5y^3}{xy}} = \sqrt{x^4y^2} = x^2y$

53. $\dfrac{\sqrt{200x^3}}{\sqrt{10x^{-1}}} = \sqrt{\dfrac{200x^3}{10x^{-1}}} = \sqrt{20x^{3-(-1)}}$

$= \sqrt{20x^4} = \sqrt{4 \cdot 5x^4} = 2x^2\sqrt{5}$

55. $\dfrac{\sqrt{48a^8b^7}}{\sqrt{3a^{-2}b^{-3}}} = \sqrt{\dfrac{48a^8b^7}{3a^{-2}b^{-3}}}$

$= \sqrt{16a^{10}b^{10}}$

$= 4a^5b^5$

57. $\dfrac{\sqrt{72xy}}{2\sqrt{2}} = \dfrac{1}{2}\sqrt{\dfrac{72xy}{2}} = \dfrac{1}{2}\sqrt{36xy}$

$= \dfrac{1}{2} \cdot 6\sqrt{xy} = 3\sqrt{xy}$

59. $\dfrac{\sqrt[3]{24x^3y^5}}{\sqrt[3]{3y^2}} = \sqrt[3]{\dfrac{24x^3y^5}{3y^2}} = \sqrt[3]{8x^3y^3} = 2xy$

61. $\dfrac{\sqrt[4]{32x^{10}y^8}}{\sqrt[4]{2x^2y^{-2}}} = \sqrt[4]{\dfrac{32x^{10}y^8}{2x^2y^{-2}}} = \sqrt[4]{16x^8y^{8-(-2)}}$

$= \sqrt[4]{16x^8y^{10}} = \sqrt[4]{16x^8y^8y^2}$

$= 2x^2y^2\sqrt[4]{y^2}$ or $2x^2y^2\sqrt{y}$

63. $\dfrac{\sqrt[3]{x^2+5x+6}}{\sqrt[3]{x+2}} = \sqrt[3]{\dfrac{x^2+5x+6}{x+2}}$

$= \sqrt[3]{\dfrac{\cancel{(x+2)}(x+3)}{\cancel{x+2}}}$

$= \sqrt[3]{x+3}$

65. $\dfrac{\sqrt[3]{a^3+b^3}}{\sqrt[3]{a+b}} = \sqrt[3]{\dfrac{a^3+b^3}{a+b}}$

$= \sqrt[3]{\dfrac{(a+b)(a^2-ab+b^2)}{a+b}}$

$= \sqrt[3]{a^2-ab+b^2}$

67. $\dfrac{\sqrt{32}}{5} + \dfrac{\sqrt{18}}{7} = \dfrac{\sqrt{16 \cdot 2}}{5} + \dfrac{\sqrt{9 \cdot 2}}{7}$

$= \dfrac{\sqrt{16} \cdot \sqrt{2}}{5} + \dfrac{\sqrt{9} \cdot \sqrt{2}}{7}$

$= \dfrac{4\sqrt{2}}{5} + \dfrac{3\sqrt{2}}{7}$

$= \dfrac{28\sqrt{2}}{35} + \dfrac{15\sqrt{2}}{35}$

$= \dfrac{28\sqrt{2} + 15\sqrt{2}}{35}$

$= \dfrac{(28+15)\sqrt{2}}{35}$

$= \dfrac{43\sqrt{2}}{35}$

69. $3x\sqrt{8xy^2} - 5y\sqrt{32x^3} + \sqrt{18x^3y^2}$

$= 3x\sqrt{4y^2 \cdot 2x} - 5y\sqrt{16x^2 \cdot 2x} + \sqrt{9x^2y^2 \cdot 2x}$

$= 3x(2y)\sqrt{2x} - 5y(4x)\sqrt{2x} + 3xy\sqrt{2x}$

$= 6xy\sqrt{2x} - 20xy\sqrt{2x} + 3xy\sqrt{2x}$

$= (6 - 20 + 3)xy\sqrt{2x}$

$= -11xy\sqrt{2x}$

71. $5\sqrt{2x^3}+\dfrac{30x^3\sqrt{24x^2}}{3x^2\sqrt{3x}}$

$=5\sqrt{2x^3}+10x\sqrt{\dfrac{24x^2}{3x}}$

$=5\sqrt{2x^3}+10x\sqrt{8x}$

$=5\sqrt{x^2\cdot 2x}+10x\sqrt{4\cdot 2x}$

$=5x\sqrt{2x}+10x(2)\sqrt{2x}$

$=5x\sqrt{2x}+20x\sqrt{2x}$

$=(5+20)x\sqrt{2x}=25x\sqrt{2x}$

73. $2x\sqrt{75xy}-\dfrac{\sqrt{81xy^2}}{\sqrt{3x^{-2}y}}$

$=2x\sqrt{75xy}-\sqrt{\dfrac{81xy^2}{3x^{-2}y}}$

$=2x\sqrt{75xy}-\sqrt{27x^3y}$

$=2x\sqrt{25\cdot 3xy}-\sqrt{9x^2\cdot 3xy}$

$=2x(5)\sqrt{3xy}-3x\sqrt{3xy}$

$=10x\sqrt{3xy}-3x\sqrt{3xy}$

$=(10-3)x\sqrt{3xy}$

$=7x\sqrt{3xy}$

75. $\dfrac{15x^4\sqrt[3]{80x^3y^2}}{5x^3\sqrt[3]{2x^2y}}-\dfrac{75\sqrt[3]{5x^3y}}{25\sqrt[3]{x^{-1}}}$

$=3x\sqrt[3]{\dfrac{80x^3y^2}{2x^2y}}-3\sqrt[3]{\dfrac{5x^3y}{x^{-1}}}$

$=3x\sqrt[3]{40xy}-3\sqrt[3]{5x^4y}$

$=3x\sqrt[3]{8\cdot 5xy}-3\sqrt[3]{x^3\cdot 5xy}$

$=3x(2)\sqrt[3]{5xy}-3x\sqrt[3]{5xy}$

$=6x\sqrt[3]{5xy}-3x\sqrt[3]{5xy}$

$=(6-3)x\sqrt[3]{5xy}=3x\sqrt[3]{5xy}$

77. $\left(\dfrac{f}{g}\right)(x)=\dfrac{\sqrt{48x^5}}{\sqrt{3x^2}}=\sqrt{\dfrac{48x^5}{3x^2}}$

$=\sqrt{16x^3}=\sqrt{16x^2\cdot x}=4x\sqrt{x}$

To get the domain, we need $x\ge 0$ and $3x^2>0$. Combining these restrictions gives us $x>0$.

Domain: $(0,\infty)$

79. $\left(\dfrac{f}{g}\right)(x)=\dfrac{\sqrt[3]{32x^6}}{\sqrt[3]{2x^2}}=\sqrt[3]{\dfrac{32x^6}{2x^2}}=\sqrt[3]{16x^4}$

$=\sqrt[3]{8x^3\cdot 2x}=2x\sqrt[3]{2x}$

Our only restriction here is that we cannot divide by 0. Thus, we need $x\ne 0$.

Domain: $(-\infty,0)\cup(0,\infty)$

81. Perimeter:

$P=2l+2w$

$=2\cdot\sqrt{125}+2\cdot 2\sqrt{20}$

$=2\cdot\sqrt{25\cdot 5}+4\sqrt{4\cdot 5}$

$=2\cdot 5\sqrt{5}+4\cdot 2\sqrt{5}$

$=10\sqrt{5}+8\sqrt{5}$

$=18\sqrt{5}$ feet

Area:

$A=lw$

$=\sqrt{125}\cdot 2\sqrt{20}$

$=2\sqrt{125\cdot 20}$

$=2\sqrt{2500}$

$=2\cdot 50$

$=100$ square feet

83. Perimeter:

$P=a+b+c$

$=\sqrt{45}+\sqrt{80}+\sqrt{125}$

$=\sqrt{9\cdot 5}+\sqrt{16\cdot 5}+\sqrt{25\cdot 5}$

$=3\sqrt{5}+4\sqrt{5}+5\sqrt{5}$

$=12\sqrt{5}$ meters

85. **a.** $f(x) = 5\sqrt{x} + 34.1$

$$f(40) - f(10) = 5\sqrt{40} + 34.1 - \left(5\sqrt{10} + 34.1\right)$$
$$= 5\sqrt{4\cdot 10} + 34.1 - 5\sqrt{10} - 34.1$$
$$= 5\cdot 2\sqrt{10} - 5\sqrt{10}$$
$$= 10\sqrt{10} - 5\sqrt{10}$$
$$= 5\sqrt{10}$$

The projected increase in the number of Americans ages 65 – 84, in millions, from 2020 to 2050 is $5\sqrt{10}$.

b. $5\sqrt{10} \approx 15.8$

This value underestimates the difference in the projected data shown in the bar graph by 2.7 million.

87. – 93. Answers will vary.

95. $\sqrt{16x} - \sqrt{9x} = \sqrt{7x}$

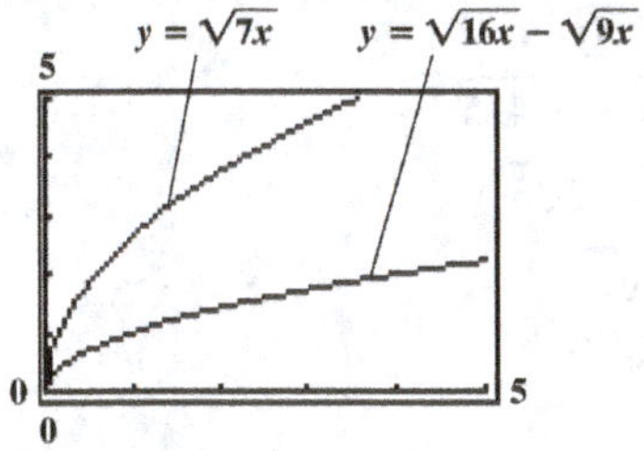

The graphs do not coincide. Correct the simplification.

$$\sqrt{16x} - \sqrt{9x} = 4\sqrt{x} - 3\sqrt{x}$$
$$= (4-3)\sqrt{x} = \sqrt{x}$$

97. makes sense

99. does not make sense; Explanations will vary. Sample explanation:

$$3\sqrt[3]{81} + 2\sqrt[3]{54} = 3\sqrt[3]{27\cdot 3} + 2\sqrt[3]{27\cdot 2}$$
$$= 3\cdot 3\sqrt[3]{3} + 2\cdot 3\sqrt[3]{2}$$
$$= 9\sqrt[3]{3} + 6\sqrt[3]{2}$$

101. false; Changes to make the statement true will vary. A sample change is: $\sqrt{5} + \sqrt{5} = 2\sqrt{5}$

103. false; Changes to make the statement true will vary. A sample change is: In order for two radical expressions to be combined, both the index and the radicand must be the same. Just because two radical expressions are completely simplified, that does not guarantee that the index and radicands match.

105. Let x = the irrational number.

$$x - \left(2\sqrt{18} - \sqrt{50}\right) = \sqrt{2}$$
$$x - 2\sqrt{18} + \sqrt{50} = \sqrt{2}$$
$$x = \sqrt{2} + 2\sqrt{18} - \sqrt{50}$$
$$x = \sqrt{2} + 2\sqrt{9\cdot 2} - \sqrt{25\cdot 2}$$
$$x = \sqrt{2} + 2\cdot 3\sqrt{2} - 5\sqrt{2}$$
$$x = \sqrt{2} + 6\sqrt{2} - 5\sqrt{2}$$
$$x = 2\sqrt{2}$$

The irrational number is $2\sqrt{2}$.

107.
$$\frac{6\sqrt{49xy}\sqrt{ab^2}}{7\sqrt{36x^{-3}y^{-5}}\sqrt{a^{-9}b^{-1}}} = \frac{6\sqrt{49xyab^2}}{7\sqrt{36x^{-3}y^{-5}a^{-9}b^{-1}}}$$
$$= \frac{6}{7}\sqrt{\frac{49x^4y^6a^{10}b^3}{36}} = \frac{6}{7}\sqrt{\frac{49x^4y^6a^{10}b^2b}{36}}$$
$$= \frac{\cancel{6}\cdot\cancel{7}x^2y^3a^5b}{\cancel{7}\cdot\cancel{6}}\sqrt{b} = x^2y^3a^5b\sqrt{b}$$

108. $2(3x-1) - 4 = 2x - (6-x)$

$$6x - 2 - 4 = 2x - 6 + x$$
$$6x - 6 = 3x - 6$$
$$3x = 0$$
$$x = 0$$

The solution set is $\{0\}$.

109. $x^2 - 8xy + 12y^2 = (x-6y)(x-2y)$

110.
$$\frac{2}{x^2+5x+6} + \frac{3x}{x^2+6x+9}$$
$$= \frac{2}{(x+3)(x+2)} + \frac{3x}{(x+3)^2}$$
$$= \frac{2}{(x+3)(x+2)}\cdot\frac{(x+3)}{(x+3)} + \frac{3x}{(x+3)^2}\cdot\frac{(x+2)}{(x+2)}$$
$$= \frac{2(x+3) + 3x(x+2)}{(x+3)^2(x+2)}$$
$$= \frac{2x + 6 + 3x^2 + 6x}{(x+3)^2(x+2)}$$
$$= \frac{3x^2 + 8x + 6}{(x+3)^2(x+2)}$$

111. **a.** $7(x+5)=7x+35$

b. $\sqrt{7}\left(x+\sqrt{5}\right)=x\sqrt{7}+\sqrt{35}$

112. **a.** $(x+5)(6x+3)=6x^2+33x+15$

b. $$\begin{aligned}\left(\sqrt{2}+5\right)\left(6\sqrt{2}+3\right)&=6\sqrt{2}^2+33\sqrt{2}+15\\&=6\cdot 2+33\sqrt{2}+15\\&=12+33\sqrt{2}+15\\&=27+33\sqrt{2}\end{aligned}$$

113. $$\begin{aligned}\frac{10y}{\sqrt[5]{4x^3y}}\cdot\frac{\sqrt[5]{8x^2y^4}}{\sqrt[5]{8x^2y^4}}&=\frac{10y\sqrt[5]{8x^2y^4}}{\sqrt[5]{4x^3y}\cdot\sqrt[5]{8x^2y^4}}\\&=\frac{10y\sqrt[5]{8x^2y^4}}{\sqrt[5]{32x^5y^5}}\\&=\frac{10y\sqrt[5]{8x^2y^4}}{2xy}\\&=\frac{5\sqrt[5]{8x^2y^4}}{x}\end{aligned}$$

Mid-Chapter Check Point – Chapter 7

1. $\sqrt{100}-\sqrt[3]{-27}=10-(-3)=10+3=13$

2. $\sqrt{8x^5y^7}=\sqrt{4x^4y^6\cdot 2xy}=2x^2y^3\sqrt{2xy}$

3. $3\sqrt[3]{4x^2}+2\sqrt[3]{4x^2}=(3+2)\sqrt[3]{4x^2}=5\sqrt[3]{4x^2}$

4. $$\begin{aligned}\left(3\sqrt[3]{4x^2}\right)\left(2\sqrt[3]{4x^2}\right)&=6\sqrt[3]{4x^2\cdot 4x^2}\\&=6\sqrt[3]{16x^4}\\&=6\sqrt[3]{8x^3\cdot 2x}\\&=6(2x)\sqrt[3]{2x}\\&=12x\sqrt[3]{2x}\end{aligned}$$

5. $$\begin{aligned}27^{2/3}+(-32)^{3/5}&=\left(\sqrt[3]{27}\right)^2+\left(\sqrt[5]{-32}\right)^3\\&=(3)^2+(-2)^3\\&=9+(-8)\\&=1\end{aligned}$$

6. $$\begin{aligned}\left(64x^3y^{1/4}\right)^{1/3}&=(64)^{1/3}\left(x^3\right)^{1/3}\left(y^{1/4}\right)^{1/3}\\&=\sqrt[3]{64}\cdot x^{3(1/3)}\cdot y^{(1/4)(1/3)}\\&=4xy^{1/12}\end{aligned}$$

7. $$\begin{aligned}5\sqrt{27}-4\sqrt{48}&=5\sqrt{9\cdot 3}-4\sqrt{16\cdot 3}\\&=5(3)\sqrt{3}-4(4)\sqrt{3}\\&=15\sqrt{3}-16\sqrt{3}\\&=(15-16)\sqrt{3}\\&=-\sqrt{3}\end{aligned}$$

8. $$\begin{aligned}\sqrt{\frac{500x^3}{4y^4}}&=\frac{\sqrt{500x^3}}{\sqrt{4y^4}}=\frac{\sqrt{100x^2\cdot 5x}}{\sqrt{4y^4}}\\&=\frac{10x\sqrt{5x}}{2y^2}=\frac{5x\sqrt{5x}}{y^2}\end{aligned}$$

9. $$\begin{aligned}\frac{x}{\sqrt[4]{x}}&=\frac{x}{x^{1/4}}=x^{1-(1/4)}\\&=x^{(4/4)-(1/4)}=x^{3/4}\\&=\sqrt[4]{x^3}\end{aligned}$$

10. $\sqrt[3]{54x^5}=\sqrt[3]{27x^3\cdot 2x^2}=3x\sqrt[3]{2x^2}$

11. $\dfrac{\sqrt[3]{160}}{\sqrt[3]{2}}=\sqrt[3]{\dfrac{160}{2}}=\sqrt[3]{80}=\sqrt[3]{8\cdot 10}=2\sqrt[3]{10}$

12. $$\begin{aligned}\sqrt[5]{\frac{x^{10}}{y^{20}}}&=\left(\frac{x^{10}}{y^{20}}\right)^{1/5}\\&=\frac{\left(x^{10}\right)^{1/5}}{\left(y^{20}\right)^{1/5}}\\&=\frac{x^{10(1/5)}}{y^{20(1/5)}}=\frac{x^2}{y^4}\end{aligned}$$

13. $$\begin{aligned}\frac{\left(x^{2/3}\right)^2}{\left(x^{1/4}\right)^3}&=\frac{x^{(2/3)\cdot 2}}{x^{(1/4)\cdot 3}}=\frac{x^{4/3}}{x^{3/4}}\\&=x^{(4/3)-(3/4)}\\&=x^{(16/12)-(9/12)}\\&=x^{7/12}\end{aligned}$$

14. $\sqrt[6]{x^6 y^4} = \left(x^6 y^4\right)^{1/6} = \left(x^6\right)^{1/6}\left(y^4\right)^{1/6}$
$= x^{6(1/6)} y^{4(1/6)} = x^1 y^{2/3}$
$= x\sqrt[3]{y^2}$

15. $\sqrt[7]{(x-2)^3} \cdot \sqrt[7]{(x-2)^6}$
$= \sqrt[7]{(x-2)^3 \cdot (x-2)^6}$
$= \sqrt[7]{(x-2)^9}$
$= \sqrt[7]{(x-2)^7 \cdot (x-2)^2}$
$= (x-2)\sqrt[7]{(x-2)^2}$

16. $\sqrt[4]{32x^{11}y^{17}} = \sqrt[4]{16x^8 y^{16} \cdot 2x^3 y}$
$= \sqrt[4]{16x^8 y^{16}} \cdot \sqrt[4]{2x^3 y}$
$= 2x^2 y^4 \sqrt[4]{2x^3 y}$

17. $4\sqrt[3]{16} + 2\sqrt[3]{54} = 4\sqrt[3]{8 \cdot 2} + 2\sqrt[3]{27 \cdot 2}$
$= 4\sqrt[3]{8} \cdot \sqrt[3]{2} + 2\sqrt[3]{27} \cdot \sqrt[3]{2}$
$= 4(2)\sqrt[3]{2} + 2(3)\sqrt[3]{2}$
$= 8\sqrt[3]{2} + 6\sqrt[3]{2}$
$= (8+6)\sqrt[3]{2}$
$= 14\sqrt[3]{2}$

18. $\dfrac{\sqrt[7]{x^4 y^9}}{\sqrt[7]{x^{-5} y^7}} = \sqrt[7]{\dfrac{x^4 y^9}{x^{-5} y^7}} = \sqrt[7]{x^9 y^2}$
$= \sqrt[7]{x^7 \cdot x^2 y^2}$
$= x\sqrt[7]{x^2 y^2}$

19. $(-125)^{-2/3} = \dfrac{1}{(-125)^{2/3}} = \dfrac{1}{\left(\sqrt[3]{-125}\right)^2}$
$= \dfrac{1}{(-5)^2} = \dfrac{1}{25}$

20. $\sqrt{2} \cdot \sqrt[3]{2} = 2^{1/2} \cdot 2^{1/3} = 2^{(1/2)+(1/3)}$
$= 2^{(3/6)+(2/6)} = 2^{5/6}$
$= \sqrt[6]{2^5} = \sqrt[6]{32}$

21. $\sqrt[3]{\dfrac{32x}{y^4}} \cdot \sqrt[3]{\dfrac{2x^2}{y^2}} = \sqrt[3]{\dfrac{32x}{y^4} \cdot \dfrac{2x^2}{y^2}}$
$= \sqrt[3]{\dfrac{64x^3}{y^6}} = \dfrac{\sqrt[3]{64x^3}}{\sqrt[3]{y^6}}$
$= \dfrac{4x}{y^2}$

22. $\sqrt{32xy^2} \cdot \sqrt{2x^3 y^5} = \sqrt{32xy^2 \cdot 2x^3 y^5}$
$= \sqrt{64x^4 y^7}$
$= \sqrt{64x^4 y^6 \cdot y}$
$= 8x^2 y^3 \sqrt{y}$

23. $4x\sqrt{6x^4 y^3} - 7y\sqrt{24x^6 y}$
$= 4x\sqrt{x^4 y^2 \cdot 6y} - 7y\sqrt{4x^6 \cdot 6y}$
$= 4x\left(x^2 y\right)\sqrt{6y} - 7y\left(2x^3\right)\sqrt{6y}$
$= 4x^3 y\sqrt{6y} - 14x^3 y\sqrt{6y}$
$= (4-14)x^3 y\sqrt{6y}$
$= -10x^3 y\sqrt{6y}$

24. $f(x) = \sqrt{30-5x}$

To find the domain, we set the radicand greater than or equal to 0.

$30 - 5x \geq 0$
$-5x \geq -30$
$x \leq 6$

Domain: $(-\infty, 6]$

25. $g(x) = \sqrt[3]{3x-15}$

The domain of a cube root is all real numbers. Since there are no other restrictions, we have

Domain: $(-\infty, \infty)$

7.5 Check Points

1. a. $\sqrt{6}\left(x+\sqrt{10}\right)=\sqrt{6}\cdot x+\sqrt{6}\sqrt{10}$

$$= x\sqrt{6}+\sqrt{60}$$
$$= x\sqrt{6}+\sqrt{4\cdot 15}$$
$$= x\sqrt{6}+2\sqrt{15}$$

b. $\sqrt[3]{y}\left(\sqrt[3]{y^2}-\sqrt[3]{7}\right)=\sqrt[3]{y}\cdot\sqrt[3]{y^2}-\sqrt[3]{y}\cdot\sqrt[3]{7}$

$$= \sqrt[3]{y^3}-\sqrt[3]{7y}$$
$$= y-\sqrt[3]{7y}$$

c. $\left(6\sqrt{5}+3\sqrt{2}\right)\left(2\sqrt{5}-4\sqrt{2}\right)=\left(6\sqrt{5}\right)\left(2\sqrt{5}\right)+\left(6\sqrt{5}\right)\left(-4\sqrt{2}\right)+\left(3\sqrt{2}\right)\left(2\sqrt{5}\right)+\left(3\sqrt{2}\right)\left(-4\sqrt{2}\right)$

$$= 12\cdot 5-24\sqrt{10}+6\sqrt{10}-12\cdot 2$$
$$= 60-24\sqrt{10}+6\sqrt{10}-24$$
$$= 36-18\sqrt{10}$$

2. a. $\left(\sqrt{5}+\sqrt{6}\right)^2=\left(\sqrt{5}\right)^2+2\cdot\sqrt{5}\cdot\sqrt{6}+\left(\sqrt{6}\right)^2$

$$= 5+2\sqrt{30}+6$$
$$= 11+2\sqrt{30}$$

b. $\left(\sqrt{6}+\sqrt{5}\right)\left(\sqrt{6}-\sqrt{5}\right)=\left(\sqrt{6}\right)^2-\left(\sqrt{5}\right)^2$

$$= 6-5$$
$$= 1$$

c. $\left(\sqrt{a}-\sqrt{7}\right)\left(\sqrt{a}+\sqrt{7}\right)=\left(\sqrt{a}\right)^2-\left(\sqrt{7}\right)^2$

$$= a-7$$

3. a. $\dfrac{\sqrt{3}}{\sqrt{7}}=\dfrac{\sqrt{3}}{\sqrt{7}}\cdot\dfrac{\sqrt{7}}{\sqrt{7}}=\dfrac{\sqrt{21}}{\sqrt{49}}=\dfrac{\sqrt{21}}{7}$

b. $\sqrt[3]{\dfrac{2}{9}}=\dfrac{\sqrt[3]{2}}{\sqrt[3]{9}}=\dfrac{\sqrt[3]{2}}{\sqrt[3]{3^2}}=\dfrac{\sqrt[3]{2}}{\sqrt[3]{3^2}}\cdot\dfrac{\sqrt[3]{3}}{\sqrt[3]{3}}=\dfrac{\sqrt[3]{6}}{\sqrt[3]{3^3}}=\dfrac{\sqrt[3]{6}}{3}$

4. a. $\sqrt{\dfrac{2x}{7y}}=\dfrac{\sqrt{2x}}{\sqrt{7y}}=\dfrac{\sqrt{2x}}{\sqrt{7y}}\cdot\dfrac{\sqrt{7y}}{\sqrt{7y}}=\dfrac{\sqrt{14xy}}{7y}$

b. $\dfrac{\sqrt[3]{x}}{\sqrt[3]{9y}}=\dfrac{\sqrt[3]{x}}{\sqrt[3]{3^2 y}}=\dfrac{\sqrt[3]{x}}{\sqrt[3]{3^2 y}}\cdot\dfrac{\sqrt[3]{3y^2}}{\sqrt[3]{3y^2}}=\dfrac{\sqrt[3]{3xy^2}}{\sqrt[3]{3^3 y^3}}=\dfrac{\sqrt[3]{3xy^2}}{3y}$

c. $$\frac{6x}{\sqrt[5]{8x^2y^4}} = \frac{6x}{\sqrt[5]{2^3x^2y^4}}$$
$$= \frac{6x}{\sqrt[5]{2^3x^2y^4}} \cdot \frac{\sqrt[5]{2^2x^3y}}{\sqrt[5]{2^2x^3y}}$$
$$= \frac{6x\sqrt[5]{2^2x^3y}}{\sqrt[5]{2^5x^5y^5}}$$
$$= \frac{6x\sqrt[5]{2^2x^3y}}{2xy}$$
$$= \frac{3\sqrt[5]{4x^3y}}{y}$$

5. $$\frac{18}{2\sqrt{3}+3} = \frac{18}{2\sqrt{3}+3} \cdot \frac{2\sqrt{3}-3}{2\sqrt{3}-3}$$
$$= \frac{36\sqrt{3}-54}{2^2\cdot 3-3^2}$$
$$= \frac{36\sqrt{3}-54}{12-9}$$
$$= \frac{36\sqrt{3}-54}{3}$$
$$= \frac{3(12\sqrt{3}-18)}{3}$$
$$= 12\sqrt{3}-18$$

6. $$\frac{3+\sqrt{7}}{\sqrt{5}-\sqrt{2}} = \frac{3+\sqrt{7}}{\sqrt{5}-\sqrt{2}} \cdot \frac{\sqrt{5}+\sqrt{2}}{\sqrt{5}+\sqrt{2}}$$
$$= \frac{3\sqrt{5}+3\sqrt{2}+\sqrt{35}+\sqrt{14}}{5-2}$$
$$= \frac{3\sqrt{5}+3\sqrt{2}+\sqrt{35}+\sqrt{14}}{3}$$

7. $$\frac{\sqrt{x+3}-\sqrt{x}}{3} = \frac{\sqrt{x+3}-\sqrt{x}}{3} \cdot \frac{\sqrt{x+3}+\sqrt{x}}{\sqrt{x+3}+\sqrt{x}}$$
$$= \frac{\left(\sqrt{x+3}\right)^2-\left(\sqrt{x}\right)^2}{3\sqrt{x+3}+3\sqrt{x}}$$
$$= \frac{x+3-x}{3\sqrt{x+3}+3\sqrt{x}}$$
$$= \frac{3}{3\left(\sqrt{x+3}+\sqrt{x}\right)}$$
$$= \frac{1}{\sqrt{x+3}+\sqrt{x}}$$

7.5 Concept and Vocabulary Check

1. 350; $-42\sqrt{10}$; $30\sqrt{10}$; -36
2. $\sqrt{10}$; $\sqrt{5}$; 10; 5; 5
3. rationalizing the denominator
4. $\sqrt{5}$
5. $\sqrt[3]{9}$
6. $7\sqrt{2}-5$
7. $3\sqrt{6}+\sqrt{5}$

7.5 Exercise Set

1. $$\sqrt{2}\left(x+\sqrt{7}\right) = \sqrt{2}\cdot x+\sqrt{2}\sqrt{7}$$
$$= x\sqrt{2}+\sqrt{14}$$

3. $$\sqrt{6}\left(7-\sqrt{6}\right) = \sqrt{6}\cdot 7-\sqrt{6}\sqrt{6}$$
$$= 7\sqrt{6}-\sqrt{36} = 7\sqrt{6}-6$$

5. $$\sqrt{3}\left(4\sqrt{6}-2\sqrt{3}\right)$$
$$= \sqrt{3}\cdot 4\sqrt{6}-\sqrt{3}\cdot 2\sqrt{3}$$
$$= 4\sqrt{18}-2\sqrt{9}$$
$$= 4\sqrt{9\cdot 2}-2\cdot 3$$
$$= 4\cdot 3\sqrt{2}-6 = 12\sqrt{2}-6$$

7. $$\sqrt[3]{2}\left(\sqrt[3]{6}+4\sqrt[3]{5}\right) = \sqrt[3]{2}\cdot\sqrt[3]{6}+\sqrt[3]{2}\cdot 4\sqrt[3]{5}$$
$$= \sqrt[3]{12}+4\sqrt[3]{10}$$

9. $$\sqrt[3]{x}\left(\sqrt[3]{16x^2}-\sqrt[3]{x}\right)$$
$$= \sqrt[3]{x}\cdot\sqrt[3]{16x^2}-\sqrt[3]{x}\cdot\sqrt[3]{x}$$
$$= \sqrt[3]{x}\cdot\sqrt[3]{8\cdot 2x^2}-\sqrt[3]{x^2}$$
$$= \sqrt[3]{8\cdot 2x^3}-\sqrt[3]{x^2}$$
$$= 2x\sqrt[3]{2}-\sqrt[3]{x^2}$$

11. $(5+\sqrt{2})(6+\sqrt{2})$
$= 5\cdot 6+5\sqrt{2}+6\sqrt{2}+\sqrt{2}\sqrt{2}$
$= 30+(5+6)\sqrt{2}+2$
$= 32+11\sqrt{2}$

13. $(6+\sqrt{5})(9-4\sqrt{5})$
$= 6\cdot 9-6\cdot 4\sqrt{5}+9\sqrt{5}-4\sqrt{5}\sqrt{5}$
$= 54-24\sqrt{5}+9\sqrt{5}-4\cdot 5$
$= 54+(-24+9)\sqrt{5}-20$
$= 34+(-15)\sqrt{5}$
$= 34-15\sqrt{5}$

15. $(6-3\sqrt{7})(2-5\sqrt{7})$
$= 6\cdot 2-6\cdot 5\sqrt{7}-2\cdot 3\sqrt{7}+3\sqrt{7}\cdot 5\sqrt{7}$
$= 12-30\sqrt{7}-6\sqrt{7}+15\cdot 7$
$= 12+(-30-6)\sqrt{7}+105$
$= 117+(-36)\sqrt{7}$
$= 117-36\sqrt{7}$

17. $(\sqrt{2}+\sqrt{7})(\sqrt{3}+\sqrt{5})$
$= \sqrt{2}\sqrt{3}+\sqrt{2}\sqrt{5}+\sqrt{7}\sqrt{3}+\sqrt{7}\sqrt{5}$
$= \sqrt{6}+\sqrt{10}+\sqrt{21}+\sqrt{35}$

19. $(\sqrt{2}-\sqrt{7})(\sqrt{3}-\sqrt{5})$
$= \sqrt{2}\sqrt{3}-\sqrt{2}\sqrt{5}-\sqrt{7}\sqrt{3}+\sqrt{7}\sqrt{5}$
$= \sqrt{6}-\sqrt{10}-\sqrt{21}+\sqrt{35}$

21. $(3\sqrt{2}-4\sqrt{3})(2\sqrt{2}+5\sqrt{3})$
$= 3\sqrt{2}(2\sqrt{2})+3\sqrt{2}(5\sqrt{3})$
$\quad -4\sqrt{3}(2\sqrt{2})-4\sqrt{3}(5\sqrt{3})$
$= 6\cdot 2+15\sqrt{6}-8\sqrt{6}-20\cdot 3$
$= 12+7\sqrt{6}-60$
$= 7\sqrt{6}-48$ or $-48+7\sqrt{6}$

23. $(\sqrt{3}+\sqrt{5})^2 = (\sqrt{3})^2+2\sqrt{3}\sqrt{5}+(\sqrt{5})^2$
$= 3+2\sqrt{15}+5$
$= 8+2\sqrt{15}$

25. $(\sqrt{3x}-\sqrt{y})^2$
$= (\sqrt{3x})^2-2\sqrt{3x}\sqrt{y}+(\sqrt{y})^2$
$= 3x-2\sqrt{3xy}+y$

27. $(\sqrt{5}+7)(\sqrt{5}-7)$
$= \sqrt{5}\sqrt{5}-7\sqrt{5}+7\sqrt{5}-7\cdot 7$
$= 5-\cancel{7\sqrt{5}}+\cancel{7\sqrt{5}}-49$
$= 5-49$
$= -44$

29. $(2-5\sqrt{3})(2+5\sqrt{3})$
$= 2\cdot 2+2\cdot 5\sqrt{3}-2\cdot 5\sqrt{3}-5\sqrt{3}\cdot 5\sqrt{3}$
$= 4+\cancel{10\sqrt{3}}-\cancel{10\sqrt{3}}-25\cdot 3$
$= 4-75$
$= -71$

31. $(3\sqrt{2}+2\sqrt{3})(3\sqrt{2}-2\sqrt{3})$
$= 3\sqrt{2}\cdot 3\sqrt{2}-3\sqrt{2}\cdot 2\sqrt{3}$
$\quad +3\sqrt{2}\cdot 2\sqrt{3}-2\sqrt{3}\cdot 2\sqrt{3}$
$= 9\cdot 2-\cancel{6\sqrt{6}}+\cancel{6\sqrt{6}}-4\cdot 3$
$= 18-12$
$= 6$

33. $(3-\sqrt{x})(2-\sqrt{x})$
$= 3\cdot 2-3\sqrt{x}-2\sqrt{x}+\sqrt{x}\sqrt{x}$
$= 6+(-3-2)\sqrt{x}+x$
$= 6+(-5)\sqrt{x}+x$
$= 6-5\sqrt{x}+x$

35. $(\sqrt[3]{x}-4)(\sqrt[3]{x}+5)$
$= \sqrt[3]{x}\sqrt[3]{x}+5\sqrt[3]{x}-4\sqrt[3]{x}-4\cdot 5$
$= \sqrt[3]{x^2}+(5-4)\sqrt[3]{x}-20$
$= \sqrt[3]{x^2}+\sqrt[3]{x}-20$

37. $\left(x+\sqrt[3]{y^2}\right)\left(2x-\sqrt[3]{y^2}\right)$

$= x\cdot 2x - x\sqrt[3]{y^2} + 2x\sqrt[3]{y^2} - \sqrt[3]{y^2}\sqrt[3]{y^2}$

$= 2x^2 + (-x+2x)\sqrt[3]{y^2} - \sqrt[3]{y^4}$

$= 2x^2 + x\sqrt[3]{y^2} - \sqrt[3]{y^3 y}$

$= 2x^2 + x\sqrt[3]{y^2} - y\sqrt[3]{y}$

39. $\dfrac{\sqrt{2}}{\sqrt{5}} = \dfrac{\sqrt{2}}{\sqrt{5}}\cdot\dfrac{\sqrt{5}}{\sqrt{5}} = \dfrac{\sqrt{2\cdot 5}}{\sqrt{5\cdot 5}} = \dfrac{\sqrt{10}}{5}$

41. $\sqrt{\dfrac{11}{x}} = \dfrac{\sqrt{11}}{\sqrt{x}} = \dfrac{\sqrt{11}}{\sqrt{x}}\cdot\dfrac{\sqrt{x}}{\sqrt{x}}$

$= \dfrac{\sqrt{11x}}{\sqrt{x^2}} = \dfrac{\sqrt{11x}}{x}$

43. $\dfrac{9}{\sqrt{3y}} = \dfrac{9}{\sqrt{3y}}\cdot\dfrac{\sqrt{3y}}{\sqrt{3y}} = \dfrac{9\sqrt{3y}}{\sqrt{3y\cdot 3y}}$

$= \dfrac{\overset{3}{\cancel{9}}\sqrt{3y}}{\underset{1}{\cancel{3}}\,y} = \dfrac{3\sqrt{3y}}{y}$

45. $\dfrac{1}{\sqrt[3]{2}} = \dfrac{1}{\sqrt[3]{2}}\cdot\dfrac{\sqrt[3]{2^2}}{\sqrt[3]{2^2}} = \dfrac{\sqrt[3]{2^2}}{\sqrt[3]{2^3}} = \dfrac{\sqrt[3]{4}}{2}$

47. $\dfrac{6}{\sqrt[3]{4}} = \dfrac{6}{\sqrt[3]{2^2}}\cdot\dfrac{\sqrt[3]{2}}{\sqrt[3]{2}} = \dfrac{6\sqrt[3]{2}}{\sqrt[3]{2^2}\sqrt[3]{2}}$

$= \dfrac{6\sqrt[3]{2}}{\sqrt[3]{2^3}} = \dfrac{6\sqrt[3]{2}}{2}$

$= 3\sqrt[3]{2}$

49. $\sqrt[3]{\dfrac{2}{3}} = \dfrac{\sqrt[3]{2}}{\sqrt[3]{3}} = \dfrac{\sqrt[3]{2}}{\sqrt[3]{3}}\cdot\dfrac{\sqrt[3]{3^2}}{\sqrt[3]{3^2}} = \dfrac{\sqrt[3]{2\cdot 3^2}}{\sqrt[3]{3^3}}$

$= \dfrac{\sqrt[3]{2\cdot 9}}{3} = \dfrac{\sqrt[3]{18}}{3}$

51. $\dfrac{4}{\sqrt[3]{x}} = \dfrac{4}{\sqrt[3]{x}}\cdot\dfrac{\sqrt[3]{x^2}}{\sqrt[3]{x^2}} = \dfrac{4\sqrt[3]{x^2}}{\sqrt[3]{x}\sqrt[3]{x^2}}$

$= \dfrac{4\sqrt[3]{x^2}}{\sqrt[3]{x^3}} = \dfrac{4\sqrt[3]{x^2}}{x}$

53. $\sqrt[3]{\dfrac{2}{y^2}} = \dfrac{\sqrt[3]{2}}{\sqrt[3]{y^2}} = \dfrac{\sqrt[3]{2}}{\sqrt[3]{y^2}}\cdot\dfrac{\sqrt[3]{y}}{\sqrt[3]{y}}$

$= \dfrac{\sqrt[3]{2y}}{\sqrt[3]{y^3}} = \dfrac{\sqrt[3]{2y}}{y}$

55. $\dfrac{7}{\sqrt[3]{2x^2}} = \dfrac{7}{\sqrt[3]{2x^2}}\cdot\dfrac{\sqrt[3]{2^2 x}}{\sqrt[3]{2^2 x}} = \dfrac{7\sqrt[3]{2^2 x}}{\sqrt[3]{2x^2}\sqrt[3]{2^2 x}}$

$= \dfrac{7\sqrt[3]{4x}}{\sqrt[3]{2^3 x^3}} = \dfrac{7\sqrt[3]{4x}}{2x}$

57. $\sqrt[3]{\dfrac{2}{xy^2}} = \dfrac{\sqrt[3]{2}}{\sqrt[3]{xy^2}} = \dfrac{\sqrt[3]{2}}{\sqrt[3]{xy^2}}\cdot\dfrac{\sqrt[3]{x^2 y}}{\sqrt[3]{x^2 y}}$

$= \dfrac{\sqrt[3]{2}\sqrt[3]{x^2 y}}{\sqrt[3]{xy^2}\sqrt[3]{x^2 y}} = \dfrac{\sqrt[3]{2x^2 y}}{\sqrt[3]{x^3 y^3}}$

$= \dfrac{\sqrt[3]{2x^2 y}}{xy}$

59. $\dfrac{3}{\sqrt[4]{x}} = \dfrac{3}{\sqrt[4]{x}}\cdot\dfrac{\sqrt[4]{x^3}}{\sqrt[4]{x^3}} = \dfrac{3\sqrt[4]{x^3}}{\sqrt[4]{xx^3}}$

$= \dfrac{3\sqrt[4]{x^3}}{\sqrt[4]{x^4}} = \dfrac{3\sqrt[4]{x^3}}{x}$

61. $\dfrac{6}{\sqrt[5]{8x^3}} = \dfrac{6}{\sqrt[5]{2^3 x^3}}\cdot\dfrac{\sqrt[5]{2^2 x^2}}{\sqrt[5]{2^2 x^2}} = \dfrac{6\sqrt[5]{4x^2}}{\sqrt[5]{2^5 x^5}}$

$= \dfrac{6\sqrt[5]{4x^2}}{2x} = \dfrac{3\sqrt[5]{4x^2}}{x}$

63. $\dfrac{2x^2 y}{\sqrt[5]{4x^2 y^4}} = \dfrac{2x^2 y}{\sqrt[5]{2^2 x^2 y^4}}\cdot\dfrac{\sqrt[5]{2^3 x^3 y}}{\sqrt[5]{2^3 x^3 y}}$

$= \dfrac{2x^2 y\sqrt[5]{8x^3 y}}{\sqrt[5]{2^5 x^5 y^5}}$

$= \dfrac{\cancel{2}x^{\cancel{2}}\cancel{y}\sqrt[5]{8x^3 y}}{\cancel{2}\cancel{x}\cancel{y}}$

$= x\sqrt[5]{8x^3 y}$

65. $\frac{9}{\sqrt{3x^2y}}=\frac{9}{\sqrt{x^2\cdot 3y}}=\frac{9}{x\sqrt{3y}}$

$=\frac{9}{x\sqrt{3y}}\cdot\frac{\sqrt{3y}}{\sqrt{3y}}$

$=\frac{9\sqrt{3y}}{x\sqrt{(3y)^2}}=\frac{9\sqrt{3y}}{x(3y)}$

$=\frac{\not{3}\cdot 3\sqrt{3y}}{\not{3}xy}=\frac{3\sqrt{3y}}{xy}$

67. $-\sqrt{\frac{75a^5}{b^3}}=-\frac{\sqrt{75a^5}}{\sqrt{b^3}}=-\frac{\sqrt{25a^4\cdot 3a}}{\sqrt{b^2\cdot b}}$

$=-\frac{5a^2\sqrt{3a}}{b\sqrt{b}}=-\frac{5a^2\sqrt{3a}}{b\sqrt{b}}\cdot\frac{\sqrt{b}}{\sqrt{b}}$

$=-\frac{5a^2\sqrt{3ab}}{b\sqrt{b^2}}=-\frac{5a^2\sqrt{3ab}}{b(b)}$

$=-\frac{5a^2\sqrt{3ab}}{b^2}$

69. $\sqrt{\frac{7m^2n^3}{14m^3n^2}}=\sqrt{\frac{n}{2m}}=\frac{\sqrt{n}}{\sqrt{2m}}$

$=\frac{\sqrt{n}}{\sqrt{2m}}\cdot\frac{\sqrt{2m}}{\sqrt{2m}}=\frac{\sqrt{2mn}}{\sqrt{(2m)^2}}$

$=\frac{\sqrt{2mn}}{2m}$

71. $\frac{3}{\sqrt[4]{x^5y^3}}=\frac{3}{\sqrt[4]{x^4\cdot xy^3}}=\frac{3}{x\sqrt[4]{xy^3}}$

$=\frac{3}{x\sqrt[4]{xy^3}}\cdot\frac{\sqrt[4]{x^3y}}{\sqrt[4]{x^3y}}$

$=\frac{3\sqrt[4]{x^3y}}{x\sqrt[4]{x^4y^4}}=\frac{3\sqrt[4]{x^3y}}{x(xy)}$

$=\frac{3\sqrt[4]{x^3y}}{x^2y}$

73. $\frac{12}{\sqrt[3]{-8x^5y^8}}=\frac{12}{\sqrt[3]{-8x^3y^6\cdot x^2y^2}}$

$=\frac{12}{-2xy^2\sqrt[3]{x^2y^2}}$

$=\frac{12}{-2xy^2\sqrt[3]{x^2y^2}}\cdot\frac{\sqrt[3]{xy}}{\sqrt[3]{xy}}$

$=\frac{12\sqrt[3]{xy}}{-2xy^2\sqrt[3]{x^3y^3}}=\frac{12\sqrt[3]{xy}}{-2xy^2(xy)}$

$=\frac{12\sqrt[3]{xy}}{-2x^2y^3}$

$=-\frac{6\sqrt[3]{xy}}{x^2y^3}$

75. $\frac{8}{\sqrt{5}+2}=\frac{8}{\sqrt{5}+2}\cdot\frac{\sqrt{5}-2}{\sqrt{5}-2}$

$=\frac{8\sqrt{5}-8\cdot 2}{\sqrt{5}\sqrt{5}-2\sqrt{5}+2\sqrt{5}-2\cdot 2}$

$=\frac{8\sqrt{5}-16}{5-\cancel{2\sqrt{5}}+\cancel{2\sqrt{5}}-4}$

$=\frac{8\sqrt{5}-16}{5-4}=\frac{8\sqrt{5}-16}{1}$

$=8\sqrt{5}-16$

77. $\frac{13}{\sqrt{11}-3}=\frac{13}{\sqrt{11}-3}\cdot\frac{\sqrt{11}+3}{\sqrt{11}+3}$

$=\frac{13\left(\sqrt{11}+3\right)}{\left(\sqrt{11}-3\right)\left(\sqrt{11}+3\right)}$

$=\frac{13\sqrt{11}+13\cdot 3}{\sqrt{11}\cdot\sqrt{11}+3\sqrt{11}-3\sqrt{11}-3\cdot 3}$

$=\frac{13\sqrt{11}+39}{11+\cancel{3\sqrt{11}}-\cancel{3\sqrt{11}}-9}$

$=\frac{13\sqrt{11}+39}{11-9}$

$=\frac{13\sqrt{11}+39}{2}$

79. $\dfrac{6}{\sqrt{5}+\sqrt{3}}$

$= \dfrac{6}{\sqrt{5}+\sqrt{3}} \cdot \dfrac{\sqrt{5}-\sqrt{3}}{\sqrt{5}-\sqrt{3}}$

$= \dfrac{6\left(\sqrt{5}-\sqrt{3}\right)}{\left(\sqrt{5}+\sqrt{3}\right)\left(\sqrt{5}-\sqrt{3}\right)}$

$= \dfrac{6\sqrt{5}-6\sqrt{3}}{\sqrt{5}\sqrt{5}-\sqrt{3}\sqrt{5}+\sqrt{3}\sqrt{5}-\sqrt{3}\sqrt{3}}$

$= \dfrac{6\sqrt{5}-6\sqrt{3}}{5-\cancel{\sqrt{15}}+\cancel{\sqrt{15}}-3}$

$= \dfrac{6\sqrt{5}-6\sqrt{3}}{5-3}$

$= \dfrac{6\sqrt{5}-6\sqrt{3}}{2}$

$= \dfrac{\cancel{2}\left(3\sqrt{5}-3\sqrt{3}\right)}{\cancel{2}}$

$= 3\sqrt{5}-3\sqrt{3}$

81. $\dfrac{\sqrt{a}}{\sqrt{a}-\sqrt{b}} = \dfrac{\sqrt{a}}{\sqrt{a}-\sqrt{b}} \cdot \dfrac{\sqrt{a}+\sqrt{b}}{\sqrt{a}+\sqrt{b}}$

$= \dfrac{\sqrt{a}\left(\sqrt{a}+\sqrt{b}\right)}{\left(\sqrt{a}-\sqrt{b}\right)\left(\sqrt{a}+\sqrt{b}\right)}$

$= \dfrac{\sqrt{a}\sqrt{a}+\sqrt{a}\sqrt{b}}{\sqrt{a}\sqrt{a}+\sqrt{a}\sqrt{b}-\sqrt{a}\sqrt{b}-\sqrt{b}\sqrt{b}}$

$= \dfrac{a+\sqrt{ab}}{a-b}$

83. $\dfrac{25}{5\sqrt{2}-3\sqrt{5}} = \dfrac{25}{5\sqrt{2}-3\sqrt{5}} \cdot \dfrac{5\sqrt{2}+3\sqrt{5}}{5\sqrt{2}+3\sqrt{5}} = \dfrac{25\left(5\sqrt{2}+3\sqrt{5}\right)}{\left(5\sqrt{2}-3\sqrt{5}\right)\left(5\sqrt{2}+3\sqrt{5}\right)}$

$= \dfrac{125\sqrt{2}+75\sqrt{5}}{5\cdot 5\sqrt{2\cdot 2}+\cancel{5\cdot 3\sqrt{2\cdot 5}}-\cancel{5\cdot 3\sqrt{2\cdot 5}}-3\cdot 3\sqrt{5\cdot 5}} = \dfrac{125\sqrt{2}+75\sqrt{5}}{25\cdot 2-9\cdot 5}$

$= \dfrac{125\sqrt{2}+75\sqrt{5}}{50-45} = \dfrac{125\sqrt{2}+75\sqrt{5}}{5} = \dfrac{\cancel{5}\left(25\sqrt{2}+15\sqrt{5}\right)}{\cancel{5}} = 25\sqrt{2}+15\sqrt{5}$

85. $$\frac{\sqrt{5}+\sqrt{3}}{\sqrt{5}-\sqrt{3}}=\frac{\sqrt{5}+\sqrt{3}}{\sqrt{5}-\sqrt{3}}\cdot\frac{\sqrt{5}+\sqrt{3}}{\sqrt{5}+\sqrt{3}}=\frac{\left(\sqrt{5}+\sqrt{3}\right)^2}{\left(\sqrt{5}-\sqrt{3}\right)\left(\sqrt{5}+\sqrt{3}\right)}$$
$$=\frac{\left(\sqrt{5}\right)^2+2\sqrt{5}\sqrt{3}+\left(\sqrt{3}\right)^2}{\sqrt{5}\cdot\sqrt{5}+\sqrt{5}\cdot\sqrt{3}-\sqrt{5}\cdot\sqrt{3}-\sqrt{3}\sqrt{3}}=\frac{5+2\sqrt{15}+3}{5+\cancel{\sqrt{15}}-\cancel{\sqrt{15}}-3}$$
$$=\frac{8+2\sqrt{15}}{5-3}=\frac{\cancel{2}\left(4+\sqrt{15}\right)}{\cancel{2}}=4+\sqrt{15}$$

87. $$\frac{\sqrt{x}+1}{\sqrt{x}+3}=\frac{\sqrt{x}+1}{\sqrt{x}+3}\cdot\frac{\sqrt{x}-3}{\sqrt{x}-3}=\frac{\sqrt{x}\cdot\sqrt{x}-3\sqrt{x}+1\sqrt{x}-3\cdot1}{\sqrt{x}\cdot\sqrt{x}-\cancel{3\sqrt{x}}+\cancel{3\sqrt{x}}-3\cdot3}$$
$$=\frac{\sqrt{x^2}+(-3+1)\sqrt{x}-3}{\sqrt{x^2}-9}=\frac{x+(-2)\sqrt{x}-3}{x-9}=\frac{x-2\sqrt{x}-3}{x-9}$$

89. $$\frac{5\sqrt{3}-3\sqrt{2}}{3\sqrt{2}-2\sqrt{3}}=\frac{5\sqrt{3}-3\sqrt{2}}{3\sqrt{2}-2\sqrt{3}}\cdot\frac{3\sqrt{2}+2\sqrt{3}}{3\sqrt{2}+2\sqrt{3}}=\frac{5\sqrt{3}\cdot3\sqrt{2}+5\sqrt{3}\cdot2\sqrt{3}-3\sqrt{2}\cdot3\sqrt{2}-3\sqrt{2}\cdot2\sqrt{3}}{3\sqrt{2}\cdot3\sqrt{2}+3\sqrt{2}\cdot2\sqrt{3}-3\sqrt{2}\cdot2\sqrt{3}-2\sqrt{3}\cdot2\sqrt{3}}$$
$$=\frac{15\sqrt{6}+10\cdot3-9\cdot2-6\sqrt{6}}{9\cdot2+\cancel{6\sqrt{6}}-\cancel{6\sqrt{6}}-4\cdot3}=\frac{15\sqrt{6}+30-18-6\sqrt{6}}{18-12}$$
$$=\frac{9\sqrt{6}+12}{6}=\frac{\cancel{3}\left(3\sqrt{6}+4\right)}{\cancel{3}\cdot2}=\frac{3\sqrt{6}+4}{2}$$

91. $$\frac{2\sqrt{x}+\sqrt{y}}{\sqrt{y}-2\sqrt{x}}=\frac{2\sqrt{x}+\sqrt{y}}{\sqrt{y}-2\sqrt{x}}\cdot\frac{\sqrt{y}+2\sqrt{x}}{\sqrt{y}+2\sqrt{x}}=\frac{2\sqrt{x}\sqrt{y}+2\sqrt{x}\cdot2\sqrt{x}+\sqrt{y}\sqrt{y}+2\sqrt{x}\sqrt{y}}{\sqrt{y}\sqrt{y}+2\sqrt{x}\sqrt{y}-2\sqrt{x}\sqrt{y}-2\sqrt{x}\cdot2\sqrt{x}}$$
$$=\frac{2\sqrt{xy}+4\sqrt{x^2}+\sqrt{y^2}+2\sqrt{xy}}{\sqrt{y^2}+\cancel{2\sqrt{xy}}-\cancel{2\sqrt{xy}}-4\sqrt{x^2}}=\frac{2\sqrt{xy}+4x+y+2\sqrt{xy}}{y-4x}=\frac{4\sqrt{xy}+4x+y}{y-4x}$$

93. $$\sqrt{\frac{3}{2}}=\frac{\sqrt{3}}{\sqrt{2}}\cdot\frac{\sqrt{3}}{\sqrt{3}}=\frac{\sqrt{3}\sqrt{3}}{\sqrt{2}\sqrt{3}}=\frac{3}{\sqrt{6}}$$

95. $$\frac{\sqrt[3]{4x}}{\sqrt[3]{y}}=\frac{\sqrt[3]{2^2x}}{\sqrt[3]{y}}\cdot\frac{\sqrt[3]{2x^2}}{\sqrt[3]{2x^2}}=\frac{\sqrt[3]{2^3x^3}}{\sqrt[3]{2x^2y}}=\frac{2x}{\sqrt[3]{2x^2y}}$$

97. $$\frac{\sqrt{x}+3}{\sqrt{x}}=\frac{\sqrt{x}+3}{\sqrt{x}}\cdot\frac{\sqrt{x}-3}{\sqrt{x}-3}=\frac{\sqrt{x}\cdot\sqrt{x}-\cancel{3\sqrt{x}}+\cancel{3\sqrt{x}}-3\cdot3}{\sqrt{x}\cdot\sqrt{x}-3\sqrt{x}}=\frac{\sqrt{x^2}-9}{\sqrt{x^2}-3\sqrt{x}}=\frac{x-9}{x-3\sqrt{x}}$$

99. $$\frac{\sqrt{a}+\sqrt{b}}{\sqrt{a}-\sqrt{b}}=\frac{\sqrt{a}+\sqrt{b}}{\sqrt{a}-\sqrt{b}}\cdot\frac{\sqrt{a}-\sqrt{b}}{\sqrt{a}-\sqrt{b}}=\frac{\sqrt{a}\cdot\sqrt{a}-\sqrt{a}\sqrt{b}+\sqrt{a}\sqrt{b}-\sqrt{b}\sqrt{b}}{\sqrt{a}\cdot\sqrt{a}-\sqrt{a}\sqrt{b}-\sqrt{a}\sqrt{b}+\sqrt{b}\sqrt{b}}$$
$$=\frac{\sqrt{a^2}-\cancel{\sqrt{ab}}+\cancel{\sqrt{ab}}-\sqrt{b^2}}{\sqrt{a^2}-\sqrt{ab}-\sqrt{ab}+\sqrt{b^2}}=\frac{a-b}{a-2\sqrt{ab}+b}$$

101. $\dfrac{\sqrt{x+5}-\sqrt{x}}{5}=\dfrac{\sqrt{x+5}-\sqrt{x}}{5}\cdot\dfrac{\sqrt{x+5}+\sqrt{x}}{\sqrt{x+5}+\sqrt{x}}=\dfrac{\left(\sqrt{x+5}\right)^2+\sqrt{x+5}\cdot\sqrt{x}-\sqrt{x+5}\cdot\sqrt{x}-\left(\sqrt{x}\right)^2}{5\left(\sqrt{x+5}+\sqrt{x}\right)}$

$=\dfrac{x+5+\sqrt{x(x+5)}-\sqrt{x(x+5)}-x}{5\left(\sqrt{x+5}+\sqrt{x}\right)}=\dfrac{5}{5\left(\sqrt{x+5}+\sqrt{x}\right)}=\dfrac{1}{\sqrt{x+5}+\sqrt{x}}$

103. $\dfrac{\sqrt{x}+\sqrt{y}}{x^2-y^2}=\dfrac{\sqrt{x}+\sqrt{y}}{x^2-y^2}\cdot\dfrac{\sqrt{x}-\sqrt{y}}{\sqrt{x}-\sqrt{y}}=\dfrac{\left(\sqrt{x}\right)^2-\cancel{\sqrt{xy}}+\cancel{\sqrt{xy}}-\left(\sqrt{y}\right)^2}{x^2\sqrt{x}-x^2\sqrt{y}-y^2\sqrt{x}+y^2\sqrt{y}}$

$=\dfrac{x-y}{x^2\left(\sqrt{x}-\sqrt{y}\right)-y^2\left(\sqrt{x}-\sqrt{y}\right)}=\dfrac{x-y}{\left(\sqrt{x}-\sqrt{y}\right)\left(x^2-y^2\right)}$

$=\dfrac{\cancel{x-y}}{\left(\sqrt{x}-\sqrt{y}\right)(x+y)\cancel{(x-y)}}=\dfrac{1}{\left(\sqrt{x}-\sqrt{y}\right)(x+y)}$

105. $\sqrt{2}+\dfrac{1}{\sqrt{2}}=\sqrt{2}+\dfrac{1}{\sqrt{2}}\cdot\dfrac{\sqrt{2}}{\sqrt{2}}$

$=\sqrt{2}+\dfrac{\sqrt{2}}{2}=\dfrac{2\sqrt{2}}{2}+\dfrac{\sqrt{2}}{2}$

$=\dfrac{2\sqrt{2}+\sqrt{2}}{2}=\dfrac{3\sqrt{2}}{2}$

107. $\sqrt[3]{25}-\dfrac{15}{\sqrt[3]{5}}=\sqrt[3]{25}-\dfrac{15}{\sqrt[3]{5}}\cdot\dfrac{\sqrt[3]{5^2}}{\sqrt[3]{5^2}}$

$=\sqrt[3]{25}-\dfrac{15\sqrt[3]{25}}{5}$

$=\sqrt[3]{25}-3\sqrt[3]{25}=-2\sqrt[3]{25}$

109. $\sqrt{6}-\sqrt{\dfrac{1}{6}}+\sqrt{\dfrac{2}{3}}$

$=\sqrt{6}-\dfrac{\sqrt{1}}{\sqrt{6}}+\dfrac{\sqrt{2}}{\sqrt{3}}$

$=\sqrt{6}-\dfrac{1}{\sqrt{6}}\cdot\dfrac{\sqrt{6}}{\sqrt{6}}+\dfrac{\sqrt{2}}{\sqrt{3}}\cdot\dfrac{\sqrt{3}}{\sqrt{3}}$

$=\sqrt{6}-\dfrac{\sqrt{6}}{6}+\dfrac{\sqrt{6}}{3}$

$=\dfrac{6\sqrt{6}}{6}-\dfrac{\sqrt{6}}{6}+\dfrac{2\sqrt{6}}{6}$

$=\dfrac{6\sqrt{6}-\sqrt{6}+2\sqrt{6}}{6}=\dfrac{7\sqrt{6}}{6}$

111. $\frac{2}{\sqrt{2}+\sqrt{3}}+\sqrt{75}-\sqrt{50}$

$=\frac{2}{\sqrt{2}+\sqrt{3}}\cdot\frac{\sqrt{2}-\sqrt{3}}{\sqrt{2}-\sqrt{3}}+\sqrt{25\cdot 3}-\sqrt{25\cdot 2}$

$=\frac{2\sqrt{2}-2\sqrt{3}}{\left(\sqrt{2}\right)^2-\left(\sqrt{3}\right)^2}+5\sqrt{3}-5\sqrt{2}$

$=\frac{2\sqrt{2}-2\sqrt{3}}{2-3}+5\sqrt{3}-5\sqrt{2}$

$=\frac{2\sqrt{2}-2\sqrt{3}}{-1}+5\sqrt{3}-5\sqrt{2}$

$=2\sqrt{3}-2\sqrt{2}+5\sqrt{3}-5\sqrt{2}$

$=7\sqrt{3}-7\sqrt{2}$

113. $f(x)=x^2-6x-4$

$f\left(3-\sqrt{13}\right)=\left(3-\sqrt{13}\right)^2-6\left(3-\sqrt{13}\right)-4$

$=9-6\sqrt{13}+13-18+6\sqrt{13}-4$

$=0$

115. $f(x)=\sqrt{9+x}$

$f\left(3\sqrt{5}\right)\cdot f\left(-3\sqrt{5}\right)=\sqrt{9+3\sqrt{5}}\cdot\sqrt{9-3\sqrt{5}}$

$=\sqrt{\left(9+3\sqrt{5}\right)\left(9-3\sqrt{5}\right)}$

$=\sqrt{9^2-\left(3\sqrt{5}\right)^2}$

$=\sqrt{81-9\cdot 5}$

$=\sqrt{81-45}$

$=\sqrt{36}$

$=6$

117. $\frac{w}{h}=\frac{2}{\sqrt{5}-1}$

$=\frac{2}{\sqrt{5}-1}\cdot\frac{\sqrt{5}+1}{\sqrt{5}+1}$

$=\frac{2\left(\sqrt{5}+1\right)}{5-1}$

$=\frac{2\left(\sqrt{5}+1\right)}{4}$

$=\frac{\sqrt{5}+1}{2}$

≈ 1.62

The ratio is approximately 1.62 to 1.

119. Perimeter $= 2l + 2w$

$$= 2\left(\sqrt{8}+1\right)+2\left(\sqrt{8}-1\right)$$
$$= 2\sqrt{8}+\not{2}+2\sqrt{8}-\not{2}$$
$$= (2+2)\sqrt{8} = 4\sqrt{8}$$
$$= 4\sqrt{4\cdot 2} = 4\cdot 2\sqrt{2} = 8\sqrt{2}$$

The perimeter is $8\sqrt{2}$ inches.

$$\text{Area} = lw = \left(\sqrt{8}+1\right)\left(\sqrt{8}-1\right)$$
$$= \left(\sqrt{8}\right)^2 - \not{\sqrt{8}} + \not{\sqrt{8}} - 1$$
$$= 8 - 1 = 7$$

The area is 7 square inches.

121. $c^2 = a^2 + b^2$

$$c^2 = \left(\sqrt{10}+\sqrt{2}\right)^2 + \left(\sqrt{10}-\sqrt{2}\right)^2$$
$$c^2 = \left(\sqrt{10}\right)^2 + 2\sqrt{10}\sqrt{2} + \left(\sqrt{2}\right)^2 + \left(\sqrt{10}\right)^2 - 2\sqrt{10}\sqrt{2} + \left(\sqrt{2}\right)^2$$
$$c^2 = 10 + 2\sqrt{20} + 2 + 10 - 2\sqrt{20} + 2$$
$$c^2 = 24$$
$$c = \sqrt{24}$$
$$c = 2\sqrt{6} \text{ inches}$$

123. – 129. Answers will vary.

131. $\left(\sqrt{x}-1\right)\left(\sqrt{x}-1\right) = x+1$

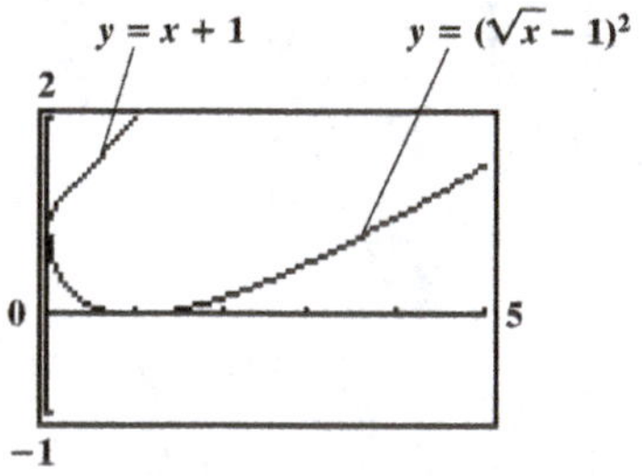

$y = x + 1$ $\quad$ $y = (\sqrt{x} - 1)^2$

2

0 $\quad$ 5

−1

The graphs do not coincide. Correct the simplification.

$$\left(\sqrt{x}-1\right)\left(\sqrt{x}-1\right)$$
$$= \left(\sqrt{x}\right)^2 - \sqrt{x} - \sqrt{x} + 1$$
$$= x - 2\sqrt{x} + 1$$

133. $\left(\sqrt{x}+1\right)^2 = x+1$

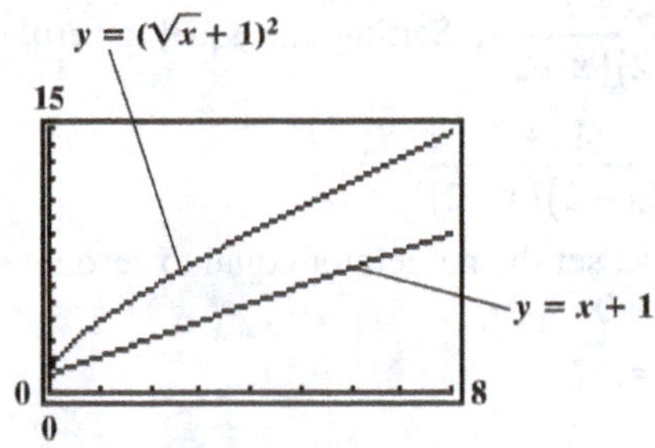

The graphs do not coincide. Correct the simplification.

$$\left(\sqrt{x}+1\right)^2 = \left(\sqrt{x}\right)^2 + 2\sqrt{x}\cdot 1 + 1^2$$
$$= x + 2\sqrt{x} + 1$$

135. makes sense

137. does not make sense; Explanations will vary. Sample explanation: Multiplying a radical expression and its conjugate will clear the expression of any radicals.

139. false; Changes to make the statement true will vary. A sample change is:

$$\frac{\sqrt{3}+7}{\sqrt{3}-2} = \frac{\sqrt{3}+7}{\sqrt{3}-2}\cdot\frac{\sqrt{3}+2}{\sqrt{3}+2}$$
$$= \frac{\left(\sqrt{3}\right)^2 + 2\sqrt{3} + 7\sqrt{3} + 14}{\left(\sqrt{3}\right)^2 - 2^2}$$
$$= \frac{3+9\sqrt{3}+14}{3-4} = \frac{17+9\sqrt{3}}{-1}$$
$$= -\left(17+9\sqrt{3}\right) = -17-9\sqrt{3}$$

141. true

143. $7\left[(2x-5)-(x+1)\right] = \left(\sqrt{7}+2\right)\left(\sqrt{7}-2\right)$

$$7\left[2x-5-x-1\right] = \left(\sqrt{7}\right)^2 - (2)^2$$
$$7(x-6) = 7-4$$
$$7x-42 = 3$$
$$7x = 45$$
$$x = \frac{45}{7}$$

The solution is $\frac{45}{7}$ and the solution set is $\left\{\frac{45}{7}\right\}$.

145. $\dfrac{1}{\sqrt{2}+\sqrt{3}+\sqrt{4}}$

$$=\frac{1}{(\sqrt{2}+\sqrt{3})+2}\cdot\frac{(\sqrt{2}+\sqrt{3})-2}{(\sqrt{2}+\sqrt{3})-2}$$

$$=\frac{\sqrt{2}+\sqrt{3}-2}{(\sqrt{2}+\sqrt{3})^2-2^2}=\frac{\sqrt{2}+\sqrt{3}-2}{2+2\sqrt{6}+3-4}$$

$$=\frac{\sqrt{2}+\sqrt{3}-2}{2\sqrt{6}+1}\cdot\frac{2\sqrt{6}-1}{2\sqrt{6}-1}$$

$$=\frac{2\sqrt{12}+2\sqrt{18}-4\sqrt{6}-\sqrt{2}-\sqrt{3}+2}{(2\sqrt{6})^2-1^2}$$

$$=\frac{4\sqrt{3}+6\sqrt{2}-4\sqrt{6}-\sqrt{2}-\sqrt{3}+2}{4\cdot 6-1}$$

$$=\frac{3\sqrt{3}+5\sqrt{2}-4\sqrt{6}+2}{24+1}$$

$$=\frac{3\sqrt{3}+5\sqrt{2}-4\sqrt{6}+2}{24-1}$$

$$=\frac{3\sqrt{3}+5\sqrt{2}-4\sqrt{6}+2}{23}$$

146. $\dfrac{2}{x-2}+\dfrac{3}{x^2-4}$

$$=\frac{2}{x-2}+\frac{3}{(x+2)(x-2)}$$

$$=\frac{2(x+2)}{(x-2)(x+2)}+\frac{3}{(x+2)(x-2)}$$

$$=\frac{2(x+2)+3}{(x-2)(x+2)}=\frac{2x+4+3}{(x-2)(x+2)}$$

$$=\frac{2x+7}{(x-2)(x+2)}\text{ or }\frac{2x+7}{x^2-4}$$

147. Using the results from Exercise 146, we know that the left side of the equation simplifies to $\dfrac{2x+7}{(x-2)(x+2)}$. Setting this equal to zero, we have $\dfrac{2x+7}{(x-2)(x+2)}=0$.

To solve, set the numerator equal to zero.

$$2x+7=0$$
$$2x=-7$$
$$x=-\frac{7}{2}$$

The solution set is $\left\{-\dfrac{7}{2}\right\}$.

148.

$\underline{-2}\rfloor$	1	0	−3	−2	5
		−2	4	−2	8
	1	−2	1	−4	13

The remainder is 13. This means that $f(-2)=13$.

149. $(\sqrt{x+4}+1)^2=(\sqrt{x+4})^2+2\sqrt{x+4}+1^2$

$$=x+4+2\sqrt{x+4}+1$$
$$=x+5+2\sqrt{x+4}$$

150. $4x^2-16x+16=4(x+4)$

$$4x^2-16x+16=4x+16$$
$$4x^2-20x=0$$
$$4x(x-5)=0$$

$x=0$ or $x-5=0$

$x=5$

The solution set is $\{0,5\}$.

151. $26-11x=16-8x+x^2$

$$0=-10+3x+x^2$$
$$0=x^2+3x-10$$
$$0=(x+5)(x-2)$$

$x+5=0$ or $x-2=0$

$x=-5$ $\quad x=2$

The solution set is $\{-5,2\}$.

7.6 Check Points

1. $\sqrt{3x+4}=8$

$$\left(\sqrt{3x+4}\right)^2=8^2$$
$$3x+4=64$$
$$3x=60$$
$$x=20$$

Check:

$$\sqrt{3x+4}=8$$
$$\sqrt{3\cdot 20+4}=8$$
$$\sqrt{60+4}=8$$
$$\sqrt{64}=8$$
$$8=8$$

The solution set is $\{20\}$.

2. $\sqrt{x-1}+7=2$

$$\sqrt{x-1}=-5$$

A principal square root cannot be negative. Thus, this equation has no solution.
The solution set is $\{\ \}$.

3. $\sqrt{6x+7}-x=2$

$$\sqrt{6x+7}=x+2$$
$$\left(\sqrt{6x+7}\right)^2=(x+2)^2$$
$$6x+7=x^2+4x+4$$
$$0=x^2-2x-3$$
$$0=(x+1)(x-3)$$

$x+1=0$ or $x-3=0$

$x=-1$ $\qquad x=3$

Check:

$$\sqrt{6x+7}-x=2$$
$$\sqrt{6(-1)+7}-(-1)=2$$
$$\sqrt{1}+1=2$$
$$1+1=2$$
$$2=2$$

$$\sqrt{6x+7}-x=2$$
$$\sqrt{6(3)+7}-(3)=2$$
$$\sqrt{25}-3=2$$
$$5-3=2$$
$$2=2$$

The solution set is $\{-1,3\}$.

4. $\sqrt{x+5}-\sqrt{x-3}=2$

$$\sqrt{x+5}=\sqrt{x-3}+2$$
$$\left(\sqrt{x+5}\right)^2=\left(\sqrt{x-3}+2\right)^2$$
$$x+5=\left(\sqrt{x-3}\right)^2+2\cdot 2\sqrt{x-3}+2^2$$
$$x+5=x-3+4\sqrt{x-3}+4$$
$$x+5=x+1+4\sqrt{x-3}$$
$$4=4\sqrt{x-3}$$
$$\frac{4}{4}=\frac{4\sqrt{x-3}}{4}$$
$$1=\sqrt{x-3}$$
$$1^2=\left(\sqrt{x-3}\right)^2$$
$$1=x-3$$
$$4=x$$

Check:

$$\sqrt{x+5}-\sqrt{x-3}=2$$
$$\sqrt{4+5}-\sqrt{4-3}=2$$
$$\sqrt{9}-\sqrt{1}=2$$
$$3-1=2$$
$$2=2$$

The solution set is $\{4\}$.

5. $(2x-3)^{\frac{1}{3}}+3=0$

$$(2x-3)^{\frac{1}{3}}=-3$$
$$\left((2x-3)^{\frac{1}{3}}\right)^3=(-3)^3$$
$$2x-3=-27$$
$$2x=-24$$
$$x=-12$$

Check:

$$(2x-3)^{\frac{1}{3}}+3=0$$
$$(2(-12)-3)^{\frac{1}{3}}+3=0$$
$$(-27)^{\frac{1}{3}}+3=0$$
$$\sqrt[3]{-27}+3=0$$
$$-3+3=0$$
$$0=0$$

The solution set is $\{-12\}$.

6. $f(x) = 3.5\sqrt{x} + 38$
$73 = 3.5\sqrt{x} + 38$
$35 = 3.5\sqrt{x}$
$\frac{35}{3.5} = \frac{3.5\sqrt{x}}{3.5}$
$10 = \sqrt{x}$
$10^2 = \left(\sqrt{x}\right)^2$
$100 = x$
The model projects that 73% of U.S. women will participate in the work force 100 years after 1960, or 2060.

7.6 Concept and Vocabulary Check

1. radical
2. extraneous
3. $2x+1$; $x^2 - 14x + 49$
4. $x+2$; $x+8-6\sqrt{x-1}$
5. $2x+3$; 8
6. true
7. false

7.6 Exercise Set

1. $\sqrt{3x-2} = 4$
$\left(\sqrt{3x-2}\right)^2 = 4^2$
$3x - 2 = 16$
$3x = 18$
$x = 6$
Check:
$\sqrt{3(6)-2} = 4$
$\sqrt{18-2} = 4$
$\sqrt{16} = 4$
$4 = 4$
The solution set is $\{6\}$.

3. $\sqrt{5x-4} - 9 = 0$
$\sqrt{5x-4} = 9$
$\left(\sqrt{5x-4}\right)^2 = 9^2$
$5x - 4 = 81$
$5x = 85$
$x = 17$
Check:
$\sqrt{5(17)-4} - 9 = 0$
$\sqrt{85-4} - 9 = 0$
$\sqrt{81} - 9 = 0$
$9 - 9 = 0$
$0 = 0$
The solution set is $\{17\}$.

5. $\sqrt{3x+7} + 10 = 4$
$\sqrt{3x+7} = -6$
Since the square root of a number is always positive, the solution set is $\varnothing$.

7. $x = \sqrt{7x+8}$
$x^2 = \left(\sqrt{7x+8}\right)^2$
$x^2 = 7x + 8$
$x^2 - 7x - 8 = 0$
$(x-8)(x+1) = 0$
Apply the zero product principle.
$x - 8 = 0$ or $x + 1 = 0$
$x = 8$ $\quad$ $x = -1$

Check:
$8 = \sqrt{7(8)+8}$ $\quad$ $-1 = \sqrt{7(-1)+8}$
$8 = \sqrt{56+8}$
$8 = \sqrt{64}$
$8 = 8$
We disregard –1 because square roots are always positive. The solution set is $\{8\}$.

9. $\sqrt{5x+1} = x+1$

$$\left(\sqrt{5x+1}\right)^2 = (x+1)^2$$
$$5x+\cancel{1} = x^2+2x+\cancel{1}$$
$$0 = x^2-3x$$
$$0 = x(x-3)$$

Apply the zero product principle.

$x=0$ or $x-3=0$

$x=3$

Both solutions check. The solution set is $\{0, 3\}$.

11. $x = \sqrt{2x-2}+1$

$$x-1 = \sqrt{2x-2}$$
$$(x-1)^2 = \left(\sqrt{2x-2}\right)^2$$
$$x^2-2x+1 = 2x-2$$
$$x^2-4x+3 = 0$$
$$(x-3)(x-1) = 0$$

Apply the zero product principle.

$x-3=0$ or $x-1=0$

$x=3$ $x=1$

Both solutions check. The solution set is $\{1, 3\}$.

13. $x-2\sqrt{x-3} = 3$

$$x-3 = 2\sqrt{x-3}$$
$$(x-3)^2 = \left(2\sqrt{x-3}\right)^2$$
$$x^2-6x+9 = 4(x-3)$$
$$x^2-6x+9 = 4x-12$$
$$x^2-10x+21 = 0$$
$$(x-7)(x-3) = 0$$

Apply the zero product principle.

$x-7=0$ or $x-3=0$

$x=7$ $x=3$

Both solutions check. The solution set is $\{3, 7\}$.

15. $\sqrt{2x-5} = \sqrt{x+4}$

$$\left(\sqrt{2x-5}\right)^2 = \left(\sqrt{x+4}\right)^2$$
$$2x-5 = x+4$$
$$x-5 = 4$$
$$x = 9$$

The solution checks. The solution set is $\{9\}$.

17. $\sqrt[3]{2x+11} = 3$

$$\left(\sqrt[3]{2x+11}\right)^3 = 3^3$$
$$2x+11 = 27$$
$$2x = 16$$
$$x = 8$$

The solution checks. The solution set is $\{8\}$.

19. $\sqrt[3]{2x-6}-4 = 0$

$$\sqrt[3]{2x-6} = 4$$
$$\left(\sqrt[3]{2x-6}\right)^3 = 4^3$$
$$2x-6 = 64$$
$$2x = 70$$
$$x = 35$$

The solution checks. The solution set is $\{35\}$.

21. $\sqrt{x-7} = 7-\sqrt{x}$

$$\left(\sqrt{x-7}\right)^2 = \left(7-\sqrt{x}\right)^2$$
$$\cancel{x}-7 = 49-14\sqrt{x}+\cancel{x}$$
$$-7 = 49-14\sqrt{x}$$
$$-56 = -14\sqrt{x}$$
$$\frac{-56}{-14} = \frac{-14\sqrt{x}}{-14}$$
$$4 = \sqrt{x}$$
$$4^2 = \left(\sqrt{x}\right)^2$$
$$16 = x$$

The solution checks. The solution set is $\{16\}$.

23. $\sqrt{x+2}+\sqrt{x-1}=3$

$$\sqrt{x+2}=3-\sqrt{x-1}$$
$$\left(\sqrt{x+2}\right)^2=\left(3-\sqrt{x-1}\right)^2$$
$$x+2=9-6\sqrt{x-1}+x-1$$
$$2=8-6\sqrt{x-1}$$
$$-6=-6\sqrt{x-1}$$
$$\frac{-6}{-6}=\frac{-6\sqrt{x-1}}{-6}$$
$$1=\sqrt{x-1}$$
$$1^2=\left(\sqrt{x-1}\right)^2$$
$$1=x-1$$
$$2=x$$

The solution checks. The solution set is $\{2\}$.

25. $2\sqrt{4x+1}-9=x-5$

$$2\sqrt{4x+1}=x+4$$
$$\left(2\sqrt{4x+1}\right)^2=(x+4)^2$$
$$2^2\left(\sqrt{4x+1}\right)^2=x^2+8x+16$$
$$4(4x+1)=x^2+8x+16$$
$$16x+4=x^2+8x+16$$
$$0=x^2-8x+12$$
$$0=(x-6)(x-2)$$
$$x-6=0 \quad \text{or} \quad x-2=0$$
$$x=6 \qquad\qquad x=2$$

Check $x=6$:

$$2\sqrt{4(6)+1}-9=6-5$$
$$2\sqrt{25}-9=1$$
$$2(5)-9=1$$
$$10-9=1$$
$$1=1$$

Check $x=2$:

$$2\sqrt{4(2)+1}-9=2-5$$
$$2\sqrt{8+1}-9=-3$$
$$2\sqrt{9}-9=-3$$
$$2(3)-9=-3$$
$$6-9=-3$$
$$-3=-3$$

Both solutions check. The solution set is $\{2,6\}$.

27. $(2x+3)^{1/3}+4=6$

$$(2x+3)^{1/3}=2$$
$$\left((2x+3)^{1/3}\right)^3=2^3$$
$$2x+3=8$$
$$2x=5$$
$$x=\frac{5}{2}$$

The solution checks. The solution set is $\left\{\frac{5}{2}\right\}$.

29. $(3x+1)^{1/4}+7=9$

$$(3x+1)^{1/4}=2$$
$$\left((3x+1)^{1/4}\right)^4=2^4$$
$$3x+1=16$$
$$3x=15$$
$$x=5$$

The solution checks. The solution set is $\{5\}$.

31. $(x+2)^{1/2}+8=4$

$$(x+2)^{1/2}=-4$$
$$\sqrt{x+2}=-4$$

The square root of a number must be positive. The solution set is $\varnothing$.

33. $\sqrt{2x-3}-\sqrt{x-2}=1$

$$\sqrt{2x-3}=\sqrt{x-2}+1$$
$$\left(\sqrt{2x-3}\right)^2=\left(\sqrt{x-2}+1\right)^2$$
$$2x-3=x-2+2\sqrt{x-2}+1$$
$$2x-3=x-1+2\sqrt{x-2}$$
$$x-2=2\sqrt{x-2}$$
$$(x-2)^2=\left(2\sqrt{x-2}\right)^2$$
$$x^2-4x+4=4(x-2)$$
$$x^2-4x+4=4x-8$$
$$x^2-8x+12=0$$
$$(x-6)(x-2)=0$$
$$x-6=0 \qquad x-2=0$$
$$x=6 \qquad\qquad x=2$$

Both solutions check. The solution set is $\{2,6\}$.

35. $3x^{1/3} = \left(x^2 + 17x\right)^{1/3}$

$$\left(3x^{1/3}\right)^3 = \left(\left(x^2 + 17x\right)^{1/3}\right)^3$$
$$3^3 x = x^2 + 17x$$
$$27x = x^2 + 17x$$
$$0 = x^2 - 10x$$
$$0 = x(x-10)$$
$$x = 0 \qquad x - 10 = 0$$
$$x = 10$$

Both solutions check. The solution set is $\{0, 10\}$.

37. $(x+8)^{1/4} = (2x)^{1/4}$

$$\left((x+8)^{1/4}\right)^4 = \left((2x)^{1/4}\right)^4$$
$$x + 8 = 2x$$
$$8 = x$$

The solution checks. The solution set is $\{8\}$.

39. $f(x) = x + \sqrt{x+5}$

$$7 = x + \sqrt{x+5}$$
$$7 - x = \sqrt{x+5}$$
$$(7-x)^2 = \left(\sqrt{x+5}\right)^2$$
$$49 - 14x + x^2 = x + 5$$
$$x^2 - 15x + 44 = 0$$
$$(x-11)(x-4) = 0$$
$$x - 11 = 0 \quad \text{or} \quad x - 4 = 0$$
$$x = 11 \qquad\qquad x = 4$$

Check $x = 11$: $11 + \sqrt{11+5} = 11 + \sqrt{16}$
$= 15 \neq 7$

Check $x = 4$: $4 + \sqrt{4+5} = 4 + \sqrt{9}$
$= 7$

Discard 11. The solution is 4.

41. $f(x) = (5x+16)^{1/3}$; $g(x) = (x-12)^{1/3}$

$$(5x+16)^{1/3} = (x-12)^{1/3}$$
$$\left[(5x+16)^{1/3}\right]^3 = \left[(x-12)^{1/3}\right]^3$$
$$5x + 16 = x - 12$$
$$4x = -28$$
$$x = -7$$

The solution is -7.

43. $r = \sqrt{\dfrac{3V}{\pi h}}$

$$r^2 = \left(\sqrt{\frac{3V}{\pi h}}\right)^2$$
$$r^2 = \frac{3V}{\pi h}$$
$$\pi r^2 h = 3V$$
$$V = \frac{\pi r^2 h}{3} \quad \text{or} \quad V = \frac{1}{3}\pi r^2 h$$

45. $t = 2\pi\sqrt{\dfrac{l}{32}}$

$$\frac{t}{2\pi} = \sqrt{\frac{l}{32}}$$
$$\left(\frac{t}{2\pi}\right)^2 = \left(\sqrt{\frac{l}{32}}\right)^2$$
$$\frac{t^2}{4\pi^2} = \frac{l}{32}$$
$$\frac{32t^2}{4\pi^2} = l$$
$$\frac{8t^2}{\pi^2} = l \quad \text{or} \quad l = \frac{8t^2}{\pi^2}$$

47. Let x = the number.

$$\sqrt{5x-4}=x-2$$
$$\left(\sqrt{5x-4}\right)^2=(x-2)^2$$
$$5x-4=x^2-4x+4$$
$$0=x^2-9x+8$$
$$0=(x-8)(x-1)$$
$$x-8=0 \quad \text{or} \quad x-1=0$$
$$x=8 \qquad\qquad x=1$$

Check $x=8$: $\sqrt{5(8)-4}=8-2$
$$\sqrt{40-4}=6$$
$$\sqrt{36}=6$$
$$6=6$$

Check $x=1$: $\sqrt{5(1)-4}=1-2$
$$\sqrt{5-4}=-1$$
$$\sqrt{1}=-1$$
$$1\neq-1$$

Discard $x=1$. The number is 8.

49. $f(x)=\sqrt{x+16}-\sqrt{x}-2$

To find the x-intercepts, set the function equal to 0 and solve for x.

$$0=\sqrt{x+16}-\sqrt{x}-2$$
$$\sqrt{x}+2=\sqrt{x+16}$$
$$\left(\sqrt{x}+2\right)^2=\left(\sqrt{x+16}\right)^2$$
$$x+4\sqrt{x}+4=x+16$$
$$4\sqrt{x}=12$$
$$\sqrt{x}=3$$
$$\left(\sqrt{x}\right)^2=3^2$$
$$x=9$$

Check $x=9$:

$$\sqrt{9+16}-\sqrt{9}-2$$
$$=\sqrt{25}-\sqrt{9}-2$$
$$=5-3-2$$
$$=0$$

The only x-intercept is 9.

51.
$$t=\frac{\sqrt{d}}{2}$$
$$1.16=\frac{\sqrt{d}}{2}$$
$$2.32=\sqrt{d}$$
$$2.32^2=\left(\sqrt{d}\right)^2$$
$$d\approx5.4$$

The vertical distance was about 5.4 feet.

53. It is represented by the point $(5.4, 1.16)$.

55. **a.** $f(x)=-4.4\sqrt{x}+38$

$f(25)=-4.4\sqrt{25}+38=16$

16% of Americans earning $25 thousand annually report fair or poor health. This underestimates the percent displayed in the graph by 1%.

b.
$$f(x)=-4.4\sqrt{x}+38$$
$$14=-4.4\sqrt{x}+38$$
$$-24=-4.4\sqrt{x}$$
$$\frac{-24}{-4.4}=\frac{-4.4\sqrt{x}}{-4.4}$$
$$\frac{24}{4.4}=\sqrt{x}$$
$$\left(\frac{24}{4.4}\right)^2=\left(\sqrt{x}\right)^2$$
$$x\approx30$$

$30 thousand is the annual income that corresponds to 14% reporting fair or poor health.

57.
$$87=29x^{1/3}$$
$$\frac{87}{29}=\frac{29x^{1/3}}{29}$$
$$3=x^{1/3}$$
$$3^3=\left(x^{1/3}\right)^3$$
$$27=x$$

A Galapagos island with an area of 27 square miles will have 87 plant species.

59.
$$365 = 0.2x^{3/2}$$
$$\frac{365}{0.2} = \frac{0.2x^{3/2}}{0.2}$$
$$1825 = x^{3/2}$$
$$1825^2 = \left(x^{3/2}\right)^2$$
$$3,330,625 = x^3$$
$$\sqrt[3]{3,330,625} = \sqrt[3]{x^3}$$
$$149.34 \approx x$$
The average distance of the Earth from the sun is approximately 149 million kilometers.

61. – 67. Answers will vary.

69. $\sqrt{x} + 3 = 5$

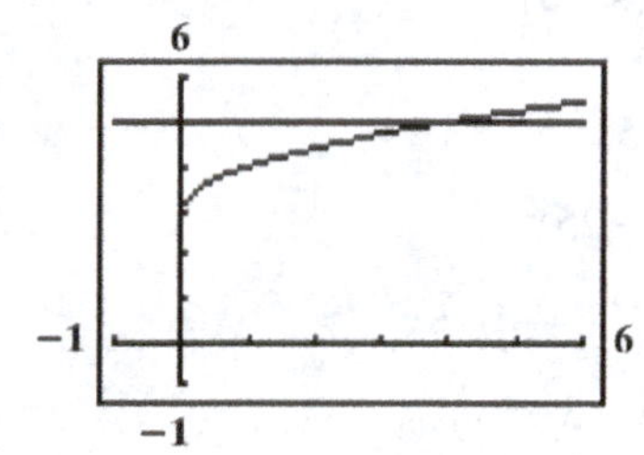

The solution set is $\{4\}$.

71. $4\sqrt{x} = x + 3$

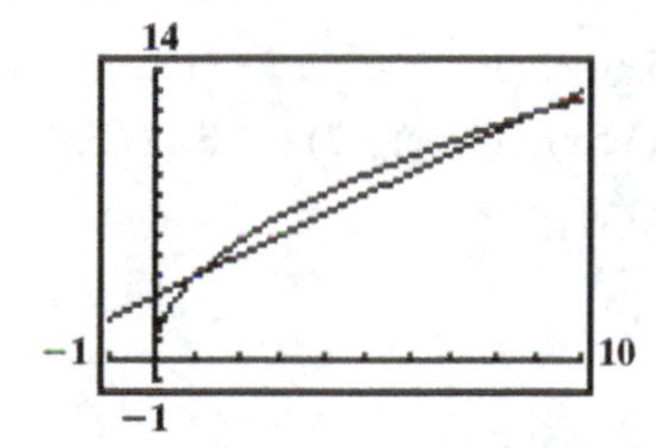

The solution set is $\{1, 9\}$.

73. does not make sense; Explanations will vary. Sample explanation: You should always substitute into the original equation. Later equations in the solution of the problem may be after extraneous roots were introduced. Substituting into such an equation will not allow you to rule out extraneous roots.

75. does not make sense; Explanations will vary. Sample explanation: Raising both sides to the first power does not introduce extraneous solutions.

77. false; Changes to make the statement true will vary. A sample change is: The first step is to square both sides, obtaining $x + 6 = x^2 + 4x + 4$.

79. true

81. $\left(\sqrt{x-7}\right)^2 + \left(\sqrt{x}\right)^2 = \left(1+\sqrt{x}\right)^2$
$$\cancel{x} - 7 + x = 1 + 2\sqrt{x} + \cancel{x}$$
$$-7 + x = 1 + 2\sqrt{x}$$
$$-8 + x = 2\sqrt{x}$$
$$(-8+x)^2 = \left(2\sqrt{x}\right)^2$$
$$64 - 16x + x^2 = 4x$$
$$x^2 - 16x + 64 = 4x$$
$$x^2 - 20x + 64 = 0$$
$$(x-16)(x-4) = 0$$
$$x - 16 = 0 \quad \text{or} \quad x - 4 = 0$$
$$x = 16 \qquad \cancel{x = 4}$$

We disregard 4. If $x = 4$, one of the legs becomes $\sqrt{4-7} = \sqrt{-3}$.

The legs of the triangle are:
$\sqrt{x-7} = \sqrt{16-7} = \sqrt{9} = 3$
$\sqrt{x} = \sqrt{16} = 4$, and
$1 + \sqrt{x} = 1 + \sqrt{16} = 1 + 4 = 5$.

83.
$$\sqrt{\sqrt{x} + \sqrt{x+9}} = 3$$
$$\left(\sqrt{\sqrt{x} + \sqrt{x+9}}\right)^2 = 3^2$$
$$\sqrt{x} + \sqrt{x+9} = 9$$
$$\sqrt{x+9} = 9 - \sqrt{x}$$
$$\left(\sqrt{x+9}\right)^2 = \left(9 - \sqrt{x}\right)^2$$
$$\cancel{x} + 9 = 81 - 18\sqrt{x} + \cancel{x}$$
$$9 = 81 - 18\sqrt{x}$$
$$-72 = -18\sqrt{x}$$
$$4 = \sqrt{x}$$
$$4^2 = \left(\sqrt{x}\right)^2$$
$$16 = x$$

Check:

$$\sqrt{\sqrt{x}+\sqrt{x+9}}=3$$
$$\sqrt{\sqrt{16}+\sqrt{16+9}}=3$$
$$\sqrt{4+\sqrt{25}}=3$$
$$\sqrt{4+5}=3$$
$$\sqrt{9}=3$$
$$3=3$$

The solution checks. The solution set is $\{16\}$.

84. $(x-4)^{2/3}=25$

$$\left((x-4)^{2/3}\right)^{3/2}=25^{3/2}$$
$$x-4=\left(\sqrt{25}\right)^3$$
$$x-4=5^3$$
$$x-4=125$$
$$x=129$$

Check:

$$(129-4)^{2/3}=25$$
$$(125)^{2/3}=25$$
$$\left(\sqrt[3]{125}\right)^2=25$$
$$5^2=25$$
$$25=25$$

The solution checks. The solution set is $\{129\}$.

85.

$\underline{-3}\rfloor$	4	−3	2	−1	−1
		−12	45	−141	426
	4	−15	47	−142	425

$$\frac{4x^4-3x^3+2x^2-x-1}{x+3}$$
$$=4x^3-15x^2+47x-142+\frac{425}{x+3}$$

86. $\dfrac{3x^2-12}{x^2+2x-8}\div\dfrac{6x+18}{x+4}$

$$=\frac{3x^2-12}{x^2+2x-8}\cdot\frac{x+4}{6x+18}$$
$$=\frac{3(x^2-4)}{(x+4)(x-2)}\cdot\frac{x+4}{6(x+3)}$$
$$=\frac{3(x+2)(x-2)}{(x-2)}\cdot\frac{1}{6(x+3)}$$
$$=\frac{3(x+2)}{1}\cdot\frac{1}{6(x+3)}$$
$$=\frac{3(x+2)}{6(x+3)}=\frac{x+2}{2(x+3)}$$

87. $y^2-6y+9-25x^2$

$$=\left(y^2-6y+9\right)-25x^2$$
$$=\left((y-3)+5x\right)\left((y-3)-5x\right)$$
$$=(y-3+5x)(y-3-5x)$$

88. $(-5+7x)-(-11-6x)=-5+7x+11+6x$

$$=6+13x$$

89. $(7-3x)(-2-5x)$

$$=(7)(-2)+(7)(-5x)+(-3x)(-2)+(-3x)(-5x)$$
$$=-14-35x+6x+15x^2$$
$$=15x^2-29x-14$$

90. $\frac{7+4\sqrt{2}}{2-5\sqrt{2}}=\frac{7+4\sqrt{2}}{2-5\sqrt{2}}\cdot\frac{2+5\sqrt{2}}{2+5\sqrt{2}}$

$$=\frac{(7+4\sqrt{2})(2+5\sqrt{2})}{(2-5\sqrt{2})(2+5\sqrt{2})}$$

$$=\frac{7\cdot 2+7\cdot 5\sqrt{2}+4\sqrt{2}\cdot 2+4\sqrt{2}\cdot 5\sqrt{2}}{(2)^2-(5\sqrt{2})^2}$$

$$=\frac{14+35\sqrt{2}+8\sqrt{2}+20\sqrt{4}}{(2)^2-(5\sqrt{2})^2}$$

$$=\frac{14+43\sqrt{2}+20\cdot 2}{4-(5)^2(\sqrt{2})^2}$$

$$=\frac{14+43\sqrt{2}+40}{4-25\cdot 2}$$

$$=\frac{54+43\sqrt{2}}{-46}$$

$$=-\frac{54+43\sqrt{2}}{46}$$

7.7 Check Points

1. a. $\sqrt{-64}=\sqrt{64\cdot -1}=\sqrt{64}\cdot\sqrt{-1}=8i$

b. $\sqrt{-11}=\sqrt{11\cdot -1}=\sqrt{11}\cdot\sqrt{-1}=i\sqrt{11}$

c. $\sqrt{-48}=\sqrt{48\cdot -1}$

$$=\sqrt{48}\cdot\sqrt{-1}$$
$$=\sqrt{16}\cdot\sqrt{3}\cdot\sqrt{-1}$$
$$=4\cdot\sqrt{3}\cdot i$$
$$=4i\sqrt{3}$$

2. a. $(5-2i)+(3+3i)=5-2i+3+3i$

$$=8+i$$

b. $(2+6i)-(12-4i)=2+6i-12+4i$

$$=-10+10i$$

3. a. $7i(2-9i)=7i\cdot 2-7i\cdot 9i$

$$=14i-63i^2$$
$$=14i-63(-1)$$
$$=63+14i$$

b. $(5+4i)(6-7i)=30-35i+24i-28i^2$

$$=30-35i+24i-28(-1)$$
$$=30+28-35i+24i$$
$$=58-11i$$

4. $\sqrt{-5}\cdot\sqrt{-7}=\sqrt{5}\sqrt{-1}\cdot\sqrt{7}\sqrt{-1}$

$$=i\sqrt{5}\cdot i\sqrt{7}$$
$$=i^2\sqrt{35}$$
$$=(-1)\sqrt{35}$$
$$=-\sqrt{35}$$

5. $\frac{6+2i}{4-3i}=\frac{6+2i}{4-3i}\cdot\frac{4+3i}{4+3i}$

$$=\frac{24+18i+8i+6i^2}{4^2-(3i)^2}$$
$$=\frac{24+18i+8i+6(-1)}{16-9i^2}$$
$$=\frac{18+26i}{16-9(-1)}$$
$$=\frac{18+26i}{25}$$
$$=\frac{18}{25}+\frac{26}{25}i$$

6. $\frac{3-2i}{4i}=\frac{3-2i}{4i}\cdot\frac{-4i}{-4i}$

$$=\frac{-12i+8i^2}{-16i^2}$$
$$=\frac{-12i+8(-1)}{-16(-1)}$$
$$=\frac{-8-12i}{16}$$
$$=-\frac{8}{16}-\frac{12}{16}i$$
$$=-\frac{1}{2}-\frac{3}{4}i$$

7. a. $i^{16}=(i^2)^8=(-1)^8=1$

b. $i^{25}=i^{24}i=(i^2)^{12}i=(-1)^{12}i=1\cdot i=i$

c. $i^{35}=i^{34}i=(i^2)^{17}i=(-1)^{17}i=(-1)i=-i$

7.7 Concept and Vocabulary Check

1. $\sqrt{-1}$; -1

2. $4i$

3. complex; imaginary; real

4. $-6i$

5. $14i$

6. 18; $-15i$; $12i$; $-10i^2$; 10

7. $2+9i$

8. $2+5i$

9. $-4i$

10. -1; 1

11. -1; -1; $-i$

7.7 Exercise Set

1. $\sqrt{-100}=\sqrt{100\cdot -1}=\sqrt{100}\cdot\sqrt{-1}=10i$

3. $\sqrt{-23}=\sqrt{23\cdot -1}=\sqrt{23}\cdot\sqrt{-1}=i\sqrt{23}$

5. $\sqrt{-18}=\sqrt{9\cdot 2\cdot -1}$
$=\sqrt{9}\cdot\sqrt{2}\cdot\sqrt{-1}$
$=3i\sqrt{2}$

7. $\sqrt{-63}=\sqrt{9\cdot 7\cdot -1}$
$=\sqrt{9}\cdot\sqrt{7}\cdot\sqrt{-1}$
$=3i\sqrt{7}$

9. $-\sqrt{-108}=-\sqrt{36\cdot 3\cdot -1}$
$=-\sqrt{36}\cdot\sqrt{3}\cdot\sqrt{-1}$
$=-6i\sqrt{3}$

11. $5+\sqrt{-36}=5+\sqrt{36\cdot -1}$
$=5+\sqrt{36}\cdot\sqrt{-1}$
$=5+6i$

13. $15+\sqrt{-3}=15+\sqrt{3\cdot -1}$
$=15+\sqrt{3}\cdot\sqrt{-1}$
$=15+i\sqrt{3}$

15. $-2-\sqrt{-18}=-2-\sqrt{9\cdot 2\cdot -1}$
$=-2-\sqrt{9}\cdot\sqrt{2}\cdot\sqrt{-1}$
$=-2-3i\sqrt{2}$

17. $(3+2i)+(5+i)$
$=3+2i+5+i=3+5+2i+i$
$=(3+5)+(2+1)i=8+3i$

19. $(7+2i)+(1-4i)$
$=7+2i+1-4i=7+1+2i-4i$
$=(7+1)+(2-4)i=8-2i$

21. $(10+7i)-(5+4i)$
$=10+7i-5-4i=10-5+7i-4i$
$=(10-5)+(7-4)i=5+3i$

23. $(9-4i)-(10+3i)$
$=9-4i-10-3i=9-10-4i-3i$
$=(9-10)+(-4-3)i$
$=-1+(-7)i=-1-7i$

25. $(3+2i)-(5-7i)$
$=3+2i-5+7i=3-5+2i+7i$
$=(3-5)+(2+7)i=-2+9i$

27. $(-5+4i)-(-13-11i)$
$=-5+4i+13+11i$
$=-5+13+4i+11i$
$=(-5+13)+(4+11)i=8+15i$

29. $8i-(14-9i)$
$=8i-14+9i=-14+8i+9i$
$=-14+(8+9)i=-14+17i$

31. $\left(2+i\sqrt{3}\right)+\left(7+4i\sqrt{3}\right)$
$=2+i\sqrt{3}+7+4i\sqrt{3}$
$=2+7+i\sqrt{3}+4i\sqrt{3}$
$=(2+7)+\left(\sqrt{3}+4\sqrt{3}\right)i=9+5i\sqrt{3}$

33. $2i(5+3i)$
$= 2i\cdot 5 + 2i\cdot 3i = 10i + 6i^2$
$= 10i + 6(-1) = -6 + 10i$

35. $3i(7i-5)$
$= 3i\cdot 7i - 3i\cdot 5 = 21i^2 - 15i$
$= 21(-1) - 15i = -21 - 15i$

37. $-7i(2-5i)$
$= -7i\cdot 2 - (-7i)5i = -14i + 35i^2$
$= -14i + 35(-1) = -35 - 14i$

39. $(3+i)(4+5i) = 12 + 15i + 4i + 5i^2$
$= 12 + 15i + 4i + 5(-1)$
$= 12 - 5 + 15i + 4i$
$= 7 + 19i$

41. $(7-5i)(2-3i)$
$= 14 - 21i - 10i + 15i^2$
$= 14 - 21i - 10i + 15(-1)$
$= 14 - 15 - 21i - 10i = -1 - 31i$

43. $(6-3i)(-2+5i)$
$= -12 + 30i + 6i - 15i^2$
$= -12 + 30i + 6i - 15(-1)$
$= -12 + 15 + 30i + 6i = 3 + 36i$

45. $(3+5i)(3-5i)$
$= 9 - \cancel{15i} + \cancel{15i} - 25i^2$
$= 9 - 25(-1) = 9 + 25$
$= 34 = 34 + 0i$

47. $(-5+3i)(-5-3i)$
$= 25 + \cancel{15i} - \cancel{15i} - 9i^2$
$= 25 - 9(-1) = 25 + 9$
$= 34 = 34 + 0i$

49. $(3-i\sqrt{2})(3+i\sqrt{2})$
$= 9 + \cancel{3i\sqrt{2}} - \cancel{3i\sqrt{2}} - 2i^2$
$= 9 - 2(-1) = 9 + 2$
$= 11 = 11 + 0i$

51. $(2+3i)^2$
$= 4 + 2\cdot 6i + 9i^2 = 4 + 12i + 9(-1)$
$= 4 - 9 + 12i = -5 + 12i$

53. $(5-2i)^2 = 25 - 2\cdot 10i + 4i^2$
$= 25 - 20i + 4(-1)$
$= 25 - 4 - 20i = 21 - 20i$

55. $\sqrt{-7}\cdot\sqrt{-2} = \sqrt{7}\sqrt{-1}\cdot\sqrt{2}\sqrt{-1}$
$= \sqrt{7}\, i\cdot\sqrt{2}\, i = \sqrt{14}\, i^2$
$= \sqrt{14}(-1) = -\sqrt{14}$
$= -\sqrt{14} + 0i$

57. $\sqrt{-9}\cdot\sqrt{-4}$
$= \sqrt{9}\sqrt{-1}\cdot\sqrt{4}\sqrt{-1} = 3i\cdot 2i = 6i^2$
$= 6(-1) = -6 = -6 + 0i$

59. $\sqrt{-7}\cdot\sqrt{-25} = \sqrt{7}\sqrt{-1}\cdot\sqrt{25}\sqrt{-1}$
$= \sqrt{7}\, i\cdot 5i = 5\sqrt{7}\, i^2$
$= 5\sqrt{7}(-1) = -5\sqrt{7}$
$= -5\sqrt{7} + 0i$

61. $\sqrt{-8}\cdot\sqrt{-3} = \sqrt{4\cdot 2}\sqrt{-1}\cdot\sqrt{3}\sqrt{-1}$
$= 2\sqrt{2}\, i\cdot\sqrt{3}\, i = 2\sqrt{6}\, i^2$
$= 2\sqrt{6}(-1) = -2\sqrt{6}$
$= -2\sqrt{6} + 0i$

63. $\dfrac{2}{3+i} = \dfrac{2}{3+i}\cdot\dfrac{3-i}{3-i} = \dfrac{6-2i}{3^2 - i^2}$
$= \dfrac{6-2i}{9-(-1)} = \dfrac{6-2i}{9+1}$
$= \dfrac{6-2i}{10} = \dfrac{6}{10} - \dfrac{2i}{10}$
$= \dfrac{3}{5} - \dfrac{1}{5}i$

65. $\dfrac{2i}{1+i} = \dfrac{2i}{1+i}\cdot\dfrac{1-i}{1-i} = \dfrac{2i - 2i^2}{1^2 - i^2}$
$= \dfrac{2i - 2(-1)}{1-(-1)} = \dfrac{2+2i}{1+1}$
$= \dfrac{2+2i}{2} = \dfrac{2}{2} + \dfrac{2i}{2} = 1 + i$

67. $\dfrac{7}{4-3i}=\dfrac{7}{4-3i}\cdot\dfrac{4+3i}{4+3i}=\dfrac{28+21i}{4^2-(3i)^2}$
$=\dfrac{28+21i}{16-9i^2}=\dfrac{28+21i}{16-9(-1)}$
$=\dfrac{28+21i}{16+9}=\dfrac{28+21i}{25}$
$=\dfrac{28}{25}+\dfrac{21}{25}i$

69. $\dfrac{6i}{3-2i}=\dfrac{6i}{3-2i}\cdot\dfrac{3+2i}{3+2i}=\dfrac{18i+12i^2}{3^2-(2i)^2}$
$=\dfrac{18i+12(-1)}{9-4i^2}=\dfrac{-12+18i}{9-4(-1)}$
$=\dfrac{-12+18i}{9+4}=\dfrac{-12+18i}{13}$
$=-\dfrac{12}{13}+\dfrac{18}{13}i$

71. $\dfrac{1+i}{1-i}=\dfrac{1+i}{1-i}\cdot\dfrac{1+i}{1+i}=\dfrac{1+2i+i^2}{1^2-i^2}$
$=\dfrac{1+2i+(-1)}{1-(-1)}=\dfrac{2i}{2}$
$=i$ or $0+i$

73. $\dfrac{2-3i}{3+i}=\dfrac{2-3i}{3+i}\cdot\dfrac{3-i}{3-i}$
$=\dfrac{6-2i-9i+3i^2}{3^2-i^2}$
$=\dfrac{6-11i+3(-1)}{9-(-1)}$
$=\dfrac{6-3-11i}{9+1}$
$=\dfrac{3-11i}{10}=\dfrac{3}{10}-\dfrac{11}{10}i$

75. $\dfrac{5-2i}{3+2i}=\dfrac{5-2i}{3+2i}\cdot\dfrac{3-2i}{3-2i}$
$=\dfrac{15-10i-6i+4i^2}{3^2-(2i)^2}$
$=\dfrac{15-16i+4(-1)}{9-4i^2}$
$=\dfrac{15-4-16i}{9-4(-1)}$
$=\dfrac{11-16i}{9+4}$
$=\dfrac{11-16i}{13}=\dfrac{11}{13}-\dfrac{16}{13}i$

77. $\dfrac{4+5i}{3-7i}=\dfrac{4+5i}{3-7i}\cdot\dfrac{3+7i}{3+7i}$
$=\dfrac{12+28i+15i+35i^2}{3^2-(7i)^2}$
$=\dfrac{12+43i+35(-1)}{9-49i^2}$
$=\dfrac{12-35+43i}{9-49(-1)}$
$=\dfrac{-23+43i}{9+49}=\dfrac{-23+43i}{58}$
$=-\dfrac{23}{58}+\dfrac{43}{58}i$

79. $\dfrac{7}{3i}=\dfrac{7}{3i}\cdot\dfrac{-3i}{-3i}=\dfrac{-21i}{-9i^2}=\dfrac{-21i}{-9(-1)}$
$=\dfrac{-21i}{9}=-\dfrac{7}{3}i$ or $0-\dfrac{7}{3}i$

81. $\dfrac{8-5i}{2i}=\dfrac{8-5i}{2i}\cdot\dfrac{-2i}{-2i}=\dfrac{-16i+10i^2}{-4i^2}$
$=\dfrac{-16i+10(-1)}{-4(-1)}=\dfrac{-10-16i}{4}$
$=-\dfrac{10}{4}-\dfrac{16}{4}i=-\dfrac{5}{2}-4i$

83. $\dfrac{4+7i}{-3i}=\dfrac{4+7i}{-3i}\cdot\dfrac{3i}{3i}=\dfrac{12i+21i^2}{-9i^2}$
$=\dfrac{12i+21(-1)}{-9(-1)}=\dfrac{-21+12i}{9}$
$=-\dfrac{21}{9}+\dfrac{12}{9}i=-\dfrac{7}{3}+\dfrac{4}{3}i$

85. $i^{10} = \left(i^2\right)^5 = (-1)^5 = -1$

87. $i^{11} = \left(i^2\right)^5 i = (-1)^5 i = -i$

89. $i^{22} = \left(i^2\right)^{11} = (-1)^{11} = -1$

91. $i^{200} = \left(i^2\right)^{100} = (-1)^{100} = 1$

93. $i^{17} = \left(i^2\right)^8 i = (-1)^8 i = i$

95. $(-i)^4 = (-1)^4 i^4 = i^4 = \left(i^2\right)^2$
$= (-1)^2 = 1$

97. $(-i)^9 = (-1)^9 i^9 = (-1)\left(i^2\right)^4 i$
$= (-1)(-1)^4 i = (-1)i$
$= -i$

99. $i^{24} + i^2 = \left(i^2\right)^{12} + (-1)$
$= (-1)^{12} + (-1)$
$= 1 + (-1) = 0$

101. $(2-3i)(1-i)-(3-i)(3+i)$
$= \left(2-2i-3i+3i^2\right)-\left(3^2-i^2\right)$
$= 2-5i+3i^2-9+i^2$
$= -7-5i+4i^2$
$= -7-5i+4(-1)$
$= -11-5i$

103. $(2+i)^2-(3-i)^2$
$= \left(4+4i+i^2\right)-\left(9-6i+i^2\right)$
$= 4+4i+i^2-9+6i-i^2$
$= -5+10i$

105. $5\sqrt{-16}+3\sqrt{-81}$
$= 5\sqrt{16}\sqrt{-1}+3\sqrt{81}\sqrt{-1}$
$= 5\cdot 4i+3\cdot 9i$
$= 20i+27i$
$= 47i$ or $0+47i$

107. $\dfrac{i^4+i^{12}}{i^8-i^7} = \dfrac{i^4+\left(i^4\right)^3}{\left(i^4\right)^2-\left(i^2\right)^3 i}$
$= \dfrac{1+1^3}{1^2-(-1)^3 i} = \dfrac{1+1}{1+i}$
$= \dfrac{2}{1+i} = \dfrac{2}{1+i}\cdot\dfrac{1-i}{1-i}$
$= \dfrac{2-2i}{1^2-i^2} = \dfrac{2-2i}{1+1}$
$= \dfrac{2-2i}{2} = 1-i$

109. $f(x) = x^2-2x+2$
$f(1+i) = (1+i)^2-2(1+i)+2$
$= 1+2i+i^2-2-2i+2$
$= 1+i^2$
$= 1-1$
$= 0$

111. $f(x) = x-3i$; $g(x) = 4x+2i$
$f(-1) = -1-3i$
$g(-1) = -4+2i$
$(fg)(-1) = (-1-3i)(-4+2i)$
$= 4-2i+12i-6i^2$
$= 4+10i-6(-1)$
$= 10+10i$

113. $f(x) = \dfrac{x^2+19}{2-x}$
$f(3i) = \dfrac{(3i)^2+19}{2-3i} = \dfrac{9i^2+19}{2-3i}$
$= \dfrac{9(-1)+19}{2-3i} = \dfrac{10}{2-3i}$
$= \dfrac{10}{2-3i}\cdot\dfrac{2+3i}{2+3i}$
$= \dfrac{20+30i}{2^2-(3i)^2} = \dfrac{20+30i}{4-9i^2}$
$= \dfrac{20+30i}{4-9(-1)} = \dfrac{20+30i}{13}$
$= \dfrac{20}{13}+\dfrac{30}{13}i$

115. $E = IR = (4-5i)(3+7i)$
$= 12 + 28i - 15i - 35i^2$
$= 12 + 13i - 35(-1)$
$= 12 + 35 + 13i = 47 + 13i$
The voltage of the circuit is $(47 + 13i)$ volts.

117. Sum:
$(5+i\sqrt{15})+(5-i\sqrt{15})$
$= 5 + \cancel{i\sqrt{15}} + 5 - \cancel{i\sqrt{15}}$
$= 5 + 5 = 10$

Product:
$(5+i\sqrt{15})(5-i\sqrt{15})$
$= 25 - \cancel{5i\sqrt{15}} + \cancel{5i\sqrt{15}} - 15i^2$
$= 25 - 15(-1) = 25 + 15 = 40$

119. – 129. Answers will vary.

131. $\sqrt{-9}+\sqrt{-16} = i\sqrt{9}+i\sqrt{16}$
$= 3i + 4i = 7i$

133. does not make sense; Explanations will vary. Sample explanation: The average of complex real numbers is never a complex imaginary number.

135. does not make sense; Explanations will vary. Sample explanation: The i in $5i$ is not a variable. It is the imaginary unit $\sqrt{-1}$.

137. false; Changes to make the statement true will vary. A sample change is: All irrational numbers are complex numbers.

139. false; Changes to make the statement true will vary. A sample change is:

$$\frac{7+3i}{5+3i} = \frac{7+3i}{5+3i}\cdot\frac{5-3i}{5-3i}$$
$$= \frac{35-21i+15i-9i^2}{5^2-(3i)^2}$$
$$= \frac{35-21i+15i-9(-1)}{25-9i^2}$$
$$= \frac{35-6i+9}{25-9(-1)}$$
$$= \frac{44-6i}{34}$$
$$= \frac{44}{34}-\frac{6}{34}i$$
$$= \frac{22}{17}-\frac{3}{17}i$$

141.
$$\frac{4}{(2+i)(3-i)} = \frac{4}{6-2i+3i-i^2}$$
$$= \frac{4}{6+i-(-1)} = \frac{4}{6+1+i}$$
$$= \frac{4}{7+i}\cdot\frac{7-i}{7-i} = \frac{28-4i}{7^2-i^2}$$
$$= \frac{28-4i}{49-(-1)} = \frac{28-4i}{50}$$
$$= \frac{28}{50}-\frac{4}{50}i = \frac{14}{25}-\frac{2}{25}i$$

143.
$$\frac{8}{1+\frac{2}{i}} = \frac{8}{\frac{i}{i}+\frac{2}{i}} = \frac{8}{\frac{2+i}{i}}$$
$$= \frac{8i}{2+i}\cdot\frac{2-i}{2-i} = \frac{8i(2-i)}{(2+i)(2-i)}$$
$$= \frac{16i-8i^2}{4-i^2} = \frac{16i-8(-1)}{4-(-1)}$$
$$= \frac{16i+8}{4+1} = \frac{8+16i}{5} = \frac{8}{5}+\frac{16i}{5}$$

144. $$\frac{\frac{x}{y^2}+\frac{1}{y}}{\frac{y}{x^2}+\frac{1}{x}} = \frac{\frac{x}{y^2}+\frac{1}{y}}{\frac{y}{x^2}+\frac{1}{x}} \cdot \frac{x^2y^2}{x^2y^2}$$
$$= \frac{\frac{x}{y^2}\cdot x^2y^2 + \frac{1}{y}\cdot x^2y^2}{\frac{y}{x^2}\cdot x^2y^2 + \frac{1}{x}\cdot x^2y^2}$$
$$= \frac{x^3+x^2y}{y^3+xy^2} = \frac{x^2(x+y)}{y^2(y+x)} = \frac{x^2}{y^2}$$

145. $$\frac{1}{x}+\frac{1}{y}=\frac{1}{z}$$
$$\frac{1}{x}\cdot xyz+\frac{1}{y}\cdot xyz=\frac{1}{z}\cdot xyz$$
$$yz+xz=xy$$
$$yz=xy-xz$$
$$yz=x(y-z)$$
$$x=\frac{yz}{y-z}$$

146. $$2x-\frac{x-3}{8}=\frac{1}{2}+\frac{x+5}{2}$$
$$8\left(2x-\frac{x-3}{8}\right)=8\left(\frac{1}{2}+\frac{x+5}{2}\right)$$
$$16x-x+3=4+4(x+5)$$
$$15x+3=4+4x+20$$
$$15x+3=4x+24$$
$$11x+3=24$$
$$11x=21$$
$$x=\frac{21}{11}$$

The solution set is $\left\{\frac{21}{11}\right\}$.

147. $$2x^2+7x-4=0$$
$$(x+4)(2x-1)=0$$
$$x+4=0 \quad \text{or} \quad 2x-1=0$$
$$x=-4 \qquad 2x=1$$
$$x=\frac{1}{2}$$

The solution set is $\left\{-4,\frac{1}{2}\right\}$.

148. $$x^2=9$$
$$x^2-9=0$$
$$(x+3)(x-3)=0$$
$$x+3=0 \quad \text{or} \quad x-3=0$$
$$x=-3 \qquad x=3$$

The solution set is $\{-3,3\}$.

149. $$3x^2=18$$
$$3(-\sqrt{6})^2=18$$
$$3(-1)^2(\sqrt{6})^2=18$$
$$3\cdot1\cdot6=18$$
$$18=18, \text{ true}$$

$-\sqrt{6}$ is a solution.

Chapter 7 Review Exercises

1. $\sqrt{81}=9$ because $9^2=81$

2. $-\sqrt{\frac{1}{100}}=-\frac{1}{10}$ because $\left(-\frac{1}{10}\right)^2=\frac{1}{100}$

3. $\sqrt[3]{-27}=-3$ because $(-3)^3=-27$

4. $\sqrt[4]{-16}$
   not a real number
   The index is even and the radicand is negative.

5. $\sqrt[5]{-32}=-2$ because $(-2)^5=-32$

6. $f(15)=\sqrt{2(15)-5}=\sqrt{30-5}$
   $=\sqrt{25}=5$
   $f(4)=\sqrt{2(4)-5}=\sqrt{8-5}=\sqrt{3}\approx1.73$
   $f\left(\frac{5}{2}\right)=\sqrt{2\left(\frac{5}{2}\right)-5}=\sqrt{5-5}$
   $=\sqrt{0}=0$
   $f(1)=\sqrt{2(1)-5}=\sqrt{2-5}=\sqrt{-3}$
   not a real number

7. $g(4)=\sqrt[3]{4(4)-8}=\sqrt[3]{16-8}=\sqrt[3]{8}=2$

$g(0)=\sqrt[3]{4(0)-8}=\sqrt[3]{-8}=-2$

$g(-14)=\sqrt[3]{4(-14)-8}=\sqrt[3]{-56-8}$

$=\sqrt[3]{-64}=-4$

8. To find the domain, set the radicand greater than or equal to zero and solve the resulting inequality.

$x-2\ge 0$

$x\ge 2$

The domain of f is $[2,\infty)$.

9. To find the domain, set the radicand greater than or equal to zero and solve the resulting inequality.

$100-4x\ge 0$

$-4x\ge -100$

$\frac{-4x}{-4}\le\frac{-100}{-4}$

$x\le 25$

The domain of g is $(-\infty,25]$.

10. $\sqrt{25x^2}=5|x|$

11. $\sqrt{(x+14)^2}=|x+14|$

12. $\sqrt{x^2-8x+16}=\sqrt{(x-4)^2}=|x-4|$

13. $\sqrt[3]{64x^3}=4x$

14. $\sqrt[4]{16x^4}=2|x|$

15. $\sqrt[5]{-32(x+7)^5}=-2(x+7)$

16. $(5xy)^{\frac{1}{3}}=\sqrt[3]{5xy}$

17. $16^{\frac{3}{2}}=\left(\sqrt{16}\right)^3=(4)^3=64$

18. $32^{\frac{4}{5}}=\left(\sqrt[5]{32}\right)^4=(2)^4=16$

19. $\sqrt{7x}=(7x)^{\frac{1}{2}}$

20. $\left(\sqrt[3]{19xy}\right)^5=(19xy)^{\frac{5}{3}}$

21. $8^{-\frac{2}{3}}=\frac{1}{8^{\frac{2}{3}}}=\frac{1}{\left(\sqrt[3]{8}\right)^2}=\frac{1}{(2)^2}=\frac{1}{4}$

22. $3x(ab)^{-\frac{4}{5}}=\frac{3x}{(ab)^{\frac{4}{5}}}$

$=\frac{3x}{\left(\sqrt[5]{ab}\right)^4}$

$=\frac{3x}{(ab)^{\frac{4}{5}}}$

$=\frac{3x}{a^{\frac{4}{5}}b^{\frac{4}{5}}}$

23. $x^{\frac{1}{3}}\cdot x^{\frac{1}{4}}=x^{\frac{1}{3}+\frac{1}{4}}=x^{\frac{4}{12}+\frac{3}{12}}=x^{\frac{7}{12}}$

24. $\frac{5^{\frac{1}{2}}}{5^{\frac{1}{3}}}=5^{\frac{1}{2}-\frac{1}{3}}=5^{\frac{3}{6}-\frac{2}{6}}=5^{\frac{1}{6}}$

25. $\left(8x^6y^3\right)^{\frac{1}{3}}=8^{\frac{1}{3}}x^{6\cdot\frac{1}{3}}y^{3\cdot\frac{1}{3}}=2x^2y$

26. $\left(x^{-\frac{2}{3}}y^{\frac{1}{4}}\right)^{\frac{1}{2}}=x^{-\frac{2}{3}\cdot\frac{1}{2}}y^{\frac{1}{4}\cdot\frac{1}{2}}$

$=x^{-\frac{1}{3}}y^{\frac{1}{8}}=\frac{y^{\frac{1}{8}}}{x^{\frac{1}{3}}}$

27. $\sqrt[3]{x^9y^{12}}=\left(x^9y^{12}\right)^{\frac{1}{3}}$

$=x^{9\cdot\frac{1}{3}}y^{12\cdot\frac{1}{3}}=x^3y^4$

28. $\sqrt[9]{x^3y^9}=\left(x^3y^9\right)^{\frac{1}{9}}=x^{3\cdot\frac{1}{9}}y^{9\cdot\frac{1}{9}}$

$=x^{\frac{1}{3}}y=y\sqrt[3]{x}$

29. $\sqrt{x}\cdot\sqrt[3]{x}=x^{\frac{1}{2}}x^{\frac{1}{3}}=x^{\frac{1}{2}+\frac{1}{3}}=x^{\frac{3}{6}+\frac{2}{6}}$

$=x^{\frac{5}{6}}=\sqrt[6]{x^5}$

30. $\dfrac{\sqrt[3]{x^2}}{\sqrt[4]{x^2}} = \dfrac{x^{\frac{2}{3}}}{x^{\frac{2}{4}}} = x^{\frac{2}{3}-\frac{1}{2}}$

$= x^{\frac{4}{6}-\frac{3}{6}} = x^{\frac{1}{6}} = \sqrt[6]{x}$

31. $\sqrt[5]{\sqrt[3]{x}} = \sqrt[5]{x^{\frac{1}{3}}} = \left(x^{\frac{1}{3}}\right)^{\frac{1}{5}} = x^{\frac{1}{3}\cdot\frac{1}{5}}$

$= x^{\frac{1}{15}} = \sqrt[15]{x}$

32. Since 2012 was 27 years after 1985, find $f(27)$.

$f(27) = 350(27)^{\frac{2}{3}} = 350\left(\sqrt[3]{27}\right)^2$

$= 350(3)^2 = 350(9) = 3150$

Expenditures were \$3150 million or \$3,150,000,000 in the year 2012.

33. $\sqrt{3x}\cdot\sqrt{7y} = \sqrt{21xy}$

34. $\sqrt[5]{7x^2}\cdot\sqrt[5]{11x} = \sqrt[5]{77x^3}$

35. $\sqrt[6]{x-5}\cdot\sqrt[6]{(x-5)^4} = \sqrt[6]{(x-5)^5}$

36. $f(x) = \sqrt{7x^2-14x+7}$

$= \sqrt{7\left(x^2-2x+1\right)}$

$= \sqrt{7(x-1)^2} = \sqrt{7}\,|x-1|$

37. $\sqrt{20x^3} = \sqrt{4\cdot 5\cdot x^2\cdot x} = \sqrt{4x^2\cdot 5x}$

$= 2x\sqrt{5x}$

38. $\sqrt[3]{54x^8y^6} = \sqrt[3]{27\cdot 2\cdot x^6\cdot x^2y^6}$

$= \sqrt[3]{27x^6y^6\cdot 2x^2}$

$= 3x^2y^2\sqrt[3]{2x^2}$

39. $\sqrt[4]{32x^3y^{11}z^5} = \sqrt[4]{16\cdot 2\cdot x^3y^8\cdot y^3\cdot z^5\cdot z}$

$= \sqrt[4]{16y^8z^4\cdot 2x^3y^3z}$

$= 2y^2z\sqrt[4]{2x^3y^3z}$

40. $\sqrt{6x^3}\cdot\sqrt{4x^2} = \sqrt{24x^5} = \sqrt{4\cdot 6\cdot x^4\cdot x}$

$= \sqrt{4x^4\cdot 6x} = 2x^2\sqrt{6x}$

41. $\sqrt[3]{4x^2y}\cdot\sqrt[3]{4xy^4} = \sqrt[3]{16x^3y^5}$

$= \sqrt[3]{8\cdot 2\cdot x^3\cdot y^3\cdot y^2}$

$= \sqrt[3]{8x^3y^3\cdot 2y^2}$

$= 2xy\sqrt[3]{2y^2}$

42. $\sqrt[5]{2x^4y^3z^4}\cdot\sqrt[5]{8xy^6z^7}$

$= \sqrt[5]{16x^5y^9z^{11}}$

$= \sqrt[5]{16\cdot x^5\cdot y^5\cdot y^4\cdot z^{10}\cdot z}$

$= \sqrt[5]{x^5y^5z^{10}\cdot 16y^4z}$

$= xyz^2\sqrt[5]{16y^4z}$

43. $\sqrt{x+1}\cdot\sqrt{x-1} = \sqrt{(x+1)(x-1)}$

$= \sqrt{x^2-1}$

44. $6\sqrt[3]{3}+2\sqrt[3]{3} = (6+2)\sqrt[3]{3} = 8\sqrt[3]{3}$

45. $5\sqrt{18}-3\sqrt{8} = 5\sqrt{9\cdot 2}-3\sqrt{4\cdot 2}$

$= 5\cdot 3\sqrt{2}-3\cdot 2\sqrt{2}$

$= 15\sqrt{2}-6\sqrt{2}$

$= (15-6)\sqrt{2} = 9\sqrt{2}$

46. $\sqrt[3]{27x^4}+\sqrt[3]{xy^6}$

$= \sqrt[3]{27x^3x}+\sqrt[3]{xy^6}$

$= 3x\sqrt[3]{x}+y^2\sqrt[3]{x}$

$= \left(3x+y^2\right)\sqrt[3]{x}$

47. $2\sqrt[3]{6}-5\sqrt[3]{48} = 2\sqrt[3]{6}-5\sqrt[3]{8\cdot 6}$

$= 2\sqrt[3]{6}-5\cdot 2\sqrt[3]{6}$

$= 2\sqrt[3]{6}-10\sqrt[3]{6}$

$= (2-10)\sqrt[3]{6} = -8\sqrt[3]{6}$

48. $\sqrt[3]{\dfrac{16}{125}} = \sqrt[3]{\dfrac{8\cdot 2}{125}} = \dfrac{2}{5}\sqrt[3]{2}$

49. $\sqrt{\dfrac{x^3}{100y^4}} = \sqrt{\dfrac{x^2\cdot x}{100y^4}}$

$= \dfrac{x}{10y^2}\sqrt{x}$ or $\dfrac{x\sqrt{x}}{10y^2}$

50. $\sqrt[4]{\frac{3y^5}{16x^{20}}} = \sqrt[4]{\frac{y^4 \cdot 3y}{16x^{20}}}$

$= \frac{y}{2x^5}\sqrt[4]{3y}$ or $\frac{y\sqrt[4]{3y}}{2x^5}$

51. $\frac{\sqrt{48}}{\sqrt{2}} = \sqrt{\frac{48}{2}} = \sqrt{24} = \sqrt{4 \cdot 6} = 2\sqrt{6}$

52. $\frac{\sqrt[3]{32}}{\sqrt[3]{2}} = \sqrt[3]{\frac{32}{2}} = \sqrt[3]{16} = \sqrt[3]{8 \cdot 2} = 2\sqrt[3]{2}$

53. $\frac{\sqrt[4]{64x^7}}{\sqrt[4]{2x^2}} = \sqrt[4]{\frac{64x^7}{2x^2}} = \sqrt[4]{32x^5}$

$= \sqrt[4]{16 \cdot 2 \cdot x^4 \cdot x}$

$= \sqrt[4]{16x^4 \cdot 2x} = 2x\sqrt[4]{2x}$

54. $\frac{\sqrt{200x^3y^2}}{\sqrt{2x^{-2}y}} = \sqrt{\frac{200x^3y^2}{2x^{-2}y}} = \sqrt{100x^5y}$

$= \sqrt{100x^4xy} = 10x^2\sqrt{xy}$

55. $\sqrt{3}\left(2\sqrt{6} + 4\sqrt{15}\right) = 2\sqrt{18} + 4\sqrt{45}$

$= 2\sqrt{9 \cdot 2} + 4\sqrt{9 \cdot 5}$

$= 2 \cdot 3\sqrt{2} + 4 \cdot 3\sqrt{5}$

$= 6\sqrt{2} + 12\sqrt{5}$

56. $\sqrt[3]{5}\left(\sqrt[3]{50} - \sqrt[3]{2}\right) = \sqrt[3]{250} - \sqrt[3]{10}$

$= \sqrt[3]{125 \cdot 2} - \sqrt[3]{10}$

$= 5\sqrt[3]{2} - \sqrt[3]{10}$

57. $\left(\sqrt{7} - 3\sqrt{5}\right)\left(\sqrt{7} + 6\sqrt{5}\right)$

$= 7 + 6\sqrt{35} - 3\sqrt{35} - 18 \cdot 5$

$= 7 + 3\sqrt{35} - 90$

$= 3\sqrt{35} - 83$ or $-83 + 3\sqrt{35}$

58. $\left(\sqrt{x} - \sqrt{11}\right)\left(\sqrt{y} - \sqrt{11}\right)$

$= \sqrt{xy} - \sqrt{11x} - \sqrt{11y} + 11$

59. $\left(\sqrt{5} + \sqrt{8}\right)^2 = 5 + 2 \cdot \sqrt{5} \cdot \sqrt{8} + 8$

$= 13 + 2\sqrt{40}$

$= 13 + 2\sqrt{4 \cdot 10}$

$= 13 + 2 \cdot 2\sqrt{10}$

$= 13 + 4\sqrt{10}$

60. $\left(2\sqrt{3} - \sqrt{10}\right)^2$

$= 4 \cdot 3 - 2 \cdot 2\sqrt{3} \cdot \sqrt{10} + 10$

$= 12 - 4\sqrt{30} + 10 = 22 - 4\sqrt{30}$

61. $\left(\sqrt{7} + \sqrt{13}\right)\left(\sqrt{7} - \sqrt{13}\right)$

$= \left(\sqrt{7}\right)^2 - \left(\sqrt{13}\right)^2 = 7 - 13 = -6$

62. $\left(7 - 3\sqrt{5}\right)\left(7 + 3\sqrt{5}\right) = 7^2 - \left(3\sqrt{5}\right)^2$

$= 49 - 9 \cdot 5$

$= 49 - 45 = 4$

63. $\frac{4}{\sqrt{6}} = \frac{4}{\sqrt{6}} \cdot \frac{\sqrt{6}}{\sqrt{6}} = \frac{4\sqrt{6}}{6} = \frac{2\sqrt{6}}{3}$

64. $\sqrt{\frac{2}{7}} = \frac{\sqrt{2}}{\sqrt{7}} = \frac{\sqrt{2}}{\sqrt{7}} \cdot \frac{\sqrt{7}}{\sqrt{7}} = \frac{\sqrt{14}}{7}$

65. $\frac{12}{\sqrt[3]{9}} = \frac{12}{\sqrt[3]{3^2}} \cdot \frac{\sqrt[3]{3}}{\sqrt[3]{3}} = \frac{12\sqrt[3]{3}}{\sqrt[3]{3^3}}$

$= \frac{12\sqrt[3]{3}}{3} = 4\sqrt[3]{3}$

66. $\sqrt{\frac{2x}{5y}} = \frac{\sqrt{2x}}{\sqrt{5y}} \cdot \frac{\sqrt{5y}}{\sqrt{5y}} = \frac{\sqrt{10xy}}{\sqrt{5^2y^2}} = \frac{\sqrt{10xy}}{5y}$

67. $\frac{14}{\sqrt[3]{2x^2}} = \frac{14}{\sqrt[3]{2x^2}} \cdot \frac{\sqrt[3]{2^2x}}{\sqrt[3]{2^2x}} = \frac{14\sqrt[3]{2^2x}}{\sqrt[3]{2^3x^3}}$

$= \frac{14\sqrt[3]{4x}}{2x} = \frac{7\sqrt[3]{4x}}{x}$

68. $\sqrt[4]{\frac{7}{3x}}=\frac{\sqrt[4]{7}}{\sqrt[4]{3x}}=\frac{\sqrt[4]{7}}{\sqrt[4]{3x}}\cdot\frac{\sqrt[4]{3^3x^3}}{\sqrt[4]{3^3x^3}}$

$=\frac{\sqrt[4]{7\cdot3^3x^3}}{\sqrt[4]{3^4x^4}}=\frac{\sqrt[4]{7\cdot27x^3}}{3x}$

$=\frac{\sqrt[4]{189x^3}}{3x}$

69. $\frac{5}{\sqrt[5]{32x^4y}}=\frac{5}{\sqrt[5]{2^5x^4y}}\cdot\frac{\sqrt[5]{xy^4}}{\sqrt[5]{xy^4}}$

$=\frac{5\sqrt[5]{xy^4}}{\sqrt[5]{2^5x^5y^5}}=\frac{5\sqrt[5]{xy^4}}{2xy}$

70. $\frac{6}{\sqrt{3}-1}=\frac{6}{\sqrt{3}-1}\cdot\frac{\sqrt{3}+1}{\sqrt{3}+1}$

$=\frac{6\left(\sqrt{3}+1\right)}{\left(\sqrt{3}\right)^2-1^2}=\frac{6\left(\sqrt{3}+1\right)}{3-1}$

$=\frac{6\left(\sqrt{3}+1\right)}{2}=3\left(\sqrt{3}+1\right)$

$=3\sqrt{3}+3$

71. $\frac{\sqrt{7}}{\sqrt{5}+\sqrt{3}}=\frac{\sqrt{7}}{\sqrt{5}+\sqrt{3}}\cdot\frac{\sqrt{5}-\sqrt{3}}{\sqrt{5}-\sqrt{3}}$

$=\frac{\sqrt{35}-\sqrt{21}}{\left(\sqrt{5}\right)^2-\left(\sqrt{3}\right)^2}$

$=\frac{\sqrt{35}-\sqrt{21}}{5-3}=\frac{\sqrt{35}-\sqrt{21}}{2}$

72. $\frac{10}{2\sqrt{5}-3\sqrt{2}}$

$=\frac{10}{2\sqrt{5}-3\sqrt{2}}\cdot\frac{2\sqrt{5}+3\sqrt{2}}{2\sqrt{5}+3\sqrt{2}}$

$=\frac{10\left(2\sqrt{5}+3\sqrt{2}\right)}{\left(2\sqrt{5}\right)^2-\left(3\sqrt{2}\right)^2}=\frac{10\left(2\sqrt{5}+3\sqrt{2}\right)}{4\cdot5-9\cdot2}$

$=\frac{10\left(2\sqrt{5}+3\sqrt{2}\right)}{20-18}=\frac{10\left(2\sqrt{5}+3\sqrt{2}\right)}{2}$

$=5\left(2\sqrt{5}+3\sqrt{2}\right)=10\sqrt{5}+15\sqrt{2}$

73. $\frac{\sqrt{x}+5}{\sqrt{x}-3}=\frac{\sqrt{x}+5}{\sqrt{x}-3}\cdot\frac{\sqrt{x}+3}{\sqrt{x}+3}$

$=\frac{x+3\sqrt{x}+5\sqrt{x}+15}{\left(\sqrt{x}\right)^2-3^2}$

$=\frac{x+8\sqrt{x}+15}{x-9}$

74. $\frac{\sqrt{7}+\sqrt{3}}{\sqrt{7}-\sqrt{3}}=\frac{\sqrt{7}+\sqrt{3}}{\sqrt{7}-\sqrt{3}}\cdot\frac{\sqrt{7}+\sqrt{3}}{\sqrt{7}+\sqrt{3}}$

$=\frac{7+2\cdot\sqrt{7}\cdot\sqrt{3}+3}{\left(\sqrt{7}\right)^2-\left(\sqrt{3}\right)^2}$

$=\frac{10+2\sqrt{21}}{7-3}=\frac{10+2\sqrt{21}}{4}$

$=\frac{2\left(5+\sqrt{21}\right)}{4}=\frac{5+\sqrt{21}}{2}$

75. $\frac{2\sqrt{3}+\sqrt{6}}{2\sqrt{6}+\sqrt{3}}=\frac{2\sqrt{3}+\sqrt{6}}{2\sqrt{6}+\sqrt{3}}\cdot\frac{2\sqrt{6}-\sqrt{3}}{2\sqrt{6}-\sqrt{3}}$

$=\frac{4\sqrt{18}-2\cdot3+2\cdot6-\sqrt{18}}{\left(2\sqrt{6}\right)^2-\left(\sqrt{3}\right)^2}$

$=\frac{3\sqrt{18}-6+12}{4\cdot6-3}=\frac{3\sqrt{9\cdot2}+6}{24-3}$

$=\frac{3\cdot3\sqrt{2}+6}{21}=\frac{9\sqrt{2}+6}{21}$

$=\frac{3\left(3\sqrt{2}+2\right)}{21}=\frac{3\sqrt{2}+2}{7}$

76. $\sqrt{\frac{2}{7}}=\frac{\sqrt{2}}{\sqrt{7}}=\frac{\sqrt{2}}{\sqrt{7}}\cdot\frac{\sqrt{2}}{\sqrt{2}}=\frac{2}{\sqrt{14}}$

77. $\frac{\sqrt[3]{3x}}{\sqrt[3]{y}}=\frac{\sqrt[3]{3x}}{\sqrt[3]{y}}\cdot\frac{\sqrt[3]{3^2x^2}}{\sqrt[3]{3^2x^2}}$

$=\frac{\sqrt[3]{3^3x^3}}{\sqrt[3]{3^2x^2y}}=\frac{3x}{\sqrt[3]{9x^2y}}$

78. $\frac{\sqrt{7}}{\sqrt{5}+\sqrt{3}}=\frac{\sqrt{7}}{\sqrt{5}+\sqrt{3}}\cdot\frac{\sqrt{7}}{\sqrt{7}}$

$=\frac{7}{\sqrt{35}+\sqrt{21}}$

79. $\dfrac{\sqrt{7}+\sqrt{3}}{\sqrt{7}-\sqrt{3}}$

$$=\frac{\sqrt{7}+\sqrt{3}}{\sqrt{7}-\sqrt{3}}\cdot\frac{\sqrt{7}-\sqrt{3}}{\sqrt{7}-\sqrt{3}}$$

$$=\frac{\left(\sqrt{7}\right)^2-\left(\sqrt{3}\right)^2}{7-2\sqrt{7}\sqrt{3}+3}=\frac{7-3}{10-2\sqrt{21}}$$

$$=\frac{4}{10-2\sqrt{21}}=\frac{4}{2\left(5-\sqrt{21}\right)}$$

$$=\frac{2}{5-\sqrt{21}}$$

80.

$$\sqrt{2x+4}=6$$
$$\left(\sqrt{2x+4}\right)^2=6^2$$
$$2x+4=36$$
$$2x=32$$
$$x=16$$

The solution checks. The solution set is $\{16\}$.

81. $\sqrt{x-5}+9=4$

$$\sqrt{x-5}=-5$$

The square root of a number is always nonnegative.
The solution set is $\varnothing$ or $\{\ \}$.

82. $\sqrt{2x-3}+x=3$

$$\sqrt{2x-3}=3-x$$
$$\left(\sqrt{2x-3}\right)^2=(3-x)^2$$
$$2x-3=9-6x+x^2$$
$$0=12-8x+x^2$$
$$0=x^2-8x+12$$
$$0=(x-6)(x-2)$$
$$x-6=0 \qquad x-2=0$$
$$x=6 \qquad x=2$$

6 is an extraneous solution. The solution set is $\{2\}$.

83. $\sqrt{x-4}+\sqrt{x+1}=5$

$$\sqrt{x-4}=5-\sqrt{x+1}$$
$$\left(\sqrt{x-4}\right)^2=\left(5-\sqrt{x+1}\right)^2$$
$$x-4=25-10\sqrt{x+1}+x+1$$
$$-30=-10\sqrt{x+1}$$
$$\frac{-30}{-10}=\frac{-10\sqrt{x+1}}{-10}$$
$$3=\sqrt{x+1}$$
$$3^2=\left(\sqrt{x+1}\right)^2$$
$$9=x+1$$
$$8=x$$

The solution checks. The solution set is $\{8\}$.

84. $\left(x^2+6x\right)^{\frac{1}{3}}+2=0$

$$\left(x^2+6x\right)^{\frac{1}{3}}=-2$$
$$\sqrt[3]{x^2+6x}=-2$$
$$\left(\sqrt[3]{x^2+6x}\right)^3=(-2)^3$$
$$x^2+6x=-8$$
$$x^2+6x+8=0$$
$$(x+4)(x+2)=0$$
$$x+4=0 \qquad x+2=0$$
$$x=-4 \qquad x=-2$$

Both solutions check. The solution set is $\{-4,-2\}$.

85. a. $f(x)=3.5\sqrt{x}+15$

$$f(30)=3.5\sqrt{30}+15$$
$$\approx 34$$

In 2010, approximately 34% of American adults were obese; This value is the same as that in the graph.

b. $f(x) = 3.5\sqrt{x} + 15$
$$f(x) = 3.5\sqrt{x} + 15$$
$$36 = 3.5\sqrt{x} + 15$$
$$21 = 3.5\sqrt{x}$$
$$\frac{21}{3.5} = \frac{3.5\sqrt{x}}{3.5}$$
$$6 = \sqrt{x}$$
$$6^2 = \left(\sqrt{x}\right)^2$$
$$36 = x$$
According to the model, 36% of adults will be obese 36 years after 1980, or 2016.

86. $20,000 = 5000\sqrt{100-x}$
$$\frac{20,000}{5000} = \frac{5000\sqrt{100-x}}{5000}$$
$$4 = \sqrt{100-x}$$
$$4^2 = \left(\sqrt{100-x}\right)^2$$
$$16 = 100 - x$$
$$-84 = -x$$
$$84 = x$$

20,000 people in the group will survive to 84 years old.

87. $\sqrt{-81} = \sqrt{81 \cdot -1} = \sqrt{81}\sqrt{-1} = 9i$

88. $\sqrt{-63} = \sqrt{9 \cdot 7 \cdot -1}$
$$= \sqrt{9}\sqrt{7}\sqrt{-1} = 3i\sqrt{7}$$

89. $-\sqrt{-8} = -\sqrt{4 \cdot 2 \cdot -1}$
$$= -\sqrt{4}\sqrt{2}\sqrt{-1} = -2i\sqrt{2}$$

90. $(7+12i)+(5-10i)$
$$= 7 + 12i + 5 - 10i = 12 + 2i$$

91. $(8-3i)-(17-7i) = 8 - 3i - 17 + 7i$
$$= -9 + 4i$$

92. $4i(3i-2) = 4i \cdot 3i - 4i \cdot 2$
$$= 12i^2 - 8i$$
$$= 12(-1) - 8i$$
$$= -12 - 8i$$

93. $(7-5i)(2+3i) = 14 + 21i - 10i - 15i^2$
$$= 14 + 11i - 15(-1)$$
$$= 14 + 11i + 15$$
$$= 29 + 11i$$

94. $(3-4i)^2 = 3^2 - 2 \cdot 3 \cdot 4i + (4i)^2$
$$= 9 - 24i + 16i^2$$
$$= 9 - 24i + 16(-1)$$
$$= 9 - 24i - 16$$
$$= -7 - 24i$$

95. $(7+8i)(7-8i) = 7^2 - (8i)^2$
$$= 49 - 64i^2$$
$$= 49 - 64(-1)$$
$$= 49 + 64$$
$$= 113 \text{ or } 113 + 0i$$

96. $\sqrt{-8} \cdot \sqrt{-3} = \sqrt{4 \cdot 2 \cdot -1} \cdot \sqrt{3 \cdot -1}$
$$= 2\sqrt{2}i \cdot \sqrt{3}i = 2\sqrt{6}i^2$$
$$= 2\sqrt{6}(-1) = -2\sqrt{6}$$
$$= -2\sqrt{6} \text{ or } -2\sqrt{6} + 0i$$

97. $\dfrac{6}{5+i} = \dfrac{6}{5+i} \cdot \dfrac{5-i}{5-i} = \dfrac{30-6i}{25-i^2}$
$$= \frac{30-6i}{25-(-1)} = \frac{30-6i}{25+1}$$
$$= \frac{30-6i}{26} = \frac{30}{26} - \frac{6}{26}i$$
$$= \frac{15}{13} - \frac{3}{13}i$$

98. $\dfrac{3+4i}{4-2i} = \dfrac{3+4i}{4-2i} \cdot \dfrac{4+2i}{4+2i}$
$$= \frac{12 + 6i + 16i + 8i^2}{16 - 4i^2}$$
$$= \frac{12 + 22i + 8(-1)}{16 - 4(-1)}$$
$$= \frac{12 + 22i - 8}{16 + 4} = \frac{4 + 22i}{20}$$
$$= \frac{4}{20} + \frac{22}{20}i = \frac{1}{5} + \frac{11}{10}i$$

99. $\frac{5+i}{3i}=\frac{5+i}{3i}\cdot\frac{i}{i}=\frac{5i+i^2}{3i^2}$
$=\frac{5i+(-1)}{3(-1)}=\frac{5i-1}{-3}$
$=\frac{-1}{-3}+\frac{5}{-3}i=\frac{1}{3}-\frac{5}{3}i$

100. $i^{16}=\left(i^2\right)^8=(-1)^8=1$

101. $i^{23}=i^{22}\cdot i=\left(i^2\right)^{11}i=(-1)^{11}i=(-1)i=-i$

Chapter 7 Test

1. a. $f(-14)=\sqrt{8-2(-14)}$
$=\sqrt{8+28}=\sqrt{36}=6$

b. To find the domain, set the radicand greater than or equal to zero and solve the resulting inequality.
$8-2x\geq 0$
$-2x\geq -8$
$x\leq 4$
The domain of f is $(-\infty,4]$.

2. $27^{-\frac{4}{3}}=\frac{1}{27^{\frac{4}{3}}}=\frac{1}{\left(\sqrt[3]{27}\right)^4}=\frac{1}{(3)^4}=\frac{1}{81}$

3. $\left(25x^{-\frac{1}{2}}y^{\frac{1}{4}}\right)^{\frac{1}{2}}=25^{\frac{1}{2}}x^{-\frac{1}{4}}y^{\frac{1}{8}}$
$=5x^{-\frac{1}{4}}y^{\frac{1}{8}}$
$=\frac{5y^{\frac{1}{8}}}{x^{\frac{1}{4}}}$

4. $\sqrt[8]{x^4}=\left(x^4\right)^{\frac{1}{8}}=x^{4\cdot\frac{1}{8}}=x^{\frac{1}{2}}=\sqrt{x}$

5. $\sqrt[4]{x}\cdot\sqrt[5]{x}=x^{\frac{1}{4}}\cdot x^{\frac{1}{5}}=x^{\frac{1}{4}+\frac{1}{5}}=x^{\frac{5}{20}+\frac{4}{20}}$
$=x^{\frac{9}{20}}=\sqrt[20]{x^9}$

6. $\sqrt{75x^2}=\sqrt{25\cdot 3x^2}=5|x|\sqrt{3}$

7. $\sqrt{x^2-10x+25}=\sqrt{(x-5)^2}$
$=|x-5|$

8. $\sqrt[3]{16x^4y^8}=\sqrt[3]{8\cdot 2\cdot x^3\cdot x\cdot y^6\cdot y^2}$
$=\sqrt[3]{8x^3y^6\cdot 2xy^2}$
$=2xy^2\sqrt[3]{2xy^2}$

9. $\sqrt[5]{-\frac{32}{x^{10}}}=\sqrt[5]{-\frac{2^5}{\left(x^2\right)^5}}=-\frac{2}{x^2}$

10. $\sqrt[3]{5x^2}\cdot\sqrt[3]{10y}=\sqrt[3]{50x^2y}$

11. $\sqrt[4]{8x^3y}\cdot\sqrt[4]{4xy^2}=\sqrt[4]{32x^4y^3}$
$=\sqrt[4]{16\cdot 2\cdot x^4\cdot y^3}$
$=\sqrt[4]{16x^4\cdot 2y^3}$
$=2x\sqrt[4]{2y^3}$

12. $3\sqrt{18}-4\sqrt{32}=3\sqrt{9\cdot 2}-4\sqrt{16\cdot 2}$
$=3\cdot 3\sqrt{2}-4\cdot 4\sqrt{2}$
$=9\sqrt{2}-16\sqrt{2}=-7\sqrt{2}$

13. $\sqrt[3]{8x^4}+\sqrt[3]{xy^6}=\sqrt[3]{8x^3\cdot x}+\sqrt[3]{xy^6}$
$=2x\sqrt[3]{x}+y^2\sqrt[3]{x}$
$=\left(2x+y^2\right)\sqrt[3]{x}$

14. $\frac{\sqrt[3]{16x^8}}{\sqrt[3]{2x^4}}=\sqrt[3]{\frac{16x^8}{2x^4}}=\sqrt[3]{8x^4}$
$=\sqrt[3]{8x^3\cdot x}=2x\sqrt[3]{x}$

15. $\sqrt{3}\left(4\sqrt{6}-\sqrt{5}\right)=\sqrt{3}\cdot 4\sqrt{6}-\sqrt{3}\cdot\sqrt{5}$
$=4\sqrt{18}-\sqrt{15}$
$=4\sqrt{9\cdot 2}-\sqrt{15}$
$=4\cdot 3\sqrt{2}-\sqrt{15}$
$=12\sqrt{2}-\sqrt{15}$

16. $\left(5\sqrt{6}-2\sqrt{2}\right)\left(\sqrt{6}+\sqrt{2}\right)$
$= 5\cdot 6+5\sqrt{12}-2\sqrt{12}-2\cdot 2$
$= 30+3\sqrt{12}-4 = 26+3\sqrt{4\cdot 3}$
$= 26+3\cdot 2\sqrt{3} = 26+6\sqrt{3}$

17. $\left(7-\sqrt{3}\right)^2 = 49-2\cdot 7\cdot\sqrt{3}+3$
$= 52-14\sqrt{3}$

18. $\sqrt{\frac{5}{x}} = \frac{\sqrt{5}}{\sqrt{x}}\cdot\frac{\sqrt{x}}{\sqrt{x}} = \frac{\sqrt{5x}}{x}$

19. $\frac{5}{\sqrt[3]{5x^2}} = \frac{5}{\sqrt[3]{5x^2}}\cdot\frac{\sqrt[3]{5^2x}}{\sqrt[3]{5^2x}} = \frac{5\sqrt[3]{5^2x}}{\sqrt[3]{5^3x^3}}$
$= \frac{5\sqrt[3]{25x}}{5x} = \frac{\sqrt[3]{25x}}{x}$

20. $\frac{\sqrt{2}-\sqrt{3}}{\sqrt{2}+\sqrt{3}} = \frac{\sqrt{2}-\sqrt{3}}{\sqrt{2}+\sqrt{3}}\cdot\frac{\sqrt{2}-\sqrt{3}}{\sqrt{2}-\sqrt{3}}$
$= \frac{2-2\sqrt{2}\sqrt{3}+3}{2-3}$
$= \frac{5-2\sqrt{6}}{-1} = -5+2\sqrt{6}$

21. $3+\sqrt{2x-3} = x$
$\sqrt{2x-3} = x-3$
$\left(\sqrt{2x-3}\right)^2 = (x-3)^2$
$2x-3 = x^2-6x+9$
$0 = x^2-8x+12$
$0 = (x-6)(x-2)$
$x-6=0 \qquad x-2=0$
$x=6 \qquad x=2$
2 is an extraneous solution. The solution set is $\{6\}$.

22. $\sqrt{x+9}-\sqrt{x-7} = 2$
$\sqrt{x+9} = 2+\sqrt{x-7}$
$\left(\sqrt{x+9}\right)^2 = \left(2+\sqrt{x-7}\right)^2$
$x+9 = 4+2\cdot 2\cdot\sqrt{x-7}+x-7$
$x+9 = 4\sqrt{x-7}+x-3$
$12 = 4\sqrt{x-7}$
$3 = \sqrt{x-7}$
$3^2 = \left(\sqrt{x-7}\right)^2$
$9 = x-7$
$16 = x$
The solution set is $\{16\}$.

23. $(11x+6)^{\frac{1}{3}}+3 = 0$
$(11x+6)^{\frac{1}{3}} = -3$
$\sqrt[3]{11x+6} = -3$
$\left(\sqrt[3]{11x+6}\right)^3 = (-3)^3$
$11x+6 = -27$
$11x = -33$
$x = -3$
The solution set is $\{-3\}$.

24. $40.4 = 2.9\sqrt{x}+20.1$
$20.3 = 2.9\sqrt{x}$
$7 = \sqrt{x}$
$7^2 = \left(\sqrt{x}\right)^2$
$49 = x$
Boys who are 49 months of age have an average height of 40.4 inches.

25. $\sqrt{-75} = \sqrt{25\cdot 3\cdot -1}$
$= \sqrt{25}\cdot\sqrt{3}\cdot\sqrt{-1} = 5i\sqrt{3}$

26. $(5-3i)-(6-9i) = 5-3i-6+9i$
$= 5-6-3i+9i$
$= -1+6i$

27. $(3-4i)(2+5i) = 6+15i-8i-20i^2$
$= 6+7i-20(-1)$
$= 6+7i+20$
$= 26+7i$

28. $\sqrt{-9}\cdot\sqrt{-4}=\sqrt{9\cdot -1}\cdot\sqrt{4\cdot -1}$
$=\sqrt{9}\cdot\sqrt{-1}\cdot\sqrt{4}\cdot\sqrt{-1}$
$=3\cdot i\cdot 2\cdot i=6i^2=6(-1)$
$=-6 \text{ or } -6+0i$

29. $\dfrac{3+i}{1-2i}=\dfrac{3+i}{1-2i}\cdot\dfrac{1+2i}{1+2i}$
$=\dfrac{3+6i+i+2i^2}{1-4i^2}$
$=\dfrac{3+7i+2(-1)}{1-4(-1)}$
$=\dfrac{3+7i-2}{1+4}$
$=\dfrac{1+7i}{5}=\dfrac{1}{5}+\dfrac{7}{5}i$

30. $i^{35}=i^{34}\cdot i=\left(i^2\right)^{17}\cdot i$
$=(-1)^{17}\cdot i$
$=(-1)i=-i$

Cumulative Review Exercises (Chapters 1 – 7)

1. $2x-y+z=-5$
$x-2y-3z=6$
$x+y-2z=1$
Add the first and third equations to eliminate y.

$2x-y+z=-5$
$x+y-2z=1$
$3x \quad -z=-4$
Multiply the third equation by 2 and add to the second equation.

$x-2y-3z=6$
$2x+2y-4z=2$
$3x \quad -7z=8$
We now have a system of two equations in two variables.

$3x-z=-4$
$3x-7z=8$
Multiply the first equation by –1 and add to the second equation.

$-3x+z=4$
$3x-7z=8$
$-6z=12$
$z=-2$
Back-substitute –2 for z to find x.

$-3x+z=4$
$-3x-2=4$
$-3x=6$
$x=-2$
Back-substitute –2 for x and z in one of the original equations to find y.

$2x-y+z=-5$
$2(-2)-y-2=-5$
$-4-y-2=-5$
$-y-6=-5$
$-y=1$
$y=-1$
The solution is $(-2,-1,-2)$ or the solution set is $\{(-2,-1,-2)\}$.

2. $3x^2-11x=4$
$3x^2-11x-4=0$
$(3x+1)(x-4)=0$
Apply the zero product principle.

$3x+1=0 \qquad x-4=0$
$3x=-1 \qquad x=4$
$x=-\dfrac{1}{3}$

The solution set is $\left\{-\dfrac{1}{3},4\right\}$.

3. $2(x+4)<5x+3(x+2)$
$2x+8<5x+3x+6$
$2x+8<8x+6$
$-6x+8<6$
$-6x<-2$
$\dfrac{-6x}{-6}>\dfrac{-2}{-6}$
$x>\dfrac{1}{3}$

The solution set is $\left(\dfrac{1}{3},\infty\right)$.

4. $$\frac{1}{x+2}+\frac{15}{x^2-4}=\frac{5}{x-2}$$

$$\frac{1}{x+2}+\frac{15}{(x+2)(x-2)}=\frac{5}{x-2}$$

So that denominators will not equal zero, x cannot equal 2 or –2. To eliminate fractions, multiply by the LCD, $(x+2)(x-2)$.

$$(x+2)(x-2)\left(\frac{1}{x+2}+\frac{15}{(x+2)(x-2)}\right)=(x+2)(x-2)\left(\frac{5}{x-2}\right)$$

$$(x+2)(x-2)\left(\frac{1}{x+2}\right)+(x+2)(x-2)\left(\frac{15}{(x+2)(x-2)}\right)=(x+2)(5)$$

$$x-2+15=5x+10$$
$$x+13=5x+10$$
$$-4x+13=10$$
$$-4x=-3$$
$$x=\frac{3}{4}$$

The solution set is $\left\{\frac{3}{4}\right\}$.

5. $\sqrt{x+2}-\sqrt{x+1}=1$

$$\sqrt{x+2}=1+\sqrt{x+1}$$
$$\left(\sqrt{x+2}\right)^2=\left(1+\sqrt{x+1}\right)^2$$
$$x+2=1+2\sqrt{x+1}+x+1$$
$$x+2=2+2\sqrt{x+1}+x$$

$$0^2=\left(2\sqrt{x+1}\right)^2$$
$$0=4(x+1)$$
$$0=4x+4$$
$$-4=4x$$
$$-1=x$$

The solution checks. The solution set is $\{-1\}$.

6. $x+2y<2$

$2y-x>4$

First consider $x+2y<2$. Replace the inequality symbol with an equal sign and we have $x+2y=2$. Solve for y to put the equation in slope-intercept form.

$$x+2y=2$$
$$2y=-x+2$$
$$y=-\frac{1}{2}x+1$$

slope $= -\frac{1}{2}$ y–intercept = 1

Now, use the origin as a test point.

$$\begin{aligned} 0+2(0) &< 2 \\ 0 &< 2 \end{aligned}$$

This is a true statement. This means that the point $(0,0)$ will fall in the shaded half-plane.

Next consider $2y-x>4$. Replace the inequality symbol with an equal sign and we have $2y-x=4$. Solve for y to put the equation in slope-intercept form.

$$\begin{aligned} 2y-x &= 4 \\ 2y &= x+4 \\ y &= \frac{1}{2}x+2 \end{aligned}$$

slope $= \frac{1}{2}$ y–intercept = 2

Now, use the origin as a test point.

$$\begin{aligned} 2(0)-0 &> 4 \\ 0 &> 4 \end{aligned}$$

This is a false statement. This means that the point $(0,0)$ will not fall in the shaded half-plane.

Next, graph each of the inequalities. The solution to the system is the intersection of the shaded half-planes.

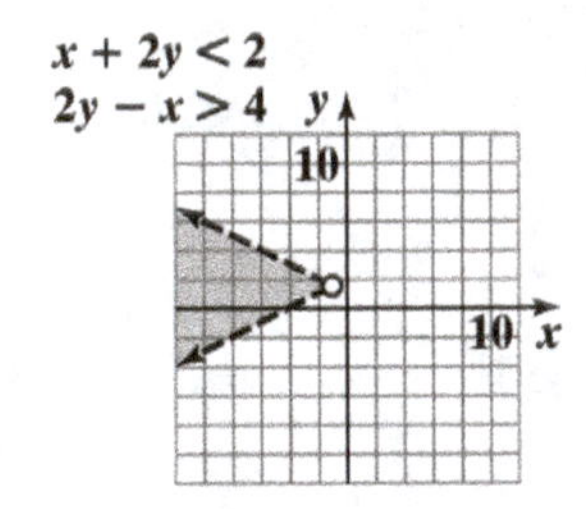

7. $\frac{8x^2}{3x^2-12} \div \frac{40}{x-2} \lim_{x\to\infty}$

$$= \frac{8x^2}{3x^2-12} \cdot \frac{x-2}{40}$$

$$= \frac{8x^2}{3(x^2-4)} \cdot \frac{x-2}{40}$$

$$= \frac{\overset{1}{\cancel{8}}x^2}{3(x+2)\cancel{(x-2)}} \cdot \frac{\cancel{x-2}}{\underset{5}{\cancel{40}}}$$

$$= \frac{x^2}{3\cdot 5(x+2)} = \frac{x^2}{15(x+2)}$$

8. $$\frac{x+\frac{1}{y}}{y+\frac{1}{x}}=\frac{xy}{xy}\cdot\frac{x+\frac{1}{y}}{y+\frac{1}{x}}=\frac{xy\cdot x+xy\cdot\frac{1}{y}}{xy\cdot y+xy\cdot\frac{1}{x}}$$
$$=\frac{x^2y+x}{xy^2+y}=\frac{x(xy+1)}{y(xy+1)}=\frac{x}{y}$$

9. $$(2x-3)(4x^2-5x-2)$$
$$=2x\cdot4x^2-2x\cdot5x-2x\cdot2-3\cdot4x^2$$
$$+3\cdot5x+3\cdot2$$
$$=8x^3-10x^2-4x-12x^2+15x+6$$
$$=8x^3-22x^2+11x+6$$

10. $$\frac{7x}{x^2-2x-15}-\frac{2}{x-5}$$
$$=\frac{7x}{(x-5)(x+3)}-\frac{2}{x-5}$$
$$=\frac{7x}{(x-5)(x+3)}-\frac{2(x+3)}{(x-5)(x+3)}$$
$$=\frac{7x-2(x+3)}{(x-5)(x+3)}=\frac{7x-2x-6}{(x-5)(x+3)}$$
$$=\frac{5x-6}{(x-5)(x+3)}$$

11. $$7(8-10)^3-7+3\div(-3)$$
$$=7(-2)^3-7+3\div(-3)$$
$$=7(-8)-7+3\div(-3)$$
$$=-56-7+(-1)=-64$$

12. $$\sqrt{80x}-5\sqrt{20x}+2\sqrt{45x}$$
$$=\sqrt{16\cdot5x}-5\sqrt{4\cdot5x}+2\sqrt{9\cdot5x}$$
$$=4\sqrt{5x}-5\cdot2\sqrt{5x}+2\cdot3\sqrt{5x}$$
$$=4\sqrt{5x}-10\sqrt{5x}+6\sqrt{5x}=0$$

13. $$\frac{\sqrt{3}-2}{2\sqrt{3}+5}=\frac{\sqrt{3}-2}{2\sqrt{3}+5}\cdot\frac{2\sqrt{3}-5}{2\sqrt{3}-5}$$
$$=\frac{2\cdot3-5\sqrt{3}-4\sqrt{3}+10}{4\cdot3-25}$$
$$=\frac{6-9\sqrt{3}+10}{12-25}$$
$$=\frac{16-9\sqrt{3}}{-13}$$
$$=-\frac{16-9\sqrt{3}}{13}$$

14.
$$\begin{array}{r}
2x^2+\ x+5 \\
x-2\,\overline{)\,2x^3-3x^2+3x-4} \\
\underline{2x^3-4x^2} \\
x^2+3x \\
\underline{x^2-2x} \\
5x-\ 4 \\
\underline{5x-10} \\
6
\end{array}$$

$$\frac{2x^3-3x^2+3x-4}{x-2}=2x^2+x+5+\frac{6}{x-2}$$

15. $$(2\sqrt{3}+5\sqrt{2})(\sqrt{3}-4\sqrt{2})$$
$$=2\cdot3-8\sqrt{6}+5\sqrt{6}-20\cdot2$$
$$=6-3\sqrt{6}-40=-34-3\sqrt{6}$$

16. $$24x^2+10x-4=2(12x^2+5x-2)$$
$$=2(3x+2)(4x-1)$$

17. $$16x^4-1=(4x^2+1)(4x^2-1)$$
$$=(4x^2+1)(2x+1)(2x-1)$$

18. Since light varies inversely as the square of the distance, we have $l = \frac{k}{d^2}$.

Use the given values to find k.

$$\begin{aligned} l &= \frac{k}{d^2} \\ 120 &= \frac{k}{10^2} \\ 120 &= \frac{k}{100} \\ 12{,}000 &= k \end{aligned}$$

The equation becomes $l = \frac{12{,}000}{d^2}$. When $d = 15$,

$l = \frac{12{,}000}{15^2} \approx 53.3$.

At a distance of 15 feet, approximately 53 lumens are provided.

19. Let x = the amount invested at 7%.
Let y = the amount invested at 9%.

$$\begin{aligned} x + \quad y &= 6000 \\ 0.07x + 0.09y &= 510 \end{aligned}$$

Solve the first equation for y.

$$\begin{aligned} x + y &= 6000 \\ y &= 6000 - x \end{aligned}$$

Substitute and solve.

$$\begin{aligned} 0.07x + 0.09(6000 - x) &= 510 \\ 0.07x + 540 - 0.09x &= 510 \\ 540 - 0.02x &= 510 \\ -0.02x &= -30 \\ x &= 1500 \end{aligned}$$

Back-substitute 1500 for x to find y.

$$\begin{aligned} y &= 6000 - x \\ y &= 6000 - 1500 = 4500 \end{aligned}$$

\$1500 was invested at 7% and \$4500 was invested at 9%.

20. Let x = the number of students enrolled last year.

$$\begin{aligned} x - 0.12x &= 2332 \\ 0.88x &= 2332 \\ x &= 2650 \end{aligned}$$

2650 students were enrolled last year.

Chapter 8
Quadratic Equations and Functions

8.1 Check Points

1. $4x^2 = 28$

   $x^2 = 7$

   Apply the square root property.

   $x = \pm\sqrt{7}$

   The solution set is $\{\pm\sqrt{7}\}$.

2. $3x^2 - 11 = 0$

   $3x^2 = 11$

   $x^2 = \frac{11}{3}$

   $x = \pm\sqrt{\frac{11}{3}}$

   $x = \pm\frac{\sqrt{11}}{\sqrt{3}}\cdot\frac{\sqrt{3}}{\sqrt{3}}$

   $x = \pm\frac{\sqrt{33}}{3}$

   The solution set is $\left\{\pm\frac{\sqrt{33}}{3}\right\}$.

3. $4x^2 + 9 = 0$

   $4x^2 = -9$

   $x^2 = -\frac{9}{4}$

   $x = \pm\sqrt{-\frac{9}{4}}$

   $x = \pm\frac{\sqrt{-9}}{\sqrt{4}}$

   $x = \pm\frac{3}{2}i$

   The solution set is $\left\{\pm\frac{3}{2}i\right\}$.

4. $(x-3)^2 = 10$

   $x - 3 = \sqrt{10}$ or $x - 3 = -\sqrt{10}$

   $x = 3 + \sqrt{10}$ $\quad$ $x = 3 - \sqrt{10}$

   The solution set is $\{3 \pm \sqrt{10}\}$.

5. **a.** $x^2 + 10x$

   The coefficient of the x-term is 10.

   Half of 10 is 5, and 5^2 is 25, which should be added to the binomial.

   The result is a perfect square trinomial.

   $x^2 + 10x + 25 = (x+5)^2$

   b. $x^2 - 3x$

   The coefficient of the x-term is -3.

   Half of -3 is $-\frac{3}{2}$, and $\left(-\frac{3}{2}\right)^2$ is $\frac{9}{4}$ which should be added to the binomial.

   The result is a perfect square trinomial.

   $x^2 - 3x + \frac{9}{4} = \left(x - \frac{3}{2}\right)^2$

   c. $x^2 + \frac{3}{4}x$

   The coefficient of the x-term is $\frac{3}{4}$.

   Half of $\frac{3}{4}$ is $\frac{3}{8}$, and $\left(\frac{3}{8}\right)^2$ is $\frac{9}{64}$ which should be added to the binomial.

   The result is a perfect square trinomial.

   $x^2 + \frac{3}{4}x + \frac{9}{64} = \left(x + \frac{3}{8}\right)^2$

6. $x^2 + 4x - 1 = 0$

   $x^2 + 4x \quad = 1$

   Half of 4 is 2, and 2^2 is 4, which should be added to both sides.

   $x^2 + 4x + 4 = 1 + 4$

   $x^2 + 4x + 4 = 5$

   $(x+2)^2 = 5$

   $x + 2 = \sqrt{5}$ or $x + 2 = -\sqrt{5}$

   $x = -2 + \sqrt{5}$ $\quad$ $x = -2 - \sqrt{5}$

   The solution set is $\{-2 \pm \sqrt{5}\}$.

7. $2x^2+3x-4=0$

$$\frac{2x^2}{2}+\frac{3x}{2}-\frac{4}{2}=\frac{0}{2}$$

$$x^2+\frac{3}{2}x-2=0$$

$$x^2+\frac{3}{2}x \quad =2$$

Half of $\frac{3}{2}$ is $\frac{3}{4}$, and $\left(\frac{3}{4}\right)^2$ is $\frac{9}{16}$, which should be added to both sides.

$$x^2+\frac{3}{2}x+\frac{9}{16}=2+\frac{9}{16}$$

$$\left(x+\frac{3}{4}\right)^2=\frac{41}{16}$$

$$x+\frac{3}{4}=\sqrt{\frac{41}{16}} \quad \text{or} \quad x+\frac{3}{4}=-\sqrt{\frac{41}{16}}$$

$$x=-\frac{3}{4}+\sqrt{\frac{41}{16}} \qquad x=-\frac{3}{4}-\sqrt{\frac{41}{16}}$$

$$x=-\frac{3}{4}+\frac{\sqrt{41}}{4} \qquad x=-\frac{3}{4}+\frac{\sqrt{41}}{4}$$

$$x=\frac{-3+\sqrt{41}}{4} \qquad x=\frac{-3+\sqrt{41}}{4}$$

The solution set is $\left\{\frac{-3\pm\sqrt{41}}{4}\right\}$.

8. $3x^2-9x+8=0$

$$\frac{3x^2}{3}-\frac{9x}{3}+\frac{8}{3}=\frac{0}{3}$$

$$x^2-3x+\frac{8}{3}=0$$

$$x^2-3x \quad =-\frac{8}{3}$$

Half of -3 is $-\frac{3}{2}$, and $\left(-\frac{3}{2}\right)^2$ is $\frac{9}{4}$, which should be added to both sides.

$$x^2-3x+\frac{9}{4}=-\frac{8}{3}+\frac{9}{4}$$

$$\left(x-\frac{3}{2}\right)^2=-\frac{32}{12}+\frac{27}{12}$$

$$\left(x-\frac{3}{2}\right)^2=-\frac{5}{12}$$

$$x-\frac{3}{2}=\pm\sqrt{-\frac{5}{12}}$$

$$x=\frac{3}{2}\pm i\sqrt{\frac{15}{36}}$$

$$x=\frac{3}{2}\pm i\frac{\sqrt{15}}{6}$$

The solution set is $\left\{\frac{3}{2}\pm i\frac{\sqrt{15}}{6}\right\}$.

9. $A=P(1+r)^t$

$$4320=3000(1+r)^2$$

$$1.44=(1+r)^2$$

$$1+r=\sqrt{1.44} \quad \text{or} \quad 1+r=-\sqrt{1.44}$$

$$1+r=1.2 \qquad 1+r=-1.2$$

$$r=0.2 \qquad r=-2.2$$

Reject -2.2 because we cannot have a negative interest rate. The solution is 0.2; the annual interest rate is 20%.

10. $x^2+20^2=50^2$

$$x^2+400=2500$$

$$x^2=2100$$

$$x=\pm\sqrt{2100}$$

$$x=\pm 10\sqrt{21}$$

$$x\approx\pm 45.8$$

The wire is attached $10\sqrt{21}$ feet, or about 45.8 feet, up the antenna.

8.1 Concept and Vocabulary Check

1. $\pm\sqrt{d}$

2. $\pm\sqrt{7}$

3. $\pm\sqrt{\frac{11}{2}}$; $\pm\frac{\sqrt{22}}{2}$

4. $\pm 3i$

5. 25

6. $\frac{9}{4}$

7. $\frac{4}{25}$

8. 9

9. $\frac{1}{9}$

8.1 Exercise Set

1. $3x^2 = 75$

$x^2 = 25$

Apply the square root property.

$x = \pm\sqrt{25}$

$x = \pm 5$

The solution set is $\{\pm 5\}$.

3. $7x^2 = 42$

$x^2 = 6$

Apply the square root property.

$x = \pm\sqrt{6}$

The solution set is $\{\pm\sqrt{6}\}$.

5. $16x^2 = 25$

$x^2 = \frac{25}{16}$

Apply the square root property.

$x = \pm\sqrt{\frac{25}{16}}$

$x = \pm\frac{5}{4}$

The solution set is $\left\{\pm\frac{5}{4}\right\}$.

7. $3x^2 - 2 = 0$

$3x^2 = 2$

$x^2 = \frac{2}{3}$

Apply the square root property.

$x = \pm\sqrt{\frac{2}{3}}$

Because the proposed solutions are opposites, rationalize both denominators at once.

$x = \pm\sqrt{\frac{2}{3}} = \pm\frac{\sqrt{2}}{\sqrt{3}} \cdot \frac{\sqrt{3}}{\sqrt{3}} = \pm\frac{\sqrt{6}}{3}$

The solution set is $\left\{\pm\frac{\sqrt{6}}{3}\right\}$.

9. $25x^2 + 16 = 0$

$25x^2 = -16$

$x^2 = -\frac{16}{25}$

Apply the square root property.

$x = \pm\sqrt{-\frac{16}{25}}$

$x = \pm\sqrt{\frac{16}{25}}\sqrt{-1}$

$x = \pm\frac{4}{5}i$

$x = 0 \pm \frac{4}{5}i = \pm\frac{4}{5}i$

The solution set is $\left\{\pm\frac{4}{5}i\right\}$.

11. $(x+7)^2 = 9$

Apply the square root property.

$x + 7 = \sqrt{9}$ or $x + 7 = -\sqrt{9}$

$x + 7 = 3$ $\quad$ $x + 7 = -3$

$x = -4$ $\quad$ $x = -10$

The solution set is $\{-10, -4\}$.

13. $(x-3)^2 = 5$

Apply the square root property.

$x - 3 = \pm\sqrt{5}$

$x = 3 \pm \sqrt{5}$

The solution set is $\{3 \pm \sqrt{5}\}$.

15. $(x+2)^2 = 8$

Apply the square root property.

$$x+2 = \pm\sqrt{8}$$
$$x+2 = \pm\sqrt{4\cdot 2}$$
$$x+2 = \pm 2\sqrt{2}$$
$$x = -2 \pm 2\sqrt{2}$$

The solution set is $\{-2 \pm 2\sqrt{2}\}$.

17. $(x-5)^2 = -9$

Apply the square root property.

$$x-5 = \pm\sqrt{-9}$$
$$x-5 = \pm 3i$$
$$x = 5 \pm 3i$$

The solution set is $\{5 \pm 3i\}$.

19. $\left(x+\frac{3}{4}\right)^2 = \frac{11}{16}$

Apply the square root property.

$$x+\frac{3}{4} = \pm\sqrt{\frac{11}{16}}$$
$$x+\frac{3}{4} = \pm\frac{\sqrt{11}}{4}$$
$$x = -\frac{3}{4} \pm \frac{\sqrt{11}}{4} = \frac{-3 \pm \sqrt{11}}{4}$$

The solution set is $\left\{\frac{-3 \pm \sqrt{11}}{4}\right\}$.

21. $x^2 - 6x + 9 = 36$

$$(x-3)^2 = 36$$

Apply the square root property.

$$x-3 = \sqrt{36} \quad \text{or} \quad x-3 = -\sqrt{36}$$
$$x-3 = 6 \qquad\qquad x-3 = -6$$
$$x = 9 \qquad\qquad x = -3$$

The solutions are 9 and -3 and the solution set is $\{-3, 9\}$.

23. $x^2 + 2x + \_\_\_\_$

Since $b = 2$, add $\left(\frac{b}{2}\right)^2 = \left(\frac{2}{2}\right)^2 = (1)^2 = 1$.

$$x^2 + 2x + 1 = (x+1)^2$$

25. $x^2 - 14x + \_\_\_\_$

Since $b = -14$, add $\left(\frac{b}{2}\right)^2 = \left(\frac{-14}{2}\right)^2 = (-7)^2 = 49$.

$$x^2 - 14x + 49 = (x-7)^2$$

27. $x^2 + 7x + \_\_\_\_$

Since $b = 7$, add $\left(\frac{b}{2}\right)^2 = \left(\frac{7}{2}\right)^2 = \frac{49}{4}$.

$$x^2 + 7x + \frac{49}{4} = \left(x+\frac{7}{2}\right)^2$$

29. $x^2 - \frac{1}{2}x + \_\_\_\_$

Since $b = -\frac{1}{2}$, add

$$\left(\frac{b}{2}\right)^2 = \left(\frac{-1}{2} \div 2\right)^2 = \left(\frac{-1}{2}\cdot\frac{1}{2}\right)^2 = \left(\frac{-1}{4}\right)^2 = \frac{1}{16}.$$

$$x^2 - \frac{1}{2}x + \frac{1}{16} = \left(x-\frac{1}{4}\right)^2$$

31. $x^2 + \frac{4}{3}x + \_\_\_\_$

Since $b = \frac{4}{3}$, add

$$\left(\frac{b}{2}\right)^2 = \left(\frac{4}{3} \div 2\right)^2 = \left(\frac{4}{3}\cdot\frac{1}{2}\right)^2 = \left(\frac{2}{3}\right)^2 = \frac{4}{9}.$$

$$x^2 + \frac{4}{3}x + \frac{4}{9} = \left(x+\frac{2}{3}\right)^2$$

33. $x^2 - \frac{9}{4}x + \_\_\_\_$

Since $b = -\frac{9}{4}$, add

$$\left(\frac{b}{2}\right)^2 = \left(-\frac{9}{4} \div 2\right)^2$$
$$= \left(-\frac{9}{4}\cdot\frac{1}{2}\right)^2 = \left(-\frac{9}{8}\right)^2 = \frac{81}{64}.$$

$$x^2 - \frac{9}{4}x + \frac{81}{64} = \left(x-\frac{9}{8}\right)^2$$

35. $x^2+4x=32$

$x^2+4x \quad =32$

Since $b=4$, add $\left(\frac{b}{2}\right)^2=\left(\frac{4}{2}\right)^2=(2)^2=4$.

$x^2+4x+4=32+4$

$(x+2)^2=36$

Apply the square root property.

$x+2=\sqrt{36}$ or $x+2=-\sqrt{36}$

$x+2=6$ $\quad$ $x+2=-6$

$x=4$ $\quad$ $x=-8$

The solution set is $\{-8,\ 4\}$.

37. $x^2+6x=-2$

$x^2+6x \quad =-2$

Since $b=6$, add $\left(\frac{b}{2}\right)^2=\left(\frac{6}{2}\right)^2=(3)^2=9$.

$x^2+6x+9=-2+9$

$(x+3)^2=7$

Apply the square root property.

$x+3=\pm\sqrt{7}$

$x=-3\pm\sqrt{7}$

The solution set is $\{-3\pm\sqrt{7}\}$.

39. $x^2-8x+1=0$

$x^2-8x \quad =-1$

Since $b=-8$, add $\left(\frac{b}{2}\right)^2=\left(\frac{-8}{2}\right)^2=(-4)^2=16$.

$x^2-8x+16=-1+16$

$(x-4)^2=15$

Apply the square root property.

$x-4=\pm\sqrt{15}$

$x=4\pm\sqrt{15}$

The solution set is $\{4\pm\sqrt{15}\}$.

41. $x^2+2x+2=0$

$x^2+2x \quad =-2$

Since $b=2$, add $\left(\frac{b}{2}\right)^2=\left(\frac{2}{2}\right)^2=(1)^2=1$.

$x^2+2x+1=-2+1$

$(x+1)^2=-1$

Apply the square root property.

$x+1=\pm\sqrt{-1}=\pm i$

$x=-1\pm i$

The solution set is $\{-1\pm i\}$.

43. $x^2+3x-1=0$

$x^2+3x \quad =1$

Since $b=3$, add $\left(\frac{b}{2}\right)^2=\left(\frac{3}{2}\right)^2=\frac{9}{4}$.

$x^2+3x+\frac{9}{4}=1+\frac{9}{4}$

$\left(x+\frac{3}{2}\right)^2=\frac{13}{4}$

Apply the square root property.

$x+\frac{3}{2}=\pm\sqrt{\frac{13}{4}}=\pm\frac{\sqrt{13}}{2}$

$x=-\frac{3}{2}\pm\frac{\sqrt{13}}{2}=\frac{-3\pm\sqrt{13}}{2}$

The solution set is $\left\{\frac{-3\pm\sqrt{13}}{2}\right\}$.

45. $x^2+\frac{4}{7}x+\frac{3}{49}=0$

$x^2+\frac{4}{7}x \quad =-\frac{3}{49}$

Since $b=\frac{4}{7}$, add $\left(\frac{1}{2}b\right)^2=\left(\frac{1}{2}\cdot\frac{4}{7}\right)^2\left(\frac{2}{7}\right)^2=\frac{4}{49}$.

$x^2+\frac{4}{7}x+\frac{4}{49}=-\frac{3}{49}+\frac{4}{49}$

$\left(x+\frac{2}{7}\right)^2=\frac{1}{49}$

Apply the square root property.

$x+\frac{2}{7}=\pm\sqrt{\frac{1}{49}}$

$x+\frac{2}{7}=\pm\frac{1}{7}$

$x=-\frac{2}{7}\pm\frac{1}{7}$

$x=-\frac{2}{7}+\frac{1}{7}=-\frac{1}{7}$ or $-\frac{2}{7}-\frac{1}{7}=-\frac{3}{7}$

The solution set is $\left\{-\frac{3}{7},-\frac{1}{7}\right\}$.

47. $x^2 + x - 1 = 0$

$x^2 + x \quad = 1$

Since $b = 1$, add $\left(\frac{b}{2}\right)^2 = \left(\frac{1}{2}\right)^2 = \frac{1}{4}$.

$x^2 + x + \frac{1}{4} = 1 + \frac{1}{4}$

$\left(x + \frac{1}{2}\right)^2 = \frac{5}{4}$

Apply the square root property.

$x + \frac{1}{2} = \pm\sqrt{\frac{5}{4}}$

$x + \frac{1}{2} = \pm\frac{\sqrt{5}}{2}$

$x = -\frac{1}{2} \pm \frac{\sqrt{5}}{2} = \frac{-1 \pm \sqrt{5}}{2}$

The solution set is $\left\{\frac{-1 \pm \sqrt{5}}{2}\right\}$.

49. $2x^2 + 3x - 5 = 0$

$x^2 + \frac{3}{2}x - \frac{5}{2} = 0$

$x^2 + \frac{3}{2}x \quad = \frac{5}{2}$

Since $b = \frac{3}{2}$, add

$\left(\frac{1}{2}b\right)^2 = \left(\frac{1}{2} \cdot \frac{3}{2}\right)^2 = \left(\frac{3}{4}\right)^2 = \frac{9}{16}$.

$x^2 + \frac{3}{2}x + \frac{9}{16} = \frac{5}{2} + \frac{9}{16}$

$\left(x + \frac{3}{4}\right)^2 = \frac{40}{16} + \frac{9}{16} = \frac{49}{16}$

Apply the square root property.

$x + \frac{3}{4} = \pm\sqrt{\frac{49}{16}} = \pm\frac{7}{4}$

$x = -\frac{3}{4} \pm \frac{7}{4}$

$x = -\frac{3}{4} + \frac{7}{4}$ or $x = -\frac{3}{4} - \frac{7}{4}$

$x = \frac{4}{4}$ $\qquad x = -\frac{10}{4}$

$x = 1$ $\qquad x = -\frac{5}{2}$

The solution set is $\left\{-\frac{5}{2}, 1\right\}$.

51. $3x^2 + 6x + 1 = 0$

$x^2 + 2x + \frac{1}{3} = 0$

$x^2 + 2x \quad = -\frac{1}{3}$

Since $b = 2$, add

$\left(\frac{b}{2}\right)^2 = \left(\frac{2}{2}\right)^2 = 1^2 = 1$.

$x^2 + 2x + 1 = -\frac{1}{3} + 1$

$(x+1)^2 = -\frac{1}{3} + \frac{3}{3} = \frac{2}{3}$

Apply the square root property.

$x + 1 = \pm\sqrt{\frac{2}{3}}$

$x + 1 = \pm\frac{\sqrt{2}}{\sqrt{3}} \cdot \frac{\sqrt{3}}{\sqrt{3}} = \pm\frac{\sqrt{6}}{3}$

$x = -1 \pm \frac{\sqrt{6}}{3} = \frac{-3 \pm \sqrt{6}}{3}$

The solution set is $\left\{\frac{-3 \pm \sqrt{6}}{3}\right\}$.

53. $3x^2 - 8x + 1 = 0$

$x^2 - \frac{8}{3}x + \frac{1}{3} = 0$

$x^2 - \frac{8}{3}x \quad = -\frac{1}{3}$

Since $b = -\frac{8}{3}$, add

$\left(\frac{1}{2}b\right)^2 = \left[\frac{1}{2}\left(-\frac{8}{3}\right)\right]^2 = \left(-\frac{4}{3}\right)^2 = \frac{16}{9}$.

$x^2 - \frac{8}{3}x + \frac{16}{9} = -\frac{1}{3} + \frac{16}{9}$

$\left(x - \frac{4}{3}\right)^2 = -\frac{3}{9} + \frac{16}{9} = \frac{13}{9}$

Apply the square root property.

$$x-\frac{4}{3}=\pm\sqrt{\frac{13}{9}}$$

$$x-\frac{4}{3}=\pm\frac{\sqrt{13}}{3}$$

$$x=\frac{4}{3}\pm\frac{\sqrt{13}}{3}=\frac{4\pm\sqrt{13}}{3}$$

The solution set is $\left\{\frac{4\pm\sqrt{13}}{3}\right\}$.

55. $8x^2-4x+1=0$

$$x^2-\frac{1}{2}x+\frac{1}{8}=0$$

$$x^2-\frac{1}{2}x \quad =-\frac{1}{8}$$

Since $b=-\frac{1}{2}$, add

$$\left(\frac{1}{2}b\right)^2=\left[\frac{1}{2}\left(-\frac{1}{2}\right)\right]^2=\left(-\frac{1}{4}\right)^2=\frac{1}{16}.$$

$$x^2-\frac{1}{2}x+\frac{1}{16}=-\frac{1}{8}+\frac{1}{16}$$

$$\left(x-\frac{1}{4}\right)^2=-\frac{2}{16}+\frac{1}{16}=-\frac{1}{16}$$

Apply the square root property.

$$x-\frac{1}{4}=\pm\sqrt{-\frac{1}{16}}$$

$$x-\frac{1}{4}=\pm\frac{1}{4}i$$

$$x=\frac{1}{4}\pm\frac{1}{4}i$$

The solution set is $\left\{\frac{1}{4}\pm\frac{1}{4}i\right\}$.

57. $2x^2-5x+7=0$

$$\frac{2x^2}{2}-\frac{5x}{2}+\frac{7}{2}=\frac{0}{2}$$

$$x^2-\frac{5}{2}x+\frac{7}{2}=0$$

$$x^2-\frac{5}{2}x \quad =-\frac{7}{2}$$

Half of $-\frac{5}{2}$ is $-\frac{5}{4}$, and $\left(-\frac{5}{4}\right)^2$ is $\frac{25}{16}$, which should be added to both sides.

$$x^2-\frac{5}{2}x+\frac{25}{16}=-\frac{7}{2}+\frac{25}{16}$$

$$\left(x-\frac{5}{4}\right)^2=-\frac{56}{16}+\frac{25}{16}$$

$$\left(x-\frac{5}{4}\right)^2=-\frac{31}{16}$$

$$x-\frac{5}{4}=\pm\sqrt{-\frac{31}{16}}$$

$$x=\frac{5}{4}\pm i\frac{\sqrt{31}}{4}$$

The solution set is $\left\{\frac{5}{4}\pm i\frac{\sqrt{31}}{4}\right\}$.

59. $g(x)=\frac{9}{25}$

$$\left(x-\frac{2}{5}\right)^2=\frac{9}{25}$$

Apply the square root property.

$$x-\frac{2}{5}=\pm\sqrt{\frac{9}{25}}=\pm\frac{3}{5}$$

$$x=\frac{2}{5}\pm\frac{3}{5}$$

$$x=\frac{2}{5}+\frac{3}{5} \quad \text{or} \quad x=\frac{2}{5}-\frac{3}{5}$$

$$x=\frac{5}{5}=1 \quad \text{or} \quad x=-\frac{1}{5}$$

The values are $-\frac{1}{5}$ and 1.

61.
$$h(x) = -125$$
$$5(x+2)^2 = -125$$
$$(x+2)^2 = -25$$
Apply the square root property.
$$x+2 = \pm\sqrt{-25}$$
$$x+2 = \pm 5i$$
$$x = -2 \pm 5i$$
The values are $-2 \pm 5i$.

63. Let x = the number.
$$3(x-2)^2 = -12$$
$$(x-2)^2 = -4$$

Apply the square root property.

$$x-2 = \pm\sqrt{-4} = \pm 2i$$
$$x = 2 \pm 2i$$

The values are $2+2i$ and $2-2i$.

65.
$$h = \frac{v^2}{2g}$$
$$2gh = v^2$$

Apply the square root property and keep only the principal square root.

$$v = \sqrt{2gh}$$

67.
$$A = P(1+r)^2$$
$$\frac{A}{P} = (1+r)^2$$

Apply the square root property and keep only the principal square root.

$$1+r = \sqrt{\frac{A}{P}}$$
$$1+r = \frac{\sqrt{A}}{\sqrt{P}} \cdot \frac{\sqrt{P}}{\sqrt{P}} = \frac{\sqrt{AP}}{P}$$
$$r = \frac{\sqrt{AP}}{P} - 1$$

69.
$$\frac{x^2}{3} + \frac{x}{9} - \frac{1}{6} = 0$$
$$3\left(\frac{x^2}{3} + \frac{x}{9} - \frac{1}{6}\right) = 3(0)$$
$$x^2 + \frac{1}{3}x - \frac{1}{2} = 0$$
$$x^2 + \frac{1}{3}x \quad = \frac{1}{2}$$

Since $b = \frac{1}{3}$, add
$$\left(\frac{1}{2}b\right)^2 = \left(\frac{1}{2} \cdot \frac{1}{3}\right)^2 = \left(\frac{1}{6}\right)^2 = \frac{1}{36}.$$

$$x^2 + \frac{1}{3}x + \frac{1}{36} = \frac{1}{2} + \frac{1}{36}$$
$$\left(x + \frac{1}{6}\right)^2 = \frac{18}{36} + \frac{1}{36} = \frac{19}{36}$$

Apply the square root property.
$$x + \frac{1}{6} = \pm\sqrt{\frac{19}{36}} = \pm\frac{\sqrt{19}}{6}$$
$$x = -\frac{1}{6} \pm \frac{\sqrt{19}}{6} = \frac{-1 \pm \sqrt{19}}{6}$$

The solution set is $\left\{\frac{-1 \pm \sqrt{19}}{6}\right\}$.

71.
$$x^2 - bx = 2b^2$$
$$x^2 - bx \quad = 2b^2$$
Since $-b$ is the linear coefficient, add
$$\left(\frac{-b}{2}\right)^2 = \frac{b^2}{4}.$$

$$x^2 - bx + \frac{b^2}{4} = 2b^2 + \frac{b^2}{4}$$
$$\left(x - \frac{b}{2}\right)^2 = \frac{8b^2}{4} + \frac{b^2}{4} = \frac{9b^2}{4}$$

Apply the square root property.
$$x - \frac{b}{2} = \pm\sqrt{\frac{9b^2}{4}} = \pm\frac{3b}{2}$$
$$x = \frac{b}{2} \pm \frac{3b}{2}$$

$$x = \frac{b}{2} + \frac{3b}{2} \quad \text{or} \quad x = \frac{b}{2} - \frac{3b}{2}$$
$$x = \frac{4b}{2} = 2b \quad \text{or} \quad x = -\frac{2b}{2} = -b$$

The solution set is $\{-b, 2b\}$.

73. a. $x^2 + 8x$

b. 16

c. $x^2 + 8x + 16$

d. $(x+4)^2$

75. $2880 = 2000(1+r)^2$

$$\frac{2880}{2000} = (1+r)^2$$
$$1.44 = (1+r)^2$$

Apply the square root property.

$$1+r = \pm\sqrt{1.44}$$
$$1+r = \pm 1.2$$
$$r = -1 \pm 1.2$$
$$r = -1 + 1.2 \quad \text{or} \quad -1 - 1.2$$
$$r = 0.2 \quad \text{or} \quad -2.2$$

Reject –2.2 because we cannot have a negative interest rate. The solution is 0.2 and the annual interest rate is 20%.

77. $1445 = 1280(1+r)^2$

$$\frac{1445}{1280} = (1+r)^2$$
$$1.12890625 = (1+r)^2$$

Apply the square root property.

$$1+r = \pm\sqrt{1.12890625}$$
$$1+r = \pm 1.0625$$
$$r = -1 \pm 1.0625$$
$$r = -1 + 1.0625 \quad \text{or} \quad -1 - 1.0625$$
$$r = 0.0625 \quad \text{or} \quad -2.0625$$

Reject –2.0625 because we cannot have a negative interest rate. The solution is 0.0625 and the annual interest rate is 6.25%.

79. a. $f(x) = 15x^2 + 340$

$$f(8) = 15(8)^2 + 340$$
$$= 1300$$

According to the model 1300 thousand students were enrolled in charter schools in 2008. This overestimates the number displayed in the graph by 23 thousand.

b. $f(x) = 15x^2 + 340$

$$3280 = 15x^2 + 340$$
$$-15x^2 = -2940$$
$$\frac{-15x^2}{-15} = \frac{-2940}{-15}$$
$$x^2 = 196$$
$$\sqrt{x^2} = \pm\sqrt{196}$$
$$x = \pm 14$$

Reject –14.

According to the model, 3280 thousand students be enrolled in charter schools 14 years after 2000, or 2014.

81. $4800 = 16t^2$

$$\frac{4800}{16} = t^2$$
$$300 = t^2$$

Apply the square root property.

$$t = \pm\sqrt{300}$$
$$t = \pm 10\sqrt{3} \approx \pm 17.3$$

Disregard –17.3 because we can't have a negative time measurement. The solution is 17.3 and we conclude that the sky diver was in a free fall for $10\sqrt{3}$ or approximately 17.3 seconds.

83.

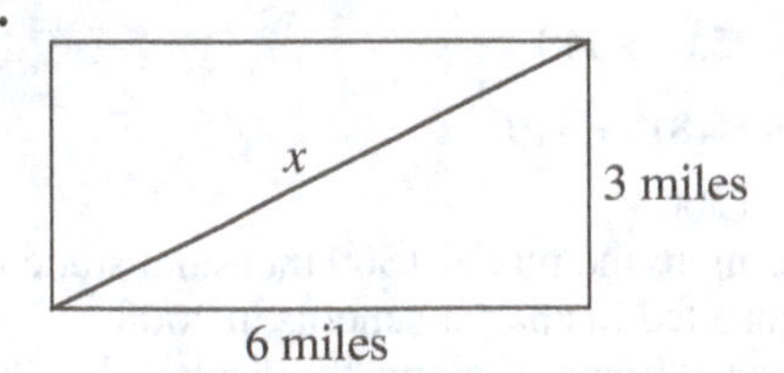

$x^2 = 6^2 + 3^2 = 36 + 9 = 45$

Apply the square root property.

$x = \pm\sqrt{45} = \pm\sqrt{9 \cdot 5} = \pm 3\sqrt{5}$

Disregard $-3\sqrt{5}$ because we can't have a negative length measurement. The solution is $3\sqrt{5}$ and we conclude that the pedestrian route is $3\sqrt{5}$ or approximately 6.7 miles long.

85. $x^2 + 10^2 = 30^2$

$x^2 + 100 = 900$

$x^2 = 800$

Apply the square root property.

$x = \pm\sqrt{800} = \pm\sqrt{400 \cdot 2} = \pm 20\sqrt{2}$

Disregard $-20\sqrt{2}$ because we can't have a negative length measurement. The solution is $20\sqrt{2}$. We conclude that the ladder reaches $20\sqrt{2}$ feet, or approximately 28.3 feet, up the house.

87.

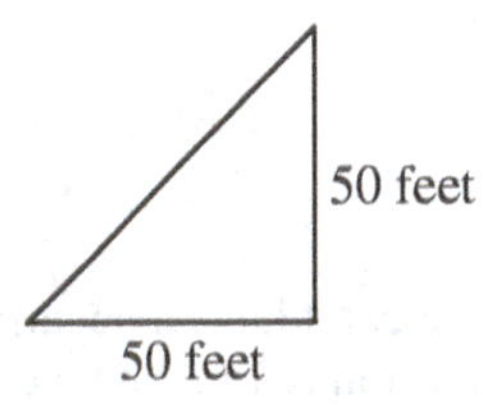

$50^2 + 50^2 = x^2$

$2500 + 2500 = x^2$

$5000 = x^2$

Apply the square root property.

$x = \pm\sqrt{5000}$

$x = \pm\sqrt{2500 \cdot 2} = \pm 50\sqrt{2} \approx \pm 70.7$

Disregard $-50\sqrt{2}$ because we cannot have a negative length measurement. The solution is $50\sqrt{2}$. We conclude that a supporting wire of $50\sqrt{2}$ feet, or approximately 70.7 feet, is required.

89. $A = lw$

$196 = (x+2+2)(x+2+2)$

$196 = (x+4)(x+4)$

$196 = (x+4)^2$

Apply the square root property.

$x + 4 = \pm\sqrt{196} = \pm 14$

$x = -4 \pm 14$

$x = -4 + 14 = 10$ or $x = -4 - 14 = -18$

Disregard -18 because we can't have a negative length measurement. The solution is 10. We conclude that the length of the original square is 10 meters.

91. – 95. Answers will vary.

97. $4 - (x+1)^2 = 0$

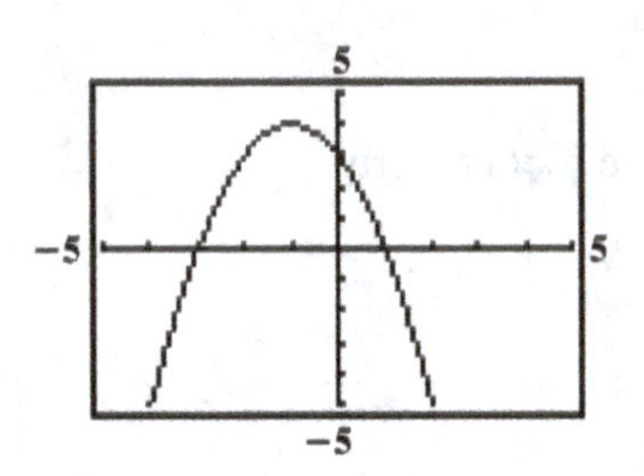

The solution set is $\{-3, 1\}$.

Check:

$x = -3$

$4 - (-3+1)^2 = 0$

$4 - (-2)^2 = 0$

$4 - 4 = 0$

$0 = 0$, true

$x = 1$

$4 - (1+1)^2 = 0$

$4 - 2^2 = 0$

$4 - 4 = 0$

$0 = 0$, true

99. Answers will vary

101. makes sense

103. does not make sense; Explanations will vary. Sample explanation: First divide both sides of the equation by 4. This makes the coefficient of x become $\frac{10}{4}$, or $\frac{5}{2}$. Thus you will add $\left(\frac{5}{2}\right)^2$, or $\frac{25}{4}$, to both sides.

105. false; Changes to make the statement true will vary. A sample change is: $(x-5)^2=12$ is equivalent to $x-5=\pm 2\sqrt{3}$.

107. false; Changes to make the statement true will vary. A sample change is: Not all quadratic equations can be solved by factoring.

109. $x^2+x+c=0$

$x^2+x \quad = -c$

Since $b=1$, add $\left(\frac{b}{2}\right)^2=\left(\frac{1}{2}\right)^2=\frac{1}{4}$.

$$x^2+x+\frac{1}{4}=-c+\frac{1}{4}$$
$$\left(x+\frac{1}{2}\right)^2=-c+\frac{1}{4}$$

Apply the square root property.

$$x+\frac{1}{2}=\pm\sqrt{-c+\frac{1}{4}}$$
$$x=-\frac{1}{2}\pm\sqrt{-c+\frac{1}{4}}$$

Simplify the solutions.

$$x=-\frac{1}{2}\pm\sqrt{\frac{-c}{1}\cdot\frac{4}{4}+\frac{1}{4}}$$
$$=-\frac{1}{2}\pm\sqrt{\frac{-4c}{4}+\frac{1}{4}}$$
$$=-\frac{1}{2}\pm\sqrt{\frac{-4c+1}{4}}$$
$$=-\frac{1}{2}\pm\frac{\sqrt{1-4c}}{2}=\frac{-1\pm\sqrt{1-4c}}{2}$$

The solutions are $\frac{-1\pm\sqrt{1-4c}}{2}$ and the solution set is $\left\{\frac{-1\pm\sqrt{1-4c}}{2}\right\}$.

111. $x^4-8x^2+15=0$

Let $u=x^2$

$$x^4-8x^2+15=0$$
$$\left(x^2\right)^2-8x^2+15=0$$
$$u^2-8u+15=0$$
$$(u-5)(u-3)=0$$

Apply the zero product principle.

$u-5=0 \qquad u-3=0$

$u=5 \qquad u=3$

Substitute x^2 for u.

$u=5 \qquad u=3$

$x^2=5 \qquad x^2=3$

Apply the square root property.

$x=\pm\sqrt{5} \qquad x=\pm\sqrt{3}$

The solution set is $\left\{\pm\sqrt{5},\pm\sqrt{3}\right\}$.

112. $4x-2-3[4-2(3-x)]$

$=4x-2-3[4-6+2x]=4x-2-3[-2+2x]$

$=4x-2+6-6x=4-2x$

113. $1-8x^3=1^3-(2x)^3$

$=(1-2x)\left(1+2x+4x^2\right)$

114.

3	1	−5	2	0	−6
		3	−6	−12	−36
	1	−2	−4	−12	−42

$$x^3-2x^2-4x-12-\frac{42}{x-3}$$

115. a. $8x^2+2x-1=0$

$(2x+1)(4x-1)=0$

$2x+1=0 \quad$ or $\quad 4x-1=0$

$2x=-1 \qquad 4x=1$

$x=-\frac{1}{2} \qquad x=\frac{1}{4}$

The solution set is $\left\{-\frac{1}{2},\frac{1}{4}\right\}$.

b. $b^2-4ac=2^2-4(8)(-1)=36$

Yes, b^2-4ac is a perfect square.

116. a. $9x^2 - 6x + 1 = 0$

$$(3x-1)^2 = 0$$
$$3x - 1 = 0$$
$$3x = 1$$
$$x = \frac{1}{3}$$

The solution set is $\left\{\frac{1}{3}\right\}$.

b. $b^2 - 4ac = (-6)^2 - 4(9)(1) = 36 - 36 = 0$

117. a.

$$3 + \frac{4}{x} = -\frac{2}{x^2}$$
$$x^2\left(3 + \frac{4}{x}\right) = x^2\left(-\frac{2}{x^2}\right)$$
$$3x^2 + 4x = -2$$
$$3x^2 + 4x + 2 = 0$$

b. $b^2 - 4ac = 4^2 - 4(3)(2) = 16 - 24 = -8$

8.2 Check Points

1. $2x^2 + 9x - 5 = 0$

$a = 2 \quad b = 9 \quad c = -5$

$$x = \frac{-b \pm \sqrt{b^2 - 4ac}}{2a}$$
$$x = \frac{-9 \pm \sqrt{9^2 - 4(2)(-5)}}{2(2)}$$
$$= \frac{-9 \pm \sqrt{81 + 40}}{4}$$
$$= \frac{-9 \pm \sqrt{121}}{4}$$
$$= \frac{-9 \pm 11}{4}$$

Evaluate the expression to obtain two solutions.

$x = \frac{-9+11}{4} = \frac{1}{2}$ or $x == \frac{-9-11}{4} = -5$

The solution set is $\left\{-5, \frac{1}{2}\right\}$.

2.

$$2x^2 = 6x - 1$$
$$2x^2 - 6x + 1 = 0$$

$a = 2 \quad b = -6 \quad c = 1$

$$x = \frac{-b \pm \sqrt{b^2 - 4ac}}{2a}$$
$$x = \frac{-(-6) \pm \sqrt{(-6)^2 - 4(2)(1)}}{2(2)}$$
$$= \frac{6 \pm \sqrt{36 - 8}}{4}$$
$$= \frac{6 \pm \sqrt{28}}{4}$$
$$= \frac{6 \pm 2\sqrt{7}}{4}$$
$$= \frac{3 \pm \sqrt{7}}{2}$$

The solution set is $\left\{\frac{3 \pm \sqrt{7}}{2}\right\}$.

3.

$$3x^2 + 5 = -6x$$
$$3x^2 + 6x + 5 = 0$$

$a = 3 \quad b = 6 \quad c = 5$

$$x = \frac{-b \pm \sqrt{b^2 - 4ac}}{2a}$$
$$x = \frac{-6 \pm \sqrt{6^2 - 4(3)(5)}}{2(3)}$$
$$= \frac{-6 \pm \sqrt{36 - 60}}{6}$$
$$= \frac{-6 \pm \sqrt{-24}}{6}$$
$$= \frac{6 \pm 2i\sqrt{6}}{6}$$
$$= \frac{6}{6} \pm \frac{2i\sqrt{6}}{6}$$
$$= 1 \pm i\frac{\sqrt{6}}{3}$$

The solution set is $\left\{1 \pm i\frac{\sqrt{6}}{3}\right\}$.

4. a. $b^2 - 4ac = 6^2 - 4(1)(9) = 0$
Since the discriminant is zero, there is one real rational solution.

b. $b^2 - 4ac = (-7)^2 - 4(2)(-4) = 81$
Since the discriminant is positive and a perfect square, there are two real rational solutions.

c. $b^2 - 4ac = (-2)^2 - 4(3)(4) = -44$
Since the discriminant is negative, there is no real solution. There are imaginary solutions that are complex conjugates.

5. a. Because the solution set is $\left\{-\frac{3}{5}, \frac{1}{4}\right\}$, we have

$$x = -\frac{3}{5} \quad \text{or} \quad x = \frac{1}{4}$$
$$5x = -3 \qquad 4x = 1$$
$$5x + 3 = 0 \qquad 4x - 1 = 0.$$

Use the zero-product principle in reverse.

$$(5x+3)(4x-1) = 0$$
$$20x^2 - 5x + 12x - 3 = 0$$
$$20x^2 + 7x - 3 = 0$$

Thus, one equation is $20x^2 + 7x - 3 = 0$. Other equations can be obtained by multiplying both sides of this equation by any nonzero real number.

b. Because the solution set is $\{-5\sqrt{2}, 5\sqrt{2}\}$, we have

$$x = -5\sqrt{2} \quad \text{or} \quad x = 5\sqrt{2}$$
$$x + 5\sqrt{2} = 0 \qquad x - 5\sqrt{2} = 0$$

Use the zero-product principle in reverse.

$$(x + 5\sqrt{2})(x - 5\sqrt{2}) = 0$$
$$x^2 - (5\sqrt{2})^2 = 0$$
$$x^2 - 25(2) = 0$$
$$x^2 - 50 = 0$$

Thus, one equation is $x^2 - 50 = 0$. Other equations can be obtained by multiplying both sides of this equation by any nonzero real number.

c. Because the solution set is $\{-7i, 7i\}$, we have

$$x = -7i \quad \text{or} \quad x = 7i$$
$$x + 7i = 0 \qquad x - 7i = 0$$

Use the zero-product principle in reverse.

$$(x+7i)(x-7i) = 0$$
$$x^2 - 7ix + 7ix - 49i^2 = 0$$
$$x^2 - 49(-1) = 0$$
$$x^2 + 49 = 0$$

Thus, one equation is $x^2 + 49 = 0$. Other equations can be obtained by multiplying both sides of this equation by any nonzero real number.

6. $P(A) = 0.01A^2 + 0.05A + 107$

$$115 = 0.01A^2 + 0.05A + 107$$
$$0 = 0.01A^2 + 0.05A - 8$$
$$a = 0.01 \quad b = 0.05 \quad c = -8$$
$$x = \frac{-b \pm \sqrt{b^2 - 4ac}}{2a}$$
$$x = \frac{-0.05 \pm \sqrt{0.05^2 - 4(0.01)(-8)}}{2(0.01)}$$
$$= \frac{-0.05 \pm \sqrt{0.3225}}{0.02}$$
$$x = \frac{-0.05 + \sqrt{0.3225}}{0.02} \quad \text{or} \quad x = \frac{-0.05 - \sqrt{0.3225}}{0.02}$$
$$x \approx 26 \qquad \cancel{x \approx -31}$$

A woman's normal systolic blood pressure is 115 mm at about 26 years of age.
This is represented by the point (26,115) on the blue graph.

8.2 Concept and Vocabulary Check

1. $\frac{-b \pm \sqrt{b^2 - 4ac}}{2a}$

2. 2; 9; -5

3. 1; -4; -1

4. $2 \pm \sqrt{2}$

5. $-1 \pm i\frac{\sqrt{6}}{2}$

6. $b^2 - 4ac$

7. no

8. two

9. the square root property

10. the quadratic formula

11. factoring and the zero–product principle

12. false

8.2 Exercise Set

1. $x^2 + 8x + 12 = 0$

$a = 1 \quad b = 8 \quad c = 12$

$$x = \frac{-8 \pm \sqrt{8^2 - 4(1)(12)}}{2(1)}$$

$$= \frac{-8 \pm \sqrt{64 - 48}}{2}$$

$$= \frac{-8 \pm \sqrt{16}}{2} = \frac{-8 \pm 4}{2}$$

Evaluate the expression to obtain two solutions.

$$x = \frac{-8 - 4}{2} = -6 \quad \text{or} \quad x = \frac{-8 + 4}{2} = -2$$

The solution set is $\{-6, -2\}$.

3. $2x^2 - 7x = -5$

$2x^2 - 7x + 5 = 0$

$a = 2 \quad b = -7 \quad c = 5$

$$x = \frac{-(-7) \pm \sqrt{(-7)^2 - 4(2)(5)}}{2(2)}$$

$$= \frac{7 \pm \sqrt{49 - 40}}{4} = \frac{7 \pm \sqrt{9}}{4} = \frac{7 \pm 3}{4}$$

Evaluate the expression to obtain two solutions.

$$x = \frac{7 - 3}{4} = 1 \quad \text{or} \quad x = \frac{7 + 3}{4} = \frac{5}{2}$$

The solution set is $\left\{1, \frac{5}{2}\right\}$.

5. $x^2 + 3x - 20 = 0$

$a = 1 \quad b = 3 \quad c = -20$

$$x = \frac{-3 \pm \sqrt{3^2 - 4(1)(-20)}}{2(1)}$$

$$= \frac{-3 \pm \sqrt{9 + 80}}{2}$$

$$= \frac{-3 \pm \sqrt{89}}{2}$$

The solution set is $\left\{\frac{-3 \pm \sqrt{89}}{2}\right\}$.

7. $3x^2 - 7x = 3$

$3x^2 - 7x - 3 = 0$

$a = 3 \quad b = -7 \quad c = -3$

$$x = \frac{-(-7) \pm \sqrt{(-7)^2 - 4(3)(-3)}}{2(3)}$$

$$= \frac{7 \pm \sqrt{49 - (-36)}}{6} = \frac{7 \pm \sqrt{85}}{6}$$

The solution set is $\left\{\frac{7 \pm \sqrt{85}}{6}\right\}$.

9. $6x^2 = 2x + 1$

$6x^2 - 2x - 1 = 0$

$a = 6 \quad b = -2 \quad c = -1$

$$x = \frac{-(-2) \pm \sqrt{(-2)^2 - 4(6)(-1)}}{2(6)}$$

$$= \frac{2 \pm \sqrt{4 + 24}}{12}$$

$$= \frac{2 \pm \sqrt{28}}{12}$$

$$= \frac{2 \pm 2\sqrt{7}}{12}$$

$$= \frac{2(1 \pm \sqrt{7})}{12}$$

$$= \frac{1 \pm \sqrt{7}}{6}$$

The solution set is $\left\{\frac{1 \pm \sqrt{7}}{6}\right\}$.

11. $$4x^2-3x=-6$$
$$4x^2-3x+6=0$$
$$a=4 \quad b=-3 \quad c=6$$

$$x=\frac{-(-3)\pm\sqrt{(-3)^2-4(4)(6)}}{2(4)}$$
$$=\frac{3\pm\sqrt{9-96}}{8}$$
$$=\frac{3\pm\sqrt{-87}}{8}$$
$$=\frac{3\pm\sqrt{87(-1)}}{8}$$
$$=\frac{3\pm i\sqrt{87}}{8}=\frac{3}{8}\pm i\frac{\sqrt{87}}{8}$$

The solution set is $\left\{\frac{3}{8}\pm i\frac{\sqrt{87}}{8}\right\}$.

13. $$x^2-4x+8=0$$
$$a=1 \quad b=-4 \quad c=8$$

$$x=\frac{-(-4)\pm\sqrt{(-4)^2-4(1)(8)}}{2(1)}$$
$$=\frac{4\pm\sqrt{16-32}}{2}$$
$$=\frac{4\pm\sqrt{-16}}{2}$$
$$=\frac{4\pm 4i}{2}=\frac{4}{2}\pm\frac{4}{2}i=2\pm 2i$$

The solution set is $\{2\pm 2i\}$.

15. $$3x^2=8x-7$$
$$3x^2-8x+7=0$$
$$a=3 \quad b=-8 \quad c=7$$

$$x=\frac{-(-8)\pm\sqrt{(-8)^2-4(3)(7)}}{2(3)}$$
$$=\frac{8\pm\sqrt{64-84}}{6}$$
$$=\frac{8\pm\sqrt{-20}}{6}$$
$$=\frac{8\pm\sqrt{4\cdot 5(-1)}}{6}$$
$$=\frac{8\pm 2i\sqrt{5}}{6}=\frac{8}{6}\pm\frac{2}{6}i\sqrt{5}=\frac{4}{3}\pm i\frac{\sqrt{5}}{3}$$

The solution set is $\left\{\frac{4}{3}\pm i\frac{\sqrt{5}}{3}\right\}$.

17. $$2x(x-2)=x+12$$
$$2x^2-4x=x+12$$
$$2x^2-5x-12=0$$
$$a=2 \quad b=-5 \quad c=-12$$

$$x=\frac{-(-5)\pm\sqrt{(-5)^2-4(2)(-12)}}{2(2)}$$
$$=\frac{5\pm\sqrt{25+96}}{4}=\frac{5\pm\sqrt{121}}{4}=\frac{5\pm 11}{4}$$

Evaluate the expression to obtain two solutions.

$$x=\frac{5-11}{4}=-\frac{3}{2} \quad\text{or}\quad x=\frac{5+11}{4}=4$$

The solution set is $\left\{-\frac{3}{2}, 4\right\}$.

19. $$x^2+8x+3=0$$
$$a=1 \quad b=8 \quad c=3$$

$$b^2-4ac=8^2-4(1)(3)=64-12=52$$

Since the discriminant is positive and not a perfect square, there are two real irrational solutions.

21. $x^2+6x+8=0$

$a=1 \quad b=6 \quad c=8$

$$\begin{aligned} b^2-4ac &= (6)^2-4(1)(8) \\ &= 36-32=4 \end{aligned}$$

Since the discriminant is greater than zero, there are two unequal real solutions. Also, since the discriminant is a perfect square, the solutions are real rational.

23. $2x^2+x+3=0$

$a=2 \quad b=1 \quad c=3$

$$b^2-4ac=1^2-4(2)(3)=1-24=-23$$

Since the discriminant is negative, there are no real solutions. There are two imaginary solutions that are complex conjugates.

25. $2x^2+6x=0$

$a=2 \quad b=6 \quad c=0$

$$\begin{aligned} b^2-4ac &= (6)^2-4(1)(0) \\ &= 36-0=36 \end{aligned}$$

Since the discriminant is greater than zero, there are two unequal real solutions. Also, since the discriminant is a perfect square, the solutions are real rational.

27. $5x^2+3=0$

$a=5 \quad b=0 \quad c=3$

$$b^2-4ac=0^2-4(5)(3)=0-60=-60$$

Since the discriminant is negative, there are no real solutions. There are two imaginary solutions that are complex conjugates.

29. $9x^2=12x-4$

$9x^2-12x+4=0$

$a=9 \quad b=-12 \quad c=4$

$$\begin{aligned} b^2-4ac &= (-12)^2-4(9)(4) \\ &= 144-144=0 \end{aligned}$$

Since the discriminant is zero, there is one repeated real rational solution.

31.

$$\begin{aligned} 3x^2-4x &= 4 \\ 3x^2-4x-4 &= 0 \\ (3x+2)(x-2) &= 0 \end{aligned}$$

$3x+2=0 \quad \text{or} \quad x-2=0$

$3x=-2 \qquad\qquad x=2$

$x=-\dfrac{2}{3}$

The solution set is $\left\{-\dfrac{2}{3},2\right\}$.

33. $x^2-2x=1$

Since $b=-2$, add $\left(\dfrac{b}{2}\right)^2=\left(\dfrac{-2}{2}\right)^2=(-1)^2=1.$

$$\begin{aligned} x^2-2x+1 &= 1+1 \\ (x-1)^2 &= 2 \end{aligned}$$

Apply the square root principle.

$$\begin{aligned} x-1 &= \pm\sqrt{2} \\ x &= 1\pm\sqrt{2} \end{aligned}$$

The solution set is $\left\{1\pm\sqrt{2}\right\}$.

35.

$$\begin{aligned} 3x^2 &= x-9 \\ 3x^2-x+9 &= 0 \end{aligned}$$

$a=3 \quad b=-1 \quad c=9$

$$\begin{aligned} x &= \frac{-(-1)\pm\sqrt{(-1)^2-4(3)(9)}}{2(3)} \\ &= \frac{1\pm\sqrt{1-108}}{6} \\ &= \frac{1\pm\sqrt{-107}}{6} \\ &= \frac{1\pm i\sqrt{107}}{6} \\ &= \frac{1}{6}\pm i\frac{\sqrt{107}}{6} \end{aligned}$$

The solution set is $\left\{\dfrac{1}{6}\pm i\dfrac{\sqrt{107}}{6}\right\}$.

37. $(2x-5)(x+1)=2$

$2x^2+2x-5x-5=2$

$2x^2-3x-7=0$

Apply the quadratic formula.

$a=2 \quad b=-3 \quad c=-7$

$$x=\frac{-(-3)\pm\sqrt{(-3)^2-4(2)(-7)}}{2(2)}$$

$$=\frac{3\pm\sqrt{9-(-56)}}{4}=\frac{3\pm\sqrt{65}}{4}$$

The solution set is $\left\{\frac{3\pm\sqrt{65}}{4}\right\}$.

39. $(3x-4)^2=16$

Apply the square root property.

$3x-4=\sqrt{16}$ or $3x-4=-\sqrt{16}$

$3x-4=4 \qquad 3x-4=-4$

$3x=8 \qquad 3x=0$

$x=\frac{8}{3} \qquad x=0$

The solution set is $\left\{0,\frac{8}{3}\right\}$.

41. $\frac{x^2}{2}+2x+\frac{2}{3}=0$

Multiply both sides of the equation by 6 to clear fractions.

$3x^2+12x+4=0$

Apply the quadratic formula.

$a=3 \quad b=12 \quad c=4$

$$x=\frac{-12\pm\sqrt{12^2-4(3)(4)}}{2(3)}$$

$$=\frac{-12\pm\sqrt{144-48}}{6}$$

$$=\frac{-12\pm\sqrt{96}}{6}$$

$$=\frac{-12\pm\sqrt{16\cdot 6}}{6}$$

$$=\frac{-12\pm 4\sqrt{6}}{6}$$

$$=\frac{2(-6\pm 2\sqrt{6})}{6}=\frac{-6\pm 2\sqrt{6}}{3}$$

The solution set is $\left\{\frac{-6\pm 2\sqrt{6}}{3}\right\}$.

43. $(3x-2)^2=10$

Apply the square root property.

$3x-2=\pm\sqrt{10}$

$3x=2\pm\sqrt{10}$

$x=\frac{2\pm\sqrt{10}}{3}$

The solution set is $\left\{\frac{2\pm\sqrt{10}}{3}\right\}$.

45. $\frac{1}{x}+\frac{1}{x+2}=\frac{1}{3}$

The LCD is $3x(x+2)$.

$$3x(x+2)\left(\frac{1}{x}+\frac{1}{x+2}\right)=3x(x+2)\left(\frac{1}{3}\right)$$

$$3(x+2)+3x=x(x+2)$$

$$3x+6+3x=x^2+2x$$

$$0=x^2-4x-6$$

Apply the quadratic formula.

$a=1 \quad b=-4 \quad c=-6$

$$x=\frac{-(-4)\pm\sqrt{(-4)^2-4(1)(-6)}}{2(1)}$$

$$=\frac{4\pm\sqrt{16-(-24)}}{2}$$

$$=\frac{4\pm\sqrt{40}}{2}=\frac{4\pm 2\sqrt{10}}{2}=2\pm\sqrt{10}$$

The solution set is $\left\{2\pm\sqrt{10}\right\}$.

47.

$$
\begin{aligned}
(2x-6)(x+2) &= 5(x-1)-12 \\
2x^2+4x-6x-12 &= 5x-5-12 \\
2x^2-2x-12 &= 5x-17 \\
2x^2-7x+5 &= 0 \\
(2x-5)(x-1) &= 0
\end{aligned}
$$

Apply the zero product principle.

$2x-5=0$ or $x-1=0$

$2x=5$ $\quad x=1$

$x=\frac{5}{2}$

The solution set is $\left\{1,\frac{5}{2}\right\}$.

49.

$$
\begin{aligned}
x^2+10 &= 2(2x-1) \\
x^2+10 &= 4x-2 \\
x^2-4x+10 &= -2 \\
x^2-4x+12 &= 0
\end{aligned}
$$

$a=1 \quad b=-4 \quad c=12$

$$
\begin{aligned}
x &= \frac{-(-4)\pm\sqrt{(-4)^2-4(1)(12)}}{2(1)} \\
&= \frac{4\pm\sqrt{16-48}}{2} \\
&= \frac{4\pm\sqrt{-32}}{2} \\
&= \frac{4\pm 4i\sqrt{2}}{2} \\
&= 4\pm 2i\sqrt{2}
\end{aligned}
$$

The solution set is $\left\{4\pm 2i\sqrt{2}\right\}$.

51. Because the solution set is $\{-3,5\}$, we have

$x=-3$ or $x=5$

$x+3=0 \quad x-5=0.$

Use the zero-product principle in reverse.

$$
\begin{aligned}
(x+3)(x-5) &= 0 \\
x^2-5x+3x-15 &= 0 \\
x^2-2x-15 &= 0
\end{aligned}
$$

53. Because the solution set is $\left\{-\frac{2}{3},\frac{1}{4}\right\}$, we have

$x=-\frac{2}{3}$ or $x=\frac{1}{4}$

$3x=-2 \quad 4x=1$

$3x+2=0 \quad 4x-1=0.$

Use the zero-product principle in reverse.

$$
\begin{aligned}
(3x+2)(4x-1) &= 0 \\
12x^2-3x+8x-2 &= 0 \\
12x^2+5x-2 &= 0
\end{aligned}
$$

55. Because the solution set is $\left\{-\sqrt{2},\sqrt{2}\right\}$, we have

$x=\sqrt{2}$ or $x=-\sqrt{2}$

$x-\sqrt{2}=0 \quad x+\sqrt{2}=0.$

Use the zero-product principle in reverse.

$$
\begin{aligned}
(x-\sqrt{2})(x+\sqrt{2}) &= 0 \\
x^2+\cancel{x\sqrt{2}}-\cancel{x\sqrt{2}}-2 &= 0 \\
x^2-2 &= 0
\end{aligned}
$$

57. Because the solution set is $\left\{-2\sqrt{5},\ 2\sqrt{5}\right\}$ we have

$x=2\sqrt{5}$ or $x=-2\sqrt{5}$

$x-2\sqrt{5}=0 \quad x+2\sqrt{5}=0.$

Use the zero-product principle in reverse.

$$
\begin{aligned}
(x-2\sqrt{5})(x+2\sqrt{5}) &= 0 \\
x^2+\cancel{2x\sqrt{5}}-\cancel{2x\sqrt{5}}-4\cdot 5 &= 0 \\
x^2-20 &= 0
\end{aligned}
$$

59. Because the solution set is $\{-6i,6i\}$, we have

$x=6i$ or $x=-6i$

$x-6i=0 \quad x+6i=0.$

Use the zero-product principle in reverse.

$$
\begin{aligned}
(x-6i)(x+6i) &= 0 \\
x^2+\cancel{6i}-\cancel{6i}-36i^2 &= 0 \\
x^2-36(-1) &= 0 \\
x^2+36 &= 0
\end{aligned}
$$

61. Because the solution set is $\{1+i,\ 1-i\}$, we have

$x = 1+i$ or $x = 1-i$

$x-(1+i)=0$ $\quad$ $x-(1-i)=0.$

Use the zero-product principle in reverse.

$$[x-(1+i)][x-(1-i)]=0$$
$$x^2 - x(1-i) - x(1+i) + (1+i)(1-i) = 0$$
$$x^2 - x + \cancel{xi} - x - \cancel{xi} + 1 - i^2 = 0$$
$$x^2 - x - x + 1 - (-1) = 0$$
$$x^2 - 2x + 2 = 0$$

63. Because the solution set is $\{1+\sqrt{2}, 1-\sqrt{2}\}$, we have

$x = 1+\sqrt{2}$ or $x = 1-\sqrt{2}$

$x-(1+\sqrt{2})=0$ $\quad$ $x-(1-\sqrt{2})=0.$

Use the zero-product principle in reverse.

$$\left(x-(1+\sqrt{2})\right)\left(x-(1-\sqrt{2})\right)=0$$
$$x^2 - x(1-\sqrt{2}) - x(1+\sqrt{2}) + (1+\sqrt{2})(1-\sqrt{2}) = 0$$
$$x^2 - x + \cancel{x\sqrt{2}} - x - \cancel{x\sqrt{2}} + 1 - 2 = 0$$
$$x^2 - 2x - 1 = 0$$

65. b. If the solutions are imaginary numbers, then the graph will not cross the x-axis.

67. a. The equation has two non-integer solutions $3 \pm \sqrt{2}$, so the graph crosses the x-axis at $3-\sqrt{2}$ and $3+\sqrt{2}$.

69. Let x = the number.

$$x^2 - (6+2x) = 0$$
$$x^2 - 2x - 6 = 0$$

Apply the quadratic formula.

$a=1 \quad b=-2 \quad c=-6$

$$x = \frac{-(-2) \pm \sqrt{(-2)^2 - 4(1)(-6)}}{2(1)}$$
$$= \frac{2 \pm \sqrt{4-(-24)}}{2}$$
$$= \frac{2 \pm \sqrt{28}}{2}$$
$$= \frac{2 \pm \sqrt{4 \cdot 7}}{2} = \frac{2 \pm 2\sqrt{7}}{2} = 1 \pm \sqrt{7}$$

Disregard $1-\sqrt{7}$ because it is negative, and we are looking for a positive number. Thus, the number is $1+\sqrt{7}$.

71.
$$\frac{1}{x^2-3x+2} = \frac{1}{x+2} + \frac{5}{x^2-4}$$
$$\frac{1}{(x-1)(x-2)} = \frac{1}{x+2} + \frac{5}{(x+2)(x-2)}$$

Multiply both sides of the equation by the least common denominator, $(x-1)(x-2)(x+2)$. This results in the following:

$$x+2 = (x-1)(x-2) + 5(x-1)$$
$$x+2 = x^2 - 2x - x + 2 + 5x - 5$$
$$x+2 = x^2 + 2x - 3$$
$$0 = x^2 + x - 5$$

Apply the quadratic formula.

$a=1 \quad b=1 \quad c=-5.$

$$x = \frac{-1 \pm \sqrt{1^2 - 4(1)(-5)}}{2(1)} = \frac{-1 \pm \sqrt{1-(-20)}}{2}$$
$$= \frac{-1 \pm \sqrt{21}}{2}$$

The solution set is $\left\{\frac{-1 \pm \sqrt{21}}{2}\right\}$.

73. $\sqrt{2}x^2+3x-2\sqrt{2}=0$

Apply the quadratic formula.

$a=\sqrt{2}\quad b=3\quad c=-2\sqrt{2}$

$$x=\frac{-3\pm\sqrt{3^2-4\left(\sqrt{2}\right)\left(-2\sqrt{2}\right)}}{2\left(\sqrt{2}\right)}$$
$$=\frac{-3\pm\sqrt{9-(-16)}}{2\sqrt{2}}$$
$$=\frac{-3\pm\sqrt{25}}{2\sqrt{2}}=\frac{-3\pm5}{2\sqrt{2}}$$

Evaluate the expression to obtain two solutions.

$$x=\frac{-3-5}{2\sqrt{2}}\quad\text{or}\quad x=\frac{-3+5}{2\sqrt{2}}$$
$$=\frac{-8}{2\sqrt{2}}\cdot\frac{\sqrt{2}}{\sqrt{2}}\qquad=\frac{2}{2\sqrt{2}}\cdot\frac{\sqrt{2}}{\sqrt{2}}$$
$$=\frac{-8\sqrt{2}}{4}\qquad=\frac{2\sqrt{2}}{4}$$
$$=-2\sqrt{2}\qquad=\frac{\sqrt{2}}{2}$$

The solution set is $\left\{-2\sqrt{2},\ \frac{\sqrt{2}}{2}\right\}$.

75. $\left|x^2+2x\right|=3$

$x^2+2x=-3$ or $x^2+2x=3$

$x^2+2x+3=0$ $\quad x^2+2x-3=0$

Apply the quadratic formula to solve $x^2+2x+3=0$.

$a=1\quad b=2\quad c=3$.

$$x=\frac{-2\pm\sqrt{2^2-4(1)(3)}}{2(1)}$$
$$=\frac{-2\pm\sqrt{4-12}}{2}$$
$$=\frac{-2\pm\sqrt{-8}}{2}$$
$$=\frac{-2\pm\sqrt{2\cdot4\cdot(-1)}}{2}$$
$$=\frac{-2\pm2i\sqrt{2}}{2}=-1\pm i\sqrt{2}$$

Apply the zero product principle to solve $x^2+2x-3=0$.

$(x+3)(x-1)=0$

$x+3=0$ or $x-1=0$

$x=-3$ or $x=1$

The solution set is $\left\{-3,\ 1,\ -1\pm i\sqrt{2}\right\}$.

77. $f(x)=0.013x^2-1.19x+28.24$

$3=0.013x^2-1.19x+28.24$

$0=0.013x^2-1.19x+25.24$

Apply the quadratic formula.

$a=0.013\quad b=-1.19\quad c=25.24$

$$x=\frac{-(-1.19)\pm\sqrt{(-1.19)^2-4(0.013)(25.24)}}{2(0.013)}$$
$$=\frac{1.19\pm\sqrt{1.4161-1.31248}}{0.026}$$
$$=\frac{1.19\pm\sqrt{0.10362}}{0.026}$$
$$\approx\frac{1.19\pm0.32190}{0.026}$$
$$\approx 58.15\text{ or }33.39$$

The solutions are approximately 33.39 and 58.15. Thus, 33 year olds and 58 year olds are expected to be in 3 fatal crashes per 100 million miles driven. The function models the actual data well.

79. Use the quadratic formula to solve $f(x)=-0.01x^2+0.7x+6.1=0$.

$a=-0.01, b=0.7, c=6.1$

$$x=\frac{-0.7\pm\sqrt{0.7^2-4(-0.01)(6.1)}}{2(-0.01)}$$
$$=\frac{-0.7\pm\sqrt{0.49+0.244}}{-0.02}$$
$$=\frac{-0.7\pm\sqrt{0.734}}{-0.02}\approx-7.8\text{ or }77.8$$

Disregard -7.8 because the distance must be positive. Thus, the maximum distance is approximately 77.8 feet. Graph (b) shows the shot's path.

81. Let x = the width of the rectangle.
Let $x+4$ = the length of the rectangle.

$$A = lw$$
$$8 = x(x+4)$$
$$0 = x^2 + 4x - 8$$

Apply the quadratic formula.
$a=1 \quad b=4 \quad c=-8$

$$x = \frac{-4 \pm \sqrt{4^2 - 4(1)(-8)}}{2(1)}$$
$$= \frac{-4 \pm \sqrt{16-(-32)}}{2}$$
$$= \frac{-4 \pm \sqrt{48}}{2} = \frac{-4 \pm 4\sqrt{3}}{2}$$
$$= -2 \pm 2\sqrt{3} \approx 1.5 \text{ or } -5.5$$

Disregard -5.5 because the width of a rectangle cannot be negative. Thus, the solution is 1.5, and the rectangle's dimensions are 1.5 meters by $1.5 + 4 = 5.5$ meters.

83. Let x = the length of the longer leg.
Let $x-1$ = the length of the shorter leg.
Let $x+7$ = the length of the hypotenuse.

$$x^2 + (x-1)^2 = (x+7)^2$$
$$x^2 + x^2 - 2x + 1 = x^2 + 14x + 49$$
$$2x^2 - 2x + 1 = x^2 + 14x + 49$$
$$x^2 - 16x - 48 = 0$$

Apply the quadratic formula.
$a=1 \quad b=-16 \quad c=-48$

$$x = \frac{-(-16) \pm \sqrt{(-16)^2 - 4(1)(-48)}}{2(1)}$$
$$= \frac{16 \pm \sqrt{256-(-192)}}{2}$$
$$= \frac{16 \pm \sqrt{448}}{2}$$
$$= \frac{16 \pm 8\sqrt{7}}{2}$$
$$= 8 \pm 4\sqrt{7} \approx 18.6 \text{ or } -2.6$$

Disregard -2.6 because the length of a leg cannot be negative. The solution is 18.6, and the lengths of the triangle's legs are approximately 18.6 inches and $18.6 - 1 = 17.6$ inches.

85. $x(20-2x) = 13$

$$20x - 2x^2 = 13$$
$$0 = 2x^2 - 20x + 13$$

Apply the quadratic formula.
$a=2 \quad b=-20 \quad c=13$

$$x = \frac{-(-20) \pm \sqrt{(-20)^2 - 4(2)(13)}}{2(2)}$$
$$= \frac{20 \pm \sqrt{400-104}}{4}$$
$$= \frac{20 \pm \sqrt{296}}{4}$$
$$= \frac{20 \pm 2\sqrt{74}}{4} = \frac{10 \pm \sqrt{74}}{2} \approx 9.3 \text{ or } 0.7$$

A gutter with depth 9.3 or 0.7 inches will have a cross-sectional area of 13 square inches.

87. Let x = the time for the first person to mow the yard alone.
Let $x+1$ = the time for the second person to mow the yard alone.

	Fractional part of job completed in 1 hour	Time working together	Fractional part of job completed in 4 hour
1st person	$\frac{1}{x}$	4	$\frac{4}{x}$
2nd person	$\frac{1}{x+1}$	4	$\frac{1}{x+1}$

$$\frac{4}{x} + \frac{4}{x+1} = 1$$
$$x(x+1)\left(\frac{4}{x} + \frac{4}{x+1}\right) = x(x+1)1$$
$$4(x+1) + 4x = x^2 + x$$
$$4x + 4 + 4x = x^2 + x$$
$$0 = x^2 - 7x - 4$$

Apply the quadratic formula.
$a = 1 \quad b = -7 \quad c = -4$

$$x = \frac{-(-7) \pm \sqrt{(-7)^2 - 4(1)(-4)}}{2(1)}$$
$$= \frac{7 \pm \sqrt{49 - (-16)}}{2}$$
$$= \frac{7 \pm \sqrt{65}}{2} \approx 7.5 \text{ or } -0.5$$

Disregard -0.5 because time cannot be negative. Thus, the solution is 7.5, and we conclude that the first person can mow the lawn alone in 7.5 hours, and the second mow the yard alone in 7.5 + 1 = 8.5 hours.

89. – 95. Answers will vary.

97. $f(x) = x(20 - 2x)$

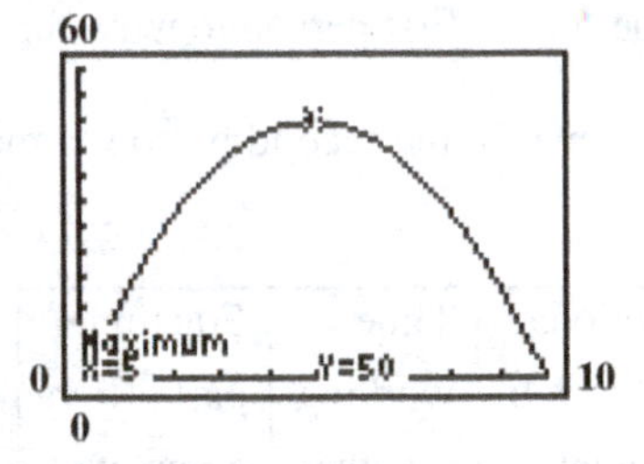

The depth of the gutter that will maximize water flow is 5 inches. This will allow for a water flow of 50 square inches. The situation described does not take full advantage of the sheets of aluminum.

99. does not make sense; Explanations will vary. Sample explanation: That is not a correct simplification because 2 is not a factor of both terms in the numerator.

101. does not make sense; Explanations will vary. Sample explanation: There are two imaginary solutions. They are not classified as irrational.

103. false; Changes to make the statement true will vary. A sample change is: The quadratic formula is developed by completing the square and the square root property.

105. true

107. The dimensions of the pool are 12 meters by 8 meters. With the tile, the dimensions will be 12 + $2x$ meters by 8 + $2x$ meters. If we take the area of the pool with the tile and subtract the area of the pool without the tile, we are left with the area of the tile only.

$$(12 + 2x)(8 + 2x) - 12(8) = 120$$
$$\cancel{96} + 24x + 16x + 4x^2 - \cancel{96} = 120$$
$$4x^2 + 40x - 120 = 0$$
$$x^2 + 10x - 30 = 0$$
$$a = 1 \quad b = 10 \quad c = -30$$

$$x = \frac{-10 \pm \sqrt{10^2 - 4(1)(-30)}}{2(1)}$$
$$= \frac{-10 \pm \sqrt{100 + 120}}{2}$$
$$= \frac{-10 \pm \sqrt{220}}{2} \approx \frac{-10 \pm 14.8}{2}$$

Evaluate the expression to obtain two solutions.

$$x = \frac{-10 + 14.8}{2} \quad \text{or} \quad x = \frac{-10 - 14.8}{2}$$
$$x = \frac{4.8}{2} \qquad x = \frac{-24.8}{2}$$
$$x = 2.4 \qquad x = -12.4$$

Disregard -12.4 because we can't have a negative width measurement. The solution is 2.4 and we conclude that the width of the uniform tile border is 2.4 meters. This is more than the 2-meter requirement, so the tile meets the zoning laws.

109. $|5x + 2| = |4 - 3x|$

$$5x + 2 = 4 - 3x \quad \text{or} \quad 5x + 2 = -(4 - 3x)$$
$$8x + 2 = 4 \qquad 5x + 2 = -4 + 3x$$
$$8x = 2 \qquad 2x + 2 = -4$$
$$x = \frac{1}{4} \qquad 2x = -6$$
$$x = -3$$

The solution set is $\left\{-3, \frac{1}{4}\right\}$.

110. $\sqrt{2x-5}-\sqrt{x-3}=1$

$$\sqrt{2x-5}=\sqrt{x-3}+1$$
$$\left(\sqrt{2x-5}\right)^2=\left(\sqrt{x-3}+1\right)^2$$
$$2x-5=x-3+2\sqrt{x-3}+1$$
$$2x-5=x-2+2\sqrt{x-3}$$
$$x-3=2\sqrt{x-3}$$

$$(x-3)^2=\left(2\sqrt{x-3}\right)^2$$
$$x^2-6x+9=4(x-3)$$
$$x^2-6x+9=4x-12$$
$$x^2-10x+21=0$$
$$(x-7)(x-3)=0$$

$x-7=0$ or $x-3=0$
$x=7$ $\qquad$ $x=3$

The solution set is $\{3,7\}$.

111. $\dfrac{5}{\sqrt{3}+x}=\dfrac{5}{\sqrt{3}+x}\cdot\dfrac{\sqrt{3}-x}{\sqrt{3}-x}=\dfrac{5\sqrt{3}-5x}{3-x^2}$

112.

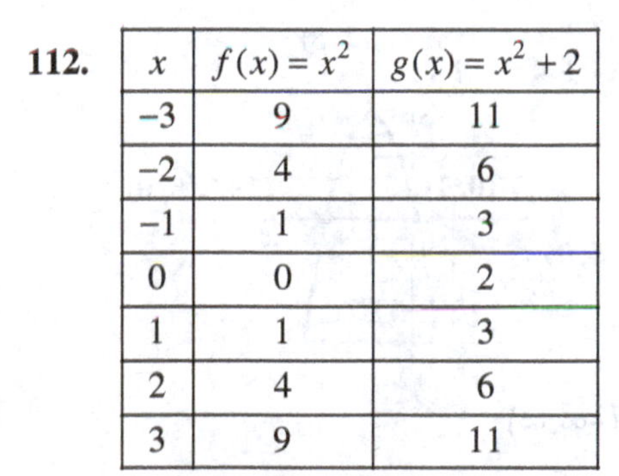

x	$f(x)=x^2$	$g(x)=x^2+2$
–3	9	11
–2	4	6
–1	1	3
0	0	2
1	1	3
2	4	6
3	9	11

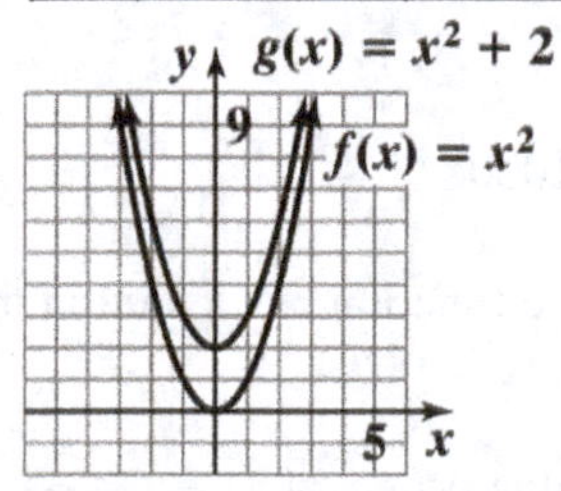

113.

x	$f(x)=x^2$	$g(x)=(x+2)^2$
–3	9	1
–2	4	0
–1	1	1
0	0	4
1	1	9
2	4	16
3	9	25

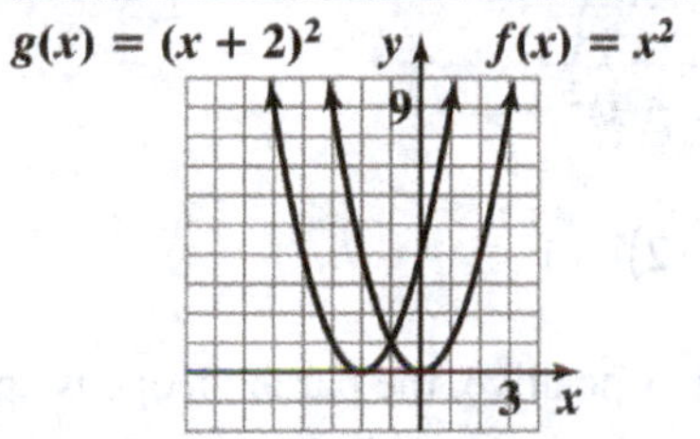

114. $f(x)=-2(x-3)^2+8$

$$0=-2(x-3)^2+8$$
$$2(x-3)^2=8$$
$$(x-3)^2=4$$
$$\sqrt{(x-3)^2}=\pm\sqrt{4}$$
$$x-3=\pm2$$
$$x=3\pm2$$

$x=3+2$ or $x=3-2$
$x=5$ $\qquad$ $x=1$

The x-intercepts are 1 and 5.

8.3 Check Points

1. $f(x)=-(x-1)^2+4$

Since $a=-1$ is negative, the parabola opens downward. The vertex of the parabola is $(h,k)=(1,4)$. Replace $f(x)$ with 0 to find x–intercepts.

$$0=-(x-1)^2+4$$
$$(x-1)^2=4$$
$$x-1=\pm\sqrt{4}$$
$$x-1=\pm2$$

$x-1=2$ or $x-1=-2$
$x=3$ $\qquad$ $x=-1$

The x–intercepts are -1 and 3.
Set $x = 0$ and solve for y to obtain the y–intercept.

$y = -(0-1)^2 + 4 = 3$

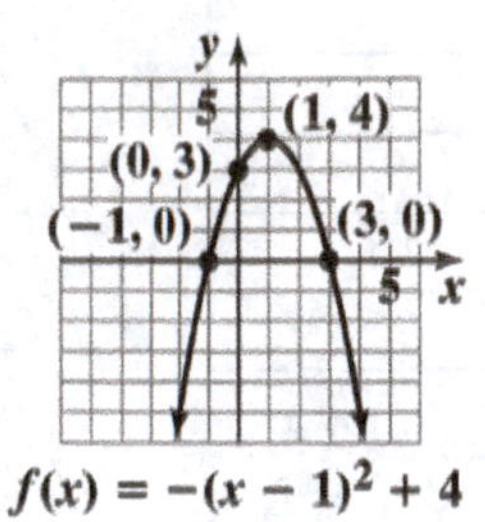

$f(x) = -(x-1)^2 + 4$

2. $f(x) = (x-2)^2 + 1$

Since $a = 1$ is positive, the parabola opens upward. The vertex of the parabola is $(h, k) = (2, 1)$.

Replace $f(x)$ with 0 to find x–intercepts.

$$0 = (x-2)^2 + 1$$
$$(x-2)^2 = -1$$
$$x - 2 = \pm\sqrt{-1}$$
$$x = 2 \pm i$$

Because this equation has no real solutions, the parabola has no x-intercepts.
Set $x = 0$ and solve for y to obtain the y–intercept.

$y = (0-2)^2 + 1 = 5$

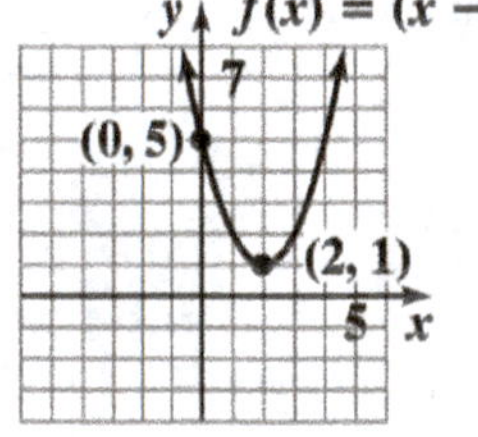

3. $f(x) = 2x^2 + 8x - 1$

The x–coordinate of the vertex of the parabola is

$-\dfrac{b}{2a} = -\dfrac{8}{2(2)} = -\dfrac{8}{4} = -2$, and the y–coordinate of the vertex of the parabola is

$f\left(-\dfrac{b}{2a}\right) = f(-2) = 2(-2)^2 + 8(-2) - 1 = -9$.

The vertex is $(-2, -9)$.

4. $f(x) = -x^2 + 4x + 1$

Since $a = -1$ is negative, the parabola opens downward. The x–coordinate of the vertex of the parabola is $-\dfrac{b}{2a} = -\dfrac{4}{2(-1)} = -\dfrac{4}{-2} = 2$ and the y–coordinate of the vertex of the parabola is

$f\left(-\dfrac{b}{2a}\right) = f(2) = -(2)^2 + 4(2) + 1 = 5.$

The vertex is $(2, 5)$.

Replace $f(x)$ with 0 to find x–intercepts.

$$0 = -x^2 + 4x + 1$$
$$x = \frac{-b \pm \sqrt{b^2 - 4ac}}{2a}$$
$$x = \frac{-4 \pm \sqrt{4^2 - 4(-1)(1)}}{2(-1)}$$
$$x = 2 \pm \sqrt{5}$$
$$x \approx -0.2 \text{ or } x \approx 4.2$$

The x–intercepts are -0.2 and 4.2.
Set $x = 0$ and solve for y to obtain the y–intercept.

$y = -0^2 + 4 \cdot 0 + 1 = 1$

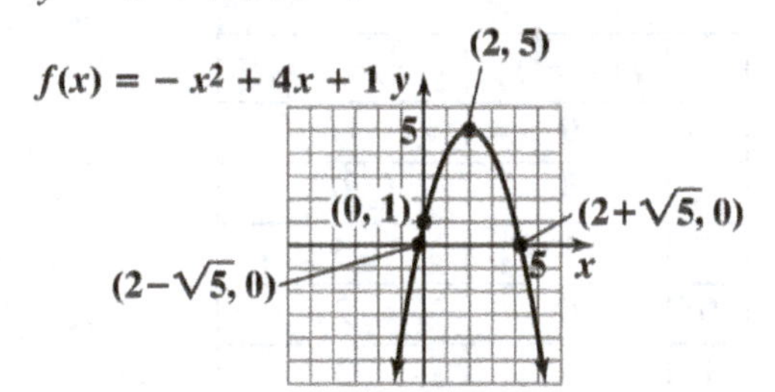

Domain: $(-\infty, \infty)$

Range: $(-\infty, 5]$

5. $f(x) = 4x^2 - 16x + 1000$

a. Because $a > 0$, the function has a minimum value.

b. The minimum value occurs at
$-\dfrac{b}{2a} = -\dfrac{-16}{2(4)} = -\dfrac{-16}{8} = 2$

The minimum of $f(x)$ is
$f(2) = 4 \cdot 2^2 - 16 \cdot 2 + 1000 = 984.$

c. Like all quadratic functions, the domain is $(-\infty, \infty)$.
Because the minimum is 984, the range includes all real numbers at or above 984.
The range is $[984, \infty)$.

6. $g(x) = -0.04x^2 + 2.1x + 6.1$
Because $a < 0$, the function has a maximum value that occurs at
$$-\frac{b}{2a} = -\frac{2.1}{2(-0.04)} = -\frac{2.1}{-0.08} = 26.25.$$
The maximum height is given by
$$g(26.25) = -0.04(26.25)^2 + 2.1(26.25) + 6.1 \approx 33.7$$
The maximum height of about 33.7 feet occurs at a horizontal distance of 26.25 feet.

7. Let the two numbers be represented by x and y, and let the product be represented by P.
We must minimize $P = xy$.
Because the difference of the two numbers is 8, then $x - y = 8$.
Solve for y in terms of x.
$$x - y = 8$$
$$-y = -x + 8$$
$$y = x - 8$$
Write P as a function of x.
$$P = xy$$
$$P(x) = x(x-8)$$
$$P(x) = x^2 - 8x$$
Because $a > 0$, the function has a minimum value that occurs at $x = -\frac{b}{2a} = -\frac{-8}{2(1)} = 4.$
Substitute to find the other number.
$$y = x - 8$$
$$y = 4 - 8$$
$$= -4$$
The two numbers are 4 and -4. The minimum product is $P = xy = (4)(-4) = -16$.

8. Let the two dimensions be represented by x and y, and let the area be represented by A.
We must maximize $A = xy$.
Because the perimeter, P, is 120, then
$$P = 2x + 2y$$
$$120 = 2x + 2y$$
Solve for y in terms of x.
$$120 = 2x + 2y$$
$$-2y = 2x - 120$$
$$y = -x + 60$$
Write A as a function of x.
$$A = xy$$
$$A(x) = x(-x + 60)$$
$$A(x) = -x^2 + 60x$$
Because $a < 0$, the function has a maximum value that occurs at $x = -\frac{b}{2a} = -\frac{60}{2(-1)} = 30.$
Substitute to find the other dimension.
$$y = -x + 60$$
$$y = -30 + 60$$
$$= 30$$
The two dimensions are 30 feet by 30 feet. The maximum area is
$A = xy = (30)(30) = 900$ square feet.

8.3 Concept and Vocabulary Check

1. parabola; upward; downward
2. lowest/minimum
3. highest/maximum
4. (h, k)
5. $-\frac{b}{2a}$; $-\frac{b}{2a}$
6. 0; solutions
7. $(f, 0)$

8.3 Exercise Set

1. The vertex of the graph is the point $(1, 1)$. This means that the equation is $h(x)=(x-1)^2+1$.

3. The vertex of the graph is the point $(1,-1)$. This means that the equation is $j(x)=(x-1)^2-1$.

5. The vertex of the graph is the point $(0,-1)$. This means that the equation is
$h(x)=(x-0)^2-1=x^2-1$.

7. The vertex of the graph is the point (1, 0). This means that the equation is
$$f(x)=(x-1)^2+0$$
$$=(x-1)^2=x^2-2x+1$$

9. $f(x)=2(x-3)^2+1$
The vertex is (3, 1).

11. $f(x)=-2(x+1)^2+5$
The vertex is $(-1,5)$.

13. $f(x)=2x^2-8x+3$
The x–coordinate of the vertex of the parabola is
$-\frac{b}{2a}=-\frac{-8}{2(2)}=-\frac{-8}{4}=2$, and the y–coordinate of the vertex of the parabola is
$$f\left(-\frac{b}{2a}\right)=f(2)=2(2)^2-8(2)+3$$
$$=2(4)-16+3$$
$$=8-16+3=-5.$$
The vertex is $(2,-5)$.

15. $f(x)=-x^2-2x+8$
The x–coordinate of the vertex of the parabola is
$-\frac{b}{2a}=-\frac{-2}{2(-1)}=-\frac{-2}{-2}=-1$,
and the y–coordinate of the vertex of the parabola is
$$f\left(-\frac{b}{2a}\right)=f(-1)$$
$$=-(-1)^2-2(-1)+8$$
$$=-1+2+8=9.$$
The vertex is $(-1,9)$.

17. $f(x)=(x-4)^2-1$

Since $a=1$ is positive, the parabola opens upward. The vertex of the parabola is $(h,k)=(4,-1)$.
Replace $f(x)$ with 0 to find x–intercepts.
$$0=(x-4)^2-1$$
$$1=(x-4)^2$$

Apply the square root property.
$$x-4=\pm\sqrt{1}=\pm1$$
$$x=4\pm1=5 \text{ or } 3$$

The x–intercepts are 5 and 3.
Set $x = 0$ and solve for y to obtain the y–intercept.
$$y=(0-4)^2-1=15$$

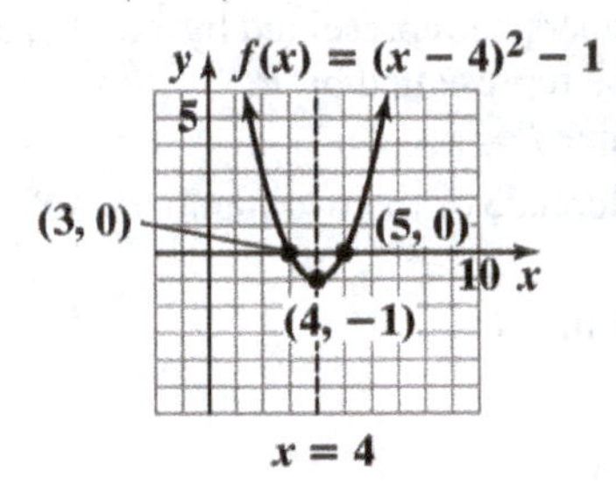

Axis of symmetry: $x=4$.
Range: $[-1,\infty)$.

19. $f(x)=(x-1)^2+2$

Since $a=1$ is positive, the parabola opens upward. The vertex of the parabola is $(h,k)=(1,2)$.
Replace $f(x)$ with 0 to find x–intercepts.
$$0=(x-1)^2+2$$
$$-2=(x-1)^2$$

Because the solutions to the equation are imaginary, there are no x–intercepts. Set $x = 0$ and solve for y to obtain the y–intercept.

$y = (0-1)^2 + 2 = (-1)^2 + 2 = 1 + 2 = 3$

The y–intercept is 3.

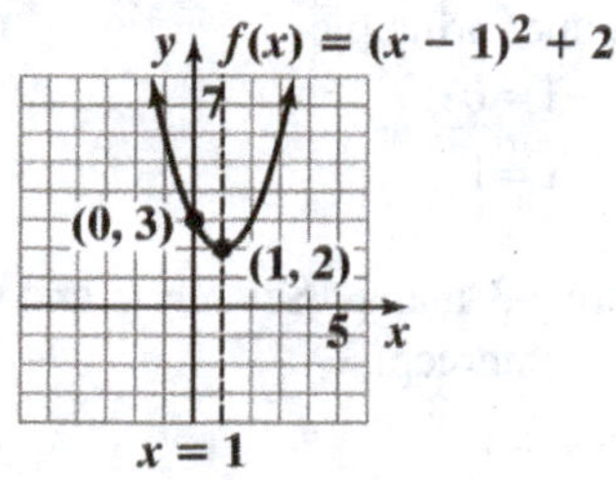

Axis of symmetry: $x = 1$.

Range: $[2, \infty)$.

21. $y - 1 = (x-3)^2$

$y = (x-3)^2 + 1$

$f(x) = (x-3)^2 + 1$

Since $a = 1$ is positive, the parabola opens upward. The vertex of the parabola is $(h, k) = (3, 1)$.

Replace $f(x)$ with 0 to find x–intercepts.

$0 = (x-3)^2 + 1$

$-1 = (x-3)^2$

Because the solutions to the equation are imaginary, there are no x–intercepts. Set $x = 0$ and solve for y to obtain the y–intercept.

$y = (0-3)^2 + 1 = (-3)^2 + 1 = 9 + 1 = 10$

The y–intercept is 10.

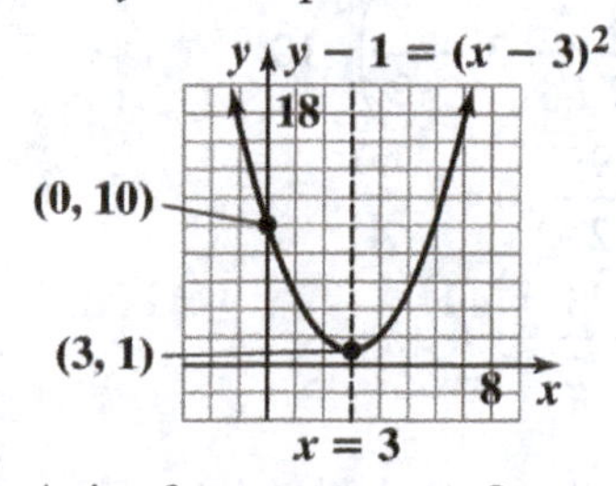

Axis of symmetry: $x = 3$.

Range: $[1, \infty)$.

23. $f(x) = 2(x+2)^2 - 1$

Since $a = 2$ is positive, the parabola opens upward. The vertex of the parabola is $(h, k) = (-2, -1)$.

Replace $f(x)$ with 0 to find x–intercepts.

$0 = 2(x+2)^2 - 1$

$1 = 2(x+2)^2$

$\frac{1}{2} = (x+2)^2$

Apply the square root property.

$x + 2 = \pm\sqrt{\frac{1}{2}}$

$x = -2 \pm \sqrt{\frac{1}{2}}$

$x \approx -2 - \sqrt{\frac{1}{2}}$ or $-2 + \sqrt{\frac{1}{2}}$

$x \approx -2.7$ or -1.3

The x–intercepts are –1.3 and –2.7.

Set $x = 0$ and solve for y to obtain the y–intercept.

$y = 2(0+2)^2 - 1$

$= 2(2)^2 - 1 = 2(4) - 1 = 8 - 1 = 7$

The y–intercept is 7.

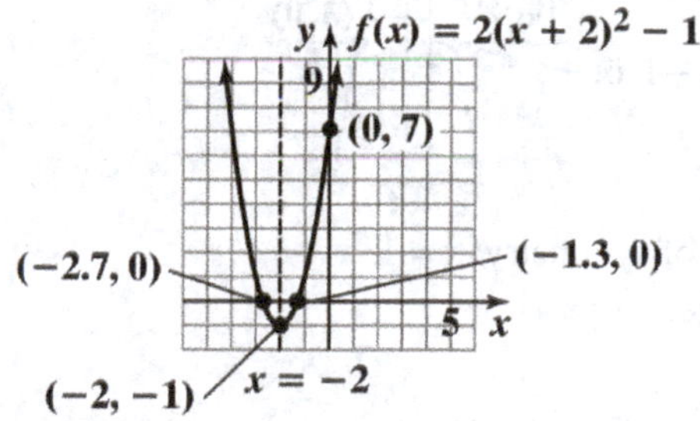

Axis of symmetry: $x = -2$.

Range: $[-1, \infty)$.

25. $f(x)=4-(x-1)^2$

$f(x)=-(x-1)^2+4$

Since $a=-1$ is negative, the parabola opens downward. The vertex of the parabola is $(h,k)=(1,4)$. Replace $f(x)$ with 0 to find x–intercepts.

$0=-(x-1)^2+4$

$-4=-(x-1)^2$

$4=(x-1)^2$

Apply the square root property.

$\sqrt{4}=x-1$ or $-\sqrt{4}=x-1$

$2=x-1 \qquad -2=x-1$

$3=x \qquad -1=x$

The x–intercepts are –1 and 3.
Set $x=0$ and solve for y to obtain the y–intercept.

$y=-(0-1)^2+4$

$=-(-1)^2+4=-1+4=3$

The y–intercept is 3.

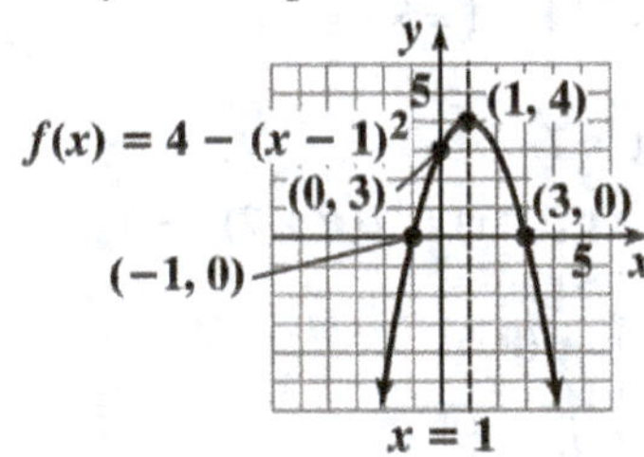

Axis of symmetry: $x=1$.
Range: $(-\infty,4]$.

27. $f(x)=x^2+2x-3$

Since $a=1$ is positive, the parabola opens upward. The x–coordinate of the vertex of the parabola is

$-\frac{b}{2a}=-\frac{2}{2(1)}=-\frac{2}{2}=-1$ and the

y–coordinate of the vertex of the parabola is

$f\left(-\frac{b}{2a}\right)=f(-1)$

$=(-1)^2+2(-1)-3$

$=1-2-3=-4.$

The vertex is $(-1,-4)$. Replace $f(x)$ with 0 to find x–intercepts.

$0=x^2+2x-3$

$0=(x+3)(x-1)$

Apply the zero product principle.

$x+3=0$ or $x-1=0$

$x=-3 \qquad x=1$

The x–intercepts are –3 and 1. Set $x=0$ and solve for y to obtain the y–intercept.

$y=(0)^2+2(0)-3=-3$

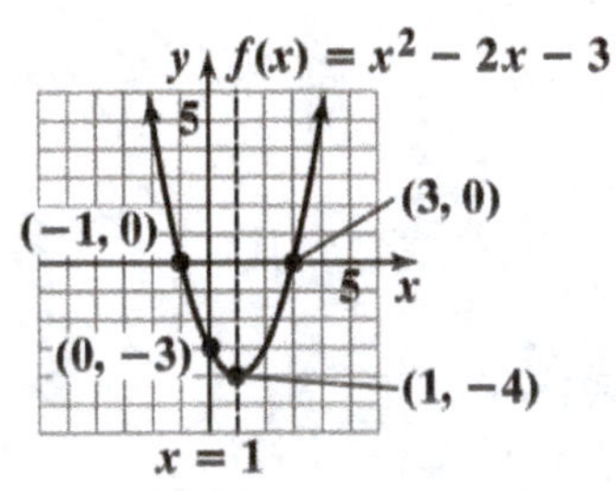

Axis of symmetry: $x=-1$.
Range: $[-4,\infty)$.

29. $f(x)=x^2+3x-10$

Since $a=1$ is positive, the parabola opens upward. The x–coordinate of the vertex of the parabola is

$-\frac{b}{2a}=-\frac{3}{2(1)}=-\frac{3}{2}$ and the

y–coordinate of the vertex of the parabola is

$f\left(-\frac{b}{2a}\right)=f\left(-\frac{3}{2}\right)$

$=\left(-\frac{3}{2}\right)^2+3\left(-\frac{3}{2}\right)-10$

$=\frac{9}{4}-\frac{9}{2}-10=-\frac{49}{4}.$

The vertex is $\left(-\frac{3}{2},-\frac{49}{4}\right)$.

Replace $f(x)$ with 0 to find the x–intercepts.

$0=x^2+3x-10$

$0=(x+5)(x-2)$

Apply the zero product principle.

$x+5=0$ or $x-2=0$

$x=-5 \qquad x=2$

The x–intercepts are –5 and 2.
Set $x = 0$ and solve for y to obtain the y–intercept.
$y = 0^2 + 3(0) - 10 = -10$

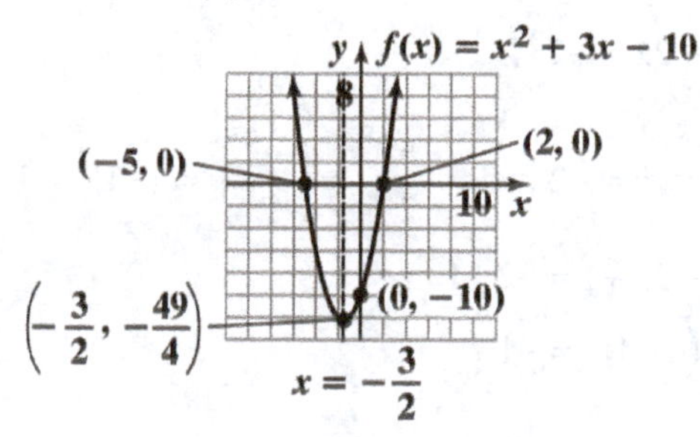

Axis of symmetry: $x = -\frac{3}{2}$.

Range: $\left[-\frac{49}{4}, \infty\right)$.

31. $f(x) = 2x - x^2 + 3$

$f(x) = -x^2 + 2x + 3$

Since $a = -1$ is negative, the parabola opens downward. The x–coordinate of the vertex of the parabola is $-\frac{b}{2a} = -\frac{2}{2(-1)} = -\frac{2}{-2} = 1$ and the y–coordinate of the vertex of the parabola is

$$f\left(-\frac{b}{2a}\right) = f(1) = -(1)^2 + 2(1) + 3$$
$$= -1 + 2 + 3 = 4.$$

The vertex is $(1, 4)$. Replace $f(x)$ with 0 to find x–intercepts.

$0 = -x^2 + 2x + 3$

$0 = x^2 - 2x - 3$

$0 = (x - 3)(x + 1)$

Apply the zero product principle.

$x - 3 = 0$ or $x + 1 = 0$

$x = 3$ $\quad$ $x = -1$

The x–intercepts are 3 and –1. Set $x = 0$ and solve for y to obtain the y–intercept. $y = -(0)^2 + 2(0) + 3 = 3$

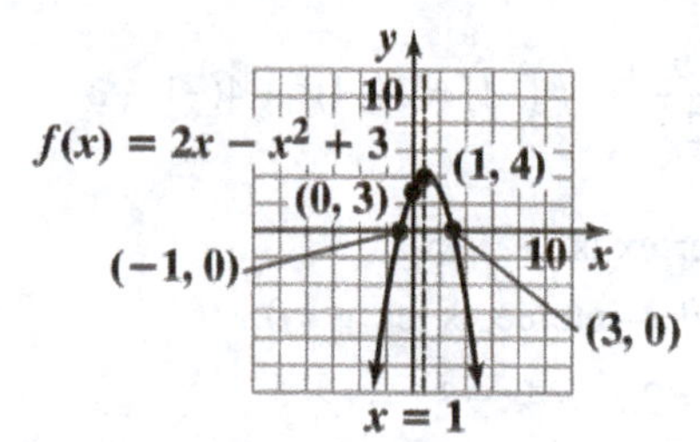

Axis of symmetry: $x = 1$.
Range: $(-\infty, 4]$.

33. $f(x) = x^2 + 6x + 3$

$$-\frac{b}{2a} = -\frac{6}{2(1)} = -3$$

$$f\left(-\frac{b}{2a}\right) = f(-3) = (-3)^2 + 6(-3) + 3 = -6$$

The vertex is $(-3, -6)$.
To find x–intercepts let $y = 0$.

$0 = x^2 + 6x + 3$

$$x = \frac{-b \pm \sqrt{b^2 - 4ac}}{2a}$$

$$x = \frac{-6 \pm \sqrt{6^2 - 4(1)(3)}}{2(1)}$$

$x = -3 \pm \sqrt{6}$

$x \approx -5.4$ or $x \approx -0.6$

The x–intercepts are –5.4 and –0.6. The y–intercept is 3.

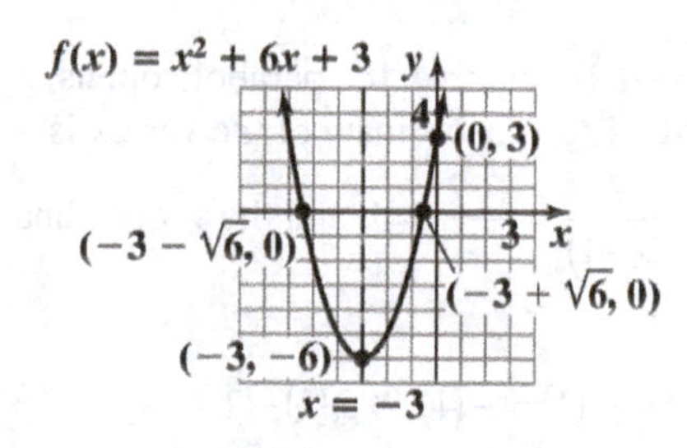

Range: $[-6, \infty)$.

35. $f(x) = 2x^2 + 4x - 3$

$-\frac{b}{2a} = -\frac{4}{2(2)} = -1$

$f\left(-\frac{b}{2a}\right) = f(-1) = 2(-1)^2 + 4(-1) - 3 = -5$

The vertex is $(-1, -5)$.
To find x–intercepts let $y = 0$.

$f(x) = 2x^2 + 4x - 3$

$x = \frac{-b \pm \sqrt{b^2 - 4ac}}{2a}$

$x = \frac{-4 \pm \sqrt{4^2 - 4(2)(-3)}}{2(2)}$

$x = -1 \pm \frac{\sqrt{10}}{2}$

$x \approx -2.6$ or $x \approx 0.6$

The x–intercepts are -2.6 and 0.6. The y–intercept is -3.

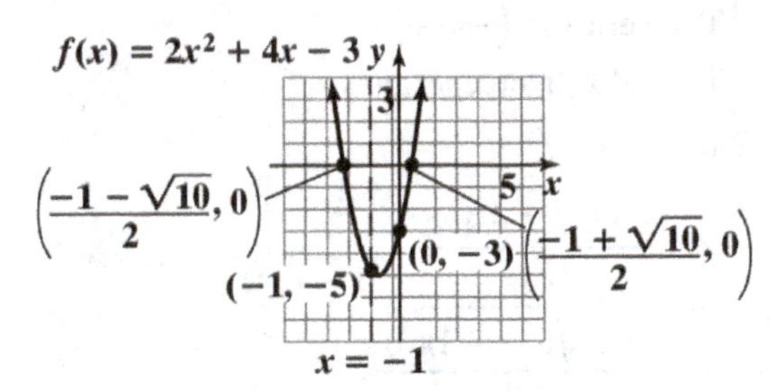

Range: $[-5, \infty)$.

37. $f(x) = 2x - x^2 - 2$

$f(x) = -x^2 + 2x - 2$

Since $a = -1$ is negative, the parabola opens downward. The x–coordinate of the vertex is

$-\frac{b}{2a} = -\frac{2}{2(-1)} = -\frac{2}{-2} = 1$ and the y–coordinate of the vertex is

$f\left(-\frac{b}{2a}\right) = f(1) = -(1)^2 + 2(1) - 2$

$= -1 + 2 - 2 = -1.$

The vertex is $(1, -1)$. Replace $f(x)$ with 0 to find x–intercepts.

$0 = -x^2 + 2x - 2$

$x^2 - 2x = -2$

Since $b = -2$, add $\left(\frac{b}{2}\right)^2 = \left(\frac{-2}{2}\right)^2 = (-1)^2 = 1$

$x^2 - 2x + 1 = -2 + 1$

$(x-1)^2 = -1$

Because the solutions to the equation are imaginary, there are no x–intercepts. Set $x = 0$ and solve for y to obtain the y–intercept. $y = 2(0) - 0^2 - 2 = -2$

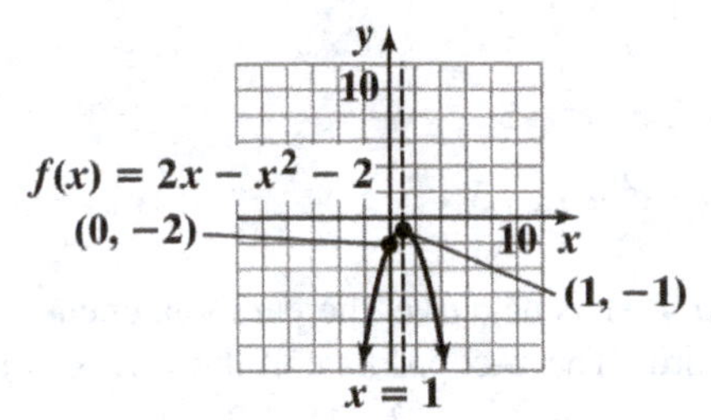

Axis of symmetry: $x = 1$.
Range: $(-\infty, -1]$.

39. $f(x) = 3x^2 - 12x - 1$

a. Since $a > 0$, the parabola opens upward and has a minimum.

b. The x–coordinate of the minimum is

$-\frac{b}{2a} = -\frac{-12}{2(3)} = -\frac{-12}{6} = 2$ and the y–coordinate of the minimum is

$f\left(-\frac{b}{2a}\right) = f(2) = 3(2)^2 - 12(2) - 1 = 12 - 24 - 1 = -1$

c. Domain: $(-\infty, \infty)$

Range: $[-13, \infty)$

41. $f(x) = -4x^2 + 8x - 3$

a. Since $a < 0$, the parabola opens downward and has a maximum.

b. The x–coordinate of the maximum is $-\frac{b}{2a} = -\frac{8}{2(-4)} = -\frac{8}{-8} = 1$ and the y–coordinate of the maximum is

$$f\left(-\frac{b}{2a}\right) = f(1) = -4(1)^2 + 8(1) - 3$$
$$= -4 + 8 - 3 = 1.$$

c. Domain: $(-\infty, \infty)$

Range: $\{y \mid y \le 1\}$ or $(-\infty, 1]$

43. $f(x) = 5x^2 - 5x$

a. Since $a > 0$, the parabola opens upward and has a minimum.

b. The x–coordinate of the minimum is $-\frac{b}{2a} = -\frac{-5}{2(5)} = -\frac{-5}{10} = \frac{1}{2}$ and the y–coordinate of the minimum is

$$f\left(-\frac{b}{2a}\right) = f\left(\frac{1}{2}\right) = 5\left(\frac{1}{2}\right)^2 - 5\left(\frac{1}{2}\right)$$
$$= 5\left(\frac{1}{4}\right) - \frac{5}{2} = \frac{5}{4} - \frac{10}{4} = -\frac{5}{4}.$$

c. Domain: $(-\infty, \infty)$

Range: $\left[-\frac{5}{4}, \infty\right)$

45. Since the parabola opens up, the vertex $(-1, -2)$ is a minimum point.
The domain is $(-\infty, \infty)$. The range is $[-2, \infty)$.

47. Since the parabola has a maximum point, it opens down. The domain is $(-\infty, \infty)$. The range is $(-\infty, -6]$.

49. $(h, k) = (5, 3)$

$$f(x) = 2(x-h)^2 + k = 2(x-5)^2 + 3$$

51. $(h, k) = (-10, -5)$

$$f(x) = 2(x-h)^2 + k$$
$$= 2[x - (-10)]^2 + (-5)$$
$$= 2(x+10)^2 - 5$$

53. Since the vertex is a maximum, the parabola opens down and $a = -3$.

$(h, k) = (-2, 4)$

$$f(x) = -3(x-h)^2 + k$$
$$= -3[x - (-2)]^2 + 4$$
$$= -3(x+2)^2 + 4$$

55. Since the vertex is a minimum, the parabola opens up and $a = 3$.

$(h, k) = (11, 0)$

$$f(x) = 3(x-h)^2 + k$$
$$= 3(x-11)^2 + 0$$
$$= 3(x-11)^2$$

57. $s(t) = -16t^2 + 64t + 160$

a. The t-coordinate of the minimum is

$$t = -\frac{b}{2a} = -\frac{64}{2(-16)} = -\frac{64}{-32} = 2.$$

The s-coordinate of the minimum is

$$s(2) = -16(2)^2 + 64(2) + 160$$
$$= -16(4) + 128 + 160$$
$$= -64 + 128 + 160 = 224$$

The ball reaches a maximum height of 224 feet 2 seconds after it is thrown.

b. $0 = -16t^2 + 64t + 160$

$$0 = t^2 - 4t - 10$$

$a = 1 \quad b = -4 \quad c = -10$

$$t = \frac{-(-4) \pm \sqrt{(-4)^2 - 4(1)(-10)}}{2(1)}$$
$$= \frac{4 \pm \sqrt{16 + 40}}{2}$$
$$= \frac{4 \pm \sqrt{56}}{2} \approx \frac{4 \pm 7.48}{2}$$

Evaluate the expression to obtain two solutions.

$x = \frac{4+7.48}{2}$ or $x = \frac{4-7.48}{2}$

$x = \frac{11.48}{2}$ $\quad$ $x = \frac{-3.48}{2}$

$x = 5.74$ $\quad$ $x = -1.74$

Disregard -1.74 because we can't have a negative time measurement. The solution is 5.74 and we conclude that the ball will hit the ground in approximately 5.7 seconds.

c. $s(0) = -16(0)^2 + 64(0) + 160$
$= -16(0) + 0 + 160 = 160$

At $t = 0$, the ball has not yet been thrown and is at a height of 160 feet. This is the height of the building.

d.

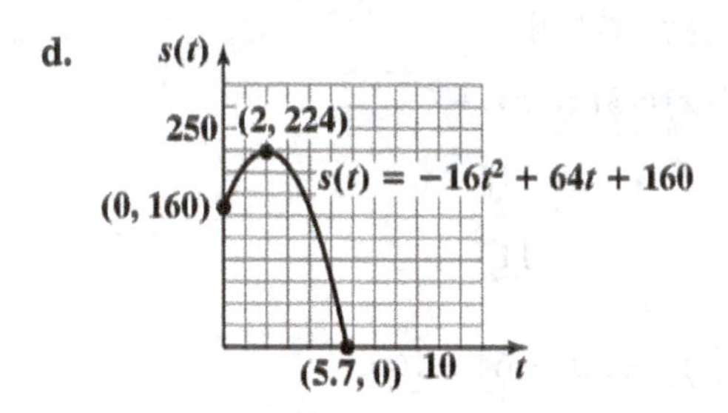

59. Let x = one of the numbers.
Let $16 - x$ = the other number.

The product is $f(x) = x(16-x)$
$= 16x - x^2 = -x^2 + 16x$

The x-coordinate of the maximum is

$x = -\frac{b}{2a} = -\frac{16}{2(-1)} = -\frac{16}{-2} = 8.$

$f(8) = -8^2 + 16(8) = -64 + 128 = 64$

The vertex is (8, 64). The maximum product is 64. This occurs when the two number are 8 and $16 - 8 = 8$.

61. Let x = one of the numbers.
Let $x - 16$ = the other number.

The product is $f(x) = x(x-16) = x^2 - 16x$.
The x-coordinate of the minimum is

$x = -\frac{b}{2a} = -\frac{16}{2(1)} = -\frac{16}{2} = -8.$

$f(-8) = (-8)^2 + 16(-8)$
$= 64 - 128 = -64$

The vertex is $(-8, -64)$. The minimum product is -64. This occurs when the two number are -8 and $-8 + 16 = 8$.

63. Maximize the area of a rectangle constructed along a river with 600 feet of fencing.

Let x = the width of the rectangle.
Let $600 - 2x$ = the length of the rectangle.
We need to maximize.

$A(x) = x(600-2x)$
$= 600x - 2x^2 = -2x^2 + 600x$

Since $a = -2$ is negative, the function opens downward and has a maximum at

$x = -\frac{b}{2a} = -\frac{600}{2(-2)} = -\frac{600}{-4} = 150.$

When the width is $x = 150$ feet, the length is $600 - 2(150) = 600 - 300 = 300$ feet.
The dimensions of the rectangular plot with maximum area are 150 feet by 300 feet. This gives an area of $150 \cdot 300 = 45{,}000$ square feet.

65. Maximize the area of a rectangle constructed with 50 yards of fencing.

Let x = the length of the rectangle. Let y = the width of the rectangle.
Since we need an equation in one variable, use the perimeter to express y in terms of x.

$2x + 2y = 50$
$2y = 50 - 2x$
$y = \frac{50-2x}{2} = 25 - x$

We need to maximize $A = xy = x(25-x)$. Rewrite A as a function of x.

$A(x) = x(25-x) = -x^2 + 25x$

Since $a = -1$ is negative, the function opens downward and has a maximum at

$$x=-\frac{b}{2a}=-\frac{25}{2(-1)}=-\frac{25}{-2}=12.5.$$

When the length x is 12.5, the width y is
$y=25-x=25-12.5=12.5.$
The dimensions of the rectangular region with maximum area are 12.5 yards by 12.5 yards. This gives an area of $12.5\cdot 12.5=156.25$ square yards.

67. Maximize the cross-sectional area of the gutter:

$$A(x)=x(20-2x)$$
$$=20x-2x^2=-2x^2+20x.$$

Since $a=-2$ is negative, the function opens downward and has a maximum at

$$x=-\frac{b}{2a}=-\frac{20}{2(-2)}=-\frac{20}{-4}=5.$$

When the height x is 5, the width is
$20-2x=20-2(5)=20-10=10.$

$$A(5)=-2(5)^2+20(5)$$
$$=-2(25)+100=-50+100=50$$

The maximum cross-sectional area is 50 square inches. This occurs when the gutter is 5 inches deep and 10 inches wide.

69. a. $C(x)=525+0.55x$

b. $P(x)=R(x)-C(x)$

$$=\left(-0.001x^2+3x\right)-(525+0.55x)$$
$$=-0.001x^2+3x-525-0.55x$$
$$=-0.001x^2+2.45x-525$$

c. Since $a=-0.001$ is negative, the function opens down and has a maximum at

$$x=-\frac{b}{2a}$$
$$=-\frac{2.45}{2(-0.001)}=-\frac{2.45}{-0.002}=1225.$$

When the number of units x is 1225, the profit is

$$P(1225)$$
$$=-0.001(1225)^2+2.45(1225)-525$$
$$=-0.001(1500625)+3001.25-525$$
$$=-1500.625+3001.25-525$$
$$=975.625$$

The store maximizes its weekly profit when 1225 roast beef sandwiches are made and sold, resulting in a profit of $975.63.

71. – 75. Answers will vary.

77. a. $y=2x^2-82x+720$

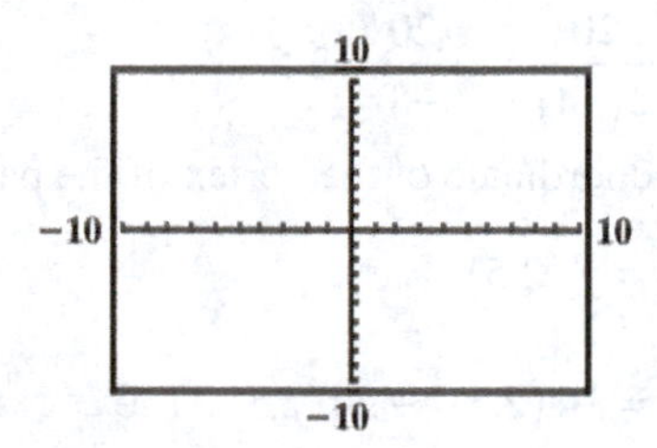

The function has no values that fall within the window.

b. $y=2x^2-82x+720$

The x–coordinate of the vertex of the parabola is $-\frac{b}{2a}=-\frac{-82}{2(2)}=-\frac{-82}{4}=20.5$ and the y–coordinate of the vertex of the parabola is

$$f\left(-\frac{b}{2a}\right)=f(20.5)$$
$$=2(20.5)^2-82(20.5)+720$$
$$=2(420.25)-1681+720$$
$$=840.5-1681+720=-120.5.$$

The vertex is $(20.5,-120.5)$.

c. Using a Ymax of 50, we have the following.

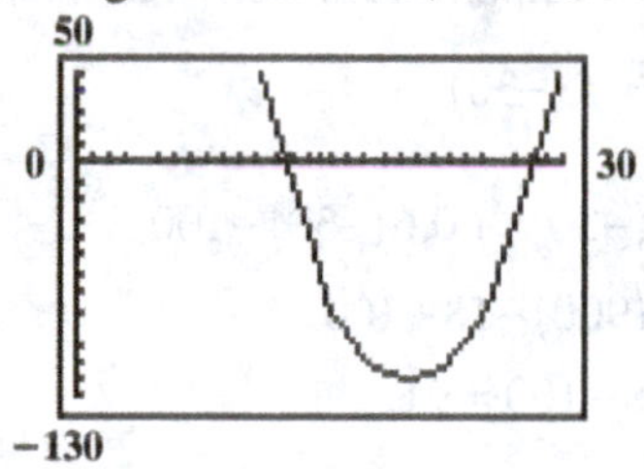

d. Answers will vary.

79. $y = -4x^2 + 20x + 160$

The x–coordinate of the vertex of the parabola is
$-\frac{b}{2a} = -\frac{20}{2(-4)} = -\frac{20}{-8} = 2.5$
and the y–coordinate of the vertex of the parabola is

$$\begin{aligned} f\left(-\frac{b}{2a}\right) &= f(2.5) \\ &= -4(2.5)^2 + 20(2.5) + 160 \\ &= -4(6.25) + 50 + 160 \\ &= -25 + 50 + 160 \\ &= 185. \end{aligned}$$

The vertex is (2.5, 185).

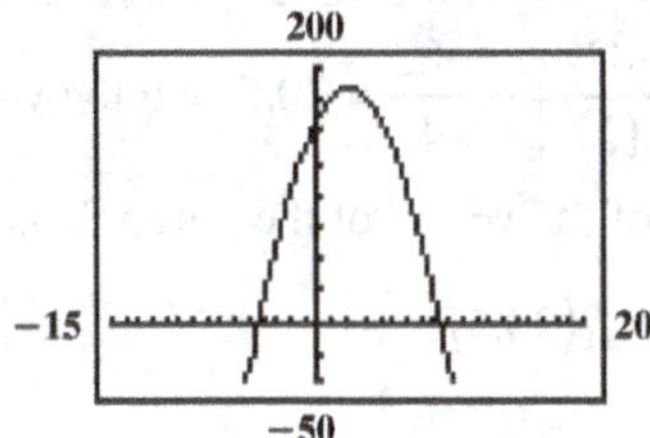

81. $y = 0.01x^2 + 0.6x + 100$

The x–coordinate of the vertex of the parabola is
$-\frac{b}{2a} = -\frac{0.6}{2(0.01)} = -\frac{0.6}{0.02} = -30$
and the y–coordinate of the vertex of the parabola is

$$\begin{aligned} f\left(-\frac{b}{2a}\right) &= f(-30) \\ &= 0.01(-30)^2 + 0.6(-30) + 100 \\ &= 0.01(900) - 18 + 100 \\ &= 9 - 18 + 100 = 91. \end{aligned}$$

The vertex is $(-30, 91)$.

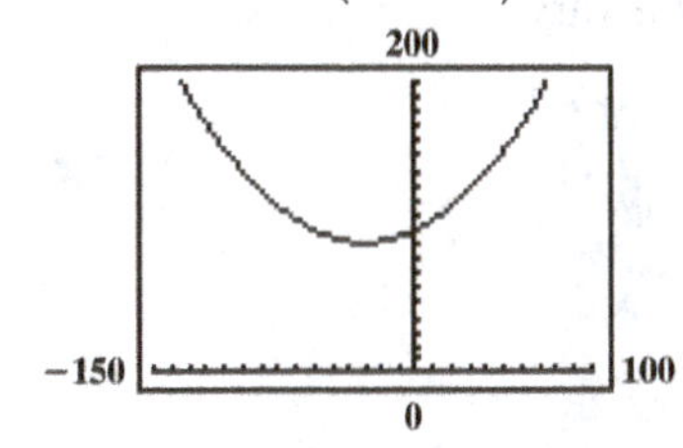

83. does not make sense; Explanations will vary. Sample explanation: If $b = 0$, then the axis of symmetry will be the y-axis.

85. does not make sense; Explanations will vary. Sample explanation: If thrown straight up, the path will be a straight line.

87. false; Changes to make the statement true will vary. A sample change is: The vertex is $(5, -1)$.

89. false; Changes to make the statement true will vary. A sample change is: The x-coordinate of the maximum is $-\frac{b}{2a} = -\frac{1}{2(-1)} = -\frac{1}{-2} = \frac{1}{2}$ and the y–coordinate of the vertex of the parabola is

$$\begin{aligned} f\left(-\frac{b}{2a}\right) &= f\left(\frac{1}{2}\right) = -\left(\frac{1}{2}\right)^2 + \frac{1}{2} + 1 \\ &= -\frac{1}{4} + \frac{1}{2} + 1 = -\frac{1}{4} + \frac{2}{4} + \frac{4}{4} = \frac{5}{4}. \end{aligned}$$

The maximum y–value is $\frac{5}{4}$.

91. $f(x) = (x-3)^2 + 2$
Since the vertex is $(3, 2)$, the axis of symmetry is the line $x = 3$. The point (6, 11) is on the parabola and lies three units to the right of the axis of symmetry. This means that the point (0, 11) will also lie on the parabola since it lies 3 units to the left of the axis of symmetry.

93. The vertex is $(h, k) = (-3, -4)$, so the equation is of the form

$$\begin{aligned} f(x) &= a(x-h)^2 + k \\ &= a\left[x - (-3)\right]^2 + (-1) \\ &= a(x+3)^2 - 1 \end{aligned}$$

Use the point $(-2, -3)$ on the graph to determine the value of a:

$$\begin{aligned} f(x) &= a(x+3)^2 - 1 \\ -3 &= a(-2+3)^2 - 1 \\ -3 &= a(1)^2 - 1 \\ -3 &= a - 1 \\ -2 &= a \end{aligned}$$

Thus, the equation of the parabola is
$f(x) = -2(x+3)^2 - 1$.

95. Let x = the number of trees over 50 that will be planted.
The function describing the annual yield per lemon tree when $x+50$ trees are planted per acre is

$$\begin{aligned} f(x) &= (x+50)(320-4x) \\ &= 320x-4x^2+16000-200x \\ &= -4x^2+120x+16000. \end{aligned}$$

This represents the number of lemon trees planted per acre multiplied by yield per tree.

The x–coordinate of the maximum is

$-\frac{b}{2a} = -\frac{120}{2(-4)} = -\frac{120}{-8} = 15$ and the y–coordinate of the vertex of the parabola is

$$f\left(-\frac{b}{2a}\right) = f(15) = -4(15)^2+120(15)+16000 = 16900$$

The maximum lemon yield is 16,900 pounds when 50 + 15 = 65 lemon trees are planted per acre.

96.

$$\frac{2}{x+5}+\frac{1}{x-5}=\frac{16}{x^2-25}$$

$$\frac{2}{x+5}+\frac{1}{x-5}=\frac{16}{(x+5)(x-5)}$$

$$(x+5)(x-5)\left(\frac{2}{x+5}+\frac{1}{x-5}\right)=(x+5)(x-5)\left(\frac{16}{(x+5)(x-5)}\right)$$

$$\begin{aligned} (x-5)(2)+(x+5)(1) &= 16 \\ 2x-10+x+5 &= 16 \\ 3x-5 &= 16 \\ 3x &= 21 \\ x &= 7 \end{aligned}$$

The solution set is $\{7\}$.

97.

$$\frac{1+\frac{2}{x}}{1-\frac{4}{x^2}} = \frac{x^2}{x^2}\cdot\frac{1+\frac{2}{x}}{1-\frac{4}{x^2}} = \frac{x^2+x\cdot 2}{x^2-4}$$

$$= \frac{x^2+2x}{(x+2)(x-2)}$$

$$= \frac{x\cancel{(x+2)}}{\cancel{(x+2)}(x-2)} = \frac{x}{x-2}$$

98. $D = \begin{vmatrix} 2 & 3 \\ 1 & -4 \end{vmatrix} = 2(-4)-1(3) = -8-3 = -11$

$D_x = \begin{vmatrix} 6 & 3 \\ 14 & -4 \end{vmatrix} = 6(-4)-14(3)$
$= -24-42 = -66$

$D_y = \begin{vmatrix} 2 & 6 \\ 1 & 14 \end{vmatrix} = 2(14)-1(6) = 28-6 = 22$

$x = \frac{D_x}{D} = \frac{-66}{-11} = 6$ $\qquad y = \frac{D_y}{D} = \frac{22}{-11} = -2$

The solution set is $\{(6,-2)\}$.

99.

$$\begin{gathered} u^2-8u-9=0 \\ (u+1)(u-9)=0 \\ u+1=0 \quad \text{or} \quad u-9=0 \\ u=-1 \qquad\qquad u=9 \end{gathered}$$

The solution set is $\{-1, 9\}$.

100.

$$\begin{gathered} 2u^2-u-10=0 \\ (u+2)(2u-5)=0 \\ u+2=0 \quad \text{or} \quad 2u-5=0 \\ u=-2 \qquad\qquad 2u=5 \\ u=\frac{5}{2} \end{gathered}$$

The solution set is $\left\{-2, \frac{5}{2}\right\}$.

101.

$$5x^{\frac{2}{3}}+11x^{\frac{1}{3}}+2=0$$

$$5\left(x^{\frac{1}{3}}\right)^2+11x^{\frac{1}{3}}+2=0$$

$$5u^2+11u+2=0$$

Mid-Chapter Check Point – Chapter 8

1. $(3x-5)^2 = 36$

Apply the square root principle.

$3x-5 = \pm\sqrt{36} = \pm 6$

$3x = 5 \pm 6$

$x = \frac{5\pm 6}{3} = \frac{11}{3}$ or $-\frac{1}{3}$

The solution set is $\left\{\frac{11}{3}, -\frac{1}{3}\right\}$.

2. $5x^2 - 2x = 7$

$5x^2 - 2x - 7 = 0$

$(5x-7)(x+1) = 0$

Apply the zero-product principle.

$5x - 7 = 0$ or $x + 1 = 0$

$5x = 7$ $\quad x = -1$

$x = \frac{7}{5}$

The solution set is $\left\{-1, \frac{7}{5}\right\}$.

3. $3x^2 - 6x - 2 = 0$

Apply the quadratic formula.

$a = 3 \quad b = -6 \quad c = -2$

$$x = \frac{-(-6) \pm \sqrt{(-6)^2 - 4(3)(-2)}}{2(3)}$$

$$= \frac{6 \pm \sqrt{36-(-24)}}{6}$$

$$= \frac{6 \pm \sqrt{60}}{6}$$

$$= \frac{6 \pm \sqrt{4 \cdot 15}}{6} = \frac{6 \pm 2\sqrt{15}}{6} = \frac{3 \pm \sqrt{15}}{3}$$

The solution set is $\left\{\frac{3 \pm \sqrt{15}}{3}\right\}$.

4. $x^2 + 6x = -2$

$x^2 + 6x + 2 = 0$

Apply the quadratic formula.

$a = 1 \quad b = 6 \quad c = 2$

$$x = \frac{-6 \pm \sqrt{6^2 - 4(1)(2)}}{2(1)}$$

$$= \frac{-6 \pm \sqrt{36-8}}{2}$$

$$= \frac{-6 \pm \sqrt{28}}{2}$$

$$= \frac{-6 \pm \sqrt{4 \cdot 7}}{2} = \frac{-6 \pm 2\sqrt{7}}{2} = -3 \pm \sqrt{7}$$

The solution set is $\left\{-3 \pm \sqrt{7}\right\}$.

5. $5x^2 + 1 = 37$

$5x^2 = 36$

$x^2 = \frac{36}{5}$

Apply the square root principle.

$$x = \pm\sqrt{\frac{36}{5}} = \pm\frac{6}{\sqrt{5}} \cdot \frac{\sqrt{5}}{\sqrt{5}} = \pm\frac{6\sqrt{5}}{5}$$

The solution set is $\left\{\pm\frac{6\sqrt{5}}{5}\right\}$.

6. $x^2 - 5x + 8 = 0$

Apply the quadratic formula.

$a = 1 \quad b = -5 \quad c = 8$

$$x = \frac{-(-5) \pm \sqrt{(-5)^2 - 4(1)(8)}}{2(1)}$$

$$= \frac{5 \pm \sqrt{25-32}}{2}$$

$$= \frac{5 \pm \sqrt{-7}}{2}$$

$$= \frac{5 \pm \sqrt{7 \cdot (-1)}}{2}$$

$$= \frac{5 \pm i\sqrt{7}}{2}$$

$$= \frac{5}{2} \pm i\frac{\sqrt{7}}{2}$$

The solution set is $\left\{\frac{5}{2} \pm \frac{\sqrt{7}}{2}i\right\}$.

7. $2x^2 + 26 = 0$

$2x^2 = -26$

$x^2 = -13$

Apply the square root principle.

$x = \pm\sqrt{-13} = \pm\sqrt{13(-1)} = \pm i\sqrt{13}$.

The solution set is $\{\pm i\sqrt{13}\}$.

8. $(2x+3)(x+2) = 10$

$2x^2 + 4x + 3x + 6 = 10$

$2x^2 + 7x - 4 = 0$

$(2x-1)(x+4) = 0$

Apply the zero-product principle.

$2x - 1 = 0$ or $x + 4 = 0$

$2x = 1$ $\quad x = -4$

$x = \frac{1}{2}$

The solution set is $\left\{-4, \frac{1}{2}\right\}$.

9. $(x+3)^2 = 24$

Apply the square root principle.

$x + 3 = \pm\sqrt{24}$

$x + 3 = \pm\sqrt{4 \cdot 6} = \pm 2\sqrt{6}$

$x = -3 \pm 2\sqrt{6}$

The solution set is $\{-3 \pm 2\sqrt{6}\}$.

10. $\frac{1}{x^2} - \frac{4}{x} + 1 = 0$

Multiply both sides of the equation by the least common denominator x^2.

$x^2\left(\frac{1}{x^2} - \frac{4}{x} + 1\right) = x^2(0)$

$1 - 4x + x^2 = 0$

$x^2 - 4x + 1 = 0$

Apply the quadratic formula.

$a = 1 \quad b = -4 \quad c = 1$

$$x = \frac{-(-4) \pm \sqrt{(-4)^2 - 4(1)(1)}}{2(1)}$$

$$= \frac{4 \pm \sqrt{16-4}}{2}$$

$$= \frac{4 \pm \sqrt{12}}{2}$$

$$= \frac{4 \pm \sqrt{4 \cdot 3}}{2} = \frac{4 \pm 2\sqrt{3}}{2} = 2 \pm \sqrt{3}$$

The solution set is $\{2 \pm \sqrt{3}\}$.

11. $x(2x-3) = -4$

$2x^2 - 3x = -4$

$2x^2 - 3x + 4 = 0$

Apply the quadratic formula.

$a = 2 \quad b = -3 \quad c = 4$

$$x = \frac{-(-3) \pm \sqrt{(-3)^2 - 4(2)(4)}}{2(2)}$$

$$= \frac{3 \pm \sqrt{9-32}}{4}$$

$$= \frac{3 \pm \sqrt{-23}}{4}$$

$$= \frac{3 \pm \sqrt{23(-1)}}{4} = \frac{3 \pm i\sqrt{23}}{4} = \frac{3}{4} \pm i\frac{\sqrt{23}}{4}$$

The solution set is $\left\{\frac{3}{4} \pm i\frac{\sqrt{23}}{4}\right\}$.

12. $\frac{x^2}{3} + \frac{x}{2} = \frac{2}{3}$

Multiply both sides of the equation by the least common denominator 6.

$6\left(\frac{x^2}{3} + \frac{x}{2}\right) = 6\left(\frac{2}{3}\right)$

$2x^2 + 3x = 4$

$2x^2 + 3x - 4 = 0$

Apply the quadratic formula.

$a = 3 \quad b = 3 \quad c = -4$

$$x = \frac{-3 \pm \sqrt{3^2 - 4(2)(-4)}}{2(2)}$$

$$= \frac{-3 \pm \sqrt{9-(-32)}}{4} = \frac{-3 \pm \sqrt{41}}{4}$$

The solution set is $\left\{\frac{-3 \pm \sqrt{41}}{4}\right\}$.

13. $\dfrac{2x}{x^2+6x+8}=\dfrac{x}{x+4}-\dfrac{2}{x+2}$

$\dfrac{2x}{(x+4)(x+2)}=\dfrac{x}{x+4}-\dfrac{2}{x+2}$

Multiply both sides of the equation by the least common denominator $(x+4)(x+2)$.

$$(x+4)(x+2)\left[\frac{2x}{(x+4)(x+2)}\right]=(x+4)(x+2)\left[\frac{x}{x+4}-\frac{2}{x+2}\right]$$

$$2x=x(x+2)-2(x+4)$$
$$2x=x^2+2x-2x-8$$
$$0=x^2-2x-8$$
$$0=(x-4)(x+2)$$

Apply the zero-product principle.

$x-4=0$ or $x+2=0$

$x=4$ $\quad x=-2$

Disregard $x=-2$ since it makes the denominator zero. Thus, the solution is 4, and the solution set is $\{4\}$.

14. $x^2+10x-3=0$

$x^2+10x \quad =3$

Since $b=10$, add $\left(\dfrac{10}{2}\right)^2=5^2=25$.

$x^2+10x+25=3+25$

$(x+5)^2=28$

Apply the square root principle.

$x+5=\pm\sqrt{28}$

$x+5=\pm\sqrt{4\cdot 7}=\pm 2\sqrt{7}$

$x=-5\pm 2\sqrt{7}$

The solution set is $\{-5\pm 2\sqrt{7}\}$.

15. $f(x)=(x-3)^2-4$

Since $a=1$ is positive, the parabola opens upward. The vertex of the parabola is $(h,k)=(3,-4)$.

Replace $f(x)$ with 0 to find x–intercepts.

$0=(x-3)^2-4$

$4=(x-3)^2$

Apply the square root property.

$x-3=\pm\sqrt{4}=\pm 2$

$x=3\pm 2=5$ or 1

The x–intercepts are 5 and 1.

Set $x=0$ and solve for y to obtain the y–intercept.

$y=(0-3)^2-4=5$

$f(x)=(x-3)^2-4$

(0, 5)
(1, 0)
(5, 0)
(3, −4)

Domain: $(-\infty,\infty)$

Range: $[-4,\infty)$

16. $g(x)=5-(x+2)^2$

$g(x)=-[x-(-2)]^2+5$

Since $a=-1$ is negative, the parabola opens downward. The vertex of the parabola is $(h,k)=(-2,5)$. Replace $g(x)$ with 0 to find x–intercepts.

$0=-(x+2)^2+5$

$-5=-(x+2)^2$

$5=(x+2)^2$

Apply the square root property.

$\sqrt{5}=x+2$ or $-\sqrt{5}=x+2$

$-2+\sqrt{5}=x$ $\quad -2-\sqrt{5}=x$

$0.24\approx x$ $\quad -4.24\approx x$

The x–intercepts are $-2+\sqrt{5}\approx 0.24$ and $-2-\sqrt{5}\approx -4.24$.

Set $x=0$ and solve for y to obtain the y–intercept.

$y=5-(0+2)^2=5-4=1$

The y–intercept is 1.

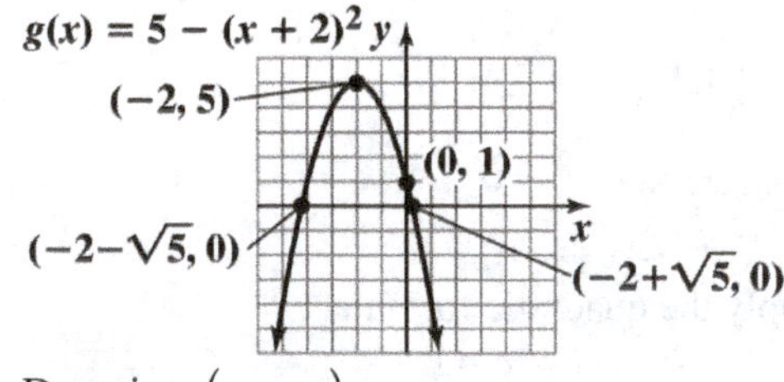

Domain: $(-\infty,\infty)$

Range: $(-\infty,5]$

17. $h(x) = -x^2 - 4x + 5$

Since $a = -1$ is negative, the parabola opens downward. The x–coordinate of the vertex of the parabola is $-\frac{b}{2a} = -\frac{-4}{2(-1)} = -\frac{-4}{-2} = -2$ and the y–coordinate of the vertex of the parabola is

$$h\left(-\frac{b}{2a}\right) = h(-2)$$
$$= -(-2)^2 - 4(-2) + 5$$
$$= -4 + 8 + 5 = 9$$

The vertex is $(-2, 9)$. Replace $h(x)$ with 0 to find the x–intercepts.

$0 = -x^2 - 4x + 5$

$0 = x^2 + 4x - 5$

$0 = (x+5)(x-1)$

Apply the zero product principle.

$x + 5 = 0$ or $x - 1 = 0$

$x = -5$ $\quad x = 1$

The x–intercepts are -5 and 1. Set $x = 0$ and solve for y to obtain the y–intercept. $y = -(0)^2 - 4(0) + 5 = 5$

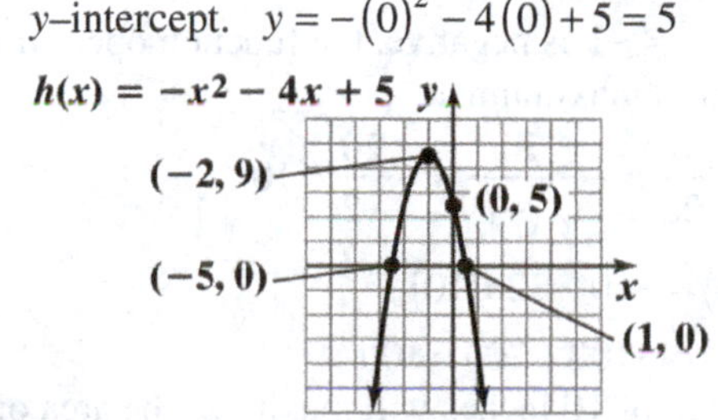

Domain: $(-\infty, \infty)$

Range: $(-\infty, 9]$

18. $f(x) = 3x^2 - 6x + 1$

Since $a = 3$, the parabola opens upward. The x–coordinate of the vertex of the parabola is $-\frac{b}{2a} = -\frac{-6}{2(3)} = -\frac{-6}{6} = 1$ and the y–coordinate of the minimum is

$$f\left(-\frac{b}{2a}\right) = f(1)$$
$$= 3(1)^2 - 6(1) + 1$$
$$= 3 - 6 + 1 = -2$$

The vertex $(1, 1)$. Replace $f(x)$ with 0 to find the x-intercepts.

$0 = 3x^2 - 6x + 1$

Apply the quadratic formula.

$a = 3 \quad b = -6 \quad c = 1$

$$x = \frac{-(-6) \pm \sqrt{(-6)^2 - 4(3)(1)}}{2(3)}$$
$$= \frac{6 \pm \sqrt{36 - 12}}{6}$$
$$= \frac{6 \pm \sqrt{24}}{6}$$
$$= \frac{6 \pm 2\sqrt{6}}{6} = \frac{3 \pm \sqrt{6}}{3} \approx 0.18 \text{ or } 1.82$$

Set $x = 0$ and solve for y to obtain the y-intercept:

$y = 3(0)^2 - 6(0) + 1 = 1$

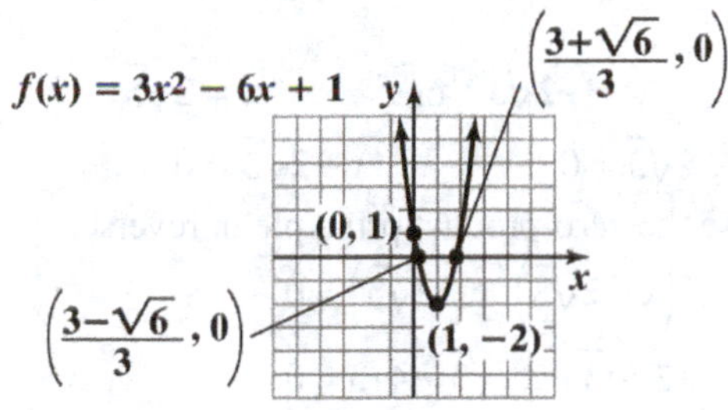

Domain: $(-\infty, \infty)$

Range: $[-2, \infty)$

19. $2x^2 + 5x + 4 = 0$

$a = 2 \quad b = 5 \quad c = 4$

$$b^2 - 4ac = 5^2 - 4(2)(4)$$
$$= 25 - 32 = -7$$

Since the discriminant is negative, there are no real solutions. There are two imaginary solutions that are complex conjugates.

20. $10x(x+4) = 15x - 15$

$10x^2 + 40x = 15x - 15$

$10x^2 - 25x + 15 = 0$

$a = 10 \quad b = -25 \quad c = 15$

$$b^2 - 4ac = (-25)^2 - 4(10)(15)$$
$$= 625 - 600 = 25$$

Since the discriminant is positive and a perfect square, there are two real rational solutions.

21. Because the solution set is $\left\{-\frac{1}{2}, \frac{3}{4}\right\}$, we have

$$x = -\frac{1}{2} \quad \text{or} \quad x = \frac{3}{4}$$
$$2x = -1 \qquad 4x = 3$$
$$2x + 1 = 0 \qquad 4x - 3 = 0$$

Use the zero-product principle in reverse.

$$(2x+1)(4x-3) = 0$$
$$8x^2 - 6x + 4x - 3 = 0$$
$$8x^2 - 2x - 3 = 0$$

22. Because the solution set is $\left\{-2\sqrt{3}, 2\sqrt{3}\right\}$ we have

$$x = -2\sqrt{3} \quad \text{or} \quad x = 2\sqrt{3}$$
$$x + 2\sqrt{3} = 0 \qquad x - 2\sqrt{3} = 0$$

Use the zero-product principle in reverse.

$$\left(x + 2\sqrt{3}\right)\left(x - 2\sqrt{3}\right) = 0$$
$$x^2 + 2x\sqrt{3} - 2x\sqrt{3} - 4 \cdot 3 = 0$$
$$x^2 - 12 = 0$$

23. $P(x) = -x^2 + 150x - 4425$

Since $a = -1$ is negative, the function opens down and has a maximum at

$$x = -\frac{b}{2a} = -\frac{150}{2(-1)} = -\frac{150}{-2} = 75.$$

$$P(75) = -75^2 + 150(75) - 4425$$
$$= -5625 + 11,250 - 4425 = 1200$$

The company will maximize its profit by manufacturing and selling 75 cabinets per day. The maximum daily profit is $1200.

24. Let x = one of the numbers.
Let $-18 - x$ = the other number.

The product is $f(x) = x(-18 - x) = -x^2 - 18x$

The x-coordinate of the maximum is

$$x = -\frac{b}{2a} = -\frac{-18}{2(-1)} = -\frac{-18}{-2} = -9.$$

$$f(-9) = -9\left[-18 - (-9)\right]$$
$$= -9(-18 + 9) = -9(-9) = 81$$

The vertex is $(-9, 81)$. The maximum product is 81. This occurs when the two number are -9 and $-18 - (-9) = -9$.

25. Let x = the measure of the height.
Let $40 - 2x$ = the measure of the base.

$$A = \frac{1}{2}bh$$
$$A(x) = \frac{1}{2}(40 - 2x)x$$
$$A(x) = -x^2 + 20x$$

Since $a = -1$ is negative, the function opens down and has a maximum at

$$x = -\frac{b}{2a} = -\frac{20}{2(-1)} = -\frac{20}{-2} = 10.$$

$$A(10) = -10^2 + 20(10)$$
$$= -100 + 200 = 100$$

A height of 10 inches will maximize the area of the triangle. The maximum area will be 100 square inches.

8.4 Check Points

1. Let $u = x^2$.

$$x^4 - 5x^2 + 6 = 0$$
$$\left(x^2\right)^2 - 5x^2 + 6 = 0$$
$$u^2 - 5u + 6 = 0$$
$$(u-3)(u-2) = 0$$

Apply the zero product principle.

$u - 3 = 0$ or $u - 2 = 0$

$u = 3$ $\quad$ $u = 2$

Replace u with x^2.

$x^2 = 3$ or $x^2 = 2$

$x = \pm\sqrt{3}$ $\quad$ $x = \pm\sqrt{2}$

The solution set is $\left\{\pm\sqrt{2}, \pm\sqrt{3}\right\}$.

2. Let $u = \sqrt{x}$.

$$x - 2\sqrt{x} - 8 = 0$$
$$\left(\sqrt{x}\right)^2 - 2\sqrt{x} - 8 = 0$$
$$u^2 - 2u - 8 = 0$$
$$(u+2)(u-4) = 0$$

Apply the zero product principle.

$u + 2 = 0$ or $u - 4 = 0$

$u = -2$ $\quad$ $u = 4$

Replace u with $\sqrt{x}$.

~~$\sqrt{x} = -2$~~ or $\sqrt{x} = 4$

$x = 16$

Disregard –2 because the square root of x cannot be a negative number. We must check 16, because both sides of the equation were raised to an even power.

Check:

$$x - 2\sqrt{x} - 8 = 0$$
$$16 - 2\sqrt{16} - 8 = 0$$
$$16 - 2 \cdot 4 - 8 = 0$$
$$16 - 8 - 8 = 0$$
$$0 = 0$$

The solution set is $\{16\}$.

3. Let $u = x^2 - 4$.

$$(x^2-4)^2 + (x^2-4) - 6 = 0$$
$$u^2 + u - 6 = 0$$
$$(u+3)(u-2) = 0$$

Apply the zero product principle and replace u.

$u + 3 = 0$ or $u - 2 = 0$

$x^2 - 4 + 3 = 0$ $\quad$ $x^2 - 4 - 2 = 0$

$x^2 - 1 = 0$ $\quad$ $x^2 - 6 = 0$

$x^2 = 1$ $\quad$ $x^2 = 6$

$x = \pm 1$ $\quad$ $x = \pm\sqrt{6}$

The solution set is $\left\{\pm\sqrt{6}, \pm 1\right\}$.

4. Let $u = x^{-1}$.

$$2x^{-2} + x^{-1} - 1 = 0$$
$$2\left(x^{-1}\right)^2 + x^{-1} - 1 = 0$$
$$2u^2 + u - 1 = 0$$
$$(u+1)(2u-1) = 0$$

Apply the zero product principle.

$u + 1 = 0$ or $2u - 1 = 0$

$u = -1$ $\quad$ $2u = 1$

$u = \frac{1}{2}$

Replace u with x^{-1}.

$u = -1$ or $u = \frac{1}{2}$

$x^{-1} = -1$ $\quad$ $x^{-1} = \frac{1}{2}$

$\left(x^{-1}\right)^{-1} = (-1)^{-1}$ $\quad$ $\left(x^{-1}\right)^{-1} = \left(\frac{1}{2}\right)^{-1}$

$x = -1$ $\quad$ $x = 2$

The solution set is $\{-1, 2\}$.

5. Let $u = x^{\frac{1}{3}}$.

$$3x^{\frac{2}{3}} - 11x^{\frac{1}{3}} - 4 = 0$$
$$3\left(x^{\frac{1}{3}}\right)^2 - 11x^{\frac{1}{3}} - 4 = 0$$
$$3u^2 - 11u - 4 = 0$$
$$(3u+1)(u-4) = 0$$

Apply the zero product principle.

$3u+1=0$ or $u-4=0$

$u = -\frac{1}{3}$ $\quad$ $u = 4$

Replace u with $x^{\frac{1}{3}}$.

$u = -\frac{1}{3}$ $\quad$ or $\quad$ $u = 4$

$x^{\frac{1}{3}} = -\frac{1}{3}$ $\quad$ $x^{\frac{1}{3}} = 4$

$\left(x^{\frac{1}{3}}\right)^3 = \left(-\frac{1}{3}\right)^3$ $\quad$ $\left(x^{\frac{1}{3}}\right)^3 = (4)^3$

$x = -\frac{1}{27}$ $\quad$ $x = 64$

The solution set is $\left\{-\frac{1}{27}, 64\right\}$.

8.4 Concept and Vocabulary Check

1. x^2; $u^2 - 13u + 36 = 0$

2. $x^{\frac{1}{2}}$ or $\sqrt{x}$; $u^2 - 2u - 8 = 0$

3. $x+3$; $u^2 + 7u - 18 = 0$

4. x^{-1}; $2u^2 - 7u + 3 = 0$

5. $x^{\frac{1}{3}}$; $u^2 + 2u - 3 = 0$

8.4 Exercise Set

1. Let $u = x^2$.

$$x^4 - 5x^2 + 4 = 0$$
$$\left(x^2\right)^2 - 5x^2 + 4 = 0$$
$$u^2 - 5u + 4 = 0$$
$$(u-4)(u-1) = 0$$

Apply the zero product principle.

$u-4=0$ or $u-1=0$

$u = 4$ $\quad$ $u = 1$

Replace u with x^2.

$x^2 = 4$ or $x^2 = 1$

$x = \pm 2$ $\quad$ $x = \pm 1$

The solution set is $\{-2, -1, 1, 2\}$.

3. Let $u = x^2$.

$$x^4 - 11x^2 + 18 = 0$$
$$\left(x^2\right)^2 - 11x^2 + 18 = 0$$
$$u^2 - 11u + 18 = 0$$
$$(u-9)(u-2) = 0$$

Apply the zero product principle.

$u-9=0$ or $u-2=0$

$u = 9$ $\quad$ $u = 2$

Replace u with x^2.

$x^2 = 9$ or $x^2 = 2$

$x = \pm 3$ $\quad$ $x = \pm\sqrt{2}$

The solution set is $\left\{-3, -\sqrt{2}, \sqrt{2}, 3\right\}$.

5. Let $u = x^2$.

$$x^4 + 2x^2 = 8$$
$$x^4 + 2x^2 - 8 = 0$$
$$\left(x^2\right)^2 + 2x^2 - 8 = 0$$
$$u^2 + 2u - 8 = 0$$
$$(u+4)(u-2) = 0$$

Apply the zero product principle.

$u+4=0$ or $u-2=0$

$u = -4$ $\quad$ $u = 2$

Replace u with x^2.

$x^2 = -4$ or $x^2 = 2$

$x = \pm\sqrt{-4}$ $\quad$ $x = \pm\sqrt{2}$

$x = \pm 2i$

The solution set is $\left\{-2i, 2i, -\sqrt{2}, \sqrt{2}\right\}$.

7. Let $u = \sqrt{x}$.

$$x + \sqrt{x} - 2 = 0$$
$$\left(\sqrt{x}\right)^2 + \sqrt{x} - 2 = 0$$
$$u^2 + u - 2 = 0$$
$$(u+2)(u-1) = 0$$

Apply the zero product principle.

$u + 2 = 0$ or $u - 1 = 0$

$u = -2$ $\quad$ $u = 1$

Replace u with $\sqrt{x}$.

~~$\sqrt{x} = -2$~~ or $\sqrt{x} = 1$

$x = 1$

Disregard –2 because the square root of x cannot be a negative number. We must check 1, because both sides of the equation were raised to an even power.

Check:

$$1 + \sqrt{1} - 2 = 0$$
$$1 + 1 - 2 = 0$$
$$2 - 2 = 0$$
$$0 = 0$$

The solution set is $\{1\}$.

9. Let $u = x^{\frac{1}{2}}$.

$$x - 4x^{\frac{1}{2}} - 21 = 0$$
$$\left(x^{\frac{1}{2}}\right)^2 - 4x^{\frac{1}{2}} - 21 = 0$$
$$u^2 - 4u - 21 = 0$$
$$(u-7)(u+3) = 0$$

Apply the zero product principle.

$u - 7 = 0$ or $u + 3 = 0$

$u = 7$ $\quad$ $u = -3$

Replace u with $x^{\frac{1}{2}}$.

$x^{\frac{1}{2}} = 7$ or $x^{\frac{1}{2}} = -3$

$\sqrt{x} = 7$ $\quad$ ~~$\sqrt{x} = -3$~~

$x = 49$

Disregard –3 because the square root of x cannot be a negative number. We must check 49, because both sides of the equation were raised to an even power.

Check:

$$49 - 4(49)^{\frac{1}{2}} - 21 = 0$$
$$49 - 4(7) - 21 = 0$$
$$49 - 28 - 21 = 0$$
$$49 - 49 = 0$$
$$0 = 0$$

The solution set is $\{49\}$.

11. Let $u = \sqrt{x}$.

$$x - 13\sqrt{x} + 40 = 0$$
$$\left(\sqrt{x}\right)^2 - 13\sqrt{x} + 40 = 0$$
$$u^2 - 13u + 40 = 0$$
$$(u-5)(u-8) = 0$$

Apply the zero product principle.

$u - 5 = 0$ or $u - 8 = 0$

$u = 5$ $\quad$ $u = 8$

Replace u with $\sqrt{x}$.

$\sqrt{x} = 5$ or $\sqrt{x} = 8$

$x = 25$ $\quad$ $x = 64$

Both solutions must be checked since both sides of the equation were raised to an even power.

$x = 25$

$$25 - 13\sqrt{25} + 40 = 0$$
$$25 - 13(5) + 40 = 0$$
$$25 - 65 + 40 = 0$$
$$65 - 65 = 0$$
$$0 = 0$$

$x = 64$

$$64 - 13\sqrt{64} + 40 = 0$$
$$64 - 13(8) + 40 = 0$$
$$64 - 104 + 40 = 0$$
$$0 = 0$$

Both solutions check. The solution set is $\{25, 64\}$.

13. Let $u = x - 5$.

$$(x-5)^2 - 4(x-5) - 21 = 0$$
$$u^2 - 4u - 21 = 0$$
$$(u-7)(u+3) = 0$$

Apply the zero product principle.

$u - 7 = 0$ or $u + 3 = 0$

$u = 7$ $\qquad u = -3$

Replace u with $x - 5$.

$x - 5 = 7$ or $x - 5 = -3$

$x = 12$ $\qquad x = 2$

The solution set is $\{2, 12\}$.

15. Let $u = x^2 - 1$.

$$(x^2-1)^2 - (x^2-1) = 2$$
$$(x^2-1)^2 - (x^2-1) - 2 = 0$$
$$u^2 - u - 2 = 0$$
$$(u-2)(u+1) = 0$$

Apply the zero product principle.

$u - 2 = 0$ or $u + 1 = 0$

$u = 2$ $\qquad u = -1$

Replace u with $x^2 - 1$.

$x^2 - 1 = 2$ or $x^2 - 1 = -1$

$x^2 = 3$ $\qquad x^2 = 0$

$x = \pm\sqrt{3}$ $\qquad x = 0$

The solution set is $\{-\sqrt{3}, 0, \sqrt{3}\}$.

17. Let $u = x^2 + 3x$.

$$(x^2+3x)^2 - 8(x^2+3x) - 20 = 0$$
$$u^2 - 8u - 20 = 0$$
$$(u-10)(u+2) = 0$$

Apply the zero product principle.

$u - 10 = 0$ or $u + 2 = 0$

$u = 10$ $\qquad u = -2$

Replace u with $x^2 + 3x$.

First, consider $u = 10$.

$$x^2 + 3x = 10$$
$$x^2 + 3x - 10 = 0$$
$$(x+5)(x-2) = 0$$

Apply the zero product principle.

$x + 5 = 0$ or $x - 2 = 0$

$x = -5$ $\qquad x = 2$

Next, consider $u = -2$.

$$x^2 + 3x = -2$$
$$x^2 + 3x + 2 = 0$$
$$(x+2)(x+1) = 0$$

Apply the zero product principle.

$x + 2 = 0$ or $x + 1 = 0$

$x = -2$ $\qquad x = -1$

The solution set is $\{-5, -2, -1, 2\}$.

19. Let $u = x^{-1}$.

$$x^{-2} - x^{-1} - 20 = 0$$
$$(x^{-1})^2 - x^{-1} - 20 = 0$$
$$u^2 - u - 20 = 0$$
$$(u-5)(u+4) = 0$$

Apply the zero product principle.

$u - 5 = 0$ or $u + 4 = 0$

$u = 5$ $\qquad u = -4$

Replace u with x^{-1}.

$x^{-1} = 5$ or $x^{-1} = -4$

$\frac{1}{x} = 5$ $\qquad \frac{1}{x} = -4$

$5x = 1$ $\qquad -4x = 1$

$x = \frac{1}{5}$ $\qquad x = -\frac{1}{4}$

The solution set is $\left\{-\frac{1}{4}, \frac{1}{5}\right\}$.

21. Let $u = x^{-1}$.

$$2x^{-2} - 7x^{-1} + 3 = 0$$
$$2\left(x^{-1}\right)^2 - 7x^{-1} + 3 = 0$$
$$2u^2 - 7u + 3 = 0$$
$$(2u - 1)(u - 3) = 0$$

Apply the zero product principle.

$2u - 1 = 0$ or $u - 3 = 0$

$2u = 1$ $\quad u = 3$

$u = \frac{1}{2}$

Replace u with x^{-1}.

$x^{-1} = \frac{1}{2}$ or $x^{-1} = 3$

$\frac{1}{x} = \frac{1}{2}$ $\quad \frac{1}{x} = 3$

$x = 2$ $\quad 3x = 1$

$x = \frac{1}{3}$

The solution set is $\left\{\frac{1}{3}, 2\right\}$.

23. Let $u = x^{-1}$.

$$x^{-2} - 4x^{-1} = 3$$
$$x^{-2} - 4x^{-1} - 3 = 0$$
$$\left(x^{-1}\right)^2 - 4x^{-1} - 3 = 0$$
$$u^2 - 4u - 3 = 0$$

$a = 1 \quad b = -4 \quad c = -3$

Use the quadratic formula.

$$u = \frac{-(-4) \pm \sqrt{(-4)^2 - 4(1)(-3)}}{2(1)}$$
$$= \frac{4 \pm \sqrt{16 + 12}}{2} = \frac{4 \pm \sqrt{28}}{2}$$
$$= \frac{4 \pm 2\sqrt{7}}{2} = \frac{2\left(2 \pm \sqrt{7}\right)}{2} = 2 \pm \sqrt{7}$$

Replace u with x^{-1}.

$$x^{-1} = 2 \pm \sqrt{7}$$
$$\frac{1}{x} = 2 \pm \sqrt{7}$$
$$\left(2 \pm \sqrt{7}\right)x = 1$$
$$x = \frac{1}{2 \pm \sqrt{7}}$$

Rationalize the denominator.

$$x = \frac{1}{2 \pm \sqrt{7}} \cdot \frac{2 \mp \sqrt{7}}{2 \mp \sqrt{7}} = \frac{2 \mp \sqrt{7}}{2^2 - \left(\sqrt{7}\right)^2}$$
$$= \frac{2 \mp \sqrt{7}}{4 - 7} = \frac{2 \mp \sqrt{7}}{-3} = \frac{-2 \pm \sqrt{7}}{3}$$

The solution set is $\left\{\frac{-2 \pm \sqrt{7}}{3}\right\}$.

25. Let $u = x^{\frac{1}{3}}$.

$$x^{\frac{2}{3}} - x^{\frac{1}{3}} - 6 = 0$$
$$\left(x^{\frac{1}{3}}\right)^2 - x^{\frac{1}{3}} - 6 = 0$$
$$u^2 - u - 6 = 0$$
$$(u - 3)(u + 2) = 0$$

Apply the zero product principle.

$u - 3 = 0$ or $u + 2 = 0$

$u = 3$ $\quad u = -2$

Replace u with $x^{\frac{1}{3}}$.

$x^{\frac{1}{3}} = 3$ or $x^{\frac{1}{3}} = -2$

$\left(x^{\frac{1}{3}}\right)^3 = 3^3$ $\quad \left(x^{\frac{1}{3}}\right)^3 = (-2)^3$

$x = 27$ $\quad x = -8$

The solution set is $\{-8, 27\}$.

27. Let $u = x^{\frac{1}{5}}$.

$$x^{\frac{2}{5}} + x^{\frac{1}{5}} - 6 = 0$$
$$\left(x^{\frac{1}{5}}\right)^2 + x^{\frac{1}{5}} - 6 = 0$$
$$u^2 + u - 6 = 0$$
$$(u+3)(u-2) = 0$$

Apply the zero product principle.

$$u+3=0 \quad \text{or} \quad u-2=0$$
$$u=-3 \qquad\qquad u=2$$

Replace u with $x^{\frac{1}{5}}$.

$$x^{\frac{1}{5}} = -3 \quad \text{or} \quad x^{\frac{1}{5}} = 2$$
$$\left(x^{\frac{1}{5}}\right)^5 = (-3)^5 \qquad \left(x^{\frac{1}{5}}\right)^5 = (2)^5$$
$$x = -243 \qquad\qquad x = 32$$

The solution set is $\{-243, 32\}$.

29. Let $u = x^{\frac{1}{4}}$.

$$2x^{\frac{1}{2}} - x^{\frac{1}{4}} = 1$$
$$2\left(x^{\frac{1}{4}}\right)^2 - x^{\frac{1}{4}} - 1 = 0$$
$$2u^2 - u - 1 = 0$$
$$(2u+1)(u-1) = 0$$
$$2u+1=0 \quad \text{or} \quad u-1=0$$
$$2u=-1 \qquad\qquad u=1$$
$$u=-\frac{1}{2}$$

Replace u with $x^{\frac{1}{4}}$.

$$x^{\frac{1}{4}} = -\frac{1}{2} \quad \text{or} \quad x^{\frac{1}{4}} = 1$$
$$\left(x^{\frac{1}{4}}\right)^4 = \left(-\frac{1}{2}\right)^4 \qquad \left(x^{\frac{1}{4}}\right)^4 = 1^4$$
$$x = \frac{1}{16} \qquad\qquad x = 1$$

Since both sides of the equations were raised to an even power, the solutions must be checked.

First, check $x = \frac{1}{16}$.

$$2\left(\frac{1}{16}\right)^{\frac{1}{2}} - \left(\frac{1}{16}\right)^{\frac{1}{4}} = 1$$
$$2\left(\frac{1}{4}\right) - \frac{1}{2} = 1$$
$$\frac{1}{2} - \frac{1}{2} = 1$$
$$0 \neq 1$$

The solution does not check, so disregard $x = \frac{1}{16}$.

Next, check $x = 1$.

$$2(1)^{\frac{1}{2}} - (1)^{\frac{1}{4}} = 1$$
$$2(1) - 1 = 1$$
$$1 = 1$$

The solution checks. The solution set is $\{1\}$.

31. Let $u = x - \frac{8}{x}$.

$$\left(x - \frac{8}{x}\right)^2 + 5\left(x - \frac{8}{x}\right) - 14 = 0$$
$$u^2 + 5u - 14 = 0$$
$$(u+7)(u-2) = 0$$
$$u+7=0 \quad \text{or} \quad u-2=0$$
$$u=-7 \qquad\qquad u=2$$

Replace u with $x - \frac{8}{x}$.

First, consider $u = -7$.

$$x - \frac{8}{x} = -7$$
$$x\left(x - \frac{8}{x}\right) = x(-7)$$
$$x^2 - 8 = -7x$$
$$x^2 + 7x - 8 = 0$$
$$(x+8)(x-1) = 0$$
$$x+8=0 \quad \text{or} \quad x-1=0$$
$$x=-8 \qquad\qquad x=1$$

Next, consider $u = 2$.

$$x - \frac{8}{x} = 2$$
$$x\left(x - \frac{8}{x}\right) = x(2)$$
$$x^2 - 8 = 2x$$
$$x^2 - 2x - 8 = 0$$
$$(x-4)(x+2) = 0$$
$$x - 4 = 0 \quad \text{or} \quad x + 2 = 0$$
$$x = 4 \qquad\qquad x = -2$$

The solution set is $\{-8, -2, 1, 4\}$.

33. $f(x) = x^4 - 5x^2 + 4$

$$y = x^4 - 5x^2 + 4$$

Set $y = 0$ to find the x–intercept(s).
$0 = x^4 - 5x^2 + 4$

Let $u = x^2$.

$$x^4 - 5x^2 + 4 = 0$$
$$\left(x^2\right)^2 - 5x^2 + 4 = 0$$
$$u^2 - 5u + 4 = 0$$
$$(u-4)(u-1) = 0$$
$$u - 1 = 0 \quad \text{or} \quad u - 4 = 0$$
$$u = 1 \qquad\qquad u = 4$$

Substitute x^2 for u.

$$x^2 = 1 \quad \text{or} \quad x^2 = 4$$
$$x = \pm 1 \qquad\qquad x = \pm 2$$

The intercepts are ± 1 and ± 2. The corresponding graph is graph c.

35. $f(x) = x^{\frac{1}{3}} + 2x^{\frac{1}{6}} - 3$

$$y = x^{\frac{1}{3}} + 2x^{\frac{1}{6}} - 3$$

Set $y = 0$ to find the x–intercept(s).
$0 = x^{\frac{1}{3}} + 2x^{\frac{1}{6}} - 3$

Let $u = x^{\frac{1}{6}}$.

$$x^{\frac{1}{3}} + 2x^{\frac{1}{6}} - 3 = 0$$
$$\left(x^{\frac{1}{6}}\right)^2 + 2x^{\frac{1}{6}} - 3 = 0$$
$$u^2 + 2u - 3 = 0$$
$$(u+3)(u-1) = 0$$
$$u + 3 = 0 \quad \text{or} \quad u - 1 = 0$$
$$u = -3 \qquad\qquad u = 1$$

Substitute $x^{\frac{1}{6}}$ for u.

$$x^{\frac{1}{6}} = -3 \quad \text{or} \quad x^{\frac{1}{6}} = 1$$
$$\left(x^{\frac{1}{6}}\right)^6 = (-3)^6 \qquad \left(x^{\frac{1}{6}}\right)^6 = (1)^6$$
$$x = 729 \qquad\qquad x = 1$$

Since both sides of the equations were raised to an even power, the solutions must be checked.

First check $x = 729$.

$$(729)^{\frac{1}{3}} + 2(729)^{\frac{1}{6}} - 3 = 0$$
$$9 + 2(3) - 3 = 0$$
$$9 + 6 - 3 = 0$$
$$12 \neq 0$$

Next check $x = 1$.

$$(1)^{\frac{1}{3}} + 2(1)^{\frac{1}{6}} - 3 = 0$$
$$1 + 2(1) - 3 = 0$$
$$1 + 2 - 3 = 0$$
$$0 = 0$$

Since 729 does not check, disregard it. The intercept is 1. The corresponding graph is graph e.

37. $f(x)=(x+2)^2-9(x+2)+20$
$y=(x+2)^2-9(x+2)+20$

Set $y=0$ to find the x–intercept(s).
$(x+2)^2-9(x+2)+20=0$

Let $u=x+2$.
$$(x+2)^2-9(x+2)+20=0$$
$$u^2-9u+20=0$$
$$(u-5)(u-4)=0$$

Apply the zero product principle.
$u-5=0$ or $u-4=0$
$u=5$ $\quad$ $u=4$

Substitute $x+2$ for u.
$x+2=5$ or $x+2=4$
$x=3$ $\quad$ $x=2$

The intercepts are 2 and 3. The corresponding graph is graph f.

39. Let $u=x^2+3x-2$
$$f(x)=-16$$
$$\left(x^2+3x-2\right)^2-10\left(x^2+3x-2\right)=-16$$
$$u^2-10u=-16$$
$$u^2-10u+16=0$$
$$(u-8)(u-2)=0$$

Apply the zero product principle.
$u-8=0$ or $u-2=0$
$u=8$ $\quad$ $u=2$

Replace u with x^2+3x-2.

First, consider $u=8$.
$$x^2+3x-2=8$$
$$x^2+3x-10=0$$
$$(x+5)(x-2)=0$$

Apply the zero product principle.
$x+5=0$ or $x-2=0$
$x=-5$ $\quad$ $x=2$

Next, consider $u=2$.
$$x^2+3x-2=2$$
$$x^2+3x-4=0$$
$$(x+4)(x-1)=0$$

Apply the zero product principle.
$x+4=0$ or $x-1=0$
$x=-4$ $\quad$ $x=1$

The solutions are -5, -4, 1, and 2.

41. Let $u=\frac{1}{x}+1$.
$$f(x)=2$$
$$3\left(\frac{1}{x}+1\right)^2+5\left(\frac{1}{x}+1\right)=2$$
$$3u^2+5u=2$$
$$3u^2+5u-2=0$$
$$(3u-1)(u+2)=0$$
$3u-1=0$ or $u+2=0$
$3u=1$ $\quad$ $u=-2$
$u=\frac{1}{3}$

Replace u with $\frac{1}{x}+1$.

First, consider $u=\frac{1}{3}$.
$$\frac{1}{x}+1=\frac{1}{3}$$
$$3x\left(\frac{1}{x}+1\right)=3x\left(\frac{1}{3}\right)$$
$$3+3x=x$$
$$2x=-3$$
$$x=-\frac{3}{2}$$

Next, consider $u = -2$.

$$\frac{1}{x}+1=-2$$
$$3x\left(\frac{1}{x}+1\right)=3x(-2)$$
$$3+3x=-6x$$
$$9x=-3$$
$$x=-\frac{3}{9}=-\frac{1}{3}$$

The solutions are $-\frac{3}{2}$ and $-\frac{1}{3}$.

43. Let $u=\sqrt{\frac{x}{x-4}}$.

$$f(x)=g(x)$$
$$\frac{x}{x-4}=13\sqrt{\frac{x}{x-4}}-36$$
$$\left(\sqrt{\frac{x}{x-4}}\right)^2=13\sqrt{\frac{x}{x-4}}-36$$
$$u^2=13u-36$$
$$u^2-13u+36=0$$
$$(u-9)(u-4)=0$$
$$u-9=0 \quad \text{or} \quad u-4=0$$
$$u=9 \qquad\qquad u=4$$

Replace u with $\sqrt{\frac{x}{x-4}}$.

First, consider $u=9$.

$$\sqrt{\frac{x}{x-4}}=9$$
$$\left(\sqrt{\frac{x}{x-4}}\right)^2=9^2$$
$$\frac{x}{x-4}=81$$
$$81(x-4)=x$$
$$81x-324=x$$
$$80x=324$$
$$x=\frac{324}{80}=\frac{81}{20}$$

Next, consider $u=4$.

$$\sqrt{\frac{x}{x-4}}=4$$
$$\left(\sqrt{\frac{x}{x-4}}\right)^2=4^2$$
$$\frac{x}{x-4}=16$$
$$16(x-4)=x$$
$$16x-64=x$$
$$15x=64$$
$$x=\frac{64}{15}$$

Since both sides of the equations were raised to an even power, the solutions must be checked. In this case, both check, so the solutions are $\frac{81}{20}$ and $\frac{64}{15}$.

45. Let $u=(x-4)^{-1}$.

$$f(x)=g(x)+12$$
$$3(x-4)^{-2}=16(x-4)^{-1}+12$$
$$3\left[(x-4)^{-1}\right]^2=16(x-4)^{-1}+12$$
$$3u^2=16u+12$$
$$3u^2-16u-12=0$$
$$(3u+2)(u-6)=0$$
$$3u+2=0 \quad \text{or} \quad u-6=0$$
$$3u=-2 \qquad\qquad u=6$$
$$u=-\frac{2}{3}$$

Replace u with $(x-4)^{-1}$.

First, consider $u=-\frac{2}{3}$.

$$(x-4)^{-1}=-\frac{2}{3}$$
$$\frac{1}{x-4}=-\frac{2}{3}$$
$$-2(x-4)=1(3)$$
$$-2x+8=3$$
$$-2x=-5$$
$$x=\frac{-5}{-2}=\frac{5}{2}$$

Next, consider $u = 6$.

$$(x-4)^{-1} = 6$$
$$\frac{1}{x-4} = 6$$
$$6(x-4) = 1$$
$$6x - 24 = 1$$
$$6x = 25$$
$$x = \frac{25}{6}$$

The solutions are $\frac{5}{2}$ and $\frac{25}{6}$.

47. $P(x) = 0.04(x+40)^2 - 3(x+40) + 104$
$$60 = 0.04(x+40)^2 - 3(x+40) + 104$$
$$0 = 0.04(x+40)^2 - 3(x+40) + 44$$

Let $u = x + 40$.
$$0.04(x+40)^2 - 3(x+40) + 44 = 0$$
$$0.04u^2 - 3u + 44 = 0$$
Solve using the quadratic formula.
$a = 0.04 \quad b = -3 \quad c = 44$

$$u = \frac{-(-3) \pm \sqrt{(-3)^2 - 4(0.04)(44)}}{2(0.04)}$$
$$= \frac{3 \pm \sqrt{9 - 7.04}}{0.08}$$
$$= \frac{3 \pm \sqrt{1.96}}{0.08} = \frac{3 \pm 1.4}{0.08} = 55 \text{ or } 20$$

Since x represents the number of years a person's age is above or below 40, $u = x + 40$ is the age we are looking for. The ages at which 60% of us feel that having a clean house is very important are 20 and 55. From the graph, we see that at 20, 58%, and at 55, 52% feel that a clean house if very important. The function models the data fairly well.

49. – 51. Answers will vary.

53. 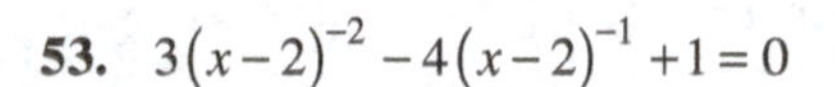
$3(x-2)^{-2} - 4(x-2)^{-1} + 1 = 0$

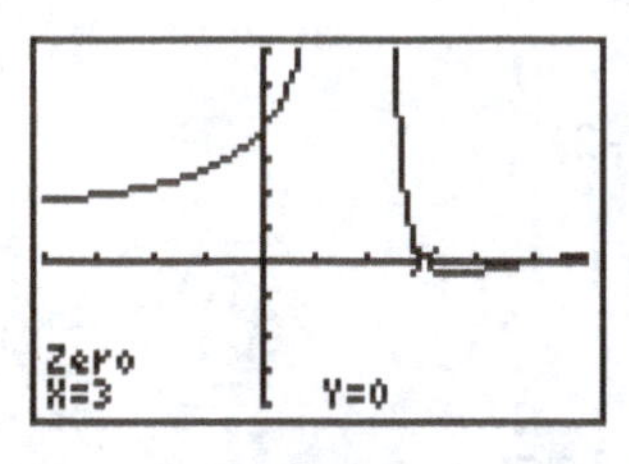

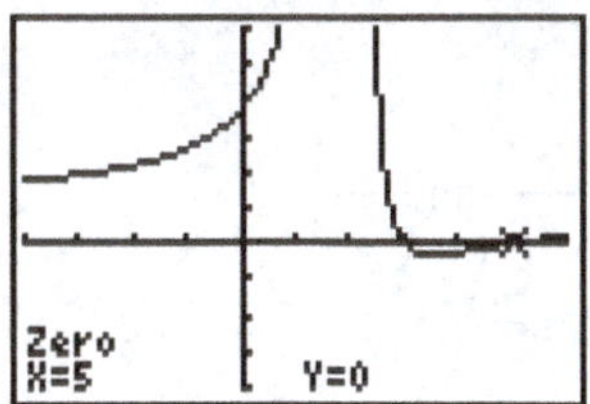

The solutions are 3 and 5. Both solutions check.

The solution set is $\{3, 5\}$.

55.
$$2x + 6\sqrt{x} = 8$$
$$2x + 6\sqrt{x} - 8 = 0$$

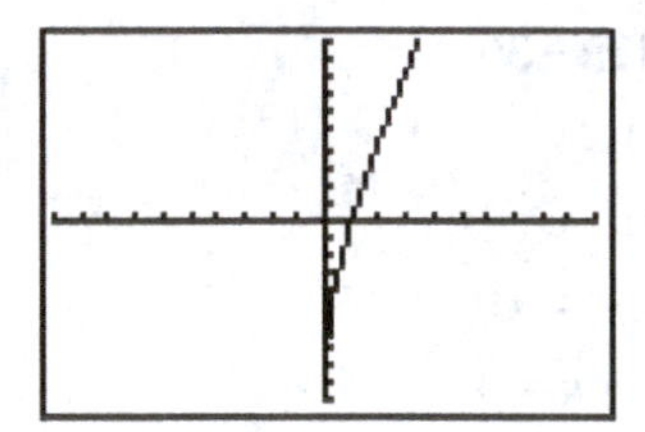

The solution is 1. The solution checks.

The solution set is $\{1\}$.

57. $(x^2 - 3x)^2 + 2(x^2 - 3x) - 24 = 0$

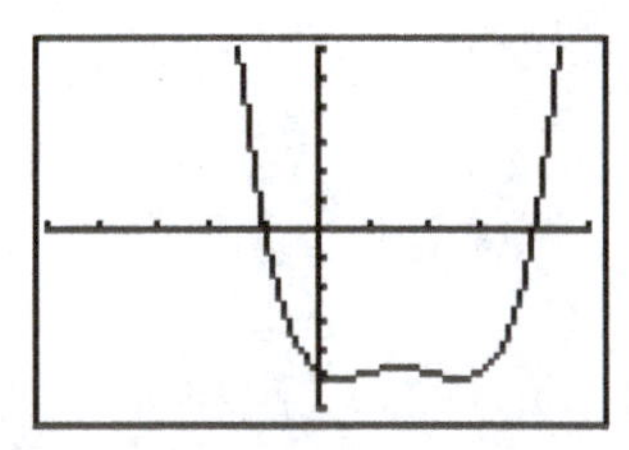

The solutions are -1 and 4.

The solution set is $\{-1, 4\}$.

59. $x^{\frac{2}{3}} - 3x^{\frac{1}{3}} + 2 = 0$

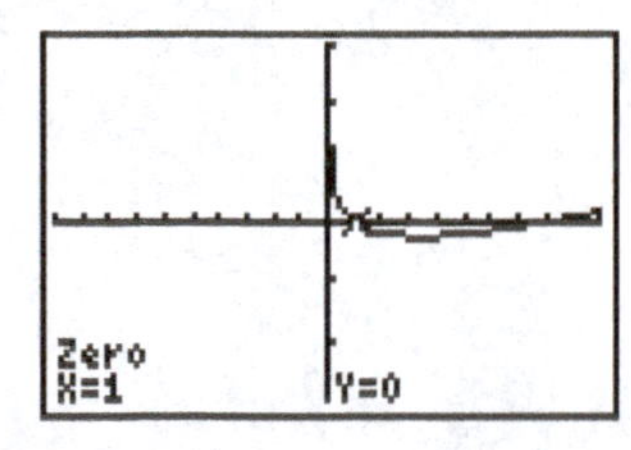

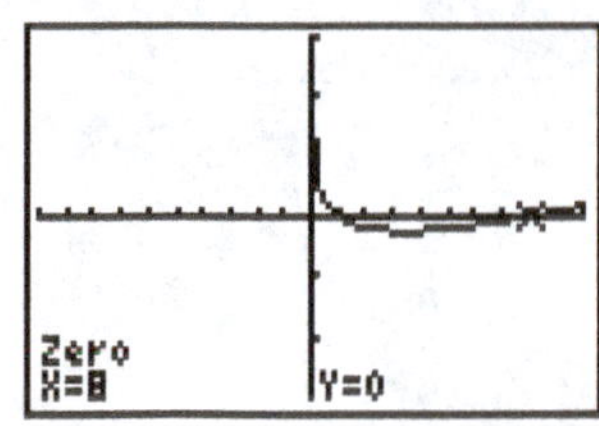

The solution set is $\{1, 8\}$.

61. makes sense

63. does not make sense; Explanations will vary. Sample explanation: Changing the order of the terms does not change the classification of this equation.

65. true

67. false; Changes to make the statement true will vary. A sample change is: To solve the equation, let $u = \sqrt{x}$.

69.
$$5x^6 + x^3 = 18$$
$$5x^6 + x^3 - 18 = 0$$

Let $u = x^3$.
$$5x^6 + x^3 - 18 = 0$$
$$5\left(x^3\right)^2 + x^3 - 18 = 0$$
$$5u^2 + u - 18 = 0$$
$$(5u - 9)(u + 2) = 0$$
$$5u - 9 = 0 \quad \text{or} \quad u + 2 = 0$$
$$5u = 9 \qquad u = -2$$
$$u = \frac{9}{5}$$

Substitute x^3 for u.
$$x^3 = \frac{9}{5} \quad \text{or} \quad x^3 = -2$$
$$x = \sqrt[3]{\frac{9}{5}} \qquad x = \sqrt[3]{-2}$$

Rationalize the denominator.
$$\sqrt[3]{\frac{9}{5}} = \frac{\sqrt[3]{9}}{\sqrt[3]{5}} \cdot \frac{\sqrt[3]{5^2}}{\sqrt[3]{5^2}} = \frac{\sqrt[3]{9 \cdot 5^2}}{\sqrt[3]{5^3}} = \frac{\sqrt[3]{225}}{5}$$

The solution set is $\left\{\sqrt[3]{-2}, \frac{\sqrt[3]{225}}{5}\right\}$.

71. $\dfrac{2x^2}{10x^3 - 2x^2} = \dfrac{2x^2}{2x^2(5x - 1)} = \dfrac{1}{5x - 1}$

72.
$$\frac{2+i}{1-i} = \frac{2+i}{1-i} \cdot \frac{1+i}{1+i} = \frac{2 + 2i + i + i^2}{1^2 - i^2}$$
$$= \frac{2 + 3i - 1}{1 - (-1)} = \frac{1 + 3i}{2} = \frac{1}{2} + \frac{3}{2}i$$

73.
$$\left[\begin{array}{cc|c} 2 & 1 & 6 \\ 1 & -2 & 8 \end{array}\right] \quad R_1 \leftrightarrow R_2$$
$$= \left[\begin{array}{cc|c} 1 & -2 & 8 \\ 2 & 1 & 6 \end{array}\right] \quad -2R_1 + R_2$$
$$= \left[\begin{array}{cc|c} 1 & -2 & 8 \\ 0 & 5 & -10 \end{array}\right] \quad \frac{1}{5}R_2$$
$$= \left[\begin{array}{cc|c} 1 & -2 & 8 \\ 0 & 1 & -2 \end{array}\right]$$

$$x - 2y = 8$$
$$y = -2$$

Back-substitute -2 for y to find x.
$$x - 2(-2) = 8$$
$$x + 4 = 8$$
$$x = 4$$

The solution set is $\{(4, -2)\}$.

74.
$$2x^2 + x = 15$$
$$2x^2 + x - 15 = 0$$
$$(x + 3)(2x - 5) = 0$$
$$x + 3 = 0 \quad \text{or} \quad 2x - 5 = 0$$
$$x = -3 \qquad 2x = 5$$
$$x = \frac{5}{2}$$

The solution set is $\left\{-3, \frac{5}{2}\right\}$.

75.

$$x^3+x^2=4x+4$$
$$x^3+x^2-4x-4=0$$
$$x^2(x+1)-4(x+1)=0$$
$$(x+1)(x^2-4)=0$$
$$(x+1)(x+2)(x-2)=0$$
$$x+1 \quad \text{or} \quad x+2=0 \quad \text{or} \quad x-2=0$$
$$x=-1 \qquad x=-2 \qquad x=2$$

The solution set is $\{-2,-1,2\}$.

76.

$$\frac{x+1}{x+3}-2=\frac{x+1}{x+3}-\frac{2(x+3)}{x+3}$$
$$=\frac{x+1}{x+3}-\frac{2x+6}{x+3}$$
$$=\frac{x+1-2x-6}{x+3}$$
$$=\frac{-x-5}{x+3}$$

8.5 Check Points

1.

$$x^2-x>20$$
$$x^2-x-20>0$$

Solve the related quadratic equation to find the boundary points.

$$x^2-x-20=0$$
$$(x+4)(x-5)=0$$
$$x+4=0 \quad \text{or} \quad x-5=0$$
$$x=-4 \qquad x=5$$

The boundary points are -4 and 5.

Interval	Test Value	Test	Conclusion
$(-\infty,-4)$	-5	$x^2-x>20$ $(-5)^2-(-5)>20$ $30>20$, true	$(-\infty,-4)$ belongs to the solution set.
$(-4,5)$	0	$x^2-x>20$ $(0)^2-(0)>20$ $0>20$, false	$(-4,5)$ does not belong to the solution set.
$(5,\infty)$	10	$x^2-x>20$ $(10)^2-(10)>20$ $90>20$, true	$(5,\infty)$ belongs to the solution set.

The solution set is $(-\infty,-4)\cup(5,\infty)$.

−8 −6 −4 −2 0 2 4 6 8 10 12

2. $2x^2 \le -6x - 1$

$2x^2 + 6x + 1 \le 0$

Solve the related quadratic equation to find the boundary points.

$2x^2 + 6x + 1 = 0$

$a = 2 \quad b = 6 \quad c = 1$

$$x = \frac{-(6) \pm \sqrt{(6)^2 - 4(2)(1)}}{2(2)}$$

$$= \frac{-6 \pm \sqrt{36 - 8}}{4}$$

$$= \frac{-6 \pm \sqrt{28}}{4}$$

$$= \frac{-6 \pm 2\sqrt{7}}{4}$$

$$= \frac{-3 \pm \sqrt{7}}{2}$$

$x = \frac{-3+\sqrt{7}}{2}$ or $x = \frac{-3-\sqrt{7}}{2}$

$x \approx -0.2 \qquad x \approx -2.8$

Interval	Test Value	Test	Conclusion
$\left(-\infty, \frac{-3-\sqrt{7}}{2}\right)$	-10	$2(-10)^2 \le -6(-10) - 1$ $200 \le 59$, false	$\left(-\infty, \frac{-3-\sqrt{7}}{2}\right)$ is not part of the solution set
$\left(\frac{-3-\sqrt{7}}{2}, \frac{-3+\sqrt{7}}{2}\right)$	-1	$2(-1)^2 \le -6(-1) - 1$ $2 \le 5$, true	$\left(\frac{-3-\sqrt{7}}{2}, \frac{-3+\sqrt{7}}{2}\right)$ is part of the solution set
$\left(\frac{-3+\sqrt{7}}{2}, \infty\right)$	0	$2(0)^2 \le -6(0) - 1$ $0 \le -1$, false	$\left(\frac{-3+\sqrt{7}}{2}, \infty\right)$ is not part of the solution set

The solution set is $\left(\frac{-3-\sqrt{7}}{2}, \frac{-3+\sqrt{7}}{2}\right)$.

−5 −4 −3 −2 −1 0 1 2 3 4 5

3. $x^3 + 3x^2 \le x + 3$

$x^3 + 3x^2 - x - 3 \le 0$

Solve the related quadratic equation to find the boundary points.

$x^3 + 3x^2 - x - 3 = 0$

$x^2(x+3) - 1(x+3) = 0$

$(x+3)(x^2 - 1) = 0$

$(x+3)(x-1)(x+1) = 0$

$x+3=0$ or $x-1=0$ or $x+1=0$

$x=-3$ $\quad$ $x=1$ $\quad$ $x=-1$

Interval	Test Value	Test	Conclusion
$(-\infty,-3)$	-5	$(-5)^3+3(-5)^2 \le (-5)+3$ $-50 \le -2$, true	$(-\infty,-3)$ is part of the solution set
$(-3,-1)$	-2	$(-2)^3+3(-2)^2 \le (-2)+3$ $4 \le 1$, false	$(-3,-1)$ is not part of the solution set
$(-1,1)$	0	$(0)^3+3(0)^2 \le (0)+3$ $0 \le 3$, true	$(-1,1)$ is part of the solution set
$(1,\infty)$	2	$(2)^3+3(2)^2 \le (2)+3$ $20 \le 5$, false	$(1,\infty)$ is not part of the solution set

The solution set is $(-\infty,-3]\cup[-1,1]$.

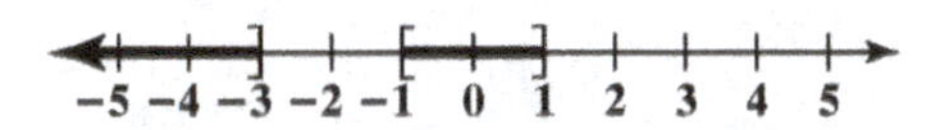

4. $\dfrac{x-5}{x+2}<0$

Find the values of x that make the numerator and denominator zero.

$x-5=0$ $\quad$ $x+2=0$

$x=5$ $\quad$ $x=-2$

The boundary points are -2 and 5.

Interval	Test Value	Test	Conclusion
$(-\infty,-2)$	-3	$\frac{-3-5}{-3+2}<0$ $8<0$, false	$(-\infty,-2)$ does not belong to the solution set.
$(-2,5)$	0	$\frac{0-5}{0+2}<0$ $-\frac{5}{2}<0$, true	$(-2,5)$ belongs to the solution set.
$(5,\infty)$	6	$\frac{6-5}{6+2}<0$ $\frac{1}{8}<0$, false	$(5,\infty)$ does not belong to the solution set.

The solution set is $(-2,5)$.

5.
$$\frac{2x}{x+1} \geq 1$$
$$\frac{2x}{x+1} - 1 \geq 0$$
$$\frac{2x}{x+1} - \frac{x+1}{x+1} \geq 0$$
$$\frac{2x - x - 1}{x+1} \geq 0$$
$$\frac{x-1}{x+1} \geq 0$$

Find the values of x that make the numerator and denominator zero.

$x - 1 = 0$ and $x + 1 = 0$

$x = 1$ $\qquad$ $x = -1$

The boundary points are -1 and 1.

Interval	Test Value	Test	Conclusion
$(-\infty, -1)$	-2	$\frac{2(-2)}{-2+1} \geq 1$; $4 \geq 1$, true	$(-\infty, -1)$ belongs to the solution set.
$(-1, 1)$	0	$\frac{2(0)}{0+1} \geq 1$; $0 \geq 1$, false	$(-1, 1)$ does not belong to the solution set.
$(1, \infty)$	2	$\frac{2(2)}{2+1} \geq 1$; $\frac{4}{3} \geq 1$, true	$(1, \infty)$ belongs to the solution set.

Exclude -1 from the solution set because it would make the denominator zero. The solution set is $(-\infty, -1) \cup [1, \infty)$.

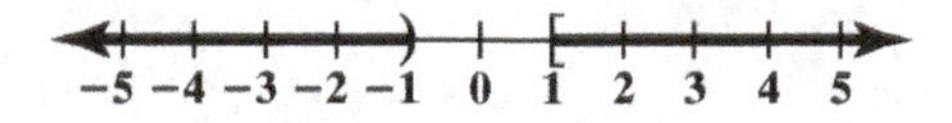

6. $s(t) = -16t^2 + 80t$

To find when the object will be more than 64 feet above the ground, solve the inequality $-16t^2 + 80t > 64$.

Solve the related quadratic equation.

$$-16t^2 + 80t = 64$$
$$-16t^2 + 80t - 64 = 0$$
$$t^2 - 5t + 4 = 0$$
$$(t-4)(t-1) = 0$$

$t - 4 = 0$ or $t - 1 = 0$

$t = 4$ $\qquad$ $t = 1$

The boundary points are 1 and 4.

Interval	Test Value	Test	Conclusion
$(0,1)$	0.5	$-16(0.5)^2+80(0.5)>64$ $36>64$, false	$(0,1)$ does not belong to the solution set.
$(1,4)$	2	$-16(2)^2+80(2)>64$ $96>64$, true	$(1,4)$ belongs to the solution set.
$(4,\infty)$	5	$-16(5)^2+80(5)>64$ $0>64$, false	$(4,\infty)$ does not belong to the solution set.

The solution set is $(1,4)$. This means that the object will be more than 64 feet above the ground between 1 and 4 seconds excluding $t=1$ and $t=4$.

8.5 Concept and Vocabulary Check

1. $x^2+8x+15=0$; boundary

2. $(-\infty,-5)$; $(-5,-3)$; $(-3,\infty)$

3. true

4. true

5. $(-\infty,-2)\cup[1,\infty)$

8.5 Exercise Set

1. $(x-4)(x+2)>0$

   Solve the related quadratic equation.

   $(x-4)(x+2)=0$

   $x-4=0$ or $x+2=0$

   $x=4$ $\quad$ $x=-2$

   The boundary points are –2 and 4.

Interval	Test Value	Test	Conclusion
$(-\infty,-2)$	–3	$(-3-4)(-3+2)>0$ $7>0$, true	$(-\infty,-2)$ belongs to the solution set.
$(-2,4)$	0	$(0-4)(0+2)>0$ $-8>0$, false	$(-2,4)$ does not belong to the solution set.
$(4,\infty)$	5	$(5-4)(5+2)>0$ $7>0$, true	$(4,\infty)$ belongs to the solution set.

The solution set is $(-\infty,-2)\cup(4,\infty)$.

3. $(x-7)(x+3) \le 0$

Solve the related quadratic equation.

$(x-7)(x+3)=0$

$x-7=0$ or $x+3=0$

$x=7$ $\quad x=-3$

The boundary points are –3 and 7.

Interval	Test Value	Test	Conclusion
$(-\infty,-3)$	–4	$(-4-7)(-4+3) \le 0$ $11 \le 0$, false	$(-\infty,-3)$ does not belong to the solution set.
$(-3,7)$	0	$(0-7)(0+3) \le 0$ $-21 \le 0$, true	$(-3,7)$ belongs to the solution set.
$(7,\infty)$	8	$(8-7)(8+3) \le 0$ $11 \le 0$, false	$(7,\infty)$ does not belong to the solution set.

The solution set is $[-3,7]$.

5. $x^2-5x+4>0$

Solve the related quadratic equation.

$x^2-5x+4=0$

$(x-4)(x-1)=0$

$x-4=0$ or $x-1=0$

$x=4$ $\quad x=1$

The boundary points are 1 and 4.

Interval	Test Value	Test	Conclusion
$(-\infty,1)$	0	$0^2-5(0)+4>0$ $4>0$, true	$(-\infty,1)$ belongs to the solution set.
$(1,4)$	2	$2^2-5(2)+4>0$ $-2>0$, false	$(1,4)$ does not belong to the solution set.
$(4,\infty)$	5	$5^2-5(5)+4>0$ $4>0$, true	$(4,\infty)$ belongs to the solution set.

The solution set is $(-\infty,1)\cup(4,\infty)$.

7. $x^2+5x+4>0$

Solve the related quadratic equation.

$$x^2+5x+4=0$$
$$(x+4)(x+1)=0$$
$$x+4=0 \quad \text{or} \quad x+1=0$$
$$x=-4 \qquad\qquad x=-1$$

The boundary points are -1 and -4.

Interval	Test Value	Test	Conclusion
$(-\infty,-4)$	-5	$(-5)^2+5(-5)+4>0$ $4>0$, true	$(-\infty,-4)$ belongs to the solution set.
$(-4,-1)$	-2	$(-2)^2+5(-2)+4>0$ $-2>0$, false	$(-4,-1)$ does not belong to the solution set.
$(-1,\infty)$	0	$0^2+5(0)+4>0$ $4>0$, true	$(-1,\infty)$ belongs to the solution set.

The solution set is $(-\infty,-4)\cup(-1,\infty)$.

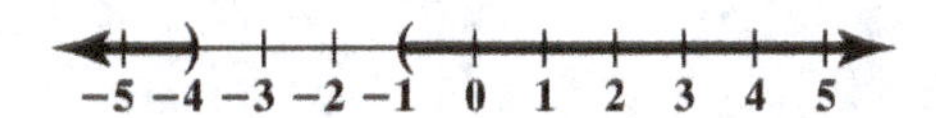

9. $x^2-6x+8\le 0$

Solve the related quadratic equation.

$$x^2-6x+8=0$$
$$(x-4)(x-2)=0$$
$$x-4=0 \quad \text{or} \quad x-2=0$$
$$x=4 \qquad\qquad x=2$$

The boundary points are 2 and 4.

Interval	Test Value	Test	Conclusion
$(-\infty,2)$	0	$0^2-6(0)+8\le 0$ $8\le 0$, false	$(-\infty,2)$ does not belong to the solution set.
$(2,4)$	3	$3^2-6(3)+8\le 0$ $-1\le 0$, true	$(2,4)$ belongs to the solution set.
$(4,\infty)$	5	$5^2-6(5)+8\le 0$ $3\le 0$, false	$(4,\infty)$ does not belong to the solution set.

The solution set is $[2,4]$.

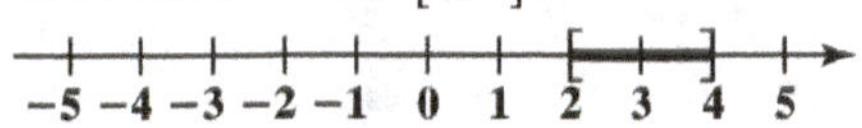

11. $3x^2+10x-8\le 0$

Solve the related quadratic equation.

$$3x^2+10x-8=0$$
$$(3x-2)(x+4)=0$$
$$3x-2=0 \quad \text{or} \quad x+4=0$$
$$3x=2 \qquad x=-4$$
$$x=\frac{2}{3}$$

The boundary points are –4 and $\frac{2}{3}$.

Interval	Test Value	Test	Conclusion
$(-\infty,-4)$	–5	$3(-5)^2+10(-5)-8\le 0$ $17\le 0$, false	$(-\infty,-4)$ does not belong to the solution set.
$\left(-4,\frac{2}{3}\right)$	0	$3(0)^2+10(0)-8\le 0$ $-8\le 0$, true	$\left(-4,\frac{2}{3}\right)$ belongs to the solution set.
$\left(\frac{2}{3},\infty\right)$	1	$3(1)^2+10(1)-8\le 0$ $5\le 0$, false	$\left(\frac{2}{3},\infty\right)$ does not belong to the solution set.

The solution set is $\left[-4,\frac{2}{3}\right]$.

–5 –4 –3 –2 –1 0 1 2 3 4 5

13. $2x^2+x<15$

$$2x^2+x-15<0$$

Solve the related quadratic equation.

$$2x^2+x-15=0$$
$$(2x-5)(x+3)=0$$
$$2x-5=0 \quad \text{or} \quad x+3=0$$
$$2x=5 \qquad x=-3$$
$$x=\frac{5}{2}$$

The boundary points are –3 and $\frac{5}{2}$.

Interval	Test Value	Test	Conclusion
$(-\infty, -3)$	-4	$2(-4)^2 + (-4) < 15$ $28 < 15$, false	$(-\infty, -3)$ does not belong to the solution set.
$\left(-3, \frac{5}{2}\right)$	0	$2(0)^2 + 0 < 15$ $0 < 15$, true	$\left(-3, \frac{5}{2}\right)$ belongs to the solution set.
$\left(\frac{5}{2}, \infty\right)$	3	$2(3)^2 + 3 < 15$ $21 < 15$, false	$\left(\frac{5}{2}, \infty\right)$ does not belong to the solution set.

The solution set is $\left(-3, \frac{5}{2}\right)$.

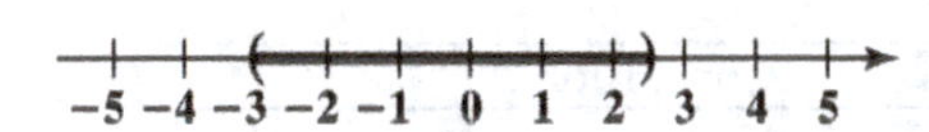

15. $4x^2 + 7x < -3$

$4x^2 + 7x + 3 < 0$

Solve the related quadratic equation.

$$4x^2 + 7x + 3 = 0$$
$$(4x+3)(x+1) = 0$$
$$4x + 3 = 0 \quad \text{or} \quad x + 1 = 0$$
$$4x = -3 \qquad x = -1$$
$$x = -\frac{3}{4}$$

The boundary points are -1 and $-\frac{3}{4}$.

Interval	Test Value	Test	Conclusion
$(-\infty, -1)$	-2	$4(-2)^2 + 7(-2) < -3$ $2 < -3$, false	$(-\infty, -1)$ does not belong to the solution set.
$\left(-1, -\frac{3}{4}\right)$	$-\frac{7}{8}$	$4\left(-\frac{7}{8}\right)^2 + 7\left(-\frac{7}{8}\right) < -3$ $-3\frac{1}{16} < -3$, true	$\left(-1, -\frac{3}{4}\right)$ belongs to the solution set.
$\left(-\frac{3}{4}, \infty\right)$	0	$4(0)^2 + 7(0) < -3$ $0 < -3$, false	$\left(-\frac{3}{4}, \infty\right)$ does not belong to the solution set.

The solution set is $\left(-1, -\frac{3}{4}\right)$.

17. $x^2 - 4x \geq 0$

Solve the related quadratic equation.

$x^2 - 4x = 0$

$x(x-4) = 0$

$x = 0$ or $x - 4 = 0$

$x = 4$

The boundary points are 0 and 4.

Interval	Test Value	Test	Conclusion
$(-\infty, 0)$	-1	$(-1)^2 - 4(-1) \geq 0$ $5 \geq 0$, true	$(-\infty, 0)$ belongs to the solution set.
$(0, 4)$	1	$(1)^2 - 4(1) \geq 0$ $-3 \geq 0$, false	$(0, 4)$ does not belong to the solution set.
$(4, \infty)$	5	$(5)^2 - 4(5) \geq 0$ $5 \geq 0$, true	$(4, \infty)$ belongs to the solution set.

The solution set is $(-\infty, 0] \cup [4, \infty)$.

19. $2x^2 + 3x > 0$

Solve the related quadratic equation.

$2x^2 + 3x = 0$

$x(2x+3) = 0$

$x = 0$ or $2x + 3 = 0$

$2x = -3$

$x = -\frac{3}{2}$

The boundary points are $-\frac{3}{2}$ and 0.

Interval	Test Value	Test	Conclusion
$\left(-\infty, -\frac{3}{2}\right)$	-2	$2(-2)^2 + 3(-2) > 0$ $2 > 0$, true	$\left(-\infty, -\frac{3}{2}\right)$ belongs to the solution set.
$\left(-\frac{3}{2}, 0\right)$	-1	$2(-1)^2 + 3(-1) > 0$ $-1 > 0$, false	$\left(-\frac{3}{2}, 0\right)$ does not belong to the solution set.
$(0, \infty)$	1	$2(1)^2 + 3(1) > 0$ $5 > 0$, true	$(0, \infty)$ belongs to the solution set.

The solution set is $\left(-\infty, -\frac{3}{2}\right) \cup (0, \infty)$.

21. $-x^2 + x \ge 0$

Solve the related quadratic equation.

$-x^2 + x = 0$
$-x(x-1) = 0$
$-x = 0 \quad \text{or} \quad x-1 = 0$
$x = 0 \qquad\qquad x = 1$

The boundary points are 0 and 1.

Interval	Test Value	Test	Conclusion
$(-\infty, 0)$	-1	$-(-1)^2 + (-1) \ge 0$ $-2 \ge 0$, false	$(-\infty, 0)$ does not belong to the solution set.
$(0,1)$	$\frac{1}{2}$	$-\left(\frac{1}{2}\right)^2 + \frac{1}{2} \ge 0$ $\frac{1}{4} \ge 0$, true	$(0,1)$ belongs to the solution set.
$(1, \infty)$	2	$-(2)^2 + 2 \ge 0$ $-2 \ge 0$, false	$(1, \infty)$ does not belong to the solution set.

The solution set is $[0,1]$.

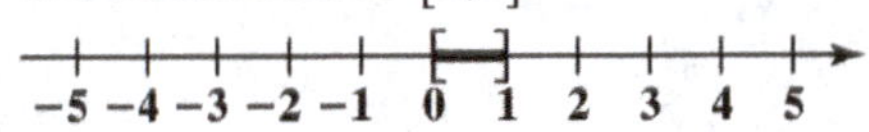

23. $x^2 \le 4x - 2$

$x^2 - 4x + 2 \le 0$

Solve the related quadratic equation, using the quadratic formula.

$x^2 - 4x + 2 = 0$
$a = 1 \quad b = -4 \quad c = 2$

$$x = \frac{-(-4) \pm \sqrt{(-4)^2 - 4(1)(2)}}{2(1)} = \frac{4 \pm \sqrt{16-8}}{2} = \frac{4 \pm \sqrt{8}}{2} = \frac{4 \pm \sqrt{4 \cdot 2}}{2}$$

$$= \frac{4 \pm 2\sqrt{2}}{2} = \frac{2\left(2 \pm \sqrt{2}\right)}{2} = 2 \pm \sqrt{2}$$

The boundary points are $2 - \sqrt{2}$ and $2 + \sqrt{2}$.

Interval	Test Value	Test	Conclusion
$\left(-\infty, 2-\sqrt{2}\right)$	0	$0^2 \le 4(0)-2$ $0 \le -2$, false	$\left(-\infty, 2-\sqrt{2}\right)$ does not belong to the solution set.
$\left(2-\sqrt{2}, 2+\sqrt{2}\right)$	2	$2^2 \le 4(2)-2$ $4 \le 6$, true	$\left(2-\sqrt{2}, 2+\sqrt{2}\right)$ belongs to the solution set.
$\left(2+\sqrt{2}, \infty\right)$	4	$4^2 \le 4(4)-2$ $16 \le 14$, false	$\left(2+\sqrt{2}, \infty\right)$ does not belong to the solution set.

The solution set is $\left[2-\sqrt{2}, 2+\sqrt{2}\right]$.

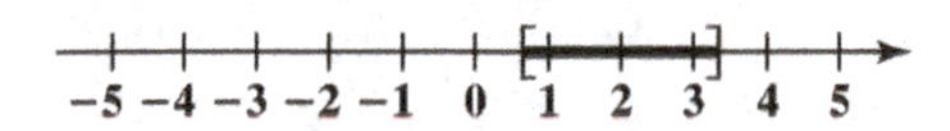

25. $3x^2 > 4x+2$

$3x^2 - 4x - 2 > 0$

Solve the related quadratic equation.

$3x^2 - 4x - 2 = 0$

$$x = \frac{-b \pm \sqrt{b^2 - 4ac}}{2a}$$

$$x = \frac{-(-4) \pm \sqrt{(-4)^2 - 4(3)(-2)}}{2(3)}$$

$$x = \frac{2 \pm \sqrt{10}}{3}$$

$x \approx -0.39$ or 1.72

Interval	Test Value	Test	Conclusion
$\left(-\infty, \frac{2-\sqrt{10}}{3}\right)$	-1	$3(-1)^2 > 4(-1)+2$ $3 > -2$, true	$\left(-\infty, \frac{2-\sqrt{10}}{3}\right)$ belongs to the solution set.
$\left(\frac{2-\sqrt{10}}{3}, \frac{2+\sqrt{10}}{3}\right)$	0	$3(0)^2 > 4(0)+2$ $0 > 2$, false	$\left(\frac{2-\sqrt{10}}{3}, \frac{2+\sqrt{10}}{3}\right)$ does not belong to the solution set.
$\left(\frac{2+\sqrt{10}}{3}, \infty\right)$	2	$3(2)^2 > 4(2)+2$ $12 > 10$, true	$\left(\frac{2+\sqrt{10}}{3}, \infty\right)$ belongs to the solution set.

The solution set is $\left(-\infty, \frac{2-\sqrt{10}}{3}\right) \cup \left(\frac{2+\sqrt{10}}{3}, \infty\right)$.

27. $2x^2 - 5x \geq 1$

$2x^2 - 5x - 1 \geq 0$

Solve the related quadratic equation.

$2x^2 - 5x - 1 = 0$

$$x = \frac{-b \pm \sqrt{b^2 - 4ac}}{2a}$$

$$x = \frac{-(-5) \pm \sqrt{(-5)^2 - 4(2)(-1)}}{2(2)}$$

$$x = \frac{5 \pm \sqrt{33}}{4}$$

$x \approx -0.19$ or 2.69

Interval	Test Value	Test	Conclusion
$\left(-\infty, \frac{5-\sqrt{33}}{4}\right)$	-1	$2(-1)^2 - 5(-1) \geq 1$ $7 \geq 1$, true	$\left(-\infty, \frac{5-\sqrt{33}}{4}\right)$ belongs to the solution set.
$\left(\frac{5-\sqrt{33}}{4}, \frac{5+\sqrt{33}}{4}\right)$	0	$2(0)^2 - 5(0) \geq 1$ $0 \geq 1$, false	$\left(\frac{5-\sqrt{33}}{4}, \frac{5+\sqrt{33}}{4}\right)$ does not belong to the solution set.
$\left(\frac{5+\sqrt{33}}{4}, \infty\right)$	3	$2(3)^2 - 5(3) \geq 1$ $3 \geq 1$, true	$\left(\frac{5+\sqrt{33}}{4}, \infty\right)$ belongs to the solution set.

The solution set is $\left(-\infty, \frac{5-\sqrt{33}}{4}\right] \cup \left[\frac{5+\sqrt{33}}{4}, \infty\right)$.

−5 −4 −3 −2 −1 0 1 2 3 4 5

29. $x^2 - 6x + 9 < 0$

Solve the related quadratic equation.

$x^2 - 6x + 9 = 0$

$(x-3)^2 = 0$

$x - 3 = 0$

$x = 3$

The boundary point is 3.

Interval	Test Value	Test	Conclusion
$(-\infty, 3)$	0	$0^2 - 6(0) + 9 < 0$ $9 < 0$, False	$(-\infty, 3)$ does not belong to the solution set.
$(3, \infty)$	4	$4^2 - 6(4) + 9 < 0$ $1 < 0$, false	$(3, \infty)$ does not belong to the solution set.

There is no solution. The solution set is $\varnothing$.

31. $(x-1)(x-2)(x-3) \geq 0$

Solve the related polynomial equation.

$(x-1)(x-2)(x-3) = 0$

$x-1=0$ or $x-2=0$ or $x-3=0$

$x=1$ $x=2$ $x=3$

The boundary points are 1, 2, and 3.

Interval	Test Value	Test	Conclusion
$(-\infty,1)$	0	$(0-1)(0-2)(0-3) \geq 0$ $-6 \geq 0$, False	$(-\infty,1)$ does not belong to the solution set.
$(1,2)$	1.5	$(1.5-1)(1.5-2)(1.5-3) \geq 0$ $0.375 \geq 0$, True	$(1,2)$ belongs to the solution set.
$(2,3)$	2.5	$(2.5-1)(2.5-2)(2.5-3) \geq 0$ $-0.375 \geq 0$, False	$(2,3)$ does not belong to the solution set.
$(3,\infty)$	4	$(4-1)(4-2)(4-3) \geq 0$ $6 \geq 0$, True	$(3,\infty)$ belongs to the solution set.

The solution set is $[1,2] \cup [3,\infty)$.

33. $x^3 + 2x^2 - x - 2 \geq 0$

Solve the related polynomial equation.

$$x^3 + 2x^2 - x - 2 = 0$$

$$x^2(x+2) - 1(x+2) = 0$$

$$(x^2-1)(x+2) = 0$$

$$(x-1)(x+1)(x+2) = 0$$

$x-1=0$ or $x+1=0$ or $x+2=0$

$x=1$ $x=-1$ $x=-2$

The boundary points are $-2, -1,$ and 1.

Interval	Test Value	Test	Conclusion
$(-\infty,-2)$	-3	$(-3)^3+2(-3)^2-(-3)-2\ge 0$ $-8\ge 0$, False	$(-\infty,-2)$ does not belong to the solution set.
$(-2,-1)$	-1.5	$(-1.5)^3+2(-1.5)^2-(-1.5)-2\ge 0$ $0.625\ge 0$, True	$(-2,-1)$ belongs to the solution set.
$(-1,1)$	0	$0^3+2(0)^2-0-2\ge 0$ $-2\ge 0$, False	$(-1,1)$ does not belong to the solution set.
$(1,\infty)$	2	$2^3+2(2)^2-2-2\ge 0$ $12\ge 0$, True	$(1,\infty)$ belongs to the solution set.

The solution set is $[-2,-1]\cup[1,\infty)$.

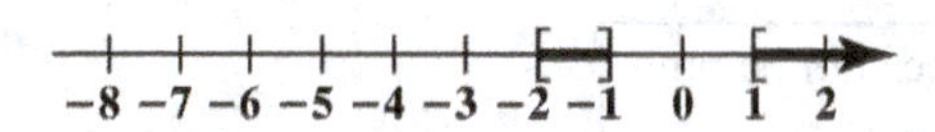

35. $x^3-3x^2-9x+27<0$

Solve the related polynomial equation.

$$x^3-3x^2-9x+27=0$$
$$x^2(x-3)-9(x-3)=0$$
$$(x^2-9)(x-3)=0$$
$$(x-3)(x+3)(x-3)=0$$
$$(x-3)^2(x+3)=0$$
$$x-3=0 \quad\text{or}\quad x+3=0$$
$$x=3 \qquad\qquad x=-3$$

The boundary points are -3 and 3.

Interval	Test Value	Test	Conclusion
$(-\infty,-3)$	-4	$(-4)^3-3(-4)^2-9(-4)+27<0$ $-49<0$, True	$(-\infty,-3)$ belongs to the solution set.
$(-3,3)$	0	$0^3-3(0)^2-9(0)+27<0$ $27<0$, False	$(-3,3)$ does not belong to the solution set.
$(3,\infty)$	4	$4^3-3(4)^2-9(4)+27<0$ $7<0$, False	$(3,\infty)$ does not belong to the solution set.

The solution set is $(-\infty,-3)$.

37. $x^3 + x^2 + 4x + 4 > 0$

Solve the related polynomial equation.

$$x^3 + x^2 + 4x + 4 = 0$$
$$x^2(x+1) + 4(x+1) = 0$$
$$(x^2 + 4)(x+1) = 0$$
$$x^2 + 4 = 0 \quad \text{or} \quad x + 1 = 0$$
$$x^2 = -4 \qquad x = -1$$
$$x = \pm\sqrt{-4}$$
$$= \pm 2i$$

The imaginary solutions will not be boundary points, so the only boundary point is -1.

Interval	Test Value	Test	Conclusion
$(-\infty, -1)$	-2	$(-2)^3 + (-2)^2 + 4(-2) + 4 > 0$ $-8 > 0$, False	$(-\infty, -1)$ does not belong to the solution set.
$(-1, \infty)$	0	$0^3 + 0^2 + 4(0) + 4 > 0$ $4 > 0$, True	$(-1, \infty)$ belongs to the solution set.

The solution set is $(-1, \infty)$.

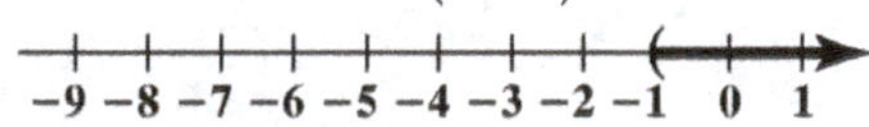

39. $x^3 \ge 9x^2$

$$x^3 - 9x^2 \ge 0$$

Solve the related polynomial equation.

$$x^3 - 9x^2 = 0$$
$$x^2(x - 9) = 0$$
$$x^2 = 0 \quad \text{or} \quad x - 9 = 0$$
$$x = \pm\sqrt{0} = 0 \qquad x = 9$$

The boundary points are 0 and 9.

Interval	Test Value	Test	Conclusion
$(-\infty, 0)$	-1	$(-1)^3 \ge 9(-1)^2$ $-1 \ge 9$, False	$(-\infty, 0)$ does not belong to the solution set.
$(0, 9)$	1	$1^3 \ge 9(1)^2$ $1 \ge 9$, False	$(0, 9)$ does not belong to the solution set.
$(9, \infty)$	10	$10^3 \ge 9(10)^2$ $1000 \ge 900$, True	$(9, \infty)$ belongs to the solution set.

The solution set is $\{0\} \cup [9, \infty)$.

−8 −6 −4 −2 0 2 4 6 8 10

41. $\dfrac{x-4}{x+3} > 0$

Find the values of x that make the numerator and denominator zero.

$x-4=0 \qquad x+3=0$

$x=4 \qquad x=-3$

The boundary points are –3 and 4. Exclude –3 from the solution set, since this would make the denominator zero.

Interval	Test Value	Test	Conclusion
$(-\infty,-3)$	-4	$\dfrac{-4-4}{-4+3} > 0$ $8 > 0$, true	$(-\infty,-3)$ belongs to the solution set.
$(-3,4)$	0	$\dfrac{0-4}{0+3} > 0$ $\dfrac{-4}{3} > 0$, false	$(-3,4)$ does not belong to the solution set.
$(4,\infty)$	5	$\dfrac{5-4}{5+3} > 0$ $\dfrac{1}{8} > 0$, true	$(4,\infty)$ belongs to the solution set.

The solution set is $(-\infty,-3)\cup(4,\infty)$.

−5 −4 −3 −2 −1 0 1 2 3 4 5

43. $\dfrac{x+3}{x+4} < 0$

Find the values of x that make the numerator and denominator zero.

$x+3=0 \qquad x+4=0$

$x=-3 \qquad x=-4$

The boundary points are –4 and –3.

Interval	Test Value	Test	Conclusion
$(-\infty,-4)$	–5	$\dfrac{-5+3}{-5+4} < 0$ $2 < 0$, false	$(-\infty,-4)$ does not belong to the solution set.
$(-4,-3)$	–3.5	$\dfrac{-3.5+3}{-3.5+4} < 0$ $-1 < 0$, true	$(-4,-3)$ belongs to the solution set.
$(-3,\infty)$	0	$\dfrac{0+3}{0+4} < 0$ $\dfrac{3}{4} < 0$, false	$(-3,\infty)$ does not belong to the solution set.

The solution set is $(-4,-3)$.

−5 −4 −3 −2 −1 0 1 2 3 4 5

45. $\dfrac{-x+2}{x-4} \geq 0$

Find the values of x that make the numerator and denominator zero.

$-x+2=0$ and $x-4=0$

$-x=-2$ $\quad x=4$

$x=2$

The boundary points are 2 and 4.

Interval	Test Value	Test	Conclusion
$(-\infty, 2)$	0	$\dfrac{-0+2}{0-4} \geq 0$ $-\dfrac{1}{2} \geq 0$, false	$(-\infty, 2)$ does not belong to the solution set.
$(2, 4)$	3	$\dfrac{-3+2}{3-4} \geq 0$ $1 \geq 0$, true	$(2, 4)$ belongs to the solution set.
$(4, \infty)$	5	$\dfrac{-5+2}{5-4} \geq 0$ $-3 \geq 0$, false	$(4, \infty)$ does not belong to the solution set.

Exclude 4 from the solution set because 4 would make the denominator zero. The solution set is $[2, 4)$

47. $\dfrac{4-2x}{3x+4} \leq 0$

Find the values of x that make the numerator and denominator zero.

$4-2x=0$ and $3x+4=0$

$-2x=-4$ $\quad 3x=-4$

$x=2$ $\quad x=-\dfrac{4}{3}$

The boundary points are $-\dfrac{4}{3}$ and 2.

Interval	Test Value	Test	Conclusion
$\left(-\infty, -\dfrac{4}{3}\right)$	−2	$\dfrac{4-2(-2)}{3(-2)+4} \leq 0$ $-4 \leq 0$, true	$\left(-\infty, -\dfrac{4}{3}\right)$ belongs to the solution set.
$\left(-\dfrac{4}{3}, 2\right)$	0	$\dfrac{4-2(0)}{3(0)+4} \leq 0$ $1 \leq 0$, false	$\left(-\dfrac{4}{3}, 2\right)$ does not belong to the solution set.
$[2, \infty)$	3	$\dfrac{4-2(3)}{3(3)+4} \leq 0$ $-\dfrac{2}{13} \leq 0$, true	$[2, \infty)$ belongs to the solution set.

Exclude $-\frac{4}{3}$ from the solution set because $-\frac{4}{3}$ would make the denominator zero. The solution set is $\left(-\infty, -\frac{4}{3}\right)\cup[2, \infty)$.

−5 −4 −3 −2 −1 0 1 2 3 4 5

49. $\frac{x}{x-3} > 0$

Find the values of x that make the numerator and denominator zero.

$x = 0$ and $x - 3 = 0$

$x = 3$

The boundary points are 0 and 3.

Interval	Test Value	Test	Conclusion
$(-\infty, 0)$	-1	$\frac{-1}{-1-3} > 0$ $\frac{1}{4} > 0$, true	$\left(-\infty, -\frac{4}{3}\right)$ belongs to the solution set.
$(0, 3)$	1	$\frac{1}{1-3} > 0$ $-\frac{1}{2} > 0$, false	$(0, 3)$ does not belong to the solution set.
$(3, \infty)$	4	$\frac{4}{4-3} > 0$ $4 > 0$, true	$(3, \infty)$ belongs to the solution set.

The solution set is $(-\infty, 0)\cup(3, \infty)$.

−3 −2 −1 0 1 2 3 4 5 6 7

51. $\frac{x+1}{x+3} < 2$

Express the inequality so that one side is zero.

$$\frac{x+1}{x+3} - 2 < 0$$

$$\frac{x+1}{x+3} - \frac{2(x+3)}{x+3} < 0$$

$$\frac{x+1-2(x+3)}{x+3} < 0$$

$$\frac{x+1-2x-6}{x+3} < 0$$

$$\frac{-x-5}{x+3} < 0$$

Find the values of x that make the numerator and denominator zero.

$$-x-5=0 \qquad x+3=0$$
$$-x=5 \qquad x=-3$$
$$x=-5$$

The boundary points are –5 and –3.

Interval	Test Value	Test	Conclusion
$(-\infty,-5)$	–6	$\frac{-6+1}{-6+3}<2$ $\frac{5}{3}<2$, true	$(-\infty,-5)$ belongs to the solution set.
$(-5,-3)$	–4	$\frac{-4+1}{-4+3}<2$ $3<2$, false	$(-5,-3)$ does not belong to the solution set.
$(-3,\infty)$	0	$\frac{0+1}{0+3}<2$ $\frac{1}{3}<2$, true	$(-3,\infty)$ belongs to the solution set.

The solution set is $(-\infty,-5)\cup(-3,\infty)$.

–6 –5 –4 –3 –2 –1 0 1 2 3 4

53. $\dfrac{x+4}{2x-1}\le 3$

Express the inequality so that one side is zero.

$$\frac{x+4}{2x-1}-3\le 0$$
$$\frac{x+4}{2x-1}-\frac{3(2x-1)}{2x-1}\le 0$$
$$\frac{x+4-3(2x-1)}{2x-1}\le 0$$
$$\frac{x+4-6x+3}{2x-1}\le 0$$
$$\frac{-5x+7}{2x-1}\le 0$$

Find the values of x that make the numerator and denominator zero.

$$-5x+7=0 \qquad 2x-1=0$$
$$-5x=-7 \qquad 2x=1$$
$$x=\frac{7}{5} \qquad x=\frac{1}{2}$$

The boundary points are $\frac{1}{2}$ and $\frac{7}{5}$.

Interval	Test Value	Test	Conclusion
$\left(-\infty, \frac{1}{2}\right)$	0	$\frac{0+4}{2(0)-1} \le 3$ $-4 \le 3$, true	$\left(-\infty, \frac{1}{2}\right)$ belongs to the solution set.
$\left(\frac{1}{2}, \frac{7}{5}\right)$	1	$\frac{1+4}{2(1)-1} \le 3$ $5 \le 3$, false	$\left(\frac{1}{2}, \frac{7}{5}\right)$ does not belong to the solution set.
$\left(\frac{7}{5}, \infty\right)$	2	$\frac{2+4}{2(2)-1} \le 3$ $2 \le 3$, true	$\left(\frac{7}{5}, \infty\right)$ belongs to the solution set.

Exclude $\frac{1}{2}$ from the solution set because $\frac{1}{2}$ would make the denominator zero. The solution set is $\left(-\infty, \frac{1}{2}\right) \cup \left[\frac{7}{5}, \infty\right)$.

−5 −4 −3 −2 −1 0 1 2 3 4 5

55. $\frac{x-2}{x+2} \le 2$

Express the inequality so that one side is zero.

$$\frac{x-2}{x+2} - 2 \le 0$$
$$\frac{x-2}{x+2} - \frac{2(x+2)}{x+2} \le 0$$
$$\frac{x-2-2(x+2)}{x+2} \le 0$$
$$\frac{x-2-2x-4}{x+2} \le 0$$
$$\frac{-x-6}{x+2} \le 0$$

Find the values of x that make the numerator and denominator zero.

$-x-6=0$ $\quad$ $x+2=0$

$-x=6$ $\quad$ $x=-2$

$x=-6$

The boundary points are -6 and -2.

Interval	Test Value	Test	Conclusion
$(-\infty, -6)$	−7	$\frac{-7-2}{-7+2} \le 2$ $\frac{9}{5} \le 2$, true	$(-\infty, -6)$ belongs to the solution set.
$(-6, -2)$	−3	$\frac{-3-2}{-3+2} \le 2$ $5 \le 2$, false	$(-6, -2)$ does not belong to the solution set.
$(-2, \infty)$	0	$\frac{0-2}{0+2} \le 2$ $-1 \le 2$, true	$(-2, \infty)$ belongs to the solution set.

Exclude −2 from the solution set because −2 would make the denominator zero. The solution set is $(-\infty, -6] \cup (-2, \infty)$.

−8 −7 −6 −5 −4 −3 −2 −1 0 1 2

57. $f(x) \geq g(x)$

$2x^2 \geq 5x - 2$

$2x^2 - 5x + 2 \geq 0$

Solve the related quadratic equation.

$2x^2 - 5x + 2 = 0$

$(2x-1)(x-2) = 0$

Apply the zero product principle.

$2x - 1 = 0$ or $x - 2 = 0$

$2x = 1$ $\quad x = 2$

$x = \frac{1}{2}$

The boundary points are $\frac{1}{2}$ and 2.

Interval	Test Value	Test	Conclusion
$\left(-\infty, \frac{1}{2}\right)$	0	$2(0)^2 \geq 5(0) - 2$ $0 \geq -2$, True	$\left(-\infty, \frac{1}{2}\right)$ belongs to the solution set.
$\left(\frac{1}{2}, 2\right)$	1	$2(1)^2 \geq 5(1) - 2$ $2 \geq 3$, False	$\left(\frac{1}{2}, 2\right)$ does not belong to the solution set.
$(2, \infty)$	3	$2(3)^2 \geq 5(3) - 2$ $18 \geq 13$, True	$(2, \infty)$ does not belong to the solution set.

The solution set is $\left(-\infty, \frac{1}{2}\right] \cup [2, \infty)$.

59. $f(x) < g(x)$

$\frac{2x}{x+1} < 1$

Express the inequality so that one side is zero.

$\frac{2x}{x+1} - 1 < 0$

$\frac{2x}{x+1} - \frac{x+1}{x+1} < 0$

$\frac{2x - x - 1}{x+1} < 0$

$\frac{x-1}{x+1} < 0$

Find the values of x that make the numerator and denominator zero.

$x - 1 = 0$ or $x + 1 = 0$

$x = 1$ $\quad x = -1$

The boundary points are -1 and 1.

Interval	Test Value	Test	Conclusion
$(-\infty,-1)$	-3	$\frac{2(-3)}{-3+1}<1$ $3<1$, false	$(-\infty,-1)$ does not belong to the solution set.
$(-1,1)$	0	$\frac{2(0)}{0+1}<1$ $0<1$, true	$(-1,1)$ belongs to the solution set.
$(1,\infty)$	2	$\frac{2(3)}{3+1}<1$ $\frac{3}{2}<1$, false	$(1,\infty)$ does not belong to the solution set.

The solution set is $(-1,1)$.

61. $|x^2+2x-36|>12$

Express the inequality without the absolute value symbol.

$x^2+2x-36<-12$ or $x^2+2x-36>12$

$x^2+2x-24<0$ $\quad$ $x^2+2x-48>0$

Solve the related quadratic equations.

$x^2+2x-24=0$ or $x^2+2x-48=0$

$(x+6)(x-4)=0$ $\quad$ $(x+8)(x-6)=0$

Apply the zero product principle.

$x+6=0$ or $x-4=0$ or $x+8=0$ or $x-6=0$

$x=-6$ $\quad$ $x=4$ $\quad$ $x=-8$ $\quad$ $x=6$

The boundary points are $-8,\ -6,\ 4$ and 6.

Test Interval	Test Number	Test	Conclusion
$(-\infty,-8)$	-9	$\left\|(-9)^2+2(-9)-36\right\|>12$ $27>12$, True	$(-\infty,-8)$ belongs to the solution set.
$(-8,-6)$	-7	$\left\|(-7)^2+2(-7)-36\right\|>12$ $1>12$, False	$(-8,-6)$ does not belong to the solution set.
$(-6,4)$	0	$\left\|0^2+2(0)-36\right\|>12$ $36>12$, True	$(-6,4)$ belongs to the solution set.
$(4,6)$	5	$\left\|5^2+2(5)-36\right\|>12$ $1>12$, False	$(4,6)$ does not belong to the solution set.
$(6,\infty)$	7	$\left\|7^2+2(7)-36\right\|>12$ $27>12$, True	$(6,\infty)$ belongs to the solution set.

The solution set is $(-\infty,-8)\cup(-6,4)\cup(6,\infty)$.

−10 −8 −6 −4 −2 0 2 4 6 8 10

63. $\frac{3}{x+3} > \frac{3}{x-2}$

Express the inequality so that one side is zero.

$$\frac{3}{x+3} - \frac{3}{x-2} > 0$$

$$\frac{3(x-2)}{(x+3)(x-2)} - \frac{3(x+3)}{(x+3)(x-2)} > 0$$

$$\frac{3x-6-3x-9}{(x+3)(x-2)} < 0$$

$$\frac{-15}{(x+3)(x-2)} < 0$$

Find the values of x that make the denominator zero.

$x+3=0$ $\quad$ $x-2=0$

$x=-3$ $\quad$ $x=2$

The boundary points are -3 and 2.

Interval	Test Value	Test	Conclusion
$(-\infty,-3)$	-4	$\frac{3}{-4+3} > \frac{3}{-4-2}$ False	$(-\infty,-3)$ does not belong to the solution set.
$(-3,2)$	0	$\frac{3}{0+3} > \frac{3}{0-2}$ True	$(-3,2)$ belongs to the solution set.
$(2,\infty)$	3	$\frac{3}{3+3} > \frac{3}{3-2}$ False	$(2,\infty)$ does not belong to the solution set.

The solution set is $(-3,2)$.

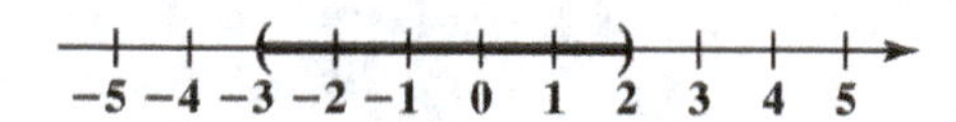

65. $\frac{x^2-x-2}{x^2-4x+3} > 0$

Find the values of x that make the numerator and denominator zero.

$x^2-x-2=0$ $\quad$ $x^2-4x+3=0$

$(x-2)(x+1)=0$ $\quad$ $(x-3)(x-1)=0$

Apply the zero product principle.

$x-2=0$ or $x+1=0$ $\quad$ $x-3=0$ or $x-1=0$

$x=2$ $\quad$ $x=-1$ $\quad$ $x=3$ $\quad$ $x=1$

The boundary points are -1, 1, 2 and 3.

Interval	Test Value	Test	Conclusion
$(-\infty,-1)$	-2	$\frac{(-2)^2-(-2)-2}{(-2)^2-4(-2)+3}>0$ $\frac{4}{15}>0$, True	$(-\infty,-1)$ belongs to the solution set.
$(-1,1)$	0	$\frac{0^2-0-2}{0^2-4(0)+3}>0$ $-\frac{2}{3}>0$, False	$(-1,1)$ does not belong to the solution set.
$(1,2)$	1.5	$\frac{1.5^2-1.5-2}{1.5^2-4(1.5)+3}>0$ $\frac{5}{3}>0$, True	$(1,2)$ belongs to the solution set.
$(2,3)$	2.5	$\frac{2.5^2-2.5-2}{2.5^2-4(2.5)+3}>0$ $-\frac{7}{3}>0$, False	$(2,3)$ does not belong to the solution set.
$(3,\infty)$	4	$\frac{4^2-4-2}{4^2-4(4)+3}>0$ $\frac{10}{3}>0$, True	$(3,\infty)$ belongs to the solution set.

The solution set is $(-\infty,-1)\cup(1,2)\cup(3,\infty)$.

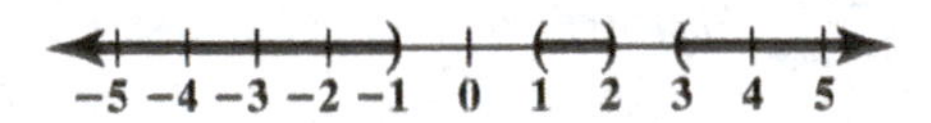

67. $2x^3+11x^2 \ge 7x+6$

$2x^3+11x^2-7x-6 \ge 0$

The graph of $f(x)=2x^3+11x^2-7x-6$ appears to cross the x-axis at -6, $-\frac{1}{2}$, and 1. Verify this numerically by substituting these values into the function.

$f(-6)=2(-6)^3+11(-6)^2-7(-6)-6=2(-216)+11(36)-(-42)-6=-432+396+42-6=0$

$f\left(-\frac{1}{2}\right)=2\left(-\frac{1}{2}\right)^3+11\left(-\frac{1}{2}\right)^2-7\left(-\frac{1}{2}\right)-6=2\left(-\frac{1}{8}\right)+11\left(\frac{1}{4}\right)-\left(-\frac{7}{2}\right)-6=-\frac{1}{4}+\frac{11}{4}+\frac{7}{2}-6=0$

$f(1)=2(1)^3+11(1)^2-7(1)-6=2(1)+11(1)-7-6=2+11-7-6=0$

Thus, the boundaries are -6, $-\frac{1}{2}$, and 1. We need to find the intervals on which $f(x)\ge 0$. These intervals are indicated on the graph where the curve is above the x-axis. Now, the curve is above the x-axis when $-6<x<-\frac{1}{2}$ and when $x>1$. Thus, the solution set is $\left[-6,-\frac{1}{2}\right]\cup[1,\infty)$.

69. $$\frac{1}{4(x+2)} \le -\frac{3}{4(x-2)}$$

$$\frac{1}{4(x+2)} + \frac{3}{4(x-2)} \le 0$$

Simplify the left side of the inequality.

$$\frac{x-2}{4(x+2)} + \frac{3(x+2)}{4(x-2)} = \frac{x-2+3x+6}{4(x+2)(x-2)} = \frac{4x+4}{4(x+2)(x-2)} = \frac{4(x+1)}{4(x+2)(x-2)} = \frac{x+1}{x^2-4}.$$

The graph of $f(x) = \frac{x+1}{x^2-4}$ crosses the x-axis at -1, and has vertical asymptotes at $x=-2$ and $x=2$. Thus, the boundaries are -2, -1, and 1. We need to find the intervals on which $f(x) \le 0$. These intervals are indicated on the graph where the curve is below the x-axis. Now, the curve is below the x-axis when $x<-2$ and when $-1<x<2$. Thus, the solution set is $(-\infty,-2)\cup[-1,2)$.

71. $s(t) = -16t^2 + 48t + 160$

To find when the height exceeds the height of the building, solve the inequality $-16t^2 + 48t + 160 > 160$.

Solve the related quadratic equation.

$$-16t^2 + 48t + 160 = 160$$
$$-16t^2 + 48t = 0$$
$$t^2 - 3t = 0$$
$$t(t-3) = 0$$

Apply the zero product principle.

$t = 0$ or $t-3=0$

$t = 3$

The boundary points are 0 and 3.

Interval	Test Value	Test	Conclusion
$(0,3)$	1	$-16(1)^2 + 48(1) + 160 > 160$ $192 > 160$, true	$(0,3)$ belongs to the solution set.
$(3,\infty)$	4	$-16(4)^2 + 48(4) + 160 > 160$ $96 > 160$, false	$(3,\infty)$ does not belong to the solution set.

The solution set is $(0,3)$. This means that the ball exceeds the height of the building between 0 and 3 seconds.

73. $f(x) = 0.0875x^2 - 0.4x + 66.6$

$g(x) = 0.0875x^2 + 1.9x + 11.6$

a. $f(35) = 0.0875(35)^2 - 0.4(35) + 66.6 \approx 160$ feet

$g(35) = 0.0875(35)^2 + 1.9(35) + 11.6 \approx 185$ feet

b. Dry pavement: graph (b)
Wet pavement: graph (a)

c. The answers to part (a) model the actual stopping distances shown in the figure extremely well. The function values and the data are identical.

d. $0.0875x^2 - 0.4x + 66.6 > 540$

$0.0875x^2 - 0.4x + 473.4 > 0$

Solve the related quadratic equation.

$0.0875x^2 - 0.4x + 473.4 = 0$

$$x = \frac{-b \pm \sqrt{b^2 - 4ac}}{2a}$$

$$x = \frac{-(-0.4) \pm \sqrt{(-0.4)^2 - 4(0.0875)(473.4)}}{2(0.0875)}$$

$x \approx -71$ or 76

Since the function's domain is $x \geq 30$, we must test the following intervals.

Interval	Test Value	Test	Conclusion
$(30, 76)$	50	$0.0875(50)^2 - 0.4(50) + 66.6 > 540$ $265.35 > 540$, False	$(30, 76)$ does not belong to the solution set.
$(76, \infty)$	100	$0.0875(100)^2 - 0.4(100) + 66.6 > 540$ $901.6 > 540$, True	$(76, \infty)$ belongs to the solution set.

On dry pavement, stopping distances will exceed 540 feet for speeds exceeding 76 miles per hour. This is represented on graph (b) to the right of point (76, 540).

75. $\overline{C}(x) = \dfrac{500,000 + 400x}{x}$

To find when the cost of producing each wheelchair does not exceed \$425, solve the inequality $\dfrac{500,000 + 400x}{x} \leq 425.$

Express the inequality so that one side is zero.

$$\frac{500,000 + 400x}{x} - 425 \leq 0$$

$$\frac{500,000 + 400x}{x} - \frac{425x}{x} \leq 0$$

$$\frac{500,000 + 400x - 425x}{x} \leq 0$$

$$\frac{500,000 - 25x}{x} \leq 0$$

Find the values of x that make the numerator and denominator zero.

$500,000 - 25x = 0 \qquad x = 0$

$500,000 = 25x$

$20,000 = x$

The boundary points are 0 and 20,000.

Interval	Test Value	Test	Conclusion
$(0, 20000)$	1	$\frac{500,000+400(1)}{1} \le 425$ $500,400 \le 425$, false	$(0, 20000)$ does not belong to the solution set.
$(20000, \infty)$	25,000	$\frac{500,000+400(25,000)}{25,000} \le 425$ $420 \le 425$, true	$(20000, \infty)$ belongs to the solution set.

The solution set is $[20000, \infty)$. This means that the company's production level will have to be at least 20,000 wheelchairs per week. The boundary corresponds to the point (20,000, 425) on the graph. When production is 20,000 or more per month, the average cost is $425 or less.

77. Let x = the length of the rectangle.
Since Perimeter $= 2(\text{length}) + 2(\text{width})$, we know

$$50 = 2x + 2(\text{width})$$
$$50 - 2x = 2(\text{width})$$
$$\text{width} = \frac{50-2x}{2} = 25 - x$$

Now, $A = (\text{length})(\text{width})$, so we have that

$$A(x) \le 114$$
$$x(25-x) \le 114$$
$$25x - x^2 \le 114$$

Solve the related equation

$$25x - x^2 = 114$$
$$0 = x^2 - 25x + 114$$
$$0 = (x-19)(x-6)$$

Apply the zero product principle.

$x - 19 = 0$ or $x - 6 = 0$

$x = 19$ $\quad x = 6$

The boundary points are 6 and 19.

Interval	Test Value	Test	Conclusion
$(0,6)$	1	$25(1) - 1^2 \le 114$ $24 \le 114$, True	$(0,6)$ belongs to the solution set.
$(6,19)$	10	$25(10) - 10^2 \le 114$ $150 \le 114$, False	$(6,19)$ does not belong to the solution set.
$(19, \infty)$	20	$25(20) - 20^2 \le 114$ $100 \le 114$, True	$(19, \infty)$ belongs to the solution set.

If the length is 6 feet, then the width is 19 feet. If the length is less than 6 feet, then the width is greater than 19 feet. Thus, if the area of the rectangle is not to exceed 114 square feet, the length of the shorter side must be 6 feet or less.

79. – 81. Answers will vary.

83. $2x^2 + 5x - 3 \le 0$
Let $y_1 = 2x^2 + 5x - 3$.

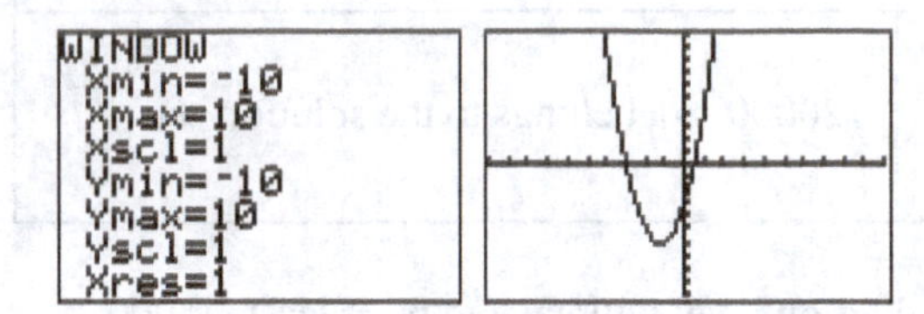

The graph is crosses the x-axis at -3 and $\frac{1}{2}$. The graph is below the x-axis when $-3 < x < \frac{1}{2}$. Thus, the solution set is $\left[-3, \frac{1}{2}\right]$.

85. $\frac{x+2}{x-3} \le 2$

$\frac{x+2}{x-3} - 2 \le 0$

Let $y_1 = \frac{x+2}{x-3} - 2$.

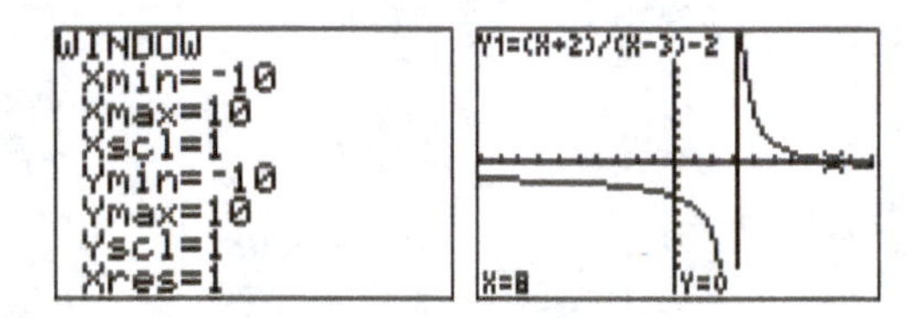

The graph is crosses the x-axis at 8. The function has a vertical asymptote at $x = 3$. The graph is below the x-axis when $x < 3$ and when $x > 8$. Thus, the solution set is $(-\infty, 3) \cup [8, \infty)$.

87. $x^3 + 2x^2 - 5x - 6 > 0$
Let $y_1 = x^3 + 2x^2 - 5x - 6$

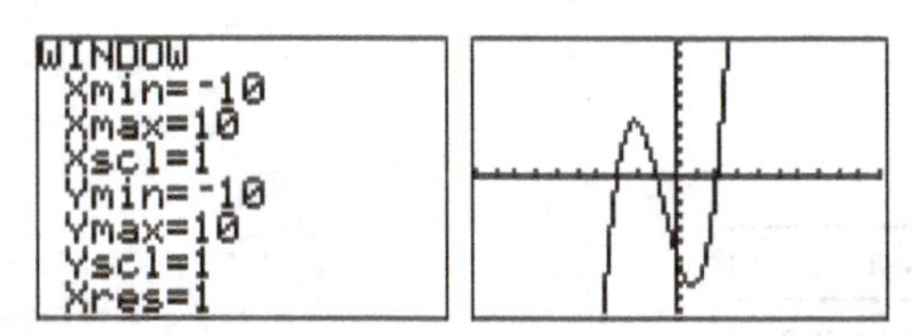

The graph is crosses the x-axis at -3, -1, and 2. The graph is above the x-axis when $-3 < x < -1$ and when $x > 2$. Thus, the solution set is $(-3, -1) \cup (2, \infty)$.

89. **a.** $f(x) = 0.1375x^2 + 0.7x + 37.8$

b. $0.1375x^2 + 0.7x + 37.8 > 446$

$0.1375x^2 + 0.7x + 408.2 > 0$

Solve the related quadratic equation.

$0.1375x^2 + 0.7x + 408.2 = 0$

$$x = \frac{-b \pm \sqrt{b^2 - 4ac}}{2a}$$

$$x = \frac{-(0.7) \pm \sqrt{(0.7)^2 - 4(0.1375)(408.2)}}{2(0.1375)}$$

$x \approx -57$ or 52

Since the function's domain must be $x \geq 0$, we must test the following intervals.

Interval	Test Value	Test	Conclusion
$(0, 52)$	10	$0.1375(10)^2 + 0.7(10) + 37.8 > 446$ $58.55 > 446$, False	$(0, 52)$ does not belong to the solution set.
$(52, \infty)$	100	$0.1375(100)^2 + 0.7(100) + 37.8 > 446$ $1482.8 > 446$, True	$(52, \infty)$ belongs to the solution set.

On wet pavement, stopping distances will exceed 446 feet for speeds exceeding 52 miles per hour.

91. does not make sense; Explanations will vary. Sample explanation: Polynomials are defined for all values.

93. does not make sense; Explanations will vary. Sample explanation: To solve this inequality you must first subtract 2 from both sides.

95. false; Changes to make the statement true will vary. A sample change is: The inequality cannot be solved by multiplying both sides by $x + 3$. We do not know if $x + 3$ is positive or negative. Thus, we would not know whether or not to reverse the order of the inequality.

97. true

99. Answers will vary. An example is $\frac{x-3}{x+4} \geq 0$.

101. $(x-2)^2 \leq 0$

Since the left hand side of the inequality is a square, we know it cannot be negative. In addition, the inequality calls for a number that is less than or equal to zero. The only possible solution is for the left hand side to equal zero. The left hand side of the inequality is zero when x is 2. Hence, the solution set is $\{2\}$.

103. $\frac{1}{(x-2)^2} > 0$

Since the denominator in the inequality is a square, we know it cannot be negative. Additionally, because the numerator is 1, the fraction will never be negative. As a result, x can be any real number except one that makes the denominator zero. Since 2 is the only value that makes the denominator zero, the solution set is $(-\infty, 2) \cup (2, \infty)$.

105. The radicand must be greater than or equal to zero:
$27 - 3x^2 \geq 0$
The inequality is true for values between 3 and –3. This means that the radicand is positive for values between 3 and –3, and the domain of the function is $[-3, 3]$.

106. $\left|\frac{x-5}{3}\right| < 8$

$$-8 < \frac{x-5}{3} < 8$$
$$-24 < x-5 < 24$$
$$-19 < x < 29$$

The solution set is $(-19, 29)$.

107. $\frac{2x+6}{x^2+8x+16} \div \frac{x^2-9}{x^2+3x-4}$

$$= \frac{2x+6}{x^2+8x+16} \cdot \frac{x^2+3x-4}{x^2-9}$$
$$= \frac{2\cancel{(x+3)}}{\cancel{(x+4)}(x+4)} \cdot \frac{\cancel{(x+4)}(x-1)}{\cancel{(x+3)}(x-3)}$$
$$= \frac{2(x-1)}{(x+4)(x-3)}$$

108. $x^4 - 16y^4$

$$= (x^2+4y^2)(x^2-4y^2)$$
$$= (x^2+4y^2)(x+2y)(x-2y)$$

109. $f(x) = 2^x$

x	$f(x) = 2^x$	(x, y)
–3	$2^{-3} = \frac{1}{8}$	$\left(-3, \frac{1}{8}\right)$
–2	$2^{-2} = \frac{1}{4}$	$\left(-2, \frac{1}{4}\right)$
–1	$2^{-1} = \frac{1}{2}$	$\left(-1, \frac{1}{2}\right)$
0	$2^0 = 1$	$(0, 1)$
1	$2^1 = 2$	$(1, 2)$
2	$2^2 = 4$	$(2, 4)$
3	$2^3 = 8$	$(3, 8)$

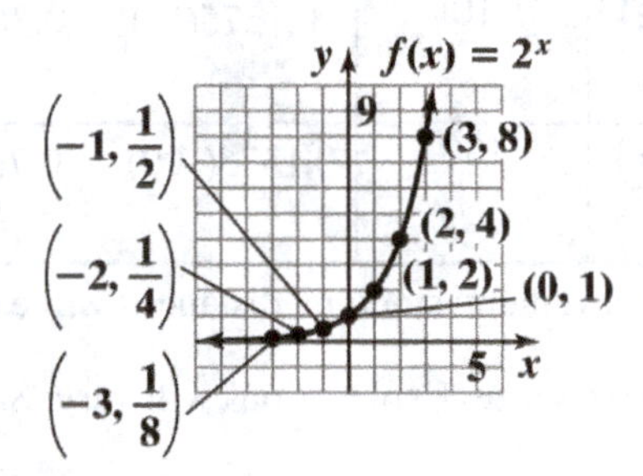

110. $f(x) = 2^{-x}$

x	$f(x) = 2^{-x}$	(x, y)
–3	$2^{-(-3)} = 8$	$(-3, 8)$
–2	$2^{-(-2)} = 4$	$(-2, 4)$
–1	$2^{-(-1)} = 2$	$(-1, 2)$
0	$2^0 = 1$	$(0, 1)$
1	$2^{-1} = \frac{1}{2}$	$\left(1, \frac{1}{2}\right)$
2	$2^{-2} = \frac{1}{4}$	$\left(2, \frac{1}{4}\right)$
3	$2^{-3} = \frac{1}{8}$	$\left(3, \frac{1}{8}\right)$

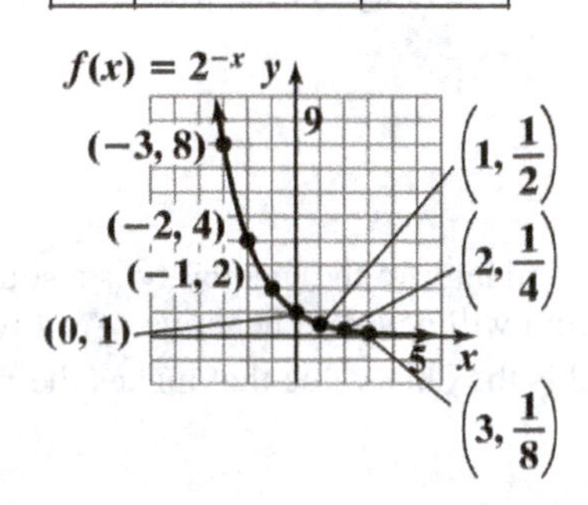

111. $f(x)=2^x+1$

x	$f(x)=2^x+1$	(x, y)
-3	$2^{-3}+1=1\frac{1}{8}$	$\left(-3,1\frac{1}{8}\right)$
-2	$2^{-2}+1=1\frac{1}{4}$	$\left(-2,1\frac{1}{4}\right)$
-1	$2^{-1}+1=1\frac{1}{2}$	$\left(-1,1\frac{1}{2}\right)$
0	$2^0+1=2$	$(0,2)$
1	$2^1+1=3$	$(1,3)$
2	$2^2+1=5$	$(2,5)$
3	$2^3+1=9$	$(3,9)$

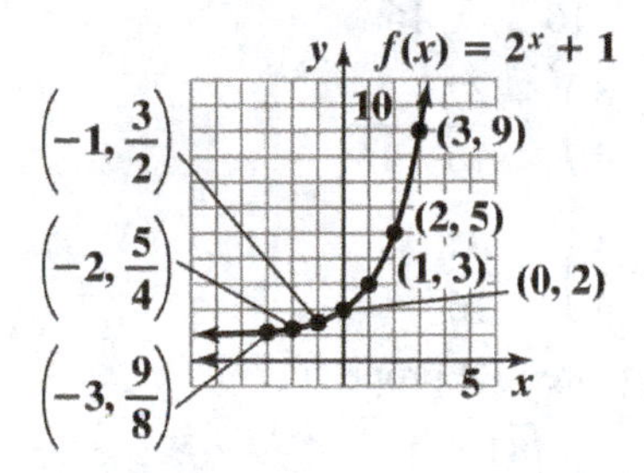

Chapter 8 Review Exercises

1. $2x^2-3=125$

$2x^2=128$

$x^2=64$

$x=\pm 8$

The solution set is $\{-8,8\}$.

2. $3x^2-150=0$

$3x^2=150$

$x^2=50$

$x=\pm\sqrt{50}$

$x=\pm\sqrt{25\cdot 2}$

$x=\pm 5\sqrt{2}$

The solution set is $\{-5\sqrt{2},5\sqrt{2}\}$.

3. $3x^2-2=0$

$3x^2=2$

$x^2=\frac{2}{3}$

$x=\pm\sqrt{\frac{2}{3}}$

Rationalize the denominator.

$x=\pm\frac{\sqrt{2}}{\sqrt{3}}\cdot\frac{\sqrt{3}}{\sqrt{3}}=\pm\frac{\sqrt{6}}{3}$

The solution set is $\left\{-\frac{\sqrt{6}}{3},\frac{\sqrt{6}}{3}\right\}$.

4. $(x-4)^2=18$

$x-4=\pm\sqrt{18}$

$x=4\pm\sqrt{9\cdot 2}$

$x=4\pm 3\sqrt{2}$

The solution set is $\{4-3\sqrt{2},4+3\sqrt{2}\}$.

5. $(x+7)^2=-36$

$x+7=\pm\sqrt{-36}$

$x=-7\pm 6i$

The solution set is $\{-7-6i,-7+6i\}$.

6. x^2+20x+ _____

Since $b = 20$, add $\left(\frac{b}{2}\right)^2=\left(\frac{20}{2}\right)^2=(10)^2=100$.

$x^2+20x+100=(x+10)^2$

7. x^2-3x+ _____

Since $b = 3$, add $\left(\frac{b}{2}\right)^2=\left(\frac{3}{2}\right)^2=\frac{9}{4}$.

$x^2-3x+\frac{9}{4}=\left(x-\frac{3}{2}\right)^2$

8. $x^2 - 12x + 27 = 0$

$x^2 - 12x \quad = -27$

Since $b = -12$, add $\left(\frac{b}{2}\right)^2 = \left(\frac{-12}{2}\right)^2 = (-6)^2 = 36$.

$$x^2 - 12x + 27 = 0$$
$$x^2 - 12x + 36 = -27 + 36$$
$$(x-6)^2 = 9$$

Apply the square root property.

$x - 6 = 3 \qquad x - 6 = -3$

$x = 9 \qquad x = 3$

The solution set is $\{3, 9\}$.

9. $x^2 - 7x - 1 = 0$

$x^2 - 7x \quad = 1$

Since $b = -7$, add $\left(\frac{b}{2}\right)^2 = \left(\frac{-7}{2}\right)^2 = \frac{49}{4}$.

$$x^2 - 7x + \frac{49}{4} = 1 + \frac{49}{4}$$
$$\left(x - \frac{7}{2}\right)^2 = \frac{4}{4} + \frac{49}{4}$$
$$\left(x - \frac{7}{2}\right)^2 = \frac{53}{4}$$

Apply the square root property.

$$x - \frac{7}{2} = \pm\sqrt{\frac{53}{4}}$$
$$x = \frac{7}{2} \pm \frac{\sqrt{53}}{2} = \frac{7 \pm \sqrt{53}}{2}$$

The solution set is $\left\{\frac{7 \pm \sqrt{53}}{2}\right\}$.

10. $2x^2 + 3x - 4 = 0$

$x^2 + \frac{3}{2}x - 2 = 0$

$x^2 + \frac{3}{2}x \quad = 2$

Since $b = \frac{3}{2}$, add

$$\left(\frac{b}{2}\right)^2 = \left(\frac{\frac{3}{2}}{2}\right)^2 = \left(\frac{3}{2} \div 2\right)^2$$
$$= \left(\frac{3}{2} \cdot \frac{1}{2}\right)^2 = \left(\frac{3}{4}\right)^2 = \frac{9}{16}.$$

$$x^2 + \frac{3}{2}x + \frac{9}{16} = 2 + \frac{9}{16}$$
$$\left(x + \frac{3}{4}\right)^2 = \frac{32}{16} + \frac{9}{16}$$
$$\left(x + \frac{3}{4}\right)^2 = \frac{41}{16}$$

Apply the square root property.

$$x + \frac{3}{4} = \pm\sqrt{\frac{41}{16}}$$
$$x = -\frac{3}{4} \pm \frac{\sqrt{41}}{4}$$
$$x = \frac{-3 \pm \sqrt{41}}{4}$$

The solution set is $\left\{\frac{-3 \pm \sqrt{41}}{4}\right\}$.

11. $A = P(1+r)^t$

$2916 = 2500(1+r)^2$

$\frac{2916}{2500} = (1+r)^2$

Apply the square root property.

$1 + r = \pm\sqrt{\frac{2916}{2500}}$

$r = -1 \pm \sqrt{1.1664}$

$r = -1 \pm 1.08$

The solutions are $-1 - 1.08 = -2.08$ and $-1 + 1.08 = 0.08$. Disregard -2.08 since we cannot have a negative interest rate. The interest rate is 0.08 or 8%.

12. $W(t) = 3t^2$

$588 = 3t^2$

$196 = t^2$

Apply the square root property.

$t^2 = 196$

$t = \pm\sqrt{196}$

$t = \pm 14$

The solutions are –14 and 14. Disregard –14, because we cannot have a negative time measurement. The fetus will weigh 588 grams after 14 weeks.

13.

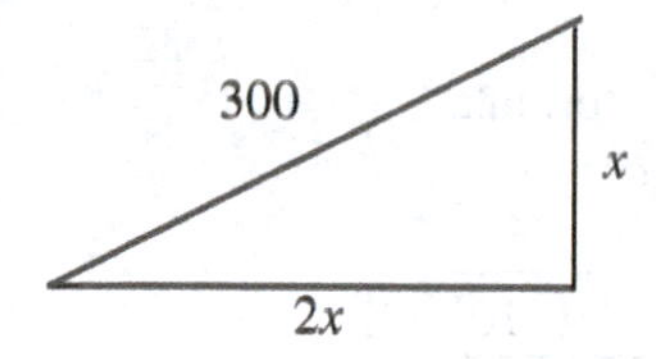

Use the Pythagorean Theorem.

$(2x)^2 + x^2 = 300^2$

$4x^2 + x^2 = 90,000$

$5x^2 = 90,000$

$x^2 = 18,000$

$x = \pm\sqrt{18,000}$

$x = \pm\sqrt{3600 \cdot 5}$

$x = \pm 60\sqrt{5}$

The solutions are $\pm 60\sqrt{5}$ meters. Disregard $-60\sqrt{5}$ meters, because we can't have a negative length measurement. Therefore, the building is $60\sqrt{5}$ meters, or approximately 134.2 meters high.

14. $x^2 = 2x + 4$

$x^2 - 2x - 4 = 0$

$a = 1 \quad b = -2 \quad c = -4$

$$x = \frac{-(-2) \pm \sqrt{(-2)^2 - 4(1)(-4)}}{2(1)}$$

$$= \frac{2 \pm \sqrt{4+16}}{2}$$

$$= \frac{2 \pm \sqrt{20}}{2}$$

$$= \frac{2 \pm \sqrt{4 \cdot 5}}{2}$$

$$= \frac{2 \pm 2\sqrt{5}}{2} = \frac{2\left(1 \pm \sqrt{5}\right)}{2} = 1 \pm \sqrt{5}$$

The solution set is $\left\{1 \pm \sqrt{5}\right\}$.

15. $x^2 - 2x + 19 = 0$

$a = 1 \quad b = -2 \quad c = 19$

$$x = \frac{-(-2) \pm \sqrt{(-2)^2 - 4(1)(19)}}{2(1)}$$

$$= \frac{2 \pm \sqrt{4-76}}{2}$$

$$= \frac{2 \pm \sqrt{-72}}{2}$$

$$= \frac{2 \pm \sqrt{-36 \cdot 2}}{2}$$

$$= \frac{2 \pm 6i\sqrt{2}}{2} = \frac{2\left(1 \pm 3i\sqrt{2}\right)}{2} = 1 \pm 3i\sqrt{2}$$

The solution set is $\left\{1 \pm 3i\sqrt{2}\right\}$.

16. $2x^2 = 3 - 4x$

$2x^2 + 4x - 3 = 0$

$a = 2 \quad b = 4 \quad c = -3$

$$x = \frac{-4 \pm \sqrt{4^2 - 4(2)(-3)}}{2(2)}$$
$$= \frac{-4 \pm \sqrt{16 + 24}}{4}$$
$$= \frac{-4 \pm \sqrt{40}}{4}$$
$$= \frac{-4 \pm \sqrt{4 \cdot 10}}{4}$$
$$= \frac{-4 \pm 2\sqrt{10}}{4}$$
$$= \frac{2\left(-2 \pm \sqrt{10}\right)}{4} = \frac{-2 \pm \sqrt{10}}{2}$$

The solution set is $\left\{\frac{-2 \pm \sqrt{10}}{2}\right\}$.

17. $x^2 - 4x + 13 = 0$

$a = 1 \quad b = -4 \quad c = 13$

Find the discriminant.

$$b^2 - 4ac = (-4)^2 - 4(1)(13)$$
$$= 16 - 52 = -36$$

Since the discriminant is negative, there are two imaginary solutions which are complex conjugates.

18. $9x^2 = 2 - 3x$

$9x^2 + 3x - 2 = 0$

$a = 9 \quad b = 3 \quad c = -2$

Find the discriminant.

$$b^2 - 4ac = 3^2 - 4(9)(-2)$$
$$= 9 + 72 = 81$$

Since the discriminant is greater than zero and a perfect square, there are two real rational solutions.

19. $2x^2 + 4x = 3$

$2x^2 + 4x - 3 = 0$

$a = 2 \quad b = 4 \quad c = -3$

Find the discriminant.

$$b^2 - 4ac = 4^2 - 4(2)(-3)$$
$$= 16 + 24 = 40$$

Since the discriminant is greater than zero but not a perfect square, there are two real irrational solutions.

20. $3x^2 - 10x - 8 = 0$

$(3x + 2)(x - 4) = 0$

Apply the zero product principle.

$3x + 2 = 0 \quad$ and $\quad x - 4 = 0$

$3x = -2 \qquad\qquad x = 4$

$x = -\frac{2}{3}$

The solution set is $\left\{-\frac{2}{3}, 4\right\}$.

21. $(2x - 3)(x + 2) = x^2 - 2x + 4$

$2x^2 + 4x - 3x - 6 = x^2 - 2x + 4$

$x^2 + 3x - 10 = 0$

Use the quadratic formula.

$a = 1 \quad b = 3 \quad c = -10$

$$x = \frac{-3 \pm \sqrt{3^2 - 4(1)(-10)}}{2(1)}$$
$$= \frac{-3 \pm \sqrt{9 - (-40)}}{2}$$
$$= \frac{-3 \pm \sqrt{49}}{2} = \frac{-3 \pm 7}{2} = -5 \text{ or } 2$$

The solution set is $\{-5, 2\}$.

22. $5x^2 - x - 1 = 0$

Use the quadratic formula.

$a = 5 \quad b = -1 \quad c = -1$

$$x = \frac{-(-1) \pm \sqrt{(-1)^2 - 4(5)(-1)}}{2(5)}$$
$$= \frac{1 \pm \sqrt{1 - (-20)}}{10} = \frac{1 \pm \sqrt{21}}{10}$$

The solution set is $\left\{\frac{1 \pm \sqrt{21}}{10}\right\}$.

23. $x^2 - 16 = 0$

$x^2 = 16$

Apply the square root principle.

$x = \pm\sqrt{16} = \pm 4$

The solution set is $\{-4, 4\}$.

24. $(x-3)^2-8=0$

$(x-3)^2=8$

Apply the square root principle.

$x-3=\pm\sqrt{8}$

$x=3\pm\sqrt{4\cdot 2}$

$x=3\pm 2\sqrt{2}$

The solution set is $\{3\pm 2\sqrt{2}\}$.

25. $3x^2-x+2=0$

Use the quadratic formula.

$a=3 \quad b=-1 \quad c=2$

$$x=\frac{-(-1)\pm\sqrt{(-1)^2-4(3)(2)}}{2(3)}$$
$$=\frac{1\pm\sqrt{1-24}}{6}$$
$$=\frac{1\pm\sqrt{-23}}{6}=\frac{1}{6}\pm i\frac{\sqrt{23}}{6}$$

The solution set is $\left\{\frac{1}{6}\pm i\frac{\sqrt{23}}{6}\right\}$.

26.
$$\frac{5}{x+1}+\frac{x-1}{4}=2$$
$$4(x+1)\left(\frac{5}{x+1}+\frac{x-1}{4}\right)=4(x+1)(2)$$
$$20+(x+1)(x-1)=8x+8$$
$$20+x^2-1=8x+8$$
$$x^2-8x+11=0$$

Use the quadratic formula.

$a=1 \quad b=-8 \quad c=11$

$$x=\frac{-(-8)\pm\sqrt{(-8)^2-4(1)(11)}}{2(1)}$$
$$=\frac{8\pm\sqrt{64-44}}{2}$$
$$=\frac{8\pm\sqrt{20}}{2}$$
$$=\frac{8\pm\sqrt{4\cdot 5}}{2}$$
$$=\frac{8\pm 2\sqrt{5}}{2}=\frac{2(4\pm\sqrt{5})}{2}=4\pm\sqrt{5}$$

The solution set is $\{4\pm\sqrt{5}\}$.

27. Because the solution set is $\left\{-\frac{1}{3},\frac{3}{5}\right\}$, we have

$x=-\frac{1}{3}$ or $x=\frac{3}{5}$

$3x=-1 \qquad 5x=3$

$3x+1=0 \qquad 5x-3=0.$

Apply the zero-product principle in reverse.

$$(3x+1)(5x-3)=0$$
$$15x^2-9x+5x-3=0$$
$$15x^2-4x-3=0$$

28. Because the solution set is $\{-9i,\ 9i\}$, we have

$x=-9i$ or $x=9i$

$x+9i=0 \qquad x-9i=0.$

Apply the zero-product principle in reverse.

$$(x+9i)(x-9i)=0$$
$$x^2-9ix+9ix-81i^2=0$$
$$x^2-81(-1)=0$$
$$x^2+81=0$$

29. Because the solution set is $\{-4\sqrt{3},4\sqrt{3}\}$, we have

$x=-4\sqrt{3}$ or $x=4\sqrt{3}$

$x+4\sqrt{3}=0 \qquad x-4\sqrt{3}=0.$

Apply the zero product principle in reverse.

$$(x+4\sqrt{3})(x-4\sqrt{3})=0$$
$$x^2-(4\sqrt{3})^2=0$$
$$x^2-16\cdot 3=0$$
$$x^2-48=0$$

30. **a.** $g(x) = 0.125x^2 + 2.3x + 27$

$g(35) = 0.125(35)^2 + 2.3(35) + 27 \approx 261$

On wet pavement, a motorcycle traveling at 35 miles per hour will require a stopping distance of 261 feet.
This answer overestimates the stopping distance shown in the graph by 1 foot.

b. $f(x) = 0.125x^2 - 0.8x + 99$

$267 = 0.125x^2 - 0.8x + 99$

$0 = 0.125x^2 - 0.8x - 168$

$$x = \frac{-b \pm \sqrt{b^2 - 4ac}}{2a}$$

$$x = \frac{-(-0.8) \pm \sqrt{(-0.8)^2 - 4(0.125)(-168)}}{2(0.125)}$$

$x \approx -33.6$ or 40

On dry pavement, a stopping distances of 267 feet will be required for a motorcycle traveling 40 miles per hour.

31. **a.** $g(35) = 0.125(35)^2 + 2.3(35) + 27 \approx 261$

This value is shown in the graph by the point (35, 261).

b. $f(40) = 0.125(40)^2 - 0.8(40) + 99 = 267$

This value is shown in the graph by the point (40, 267).

32. $0 = -16t^2 + 140t + 3$

Apply the Pythagorean Theorem.
$a = -16 \quad b = 140 \quad c = 3$

$$= \frac{-140 \pm \sqrt{19,600 + 192}}{-32}$$

$$= \frac{-140 \pm \sqrt{19,792}}{-32} \approx \frac{-140 \pm 140.7}{-32}$$

$$\approx \frac{-140 - 140.7}{-32} \text{ or } \frac{-140 + 140.7}{-32}$$

$$\approx \frac{-280.7}{-32} \text{ or } \frac{0.7}{-32}$$

≈ 8.8 or -0.02

Disregard –0.02 because we cannot have a negative time measurement. The solution is approximately 8.8. The ball will hit the ground in about 8.8 seconds.

33. $f(x) = -(x+1)^2 + 4$

Since $a = -1$ is negative, the parabola opens downward. The vertex of the parabola is $(h, k) = (-1, 4)$ and the axis of symmetry is $x = -1$. Replace $f(x)$ with 0 to find x–intercepts.

$0 = -(x+1)^2 + 4$

$(x+1)^2 = 4$

Apply the square root property.

$x + 1 = \sqrt{4}$ or $x + 1 = -\sqrt{4}$

$x + 1 = 2 \qquad x + 1 = -2$

$x = 1 \qquad x = -3$

The x–intercepts are 1 and –3. Set $x = 0$ and solve for y to obtain the y–intercept.

$y = -(0+1)^2 + 4$

$y = -(1)^2 + 4$

$y = -1 + 4 = 3$

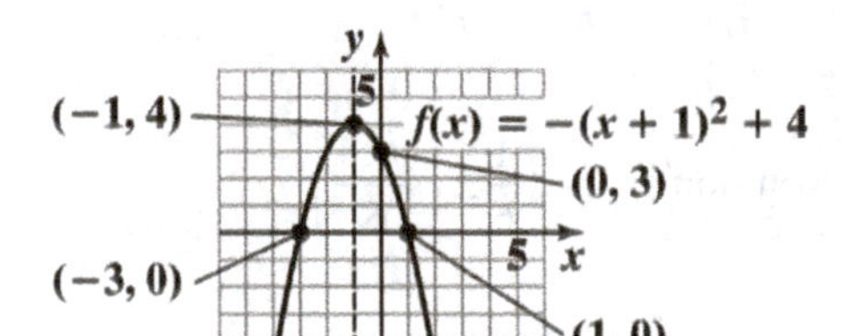

Axis of symmetry: $x = -1$.

34. $f(x) = (x+4)^2 - 2$

Since $a = 1$ is positive, the parabola opens upward. The vertex of the parabola is $(h, k) = (-4, -2)$ and the axis of symmetry is $x = -4$. Replace $f(x)$ with 0 to find x–intercepts.

$0 = (x+4)^2 - 2$

$2 = (x+4)^2$

Apply the square root property.

$x + 4 = \sqrt{2}$ or $x + 4 = -\sqrt{2}$

$x = -4 + \sqrt{2} \qquad x = -4 - \sqrt{2}$

The x–intercepts are $-4-\sqrt{2}$ and $-4+\sqrt{2}$. Set $x=0$ and solve for y to obtain the y–intercept.

$y=(0+4)^2-2$

$y=4^2-2$

$y=16-2$

$y=14$

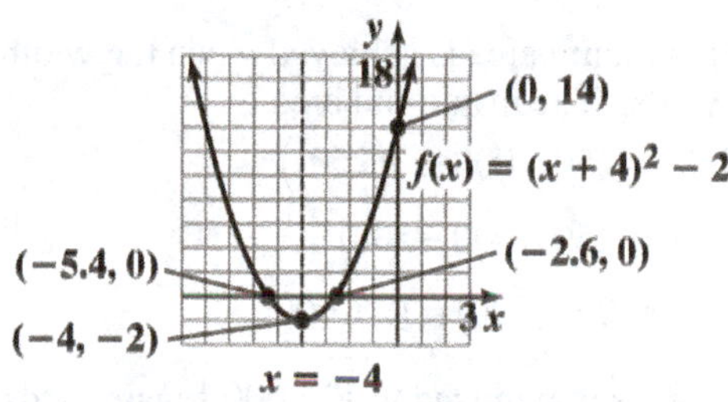

Axis of symmetry: $x=-4$.

35. $f(x)=-x^2+2x+3$

Since $a=-1$ is negative, the parabola opens downward. The x–coordinate of the vertex of the parabola is $-\frac{b}{2a}=-\frac{2}{2(-1)}=-\frac{2}{-2}=1$ and the y–coordinate of the vertex of the parabola is

$$f\left(-\frac{b}{2a}\right)=f(1)$$
$$=-1^2+2(1)+3$$
$$=-1+2+3=4.$$

The vertex is $(1, 4)$. Replace $f(x)$ with 0 to find x–intercepts.

$0=-x^2+2x+3$

$0=x^2-2x-3$

$0=(x-3)(x+1)$

Apply the zero product principle.

$x-3=0$ or $x+1=0$

$x=3$ $\quad$ $x=-1$

The x–intercepts are -1 and 3. Set $x=0$ and solve for y to obtain the y–intercept.

$y=-0^2+2(0)+3$

$y=0+0+3$

$y=3$

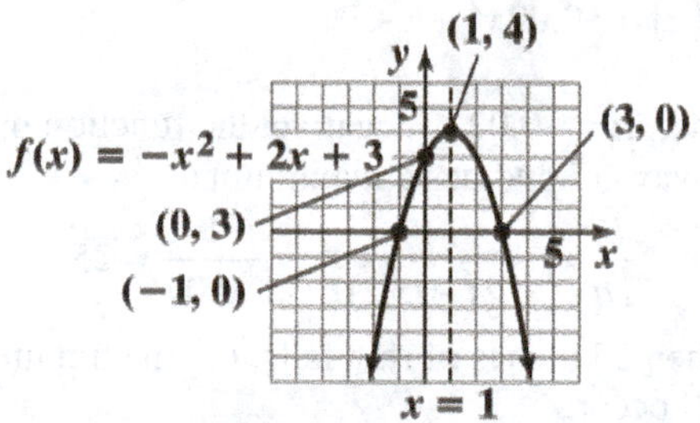

Axis of symmetry: $x=1$.

36. $f(x)=2x^2-4x-6$

Since $a=2$ is positive, the parabola opens upward. The x–coordinate of the vertex of the parabola is $-\frac{b}{2a}=-\frac{-4}{2(2)}=-\frac{-4}{4}=1$ and the y–coordinate of the vertex of the parabola is

$$f\left(-\frac{b}{2a}\right)=f(1)$$
$$=2(1)^2-4(1)-6$$
$$=2(1)-4-6$$
$$=2-4-6=-8.$$

The vertex is $(1,-8)$. Replace $f(x)$ with 0 to find x–intercepts.

$0=2x^2-4x-6$

$0=x^2-2x-3$

$0=(x-3)(x+1)$

Apply the zero product principle.

$x-3=0$ or $x+1=0$

$x=3$ $\quad$ $x=-1$

The x–intercepts are -1 and 3. Set $x=0$ and solve for y to obtain the y–intercept.

$y=2(0)^2-4(0)-6$

$y=2(0)-0-6$

$y=0-0-6=-6$

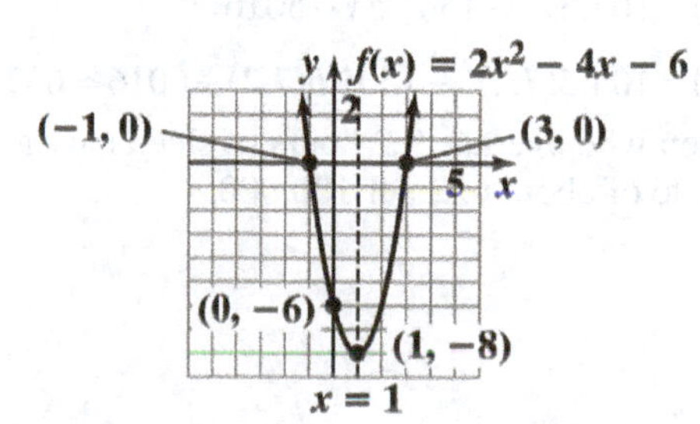

Axis of symmetry: $x=1$.

37. $f(x) = -0.02x^2 + x + 1$

Since $a = -0.02$ is negative, the function opens downward and has a maximum at

$$x = -\frac{b}{2a} = -\frac{1}{2(-0.02)} = -\frac{1}{-0.04} = 25.$$

When 25 inches of rain falls, the maximum growth will occur.
The maximum growth is

$$\begin{aligned} f(25) &= -0.02(25)^2 + 25 + 1 \\ &= -0.02(625) + 25 + 1 \\ &= -12.5 + 25 + 1 = 13.5. \end{aligned}$$

A maximum yearly growth of 13.5 inches occurs when 25 inches of rain falls per year.

38. $s(t) = -16t^2 + 400t + 40$

Since $a = -16$ is negative, the function opens downward and has a maximum at

$$x = -\frac{b}{2a} = -\frac{400}{2(-16)} = -\frac{400}{-32} = 12.5.$$

At 12.5 seconds, the rocket reaches its maximum height.
The maximum height is

$$\begin{aligned} s(12.5) &= -16(12.5)^2 + 400(12.5) + 40 \\ &= -16(156.25) + 5000 + 40 \\ &= -2500 + 5000 + 40 = 2540. \end{aligned}$$

The rocket reaches a maximum height of 2540 feet in 12.5 seconds.

39. $f(x) = 104.5x^2 - 1501.5x + 6016$

Since a is positive, the function opens upward and has a minimum at $x = -\frac{b}{2a} = -\frac{-1501.5}{2(104.5)} \approx 7.2.$

At 7.2 hours, the death rate reaches its minimum.
The minimum death rate is

$$f(x) = 104.5x^2 - 1501.5x + 6016$$

$$f(7.2) = 104.5(7.2)^2 - 1501.5(7.2) + 6016 \approx 622$$

U.S. men who average 7.2 hours of sleep have a death rate of about 622 per 100,000.

40. Maximize the area using $A = lw$.

$$A(x) = x(1000 - 2x)$$

$$A(x) = -2x^2 + 1000x$$

Since $a = -2$ is negative, the function opens downward and has a maximum at

$$x = -\frac{b}{2a} = -\frac{1000}{2(-2)} = -\frac{1000}{-4} = 250.$$

The maximum area is achieved when the width is 250 yards. The maximum area is

$$\begin{aligned} A(250) &= 250(1000 - 2(250)) \\ &= 250(1000 - 500) \\ &= 250(500) = 125,000. \end{aligned}$$

The area is maximized at 125,000 square yards when the width is 250 yards and the length is $1000 - 2 \cdot 250 = 500$ yards.

41. Let x = one of the numbers.
Let $14 + x$ = the other number.

We need to minimize the function

$$\begin{aligned} P(x) &= x(14 + x) \\ &= 14x + x^2 \\ &= x^2 + 14x. \end{aligned}$$

The minimum is at $x = -\frac{b}{2a} = -\frac{14}{2(1)} = -\frac{14}{2} = -7.$

The other number is $14 + x = 14 + (-7) = 7.$

The numbers which minimize the product are 7 and -7. The minimum product is $-7 \cdot 7 = -49.$

42. Let $u = x^2$.

$$x^4 - 6x^2 + 8 = 0$$

$$\left(x^2\right)^2 - 6x^2 + 8 = 0$$

$$u^2 - 6u + 8 = 0$$

$$(u - 4)(u - 2) = 0$$

$$u - 4 = 0 \quad \text{or} \quad u - 2 = 0$$

$$u = 4 \qquad\qquad u = 2$$

Replace u by x^2.

$$x^2 = 4 \quad \text{or} \quad x^2 = 2$$

$$x = \pm 2 \qquad\qquad x = \pm\sqrt{2}$$

The solution set is $\left\{-2, -\sqrt{2}, \sqrt{2}, 2\right\}$.

43. Let $u = \sqrt{x}$.

$$x + 7\sqrt{x} - 8 = 0$$
$$\left(\sqrt{x}\right)^2 + 7\sqrt{x} - 8 = 0$$
$$u^2 + 7u - 8 = 0$$
$$(u+8)(u-1) = 0$$

Apply the zero product principle.

$u + 8 = 0$ or $u - 1 = 0$

$u = -8$ $\quad u = 1$

Replace u by $\sqrt{x}$.

~~$\sqrt{x} = -8$~~ or $\sqrt{x} = 1$

$x = 1$

Disregard –8 because the square root of x cannot be a negative number.

We must check 1, because both sides of the equation were raised to an even power.

Check:

$$1 + 7\sqrt{1} - 8 = 0$$
$$1 + 7(1) - 8 = 0$$
$$1 + 7 - 8 = 0$$
$$8 - 8 = 0$$
$$0 = 0$$

The solution set is $\{1\}$.

44. Let $u = x^2 + 2x$.

$$\left(x^2 + 2x\right)^2 - 14\left(x^2 + 2x\right) = 15$$
$$\left(x^2 + 2x\right)^2 - 14\left(x^2 + 2x\right) - 15 = 0$$
$$u^2 - 14u - 15 = 0$$
$$(u - 15)(u + 1) = 0$$

$u - 15 = 0$ or $u + 1 = 0$

$u = 15$ $\quad u = -1$

Replace u by $x^2 + 2x$.

First, consider $u = 15$.

$$x^2 + 2x = 15$$
$$x^2 + 2x - 15 = 0$$
$$(x + 5)(x - 3) = 0$$

$x + 5 = 0$ or $x - 3 = 0$

$x = -5$ $\quad x = 3$

Next, consider $u = -1$.

$$x^2 + 2x = -1$$
$$x^2 + 2x + 1 = 0$$
$$(x + 1)^2 = 0$$
$$x + 1 = 0$$
$$x = -1$$

The solution set is $\{-5, -1, 3\}$.

45. Let $u = x^{-1}$.

$$x^{-2} + x^{-1} - 56 = 0$$
$$\left(x^{-1}\right)^2 + x^{-1} - 56 = 0$$
$$u^2 + u - 56 = 0$$
$$(u + 8)(u - 7) = 0$$

$u + 8 = 0$ or $u - 7 = 0$

$u = -8$ $\quad u = 7$

Replace u by x^{-1}.

$x^{-1} = -8$ or $x^{-1} = 7$

$\frac{1}{x} = -8$ $\quad \frac{1}{x} = 7$

$-8x = 1$ $\quad 7x = 1$

$x = -\frac{1}{8}$ $\quad x = \frac{1}{7}$

The solution set is $\left\{-\frac{1}{8}, \frac{1}{7}\right\}$.

46. Let $u = x^{\frac{1}{3}}$.

$$x^{\frac{2}{3}} - x^{\frac{1}{3}} - 12 = 0$$
$$\left(x^{\frac{1}{3}}\right)^2 - x^{\frac{1}{3}} - 12 = 0$$
$$u^2 - u - 12 = 0$$
$$(u-4)(u+3) = 0$$
$$u - 4 = 0 \quad \text{or} \quad u + 3 = 0$$
$$u = 4 \qquad\qquad u = -3$$

Replace u by $x^{\frac{1}{3}}$.

$$x^{\frac{1}{3}} = 4 \quad \text{or} \quad x^{\frac{1}{3}} = -3$$
$$\left(x^{\frac{1}{3}}\right)^3 = 4^3 \qquad \left(x^{\frac{1}{3}}\right)^3 = (-3)^3$$
$$x = 64 \qquad\qquad x = -27$$

The solution set is $\{-27, 64\}$.

47. Let $u = x^{\frac{1}{4}}$.

$$x^{\frac{1}{2}} + 3x^{\frac{1}{4}} - 10 = 0$$
$$\left(x^{\frac{1}{4}}\right)^2 + 3x^{\frac{1}{4}} - 10 = 0$$
$$u^2 + 3u - 10 = 0$$
$$(u+5)(u-2) = 0$$
$$u + 5 = 0 \quad \text{or} \quad u - 2 = 0$$
$$u = -5 \qquad\qquad u = 2$$

Replace u by $x^{\frac{1}{4}}$.

$$x^{\frac{1}{4}} = -5 \quad \text{or} \quad x^{\frac{1}{4}} = 2$$
$$\sqrt[4]{x} = -5 \text{ (crossed out)} \qquad \left(x^{\frac{1}{4}}\right)^4 = 2^4$$
$$x = 16$$

Disregard –5 because the fourth root of x cannot be a negative number.
We must check 16, because both sides of the equation were raised to an even power.

Check $x = 16$.

$$16^{\frac{1}{2}} + 3(16)^{\frac{1}{4}} - 10 = 0$$
$$4 + 3(2) - 10 = 0$$
$$4 + 6 - 10 = 0$$
$$10 - 10 = 0$$
$$0 = 0$$

The solution checks. The solution set is $\{16\}$.

48. $2x^2+5x-3<0$

Solve the related quadratic equation.

$$2x^2+5x-3=0$$
$$(2x-1)(x+3)=0$$
$$2x-1=0 \quad \text{or} \quad x+3=0$$
$$2x=1 \qquad x=-3$$
$$x=\frac{1}{2}$$

The boundary points are -3 and $\frac{1}{2}$.

Interval	Test Value	Test	Conclusion
$(-\infty,-3)$	-4	$2(-4)^2+5(-4)-3<0$ false	$(-\infty,-3)$ does not belong to the solution set.
$\left(-3,\frac{1}{2}\right)$	0	$2(0)^2+5(0)-3<0$ true	$\left(-3,\frac{1}{2}\right)$ belongs to the solution set.
$\left(\frac{1}{2},\infty\right)$	1	$2(1)^2+5(1)-3<0$ false	$\left(\frac{1}{2},\infty\right)$ does not belong to the solution set.

The solution set is $\left(-3,\frac{1}{2}\right)$.

−5 −4 −3 −2 −1 0 1 2 3 4 5

49. $2x^2+9x+4\ge 0$

Solve the related quadratic equation.

$$2x^2+9x+4=0$$
$$(2x+1)(x+4)=0$$
$$2x+1=0 \quad \text{or} \quad x+4=0$$
$$2x=-1 \qquad x=-4$$
$$x=-\tfrac{1}{2}$$

The boundary points are -4 and $-\frac{1}{2}$.

Interval	Test Value	Test	Conclusion
$(-\infty,-4]$	-5	$2(-5)^2+9(-5)+4\ge 0$ true	$(-\infty,-4]$ belongs to the solution set.
$\left[-4,-\frac{1}{2}\right]$	-1	$2(-1)^2+9(-1)+4\ge 0$ false	$\left[-4,-\frac{1}{2}\right]$ does not belong to the solution set.
$\left[-\frac{1}{2},\infty\right)$	0	$2(0)^2+9(0)+4\ge 0$ true	$\left[-\frac{1}{2},\infty\right)$ belongs to the solution set.

The solution set is $(-\infty,-4]\cup\left[-\frac{1}{2},\infty\right)$.

−5 −4 −3 −2 −1 0 1 2 3 4 5

50. $x^3 + 2x^2 > 3x$

Solve the related polynomial equation.

$$x^3 + 2x^2 = 3x$$
$$x^3 + 2x^2 - 3x = 0$$
$$x(x^2 + 2x - 3) = 0$$
$$x(x+3)(x-1) = 0$$
$$x = 0 \quad \text{or} \quad x+3=0 \quad \text{or} \quad x-1=0$$
$$x=-3 \qquad\qquad x=1$$

The boundary points are -3, 0, and 1.

Interval	Test Value	Test	Conclusion
$(-\infty,-3)$	-4	$(-4)^3 + 2(-4)^2 > 3(-4)$ false	$(-\infty,-3)$ does not belong to the solution set.
$(-3,0)$	-2	$(-2)^3 + 2(-2)^2 > 3(-2)$ true	$(-3,0)$ belongs to the solution set.
$(0,1)$	0.5	$0.5^3 + 2(0.5)^2 > 3(0.5)$ false	$(0,1)$ does not belong to the solution set.
$(1,\infty)$	2	$2^3 + 2(2)^2 > 3(2)$ true	$(1,\infty)$ belongs to the solution set.

The solution set is $(-3,0)\cup(1,\infty)$.

−4 −3 −2 −1 0 1 2 3 4 5 6

51. $\dfrac{x-6}{x+2} > 0$

Find the values of x that make the numerator and denominator zero.

$$x-6=0 \qquad x+2=0$$
$$x=6 \qquad x=-2$$

The boundary points are -2 and 6.

Interval	Test Value	Test	Conclusion
$(-\infty,-2)$	-3	$\dfrac{-3-6}{-3+2} > 0$ true	$(-\infty,-2)$ belongs to the solution set.
$(-2,6)$	0	$\dfrac{0-6}{0+2} > 0$ false	$(-2,6)$ does not belong to the solution set.
$(6,\infty)$	7	$\dfrac{7-6}{7+2} > 0$ true	$(6,\infty)$ belongs to the solution set.

The solution set is $(-\infty,-2)\cup(6,\infty)$.

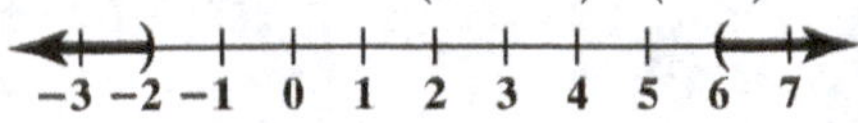

52. $\dfrac{x+3}{x-4} \le 5$

Express the inequality so that one side is zero.

$$\frac{x+3}{x-4} - 5 \le 0$$

$$\frac{x+3}{x-4} - \frac{5(x-4)}{x-4} \le 0$$

$$\frac{x+3-5(x-4)}{x-4} \le 0$$

$$\frac{x+3-5x+20}{x-4} \le 0$$

$$\frac{-4x+23}{x-4} \le 0$$

Find the values of x that make the numerator and denominator zero.

$-4x+23=0$ and $x-4=0$

$-4x=-23$ $\qquad x=4$

$x=\dfrac{23}{4}$

The boundary points are 4 and $\dfrac{23}{4}$. Exclude 4 from the solution set, since this would make the denominator zero.

Interval	Test Value	Test	Conclusion
$(-\infty, 4)$	0	$\frac{0+3}{0-4} \le 5$ $\frac{3}{-4} \le 5$, true	$(-\infty, 4)$ belongs to the solution set.
$\left(4, \frac{23}{4}\right]$	5	$\frac{5+3}{5-4} \le 5$ $8 \le 5$, false	$\left(4, \frac{23}{4}\right]$ does not belong to the solution set.
$\left[\frac{23}{4}, \infty\right)$	6	$\frac{6+3}{6-4} \le 5$ $\frac{9}{2} \le 5$, true	$\left[\frac{23}{4}, \infty\right)$ belongs to the solution set.

The solution set is $(-\infty, 4) \cup \left[\dfrac{23}{4}, \infty\right)$.

−3 −2 −1 0 1 2 3 4 5 6 7

53. $s(t) = -16t^2 + 48t$

To find when the height is more than 32 feet above the ground, solve the inequality $-16t^2 + 48t > 32$.

Solve the related quadratic equation.

$$-16t^2 + 48t = 32$$

$$-16t^2 + 48t - 32 = 0$$

$$t^2 - 3t + 2 = 0$$
$$(t-2)(t-1) = 0$$
$$t - 2 = 0 \quad \text{or} \quad t - 1 = 0$$
$$t = 2 \qquad\qquad t = 1$$

The boundary points are 1 and 2.

Interval	Test Value	Test	Conclusion
$(0,1)$	0.5	$-16(0.5)^2 + 48(0.5) > 32$ $20 > 32$, false	$(0,1)$ does not belong to the solution set.
$(1,2)$	1.5	$-16(1.5)^2 + 48(1.5) > 32$ $36 > 32$, true	$(1,2)$ belongs to the solution set.
$(2,\infty)$	3	$-16(3)^2 + 48(3) > 32$ $0 > 32$, false	$(2,\infty)$ does not belong to the solution set.

The solution set is $(1,2)$. This means that the ball will be more than 32 feet above the graph between 1 and 2 seconds.

54. a. $H(0) = \frac{15}{8}(0)^2 - 30(0) + 200 = \frac{15}{8}(0) - 0 + 200 = 0 - 0 + 200 = 200$

The heart rate is 200 beats per minute immediately following the workout.

b.

$$\frac{15}{8}x^2 - 30x + 200 > 110$$
$$\frac{15}{8}x^2 - 30x + 90 > 0$$
$$\frac{8}{15}\left(\frac{15}{8}x^2 - 30x + 90\right) > \frac{8}{15}(0)$$
$$x^2 - \frac{8}{15}(30x) + \frac{8}{15}(90) > 0$$

$$x^2 - 16x + 48 > 0$$
$$(x-12)(x-4) > 0$$

Apply the zero product principle.

$$x - 12 = 0 \quad \text{or} \quad x - 4 = 0$$
$$x = 12 \qquad\qquad x = 4$$

The boundary points are 4 and 12.

Interval	Test Value	Test	Conclusion
$(0,4)$	1	$\frac{15}{8}(1)^2-30(1)+200>110$ $171\frac{7}{8}>110$, true	$(0,4)$ belongs to the solution set.
$(4,12)$	5	$\frac{15}{8}(5)^2-30(5)+200>110$ $96\frac{7}{8}>110$, false	$(4,12)$ does not belong to the solution set.
$(12,\infty)$	13	$\frac{15}{8}(13)^2-30(13)+200>110$ $126\frac{7}{8}>110$, true	$(12,\infty)$ does not belong to the solution set.

The solution set is $(0,4)\cup(12,\infty)$. This means that the heart rate exceeds 110 beats per minute between 0 and 4 minutes after the workout and more than 12 minutes after the workout. Between 0 and 4 minutes provides a more realistic answer since it is unlikely that the heart rate will begin to climb again without further exertion. Model breakdown occurs for the interval $(12,\infty)$.

Chapter 8 Test

1.
$$2x^2-5=0$$
$$2x^2=5$$
$$x^2=\frac{5}{2}$$
$$x=\pm\sqrt{\frac{5}{2}}$$

Rationalize the denominators.

$$x=\pm\frac{\sqrt{5}}{\sqrt{2}}\cdot\frac{\sqrt{2}}{\sqrt{2}}=\pm\frac{\sqrt{10}}{2}$$

The solution set is $\left\{\pm\frac{\sqrt{10}}{2}\right\}$.

2.
$$(x-3)^2=20$$
$$x-3=\pm\sqrt{20}$$
$$x=3\pm\sqrt{4\cdot 5}$$
$$x=3\pm2\sqrt{5}$$

The solution set is $\left\{3\pm2\sqrt{5}\right\}$.

3. x^2-16x+____

Since $b=-16$, add $\left(\frac{b}{2}\right)^2=\left(\frac{-16}{2}\right)^2=(-8)^2=64$.

$x^2-16x+64=(x-8)^2$

4. $x^2+\frac{2}{5}x+$____

Since $b=\frac{2}{5}$, add $\left(\frac{1}{2}b\right)^2=\left(\frac{1}{2}\cdot\frac{2}{5}\right)^2=\left(\frac{1}{5}\right)^2=\frac{1}{25}$.

$$x^2+\frac{2}{5}x+\frac{1}{25}=\left(x+\frac{1}{5}\right)^2$$

5. $x^2-6x+7=0$

$x^2-6x\quad=-7$

Since $b=-6$, add $\left(\frac{b}{2}\right)^2=\left(\frac{-6}{2}\right)^2=(-3)^2=9$.

$$x^2-6x+9=-7+9$$
$$(x-3)^2=2$$

Apply the square root property.

$$x-3=\pm\sqrt{2}$$
$$x=3\pm\sqrt{2}$$

The solution set is $\{3\pm\sqrt{2}\}$.

6. Use the Pythagorean Theorem.

$$50^2+50^2=x^2$$
$$2500+2500=x^2$$
$$5000=x^2$$
$$\pm\sqrt{5000}=x$$
$$\pm\sqrt{2500\cdot 2}=x$$
$$\pm 50\sqrt{2}=x$$

The solutions are $\pm 50\sqrt{2}$ feet. Disregard $-50\sqrt{2}$ feet because we can't have a negative length measurement. The width of the pond is $50\sqrt{2}$ feet.

7. $3x^2+4x-2=0$

$a=3\quad b=4\quad c=-2$

Find the discriminant.

$$b^2-4ac=4^2-4(3)(-2)$$
$$=16+24=40$$

Since the discriminant is greater than zero but not a perfect square, there are two real irrational solutions.

8. $x^2=4x-8$

$x^2-4x+8=0$

$a=1\quad b=-4\quad c=8$

Find the discriminant.

$$b^2-4ac=(-4)^2-4(1)(8)$$
$$=16-32=-16$$

Since the discriminant is negative, there are two imaginary solutions which are complex conjugates.

9. $2x^2+9x=5$

$$2x^2+9x-5=0$$
$$(2x-1)(x+5)=0$$

Apply the zero-product principle.

$2x-1=0$ and $x+5=0$

$2x=1 \qquad x=-5$

$x=\frac{1}{2}$

The solution set is $\left\{-5,\frac{1}{2}\right\}$.

10. $x^2+8x+5=0$

Solve using the quadratic formula.

$a=1 \quad b=8 \quad c=5$

$$x=\frac{-8\pm\sqrt{8^2-4(1)(5)}}{2(1)}$$
$$=\frac{-8\pm\sqrt{64-20}}{2}$$
$$=\frac{-8\pm\sqrt{44}}{2}$$
$$=\frac{-8\pm\sqrt{4\cdot 11}}{2}$$
$$=\frac{-8\pm 2\sqrt{11}}{2}$$
$$=\frac{2\left(-4\pm\sqrt{11}\right)}{2}=-4\pm\sqrt{11}$$

The solution set is $\left\{-4\pm\sqrt{11}\right\}$.

11. $(x+2)^2+25=0$

$(x+2)^2=-25$

Apply the square root principle.

$x+2=\pm\sqrt{-25}$

$x=-2\pm 5i$

The solution set is $\{-2\pm 5i\}$.

12. $2x^2-6x+5=0$

$a=2 \quad b=-6 \quad c=5$

$$x=\frac{-(-6)\pm\sqrt{(-6)^2-4(2)(5)}}{2(2)}$$
$$=\frac{6\pm\sqrt{36-40}}{4}$$
$$=\frac{6\pm\sqrt{-4}}{4}$$
$$=\frac{6\pm 2i}{4}=\frac{6}{4}\pm\frac{2}{4}i=\frac{3}{2}\pm\frac{1}{2}i$$

The solution set is $\left\{\frac{3}{2}\pm\frac{1}{2}i\right\}$.

13. Because the solution set is $\{-3,7\}$, we have

$x=-3$ or $x=7$

$x+3=0 \qquad x-7=0$

Apply the zero-product principle in reverse.

$(x+3)(x-7)=0$

$x^2-7x+3x-21=0$

$x^2-4x-21=0$

14. Because the solution set is $\{-10i,10i\}$, we have

$x=-10i$ or $x=10i$

$x+10i=0 \qquad x-10i=0$

Apply the zero-product principle in reverse.

$(x+10i)(x-10i)=0$

$x^2-100i^2=0$

$x^2-100(-1)=0$

$x^2+100=0$

15. a. 2011 is 8 years after 2003.

$f(x)=1.7x^2+6x+26$

$f(8)=1.7(8)^2+6(8)+26$

$=182.8$

≈ 183

According to the function, in 2011 there were 183 "Bicycle Friendly" communities. This overestimates the number shown in the graph by 3.

b. $f(x)=1.7x^2+6x+26$

$826=1.7x^2+6x+26$

$0=1.7x^2+6x-800$

$$x=\frac{-b\pm\sqrt{b^2-4ac}}{2a}$$
$$x=\frac{-(6)\pm\sqrt{(6)^2-4(1.7)(-800)}}{2(1.7)}$$
$$=\frac{-6\pm\sqrt{36+5440}}{2(1.7)}$$
$$=\frac{-6\pm\sqrt{5476}}{3.4}$$
$$=\frac{-6\pm 74}{3.4}$$

$x=20$ or $-23\frac{9}{17}$

According to the function, there will be 826 "Bicycle Friendly" communities 20 years after 2003, or 2023.

16. $f(x)=(x+1)^2+4$

Since $a=1$ is positive, the parabola opens upward. The vertex of the parabola is $(h,k)=(-1,4)$ and the axis of symmetry is $x=-1$. Replace $f(x)$ with 0 to find x–intercepts.

$$0=(x+1)^2+4$$
$$-4=(x+1)^2$$

This will be result in complex solutions. As a result, there are no x–intercepts. Set $x=0$ and solve for y to obtain the y–intercept.

$$y=(0+1)^2+4=1+4=5$$

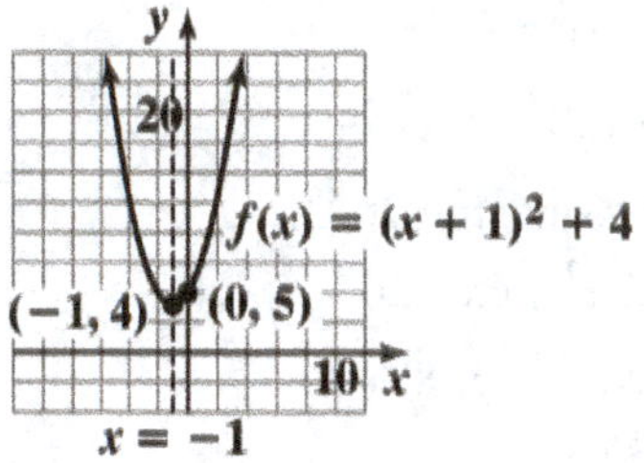

Axis of symmetry: $x=-1$.

17. $f(x)=x^2-2x-3$

Since $a=1$ is positive, the parabola opens upward. The x–coordinate of the vertex of the parabola is $-\frac{b}{2a}=-\frac{-2}{2(1)}=-\frac{-2}{2}=1$ and the y–coordinate of the vertex of the parabola is

$$f\left(-\frac{b}{2a}\right)=f(1)$$
$$=1^2-2(1)-3$$
$$=1-2-3$$
$$=-4.$$

The vertex is $(1,-4)$. Replace $f(x)$ with 0 to find x–intercepts.

$$0=x^2-2x-3$$
$$0=(x-3)(x+1)$$

Apply the zero-product principle.

$x-3=0$ or $x+1=0$

$x=3$ $\quad$ $x=-1$

The x–intercepts are –1 and 3. Set $x=0$ and solve for y to obtain the y–intercept.

$$y=0^2-2(0)-3=-3$$

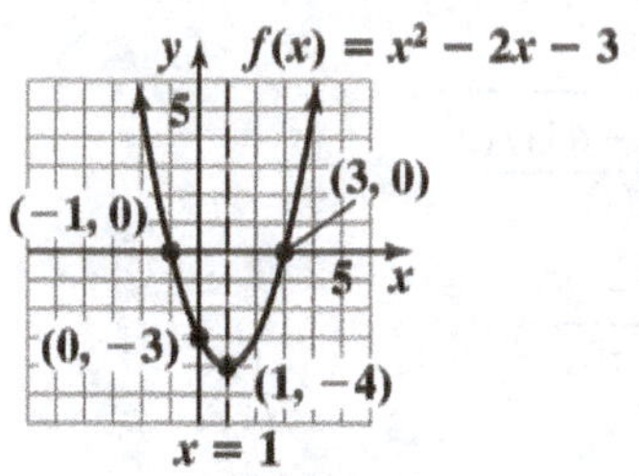

Axis of symmetry: $x=1$.

18. $s(t)=-16t^2+64t+5$

Since $a=-16$ is negative, the function opens downward and has a maximum at $x=-\frac{b}{2a}=-\frac{64}{2(-16)}=-\frac{64}{-32}=2.$

The ball reaches its maximum height in two seconds. The maximum height is

$$s(2)=-16(2)^2+64(2)+5$$
$$=-16(4)+128+5$$
$$=-64+128+5=69.$$

The baseball reaches a maximum height of 69 feet after 2 seconds.

19. $0=-16t^2+64t+5$

Solve using the quadratic formula.

$a=-16 \quad b=64 \quad c=5$

$$x=\frac{-64\pm\sqrt{64^2-4(-16)(5)}}{2(-16)}$$
$$=\frac{-64\pm\sqrt{4096+320}}{-32}$$
$$=\frac{-64\pm\sqrt{4416}}{-32}$$
$$\approx 4.1 \text{ or } -0.1$$

Disregard –0.1 since we cannot have a negative time measurement. The solution is 4.1 and we conclude that the baseball hits the ground in approximately 4.1 seconds.

20. $f(x) = -x^2 + 46x - 360$

Since $a = -1$ is negative, the function opens downward and has a maximum at $x = -\frac{b}{2a} = -\frac{46}{2(-1)} = -\frac{46}{-2} = 23.$

$f(23) = -23^2 + 46(23) - 360 = 169$

Profit is maximized when 23 computers are manufactured. This produces a profit of \$169 hundreds or \$16,900.

21. Let $u = 2x - 5.$

$$(2x-5)^2 + 4(2x-5) + 3 = 0$$
$$u^2 + 4u + 3 = 0$$
$$(u+3)(u+1) = 0$$
$$u+3 = 0 \quad \text{or} \quad u+1 = 0$$
$$u = -3 \qquad\qquad u = -1$$

Replace u by $2x-5.$

$$2x-5 = -3 \quad \text{or} \quad 2x-5 = -1$$
$$2x = 2 \qquad\qquad 2x = 4$$
$$x = 1 \qquad\qquad x = 2$$

The solution set is $\{1, 2\}$.

22. Let $u = x^2.$

$$x^4 - 13x^2 + 36 = 0$$
$$\left(x^2\right)^2 - 13x^2 + 36 = 0$$
$$u^2 - 13u + 36 = 0$$
$$(u-9)(u-4) = 0$$
$$u-9 = 0 \quad \text{or} \quad u-4 = 0$$
$$u = 9 \qquad\qquad u = 4$$

Replace u by $x^2.$

$$x^2 = 9 \quad \text{or} \quad x^2 = 4$$
$$x = \pm 3 \qquad\qquad x = \pm 2$$

The solution set is $\{-3, -2, 2, 3\}$.

23. Let $u = x^{1/3}$.

$$x^{2/3} - 9x^{1/3} + 8 = 0$$
$$\left(x^{1/3}\right)^2 - 9x^{1/3} + 8 = 0$$
$$u^2 - 9u + 8 = 0$$
$$(u-8)(u-1) = 0$$
$$u - 8 = 0 \quad \text{or} \quad u - 1 = 0$$
$$u = 8 \qquad\qquad u = 1$$

Replace u by $x^{1/3}$.

$$x^{1/3} = 8 \qquad \text{or} \qquad x^{1/3} = 1$$
$$x = 8^3 = 512 \qquad\qquad x = 1^3 = 1$$

The solution set is $\{1,\ 512\}$.

24. $x^2 - x - 12 < 0$

Solve the related quadratic equation.

$$x^2 - x - 12 = 0$$
$$(x-4)(x+3) = 0$$
$$x - 4 = 0 \quad \text{or} \quad x + 3 = 0$$
$$x = 4 \qquad\qquad x = -3$$

The boundary points are -3 and 4.

Interval	Test Value	Test	Conclusion
$(-\infty, -3)$	-4	$(-4)^2 - (-4) - 12 < 0$ $8 < 0$, false	$(-\infty, -3)$ does not belong to the solution set.
$(-3, 4)$	0	$0^2 - 0 - 12 < 0$ $-12 < 0$, true	$(-3, 4)$ belongs to the solution set.
$(4, \infty)$	5	$5^2 - 5 - 12 < 0$ $8 < 0$, false	$(4, \infty)$ does not belong to the solution set.

The solution set is $(-3, 4)$.

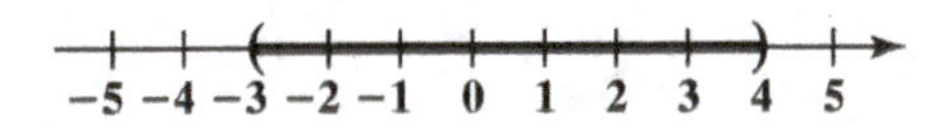

25. $\dfrac{2x+1}{x-3} \le 3$

Express the inequality so that one side is zero.

$$\frac{2x+1}{x-3} - 3 \le 0$$

$$\frac{2x+1}{x-3} - \frac{3(x-3)}{x-3} \le 0$$

$$\frac{2x+1-3(x-3)}{x-3} \le 0$$

$$\frac{2x+1-3x+9}{x-3} \le 0$$

$$\frac{-x+10}{x-3} \le 0$$

Find the values of x that make the numerator and denominator zero.

$-x+10=0$ and $x-3=0$

$-x=-10$ $\qquad x=3$

$x=10$

The boundary points are 3 and 10. Exclude 3 from the solution set, since this would make the denominator zero.

Interval	Test Value	Test	Conclusion
$(-\infty, 3)$	0	$\frac{2(0)+1}{0-3} \le 3$ $-\frac{1}{3} \le 3$, true	$(-\infty, 3)$ belongs to the solution set.
$(3, 10]$	4	$\frac{2(4)+1}{4-3} \le 3$ $9 \le 3$, false	$(3, 10]$ does not belong to the solution set.
$[10, \infty)$	11	$\frac{2(11)+1}{11-3} \le 3$ $\frac{23}{8} \le 3$, true	$[10, \infty)$ belongs to the solution set.

The solution set is $(-\infty, 3) \cup [10, \infty)$.

−4 −2 0 2 4 6 8 10 12 14 16

Cumulative Review Exercises (Chapters 1 – 8)

1. $9(x-1)=1+3(x-2)$

$$9x-9=1+3x-6$$
$$9x-9=3x-5$$
$$6x-9=-5$$
$$6x=4$$
$$x=\frac{4}{6}=\frac{2}{3}$$

The solution set is $\left\{\frac{2}{3}\right\}$.

2. $3x+4y=-7$
$x-2y=-9$

Multiply the second equation by 2 and add the result to the first equation.

$$3x+4y=-7$$
$$2x-4y=-18$$
$$5x\quad=-25$$
$$x=-5$$

Back substitute into the second equation.

$$-5-2y=-9$$
$$-2y=-4$$
$$y=2$$

The solution set is $\{(-5,2)\}$.

3. $x-y+3z=-9$
$2x+3y-z=16$
$5x+2y-z=15$

Multiply the second equation by 3 and add to the first equation to eliminate z.

$$x-y+3z=-9$$
$$6x+9y-3z=48$$
$$7x+8y\quad=39$$

Multiply the second equation by -1 and add to the third equation.

$$-2x-3y+z=-16$$
$$5x+2y-z=15$$
$$3x-y\quad=-1$$

We now have a system of two equations in two variables.

$$7x+8y=39$$
$$3x-y=-1$$

Multiply the second equation by 8 and add to the second equation.

$$7x+8y=39$$
$$24x-8y=-8$$
$$31x\quad=31$$
$$x=1$$

Back-substitute 1 for x to find y.

$$3x-y=-1$$
$$3(1)-y=-1$$
$$3-y=-1$$
$$-y=-4$$
$$y=4$$

Back-substitute 1 for x and 4 for y to find z.

$$x-y+3z=-9$$
$$1-4+3z=-9$$
$$-3+3z=-9$$
$$3z=-6$$
$$z=-2$$

The solution set is $\{(1,4,-2)\}$.

4. $7x+18\le 9x-2$

$$-2x+18\le -2$$
$$-2x\le -20$$
$$\frac{-2x}{-2}\ge\frac{-20}{-2}$$
$$x\ge 10$$

The solution set is $[10,\infty)$.

5. $4x-3<13$ and $-3x-4\ge 8$

$4x<16$ $\qquad$ $-3x\ge 12$

$x<4$ $\qquad$ $x\le -4$

For a value to be in the solution set, it must be both less than 4 and less than or equal to -4. Now only values that are less than or equal to -4 meet both conditions. Therefore, the solution set is $(-\infty,-4]$.

6. $2x+4>8$ or $x-7\geq 3$

$$2x>4 \qquad x\geq 10$$
$$x>2$$

For a value to be in the solution set, it must satisfy either of the conditions. Now, all numbers that are greater than or equal to 10 are also greater than 2. Therefore, the solution set is $(2,\infty)$.

7. $|2x-1|<5$

$$-5<2x-1<5$$
$$-4<2x<6$$
$$-2<x<3$$

The solution set is $(-2,3)$.

8. $\left|\frac{2}{3}x-4\right|=2$

$$\frac{2}{3}x-4=-2 \quad \text{or} \quad \frac{2}{3}x-4=2$$
$$3\left(\frac{2}{3}x-4\right)=3(-2) \qquad 3\left(\frac{2}{3}x-4\right)=3(2)$$
$$2x-12=-6 \qquad 2x-12=6$$
$$2x=6 \qquad 2x=18$$
$$x=3 \qquad x=9$$

The solution set is $\{3,9\}$.

9.

$$\frac{4}{x-3}-\frac{6}{x+3}=\frac{24}{x^2-9}$$
$$\frac{4}{x-3}-\frac{6}{x+3}=\frac{24}{(x-3)(x+3)}$$
$$(x-3)(x+3)\left[\frac{4}{x-3}-\frac{6}{x+3}\right]=(x-3)(x+3)\left[\frac{24}{(x-3)(x+3)}\right]$$
$$4(x+3)-6(x-3)=24$$
$$4x+12-6x+18=24$$
$$-2x+30=24$$
$$-2x=-6$$
$$x=3$$

Disregard 3 since it causes a result of 0 in the denominator of a fraction.
Thus, the equation has no solution. The solution set is $\varnothing$.

10. $\sqrt{x+4}-\sqrt{x-3}=1$

$$\begin{aligned}
\sqrt{x+4} &= \sqrt{x-3}+1 \\
\left(\sqrt{x+4}\right)^2 &= \left(\sqrt{x-3}+1\right)^2 \\
x+4 &= (x-3)+2\sqrt{x-3}+1 \\
x+4 &= x-2+2\sqrt{x-3} \\
6 &= 2\sqrt{x-3} \\
3 &= \sqrt{x-3} \\
(3)^2 &= \left(\sqrt{x-3}\right)^2 \\
9 &= x-3 \\
12 &= x
\end{aligned}$$

Since we square both sides of the equation, we must check to make sure 12 is not extraneous:

$$\begin{aligned}
\sqrt{12+4}-\sqrt{12-3} &= 1 \\
\sqrt{16}-\sqrt{9} &= 1 \\
4-3 &= 1 \\
1 &= 1 \quad \text{True}
\end{aligned}$$

Thus, the solution set is $\{12\}$.

11. $2x^2 = 5-4x$

$$2x^2+4x-5=0$$

Apply the quadratic formula:

$a=2 \quad b=4 \quad c=-5$

$$\begin{aligned}
x &= \frac{-4\pm\sqrt{4^2-4(2)(-5)}}{2(2)} \\
&= \frac{-4\pm\sqrt{56}}{4} \\
&= \frac{-4\pm\sqrt{4\cdot 14}}{4} \\
&= \frac{-4\pm 2\sqrt{14}}{4} = \frac{-2\pm\sqrt{14}}{2}
\end{aligned}$$

The solution set is $\left\{\frac{-2\pm\sqrt{14}}{2}\right\}$.

12. $x^{\frac{2}{3}} - 5x^{\frac{1}{3}} + 6 = 0$

$\left(x^{\frac{1}{3}}\right)^2 - 5x^{\frac{1}{3}} + 6 = 0$

Let $u = x^{\frac{1}{3}}$.

$u^2 - 5u + 6 = 0$

$(u-3)(u-2) = 0$

$u-3=0$ or $u-2=0$

$u = 3$ $\quad u = 2$

Substitute $x^{\frac{1}{3}}$ back in for u.

$x^{\frac{1}{3}} = 3$ or $x^{\frac{1}{3}} = 2$

$\left(x^{\frac{1}{3}}\right)^3 = 3^3$ $\quad \left(x^{\frac{1}{3}}\right)^3 = 2^3$

$x = 27$ $\quad x = 8$

The solution set is $\{8, 27\}$.

13. $2x^2 + x - 6 \le 0$

Solve the related quadratic equation.

$2x^2 + x - 6 = 0$

$(2x-3)(x+2) = 0$

$2x-3=0$ or $x+2=0$

$2x = 3$ $\quad x = -2$

$x = \frac{3}{2}$

The boundary points are -2 and $\frac{3}{2}$.

Interval	Test Value	Test	Conclusion
$(-\infty, -2)$	-3	$2(-3)^2 + (-3) - 6 \le 0$ $9 \le 0$, false	$(-\infty, -2)$ does not belong to the solution set.
$\left(-2, \frac{3}{2}\right)$	0	$2(0)^2 + 0 - 6 \le 0$ $-6 \le 0$, true	$\left(-2, \frac{3}{2}\right)$ belongs to the solution set.
$\left(\frac{3}{2}, \infty\right)$	2	$2(2)^2 + 2 - 6 \le 0$ $4 \le 0$, false	$\left(\frac{3}{2}, \infty\right)$ does not belong to the solution set.

The solution set is $\left[-2, \frac{3}{2}\right]$.

14. $x-3y=6$

$$-3y=-x+6$$
$$y=\frac{-x+6}{-3}$$
$$y=\frac{1}{3}x-2$$

The slope is $m=\frac{1}{3}$ and the y-intercept is $b=-2$.

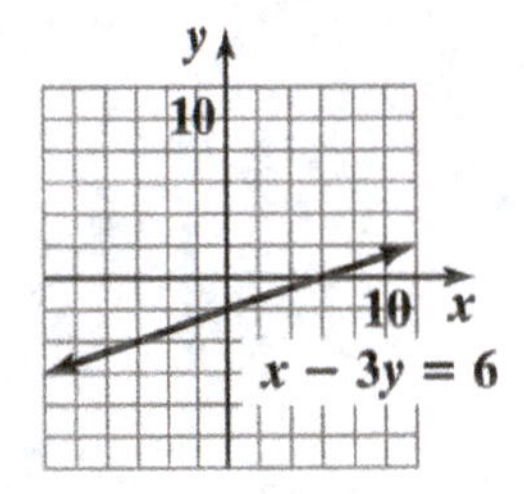

15. $f(x)=\frac{1}{2}x-1$

This is a linear function with slope $m=\frac{1}{2}$ and y-intercept $b=-1$.

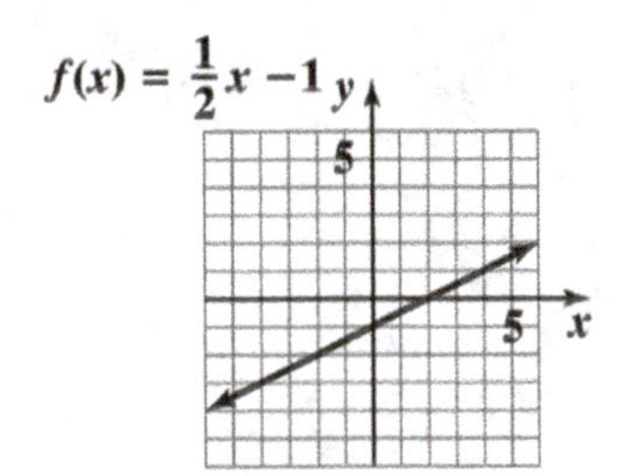

16. $3x-2y>-6$

First, graph the equation $3x-2y=-6$ as a dashed line.

$$3x-2y=-6$$
$$-2y=-3x-6$$
$$y=\frac{-3x-6}{-2}$$
$$y=\frac{3}{2}x+3$$

The slope is $m=\frac{3}{2}$ and the y-intercept is $b=3$.
Next, use the origin as a test point.

$$3(0)-2(0)>-6$$
$$0>-6$$

This is a true statement. This means that the point, $(0,0)$, will fall in the shaded half-plane.

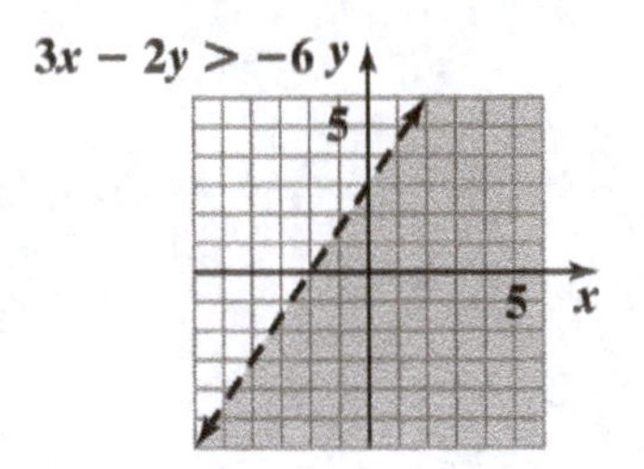

17. $f(x)=-2(x-3)^2+2$

Since $a=-2$ is negative, the parabola opens downward. The vertex of the parabola is $(h,k)=(3,2)$. Replace $f(x)$ with 0 to find x–intercepts.

$$0=-2(x-3)^2+2$$
$$2(x-3)^2=2$$
$$(x-3)^2=1$$
$$x-3=\pm1$$
$$x=3\pm1$$
$$x=3-1 \text{ or } 3+1$$
$$x=2 \text{ or } 4$$

The x–intercepts are 2 and 4.
Set $x=0$ to obtain the y–intercept.

$$f(0)=-2(0-3)^2+2$$
$$=-2(-3)^2+2$$
$$=-2(9)+2=-18+2=-16$$

The y-intercept is -16.

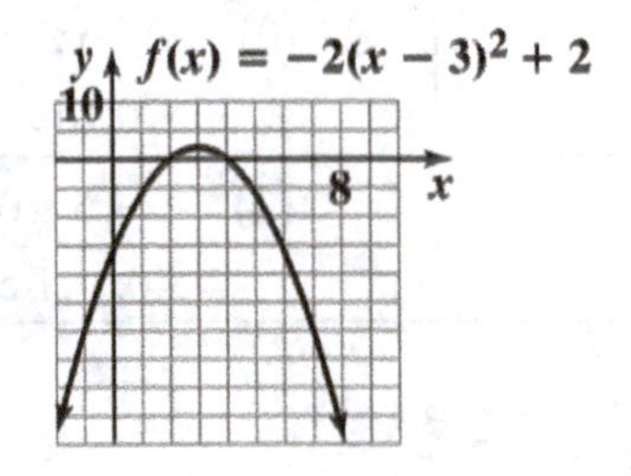

18. $4[2x-6(x-y)]=4(2x-6x+6y)$
$=4(-4x+6y)$
$=-16x+24y$

19. $(-5x^3y^2)(4x^4y^{-6})=-20x^{3+4}y^{2+(-6)}$
$=-20x^7y^{-4}$
$=-\frac{20x^7}{y^4}$

20. $(8x^2-9xy-11y^2)-(7x^2-4xy+5y^2)$
$=8x^2-9xy-11y^2-7x^2+4xy-5y^2$
$=x^2-5xy-16y^2$

21. $(3x-1)(2x+5)=6x^2+15x-2x-5$
$=6x^2+13x-5$

22. $(3x^2-4y)^2$
$=(3x^2)^2-2(3x^2)(4y)+(4y)^2$
$=9x^4-24x^2y+16y^2$

23. $\frac{3x}{x+5}-\frac{2}{x^2+7x+10}$
$=\frac{3x}{x+5}-\frac{2}{(x+5)(x+2)}$
$=\frac{3x}{x+5}\cdot\frac{(x+2)}{(x+2)}-\frac{2}{(x+5)(x+2)}$
$=\frac{3x^2+6x}{(x+5)(x+2)}-\frac{2}{(x+5)(x+2)}$
$=\frac{3x^2+6x-2}{(x+5)(x+2)}$

24. $\frac{1-\frac{9}{x^2}}{1+\frac{3}{x}}=\frac{1-\frac{9}{x^2}}{1+\frac{3}{x}}\cdot\frac{x^2}{x^2}$
$=\frac{x^2-9}{x^2+3x}$
$=\frac{(x-3)(x+3)}{x(x+3)}=\frac{x-3}{x}$

25. $\frac{x^2-6x+8}{3x+9}\div\frac{x^2-4}{x+3}$
$=\frac{x^2-6x+8}{3x+9}\cdot\frac{x+3}{x^2-4}$
$=\frac{(x-4)\cancel{(x-2)}}{3\cancel{(x+3)}}\cdot\frac{\cancel{x+3}}{\cancel{(x-2)}(x+2)}$
$=\frac{x-4}{3(x+2)}$ or $\frac{x-4}{3x+6}$

26. $\sqrt{5xy}\cdot\sqrt{10x^2y}=\sqrt{50x^3y^2}$
$=\sqrt{25\cdot2\cdot x^2\cdot x\cdot y^2}$
$=5xy\sqrt{2x}$

27. $4\sqrt{72}-3\sqrt{50}=4\sqrt{36\cdot2}-3\sqrt{25\cdot2}$
$=4\cdot6\sqrt{2}-3\cdot5\sqrt{2}$
$=24\sqrt{2}-15\sqrt{2}$
$=9\sqrt{2}$

28. $(5+3i)(7-3i)=35-15i+21i-9i^2$
$=35+6i-9(-1)$
$=35+6i+9$
$=44+6i$

29. $81x^4-1=(9x^2+1)(9x^2-1)$
$=(9x^2+1)(3x+1)(3x-1)$

30. $24x^3-22x^2+4x=2x(12x^2-11x+2)$
$=2x(4x-1)(3x-2)$

31. $x^3+27y^3=x^3+(3y)^3$
$=(x+3y)[x^2-x(3y)+(3y)^2]$
$=(x+3y)(x^2-3xy+9y^2)$

32. $(f-g)(x)=f(x)-g(x)$
$=(x^2+3x-15)-(x-2)$
$=x^2+3x-15-x+2$
$=x^2+2x-13$

$(f-g)(5)=5^2+2\cdot5-13$
$=25+10-13=22$

33. $\left(\frac{f}{g}\right)(x) = \frac{f(x)}{g(x)} = \frac{x^2+3x-15}{x-2}$

The domain of $\frac{f}{g}$ is $(-\infty, 2)\cup(2, \infty)$.

34. $f(x) = x^2+3x-15$

$$\begin{aligned}\frac{f(a+h)-f(a)}{h} &= \frac{\left((a+h)^2+3(a+h)-15\right)-\left(a^2+3a-15\right)}{h}\\ &= \frac{a^2+2ah+h^2+3a+3h-15-a^2-3a+15}{h}\\ &= \frac{2ah+h^2+3h}{h}\\ &= 2a+h+3\end{aligned}$$

35.

$$\begin{array}{r|rrrr} -2 & 3 & -1 & 4 & 8\\ & & -6 & 14 & -36\\ \hline & 3 & -7 & 18 & -28\end{array}$$

$$\left(3x^3-x^2+4x+8\right)\div(x+2) = 3x^2-7x+18-\frac{28}{x+2}$$

36.

$$\begin{aligned}I &= \frac{R}{R+r}\\ I(R+r) &= R\\ IR+Ir &= R\\ Ir &= R-IR\\ Ir &= (1-I)R\\ \frac{Ir}{1-I} &= R \text{ or } R = -\frac{Ir}{I-1}\end{aligned}$$

37. $3x+y=9$

$y=-3x+9$

The line whose equation we want to find has a slope of $m=-3$, the same as that of the line above. Using this slope with the point through which the line passes, $(-2, 5)$, we first find the point-slope equation and then put it in slope-intercept form.

$$\begin{aligned}y-y_1 &= m(x-x_1)\\ y-5 &= -3(x-(-2))\\ y-5 &= -3(x+2)\\ y-5 &= -3x-6\\ y &= -3x-1\end{aligned}$$

The slope-intercept equation of the line through $(-2, 5)$ and parallel to $3x+y=9$ is $y=-3x-1$.

38. $\begin{vmatrix} -2 & -4 \\ 5 & 7 \end{vmatrix} = -2(7)-5(-4)$

$= -14+20$

$= 6$

39. Let x = the computer's original price.

$$x-0.30x = 434$$
$$0.70x = 434$$
$$x = \frac{434}{0.70} = 620$$

The original price of the computer was \$620.

40. Let x = the width of the rectangle.
Let $3x + 1$ = the length of the rectangle.

$$x(3x+1) = 52$$
$$3x^2 + x - 52 = 0$$
$$(3x+13)(x-4) = 0$$
$$3x+13 = 0 \quad \text{or} \quad x-4 = 0$$
$$x = -\frac{13}{3} \qquad x = 4$$

Disregard $-\frac{13}{3}$ because the width of a rectangle cannot be negative. If $x = 4$, then $3x+1 = 3(4)+1 = 13$. Thus, the length of the rectangle is 13 yards and the width is 4 yards.

41. Let x = the amount invested at 12%.
Let $4000 - x$ = the amount invested at 14%.

$$0.12x + 0.14(4000-x) = 508$$
$$0.12x + 560 - 0.14x = 508$$
$$-0.02x + 560 = 508$$
$$-0.02x = -52$$
$$x = 2600$$
$$4000 - x = 4000 - 2600 = 1400$$

Thus, \$2600 was invested at 12% and \$1400 was invested at 14%.

42. Because I varies inversely as R, we have the following for a constant k:

$$I = \frac{k}{R}$$

Use the fact that $I = 5$ when $R = 22$ to find k:

$$5 = \frac{k}{22}$$
$$k = 22 \cdot 5 = 110$$

Thus, the equation relating I and R is $I = \frac{110}{R}$.

If $R = 10$, then $I = \frac{110}{10} = 11$.

A current of 11 amperes is required when the resistance is 10 ohms.

Chapter 9
Exponential and Logarithmic Functions

9.1 Check Points

1. $f(x) = 42.2(1.56)^x$

$f(3) = 42.2(1.56)^3 \approx 160.20876 \approx 160$

The average amount spent after three hours at a mall is \$160. This overestimates the amount shown in the figure by \$11.

2. $f(x) = 3^x$

x	$f(x) = 3^x$	(x, y)
-3	$3^{-3} = \frac{1}{27}$	$\left(-3, \frac{1}{27}\right)$
-2	$3^{-2} = \frac{1}{9}$	$\left(-2, \frac{1}{9}\right)$
-1	$3^{-1} = \frac{1}{3}$	$\left(-1, \frac{1}{3}\right)$
0	$3^0 = 1$	$(0, 1)$
1	$3^1 = 3$	$(1, 3)$
2	$3^2 = 9$	$(2, 9)$
3	$3^3 = 27$	$(3, 27)$

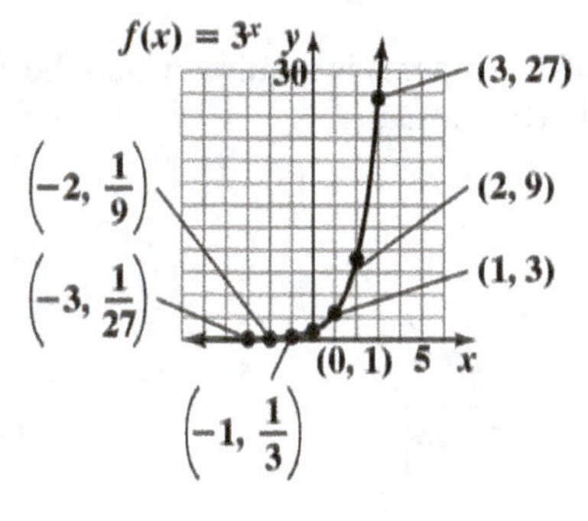

3. $f(x) = \left(\frac{1}{3}\right)^x = 3^{-x}$

x	$f(x) = \left(\frac{1}{3}\right)^x = 3^{-x}$	(x, y)
-3	$3^{-(-3)} = 27$	$(-3, 27)$
-2	$3^{-(-2)} = 9$	$(-2, 9)$
-1	$3^{-(-1)} = 3$	$(-1, 3)$
0	$3^0 = 1$	$(0, 1)$
1	$3^{-1} = \frac{1}{3}$	$\left(1, \frac{1}{3}\right)$
2	$3^{-2} = \frac{1}{9}$	$\left(2, \frac{1}{9}\right)$
3	$3^{-3} = \frac{1}{27}$	$\left(3, \frac{1}{27}\right)$

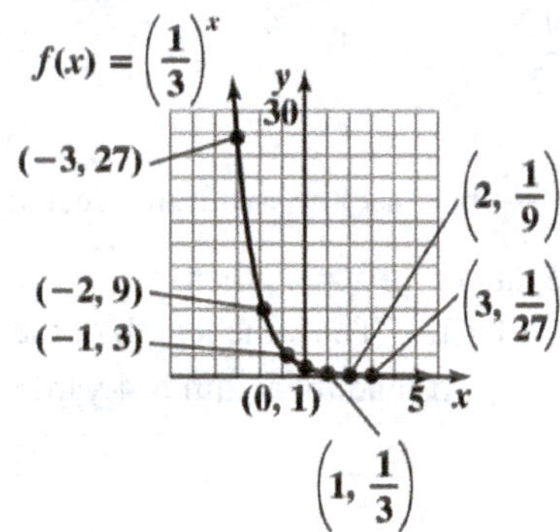

4. $f(x) = 3^x$ and $g(x) = 3^{x-1}$

x	$f(x) = 3^x$	$g(x) = 3^{x-1}$
-2	$3^{-2} = \frac{1}{9}$	$3^{-2-1} = 3^{-3} = \frac{1}{27}$
-1	$3^{-1} = \frac{1}{3}$	$3^{-1-1} = 3^{-2} = \frac{1}{9}$
0	$3^0 = 1$	$3^{0-1} = 3^{-1} = \frac{1}{3}$
1	$3^1 = 3$	$3^{1-1} = 3^0 = 1$
2	$3^2 = 9$	$3^{2-1} = 3^1 = 3$

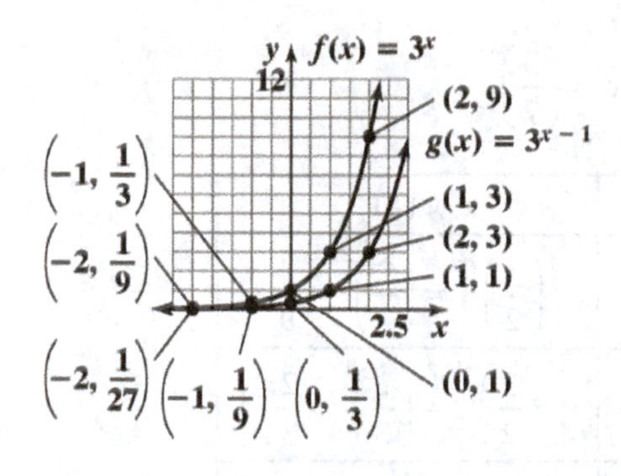

The graph of $g(x)=3^{x-1}$ is the graph of $f(x)=3^x$ shifted 1 unit to the right.

5. $f(x)=2^x$ and $g(x)=2^x+3$

x	$f(x)=2^x$	$g(x)=2^x+3$
−2	$2^{-2}=\frac{1}{4}$	$2^{-2}+3=3\frac{1}{4}$
−1	$2^{-1}=\frac{1}{2}$	$2^{-1}+3=3\frac{1}{2}$
0	$2^0=1$	$2^0+3=4$
1	$2^1=2$	$2^1+3=5$
2	$2^2=4$	$2^2+3=7$

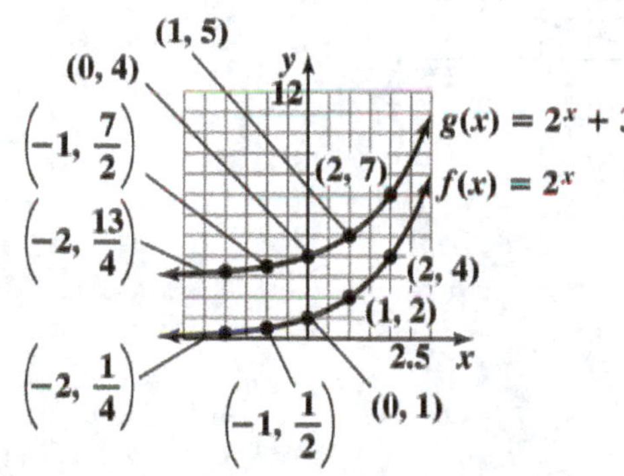

The graph of $g(x)=2^x+3$ is the graph of $f(x)=2^x$ shifted up 3 units.

6. 2012 is 34 years after 1978.

$f(x)=1066e^{0.042x}$

$f(34)=1066e^{0.042(34)}\approx 4446$

In 2012 the gray wolf population of the Western Great Lakes is projected to be about 4446.

7. a. $A=P\left(1+\frac{r}{n}\right)^{nt}$

$=\$10,000\left(1+\frac{0.08}{4}\right)^{4\cdot 5}\approx \$14,859.47$

b. $A=Pe^{rt}=\$10,000e^{0.08(5)}\approx \$14,918.25$

9.1 Concept and Vocabulary Check

1. b^x; $(-\infty,\infty)$; $(0,\infty)$
2. x; $y=0$; horizontal
3. e; natural; 2.72
4. A; P; r; n;
5. semiannually; quarterly; continuous

9.1 Exercise Set

1. $2^{3.4}\approx 10.556$

3. $3^{\sqrt{5}}\approx 11.665$

5. $4^{-1.5}=0.125$

7. $e^{2.3}\approx 9.974$

9. $e^{-0.95}\approx 0.387$

11. $f(x)=3^x$

x	$f(x)$
−2	$3^{-2}=\frac{1}{3^2}=\frac{1}{9}$
−1	$3^{-1}=\frac{1}{3^1}=\frac{1}{3}$
0	$3^0=1$
1	$3^1=3$
2	$3^2=9$

This function matches graph (d).

13. $f(x) = 3^x - 1$

x	$f(x)$
-2	$3^{-2} - 1 = \frac{1}{3^2} - 1 = \frac{1}{9} - 1 = -\frac{8}{9}$
-1	$3^{-1} - 1 = \frac{1}{3^1} - 1 = \frac{1}{3} - 1 = -\frac{2}{3}$
0	$3^0 - 1 = 1 - 1 = 0$
1	$3^1 - 1 = 3 - 1 = 2$
2	$3^2 - 1 = 9 - 1 = 8$

This function matches graph (e).

15. $f(x) = 3^{-x}$

x	$f(x)$
-2	$3^{-(-2)} = 3^2 = 9$
-1	$3^{-(-1)} = 3^1 = 3$
0	$3^{-(0)} = 3^0 = 1$
1	$3^{-(1)} = 3^{-1} = \frac{1}{3}$
2	$3^{-(2)} = 3^{-2} = \frac{1}{3^2} = \frac{1}{9}$

This function matches graph (f).

17. $f(x) = 4^x$

x	$f(x)$
-2	$4^{-2} = \frac{1}{4^2} = \frac{1}{16}$
-1	$4^{-1} = \frac{1}{4^1} = \frac{1}{4}$
0	$4^0 = 1$
1	$4^1 = 4$
2	$4^2 = 16$

$f(x) = 4^x$

19. $g(x) = \left(\frac{3}{2}\right)^x$

x	$g(x)$
-2	$\left(\frac{3}{2}\right)^{-2} = \left(\frac{2}{3}\right)^2 = \frac{4}{9}$
-1	$\left(\frac{3}{2}\right)^{-1} = \left(\frac{2}{3}\right)^1 = \frac{2}{3}$
0	$\left(\frac{3}{2}\right)^0 = 1$
1	$\left(\frac{3}{2}\right)^1 = \frac{3}{2}$
2	$\left(\frac{3}{2}\right)^2 = \frac{9}{4}$

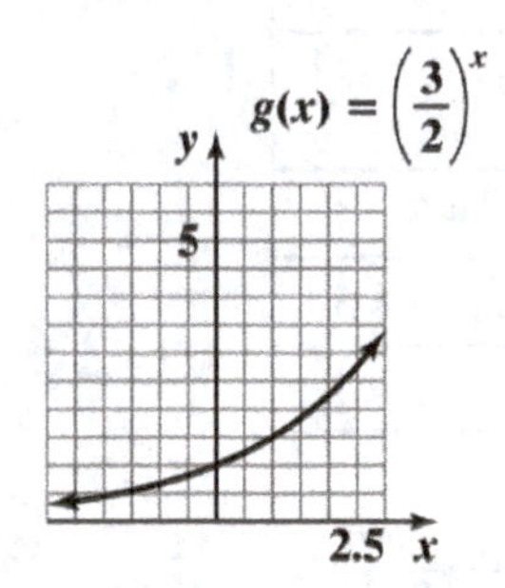

21. $h(x)=\left(\frac{1}{2}\right)^x$

x	$h(x)$
−2	$\left(\frac{1}{2}\right)^{-2}=\left(\frac{2}{1}\right)^2=\frac{4}{1}=4$
−1	$\left(\frac{1}{2}\right)^{-1}=\left(\frac{2}{1}\right)^1=\frac{2}{1}=2$
0	$\left(\frac{1}{2}\right)^0=1$
1	$\left(\frac{1}{2}\right)^1=\frac{1}{2}$
2	$\left(\frac{1}{2}\right)^2=\frac{1}{4}$

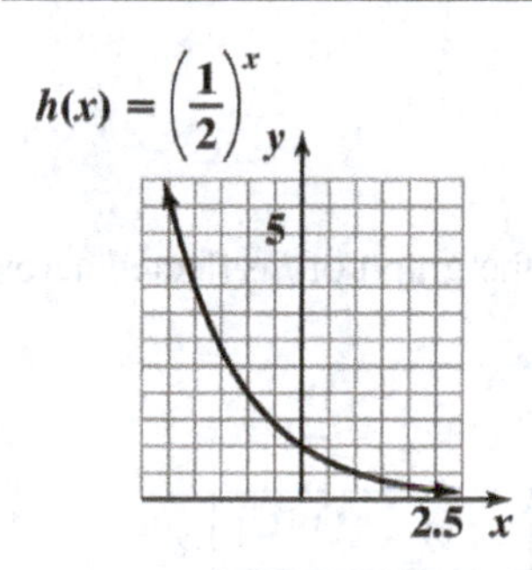

23. $f(x)=(0.6)^x=\left(\frac{6}{10}\right)^x=\left(\frac{3}{5}\right)^x$

x	$f(x)$
−2	$\left(\frac{3}{5}\right)^{-2}=\left(\frac{5}{3}\right)^2=\frac{25}{9}$
−1	$\left(\frac{3}{5}\right)^{-1}=\left(\frac{5}{3}\right)^1=\frac{5}{3}$
0	$\left(\frac{3}{5}\right)^0=1$
1	$\left(\frac{3}{5}\right)^1=\frac{3}{5}$
2	$\left(\frac{3}{5}\right)^2=\frac{9}{25}$

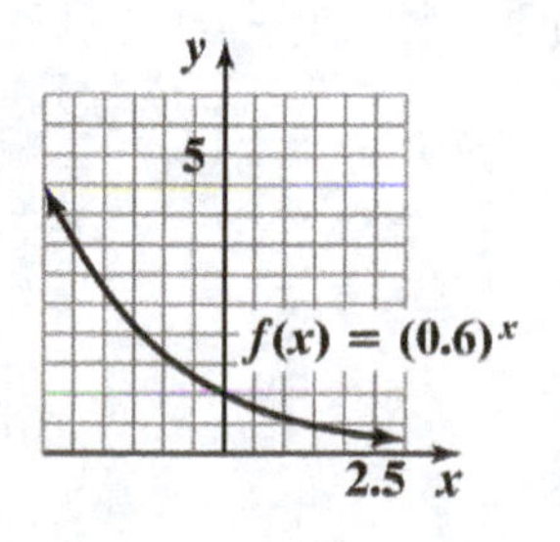

25. $f(x)=2^x$ and $g(x)=2^{x+1}$

x	$f(x)=2^x$	$g(x)=2^{x+1}$
−2	$\frac{1}{4}$	$\frac{1}{2}$
−1	$\frac{1}{2}$	1
0	1	2
1	2	4
2	4	8

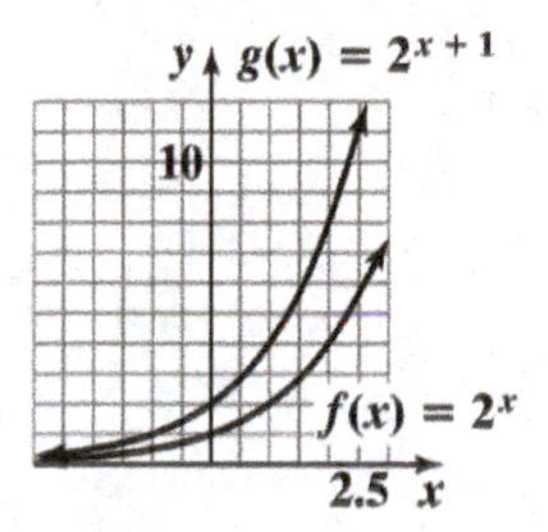

The graph of g is the graph of f shifted 1 unit to the left.

27. $f(x)=2^x$ and $g(x)=2^{x-2}$

x	$f(x)=2^x$	$g(x)=2^{x-2}$
−2	$\frac{1}{4}$	$\frac{1}{16}$
−1	$\frac{1}{2}$	$\frac{1}{8}$
0	1	$\frac{1}{2}$
1	2	1
2	4	2

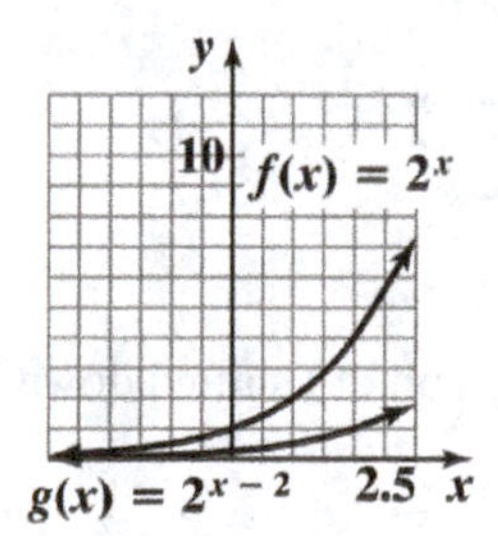

The graph of g is the graph of f shifted 2 units to the right.

29. $f(x)=2^x$ and $g(x)=2^x+1$

x	$f(x)=2^x$	$g(x)=2^x+1$
−2	$\frac{1}{4}$	$\frac{5}{4}$
−1	$\frac{1}{2}$	$\frac{3}{2}$
0	1	2
1	2	3
2	4	5

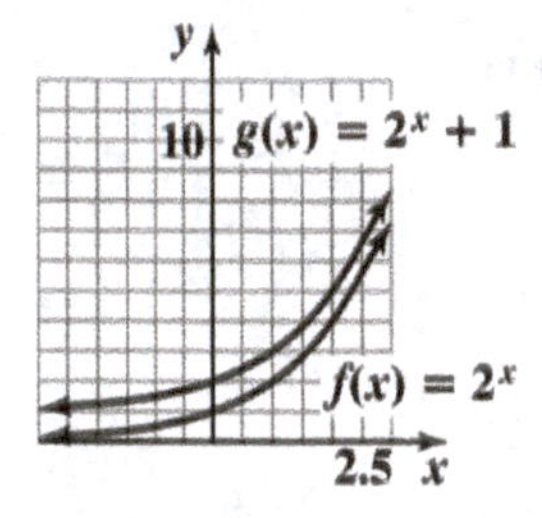

The graph of g is the graph of f shifted up 1 unit.

31. $f(x)=2^x$ and $g(x)=2^x-2$

x	$f(x)=2^x$	$g(x)=2^x-2$
−2	$\frac{1}{4}$	$-\frac{7}{4}$
−1	$\frac{1}{2}$	$-\frac{3}{2}$
0	1	−1
1	2	0
2	4	2

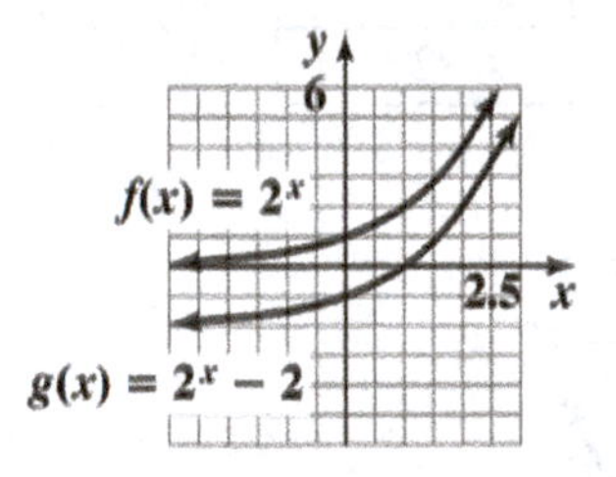

The graph of g is the graph of f shifted down 2 units.

33. $f(x)=3^x$ and $g(x)=-3^x$

x	$f(x)=3^x$	$g(x)=-3^x$
−2	$\frac{1}{9}$	$-\frac{1}{9}$
−1	$\frac{1}{3}$	$-\frac{1}{3}$
0	1	−1
1	3	−3
2	9	−9

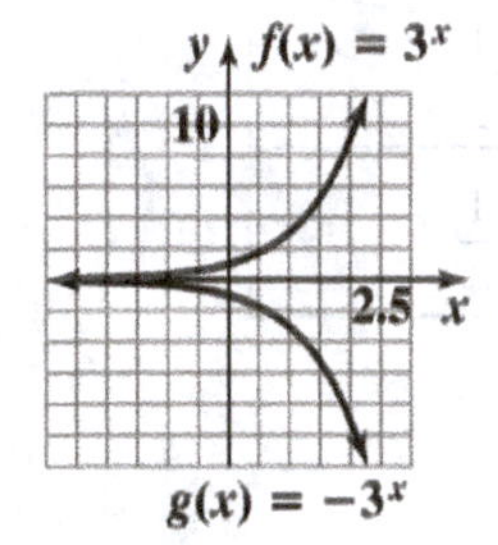

The graph of g is the graph of f reflected across the x–axis.

35. $f(x)=2^x$ and $g(x)=2^{x+1}-1$

x	$f(x)=2^x$	$g(x)=2^{x+1}-1$
−2	$\frac{1}{4}$	$-\frac{1}{2}$
−1	$\frac{1}{2}$	0
0	1	1
1	2	3
2	4	7

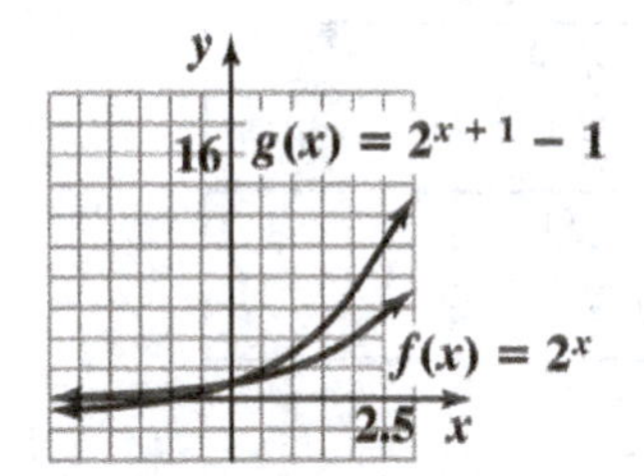

The graph of g is the graph of f shifted 1 unit down and 1 unit to the left.

37. $f(x) = 3^x$ and $g(x) = \frac{1}{3} \cdot 3^x$

x	$f(x) = 3^x$	$g(x) = \frac{1}{3} \cdot 3^x$
−2	$\frac{1}{9}$	$\frac{1}{27}$
−1	$\frac{1}{3}$	$\frac{1}{9}$
0	1	$\frac{1}{3}$
1	3	1
2	9	3

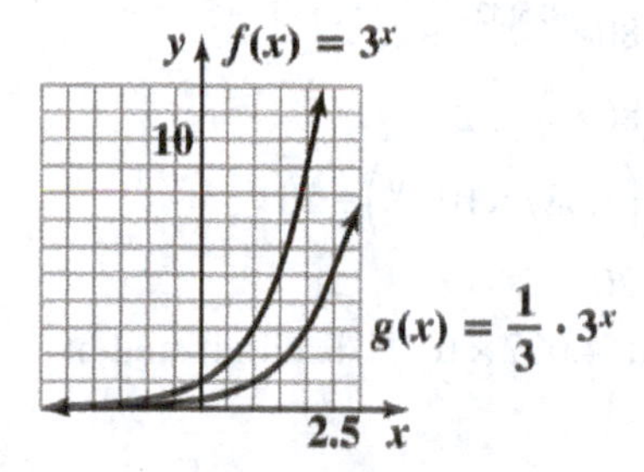

The graph of g is the graph of f compressed vertically by a factor of $\frac{1}{3}$.

39. **a.** $A = 10,000\left(1+\frac{0.055}{2}\right)^{2(5)} \approx 13,116.51$

The balance in the account is \$13,116.51 after 5 years of semiannual compounding.

b. $A = 10,000\left(1+\frac{0.055}{12}\right)^{12(5)} \approx 13,157.04$

The balance in the account is \$13,157.04 after 5 years of monthly compounding.

c. $A = Pe^{rt} = 10,000e^{0.055(5)} \approx 13,165.31$

The balance in the account is \$13,165.31 after 5 years of continuous compounding.

41. Monthly Compounding

$A = 12,000\left(1+\frac{0.07}{12}\right)^{12(3)} \approx 14,795.11$

Continuous Compounding

$A = 12,000e^{0.0685(3)} \approx 14737.67$

Monthly compounding at 7% yields the greatest return.

43. Domain: $(-\infty, \infty)$
Range: $(-2, \infty)$

45. Domain: $\{x \mid x \text{ is a real number}\}$ or $(-\infty, \infty)$
Range: $(1, \infty)$

47. Domain: $(-\infty, \infty)$
Range: $(0, \infty)$

49.

x	$f(x) = 2^x$	$g(x) = 2^{-x}$
−2	$\frac{1}{4}$	4
−1	$\frac{1}{2}$	2
0	1	1
1	2	$\frac{1}{2}$
2	4	$\frac{1}{4}$

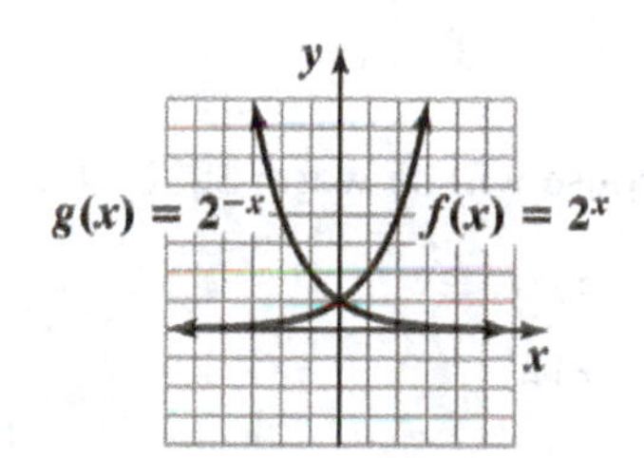

The point of intersection is $(0,1)$.

51.

x	$y = 2^x$
−2	$\frac{1}{4}$
−1	$\frac{1}{2}$
0	1
1	2
2	4

y	$x = 2^y$
−2	$\frac{1}{4}$
−1	$\frac{1}{2}$
0	1
1	2
2	4

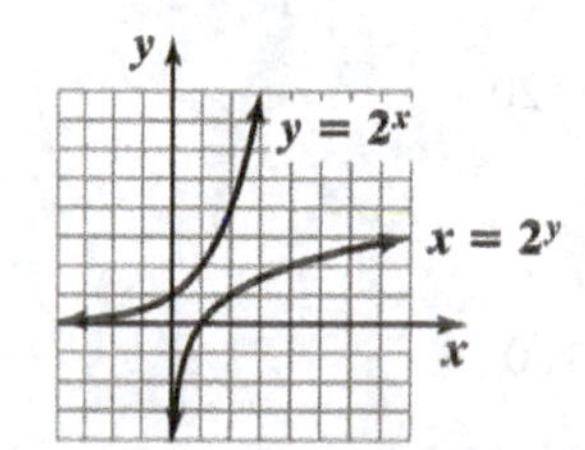

53. a. $f(0) = 574(1.026)^0$
$= 574(1) = 574$

India's population in 1974 was 574 million.

b. $f(27) = 574(1.026)^{27} \approx 1148$

India's population in 2001 will be 1148 million.

c. Since $2028 - 1974 = 54$, find
$f(54) = 574(1.026)^{54} \approx 2295$.
India's population in 2028 will be 2295 million.

d. $2055 - 1974 = 81$, find
$f(54) = 574(1.026)^{81} \approx 4590$.
India's population in 2055 will be 4590 million.

e. India's population appears to be doubling every 27 years.

55. $S = 465,000(1+0.06)^{10}$
$= 465,000(1.06)^{10} \approx 832,744$

In 10 years, the house will be worth $832,744.

57. a. $f(x) = 19x + 127$
$f(14) = 19(14) + 127$
$= 393$
According to the linear model, there were about 393 million active Facebook users in February 2010.

b. $g(x) = 152.6e^{0.0667x}$
$g(14) = 152.6e^{0.0667(14)}$
≈ 388
According to the exponential model, there were about 388 million active Facebook users in February 2010.

c. The linear model is the better model for the data in February 2010.

59. a. $f(0) = 80e^{-0.5(0)} + 20$
$= 80e^0 + 20$
$= 80(1) + 20$
$= 80 + 20 = 100$

100% of information is remembered at the moment it is first learned.

b. $f(1) = 80e^{-0.5(1)} + 20$
$= 80e^{-0.5} + 20 \approx 68.522$

About 68.5% of information is remembered after one week.

c. $f(4) = 80e^{-0.5(4)} + 20$
$= 80e^{-2} + 20$
$= 10.827 + 20 = 30.827$

Approximately 30.8% of information is remembered after four weeks.

d. $f(52) = 80e^{-0.5(52)} + 20$
$= 80e^{-26} + 20$
$= (4.087 \times 10^{-10}) + 20$
≈ 20
(Note that 4.087×10^{-10} is eliminated in rounding.)

Approximately 20% of information is remembered after one year.

61. $f(30) = \dfrac{90}{1+270e^{-0.122(30)}}$
$= \dfrac{90}{1+270e^{-3.66}}$
$= \dfrac{90}{1+6.948} = \dfrac{90}{7.948} \approx 11.3$

Approximately 11.3% of 30-year-olds have some coronary heart disease.

63. a. $N(0) = \dfrac{30,000}{1+20e^{-1.5(0)}}$
$= \dfrac{30,000}{1+20e^0}$
$= \dfrac{30,000}{1+20(1)}$
$= \dfrac{30,000}{1+20}$
$= \dfrac{30,000}{21} \approx 1428.6$
Approximately 1429 people became ill with the flu when the epidemic began.

b. $N(3) = \dfrac{30,000}{1+20e^{-1.5(3)}}$

$= \dfrac{30,000}{1+20e^{-4.5}} \approx 24,546$

Approximately 24,546 people became ill with the flu by the end of the third week.

c. The epidemic cannot grow indefinitely because there are a limited number of people that can become ill. Because there are 30,000 people in the town, the limit is 30,000.

65. – 67. Answers will vary

69. a. $Q(t) = 10,000\left(1+\dfrac{0.05}{4}\right)^{4t}$

$M(t) = 10,000\left(1+\dfrac{0.045}{12}\right)^{12t}$

b.

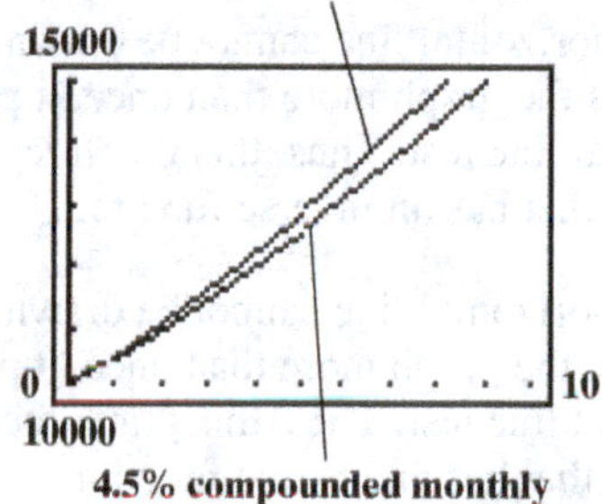

The bank paying 5% compounded quarterly offers a better return.

71. does not make sense; Explanations will vary. Sample explanation: The horizontal asymptote is $y = 0$.

73. does not make sense; Explanations will vary. Sample explanation: An exponential function would be a better choice.

75. false; Changes to make the statement true will vary. A sample change is: The amount of money will not increase without bound.

77. false; Changes to make the statement true will vary. A sample change is: $f(a+b) = 2^{a+b} = 2^a \cdot 2^b$, not $f(a) + f(b) = 2^a + 2^b$.

79. Graph (a) is $y = \left(\dfrac{1}{3}\right)^x$.

Graph (b) is $y = \left(\dfrac{1}{5}\right)^x$.

Graph (c) is $y = 5^x$.

Graph (d) is $y = 3^x$.

Answers will vary. One possibility follows: A base between 0 and 1 will rise to the left and a base greater than 1 will rise to the right.

81.

$$D = \frac{ab}{a+b}$$
$$D(a+b) = ab$$
$$Da + Db = ab$$
$$Da = ab - Db$$
$$Da = (a-D)b$$
$$b = \frac{Da}{a-D}$$

82. $\begin{vmatrix} 3 & -2 \\ 7 & -5 \end{vmatrix} = 3(-5) - 7(-2) = -15 + 14 = -1$

83.

$$x(x-3) = 10$$
$$x^2 - 3x = 10$$
$$x^2 - 3x - 10 = 0$$
$$(x-5)(x+2) = 0$$

Apply the zero product principle.

$x - 5 = 0$ or $x + 2 = 0$

$x = 5$ $\qquad x = -2$

The solutions are 5 and –2, and the solution set is $\{-2, 5\}$.

84. a. $f(x) = 3x - 4$

$f(5) = 3(5) - 4 = 11$

b. $g(x) = x^2 + 6$

$g(f(5)) = g(11) = (11)^2 + 6 = 127$

85. $3\left(\dfrac{x-2}{3}\right) + 2 = x - 2 + 2 = x$

86.
$$x = 7y - 5$$
$$x + 5 = 7y$$
$$\frac{x+5}{7} = y$$
$$y = \frac{x+5}{7}$$

9.2 Check Points

1. a. $(f \circ g)(x) = f(g(x))$
$= f(x^2 - 1)$
$= 5(x^2 - 1) + 6$
$= 5x^2 - 5 + 6$
$= 5x^2 + 1$

b. $(g \circ f)(x) = g(f(x))$
$= g(5x + 6)$
$= (5x+6)^2 - 1$
$= 25x^2 + 60x + 36 - 1$
$= 25x^2 + 60x + 35$

2. $f(g(x)) = 7\left(\frac{x}{7}\right) = x$

$g(f(x)) = \frac{7x}{7} = x$

3. $f(g(x)) = 4\left(\frac{x+7}{4}\right) - 7 = x + 7 - 7 = x$

$g(f(x)) = \frac{(4x-7)+7}{4} = \frac{4x}{4} = x$

4. $f(x) = 2x + 7$
$y = 2x + 7$
Interchange x and y and solve for y.
$x = 2y + 7$
$x - 7 = 2y$
$\frac{x-7}{2} = y$
$f^{-1}(x) = \frac{x-7}{2}$

5. $f(x) = 4x^3 - 1$
$y = 4x^3 - 1$
Interchange x and y and solve for y.
$x = 4y^3 - 1$
$x + 1 = 4y^3$
$\frac{x+1}{4} = y^3$
$\sqrt[3]{\frac{x+1}{4}} = \sqrt[3]{y^3}$
$\sqrt[3]{\frac{x+1}{4}} = y$
$f^{-1}(x) = \sqrt[3]{\frac{x+1}{4}}$

6. a. Since a horizontal line can be drawn that intersects the graph more than once, it fails the horizontal line test. Thus, this graph does not represent a function that has an inverse function.

b. Since a horizontal line cannot be drawn that intersects the graph more than once, it passes the horizontal line test. Thus, this graph represents a function that has an inverse function.

c. Since a horizontal line cannot be drawn that intersects the graph more than once, it passes the horizontal line test. Thus, this graph represents a function that has an inverse function.

7. Since f has a line segment from $(-2, -2)$ to $(-1, 0)$, then f^{-1} has a line segment from $(-2, -2)$ to $(0, -1)$.
Since f has a line segment from $(-1, 0)$ to $(1, 2)$, then f^{-1} has a line segment from $(0, -1)$ to $(2, 1)$.

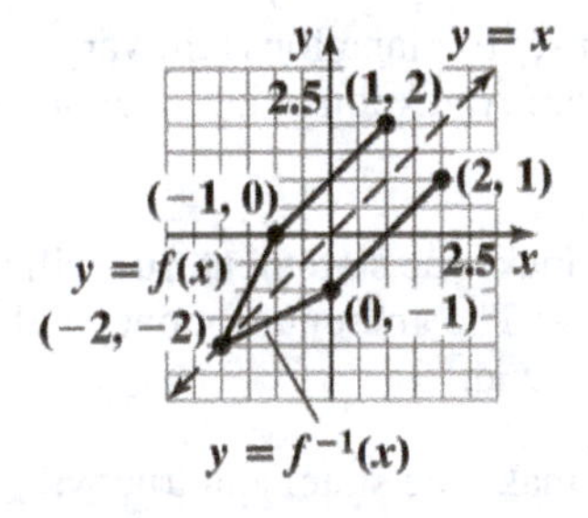

9.2 Concept and Vocabulary Check

1. composition; $f(g(x))$

2. f; $g(x)$

3. composition; $g(f(x))$

4. g; $f(x)$

5. false

6. false

7. inverse

8. x; x

9. horizontal; one-to-one

10. $y = x$

9.2 Exercise Set

1. a. $(f \circ g)(x) = f(g(x)) = f(x+7) = 2(x+7) = 2x+14$

b. $(g \circ f)(x) = g(f(x)) = g(2x) = 2x+7$

c. $(f \circ g)(2) = 2(2)+14 = 4+14 = 18$

3. a. $(f \circ g)(x) = f(g(x)) = f(2x+1) = (2x+1)+4 = 2x+5$

b. $(g \circ f)(x) = g(f(x)) = g(x+4) = 2(x+4)+1 = 2x+8+1 = 2x+9$

c. $(f \circ g)(2) = 2(2)+5 = 4+5 = 9$

5. a. $(f \circ g)(x) = f(g(x)) = f(5x^2-2) = 4(5x^2-2)-3 = 20x^2-8-3 = 20x^2-11$

b. $(g \circ f)(x) = g(f(x)) = g(4x-3) = 5(4x-3)^2-2 = 5(16x^2-24x+9)-2 = 80x^2-120x+45-2 = 80x^2-120x+43$

c. $(f \circ g)(2) = 20(2)^2-11 = 20(4)-11 = 80-11 = 69$

7. a. $(f \circ g)(x) = f(g(x)) = f(x^2-2) = (x^2-2)^2+2 = x^4-4x^2+4+2 = x^4-4x^2+6$

b. $(g \circ f)(x) = g(f(x)) = g(x^2+2) = (x^2+2)^2-2 = x^4+4x^2+4-2 = x^4+4x^2+2$

c. $(f \circ g)(2) = 2^4-4(2)^2+6 = 16-4(4)+6 = 16-16+6 = 6$

9. a. $(f \circ g)(x) = f(g(x)) = f(x-1) = \sqrt{x-1}$

b. $(g \circ f)(x) = g(f(x)) = g(\sqrt{x}) = \sqrt{x}-1$

c. $(f \circ g)(2) = \sqrt{2-1} = \sqrt{1} = 1$

11. a. $(f\circ g)(x)=f(g(x))$
$$=f\left(\frac{x+3}{2}\right)$$
$$=2\left(\frac{x+3}{2}\right)-3$$
$$=x+3-3=x$$

b. $(g\circ f)(x)=g(f(x))$
$$=g(2x-3)$$
$$=\frac{(2x-3)+3}{2}$$
$$=\frac{2x-3+3}{2}=\frac{2x}{2}=x$$

c. $(f\circ g)(2)=2$

13. a. $(f\circ g)(x)=f(g(x))$
$$=f\left(\frac{1}{x}\right)$$
$$=\frac{1}{\frac{1}{x}}=1\cdot\frac{x}{1}=x$$

b. $(g\circ f)(x)=g(f(x))$
$$=g\left(\frac{1}{x}\right)$$
$$=\frac{1}{\frac{1}{x}}=1\cdot\frac{x}{1}=x$$

c. $(f\circ g)(2)=2$

15. $f(g(x))=f\left(\frac{x}{4}\right)=4\left(\frac{x}{4}\right)=x$

$g(f(x))=g(4x)=\frac{4x}{4}=x$

The functions are inverses.

17. $f(g(x))=f\left(\frac{x-8}{3}\right)$
$$=3\left(\frac{x-8}{3}\right)+8$$
$$=x-8+8=x$$

$g(f(x))=g(3x+8)$
$$=\frac{(3x+8)-8}{3}$$
$$=\frac{3x+8-8}{3}=\frac{3x}{3}=x$$

The functions are inverses.

19. $f(g(x))=f\left(\frac{x+5}{9}\right)$
$$=5\left(\frac{x+5}{9}\right)-9$$
$$=\frac{5x+25}{9}-\frac{81}{9}$$
$$=\frac{5x+25-81}{9}=\frac{5x-56}{9}$$

$g(f(x))=g(5x-9)$
$$=\frac{(5x-9)+5}{9}=\frac{5x-4}{9}$$

Since $f(g(x))\neq g(f(x))\neq x$, we conclude the functions are not inverses.

21. $f(g(x))=f\left(\frac{3}{x}+4\right)$
$$=\frac{3}{\left(\frac{3}{x}+4\right)-4}$$
$$=\frac{3}{\frac{3}{x}+4-4}=\frac{3}{\frac{3}{x}}=3\cdot\frac{x}{3}=x$$

$g(f(x))=g\left(\frac{3}{x-4}\right)$
$$=\frac{3}{\frac{3}{x-4}}+4$$
$$=3\cdot\frac{x-4}{3}+4=x-4+4=x$$

The functions are inverses.

23. $f(g(x)) = f(-x) = -(-x) = x$

$g(f(x)) = g(-x) = -(-x) = x$

The functions are inverses.

25. a. $f(x) = x+3$

$y = x+3$

Interchange x and y and solve for y.

$x = y+3$

$x-3 = y$

$f^{-1}(x) = x-3$

b. $f(f^{-1}(x)) = f(x-3)$

$= (x-3)+3$

$= x-3+3 = x$

$f^{-1}(f(x)) = f(x+3)$

$= (x+3)-3$

$= x+3-3 = x$

27. a. $f(x) = 2x$

$y = 2x$

Interchange x and y and solve for y.

$x = 2y$

$\frac{x}{2} = y$

$f^{-1}(x) = \frac{x}{2}$

b. $f(f^{-1}(x)) = f\left(\frac{x}{2}\right) = 2\left(\frac{x}{2}\right) = x$

$f^{-1}(f(x)) = f(2x) = \frac{2x}{2} = x$

29. a. $f(x) = 2x+3$

$y = 2x+3$

Interchange x and y and solve for y.

$x = 2y+3$

$x-3 = 2y$

$\frac{x-3}{2} = y$

$f^{-1}(x) = \frac{x-3}{2}$

b. $f(f^{-1}(x)) = f\left(\frac{x-3}{2}\right)$

$= 2\left(\frac{x-3}{2}\right)+3$

$= x-3+3 = x$

$f^{-1}(f(x)) = f^{-1}(2x+3)$

$= \frac{(2x+3)-3}{2}$

$= \frac{2x+3-3}{2} = \frac{2x}{2} = x$

31. a. $f(x) = x^3+2$

$y = x^3+2$

Interchange x and y and solve for y.

$x = y^3+2$

$x-2 = y^3$

$\sqrt[3]{x-2} = y$

$f^{-1}(x) = \sqrt[3]{x-2}$

b. $f(f^{-1}(x)) = f\left(\sqrt[3]{x-2}\right)$

$= \left(\sqrt[3]{x-2}\right)^3+2$

$= x-2+2 = x$

$f^{-1}(f(x)) = f^{-1}\left(x^3+2\right)$

$= \sqrt[3]{\left(x^3+2\right)-2}$

$= \sqrt[3]{x^3+2-2}$

$= \sqrt[3]{x^3} = x$

33. a. $f(x) = (x+2)^3$

$y = (x+2)^3$

Interchange x and y and solve for y.

$x = (y+2)^3$

$\sqrt[3]{x} = \sqrt[3]{(y+2)^3}$

$\sqrt[3]{x} = y+2$

$\sqrt[3]{x}-2 = y$

$f^{-1}(x) = \sqrt[3]{x}-2$

b. $f\left(f^{-1}(x)\right) = f\left(\sqrt[3]{x} - 2\right)$
$= \left(\left(\sqrt[3]{x} - 2\right) + 2\right)^3$
$= \left(\sqrt[3]{x} - 2 + 2\right)^3$
$= \left(\sqrt[3]{x}\right)^3 = x$

$f^{-1}\left(f(x)\right) = f^{-1}\left((x+2)^3\right)$
$= \sqrt[3]{(x+2)^3} - 2$
$= x + 2 - 2 = x$

35. a. $f(x) = \dfrac{1}{x}$
$y = \dfrac{1}{x}$

Interchange x and y and solve for y.

$x = \dfrac{1}{y}$
$xy = 1$
$y = \dfrac{1}{x}$
$f^{-1}(x) = \dfrac{1}{x}$

b. $f\left(f^{-1}(x)\right) = f\left(\dfrac{1}{x}\right)$
$= \dfrac{1}{\frac{1}{x}} = 1 \cdot \dfrac{x}{1} = x$

$f^{-1}\left(f(x)\right) = f^{-1}\left(\dfrac{1}{x}\right)$
$= \dfrac{1}{\frac{1}{x}} = 1 \cdot \dfrac{x}{1} = x$

37. a. $f(x) = \sqrt{x}$
$y = \sqrt{x}$

Interchange x and y and solve for y.

$x = \sqrt{y}$
$x^2 = y$
$f^{-1}(x) = x^2, \quad x \ge 0$

b. $f\left(f^{-1}(x)\right) = f\left(x^2\right) = \sqrt{x^2} = x$
$f^{-1}\left(f(x)\right) = f^{-1}\left(\sqrt{x}\right)$
$= \left(\sqrt{x}\right)^2 = x$

39. a. $f(x) = x^2 + 1$
$y = x^2 + 1$

Interchange x and y and solve for y.

$x = y^2 + 1$
$x - 1 = y^2$
$\sqrt{x-1} = y$
$f^{-1}(x) = \sqrt{x-1}$

b. $f\left(f^{-1}(x)\right) = f\left(\sqrt{x-1}\right)$
$= \left(\sqrt{x-1}\right)^2 + 1$
$= x - 1 + 1 = x$

$f^{-1}\left(f(x)\right) = f^{-1}\left(x^2 + 1\right)$
$= \sqrt{\left(x^2+1\right) - 1}$
$= \sqrt{x^2 + 1 - 1}$
$= \sqrt{x^2} = x$

41. a. $f(x) = \dfrac{2x+1}{x-3}$
$y = \dfrac{2x+1}{x-3}$

Interchange x and y and solve for y.

$x = \dfrac{2y+1}{y-3}$
$x(y-3) = 2y + 1$
$xy - 3x = 2y + 1$
$xy - 2y = 3x + 1$
$(x-2)y = 3x + 1$
$y = \dfrac{3x+1}{x-2}$

$f^{-1}(x) = \dfrac{3x+1}{x-2}$

b. $f\left(f^{-1}(x)\right)=f\left(\dfrac{3x+1}{x-2}\right)$

$$=\frac{2\left(\frac{3x+1}{x-2}\right)+1}{\left(\frac{3x+1}{x-2}\right)-3}$$

$$=\frac{x-2}{x-2}\cdot\frac{2\left(\frac{3x+1}{x-2}\right)+1}{\left(\frac{3x+1}{x-2}\right)-3}$$

$$=\frac{2(3x+1)+1(x-2)}{(3x+1)-3(x-2)}$$

$$=\frac{6x+2+x-2}{3x+1-3x+6}$$

$$=\frac{7x}{7}$$

$$=x$$

$f^{-1}\left(f(x)\right)=f^{-1}\left(\dfrac{2x+1}{x-3}\right)$

$$=\frac{3\left(\frac{2x+1}{x-3}\right)+1}{\left(\frac{2x+1}{x-3}\right)-2}$$

$$=\frac{x-3}{x-3}\cdot\frac{3\left(\frac{2x+1}{x-3}\right)+1}{\left(\frac{2x+1}{x-3}\right)-2}$$

$$=\frac{3(2x+1)+1(x-3)}{(2x+1)-2(x-3)}$$

$$=\frac{6x+3+x-3}{2x+1-2x+6}$$

$$=\frac{7x}{7}$$

$$=x$$

43. a. $f(x)=\sqrt[3]{x-4}+3$

$y=\sqrt[3]{x-4}+3$

Interchange x and y and solve for y.

$$x=\sqrt[3]{y-4}+3$$

$$x-3=\sqrt[3]{y-4}$$

$$(x-3)^3=y-4$$

$$(x-3)^3+4=y$$

$$f^{-1}(x)=(x-3)^3+4$$

b. $f\left(f^{-1}(x)\right)=f\left((x-3)^3+4\right)$

$$=\sqrt[3]{\left((x-3)^3+4\right)-4}+3$$

$$=\sqrt[3]{(x-3)^3+4-4}+3$$

$$=\sqrt[3]{(x-3)^3}+3$$

$$=x-3+3$$

$$=x$$

$f^{-1}\left(f(x)\right)=f^{-1}\left(\sqrt[3]{x-4}+3\right)$

$$=\left(\left(\sqrt[3]{x-4}+3\right)-3\right)^3+4$$

$$=\left(\sqrt[3]{x-4}+3-3\right)^3+4$$

$$=\left(\sqrt[3]{x-4}\right)^3+4$$

$$=x-4+4$$

$$=x$$

45. The graph does not satisfy the horizontal line test so the function does not have an inverse.

47. The graph does not satisfy the horizontal line test so the function does not have an inverse.

49. The graph satisfies the horizontal line test so the function has an inverse.

51.

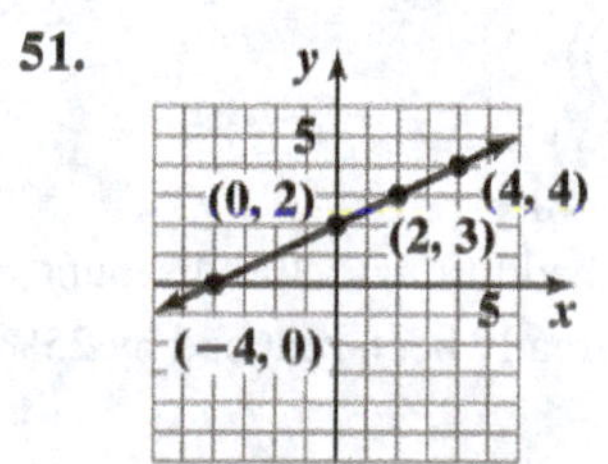

53.

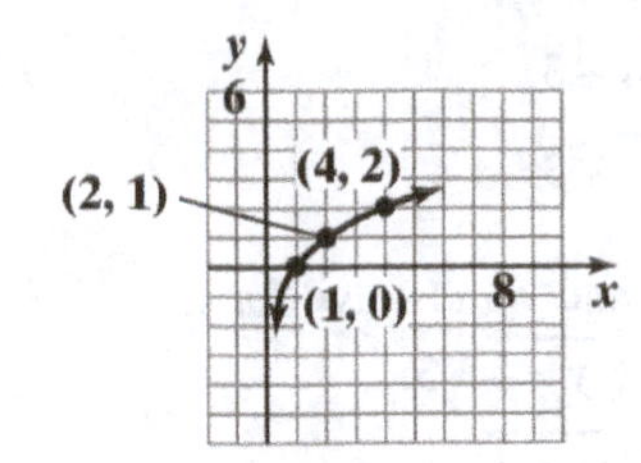

55. $f(g(1)) = f(1) = 5$

57. $(g \circ f)(-1) = g(f(-1)) = g(1) = 1$

59. $f^{-1}(g(10)) = f^{-1}(-1) = 2$, since $f(2) = -1$.

61. $(f \circ g)(-1) = f(g(-1)) = f(-3) = 1$

63. $(g \circ f)(0) = g(f(0)) = g(2) = -6$

65.
$$\begin{aligned}(f \circ g)(0) &= f(g(0)) \\ &= f(4 \cdot 0 - 1) \\ &= f(-1) = 2(-1) - 5 = -7\end{aligned}$$

67. Let $f^{-1}(1) = x$. Then
$$\begin{aligned}f(x) &= 1 \\ 2x - 5 &= 1 \\ 2x &= 6 \\ x &= 3\end{aligned}$$
Thus, $f^{-1}(1) = 3$

69.
$$\begin{aligned}g(f[h(1)]) &= g(f[1^2 + 1 + 2]) \\ &= g(f(4)) \\ &= g(2 \cdot 4 - 5) \\ &= g(3) \\ &= 4 \cdot 3 - 1 = 11\end{aligned}$$

71. a. f represents the price after a \$400 discount; g represents the price after a 25% discount (75% of the regular price).

b.
$$\begin{aligned}(f \circ g)(x) &= f(g(x)) \\ &= f(0.75x) \\ &= 0.75x - 400\end{aligned}$$
$f \circ g$ represents and additional \$400 discount on a price that has already been reduced by 25%.

c.
$$\begin{aligned}(g \circ f)(x) &= g(f(x)) \\ &= g(x - 400) \\ &= 0.75(x - 400) \\ &= 0.75x - 300\end{aligned}$$
$g \circ f$ represents an additional 25% discount on a price that has already been reduced by \$400.

d. $0.75x - 400 < 0.75x - 300$, so $f \circ g$ models the greater discount. It has a savings of \$100 over $g \circ f$.

e.
$$\begin{aligned}f(x) &= x - 400 \\ y &= x - 400\end{aligned}$$

Interchange x and y and solve for y.
$$\begin{aligned}x &= y - 400 \\ x + 400 &= y \\ f^{-1}(x) &= x + 400\end{aligned}$$
f^{-1} represents the regular price of the computer, since the value of x here is the price after a \$400 discount.

73. a. f: {(U.S., 1%), (U.K., 8%), (Italy, 5%), (France, 5%), (Holland, 30%)}

b. Inverse: {(1%, U.S.), (8%, U.K.), (5%, Italy), (5%, France), (30%, Holland)}
The inverse is not a function because the input 5% is associated with two outputs, Italy and France.

75. a. We know that f has an inverse because no horizontal line intersects the graph of f in more than one point.

b. $f^{-1}(0.25)$, or approximately 15, represents the number of people who must be in a room so that the probability of two sharing a birthday would be 0.25; $f^{-1}(0.5)$, or approximately 23, represents the number of people who must be in a room so that the probability of 2 sharing a birthday would be 0.5;

$f^{-1}(0.7)$, or approximately 30, represents the number of people who must be in a room so that the probability of two sharing a birthday would be 0.70.

77. $f(g(x)) = f\left(\frac{5}{9}(x-32)\right)$

$= \frac{9}{5}\left[\frac{5}{9}(x-32)\right]+32$

$= (x-32)+32$

$= x$

$g(f(x)) = g\left(\frac{9}{5}x+32\right)$

$= \frac{5}{9}\left[\left(\frac{9}{5}x+32\right)-32\right]$

$= \frac{5}{9}\left(\frac{9}{5}x\right)$

$= x$

Since $f(g(x)) = x$ and $g(f(x)) = x$, then f and g are inverses.

79. – 83. Answers will vary

85. $f(x) = \sqrt[3]{2-x}$

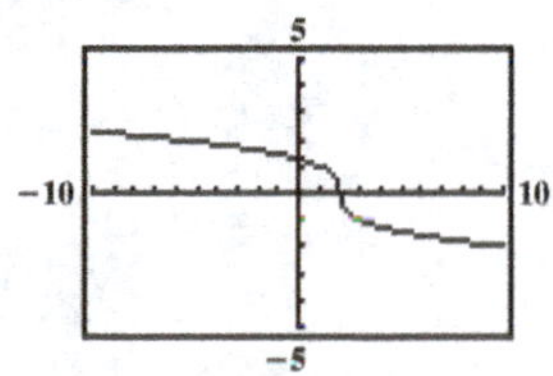

f has an inverse function because it passes the horizontal line test.

87. $f(x) = \frac{x^4}{4}$

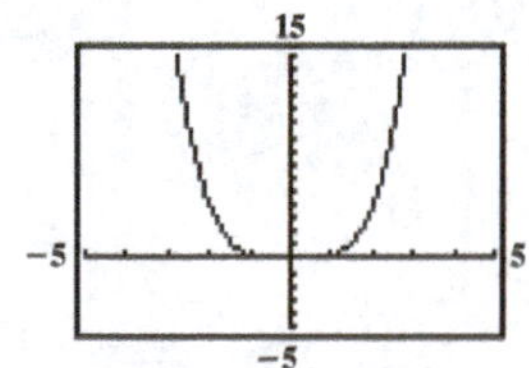

f does not have an inverse function because it does not pass the horizontal line test.

89. $f(x) = (x-1)^3$

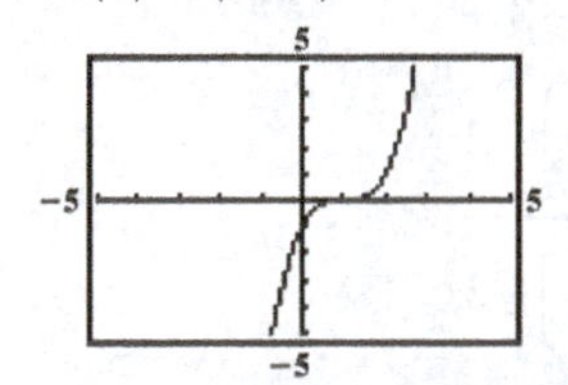

f has an inverse function because it passes the horizontal line test.

91. $f(x) = x^3 + x + 1$

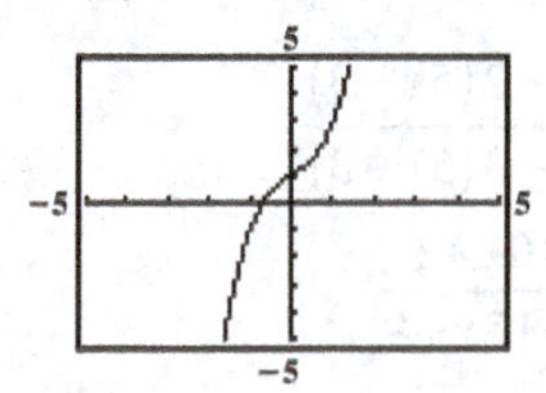

f has an inverse function because it passes the horizontal line test.

93. $f(x) = 4x+4$

$g(x) = 0.25x-1$

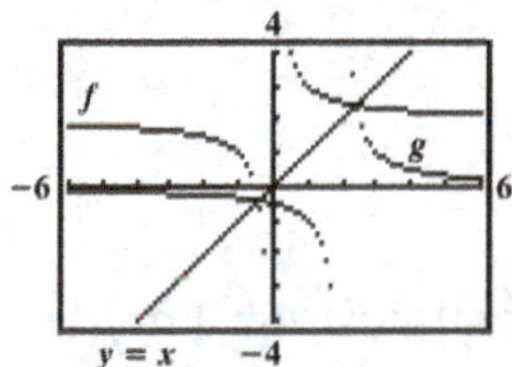

f and g are inverses because they are reflections of each other over the line $y = x$.

95. does not make sense; Explanations will vary. Sample explanation: The diagram illustrates $g(f(x))$.

97. makes sense

99. false; Changes to make the statement true will vary. A sample change is: The inverse is $\{(4,1),(7,2)\}$.

101. true

103. Answers will vary. One example is $f(x) = \sqrt{x+5}$ and $g(x) = 3x^2$.

105. $f\left(f(x)\right)=\dfrac{3\left(f(x)\right)-2}{5\left(f(x)\right)-3}$

$$=\frac{3\left(\frac{3x-2}{5x-3}\right)-2}{5\left(\frac{3x-2}{5x-3}\right)-3}$$

$$=\frac{3\left(\frac{3x-2}{5x-3}\right)-2}{5\left(\frac{3x-2}{5x-3}\right)-3}\cdot\frac{5x-3}{5x-3}$$

$$=\frac{3(3x-2)-2(5x-3)}{5(3x-2)-3(5x-3)}$$

$$=\frac{9x-6-10x+6}{15x-10-15x+9}$$

$$=\frac{-x}{-1}=x$$

Since $f\left(f(x)\right)=x$, *f* is it own inverse.

106. $f(x)=m_1x+b_1$

$g(x)=m_2x+b_2$

First find $(f\circ g)(x)$.

$$(f\circ g)(x)=f\left(g(x)\right)=f\left(m_2x+b_2\right)$$
$$=m_1\left(m_2x+b_2\right)+b_1$$
$$=m_1m_2x+m_1b_2+b_1$$

The slope of the composite function is m_1m_2. The slope of *f* is m_1 and the slope of *g* is m_2, thus the product m_1m_2 is the same as the slope of the composite function.

107. $\dfrac{4.3\times10^5}{8.6\times10^{-4}}=\dfrac{4.3}{8.6}\times\dfrac{10^5}{10^{-4}}=0.5\times10^9$

$$=5\times10^{-1}\times10^9=5\times10^8$$

108. $f(x) = x^2 - 4x + 3$

The x–coordinate of the vertex is $-\frac{b}{2a} = -\frac{-4}{2(1)} = -\frac{-4}{2} = 2$ and the y–coordinate of the vertex of the parabola is

$$f\left(-\frac{b}{2a}\right) = f(2) = 2^2 - 4(2) + 3$$
$$= 4 - 8 + 3 = -1.$$

The x–intercepts are 1 and 3.

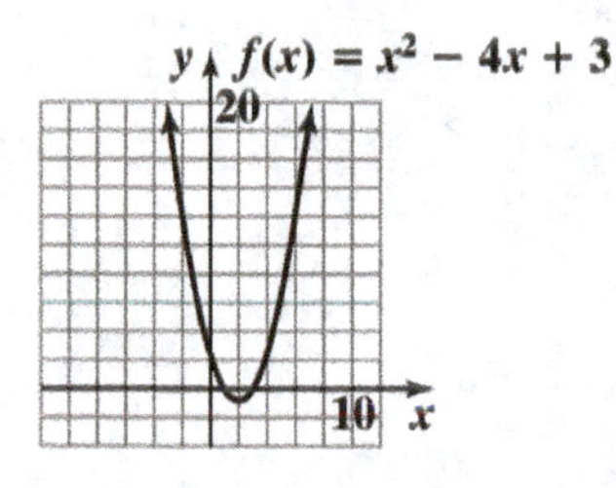

109. $\sqrt{x+4} - \sqrt{x-1} = 1$

$$\sqrt{x+4} = \sqrt{x-1} + 1$$
$$x + 4 = x - 1 + 2\sqrt{x-1} + 1$$
$$x + 4 = x + 2\sqrt{x-1}$$
$$4 = 2\sqrt{x-1}$$
$$2 = \sqrt{x-1}$$
$$4 = x - 1$$
$$5 = x$$

110. There is no method for solving $x = 2^y$ for y.

111. 25 requires a power of $\frac{1}{2}$ to obtain 5.

$$25^{\frac{1}{2}} = 5$$

112. $(x-3)^2 > 0$

Solve the related quadratic equation to find the boundary points.

$$(x-3)^2 = 0$$
$$x - 3 = 0$$
$$x = 3$$

Interval	Test Value	Test	Conclusion
$(-\infty, 3)$	0	$(0-3)^2 > 0$ $9 > 0$, true	$(-\infty, 3)$ is part of the solution set
$(3, \infty)$	5	$(5-3)^2 > 0$ $4 > 0$, true	$(3, \infty)$ is part of the solution set

3 does not satisfy the inequality.

The solution set is $(-\infty, 3) \cup (3, \infty)$ or $\{x \mid x \neq 3\}$.

9.3 Check Points

1. **a.** $3 = \log_7 x$
   $7^3 = x$

   b. $2 = \log_b 25$
   $b^2 = 25$

   c. $\log_4 26 = y$
   $4^y = 26$

2. **a.** $2^5 = x$
   $5 = \log_2 x$

   b. $b^3 = 27$
   $3 = \log_b 27$

   c. $e^y = 33$
   $y = \log_e 33$

3. **a.** $\log_{10} 100 = 2$ because $10^2 = 100$.

   b. $\log_3 3 = 1$ because $3^1 = 3$.

   c. $\log_{36} 6 = \frac{1}{2}$ because $36^{\frac{1}{2}} = \sqrt{36} = 6$.

4. **a.** Because $\log_b b = 1$, we conclude $\log_9 9 = 1$.

   b. Because $\log_b 1 = 0$, we conclude $\log_8 1 = 0$.

5. **a.** Because $\log_b b^x = x$, we conclude $\log_7 7^8 = 8$.

   b. Because $b^{\log_b x} = x$, we conclude $3^{\log_3 17} = 17$.

6. Set up a table of coordinates for $f(x) = 3^x$.

x	-2	-1	0	1	2	3
$f(x) = 3^x$	$\frac{1}{9}$	$\frac{1}{3}$	1	3	9	27

Reverse these coordinates to obtain the coordinates of $g(x) = \log_3 x$.

x	$\frac{1}{9}$	$\frac{1}{3}$	1	3	9	27
$g(x) = \log_3 x$	-2	-1	0	1	2	3

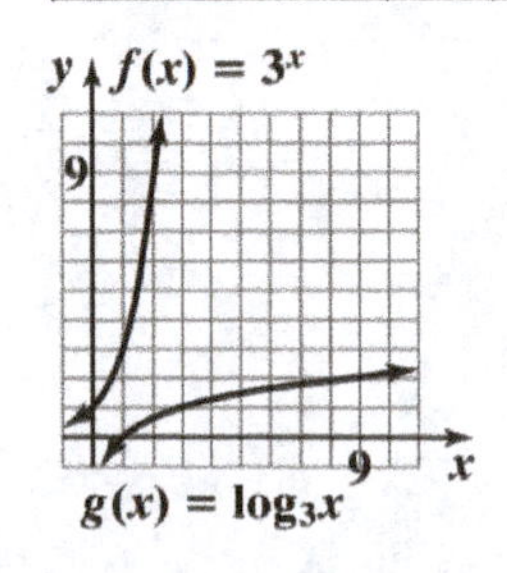

7. $x - 5 > 0$
   $x > 5$
   The domain of h is $(5, \infty)$.

8. $f(x) = 29 + 48.8\log(x+1)$
   $f(10) = 29 + 48.8\log(10+1)$
   $= 29 + 48.8\log 11$
   ≈ 80
   A 10-year-old boy has attained approximately 80% of his adult height.

9. $R = \log\frac{I}{I_0}$
   $R = \log\frac{10,000I_0}{I_0}$
   $= \log 10,000$
   $= \log 10^4$
   $= 4$
   The magnitude on the Richter Scale is 4.

10. **a.** The domain of f consists of all x for which $4 - x > 0$.
    $4 - x > 0$
    $-x > -4$
    $x < 4$
    The domain of f is $(-\infty, 4)$.

    b. The domain of g consists of all x for which $x^2 > 0$. It follows that the domain is all real numbers except 0.
    The domain of g is $(-\infty, 0) \cup (0, \infty)$.

11. $f(x) = 13.4\ln x - 11.6$
    $f(30) = 13.4\ln 30 - 11.6$
    ≈ 34
    The temperature increase after 30 minutes will be 34°. The function models the actual increase shown in the figure extremely well.

9.3 Concept and Vocabulary Check

1. $b^y = x$

2. logarithmic; b

3. 1

4. 0

5. x

6. x

7. $(0,\infty)$; $(-\infty,\infty)$

8. y; $x=0$; vertical

9. $5-x>0$

10. common; $\log x$

11. natural; $\ln x$

9.3 Exercise Set

1. $4=\log_2 16$

$2^4=16$

3. $2=\log_3 x$

$3^2=x$

5. $5=\log_b 32$

$b^5=32$

7. $\log_6 216=y$

$6^y=216$

9. $2^3=8$

$\log_2 8=3$

11. $2^{-4}=\frac{1}{16}$

$\log_2 \frac{1}{16}=-4$

13. $\sqrt[3]{8}=2$

$8^{\frac{1}{3}}=2$

$\log_8 2=\frac{1}{3}$

15. $13^2=x$

$\log_{13} x=2$

17. $b^3=1000$

$\log_b 1000=3$

19. $7^y=200$

$\log_7 200=y$

21. $\log_4 16=y$

$4^y=16$

$4^y=4^2$

$y=2$

23. $\log_2 64=y$

$2^y=64$

$2^y=2^6$

$y=6$

25. $\log_5 \frac{1}{5}=y$

$5^y=\frac{1}{5}$

$5^y=5^{-1}$

$y=-1$

27. $\log_2 \frac{1}{8}=y$

$2^y=\frac{1}{8}$

$2^y=\frac{1}{2^3}$

$2^y=2^{-3}$

$y=-3$

29. $\log_7 \sqrt{7}=y$

$7^y=\sqrt{7}$

$7^y=7^{\frac{1}{2}}$

$y=\frac{1}{2}$

31. $\log_2 \frac{1}{\sqrt{2}}=y$

$2^y=\frac{1}{\sqrt{2}}$

$2^y=\frac{1}{2^{\frac{1}{2}}}$

$2^y=2^{-\frac{1}{2}}$

$y=-\frac{1}{2}$

33. $\log_{64} 8 = y$

$64^y = 8$

$64^y = 64^{\frac{1}{2}}$

$y = \frac{1}{2}$

35. $\log_5 5 = y$

$5^y = 5^1$

$y = 1$

37. $\log_4 1 = y$

$4^y = 1$

$4^y = 4^0$

$y = 0$

39. $\log_5 5^7 = y$

$5^y = 5^7$

$y = 7$

41. Since $b^{\log_b x} = x$, $8^{\log_8 19} = 19$.

43. $f(x) = 4^x$

$g(x) = \log_4 x$

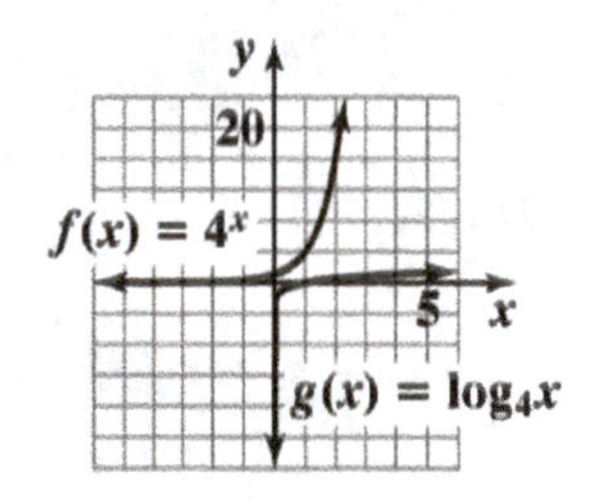

45. $f(x) = \left(\frac{1}{2}\right)^x$

$g(x) = \log_{\frac{1}{2}} x$

$f(x) = \left(\frac{1}{2}\right)^x$

y

10

5

x

$g(x) = \log_{1/2} x$

47. $f(x) = \log_5 (x+4)$

$x + 4 > 0$

$x > -4$

The domain of f is $(-4, \infty)$.

49. $f(x) = \log_5 (2 - x)$

$2 - x > 0$

$-x > -2$

$x < 2$

The domain of f is $(-\infty, 2)$.

51. $f(x) = \ln (x-2)^2$

The domain of g is all real numbers for which $(x-2)^2 > 0$. The only number that must be excluded is 2. The domain of f is $(-\infty, 2) \cup (2, \infty)$.

53. $\log 100 = y$

$10^y = 100$

$10^y = 10^2$

$y = 2$

55. $\log 10^7 = y$

$10^y = 10^7$

$y = 7$

57. Since $10^{\log x} = x$, $10^{\log 33} = 33$.

59. $\ln 1 = y$

$e^y = 1$

$e^y = e^0$

$y = 0$

61. Since $\ln e^x = x$, $\ln e^6 = 6$.

63. $\ln \frac{1}{e^6} = \ln e^{-6}$

Since $\ln e^x = x$, $\ln e^{-6} = -6$.

65. Since $e^{\ln x} = x$, $e^{\ln 125} = 125$.

67. Since $\ln e^x = x$, $\ln e^{9x} = 9x$.

69. Since $e^{\ln x} = x$, $e^{\ln 5x^2} = 5x^2$.

71. Since $10^{\log x} = x$, $10^{\log\sqrt{x}} = \sqrt{x}$.

73.
$$\log_3(x-1) = 2$$
$$3^2 = x-1$$
$$9 = x-1$$
$$10 = x$$

The solution set is $\{10\}$.

75.
$$\log_4 x = -3$$
$$4^{-3} = x$$
$$x = \frac{1}{4^3} = \frac{1}{64}$$

The solution set is $\left\{\frac{1}{64}\right\}$.

77. $\log_3(\log_7 7) = \log_3 1 = 0$

79.
$$\log_2(\log_3 81) = \log_2(\log_3 3^4)$$
$$= \log_2 4 = \log_2 2^2 = 2$$

81. **(d)** The graph is similar to that of $y = \ln x$, but shifted left 2 units.

83. **(c)** The graph is similar to that of $y = \ln x$, but shifted up 2 units.

85. **(b)** The graph is similar to that of $y = \ln x$, but reflected across the *y*-axis and then shifted right 1 unit.

87.
$$f(13) = 62 + 35\log(13-4)$$
$$= 62 + 35\log(9) \approx 95.4$$

A 13-year-old girl is approximately 95.4% of her adult height.

89. **a.** 2008 is 39 years after 1969.
$$f(x) = -7.52\ln x + 53$$
$$f(39) = -7.52\ln 39 + 53$$
$$\approx 25.5$$
According to the function, 25.5% of first-year college men expressed antifeminist views in 2008. This underestimates the value in the graph by 0.7%.

b. 2015 is 46 years after 1969.
$$f(x) = -7.52\ln x + 53$$
$$f(46) = -7.52\ln 46 + 53$$
$$\approx 24.2$$
According to the function, 24.2% of first-year college men will express antifeminist views in 2015.

91.
$$D = 10\log\left(10^{12}\left(6.3\times10^6\right)\right)$$
$$= 10\log\left(6.3\times10^{18}\right) \approx 188.0$$

The decibel level of a blue whale is approximately 188 decibels. At close range, the sound could rupture the human ear drum.

93. **a.** The original exam was at time, $t = 0$.

$$f(0) = 88 - 15\ln(0+1)$$
$$= 88 - 15\ln(1) \approx 88$$

The average score on the original exam was 88.

b.
$$f(2) = 88 - 15\ln(2+1)$$
$$= 88 - 15\ln(3) \approx 71.5$$

$$f(4) = 88 - 15\ln(4+1)$$
$$= 88 - 15\ln(5) \approx 63.9$$

$$f(6) = 88 - 15\ln(6+1)$$
$$= 88 - 15\ln(7) \approx 58.8$$

$$f(8) = 88 - 15\ln(8+1)$$
$$= 88 - 15\ln(9) \approx 55.0$$

$$f(10) = 88 - 15\ln(10+1)$$
$$= 88 - 15\ln(11) \approx 52.0$$

$$f(12) = 88 - 15\ln(12+1)$$
$$= 88 - 15\ln(13) \approx 49.5$$

The average score for the tests is as follows:

2 months:	71.5
4 months:	63.9
6 months:	58.8
8 months:	55.0
10 months:	52.0
12 months:	49.5

c.

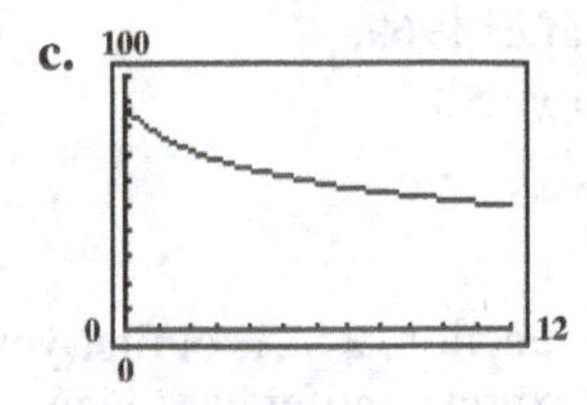

The students remembered less of the material over time.

95. – 101. Answers will vary.

103. $f(x) = \ln x \qquad g(x) = \ln x + 3$

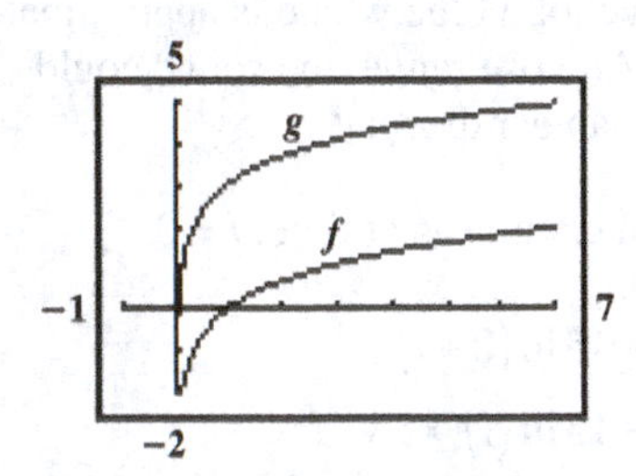

The graph of g is the graph of f shifted up 3 units.

105. $f(x) = \log x \qquad g(x) = \log(x-2)+1$

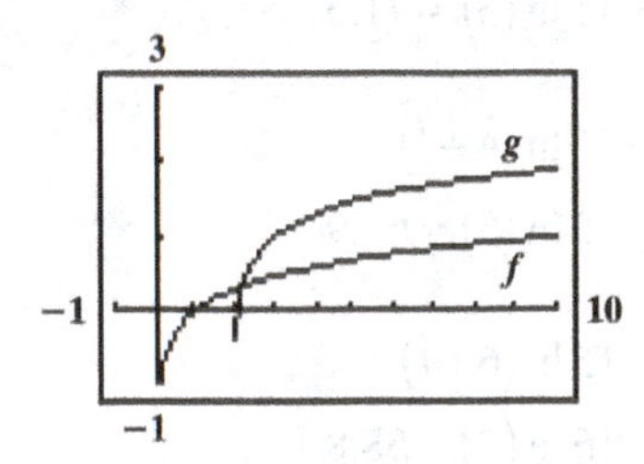

The graph of g is the graph of f shifted 2 units to the right and 1 unit up.

107. a. $f(x) = \ln(3x)$

$g(x) = \ln 3 + \ln x$

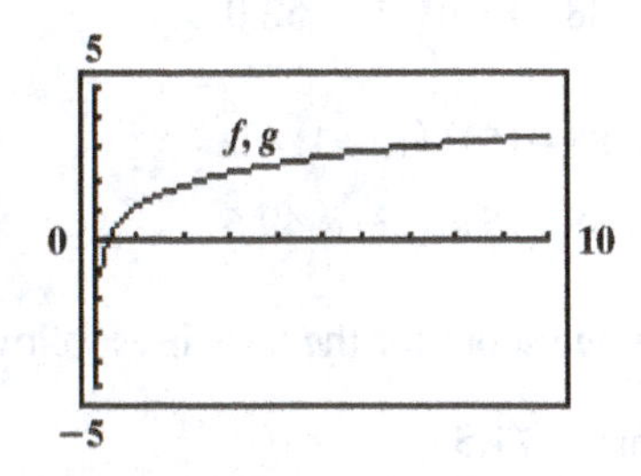

The graphs coincide.

b. $f(x) = \log(5x^2)$

$g(x) = \log 5 + \log x^2$

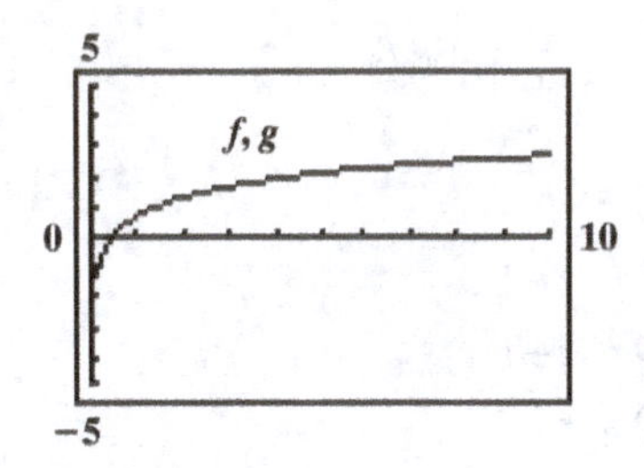

The graphs coincide.

c. $f(x) = \ln(2x^3)$

$g(x) = \ln 2 + \ln x^3$

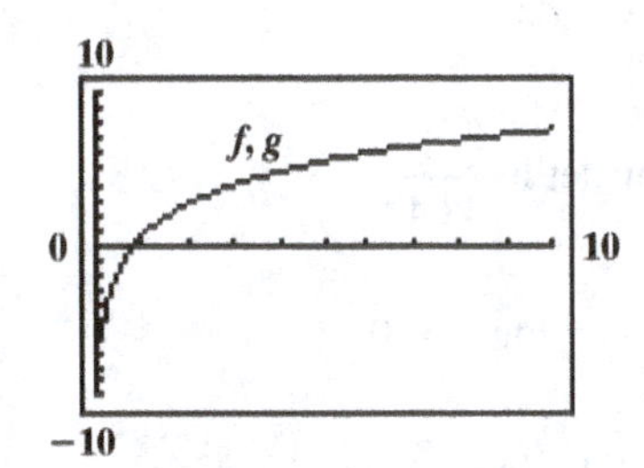

The graphs coincide.

d. In each case, the function, f, is equivalent to g. This means that $\log_b(MN) = \log_b M + \log_b N$.

e. The logarithm of a product is equal to <u>the sum of the logarithms of the factors</u>.

109. makes sense

111. makes sense

113. false; Changes to make the statement true will vary. A sample change is: $\dfrac{\log_2 8}{\log_2 4} = \dfrac{3}{2}$

115. false; Changes to make the statement true will vary. A sample change is: The domain of $f(x) = \log_2 x$ is $(0, \infty)$.

117. To evaluate $\dfrac{\log_3 81 - \log_\pi 1}{\log_{2\sqrt{2}} 8 - \log 0.001}$, consider each of the terms independently.

$$\begin{aligned} \log_3 81 &= y \\ 3^y &= 81 \\ 3^y &= 3^4 \\ y &= 4 \end{aligned} \qquad \begin{aligned} \log_\pi 1 &= y \\ \pi^y &= 1 \\ \pi^y &= \pi^0 \\ y &= 0 \end{aligned}$$

$$\begin{aligned} \log_{2\sqrt{2}} 8 &= y \\ \left(2\sqrt{2}\right)^y &= 8 \\ \left(2^1 2^{\frac{1}{2}}\right)^y &= 2^3 \\ \left(2^{\frac{3}{2}}\right)^y &= 2^3 \\ (2)^{\frac{3}{2}y} &= 2^3 \\ \frac{3}{2}y &= 3 \\ y &= 3 \cdot \frac{2}{3} = 2 \end{aligned} \qquad \begin{aligned} \log 0.001 &= y \\ 10^y &= 0.001 \\ 10^y &= 10^{-3} \\ y &= -3 \end{aligned}$$

$$\frac{\log_3 81 - \log_\pi 1}{\log_{2\sqrt{2}} 8 - \log 0.001} = \frac{4-0}{2-(-3)} = \frac{4}{2+3} = \frac{4}{5}$$

119. To determine which expression represents a greater number, rewrite the expressions in exponential notation.

$$\begin{aligned} \log_4 60 &= x \\ 4^x &= 60 \end{aligned} \qquad \begin{aligned} \log_3 40 &= y \\ 3^y &= 40 \end{aligned}$$

First consider $4^x = 60$. We know that $4^2 = 16$ and $4^3 = 64$, so x falls between 2 and 3 and is much closer to 3. Next consider $3^y = 40$. We know that $3^3 = 27$. This means that y is greater than 3, and therefore greater than x. This means that $\log_3 40$ represents the greater number.

120. Rewrite the equations in $Ax + By = C$ form.

$$\begin{aligned} 2x + 5y &= 11 \\ 3x - 2y &= -12 \end{aligned}$$

Multiply the first equation by 2 and the second equation by 5 and solve by addition.

$$\begin{aligned} 4x + 10y &= 22 \\ 15x - 10y &= -60 \\ \hline 19x &= -38 \\ x &= -2 \end{aligned}$$

Back-substitute -2 for x to find y.

$$\begin{aligned} 2(-2) + 5y &= 11 \\ -4 + 5y &= 11 \\ 5y &= 15 \\ y &= 3 \end{aligned}$$

The solution is $(-2, 3)$ and the solution set is $\{(-2, 3)\}$.

121. $$\begin{aligned} 6x^2 - 8xy + 2y^2 &= 2\left(3x^2 - 4xy + y^2\right) \\ &= 2(3x - y)(x - y) \end{aligned}$$

122. $$\begin{aligned} x + 3 &\le -4 \\ x &\le -7 \end{aligned} \quad \text{or} \quad \begin{aligned} 2 - 7x &\le 16 \\ -7x &\le 14 \\ x &\ge -2 \end{aligned}$$

The solution set is $\{x \mid x \le -7 \text{ or } x \ge -2\}$ or $(-\infty, -7] \cup [-2, \infty)$.

123. **a.** $\log_2 32 = \log_2 2^5 = 5$

b. $\log_2 8 + \log_2 4 = \log_2 2^3 + \log_2 2^2 = 3 + 2 = 5$

c. $\log_2 (8 \cdot 4) = \log_2 8 + \log_2 4$

124. **a.** $\log_2 16 = \log_2 2^4 = 4$

b. $\log_2 32 - \log_2 2 = \log_2 2^5 - \log_2 2 = 5 - 1 = 4$

c. $\log_2 \left(\frac{32}{2}\right) = \log_2 32 - \log_2 2$

125. **a.** $\log_3 81 = \log_3 3^4 = 4$

b. $2\log_3 9 = 2\log_3 3^2 = 2 \cdot 2 = 4$

c. $\log_3 9^2 = 2\log_3 9$

9.4 Check Points

1. **a.** $\log_6(7 \cdot 11) = \log_6 7 + \log_6 11$

b.
$$\begin{aligned}\log(100x) &= \log 100 + \log x \\ &= \log 10^2 + \log x \\ &= 2 + \log x\end{aligned}$$

2. **a.** $\log_8\left(\frac{23}{x}\right) = \log_8 23 - \log_8 x$

b.
$$\begin{aligned}\ln\left(\frac{e^5}{11}\right) &= \ln e^5 - \ln 11 \\ &= 5 - \ln 11\end{aligned}$$

3. **a.** $\log_6 8^9 = 9\log_6 8$

b. $\ln \sqrt[3]{x} = \ln x^{\frac{1}{3}} = \frac{1}{3}\ln x$

c. $\log(x+4)^2 = 2\log(x+4)$

4. **a.**
$$\begin{aligned}\log_b\left(x^4\sqrt[3]{y}\right) &= \log_b x^4 + \log_b \sqrt[3]{y} \\ &= \log_b x^4 + \log_b y^{\frac{1}{3}} \\ &= 4\log_b x + \frac{1}{3}\log_b y\end{aligned}$$

b.
$$\begin{aligned}\log_5\left(\frac{\sqrt{x}}{25y^3}\right) &= \log_5\left(\frac{x^{\frac{1}{2}}}{25y^3}\right) \\ &= \log_5 x^{\frac{1}{2}} - \log_5\left(25y^3\right) \\ &= \log_5 x^{\frac{1}{2}} - \left(\log_5 25 + \log_5 y^3\right) \\ &= \log_5 x^{\frac{1}{2}} - \log_5 25 - \log_5 y^3 \\ &= \frac{1}{2}\log_5 x - 2 - 3\log_5 y\end{aligned}$$

5. a. $\log 25 + \log 4 = \log(25 \cdot 4)$
$= \log 100$
$= 2$

b. $\log(7x+6) - \log x = \log\left(\frac{7x+6}{x}\right)$

6. a. $2\ln x + \frac{1}{3}\ln(x+5) = \ln x^2 + \ln(x+5)^{\frac{1}{3}}$
$= \ln x^2 + \ln \sqrt[3]{x+5}$
$= \ln\left(x^2 \sqrt[3]{x+5}\right)$

b. $2\log(x-3) - \log x = \log(x-3)^2 - \log x$
$= \log \frac{(x-3)^2}{x}$

c. $\frac{1}{4}\log_b x - 2\log_b 5 - 10\log_b y = \log_b x^{\frac{1}{4}} - \log_b 5^2 - \log_b y^{10}$
$= \log_b x^{\frac{1}{4}} - \left(\log_b 5^2 + \log_b y^{10}\right)$
$= \log_b \sqrt[4]{x} - \left(\log_b 25 + \log_b y^{10}\right)$
$= \log_b \sqrt[4]{x} - \left(\log_b 25y^{10}\right)$
$= \log_b\left(\frac{\sqrt[4]{x}}{25y^{10}}\right)$

7. $\log_7 2506 = \frac{\log 2506}{\log 7} \approx 4.02$

8. $\log_7 2506 = \frac{\ln 2506}{\ln 7} \approx 4.02$

9.4 Concept and Vocabulary Check

1. $\log_b M + \log_b N$; sum

2. $\log_b M - \log_b N$; difference

3. $p\log_b M$; product

4. $\frac{\log_a M}{\log_a b}$

9.4 Exercise Set

1. $\log_5(7\cdot 3)=\log_5 7+\log_5 3$

3. $\log_7(7x)=\log_7 7+\log_7 x=1+\log_7 x$

5. $\log(1000x)=\log 1000+\log x$
$=3+\log x$

7. $\log_7\left(\frac{7}{x}\right)=\log_7 7-\log_7 x=1-\log_7 x$

9. $\log\left(\frac{x}{100}\right)=\log x-\log 100=\log x-2$

11. $\log_4\left(\frac{64}{y}\right)=\log_4 64-\log_4 y$
$=3-\log_4 y$

13. $\ln\left(\frac{e^2}{5}\right)=\ln e^2-\ln 5=2-\ln 5$

15. $\log_b x^3=3\log_b x$

17. $\log N^{-6}=-6\log N$

19. $\ln\sqrt[5]{x}=\ln x^{\frac{1}{5}}=\frac{1}{5}\ln x$

21. $\log_b x^2y=\log_b x^2+\log_b y$
$=2\log_b x+\log_b y$

23. $\log_4\left(\frac{\sqrt{x}}{64}\right)=\log_4\sqrt{x}-\log_4 64$
$=\log_4 x^{\frac{1}{2}}-3$
$=\frac{1}{2}\log_4 x\ -3$

25. $\log_6\left(\frac{36}{\sqrt{x+1}}\right)=\log_6 36-\log_6\sqrt{x+1}$
$=2-\log_6(x+1)^{\frac{1}{2}}$
$=2-\frac{1}{2}\log_6(x+1)$

27. $\log_b\left(\frac{x^2y}{z^2}\right)=\log_b x^2y-\log_b z^2$
$=\log_b x^2+\log_b y-2\log_b z$
$=2\log_b x+\log_b y-2\log_b z$

29. $\log\sqrt{100x}=\log(100x)^{\frac{1}{2}}$
$=\frac{1}{2}\log(100x)$
$=\frac{1}{2}(\log 100+\log x)$
$=\frac{1}{2}(2+\log x)$
$=1+\frac{1}{2}\log x$

31. $\log\sqrt[3]{\frac{x}{y}}=\log\left(\frac{x}{y}\right)^{\frac{1}{3}}$
$=\frac{1}{3}\log\left(\frac{x}{y}\right)$
$=\frac{1}{3}(\log x-\log y)$
$=\frac{1}{3}\log x-\frac{1}{3}\log y$

33. $\log_b\left(\frac{\sqrt{x}y^3}{z^3}\right)$
$=\log_b\left(\frac{x^{\frac{1}{2}}y^3}{z^3}\right)$
$=\log_b\left(x^{\frac{1}{2}}y^3\right)-\log_b z^3$
$=\log_b x^{\frac{1}{2}}+\log_b y^3-\log_b z^3$
$=\frac{1}{2}\log_b x+3\log_b y-3\log_b z$

35. $\log_5 \sqrt[3]{\frac{x^2 y}{25}}$
$= \log_5 \left(\frac{x^2 y}{25}\right)^{\frac{1}{3}}$
$= \frac{1}{3}\log_5 \left(\frac{x^2 y}{25}\right)$
$= \frac{1}{3}\left(\log_5 \left[x^2 y\right] - \log_5 25\right)$
$= \frac{1}{3}\left(\log_5 x^2 + \log_5 y - \log_5 5^2\right)$
$= \frac{1}{3}\left(2\log_5 x + \log_5 y - 2\right)$
$= \frac{2}{3}\log_5 x + \frac{1}{3}\log_5 y - \frac{2}{3}$

37. $\log 5 + \log 2 = \log(5 \cdot 2) = \log 10 = 1$

39. $\ln x + \ln 7 = \ln(x \cdot 7) = \ln(7x)$

41. $\log_2 96 - \log_2 3 = \log_2 \frac{96}{3} = \log_2 32 = 5$

43. $\log(2x+5) - \log x = \log\left(\frac{2x+5}{x}\right)$

45. $\log x + 3\log y = \log x + \log y^3$
$= \log\left(xy^3\right)$

47. $\frac{1}{2}\ln x + \ln y = \ln x^{\frac{1}{2}} + \ln y$
$= \ln\left(x^{\frac{1}{2}} y\right) = \ln\left(y\sqrt{x}\right)$

49. $2\log_b x + 3\log_b y = \log_b x^2 + \log_b y^3$
$= \log_b \left(x^2 y^3\right)$

51. $5\ln x - 2\ln y = \ln x^5 - \ln y^2$
$= \ln\left(\frac{x^5}{y^2}\right)$

53. $3\ln x - \frac{1}{3}\ln y = \ln x^3 - \ln y^{\frac{1}{3}}$
$= \ln\left(\frac{x^3}{y^{\frac{1}{3}}}\right) = \ln\left(\frac{x^3}{\sqrt[3]{y}}\right)$

55. $4\ln(x+6) - 3\ln x = \ln(x+6)^4 - \ln x^3$
$= \ln\left[\frac{(x+6)^4}{x^3}\right]$

57. $3\ln x + 5\ln y - 6\ln z$
$= \ln x^3 + \ln y^5 - \ln z^6$
$= \ln\left(x^3 y^5\right) - \ln z^6 = \ln\left(\frac{x^3 y^5}{z^6}\right)$

59. $\frac{1}{2}\left(\log_5 x + \log_5 y\right) - 2\log_5 (x+1)$
$= \frac{1}{2}\log_5 (xy) - \log_5 (x+1)^2$
$= \log_5 (xy)^{\frac{1}{2}} - \log_5 (x+1)^2$
$= \log_5 \sqrt{xy} - \log_5 (x+1)^2$
$= \log_5 \left[\frac{\sqrt{xy}}{(x+1)^2}\right]$

61. $\log_5 13 = \frac{\log 13}{\log 5} \approx 1.5937$

63. $\log_{14} 87.5 = \frac{\log 87.5}{\log 14} \approx 1.6944$

65. $\log_{0.1} 17 = \frac{\log 17}{\log 0.1} \approx -1.2304$

67. $\log_\pi 63 = \frac{\log 63}{\log \pi} \approx 3.6193$

69. $\log_b \frac{3}{2} = \log_b 3 - \log_b 2 = C - A$

71. $\log_b 8 = \log_b 2^3 = 3\log_b 2 = 3A$

73. $\log_b \sqrt{\frac{2}{27}} = \log_b \left(\frac{2}{27}\right)^{\frac{1}{2}}$

$= \frac{1}{2}\log_b \left(\frac{2}{3^3}\right)$

$= \frac{1}{2}\left(\log_b 2 - \log_b 3^3\right)$

$= \frac{1}{2}\left(\log_b 2 - 3\log_b 3\right)$

$= \frac{1}{2}\log_b 2 - \frac{3}{2}\log_b 3$

$= \frac{1}{2}A - \frac{3}{2}C$

75. false; Changes to make the statement true will vary. A sample change is: $\ln e = 1$.

77. false; Changes to make the statement true will vary. A sample change is: $\log_4 (2x)^3 = 3\log_4 (2x)$.

79. true

81. true

83. false; Changes to make the statement true will vary. A sample change is: $\log(x+3) - \log(2x) = \log\left(\frac{x+3}{2x}\right)$.

85. true

87. true

89. **a.** $\log_3 9 = 2$

b. $\log_3 x + 4\log_3 y - 2 = \log_3 x + 4\log_3 y - \log_3 9$

$= \log_3 x + \log_3 y^4 - \log_3 9$

$= \log_3 \frac{xy^4}{9}$

91. **a.** $\log_{25} 5 = \frac{1}{2}$

b. $\log_{25} x + \log_{25}(x^2 - 1) - \log_{25}(x+1) - \frac{1}{2} = \log_{25} x + \log_{25}(x^2 - 1) - \log_{25}(x+1) - \log_{25} 5$

$= \log_{25} \frac{x(x^2 - 1)}{5(x+1)}$

$= \log_{25} \frac{x(x-1)(x+1)}{5(x+1)}$

$= \log_{25} \frac{x(x-1)}{5}$

93. **a.** $D = 10(\log I - \log I_0)$

$= 10\log\frac{I}{I_0}$

b. $D = 10\log\frac{I}{I_0}$

$= 10\log\frac{100I_0}{I_0}$

$= 10\log 100$

$= 10 \cdot 2$

$= 20$

A sound that has an intensity 100 times the intensity of a softer sound, it is 20 decibels louder than the softer sound.

95. – 101. Answers will vary.

103. **a.** $y = \log_3 x = \frac{\log x}{\log 3}$

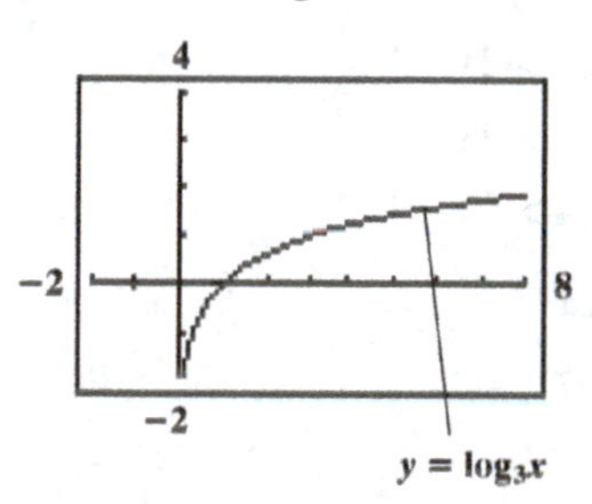

b. $y = 2 + \log_3 x$

$y = \log_3 (x+2)$

$y = -\log_3 x$

$y = \log_3 x$

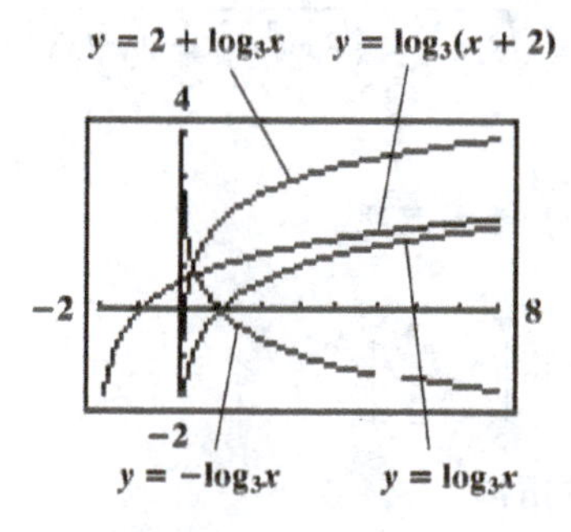

The graph of $y = 2 + \log_3 x$ is the graph of $y = \log_3 x$ shifted up two units.
The graph of $y = \log_3 (x+2)$ is the graph of $y = \log_3 x$ shifted 2 units to the left.
The graph of $y = -\log_3 x$ is the graph of $y = \log_3 x$ reflected about the x–axis.

105. $y = \log_3 x$

$y = \log_{25} x$

$y = \log_{100} x$

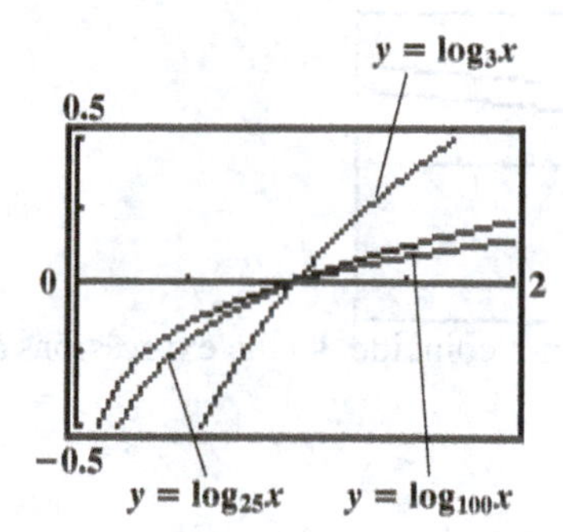

a. $y = \log_{100} x$ is on top. $y = \log_3 x$ is on the bottom.

b. $y = \log_3 x$ is on top. $y = \log_{100} x$ is on the bottom.

c. If $y = \log_b x$ is graphed for two different values of b, the graph of the one with the larger base will be on top in the interval (0, 1) and the one with the smaller base will be on top in the interval $(1, \infty)$. Likewise, if $y = \log_b x$ is graphed for two different values of b, the graph of the one with the smaller base will be on the bottom in the interval (0, 1) and the one with the larger base will be on the bottom in the interval $(1, \infty)$.

107. Answers will vary. One example follows. To disprove the statement $\log\frac{x}{y} = \frac{\log x}{\log y}$, let $y = 3$.

Graph $y = \log\frac{x}{3}$ and $y = \frac{\log x}{\log 3}$.

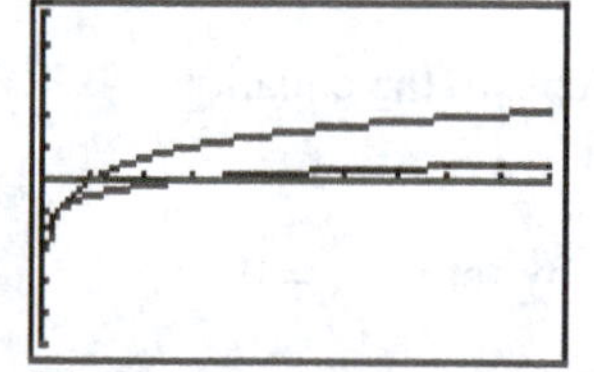

The graphs do not coincide, so the expressions are not equivalent.

109. Answers will vary. One example follows. To disprove the statement $\ln(xy) = (\ln x)(\ln y)$, let $y = 3$.
Graph $y = \ln(x \cdot 3)$ and $y = (\ln x)(\ln 3)$.

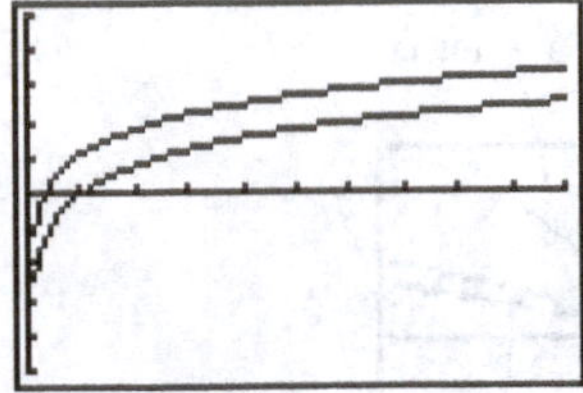

The graphs do not coincide, so the expressions are not equivalent.

111. makes sense

113. makes sense

115. true

117. false; Changes to make the statement true will vary. A sample change is: $\log_b(x^3 + y^3)$ cannot be simplified. If we were taking the logarithm of a product and not a sum, we would have been able to simplify as follows.

$$\log_b(x^3 y^3) = \log_b x^3 + \log_b y^3$$
$$= 3\log_b x + 3\log_b y$$

119. Recall that when a logarithm is written without a base, the base is 10.

$$\log e = \frac{\ln e}{\ln 10} = \frac{1}{\ln 10}$$

121. $e^{\ln 8x^5 - \ln 2x^2} = \frac{e^{\ln 8x^5}}{e^{\ln 2x^2}} = \frac{8x^5}{2x^2} = 4x^3$

122. $5x - 2y > 10$

First, find the intercepts to the equation $5x - 2y = 10$.

Find the x–intercept by setting $y = 0$.

$$5x - 2(0) = 10$$
$$5x = 10$$
$$x = 2$$

Find the y–intercept by setting $x = 0$.

$$5(0) - 2y = 10$$
$$-2y = 10$$
$$y = -5$$

Next, use the origin as a test point.

$$5(0) - 2(0) > 10$$
$$0 - 0 > 10$$
$$0 > 10$$

This is a false statement. This means that the origin will not fall in the shaded half-plane.

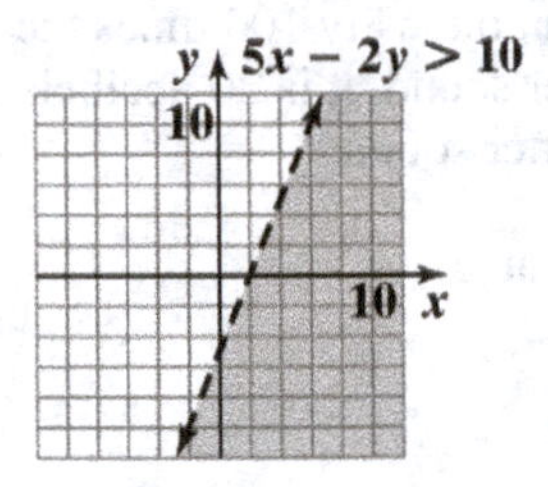

123.
$$x - 2(3x - 2) > 2x - 3$$
$$x - 6x + 4 > 2x - 3$$
$$-5x + 4 > 2x - 3$$
$$-7x + 4 > -3$$
$$-7x > -7$$
$$x < 1$$

The solution set is $(-\infty, 1)$.

124. $\frac{\sqrt[3]{40x^2y^6}}{\sqrt[3]{5xy}} = \sqrt[3]{\frac{40x^2y^6}{5xy}}$
$= \sqrt[3]{8xy^5} = \sqrt[3]{8xy^3y^2} = 2y\sqrt[3]{xy^2}$

125. $16^{\frac{3}{2}} = \left(\sqrt{16}\right)^3 = 4^3 = 64$

126.
$$3\ln(2x) = 3\ln\left(2 \cdot \frac{e^4}{2}\right)$$
$$= 3\ln e^4$$
$$= 3 \cdot 4$$
$$= 12$$

127.
$$\frac{x+2}{4x+3}=\frac{1}{x}$$
$$x(4x+3)\left(\frac{x+2}{4x+3}\right)=x(4x+3)\left(\frac{1}{x}\right)$$
$$x(x+2)=4x+3$$
$$x^2+2x=4x+3$$
$$x^2-2x-3=0$$
$$(x+1)(x-3)=0$$
$$x+1=0 \quad \text{or} \quad x-3=0$$
$$x=-1 \qquad x=3$$
The solution set is $\{-1,3\}$.

Mid-Chapter Check Points – Chapter 9

1. $(f\circ g)(x)=f(g(x))=f(4x-5)$
$$=3(4x-5)+2$$
$$=12x-15+2=12x-13$$
$(g\circ f)(x)=g(f(x))=g(3x+2)$
$$=4(3x+2)-5$$
$$=12x+8-5=12x+3$$
Since $(f\circ g)(x)\neq(g\circ f)(x)\neq x$, we conclude the functions are not inverses.

2. $(f\circ g)(x)=f(g(x))$
$$=f\left(\frac{x^3-5}{7}\right)$$
$$=\sqrt[3]{7\left(\frac{x^3-5}{7}\right)+5}$$
$$=\sqrt[3]{(x^3-5)+5}=\sqrt[3]{x^3}=x$$

$(g\circ f)(x)=g(f(x))$
$$=g\left(\sqrt[3]{7x+5}\right)$$
$$=\frac{\left(\sqrt[3]{7x+5}\right)^3-5}{7}$$
$$=\frac{(7x+5)-5}{7}=\frac{7x}{7}=x$$

The functions are inverses.

3. $(f\circ g)(x)=f(g(x))$
$$=f(5^x)=\log_5 5^x=x$$
$(g\circ f)(x)=g(f(x))$
$$=g(\log_5 x)=5^{\log_5 x}=x$$

The functions are inverses.

4. a. $(f\circ g)(6)=f(g(6))$
$$=f\left(\sqrt{6+3}\right)$$
$$=f\left(\sqrt{9}\right)$$
$$=f(3)=\frac{3-1}{3}=\frac{2}{3}$$

b. $(g\circ f)(-1)=g(f(-1))$
$$=g\left(\frac{-1-1}{-1}\right)$$
$$=g(2)$$
$$=\sqrt{2+3}=\sqrt{5}$$

c. $(f\circ f)(5)=f(f(5))$
$$=f\left(\frac{5-1}{5}\right)$$
$$=f\left(\frac{4}{5}\right)$$
$$=\frac{\frac{4}{5}-1}{\frac{4}{5}}=\frac{-\frac{1}{5}}{\frac{4}{5}}$$
$$=-\frac{1}{5}\cdot\frac{5}{4}=-\frac{1}{4}$$

d. $(g\circ g)(-2)=g(g(-2))$
$$=g\left(\sqrt{-2+3}\right)$$
$$=g\left(\sqrt{1}\right)=g(1)$$
$$=\sqrt{1+3}=\sqrt{4}=2$$

5. $f(x)=\dfrac{2x+5}{4}$

$$y=\frac{2x+5}{4}$$

Interchange x and y and solve for y.

$$x=\frac{2y+5}{4}$$
$$4x=2y+5$$
$$4x-5=2y$$
$$\frac{4x-5}{2}=y$$
$$f^{-1}(x)=\frac{4x-5}{2}$$

6. $f(x)=10x^3-7$

$$y=10x^3-7$$

Interchange x and y and solve for y.

$$x=10y^3-7$$
$$x+7=10y^3$$
$$\frac{x+7}{10}=y^3$$
$$\sqrt[3]{\frac{x+7}{10}}=y$$
$$f^{-1}(x)=\sqrt[3]{\frac{x+7}{10}}$$

7. Interchange the x and y coordinates of each ordered pair in the set: $f^{-1}=\{(5,2),(-7,10),(-10,11)\}$.

8. The graph passes the vertical line test but fails the horizontal line test. Thus, the graph represents a function, but its inverse is not a function.

9. The graph passes both the vertical line test and the horizontal line test. Thus, the graph represents a function, and its inverse is a function as well.

10. The graph fails the vertical line test, so it does not represent a function.

11.

x	$f(x)=2^x-3$
-2	$-\frac{11}{4}=-2.75$
-1	$-\frac{5}{2}=-2.5$
0	-3
1	-1
2	1
3	5

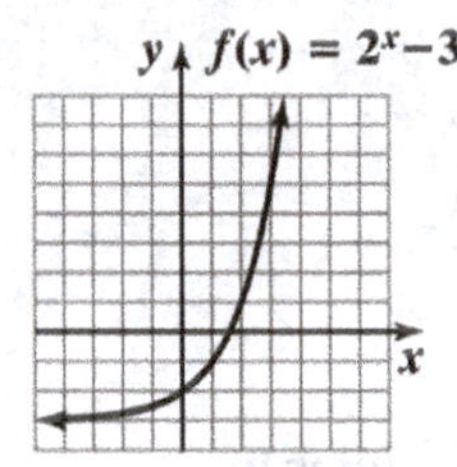

Domain: $(-\infty,\infty)$

Range: $(-3,\infty)$

12.

x	$f(x)=\left(\frac{1}{3}\right)^x$
-2	9
-1	3
0	1
1	$\frac{1}{3}$
2	$\frac{1}{9}$

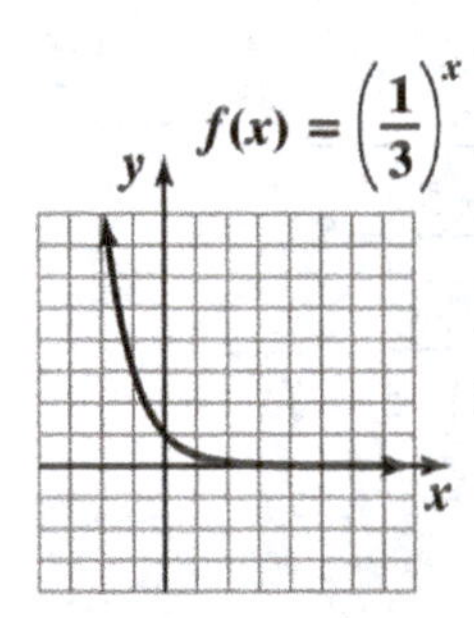

Domain: $(-\infty,\infty)$

Range: $(0,\infty)$

13.

x	$f(x)=\log_2 x$
$\frac{1}{4}$	-2
$\frac{1}{2}$	-1
1	0
2	1
4	2

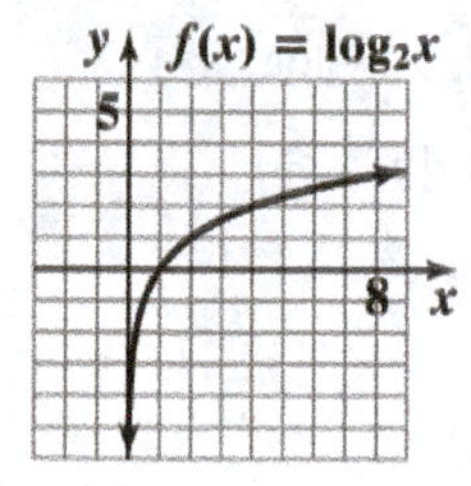

Domain: $(0,\infty)$

Range: $(-\infty,\infty)$

14.

x	$f(x)=\log_2 x+1$
$\frac{1}{4}$	-1
$\frac{1}{2}$	0
1	1
2	2
4	3

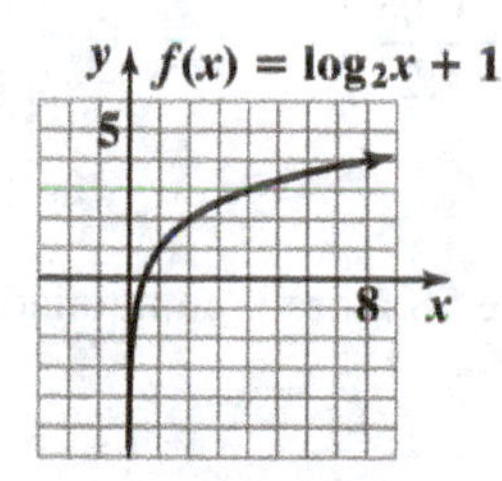

Domain: $(0,\infty)$

Range: $(-\infty,\infty)$

15. $f(x)=\log_3(x+6)$

The argument of the logarithm must be positive:

$x+6>0$

$x>-6$

Domain: $(-6,\infty)$

16. $f(x)=\log_3 x+6$

The argument of the logarithm must be positive:

$x>0$

Domain: $(0,\infty)$

17. $\log_3(x+6)^2$

The argument of the logarithm must be positive.

Now $(x+6)^2$ is always positive, except when $x=-6$.

Domain: $(-\infty-6)\cup(-6,\infty)$

18. $f(x)=3^{x+6}$

Domain: $(-\infty,\infty)$

19. $\log_2 8+\log_5 25=\log_2 2^3+\log_5 5^2$

$=3+2=5$

20. $\log_3\frac{1}{9}=\log_3\frac{1}{3^2}=\log_3 3^{-2}=-2$

21. Let $\log_{100}10=y$

$$100^y=10$$
$$\left(10^2\right)^y=10^1$$
$$10^{2y}=10^1$$
$$2y=1$$
$$y=\frac{1}{2}$$

22. $\log\sqrt[3]{10}=\log 10^{\frac{1}{3}}=\frac{1}{3}$

23. $\log_2(\log_3 81)=\log_2\left(\log_3 3^4\right)$

$=\log_2 4=\log_2 2^2=2$

24. $\log_3\left(\log_2 \frac{1}{8}\right) = \log_3\left(\log_2 \frac{1}{2^3}\right)$
$= \log_3\left(\log_2 2^{-3}\right)$
$= \log_3(-3)$
$=$ not possible
This expression is impossible to evaluate.

25. $6^{\log_6 5} = 5$

26. $\ln e^{\sqrt{7}} = \sqrt{7}$

27. $10^{\log 13} = 13$

28. $\log_{100} 0.1 = y$
$100^y = 0.1$
$\left(10^2\right)^y = \frac{1}{10}$
$10^{2y} = 10^{-1}$
$2y = -1$
$y = -\frac{1}{2}$

29. $\log_\pi \pi^{\sqrt{\pi}} = \sqrt{\pi}$

30. $\log\left(\frac{\sqrt{xy}}{1000}\right) = \log\left(\sqrt{xy}\right) - \log 1000$
$= \log(xy)^{\frac{1}{2}} - \log 10^3$
$= \frac{1}{2}\log(xy) - 3$
$= \frac{1}{2}(\log x + \log y) - 3$
$= \frac{1}{2}\log x + \frac{1}{2}\log y - 3$

31. $\ln\left(e^{19}x^{20}\right) = \ln e^{19} + \ln x^{20}$
$= 19 + 20\ln x$

32. $8\log_7 x - \frac{1}{3}\log_7 y = \log_7 x^8 - \log_7 y^{\frac{1}{3}}$
$= \log_7\left(\frac{x^8}{y^{\frac{1}{3}}}\right)$
$= \log_7\left(\frac{x^8}{\sqrt[3]{y}}\right)$

33. $7\log_5 x + 2\log_5 x = \log_5 x^7 + \log_5 x^2$
$= \log_5\left(x^7 \cdot x^2\right)$
$= \log_5 x^9$

34. $\frac{1}{2}\ln x - 3\ln y - \ln(z-2)$
$= \ln x^{\frac{1}{2}} - \ln y^3 - \ln(z-2)$
$= \ln\sqrt{x} - \left[\ln y^3 + \ln(z-2)\right]$
$= \ln\sqrt{x} - \ln\left[y^3(z-2)\right]$
$= \ln\left[\frac{\sqrt{x}}{y^3(z-2)}\right]$

35. Continuously: $A = 8000e^{0.08(3)}$
$\approx 10{,}170$

Monthly: $A = 8000\left(1+\frac{0.08}{12}\right)^{12\cdot 3}$
$\approx 10{,}162$
$10{,}170 - 10{,}162 = 8$
Interest returned will be \$8 more if compounded continuously.

9.5 Check Points

1. a. $5^{3x-6} = 125$

$$5^{3x-6} = 5^3$$
$$3x-6=3$$
$$3x=9$$
$$x=3$$

The solution set is $\{3\}$.

b. $4^x = 32$

$$\left(2^2\right)^x = 2^5$$
$$2^{2x} = 2^5$$
$$2x = 5$$
$$x = \frac{5}{2}$$

The solution set is $\left\{\frac{5}{2}\right\}$.

2. a. Take the natural log of both sides of the equation.

$$5^x = 134$$
$$\ln 5^x = \ln 134$$
$$x \ln 5 = \ln 134$$
$$x = \frac{\ln 134}{\ln 5}$$
$$x \approx 3.04$$

The solution set is $\left\{\frac{\ln 134}{\ln 5} \approx 3.04\right\}$.

b. Take the common log of both sides of the equation.

$$10^x = 8000$$
$$\log 10^x = \log 8000$$
$$x = \log 8000$$
$$x \approx 3.90$$

The solution set is $\{\log 8000 \approx 3.90\}$.

3. Isolate the exponential expression then take the natural log of both sides of the equation.

$$7e^{2x} - 5 = 58$$
$$7e^{2x} = 63$$
$$e^{2x} = 9$$
$$\ln e^{2x} = \ln 9$$
$$2x = \ln 9$$
$$x = \frac{\ln 9}{2}$$
$$x = \frac{\ln 3^2}{2}$$
$$x = \frac{2 \ln 3}{2}$$
$$x = \ln 3$$
$$x \approx 1.10$$

The solution set is $\{\ln 3 \approx 1.10\}$.

4. a. $\log_2 (x-4) = 3$

$$\log_2 (x-4) = 3$$
$$2^3 = x-4$$
$$8 = x-4$$
$$12 = x$$

12 checks. The solution set is $\{12\}$.

b. $4 \ln(3x) = 8$

$$\ln(3x) = 2$$
$$e^2 = 3x$$
$$\frac{e^2}{3} = x$$

$\frac{e^2}{3}$ checks. The solution set is $\left\{\frac{e^2}{3}\right\}$.

5. $\log x + \log(x-3) = 1$

$$\log(x^2 - 3x) = 1$$
$$10^1 = x^2 - 3x$$
$$0 = x^2 - 3x - 10$$
$$0 = (x+2)(x-5)$$

$x+2=0$ or $x-5=0$

$x=-2$ $\qquad$ $x=5$

The number -2 does not check. The solution set is $\{5\}$.

6. $\ln(x-3) = \ln(7x-23) - \ln(x+1)$

$$\ln(x-3) = \ln\left(\frac{7x-23}{x+1}\right)$$

$$x-3 = \frac{7x-23}{x+1}$$

$$(x+1)(x-3) = (x+1)\frac{7x-23}{x+1}$$

$$x^2 - 2x - 3 = 7x - 23$$

$$x^2 - 9x + 20 = 0$$

$$(x-4)(x-5) = 0$$

$$x-4=0 \quad \text{or} \quad x-5=0$$

$$x=4 \qquad\qquad x=5$$

Both numbers check. The solution set is $\{4, 5\}$.

7. $R = 6e^{12.77x}$

$$6e^{12.77x} = 7$$

$$e^{12.77x} = \frac{7}{6}$$

$$\ln e^{12.77x} = \ln\frac{7}{6}$$

$$12.77x = \ln\frac{7}{6}$$

$$x = \frac{\ln\frac{7}{6}}{12.77}$$

$$x \approx 0.01$$

For a blood alcohol concentration of 0.01, the risk of a car accident is 7%.

8. $A = P\left(1+\frac{r}{n}\right)^{nt}$

$$3600 = 1000\left(1+\frac{0.08}{4}\right)^{4t}$$

$$3.6 = 1.02^{4t}$$

$$1.02^{4t} = 3.6$$

$$\ln 1.02^{4t} = \ln 3.6$$

$$4t \ln 1.02 = \ln 3.6$$

$$t = \frac{\ln 3.6}{4 \ln 1.02}$$

$$t \approx 16.2$$

After approximately 16.2 years, the $1000 will grow to $3600.

9. $g(x) = -7\ln x + 59$

$$40 = -7\ln x + 59$$

$$-19 = -7\ln x$$

$$\frac{-19}{-7} = \frac{-7\ln x}{-7}$$

$$\frac{19}{7} = \ln x$$

$$x = e^{\frac{19}{7}}$$

$$x \approx 15$$

Approximately 15 years after 1979, in the year 1994, 40% of first-year college men opposed homosexual relationships.

9.5 Concept and Vocabulary Check

1. $M = N$
2. $4x - 1$
3. $\dfrac{\ln 20}{\ln 9}$
4. $\ln 6$
5. 5^3
6. $(x^2 + x)$
7. $\dfrac{7x-23}{x+1}$
8. false
9. true
10. false

9.5 Exercise Set

1. $2^x = 64$

$$2^x = 2^6$$

$$x = 6$$

The solution set is $\{6\}$.

3. $5^x = 125$

$5^x = 5^3$

$x = 3$

The solution set is $\{3\}$.

5. $2^{2x-1} = 32$

$2^{2x-1} = 2^5$

$2x - 1 = 5$

$2x = 6$

$x = 3$

The solution set is $\{3\}$.

7. $4^{2x-1} = 64$

$4^{2x-1} = 4^3$

$2x - 1 = 3$

$2x = 4$

$x = 2$

The solution set is $\{2\}$.

9. $32^x = 8$

$(2^5)^x = 2^3$

$2^{5x} = 2^3$

$5x = 3$

$x = \frac{3}{5}$

The solution set is $\left\{\frac{3}{5}\right\}$.

11. $9^x = 27$

$(3^2)^x = 3^3$

$3^{2x} = 3^3$

$2x = 3$

$x = \frac{3}{2}$

The solution set is $\left\{\frac{3}{2}\right\}$.

13. $3^{1-x} = \frac{1}{27}$

$3^{1-x} = \frac{1}{3^3}$

$3^{1-x} = 3^{-3}$

$1 - x = -3$

$-x = -4$

$x = 4$

The solution set is $\{4\}$.

15. $6^{\frac{x-3}{4}} = \sqrt{6}$

$6^{\frac{x-3}{4}} = 6^{\frac{1}{2}}$

$\frac{x-3}{4} = \frac{1}{2}$

$2(x-3) = 4(1)$

$2x - 6 = 4$

$2x = 10$

$x = 5$

The solution set is $\{5\}$.

17. $4^x = \frac{1}{\sqrt{2}}$

$(2^2)^x = \frac{1}{2^{\frac{1}{2}}}$

$2^{2x} = 2^{-\frac{1}{2}}$

$2x = -\frac{1}{2}$

$x = \frac{1}{2}\left(-\frac{1}{2}\right) = -\frac{1}{4}$

The solution set is $\left\{-\frac{1}{4}\right\}$.

19. $e^x = 5.7$

$\ln e^x = \ln 5.7$

$x = \ln 5.7 \approx 1.74$

The solution set is $\{\ln 5.7 \approx 1.74\}$.

21.
$$10^x = 3.91$$
$$\log 10^x = \log 3.91$$
$$x = \log 3.91$$
$$x \approx 0.59$$

The solution set is $\{\log 3.91 \approx 0.59\}$.

23.
$$5^x = 17$$
$$\ln 5^x = \ln 17$$
$$x \ln 5 = \ln 17$$
$$x = \frac{\ln 17}{\ln 5} \approx 1.76$$

The solution set is $\left\{\frac{\ln 17}{\ln 5} \approx 1.76\right\}$.

25.
$$5e^x = 25$$
$$e^x = 5$$
$$\ln e^x = \ln 5$$
$$x = \ln 5 \approx 1.61$$

The solution set is $\{\ln 5 \approx 1.61\}$.

27.
$$3e^{5x} = 1977$$
$$e^{5x} = 659$$
$$\ln e^{5x} = \ln 659$$
$$5x = \ln 659$$
$$x = \frac{\ln 659}{5} \approx 1.30$$

The solution set is $\left\{\frac{\ln 659}{5} \approx 1.30\right\}$.

29.
$$e^{0.7x} = 13$$
$$\ln e^{0.7x} = \ln 13$$
$$0.7x = \ln 13$$
$$x = \frac{\ln 13}{0.7} \approx 3.66$$

The solution set is $\left\{\frac{\ln 13}{0.7} \approx 3.66\right\}$.

31.
$$1250e^{0.055x} = 3750$$
$$e^{0.055x} = 3$$
$$\ln e^{0.055x} = \ln 3$$
$$0.055x = \ln 3$$
$$x = \frac{\ln 3}{0.055} \approx 19.97$$

The solution set is $\left\{\frac{\ln 3}{0.055} \approx 19.97\right\}$.

33.
$$30 - (1.4)^x = 0$$
$$-1.4^x = -30$$
$$1.4^x = 30$$
$$\ln 1.4^x = \ln 30$$
$$x \ln 1.4 = \ln 30$$
$$x = \frac{\ln 30}{\ln 1.4} \approx 10.11$$

The solution set is $\left\{\frac{\ln 30}{\ln 1.4} \approx 10.11\right\}$.

35.
$$e^{1-5x} = 793$$
$$\ln e^{1-5x} = \ln 793$$
$$1 - 5x = \ln 793$$
$$-5x = \ln 793 - 1$$
$$x = \frac{-(\ln 793 - 1)}{5}$$
$$x = \frac{1 - \ln 793}{5} \approx -1.14$$

The solution set is $\left\{\frac{1 - \ln 793}{5} \approx -1.14\right\}$.

37.
$$7^{x+2} = 410$$
$$\ln 7^{x+2} = \ln 410$$
$$(x+2)\ln 7 = \ln 410$$
$$x + 2 = \frac{\ln 410}{\ln 7}$$
$$x = \frac{\ln 410}{\ln 7} - 2 \approx 1.09$$

The solution set is $\left\{\frac{\ln 410}{\ln 7} - 2 \approx 1.09\right\}$.

39.
$$2^{x+1}=5^x$$
$$\ln 2^{x+1}=\ln 5^x$$
$$(x+1)\ln 2=x\ln 5$$
$$x\ln 2+\ln 2=x\ln 5$$
$$x\ln 2=x\ln 5-\ln 2$$
$$x\ln 2-x\ln 5=-\ln 2$$
$$x(\ln 2-\ln 5)=-\ln 2$$
$$x=\frac{-\ln 2}{\ln 2-\ln 5}$$
$$x=\frac{\ln 2}{\ln 5-\ln 2}\approx 0.76$$

The solution set is $\left\{\frac{\ln 2}{\ln 5-\ln 2}\approx 0.76\right\}$.

41. $\log_3 x=4$
$$x=3^4$$
$$x=81$$

The solution set is $\{81\}$.

43. $\log_2 x=-4$
$$x=2^{-4}$$
$$x=\frac{1}{2^4}=\frac{1}{16}$$

The solution set is $\left\{\frac{1}{16}\right\}$.

45. $\log_9 x=\frac{1}{2}$
$$x=9^{\frac{1}{2}}$$
$$x=\sqrt{9}=3$$

The solution set is $\{3\}$.

47. $\log x=2$
$$x=10^2$$
$$x=100$$

The solution set is $\{100\}$.

49. $\log_4(x+5)=3$
$$x+5=4^3$$
$$x+5=64$$
$$x=59$$

The solution set is $\{59\}$.

51. $\log_3(x-4)=-3$
$$x-4=3^{-3}$$
$$x-4=\frac{1}{3^3}$$
$$x-4=\frac{1}{27}$$
$$x=\frac{1}{27}+4$$
$$x=\frac{1}{27}+\frac{108}{27}=\frac{109}{27}$$

The solution set is $\left\{\frac{109}{27}\right\}$.

53. $\log_4(3x+2)=3$
$$3x+2=4^3$$
$$3x+2=64$$
$$3x=62$$
$$x=\frac{62}{3}$$

The solution set is $\left\{\frac{62}{3}\right\}$.

55. $\ln x=2$
$$e^{\ln x}=e^2$$
$$x=e^2\approx 7.39$$

The solution set is $\left\{e^2\approx 7.39\right\}$.

57. $\ln x=-3$
$$x=e^{-3}=\frac{1}{e^3}$$

The solution set is $\left\{e^{-3}=\frac{1}{e^3}\approx 0.05\right\}$.

59. $5\ln(2x)=20$
$$\ln(2x)=4$$
$$e^{\ln(2x)}=e^4$$
$$2x=e^4$$
$$x=\frac{e^4}{2}\approx 27.30$$

The solution set is $\left\{\frac{e^4}{2}\approx 27.30\right\}$.

61. $6+2\ln x=5$

$$2\ln x=-1$$

$$e^{\ln x}=e^{-\frac{1}{2}}$$

$$x=e^{-\frac{1}{2}}\approx 0.61$$

The solution set is $\left\{e^{-\frac{1}{2}}\approx 0.61\right\}$.

63. $\ln\sqrt{x+3}=1$

$$\ln(x+3)^{\frac{1}{2}}=1$$

$$\frac{1}{2}\ln(x+3)=1$$

$$\ln(x+3)=2$$

$$e^{\ln(x+3)}=e^2$$

$$x+3=e^2$$

$$x=e^2-3\approx 4.39$$

The solution set is $\{e^2-3\approx 4.39\}$.

65. $\log_5 x+\log_5(4x-1)=1$

$$\log_5(x(4x-1))=1$$

$$x(4x-1)=5^1$$

$$4x^2-x=5$$

$$4x^2-x-5=0$$

$$(4x-5)(x+1)=0$$

$$4x-5=0 \quad\text{and}\quad x+1=0$$

$$4x=5 \qquad x=-1$$

$$x=\frac{5}{4}$$

We disregard -1 because it would result in taking the logarithm of a negative number in the original equation. The solution set is $\left\{\frac{5}{4}\right\}$.

67. $\log_3(x-5)+\log_3(x+3)=2$

$$\log_3((x-5)(x+3))=2$$

$$(x-5)(x+3)=3^2$$

$$x^2-2x-15=9$$

$$x^2-2x-24=0$$

$$(x-6)(x+4)=0$$

$$x-6=0 \quad\text{and}\quad x+4=0$$

$$x=6 \qquad x=-4$$

We disregard -4 because it would result in taking the logarithm of a negative number in the original equation. The solution set is $\{6\}$.

69. $\log_2(x+2)-\log_2(x-5)=3$

$$\log_2\frac{x+2}{x-5}=3$$

$$\frac{x+2}{x-5}=2^3$$

$$\frac{x+2}{x-5}=8$$

$$x+2=8(x-5)$$

$$x+2=8x-40$$

$$-7x+2=-40$$

$$-7x=-42$$

$$x=6$$

The solution set is $\{6\}$.

71. $\log(3x-5)-\log(5x)=2$

$$\log\frac{3x-5}{5x}=2$$

$$\frac{3x-5}{5x}=10^2$$

$$\frac{3x-5}{5x}=100$$

$$3x-5=500x$$

$$-5=497x$$

$$-\frac{5}{497}=x$$

We disregard $-\frac{5}{497}$ because it would result in taking the logarithm of a negative number in the original equation. Therefore, the equation has no solution. The solution set is $\varnothing$ or $\{\ \}$.

73. $\ln(x+1)-\ln x=1$

$$\ln\left(\frac{x+1}{x}\right)=1$$
$$\frac{x+1}{x}=e^1$$
$$x+1=ex$$
$$1=ex-x$$
$$1=(e-1)x$$
$$x=\frac{1}{e-1}\approx 0.58$$

The solution set is $\left\{\frac{1}{e-1}\approx 0.58\right\}$.

75. $\log_3(x+4)=\log_3 7$

$$x+4=7$$
$$x=3$$

The solution set is $\{3\}$.

77. $\log(x+4)=\log x+\log 4$

$$\log(x+4)=\log 4x$$
$$x+4=4x$$
$$4=3x$$
$$x=\frac{4}{3}$$

The solution set is $\left\{\frac{4}{3}\right\}$.

79. $\log(3x-3)=\log(x+1)+\log 4$

$$\log(3x-3)=\log(4x+4)$$
$$3x-3=4x+4$$
$$-7=x$$

This value is rejected. The solution set is $\{\ \}$.

81. $2\log x=\log 25$

$$\log x^2=\log 25$$
$$x^2=25$$
$$x=\pm 5$$

-5 is rejected. The solution set is $\{5\}$.

83. $\log(x+4)-\log 2=\log(5x+1)$

$$\log\frac{x+4}{2}=\log(5x+1)$$
$$\frac{x+4}{2}=5x+1$$
$$x+4=10x+2$$
$$-9x=-2$$
$$x=\frac{2}{9}$$

The solution set is $\left\{\frac{2}{9}\right\}$.

85. $2\log x-\log 7=\log 112$

$$\log x^2-\log 7=\log 112$$
$$\log\frac{x^2}{7}=\log 112$$
$$\frac{x^2}{7}=112$$
$$x^2=784$$
$$x=\pm 28$$

-28 is rejected. The solution set is $\{28\}$.

87. $\log x+\log(x+3)=\log 10$

$$\log(x^2+3x)=\log 10$$
$$x^2+3x=10$$
$$x^2+3x-10=0$$
$$(x+5)(x-2)=0$$
$$x=-5 \text{ or } x=2$$

-5 is rejected. The solution set is $\{2\}$.

89. $\ln(x-4)+\ln(x+1)=\ln(x-8)$

$$\ln(x^2-3x-4)=\ln(x-8)$$
$$x^2-3x-4=x-8$$
$$x^2-4x+4=0$$
$$(x-2)(x-2)=0$$
$$x=2$$

2 is rejected. The solution set is $\{\ \}$.

91.
$$5^{2x} \cdot 5^{4x} = 125$$
$$5^{2x+4x} = 5^3$$
$$5^{6x} = 5^3$$
$$6x = 3$$
$$x = \frac{1}{2}$$

The solution set is $\left\{\frac{1}{2}\right\}$.

93.
$$3^{x^2} = 45$$
$$\ln 3^{x^2} = \ln 45$$
$$x^2 \ln 3 = \ln 45$$
$$x^2 = \frac{\ln 45}{\ln 3}$$
$$x = \pm\sqrt{\frac{\ln 45}{\ln 3}} \approx \pm 1.86$$

The solution set is $\left\{\pm\sqrt{\frac{\ln 45}{\ln 3}} \approx \pm 1.86\right\}$.

95.
$$\log_2(x-6) + \log_2(x-4) - \log_2 x = 2$$
$$\log_2\left[(x-6)(x-4)\right] - \log_2 x = 2$$
$$\log_2\left[\frac{(x-6)(x-4)}{x}\right] = 2$$
$$\log_2\left(\frac{x^2-10x+24}{x}\right) = 2$$
$$\frac{x^2-10x+24}{x} = 2^2$$
$$\frac{x^2-10x+24}{x} = 4$$
$$x^2-10x+24 = 4x$$
$$x^2-14x+24 = 0$$
$$(x-12)(x-2) = 0$$

Apply the zero-product property:

$x-12=0$ or $x-2=0$

$x=12$ $\quad$ $x=2$

We disregard 2 because it would result in taking the logarithm of a negative number in the original equation. The solution set is $\{12\}$.

97.
$$5^{x^2-12} = 25^{2x}$$
$$5^{x^2-12} = \left(5^2\right)^{2x}$$
$$5^{x^2-12} = 5^{4x}$$
$$x^2 - 12 = 4x$$
$$x^2 - 4x - 12 = 0$$
$$(x-6)(x+2) = 0$$

$x-6=0$ or $x+2=0$

$x=6$ $\quad$ $x=-2$

The solution set is $\{-2,\ 6\}$.

99. a. 2005 is 0 years after 2005.

$$A = 36.1e^{0.0126t}$$
$$A = 36.1e^{0.0126(0)} = 36.1$$

The population of California was 36.1 million in 2005.

b.
$$A = 36.1e^{0.0126t}$$
$$40 = 36.1e^{0.0126t}$$
$$\frac{40}{36.1} = e^{0.0126t}$$
$$\ln\frac{40}{36.1} = \ln e^{0.0126t}$$
$$0.0126t = \ln\frac{40}{36.1}$$
$$t = \frac{\ln\frac{40}{36.1}}{0.0126} \approx 8$$

The population of California will reach 40 million about 8 years after 2005, or 2013.

101.
$$f(x) = 20(0.975)^x$$
$$1 = 20(0.975)^x$$
$$\frac{1}{20} = 0.975^x$$
$$\ln\frac{1}{20} = \ln 0.975^x$$
$$\ln\frac{1}{20} = x\ln 0.975$$
$$x = \frac{\ln\frac{1}{20}}{\ln 0.975}$$
$$x \approx 118$$

There is 1% of surface sunlight at 118 feet. This is represented by the point (118,1).

103.

$$20000 = 12500\left(1+\frac{0.0575}{4}\right)^{4t}$$
$$20000 = 12500(1+0.014375)^{4t}$$
$$20000 = 12500(1.014375)^{4t}$$
$$\frac{20000}{12500} = (1.014375)^{4t}$$
$$1.6 = (1.014375)^{4t}$$
$$\ln 1.6 = \ln(1.014375)^{4t}$$
$$\ln 1.6 = 4t\ln 1.014375$$
$$\frac{4t\ln 1.014375}{4\ln 1.014375} = \frac{\ln 1.6}{4\ln 1.014375}$$
$$t = \frac{\ln 1.6}{4\ln 1.014375} \approx 8.2$$

It will take approximately 8.2 years.

105.

$$1400 = 1000\left(1+\frac{r}{360}\right)^{360(2)}$$
$$\frac{1400}{1000} = \left(1+\frac{r}{360}\right)^{720}$$
$$1.4 = \left(1+\frac{r}{360}\right)^{720}$$
$$\ln 1.4 = \ln\left(1+\frac{r}{360}\right)^{720}$$
$$\ln 1.4 = 720\ln\left(1+\frac{r}{360}\right)$$
$$\frac{\ln 1.4}{720} = \ln\left(1+\frac{r}{360}\right)$$
$$e^{\frac{\ln 1.4}{720}} = e^{\ln\left(1+\frac{r}{360}\right)}$$
$$e^{\frac{\ln 1.4}{720}} = 1+\frac{r}{360}$$
$$1+\frac{r}{360} = e^{\frac{\ln 1.4}{720}}$$
$$\frac{r}{360} = e^{\frac{\ln 1.4}{720}} - 1$$
$$r = 360\left(e^{\frac{\ln 1.4}{720}} - 1\right) \approx 0.168$$

The annual interest rate is approximately 16.8%.

107.

$$16000 = 8000e^{0.08t}$$
$$\frac{16000}{8000} = e^{0.08t}$$
$$2 = e^{0.08t}$$
$$\ln 2 = \ln e^{0.08t}$$
$$\ln 2 = 0.08t$$
$$t = \frac{\ln 2}{0.08} \approx 8.7$$

It will take approximately 8.7 years to double the money.

109.

$$7050 = 2350e^{r7}$$
$$\frac{7050}{2350} = e^{7r}$$
$$3 = e^{7r}$$
$$\ln 3 = \ln e^{7r}$$
$$\ln 3 = 7r$$
$$r = \frac{\ln 3}{7} \approx 0.157$$

The annual interest rate would have to be 15.7% to triple the money.

111. a. 2009 is 3 years after 2006.

$$f(x) = 1.2\ln x + 15.7$$
$$f(3) = 1.2\ln 3 + 15.7 \approx 17.0$$

According to the function, 17.0% of the of the U.S. gross domestic product went toward healthcare in 2009. This underestimates the value shown in the graph by 0.3%.

b.

$$f(x) = 1.2\ln x + 15.7$$
$$18.5 = 1.2\ln x + 15.7$$
$$2.8 = 1.2\ln x$$
$$\frac{2.8}{1.2} = \frac{1.2\ln x}{1.2}$$
$$\frac{2.8}{1.2} = \ln x$$
$$x = e^{\frac{2.8}{1.2}}$$
$$x \approx 10$$

If the trend continues, 18.5% of the U.S. gross domestic product will go toward healthcare 10 years after 2006, or 2016.

113.
$$50 = 95 - 30\log_2 x$$
$$-45 = -30\log_2 x$$
$$\frac{-45}{-30} = \log_2 x$$
$$\log_2 x = \frac{3}{2}$$
$$x = 2^{\frac{3}{2}} \approx 2.8$$

After approximately 2.8 days, only half the students recall the important features of the lecture. This is represented by the point (2.8, 50).

115. a.
$$\text{pH} = -\log x$$
$$5.6 = -\log x$$
$$-5.6 = \log x$$
$$x = 10^{-5.6}$$

The hydrogen ion concentration is $10^{-5.6}$ mole per liter.

b.
$$\text{pH} = -\log x$$
$$2.4 = -\log x$$
$$-2.4 = \log x$$
$$x = 10^{-2.4}$$

The hydrogen ion concentration is $10^{-2.4}$ mole per liter.

c. $\frac{10^{-2.4}}{10^{-5.6}} = 10^{-2.4-(-5.6)} = 10^{3.2}$

The concentration of the acidic rainfall in part (b) is $10^{3.2}$ times greater than the normal rainfall in part (a).

117. –121. Answers will vary.

123. $2^{x+1} = 8$

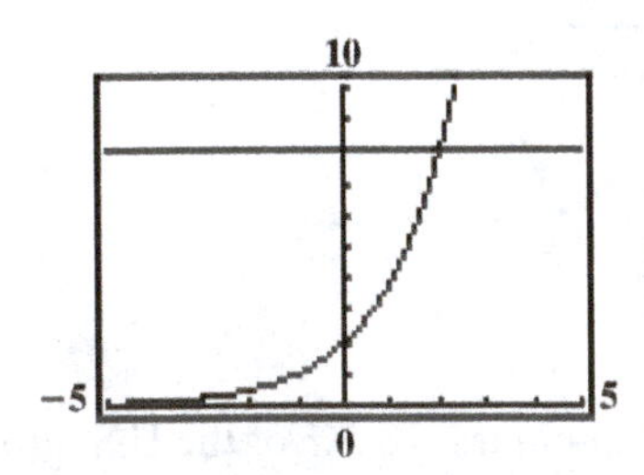

The solution is 2, and the solution set is $\{2\}$.
Verify the solution algebraically:
$$2^{2+1} = 8$$
$$2^3 = 8$$
$$8 = 8$$

125. $\log_3(4x-7) = 2$

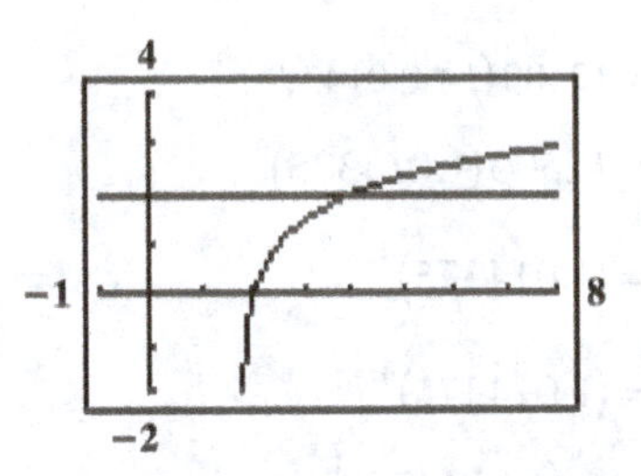

The solution is 4, and the solution set is $\{4\}$.

Verify the solution algebraically:
$$\log_3(4\cdot4-7) = 2$$
$$\log_3(16-7) = 2$$
$$\log_3 9 = 2$$
$$\log_3 3^2 = 2$$
$$2 = 2$$

127. $\log(x+3) + \log x = 1$

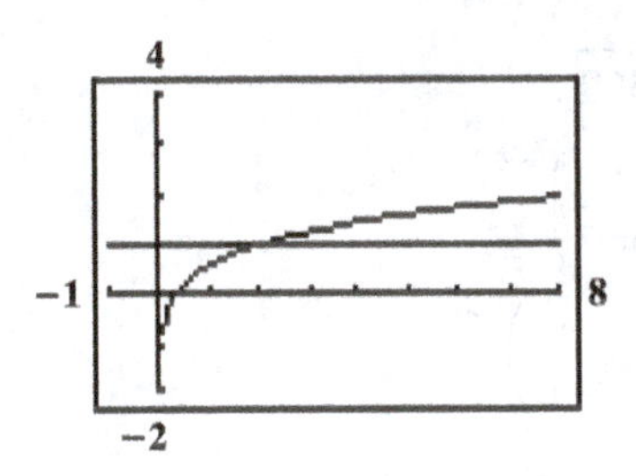

The solution is 2, and the solution set is $\{2\}$.

Verify the solution algebraically:
$$\log(2+3) + \log 2 = 1$$
$$\log 5 + \log 2 = 1$$
$$\log(5\cdot2) = 1$$
$$\log 10 = 1$$
$$1 = 1$$

129. $3^x = 2x+3$

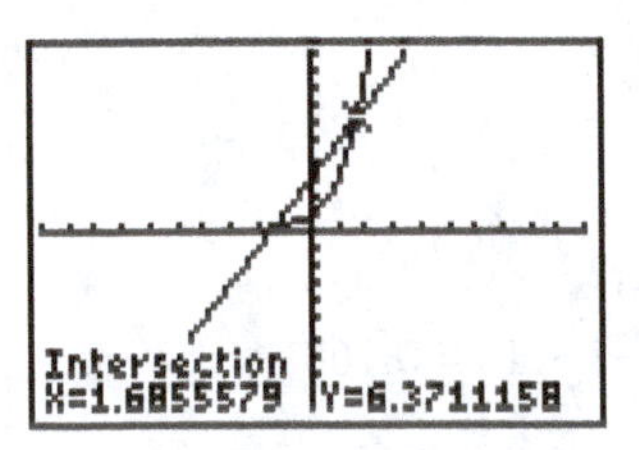

The solution set is $\{-1.39, 1.69\}$.
The solutions check algebraically.

131. $f(x) = 0.48\ln(x+1) + 27$

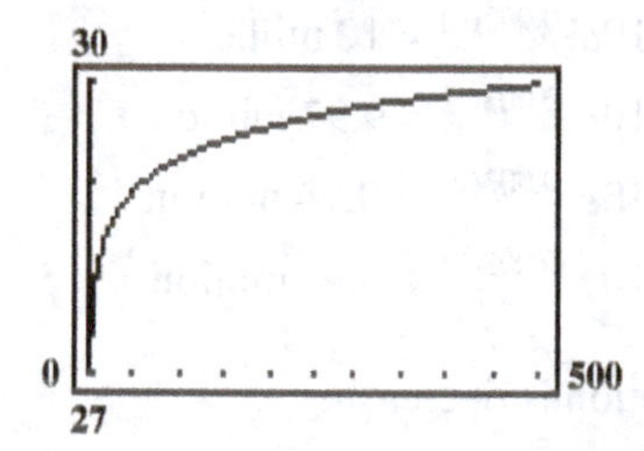

The barometric air pressure increases as the distance from the eye increases. It increases quickly at first, and the more slowly over time.

133. $P(t) = 145e^{-0.092t}$

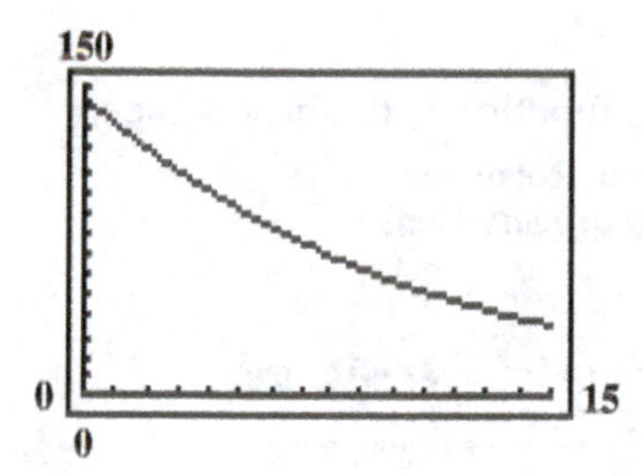

The runner's pulse will be 70 beats per minute after approximately 7.9 minutes.

Verifying algebraically:

$$P(7.9) = 145e^{-0.092(7.9)}$$
$$= 145e^{-07268} \approx 70$$

135. does not make sense; Explanations will vary. Sample explanation: $2^x = 15$ requires logarithms. $2^x = 16$ can be solved by rewriting 16 as 2^4.

$$2^x = 15$$
$$\ln 2^x = \ln 15$$
$$x \ln 2 = \ln 15$$
$$x = \frac{\ln 15}{\ln 2}$$

$$2^x = 16$$
$$2^x = 2^4$$
$$x = 4$$

137. makes sense

139. false; Changes to make the statement true will vary. A sample change is: If $\log(x+3) = 2$, then $10^2 = x+3$.

141. true

143. $A_{4000} = 4000\left(1+\frac{0.03}{1}\right)^{1t}$

$A_{2000} = 2000\left(1+\frac{0.05}{1}\right)^{1t}$

Set the right hand sides of the equations equal and solve for t.

$$4000\left(1+\frac{0.03}{1}\right)^{1t} = 2000\left(1+\frac{0.05}{1}\right)^{1t}$$
$$4000(1+0.03)^t = 2000(1+0.05)^t$$
$$2(1.03)^t = (1.05)^t$$
$$\ln 2(1.03)^t = \ln(1.05)^t$$
$$\ln 2 + \ln(1.03)^t = t\ln(1.05)$$
$$\ln 2 + t\ln(1.03) = t\ln(1.05)$$
$$\ln 2 = t\ln(1.05) - t\ln(1.03)$$
$$\ln 2 = t(\ln(1.05) - \ln(1.03))$$
$$t = \frac{\ln 2}{\ln(1.05) - \ln(1.03)}$$
$$t \approx 36.0$$

In approximately 36 years, the two accounts will have the same balance.

145. $(\log x)(2\log x + 1) = 6$

Let $u = \log x$.

$$(u)(2u+1) = 6$$
$$2u^2 + u = 6$$
$$2u^2 + u - 6 = 0$$
$$(2u-3)(u+2) = 0$$
$$2u - 3 = 0 \quad \text{or} \quad u + 2 = 0$$
$$2u = 3 \qquad u = -2$$
$$u = \frac{3}{2}$$

Substitute $\log x$ for u.

$$u = \frac{3}{2} \quad \text{or} \quad u = -2$$
$$\log x = \frac{3}{2} \qquad \log x = -2$$
$$x = 10^{\frac{3}{2}} \qquad x = 10^{-2}$$

The solution set is $\left\{10^{-2}, 10^{\frac{3}{2}}\right\}$.

147. $\sqrt{2x-1} - \sqrt{x-1} = 1$

$$\sqrt{2x-1} = 1 + \sqrt{x-1}$$
$$\left(\sqrt{2x-1}\right)^2 = \left(1+\sqrt{x-1}\right)^2$$
$$2x-1 = 1 + 2\sqrt{x-1} + x - 1$$
$$2x-1 = 2\sqrt{x-1} + x$$
$$x-1 = 2\sqrt{x-1}$$
$$x-1 = 2\sqrt{x-1}$$
$$(x-1)^2 = \left(2\sqrt{x-1}\right)^2$$
$$x^2 - 2x + 1 = 4(x-1)$$
$$x^2 - 2x + 1 = 4x - 4$$
$$x^2 - 6x + 5 = 0$$
$$(x-1)(x-5) = 0$$
$$x-1 = 0 \quad \text{or} \quad x-5 = 0$$
$$x = 1 \qquad x = 5$$

solution set: $\{1, 5\}$

148. $\dfrac{3}{x+1} - \dfrac{5}{x} = \dfrac{19}{x^2+x}$

$$\frac{3}{x+1} - \frac{5}{x} = \frac{19}{x(x+1)}$$
$$x(x+1)\left(\frac{3}{x+1} - \frac{5}{x}\right) = x(x+1)\left(\frac{19}{x(x+1)}\right)$$
$$x(3) - 5(x+1) = 19$$
$$3x - 5x - 5 = 19$$
$$-2x - 5 = 19$$
$$-2x = 24$$
$$x = -12$$

The solution set is $\{-12\}$.

149. $\left(-2x^3y^{-2}\right)^{-4} = \left(-\dfrac{2x^3}{y^2}\right)^{-4} = \left(-\dfrac{y^2}{2x^3}\right)^4 = \dfrac{y^8}{16x^{12}}$

150. $A = 10e^{-0.003t}$

a. 2006: $A = 10e^{-0.003(0)} = 10$ million

2007: $A = 10e^{-0.003(1)} \approx 9.97$ million

2008: $A = 10e^{-0.003(2)} \approx 9.94$ million

2009: $A = 10e^{-0.003(3)} \approx 9.91$ million

b. The population is decreasing.

151. a. $e^{\ln 3} = 3$

b.
$$e^{\ln 3} = 3$$
$$\left(e^{\ln 3}\right)^x = 3^x$$
$$e^{(\ln 3)x} = 3^x$$

152. An exponential function is the best choice.

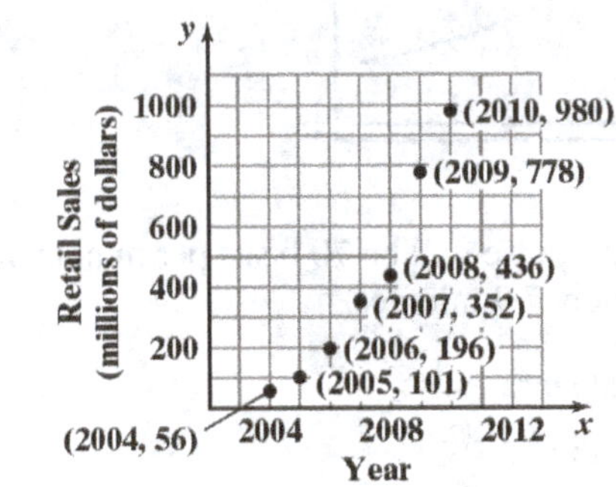

9.6 Check Points

1. a. $A_0 = 643$. Since 2006 is 16 years after 1990, when $t = 16$, $A = 906$.

$$A = A_0e^{kt}$$
$$906 = 643e^{k(16)}$$
$$\frac{906}{643} = e^{16k}$$
$$\ln\left(\frac{906}{643}\right) = \ln e^{16k}$$
$$\ln\left(\frac{906}{643}\right) = 16k$$
$$k = \frac{\ln\left(\frac{906}{643}\right)}{16} \approx 0.021$$

Thus, the growth function is $A = 643e^{0.021t}$.

b.
$$A = 643e^{0.021t}$$
$$2000 = 643e^{0.021t}$$
$$\frac{2000}{643} = e^{0.021t}$$
$$\ln\left(\frac{2000}{643}\right) = \ln e^{0.021t}$$
$$\ln\left(\frac{2000}{643}\right) = 0.021t$$
$$t = \frac{\ln\left(\frac{2000}{643}\right)}{0.021} \approx 54$$

Africa's population will reach 2000 million approximately 54 years after 1990, or 2044.

2. a.
$$A = A_0e^{kt}$$
$$\frac{A_0}{2} = A_0e^{k28}$$
$$\frac{1}{2} = e^{28k}$$
$$\ln\left(\frac{1}{2}\right) = \ln e^{28k}$$
$$\ln\left(\frac{1}{2}\right) = 28k$$
$$k = \frac{\ln\left(\frac{1}{2}\right)}{28} \approx -0.0248$$

Thus, the decay model is $A = A_0e^{-0.0248t}$.

b.
$$A = A_0e^{-0.0248t}$$
$$10 = 60e^{-0.0248t}$$
$$\frac{10}{60} = e^{-0.0248t}$$
$$\ln\left(\frac{10}{60}\right) = \ln e^{-0.0248t}$$
$$\ln\left(\frac{10}{60}\right) = -0.0248t$$
$$t = \frac{\ln\left(\frac{10}{60}\right)}{-0.0248} \approx 72.2$$

It will take about 72.2 years to decay to a level of 10 grams.

3. A logarithmic function would be a good choice for modeling the data.

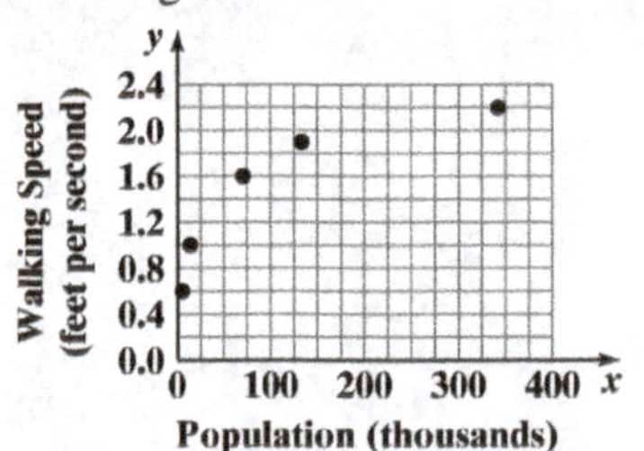

4. An exponential function would be a good choice for modeling the data although model choices may vary.

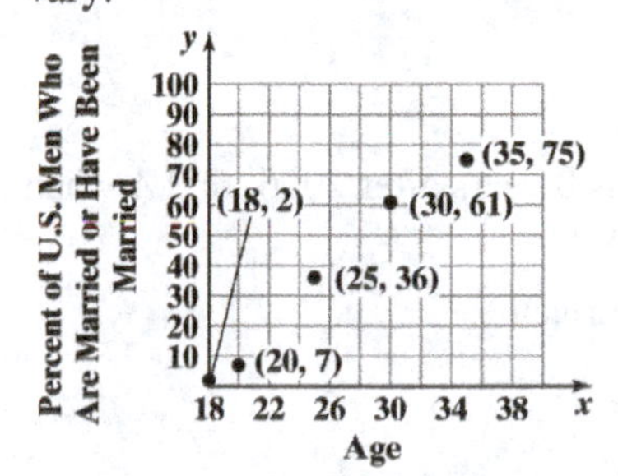

5. a. 1970 is 21 years after 1949.
$$f(x) = 0.074x + 2.294$$
$$f(21) = 0.074(21) + 2.294$$
$$\approx 3.8$$
$$g(x) = 2.577(1.017)^x$$
$$g(21) = 2.577(1.017)^{21}$$
$$\approx 3.7$$

The exponential function g serves as a better model for 1970.

b. 2050 is 101 years after 1949.
$$f(x) = 0.074x + 2.294$$
$$f(101) = 0.074(101) + 2.294$$
$$\approx 9.8$$
$$g(x) = 2.577(1.017)^x$$
$$g(101) = 2.577(1.017)^{101}$$
$$\approx 14.1$$

The linear function f serves as a better model for 2050.

6.
$$y = 4(7.8)^x$$
$$= 4e^{(\ln 7.8)x}$$

Rounded to three decimal places:
$$y = 4e^{(\ln 7.8)x}$$
$$= 4e^{2.054x}$$

9.6 Concept and Vocabulary Check

1. >0; <0

2. A_0; A

3. logarithmic

4. exponential

5. linear

6. $\ln 5$

9.6 Exercise Set

1. Since 2006 is 0 years after 2006, find A when $t = 0$:

$$A = 127.5e^{0.001t}$$
$$A = 127.5e^{0.001(0)}$$
$$A = 127.5e^{0}$$
$$A = 127.5(1)$$
$$A = 127.5$$

In 2006, the population of Japan was 127.5 million.

3. Iraq has the greatest growth rate at 2.7% per year.

5. Substitute $A = 1238$ into the model for India and solve for t:

$$1238 = 1095.4e^{0.014t}$$
$$\frac{1238}{1095.4} = e^{0.014t}$$
$$\ln\frac{1238}{1095.4} = \ln e^{0.014t}$$
$$\ln\frac{1238}{1095.4} = 0.014t$$
$$t = \frac{\ln\frac{1238}{1095.4}}{0.014} \approx 9$$

The population of India will be 1238 million approximately 9 years after 2006, or 2015.

7. a. $A_0 = 6.04$. Since 2050 is 50 years after 2000, when $t = 50$, $A = 10$.

$$A = A_0e^{kt}$$
$$10 = 6.04e^{k(50)}$$
$$\frac{10}{6.04} = e^{50k}$$
$$\ln\left(\frac{10}{6.04}\right) = \ln e^{50k}$$
$$\ln\left(\frac{10}{6.04}\right) = 50k$$
$$k = \frac{\ln\left(\frac{10}{6.04}\right)}{50} \approx 0.01$$

Thus, the growth function is $A = 6.04e^{0.01t}$.

b. $9 = 6.04e^{0.01t}$

$$\frac{9}{6.04} = e^{0.01t}$$
$$\ln\left(\frac{9}{6.04}\right) = \ln e^{0.01t}$$
$$\ln\left(\frac{9}{6.04}\right) = 0.01t$$
$$t = \frac{\ln\left(\frac{9}{6.04}\right)}{0.01} \approx 40$$

Now, 2000 + 40 = 2040, so the population will be 9 million in approximately the year 2040.

9. Since 2025 is 18 years after 2007, find A when $t = 18$:

$$A = 91.1e^{0.0147t}$$
$$A = 91.1e^{0.0147(18)}$$
$$A \approx 118.7$$

In 2025, the population of the Philippines will be 118.7 million.

11. $A_0 = 44.4$ and $A = 55.2$. Since 2025 is 18 years after 2007, $t = 18$

$$A = A_0 e^{kt}$$
$$55.2 = 44.4e^{k(18)}$$
$$\frac{55.2}{44.4} = e^{18k}$$
$$18k = \ln\left(\frac{55.2}{44.4}\right)$$
$$k = \frac{\ln\left(\frac{55.2}{44.4}\right)}{18}$$
$$k \approx 0.0121$$

13. $A_0 = 44.0$ and $A = 40.0$. Since 2025 is 18 years after 2007, $t = 18$

$$A = A_0 e^{kt}$$
$$40.0 = 44.0e^{k(18)}$$
$$\frac{40.0}{44.0} = e^{18k}$$
$$18k = \ln\left(\frac{40.0}{44.0}\right)$$
$$k = \frac{\ln\left(\frac{40.0}{44.0}\right)}{18}$$
$$k \approx -0.0053$$

15. $A = 16e^{-0.000121t}$

$$A = 16e^{-0.000121(5715)}$$
$$A = 16e^{-0.691515}$$
$$A \approx 8.01$$

Approximately 8 grams of carbon-14 will be present in 5715 years.

17. After 10 seconds, there will be $16 \cdot \frac{1}{2} = 8$ grams present. After 20 seconds, there will be $8 \cdot \frac{1}{2} = 4$ grams present. After 30 seconds, there will be $4 \cdot \frac{1}{2} = 2$ grams present. After 40 seconds, there will be $2 \cdot \frac{1}{2} = 1$ grams present. After 50 seconds, there will be $1 \cdot \frac{1}{2} = \frac{1}{2}$ gram present.

19.

$$A = A_0 e^{-0.000121t}$$
$$15 = 100e^{-0.000121t}$$
$$\frac{15}{100} = e^{-0.000121t}$$
$$\ln 0.15 = \ln e^{-0.000121t}$$
$$\ln 0.15 = -0.000121t$$
$$t = \frac{\ln 0.15}{-0.000121} \approx 15,679$$

The paintings are approximately 15,679 years old.

21. a.

$$\frac{1}{2} = 1e^{k1.31}$$
$$\ln\frac{1}{2} = \ln e^{1.31k}$$
$$\ln\frac{1}{2} = 1.31k$$
$$k = \frac{\ln\frac{1}{2}}{1.31} \approx -0.52912$$

The exponential model is given by $A = A_0 e^{-0.52912t}$.

b.

$$A = A_0 e^{-0.52912t}$$
$$0.945A_0 = A_0 e^{-0.52912t}$$
$$0.945 = e^{-0.52912t}$$
$$\ln 0.945 = \ln e^{-0.52912t}$$
$$\ln 0.945 = -0.52912t$$
$$t = \frac{\ln 0.945}{-0.52912} \approx 0.1069$$

The age of the dinosaur bones is approximately 0.1069 billion or 106,900,000 years old.

23.

$$2A_0 = A_0 e^{kt}$$
$$2 = e^{kt}$$
$$\ln 2 = \ln e^{kt}$$
$$\ln 2 = kt$$
$$t = \frac{\ln 2}{k}$$

The population will double in $t = \frac{\ln 2}{k}$ years.

25. $A = 4.1e^{0.01t}$

a. $k = 0.01$, so New Zealand's growth rate is 1%.

b.
$$A = 4.1e^{0.01t}$$
$$2 \cdot 4.1 = 4.1e^{0.01t}$$
$$2 = e^{0.01t}$$
$$\ln 2 = \ln e^{0.01t}$$
$$\ln 2 = 0.01t$$
$$t = \frac{\ln 2}{0.01} \approx 69$$
New Zealand's population will double in approximately 69 years.

27. a.

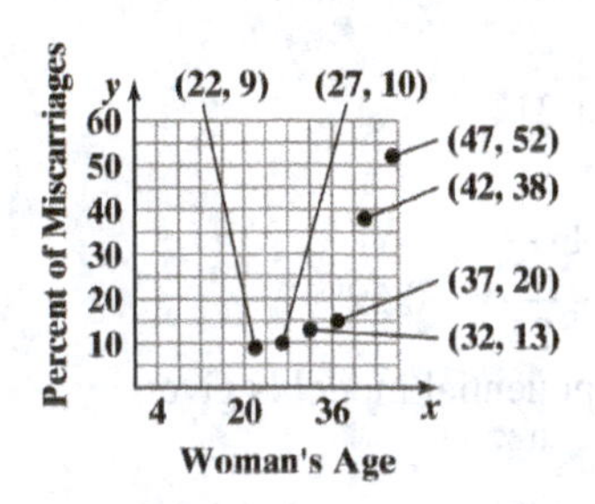

b. An exponential function appears to be the best choice for modeling the data.

29. a.

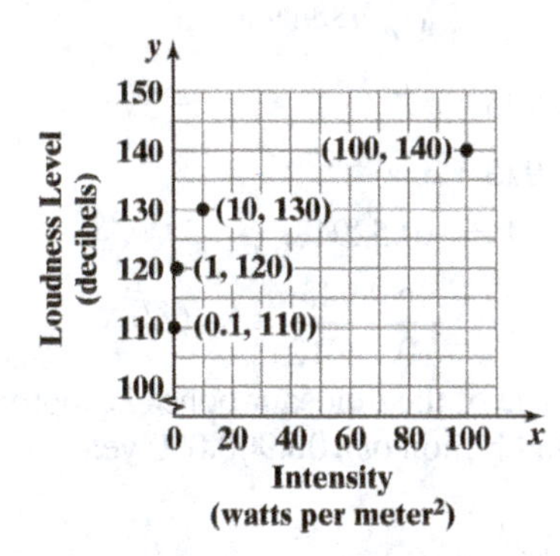

b. A logarithmic function appears to be the best choice for modeling the data.

31. a.

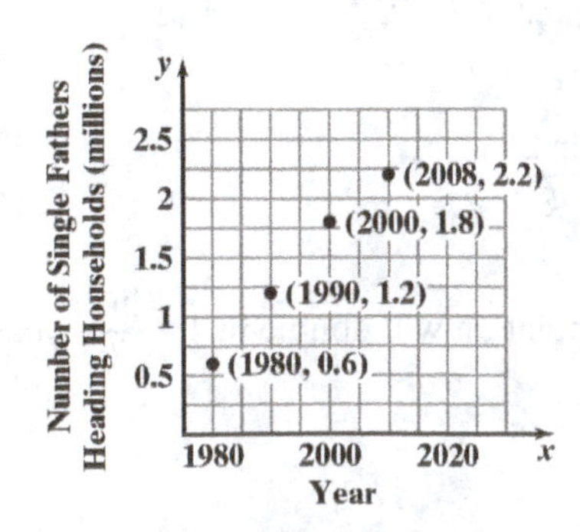

b. A linear function appears to be the best choice for modeling the data.

33.
$$y = 100(4.6)^x$$
$$y = 100e^{(\ln 4.6)x}$$
$$y = 100e^{1.526x}$$

35.
$$y = 2.5(0.7)^x$$
$$y = 2.5e^{(\ln 0.7)x}$$
$$y = 2.5e^{-0.357x}$$

37. – 43. Answers will vary.

45. a. The exponential model is $y = 200.9(1.011)^x$. Since $r \approx 0.999$ is very close to 1, the model fits the data well.

b.
$$y = 200.9(1.011)^x$$
$$y = 200.9e^{(\ln 1.011)x}$$
$$y = 200.9e^{0.0109x}$$
Since $k = .0109$, the population of the United States is increasing by about 1% each year.

47. The linear model is $y = 2.674x + 197.756$. Since $r \approx 0.997$ is close to 1, the model fits the data very well.

49. Using r, the model of best fit is the exponential model $y = 200.9(1.011)^x$.
The model of second best fit is the linear model $y = 2.674x + 197.756$.

Using the exponential model:
$$352 = 200.9(1.011)^x$$
$$\frac{352}{200.9} = (1.011)^x$$
$$\ln\left(\frac{352}{200.9}\right) = \ln(1.011)^x$$
$$\ln\left(\frac{352}{200.9}\right) = x\ln(1.011)$$
$$x = \frac{\ln\left(\frac{352}{200.9}\right)}{\ln(1.011)} \approx 51$$
$1969 + 51 = 2020$

Using the linear model:

$$y = 2.674x + 197.756$$
$$352 = 2.674x + 197.756$$
$$154.244 = 2.674x$$
$$x = \frac{154.244}{2.674} \approx 58$$
$$1969 + 58 = 2027$$

According to the exponential model, the U.S. population will reach 352 million around the year 2020. According to the linear model, the U.S. population will reach 352 million around the year 2027. Explanations will vary.

51. Models and predictions will vary. Sample models are provided

Exercise 27: $y = 1.402(1.078)^x$

Exercise 28: $y = 2896.7(1.056)^x$

Exercise 29: $y = 120 + 4.343\ln x$

Exercise 30: $y = -11.629 + 13.424\ln x$

Exercise 31: $y = 0.058x + 0.558$ (where x is the number of years after 1979)

Exercise 32: $y = 5.3x + 9.5$ (where x is the number of years after 2003)

53. does not make sense; Explanations will vary. Sample explanation: This is not necessarily so. Growth rate measures how fast a population is growing relative to that population. It does not indicate how the size of a population compares to the size of another population.

55. makes sense

57. true

59. true

61. $$\frac{x^2-9}{2x^2+7x+3} \div \frac{x^2-3x}{2x^2+11x+5}$$
$$= \frac{x^2-9}{2x^2+7x+3} \cdot \frac{2x^2+11x+5}{x^2-3x}$$
$$= \frac{(x+3)(x-3)}{(2x+1)(x+3)} \cdot \frac{(2x+1)(x+5)}{x(x-3)}$$
$$= \frac{x+5}{x}$$

62. $$x^{\frac{2}{3}} + 2x^{\frac{1}{3}} - 3 = 0$$

Let $t = x^{\frac{1}{3}}$.

$$\left(x^{\frac{1}{3}}\right)^2 + 2x^{\frac{1}{3}} - 3 = 0$$
$$t^2 + 2t - 3 = 0$$
$$(t+3)(t-1) = 0$$
$$t + 3 = 0 \quad \text{or} \quad t - 1 = 0$$
$$t = -3 \qquad\qquad t = 1$$

Substitute $x^{\frac{1}{3}}$ for t.

$$x^{\frac{1}{3}} = -3 \quad \text{or} \quad x^{\frac{1}{3}} = 1$$
$$\left(x^{\frac{1}{3}}\right)^3 = (-3)^3 \qquad \left(x^{\frac{1}{3}}\right)^3 = (1)^3$$
$$x = -27 \qquad\qquad x = 1$$

The solution set is $\{-27, 1\}$.

63. $$6\sqrt{2} - 2\sqrt{50} + 3\sqrt{98}$$
$$= 6\sqrt{2} - 2\sqrt{25\cdot 2} + 3\sqrt{49\cdot 2}$$
$$= 6\sqrt{2} - 2\cdot 5\sqrt{2} + 3\cdot 7\sqrt{2}$$
$$= 6\sqrt{2} - 10\sqrt{2} + 21\sqrt{2} = 17\sqrt{2}$$

64. $$\sqrt{(x_2-x_1)^2 + (y_2-y_1)^2} = \sqrt{(1-7)^2 + (-1-2)^2}$$
$$= \sqrt{(-6)^2 + (-3)^2}$$
$$= \sqrt{36+9}$$
$$= \sqrt{45}$$
$$= \sqrt{9\cdot 5}$$
$$= 3\sqrt{5}$$

65. $$\left(\frac{x_1+x_2}{2}, \frac{y_1+y_2}{2}\right) = \left(\frac{7+1}{2}, \frac{2+(-1)}{2}\right)$$
$$= \left(4, \frac{1}{2}\right)$$

66. Graph of Circle:

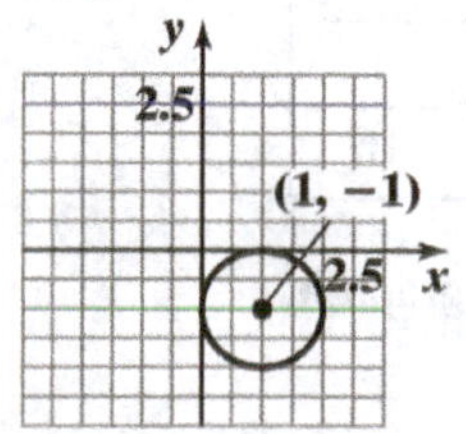

Chapter 9 Review Exercises

1. $f(x) = 4^x$

x	$f(x)$
-2	$4^{-2} = \frac{1}{4^2} = \frac{1}{16}$
-1	$4^{-1} = \frac{1}{4^1} = \frac{1}{4}$
0	$4^0 = 1$
1	$4^1 = 4$
2	$4^2 = 16$

The coordinates match graph **d**.

2. $f(x) = 4^{-x}$

x	$f(x)$
-2	$4^{-(-2)} = 4^2 = 16$
-1	$4^{-(-1)} = 4^1 = 4$
0	$4^{-0} = 4^0 = 1$
1	$4^{-1} = \frac{1}{4^1} = \frac{1}{4}$
2	$4^{-2} = \frac{1}{4^2} = \frac{1}{16}$

The coordinates match graph **a.**

3. $f(x) = -4^{-x}$

x	$f(x)$
-2	$-4^{-(-2)} = -4^2 = -16$
-1	$-4^{-(-1)} = -4^1 = -4$
0	$-4^{-0} = -4^0 = -1$
1	$-4^{-1} = -\frac{1}{4^1} = -\frac{1}{4}$
2	$-4^{-2} = -\frac{1}{4^2} = -\frac{1}{16}$

The coordinates match graph **b.**

4. $f(x) = -4^{-x} + 3$

x	$f(x)$
-2	$-4^{-(-2)} + 3 = -4^2 + 3 = -16 + 3 = -13$
-1	$-4^{-(-1)} + 3 = -4^1 + 3 = -4 + 3 = -1$
0	$-4^{-0} + 3 = -4^0 + 3 = -1 + 3 = 2$
1	$-4^{-1} + 3 = -\frac{1}{4^1} + 3 = -\frac{1}{4} + 3 = \frac{11}{4}$
2	$-4^{-2} + 3 = -\frac{1}{4^2} + 3 = -\frac{1}{16} + 3 = \frac{47}{16}$

The coordinates match graph **c.**

5. $f(x) = 2^x$ and $g(x) = 2^{x-1}$

x	$f(x)$	$g(x)$
-2	$\frac{1}{4}$	$\frac{1}{8}$
-1	$\frac{1}{2}$	$\frac{1}{4}$
0	1	$\frac{1}{2}$
1	2	1
2	4	2

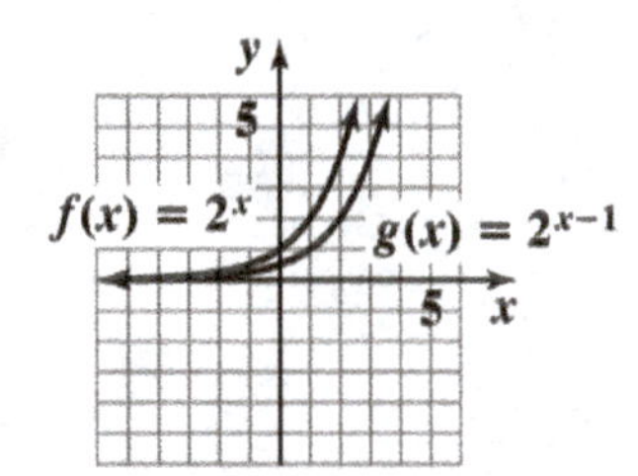

The graph of g is the graph of f shifted 1 unit to the right.

6. $f(x)=2^x$ and $g(x)=\left(\frac{1}{2}\right)^x$

x	$f(x)$	$g(x)$
−2	$\frac{1}{4}$	4
−1	$\frac{1}{2}$	2
0	1	1
1	2	$\frac{1}{2}$
2	4	$\frac{1}{4}$

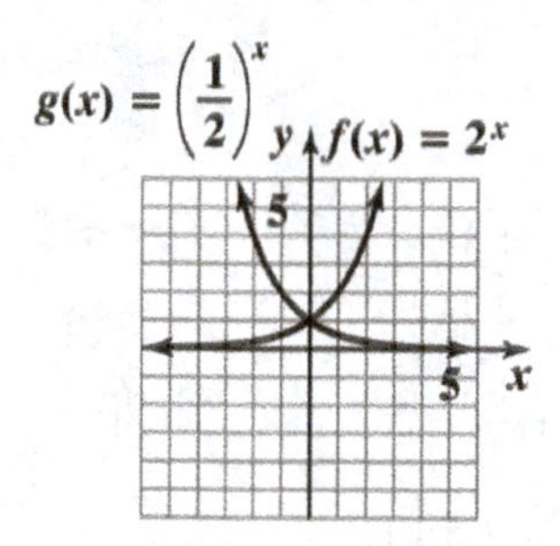

The graph of g is the graph of f reflected across the y–axis.

7. $f(x)=3^x$ and $g(x)=3^x-1$

x	$f(x)$	$g(x)$
−2	$\frac{1}{9}$	$-\frac{8}{9}$
−1	$\frac{1}{3}$	$-\frac{2}{3}$
0	1	0
1	3	2
2	9	8

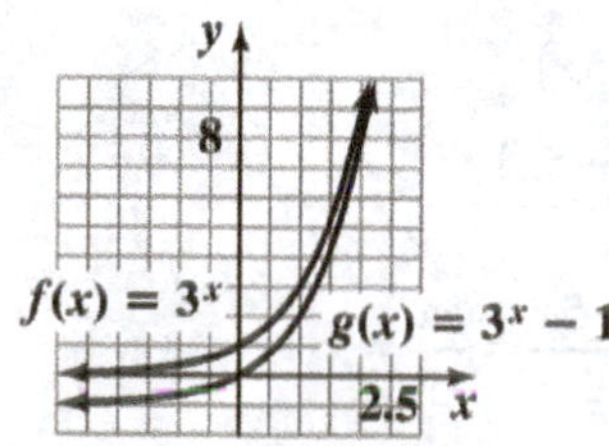

The graph of g is the graph of f shifted down 1 unit.

8. $f(x)=3^x$ and $g(x)=-3^x$

x	$f(x)$	$g(x)$
−2	$\frac{1}{9}$	$-\frac{1}{9}$
−1	$\frac{1}{3}$	$-\frac{1}{3}$
0	1	-1
1	3	-3
2	9	-9

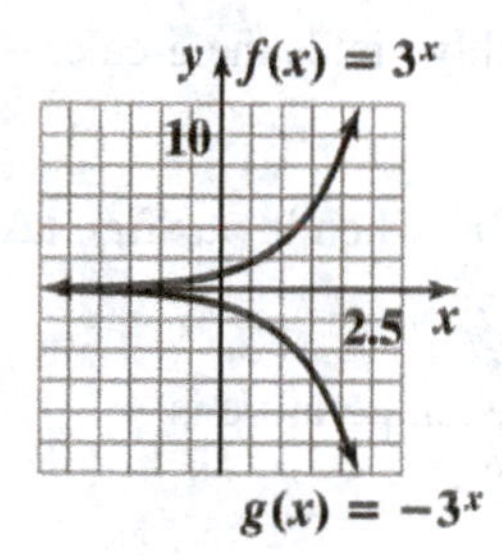

The graph of g is the graph of f reflected across the x–axis.

9. 5.5% Compounded Semiannually:

$$A=5000\left(1+\frac{0.055}{2}\right)^{2\cdot 5}$$
$$=5000(1+0.0275)^{10}$$
$$=5000(1.0275)^{10}\approx 6558.26$$

5.25% Compounded Monthly:

$$A=5000\left(1+\frac{0.0525}{12}\right)^{12\cdot 5}$$
$$=5000(1+0.004375)^{60}$$
$$=5000(1.004375)^{60}\approx 6497.16$$

5.5% compounded semiannually yields the greater return.

10. 7.0% Compounded Monthly:

$$A = 14000\left(1+\frac{0.07}{12}\right)^{12\cdot 10}$$
$$= 14000\left(1+\frac{7}{1200}\right)^{120}$$
$$= 14000\left(\frac{1207}{1200}\right)^{120} \approx 28135.26$$

6.85% Compounded Continuously:

$$A = 14000e^{0.0685\cdot 10}$$
$$= 14000e^{0.685} \approx 27772.81$$

7.0% compounded monthly yields the greater return.

11. a. The coffee was $200°F$ when it was first taken out of the microwave.

b. After 20 minutes, the temperature is approximately $119°F$.

c. The coffee will cool to a low of $70°F$. This means that the temperature of the room is $70°F$.

12. a.
$$(f \circ g)(x) = f(g(x))$$
$$= f(4x-1)$$
$$= (4x-1)^2 + 3$$
$$= 16x^2 - 8x + 1 + 3$$
$$= 16x^2 - 8x + 4$$

b.
$$(g \circ f)(x) = g(f(x))$$
$$= g(x^2+3)$$
$$= 4(x^2+3) - 1$$
$$= 4x^2 + 12 - 1$$
$$= 4x^2 + 11$$

c.
$$(f \circ g)(3) = 16(3)^2 - 8(3) + 4$$
$$= 16(9) - 24 + 4$$
$$= 144 - 24 + 4$$
$$= 124$$

13. a.
$$(f \circ g)(x) = f(g(x))$$
$$= f(x+1) = \sqrt{x+1}$$

b.
$$(g \circ f)(x) = g(f(x))$$
$$= g(\sqrt{x}) = \sqrt{x} + 1$$

c. $(f \circ g)(3) = \sqrt{3+1} = \sqrt{4} = 2$

14. $f(x) = \frac{3}{5}x + \frac{1}{2}$ and $g(x) = \frac{5}{3}x - 2$

$$f(g(x)) = f\left(\frac{5}{3}x - 2\right)$$
$$= \frac{3}{5}\left(\frac{5}{3}x - 2\right) + \frac{1}{2}$$
$$= \frac{3}{5}\left(\frac{5}{3}x\right) - \left(\frac{3}{5}\right)2 + \frac{1}{2}$$
$$= x - \frac{6}{5} + \frac{1}{2}$$
$$= x - \frac{7}{10}$$

$$g(f(x)) = g\left(\frac{3}{5}x + \frac{1}{2}\right)$$
$$= \frac{5}{3}\left(\frac{3}{5}x + \frac{1}{2}\right) - 2$$
$$= \frac{5}{3}\left(\frac{3}{5}x\right) + \left(\frac{5}{3}\right)\frac{1}{2} - 2$$
$$= x + \frac{5}{6} - 2$$
$$= x - \frac{7}{6}$$

The functions are not inverses.

15. $f(x) = 2 - 5x$ and $g(x) = \frac{2-x}{5}$

$$f(g(x)) = f\left(\frac{2-x}{5}\right)$$
$$= 2 - 5\left(\frac{2-x}{5}\right)$$
$$= 2 - (2-x) = 2 - 2 + x = x$$

$$g(f(x)) = g(2-5x)$$
$$= \frac{2-(2-5x)}{5}$$
$$= \frac{2-2+5x}{5} = \frac{5x}{5} = x$$

The functions are inverses.

16. a. $f(x) = 4x - 3$

$y = 4x - 3$

Interchange x and y and solve for y.

$$x = 4y - 3$$
$$x + 3 = 4y$$
$$\frac{x+3}{4} = y$$

$$f^{-1}(x) = \frac{x+3}{4}$$

b. $f\left(f^{-1}(x)\right) = f\left(\frac{x+3}{4}\right)$

$$= 4\left(\frac{x+3}{4}\right) - 3$$
$$= x + 3 - 3 = x$$

$$f^{-1}(f(x)) = f(4x - 3)$$
$$= \frac{(4x-3)+3}{4}$$
$$= \frac{4x-3+3}{4} = \frac{4x}{4} = x$$

17. a. $f(x) = \sqrt{x+2}$

$y = \sqrt{x+2}$

Interchange x and y and solve for y.

$$x = \sqrt{y+2}$$
$$x^2 = y + 2$$
$$x^2 - 2 = y$$

$$f^{-1}(x) = x^2 - 2 \text{ for } x \geq 0$$

b. $f\left(f^{-1}(x)\right) = f\left(x^2 - 2\right)$

$$= \sqrt{\left(x^2 - 2\right) + 2}$$
$$= \sqrt{x^2 - 2 + 2}$$
$$= \sqrt{x^2} = x$$

$$f^{-1}(f(x)) = f\left(\sqrt{x+2}\right)$$
$$= \left(\sqrt{x+2}\right)^2 - 2$$
$$= x + 2 - 2 = x$$

18. a. $f(x) = 8x^3 + 1$

$y = 8x^3 + 1$

Interchange x and y and solve for y.

$$x = 8y^3 + 1$$
$$x - 1 = 8y^3$$
$$\frac{x-1}{8} = y^3$$
$$\sqrt[3]{\frac{x-1}{8}} = y$$
$$\frac{\sqrt[3]{x-1}}{2} = y$$

$$f^{-1}(x) = \frac{\sqrt[3]{x-1}}{2}$$

b. $f\left(f^{-1}(x)\right) = f\left(\frac{\sqrt[3]{x-1}}{2}\right)$

$$= 8\left(\frac{\sqrt[3]{x-1}}{2}\right)^3 + 1$$
$$= 8\left(\frac{x-1}{8}\right) + 1$$
$$= x - 1 + 1 = x$$

$$f^{-1}(f(x)) = f\left(8x^3 + 1\right)$$
$$= \frac{\sqrt[3]{\left(8x^3 + 1\right) - 1}}{2}$$
$$= \frac{\sqrt[3]{8x^3 + 1 - 1}}{2}$$
$$= \frac{\sqrt[3]{8x^3}}{2} = \frac{2x}{2} = x$$

19. Since the graph satisfies the horizontal line test, it has an inverse function.

20. Since the graph does not satisfy the horizontal line test, it does not have an inverse function.

21. Since the graph satisfies the horizontal line test, it has an inverse function.

22. Since the graph does not satisfy the horizontal line test, it does not have an inverse function.

23. Since the points $(-3,-1),(0,0)$ and $(2,4)$ lie on the graph of the function, the points $(-1,-3)$, $(0,0)$ and $(4,2)$ lie on the inverse function.

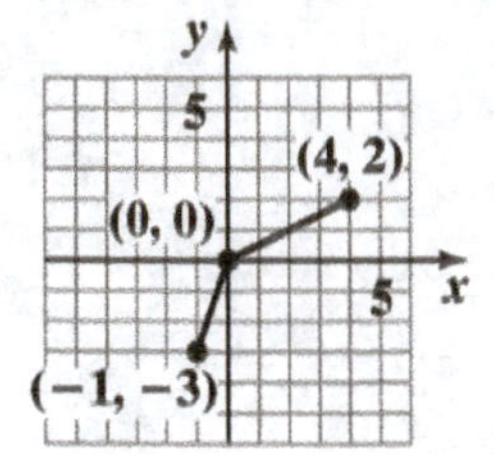

24. $\frac{1}{2}=\log_{49}7$

$49^{\frac{1}{2}}=7$

25. $3=\log_4 x$

$4^3=x$

26. $\log_3 81=y$

$3^y=81$

27. $6^3=216$

$\log_6 216=3$

28. $b^4=625$

$\log_b 625=4$

29. $13^y=874$

$\log_{13}874=y$

30. $\log_4 64=\log_4 4^3=3$ because $\log_b b^x=x$.

31. $\log_5\frac{1}{25}=\log_5\frac{1}{5^2}=\log_5 5^{-2}=-2$ because $\log_b b^x=x$.

32. $\log_3(-9)$

This logarithm cannot be evaluated because –9 is not in the domain of $y=\log_3 x$.

33. $\log_{16}4=y$

$16^y=4$

$(4^2)^y=4$

$4^{2y}=4^1$

$2y=1$

$y=\frac{1}{2}$

34. $\log_{17}17=1$ because $17^1=17$.

35. $\log_3 3^8=8$ because $\log_b b^x=x$.

36. Because $\ln e^x=x$, we conclude that $\ln e^5=5$.

37. $\log_3\frac{1}{\sqrt{3}}=\log_3\frac{1}{3^{\frac{1}{2}}}=\log_3 3^{-\frac{1}{2}}=-\frac{1}{2}$ because $\log_b b^x=x$.

38. $\ln\frac{1}{e^2}=\ln e^{-2}=-2$ because $\log_b b^x=x$.

39. $\log\frac{1}{1000}=\log\frac{1}{10^3}=\log 10^{-3}=-3$ because $\log_b b^x=x$.

40. Recall that $\log_b b=1$ and $\log_b 1=0$ for all $b>0$, $b\neq 1$. Therefore, $\log_3(\log_8 8)=\log_3 1=0$.

41. $f(x)=2^x$; $g(x)=\log_2 x$

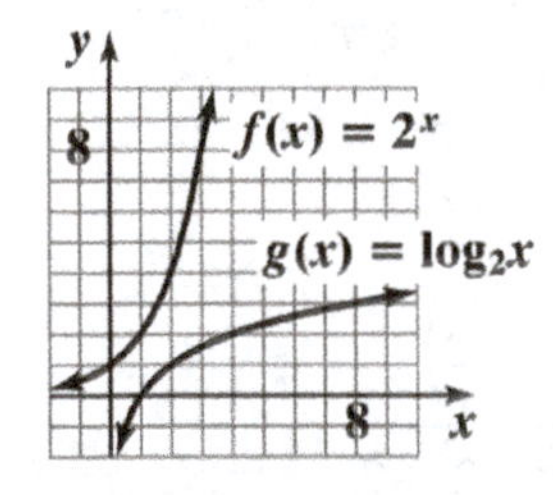

Domain of *f*: $\{x\,|\,x\text{ is a real number}\}$ or $(-\infty,\infty)$

Range of *f*: $\{y\,|\,y>0\}$ or $(0,\infty)$

Domain of *g*: $\{x\,|\,x>0\}$ or $(0,\infty)$

Range of *g*: $\{y\,|\,y\text{ is a real number}\}$ or $(-\infty,\infty)$

42. $f(x)=\left(\frac{1}{3}\right)^x$; $g(x)=\log_{\frac{1}{3}} x$

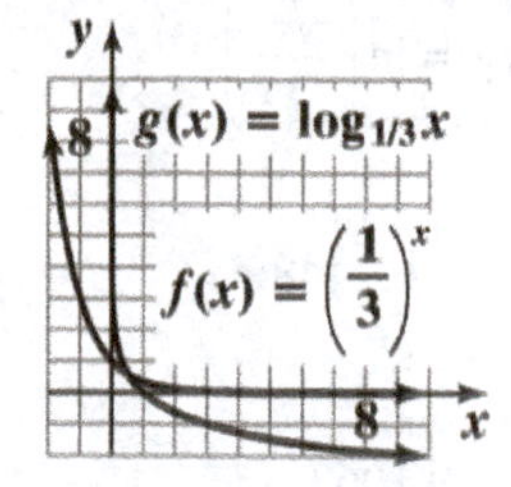

Domain of *f*: $\{x \mid x \text{ is a real number}\}$ or $(-\infty,\infty)$
Range of *f*: $\{y \mid y>0\}$ or $(0,\infty)$

Domain of *g*: $\{x \mid x>0\}$ or $(0,\infty)$
Range of *g*: $\{y \mid y \text{ is a real number}\}$ or $(-\infty,\infty)$

43. $f(x)=\log_8(x+5)$

$$x+5>0$$
$$x>-5$$

The domain of *f* is $\{x|x>-5\}$ or $(-5,\infty)$.

44. $f(x)=\log(3-x)$

$$3-x>0$$
$$-x>-3$$
$$x<3$$

The domain of *f* is $\{x|x<3\}$ or $(-\infty,3)$.

45. $f(x)=\ln(x-1)^2$
The domain of *g* is all real numbers for which $(x-1)^2>0$. The only number that must be excluded is 1. The domain of *f* is $\{x|x\neq 1\}$ or $(-\infty,1)\cup(1,\infty)$.

46. Since $\ln e^x=x$, $\ln e^{6x}=6x$.

47. Since $e^{\ln x}=x$, $e^{\ln\sqrt{x}}=\sqrt{x}$.

48. Since $10^{\log x}=x$, $10^{\log 4x^2}=4x^2$.

49.
$$R=\log\frac{I}{I_0}$$
$$R=\log\frac{1000I_0}{I_0}$$
$$R=\log 1000$$
$$10^R=1000$$
$$10^R=10^3$$
$$R=3$$
The magnitude on the Richter scale is 3.

50. a. $f(0)=76-18\log(0+1)$
$=76-18\log 1=76-18(0)=76$

The average score when the exam was first given was 76.

b. $f(2)=76-18\log(2+1)$
$=76-18\log(3)\approx 67.4$

$f(4)=76-18\log(4+1)$
$=76-18\log(5)\approx 63.4$

$f(6)=76-18\log(6+1)$
$=76-18\log(7)\approx 60.8$

$f(8)=76-18\log(8+1)$
$=76-18\log(9)\approx 58.8$

$f(12)=76-18\log(12+1)$
$=76-18\log(13)\approx 55.9$

The average scores were as follows:

2 months	67.4
4 months	63.4
6 months	60.8
8 months	58.8
12 months	55.9.

c.

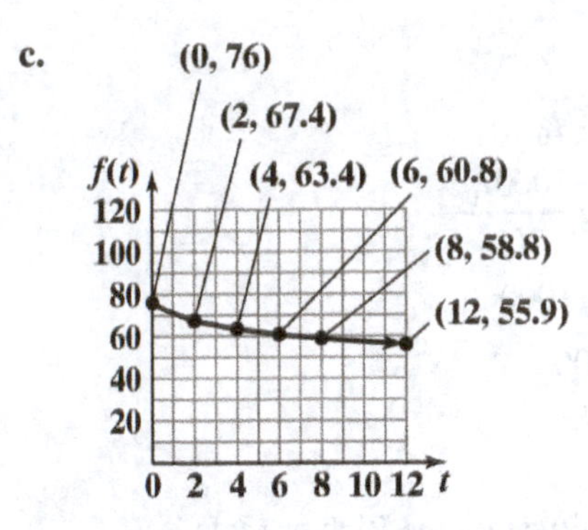

The students retain less material over time.

51. $t = \frac{1}{0.06}\ln\left(\frac{12}{12-5}\right)$
$= \frac{1}{0.06}\ln\left(\frac{12}{7}\right) \approx 9.0$

It will take approximately 9 weeks for the man to run 5 miles per hour.

52. $\log_6\left(36x^3\right) = \log_6 36 + \log_6 x^3 = 2 + 3\log_6 x$

53. $\log_4 \frac{\sqrt{x}}{64} = \log_4 \sqrt{x} - \log_4 64$
$= \log_4 x^{\frac{1}{2}} - 3 = \frac{1}{2}\log_4 x - 3$

54. $\log_2\left(\frac{xy^2}{64}\right) = \log_2 xy^2 - \log_2 64$
$= \log_2 x + \log_2 y^2 - 6$
$= \log_2 x + 2\log_2 y - 6$

55. $\ln\sqrt[3]{\frac{x}{e}} = \ln\left(\frac{x}{e}\right)^{\frac{1}{3}} = \frac{1}{3}\ln\left(\frac{x}{e}\right)$
$= \frac{1}{3}(\ln x - \ln e)$
$= \frac{1}{3}(\ln x - 1) = \frac{1}{3}\ln x - \frac{1}{3}$

56. $\log_b 7 + \log_b 3 = \log_b (7\cdot 3) = \log_b 21$

57. $\log 3 - 3\log x = \log 3 - \log x^3 = \log\frac{3}{x^3}$

58. $3\ln x + 4\ln y = \ln x^3 + \ln y^4 = \ln\left(x^3 y^4\right)$

59. $\frac{1}{2}\ln x - \ln y = \ln x^{\frac{1}{2}} - \ln y$
$= \ln\sqrt{x} - \ln y = \ln\left(\frac{\sqrt{x}}{y}\right)$

60. $\log_6 72{,}348 = \frac{\log 72{,}348}{\log 6} \approx 6.2448$

61. $\log_4 0.863 = \frac{\log 0.863}{\log 4} \approx -0.1063$

62. true

63. false; Changes to make the statement true will vary. A sample change is:
$\log(x+9) - \log(x+1) = \log\left(\frac{x+9}{x+1}\right)$.

64. false; Changes to make the statement true will vary. A sample change is: $\log_2 x^4 = 4\log_2 x$.

65. true

66. $2^{4x-2} = 64$
$2^{4x-2} = 2^6$
$4x - 2 = 6$
$4x = 8$
$x = 2$

The solution set is $\{2\}$.

67. $125^x = 25$
$\left(5^3\right)^x = 5^2$
$5^{3x} = 5^2$
$3x = 2$
$x = \frac{2}{3}$

The solution set is $\left\{\frac{2}{3}\right\}$.

68. $9^x = \frac{1}{27}$

$(3^2)^x = 3^{-3}$

$3^{2x} = 3^{-3}$

$2x = -3$

$x = -\frac{3}{2}$

The solution set is $\left\{-\frac{3}{2}\right\}$.

69. $8^x = 12{,}143$

$\ln 8^x = \ln 12{,}143$

$x \ln 8 = \ln 12{,}143$

$x = \frac{\ln 12{,}143}{\ln 8} \approx 4.52$

The solution set is $\left\{\frac{\ln 12{,}143}{\ln 8} \approx 4.52\right\}$.

70. $9e^{5x} = 1269$

$e^{5x} = \frac{1269}{9}$

$\ln e^{5x} = \ln 141$

$5x = \ln 141$

$x = \frac{\ln 141}{5} \approx 0.99$

The solution set is $\left\{\frac{\ln 141}{5} \approx 0.99\right\}$.

71. $30e^{0.045x} = 90$

$e^{0.045x} = \frac{90}{30}$

$\ln e^{0.045x} = \ln 3$

$0.045x = \ln 3$

$x = \frac{\ln 3}{0.045} \approx 24.41$

The solution set is $\left\{\frac{\ln 3}{0.045} \approx 24.41\right\}$.

72. $\log_5 x = -3$

$x = 5^{-3}$

$x = \frac{1}{125}$

The solution set is $\left\{\frac{1}{125}\right\}$.

73. $\log x = 2$

$x = 10^2$

$x = 100$

The solution set is $\{100\}$.

74. $\log_4(3x-5) = 3$

$3x - 5 = 4^3$

$3x - 5 = 64$

$3x = 69$

$x = 23$

The solution set is $\{23\}$.

75. $\ln x = -1$

$x = e^{-1}$

$x = \frac{1}{e}$

The solution set is $\left\{\frac{1}{e}\right\}$.

76. $3 + 4\ln(2x) = 15$

$4\ln(2x) = 12$

$\ln(2x) = 3$

$2x = e^3$

$x = \frac{e^3}{2}$

The solution set is $\left\{\frac{e^3}{2}\right\}$.

77. $\log_2(x+3) + \log_2(x-3) = 4$

$\log_2((x+3)(x-3)) = 4$

$\log_2(x^2-9) = 4$

$x^2 - 9 = 2^4$

$x^2 - 9 = 16$

$x^2 = 25$

$x = \pm 5$

We disregard -5 because it would result in taking the logarithm of a negative number in the original equation. The solution set is $\{5\}$.

78. $\log_3(x-1)-\log_3(x+2)=2$

$$\log_3\frac{x-1}{x+2}=2$$
$$\frac{x-1}{x+2}=3^2$$
$$\frac{x-1}{x+2}=9$$
$$x-1=9(x+2)$$
$$x-1=9x+18$$
$$-8x-1=18$$
$$-8x=19$$
$$x=-\frac{19}{8}$$

We disregard $-\frac{19}{8}$ because it would result in taking the logarithm of a negative number in the original equation. There is no solution. The solution set is $\varnothing$ or $\{\ \}$.

79. $\log_4(3x-5)=\log_4 3$

$$3x-5=3$$
$$3x=8$$
$$x=\frac{8}{3}$$

The solution set is $\left\{\frac{8}{3}\right\}$.

80. $\ln(x+4)-\ln(x+1)=\ln x$

$$\ln\frac{x+4}{x+1}=\ln x$$
$$\frac{x+4}{x+1}=x$$
$$(x+1)\frac{x+4}{x+1}=x(x+1)$$
$$x+4=x^2+x$$
$$x^2=4$$
$$x=\pm 2$$

We disregard –2 because it would result in taking the logarithm of a negative number in the original equation. The solution set is $\{2\}$.

81. $\log_6(2x+1)=\log_6(x-3)+\log_6(x+5)$

$$\log_6(2x+1)=\log_6(x^2+2x-15)$$
$$2x+1=x^2+2x-15$$
$$x^2=16$$
$$x=\pm 4$$

We disregard –4 because it would result in taking the logarithm of a negative number in the original equation. The solution set is $\{4\}$.

82. $P(x)=14.7e^{-0.21x}$

$$4.6=14.7e^{-0.21x}$$
$$\frac{4.6}{14.7}=e^{-0.21x}$$
$$\ln\frac{4.6}{14.7}=\ln e^{-0.21x}$$
$$\ln\frac{4.6}{14.7}=-0.21x$$
$$t=\frac{\ln\frac{4.6}{14.7}}{-0.21}\approx 5.5$$

The peak of Mt. Everest is about 5.5 miles above sea level.

83. $f(t)=364(1.005)^t$

$$560=364(1.005)^t$$
$$\frac{560}{364}=(1.005)^t$$
$$\ln\frac{560}{364}=\ln(1.005)^t$$
$$\ln\frac{560}{364}=t\ln 1.005$$
$$t=\frac{\ln\frac{560}{364}}{\ln 1.005}\approx 86.4$$

The carbon dioxide concentration will be double the pre-industrial level approximately 86 years after the year 2000 in the year 2086.

84. $W(x) = 0.37\ln x + 0.05$

$$3.38 = 0.37\ln x + 0.05$$
$$3.33 = 0.37\ln x$$
$$9 = \ln x$$
$$e^9 = e^{\ln x}$$
$$x = e^9 \approx 8103$$

The population of New Your City is approximately 8103 thousand, or 8,103,000.

85. $20,000 = 12,500\left(1 + \frac{0.065}{4}\right)^{4t}$

$$20,000 = 12,500(1 + 0.01625)^{4t}$$
$$20,000 = 12,500(1.01625)^{4t}$$
$$1.6 = (1.01625)^{4t}$$
$$\ln 1.6 = \ln(1.01625)^{4t}$$
$$\ln 1.6 = 4t\ln 1.01625$$
$$\frac{\ln 1.6}{4\ln 1.01625} = \frac{4t\ln 1.01625}{4\ln 1.01625}$$
$$t = \frac{\ln 1.6}{4\ln 1.01625} \approx 7.3$$

It will take approximately 7.3 years.

86. $3(50,000) = 50,000e^{0.075t}$

$$\frac{3(50,000)}{50,000} = e^{0.075t}$$
$$3 = e^{0.075t}$$
$$\ln 3 = \ln e^{0.075t}$$
$$\ln 3 = 0.075t$$
$$t = \frac{\ln 3}{0.075} \approx 14.6$$

The money will triple in approximately 14.6 years.

87. $3 = e^{r5}$

$$\ln 3 = \ln e^{5r}$$
$$\ln 3 = 5r$$
$$r = \frac{\ln 3}{5} \approx 0.220$$

The money will triple in 5 years if the interest rate is approximately 22%.

88. **a.** $t = 2008 - 1990 = 18$

$$A = 22.4e^{kt}$$
$$46.9 = 22.4e^{k(18)}$$
$$\frac{46.9}{22.4} = e^{18k}$$
$$\ln\frac{46.9}{22.4} = \ln e^{18k}$$
$$\ln\frac{46.9}{22.4} = 18k$$
$$k = \frac{\ln\frac{46.9}{22.4}}{18} \approx 0.041$$

b. Note that 2015 is 25 years after 1990, find A for $t = 25$.

$$A = 22.4e^{0.041t}$$
$$= 22.4e^{0.041(25)} \approx 62.4$$

The population will be about 62.4 million the year 2015.

c.

$$68 = 22.4e^{0.041t}$$
$$\frac{68}{22.4} = e^{0.041t}$$
$$\ln\frac{68}{22.4} = \ln e^{0.041t}$$
$$\ln\frac{68}{22.4} = 0.041t$$
$$t = \frac{\ln\frac{68}{22.4}}{0.041} \approx 27$$

The Hispanic resident population will reach 68 million approximately 27 years after 1990, or 2017.

89. Find k:

$$A = A_0e^{kt}$$
$$\frac{A_0}{2} = A_0e^{k\cdot 140}$$
$$\frac{1}{2} = e^{140k}$$
$$\ln\frac{1}{2} = \ln e^{140k}$$
$$\ln\frac{1}{2} = 140k$$
$$k = \frac{\ln\frac{1}{2}}{140} \approx -0.00495$$

Thus, $A = A_0 e^{-0.00495t}$.

$$0.20A_0 = A_0 e^{-0.00495t}$$
$$0.20 = e^{-0.00495t}$$
$$\ln 0.20 = \ln e^{-0.00495t}$$
$$\ln 0.20 = -0.00495t$$
$$t = \frac{\ln 0.20}{-0.00495} \approx 325$$

It will take 325 days for polonium-210 to decay to 20% of its original amount.

90. a.

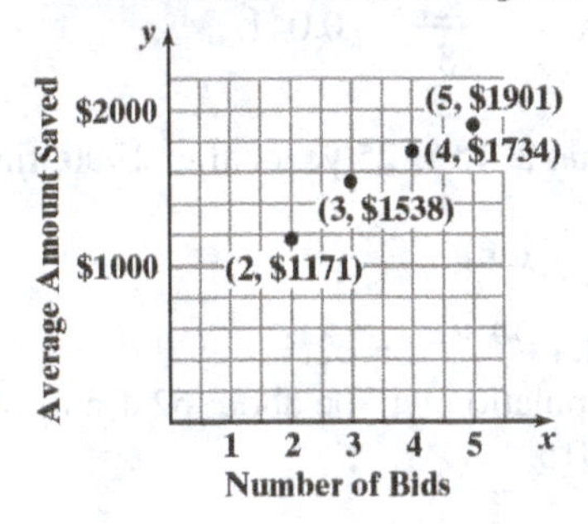

b. A logarithmic function appears to be the best choice for modeling the data.

91. a.

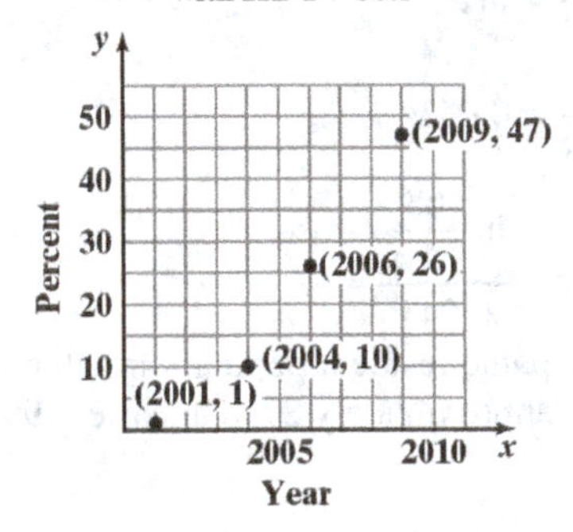

b. An exponential function appears to be the best choice for modeling the data.

92. $y = 73(2.6)^x$

$$y = 73e^{(\ln 2.6)x}$$
$$y = 73e^{0.956x}$$

93. $y = 6.5(0.43)^x$

$$y = 6.5e^{(\ln 0.43)x}$$
$$y = 6.5e^{-0.844x}$$

Chapter 9 Test

1. $f(x) = 2^x$

$g(x) = 2^{x+1}$

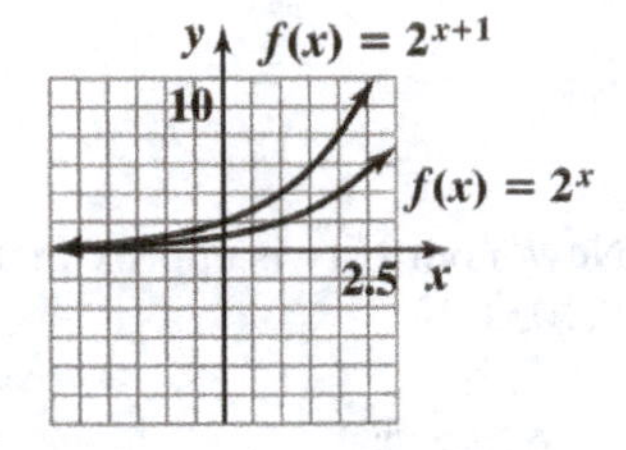

2. Semiannual Compounding:

$$A = 3000\left(1 + \frac{0.065}{2}\right)^{2(10)}$$
$$= 3000(1.0325)^{20} \approx 5687.51$$

Continuous Compounding:

$$A = 3000e^{0.06(10)} = 3000e^{0.6} \approx 5466.36$$

Semiannual compounding at 6.5% yields a greater return. The difference in the yields is $221.

3. $f(x) = x^2 + x$ and $g(x) = 3x - 1$

$$(f \circ g)(x) = f(g(x)) = f(3x-1)$$
$$= (3x-1)^2 + (3x-1)$$
$$= 9x^2 - 6x + 1 + 3x - 1$$
$$= 9x^2 - 3x$$

$$(g \circ f)(x) = g(f(x))$$
$$= g(x^2 + x)$$
$$= 3(x^2 + x) - 1$$
$$= 3x^2 + 3x - 1$$

4. $f(x) = 5x - 7$

$y = 5x - 7$

Interchange x and y and solve for y.

$$x = 5y - 7$$
$$x + 7 = 5y$$
$$\frac{x+7}{5} = y$$

$$f^{-1}(x) = \frac{x+7}{5}$$

5. a. The function passes the horizontal line test (i.e., no horizontal line intersects the graph of *f* in more than one point), so we know its inverse is a function.

b. $f(80) = 2000$

c. $f^{-1}(2000)$ represents the income, \$80 thousand, of a family that gives \$2000 to charity.

6. $\log_5 125 = 3$

$5^3 = 125$

7. $\sqrt{36} = 6$

$36^{\frac{1}{2}} = 6$

$\log_{36} 6 = \frac{1}{2}$

8. $f(x) = 3^x$

$g(x) = \log_3 x$

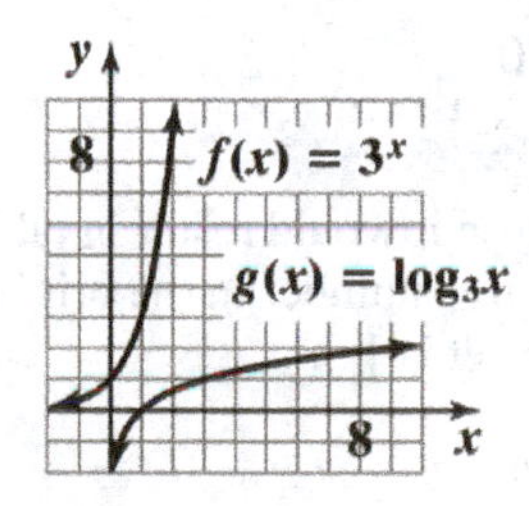

Domain of *f*: $(-\infty, \infty)$.

Range of *f*: $(0, \infty)$

Domain of *g*: $(0, \infty)$

Range of *g*: $(-\infty, \infty)$.

9. Since $\ln e^x = x$, $\ln e^{5x} = 5x$.

10. $\log_b b = 1$ because $b^1 = b$.

11. $\log_6 1 = 0$ because $6^0 = 1$.

12. $f(x) = \log_5 (x-7)$

$x - 7 > 0$

$x > 7$

The domain of *f* is $(7, \infty)$.

13. $D = 10\log\frac{I}{I_0}$

$D = 10\log\frac{10^{12} I_0}{I_0}$

$= 10\log 10^{12} = 10(12) = 120$

The sound has a loudness of 120 decibels.

14. $\log_4 (64x^5) = \log_4 64 + \log_4 x^5$

$= 3 + 5\log_4 x$

15. $\log_3 \frac{\sqrt[3]{x}}{81} = \log_3 \sqrt[3]{x} - \log_3 81$

$= \log_3 x^{\frac{1}{3}} - 4 = \frac{1}{3}\log_3 x - 4$

16. $6\log x + 2\log y = \log x^6 + \log y^2$

$= \log(x^6 y^2)$

17. $\ln 7 - 3\ln x = \ln 7 - \ln x^3 = \ln\left(\frac{7}{x^3}\right)$

18. $\log_{15} 71 = \frac{\ln 71}{\ln 15} \approx 1.5741$

19. $3^{x-2} = 81$

$3^{x-2} = 3^4$

$x - 2 = 4$

$x = 6$

The solution set is $\{6\}$.

20. $5^x = 1.4$

$\ln 5^x = \ln 1.4$

$x\ln 5 = \ln 1.4$

$x = \frac{\ln 1.4}{\ln 5} \approx 0.21$

The solution set is $\left\{\frac{\ln 1.4}{\ln 5} \approx 0.21\right\}$.

21. $400e^{0.005x} = 1600$

$$e^{0.005x} = \frac{1600}{400}$$

$$\ln e^{0.005x} = \ln 4$$
$$0.005x = \ln 4$$
$$x = \frac{\ln 4}{0.005} \approx 277.26$$

The solution set is $\left\{\frac{\ln 4}{0.005} \approx 277.26\right\}$.

22. $\log_{25} x = \frac{1}{2}$

$$x = 25^{\frac{1}{2}} = \sqrt{25} = 5$$

The solution set is $\{5\}$.

23. $\log_6 (4x-1) = 3$

$$4x-1 = 6^3$$
$$4x-1 = 216$$
$$4x = 217$$
$$x = \frac{217}{4}$$

The solution set is $\left\{\frac{217}{4}\right\}$.

24. $2\ln(3x) = 8$

$$\ln(3x) = \frac{8}{2}$$
$$e^{\ln(3x)} = e^4$$
$$3x = e^4$$
$$x = \frac{e^4}{3}$$

The solution set is $\left\{\frac{e^4}{3}\right\}$.

25. $\log x + \log(x+15) = 2$

$$\log(x(x+15)) = 2$$
$$x(x+15) = 10^2$$
$$x^2 + 15 = 100$$
$$x^2 + 15 - 100 = 0$$
$$(x+20)(x-5) = 0$$
$$x+20 = 0 \quad \text{or} \quad x-5 = 0$$
$$x = -20 \qquad\qquad x = 5$$

We disregard -20 because it would result in taking the logarithm of a negative number in the original equation. The solution set is $\{5\}$

26. $\ln(x-4) - \ln(x+1) = \ln 6$

$$\ln\left(\frac{x-4}{x+1}\right) = \ln 6$$
$$\frac{x-4}{x+1} = 6$$
$$(x+1)\frac{x-4}{x+1} = 6(x+1)$$
$$x-4 = 6x+6$$
$$-5x = 10$$
$$x = -2$$

We disregard -2 because it would result in taking the logarithm of a negative number in the original equation. The solution set is $\{\ \}$.

27. **a.** $P(0) = 82.4e^{-0.002(0)}$

$$= 82.4e^0 = 82.4(1) = 82.4$$

In 2006, the population of Germany was 82.4 million.

b. The population of Germany is decreasing. The growth rate, $k = -0.002$, is negative.

c. $80.6 = 82.4e^{-0.002t}$

$$\frac{80.6}{82.4} = e^{-0.002t}$$
$$\ln\frac{80.6}{82.4} = \ln e^{-0.002t}$$
$$\ln\frac{80.6}{82.4} = -0.002t$$
$$t = \frac{\ln\frac{80.6}{82.4}}{-0.002} \approx 11$$

The population of Germany will be 80.6 million approximately 11 years after 2006, or 2017.

28.
$$8000 = 4000\left(1+\frac{0.05}{4}\right)^{4t}$$
$$\frac{8000}{4000} = (1+0.0125)^{4t}$$
$$2 = (1.0125)^{4t}$$
$$\ln 2 = \ln(1.0125)^{4t}$$
$$\ln 2 = 4t\ln(1.0125)$$
$$\frac{\ln 2}{4\ln(1.0125)} = \frac{4t\ln(1.0125)}{4\ln(1.0125)}$$
$$t = \frac{\ln 2}{4\ln(1.0125)} \approx 13.9$$

It will take approximately 13.9 years for the money to grow to \$8000.

29.
$$2 = 1e^{r10}$$
$$2 = e^{10r}$$
$$\ln 2 = \ln e^{10r}$$
$$\ln 2 = 10r$$
$$r = \frac{\ln 2}{10} \approx 0.069$$

The money will double in 10 years with an interest rate of approximately 6.9%.

30. Substitute $A_0 = 509$, $A = 729$, and $t = 2000-1990 = 10$ into the general growth function to determine the growth rate k:
$$A = A_0e^{kt}$$
$$729 = 509e^{k(10)}$$
$$\frac{729}{509} = e^{10k}$$
$$\ln\frac{729}{509} = \ln e^{10k}$$
$$\ln\frac{729}{509} = 10k$$
$$k = \frac{\ln\frac{729}{509}}{10} \approx 0.036$$

The exponential growth function is $A = 509e^{0.036t}$.

31.
$$A = A_0e^{-0.000121t}$$
$$5 = 100e^{-0.000121t}$$
$$\frac{5}{100} = e^{-0.000121t}$$
$$\ln 0.05 = \ln e^{-0.000121t}$$
$$\ln 0.05 = -0.000121t$$
$$t = \frac{\ln 0.05}{-0.000121} \approx 24758$$

The man died approximately 24,758 years ago.

32. Plot the ordered pairs.

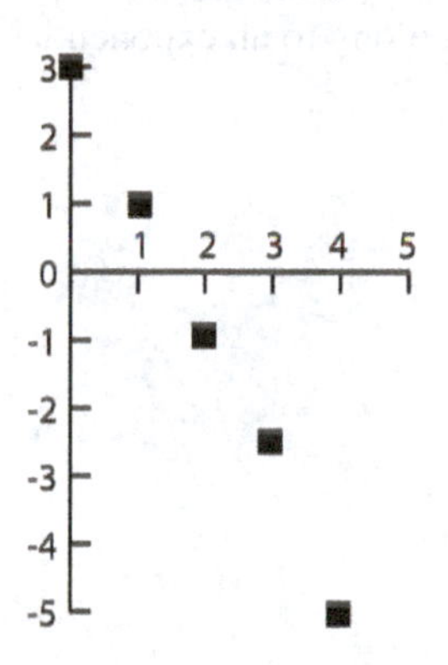

The values appear to belong to a linear function.

33. Plot the ordered pairs.

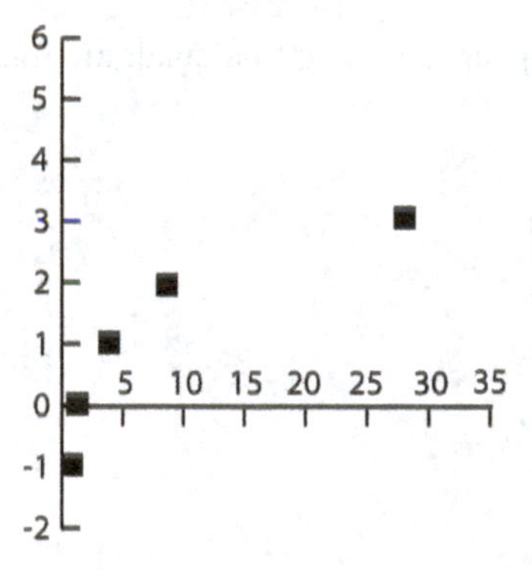

The values appear to belong to a logarithmic function.

34. Plot the ordered pairs.

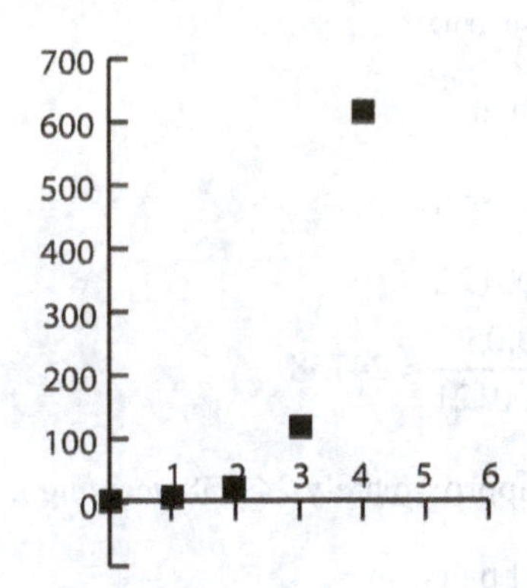

The values appear to belong to an exponential function.

35. Plot the ordered pairs.

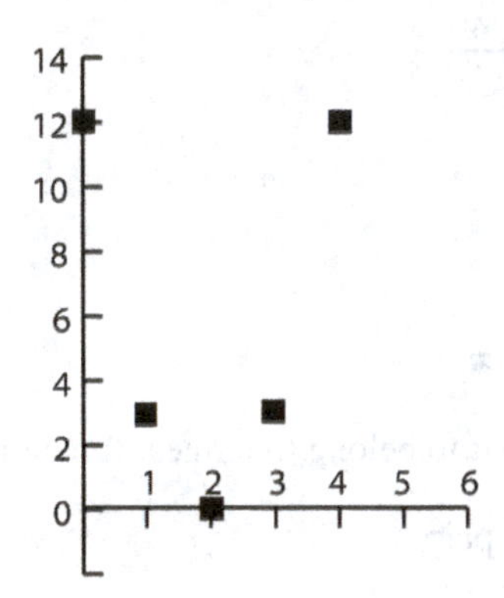

The values appear to belong to a quadratic function.

36. $y = 96(0.38)^x$

$y = 96e^{(\ln 0.38)x}$

$y = 96e^{-0.968x}$

Cumulative Review Exercises (Chapters 1 – 9)

1.
$$\begin{aligned} 8-(4x-5) &= x-7 \\ 8-4x+5 &= x-7 \\ 13-4x &= x-7 \\ 13 &= 5x-7 \\ 20 &= 5x \\ 4 &= x \end{aligned}$$

The solution set is $\{4\}$.

2.
$$\begin{aligned} 5x+4y &= 22 \\ 3x-8y &= -18 \end{aligned}$$

Multiply the first equation by 2 and solve by addition.

$$\begin{aligned} 10x+8y &= 44 \\ 3x-8y &= -18 \\ \hline 13x &= 26 \\ x &= 2 \end{aligned}$$

Back-substitute 2 for x to find y.

$$\begin{aligned} 5(2)+4y &= 22 \\ 10+4y &= 22 \\ 4y &= 12 \\ y &= 3 \end{aligned}$$

The solution set is $\{(2,3)\}$.

3.
$$\begin{aligned} -3x+2y+4z &= 6 \\ 7x-y+3z &= 23 \\ 2x+3y+z &= 7 \end{aligned}$$

Multiply the second equation by 2 and add to the first equation to eliminate y.

$$\begin{aligned} -3x+2y+4z &= 6 \\ 14x-2y+6z &= 46 \\ \hline 11x+10z &= 52 \end{aligned}$$

Multiply the second equation by 3 and add to the third equation to eliminate y.

$$\begin{aligned} 21x-3y+9z &= 69 \\ 2x+3y+z &= 7 \\ \hline 23x+10z &= 76 \end{aligned}$$

The system of two variables in two equations is:

$11x+10z=52$

$23x+10z=76$

Multiply the first equation by –1 and add to the second equation.

$$\begin{aligned}-11x-10z&=-52\\ \underline{23x+10z}&=\underline{\ 76}\\ 12x&=24\\ x&=2\end{aligned}$$

Back-substitute 2 for x to find z.

$$\begin{aligned}11(2)+10z&=52\\ 22+10z&=52\\ 10z&=30\\ z&=3\end{aligned}$$

Back-substitute 2 for x and 3 for z to find y.

$$\begin{aligned}-3(2)+2y+4(3)&=6\\ -6+2y+12&=6\\ 2y&=0\\ y&=0\end{aligned}$$

The solution is {(2, 0, 3)}.

4. $|x-1|>3$

$x-1<-3$ or $x-1>3$

$x<-2$ $\quad$ $x>4$

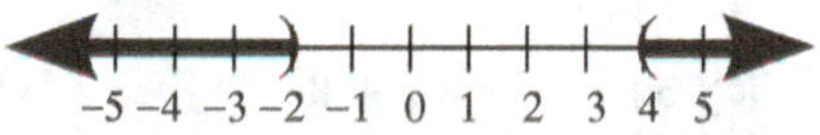

The solution set is $(-\infty,-2)\cup(4,\infty)$.

5. $\sqrt{x+4}-\sqrt{x-4}=2$

$$\begin{aligned}\sqrt{x+4}&=2+\sqrt{x-4}\\ \left(\sqrt{x+4}\right)^2&=\left(2+\sqrt{x-4}\right)^2\\ x+4&=4+4\sqrt{x-4}+x-4\\ \not{x}+4&=4\sqrt{x-4}+\not{x}\\ 4&=4\sqrt{x-4}\\ 1&=\sqrt{x-4}\\ 1^2&=\left(\sqrt{x-4}\right)^2\\ 1&=x-4\\ 5&=x\end{aligned}$$

The solution set is $\{5\}$.

6. $x-4\ge 0$ and $-3x\le -6$

$x\ge 4$ $\quad$ $x\ge 2$

For a value to be in the solution set, it must satisfy both of the conditions $x\ge 4$ and $x\ge 2$. Now any value that is 4 or larger is also larger than 2. But values between 2 and 4 do not satisfy both conditions. Therefore, only values that are 4 or larger will be in the solution set. Thus, the solution set is $[4,\infty)$.

7. $2x^2=3x-2$

$2x^2-3x+2=0$

Solve using the quadratic formula.

$a=2$ $\quad$ $b=-3$ $\quad$ $c=2$

$$\begin{aligned}x&=\frac{-(-3)\pm\sqrt{(-3)^2-4(2)(2)}}{2(2)}\\ &=\frac{3\pm\sqrt{9-16}}{4}\\ &=\frac{3\pm\sqrt{-7}}{4}=\frac{3\pm i\sqrt{7}}{4}=\frac{3}{4}\pm\frac{\sqrt{7}}{4}i\end{aligned}$$

The solutions are $\frac{3}{4}\pm\frac{\sqrt{7}}{4}i$, and the solution set is $\left\{\frac{3}{4}-\frac{\sqrt{7}}{4}i,\frac{3}{4}+\frac{\sqrt{7}}{4}i\right\}$.

8. $3x=15+5y$

Find the x–intercept by setting $y=0$ and solving.

$$\begin{aligned}3x&=15+\cancel{5(0)}\\ 3x&=15\\ x&=5\end{aligned}$$

Find the y–intercept by setting $x=0$ and solving.

$$\begin{aligned}\cancel{3(0)}&=15+5y\\ 0&=15+5y\\ -15&=5y\\ -3&=y\end{aligned}$$

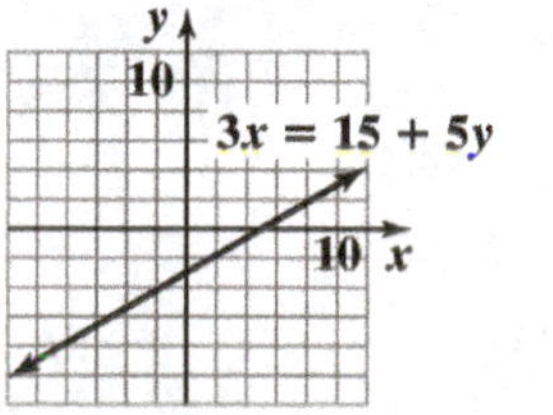

9. $2x-3y>6$

First, find the intercepts to the equation $2x-3y=6$.

Find the x–intercept by setting $y = 0$ and solving.

$$2x-3(0)=6$$
$$2x=6$$
$$x=3$$

Find the y–intercept by setting $x = 0$ and solving.

$$2(0)-3y=6$$
$$-3y=6$$
$$y=-2$$

Next, use the origin as a test point.

$$2(0)-3(0)>6$$
$$0>6$$

This is a false statement. This means that the origin will not fall in the shaded half-plane.

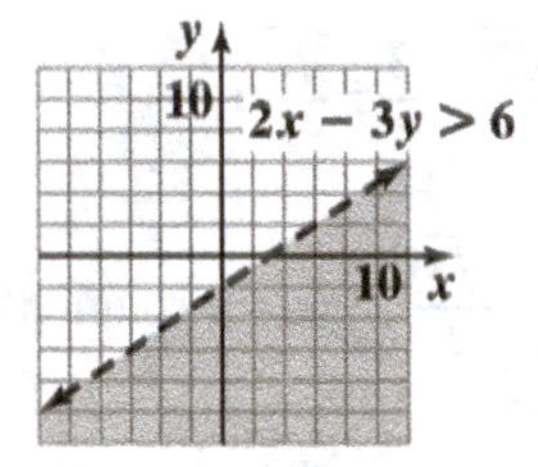

10. $f(x)=-\dfrac{1}{2}x+1$

$m=-\dfrac{1}{2}$; $y-\text{intercept}=1$

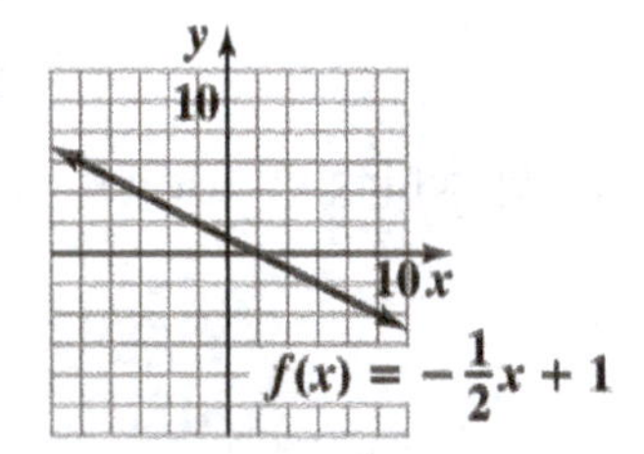

11. $f(x)=x^2+6x+8$

Since $a=1$ is positive, the parabola opens upward. The x–coordinate of the vertex of the parabola is $-\dfrac{b}{2a}=-\dfrac{6}{2(1)}=-\dfrac{6}{2}=-3$ and the y–coordinate of the vertex of the parabola is

$$f\left(-\frac{b}{2a}\right)=f(-3)=(-3)^2+6(-3)+8$$
$$=9-18+8=-1.$$

The vertex is $(-3,-1)$. Replace $f(x)$ with 0 to find x–intercepts.

$$0=x^2+6x+8$$
$$0=(x+4)(x+2)$$
$$x+4=0 \quad \text{or} \quad x+2=0$$
$$x=-4 \qquad\qquad x=-2$$

The x–intercepts are –4 and –2. Set $x = 0$ and solve for y to obtain the y–intercept.

$$y=0^2+6(0)+8$$
$$y=0+0+8$$
$$y=8$$

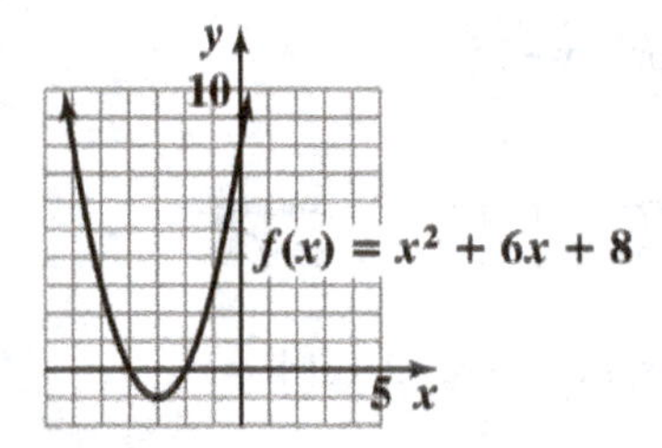

12. $f(x)=(x-3)^2-4$

Since $a=1$ is positive, the parabola opens upward. The vertex of the parabola is $(h,k)=(3,-4)$ and the axis of symmetry is $x=3$. Replace $f(x)$ with 0 to find x–intercepts.

$$0=(x-3)^2-4$$
$$4=(x-3)^2$$

Apply the square root property.

$$x-3=-2 \quad \text{or} \quad x-3=2$$
$$x=1 \qquad\qquad x=5$$

The x–intercepts are 1 and 5.

Set $x = 0$ and solve for y to obtain the y–intercept.

$y = (0-3)^2 - 4$
$y = (-3)^2 - 4$
$y = 9 - 4$
$y = 5$

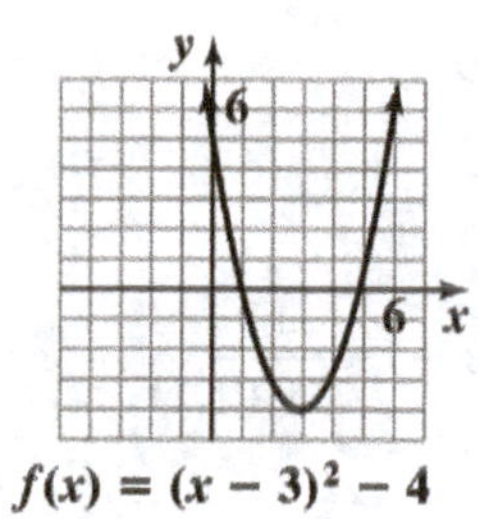

$f(x) = (x-3)^2 - 4$

13. $\begin{vmatrix} 3 & 1 & 0 \\ 0 & 5 & -6 \\ -2 & -1 & 0 \end{vmatrix}$

$= 3\begin{vmatrix} 5 & -6 \\ -1 & 0 \end{vmatrix} - 0\begin{vmatrix} 1 & 0 \\ -1 & 0 \end{vmatrix} + (-2)\begin{vmatrix} 1 & 0 \\ 5 & -6 \end{vmatrix}$

$= 3(5(0)-(-1)(-6)) + (-2)(1(-6)-5(0))$

$= 3(-6) + (-2)(-6) = -18 + 12 = -6$

14. $A = \dfrac{cd}{c+d}$

$A(c+d) = cd$
$Ac + Ad = cd$
$Ac - cd = -Ad$
$c(A-d) = -Ad$
$c = -\dfrac{Ad}{A-d}$ or $\dfrac{Ad}{d-A}$

15. $f(g(x)) = [g(x)]^2 + 3[g(x)] - 15$
$= (x-2)^2 + 3(x-2) - 15$
$= x^2 - 4x + 4 + 3x - 6 - 15$
$= x^2 - x - 17$

16. $g(f(x)) = f(x) - 2$
$= (x^2 + 3x - 15) - 2$
$= x^2 + 3x - 17$

17. $g(a+h) - g(a) = ((a+h)-2) - (a-2)$
$= a + h - 2 - a + 2$
$= h$

18. $f(x) = 7x - 3$
$y = 7x - 3$

Interchange x and y, and solve for y.

$x = 7y - 3$
$x + 3 = 7y$
$\dfrac{x+3}{7} = y$

Thus, $f^{-1}(x) = \dfrac{x+3}{7}$.

19. $f(x) = \dfrac{x-2}{x^2 - 3x + 2}$

Since a denominator cannot equal zero, exclude from the domain all values which make $x^2 - 3x + 2 = 0$.

$x^2 - 3x + 2 = 0$
$(x-2)(x-1) = 0$
$x - 2 = 0$ or $x - 1 = 0$
$x = 2$ $\quad$ $x = 1$

The domain of f is $(-\infty, 1) \cup (1, 2) \cup (2, \infty)$.

20. $f(x) = \ln(2x - 8)$

To find the domain, find all values of x for which $2x - 8$ is greater than zero.

$2x - 8 > 0$
$2x > 8$
$x > 4$

The domain of f is $(4, \infty)$.

21. First, solve for y to obtain the slope of the line whose equation is $2x + y = 10$.

$$2x + y = 10$$
$$y = -2x + 10$$

The slope is –2. The line we want to find is perpendicular to this line, so we know the slope will be $\frac{1}{2}$.

Using the point, $(-2, 4)$, and the slope, $\frac{1}{2}$, we can write the equation in point-slope form.

$$y - y_1 = m(x - x_1)$$
$$y - 4 = \frac{1}{2}(x - (-2))$$
$$y - 4 = \frac{1}{2}(x + 2)$$

Solve for y to obtain slope-intercept form.

$$y - 4 = \frac{1}{2}(x + 2)$$
$$y - 4 = \frac{1}{2}x + 1$$
$$y = \frac{1}{2}x + 5$$
$$f(x) = \frac{1}{2}x + 5$$

22. $\dfrac{-5x^3y^7}{15x^4y^{-2}} = \dfrac{-y^7y^2}{3x} = \dfrac{-y^9}{3x} = -\dfrac{y^9}{3x}$

23. $(4x^2 - 5y)^2$

$$= (4x^2)^2 + 2(4x^2)(-5y) + (-5y)^2$$
$$= 16x^4 - 40x^2y + 25y^2$$

24.

$$\begin{array}{r|l} & x^2 - 5x + 1 \\ 5x+1 & 5x^3 - 24x^2 + 0x + 9 \\ & \underline{5x^3 + \quad x^2} \\ & -25x^2 + 0x \\ & \underline{-25x^2 - 5x} \\ & 5x + 9 \\ & \underline{5x + 1} \\ & 8 \end{array}$$

$$\frac{5x^3 - 24x^2 + 9}{5x + 1} = x^2 - 5x + 1 + \frac{8}{5x + 1}$$

25.

$$\frac{\sqrt[3]{32xy^{10}}}{\sqrt[3]{2xy^2}} = \sqrt[3]{\frac{32xy^{10}}{2xy^2}}$$
$$= \sqrt[3]{16y^8}$$
$$= \sqrt[3]{8 \cdot 2y^6y^2} = 2y^2\sqrt[3]{2y^2}$$

26.

$$\frac{x+2}{x^2 - 6x + 8} + \frac{3x - 8}{x^2 - 5x + 6}$$
$$= \frac{x+2}{(x-4)(x-2)} + \frac{3x-8}{(x-2)(x-3)}$$
$$= \frac{(x+2)(x-3)}{(x-4)(x-2)(x-3)} + \frac{(3x-8)(x-4)}{(x-4)(x-2)(x-3)}$$
$$= \frac{x^2 - 3x + 2x - 6 + 3x^2 - 12x - 8x + 32}{(x-4)(x-2)(x-3)}$$
$$= \frac{4x^2 - 21x + 26}{(x-4)(x-2)(x-3)}$$
$$= \frac{(4x-13)\cancel{(x-2)}}{(x-4)\cancel{(x-2)}(x-3)} = \frac{4x-13}{(x-4)(x-3)}$$

27. $x^4 - 4x^3 + 8x - 32$

$$= x^3(x-4) + 8(x-4)$$
$$= (x-4)(x^3 + 8)$$
$$= (x-4)(x+2)(x^2 - 2x + 4)$$

28. $2x^2 + 12xy + 18y^2$

$$= 2(x^2 + 6xy + 9y^2) = 2(x + 3y)^2$$

29. $2\ln x - \frac{1}{2}\ln y = \ln x^2 - \ln y^{\frac{1}{2}}$

$$= \ln\left(\frac{x^2}{y^{\frac{1}{2}}}\right)$$

$$= \ln\left(\frac{x^2}{\sqrt{y}}\right)$$

30. Let x = the width of the carpet.
Let $2x + 4$ = the length of the carpet.

$$x(2x+4) = 48$$
$$2x^2 + 4x = 48$$
$$2x^2 + 4x - 48 = 0$$
$$x^2 + 2x - 24 = 0$$
$$(x+6)(x-4) = 0$$
$$x+6=0 \quad \text{and} \quad x-4=0$$
$$x=-6 \qquad x=4$$

We disregard –6 because we can't have a negative length measurement. The width of the carpet is 4 feet and the length of the carpet is
$2x+4 = 2(4)+4 = 8+4 = 12$ feet.

31. Let x = time it takes when working together.

	Part done in 1 hour	Time Working Together	Part done in x hours
You	$\frac{1}{2}$	x	$\frac{x}{2}$
Your Sister	$\frac{1}{3}$	x	$\frac{x}{3}$

$$\frac{x}{2} + \frac{x}{3} = 1$$
$$6\left(\frac{x}{2} + \frac{x}{3}\right) = 6(1)$$
$$6\left(\frac{x}{2}\right) + 6\left(\frac{x}{3}\right) = 6$$
$$3x + 2x = 6$$
$$5x = 6$$
$$x = \frac{6}{5}$$

If you and your sister work together, it will take $\frac{6}{5}$ hours, or 1 hour and 12 minutes, to clean the house.

32. Let x = the rate of the current.

	distance d	rate r	time $t = \frac{d}{r}$
with the current	20	$15+x$	$\frac{20}{15+x}$
against the current	10	$15-x$	$\frac{10}{15-x}$

$$\frac{20}{15+x} = \frac{10}{15-x}$$
$$20(15-x) = 10(15+x)$$
$$300 - 20x = 150 + 10x$$
$$300 = 150 + 30x$$
$$150 = 30x$$
$$5 = x$$

The rate of the current is 5 miles per hour.

33.
$$A = Pe^{rt}$$
$$18,000 = 6000e^{r(10)}$$
$$3 = e^{10r}$$
$$\ln 3 = \ln e^{10r}$$
$$\ln 3 = 10r$$
$$r = \frac{\ln 3}{10} \approx 0.11$$

An interest rate of approximately 11% compounded continuously would be required for \$6000 to grow to \$18,000 in 10 years.

Chapter 10
Conic Sections and Systems of Nonlinear Equations

10.1 Check Points

1. $d = \sqrt{(x_2 - x_1)^2 + (y_2 - y_1)^2}$

$d = \sqrt{(2-(-1))^2 + (3-(-3))^2}$

$= \sqrt{3^2 + 6^2} = \sqrt{9+36}$

$= \sqrt{45}$

$= 3\sqrt{5}$

≈ 6.71 units

2. Midpoint $= \left(\frac{x_1 + x_2}{2}, \frac{y_1 + y_2}{2}\right)$

$= \left(\frac{1+7}{2}, \frac{2+(-3)}{2}\right)$

$= \left(\frac{8}{2}, \frac{-1}{2}\right) = \left(4, -\frac{1}{2}\right)$

3. $(x-h)^2 + (y-k)^2 = r^2$

$(x-0)^2 + (y-0)^2 = 4^2$

$x^2 + y^2 = 16$

4. $(x-h)^2 + (y-k)^2 = r^2$

$(x-5)^2 + (y-(-6))^2 = 10^2$

$(x-5)^2 + (y+6)^2 = 100$

5. $(x+3)^2 + (y-1)^2 = 4$

$(x-(-3))^2 + (y-1)^2 = 2^2$

The center is $(-3, 1)$ and the radius is 2 units.

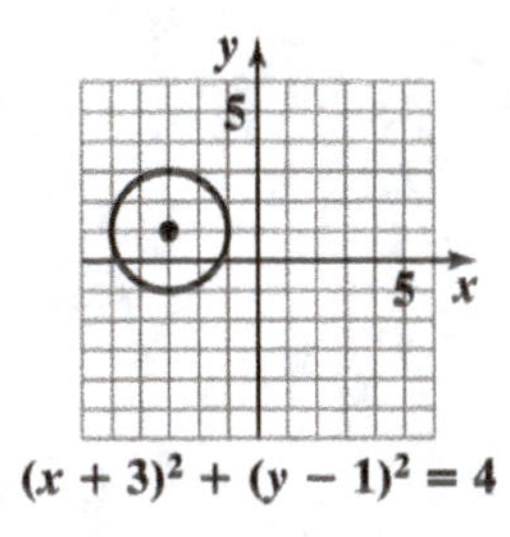

$(x + 3)^2 + (y - 1)^2 = 4$

6. $x^2 + y^2 + 4x - 4y - 1 = 0$

$(x^2 + 4x \quad) + (y^2 - 4y \quad) = 1$

Complete the squares.

$\left(\frac{b}{2}\right)^2 = \left(\frac{4}{2}\right)^2 = (2)^2 = 4$

$\left(\frac{b}{2}\right)^2 = \left(\frac{-4}{2}\right)^2 = (-2)^2 = 4$

$(x^2 + 4x + 4) + (y^2 - 4y + 4) = 1 + 4 + 4$

$(x+2)^2 + (y-2)^2 = 9$

$(x-(-2))^2 + (y-2)^2 = 3^2$

The center is $(-2, 2)$ and the radius is 3 units.

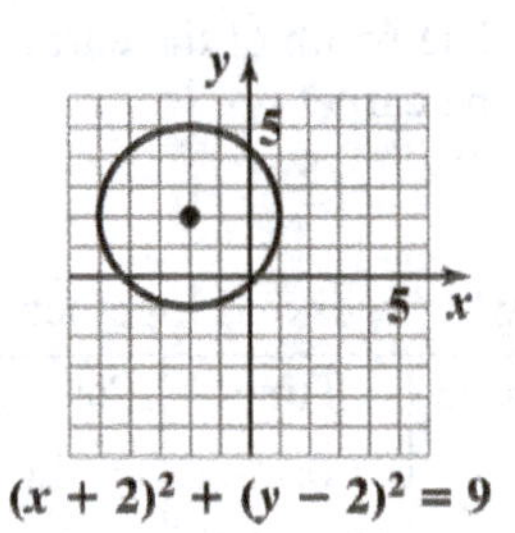

$(x + 2)^2 + (y - 2)^2 = 9$

10.1 Concept and Vocabulary Check

1. $\sqrt{(x_2 - x_1)^2 + (y_2 - y_1)^2}$

2. $\left(\frac{x_1 + x_2}{2}, \frac{y_1 + y_2}{2}\right)$

3. circle; center; radius

4. $(x-h)^2 + (y-k)^2 = r^2$

5. general

6. 4; 16

10.1 Exercise Set

1. $$d=\sqrt{(14-2)^2+(8-3)^2}$$
$$=\sqrt{12^2+5^2}=\sqrt{144+25}$$
$$=\sqrt{169}=13$$

3. $$d=\sqrt{(6-4)^2+(3-1)^2}$$
$$=\sqrt{2^2+2^2}=\sqrt{4+4}$$
$$=\sqrt{8}=\sqrt{4\cdot 2}=2\sqrt{2}\approx 2.83$$

5. $$d=\sqrt{(-3-0)^2+(4-0)^2}$$
$$=\sqrt{(-3)^2+4^2}=\sqrt{9+16}$$
$$=\sqrt{25}=5$$

7. $$d=\sqrt{(3-(-2))^2+(-4-(-6))^2}$$
$$=\sqrt{5^2+2^2}=\sqrt{25+4}$$
$$=\sqrt{29}\approx 5.39$$

9. $$d=\sqrt{(4-0)^2+(1-(-3))^2}$$
$$=\sqrt{4^2+4^2}=\sqrt{16+16}$$
$$=\sqrt{32}=\sqrt{16\cdot 2}=4\sqrt{2}\approx 5.66$$

11. $$d=\sqrt{(3.5-(-0.5))^2+(8.2-6.2)^2}$$
$$=\sqrt{4^2+2^2}=\sqrt{16+4}$$
$$=\sqrt{20}=\sqrt{4\cdot 5}=2\sqrt{5}\approx 4.47$$

13. $$d=\sqrt{(\sqrt{5}-0)^2+(0-(-\sqrt{3}))^2}$$
$$=\sqrt{(\sqrt{5})^2+(\sqrt{3})^2}=\sqrt{5+3}$$
$$=\sqrt{8}=\sqrt{4\cdot 2}=2\sqrt{2}\approx 2.83$$

15. $$d=\sqrt{(3\sqrt{3}-(-\sqrt{3}))^2+(\sqrt{5}-4\sqrt{5})^2}$$
$$=\sqrt{(4\sqrt{3})^2+(-3\sqrt{5})^2}$$
$$=\sqrt{16\cdot 3+9\cdot 5}=\sqrt{48+45}$$
$$=\sqrt{93}\approx 9.64$$

17. $$d=\sqrt{\left(\frac{7}{3}-\frac{1}{3}\right)^2+\left(\frac{1}{5}-\frac{6}{5}\right)^2}$$
$$=\sqrt{\left(\frac{6}{3}\right)^2+\left(-\frac{5}{5}\right)^2}$$
$$=\sqrt{2^2+(-1)^2}=\sqrt{4+1}$$
$$=\sqrt{5}\approx 2.24$$

19. $$\text{Midpoint}=\left(\frac{6+2}{2},\frac{8+4}{2}\right)$$
$$=\left(\frac{8}{2},\frac{12}{2}\right)=(4,6)$$

21. $$\text{Midpoint}=\left(\frac{-2+(-6)}{2},\frac{-8+(-2)}{2}\right)$$
$$=\left(\frac{-8}{2},\frac{-10}{2}\right)=(-4,-5)$$

23. $$\text{Midpoint}=\left(\frac{-3+6}{2},\frac{-4+(-8)}{2}\right)$$
$$=\left(\frac{3}{2},\frac{-12}{2}\right)=\left(\frac{3}{2},-6\right)$$

25. $$\text{Midpoint}=\left(\frac{-\frac{7}{2}+\left(-\frac{5}{2}\right)}{2},\frac{\frac{3}{2}+\left(-\frac{11}{2}\right)}{2}\right)$$
$$=\left(\frac{-\frac{12}{2}}{2},\frac{-\frac{8}{2}}{2}\right)$$
$$=\left(-\frac{12}{2}\cdot\frac{1}{2},-\frac{8}{2}\cdot\frac{1}{2}\right)$$
$$=\left(-\frac{12}{4},-\frac{8}{4}\right)=(-3,-2)$$

27. $$\text{Midpoint}=\left(\frac{8+(-6)}{2},\frac{3\sqrt{5}+7\sqrt{5}}{2}\right)$$
$$=\left(\frac{2}{2},\frac{10\sqrt{5}}{2}\right)=\left(1,5\sqrt{5}\right)$$

29. $\text{Midpoint} = \left(\frac{\sqrt{18}+\sqrt{2}}{2}, \frac{-4+4}{2}\right)$
$= \left(\frac{\sqrt{9\cdot 2}+\sqrt{2}}{2}, \frac{0}{2}\right)$
$= \left(\frac{3\sqrt{2}+\sqrt{2}}{2}, 0\right)$
$= \left(\frac{4\sqrt{2}}{2}, 0\right) = \left(2\sqrt{2}, 0\right)$

31. $(x-h)^2+(y-k)^2=r^2$
$(x-0)^2+(y-0)^2=7^2$
$x^2+y^2=49$

33. $(x-h)^2+(y-k)^2=r^2$
$(x-3)^2+(y-2)^2=5^2$
$(x-3)^2+(y-2)^2=25$

35. $(x-h)^2+(y-k)^2=r^2$
$(x-(-1))^2+(y-4)^2=2^2$
$(x+1)^2+(y-4)^2=4$

37. $(x-h)^2+(y-k)^2=r^2$
$(x-(-3))^2+(y-(-1))^2=\left(\sqrt{3}\right)^2$
$(x+3)^2+(y+1)^2=3$

39. $(x-h)^2+(y-k)^2=r^2$
$(x-(-4))^2+(y-0)^2=10^2$
$(x+4)^2+y^2=100$

41. $x^2+y^2=16$
$(x-0)^2+(y-0)^2=4^2$
The center is $(0,0)$ and the radius is 4 units.

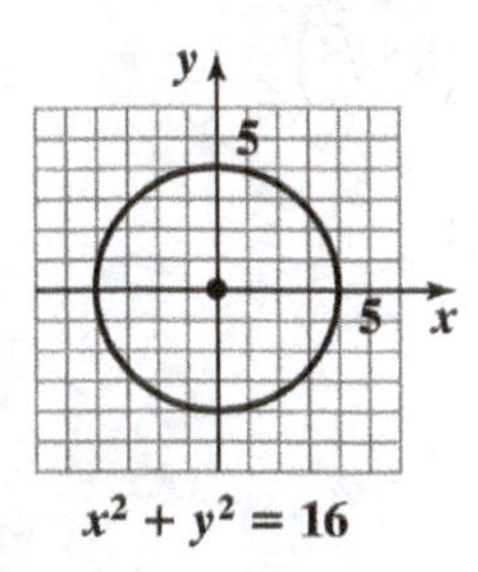

$x^2 + y^2 = 16$

43. $(x-3)^2+(y-1)^2=36$
$(x-3)^2+(y-1)^2=6^2$
The center is $(3,1)$ and the radius is 6 units.

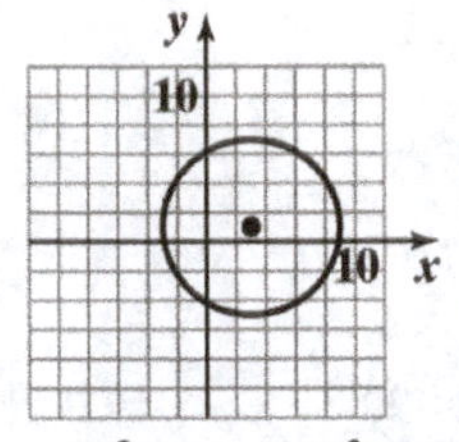

$(x-3)^2+(y-1)^2=36$

45. $(x+3)^2+(y-2)^2=4$
$(x-(-3))^2+(y-2)^2=2^2$
The center is $(-3,2)$ and the radius is 2 units.

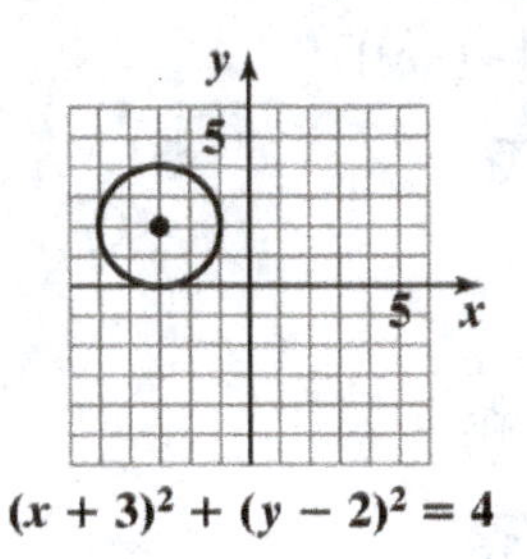

$(x+3)^2+(y-2)^2=4$

47. $(x+2)^2+(y+2)^2=4$
$(x-(-2))^2+(y-(-2))^2=2^2$
The center is $(-2,-2)$ and the radius is 2 units.

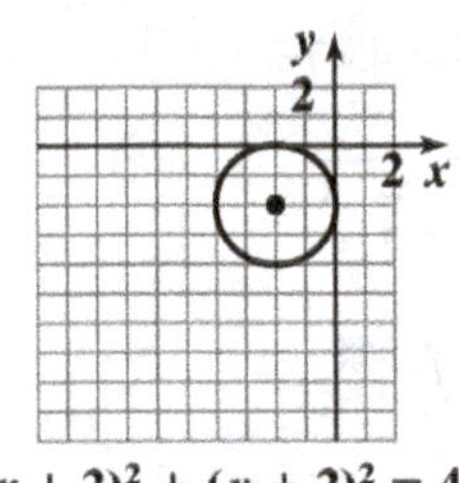

$(x+2)^2+(y+2)^2=4$

49. $x^2+y^2+6x+2y+6=0$

$(x^2+6x\quad)+(y^2+2y\quad)=-6$

Complete the squares.

$\left(\frac{b}{2}\right)^2=\left(\frac{6}{2}\right)^2=(3)^2=9$

$\left(\frac{b}{2}\right)^2=\left(\frac{2}{2}\right)^2=(1)^2=1$

$(x^2+6x+9)+(y^2+2y+1)=-6+9+1$

$(x+3)^2+(y+1)^2=4$

The center is $(-3,-1)$ and the radius is 2 units.

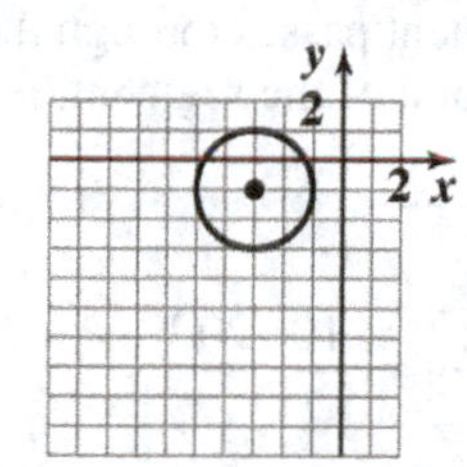

$x^2+y^2+6x+2y+6=0$

51. $x^2+y^2-10x-6y-30=0$

$(x^2-10x\quad)+(y^2-6y\quad)=30$

Complete the squares.

$\left(\frac{b}{2}\right)^2=\left(\frac{-10}{2}\right)^2=(-5)^2=25$

$\left(\frac{b}{2}\right)^2=\left(\frac{-6}{2}\right)^2=(-3)^2=9$

$(x^2-10x+25)+(y^2-6y+9)=30+25+9$

$(x-5)^2+(y-3)^2=64$

The center is $(5,3)$ and the radius is 8 units.

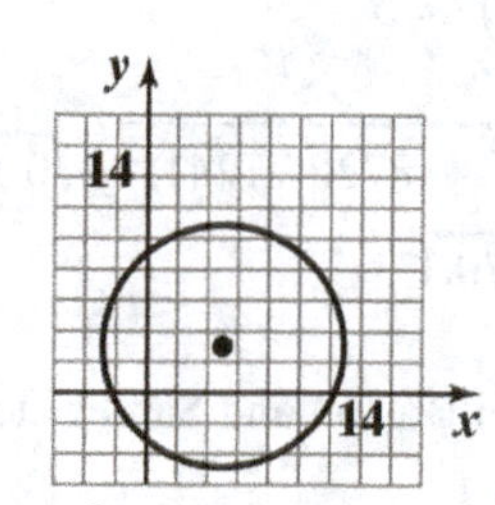

$x^2+y^2-10x-6y-30=0$

53. $x^2+y^2+8x-2y-8=0$

$(x^2+8x\quad)+(y^2-2y\quad)=8$

Complete the squares.

$\left(\frac{b}{2}\right)^2=\left(\frac{8}{2}\right)^2=(4)^2=16$

$\left(\frac{b}{2}\right)^2=\left(\frac{-2}{2}\right)^2=(-1)^2=1$

$(x^2+8x+16)+(y^2-2y+1)=8+16+1$

$(x+4)^2+(y-1)^2=25$

The center is $(-4,1)$ and the radius is 5 units.

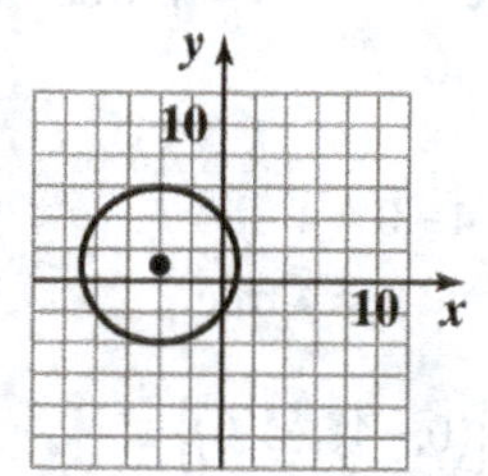

$x^2+y^2+8x-2y-8=0$

55. $x^2-2x+y^2-15=0$

$(x^2-2x\quad)+y^2=15$

Complete the square.

$\left(\frac{b}{2}\right)^2=\left(\frac{-2}{2}\right)^2=(-1)^2=1$

$(x^2-2x+1)+y^2=15+1$

$(x-1)^2+y^2=16$

The center is $(1,0)$ and the radius is 4 units.

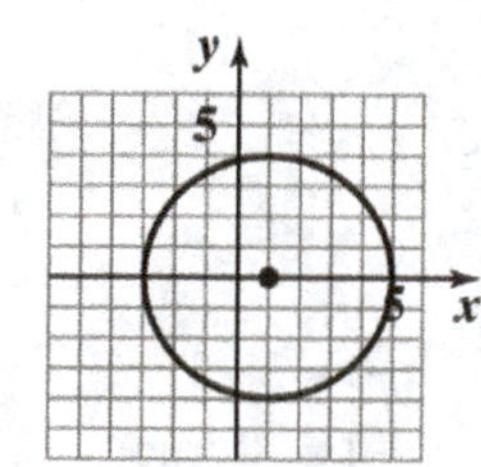

$x^2-2x+y^2-15=0$

57.

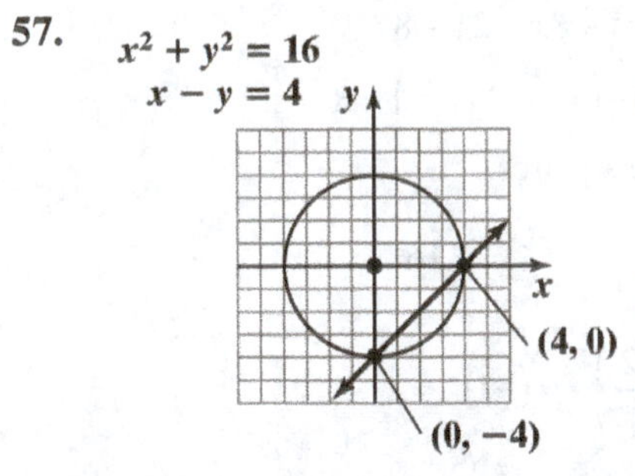

Intersection points: $(0,-4)$ and $(4,0)$

Check $(0,-4)$:

$0^2 + (-4)^2 = 16 \qquad 0-(-4) = 4$

$16 = 16$ true $\qquad 4 = 4$ true

Check $(4,0)$:

$4^2 + 0^2 = 16 \qquad 4 - 0 = 4$

$16 = 16$ true $\qquad 4 = 4$ true

The solution set is $\{(0,-4),(4,0)\}$.

59.

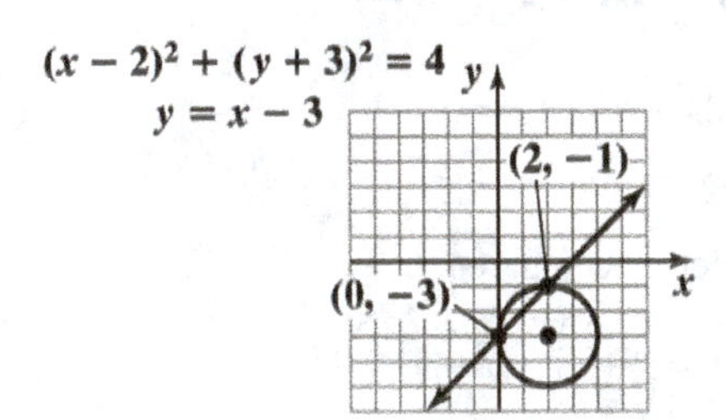

Intersection points: $(0,-3)$ and $(2,-1)$

Check $(0,-3)$:

$(0-2)^2 + (-3+3)^2 = 9 \qquad -3 = 0-3$

$(-2)^2 + 0^2 = 4 \qquad -3 = -3$ true

$4 = 4$

true

Check $(2,-1)$:

$(2-2)^2 + (-1+3)^2 = 4 \qquad -1 = 2-3$

$0^2 + 2^2 = 4 \qquad -1 = -1$ true

$4 = 4$

true

The solution set is $\{(0,-3),(2,-1)\}$.

61. From the graph we can see that the center of the circle is at $(2,-1)$ and the radius is 2 units. Therefore, the equation is

$$(x-2)^2 + (y-(-1))^2 = 2^2$$
$$(x-2)^2 + (y+1)^2 = 4$$

63. From the graph we can see that the center of the circle is at $(-3,-2)$ and the radius is 1 unit. Therefore, the equation is

$$(x-(-3))^2 + (y-(-2))^2 = 1^2$$
$$(x+3)^2 + (y+2)^2 = 1$$

65. a. Since the line segment passes through the center, the center is the midpoint of the segment.

$$M = \left(\frac{x_1 + x_2}{2}, \frac{y_1 + y_2}{2}\right)$$
$$= \left(\frac{3+7}{2}, \frac{9+11}{2}\right) = \left(\frac{10}{2}, \frac{20}{2}\right)$$
$$= (5,10)$$

The center is $(5,10)$.

b. The radius is the distance from the center to one of the points on the circle. Using the point $(3,9)$, we get:

$$d = \sqrt{(5-3)^2 + (10-9)^2}$$
$$= \sqrt{2^2 + 1^2} = \sqrt{4+1}$$
$$= \sqrt{5}$$

The radius is $\sqrt{5}$ units.

c. $(x-5)^2 + (y-10)^2 = (\sqrt{5})^2$

$(x-5)^2 + (y-10)^2 = 5$

67. $d = \sqrt{(8495-4422)^2 + (8720-1241)^2} \cdot \sqrt{0.1}$

$d = \sqrt{72,524,770} \cdot \sqrt{0.1}$

$d \approx 2693$

The distance between Boston and San Francisco is about 2693 miles.

69. If we place L.A. at the origin, then we want the equation of a circle with center at $(-2.4, -2.7)$ and radius 30.

$$(x-(-2.4))^2+(y-(-2.7))^2=30^2$$
$$(x+2.4)^2+(y+2.7)^2=900$$

71. – 77. Answers will vary.

79. $(y+1)^2=36-(x-3)^2$

$$y+1=\pm\sqrt{36-(x-3)^2}$$
$$y=-1\pm\sqrt{36-(x-3)^2}$$

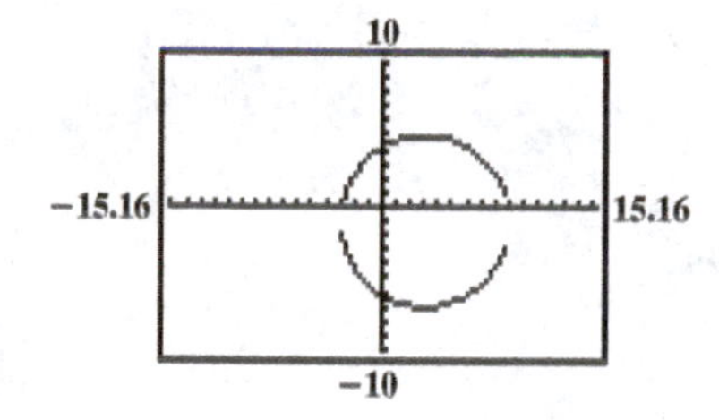

81. makes sense

83. does not make sense; Explanations will vary. Sample explanation: Because of the negative on the constant 4, this is not the equation of a circle.

85. false; Changes to make the statement true will vary. A sample change is: The circle has a radius of 4.

87. false; Changes to make the statement true will vary. A sample change is: Because the variables are not squared, this is a linear equation, not a circle.

89. Distance from A to B:

$$d=\sqrt{(3-1)^2+((3+d)-(1+d))^2}$$
$$=\sqrt{(2)^2+(3+d-1-d)^2}=\sqrt{4+(2)^2}$$
$$=\sqrt{4+4}=\sqrt{8}=\sqrt{4\cdot 2}=2\sqrt{2}$$

Distance from B to C:

$$d=\sqrt{(6-3)^2+((6+d)-(3+d))^2}$$
$$=\sqrt{(3)^2+(6+d-3-d)^2}=\sqrt{9+(3)^2}$$
$$=\sqrt{9+9}=\sqrt{18}=\sqrt{9\cdot 2}=3\sqrt{2}$$

Distance from A to C.

$$d=\sqrt{(6-1)^2+((6+d)-(1+d))^2}$$
$$=\sqrt{(5)^2+(6+d-1-d)^2}=\sqrt{25+(5)^2}$$
$$=\sqrt{25+25}=\sqrt{50}=\sqrt{25\cdot 2}=5\sqrt{2}$$

If the points are collinear, $d_{AB}+d_{BC}=d_{AC}$.

$$d_{AB}+d_{BC}=2\sqrt{2}+3\sqrt{2}=5\sqrt{2}$$

Since this is the same as the distance from A to C, we know that the points are collinear.

91. $d=\sqrt{(x_2-x_1)^2+(y_2-y_1)^2}$

$$5=\sqrt{(x-2)^2+(2-(-1))^2}$$
$$5^2=(x-2)^2+3^2$$
$$25=(x-2)^2+9$$
$$16=(x-2)^2$$
$$\pm 4=x-2$$
$$2\pm 4=x$$

Therefore, $x=2-4=-2$ or $x=2+4=6$. There are two points with y-coordinate 2 whose distance is 5 units from the point $(2,-1)$. The two points are $(-2,2)$ and $(6,2)$.

93. The center of the circle with equation, $x^2+y^2=25$, is the point (0,0). First, find the slope of the line going through the center and the point, (3, –4).

$$m=\frac{y_2-y_1}{x_2-x_1}=\frac{-4-0}{3-0}=-\frac{4}{3}$$

Since the tangent line is perpendicular to the line going through the center and the point, (3, –4), we know that its slope will be $\frac{3}{4}$. We can now write the point-slope equation of the line.

$$y-(-4)=\frac{3}{4}(x-3)$$
$$y+4=\frac{3}{4}(x-3)$$

94. $f\left(g\left(x\right)\right) = f\left(3x+4\right) = \left(3x+4\right)^2 - 2$
$= 9x^2 + 24x + 16 - 2$
$= 9x^2 + 24x + 14$

$g\left(f\left(x\right)\right) = g\left(x^2 - 2\right) = 3\left(x^2 - 2\right) + 4$
$= 3x^2 - 6 + 4 = 3x^2 - 2$

95. $2x = \sqrt{7x-3} + 3$
$2x - 3 = \sqrt{7x-3}$
$\left(2x-3\right)^2 = 7x - 3$
$4x^2 - 12x + 9 = 7x - 3$
$4x^2 - 19x + 12 = 0$
$\left(4x-3\right)\left(x-4\right) = 0$
$4x - 3 = 0$ or $x - 4 = 0$
$4x = 3$ $\quad x = 4$
$x = \frac{3}{4}$

The solution $\frac{3}{4}$ does not check. The solution set is $\{4\}$.

96. $|2x-5| < 10$
$-10 < 2x - 5 < 10$
$-10 + 5 < 2x - 5 + 5 < 10 + 5$
$-5 < 2x < 15$
$-\frac{5}{2} < x < \frac{15}{2}$

The solution set is $\left\{x \middle| -\frac{5}{2} < x < \frac{15}{2}\right\}$ or $\left(-\frac{5}{2}, \frac{15}{2}\right)$.

97. $\frac{x^2}{9} + \frac{y^2}{4} = 1$
$\frac{x^2}{9} + \frac{0^2}{4} = 1$
$\frac{x^2}{9} = 1$
$x^2 = 9$
$x = \pm 3$
The x-intercepts are -3 and 3.

98. $\frac{x^2}{9} + \frac{y^2}{4} = 1$
$\frac{0^2}{9} + \frac{y^2}{4} = 1$
$\frac{y^2}{4} = 1$
$y^2 = 4$
$x = \pm 2$
The y-intercepts are -2 and 2.

99. $25x^2 + 16y^2 = 400$
$\frac{25x^2}{400} + \frac{16y^2}{400} = \frac{400}{400}$
$\frac{x^2}{16} + \frac{y^2}{25} = 1$

10.2 Check Points

1. $\frac{x^2}{36} + \frac{y^2}{9} = 1$

Because the denominator of the x^2 – term is greater than the denominator of the y^2 – term, the major axis is horizontal. Since $a^2 = 36$, $a = 6$ and the vertices are $(-6, 0)$ and $(6, 0)$. Since $b^2 = 9$, $b = 3$ and endpoints of the minor axis are $(0, -3)$ and $(0, 3)$.

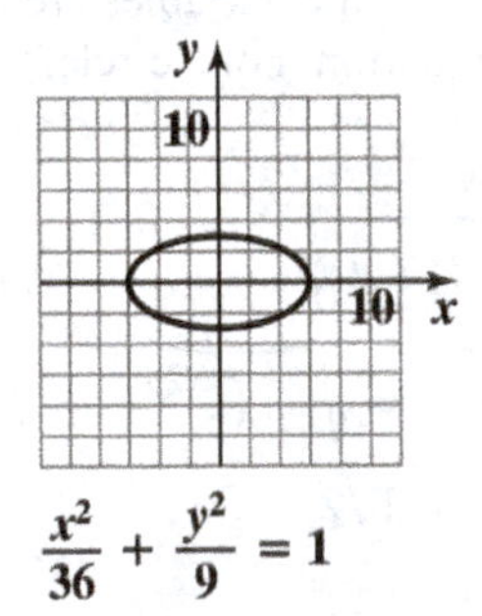

$\frac{x^2}{36} + \frac{y^2}{9} = 1$

2. $16x^2 + 9y^2 = 144$

$$\frac{16x^2}{144} + \frac{9y^2}{144} = \frac{144}{144}$$

$$\frac{x^2}{9} + \frac{y^2}{16} = 1$$

Because the denominator of the y^2 – term is greater than the denominator of the x^2 – term, the major axis is vertical. Since $a^2 = 16$, $a = 4$ and the vertices are $(0,-4)$ and $(0,4)$. Since $b^2 = 9$, $b = 3$ and endpoints of the minor axis are $(-3,0)$ and $(3,0)$.

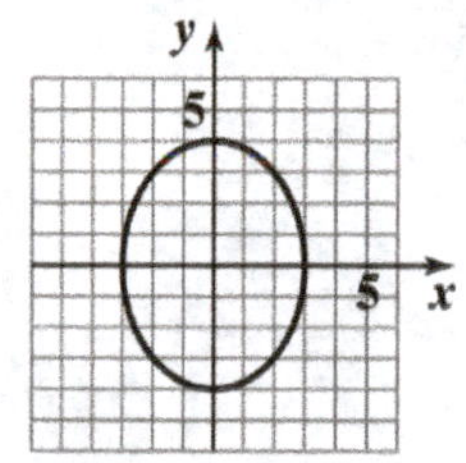

$16x^2 + 9y^2 = 144$

3. $\frac{(x+1)^2}{9} + \frac{(y-2)^2}{4} = 1$

The center of the ellipse is $(-1,2)$. Because the denominator of the x^2 – term is greater than the denominator of the y^2 – term, the major axis is horizontal. Since $a^2 = 9$, $a = 3$ and the vertices lie 3 units to the right and left of the center. Since $b^2 = 4$, $b = 2$ and endpoints of the minor axis lie 2 units above and below the center.

Center	Vertices	Endpoints Minor Axis
$(-1,2)$	$(-1-3,2)$ $=(-4,2)$	$(-1,2-2)$ $=(-1,0)$
	$(-1+3,2)$ $=(2,2)$	$(-1,2+2)$ $=(-1,4)$

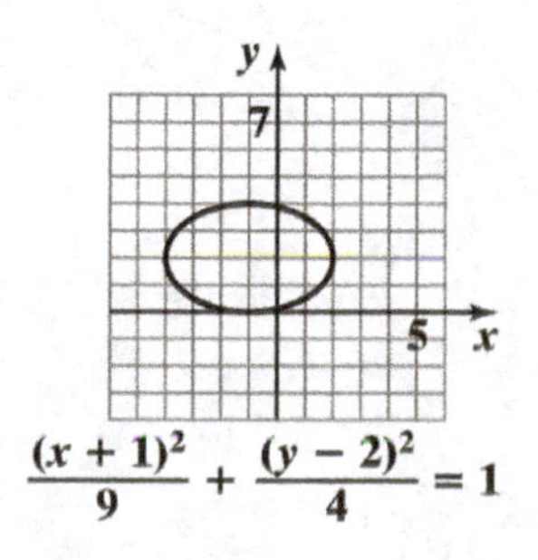

$\frac{(x+1)^2}{9} + \frac{(y-2)^2}{4} = 1$

4. Using the equation $\frac{x^2}{a^2} + \frac{y^2}{b^2} = 1$ the archway can be expressed as $\frac{x^2}{20^2} + \frac{y^2}{10^2} = 1$ or $\frac{x^2}{400} + \frac{y^2}{100} = 1$.

Since the truck is 12 feet wide, we need to determine the height of the archway at $\frac{12}{2} = 6$ feet from the center.

Substitute 6 for x to find the height y.

$$\frac{6^2}{400} + \frac{y^2}{100} = 1$$

$$\frac{36}{400} + \frac{y^2}{100} = 1$$

$$400\left(\frac{36}{400} + \frac{y^2}{100}\right) = 400(1)$$

$$36 + 4y^2 = 400$$

$$4y^2 = 364$$

$$y^2 = 91$$

$$y = \sqrt{91} \approx 9.54$$

The height of the archway 6 feet from the center is approximately 9.54 feet. Since the truck is 9 feet high, the truck will clear the archway.

10.2 Concept and Vocabulary Check

1. ellipse; foci; center
2. 25; -5; 5; $(-5,0)$; $(5,0)$; 9; -3; 3; $(0,-3)$; $(0,3)$
3. 25; -5; 5; $(0,-5)$; $(0,5)$; 9; -3; 3; $(-3,0)$; $(3,0)$
4. $(-1,4)$
5. $(-2,2)$; $(8,-2)$

10.2 Exercise Set

1. $\frac{x^2}{16}+\frac{y^2}{4}=1$

Because the denominator of the x^2 – term is greater than the denominator of the y^2 – term, the major axis is horizontal. Since $a^2=16$, $a=4$ and the vertices are $(-4,0)$ and $(4,0)$. Since $b^2=4$, $b=2$ and endpoints of the minor axis are $(0,-2)$ and $(0,2)$.

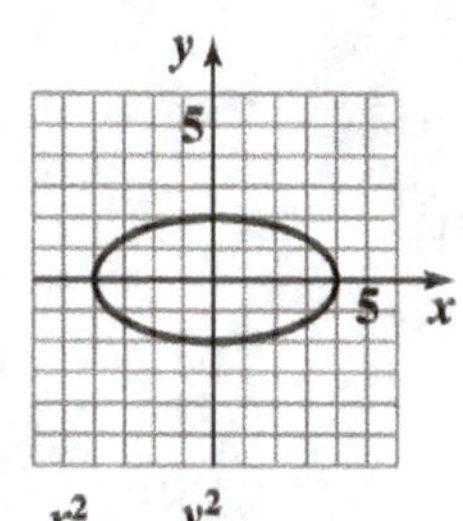

3. $\frac{x^2}{9}+\frac{y^2}{36}=1$

Because the denominator of the y^2 – term is greater than the denominator of the x^2 – term, the major axis is vertical. Since $a^2=36$, $a=6$ and the vertices are $(0,-6)$ and $(0,6)$. Since $b^2=9$, $b=3$ and endpoints of the minor axis are $(-3,0)$ and $(3,0)$.

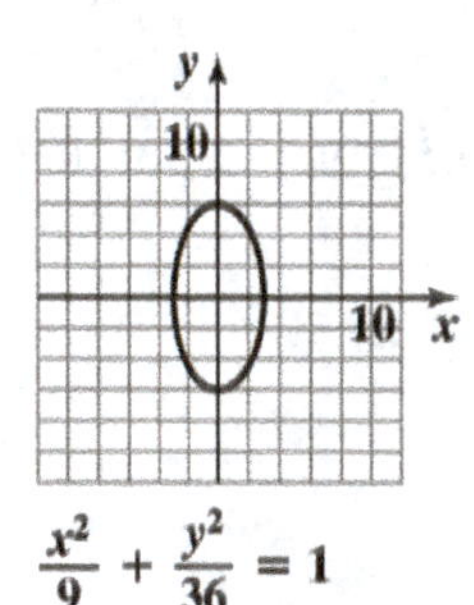

5. $\frac{x^2}{25}+\frac{y^2}{64}=1$

Because the denominator of the y^2 – term is greater than the denominator of the x^2 – term, the major axis is vertical. Since $a^2=64$, $a=8$ and the vertices are $(0,-8)$ and $(0,8)$. Since $b^2=25$, $b=5$ and endpoints of the minor axis are $(-5,0)$ and $(5,0)$.

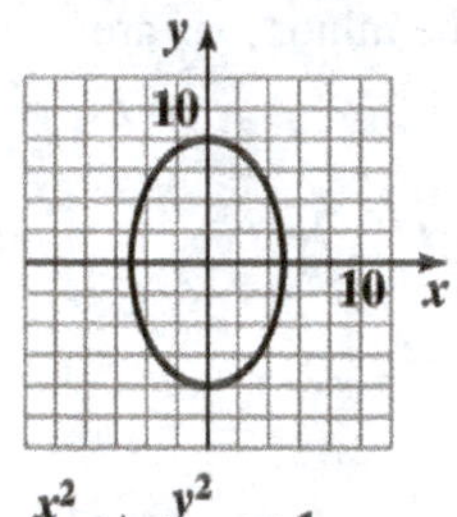

7. $\frac{x^2}{49}+\frac{y^2}{81}=1$

Because the denominator of the y^2 – term is greater than the denominator of the x^2 – term, the major axis is vertical. Since $a^2=81$, $a=9$ and the vertices are $(0,-9)$ and $(0,9)$. Since $b^2=49$, $b=7$ and endpoints of the minor axis are $(-7,0)$ and $(7,0)$.

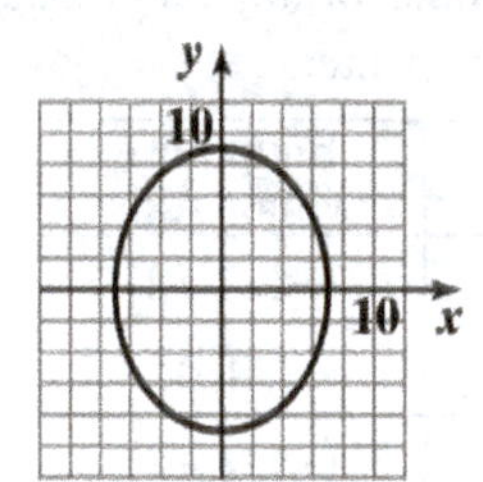

9. $25x^2 + 4y^2 = 100$

$$\frac{25x^2}{100} + \frac{4y^2}{100} = \frac{100}{100}$$

$$\frac{x^2}{4} + \frac{y^2}{25} = 1$$

Because the denominator of the y^2 – term is greater than the denominator of the x^2 – term, the major axis is vertical. Since $a^2 = 25$, $a = 5$ and the vertices are $(0,-5)$ and $(0,5)$. Since $b^2 = 4$, $b = 2$ and endpoints of the minor axis are $(-2,0)$ and $(2,0)$.

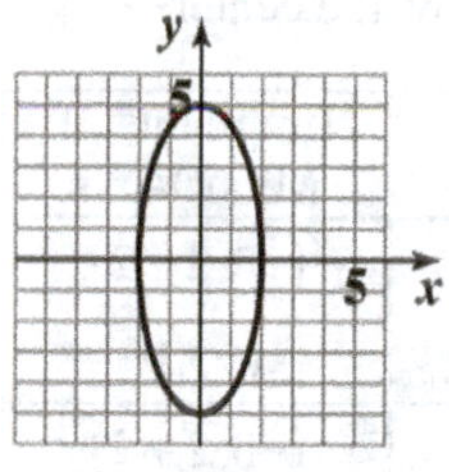

$25x^2 + 4y^2 = 100$

11. $4x^2 + 16y^2 = 64$

$$\frac{4x^2}{64} + \frac{16y^2}{64} = \frac{64}{64}$$

$$\frac{x^2}{16} + \frac{y^2}{4} = 1$$

Because the denominator of the x^2 – term is greater than the denominator of the y^2 – term, the major axis is horizontal. Since $a^2 = 16$, $a = 4$ and the vertices are $(-4,0)$ and $(4,0)$. Since $b^2 = 4$, $b = 2$ and endpoints of the minor axis are $(0,-2)$ and $(0,2)$.

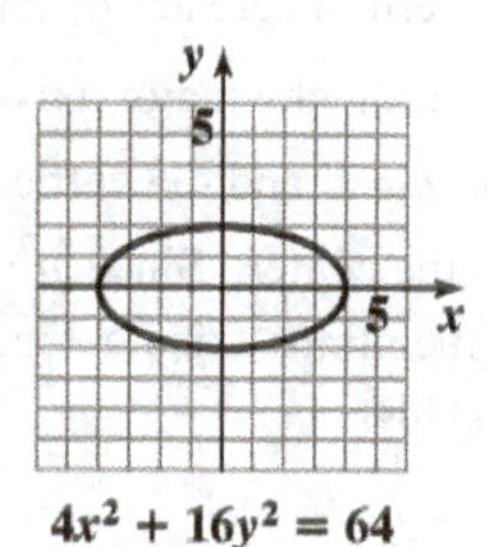

$4x^2 + 16y^2 = 64$

13. $25x^2 + 9y^2 = 225$

$$\frac{25x^2}{225} + \frac{9y^2}{225} = \frac{225}{225}$$

$$\frac{x^2}{9} + \frac{y^2}{25} = 1$$

Because the denominator of the y^2 – term is greater than the denominator of the x^2 – term, the major axis is vertical. Since $a^2 = 25$, $a = 5$ and the vertices are $(0,-5)$ and $(0,5)$. Since $b^2 = 9$, $b = 3$ and endpoints of the minor axis are $(-3,0)$ and $(3,0)$.

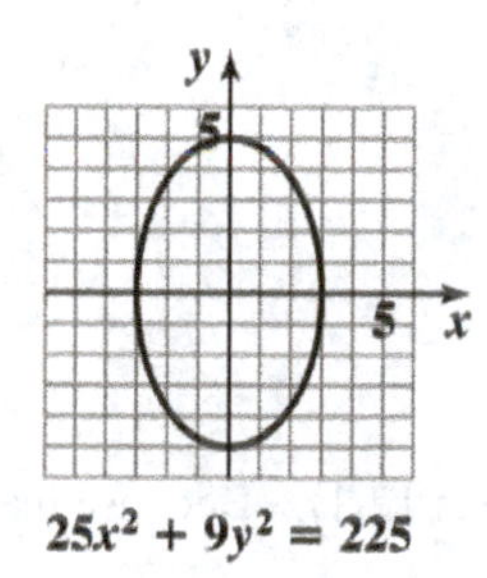

$25x^2 + 9y^2 = 225$

15. $x^2 + 2y^2 = 8$

$$\frac{x^2}{8} + \frac{2y^2}{8} = \frac{8}{8}$$

$$\frac{x^2}{8} + \frac{y^2}{4} = 1$$

Because the denominator of the x^2 – term is greater than the denominator of the y^2 – term, the major axis is horizontal. Since $a^2 = 8$, $a = \sqrt{8} = 2\sqrt{2}$ and the vertices are $(-2\sqrt{2},0)$ and $(2\sqrt{2},0)$. Since $b^2 = 4$, $b = 2$ and endpoints of the minor axis are $(0,-2)$ and $(0,2)$.

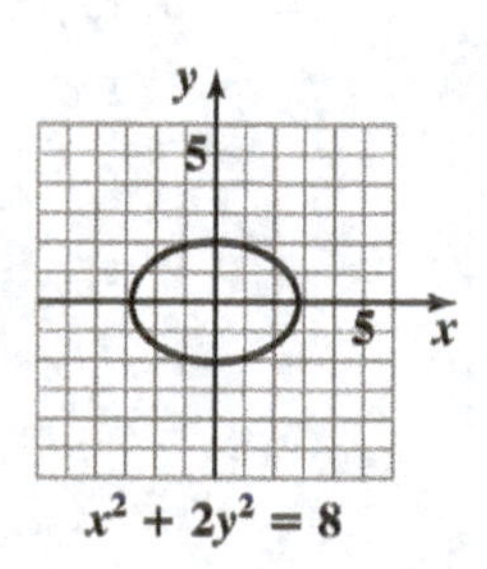

$x^2 + 2y^2 = 8$

17. From the graph, we see that the center of the ellipse is the origin, the major axis is horizontal with $a = 2$, and $b = 1$.

$$\frac{x^2}{2^2}+\frac{y^2}{1^2}=1$$
$$\frac{x^2}{4}+\frac{y^2}{1}=1$$

19. From the graph, we see that the center of the ellipse is the origin, the major axis is vertical with $a = 2$, and $b = 1$.

$$\frac{x^2}{1^2}+\frac{y^2}{2^2}=1$$
$$\frac{x^2}{1}+\frac{y^2}{4}=1$$

21. $\frac{(x-2)^2}{9}+\frac{(y-1)^2}{4}=1$

The center of the ellipse is $(2,1)$. Because the denominator of the x^2 – term is greater than the denominator of the y^2 – term, the major axis is horizontal. Since $a^2 = 9$, $a = 3$ and the vertices lie 3 units to the left and right of the center. Since $b^2 = 4$, $b = 2$ and endpoints of the minor axis lie two units above and below the center.

Center	Vertices	Endpoints of Minor Axis
$(2,1)$	$(2-3,1)$ $=(-1,1)$	$(2,1-2)$ $=(2,-1)$
	$(2+3,1)$ $=(5,1)$	$(2,1+2)$ $=(2,3)$

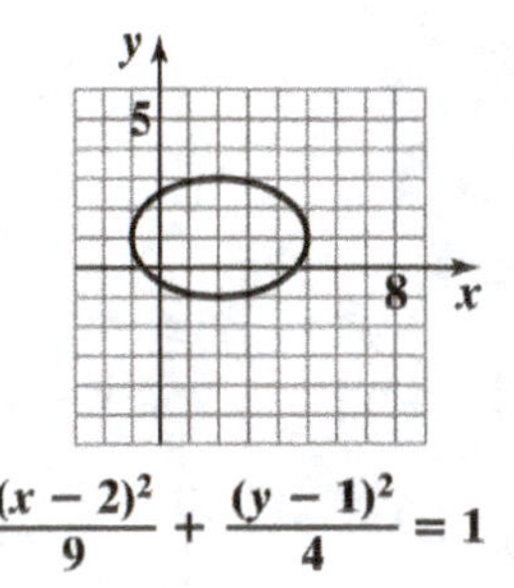

$\frac{(x-2)^2}{9}+\frac{(y-1)^2}{4}=1$

23. $(x+3)^2+4(y-2)^2=16$

$$\frac{(x+3)^2}{16}+\frac{4(y-2)^2}{16}=\frac{16}{16}$$
$$\frac{(x+3)^2}{16}+\frac{(y-2)^2}{4}=1$$

The center of the ellipse is $(-3,2)$. Because the denominator of the x^2 – term is greater than the denominator of the y^2 – term, the major axis is horizontal. Since $a^2 = 16$, $a = 4$ and the vertices lie 4 units to the left and right of the center. Since $b^2 = 4$, $b = 2$ and endpoints of the minor axis lie two units above and below the center.

Center	Vertices	Endpoints of Minor Axis
$(-3,2)$	$(-3-4,2)$ $=(-7,2)$	$(-3,2-2)$ $=(-3,0)$
	$(-3+4,2)$ $=(1,2)$	$(-3,2+2)$ $=(-3,4)$

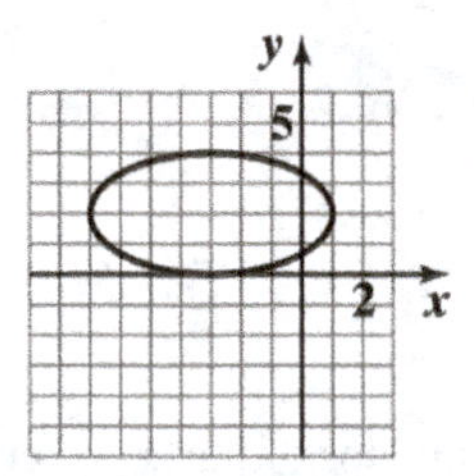

$(x+3)^2+4(y-2)^2=16$

25. $\frac{(x-4)^2}{9}+\frac{(y+2)^2}{25}=1$

The center of the ellipse is $(4,-2)$. Because the denominator of the y^2 – term is greater than the denominator of the x^2 – term, the major axis is vertical. Since $a^2 = 25$, $a = 5$ and the vertices lie 5 units above and below the center. Since $b^2 = 9$, $b = 3$ and endpoints of the minor axis lie 3 units to the right and left of the center.

Center	Vertices	Endpoints Minor Axis
$(4,-2)$	$(4,-2-5)$ $=(4,-7)$	$(4-3,-2)$ $=(1,-2)$
	$(4,-2+5)$ $=(4,3)$	$(4+3,-2)$ $=(7,-2)$

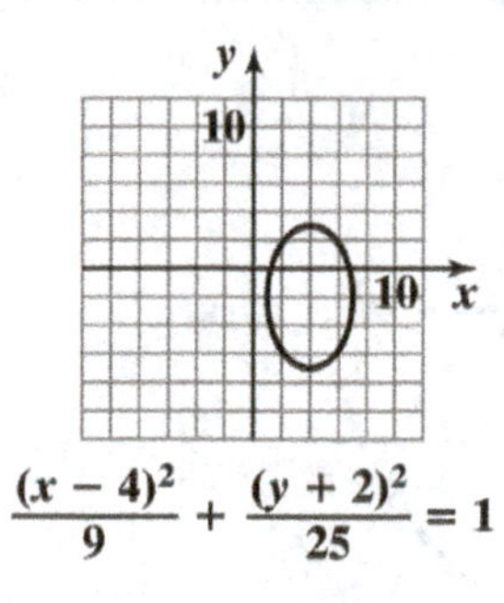

$$\frac{(x-4)^2}{9}+\frac{(y+2)^2}{25}=1$$

27. $\frac{x^2}{25}+\frac{(y-2)^2}{36}=1$

The center of the ellipse is $(0,2)$. Because the denominator of the y^2 – term is greater than the denominator of the x^2 – term, the major axis is vertical. Since $a^2=36$, $a=6$ and the vertices lie 6 units above and below the center. Since $b^2=25$, $b=5$ and endpoints of the minor axis lie 5 units to the left and right of the center.

Center	Vertices	Endpoint Minor Axis
$(0,2)$	$(0,2-6)$ $=(0,-4)$	$(0-5,2)$ $=(-5,2)$
	$(0,2+6)$ $=(0,8)$	$(0+5,2)$ $=(5,2)$

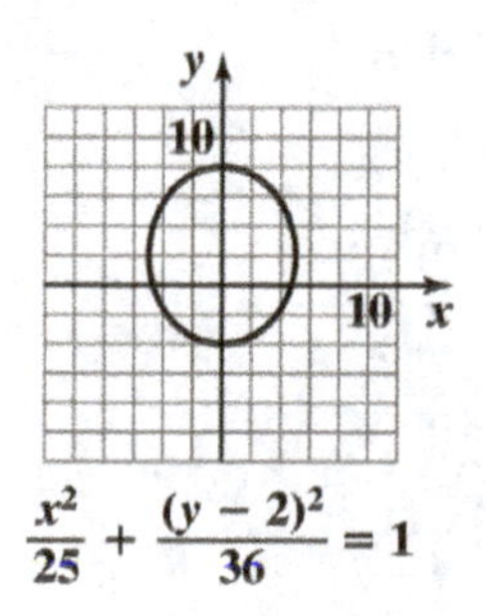

$$\frac{x^2}{25}+\frac{(y-2)^2}{36}=1$$

29. $\frac{(x+3)^2}{9}+(y-2)^2=1$

$$\frac{(x+3)^2}{9}+\frac{(y-2)^2}{1}=1$$

The center of the ellipse is $(-3,2)$. Because the denominator of the x^2 – term is greater than the denominator of the y^2 – term, the major axis is horizontal. Since $a^2=9$, $a=3$ and the vertices lie 3 units to the left and right of the center. Since $b^2=1$, $b=1$ and endpoints of the minor axis lie two units above and below the center.

Center	Vertices	Endpoints of Minor Axis
$(-3,2)$	$(-3+3,2)$ $=(0,2)$	$(-3,2-1)$ $=(-3,1)$
	$(-3-3,2)$ $=(-6,2)$	$(-3,2+1)$ $=(-3,3)$

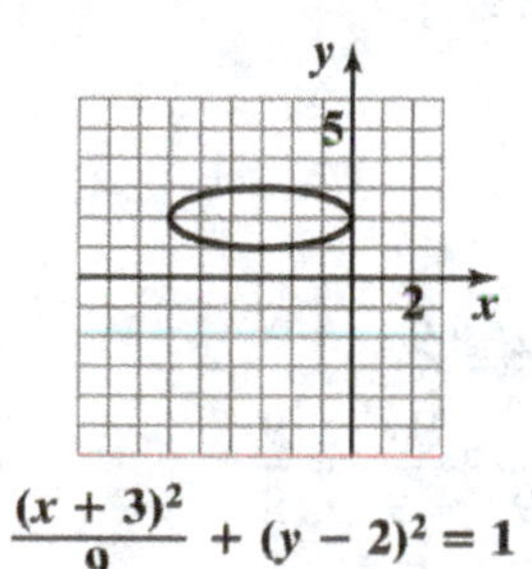

$$\frac{(x+3)^2}{9}+(y-2)^2=1$$

31. $9(x-1)^2+4(y+3)^2=36$

$$\frac{9(x-1)^2}{36}+\frac{4(y+3)^2}{36}=\frac{36}{36}$$

$$\frac{(x-1)^2}{4}+\frac{(y+3)^2}{9}=1$$

The center of the ellipse is $(1,-3)$. Because the denominator of the y^2 – term is greater than the denominator of the x^2 – term, the major axis is vertical. Since $a^2=9$, $a=3$ and the vertices lie 3 units above and below the center. Since $b^2=4$, $b=2$ and endpoints of the minor axis lie 2 units to the right and left of the center.

Center	Vertices	Endpoints of Minor Axis
$(1,-3)$	$(1,-3-3)$ $=(1,-6)$	$(1-2,-3)$ $=(-1,-3)$
	$(1,-3+3)$ $=(1,0)$	$(1+2,-3)$ $=(3,-3)$

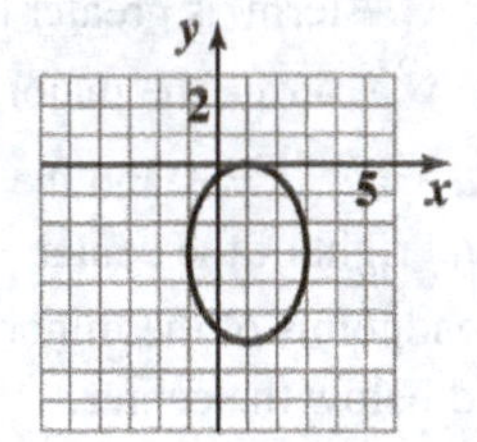

$9(x-1)^2+4(y+3)^2=36$

33. From the graph we see that the center of the ellipse is $(h,k)=(-1,1)$. We also see that the major axis is horizontal. The length of the major axis is 4 units, so $a=2$. The length of the minor axis is 2 units so $b=1$. Therefore, the equation of the ellipse is

$$\frac{(x-(-1))^2}{2^2}+\frac{(y-1)^2}{1^2}=1$$

$$\frac{(x+1)^2}{4}+\frac{(y-1)^2}{1}=1$$

35. $x^2+y^2=1$ and $x^2+9y^2=9$

$$\frac{x^2}{9}+\frac{9y^2}{9}=\frac{9}{9}$$

$$\frac{x^2}{9}+\frac{y^2}{1}=1$$

The first equation is that of a circle with center at the origin and $r=1$. The second equation is that of an ellipse with center at the origin, horizontal major axis of length 6 units ($a^2=9$, so $a=3$), and vertical minor axis of length 2 units ($b^2=1$, so $b=1$).

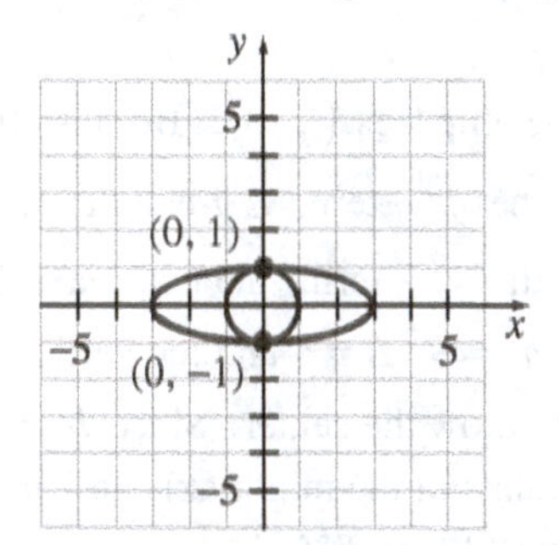

Check each intersection point.

The solutions set is $\{(0,-1),(0,1)\}$.

37. $\frac{x^2}{25}+\frac{y^2}{9}=1$ and $y=3$

The first equation is for an ellipse centered at the origin with horizontal major axis of length 10 units ($a^2=25$, so $a=5$) and vertical minor axis of length 6 units ($b^2=9$, so $b=3$). The second equation is for a horizontal line with a y-intercept of 3.

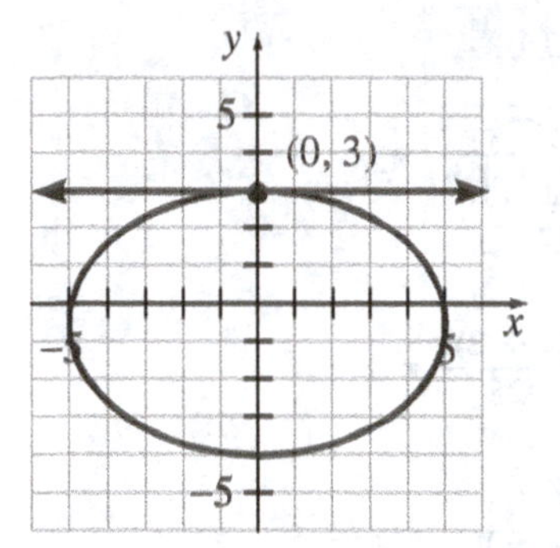

Check the intersection point.

The solution set is $\{(0,3)\}$.

39. $4x^2+y^2=4$ and $2x-y=2$

$$\frac{4x^2}{4}+\frac{y^2}{4}=\frac{4}{4} \qquad -y=-2x+2$$

$$y=2x-2$$

$$\frac{x^2}{1}+\frac{y^2}{4}=1$$

The first equation is for an ellipse centered at the origin with vertical major axis of length 4 units ($a^2=4$, so $a=2$) and horizontal minor axis of length 2 units ($b^2=1$, so $b=1$). The second equation is for a line with slope 2 and y-intercept -2.

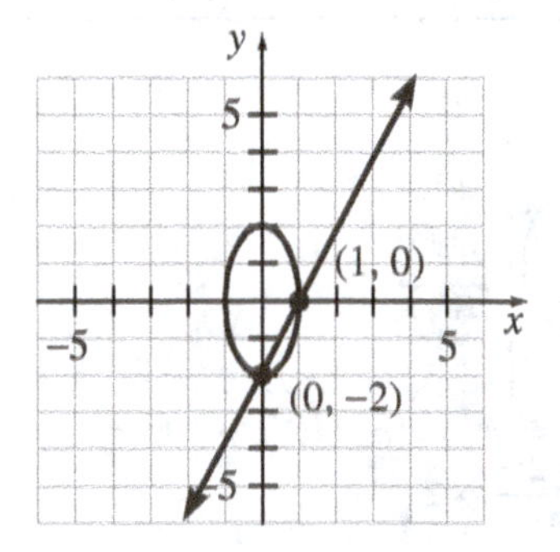

Check the intersection points.

The solution set is $\{(0,-2),(1,0)\}$.

41.

$$y = -\sqrt{16-4x^2}$$
$$y^2 = \left(-\sqrt{16-4x^2}\right)^2$$
$$y^2 = 16-4x^2$$
$$4x^2 + y^2 = 16$$
$$\frac{x^2}{4}+\frac{y^2}{16}=1$$

We want to graph the bottom half of an ellipse centered at the origin with a vertical major axis of length 8 units ($a^2 = 16$, so $a = 4$) and horizontal minor axis of length 4 units ($b^2 = 4$, so $b = 2$).

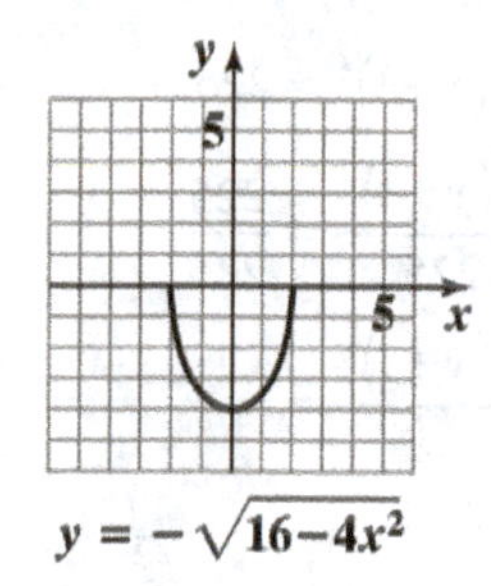

$y = -\sqrt{16-4x^2}$

43. From the figure, we see that the major axis is horizontal with $a = 15$, and $b = 10$.

$$\frac{x^2}{15^2}+\frac{y^2}{10^2}=1$$
$$\frac{x^2}{225}+\frac{y^2}{100}=1$$

Since the truck is 8 feet wide, we need to determine the height of the archway at $\frac{8}{2} = 4$ feet from the center.

$$\frac{4^2}{225}+\frac{y^2}{100}=1$$
$$\frac{16}{225}+\frac{y^2}{100}=1$$
$$900\left(\frac{16}{225}+\frac{y^2}{100}\right)=900(1)$$
$$4(16)+9y^2 = 900$$
$$64+9y^2 = 900$$
$$9y^2 = 836$$
$$y^2 = \frac{836}{9}$$
$$y = \sqrt{\frac{836}{9}} \approx 9.64$$

The height of the archway 4 feet from the center is approximately 9.64 feet. Since the truck is 7 feet high, the truck will clear the archway.

45. a.

$$\frac{x^2}{48^2}+\frac{y^2}{23^2}=1$$
$$\frac{x^2}{2304}+\frac{y^2}{529}=1$$

b.

$$c^2 = a^2 - b^2$$
$$c^2 = 48^2 - 23^2$$
$$c^2 = 2304 - 529$$
$$c^2 = 1775$$
$$c = \sqrt{1775} \approx 42.1$$

The desk was situated approximately 42 feet from the center of the ellipse.

47. – 51. Answers will vary.

53. Answers will vary. For example, consider Exercise 21.

$$\frac{(x-2)^2}{9}+\frac{(y-1)^2}{4}=1$$
$$\frac{(y-1)^2}{4}=1-\frac{(x-2)^2}{9}$$
$$(y-1)^2=4\left(1-\frac{(x-2)^2}{9}\right)$$
$$(y-1)^2=4-\frac{4(x-2)^2}{9}$$
$$y-1=\pm\sqrt{4-\frac{4(x-2)^2}{9}}$$
$$y=1\pm\sqrt{4-\frac{4(x-2)^2}{9}}$$

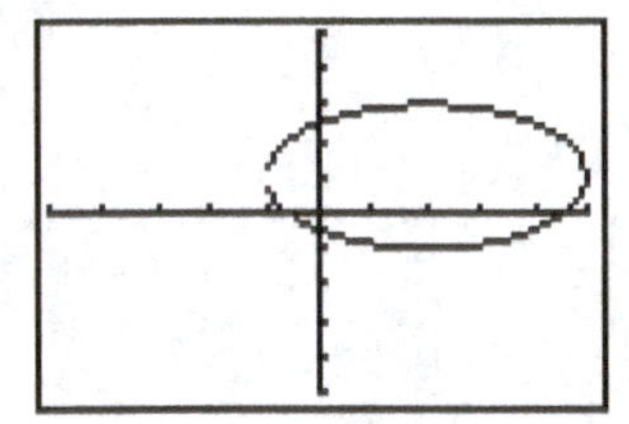

55. does not make sense; Explanations will vary. Sample explanation: An ellipse is symmetrical about both axes.

57. makes sense

59. false; Changes to make the statement true will vary. A sample change is: Ellipses are not functions. They do not pass the vertical line test.

61. true

63.
$$9x^2+25y^2-36x+50y-164=0$$
$$(9x^2-36x\quad)+(25y^2+50y\quad)=164$$
$$9(x^2-4x\quad)+25(y^2+2y\quad)=164$$

Complete the squares.

$$\left(\frac{b}{2}\right)^2=\left(\frac{-4}{2}\right)^2=(-2)^2=4$$
$$\left(\frac{b}{2}\right)^2=\left(\frac{2}{2}\right)^2=(1)^2=1$$

$$9(x^2-4x+4)+25(y^2+2y+1)=164+9(4)+25(1)$$
$$9(x-2)^2+25(y+1)^2=164+36+25$$
$$9(x-2)^2+25(y+1)^2=225$$
$$\frac{9(x-2)^2}{225}+\frac{25(y+1)^2}{225}=\frac{225}{225}$$
$$\frac{(x-2)^2}{25}+\frac{(y+1)^2}{9}=1$$

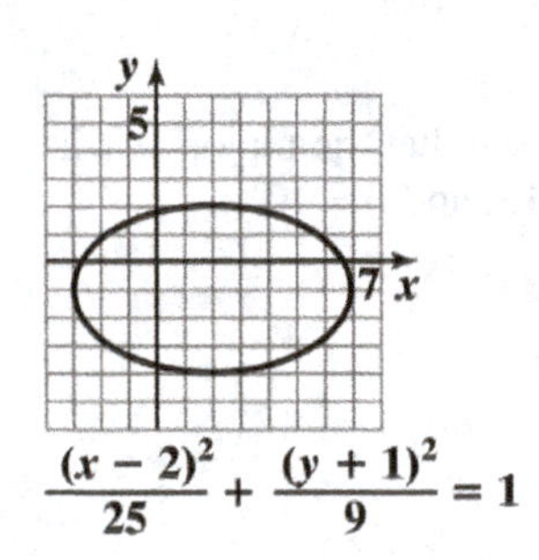

$$\frac{(x-2)^2}{25}+\frac{(y+1)^2}{9}=1$$

65. **a.** The perigee is 5000 – 16 – 4000 = 984 miles above the Earth.

b. The apogee is 5000 + 16 – 4000 = 1016 miles above the Earth.

67.
$$x^3+2x^2-4x-8=x^2(x+2)-4(x+2)$$
$$=(x+2)(x^2-4)$$
$$=(x+2)(x+2)(x-2)$$
$$=(x+2)^2(x-2)$$

68. $\sqrt[3]{40x^4y^7}=\sqrt[3]{8\cdot 5x^3xy^6y}=2xy^2\sqrt[3]{5xy}$

69.
$$\frac{2}{x+2}+\frac{4}{x-2}=\frac{x-1}{x^2-4}$$
$$\frac{2}{x+2}+\frac{4}{x-2}=\frac{x-1}{(x+2)(x-2)}$$
$$(x+2)(x-2)\left(\frac{2}{x+2}+\frac{4}{x-2}\right)=(x+2)(x-2)\left(\frac{x-1}{(x+2)(x-2)}\right)$$
$$2(x-2)+4(x+2)=x-1$$
$$2x-4+4x+8=x-1$$
$$6x+4=x-1$$
$$5x=-5$$
$$x=-1$$
The solution set is $\{-1\}$.

70. $4x^2-9y^2=36$
$$\frac{4x^2}{36}-\frac{9y^2}{36}=\frac{36}{36}$$
$$\frac{x^2}{9}-\frac{y^2}{4}=1$$
The terms are separated by subtraction rather than by addition.

71. $\frac{x^2}{16}-\frac{y^2}{9}=1$

a. Substitute 0 for y.
$$\frac{x^2}{16}-\frac{0^2}{9}=1$$
$$\frac{x^2}{16}=1$$
$$x^2=16$$
$$x=\pm 4$$
The x-intercepts are -4 and 4.

b. $\frac{0^2}{16}-\frac{y^2}{9}=1$
$$-\frac{y^2}{9}=1$$
$$y^2=-9$$
The equation $y^2=-9$ has no real solutions.

72. $\frac{y^2}{9}-\frac{x^2}{16}=1$

a. Substitute 0 for x.

$$\frac{y^2}{9}-\frac{0^2}{16}=1$$
$$\frac{y^2}{9}=1$$
$$y^2=9$$
$$y=\pm 3$$

The y-intercepts are -3 and 3.

b. $$\frac{0^2}{9}-\frac{x^2}{16}=1$$
$$-\frac{x^2}{16}=1$$
$$x^2=-16$$

The equation $x^2=-16$ has no real solutions.

10.3 Check Points

1. **a.** Since the x^2 – term is positive, the transverse axis lies along the x–axis. Also, since $a^2=25$ and $a=5$, the vertices are $(-5,0)$ and $(5,0)$.

   b. Since the y^2 – term is positive, the transverse axis lies along the y–axis. Also, since $a^2=25$ and $a=5$, the vertices are $(0,-5)$ and $(0,5)$.

2. $\frac{x^2}{36}-\frac{y^2}{9}=1$

   Since the x^2 – term is positive, the transverse axis lies along the x–axis. Also, since $a^2=36$ and $a=6$, the vertices are $(-6,0)$ and $(6,0)$. Construct a rectangle using –6 and 6 on the x–axis, and –3 and 3 on the y–axis. Draw extended diagonals to obtain the asymptotes. Draw the two branches of the hyperbola by starting at each vertex and approaching the asymptotes.

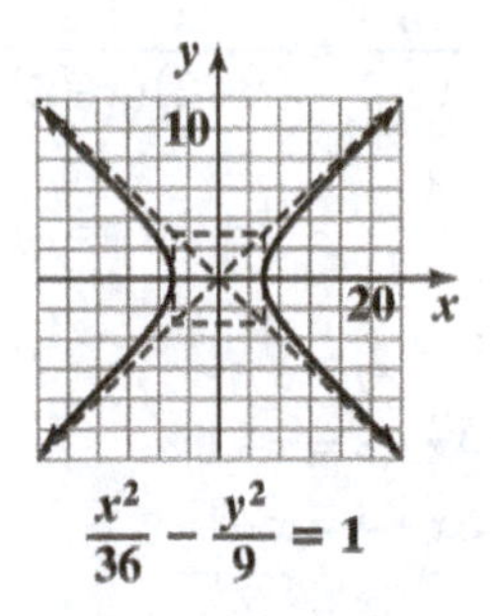

$\frac{x^2}{36}-\frac{y^2}{9}=1$

3. First write the equation in standard form.

$$y^2-4x^2=4$$
$$\frac{y^2}{4}-\frac{4x^2}{4}=\frac{4}{4}$$
$$\frac{y^2}{4}-\frac{x^2}{1}=1$$

   The equation is in the form $\frac{y^2}{a^2}-\frac{x^2}{b^2}=1$ with $a^2=4$ and $b^2=1$. We know the transverse axis lies on the y-axis and the vertices are $(0,-2)$ and $(0,2)$. Because $a^2=4$ and $b^2=1$, $a=2$ and $b=1$. Construct a rectangle using –2 and 2 on the y–axis, and –1 and 1 on the x–axis. Draw extended diagonals to obtain the asymptotes. Draw the two branches of the hyperbola by starting at each vertex and approaching the asymptotes.

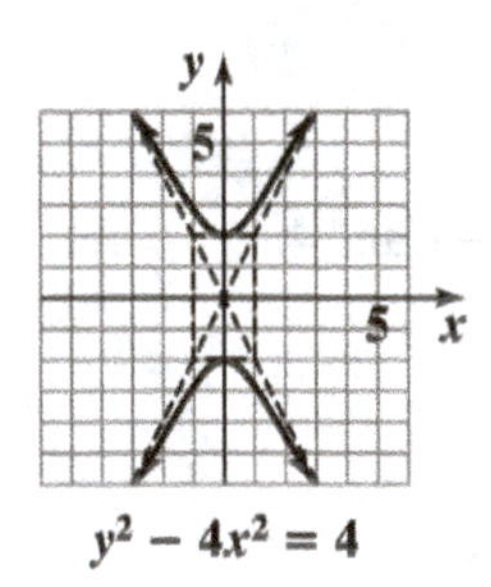

$y^2-4x^2=4$

10.3 Concept and Vocabulary Check

1. hyperbola; vertices; transverse
2. $(-5,0)$; $(5,0)$
3. $(0,-5)$; $(0,5)$
4. asymptotes; center

5. dividing; 36
6. ellipse
7. hyperbola
8. neither
9. hyperbola
10. ellipse

10.3 Exercise Set

1. Since the x^2 – term is positive, the transverse axis lies along the x–axis. Also, since $a^2 = 4$ and $a = 2$, the vertices are $(-2,0)$ and $(2,0)$. This corresponds to graph (b).

3. Since the y^2 – term is positive, the transverse axis lies along the y–axis. Also, since $a^2 = 4$ and $a = 2$, the vertices are $(0,-2)$ and $(0,2)$. This corresponds to graph (a).

5. $\frac{x^2}{9} - \frac{y^2}{25} = 1$

   The equation is in the form $\frac{x^2}{a^2} - \frac{y^2}{b^2} = 1$ with $a^2 = 9$, and $b^2 = 25$. We know the transverse axis lies on the x-axis and the vertices are $(-3,0)$ and $(3,0)$. Because $a^2 = 9$ and $b^2 = 25$, $a = 3$ and $b = 5$. Construct a rectangle using –3 and 3 on the x–axis, and –5 and 5 on the y–axis. Draw extended diagonals to obtain the asymptotes. Graph the hyperbola.

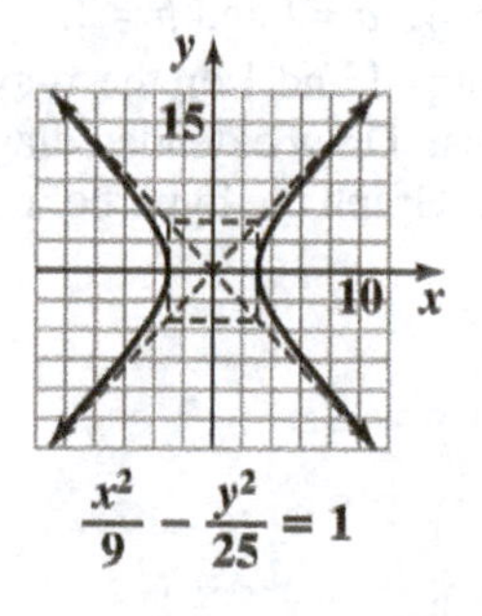

$\frac{x^2}{9} - \frac{y^2}{25} = 1$

7. $\frac{x^2}{100} - \frac{y^2}{64} = 1$

   The equation is in the form $\frac{x^2}{a^2} - \frac{y^2}{b^2} = 1$ with $a^2 = 100$, and $b^2 = 64$. We know the transverse axis lies on the x-axis and the vertices are $(-10,0)$ and $(10,0)$. Because $a^2 = 100$ and $b^2 = 64$, $a = 10$ and $b = 8$. Construct a rectangle using –10 and 10 on the x–axis, and –8 and 8 on the y–axis. Draw extended diagonals to obtain the asymptotes. Graph the hyperbola.

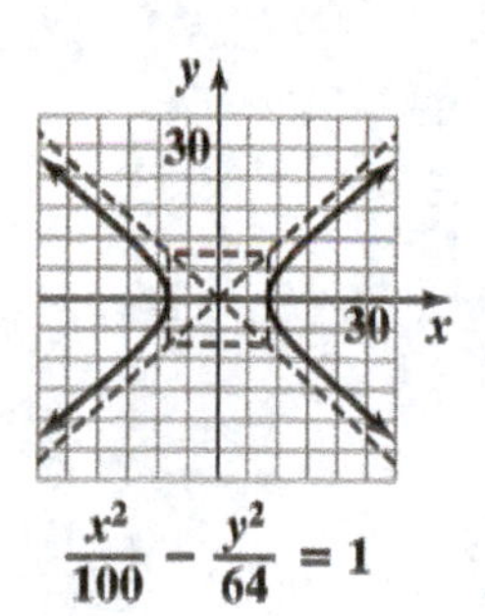

$\frac{x^2}{100} - \frac{y^2}{64} = 1$

9. $\frac{x^2}{16} - \frac{y^2}{36} = 1$

   The equation is in the form $\frac{x^2}{a^2} - \frac{y^2}{b^2} = 1$ with $a^2 = 16$, and $b^2 = 36$. We know the transverse axis lies on the x-axis and the vertices are $(-4,0)$ and $(4,0)$. Because $a^2 = 16$ and $b^2 = 36$, $a = 4$ and $b = 6$. Construct a rectangle using –4 and 4 on the x–axis, and –6 and 6 on the y–axis. Draw extended diagonals to obtain the asymptotes. Graph the hyperbola.

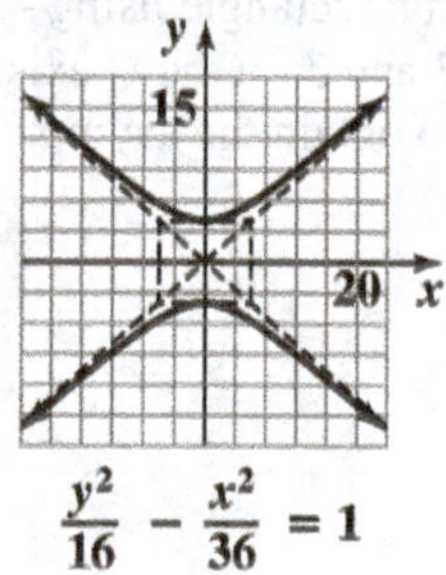

$\frac{y^2}{16} - \frac{x^2}{36} = 1$

11. $\frac{y^2}{36}-\frac{x^2}{25}=1$

The equation is in the form $\frac{y^2}{a^2}-\frac{x^2}{b^2}=1$ with $a^2=36$, and $b^2=25$. We know the transverse axis lies on the y-axis and the vertices are $(0,-6)$ and $(0,6)$. Because $a^2=36$ and $b^2=25$, $a=6$ and $b=5$. Construct a rectangle using –5 and 5 on the x–axis, and –6 and 6 on the y–axis. Draw extended diagonals to obtain the asymptotes. Graph the hyperbola.

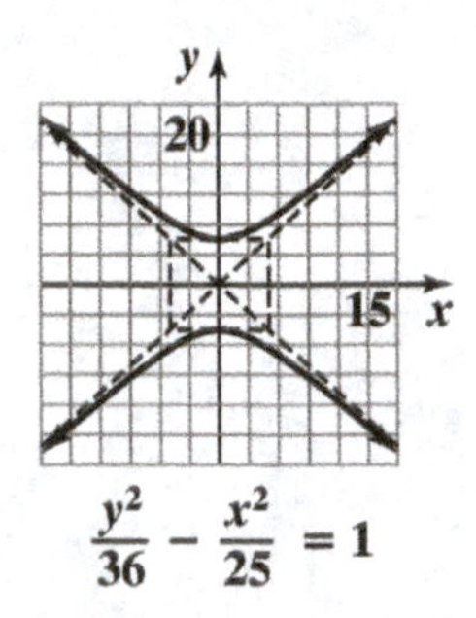

$\frac{y^2}{36}-\frac{x^2}{25}=1$

13. $9x^2-4y^2=36$

$\frac{9x^2}{36}-\frac{4y^2}{36}=\frac{36}{36}$

$\frac{x^2}{4}-\frac{y^2}{9}=1$

The equation is in the form $\frac{x^2}{a^2}-\frac{y^2}{b^2}=1$ with $a^2=4$ and $b^2=9$. We know the transverse axis lies on the x-axis and the vertices are $(-2,0)$ and $(2,0)$. Because $a^2=4$ and $b^2=9$, $a=2$ and $b=3$. Construct a rectangle using –2 and 2 on the x–axis, and –3 and 3 on the y–axis. Draw extended diagonals to obtain the asymptotes. Graph the hyperbola.

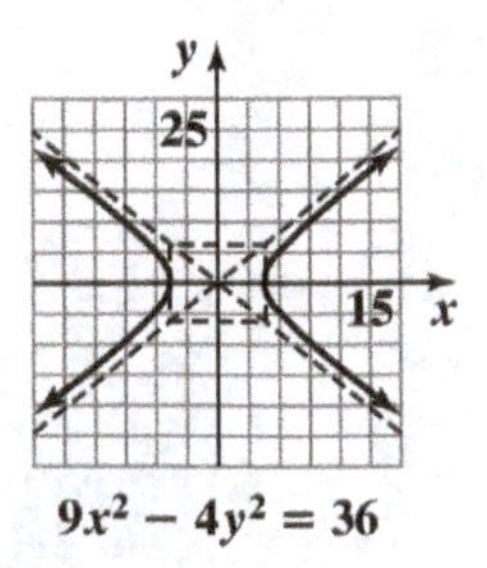

$9x^2-4y^2=36$

15. $9y^2-25x^2=225$

$\frac{9y^2}{225}-\frac{25x^2}{225}=\frac{225}{225}$

$\frac{y^2}{25}-\frac{x^2}{9}=1$

The equation is in the form $\frac{y^2}{a^2}-\frac{x^2}{b^2}=1$ with $a^2=25$ and $b^2=9$. We know the transverse axis lies on the y-axis and the vertices are $(0,-5)$ and $(0,5)$. Because $a^2=25$ and $b^2=9$, $a=5$ and $b=3$. Construct a rectangle using –3 and 3 on the x–axis, and –5 and 5 on the y–axis. Draw extended diagonals to obtain the asymptotes. Graph the hyperbola.

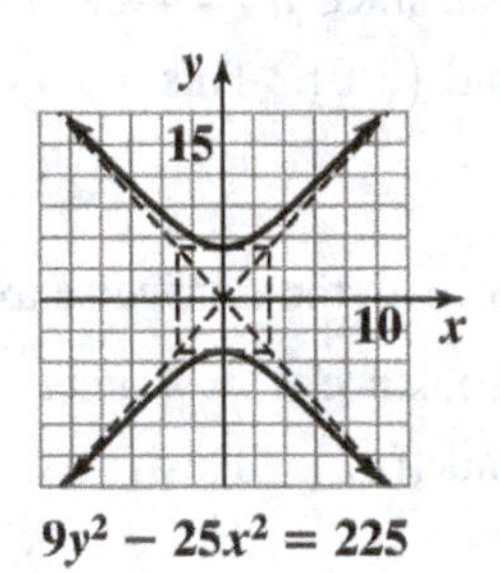

$9y^2-25x^2=225$

17. $4x^2=4+y^2$

$4x^2-y^2=4$

$\frac{4x^2}{4}-\frac{y^2}{4}=\frac{4}{4}$

$\frac{x^2}{1}-\frac{y^2}{4}=1$

The equation is in the form $\frac{x^2}{a^2}-\frac{y^2}{b^2}=1$ with $a^2=1$ and $b^2=4$. We know the transverse axis lies on the x-axis and the vertices are $(-1,0)$ and $(1,0)$.

Because $a^2=1$ and $b^2=4$, $a=1$ and $b=2$. Construct a rectangle using –1 and 1 on the x–axis, and –2 and 2 on the y–axis. Draw extended diagonals to obtain the asymptotes. Graph the hyperbola.

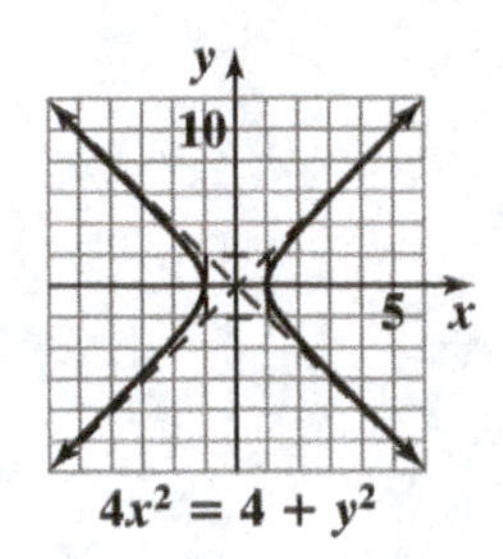

$4x^2=4+y^2$

19. The graph shows that the transverse axis lies along the x–axis and the vertices are $(-3,0)$ and $(3,0)$. This means that $a = 3$. We also see that $b = 5$.

$$\frac{x^2}{a^2}-\frac{y^2}{b^2}=1$$
$$\frac{x^2}{3^2}-\frac{y^2}{5^2}=1$$
$$\frac{x^2}{9}-\frac{y^2}{25}=1$$

21. The graph shows that the transverse axis lies along the y–axis and the vertices are $(0,-2)$ and $(0,2)$. This means that $a = 2$. We also see that $b = 3$.

$$\frac{y^2}{a^2}-\frac{x^2}{b^2}=1$$
$$\frac{y^2}{2^2}-\frac{x^2}{3^2}=1$$
$$\frac{y^2}{4}-\frac{x^2}{9}=1$$

23. $\frac{x^2}{9}-\frac{y^2}{16}=1$

The equation is for a hyperbola in standard form with the transverse axis on the x-axis. We have $a^2=9$ and $b^2=16$, so $a=3$ and $b=4$. Therefore, the vertices are at $(\pm a,0)$ or $(\pm 3,0)$. Using a dashed line, we construct a rectangle using the ± 3 on the x-axis and ± 4 on the y-axis. Then use dashed lines to draw extended diagonals for the rectangle. These represent the asymptotes of the graph.

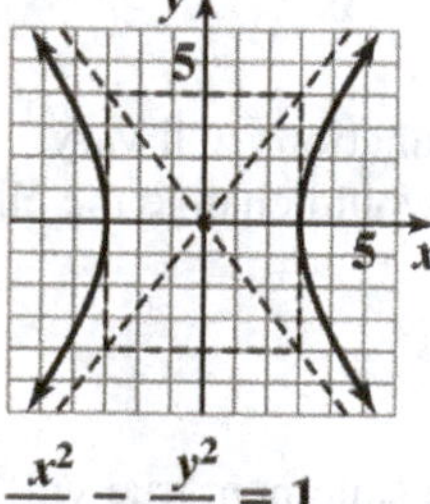

From the graph we determine the following:
Domain: $(-\infty,-3]\cup[3,\infty)$
Range: $(-\infty,\infty)$

25. $\frac{x^2}{9}+\frac{y^2}{16}=1$

The equation is for an ellipse in standard form with major axis along the y-axis. We have $a^2=16$ and $b^2=9$, so $a=4$ and $b=3$. Therefore, the vertices are $(0,\pm a)$ or $(0,\pm 4)$. The endpoints of the minor axis are $(\pm b,0)$ or $(\pm 3,0)$.

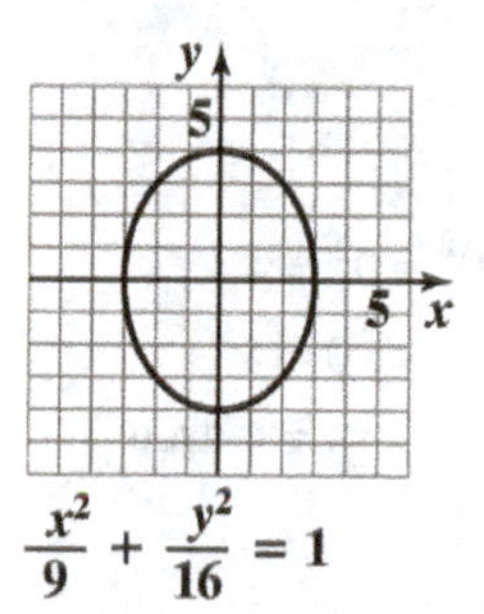

From the graph we determine the following:
Domain: $[-3,3]$
Range: $[-4,4]$

27. $\frac{y^2}{16}-\frac{x^2}{9}=1$

The equation is in standard form with the transverse axis on the y-axis. We have $a^2=16$ and $b^2=9$, so $a=4$ and $b=3$. Therefore, the vertices are at $(0,\pm a)$ or $(0,\pm 4)$. Using a dashed line, we construct a rectangle using the ± 4 on the y-axis and ± 3 on the x-axis. Then use dashed lines to draw extended diagonals for the rectangle. These represent the asymptotes of the graph.

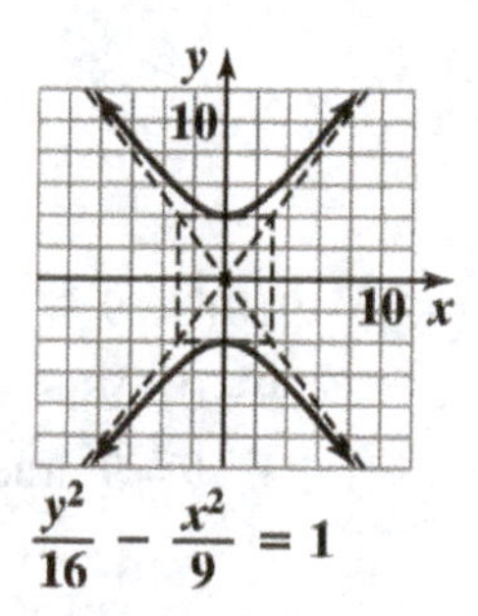

From the graph we determine the following:
Domain: $(-\infty,\infty)$
Range: $(-\infty,-4]\cup[4,\infty)$

29. $x^2 - y^2 = 4$
$x^2 + y^2 = 4$

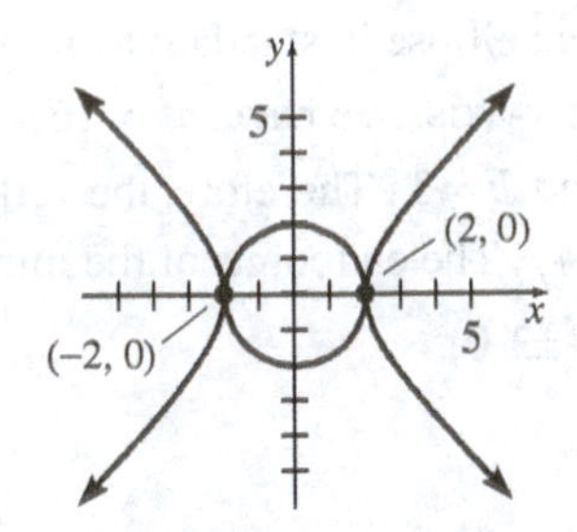

Check $(-2,0)$:

$(-2)^2 - 0^2 = 4$ $\quad$ $(-2)^2 + 0^2 = 4$
$4 - 0 = 4$ $\quad$ $4 + 0 = 4$
$4 = 4$ true $\quad$ $4 = 4$ true

Check $(2,0)$:

$(2)^2 - 0^2 = 4$ $\quad$ $(2)^2 + 0^2 = 4$
$4 - 0 = 4$ $\quad$ $4 + 0 = 4$
$4 = 4$ true $\quad$ $4 = 4$ true

The solution set is $\{(-2,0),(2,0)\}$.

31. $9x^2 + y^2 = 9$ or $\frac{x^2}{1} + \frac{y^2}{9} = 1$
$y^2 - 9x^2 = 9$ $\quad$ $\frac{y^2}{9} - \frac{x^2}{1} = 1$

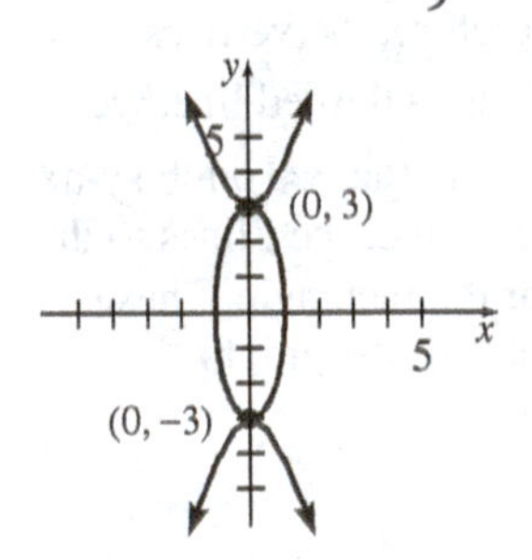

Check $(0,-3)$:

$9(0)^2 + (-3)^2 = 9$ $\quad$ $(-3)^2 - 9(0)^2 = 9$
$0 + 9 = 9$ $\quad$ $9 - 0 = 9$
$9 = 9$ true $\quad$ $9 = 9$ true

Check $(0,3)$:

$9(0)^2 + (3)^2 = 9$ $\quad$ $(3)^2 - 9(0)^2 = 9$
$0 + 9 = 9$ $\quad$ $9 - 0 = 9$
$9 = 9$ true $\quad$ $9 = 9$ true

The solution set is $\{(0,-3),(0,3)\}$.

33.
$$625y^2 - 400x^2 = 250{,}000$$
$$\frac{625y^2}{250{,}000} - \frac{400x^2}{250{,}000} = \frac{250{,}000}{250{,}000}$$
$$\frac{y^2}{400} - \frac{x^2}{625} = 1$$

Since the houses at the vertices of the hyperbola will be closest, find the distance between the vertices. Since $a^2 = 400$, $a = 20$. The houses are $20 + 20 = 40$ yards apart.

35. – 39. Answers will vary.

41. $\frac{x^2}{4} - \frac{y^2}{9} = 0$

Solve the equation for y.

$$\frac{x^2}{4} = \frac{y^2}{9}$$
$$9x^2 = 4y^2$$
$$\frac{9}{4}x^2 = y^2$$
$$\pm\sqrt{\frac{9}{4}x^2} = y$$
$$\pm\frac{3}{2}x = y$$

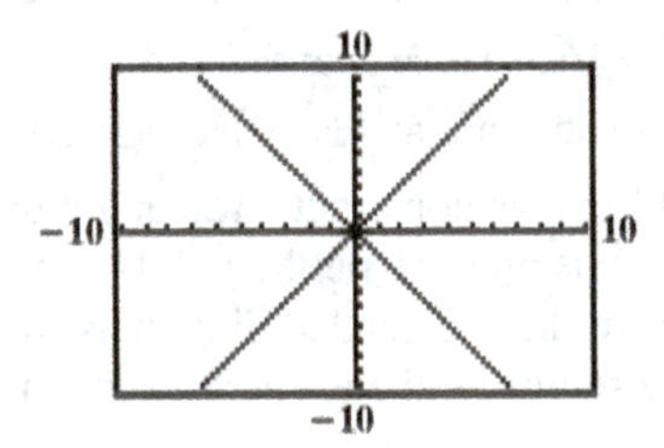

The graph is not a hyperbola. The graph is two lines.

43. does not make sense; Explanations will vary. Sample explanation: This would change the ellipse to a hyperbola.

45. makes sense

47. false; Changes to make the statement true will vary. A sample change is: If a hyperbola has a transverse axis along the x–axis and one of the branches is removed, the remaining branch does not define a function of x.

49. true

51. $\dfrac{(x-2)^2}{16}-\dfrac{(y-3)^2}{9}=1$

This is the equation of a hyperbola with center $(2,3)$. The transverse axis is horizontal and the vertices lie 4 units to the right and left of $(2,3)$ at $(2-4,3)=(-2,3)$ and $(2+4,3)=(6,3)$. Construct two sides of a rectangle using –2 and 6 on the x–axis. The remaining two sides of the rectangle are constructed 3 units above and 3 units below the center at $3-3=0$ and $3+3=6$. Draw extended diagonals to obtain the asymptotes.

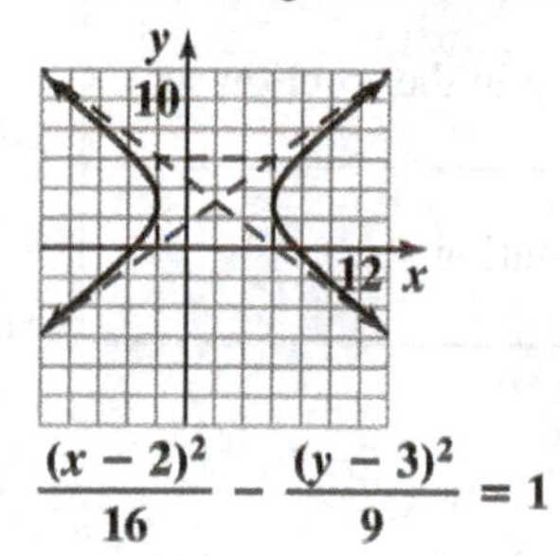

$$\frac{(x-2)^2}{16}-\frac{(y-3)^2}{9}=1$$

53. 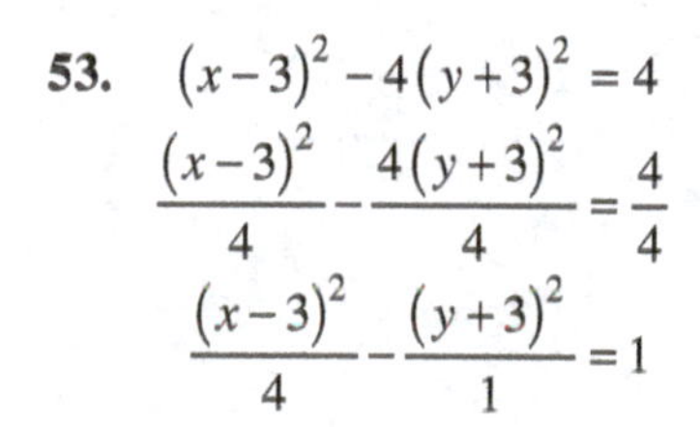

$$(x-3)^2-4(y+3)^2=4$$
$$\frac{(x-3)^2}{4}-\frac{4(y+3)^2}{4}=\frac{4}{4}$$
$$\frac{(x-3)^2}{4}-\frac{(y+3)^2}{1}=1$$

This is the equation of a hyperbola with center $(3,-3)$. The transverse axis is horizontal and the vertices lie 2 units to the right and left of $(3,-3)$ at $(3-2,-3)=(1,-3)$ and $(3+2,-3)=(5,-3)$ Construct two sides of a rectangle using 1 and 5 on the x–axis. The remaining two sides of the rectangle are constructed 1 unit above and below center at $-3-1=-4$ and $-3+1=-2$. Draw extended diagonals to obtain the asymptotes.

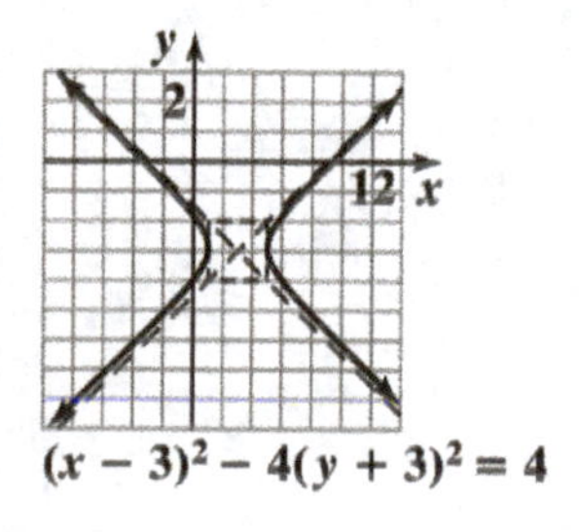

$$(x-3)^2-4(y+3)^2=4$$

55. Since the vertices are $(6,0)$ and $(-6,0)$, we know that the transverse axis lies along the x–axis and $a=6$. Use the equation of the asymptote, $y=4x$, to find b.

$$y=4x=4(6)=24$$

This means that $b=\pm 24$.

$$\frac{x^2}{6^2}-\frac{y^2}{24^2}=1$$
$$\frac{x^2}{36}-\frac{y^2}{576}=1$$

57. $y=-x^2-4x+5$

The x–coordinate of the vertex is $-\dfrac{b}{2a}=-\dfrac{-4}{2(-1)}=-\dfrac{-4}{-2}=-2$ and the y–coordinate of the vertex is

$$f\left(-\frac{b}{2a}\right)=f(-2)=-(-2)^2-4(-2)+5$$
$$=-4+8+5=9.$$

The vertex is at (–2, 9). The x–intercepts are –5 and 1. The y–intercept is 5.

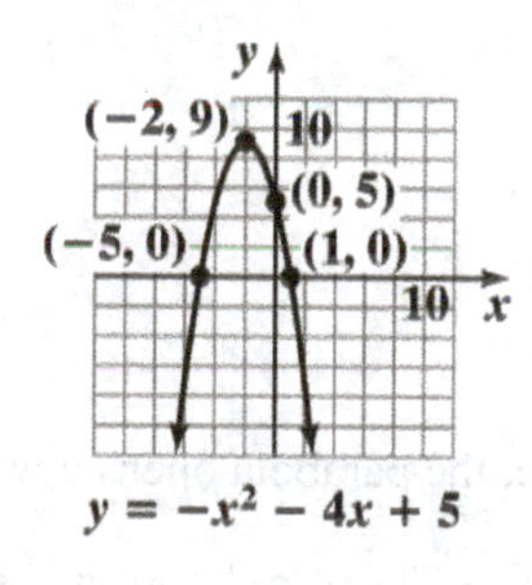

$$y=-x^2-4x+5$$

58. $3x^2 - 11x - 4 \geq 0$

$$3x^2 - 11x - 4 = 0$$
$$(3x+1)(x-4) = 0$$
$$3x+1=0 \quad \text{or} \quad x-4=0$$
$$3x=-1 \qquad x=4$$
$$x=-\frac{1}{3}$$

Interval	Test Value	Substitution	Conclusion
$\left(-\infty, -\frac{1}{3}\right]$	-1	$3(-1)^2 - 11(-1) - 4 \geq 0$ $10 \geq 0$, true	$\left(-\infty, -\frac{1}{3}\right]$ belongs in the solution set.
$\left[-\frac{1}{3}, 4\right]$	0	$3(0)^2 - 11(0) - 4 \geq 0$ $-4 \geq 0$, false	$\left[-\frac{1}{3}, 4\right]$ does not belong in the solution set.
$[4, \infty)$	5	$3(5)^2 - 11(5) - 4 \geq 0$ $16 \geq 0$, true	$[4, \infty)$ belongs in the solution set.

The solution set is $\left(-\infty, -\frac{1}{3}\right] \cup [4, \infty)$.

59. $\log_4(3x+1) = 3$

$$3x+1 = 4^3$$
$$3x+1 = 64$$
$$3x = 63$$
$$x = 21$$

The solution set is $\{21\}$.

60. $y = x^2 + 4x - 5$

Since $a = 1$ is positive, the parabola opens upward. The x-coordinate of the vertex is $x = -\frac{b}{2a} = -\frac{4}{2(1)} = -2$. The y-coordinate of the vertex is $y = (-2)^2 + 4(-2) - 5 = -9$.
Vertex: $(-2, -9)$.

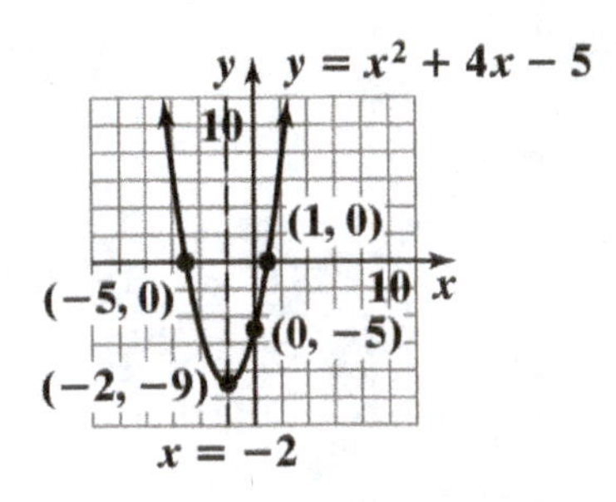

61. $y = -3(x-1)^2 + 2$

Since $a = -3$ is negative, the parabola opens downward. The vertex of the parabola is $(h,k) = (1,2)$.
The y–intercept is -1.

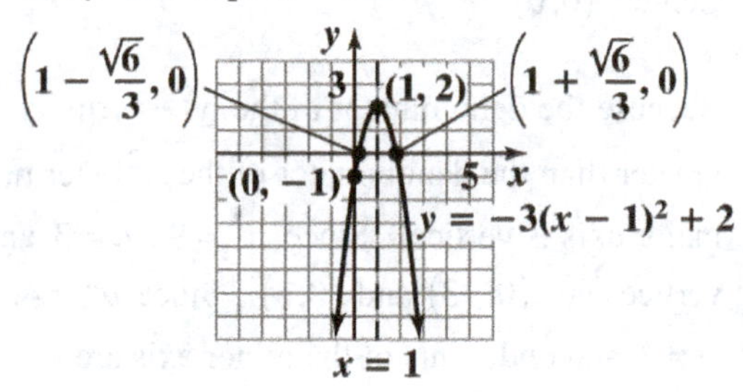

62. Find the y-intercepts.

$$x = -3(y-1)^2 + 2$$
$$0 = -3(y-1)^2 + 2$$
$$3(y-1)^2 = 2$$
$$(y-1)^2 = \frac{2}{3}$$
$$y - 1 = \pm\sqrt{\frac{2}{3}}$$
$$y = 1 \pm \sqrt{\frac{2}{3}}$$
$$y \approx 0.2 \text{ or } 1.8$$

The y-intercepts are approximately 0.2 and 1.8.

Mid-Chapter 10 Check Point

1. $x^2 + y^2 = 9$

Center: $(0,0)$
Radius: $r = \sqrt{9} = 3$

We plot points that are 3 units to the left, right, above, and below the center. These points are $(-3,0)$, $(3,0)$, $(0,3)$ and $(0,-3)$.

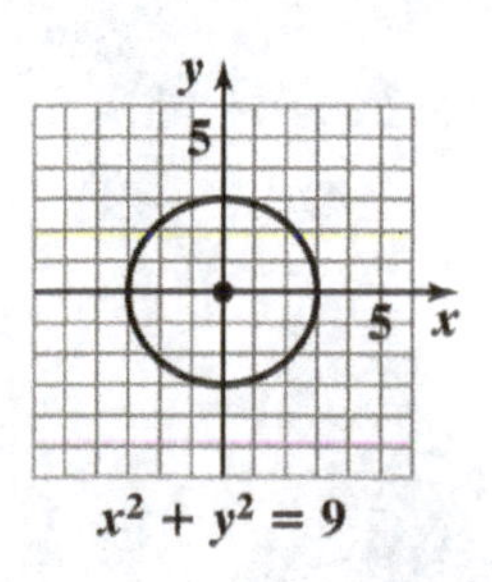

2. $(x-3)^2 + (y+2)^2 = 25$

Center: $(3,-2)$
Radius: $r = \sqrt{25} = 5$

We plot the points that are 5 units to the left, right, above and below the center.

These points are $(-2,-2)$, $(8,-2)$, $(3,3)$, and $(3,-7)$.

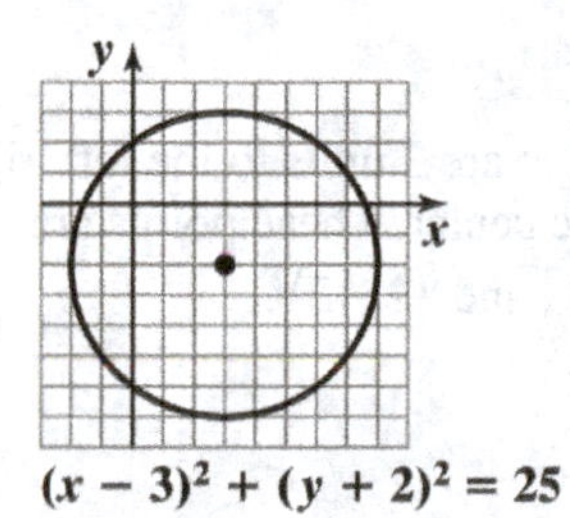

3. $x^2 + (y-1)^2 = 4$

Center: $(0,1)$
Radius: $r = \sqrt{4} = 2$

We plot the points that are 2 units to the left, right, above, and below the center. These points are $(-2,1)$, $(2,1)$, $(0,3)$, and $(0,-1)$.

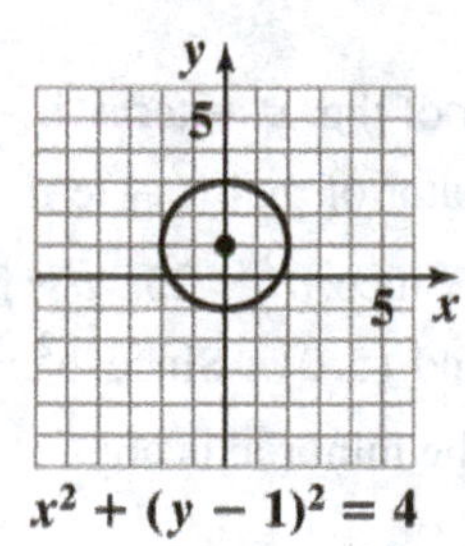

4. $x^2+y^2-4x-2y-4=0$

Complete the square in both x and y to get the equation in standard form.

$$\left(x^2-4x\right)+\left(y^2-2y\right)=4$$
$$\left(x^2-4x+4\right)+\left(y^2-2y+1\right)=4+4+1$$
$$(x-2)^2+(y-1)^2=9$$

Center: $(2,1)$

Radius: $r=\sqrt{9}=3$

We plot the points that are 3 units to the left, right, above, and below the center. These points are $(-1,1)$, $(5,1)$, $(2,4)$, and $(2,-2)$.

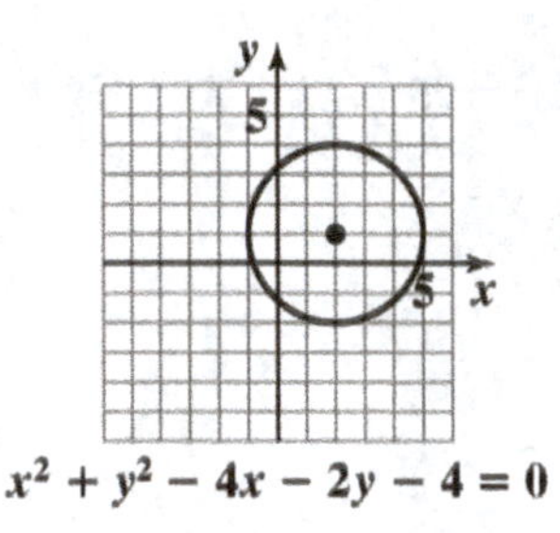

$x^2+y^2-4x-2y-4=0$

5. $\dfrac{x^2}{25}+\dfrac{y^2}{4}=1$

Center: $(0,0)$

Because the denominator of the x^2 – term is greater than the denominator of the y^2 – term, the major axis is horizontal. Since $a^2=25$, $a=5$ and the vertices are $(-5,0)$ and $(5,0)$. Since $b^2=4$, $b=2$ and endpoints of the minor axis are $(0,-2)$ and $(0,2)$.

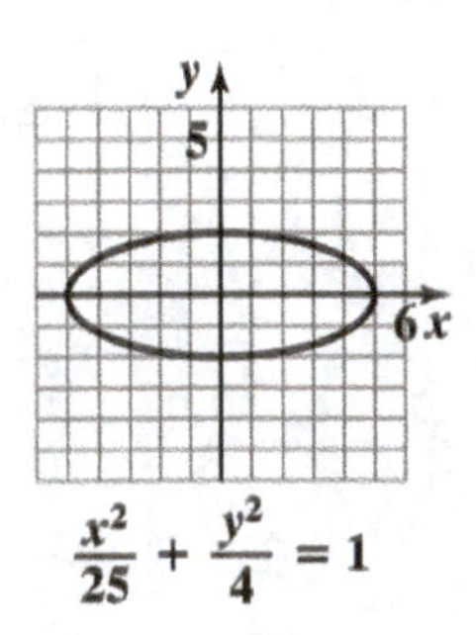

$\dfrac{x^2}{25}+\dfrac{y^2}{4}=1$

6. $9x^2+4y^2=36$

Divide both sides by 36 to get the standard form:
$\dfrac{x^2}{4}+\dfrac{y^2}{9}=1$

Center: $(0,0)$

Because the denominator of the y^2 – term is greater than the denominator of the x^2 – term, the major axis is vertical. Since $a^2=9$, $a=3$ and the vertices are $(0,-3)$ and $(0,3)$. Since $b^2=4$, $b=2$ and endpoints of the minor axis are $(-2,0)$ and $(2,0)$.

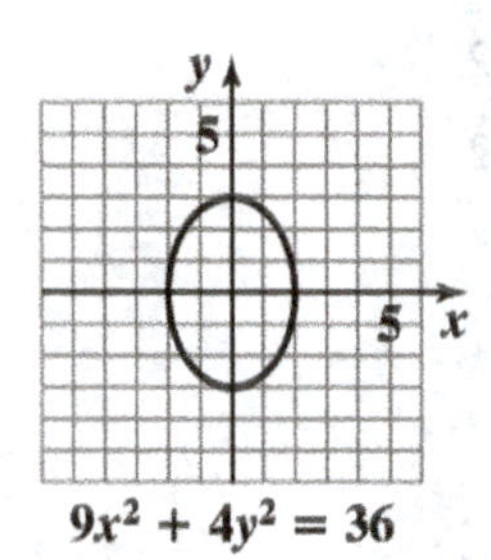

$9x^2+4y^2=36$

7. $\dfrac{(x-2)^2}{16}+\dfrac{(y+1)^2}{25}=1$

Center: $(2,-1)$

Because the denominator of the y^2 – term is greater than the denominator of the x^2 – term, the major axis is vertical. We have $a^2=25$ and $b^2=16$, so $a=5$ and $b=4$. The vertices lie 5 units above and below the center. The endpoints of the minor axis lie 4 units to the left and right of the center.

Vertices: $(2,4)$ and $(2,-6)$

Minor endpoints: $(-2,-1)$ and $(6,-1)$

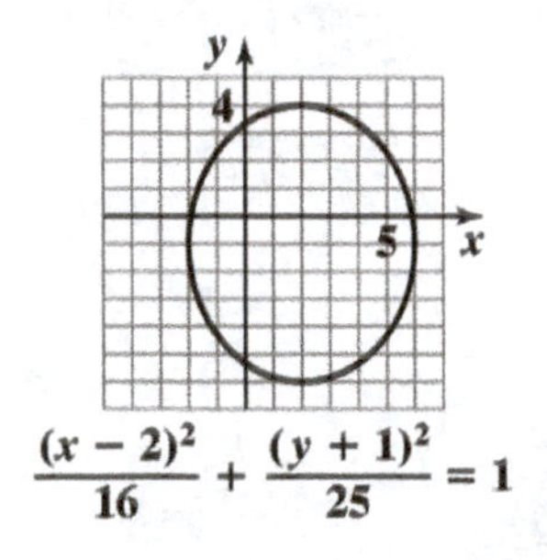

$\dfrac{(x-2)^2}{16}+\dfrac{(y+1)^2}{25}=1$

8. $\frac{(x+2)^2}{25}+\frac{(y-1)^2}{16}=1$

Center: $(-2,1)$

Because the denominator of the x^2 – term is greater than the denominator of the y^2 – term, the major axis is horizontal. We have $a^2 = 25$ and $b^2 = 16$, so $a = 5$ and $b = 4$. The vertices lie 5 units to the left and right of the center. The endpoints of the minor axis lie 4 units above and below the center.

Vertices: $(-7,1)$ and $(3,1)$

Minor endpoints: $(-2,5)$ and $(-2,-3)$

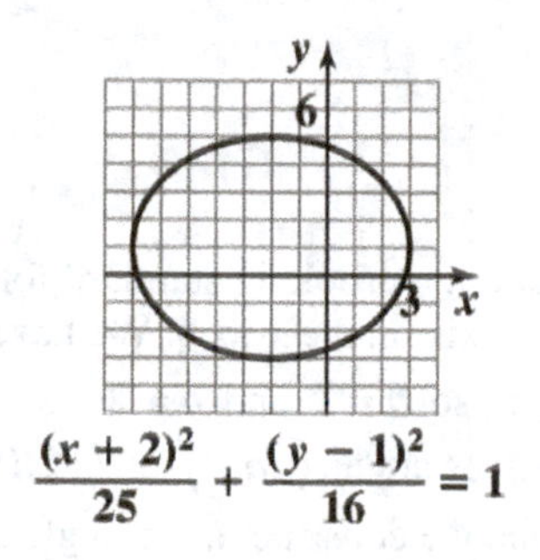

$\frac{(x+2)^2}{25}+\frac{(y-1)^2}{16}=1$

9. $\frac{x^2}{9}-y^2=1$

The equation is for a hyperbola in standard form with the transverse axis on the x-axis. We have $a^2 = 9$ and $b^2 = 1$, so $a = 3$ and $b = 1$. Therefore, the vertices are at $(\pm a, 0)$ or $(\pm 3, 0)$. Using a dashed line, we construct a rectangle using the ± 3 on the x-axis and ± 1 on the y-axis. Then use dashed lines to draw extended diagonals for the rectangle. These represent the asymptotes of the graph.

Graph the hyperbola.

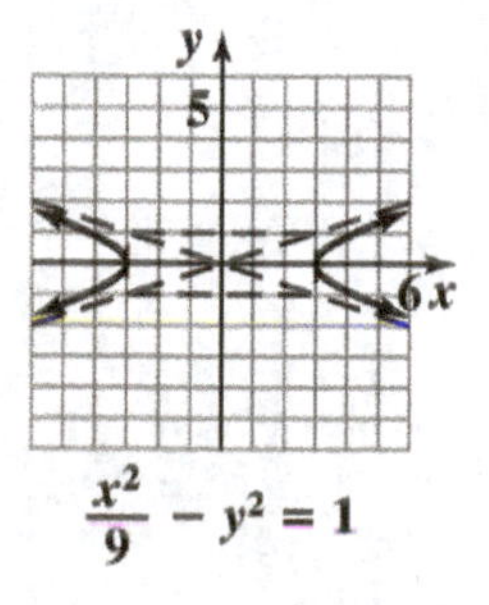

$\frac{x^2}{9}-y^2=1$

10. $\frac{y^2}{9}-x^2=1$

The equation is in the form $\frac{y^2}{a^2}-\frac{x^2}{b^2}=1$ with $a^2 = 9$, and $b^2 = 1$. We know the transverse axis lies on the y-axis and the vertices are $(0,-3)$ and $(0,3)$. Because $a^2 = 9$ and $b^2 = 1$, $a = 3$ and $b = 1$. Construct a rectangle using –1 and 1 on the x–axis, and –3 and 3 on the y–axis. Draw extended diagonals to obtain the asymptotes.

Graph the hyperbola.

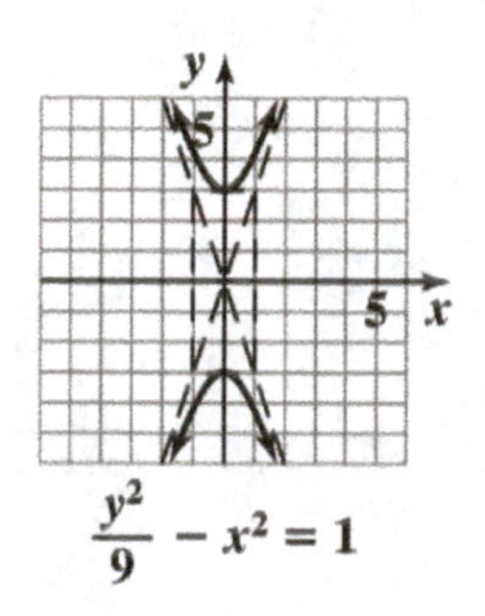

$\frac{y^2}{9}-x^2=1$

11. $y^2-4x^2=16$

$\frac{y^2}{16}-\frac{x^2}{4}=1$

The equation is in the form $\frac{y^2}{a^2}-\frac{x^2}{b^2}=1$ with $a^2 = 16$, and $b^2 = 4$. We know the transverse axis lies on the y-axis and the vertices are $(0,-4)$ and $(0,4)$. Because $a^2 = 16$ and $b^2 = 4$, $a = 4$ and $b = 2$. Construct a rectangle using –2 and 2 on the x–axis, and –4 and 4 on the y–axis. Draw extended diagonals to obtain the asymptotes. Graph the hyperbola.

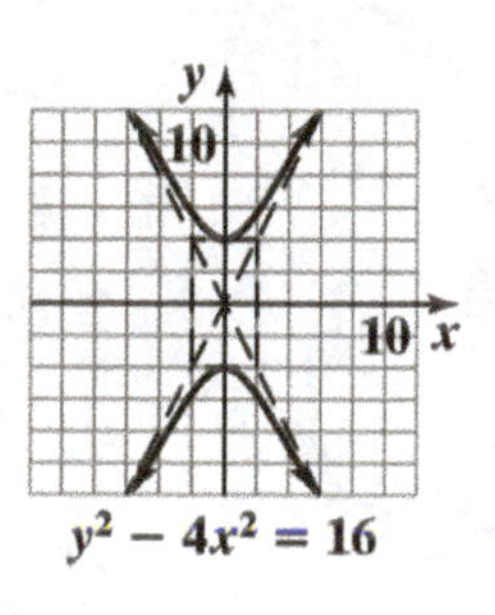

$y^2-4x^2=16$

12. $4x^2 - 49y^2 = 196$

$$\frac{x^2}{49} - \frac{y^2}{4} = 1$$

The equation is for a hyperbola in standard form with the transverse axis on the x-axis. We have $a^2 = 49$ and $b^2 = 4$, so $a = 7$ and $b = 2$. Therefore, the vertices are at $(\pm a, 0)$ or $(\pm 7, 0)$. Using a dashed line, we construct a rectangle using the ± 7 on the x-axis and ± 2 on the y-axis. Then use dashed lines to draw extended diagonals for the rectangle. These represent the asymptotes of the graph.

Graph the hyperbola.

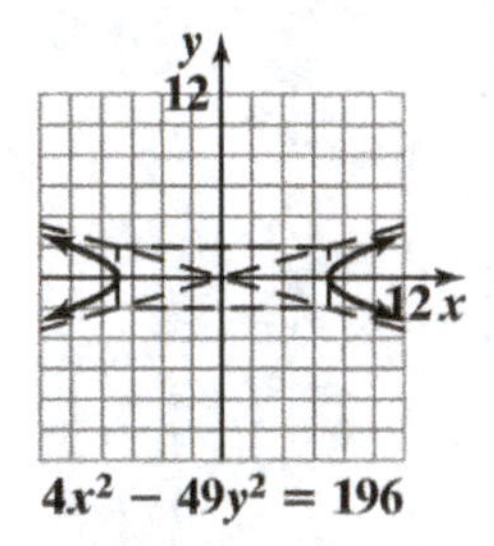

13. $x^2 + y^2 = 4$

This is the equation of a circle centered at the origin with radius $r = \sqrt{4} = 2$.

We can plot points that are 2 units to the left, right, above, and below the origin and then graph the circle. The points are $(-2,0)$, $(2,0)$, $(0,2)$, and $(0,-2)$.

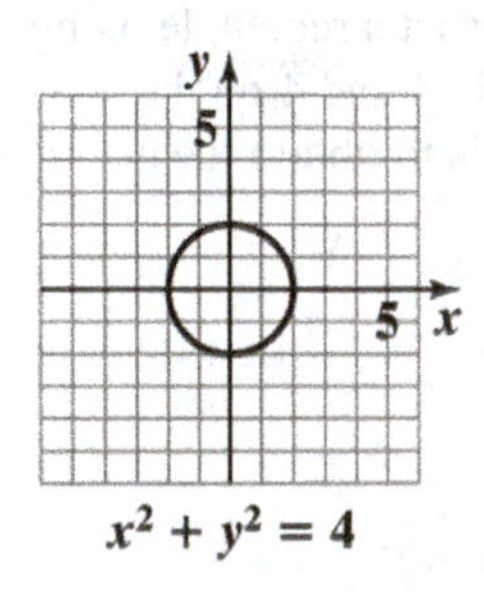

14. $x + y = 4$

$y = -x + 4$

This is the equation of a line with slope $m = -1$ and a y-intercept of 4. We can plot the point $(0,4)$, use the slope to get an additional point, connect the points with a straight line and then extend the line to represent the graph of the equation.

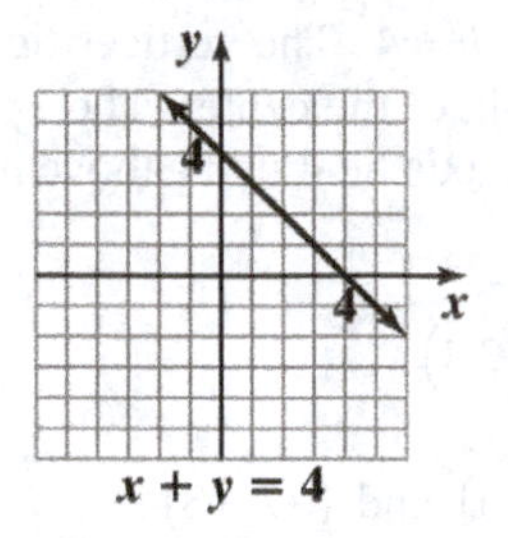

15. $x^2 - y^2 = 4$

$$\frac{x^2}{4} - \frac{y^2}{4} = 1$$

The equation is for a hyperbola in standard form with the transverse axis on the x-axis. We have $a^2 = 4$ and $b^2 = 4$, so $a = 2$ and $b = 2$. Therefore, the vertices are at $(\pm a, 0)$ or $(\pm 2, 0)$. Using a dashed line, we construct a rectangle using the ± 2 on the x-axis and ± 2 on the y-axis. Then use dashed lines to draw extended diagonals for the rectangle. These represent the asymptotes of the graph.

Graph the hyperbola.

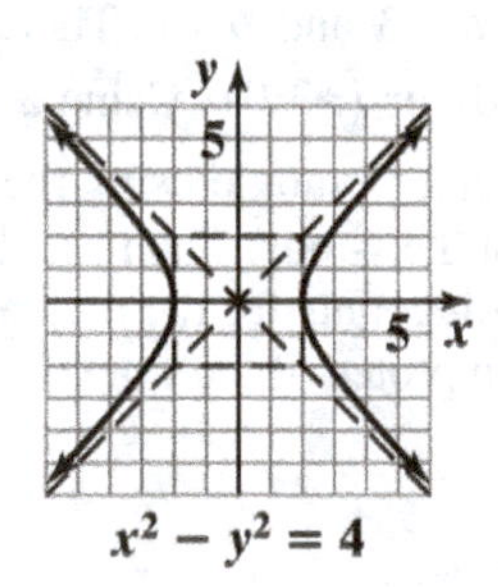

16. $x^2+4y^2=4$

$$\frac{x^2}{4}+\frac{y^2}{1}=1$$

Center: $(0,0)$

Because the denominator of the x^2 – term is greater than the denominator of the y^2 – term, the major axis is horizontal. We have $a^2=4$ and $b^2=1$, so $a=2$ and $b=1$. The vertices lie 2 units to the left and right of the center. The endpoints of the minor axis lie 1 unit above and below the center.

Vertices: $(-2,0)$ and $(2,0)$

Minor endpoints: $(0,-1)$ and $(0,1)$

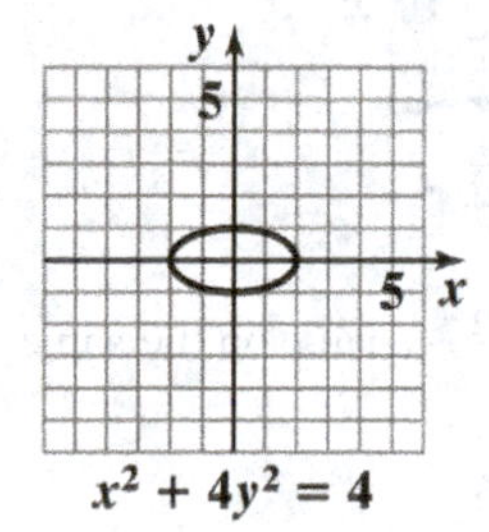

$x^2+4y^2=4$

17. $(x+1)^2+(y-1)^2=4$

Center: $(-1,1)$

Radius: $r=\sqrt{4}=2$

We plot the points that are 2 units to the left, right, above and below the center.

These points are $(-3,1)$, $(1,1)$, $(-1,3)$, and $(-1,-1)$.

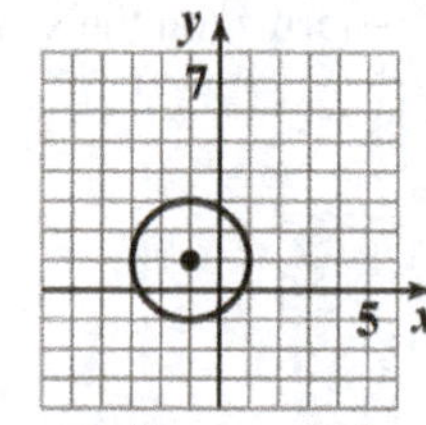

$(x+1)^2+(y-1)^2=4$

18. $x^2+4(y-1)^2=4$

$$\frac{x^2}{4}+\frac{(y-1)^2}{1}=1$$

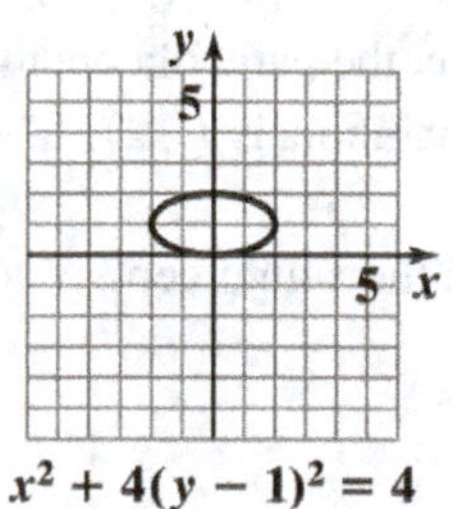

$x^2+4(y-1)^2=4$

19. $d=\sqrt{(-2-2)^2+(2-(-2))^2}$

$=\sqrt{(-4)^2+(4)^2}$

$=\sqrt{16+16}=\sqrt{32}$

$=\sqrt{16\cdot 2}$

$=4\sqrt{2}\approx 5.66$ units

$$\left(\frac{2+(-2)}{2},\frac{-2+2}{2}\right)=\left(\frac{0}{2},\frac{0}{2}\right)=(0,0)$$

The length of the segment is $4\sqrt{2}\approx 5.66$ units and the midpoint is the origin, $(0,0)$.

20. $d=\sqrt{(-10-(-5))^2+(14-8)^2}$

$=\sqrt{(-5)^2+(6)^2}$

$=\sqrt{25+36}$

$=\sqrt{61}\approx 7.81$ units

$$\left(\frac{-5+(-10)}{2},\frac{8+14}{2}\right)=\left(\frac{-15}{2},\frac{22}{2}\right)$$

$$=\left(-\frac{15}{2},11\right)$$

The length of the segment is $\sqrt{61}\approx 7.81$ units and the midpoint is $\left(-\frac{15}{2},11\right)$.

10.4 Check Points

1. $x = -(y-2)^2 + 1$

   This is a parabola of the form $x = a(y-k)^2 + h$. Since $a = -1$ is negative, the parabola opens to the left. The vertex of the parabola is $(1, 2)$. The axis of symmetry is $y = 2$.
   Replace y with 0 to find the x–intercept.

   $$x = -(y-2)^2 + 1$$
   $$= -(0-2)^2 + 1$$
   $$= -4 + 1$$
   $$= -3$$

   Replace x with 0 to find the y–intercepts.

   $$x = -(y-2)^2 + 1$$
   $$0 = -(y-2)^2 + 1$$
   $$0 = -(y^2 - 4y + 4) + 1$$
   $$0 = -y^2 + 4y - 4 + 1$$
   $$0 = -y^2 + 4y - 3$$
   $$0 = y^2 - 4y + 3$$
   $$0 = (y-1)(y-3)$$
   $$y - 1 = 0 \quad \text{or} \quad y - 3 = 0$$
   $$y = 1 \qquad\qquad y = 3$$

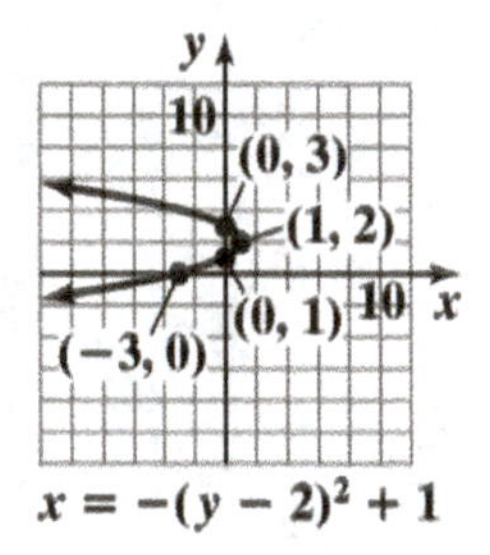

2. $x = y^2 + 8y + 7$

   This is a parabola of the form $x = ay^2 + by + c$. Since $a = 1$ is positive, the parabola opens to the right.
   The y–coordinate of the vertex is

   $$y = -\frac{b}{2a} = -\frac{8}{2(1)} = -\frac{8}{2} = -4.$$

   The x–coordinate of the vertex is
   $x = y^2 + 8y + 7 = (-4)^2 + 8(-4) + 7 = -9.$

   The vertex of the parabola is $(-9, -4)$. The axis of symmetry is $y = -4$.

   Replace y with 0 to find the x–intercept.
   $x = y^2 + 8y + 7 = 0^2 + 8(0) + 7 = 7$

   Replace x with 0 to find the y–intercepts.

   $$x = y^2 + 8y + 7$$
   $$0 = y^2 + 8y + 7$$
   $$0 = (y+7)(y+1)$$
   $$y + 7 = 0 \quad \text{or} \quad y + 1 = 0$$
   $$y = -7 \qquad\qquad y = -1$$

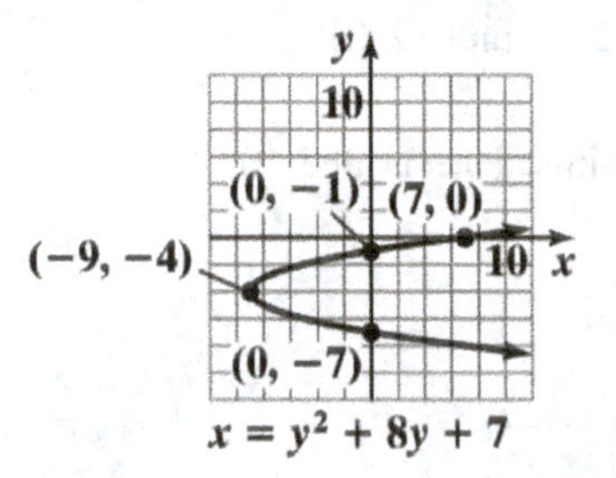

3. **a.** Collect x^2 – and y^2 – terms on the same side of the equation.

   $$x^2 = 4y^2 + 16$$
   $$x^2 - 4y^2 = 16$$

   Because x^2 and y^2 have opposite signs, the equation's graph is a hyperbola.

   b. Collect x^2 – and y^2 – terms on the same side of the equation.

   $$x^2 = 16 - 4y^2$$
   $$x^2 + 4y^2 = 16$$

   Because x^2 and y^2 have different positive coefficients, the equation's graph is an ellipse.

   c. Collect x^2 – and y^2 – terms on the same side of the equation.

   $$4x^2 = 16 - 4y^2$$
   $$4x^2 + 4y^2 = 16$$
   or
   $$x^2 + y^2 = 4$$

   Because x^2 and y^2 have equal positive coefficients, the equation's graph is a circle.

d. $x = -4y^2 + 16y$

Since only one variable is squared, the graph of the equation is a parabola. Furthermore, this parabola is in the form $x = ay^2 + by + c$, with a negative coefficient of the y^2 – term. Thus this horizontal parabola opens to the left.

10.4 Concept and Vocabulary Check

1. parabola; focus
2. >0
3. <0
4. to the left
5. to the right
6. upward
7. downward
8. downward
9. to the right
10. to the left
11. downward
12. upward
13. (h,k)
14. $-\dfrac{b}{2a}$
15. conic;

    a. hyperbola

    b. ellipse

    c. circle

    d. parabola

10.4 Exercise Set

1. Since $a = 1$, the parabola opens to the right. The vertex of the parabola is $(-1, 2)$. The equation's graph is b.

3. Since $a = 1$, the parabola opens to the right. The vertex of the parabola is $(1, -2)$. The equation's graph is f.

5. Since $a = -1$, the parabola opens to the left. The vertex of the parabola is $(1, 2)$. The equation's graph is a.

7. $x = 2y^2$

$x = 2(y-0)^2 + 0$

The vertex is the point $(0,0)$.

9. $x = (y-2)^2 + 3$

The vertex is the point $(3,2)$.

11. $x = -4(y+2)^2 - 1$

The vertex is the point $(-1,-2)$.

13. $x = 2(y-6)^2$

$x = 2(y-6)^2 + 0$

The vertex is the point $(0,6)$.

15. $x = y^2 - 6y + 6$

The y–coordinate of the vertex is

$-\dfrac{b}{2a} = -\dfrac{-6}{2(1)} = -\dfrac{-6}{2} = 3.$

The x–coordinate of the vertex is

$f(3) = 3^2 - 6(3) + 6 = 9 - 18 + 6 = -3.$

The vertex is the point $(-3,3)$.

17. $x = 3y^2 + 6y + 7$

The y–coordinate of the vertex is

$-\dfrac{b}{2a} = -\dfrac{6}{2(3)} = -\dfrac{6}{6} = -1.$

The x–coordinate of the vertex is

$f(-1) = 3(-1)^2 + 6(-1) + 7$

$= 3(1) - 6 + 7 = 3 - 6 + 7 = 4.$

The vertex is the point $(4,-1)$.

19. $x=(y-2)^2-4$

This is a parabola of the form $x=a(y-k)^2+h$. Since $a=1$ is positive, the parabola opens to the right. The vertex of the parabola is $(-4,2)$. The axis of symmetry is $y=2$. Replace y with 0 to find the x–intercept.

$x=(0-2)^2-4=4-4=0$

The x–intercept is 0. Replace x with 0 to find the y–intercepts.

$0=(y-2)^2-4$

$0=y^2-4y+4-4$

$0=y^2-4y$

$0=y(y-4)$

Apply the zero product principle.

$y=0$ and $y-4=0$

$y=4$

The y–intercepts are 0 and 4.

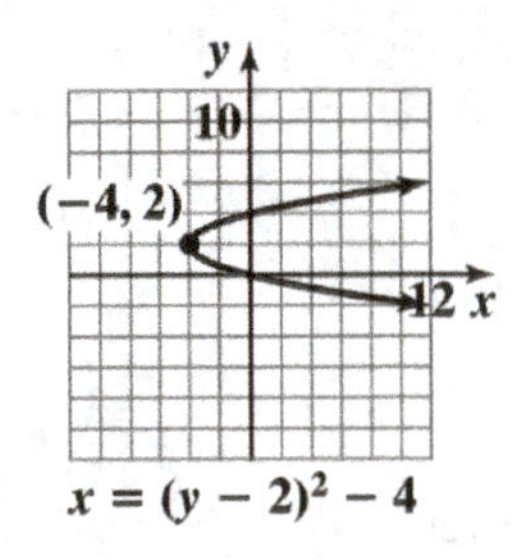

$x=(y-2)^2-4$

21. $x=(y-3)^2-5$

This is a parabola of the form $x=a(y-k)^2+h$. Since $a=1$ is positive, the parabola opens to the right. The vertex of the parabola is $(-5,3)$. The axis of symmetry is $y=3$. Replace y with 0 to find the x–intercept.

$x=(0-3)^2-5=(-3)^2-5=9-5=4$

The x–intercept is 0. Replace x with 0 to find the y–intercepts.

$0=(y-3)^2-5$

$0=y^2-6y+9-5$

$0=y^2-6y+4$

Solve using the quadratic formula.

$$x=\frac{-b\pm\sqrt{b^2-4ac}}{2a}$$

$$=\frac{-(-6)\pm\sqrt{(-6)^2-4(1)4}}{2(1)}$$

$$=\frac{6\pm\sqrt{36-16}}{2}=\frac{6\pm\sqrt{20}}{2}$$

$$=\frac{6\pm2\sqrt{5}}{2}=3\pm\sqrt{5}$$

The y–intercepts are $3-\sqrt{5}$ and $3+\sqrt{5}$.

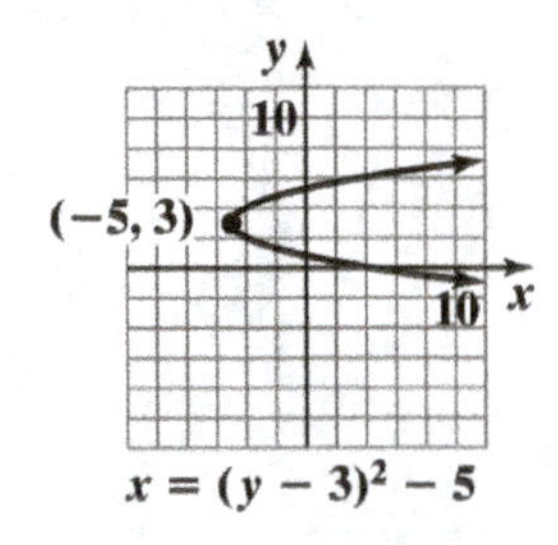

$x=(y-3)^2-5$

23. $x=-(y-5)^2+4$

This is a parabola of the form $x=a(y-k)^2+h$. Since $a=-1$ is negative, the parabola opens to the left. The vertex of the parabola is $(4,5)$. The axis of symmetry is $y=5$. Replace y with 0 to find the x–intercept.

$x=-(0-5)^2+4=-(-5)^2+4$

$=-25+4=-21$

The x–intercept is 0. Replace x with 0 to find the y–intercepts.

$0=-(y-5)^2+4$

$0=-(y^2-10y+25)+4$

$0=-y^2+10y-25+4$

$0=-y^2+10y-21$

$0=y^2-10y+21$

$0=(y-7)(y-3)$

$y-7=0$ or $y-3=0$

$y=7$ $\quad y=3$

The y–intercepts are 3 and 7 .

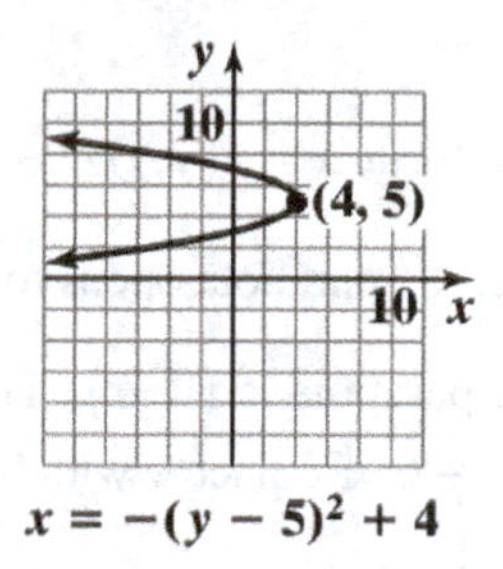

$x = -(y-5)^2 + 4$

25. $x = (y-4)^2 + 1$

This is a parabola of the form $x = a(y-k)^2 + h$. Since $a = 1$ is positive, the parabola opens to the right. The vertex of the parabola is $(1, 4)$. The axis of symmetry is $y = 4$. Replace y with 0 to find the x–intercept.

$$\begin{aligned} x &= (0-4)^2 + 1 \\ &= (-4)^2 + 1 \\ &= 16 + 1 \\ &= 17 \end{aligned}$$

The x–intercept is 0. Replace x with 0 to find the y–intercepts.

$$0 = (y-4)^2 + 1$$
$$0 = y^2 - 8y + 16 + 1$$
$$0 = y^2 - 8y + 17$$

Solve using the quadratic formula.

$$\begin{aligned} y &= \frac{-b \pm \sqrt{b^2 - 4ac}}{2a} \\ &= \frac{-(-8) \pm \sqrt{(-8)^2 - 4(1)(17)}}{2(1)} \\ &= \frac{8 \pm \sqrt{64 - 68}}{2} \\ &= \frac{8 \pm \sqrt{-4}}{2} \\ &= \frac{8 \pm 2i}{2} \\ &= 4 \pm i \end{aligned}$$

The solutions are complex, so there are no y–intercepts.

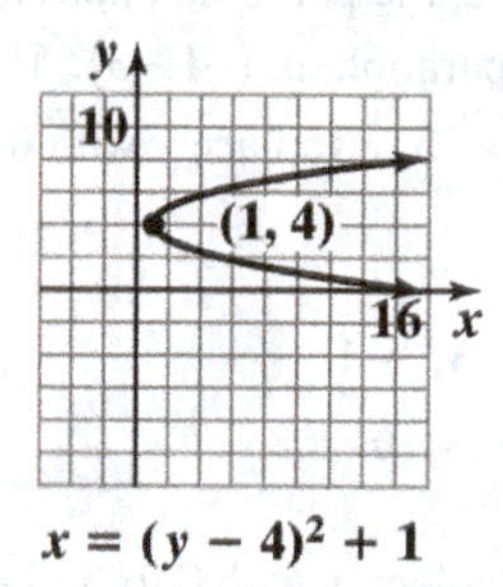

$x = (y-4)^2 + 1$

27. $x = -3(y-5)^2 + 3$

This is a parabola of the form $x = a(y-k)^2 + h$. Since $a = -3$ is negative, the parabola opens to the left. The vertex of the parabola is $(3, 5)$. The axis of symmetry is $y = 5$. Replace y with 0 to find the x–intercept.

$$\begin{aligned} x &= -3(0-5)^2 + 3 = -3(-5)^2 + 3 \\ &= -3(25) + 3 = -75 + 3 = -72 \end{aligned}$$

The x–intercept is 0. Replace x with 0 to find the y–intercepts.

$$0 = -3(y-5)^2 + 3$$
$$0 = -3(y^2 - 10y + 25) + 3$$
$$0 = -3y^2 + 30y - 75 + 3$$
$$0 = -3y^2 + 30y - 72$$
$$0 = y^2 - 10y + 24$$
$$0 = (y-6)(y-4)$$

$y - 6 = 0$ or $y - 4 = 0$

$y = 6$ $\quad$ $y = 4$

The y–intercepts are 4 and 6.

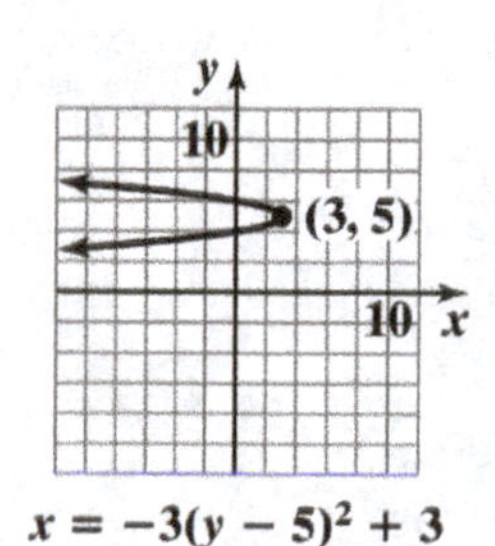

$x = -3(y-5)^2 + 3$

29. $x = -2(y+3)^2 - 1$

This is a parabola of the form $x = a(y-k)^2 + h$. Since $a = -2$ is negative, the parabola opens to the left. The vertex of the parabola is $(-1,-3)$. The axis of symmetry is $y = -3$. Replace y with 0 to find the x–intercept.

$$x = -2(0+3)^2 - 1 = -2(3)^2 - 1$$
$$= -2(9) - 1 = -18 - 1 = -19$$

The x–intercept is 0. Replace x with 0 to find the y–intercepts.

$$0 = -2(y+3)^2 - 1$$
$$0 = -2(y^2 + 6x + 9) - 1$$
$$0 = -2y^2 - 12x - 18 - 1$$
$$0 = -2y^2 - 12x - 19$$
$$0 = 2y^2 + 12x + 19$$

Solve using the quadratic formula.

$$y = \frac{-12 \pm \sqrt{12^2 - 4(2)(19)}}{2(2)}$$
$$= \frac{-12 \pm \sqrt{144 - 152}}{4}$$
$$= \frac{-12 \pm \sqrt{-8}}{4}$$

Since the solutions will be complex, there are no y–intercepts.

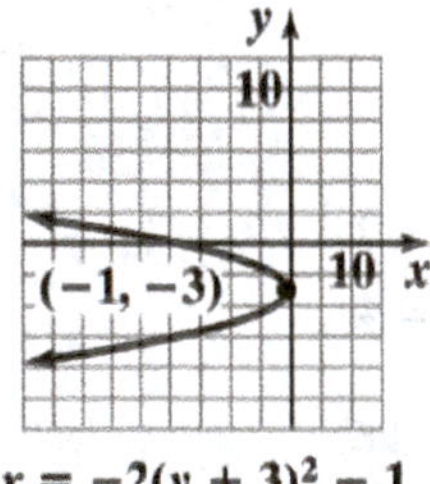

$x = -2(y + 3)^2 - 1$

31. $x = \frac{1}{2}(y+2)^2 + 1$

This is a parabola of the form $x = a(y-k)^2 + h$. Since $a = \frac{1}{2}$ is positive, the parabola opens to the right. The vertex of the parabola is $(1,-2)$. The axis of symmetry is $y = -2$. Replace y with 0 to find the x–intercept.

$$x = \frac{1}{2}(0+2)^2 + 1$$
$$= \frac{1}{2}(4) + 1 = 2 + 1 = 3$$

The x–intercept is 3. Replace x with 0 to find the y–intercepts.

$$0 = \frac{1}{2}(y+2)^2 + 1$$
$$0 = \frac{1}{2}(y^2 + 4y + 4) + 1$$
$$0 = \frac{1}{2}y^2 + 2y + 2 + 1$$
$$0 = \frac{1}{2}y^2 + 2y + 3$$
$$0 = y^2 + 4y + 6$$

Solve using the quadratic formula.

$$y = \frac{-4 \pm \sqrt{4^2 - 4(1)(6)}}{2(1)}$$
$$= \frac{-4 \pm \sqrt{16 - 24}}{2}$$
$$= \frac{-4 \pm \sqrt{-8}}{2}$$

Since the solutions will be complex, there are no y–intercepts.

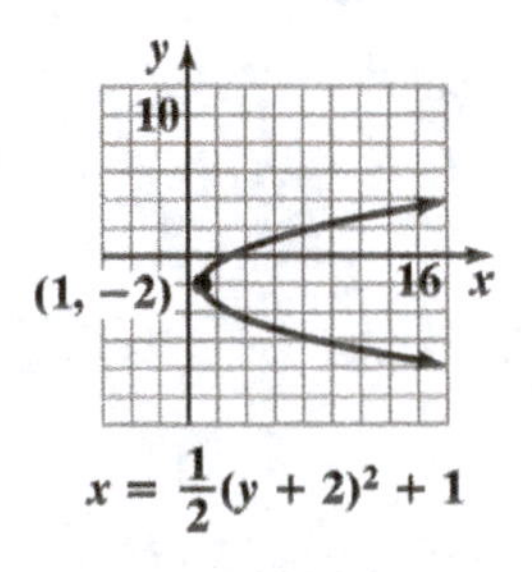

$x = \frac{1}{2}(y + 2)^2 + 1$

33. $x = y^2 + 2y - 3$

This is a parabola of the form $x = ay^2 + by + c$. Since $a = 1$ is positive, the parabola opens to the right. The y–coordinate of the vertex is $-\frac{b}{2a} = -\frac{2}{2(1)} = -\frac{2}{2} = -1$. The x–coordinate of the vertex is $x = (-1)^2 + 2(-1) - 3 = 1 - 2 - 3 = -4$.

The vertex of the parabola is $(-4, -1)$. The axis of symmetry is $y = -1$. Replace y with 0 to find the x–intercept.

$x = 0^2 + 2(0) - 3 = 0 + 0 - 3 = -3$

The x–intercept is –3. Replace x with 0 to find the y–intercepts.

$0 = y^2 + 2y - 3$

$0 = (y+3)(y-1)$

$y + 3 = 0$ or $y - 1 = 0$

$y = -3$ $\quad$ $y = 1$

The y–intercepts are –3 and 1.

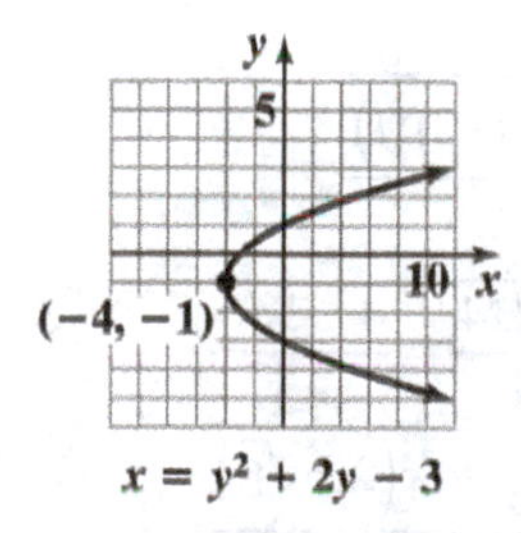

35. $x = -y^2 - 4y + 5$

This is a parabola of the form $x = ay^2 + by + c$. Since $a = -1$ is negative, the parabola opens to the left. The y–coordinate of the vertex is $-\frac{b}{2a} = -\frac{-4}{2(-1)} = -\frac{-4}{-2} = -2$. The x–coordinate of the vertex is

$x = -(-2)^2 - 4(-2) + 5 = -4 + 8 + 5 = 9$.

The vertex of the parabola is $(9, -2)$. The axis of symmetry is $y = -2$. Replace y with 0 to find the x–intercept.

$x = -0^2 - 4(0) + 5 = 0 - 0 + 5 = 5$

The x–intercept is 5. Replace x with 0 to find the y–intercepts.

$0 = -y^2 - 4y + 5$

$0 = y^2 + 4y - 5$

$0 = (y+5)(y-1)$

$y + 5 = 0$ or $y - 1 = 0$

$y = -5$ $\quad$ $y = 1$

The y–intercepts are –5 and 1.

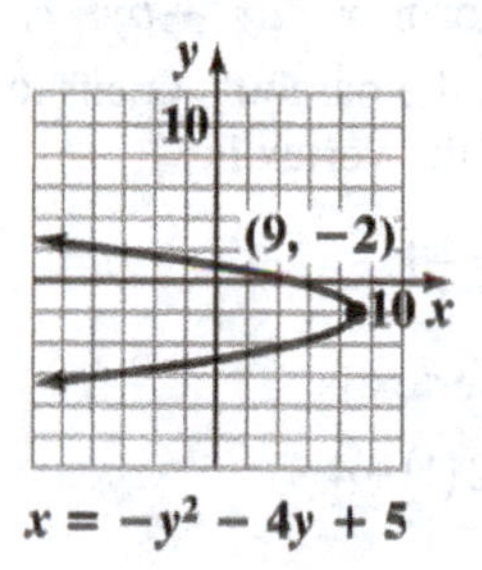

37. $x = y^2 + 6y$

This is a parabola of the form $x = ay^2 + by + c$. Since $a = 1$ is positive, the parabola opens to the right. The y–coordinate of the vertex is $-\frac{b}{2a} = -\frac{6}{2(1)} = -\frac{6}{2} = -3$.

The x–coordinate of the vertex is $x = (-3)^2 + 6(-3) = 9 - 18 = -9$.

The vertex of the parabola is $(-9, -3)$. The axis of symmetry is $y = -3$. Replace y with 0 to find the x–intercept.

$x = 0^2 + 6(0) = 0$

The x–intercept is 0. Replace x with 0 to find the y–intercepts.

$0 = y^2 + 6y$

$0 = y(y+6)$

$y = 0$ or $y + 6 = 0$

$y = -6$

The y–intercepts are –6 and 0.

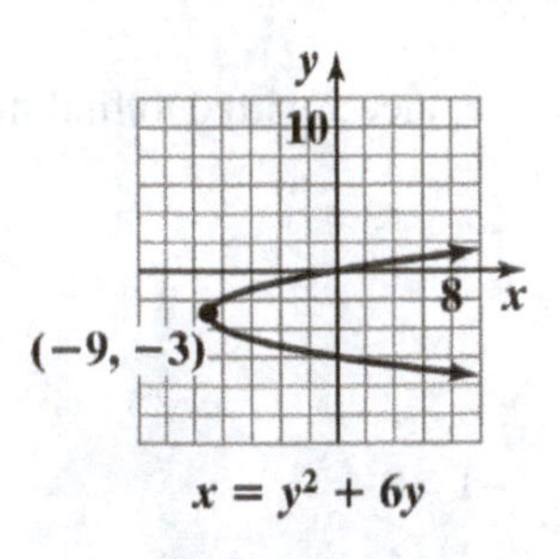

39. $x = -2y^2 - 4y$

This is a parabola of the form $x = ay^2 + by + c$. Since $a = -2$ is negative, the parabola opens to the left. The y–coordinate of the vertex is $-\frac{b}{2a} = -\frac{-4}{2(-2)} = -\frac{-4}{-4} = -1.$

The x–coordinate of the vertex is

$$x = -2(-1)^2 - 4(-1) = -2(1) + 4$$
$$= -2 + 4 = 2.$$

The vertex of the parabola is $(2, -1)$. The axis of symmetry is $y = -1$. Replace y with 0 to find the x–intercept.

$$x = -2(0)^2 - 4(0) = -2(0) - 0 = 0$$

The x–intercept is 0. Replace x with 0 to find the y–intercepts.

$$0 = -2y^2 - 4y$$
$$0 = y^2 + 2y$$
$$0 = y(y+2)$$
$$y = 0 \quad \text{or} \quad y + 2 = 0$$
$$y = -2$$

The y–intercepts are –2 and 0.

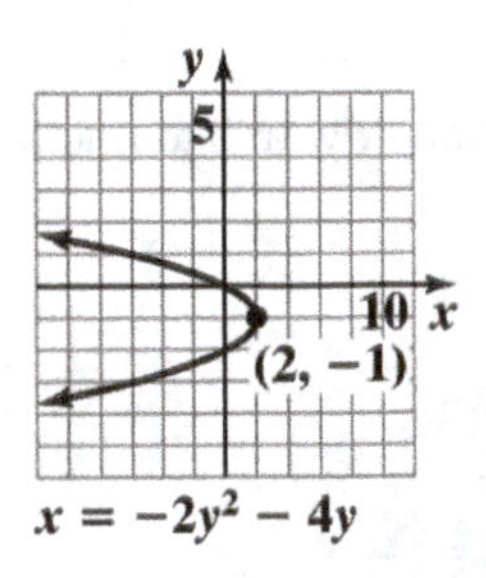

41. $x = -2y^2 - 4y + 1$

This is a parabola of the form $x = ay^2 + by + c$. Since $a = -2$ is negative, the parabola opens to the left. The y–coordinate of the vertex is $-\frac{b}{2a} = -\frac{-4}{2(-2)} = -\frac{-4}{-4} = -1.$

The x–coordinate of the vertex is

$$x = -2(-1)^2 - 4(-1) + 1 = -2(1) + 4 + 1$$
$$= -2 + 4 + 1 = 3$$

The vertex of the parabola is $(3, -1)$. The axis of symmetry is $y = -1$. Replace y with 0 to find the x–intercept.

$$x = -2(0)^2 - 4(0) + 1 = -2(0) - 0 + 1$$
$$= 0 - 0 + 1 = 1$$

The x–intercept is 1. Replace x with 0 to find the y–intercepts.

$$0 = -2y^2 - 4y + 1$$

Solve using the quadratic formula.

$$y = \frac{-b \pm \sqrt{b^2 - 4ac}}{2a}$$
$$= \frac{-(-4) \pm \sqrt{(-4)^2 - 4(-2)(1)}}{2(-2)}$$
$$= \frac{4 \pm \sqrt{16+8}}{-4} = \frac{4 \pm \sqrt{24}}{-4} = \frac{4 \pm 2\sqrt{6}}{-4}$$
$$= \frac{2(2 \pm \sqrt{6})}{-4} = \frac{2 \pm \sqrt{6}}{-2} = \frac{-(2 \pm \sqrt{6})}{2}$$
$$= \frac{-2 \pm \sqrt{6}}{2}$$

The y–intercepts are $\frac{-2 \pm \sqrt{6}}{2}$.

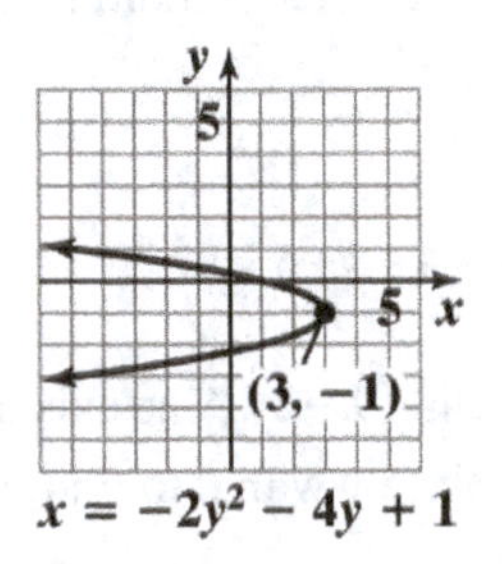

43. a. Since the squared term is y, the parabola is horizontal.

b. Since $a = 2$ is positive, the parabola opens to the right.

c. The vertex is the point $(2,1)$.

45. a. Since the squared term is x, the parabola is vertical.

b. Since $a = 2$ is positive, the parabola opens up.

c. The vertex is the point $(1,2)$.

47. a. Since the squared term is x, the parabola is vertical.

b. Since $a = -1$ is negative, the parabola opens down.

c. The vertex is the point $(-3,4)$.

49. a. Since the squared term is y, the parabola is horizontal.

b. Since $a = -1$ is negative, the parabola opens to the left.

c. The vertex is the point $(4,-3)$.

51. a. Since the squared term is x, the parabola is vertical.

b. Since $a = 1$ is positive, the parabola opens up.

c. The x–coordinate of the vertex is

$$-\frac{b}{2a} = -\frac{-4}{2(1)} = -\frac{-4}{2} = 2.$$

The y–coordinate of the vertex is

$$f(2) = 2^2 - 4(2) - 1$$
$$= 4 - 8 - 1 = -5.$$

The vertex is the point $(2,-5)$.

53. a. Since the squared term is y, the parabola is horizontal.

b. Since $a = -1$ is negative, the parabola opens to the left.

c. The y–coordinate of the vertex is

$$-\frac{b}{2a} = -\frac{4}{2(-1)} = -\frac{4}{-2} = 2.$$

The x–coordinate of the vertex is

$$f(2) = -(2)^2 + 4(2) + 1$$
$$= -4 + 8 + 1 = 5.$$

The vertex is the point $(5,2)$.

55. $x - 7 - 8y = y^2$

Since only one variable is squared, the graph of the equation is a parabola.

57. $4x^2 = 36 - y^2$

$4x^2 + y^2 = 36$

Because x^2 and y^2 have different positive coefficients, the equation's graph is an ellipse.

59. $x^2 = 36 + 4y^2$

$x^2 - 4y^2 = 36$

Because x^2 and y^2 have opposite signs, the equation's graph is a hyperbola.

61. $3x^2 = 12 - 3y^2$

$3x^2 + 3y^2 = 12$

Because x^2 and y^2 have the same positive coefficient, the equation's graph is a circle.

63. $3x^2 = 12 + 3y^2$

$3x^2 - 3y^2 = 12$

Because x^2 and y^2 have opposite signs, the equation's graph is a hyperbola.

65. $x^2 - 4y^2 = 16$

Because x^2 and y^2 have opposite signs, the equation's graph is a hyperbola.

$$\frac{x^2}{16} - \frac{4y^2}{16} = \frac{16}{16}$$
$$\frac{x^2}{16} - \frac{y^2}{4} = 1$$

The equation is in the form $\frac{x^2}{a^2} - \frac{y^2}{b^2} = 1$ with $a^2 = 16$, and $b^2 = 4$. We know the transverse axis lies on the x-axis and the vertices are $(-4,0)$ and $(4,0)$. Because $a^2 = 16$ and $b^2 = 4$, $a = 4$ and $b = 2$. Construct a rectangle using –4 and 4 on the x–axis, and –2 and 2 on the y–axis. Draw extended diagonals to obtain the asymptotes.

Graph the hyperbola.

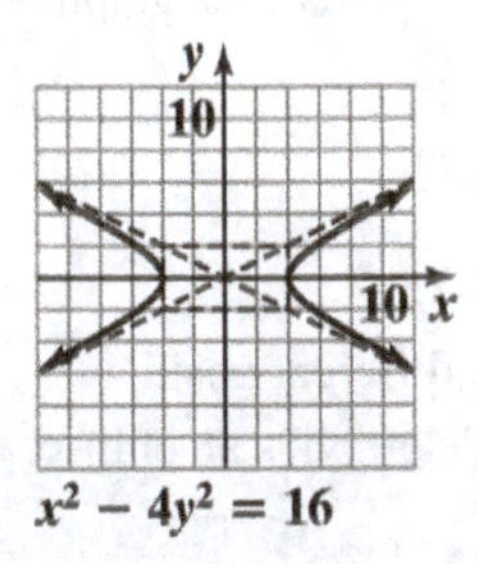

67. $4x^2 + 4y^2 = 16$

Because x^2 and y^2 have the same positive coefficient, the equation's graph is a circle.

$$\frac{4x^2}{4} + \frac{4y^2}{4} = \frac{16}{4}$$
$$x^2 + y^2 = 4$$

The center is $(0,0)$ and the radius is 2 units.

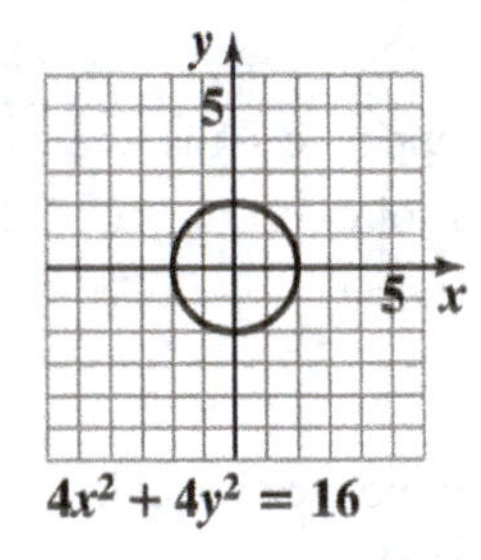

69. $x^2 + 4y^2 = 16$

Because x^2 and y^2 have different positive coefficients, the equation's graph is an ellipse.

$$\frac{x^2}{16} + \frac{4y^2}{16} = \frac{16}{16}$$
$$\frac{x^2}{16} + \frac{y^2}{4} = 1$$

The vertices are $(-4,0)$ and $(4,0)$. The endpoints of the minor axis are $(0,-2)$ and $(0,2)$.

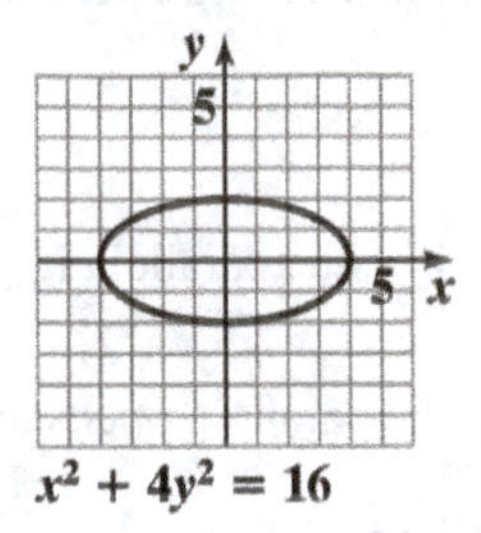

71. $x = (y-1)^2 - 4$

Since only one variable is squared, the graph of the equation is a parabola.

This is a parabola of the form $x = a(y-k)^2 + h$. Since $a = 1$ is positive, the parabola opens to the right. The vertex of the parabola is $(-4,1)$. The axis of symmetry is $y = 1$. Replace y with 0 to find the x–intercept.

$$x = (0-1)^2 - 4 = (-1)^2 - 4 = 1 - 4 = -3$$

The x–intercept is -3. Replace x with 0 to find the y–intercepts.

$$0 = (y-1)^2 - 4$$
$$0 = y^2 - 2y + 1 - 4$$
$$0 = y^2 - 2y - 3$$
$$0 = (y-3)(y+1)$$
$$y - 3 = 0 \quad \text{or} \quad y + 1 = 0$$
$$y = 3 \qquad\qquad y = -1$$

The y–intercepts are -1 and 3.

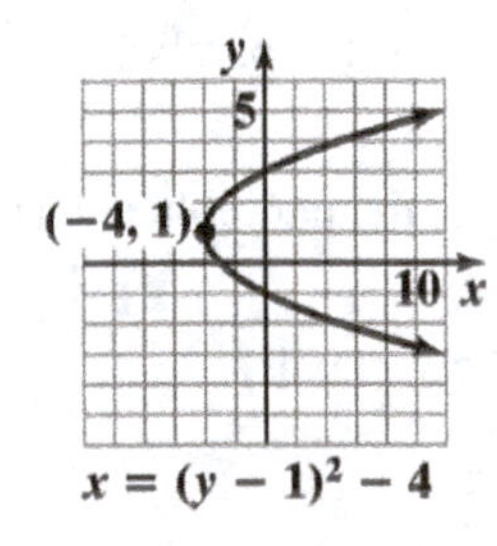

73. $(x-2)^2 + (y+1)^2 = 16$

Because x^2 and y^2 have the same positive coefficient, the equation's graph is a circle.

The center is $(2,-1)$ and the radius is 4 units.

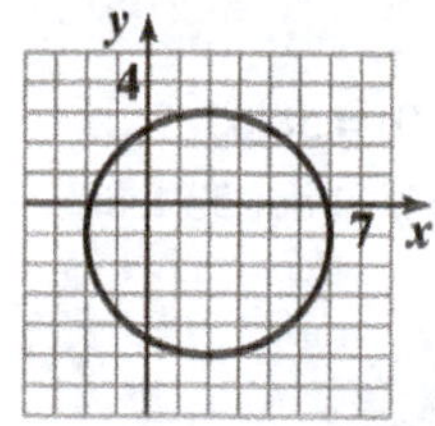

$(x-2)^2+(y+1)^2=16$

75. The y-coordinate of the vertex is
$y=-\frac{b}{2a}=-\frac{6}{2(1)}=-3$.

The x-coordinate of the vertex is
$x=(-3)^2+6(-3)+5=-4$.

The vertex is $(-4,-3)$.

Since the squared term is y and $a>0$, the graph opens to the right.

Domain: $[-4,\infty)$

Range: $(-\infty,\infty)$

The relation is not a function.

77. The x-coordinate of the vertex is
$x=-\frac{b}{2a}=-\frac{(4)}{2(-1)}=2$.

The y-coordinate of the vertex is
$y=-(2)^2+4(2)-3=1$.

The vertex is $(2,1)$.

Since the squared term is x and $a<0$, the graph opens down.

Domain: $(-\infty,\infty)$

Range: $(-\infty,1]$

The relation is a function.

79. The equation is in the form $x=a(y-k)^2+h$
From the equation, we can see that the vertex is $(3,1)$.

Since the squared term is y and $a<0$, the graph opens to the left.

Domain: $(-\infty,3]$

Range: $(-\infty,\infty)$

The relation is not a function.

81. $x=(y-2)^2-4$

$y=-\frac{1}{2}x$

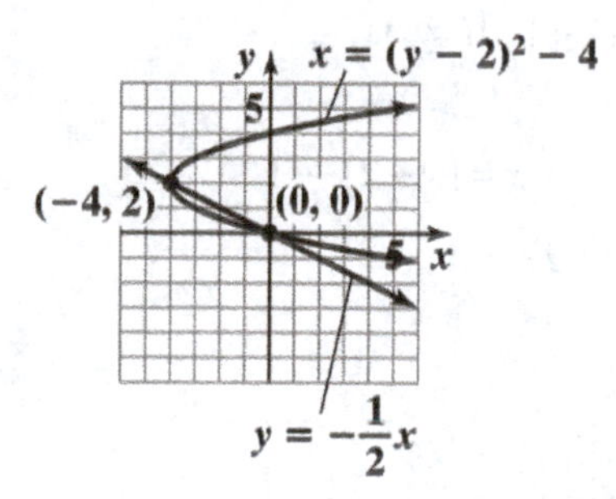

Check $(-4,2)$:

$-4=(2-2)^2-4$
$-4=0-4$
$-4=-4$ true

$2=-\frac{1}{2}(-4)$
$2=2$ true

Check $(0,0)$:

$0=(0-2)^2-4$
$0=4-4$
$0=0$ true

$0=-\frac{1}{2}(0)$
$0=0$ true

The solution set is $\{(-4,2),(0,0)\}$.

83. $x = y^2 - 3$
$x = y^2 - 3y$

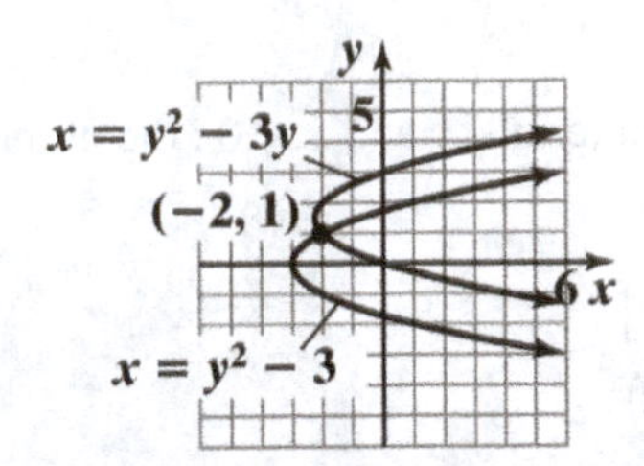

Check $(-2,1)$:

$-2 = (1)^2 - 3 \qquad -2 = (1)^2 - 3(1)$
$-2 = 1-3 \qquad -2 = 1-3$
$-2 = -2$ true $\qquad -2 = -2$ true

The solution set is $\{(-2,1)\}$.

85. $x = (y+2)^2 - 1$
$(x-2)^2 + (y+2)^2 = 1$

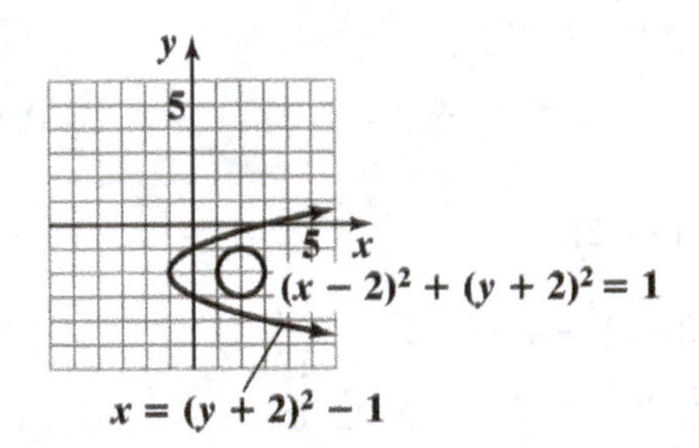

The two graphs do not cross. Therefore, the solution set is the empty set, $\{\ \}$ or $\varnothing$.

87. a. $y = ax^2$
$316 = a(1750)^2$
$316 = a(3062500)$
$\frac{316}{3062500} = a$
$0.0001032 = a$

The equation is $y = 0.0001032x^2$.

b. To find the height of the cable 1000 feet from the tower, find y when $x = 1750 - 1000 = 750$.

$y = 0.0001032(750)^2$
$= 0.0001032(562,500) = 58.05$

The height of the cable is about 58 feet.

89. a. $y = ax^2$
$2 = a(6)^2$
$2 = a(36)$
$\frac{2}{36} = a$
$\frac{1}{18} = a$

The equation is $y = \frac{1}{18}x^2$.

b. $a = \frac{1}{4p}$
$\frac{1}{18} = \frac{1}{4p}$
$4p = 18$
$p = \frac{18}{4} = 4.5$

The receiver should be placed 4.5 feet from the base of the dish.

91. a. ellipse

b. $x^2 + 4y^2 = 4$

93. – 99. Answers will vary.

101.

$$y^2 + 2y - 6x + 13 = 0$$
$$y^2 + 2y + (-6x + 13) = 0$$
$$a = 1 \qquad b = 2 \qquad c = -6x + 13$$

$$y = \frac{-2 \pm \sqrt{2^2 - 4(1)(-6x+13)}}{2(1)}$$
$$= \frac{-2 \pm \sqrt{4 - 4(-6x+13)}}{2}$$
$$= \frac{-2 \pm \sqrt{4 + 24x - 52}}{2}$$
$$= \frac{-2 \pm \sqrt{24x - 48}}{2} = \frac{-2 \pm \sqrt{4(6x-12)}}{2}$$
$$= \frac{-2 \pm 2\sqrt{6x-12}}{2}$$
$$= -1 \pm \sqrt{6x-12}$$

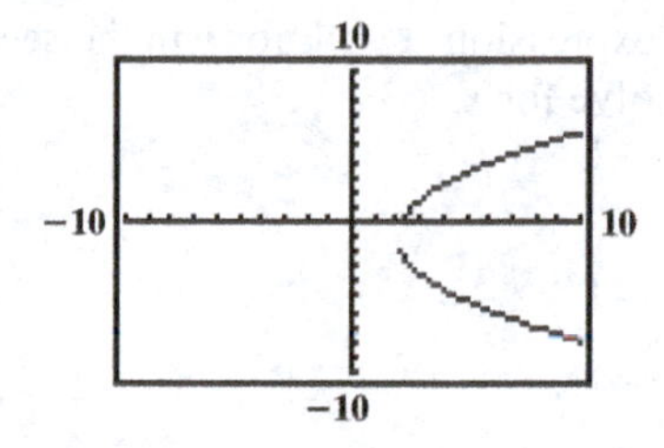

103. Answers will vary. For example, consider Exercise 19.

$$x = (y-2)^2 - 4$$
$$x + 4 = (y-2)^2$$
$$\pm\sqrt{x+4} = y - 2$$
$$2 \pm \sqrt{x+4} = y$$

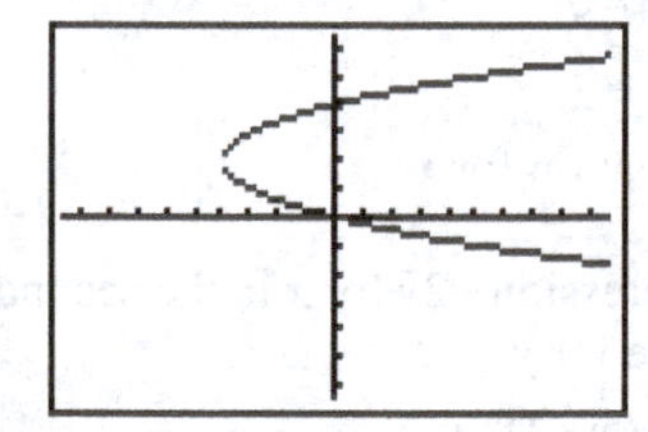

105. makes sense

107. makes sense

109. true

111. false; Changes to make the statement true will vary. A sample change is: $x = a(y-k)+h$ is not a parabola. There is no squared variable.

113.

$$y = ax^2$$
$$-50 = a(100)^2$$
$$-50 = a(10{,}000)$$
$$a = -\frac{50}{10{,}000} = -\frac{1}{200}$$

The equation of the parabola is $y = -\frac{1}{200}x^2$.

$$y = -\frac{1}{200}(30)^2 = -\frac{1}{200}(900) = -4.5$$

The height of the arch is 50 feet. The archway comes down 4.5 feet, so the height of the arch 30 feet from the center is 50 – 4.5 = 45.5 feet.

114. $f(x) = 2^{1-x}$

x	$f(x)$
−2	8
−1	4
0	2
1	1
2	$\frac{1}{2}$

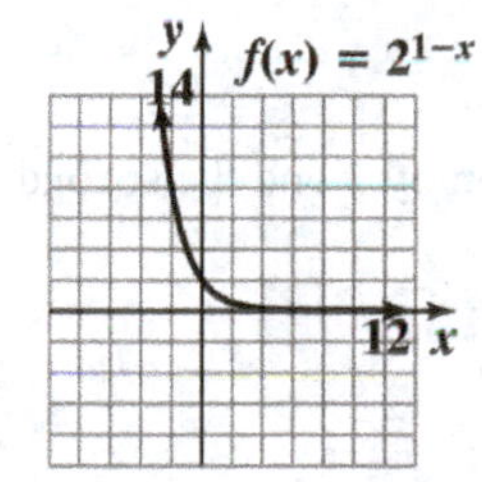

115. $f(x) = \frac{1}{3}x - 5$

$$y = \frac{1}{3}x - 5$$

Interchange x and y and solve for y.

$$x = \frac{1}{3}y - 5$$
$$x + 5 = \frac{1}{3}y$$
$$3x + 15 = y$$
$$f^{-1}(x) = 3x + 15$$

116. $(x+1)^2+(x+3)^2=4$

$x^2+2x+1+x^2+6x+9=4$

$2x^2+8x+10=4$

$2x^2+8x+6=0$

$x^2+4x+3=0$

$(x+1)(x+3)=0$

$x+1=0$ or $x+3=0$

$x=-1$ $\quad$ $x=-3$

The solution set is $\{-3,-1\}$.

117. $4x+3y=4$

$4x+3\overbrace{(2x-7)}^{y}=4$

$4x+6x-21=4$

$10x-21=4$

$10x=25$

$x=2.5$

Substitute to find y.

$y=2x-7$

$y=2(2.5)-7=-2$

The solution is $(2.5,-2)$.

The solution set is $\{(2.5,-2)\}$.

118. $2x+4y=-4$

$3x+5y=-3$

Multiply the first equation by 3 and the second equation by -2.

$6x+12y=-12$

$-6x-10y=\ \ 6$

$2y=-6$

$y=-3$

Back-substitute to find x.

$2x+4y=-4$

$2x+4(-3)=-4$

$2x-12=-4$

$2x=8$

$x=4$

The solution is $(4,-3)$.

The solution set is $\{(4,-3)\}$.

119. $x^2=2(3x-9)+10$

$x^2=6x-18+10$

$x^2=6x-8$

$x^2-6x+8=0$

$(x-2)(x-4)=0$

$x-2=0$ or $x-4=0$

$x=2$ $\quad$ $x=4$

The solution set is $\{2,4\}$.

10.5 Check Points

1. $x^2=y-1$

$4x-y=-1$

Solve the first equation for y.

$y=x^2+1$

Substitute the expression x^2+1 for y in the second equation and solve for x.

$4x-y\ =-1$

$4x-\overbrace{(x^2+1)}^{y}=-1$

$4x-x^2-1=-1$

$x^2-4x=0$

$x(x-4)=0$

$x=0$ or $x-4=0$

$x=4$

If $x=0$, $y=(0)^2+1=1$.

If $x=4$, $y=(4)^2+1=17$.

The solution set is $\{(0, 1), (4, 17)\}$.

2. $x+2y=0$

$(x-1)^2+(y-1)^2=5$

Solve the first equation for x.

$x=-2y$

Substitute the expression $-2y$ for x in the second equation and solve for y.

$(x-1)^2+(y-1)^2=5$

$(\overbrace{-2y}^{x}-1)^2+(y-1)^2=5$

$4y^2+4y+1+y^2-2y+1=5$

$5y^2+2y-3=0$

$(5y-3)(y+1)=0$

$5y-3=0$ or $y+1=0$

$y=\frac{3}{5}$ or $y=-1$

If $y=\frac{3}{5}$, $x=-2\left(\frac{3}{5}\right)=-\frac{6}{5}$.

If $y=-1$, $x=-2(-1)=2$.

The solution set is $\left\{\left(-\frac{6}{5}, \frac{3}{5}\right), (2, -1)\right\}$.

3. $3x^2+2y^2=35$

$4x^2+3y^2=48$

Eliminate the y^2-term by multiplying the first equation by –3 and the second equation by 2. Add the resulting equations.

$$\begin{aligned} -9x^2-6y^2&=-105\\ 8x^2+6y^2&=96\\ \hline -x^2&=-9\\ x^2&=9\\ x&=\pm 3 \end{aligned}$$

If $x=3$,

$$\begin{aligned} 3(3)^2+2y^2&=35\\ y^2&=4\\ y&=\pm 2 \end{aligned}$$

If $x=-3$,

$$\begin{aligned} 3(-3)^2+2y^2&=35\\ y^2&=4\\ y&=\pm 2 \end{aligned}$$

The solution set is $\{(3,2),(3,-2),(-3,2),(-3,-2)\}$.

4. $y=x^2+5$

$x^2+y^2=25$

Arrange the first equation so that variable terms appear on the left, and constants appear on the right. Add the resulting equations to eliminate the x^2-terms and solve for y.

$$\begin{aligned} -x^2+y&=5\\ x^2+y^2&=25\\ \hline y^2+y&=30 \end{aligned}$$

$y^2+y-30=0$

$(y+6)(y-5)=0$

$y+6=0$ or $y-5=0$

$y=-6$ or $y=5$

If $y=-6$,

$$\begin{aligned} x^2+(-6)^2&=25\\ x^2&=-11 \end{aligned}$$

When $y=-6$ there is no real solution.

If $y=5$,

$$\begin{aligned} x^2+(5)^2&=25\\ x^2&=0\\ x&=0 \end{aligned}$$

The solution set is $\{(0, 5)\}$.

5. $2x+2y=20$

$xy=21$

Solve the second equation for x.

$x=\frac{21}{y}$

Substitute the expression $\frac{21}{y}$ for x in the first equation and solve for y.

$$\begin{aligned} 2x+2y&=20\\ 2\left(\frac{21}{y}\right)+2y&=20\\ \frac{42}{y}+2y&=20 \end{aligned}$$

$y^2-10y+21=0$

$(y-7)(y-3)=0$

$y-7=0$ or $y-3=0$

$y=7$ or $y=3$

If $y=7$, $x=\frac{21}{7}=3$.

If $y=3$, $x=\frac{21}{3}=7$.

The dimensions are 7 feet by 3 feet.

10.5 Concept and Vocabulary Check

1. nonlinear

2. $\{(-4,3),(0,-1)\}$

3. 3

4. $\{(2,\sqrt{3}),(2,-\sqrt{3}),(-2,\sqrt{3}),(-2,-\sqrt{3})\}$

5. -1; $y^2+y=6$

6. $\frac{4}{x}$; $\frac{4}{x}$; y

10.5 Exercise Set

1. $x+y=2$
   $y=x^2-4$

   Substitute x^2+4 for y in the first equation and solve for x.

   $x+(x^2-4)=2$
   $x+x^2-4=2$
   $x^2+x-6=0$
   $(x+3)(x-2)=0$
   $x+3=0$ or $x-2=0$
   $x=-3$ $\quad x=2$

   Substitute -3 and 2 for x in the second equation to find y.

   $x=-3$ or $x=2$
   $y=(-3)^2-4$ $\quad y=2^2-4$
   $y=9-4$ $\quad y=4-4$
   $y=5$ $\quad y=0$

   The solution set is $\{(-3,5),\ (2,0)\}$.

3. $x+y=2$
   $y=x^2-4x+4$

   Substitute x^2-4x+4 for y in the first equation and solve for x.

   $x+x^2-4x+4=2$
   $x^2-3x+4=2$
   $x^2-3x+2=0$
   $(x-2)(x-1)=0$
   $x-2=0$ or $x-1=0$
   $x=2$ $\quad x=1$

   Substitute 1 and 2 for x to find y.

   $x=2$ or $x=1$
   $x+y=2$ $\quad x+y=2$
   $2+y=2$ $\quad 1+y=2$
   $y=0$ $\quad y=1$

   The solution set is $\{(1,1),(2,0)\}$.

5. $y=x^2-4x-10$
   $y=-x^2-2x+14$

   Substitute $-x^2-2x+14$ for y in the first equation and solve for x.

   $-x^2-2x+14=x^2-4x-10$
   $0=2x^2-2x-24$

   $0=x^2-x-12$
   $0=(x-4)(x+3)$
   $x-4=0$ or $x+3=0$
   $x=4$ $\quad x=-3$

   Substitute -3 and 4 for x to find y.

   $x=4$
   $y=4^2-4(4)-10$
   $=16-16-10=-10$

   $x=-3$
   $y=(-3)^2-4(-3)-10$
   $=9+12-10=11$

   The solution set is $\{(-3,11),(4,-10)\}$.

7. $x^2 + y^2 = 25$
$x - y = 1$

Solve the second equation for x.
$x - y = 1$
$x = y + 1$

Substitute $y+1$ for x to find y.

$$x^2 + y^2 = 25$$
$$(y+1)^2 + y^2 = 25$$
$$y^2 + 2y + 1 + y^2 = 25$$
$$2y^2 + 2y + 1 = 25$$
$$2y^2 + 2y - 24 = 0$$
$$y^2 + y - 12 = 0$$
$$(y+4)(y-3) = 0$$
$$y + 4 = 0 \quad \text{or} \quad y - 3 = 0$$
$$y = -4 \qquad y = 3$$

Substitute –4 and 3 for y to find x.

$$y = -4 \qquad y = 3$$
$$x = -4 + 1 \qquad x = 3 + 1$$
$$x = -3 \qquad x = 4$$

The solution set is $\{(-3,-4),\ (4,3)\}$.

9. $xy = 6$
$2x - y = 1$

Solve the first equation for y.
$xy = 6$
$y = \frac{6}{x}$

Substitute $\frac{6}{x}$ for y in the second equation and solve for x.

$$2x - \frac{6}{x} = 1$$
$$x\left(2x - \frac{6}{x}\right) = x(1)$$
$$2x^2 - 6 = x$$
$$2x^2 - x - 6 = 0$$
$$(2x+3)(x-2) = 0$$
$$x - 2 = 0 \quad \text{or} \quad 2x + 3 = 0$$
$$x = 2 \qquad 2x = -3$$
$$x = -\tfrac{3}{2}$$

Substitute 2 and $-\frac{3}{2}$ for x to find y.

$$x = 2 \quad \text{or} \quad x = -\frac{3}{2}$$
$$2y = 6 \qquad -\frac{3}{2}y = 6$$
$$y = 3 \qquad -\frac{2}{3}\left(-\frac{3}{2}\right)y = \left(-\frac{2}{3}\right)6$$
$$y = -4$$

The solution set is $\{(2,3),(-\tfrac{3}{2},-4)\}$.

11. $y^2 = x^2 - 9$
$2y = x - 3$

Solve the second equation for x.
$2y = x - 3$
$2y + 3 = x$

Substitute $2y+3$ for x to find y.

$$y^2 = (2y+3)^2 - 9$$
$$y^2 = 4y^2 + 12y + 9 - 9$$
$$y^2 = 4y^2 + 12y$$
$$0 = 3y^2 + 12y$$
$$0 = 3y(y+4)$$
$$3y = 0 \quad \text{or} \quad y + 4 = 0$$
$$y = 0 \qquad y = -4$$

Substitute –4 and 0 for y to find x.

$$y = 0 \quad \text{or} \quad y = -4$$
$$2(0) + 3 = x \qquad 2(-4) + 3 = x$$
$$3 = x \qquad -8 + 3 = x$$
$$-5 = x$$

The solution set is $\{(-5,-4),\ (3,0)\}$.

13. $xy = 3$
$x^2 + y^2 = 10$

Solve the first equation for y.
$xy = 3$
$y = \frac{3}{x}$

Substitute $\frac{3}{x}$ for y to find x.

$$x^2 + \left(\frac{3}{x}\right)^2 = 10$$
$$x^2 + \frac{9}{x^2} = 10$$
$$x^2\left(x^2 + \frac{9}{x^2}\right) = x^2(10)$$
$$x^4 + 9 = 10x^2$$
$$x^4 - 10x^2 + 9 = 0$$
$$(x^2 - 9)(x^2 - 1) = 0$$
$$(x+3)(x-3)(x+1)(x-1) = 0$$

Apply the zero-product principle.
$x + 3 = 0 \qquad x - 3 = 0$
$x = -3 \qquad x = 3$

$x + 1 = 0 \qquad x - 1 = 0$
$x = -1 \qquad x = 1$

Substitute ± 1 and ± 3 for x to find y.
$x = -3 \qquad x = 3$
$y = \frac{3}{-3} \qquad y = \frac{3}{3}$
$y = -1 \qquad y = 1$

$x = -1 \qquad x = 1$
$y = \frac{3}{-1} \qquad y = \frac{3}{1}$
$y = -3 \qquad y = 3$

The solution set is $\{(-3,-1),(-1,-3),(1,3),(3,1)\}$.

15. $x + y = 1$
$x^2 + xy - y^2 = -5$

Solve the first equation for y.
$x + y = 1$
$y = -x + 1$

Substitute $-x+1$ for y and solve for x.

$$x^2 + x(-x+1) - (-x+1)^2 = -5$$
$$x^2 - x^2 + x - (x^2 - 2x + 1) = -5$$
$$x^2 - x^2 + x - x^2 + 2x - 1 = -5$$
$$-x^2 + 3x - 1 = -5$$
$$-x^2 + 3x + 4 = 0$$
$$x^2 - 3x - 4 = 0$$
$$(x-4)(x+1) = 0$$
$$x - 4 = 0 \quad \text{or} \quad x + 1 = 0$$
$$x = 4 \qquad\qquad x = -1$$

Substitute -1 and 4 for x to find y.
$x = 4 \quad$ or $\quad x = -1$
$y = -4 + 1 \qquad y = -(-1) + 1$
$y = -3 \qquad y = 1 + 1$
$y = 2$

The solution set is $\{(4,-3),\ (-1,2)\}$.

17. $x + y = 1$
$(x-1)^2 + (y+2)^2 = 10$

Solve the first equation for y.
$x + y = 1$
$y = -x + 1$

Substitute $-x+1$ for y to find x.

$$(x-1)^2 + ((-x+1)+2)^2 = 10$$
$$(x-1)^2 + (-x+1+2)^2 = 10$$
$$(x-1)^2 + (-x+3)^2 = 10$$
$$x^2 - 2x + 1 + x^2 - 6x + 9 = 10$$
$$2x^2 - 8x + 10 = 10$$
$$2x^2 - 8x = 0$$
$$2x(x-4) = 0$$
$$2x = 0 \quad \text{or} \quad x - 4 = 0$$
$$x = 0 \qquad\qquad x = 4$$

Substitute 0 and 4 for x to find y.

$x = 0$ or $x = 4$

$y = -0+1$ $\quad$ $y = -4+1$

$y = 1$ $\quad$ $y = -3$

The solution set is $\{(0,1), (4,-3)\}$.

19. Solve the system by addition.

$$\begin{aligned} x^2 + y^2 &= 13 \\ x^2 - y^2 &= 5 \\ \hline 2x^2 &= 18 \\ x^2 &= 9 \\ x &= \pm 3 \end{aligned}$$

Substitute ± 3 for x to find y.

$$\begin{aligned} x &= \pm 3 \\ (\pm 3)^2 + y^2 &= 13 \\ 9 + y^2 &= 13 \\ y^2 &= 4 \\ y &= \pm 2 \end{aligned}$$

The solution set is $\{(-3,-2),(-3,2),(3,-2),(3,2)\}$.

21. $x^2 - 4y^2 = -7$

$3x^2 + y^2 = 31$

Multiply the first equation by –3 and add to the second equation.

$$\begin{aligned} -3x^2 + 12y^2 &= 21 \\ 3x^2 + y^2 &= 31 \\ \hline 13y^2 &= 52 \\ y^2 &= 4 \\ y &= \pm 2 \end{aligned}$$

Substitute –2 and 2 for y to find x.

$$\begin{aligned} y &= \pm 2 \\ 3x^2 + (\pm 2)^2 &= 31 \\ 3x^2 + 4 &= 31 \\ 3x^2 &= 27 \\ x^2 &= 9 \\ x &= \pm 3 \end{aligned}$$

The solution set is $\{(-3,-2),(-3,2),(3,-2),(3,2)\}$.

23. $3x^2 + 4y^2 - 16 = 0$

$2x^2 - 3y^2 - 5 = 0$

Multiply the first equation by 3 and the second equation by 4 and solve by addition.

$$\begin{aligned} 9x^2 + 12y^2 - 48 &= 0 \\ 8x^2 - 12y^2 - 20 &= 0 \\ \hline 17x^2 - 68 &= 0 \\ 17x^2 &= 68 \\ x^2 &= 4 \\ x &= \pm 2 \end{aligned}$$

Substitute ± 2 for x to find y.

$$\begin{aligned} 2(\pm 2)^2 - 3y^2 - 5 &= 0 \\ 2(4) - 3y^2 - 5 &= 0 \\ 8 - 3y^2 - 5 &= 0 \\ 3 - 3y^2 &= 0 \\ 3 &= 3y^2 \\ 1 &= y^2 \\ \pm 1 &= y \end{aligned}$$

The solution set is $\{(-2,-1),(-2,1),(2,-1),(2,1)\}$.

25. $x^2 + y^2 = 25$

$(x-8)^2 + y^2 = 41$

Multiply the first equation by –1 and solve by addition.

$$\begin{aligned} -x^2 \qquad - y^2 &= -25 \\ (x-8)^2 + y^2 &= 41 \\ \hline -x^2 + (x-8)^2 &= 16 \\ -x^2 + x^2 - 16x + 64 &= 16 \\ -16x + 64 &= 16 \\ -16x &= -48 \\ x &= 3 \end{aligned}$$

Substitute 3 for x to find y.

$$\begin{aligned} x &= 3 \\ 3^2 + y^2 &= 25 \\ 9 + y^2 &= 25 \\ y^2 &= 16 \\ y &= \pm 4 \end{aligned}$$

The solution set is $\{(3,-4), (3,4)\}$.

27. $y^2 - x = 4$
$x^2 + y^2 = 4$

Multiply the first equation by –1 and solve by addition.

$$\begin{aligned} -y^2 + x &= -4 \\ x^2 + y^2 &= 4 \\ \hline x^2 + x &= 0 \\ x(x+1) &= 0 \end{aligned}$$

Apply the zero-product principle.
$x = 0$ or $x + 1 = 0$
$x = -1$

Substitute –1 and 0 for x to find y.

$x = 0$: $y^2 - 0 = 4$, $y^2 = 4$, $y = \pm 2$

or

$x = -1$: $y^2 - (-1) = 4$, $y^2 + 1 = 4$, $y^2 = 3$, $y = \pm\sqrt{3}$

The solution set is $\{(-1, -\sqrt{3}), (-1, \sqrt{3}), (0, -2), (0, 2)\}$.

29. $3x^2 + 4y^2 = 16$
$2x^2 - 3y^2 = 5$

Multiply the first equation by –2 and the second equation by 3 and solve by addition.

$$\begin{aligned} -6x^2 - 8y^2 &= -32 \\ 6x^2 - 9y^2 &= 15 \\ \hline -17y^2 &= -17 \\ y^2 &= 1 \\ y &= \pm 1 \end{aligned}$$

Substitute ± 1 for y to find x.

$$\begin{aligned} y &= \pm 1 \\ 3x^2 + 4(\pm 1)^2 &= 16 \\ 3x^2 + 4(1) &= 16 \\ 3x^2 + 4 &= 16 \\ 3x^2 &= 12 \\ x^2 &= 4 \\ x &= \pm 2 \end{aligned}$$

The solution set is $\{(-2, -1), (-2, 1), (2, -1), (2, 1)\}$.

31. $2x^2 + y^2 = 18$
$xy = 4$

Solve the second equation for y.

$xy = 4 \rightarrow y = \frac{4}{x}$

Substitute $\frac{4}{x}$ for y in the second equation and solve for x.

$$\begin{aligned} 2x^2 + \left(\frac{4}{x}\right)^2 &= 18 \\ 2x^2 + \frac{16}{x^2} &= 18 \\ x^2\left(2x^2 + \frac{16}{x^2}\right) &= x^2(18) \\ 2x^4 + 16 &= 18x^2 \\ 2x^4 - 18x^2 + 16 &= 0 \\ x^4 - 9x^2 + 8 &= 0 \\ (x^2 - 8)(x^2 - 1) &= 0 \\ (x^2 - 8)(x+1)(x-1) &= 0 \end{aligned}$$

Apply the zero product principle.

$x^2 - 8 = 0$, $x^2 = 8$, $x = \pm\sqrt{8}$, $x = \pm 2\sqrt{2}$

$x + 1 = 0$, $x = -1$

$x - 1 = 0$, $x = 1$

Substitute $\pm 2\sqrt{2}$ and ± 1 for x to find y.

$x = 1$: $y = \frac{4}{1}$, $y = 4$

$x = -1$: $y = \frac{4}{-1}$, $y = -4$

$x = 2\sqrt{2}$: $y = \frac{4}{2\sqrt{2}}$, $y = \frac{2}{\sqrt{2}} \cdot \frac{\sqrt{2}}{\sqrt{2}}$

$x = -2\sqrt{2}$: $y = \frac{4}{-2\sqrt{2}}$, $y = -\frac{2}{\sqrt{2}} \cdot \frac{\sqrt{2}}{\sqrt{2}}$

$y = \frac{2\sqrt{2}}{2}$ $\quad y = -\frac{2\sqrt{2}}{2}$

$y = \sqrt{2}$ $\quad y = -\sqrt{2}$

The solution set is $\{(-2\sqrt{2}, -\sqrt{2}), (-1,-4), (1,4), (2\sqrt{2}, \sqrt{2})\}$.

33. $x^2 + 4y^2 = 20$
$x + 2y = 6$

Solve the second equation for x.

$x + 2y = 6$
$x = 6 - 2y$

Substitute $6 - 2y$ for x to find y.

$$(6-2y)^2 + 4y^2 = 20$$
$$36 - 24y + 4y^2 + 4y^2 = 20$$
$$36 - 24y + 8y^2 = 20$$
$$8y^2 - 24y + 16 = 0$$
$$y^2 - 3y + 2 = 0$$
$$(y-2)(y-1) = 0$$
$$y - 2 = 0 \quad \text{or} \quad y - 1 = 0$$
$$y = 2 \qquad\qquad y = 1$$

Substitute 1 and 2 for y to find x.

$y = 2$ or $y = 1$

$x = 6 - 2(2)$ $\quad x = 6 - 2(1)$

$x = 6 - 4$ $\quad x = 6 - 2$

$x = 2$ $\quad x = 4$

The solution set is $\{(2,2), (4,1)\}$.

35. Eliminate y by adding the two equations.

$$x^3 + y = 0$$
$$\underline{x^2 - y = 0}$$
$$x^3 + x^2 = 0$$
$$x^2(x+1) = 0$$
$$x = 0 \quad \text{or} \quad x = -1$$

Substitute –1 and 0 for x to find y.

$x = 0$ or $x = -1$

$0^2 - y = 0$ $\quad (-1)^2 - y = 0$

$-y = 0$ $\quad 1 - y = 0$

$y = 0$ $\quad -y = -1$

$y = 1$

The solution set is $\{(-1,1), (0,0)\}$.

37. $x^2 + (y-2)^2 = 4$
$x^2 - 2y = 0$

Solve the second equation for x^2.

$x^2 - 2y = 0$
$x^2 = 2y$

Substitute $2y$ for x^2 in the first equation and solve for y.

$$2y + (y-2)^2 = 4$$
$$2y + y^2 - 4y + 4 = 4$$
$$y^2 - 2y + 4 = 4$$
$$y^2 - 2y = 0$$
$$y(y-2) = 0$$
$$y = 0 \quad \text{or} \quad y - 2 = 0$$
$$y = 2$$

Substitute 0 and 2 for y to find x.

$y = 0$ or $y = 2$

$x^2 = 2(0)$ $\quad x^2 = 2(2)$

$x^2 = 0$ $\quad x^2 = 4$

$x = 0$ $\quad x = \pm 2$

The solution set is $\{(0,0), (-2,2), (2,2)\}$.

39. $y = (x+3)^2$
$x + 2y = -2$

Substitute $(x+3)^2$ for y in the second equation.

$$x + 2(x+3)^2 = -2$$
$$x + 2(x^2 + 6x + 9) = -2$$
$$x + 2x^2 + 12x + 18 = -2$$
$$2x^2 + 13x + 18 = -2$$
$$2x^2 + 13x + 20 = 0$$
$$(2x+5)(x+4) = 0$$
$$2x + 5 = 0 \quad \text{or} \quad x + 4 = 0$$
$$x = -\tfrac{5}{2} \qquad x = -4$$

Substitute $-\frac{5}{2}$ and -4 for x to find y.

$x = -\frac{5}{2}$ or $x = -4$

$$-\tfrac{5}{2} + 2y = -2 \qquad -4 + 2y = -2$$
$$-5 + 4y = -4 \qquad 2y = 2$$
$$4y = 1 \qquad y = 1$$
$$y = \frac{1}{4}$$

The solution set is $\left\{(-4, 1), \left(-\frac{5}{2}, \frac{1}{4}\right)\right\}$.

41. $x^2 + y^2 + 3y = 22$
$2x + y = -1$

Solve the second equation for y.

$$2x + y = -1$$
$$y = -2x - 1$$

Substitute $-2x - 1$ for y to find x.

$$x^2 + (-2x-1)^2 + 3(-2x-1) = 22$$
$$x^2 + 4x^2 + 4x + 1 - 6x - 3 = 22$$
$$5x^2 - 2x - 2 = 22$$
$$5x^2 - 2x - 24 = 0$$
$$(5x - 12)(x + 2) = 0$$
$$5x - 12 = 0 \quad \text{or} \quad x + 2 = 0$$
$$5x = 12 \qquad x = -2$$
$$x = \frac{12}{5}$$

Substitute -2 and $\frac{12}{5}$ for x to find y.

$x = \frac{12}{5}$ or $x = -2$

$$y = -2\left(\frac{12}{5}\right) - 1 \qquad y = -2(-2) - 1$$
$$y = -\frac{24}{5} - \frac{5}{5} \qquad y = 4 - 1$$
$$y = -\frac{29}{5} \qquad y = 3$$

The solution set is $\left\{\left(\frac{12}{5}, -\frac{29}{5}\right), (-2, 3)\right\}$.

43. Let x = one of the numbers.
Let y = the other number.

$$x + y = 10$$
$$xy = 24$$

Solve the second equation for y.

$$xy = 24$$
$$y = \frac{24}{x}$$

Substitute $\frac{24}{x}$ for y in the first equation and solve for x.

$$x + \frac{24}{x} = 10$$
$$x\left(x + \frac{24}{x}\right) = x(10)$$
$$x^2 + 24 = 10x$$
$$x^2 - 10x + 24 = 0$$
$$(x-6)(x-4) = 0$$
$$x - 6 = 0 \quad \text{or} \quad x - 4 = 0$$
$$x = 6 \qquad x = 4$$

Substitute 6 and 4 for x to find y.

$x = 6$ or $x = 4$

$$y = \frac{24}{6} \qquad y = \frac{24}{4}$$
$$y = 4 \qquad y = 6$$

The numbers are 4 and 6.

45. Let x = one of the numbers.
Let y = the other number.

$$\begin{aligned} x^2 - y^2 &= 3 \\ 2x^2 + y^2 &= 9 \\ \hline 3x^2 &= 12 \\ x^2 &= 4 \\ x &= \pm 2 \end{aligned}$$

Substitute ± 2 for x to find y.

$$\begin{aligned} x &= \pm 2 \\ (\pm 2)^2 - y^2 &= 3 \\ 4 - y^2 &= 3 \\ -y^2 &= -1 \\ y^2 &= 1 \\ y &= \pm 1 \end{aligned}$$

The numbers are either 2 and –1, 2 and 1, –2 and –1, or –2 and 1.

47. $2x^2 + xy = 6$
$x^2 + 2xy = 0$

Multiply the first equation by -2 and add the two equations.

$$\begin{aligned} -4x^2 - 2xy &= -12 \\ x^2 + 2xy &= 0 \\ \hline -3x^2 &= -12 \\ x^2 &= 4 \\ x &= \pm 2 \end{aligned}$$

Back-substitute these values for x in the second equation and solve for y.

For $x = -2$: $(-2)^2 + 2(-2)y = 0$
$4 - 4y = 0$
$-4y = -4$
$y = 1$

For $x = 2$: $(2)^2 + 2(2)y = 0$
$4 + 4y = 0$
$4y = -4$
$y = -1$

The solution set is $\{(-2,1),(2,-1)\}$.

49. $-4x + y = 12$
$y = x^3 + 3x^2$

Substitute $x^3 + 3x^2$ for y in the first equation and solve for x.

$$\begin{aligned} -4x + (x^3 + 3x^2) &= 12 \\ x^3 + 3x^2 - 4x - 12 &= 0 \\ x^2(x+3) - 4(x+3) &= 0 \\ (x+3)(x^2 - 4) &= 0 \\ (x+3)(x-2)(x+2) &= 0 \end{aligned}$$

$x = -3$, $x = 2$, or $x = -2$

Substitute these values for x in the second equation and solve for y.

For $x = -3$: $y = (-3)^3 + 3(-3)^2$
$= -27 + 27$
$= 0$

For $x = 2$: $y = (2)^3 + 3(2)^2$
$= 8 + 12$
$= 20$

For $x = -2$: $y = (-2)^3 + 3(-2)^2$
$= -8 + 12$
$= 4$

The solution set is $\{(-3,0),(2,20),(-2,4)\}$.

51. $\frac{3}{x^2} + \frac{1}{y^2} = 7$
$\frac{5}{x^2} - \frac{2}{y^2} = -3$

Multiply the first equation by 2 and add the equations.

$$\begin{aligned} \frac{6}{x^2} + \frac{2}{y^2} &= 14 \\ \frac{5}{x^2} - \frac{2}{y^2} &= -3 \\ \hline \frac{11}{x^2} &= 11 \\ x^2 &= 1 \\ x &= \pm 1 \end{aligned}$$

Back-substitute these values for x in the first equation and solve for y.

For $x = -1$:

$$\frac{3}{(-1)^2}+\frac{1}{y^2}=7$$
$$3+\frac{1}{y^2}=7$$
$$\frac{1}{y^2}=4$$
$$y^2=\frac{1}{4}$$
$$y=\pm\frac{1}{2}$$

For $x = 1$:

$$\frac{3}{(1)^2}+\frac{1}{y^2}=7$$
$$3+\frac{1}{y^2}=7$$
$$\frac{1}{y^2}=4$$
$$y^2=\frac{1}{4}$$
$$y=\pm\frac{1}{2}$$

The solution set is

$$\left\{\left(-1,-\frac{1}{2}\right),\left(-1,\frac{1}{2}\right),\left(1,-\frac{1}{2}\right),\left(1,\frac{1}{2}\right)\right\}.$$

53. Answers will vary. One example:

Circle: $x^2+y^2=9$

Ellipse: $\frac{x^2}{9}+\frac{y^2}{49}=1$

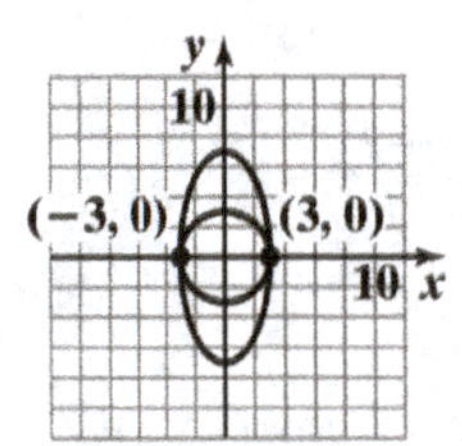

Circle : $x^2+y^2=9$

Ellipse : $\frac{x^2}{9}+\frac{y^2}{49}=1$

Solutions: $(-3,0)$ and $(3,0)$.

55. $16x^2+4y^2=64$

$$y=x^2-4$$

Solve the second equation for x^2.

$$y=x^2-4$$
$$y+4=x^2$$

Substitute $y+4$ for x^2 in the first equation and solve for y.

$$16(y+4)+4y^2=64$$
$$16y+64+4y^2=64$$
$$16y+4y^2=0$$
$$4y(4+y)=0$$
$$4y=0 \quad \text{or} \quad 4+y=0$$
$$y=0 \qquad\qquad y=-4$$

Substitute 0 and 4 for y to find x.

$$y=0 \quad \text{or} \quad y=-4$$
$$0=x^2-4 \qquad -4=x^2-4$$
$$4=x^2 \qquad 0=x^2$$
$$\pm2=x \qquad 0=x$$

The comet intersects the planet's orbit at the points $(2,0),(-2,0)$ and $(0,-4)$.

57. Let x = the length of the rectangle.
Let y = the width of the rectangle.

Perimeter: $2x+2y=36$

Area: $xy=77$

Solve the second equation for y.

$$xy=77$$
$$y=\frac{77}{x}$$

Substitute $\frac{77}{x}$ for y in the first equation and solve for x.

$$2x+2\left(\frac{77}{x}\right)=36$$
$$2x+\frac{154}{x}=36$$
$$x\left(2x+\frac{154}{x}\right)=x(36)$$
$$2x^2+154=36x$$
$$2x^2-36x+154=0$$
$$x^2-18x+77=0$$
$$(x-7)(x-11)=0$$
$$x-7=0 \quad \text{or} \quad x-11=0$$
$$x=7 \qquad\qquad x=11$$

Substitute 7 and 11 for x to find y.

$x = 7$ or $x = 11$

$y = \frac{77}{7}$ $\quad y = \frac{77}{11}$

$y = 11$ $\quad y = 7$

The dimensions of the rectangle are 7 feet by 11 feet.

59. Let x = the length of the screen.
Let y = the width of the screen.

$$x^2 + y^2 = 10^2$$
$$xy = 48$$

Solve the second equation for y.

$$xy = 48$$
$$y = \frac{48}{x}$$

Substitute $\frac{48}{x}$ for y to find x.

$$x^2 + \left(\frac{48}{x}\right)^2 = 10^2$$
$$x^2 + \frac{2304}{x^2} = 100$$
$$x^2\left(x^2 + \frac{2304}{x^2}\right) = x^2(100)$$
$$x^4 + 2304 = 100x^2$$
$$x^4 - 100x^2 + 2304 = 0$$
$$(x^2 - 64)(x^2 - 36) = 0$$
$$(x+8)(x-8)(x+6)(x-6) = 0$$

Apply the zero product principle.

$x + 8 = 0$ $\quad x - 8 = 0$

$x = -8$ $\quad x = 8$

$x + 6 = 0$ $\quad x + 6 = 0$

$x = -6$ $\quad x = -6$

Disregard –8 and –6 because we cannot have a negative length. Substitute 8 and 6 for x to find y.

$x = 8$ or $x = 6$

$y = \frac{48}{8}$ $\quad y = \frac{48}{6}$

$y = 6$ $\quad y = 8$

The dimensions of the screen are 8 inches by 6 inches.

61. $x^2 - y^2 = 21$
$4x + 2y = 24$

Solve for y in the second equation.

$$4x + 2y = 24$$
$$2y = 24 - 4x$$
$$y = 12 - 2x$$

Substitute $12 - 2x$ for y and solve for x.

$$x^2 - (12 - 2x)^2 = 21$$
$$x^2 - (144 - 48x + 4x^2) = 21$$
$$x^2 - 144 + 48x - 4x^2 = 21$$
$$-3x^2 + 48x - 144 = 21$$
$$-3x^2 + 48x - 165 = 0$$
$$x^2 - 16x + 55 = 0$$
$$(x-5)(x-11) = 0$$

$x - 5 = 0$ or $x - 11 = 0$

$x = 5$ $\quad x = 11$

Substitute 5 and 11 for x to find y.

$x = 5$ or $x = 11$

$y = 12 - 2(5)$ $\quad y = 12 - 2(11)$

$y = 12 - 10$ $\quad y = 12 - 22$

$y = 2$ $\quad y = -10$

Disregard –10 because we can't have a negative length measurement. The larger square is 5 meters by 5 meters and the smaller square to be cut out is 2 meters by 2 meters.

63. **a.** It appears from the graphs that the percentage of white-collar workers was the same as blue-collar workers between the 1940s and 1960s.

b. $0.5x - y = -18$
$y = -0.004x^2 + 0.23x + 41$

Substitute $-0.004x^2 + 0.23x + 41$ in the first equation and solve for x.

$$0.5x - y = -18$$
$$0.5x - \overbrace{(-0.004x^2 + 0.23x + 41)}^{y} = -18$$
$$0.5x + 0.004x^2 - 0.23x - 41 = -18$$
$$0.004x^2 + 0.27x - 23 = 0$$

Use the quadratic formula.

$$x = \frac{-b \pm \sqrt{b^2 - 4ac}}{2a}$$

$$x = \frac{-(0.27) \pm \sqrt{0.27^2 - 4(0.004)(-23)}}{2(0.004)}$$

$x \approx 49$ or -117

The percentage of white-collar workers was the same as blue-collar workers 49 years after 1900, or 1949.

Let $x = 49$ and solve for y in the white-collar model.

$$0.5x - y = -18$$
$$0.5(49) - y = -18$$
$$24.5 - y = -18$$
$$-y = -42.5$$
$$y \approx 43$$

The percentage of white-collar workers in 1949 was about 43%

Let $x = 49$ and solve for y in the blue-collar model.

$y = -0.004(49)^2 + 0.23(49) + 41 \approx 43\%$

The percentage of blue-collar workers in 1949 was about 43%.

c. According to the graph, the percentage of white-collar workers was the same as farmers in 1920. The percentages of white-collar workers and farmers in 1920 were both 28%.

d.
$$0.5x - y = -18$$
$$\underline{0.4x + y = \ 35}$$
$$0.9x \quad = 17$$
$$x = \frac{17}{0.9}$$
$$x \approx 19$$

According to the models, the percentage of white-collar workers was the same as farmers 19 years after 1900, or 1919.

Let $x = 19$ and solve for y in the white-collar model.

$$0.5x - y = -18$$
$$0.5(19) - y = -18$$
$$9.5 - y = -18$$
$$-y = -27.5$$
$$y = 27.5$$

The percentage of white-collar workers in 1919 was about 27.5%

Let $x = 19$ and solve for y in the farming model.

$$0.4x + y = 35$$
$$0.4(19) + y = 35$$
$$0.4(19) + y = 35$$
$$7.6 + y = 35$$
$$y = 27.4$$

The percentage of farm workers in 1919 was about 27.4%

These answers model the actual data from part c (the graph) fairly well.

65. – 67. Answers will vary.

69. Answers will vary. For example, consider the following system.

$y = x$

$y = x^2 + 5$

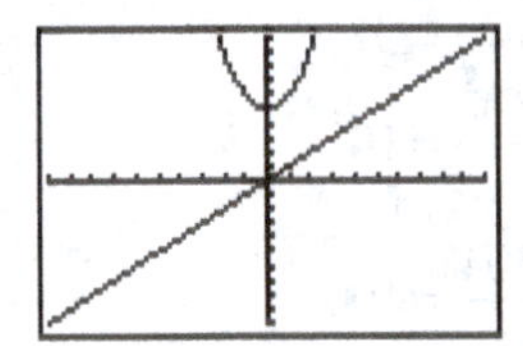

Since the graphs do not intersect, there are no ordered pairs that are real numbers in the solution set.

71. does not make sense; Explanations will vary. Sample explanation: Since the orbits of earth and Mars do not intersect, their system of equations will have no solution.

73. makes sense

75. true

77. false; Changes to make the statement true will vary. A sample change is: It is possible that a system of two equations in two variables whose graphs represent a parabola and a circle to have one real ordered-pair solution. This will occur if the graphs intersect in a single point.

79. $\log_y x = 3$
$\log_y (4x) = 5$

Rewrite the equations.
$y^3 = x$
$y^5 = 4x$

Substitute y^3 for x in the second equation and solve for y.

$$y^5 = 4y^3$$
$$y^5 - 4y^3 = 0$$
$$y^3\left(y^2 - 4\right) = 0$$
$$y^3\left(y+2\right)\left(y-2\right) = 0$$

Apply the zero product principle.

$y^3 = 0 \qquad y+2=0 \qquad y-2=0$
$y = 0 \qquad y=-2 \qquad y=2$

Disregard 0 and –2 because the base of a logarithm must be greater than zero.

Substitute 2 for y to find x.
$y^3 = x$
$2^3 = x$
$8 = x$

The solution is $(8, 2)$ and the solution set is $\{(8, 2)\}$.

81. $3x - 2y \le 6$

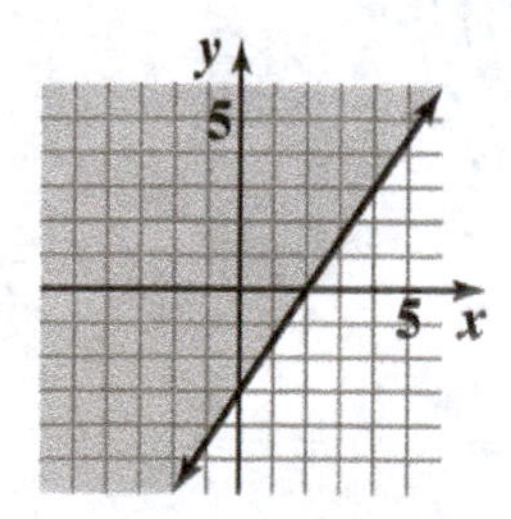

82. $m = \frac{5-(-3)}{1-(-2)} = \frac{5+3}{1+2} = \frac{8}{3}$

The slope is $\frac{8}{3}$.

83.

$$\begin{array}{r} 2x^2 - 4x + 3 \\ 3x - 2 \\ \hline 6x^3 - 12x^2 + 9x \\ -4x^2 + 8x - 6 \\ \hline 6x^3 - 16x^2 + 17x - 6 \end{array}$$

84. For $n = 1$; $\frac{(-1)^n}{3^n - 1} = \frac{(-1)^1}{3^1 - 1} = \frac{-1}{3-1} = -\frac{1}{2}$

For $n = 2$; $\frac{(-1)^n}{3^n - 1} = \frac{(-1)^2}{3^2 - 1} = \frac{1}{9-1} = \frac{1}{8}$

For $n = 3$; $\frac{(-1)^n}{3^n - 1} = \frac{(-1)^3}{3^3 - 1} = \frac{-1}{27-1} = -\frac{1}{26}$

For $n = 4$; $\frac{(-1)^n}{3^n - 1} = \frac{(-1)^4}{3^4 - 1} = \frac{1}{81-1} = \frac{1}{80}$

85. $5 \cdot 4 \cdot 3 \cdot 2 \cdot 1 = 120$

86. For $n = 1$; $n^2 + 1 = 1^2 + 1 = 1 + 1 = 2$

For $n = 2$; $n^2 + 1 = 2^2 + 1 = 4 + 1 = 5$

For $n = 3$; $n^2 + 1 = 3^2 + 1 = 9 + 1 = 10$

For $n = 4$; $n^2 + 1 = 4^2 + 1 = 16 + 1 = 17$

For $n = 5$; $n^2 + 1 = 5^2 + 1 = 25 + 1 = 26$

For $n = 6$; $n^2 + 1 = 6^2 + 1 = 36 + 1 = 37$

$2 + 5 + 10 + 17 + 26 + 37 = 97$

Chapter 10 Review Exercises

1. $d = \sqrt{\left(3-(-2)\right)^2 + \left(9-(-3)\right)^2}$

$= \sqrt{(3+2)^2 + (9+3)^2}$

$= \sqrt{5^2 + 12^2} = \sqrt{25 + 144}$

$= \sqrt{169} = 13$

2. $d = \sqrt{\left(-2-(-4)\right)^2 + (5-3)^2}$

$= \sqrt{(-2+4)^2 + 2^2} = \sqrt{2^2 + 4}$

$= \sqrt{4+4} = \sqrt{8} = \sqrt{4 \cdot 2} = 2\sqrt{2} \approx 2.83$

3. Midpoint $= \left(\frac{2+(-12)}{2}, \frac{6+4}{2}\right)$

$= \left(\frac{-10}{2}, \frac{10}{2}\right) = (-5, 5)$

4. Midpoint $= \left(\frac{4+(-15)}{2}, \frac{-6+2}{2}\right)$

$= \left(\frac{-11}{2}, \frac{-4}{2}\right) = \left(-\frac{11}{2}, -2\right)$

5. $(x-0)^2 + (y-0)^2 = 3^2$

$x^2 + y^2 = 9$

6. $(x-(-2))^2 + (y-4)^2 = 6^2$

$(x+2)^2 + (y-4)^2 = 36$

7. $x^2 + y^2 = 1$

$(x-0)^2 + (y-0)^2 = 1^2$

The center is $(0,0)$ and the radius is 1 unit.

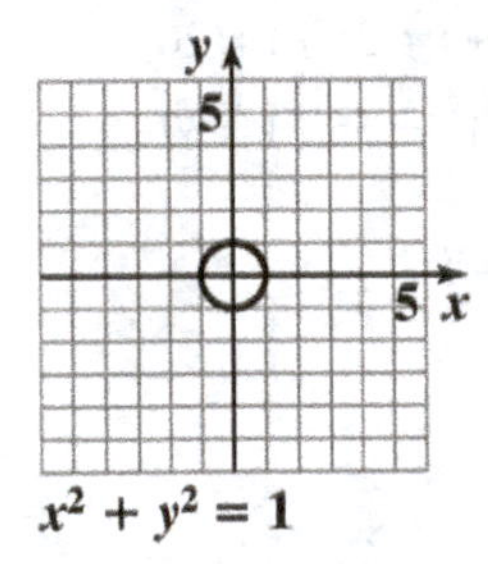

8. $(x+2)^2 + (y-3)^2 = 9$

$(x-(-2))^2 + (y-3)^2 = 3^2$

The center is $(-2, 3)$ and the radius is 3 units.

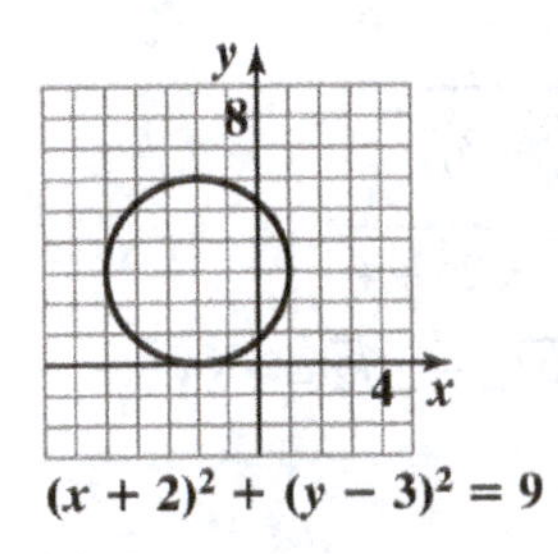

9. $x^2 + y^2 - 4x + 2y - 4 = 0$

$(x^2 - 4x \quad) + (y^2 + 2y \quad) = 4$

Complete the squares.

$\left(\frac{b}{2}\right)^2 = \left(\frac{-4}{2}\right)^2 = (-2)^2 = 4$

$\left(\frac{b}{2}\right)^2 = \left(\frac{2}{2}\right)^2 = 1^2 = 1$

$(x^2 - 4x + 4) + (y^2 + 2y + 1) = 4 + 4 + 1$

$(x-2)^2 + (y+1)^2 = 9$

$(x-2)^2 + (y-(-1))^2 = 3^2$

The center is $(2, -1)$ and the radius is 3 units.

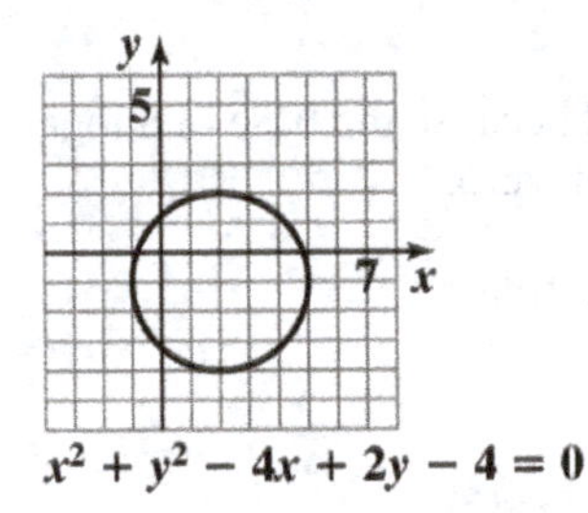

10. $x^2 + y^2 - 4y = 0$

$x^2 + (y^2 - 4y \quad) = 0$

Complete the square.

$\left(\frac{b}{2}\right)^2 = \left(\frac{-4}{2}\right)^2 = (-2)^2 = 4$

$x^2 + (y^2 - 4y + 4) = 0 + 4$

$(x-0)^2 + (y-2)^2 = 4$

$(x-0)^2 + (y-2)^2 = 2^2$

The center is $(0, 2)$ and the radius is 2 units.

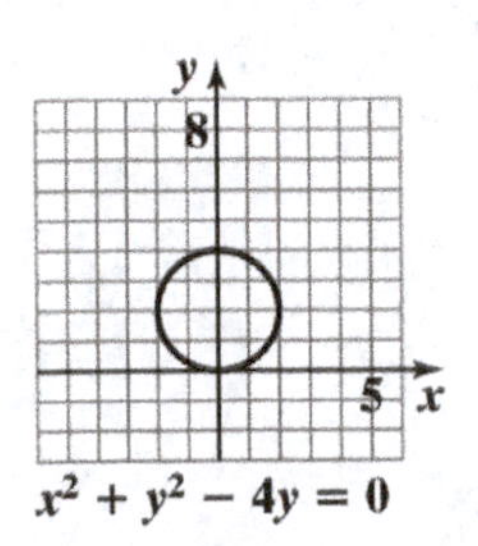

11. $\dfrac{x^2}{36}+\dfrac{y^2}{25}=1$

Because the denominator of the x^2 – term is greater than the denominator of the y^2 – term, the major axis is horizontal. Since $a^2=36$, $a=6$ and the vertices are $(-6,0)$ and $(6,0)$. Since $b^2=25$, $b=5$ and endpoints of the minor axis are $(0,-5)$ and $(0,5)$.

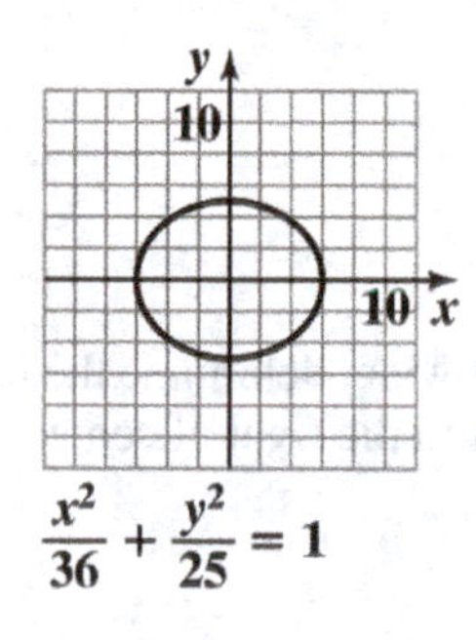

$\dfrac{x^2}{36}+\dfrac{y^2}{25}=1$

12. $\dfrac{x^2}{25}+\dfrac{y^2}{16}=1$

Because the denominator of the x^2 – term is greater than the denominator of the y^2 – term, the major axis is horizontal. Since $a^2=25$, $a=5$ and the vertices are $(-5,0)$ and $(5,0)$. Since $b^2=16$, $b=4$ and endpoints of the minor axis are $(0,-4)$ and $(0,4)$.

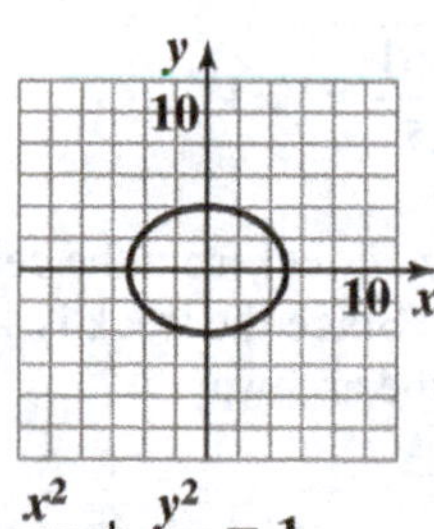

$\dfrac{x^2}{25}+\dfrac{y^2}{16}=1$

13. $4x^2+y^2=16$

$\dfrac{4x^2}{16}+\dfrac{y^2}{16}=\dfrac{16}{16}$

$\dfrac{x^2}{4}+\dfrac{y^2}{16}=1$

Because the denominator of the y^2 – term is greater than the denominator of the x^2 – term, the major axis is vertical. Since $a^2=16$, $a=4$ and the vertices are $(0,-4)$ and $(0,4)$. Since $b^2=4$, $b=2$ and endpoints of the minor axis are $(-2,0)$ and $(2,0)$.

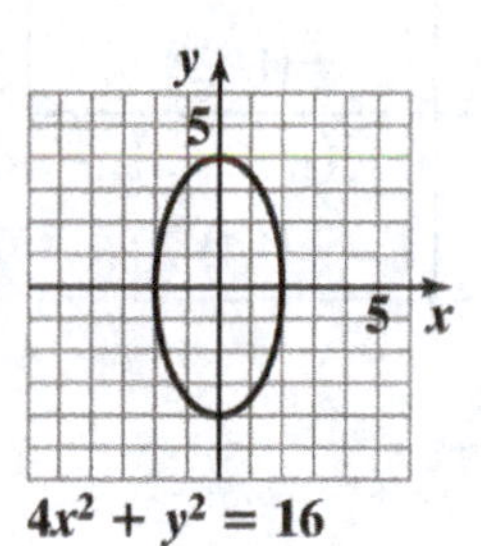

$4x^2+y^2=16$

14. $4x^2+9y^2=36$

$\dfrac{4x^2}{36}+\dfrac{9y^2}{36}=\dfrac{36}{36}$

$\dfrac{x^2}{9}+\dfrac{y^2}{4}=1$

Because the denominator of the x^2 – term is greater than the denominator of the y^2 – term, the major axis is horizontal. Since $a^2=9$, $a=3$ and the vertices are $(-3,0)$ and $(3,0)$. Since $b^2=4$, $b=2$ and endpoints of the minor axis are $(0,-2)$ and $(0,2)$.

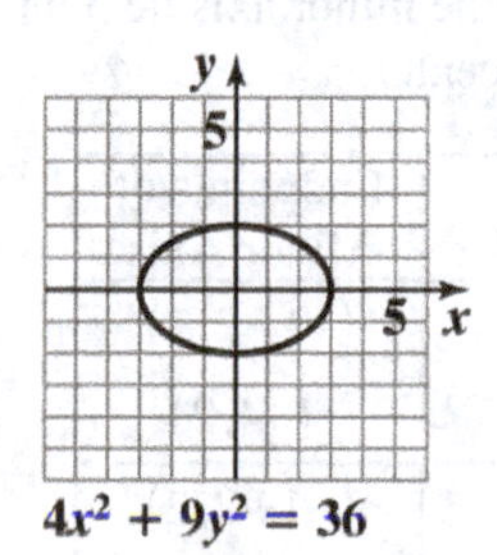

$4x^2+9y^2=36$

15. $\frac{(x-1)^2}{16}+\frac{(y+2)^2}{9}=1$

The center of the ellipse is $(1,-2)$. Because the denominator of the x^2 – term is greater than the denominator of the y^2 – term, the major axis is horizontal. Since $a^2=16$, $a=4$ and the vertices lie 4 units to the left and right of the center. Since $b^2=9$, $b=3$ and endpoints of the minor axis lie 3 units above and below the center.

Center	Vertices	Endpoints of Minor Axis
$(1,-2)$	$(1-4,-2)$ $=(-3,-2)$	$(1,-2-3)$ $=(1,-5)$
	$(1+4,-2)$ $=(5,-2)$	$(1,-2+3)$ $=(1,1)$

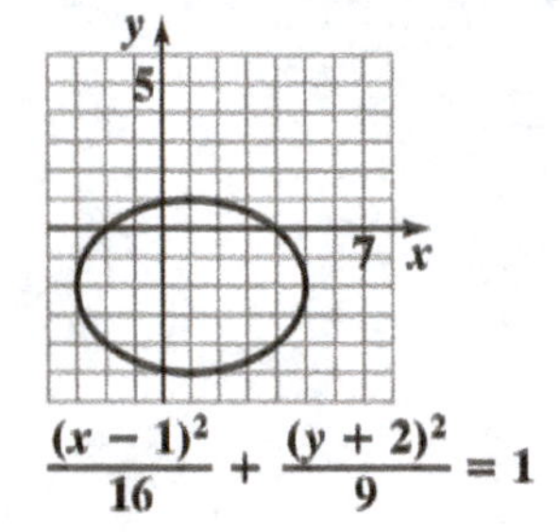

16. $\frac{(x+1)^2}{9}+\frac{(y-2)^2}{16}=1$

The center of the ellipse is $(-1,2)$. Because the denominator of the y^2 – term is greater than the denominator of the x^2 – term the major axis is vertical. Since $a^2=16$, $a=4$ and the vertices lie 4 units above and below the center. Since $b^2=9$, $b=3$ and endpoints of the minor axis lie 3 units to the left and right of the center.

Center	Vertices	Endpoints of Minor Axis
$(-1,2)$	$(-1,2-4)$ $=(-1,-2)$	$(-1-3,2)$ $=(-4,2)$
	$(-1,2+4)$ $=(-1,6)$	$(-1+3,2)$ $=(2,2)$

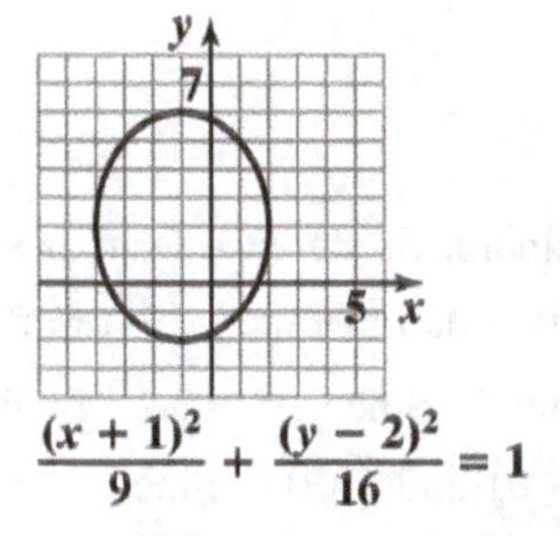

17. From the figure, we see that the major axis is horizontal with $a = 25$, and $b = 15$.

$$\frac{x^2}{25^2}+\frac{y^2}{15^2}=1$$
$$\frac{x^2}{625}+\frac{y^2}{225}=1$$

Since the truck is 14 feet wide, determine the height of the archway at 14 feet to the right of center.

$$\frac{14^2}{625}+\frac{y^2}{225}=1$$
$$\frac{196}{625}+\frac{y^2}{225}=1$$

$$5625\left(\frac{196}{625}+\frac{y^2}{225}\right)=5625(1)$$
$$9(196)+25y^2=5625$$
$$1764+25y^2=5625$$
$$25y^2=3861$$
$$y^2=\frac{3861}{25}$$
$$y=\sqrt{\frac{3861}{25}}\approx 12.43$$

The height of the archway 14 feet from the center is approximately 12.43 feet. Since the truck is 12 feet high, the truck will clear the archway.

18. $\frac{x^2}{16} - y^2 = 1$

$\frac{x^2}{16} - \frac{y^2}{1} = 1$

The equation is in the form $\frac{x^2}{a^2} - \frac{y^2}{b^2} = 1$ with $a^2 = 16$, and $b^2 = 1$. We know the transverse axis lies on the x-axis and the vertices are $(-4,0)$ and $(4,0)$. Because $a^2 = 16$ and $b^2 = 1$, $a = 4$ and $b = 1$. Construct a rectangle using –4 and 4 on the x–axis, and –1 and 1 on the y–axis. Draw extended diagonals to obtain the asymptotes. Graph the hyperbola.

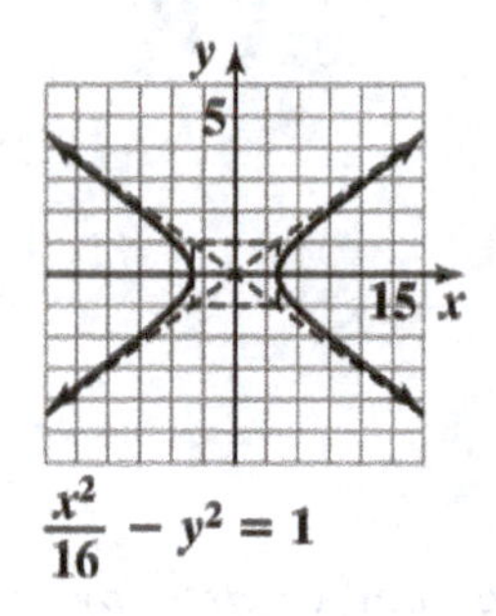

$\frac{x^2}{16} - y^2 = 1$

19. $\frac{y^2}{16} - x^2 = 1$

$\frac{y^2}{16} - \frac{x^2}{1} = 1$

The equation is in the form $\frac{y^2}{a^2} - \frac{x^2}{b^2} = 1$ with $a^2 = 16$, and $b^2 = 1$. We know the transverse axis lies on the y-axis and the vertices are $(0,-4)$ and $(0,4)$. Because $a^2 = 16$ and $b^2 = 1$, $a = 4$ and $b = 1$. Construct a rectangle using –1 and 1 on the x–axis, and –4 and 4 on the y–axis. Draw extended diagonals to obtain the asymptotes. Graph the hyperbola.

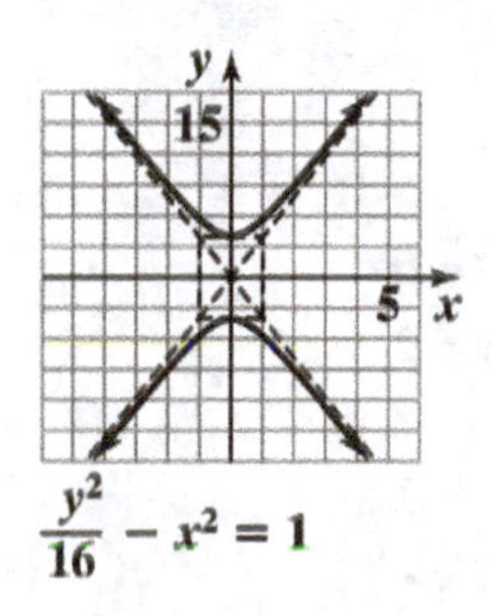

$\frac{y^2}{16} - x^2 = 1$

20. $9x^2 - 16y^2 = 144$

$\frac{9x^2}{144} - \frac{16y^2}{144} = \frac{144}{144}$

$\frac{x^2}{16} - \frac{y^2}{9} = 1$

The equation is in the form $\frac{x^2}{a^2} - \frac{y^2}{b^2} = 1$ with $a^2 = 16$, and $b^2 = 9$. We know the transverse axis lies on the x-axis and the vertices are $(-4,0)$ and $(4,0)$. Because $a^2 = 16$ and $b^2 = 9$, $a = 4$ and $b = 3$. Construct a rectangle using –4 and 4 on the x–axis, and –3 and 3 on the y–axis. Draw extended diagonals to obtain the asymptotes. Graph the hyperbola.

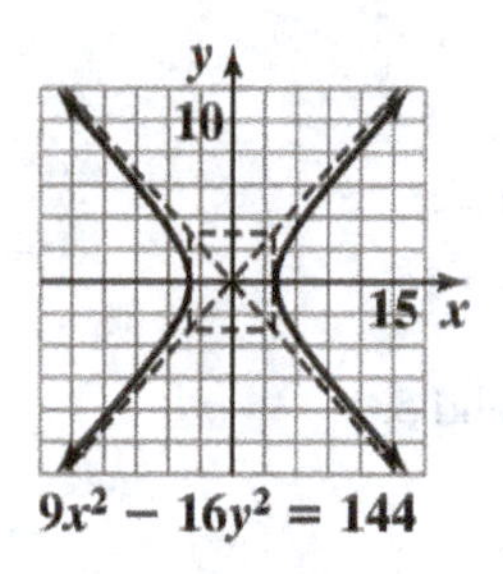

$9x^2 - 16y^2 = 144$

21. $4y^2 - x^2 = 16$

$\frac{4y^2}{16} - \frac{x^2}{16} = \frac{16}{16}$

$\frac{y^2}{4} - \frac{x^2}{16} = 1$

The equation is in the form $\frac{y^2}{a^2} - \frac{x^2}{b^2} = 1$ with $a^2 = 4$, and $b^2 = 16$. We know the transverse axis lies on the y-axis and the vertices are $(0,-2)$ and $(0,2)$. Because $a^2 = 4$ and $b^2 = 16$, $a = 2$ and $b = 4$. Construct a rectangle using –2 and 2 on the y–axis, and –4 and 4 on the x–axis. Draw extended diagonals to obtain the asymptotes. Graph the hyperbola.

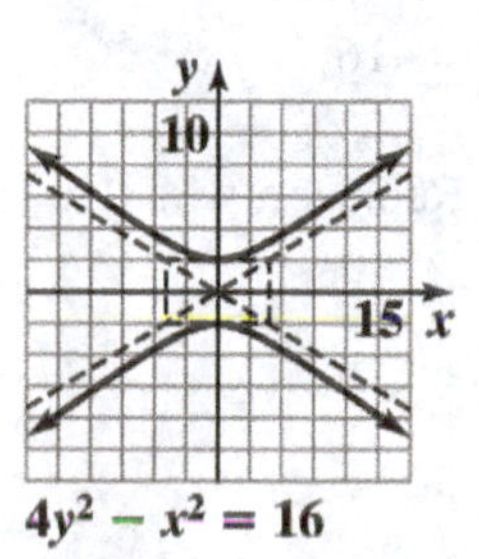

$4y^2 - x^2 = 16$

22. $x=(y-3)^2-4$

This is a parabola of the form $x=a(y-k)^2+h$. Since $a=1$ is positive, the parabola opens to the right. The vertex of the parabola is $(-4,3)$. The axis of symmetry is $y=3$. Replace y with 0 to find the x–intercept.

$x=(0-3)^2-4=(-3)^2-4=9-4=5$

The x–intercept is 5. Replace x with 0 to find the y–intercepts.

$0=(y-3)^2-4$

$0=y^2-6y+9-4$

$0=y^2-6y+5$

$0=(y-5)(y-1)$

$y-5=0$ or $y-1=0$

$y=5$ $\qquad y=1$

The y–intercepts are 1 and 5.

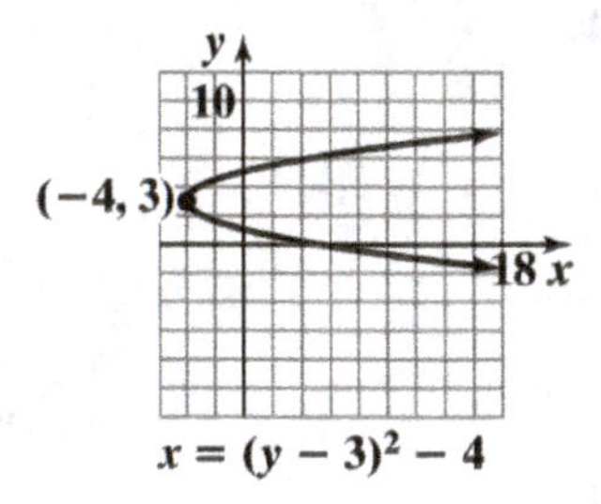

23. $x=-2(y+3)^2+2$

This is a parabola of the form $x=a(y-k)^2+h$. Since $a=-2$ is negative, the parabola opens to the left. The vertex of the parabola is $(2,-3)$. The axis of symmetry is $y=-3$. Replace y with 0 to find the x–intercept.

$x=-2(0+3)^2+2=-2(3)^2+2$

$=-2(9)+2=-18+2=-16$

The x–intercept is -16. Replace x with 0 to find the y–intercepts.

$0=-2(y+3)^2+2$

$0=-2(y^2+6y+9)+2$

$0=-2y^2-12y-18+2$

$0=-2y^2-12y-16$

$0=y^2+6y+8$

$0=(y+4)(y+2)$

$y+4=0$ or $y+2=0$

$y=-4$ $\qquad y=-2$

The y–intercepts are –4 and –2.

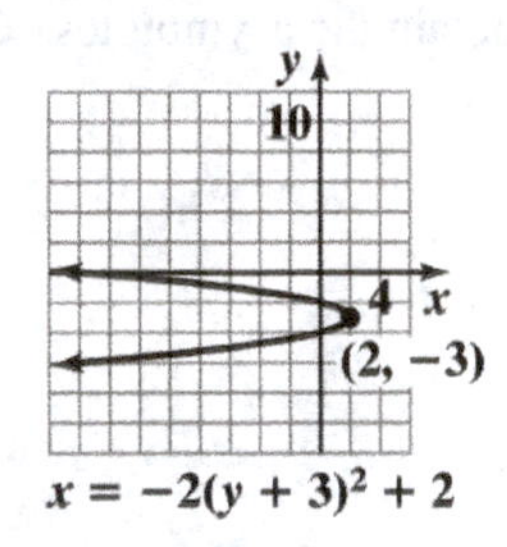

24. $x=y^2-8y+12$

This is a parabola of the form $x=ay^2+by+c$. Since $a=1$ is positive, the parabola opens to the right. The y–coordinate of the vertex is $-\frac{b}{2a}=-\frac{-8}{2(1)}=-\frac{-8}{2}=4$. The x–coordinate of the vertex is

$x=4^2-8(4)+12=16-32+12$

$=16-32+12=-4.$

The vertex of the parabola is $(-4,4)$. The axis of symmetry is $y=4$.

Replace y with 0 to find the x–intercept.

$x=0^2-8(0)+12=12$

The x–intercept is 12. Replace x with 0 to find the y–intercepts.

$0=y^2-8y+12$

$0=(y-6)(y-2)$

$y-6=0$ or $y-2=0$

$y=6$ $\qquad y=2$

The y–intercepts are 2 and 6.

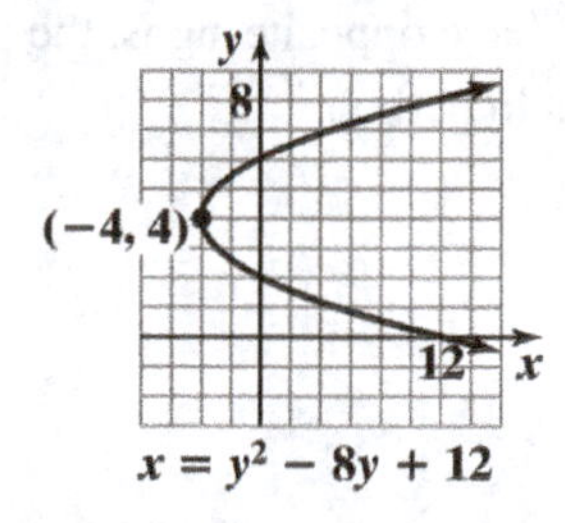

25. $x = -y^2 - 4y + 6$

This is a parabola of the form $x = ay^2 + by + c$. Since $a = -1$ is negative, the parabola opens to the left. The y–coordinate of the vertex is $-\frac{b}{2a} = -\frac{-4}{2(-1)} = -\frac{-4}{-2} = -2.$ The x–coordinate of the vertex is

$$x = -(-2)^2 - 4(-2) + 6$$
$$= -4 + 8 + 6 = 10.$$

The vertex of the parabola is $(10, -2)$. The axis of symmetry is $y = -2$.

Replace y with 0 to find the x–intercept.
$x = -0^2 - 4(0) + 6 = 0^2 - 0 + 6 = 6$

The x–intercept is 6. Replace x with 0 to find the y–intercepts.
$0 = -y^2 - 4y + 6$

Solve using the quadratic formula.

$$y = \frac{-b \pm \sqrt{b^2 - 4ac}}{2a}$$
$$= \frac{-(-4) \pm \sqrt{(-4)^2 - 4(-1)(6)}}{2(-1)} = \frac{4 \pm \sqrt{16 + 24}}{-2}$$
$$= \frac{4 \pm \sqrt{40}}{-2} = \frac{4 \pm 2\sqrt{10}}{-2} = -2 \pm \sqrt{10}$$

The y–intercepts are $-2 \pm \sqrt{10}$.

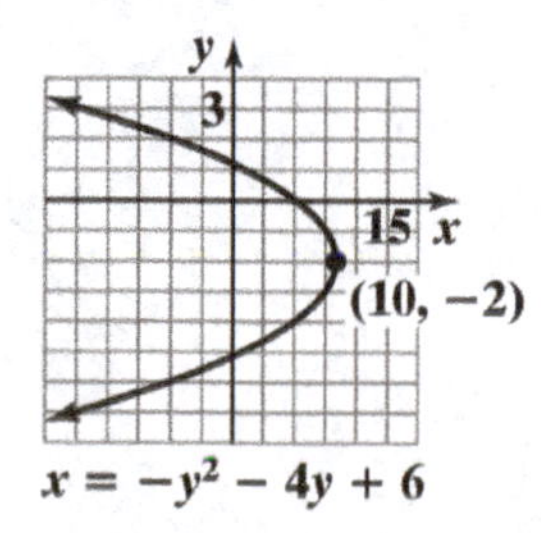

26. $x + 8y = y^2 + 10$

Since only one variable is squared, the graph of the equation is a parabola.

27.
$$16x^2 = 32 - y^2$$
$$16x^2 + y^2 = 32$$

Because x^2 and y^2 have different positive coefficients, the equation's graph is an ellipse.

28.
$$x^2 = 25 + 25y^2$$
$$x^2 - 25y^2 = 25$$

Because x^2 and y^2 have opposite signs, the equation's graph is a hyperbola.

29.
$$x^2 = 4 - y^2$$
$$x^2 + y^2 = 4$$

Because x^2 and y^2 have the same positive coefficient, the equation's graph is a circle.

30.
$$36y^2 = 576 + 16x^2$$
$$36y^2 - 16x^2 = 576$$

Because x^2 and y^2 have opposite signs, the equation's graph is a hyperbola.

31. $\frac{(x+3)^2}{9} + \frac{(y-4)^2}{25} = 1$

Because x^2 and y^2 have different positive coefficients, the equation's graph is an ellipse.

32. $y = x^2 + 6x + 9$

Since only one variable is squared, the graph of the equation is a parabola.

33. $5x^2 + 5y^2 = 180$

Because x^2 and y^2 have the same positive coefficient, the equation's graph is a circle.

Divide both sides of the equation by 5.
$x^2 + y^2 = 36$

The center is (0, 0) and the radius is 6 units.

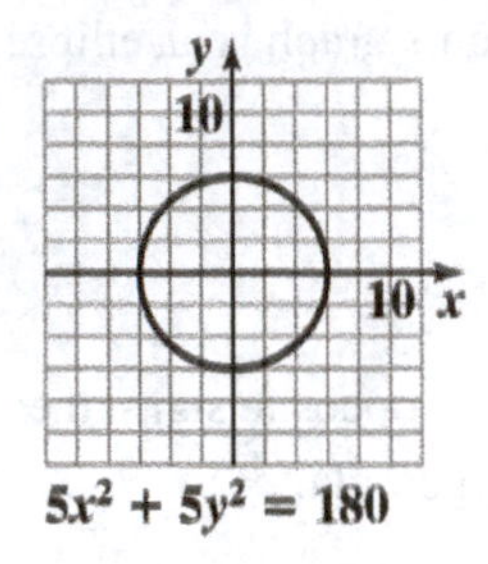

34. $4x^2 + 9y^2 = 36$

Because x^2 and y^2 have different positive coefficients, the equation's graph is an ellipse.

$$\frac{4x^2}{36} + \frac{9y^2}{36} = \frac{36}{36}$$

$$\frac{x^2}{9} + \frac{y^2}{4} = 1$$

Because the denominator of the x^2 – term is greater than the denominator of the y^2 – term, the major axis is horizontal. Since $a^2 = 9$, $a = 3$ and the vertices are $(-3, 0)$ and $(3, 0)$. Since $b^2 = 4$, $b = 2$ and endpoints of the minor axis are $(0, -2)$ and $(0, 2)$.

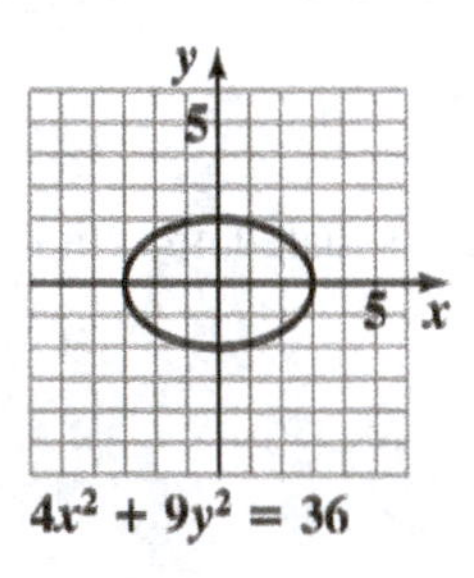

35. $4x^2 - 9y^2 = 36$

Because x^2 and y^2 have opposite signs, the equation's graph is a hyperbola.

$$\frac{4x^2}{36} - \frac{9y^2}{36} = \frac{36}{36}$$

$$\frac{x^2}{9} - \frac{y^2}{4} = 1$$

The equation is in the form $\frac{x^2}{a^2} - \frac{y^2}{b^2} = 1$ with $a^2 = 9$, and $b^2 = 4$. We know the transverse axis lies on the x-axis and the vertices are $(-3, 0)$ and $(3, 0)$. Because $a^2 = 9$ and $b^2 = 4$, $a = 3$ and $b = 2$. Construct a rectangle using –3 and 3 on the x–axis, and –2 and 2 on the y–axis. Draw extended diagonals to obtain the asymptotes. Graph the hyperbola.

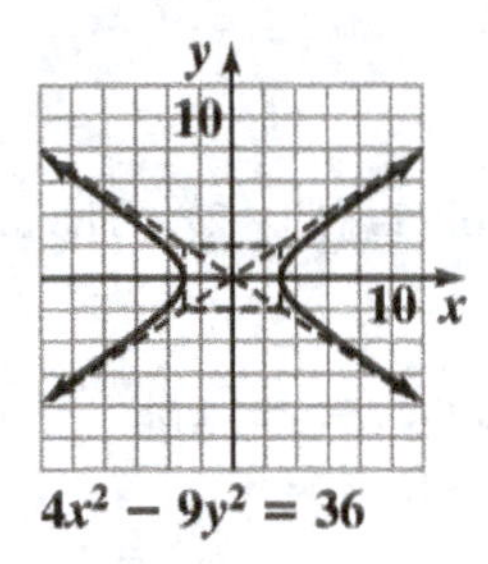

36. $\frac{x^2}{25} + \frac{y^2}{1} = 1$

Because x^2 and y^2 have different positive coefficients, the equation's graph is an ellipse.

Because the denominator of the x^2 – term is greater than the denominator of the y^2 – term, the major axis is horizontal. Since $a^2 = 25$, $a = 5$ and the vertices are $(-5, 0)$ and $(5, 0)$. Since $b^2 = 1$, $b = 1$ and endpoints of the minor axis are $(0, -1)$ and $(0, 1)$.

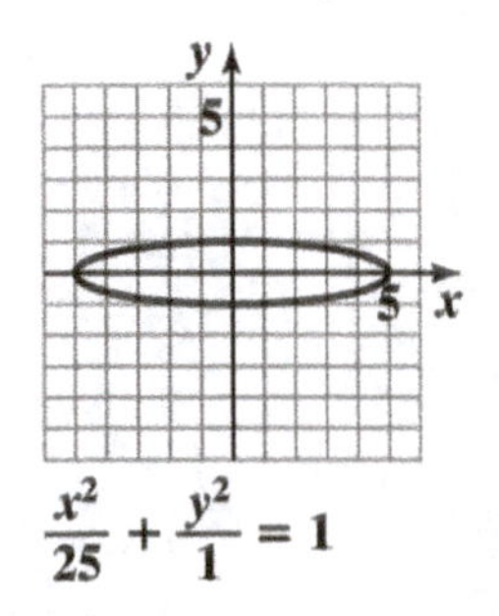

37. $x+3=-y^2+2y$
$$x=-y^2+2y-3$$

Since only one variable is squared, the graph of the equation is a parabola.

This is a parabola of the form $x=ay^2+by+c$. Since $a=-1$ is negative, the parabola opens to the left. The y–coordinate of the vertex is $-\frac{b}{2a}=-\frac{2}{2(-1)}=-\frac{2}{-2}=1$. The x–coordinate of the vertex is $x=-1^2+2(1)-3=-1+2-3=-2$.

The vertex of the parabola is $(-2,1)$.

Replace y with 0 to find the x–intercept.
$$x=-0^2+2(0)-3=0+0-3=-3$$

The x–intercept is -3. Replace x with 0 to find the y–intercepts.
$$0=-y^2+2y-3$$

Solve using the quadratic formula.
$$y=\frac{-2\pm\sqrt{2^2-4(-1)(-3)}}{2(-1)}$$
$$=\frac{-2\pm\sqrt{4-12}}{-2}=\frac{-2\pm\sqrt{-8}}{-2}$$

We do not need to simplify further. The solutions are complex and there are no y–intercepts.

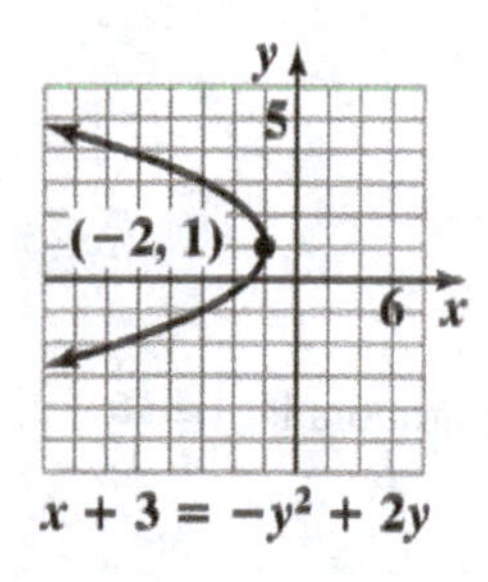

$x+3=-y^2+2y$

38. $y-3=x^2-2x$
$$y=x^2-2x+3$$

Since only one variable is squared, the graph of the equation is a parabola.

This is a parabola of the form $y=ax^2+bx+c$. Since $a=1$ is positive, the parabola opens to the right. The x–coordinate of the vertex is $-\frac{b}{2a}=-\frac{-2}{2(1)}=-\frac{-2}{2}=1$. The y–coordinate of the vertex is $y=1^2-2(1)+3=1-2+3=2$.

The vertex of the parabola is $(1,2)$.

Replace x with 0 to find the y–intercept.
$$y=0^2-2(0)+3=0-0+3=3$$

The y–intercept is 3. Replace y with 0 to find the x–intercepts.
$$0=x^2-2x+3$$

Solve using the quadratic formula.
$$x=\frac{-2\pm\sqrt{2^2-4(-1)(-3)}}{2(-1)}$$
$$=\frac{-2\pm\sqrt{4-12}}{-2}=\frac{-2\pm\sqrt{-8}}{-2}$$

We do not need to simplify further. The solutions are complex and there are no x–intercepts.

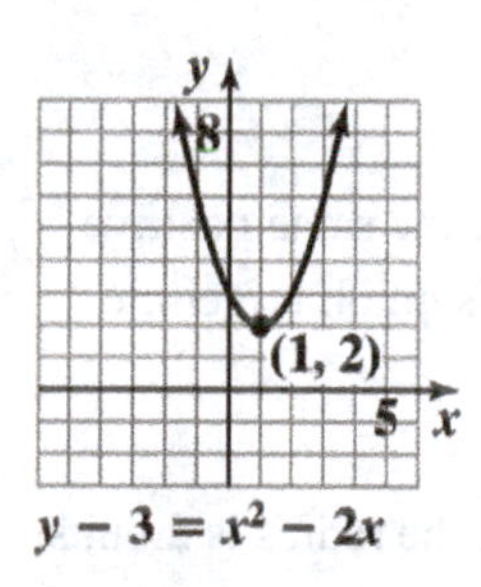

$y-3=x^2-2x$

39. $\frac{(x+2)^2}{16}+\frac{(y-5)^2}{4}=1$

Because x^2 and y^2 have different positive coefficients, the equation's graph is an ellipse.

The center of the ellipse is $(-2,5)$. Because the denominator of the x^2 – term is greater than the denominator of the y^2 – term, the major axis is horizontal. Since $a^2=16$, $a=4$ and the vertices lie 4 units to the left and right of the center. Since $b^2=4$, $b=2$ and endpoints of the minor axis lie two units above and below the center.

Center	Vertices	Endpoints of Minor Axis
$(-2,5)$	$(-2-4,5)$ $=(-6,5)$	$(-2,5-2)$ $=(-2,3)$
	$(-2+4,5)$ $=(2,5)$	$(-2,5+2)$ $=(-2,7)$

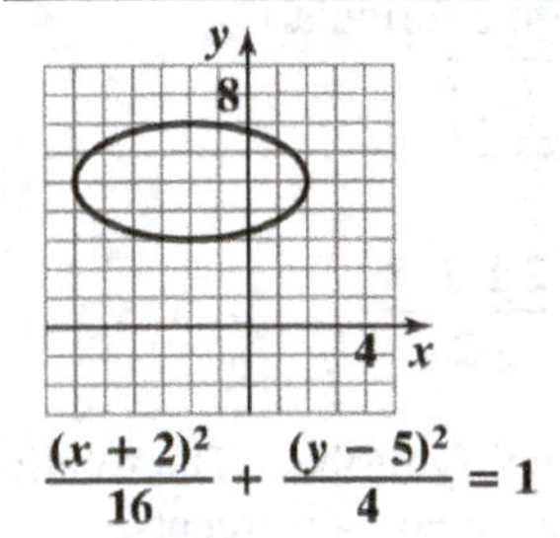

40. $(x-3)^2+(y+2)^2=4$

Because x^2 and y^2 have the same positive coefficient, the equation's graph is a circle.

$(x-3)^2+(y+2)^2=4$

The center is $(3,-2)$ and the radius is 2 units.

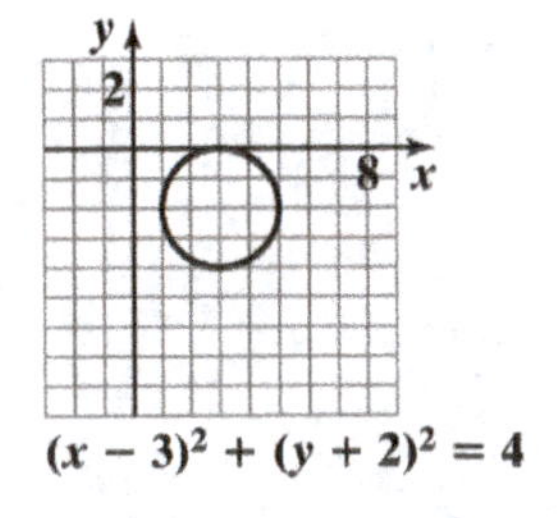

41. $x^2+y^2+6x-2y+6=0$

Because x^2 and y^2 have the same positive coefficient, the equation's graph is a circle.

$$x^2+y^2+6x-2y+6=0$$

$$(x^2+6x\quad)+(y^2-2y\quad)=-6$$

Complete the squares.

$$\left(\frac{b}{2}\right)^2=\left(\frac{6}{2}\right)^2=(3)^2=9$$

$$\left(\frac{b}{2}\right)^2=\left(\frac{-2}{2}\right)^2=(-1)^2=1$$

$$(x^2+6x+9)+(y^2-2y+1)=-6+9+1$$

$$(x+3)^2+(y-1)^2=4$$

The center is $(-3,1)$ and the radius is 2 units.

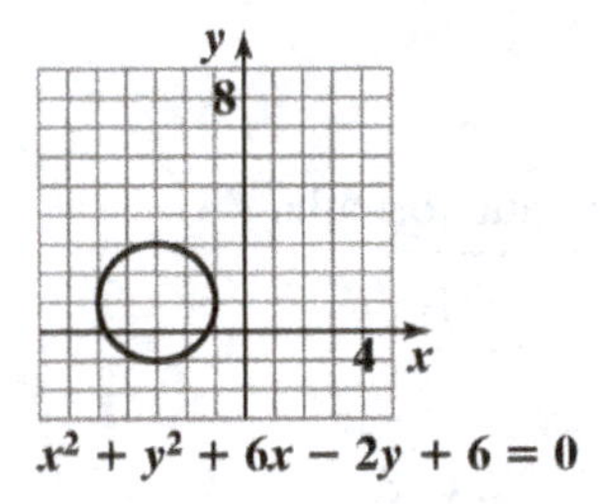

42. a. Using the point (6, 3), substitute for x and y to find a in $y=ax^2$.

$$3=a(6)^2$$
$$3=a(36)$$
$$a=\frac{3}{36}=\frac{1}{12}$$

The equation for the parabola is $y=\frac{1}{12}x^2$.

b. $a=\frac{1}{4p}$

$$\frac{1}{12}=\frac{1}{4p}$$
$$4p=12$$
$$p=3$$

The light source should be placed at the point (0, 3). This is the point 3 inches above the vertex.

43. $5y = x^2 - 1$
$x - y = 1$

Solve the second equation for *y*.
$x - y = 1$
$-y = -x + 1$
$y = x - 1$

Substitute $x - 1$ for *y* in the first equation.
$5(x-1) = x^2 - 1$
$5x - 5 = x^2 - 1$
$0 = x^2 - 5x + 4$
$0 = (x-4)(x-1)$
$x - 4 = 0$ or $x - 1 = 0$
$x = 4$ $\quad x = 1$

Back-substitute 1 and 4 for *x* to find *y*.

$x = 4$	or	$x = 1$
$y = x - 1$		$y = x - 1$
$y = 4 - 1$		$y = 1 - 1$
$y = 3$		$y = 0$

The solution set is $\{(1,0),(4,3)\}$.

44. $y = x^2 + 2x + 1$
$x + y = 1$

Solve the second equation for *y*.
$x + y = 1$
$y = -x + 1$

Substitute $-x + 1$ for *y* in the first equation.
$-x + 1 = x^2 + 2x + 1$
$0 = x^2 + 3x$
$0 = x(x+3)$
$x = 0$ or $x + 3 = 0$
$x = -3$

Back-substitute –3 and 0 for *x* to find *y*.

$x = 0$	or	$x = -3$
$y = -x + 1$		$y = -x + 1$
$y = -0 + 1$		$y = -(-3) + 1$
$y = 1$		$y = 3 + 1$
		$y = 4$

The solution set is $\{(-3,4),(0,1)\}$.

45. $x^2 + y^2 = 2$
$x + y = 0$

Solve the second equation for *y*.
$x + y = 0$
$y = -x$

Substitute $-x$ for *y* in the first equation.
$x^2 + (-x)^2 = 2$
$x^2 + x^2 = 2$
$2x^2 = 2$
$x^2 = 1$
$x = \pm 1$

Back-substitute –1 and 1 for *x* to find *y*.

$x = -1$	or	$x = 1$
$y = -x$		$y = -x$
$y = -(-1)$		$y = -1$
$y = 1$		

The solution set is $\{(-1,1),\ (1,-1)\}$.

46. $2x^2 + y^2 = 24$
$x^2 + y^2 = 15$

Multiple the second equation by –1 and add to the first equation.

$$\begin{array}{r} 2x^2 + y^2 = 24 \\ \underline{-x^2 - y^2 = -15} \\ x^2 = 9 \\ x = \pm 3 \end{array}$$

Back-substitute –3 and 3 for *x* to find *y*.
$x = \pm 3$
$(\pm 3)^2 + y^2 = 15$
$9 + y^2 = 15$
$y^2 = 6$
$y = \pm\sqrt{6}$

The solution set is $\{(-3,-\sqrt{6}),$
$(-3,\sqrt{6}),(3,-\sqrt{6}),\ (3,\sqrt{6})\}$.

47. $xy-4=0$
$y-x=0$

Solve the second equation for y.

$$y-x=0$$
$$y=x$$

Substitute x for y in the first equation and solve for x.

$$x(x)-4=0$$
$$x^2-4=0$$
$$(x+2)(x-2)=0$$
$$x+2=0 \quad \text{or} \quad x-2=0$$
$$x=-2 \qquad x=2$$

Back-substitute –2 and 2 for x to find y.

$$x=-2 \quad \text{or} \quad x=2$$
$$y=x \qquad y=x$$
$$y=-2 \qquad y=2$$

The solution set is $\{(-2,-2),\ (2,2)\}$.

48. $y^2=4x$
$x-2y+3=0$

Solve the second equation for x.

$$x-2y+3=0$$
$$x=2y-3$$

Substitute $2y-3$ for x in the first equation and solve for y.

$$y^2=4(2y-3)$$
$$y^2=8y-12$$
$$y^2-8y+12=0$$
$$(y-6)(y-2)=0$$
$$y-6=0 \quad \text{or} \quad y-2=0$$
$$y=6 \qquad y=2$$

Back-substitute 2 and 6 for y to find x.

$$y=6 \quad \text{or} \quad y=2$$
$$x=2y-3 \qquad x=2y-3$$
$$x=2(6)-3 \qquad x=2(2)-3$$
$$x=12-3 \qquad x=4-3$$
$$x=9 \qquad x=1$$

The solution set is $\{(1,2),(9,6)\}$.

49. $x^2+y^2=10$
$y=x+2$

Substitute $x+2$ for y in the first equation and solve for x.

$$x^2+(x+2)^2=10$$
$$x^2+x^2+4x+4=10$$
$$2x^2+4x+4=10$$
$$2x^2+4x-6=0$$
$$x^2+2x-3=0$$
$$(x+3)(x-1)=0$$
$$x+3=0 \quad \text{or} \quad x-1=0$$
$$x=-3 \qquad x=1$$

Back-substitute –3 and 1 for x to find y.

$$x=-3 \quad \text{or} \quad x=1$$
$$y=x+2 \qquad y=x+2$$
$$y=-3+2 \qquad y=1+2$$
$$y=-1 \qquad y=3$$

The solution set is $\{(-3,-1),\ (1,3)\}$.

50. $xy=1$
$y=2x+1$

Substitute $2x+1$ for y in the first equation and solve for x.

$$x(2x+1)=1$$
$$2x^2+x=1$$
$$2x^2+x-1=0$$
$$(2x-1)(x+1)=0$$
$$2x-1=0 \quad \text{or} \quad x+1=0$$
$$2x=1 \qquad x=-1$$
$$x=\frac{1}{2}$$

Back-substitute –1 and $\frac{1}{2}$ for x to find y.

$x=-1 \quad \text{or} \quad x=\frac{1}{2}$

$y=2x+1 \qquad y=2x+1$

$y=2(-1)+1 \qquad y=2\left(\frac{1}{2}\right)+1$

$y=-2+1 \qquad y=1+1$

$y=-1 \qquad y=2$

The solution set is $\left\{(-1,-1),\left(\frac{1}{2},2\right)\right\}$.

51. $x+y+1=0$

$x^2+y^2+6y-x=-5$

Solve for y in the first equation.

$x+y+1=0$

$y=-x-1$

Substitute $-x-1$ for y in the second equation and solve for x.

$$x^2+(-x-1)^2+6(-x-1)-x=-5$$
$$x^2+x^2+2x+1-6x-6-x=-5$$
$$2x^2-5x-5=-5$$
$$2x^2-5x=0$$
$$x(2x-5)=0$$

$x=0 \quad \text{or} \quad 2x-5=0$

$2x=5$

$x=\frac{5}{2}$

Back-substitute 0 and $\frac{5}{2}$ for x to find y.

$x=0 \quad \text{or} \quad x=\frac{5}{2}$

$y=-x-1 \qquad y=-x-1$

$y=-0-1 \qquad y=-\frac{5}{2}-1$

$y=-1 \qquad y=-\frac{7}{2}$

The solution set is $\left\{(0,-1),\left(\frac{5}{2},-\frac{7}{2}\right)\right\}$.

52. $x^2+y^2=13$

$x^2-y=7$

Solve for x^2 in the second equation.

$x^2-y=7$

$x^2=y+7$

Substitute $y+7$ for x^2 in the first equation and solve for y.

$$(y+7)+y^2=13$$
$$y^2+y+7=13$$
$$y^2+y-6=0$$
$$(y+3)(y-2)=0$$

Back-substitute -3 and 2 for y to find x.

$y=-3 \quad \text{or} \quad y=2$

$x^2=y+7 \qquad x^2=y+7$

$x^2=-3+7 \qquad x^2=2+7$

$x^2=4 \qquad x^2=9$

$x=\pm 2 \qquad x=\pm 3$

The solution set is $\{(-3,2),(-2,-3),(2,-3),(3,2)\}$.

53. $2x^2+3y^2=21$

$3x^2-4y^2=23$

Multiply the first equation by 4 and the second equation by 3.

$$8x^2+12y^2=84$$
$$\underline{9x^2-12y^2=69}$$
$$17x^2=153$$
$$x^2=9$$
$$x=\pm 3$$

Back-substitute ± 3 for x to find y.

$x=\pm 3$

$$2(\pm 3)^2+3y^2=21$$
$$2(9)+3y^2=21$$
$$18+3y^2=21$$
$$3y^2=3$$
$$y^2=1$$
$$y=\pm 1$$

The solution set is $\{(-3,-1),(-3,1),(3,-1),(3,1)\}$.

54. Let x = the length of the rectangle.
Let y = the width of the rectangle.

$$2x + 2y = 26$$
$$xy = 40$$

Solve the first equation for y.

$$2x + 2y = 26$$
$$x + y = 13$$
$$y = 13 - x$$

Substitute $13 - x$ for y in the second equation.

$$x(13 - x) = 40$$
$$13x - x^2 = 40$$
$$0 = x^2 - 13x + 40$$
$$0 = (x-8)(x-5)$$

$x - 8 = 0$ or $x - 5 = 0$

$x = 8$ $\qquad$ $x = 5$

Back-substitute 5 and 8 for x to find y.

$x = 8$ or $x = 5$

$y = 13 - 8$ $\qquad$ $y = 13 - 5$

$y = 5$ $\qquad$ $y = 8$

The solutions are the same. The dimensions are 8 meters by 5 meters.

55.

$$2x + y = 8$$
$$xy = 6$$

Solve the first equation for y.

$$2x + y = 8$$
$$y = -2x + 8$$

Substitute $-2x + 8$ for y in the second equation.

$$x(-2x + 8) = 6$$
$$-2x^2 + 8x = 6$$
$$-2x^2 + 8x - 6 = 0$$
$$x^2 - 4x + 3 = 0$$
$$(x-3)(x-1) = 0$$

$x - 3 = 0$ or $x - 1 = 0$

$x = 3$ $\qquad$ $x = 1$

Back-substitute 1 and 3 for x to find y.

$x = 3$ or $x = 1$

$y = -2x + 8$ $\qquad$ $y = -2x + 8$

$y = -2(3) + 8$ $\qquad$ $y = -2(1) + 8$

$y = -6 + 8$ $\qquad$ $y = -2 + 8$

$y = 2$ $\qquad$ $y = 6$

The solutions are the points $(1, 6)$ and $(3, 2)$.

56. Using the formula for the area, we have $x^2 + y^2 = 2900$. Since there are 240 feet of fencing available, we have

$$x + (x + y) + y + y + (x - y) + x = 240$$
$$x + x + y + y + y + x - y + x = 240$$
$$4x + 2y = 240.$$

The system of two variables in two equations is as follows.

$$x^2 + y^2 = 2900$$
$$4x + 2y = 240$$

Solve the second equation for y.

$$4x + 2y = 240$$
$$2y = -4x + 240$$
$$y = -2x + 120$$

Substitute $-2x + 120$ for y to find x.

$$x^2 + (-2x + 120)^2 = 2900$$
$$x^2 + 4x^2 - 480x + 14400 = 2900$$
$$5x^2 - 480x + 11500 = 0$$
$$x^2 - 96x + 2300 = 0$$
$$(x - 50)(x - 46) = 0$$

$x - 50 = 0$ or $x - 46 = 0$

$x = 50$ $\qquad$ $x = 46$

Back-substitute 46 and 50 for x to find y.

$x = 50$ $\qquad$ $x = 46$

$y = -2x + 120$ $\qquad$ $y = -2x + 120$

$y = -2(50) + 120$ $\qquad$ $y = -2(46) + 120$

$y = -100 + 120$ $\qquad$ $y = -92 + 120$

$y = 20$ $\qquad$ $y = 28$

The solutions are $x = 50$ feet and $y = 20$ feet or $x = 46$ feet and $y = 28$ feet.

Chapter 10 Test

1. $d = \sqrt{(2-(-1))^2 + (-3-5)^2}$
$= \sqrt{(3)^2 + (-8)^2}$
$= \sqrt{9+64} = \sqrt{73} \approx 8.54$
The distance between the points is $\sqrt{73}$ or 8.54 units.

2. Midpoint $= \left(\frac{-5+12}{2}, \frac{-2+(-6)}{2}\right)$
$= \left(\frac{7}{2}, \frac{-8}{2}\right) = \left(\frac{7}{2}, -4\right)$

The midpoint is $\left(\frac{7}{2}, -4\right)$.

3. $(x-3)^2 + (y-(-2))^2 = 5^2$
$(x-3)^2 + (y+2)^2 = 25$

4. $(x-5)^2 + (y+3)^2 = 49$
$(x-5)^2 + (y-(-3))^2 = 7^2$
The center is $(5,-3)$ and the radius is 7 units.

5. $x^2 + y^2 + 4x - 6y - 3 = 0$
$(x^2 + 4x \quad) + (y^2 - 6y \quad) = 3$
Complete the squares.
$\left(\frac{b}{2}\right)^2 = \left(\frac{4}{2}\right)^2 = (2)^2 = 4$
$\left(\frac{b}{2}\right)^2 = \left(\frac{-6}{2}\right)^2 = (-3)^2 = 9$

$(x^2 + 4x + 4) + (y^2 - 6y + 9) = 3 + 4 + 9$
$(x+2)^2 + (y-3)^2 = 16$
$(x-(-2))^2 + (y-3)^2 = 4^2$
The center is $(-2,3)$ and the radius is 4 units.

6. $x = -2(y+3)^2 + 7$
$x = -2(y-(-3))^2 + 7$

The vertex of the parabola is $(7,-3)$.

7. $x = y^2 + 10y + 23$

The y–coordinate of the vertex is
$-\frac{b}{2a} = -\frac{10}{2(1)} = -\frac{10}{2} = -5.$

The x–coordinate of the vertex is
$x = (-5)^2 + 10(-5) + 23$
$= 25 - 50 + 23$
$= 25 - 50 + 23 = -2.$

The vertex of the parabola is $(-2,-5)$.

8. $\frac{x^2}{4} - \frac{y^2}{9} = 1$

Because x^2 and y^2 have opposite signs, the equation's graph is a hyperbola.

The equation is in the form $\frac{x^2}{a^2} - \frac{y^2}{b^2} = 1$ with $a^2 = 4$, and $b^2 = 9$. We know the transverse axis lies on the x-axis and the vertices are $(-2,0)$ and $(2,0)$. Because $a^2 = 4$ and $b^2 = 9$, $a = 2$ and $b = 3$. Construct a rectangle using –2 and 2 on the x–axis, and –3 and 3 on the y–axis. Draw extended diagonals to obtain the asymptotes. Graph the hyperbola.

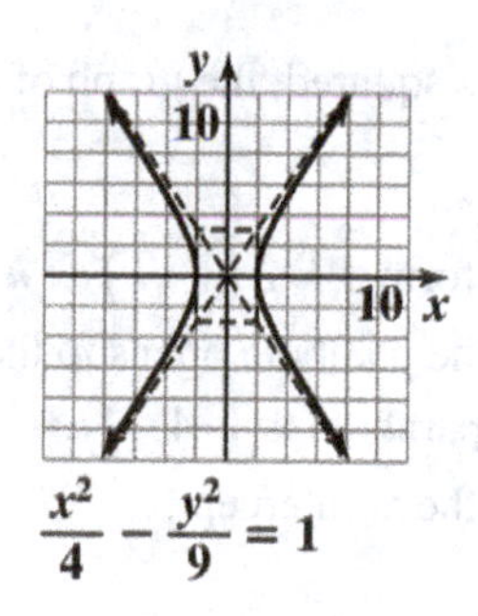

$\frac{x^2}{4} - \frac{y^2}{9} = 1$

9. $4x^2 + 9y^2 = 36$

Because x^2 and y^2 have different positive coefficients, the equation's graph is an ellipse.

$$\frac{4x^2}{36} + \frac{9y^2}{36} = \frac{36}{36}$$
$$\frac{x^2}{9} + \frac{y^2}{4} = 1$$

Because the denominator of the x^2 – term is greater than the denominator of the y^2 – term, the major axis is horizontal.

Since $a^2 = 9$, $a = 3$ and the vertices are $(-3, 0)$ and $(3, 0)$. Since $b^2 = 4$, $b = 2$ and endpoints of the minor axis are $(0, -2)$ and $(0, 2)$.

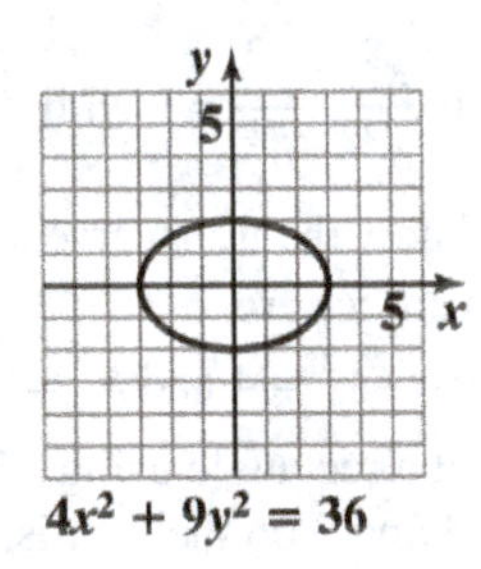

$4x^2 + 9y^2 = 36$

10. $x = (y+1)^2 - 4$

Since only one variable is squared, the graph of the equation is a parabola.

This is a parabola of the form $x = a(y-k)^2 + h$. Since $a = 1$ is positive, the parabola opens to the right. The vertex of the parabola is $(-4, -1)$. Replace y with 0 to find the x–intercept.

$$x = (0+1)^2 - 4 = (1)^2 - 4 = 1 - 4 = -3.$$

The x–intercept is -3. Replace x with 0 to find the y–intercepts.

$$0 = (y+1)^2 - 4$$
$$0 = y^2 + 2y + 1 - 4$$
$$0 = y^2 + 2y - 3$$
$$0 = (y+3)(y-1)$$

$$y + 3 = 0 \quad \text{or} \quad y - 1 = 0$$
$$y = -3 \qquad\qquad y = 1$$

The y–intercepts are -3 and 1.

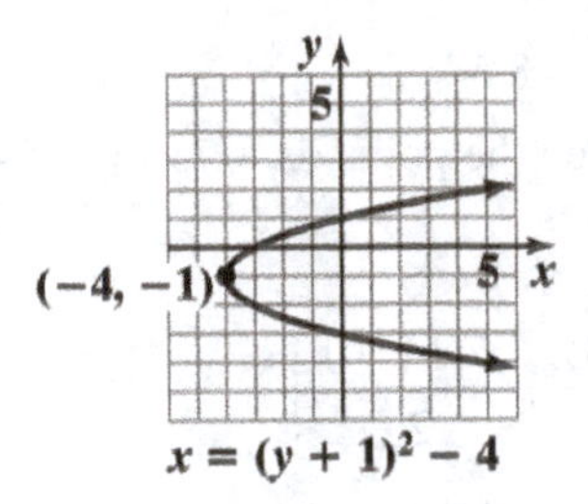

$x = (y + 1)^2 - 4$

11. $16x^2 + y^2 = 16$

Because x^2 and y^2 have different positive coefficients, the equation's graph is an ellipse.

$$\frac{16x^2}{16} + \frac{y^2}{16} = \frac{16}{16}$$
$$\frac{x^2}{1} + \frac{y^2}{16} = 1$$

Because the denominator of the y^2 – term is greater than the denominator of the x^2 – term, the major axis is vertical. Since $a^2 = 16$, $a = 4$ and the vertices are $(0, -4)$ and $(0, 4)$. Since $b^2 = 1$, $b = 1$ and endpoints of the minor axis are $(-1, 0)$ and $(1, 0)$.

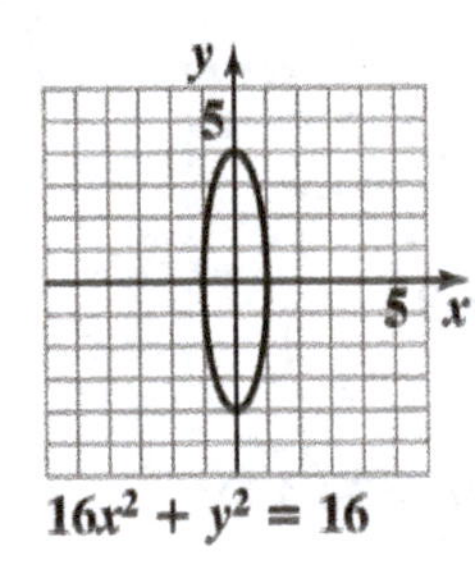

$16x^2 + y^2 = 16$

12. $25y^2 = 9x^2 + 225$

$25y^2 - 9x^2 = 225$

Because x^2 and y^2 have opposite signs, the equation's graph is a hyperbola.

$$\frac{25y^2}{225} - \frac{9x^2}{225} = \frac{225}{225}$$

$$\frac{y^2}{9} - \frac{x^2}{25} = 1$$

The equation is in the form $\frac{y^2}{a^2} - \frac{x^2}{b^2} = 1$ with $a^2 = 9$, and $b^2 = 25$. We know the transverse axis lies on the y-axis and the vertices are $(0,-3)$ and $(0,3)$. Because $a^2 = 9$ and $b^2 = 25$, $a = 3$ and $b = 5$. Construct a rectangle using –5 and 5 on the x–axis, and –3 and 3 on the y–axis. Draw extended diagonals to obtain the asymptotes. Graph the hyperbola.

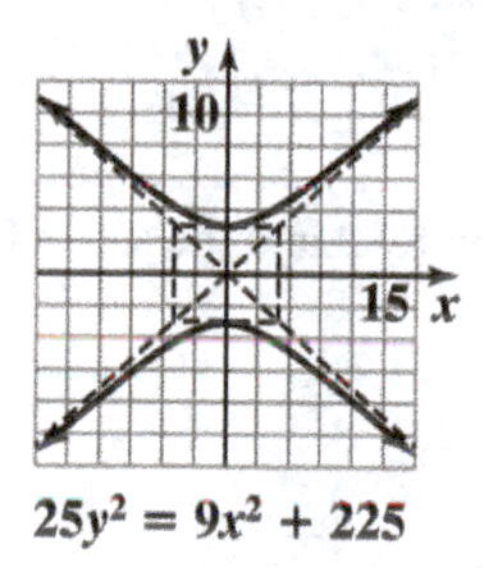

$25y^2 = 9x^2 + 225$

13. $x = -y^2 + 6y$

Since only one variable is squared, the graph of the equation is a parabola.

This is a parabola of the form $x = ay^2 + by + c$. Since $a = -1$ is negative, the parabola opens to the left. The y–coordinate of the vertex is $-\frac{b}{2a} = -\frac{6}{2(-1)} = -\frac{6}{-2} = 3$. The x–coordinate of the vertex is $x = -3^2 + 6(3) = -9 + 18 = 9$.

The vertex of the parabola is $(9,3)$.

Replace y with 0 to find the x–intercept

$x = -0^2 + 6(0) = 0 + 0 = 0$

The x–intercept is 0. Replace x with 0 to find the y–intercepts.

$0 = -y^2 + 6y$

$0 = -y(y-6)$

$-y = 0$ or $y - 6 = 0$

$y = 0$ $\quad$ $y = 6$

The y–intercepts are 0 and 6.

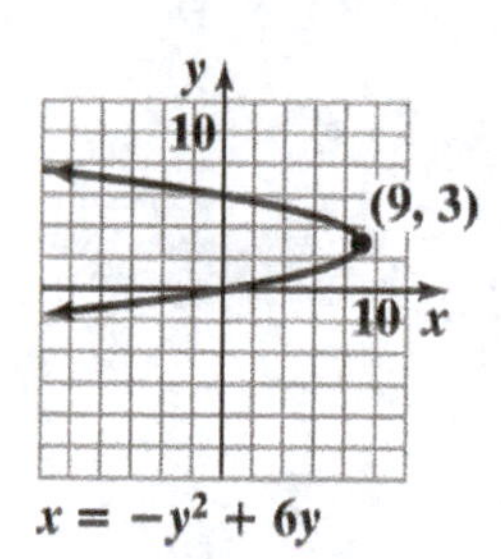

$x = -y^2 + 6y$

14. $\frac{(x-2)^2}{16} + \frac{(y+3)^2}{9} = 1$

Because x^2 and y^2 have different positive coefficients, the equation's graph is an ellipse.

The center of the ellipse is $(2,-3)$. Because the denominator of the x^2 – term is greater than the denominator of the y^2 – term, the major axis is horizontal. Since $a^2 = 16$, $a = 4$ and the vertices lie 4 units to the left and right of the center. Since $b^2 = 9$, $b = 3$ and endpoints of the minor axis lie 3 units above and below the center.

Center	Vertices	Endpoints of Minor Axis
$(2,-3)$	$(2-4,-3)$ $=(-2,-3)$	$(2,-3-3)$ $=(2,-6)$
	$(2+4,-3)$ $=(6,-3)$	$(2,-3+3)$ $=(2,0)$

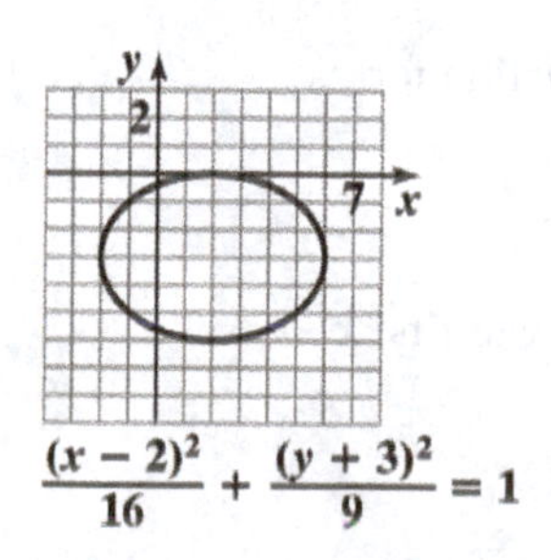

$\frac{(x-2)^2}{16} + \frac{(y+3)^2}{9} = 1$

15. $(x+1)^2+(y+2)^2=9$

Because x^2 and y^2 have the same positive coefficient, the equation's graph is a circle.

The center of the circle is $(-1,-2)$ and the radius is 3.

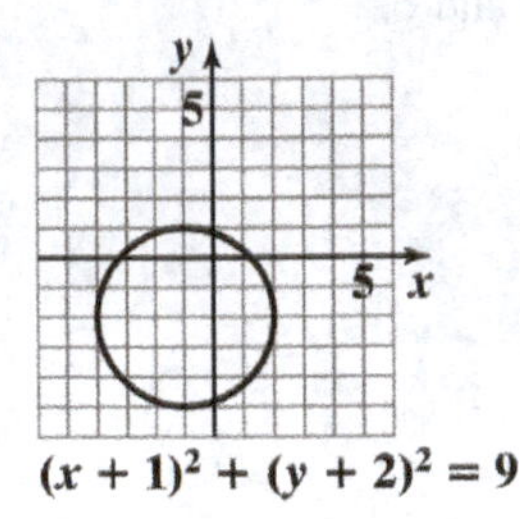

$(x+1)^2+(y+2)^2=9$

16. $\frac{x^2}{4}+\frac{y^2}{4}=1$

$4\left(\frac{x^2}{4}+\frac{y^2}{4}\right)=4(1)$

$x^2+y^2=4$

Because x^2 and y^2 have the same positive coefficient, the equation's graph is a circle.

The circle has center $(0,0)$ and radius 2.

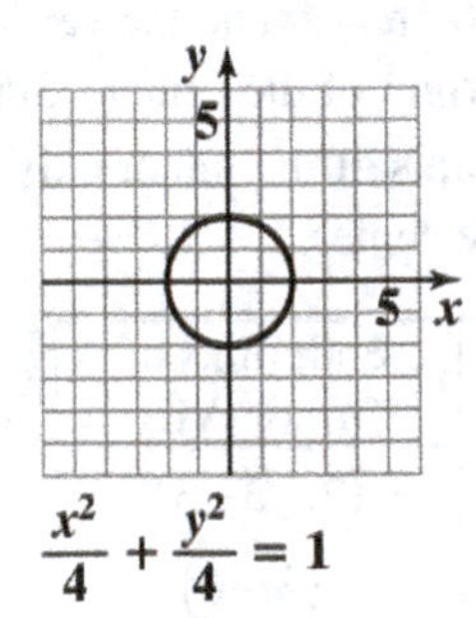

$\frac{x^2}{4}+\frac{y^2}{4}=1$

17. $x^2+y^2=25$
$x+y=1$

Solve the second equation for y.
$x+y=1$
$y=-x+1$

Substitute $-x+1$ for y to find x.

$$x^2+(-x+1)^2=25$$
$$x^2+x^2-2x+1=25$$
$$2x^2-2x+1=25$$
$$2x^2-2x-24=0$$
$$x^2-x-12=0$$
$$(x-4)(x+3)=0$$
$x-4=0$ or $x+3=0$
$x=4$ $\qquad x=-3$

Back-substitute –3 and 4 for x to find y.

$x=4$ or $x=-3$

$y=-x+1$ $\qquad y=-x+1$
$y=-4+1$ $\qquad y=-(-3)+1$
$y=-3$ $\qquad y=3+1$
$\qquad y=4$

The solution set is $\{(-3,4),\ (4,-3)\}$.

18. $2x^2-5y^2=-2$
$3x^2+2y^2=35$

Multiply the first equation by 2 and the second equation by 5.

$$4x^2-10y^2=-4$$
$$15x^2+10y^2=175$$
$$19x^2=171$$
$$x^2=9$$
$$x=\pm 3$$

In this case, we can back-substitute 9 for x^2 to find y.

$$x^2=9$$
$$2x^2-5y^2=-2$$
$$2(9)-5y^2=-2$$
$$18-5y^2=-2$$
$$-5y^2=-20$$
$$y^2=4$$
$$y=\pm 2$$

We have $x=\pm 3$ and $y=\pm 2$, the solution set is $\{(-3,-2),(-3,2),(3,-2),(3,2)\}$.

19. $2x + y = 39$
$xy = 180$

Solve the first equation for y.
$2x + y = 39$
$y = 39 - 2x$

Substitute $39 - 2x$ for y to find x.
$x(39 - 2x) = 180$
$39x - 2x^2 = 180$
$0 = 2x^2 - 39x + 180$
$0 = (2x - 15)(x - 12)$
$2x - 15 = 0$ or $x - 12 = 0$
$2x = 15$ $\quad x = 12$
$x = \frac{15}{2}$

Back-substitute $\frac{15}{2}$ and 12 for x to find y.

$x = \frac{15}{2}$ or $x = 12$
$y = 39 - 2x$ $\quad y = 39 - 2x$
$y = 39 - 2\left(\frac{15}{2}\right)$ $\quad y = 39 - 2(12)$
$y = 39 - 15$ $\quad y = 39 - 24$
$y = 24$ $\quad y = 15$

The dimensions are 15 feet by 12 feet or 24 feet by $\frac{15}{2}$ or 7.5 feet.

20. Let x = the length of the rectangle.
Let y = the width of the rectangle.

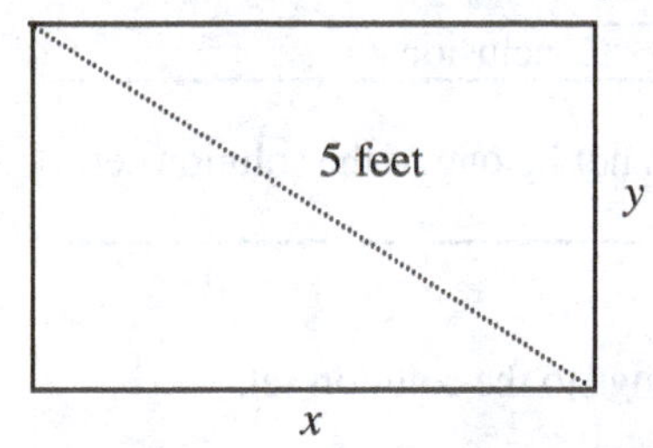

Using the Pythagorean Theorem, we obtain $x^2 + y^2 = 5^2$. Since the perimeter is 14 feet, we have $2x + 2y = 14$.

The system of two equations in two variables is as follows.
$x^2 + y^2 = 25$
$2x + 2y = 14$

Solve the second equation for y.
$2x + 2y = 14$
$2y = 14 - 2x$
$y = 7 - x$

Substitute $7 - x$ for y to find x.
$x^2 + (7 - x)^2 = 25$
$x^2 + 49 - 14x + x^2 = 25$
$2x^2 - 14x + 49 = 25$
$2x^2 - 14x + 24 = 0$
$x^2 - 7x + 12 = 0$
$(x - 4)(x - 3) = 0$
$x - 4 = 0$ or $x - 3 = 0$
$x = 4$ $\quad x = 3$

Back-substitute 3 and 4 for x to find y.
$x = 4$ or $x = 3$
$y = 7 - x$ $\quad y = 7 - x$
$y = 7 - 4$ $\quad y = 7 - 3$
$y = 3$ $\quad y = 4$

The solutions are the same. The dimensions are 4 feet by 3 feet.

Cumulative Review Exercises

(Chapters 1-10)

1. $3x + 7 > 4$ or $6 - x < 1$
$3x > -3$ $\quad -x < -5$
$x > -1$ $\quad x > 5$

The solution set is $(-1, \infty)$.

2. $x(2x - 7) = 4$
$2x^2 - 7x = 4$
$2x^2 - 7x - 4 = 0$
$(2x + 1)(x - 4) = 0$
$2x + 1 = 0$ or $x - 4 = 0$
$2x = -1$ $\quad x = 4$
$x = -\frac{1}{2}$

The solution set is $\left\{-\frac{1}{2}, 4\right\}$.

3. $\dfrac{5}{x-3}=1+\dfrac{30}{x^2-9}$

$\dfrac{5}{x-3}=1+\dfrac{30}{(x+3)(x-3)}$

Multiply both sides of the equation by the LCD, $(x+3)(x-3)$.

$$(x+3)(x-3)\left(\frac{5}{x-3}\right)=(x+3)(x-3)\left(1+\frac{30}{(x+3)(x-3)}\right)$$
$$(x+3)(5)=(x+3)(x-3)+30$$
$$5x+15=x^2-9+30$$
$$15=x^2-5x+21$$
$$0=x^2-5x+6$$
$$0=(x-3)(x-2)$$
$$x-3=0 \quad \text{or} \quad x-2=0$$
$$x=3 \qquad\qquad x=2$$

Disregard 3 because it would make the denominator zero. The solution set is $\{2\}$.

4. $3x^2+8x+5<0$

Solve the related quadratic equation.

$$3x^2+8x+5=0$$
$$(3x+5)(x+1)=0$$
$$3x+5=0 \quad \text{or} \quad x+1=0$$
$$3x=-5 \qquad\qquad x=-1$$
$$x=-\frac{5}{3}$$

The boundary points are $-\dfrac{5}{3}$ and -1.

Interval	Test Value	Test	Conclusion
$\left(-\infty,-\frac{5}{3}\right)$	-2	$3(-2)^2+8(-2)+5<0$ $1<0$, false	$\left(-\infty,-\frac{5}{3}\right)$ does not belong to the solution set.
$\left(-\frac{5}{3},-1\right)$	$-\frac{4}{3}$	$3\left(-\frac{4}{3}\right)^2+8\left(-\frac{4}{3}\right)+5<0$ $-\frac{1}{3}<0$, true	$\left(-\frac{5}{3},-1\right)$ belongs to the solution set.
$(-1,\infty)$	0	$3(0)^2+8(0)+5<0$ $5<0$, false	$(-1,\infty)$ does not belong to the solution set.

The solution set is $\left(-\dfrac{5}{3},-1\right)$ or $\left\{x\,\middle|\,-\dfrac{5}{3}<x<-1\right\}$.

5. $3^{2x-1} = 81$

$$3^{2x-1} = 3^4$$
$$2x - 1 = 4$$
$$2x = 5$$
$$x = \frac{5}{2}$$

The solution set is $\left\{\frac{5}{2}\right\}$.

6. $30e^{0.7x} = 240$

$$e^{0.7x} = 80$$
$$\ln e^{0.7x} = \ln 8$$
$$0.7x = \ln 8$$
$$x = \frac{\ln 8}{0.7} = \frac{2.08}{0.7} \approx 2.97$$

The solution set is $\left\{\frac{\ln 8}{0.7} \approx 2.97\right\}$.

7. $3x^2 + 4y^2 = 39$

$$5x^2 - 2y^2 = -13$$

Multiply the second equation by 2 and add to the first equation.

$$3x^2 + 4y^2 = 39$$
$$\underline{10x^2 - 4y^2 = -26}$$
$$13x^2 = 13$$
$$x^2 = 1$$
$$x = \pm 1$$

We can back-substitute 1 for x^2 to find y.

$$x^2 = 1$$
$$3x^2 + 4y^2 = 39$$
$$3(1) + 4y^2 = 39$$
$$3 + 4y^2 = 39$$
$$4y^2 = 36$$
$$y^2 = 9$$
$$y = \pm 3$$

We have $x = \pm 1$ and $y = \pm 3$, the solution set is $\{(-1,-3),(-1,3),(1,-3),(1,3)\}$.

8. $f(x) = -\frac{2}{3}x + 4$

$$y = -\frac{2}{3}x + 4$$

The y–intercept is 4 and the slope is $-\frac{2}{3}$. We can write the slope as $m = \frac{-2}{3} = \frac{\text{rise}}{\text{run}}$ and use the intercept and the slope to graph the function.

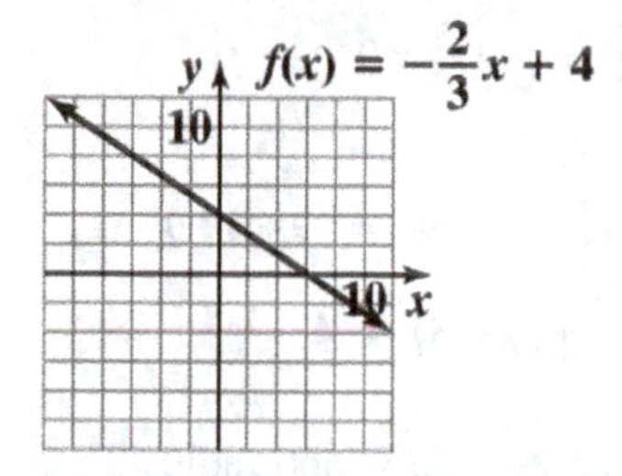

9. $3x - y > 6$

First, find the intercepts to the equation $3x - y = 6$.

Find the x–intercept by setting y equal to zero.

$$3x - 0 = 6$$
$$3x = 6$$
$$x = 2$$

Find the y–intercept by setting x equal to zero.

$$3(0) - y = 6$$
$$-y = 6$$
$$y = -6$$

Next, use the origin as a test point.

$$3(0) - 0 > 6$$
$$0 - 0 > 6$$
$$0 > 6$$

This is a false statement. This means that the origin will not fall in the shaded half-plane.

$3x - y > 6$

10. $x^2+y^2+4x-6y+9=0$

Because x^2 and y^2 have the same positive coefficient, the equation's graph is a circle.

$(x^2+4x\quad)+(y^2-6y\quad)=-9$

Complete the squares.

$$\left(\frac{b}{2}\right)^2=\left(\frac{4}{2}\right)^2=(2)^2=4$$

$$\left(\frac{b}{2}\right)^2=\left(\frac{-6}{2}\right)^2=(-3)^2=9$$

$$(x^2+4x+4)+(y^2-6y+9)=-9+4+9$$

$$(x+2)^2+(y-3)^2=4$$

The circle has center $(-2,3)$ and radius 2.

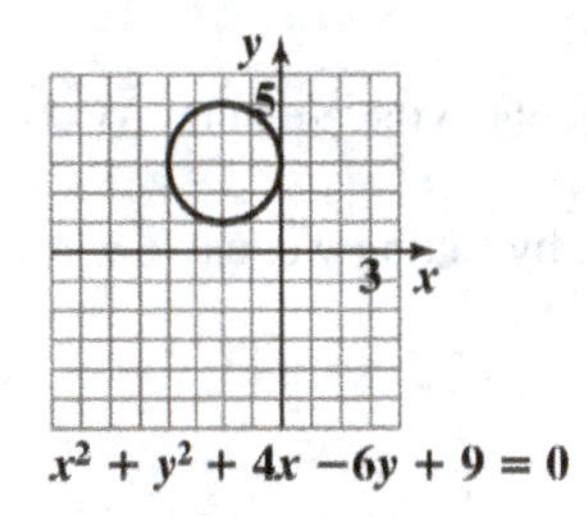

11. $9x^2-4y^2=36$

Because x^2 and y^2 have opposite signs, the equation's graph is a hyperbola.

$$\frac{9x^2}{36}-\frac{4y^2}{36}=\frac{36}{36}$$

$$\frac{x^2}{4}-\frac{y^2}{9}=1$$

The equation is in the form $\frac{x^2}{a^2}-\frac{y^2}{b^2}=1$ with $a^2=4$, and $b^2=9$. We know the transverse axis lies on the x-axis and the vertices are $(-2,0)$ and $(2,0)$. Because $a^2=4$ and $b^2=9$, $a=2$ and $b=3$. Construct a rectangle using −2 and 2 on the x–axis, and −3 and 3 on the y–axis. Draw extended diagonals to obtain the asymptotes.

Graph the hyperbola.

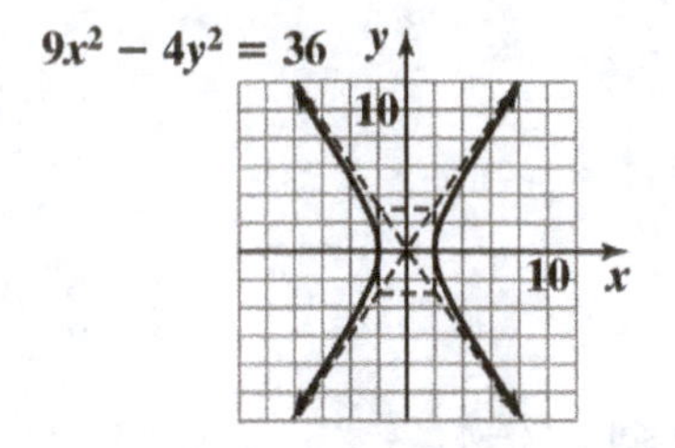

12. $-2(3^2-12)^3-45\div9-3$

$=-2(9-12)^3-45\div9-3$

$=-2(-3)^3-45\div9-3$

$=-2(-27)-45\div9-3$

$=54-5-3=46$

13. $(3x^3-19x^2+17x+4)\div(3x-4)$

Rewrite the polynomials in descending order and divide.

$$\begin{array}{r} x^2-5x-1 \\ 3x-4\overline{)3x^3-19x^2+17x+4} \\ \underline{3x^3-4x^2\qquad\qquad} \\ -15x^2+17x\quad \\ \underline{-15x^2+20x\quad} \\ -3x+4 \\ \underline{-3x+4} \\ 0 \end{array}$$

$$\frac{3x^3-19x^2+17x+4}{3x-4}=x^2-5x-1$$

14. $\sqrt[3]{4x^2y^5}\cdot\sqrt[3]{4xy^2}$

$=\sqrt[3]{4x^2y^5\,4xy^2}=\sqrt[3]{16x^3y^7}$

$=\sqrt[3]{8\cdot2x^3y^6y}=2xy^2\sqrt[3]{2y}$

15. $(2+3i)(4-i)$

$=8-2i+12i-3i^2=8+10i-3(-1)$

$=8+10i+3=11+10i$

16. $12x^3-36x^2+27x=3x(4x^2-12x+9)$

$=3x(2x-3)^2$

17. $x^3 - 2x^2 - 9x + 18$

$= x^2(x-2) - 9(x-2)$

$= (x-2)(x^2-9)$

$= (x-2)(x+3)(x-3)$

18. Since the radicand must be positive, the domain will exclude all values of x which make the radicand less than zero.

$6 - 3x \geq 0$

$-3x \geq -6$

$x \leq 2$

The domain of $f = \{x | x \leq 2\}$ or $(-\infty, 2]$.

19. $\frac{1-\sqrt{x}}{1+\sqrt{x}} = \frac{1-\sqrt{x}}{1+\sqrt{x}} \cdot \frac{1-\sqrt{x}}{1-\sqrt{x}}$

$= \frac{(1-\sqrt{x})^2}{1^2 - (\sqrt{x})^2}$

$= \frac{(1-\sqrt{x})^2}{1-x}$ or $\frac{1-2\sqrt{x}+x}{1-x}$

20. $\frac{1}{3}\ln x + 7\ln y = \ln x^{\frac{1}{3}} + \ln y^7 = \ln\left(x^{\frac{1}{3}} y^7\right)$

21. $(3x^3 - 5x^2 + 2x - 1) \div (x-2)$

$$\begin{array}{r|rrrr} 2 & 3 & -5 & 2 & -1 \\ & & 6 & 2 & 8 \\ \hline & 3 & 1 & 4 & 7 \end{array}$$

$(3x^3 - 5x^2 + 2x - 1) \div (x-2)$

$= 3x^2 + x + 4 + \frac{7}{x-2}$

22. $x = -2\sqrt{3}$ or $x = 2\sqrt{3}$

$x + 2\sqrt{3} = 0$ $\quad x - 2\sqrt{3} = 0$

Multiply the factors to obtain the polynomial.

$(x+2\sqrt{3})(x-2\sqrt{3}) = 0$

$x^2 - (2\sqrt{3})^2 = 0$

$x^2 - 4 \cdot 3 = 0$

$x^2 - 12 = 0$

23. Let x = the rate of the slower car.

	r	t	d
Fast	$x + 10$	2	$2(x + 10)$
Slow	x	2	$2x$

$2(x+10) + 2x = 180$

$2x + 20 + 2x = 180$

$4x + 20 = 180$

$4x = 160$

$x = 40$

The rate of the slower car is 40 miles per hour and the rate of the faster car is 40 + 10 = 50 miles per hour.

24. Let x = the number of miles driven in a day.

$C_R = 39 + 0.16x$

$C_A = 25 + 0.24x$

Set the costs equal.

$39 + 0.16x = 25 + 0.24x$

$39 = 25 + 0.08x$

$14 = 0.08x$

$\frac{14}{0.08} = x$

$x = 175$

The cost is the same when renting from either company when 175 miles are driven in a day.

$C_R = 39 + 0.16(175) = 39 + 28 = 67$

When 175 miles are driven, the cost is $67.

25. Let x = the number of apples.
Let y = the number of bananas.

$$3x+2y=354$$
$$2x+3y=381$$

Multiply the first equation by –3 and the second equation by 2 and solve by addition.

$$\begin{array}{r} -9x-6y=-1062 \\ 4x+6y=762 \\ \hline -5x=-300 \\ x=60 \end{array}$$

Back-substitute 60 for x to find y.

$$3(60)+2y=354$$
$$180+2y=354$$
$$2y=174$$
$$y=87$$

There are 60 calories in an apple and 87 calories in a banana.

Chapter 11
Sequences, Series, and the Binomial Theorem

11.1 Check Points

1. a. $a_n = 2n+5$

$a_1 = 2(1)+5 = 7$

$a_2 = 2(2)+5 = 9$

$a_3 = 2(3)+5 = 11$

$a_4 = 2(4)+5 = 13$

The first four terms are 7, 9, 11, and 13.

b. $a_n = \dfrac{(-1)^n}{2^n+1}$

$a_1 = \dfrac{(-1)^1}{2^1+1} == \dfrac{-1}{3} - \dfrac{1}{3}$

$a_2 = \dfrac{(-1)^2}{2^2+1} = \dfrac{1}{5}$

$a_3 = \dfrac{(-1)^3}{2^3+1} = \dfrac{-1}{9} = -\dfrac{1}{9}$

$a_4 = \dfrac{(-1)^4}{2^4+1} = \dfrac{1}{17}$

The first four terms are $-\frac{1}{3}$, $\frac{1}{5}$, $-\frac{1}{9}$, and $\frac{1}{17}$.

2. $a_n = \dfrac{20}{(n+1)!}$

$a_1 = \dfrac{20}{(1+1)!} = \dfrac{20}{2!} = 10$

$a_2 = \dfrac{20}{(2+1)!} = \dfrac{20}{3!} = \dfrac{20}{6} = \dfrac{10}{3}$

$a_3 = \dfrac{20}{(3+1)!} = \dfrac{20}{4!} = \dfrac{20}{24} = \dfrac{5}{6}$

$a_4 = \dfrac{20}{(4+1)!} = \dfrac{20}{5!} = \dfrac{20}{120} = \dfrac{1}{6}$

The first four terms are 10, $\frac{10}{3}$, $\frac{5}{6}$, and $\frac{1}{6}$.

3. a. $\sum_{i=1}^{6} 2i^2$

$= 2(1)^2 + 2(2)^2 + 2(3)^2$

$\quad + 2(4)^2 + 2(5)^2 + 2(6)^2$

$= 2+8+18+32+50+72$

$= 182$

b. $\sum_{k=3}^{5} \left(2^k - 3\right)$

$= \left(2^3-3\right) + \left(2^4-3\right) + \left(2^5-3\right)$

$= (8-3)+(16-3)+(32-3)$

$= 5+13+29$

$= 47$

c. $\sum_{i=1}^{5} 4 = 4+4+4+4+4 = 20$

4. a. The sum has nine terms, each of the form i^2, starting at $i=1$ and ending at $i=9$.

$1^2+2^2+3^2+\cdots+9^2 = \sum_{i=1}^{9} i^2$

b. The sum has n terms, each of the form $\dfrac{1}{2^{i-1}}$, starting at $i=1$ and ending at $i=n$.

$1+\dfrac{1}{2}+\dfrac{1}{4}+\dfrac{1}{8}+\cdots+\dfrac{1}{2^{n-1}} = \sum_{i=1}^{n} \dfrac{1}{2^{i-1}}$

11.1 Concept and Vocabulary Check

1. sequence; integers; terms
2. general
3. -1
4. $-\dfrac{1}{3}$
5. factorial; 5; 1; 1
6. a_1; a_2; a_3; a_n; index; upper limit; lower limit

11.1 Exercise Set

1. $a_n = 3n + 2$

$a_1 = 3(1) + 2 = 3 + 2 = 5$

$a_2 = 3(2) + 2 = 6 + 2 = 8$

$a_3 = 3(3) + 2 = 9 + 2 = 11$

$a_4 = 3(4) + 2 = 12 + 2 = 14$

The first four terms are 5, 8, 11, 14.

3. $a_n = 3^n$

$a_1 = 3^1 = 3$

$a_2 = 3^2 = 9$

$a_3 = 3^3 = 27$

$a_4 = 3^4 = 81$

The first four terms are 3, 9, 27, 81.

5. $a_n = (-3)^n$

$a_1 = (-3)^1 = -3$

$a_2 = (-3)^2 = 9$

$a_3 = (-3)^3 = -27$

$a_4 = (-3)^4 = 81$

The first four terms are –3, 9, –27, 81.

7. $a_n = (-1)^n (n + 3)$

$a_1 = (-1)^1 (1 + 3) = -1(4) = -4$

$a_2 = (-1)^2 (2 + 3) = 1(5) = 5$

$a_3 = (-1)^3 (3 + 3) = -1(6) = -6$

$a_4 = (-1)^4 (4 + 3) = 1(7) = 7$

The first four terms are –4, 5, –6, 7.

9. $a_n = \dfrac{2n}{n + 4}$

$a_1 = \dfrac{2(1)}{1 + 4} = \dfrac{2}{5}$

$a_2 = \dfrac{2(2)}{2 + 4} = \dfrac{4}{6} = \dfrac{2}{3}$

$a_3 = \dfrac{2(3)}{3 + 4} = \dfrac{6}{7}$

$a_4 = \dfrac{2(4)}{4 + 4} = \dfrac{8}{8} = 1$

The first four terms are $\dfrac{2}{5}, \dfrac{2}{3}, \dfrac{6}{7}, 1$.

11. $a_n = \dfrac{(-1)^{n+1}}{2^n - 1}$

$a_1 = \dfrac{(-1)^{1+1}}{2^1 - 1} = \dfrac{(-1)^2}{2 - 1} = \dfrac{1}{1} = 1$

$a_2 = \dfrac{(-1)^{2+1}}{2^2 - 1} = \dfrac{(-1)^3}{4 - 1} = \dfrac{-1}{3} = -\dfrac{1}{3}$

$a_3 = \dfrac{(-1)^{3+1}}{2^3 - 1} = \dfrac{(-1)^4}{8 - 1} = \dfrac{1}{7}$

$a_4 = \dfrac{(-1)^{4+1}}{2^4 - 1} = \dfrac{(-1)^5}{16 - 1} = \dfrac{-1}{15} = -\dfrac{1}{15}$

The first four terms are $1, -\dfrac{1}{3}, \dfrac{1}{7}, -\dfrac{1}{15}$.

13. $a_n = \dfrac{n^2}{n!}$

$a_1 = \dfrac{1^2}{1!} = \dfrac{1}{1} = 1$

$a_2 = \dfrac{2^2}{2!} = \dfrac{4}{2 \cdot 1} = \dfrac{4}{2} = 2$

$a_3 = \dfrac{3^2}{3!} = \dfrac{9}{3 \cdot 2 \cdot 1} = \dfrac{3}{2}$

$a_4 = \dfrac{4^2}{4!} = \dfrac{16}{4 \cdot 3 \cdot 2 \cdot 1} = \dfrac{2}{3}$

The first four terms are $1, 2, \dfrac{3}{2}, \dfrac{2}{3}$.

15. $a_n = 2(n + 1)!$

$a_1 = 2(1 + 1)! = 2 \cdot 2! = 2 \cdot 2 \cdot 1 = 4$

$a_2 = 2(2 + 1)! = 2 \cdot 3! = 2 \cdot 3 \cdot 2 \cdot 1 = 12$

$a_3 = 2(3 + 1)! = 2 \cdot 4! = 2 \cdot 4 \cdot 3 \cdot 2 \cdot 1$
$= 48$

$a_4 = 2(4 + 1)! = 2 \cdot 5! = 2 \cdot 5 \cdot 4 \cdot 3 \cdot 2 \cdot 1$
$= 240$

17. $\sum_{i=1}^{6} 5i = 5(1)+5(2)+5(3)+5(4)+5(5)+5(6) = 5+10+15+20+25+30 = 105$

19. $\sum_{i=1}^{4} 2i^2 = 2(1)^2+2(2)^2+2(3)^2+2(4)^2 = 2(1)+2(4)+2(9)+2(16)$

$= 2+8+18+32 = 60$

21. $\sum_{k=1}^{5} k(k+4) = 1(1+4)+2(2+4)+3(3+4)+4(4+4)+5(5+4)$

$= 1(5)+2(6)+3(7)+4(8)+5(9) = 5+12+21+32+45 = 115$

23. $\sum_{i=1}^{4}\left(-\frac{1}{2}\right)^i = \left(-\frac{1}{2}\right)^1+\left(-\frac{1}{2}\right)^2+\left(-\frac{1}{2}\right)^3+\left(-\frac{1}{2}\right)^4 = -\frac{1}{2}+\frac{1}{4}+\left(-\frac{1}{8}\right)+\frac{1}{16}$

$= -\frac{1}{2}\cdot\frac{8}{8}+\frac{1}{4}\cdot\frac{4}{4}+\left(-\frac{1}{8}\right)\frac{2}{2}+\frac{1}{16} = -\frac{8}{16}+\frac{4}{16}-\frac{2}{16}+\frac{1}{16}$

$= \frac{-8+4-2+1}{16} = -\frac{5}{16}$

25. $\sum_{i=5}^{9} 11 = 11+11+11+11+11 = 55$

27. $\sum_{i=0}^{4}\frac{(-1)^i}{i!} = \frac{(-1)^0}{0!}+\frac{(-1)^1}{1!}+\frac{(-1)^2}{2!}+\frac{(-1)^3}{3!}+\frac{(-1)^4}{4!} = \frac{1}{1}+\frac{-1}{1}+\frac{1}{2\cdot 1}+\frac{-1}{3\cdot 2\cdot 1}+\frac{1}{4\cdot 3\cdot 2\cdot 1}$

$= 1-1+\frac{1}{2}-\frac{1}{6}+\frac{1}{24} = \frac{1}{2}\cdot\frac{12}{12}-\frac{1}{6}\cdot\frac{4}{4}+\frac{1}{24} = \frac{12}{24}-\frac{4}{24}+\frac{1}{24} = \frac{12-4+1}{24} = \frac{9}{24} = \frac{3}{8}$

29. $\sum_{i=1}^{5}\frac{i!}{(i-1)!} = \frac{1!}{(1-1)!}+\frac{2!}{(2-1)!}+\frac{3!}{(3-1)!}+\frac{4!}{(4-1)!}+\frac{5!}{(5-1)!} = \frac{1!}{0!}+\frac{2!}{1!}+\frac{3!}{2!}+\frac{4!}{3!}+\frac{5!}{4!}$

$= \frac{1}{1}+\frac{2\cdot \cancel{1!}}{\cancel{1!}}+\frac{3\cdot \cancel{2!}}{\cancel{2!}}+\frac{4\cdot \cancel{3!}}{\cancel{3!}}+\frac{5\cdot \cancel{4!}}{\cancel{4!}} = 1+2+3+4+5 = 15$

31. $1^2+2^2+3^2+\ldots+15^2 = \sum_{i=1}^{15} i^2$

33. $2+2^2+2^3+\ldots+2^{11} = \sum_{i=1}^{11} 2^i$

35. $1+2+3+\ldots+30 = \sum_{i=1}^{30} i$

37. $\frac{1}{2}+\frac{2}{3}+\frac{3}{4}+\ldots+\frac{14}{14+1} = \sum_{i=1}^{14}\frac{i}{i+1}$

39. $4+\frac{4^2}{2}+\frac{4^3}{3}+...+\frac{4^n}{n}=\sum_{i=1}^{n}\frac{4^i}{i}$

41. $1+3+5+...+(2n-1)=\sum_{i=1}^{n}(2i-1)$

43. $5+7+9+11+...+31=\sum_{k=2}^{15}(2k+1)$ or

$$=\sum_{k=1}^{14}(2k+3)$$

45. $a+ar+ar^2+...+ar^{12}=\sum_{k=0}^{12}ar^k$

47. $a+(a+d)+(a+2d)+...+a(a+nd)$

$$=\sum_{k=0}^{n}(a+kd)$$

49. $\sum_{i=1}^{5}(a_i^2+1)=((-4)^2+1)+((-2)^2+1)+((0)^2+1)+((2)^2+1)+((4)^2+1)$

$$=17+5+1+5+17$$
$$=45$$

51. $\sum_{i=1}^{5}(2a_i+b_i)=(2(-4)+4)+(2(-2)+2)+(2(0)+0)+(2(2)+(-2))+(2(4)+(-4))$

$$=-4+(-2)+0+2+4=0$$

53. $\sum_{i=4}^{5}\left(\frac{a_i}{b_i}\right)^2=\left(\frac{2}{-2}\right)^2+\left(\frac{4}{-4}\right)^2=(-1)^2+(-1)^2=1+1=2$

55. $\sum_{i=1}^{5}a_i^2+\sum_{i=1}^{5}b_i^2=((-4)^2+(-2)^2+0^2+2^2+4^2)+(4^2+2^2+0^2+(-2)^2+(-4)^2)$

$$=(16+4+0+4+16)+(16+4+0+4+16)$$
$$=80$$

57. a. $\sum_{i=1}^{8}a_i=100+120+145+170+200+220+260+300=1515$

A total of 1515 thousand, or 1,515,000, autism cases were diagnosed in the United States from 2001 through 2008.

b. $\sum_{i=1}^{8}a_i=(28\cdot1+63)+(28\cdot2+63)+(28\cdot3+63)+(28\cdot4+63)+(28\cdot5+63)+(28\cdot6+63)+(28\cdot7+63)+(28\cdot8+63)$

$$=91+119+147+175+203+231+259+287$$
$$=1512$$

The model underestimates the actual sum by 3 thousand.

59. $a_{20} = 6000\left(1+\frac{0.06}{4}\right)^{20} = 6000(1+0.015)^{20} = 6000(1.015)^{20} = 8081.13$

The balance in the account after 5 years if \$8081.13.

61. – 65. Answers will vary.

67. Answers will vary.

69. $a_n = \frac{n}{n+1}$;

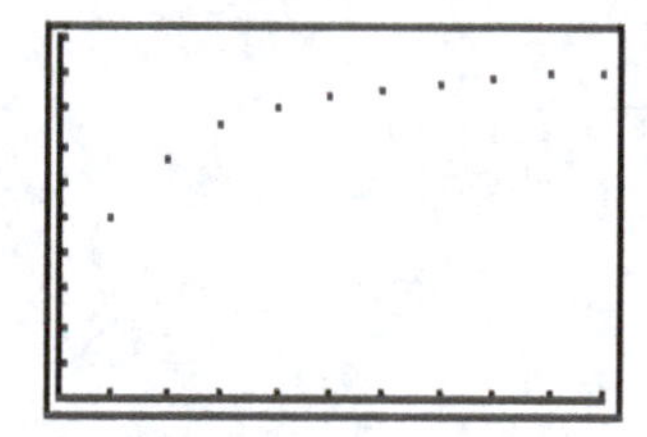

As n gets larger, the terms get closer to 1.

71. $a_n = \frac{2n^2+5n-7}{n^3}$

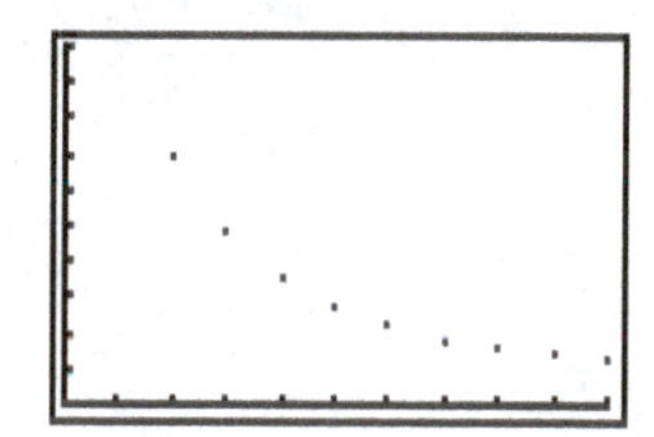

As n gets larger, the terms get closer to 0.

73. does not make sense; Explanations will vary. Sample explanation: There is nothing that implies that there is a negative number of sheep.

75. makes sense

77. false; Changes to make the statement true will vary. A sample change is:

$$\sum_{i=1}^{2}(-1)^i 2^i = (-1)^1 2^1 + (-1)^2 2^2 = -1(2)+1(4) = -2+4 = 2$$

79. true

81. $a_n = \frac{1}{n}$

83. $a_n = (-1)^n$

85. $a_n = \frac{n+2}{n+1}$

87. $a_n = \frac{(n+1)^2}{n}$

89. $\frac{600!}{599!} = \frac{600 \cdot \cancel{599!}}{\cancel{599!}} = 600$

91. $\frac{n!}{(n-3)!} = \frac{n(n-1)(n-2)\cancel{(n-3)!}}{\cancel{(n-3)!}}$

$= n(n-1)(n-2)$

$= n(n^2 - 3n + 2)$

$= n^3 - 3n^2 + 2n$

93. $\sum_{i=2}^{4} 2i \log x = 2(2)\log x + 2(3)\log x + 2(4)\log x$

$= 4\log x + 6\log x + 8\log x$

$= \log x^4 + \log x^6 + \log x^8$

$= \log\left(x^4 \cdot x^6 \cdot x^8\right) = \log x^{18}$

95. $\sqrt[3]{40x^4y^7} = \sqrt[3]{8 \cdot 5x^3xy^6y} = 2xy^2\sqrt[3]{5xy}$

96. $27x^3 - 8 = (3x-2)\left(9x^2 + 6x + 4\right)$

97.

$$\frac{6}{x} + \frac{6}{x+2} = \frac{5}{2}$$

$$2x(x+2)\left(\frac{6}{x} + \frac{6}{x+2}\right) = 2x(x+2)\left(\frac{5}{2}\right)$$

$$2(x+2)(6) + 2x(6) = x(x+2)(5)$$

$$12(x+2) + 12x = 5x(x+2)$$

$$12x + 24 + 12x = 5x^2 + 10x$$

$$24x + 24 = 5x^2 + 10x$$

$$0 = 5x^2 - 14x - 24$$

$$0 = (5x+6)(x-4)$$

Apply the zero product principle.

$5x + 6 = 0$ or $x - 4 = 0$

$5x = -6$ $\quad x = 4$

$x = -\frac{6}{5}$

The solution set is $\left\{-\frac{6}{5}, 4\right\}$.

98. $a_2 - a_1 = 3 - 8 = -5$
$a_3 - a_2 = -2 - 3 = -5$
$a_4 - a_3 = -7 - (-2) = -5$
$a_5 - a_4 = -12 - (-7) = -5$
The difference between consecutive terms is always -5.

99. $a_2 - a_1 = (4(2) - 3) - (4(1) - 3) = 4$
$a_3 - a_2 = (4(3) - 3) - (4(2) - 3) = 4$
$a_4 - a_3 = (4(4) - 3) - (4(3) - 3) = 4$
$a_5 - a_4 = (4(5) - 3) - (4(4) - 3) = 4$
The difference between consecutive terms is always 4.

100. $a_n = 4 + (n-1)(-7)$
$a_8 = 4 + (8-1)(-7) = 4 + (7)(-7) = 4 - 49 = -45$

11.2 Check Points

1. $a_1 = 100$
$a_2 = 100 + (-30) = 70$
$a_3 = 70 + (-30) = 40$
$a_4 = 40 + (-30) = 10$
$a_5 = 10 + (-30) = -20$
$a_6 = -20 + (-30) = -50$

2. $a_1 = 6,\ d = -5$
To find the ninth term, a_9, replace n in the formula with 9, a_1 with 6, and d with –5.
$a_n = a_1 + (n-1)d$
$a_9 = 6 + (9-1)(-5)$
$= 6 + 8(-5)$
$= 6 + (-40)$
$= -34$

3. a. $a_n = a_1 + (n-1)d$
$= 32 + (n-1)0.7$
$= 0.7n + 31.3$

b. 2024 is 21 years after 2003.
$a_n = 0.7n + 31.3$
$a_{21} = 0.7(21) + 31.3 = 46$
In 2024 Americans will average 46 car meals.

4. 3, 6, 9, 12, ...
To find the sum of the first 15 terms, S_{15}, replace n in the formula with 15.

$$S_n = \frac{n}{2}(a_1 + a_n)$$
$$S_{15} = \frac{15}{2}(a_1 + a_{15})$$

Use the formula for the general term of a sequence to find a_{15}. The common difference, *d*, is 3, and the first term, a_1, is 3.

$$\begin{aligned} a_n &= a_1 + (n-1)d \\ a_{15} &= 3 + (15-1)(3) \\ &= 3 + 14(3) \\ &= 3 + 42 \\ &= 45 \end{aligned}$$

Thus, $S_{15} = \frac{15}{2}(3+45) = \frac{15}{2}(48) = 360$.

5. $\sum_{i=1}^{30}(6i-11) = (6\cdot 1-11)+(6\cdot 2-11)+(6\cdot 3-11)+\ldots+(6\cdot 30-11) = -5+1+7+\ldots+169$

The first term, a_1, is –5; the common difference, *d*, is $1-(-5)=6$; the last term, a_{30}, is 169. Substitute $n = 30$, $a_1 = -5$, and $a_{30} = 169$ in the formula $S_n = \frac{n}{2}(a_1 + a_n)$.

$$S_{30} = \frac{30}{2}(-5+169) = 15(164) = 2460$$

Thus, $\sum_{i=1}^{30}(6i-11) = 2460$

6. $a_n = 1800n + 64{,}130$

$$\begin{aligned} a_1 &= 1800(1) + 64{,}130 = 65{,}930 \\ a_{10} &= 1800(10) + 64{,}130 = 82{,}130 \end{aligned}$$

$$S_n = \frac{n}{2}(a_1 + a_n)$$

$$\begin{aligned} S_{10} &= \frac{10}{2}(a_1 + a_{10}) \\ &= 5(65{,}930 + 82{,}130) \\ &= 5(148{,}060) \\ &= \$740{,}300 \end{aligned}$$

It would cost \$7400 for the ten-year period beginning in 2014.

11.2 Concept and Vocabulary Check

1. arithmetic; common difference

2. $a_1 + (n-1)d$; first term; common difference

3. $\frac{n}{2}(a_1 + a_2)$; first term; *n*th term

4. 2; 116

5. 8; 13; 18; 5

11.2 Exercise Set

1. Since $6-2=4,\ d=4.$

3. Since $-2-(-7)=5,\ d=5.$

5. Since $711-714=-3,\ d=-3.$

7.
$$a_1=200$$
$$a_2=200+20=220$$
$$a_3=220+20=240$$
$$a_4=240+20=260$$
$$a_5=260+20=280$$
$$a_6=280+20=300$$

9.
$$a_1=-7$$
$$a_2=-7+4=-3$$
$$a_3=-3+4=1$$
$$a_4=1+4=5$$
$$a_5=5+4=9$$
$$a_6=9+4=13$$

11.
$$a_1=300$$
$$a_2=300-90=210$$
$$a_3=210-90=120$$
$$a_4=120-90=30$$
$$a_5=30-90=-60$$
$$a_6=-60-90=-150$$

13.
$$a_1=\frac{5}{2}$$
$$a_2=\frac{5}{2}-\frac{1}{2}=\frac{4}{2}=2$$
$$a_3=\frac{4}{2}-\frac{1}{2}=\frac{3}{2}$$
$$a_4=\frac{3}{2}-\frac{1}{2}=\frac{2}{2}=1$$
$$a_5=1-\frac{1}{2}=\frac{1}{2}$$
$$a_6=\frac{1}{2}-\frac{1}{2}=0$$

15. $a_1 = -0.4$
$a_2 = -0.4 - 1.6 = -2$
$a_3 = -2 - 1.6 = -3.6$
$a_4 = -3.6 - 1.6 = -5.2$
$a_5 = -5.2 - 1.6 = -6.8$
$a_6 = -6.8 - 1.6 = -8.4$

17. $a_6 = 13 + (6-1)4 = 13 + (5)4$
$= 13 + 20 = 33$

19. $a_{50} = 7 + (50-1)5 = 7 + (49)5$
$= 7 + 245 = 252$

21. $a_{200} = -40 + (200-1)5 = -40 + (199)5$
$= -40 + 995 = 955$

23. $a_{60} = 35 + (60-1)(-3) = 35 + (59)(-3)$
$= 35 + (-177) = -142$

25. $a_n = a_1 + (n-1)d = 1 + (n-1)4$
$= 1 + 4n - 4 = 4n - 3$

$a_{20} = 4(20) - 3 = 80 - 3 = 77$

27. $a_n = a_1 + (n-1)d = 7 + (n-1)(-4)$
$= 7 - 4n + 4 = 11 - 4n$

$a_{20} = 11 - 4(20) = 11 - 80 = -69$

29. $a_n = a_1 + (n-1)d = -20 + (n-1)(-4)$
$= -20 - 4n + 4 = -4n - 16$

$a_{20} = -4(20) - 16 = -80 - 16 = -96$

31. $a_n = a_1 + (n-1)d = -\frac{1}{3} + (n-1)\left(\frac{1}{3}\right)$
$= -\frac{1}{3} + \frac{1}{3}n - \frac{1}{3} = \frac{1}{3}n - \frac{2}{3}$

$a_{20} = \frac{1}{3}(20) - \frac{2}{3} = \frac{20}{3} - \frac{2}{3} = \frac{18}{3} = 6$

33. $a_n = a_1 + (n-1)d = 4 + (n-1)(-0.3)$
$= 4 - 0.3n + 0.3 = 4.3 - 0.3n$

$a_{20} = 4.3 - 0.3(20) = 4.3 - 6 = -1.7$

35. First find a_{20}.

$$a_{20} = 4+(20-1)6 = 4+(19)6$$
$$= 4+114 = 118$$

$$S_{20} = \frac{20}{2}(4+118) = 10(122) = 1220$$

37. First find a_{50}.

$$a_{50} = -10+(50-1)4 = -10+(49)4$$
$$= -10+196 = 186$$

$$S_{50} = \frac{50}{2}(-10+186) = 25(176) = 4400$$

39. First find a_{100}.

$$a_{100} = 1+(100-1)1 = 1+(99)1$$
$$= 1+99 = 100$$

$$S_{100} = \frac{100}{2}(1+100) = 50(101) = 5050$$

41. First find a_{60}.

$$a_{60} = 2+(60-1)2 = 2+(59)2$$
$$= 2+118 = 120$$

$$S_{60} = \frac{60}{2}(2+120) = 30(122) = 3660$$

43. The even integers between 21 and 45 start with 22 and end with 44.

$$44 = 22+(n-1)2$$
$$22 = 2(n-1)$$
$$11 = n-1$$
$$12 = n$$

$$S_{12} = \frac{12}{2}(22+44) = 6(66) = 396$$

45. $$\sum_{i=1}^{17}(5i+3) = (5(1)+3)+(5(2)+3)+(5(3)+3)+...+(5(17)+3)$$
$$= (5+3)+(10+3)+(15+3)+...+(85+3) = 8+13+18+...+88$$

$$S_{17} = \frac{17}{2}(8+88) = \frac{17}{2}(96) = 17(48) = 816$$

47. $\sum_{i=1}^{30}(-3i+5)=(-3(1)+5)+(-3(2)+5)+(-3(3)+5)+...+(-3(30)+5)$

$$=(-3+5)+(-6+5)+(-9+5)+...+(-90+5)=2+(-1)+(-4)+...+(-85)$$

$$S_{30}=\frac{30}{2}(2+(-85))=15(-83)=-1245$$

49. $\sum_{i=1}^{100}4i=4(1)+4(2)+4(3)+...+4(100)=4+8+12+...+400$

$$S_{100}=\frac{100}{2}(4+400)=50(404)=20,200$$

51. First find a_{14} and b_{12}:

$$a_{14}=a_1+(n-1)d$$
$$=1+(14-1)(-3-1)=-51$$

$$b_{12}=b_1+(n-1)d$$
$$=3+(12-1)(8-3)=58$$

So, $a_{14}+b_{12}=-51+58=7$.

53. $a_n=a_1+(n-1)d$

$$-83=1+(n-1)(-3-1)$$
$$-83=1+-4(n-1)$$
$$-84=-4n+4$$
$$-88=-4n$$
$$n=22$$

There are 22 terms.

55. $S_n=\frac{n}{2}(a_1+a_n)$

For $\{a_n\}$: $S_{14}=\frac{14}{2}(a_1+a_{14})=7(1+(-51))=-350$

For $\{b_n\}$: $S_{14}=\frac{14}{2}(b_1+b_{14})=7(3+68)=497$

So $\sum_{n=1}^{14}b_n-\sum_{n=1}^{14}a_n=497-(-350)=847$

57. Two points on the graph are $(1, 1)$ and $(2, -3)$.
Finding the slope of the line;

$$m = \frac{y_2 - y_1}{x_2 - x_1} = \frac{-3-1}{2-1} = \frac{-4}{1} = -4$$

Using the point-slope form of an equation of a line;

$$y - y_2 = m(x - x_2)$$
$$y - 1 = -4(x-1)$$
$$y - 1 = -4x + 4$$
$$y = -4x + 5$$

Thus, $f(x) = -4x + 5$.

59. Using $a_n = a_1 + (n-1)d$ and $a_2 = 4$:

$$a_2 = a_1 + (2-1)d$$
$$4 = a_1 + d$$

And since $a_6 = 16$:

$$a_6 = a_1 + (6-1)d$$
$$16 = a_1 + 5d$$

The system of equations is

$$4 = a_1 + d$$
$$16 = a_1 + 5d$$

Solving the first equation for a_1:

$$a_1 = 4 - d$$

Substituting the value into the second equation and solving for d:

$$16 = (4-d) + 5d$$
$$16 = 4 + 4d$$
$$12 = 4d$$
$$3 = d$$

Then $a_n = a_1 + (n-1)d$

$$a_n = 1 + (n-1)3$$
$$a_n = 1 + 3n - 3$$
$$a_n = 3n - 2$$

61. a. $a_n = a_1 + (n-1)d$

$$a_n = 11.0 + (n-1)0.5$$
$$= 11.0 + 0.5n - 0.5$$
$$= 0.5n + 10.5$$

b. $a_n = 0.5n + 10.5$

$$= 0.5(50) + 10.5$$
$$= 35.5$$

The percentage is projected to be 35.5% in 2019.

63. Company *A*

$$a_n = 24000 + (n-1)1600$$
$$= 24000 + 1600n - 1600$$
$$= 1600n + 22400$$

$$a_{10} = 1600(10) + 22400$$
$$= 16000 + 22400 = 38400$$

Company *B*

$$a_n = 28000 + (n-1)1000$$
$$= 28000 + 1000n - 1000$$
$$= 1000n + 27000$$

$$a_{10} = 1000(10) + 27000$$
$$= 10000 + 27000 = 37000$$

Company *A* will pay \$1400 more in year 10.

65. a. Total cost:
$\$5836 + \$6185 + \$6585 + \$7020 = \$25,626$

b. $a_1 = 395(1) + 5419 = 5814$

$$a_4 = 395(4) + 5419 = 6999$$

$$S_n = \frac{n}{2}(a_1 + a_n)$$

$$S_4 = \frac{4}{2}(5814 + 6999) = 2(12,813) = \$25,626$$

The model gives actual sum of \$25,626.

67. Answers will vary.

69. Company *A*

$$a_n = 19,000 + (n-1)2600$$
$$= 19,000 + 2600n - 2600$$
$$= 2600n + 16,400$$

$$a_{10} = 2600(10) + 16400$$
$$= 26,000 + 16,400 = 42,400$$

$$S_n = \frac{n}{2}(a_1 + a_{10})$$
$$S_{10} = \frac{10}{2}(19,000 + 42,400)$$
$$= 5(61,400) = \$307,000$$

Company *B*

$$a_n = 27,000 + (n-1)1200$$
$$= 27,000 + 1200n - 1200$$
$$= 1200n + 25,800$$

$$a_{10} = 1200(10) + 25,800$$
$$= 12,000 + 25,800$$
$$= 37,800$$

$$S_n = \frac{n}{2}(a_1 + a_{10})$$
$$S_{10} = \frac{10}{2}(27,000 + 37,800)$$
$$= 5(64,800) = \$324,000$$

Company *B* pays the greater total amount.

71. $a_{38} = a_1 + (n-1)d$

$$= 20 + (38-1)(3)$$
$$= 20 + 37(3) = 131$$

$$S_{38} = \frac{n}{2}(a_1 + a_{38})$$
$$= \frac{38}{2}(20 + 131)$$
$$= 19(151)$$
$$= 2869$$

There are 2869 seats in this section of the stadium.

73. – 75. Answers will vary.

77. Answers will vary. For example, consider Exercise 45.

$$\sum_{i=1}^{17}(5i+3)$$

```
sum(seq(5I+3,I,1
,17))
                816
```

79. makes sense

81. makes sense

83. false; Changes to make the statement true will vary. A sample change is: The common difference is –2.

85. true

87. From the sequence, we see that $a_1 = 21700$ and $d = 23172 - 21700 = 1472$.

We know that $a_n = a_1 + (n-1)d$. We can substitute what we know to find *n*.

$$314,628 = 21,700 + (n-1)1472$$
$$292,928 = (n-1)1472$$
$$\frac{292,928}{1472} = \frac{(n-1)1472}{1472}$$
$$199 = n-1$$
$$200 = n$$

314,628 is the 200th term of the sequence.

89. $1 + 3 + 5 + ... + (2n-1)$

$$S_n = \frac{n}{2}(a_1 + a_n) = \frac{n}{2}(1 + (2n-1))$$
$$= \frac{n}{2}(1 + 2n - 1) = \frac{n}{2}(2n)$$
$$= n(n) = n^2$$

90. $\log(x^2 - 25) - \log(x+5) = 3$

$$\log\left(\frac{x^2-25}{x+5}\right) = 3$$

$$\log\left(\frac{(x-5)(x+5)}{x+5}\right) = 3$$

$$\log(x-5) = 3$$

$$x - 5 = 10^3$$

$$x - 5 = 1000$$

$$x = 1005$$

The solution set is $\{1005\}$.

91. $x^2 + 3x \le 10$

Solve the related quadratic equation.

$$x^2 + 3x - 10 = 0$$

$$(x+5)(x-2) = 0$$

Apply the zero product principle.

$x + 5 = 0$ or $x - 2 = 0$

$x = -5$ $\quad$ $x = 2$

The boundary points are -5 and 2.

Interval	Test Value	Test	Conclusion
$(-\infty, -5]$	-6	$(-6)^2 + 3(-6) \le 10$ $18 \le 10$, false	$(-\infty, -5]$ does not belong to the solution set.
$[-5, 2]$	0	$0^2 + 3(0) \le 10$ $0 \le 10$, true	$[-5, 2]$ belongs to the solution set.
$[2, \infty)$	3	$3^2 + 3(3) \le 10$ $18 \le 10$, false	$[2, \infty)$ does not belong to the solution set.

The solution set is $[-5, 2]$.

92. $A = \dfrac{Pt}{P+t}$

$$A(P+t) = Pt$$

$$AP + At = Pt$$

$$AP - Pt = -At$$

$$P(A-t) = -At$$

$$P = -\frac{At}{A-t} \text{ or } \frac{At}{t-A}$$

93. $\frac{a_2}{a_1} = \frac{-2}{1} = -2$

$\frac{a_3}{a_2} = \frac{4}{-2} = -2$

$\frac{a_4}{a_3} = \frac{-8}{4} = -2$

$\frac{a_5}{a_4} = \frac{16}{-8} = -2$

The ratio of a term to the term that directly precedes it is always −2.

94. $\frac{a_2}{a_1} = \frac{3 \cdot 5^2}{3 \cdot 5^1} = 5$

$\frac{a_3}{a_2} = \frac{3 \cdot 5^3}{3 \cdot 5^2} = 5$

$\frac{a_4}{a_3} = \frac{3 \cdot 5^4}{3 \cdot 5^3} = 5$

$\frac{a_5}{a_4} = \frac{3 \cdot 5^5}{3 \cdot 5^4} = 5$

The ratio of a term to the term that directly precedes it is always 5.

95. $a_n = a_1 3^{n-1}$

$a_7 = 11 \cdot 3^{7-1} = 11 \cdot 3^6 = 11 \cdot 729 = 8019$

11.3 Check Points

1. $a_1 = 12,\ r = \frac{1}{2}$

$a_2 = 12\left(\frac{1}{2}\right)^1 = 6$

$a_3 = 12\left(\frac{1}{2}\right)^2 = \frac{12}{4} = 3$

$a_4 = 12\left(\frac{1}{2}\right)^3 = \frac{12}{8} = \frac{3}{2}$

$a_5 = 12\left(\frac{1}{2}\right)^4 = \frac{12}{16} = \frac{3}{4}$

$a_6 = 12\left(\frac{1}{2}\right)^5 = \frac{12}{32} = \frac{3}{8}$

The first six terms are $12,\ 6,\ 3,\ \frac{3}{2},\ \frac{3}{4}$, and $\frac{3}{8}$.

2. $a_1 = 5,\ r = -3$

$a_n = a_1 r^{n-1}$

$a_7 = 5(-3)^{7-1} = 5(-3)^6 = 5(729) = 3645$

The seventh term is 3645.

3. 3, 6, 12, 24, 48, ...

$r = \frac{6}{3} = 2,\ a_1 = 3$

$a_n = 3(2)^{n-1}$

$a_8 = 3(2)^{8-1} = 3(2)^7 = 3(128) = 384$

The eighth term is 384.

4. $a_1 = 2,\ r = \frac{-6}{2} = -3$

$S_n = \frac{a_1(1-r^n)}{1-r}$

$S_9 = \frac{2\left(1-(-3)^9\right)}{1-(-3)} = \frac{2(19,684)}{4} = 9842$

The sum of the first nine terms is 9842.

5. $\sum_{i=1}^{8} 2 \cdot 3^i$

$a_1 = 2 \cdot (3)^1 = 6,\ r = 3$

$S_n = \frac{a_1(1-r^n)}{1-r}$

$S_8 = \frac{6\left(1-3^8\right)}{1-3} = \frac{6(-6560)}{-2} = 19,680$

Thus, $\sum_{i=1}^{8} 2 \cdot 3^i = 19,680$.

6. $a_1 = 30,000,\ r = 1.06$

$S_n = \frac{a_1(1-r^n)}{1-r}$

$S_{30} = \frac{30,000\left(1-(1.06)^{30}\right)}{1-1.06} \approx 2,371,746$

The total lifetime salary is \$2,371,746.

7. a. $A = \frac{P\left[\left(1+\frac{r}{n}\right)^{nt}-1\right]}{\frac{r}{n}}$

$P = 100,\ r = 0.095,\ n = 12,\ t = 35$

$$A = \frac{100\left[\left(1+\frac{0.095}{12}\right)^{12\cdot 35}-1\right]}{\frac{0.095}{12}} \approx 333{,}946$$

The value of the IRA will be $333,946.

b. Interest = Value of IRA − Total deposits

$\approx \$333{,}946 - \$100\cdot 12\cdot 35$

$\approx \$333{,}946 - \$42{,}000$

$\approx \$291{,}946$

8. $3+2+\frac{4}{3}+\frac{8}{9}+\cdots$

$a_1 = 3,\ r = \frac{2}{3}$

$S = \frac{a_1}{1-r}$

$S = \frac{3}{1-\frac{2}{3}} = \frac{3}{\frac{1}{3}} = 9$

The sum of this infinite geometric series is 9.

9. $0.\overline{9} = 0.9999\cdots = \frac{9}{10}+\frac{9}{100}+\frac{9}{1000}+\cdots$

$a_1 = \frac{9}{10},\ r = \frac{1}{10}$

$$S = \frac{\frac{9}{10}}{1-\frac{1}{10}} = \frac{\frac{9}{10}}{\frac{9}{10}} = 1$$

An equivalent fraction for $0.\overline{9}$ is 1.

10. $a_1 = 1000(0.8) = 800,\ r = 0.8$

$S = \frac{800}{1-0.8} = 4000$

The total amount spent is $4000.

11.3 Concept and Vocabulary Check

1. geometric; common ratio

2. $a_1 r^{n-1}$; first term; common ratio

3. $\frac{a_1(1-r^n)}{1-r}$; first term; common ratio

4. annuity; P; r; n

5. infinite geometric series; 1; $\frac{a_1}{1-r}$; $|r| \geq 1$

6. 2; 4; 8; 16; 2

7. arithmetic

8. geometric

9. geometric

10. arithmetic

11.3 Exercise Set

1. $r = \frac{a_2}{a_1} = \frac{15}{5} = 3$

3. $r = \frac{a_2}{a_1} = \frac{30}{-15} = -2$

5. $r = \frac{a_2}{a_1} = \frac{\frac{9}{2}}{3} = \frac{9}{2}\cdot\frac{1}{3} = \frac{3}{2}$

7. $r = \frac{a_2}{a_1} = \frac{0.04}{-0.4} = -0.1$

9. The first term is 2.
The second term is $2\cdot 3 = 6$.
The third term is $6\cdot 3 = 18$.
The fourth term is $18\cdot 3 = 54$.
The fifth term is $54\cdot 3 = 162$.

11. The first term is 20.

The second term is $20\cdot\frac{1}{2} = 10$.

The third term is $10\cdot\frac{1}{2} = 5$.

The fourth term is $5\cdot\frac{1}{2} = \frac{5}{2}$.

The fifth term is $\frac{5}{2}\cdot\frac{1}{2} = \frac{5}{4}$.

13. The first term is –4.
The second term is $-4(-10)=40$.
The third term is $40(-10)=-400$.
The fourth term is $-400(-10)=4000$.
The fifth term is $4000(-10)=-40{,}000$.

15. The first term is $-\frac{1}{4}$.
The second term is $-\frac{1}{4}(-2)=\frac{1}{2}$.
The third term is $\frac{1}{2}(-2)=-1$.
The fourth term is $-1(-2)=2$.
The fifth term is $2(-2)=-4$.

17. $a_8=6(2)^{8-1}=6(2)^7=6(128)=768$

19. $a_{12}=5(-2)^{12-1}=5(-2)^{11}$
$=5(-2048)=-10{,}240$

21. $a_6=6400\left(-\frac{1}{2}\right)^{6-1}=6400\left(-\frac{1}{2}\right)^5$
$=-200$

23. $a_8=1{,}000{,}000(0.1)^{8-1}$
$=1{,}000{,}000(0.1)^7$
$=1{,}000{,}000(0.0000001)=0.1$

25. $r=\frac{a_2}{a_1}=\frac{12}{3}=4$
$a_n=a_1r^{n-1}=3(4)^{n-1}$
$a_7=3(4)^{7-1}=3(4)^6$
$=3(4096)=12{,}288$

27. $r=\frac{a_2}{a_1}=\frac{6}{18}=\frac{1}{3}$
$a_n=a_1r^{n-1}=18\left(\frac{1}{3}\right)^{n-1}$
$a_7=18\left(\frac{1}{3}\right)^{7-1}=18\left(\frac{1}{3}\right)^6$
$=18\left(\frac{1}{729}\right)=\frac{18}{729}=\frac{2}{81}$

29. $r=\frac{a_2}{a_1}=\frac{-3}{1.5}=-2$
$a_n=a_1r^{n-1}=1.5(-2)^{n-1}$
$a_7=1.5(-2)^{7-1}=1.5(-2)^6$
$=1.5(64)=96$

31. $r=\frac{a_2}{a_1}=\frac{-0.004}{0.0004}=-10$
$a_n=a_1r^{n-1}=0.0004(-10)^{n-1}$
$a_7=0.0004(-10)^{7-1}=0.0004(-10)^6$
$=0.0004(1000000)=400$

33. $r=\frac{a_2}{a_1}=\frac{6}{2}=3$

$$S_{12}=\frac{2\left(1-3^{12}\right)}{1-3}=\frac{2(1-531{,}441)}{-2}$$
$$=\frac{2(-531{,}440)}{-2}=\frac{-1{,}062{,}880}{-2}$$
$$=531{,}440$$

35. $r=\frac{a_2}{a_1}=\frac{-6}{3}=-2$

$$S_{11}=\frac{a_1\left(1-r^n\right)}{1-r}=\frac{3\left(1-(-2)^{11}\right)}{1-(-2)}$$
$$=\frac{\cancel{3}\left(1-(-2048)\right)}{\cancel{3}}=2049$$

37. $r=\frac{a_2}{a_1}=\frac{3}{-\frac{3}{2}}=3\div\left(-\frac{3}{2}\right)$
$=3\cdot\left(-\frac{2}{3}\right)=-2$

$$S_{14}=\frac{a_1\left(1-r^n\right)}{1-r}=\frac{-\frac{3}{2}\left(1-(-2)^{14}\right)}{1-(-2)}$$
$$=\frac{-\frac{3}{2}(1-(16,384))}{3}=\frac{-\frac{3}{2}(-16,383)}{3}$$
$$=-\frac{3}{2}(-16,383)\div 3=\frac{49,149}{2}\cdot\frac{1}{3}$$
$$=\frac{16,383}{2}$$

39. $\displaystyle\sum_{i=1}^{8}3^i=\frac{3\left(1-3^8\right)}{1-3}=\frac{3(1-6561)}{-2}$
$$=\frac{3(-6560)}{-2}=\frac{-19,680}{-2}=9840$$

41. $\displaystyle\sum_{i=1}^{10}5\cdot 2^i=\frac{10\left(1-2^{10}\right)}{1-2}=\frac{10(1-1024)}{-1}$
$$=\frac{10(-1023)}{-1}=10,230$$

43. $\displaystyle\sum_{i=1}^{6}\left(\frac{1}{2}\right)^{i+1}=\frac{\frac{1}{4}\left(1-\left(\frac{1}{2}\right)^6\right)}{1-\frac{1}{2}}=\frac{\frac{1}{4}\left(1-\frac{1}{64}\right)}{\frac{1}{2}}$
$$=\frac{\frac{1}{4}\left(\frac{64}{64}-\frac{1}{64}\right)}{\frac{1}{2}}=\frac{\frac{1}{4}\left(\frac{63}{64}\right)}{\frac{1}{2}}$$
$$=\frac{1}{4}\left(\frac{63}{64}\right)\div\frac{1}{2}=\frac{1}{\cancel{4}_2}\left(\frac{63}{64}\right)\cdot\frac{\cancel{2}}{1}$$
$$=\frac{63}{128}$$

45. $r=\dfrac{a_2}{a_1}=\dfrac{\frac{1}{3}}{1}=\dfrac{1}{3}$
$$S=\frac{a_1}{1-r}=\frac{1}{1-\frac{1}{3}}=\frac{1}{\frac{2}{3}}=1\div\frac{2}{3}=1\cdot\frac{3}{2}=\frac{3}{2}$$

47. $r=\dfrac{a_2}{a_1}=\dfrac{\frac{3}{4}}{3}=\dfrac{3}{4}\div 3=\dfrac{3}{4}\cdot\dfrac{1}{3}=\dfrac{1}{4}$
$$S=\frac{a_1}{1-r}=\frac{3}{1-\frac{1}{4}}=\frac{3}{\frac{3}{4}}=3\div\frac{3}{4}$$
$$=3\cdot\frac{4}{3}=\frac{12}{3}=4$$

49. $r=\dfrac{a_2}{a_1}=\dfrac{-\frac{1}{2}}{1}=-\dfrac{1}{2}$
$$S=\frac{a_1}{1-r}=\frac{1}{1-\left(-\frac{1}{2}\right)}=\frac{1}{\frac{3}{2}}=1\div\frac{3}{2}$$
$$=1\cdot\frac{2}{3}=\frac{2}{3}$$

51. $r=-0.3$
$$a_1=26(-0.3)^{1-1}=26(-0.3)^0$$
$$=26(1)=26$$
$$S=\frac{26}{1-(-0.3)}=\frac{26}{1.3}=20$$

53. $0.\overline{5}=\dfrac{a_1}{1-r}=\dfrac{\frac{5}{10}}{1-\frac{1}{10}}=\dfrac{\frac{5}{10}}{\frac{9}{10}}=\dfrac{5}{10}\div\dfrac{9}{10}$
$$=\frac{5}{10}\cdot\frac{10}{9}=\frac{5}{9}$$

55. $0.\overline{47}=\dfrac{a_1}{1-r}=\dfrac{\frac{47}{100}}{1-\frac{1}{100}}=\dfrac{\frac{47}{100}}{\frac{99}{100}}$
$$=\frac{47}{100}\div\frac{99}{100}=\frac{47}{100}\cdot\frac{100}{99}=\frac{47}{99}$$

57. $0.\overline{257}=\dfrac{a_1}{1-r}=\dfrac{\frac{257}{1000}}{1-\frac{1}{1000}}=\dfrac{\frac{257}{1000}}{\frac{999}{1000}}$
$=\dfrac{257}{1000}\div\dfrac{999}{1000}=\dfrac{257}{1000}\cdot\dfrac{1000}{999}$
$=\dfrac{257}{999}$

59. The sequence is arithmetic with common difference $d=1$.

61. The sequence is geometric with common ratio $r=2$.

63. The sequence is neither arithmetic nor geometric.

65. First find a_{10} and b_{10}:

$$a_{10}=a_1r^{n-1}$$
$$=(-5)\left(\frac{10}{-5}\right)^{10-1}=(-5)(-2)^9=2560$$
$$b_{10}=b_1+(n-1)d$$
$$=10+(10-1)(-5-10)$$
$$=10+(9)(-15)=-125$$

So, $a_{10}+b_{10}=2560+(-125)=2435$.

67. From Exercise 65, $a_{10}=2560$ and $b_{10}=-125$.

For $\{a_n\}$,
$$r=\frac{10}{-5}=-2$$
$$S_{10}=\frac{a_1(1-r^n)}{1-r}=\frac{(-5)\left(1-(-2)^{10}\right)}{1-(-2)}$$
$$=\frac{(-5)(-1023)}{3}=1705$$

For $\{b_n\}$,
$$S_n=\frac{n}{2}(b_1+b_n)=\frac{10}{2}(10+(-125))$$
$$=5(-115)=-575$$

So, $\displaystyle\sum_{n=1}^{10}a_n-\sum_{n=1}^{10}b_n=1705-(-575)=2280$

69. For $\{a_n\}$,
$$S_6=\frac{a_1(1-r^n)}{1-r}=\frac{(-5)\left(1-(-2)^6\right)}{1-(-2)}$$
$$=\frac{(-5)(-63)}{3}=105$$

For $\{c_n\}$,
$$S=\frac{a_1}{1-r}=\frac{-2}{1-\frac{1}{-2}}=\frac{-2}{\frac{3}{2}}=-\frac{4}{3}$$

So, $S_6\cdot S=105\left(-\dfrac{4}{3}\right)=-140$

71. It is given that $a_4=27$. Using the formula $a_n=a_1r^{n-1}$ when $n=4$ we have:
$$27=8r^{4-1}$$
$$\frac{27}{8}=r^3$$
$$r=\sqrt[3]{\frac{27}{8}}=\frac{3}{2}$$

Then
$$a_n=a_1r^{n-1}$$
$$a_2=8\left(\frac{3}{2}\right)^{2-1}=8\left(\frac{3}{2}\right)=12$$
$$a_3=8\left(\frac{3}{2}\right)^{3-1}=8\left(\frac{3}{2}\right)^2=8\left(\frac{9}{4}\right)=18$$

73. Find the total value of the lump-sum investment.
$A=P(1+r)^t=30,000(1+0.05)^{20}\approx 79,599$
Find the total value of the annuity.
$$A=\frac{P\left[\left(1+\frac{r}{n}\right)^{nt}-1\right]}{\frac{r}{n}}=\frac{1500\left[\left(1+\frac{0.05}{1}\right)^{20}-1\right]}{\frac{0.05}{1}}\approx 49,599$$

$\$79,599-\$49,599=\$30,000$

You will have $30,000 more from the lump-sum investment.

75. $r = \frac{a_2}{a_1} = \frac{2}{1} = 2$

$a_{15} = 1(2)^{15-1} = (2)^{14} = 16,384$

On the fifteenth day, you will put aside $16,384 for savings.

77. $r = 1.04$

$$\begin{aligned} a_7 &= 3,000,000(1.04)^{7-1} \\ &= 3,000,000(1.04)^6 \\ &= 3,000,000(1.265319) \\ &= 3,795,957 \end{aligned}$$

The athlete's salary for year 7 will be $3,795,957.

79. a. $r_{2003 \text{ to } 2004} = \frac{35.89}{35.48} \approx 1.01$

$r_{2004 \text{ to } 2005} = \frac{36.13}{35.89} \approx 1.01$

$r_{2005 \text{ to } 2006} = \frac{36.46}{36.13} \approx 1.01$

r is approximately 1.01 for each division.

b. $a_n = a_1 r^{n-1}$

$a_n = 35.48(1.01)^{n-1}$

c. Since year 2010 is the 8th term, find a_8.

$a_n = 35.48(1.01)^{n-1}$

$a_8 = 35.48(1.01)^{8-1} \approx 38.04$

The population of California will be approximately 38.04 million in 2010.

81. $r = \frac{a_2}{a_1} = \frac{2}{1} = 2$

$$\begin{aligned} S_{15} &= \frac{a_1(1-r^n)}{1-r} = \frac{1(1-(2)^{15})}{1-2} \\ &= \frac{(1-32,768)}{-1} = \frac{(-32,767)}{-1} = 32,767 \end{aligned}$$

Your savings will be $32,767 over the 15 days.

83. $r = 1.05$

$$S_{20} = \frac{a_1(1-r^n)}{1-r} = \frac{24,000\left(1-(1.05)^{20}\right)}{1-1.05}$$
$$= \frac{24,000(1-2.6533)}{-0.05}$$
$$= \frac{24,000(-1.6533)}{-0.05} = 793,583$$

The total lifetime salary over the 20 years is $793,583.

85. $r = 0.9$

$$S_{10} = \frac{a_1(1-r^n)}{1-r} = \frac{20\left(1-(0.9)^{10}\right)}{1-0.9}$$
$$= \frac{20(1-0.348678)}{0.1}$$
$$= \frac{20(0.651322)}{0.1} = 130.264$$

After 10 swings, the pendulum covers a distance of approximately 130.26 inches.

87. a. $$A = \frac{P\left[\left(1+\frac{r}{n}\right)^{nt} - 1\right]}{\frac{r}{n}} = \frac{2000\left[\left(1+\frac{0.075}{1}\right)^{5} - 1\right]}{\frac{0.075}{1}} \approx \$11,617$$

b. $\$11,617 - 5\times\$2000 = \$1617$

89. a. $$A = \frac{P\left[\left(1+\frac{r}{n}\right)^{nt} - 1\right]}{\frac{r}{n}} = \frac{50\left[\left(1+\frac{0.055}{12}\right)^{12\times 40} - 1\right]}{\frac{0.055}{12}} \approx \$87,052$$

b. $\$87,052 - \$50\cdot 12\cdot 40 = \$63,052$

91. a. $$A = \frac{P\left[\left(1+\frac{r}{n}\right)^{nt} - 1\right]}{\frac{r}{n}} = \frac{10,000\left[\left(1+\frac{0.105}{4}\right)^{4\times 10} - 1\right]}{\frac{0.105}{4}} \approx \$693,031$$

b. $\$693,031 - \$10,000\cdot 4\cdot 10 = \$293,031$

93. $r = 60\% = 0.6$

$a_1 = 6(.6) = 3.6$

$$S = \frac{3.6}{1-0.6} = \frac{3.6}{0.4} = 9$$

The total economic impact of the factory will be $9 million per year.

95. $r = \dfrac{1}{4}$

$$S = \frac{\frac{1}{4}}{1-\frac{1}{4}} = \frac{\frac{1}{4}}{\frac{3}{4}} = \frac{1}{4} \div \frac{3}{4} = \frac{1}{4} \cdot \frac{4}{3} = \frac{1}{3}$$

Eventually $\frac{1}{3}$ of the largest square will be shaded.

97. – 103. Answers will vary.

105. Answers will vary. For example, consider Exercise 25.

$a_n = 3(4)^{n-1}$

```
seq(3(4)^(N-1),N
,7,7)
                {12288}
```

107. $f(x) = \dfrac{2\left[1-\left(\frac{1}{3}\right)^x\right]}{1-\frac{1}{3}}$

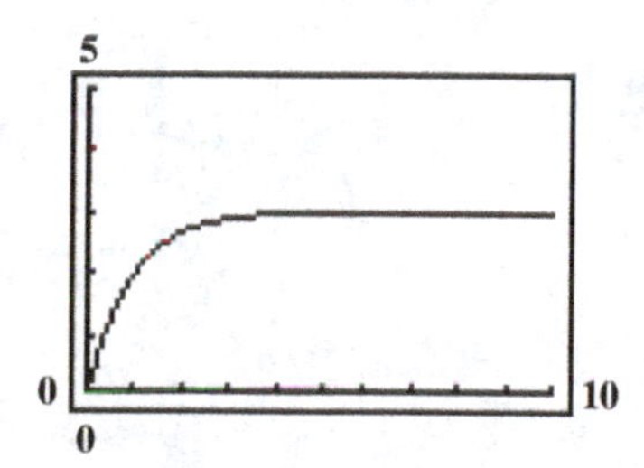

$$S = \frac{2}{1-\frac{1}{3}} = \frac{2}{\frac{2}{3}} = 2 \div \frac{2}{3} = 2 \cdot \frac{3}{2} = 3$$

The sum of the series and the asymptote of the function are both 3.

109. makes sense

111. makes sense

113. false; Changes to make the statement true will vary. A sample change is: The sequence is not geometric. There is not a common ratio.

115. false; Changes to make the statement true will vary. A sample change is: The sum of the sequence is $\dfrac{10}{1-\left(-\frac{1}{2}\right)}$.

117.
$$\begin{aligned} S &= \frac{a_1}{1-r} \\ 20,000 &= \frac{x}{1-0.9} \\ 20,000 &= \frac{x}{0.1} \\ 20,000(0.1) &= x \\ 2000 &= x \end{aligned}$$

To keep 20,000 flies in the population, 2000 flies should be released each day.

119.
$$\begin{aligned} \sqrt{28} - 3\sqrt{7} + \sqrt{63} &= \sqrt{4\cdot 7} - 3\sqrt{7} + \sqrt{9\cdot 7} \\ &= 2\sqrt{7} - 3\sqrt{7} + 3\sqrt{7} \\ &= 2\sqrt{7} \end{aligned}$$

120.
$$2x^2 = 4 - x$$

$2x^2 + x - 4 = 0$
$a = 2 \quad b = 1 \quad c = -4$
Solve using the quadratic formula.

$$\begin{aligned} x &= \frac{-1 \pm \sqrt{1^2 - 4(2)(-4)}}{2(2)} \\ &= \frac{-1 \pm \sqrt{1+32}}{4} = \frac{-1 \pm \sqrt{33}}{4} \end{aligned}$$

The solution set is $\left\{ \frac{-1 \pm \sqrt{33}}{4} \right\}$.

121.
$$\begin{aligned} \frac{6}{\sqrt{3} - \sqrt{5}} &= \frac{6}{\sqrt{3} - \sqrt{5}} \cdot \frac{\sqrt{3} + \sqrt{5}}{\sqrt{3} + \sqrt{5}} \\ &= \frac{6\left(\sqrt{3} + \sqrt{5}\right)}{3 - 5} \\ &= \frac{6\left(\sqrt{3} + \sqrt{5}\right)}{-2} \\ &= -3\left(\sqrt{3} + \sqrt{5}\right) \end{aligned}$$

122. The exponents begin with the exponent on $a + b$ and decrease by 1 in each successive term.

123. The exponents begin with 0, increase by 1 in each successive term, and end with the exponent on $a + b$.

124. The sum of the exponents is the exponent on $a + b$.

Mid-Chapter Check Point – Chapter 11

1. $a_n = (-1)^{n+1}\dfrac{n}{(n-1)!}$

$a_1 = (-1)^{1+1}\dfrac{1}{(1-1)!} = (-1)^2\dfrac{1}{0!} = 1\cdot 1 = 1$

$a_2 = (-1)^{2+1}\dfrac{2}{(2-1)!} = (-1)^3\dfrac{2}{1!} = (-1)(2) = -2$

$a_3 = (-1)^{3+1}\dfrac{3}{(3-1)!} = (-1)^4\dfrac{3}{2!} = 1\cdot\dfrac{3}{2} = \dfrac{3}{2}$

$a_4 = (-1)^{4+1}\dfrac{4}{(4-1)!} = (-1)^5\dfrac{4}{3!} = (-1)\dfrac{4}{6} = -\dfrac{2}{3}$

$a_5 = (-1)^{5+1}\dfrac{5}{(5-1)!} = (-1)^6\dfrac{5}{4!} = 1\cdot\dfrac{5}{24} = \dfrac{5}{24}$

2. Using $a_n = a_1 + (n-1)d$;

$a_1 = 5$

$a_2 = 5 + (2-1)(-3) = 5 + 1(-3) = 5 - 3 = 2$

$a_3 = 5 + (3-1)(-3) = 5 + 2(-3) = 5 - 6 = -1$

$a_4 = 5 + (4-1)(-3) = 5 + 3(-3) = 5 - 9 = -4$

$a_5 = 5 + (5-1)(-3) = 5 + 4(-3) = 5 - 12 = -7$

3. Using $a_n = a_1 r^{n-1}$;

$a_1 = 5$

$a_2 = 5(-3)^{2-1} = 5(-3)^1 = 5(-3) = -15$

$a_3 = 5(-3)^{3-1} = 5(-3)^2 = 5(9) = 45$

$a_4 = 5(-3)^{4-1} = 5(-3)^3 = 5(-27) = -135$

$a_5 = 5(-3)^{5-1} = 5(-3)^4 = 5(81) = 405$

4. $d = a_2 - a_1 = 6 - 2 = 4$

$$\begin{aligned} a_n &= a_1 + (n-1)d \\ &= 2 + (n-1)4 \\ &= 2 + 4n - 4 \\ &= 4n - 2 \end{aligned}$$

$a_{20} = 4(20) - 2 = 78$

5. $r = \dfrac{a_2}{a_1} = \dfrac{6}{3} = 2$

$a_n = a_1 r^{n-1} = 3(2)^{n-1}$

$a_{10} = 3(2)^{10-1} = 3(2)^9 = 1536$

6. $d = a_2 - a_1 = 1 - \frac{3}{2} = -\frac{1}{2}$

$$a_n = a_1 + (n-1)d = \frac{3}{2} + (n-1)\left(-\frac{1}{2}\right)$$
$$= \frac{3}{2} - \frac{1}{2}n + \frac{1}{2} = -\frac{1}{2}n + 2$$
$$a_{30} = -\frac{1}{2}(30) + 2 = -15 + 2 = -13$$

7. First find r;

$$r = \frac{a_2}{a_1} = \frac{10}{5} = 2$$
$$S_{10} = \frac{a_1(1-r^n)}{1-r} = \frac{5(1-(2)^{10})}{1-(2)}$$
$$= \frac{5(1-1024)}{-1} = \frac{5(-1023)}{-1} = 5115$$

8. First find a_{10};

$$d = a_2 - a_1 = 0 - (-2) = 2$$
$$a_{50} = a_1 + (n-1)d$$
$$= -2 + (50-1)(2) = -2 + 49(2) = 96$$
$$S_{50} = \frac{n}{2}(a_1 + a_n) = \frac{50}{2}(-2 + 96) = 25(94)$$
$$= 2350$$

9. First find r;

$$r = \frac{a_2}{a_1} = \frac{40}{-20} = -2$$
$$S_{10} = \frac{a_1(1-r^n)}{1-r} = \frac{-20(1-(-2)^{10})}{1-(-2)}$$
$$= \frac{-20(-1023)}{3} = \frac{20460}{3} = 6820$$

10. First find a_{100};

$$d = a_2 - a_1 = -2 - 4 = -6$$
$$a_{100} = a_1 + (n-1)d = 4 + (100-1)(-6)$$
$$= 4 + 99(-6) = -590$$
$$S_{100} = \frac{n}{2}(a_1 + a_n) = \frac{100}{2}(4 - 590)$$
$$= 50(-586)$$
$$= -29,300$$

11. $\sum_{i=1}^{4}(i+4)(i-1) = (1+4)(1-1) + (2+4)(2-1) + (3+4)(3-1) + (4+4)(4-1)$

$$= 5(0) + 6(1) + 7(2) + 8(3) = 0 + 6 + 14 + 24 = 44$$

12. $\sum_{i=1}^{50}(3i-2)=(3\cdot 1-2)+(3\cdot 2-2)+(3\cdot 3-2)+...+(3\cdot 50-2)$

$=(3-2)+(6-2)+(9-2)+...+(150-2)$

$=1+4+7+...+148$

The sum of this arithmetic sequence is given by $S_n=\frac{n}{2}(a_1+a_n)$;

$S_{50}=\frac{50}{2}(1+148)=25(149)=3725$

13. $\sum_{i=1}^{6}\left(\frac{3}{2}\right)^i$

$=\left(\frac{3}{2}\right)^1+\left(\frac{3}{2}\right)^2+\left(\frac{3}{2}\right)^3+\left(\frac{3}{2}\right)^4+\left(\frac{3}{2}\right)^5+\left(\frac{3}{2}\right)^6$

$=\frac{3}{2}+\frac{9}{4}+\frac{27}{8}+\frac{81}{16}+\frac{243}{32}+\frac{729}{64}=\frac{1995}{64}$

14. $\sum_{i=1}^{\infty}\left(-\frac{2}{5}\right)^{i-1}$

$=\left(-\frac{2}{5}\right)^{1-1}+\left(-\frac{2}{5}\right)^{2-1}+\left(-\frac{2}{5}\right)^{3-1}+...$

$=\left(-\frac{2}{5}\right)^{0}+\left(-\frac{2}{5}\right)^{1}+\left(-\frac{2}{5}\right)^{2}+...$

$=1+\left(-\frac{2}{5}\right)+\frac{4}{25}+...$

This is an infinite geometric sequence with $r=\frac{a_2}{a_1}=\frac{-\frac{2}{5}}{1}=-\frac{2}{5}$.

Using $S=\frac{a_1}{1-r}=\frac{1}{1-\left(-\frac{2}{5}\right)}=\frac{1}{\frac{7}{5}}=\frac{5}{7}$

15. $0.\overline{45}=\frac{a_1}{1-r}=\frac{\frac{45}{100}}{1-\frac{1}{100}}=\frac{\frac{45}{100}}{\frac{99}{100}}$

$=\frac{45}{100}\div\frac{99}{100}=\frac{45}{100}\cdot\frac{100}{99}=\frac{45}{99}=\frac{5}{11}$

16. Answers will vary. An example is $\sum_{i=1}^{18}\frac{i}{i+2}$.

17. The arithmetic sequence is 16, 48, 80, 112, ….

First find a_{15} where $d = a_2 - a_1 = 48 - 16 = 32$.

$$\begin{aligned} a_{15} &= a_1 + (n-1)d = 16 + (15-1)(32) \\ &= 16 + 14(32) \\ &= 16 + 448 = 464 \end{aligned}$$

The distance the skydiver falls during the 15th second is 464 feet.

$$\begin{aligned} S_{15} &= \frac{n}{2}(a_1 + a_n) = \frac{15}{2}(16 + 464) \\ &= 7.5(480) = 3600 \end{aligned}$$

The total distance the skydiver falls in 15 seconds is 3600 feet.

18. $$\begin{aligned} A &= P(1+r)^t \\ &= 500,000(1+0.10)^8 \\ &\approx 1,071,794 \end{aligned}$$

The value of the house after 8 years will be $1,071,794.

11.4 Check Points

1. a. $\binom{6}{3} = \frac{6!}{3!(6-3)!} = \frac{6!}{3!3!} = \frac{5\cdot 4}{1} = 20$

b. $\binom{6}{0} = \frac{6!}{0!(6-0)!} = \frac{6!}{6!} = 1$

c. $\binom{8}{2} = \frac{8!}{2!(8-2)!} = \frac{8!}{2!6!} = \frac{8\cdot 7}{2} = 28$

d. $\binom{3}{3} = \frac{3!}{3!(3-3)!} = \frac{3!}{3!0!} = \frac{3!}{3!} = 1$

2. $(x+1)^4 = \binom{4}{0}x^4 + \binom{4}{1}x^3 + \binom{4}{2}x^2 + \binom{4}{3}x + \binom{4}{4} = x^4 + 4x^3 + 6x^2 + 4x + 1$

3. $$\begin{aligned} (x-2y)^5 &= \binom{5}{0}x^5(-2y)^0 + \binom{5}{1}x^4(-2y)^1 + \binom{5}{2}x^3(-2y)^2 + \binom{5}{3}x^2(-2y)^3 + \binom{5}{4}x(-2y)^4 + \binom{5}{5}x^0(-2y)^5 \\ &= x^5 - 5x^4(2y) + 10x^3(4y^2) - 10x^2(8y^3) + 5x(16y^4) - 32y^5 \\ &= x^5 - 10x^4y + 40x^3y^2 - 80x^2y^3 + 80xy^4 - 32y^5 \end{aligned}$$

4. $(2x+y)^9$

fifth term $= \binom{9}{4}(2x)^5y^4 = \frac{9!}{4!5!}(32x^5)y^4 = 4032x^5y^4$

11.4 Concept and Vocabulary Check

1. binomial

2. $\dfrac{8!}{2!6!}$

3. $\dfrac{m!}{r!(n-r)!}$

4. $\binom{5}{1}$; $\binom{5}{2}$; $\binom{5}{3}$; $\binom{5}{4}$; $\binom{5}{5}$

5. $\binom{n}{1}$; $\binom{n}{2}$; $\binom{n}{3}$; $\binom{n}{n}$; n

6. Binomial

7. $a^{n-r}b^{r}$

11.4 Exercise Set

1. $\binom{8}{3}=\dfrac{8!}{3!(8-3)!}=\dfrac{8!}{3!5!}=\dfrac{8\cdot 7\cdot \cancel{6}\cdot \cancel{5!}}{\cancel{3\cdot 2}\cdot 1\cdot \cancel{5!}}=56$

3. $\binom{12}{1}=\dfrac{12!}{1!(12-1)!}=\dfrac{12\cdot \cancel{11!}}{1\cdot \cancel{11!}}=12$

5. $\binom{6}{6}=\dfrac{\cancel{6!}}{\cancel{6!}(6-6)!}=\dfrac{1}{0!}=\dfrac{1}{1}=1$

7. $\binom{100}{2}=\dfrac{100!}{2!(100-2)!}=\dfrac{100\cdot 99\cdot \cancel{98!}}{2\cdot 1\cdot \cancel{98!}}$
$=4950$

9. Applying the Binomial Theorem to $(x+2)^3$, we have $a=x$, $b=2$, and $n=3$.

$$(x+2)^3=\binom{3}{0}x^3+\binom{3}{1}x^2(2)+\binom{3}{2}x(2)^2+\binom{3}{3}2^3$$
$$=\frac{3!}{0!(3-0)!}x^3+\frac{3!}{1!(3-1)!}2x^2+\frac{3!}{2!(3-2)!}4x+\frac{3!}{3!(3-3)!}8$$
$$=\frac{\cancel{3!}}{1\cdot \cancel{3!}}x^3+\frac{3\cdot \cancel{2!}}{1\cdot \cancel{2!}}2x^2+\frac{3\cdot \cancel{2!}}{\cancel{2!}1!}4x+\frac{\cancel{3!}}{\cancel{3!}0!}8=x^3+3\left(2x^2\right)+3(4x)+1(8)$$
$$=x^3+6x^2+12x+8$$

11. Applying the Binomial Theorem to $(3x+y)^3$, we have $a=3x$, $b=y$, and $n=3$.

$$(3x+y)^3=\binom{3}{0}(3x)^3+\binom{3}{1}(3x)^2y+\binom{3}{2}(3x)y^2+\binom{3}{3}y^3$$
$$=\frac{3!}{0!(3-0)!}27x^3+\frac{3!}{1!(3-1)!}9x^2y+\frac{3!}{2!(3-2)!}3xy^2+\frac{3!}{3!(3-3)!}y^3$$
$$=\frac{\cancel{3!}}{1\cdot\cancel{3!}}27x^3+\frac{3\cdot\cancel{2!}}{1\cdot\cancel{2!}}9x^2y+\frac{3\cdot\cancel{2!}}{\cancel{2!}1!}3xy^2+\frac{\cancel{3!}}{\cancel{3!}0!}y^3=27x^3+3\left(9x^2y\right)+3\left(3xy^2\right)+1\left(y^3\right)$$
$$=27x^3+27x^2y+9xy^2+y^3$$

13. Applying the Binomial Theorem to $(5x-1)^3$, we have $a=5x$, $b=-1$, and $n=3$.

$$(5x-1)^3=\binom{3}{0}(5x)^3+\binom{3}{1}(5x)^2(-1)+\binom{3}{2}(5x)(-1)^2+\binom{3}{3}(-1)^3$$
$$=\frac{3!}{0!(3-0)!}125x^3-\frac{3!}{1!(3-1)!}25x^2+\frac{3!}{2!(3-2)!}5x(1)-\frac{3!}{3!(3-3)!}$$
$$=\frac{\cancel{3!}}{1\cdot\cancel{3!}}125x^3-\frac{3\cdot\cancel{2!}}{1\cdot\cancel{2!}}25x^2+\frac{3\cdot\cancel{2!}}{\cancel{2!}1!}5x-\frac{\cancel{3!}}{\cancel{3!}0!}=125x^3-3\left(25x^2\right)+3(5x)-1$$
$$=125x^3-75x^2+15x-1$$

15. Applying the Binomial Theorem to $(2x+1)^4$, we have $a=2x$, $b=1$, and $n=4$.

$$(2x+1)^4=\binom{4}{0}(2x)^4+\binom{4}{1}(2x)^3+\binom{4}{2}(2x)^2+\binom{4}{3}2x+\binom{4}{4}$$
$$=\frac{4!}{0!(4-0)!}16x^4+\frac{4!}{1!(4-1)!}8x^3\cdot1+\frac{4!}{2!(4-2)!}4x^2\cdot1^2+\frac{4!}{3!(4-3)!}2x\cdot1^3+\frac{4!}{4!(4-4)!}\cdot1^4$$
$$=\frac{\cancel{4!}}{0!\cancel{4!}}16x^4+\frac{4!}{1!3!}8x^3\cdot1+\frac{4!}{2!2!}4x^2\cdot1+\frac{4!}{3!1!}2x\cdot1+\frac{\cancel{4!}}{\cancel{4!}0!}\cdot1$$
$$=1\left(16x^4\right)+\frac{4\cdot\cancel{3!}}{1\cdot\cancel{3!}}8x^3+\frac{4\cdot3\cdot\cancel{2!}}{2\cdot1\cdot\cancel{2!}}4x^2+\frac{4\cdot\cancel{3!}}{\cancel{3!}\cdot1}2x+1=16x^4+4\left(8x^3\right)+6\left(4x^2\right)+4(2x)+1$$
$$=16x^4+32x^3+24x^2+8x+1$$

17. Applying the Binomial Theorem to $\left(x^2+2y\right)^4$, we have $a=x^2$, $b=2y$, and $n=4$.

$$\left(x^2+2y\right)^4=\binom{4}{0}\left(x^2\right)^4+\binom{4}{1}\left(x^2\right)^3(2y)+\binom{4}{2}\left(x^2\right)^2(2y)^2+\binom{4}{3}x^2(2y)^3+\binom{4}{4}(2y)^4$$
$$=\frac{4!}{0!(4-0)!}x^8+\frac{4!}{1!(4-1)!}2x^6y+\frac{4!}{2!(4-2)!}4x^4y^2+\frac{4!}{3!(4-3)!}8x^2y^3+\frac{4!}{4!(4-4)!}16y^4$$
$$=\frac{\cancel{4!}}{0!\cancel{4!}}x^8+\frac{4!}{1!3!}2x^6y+\frac{4!}{2!2!}4x^4y^2+\frac{4!}{3!1!}8x^2y^3+\frac{\cancel{4!}}{\cancel{4!}0!}16y^4$$
$$=1\left(x^8\right)+\frac{4\cdot\cancel{3!}}{1\cdot\cancel{3!}}2x^6y+\frac{4\cdot3\cdot\cancel{2!}}{2\cdot1\cdot\cancel{2!}}4x^4y^2+\frac{4\cdot\cancel{3!}}{\cancel{3!}\cdot1}8x^2y^3+16y^4$$
$$=x^8+4\left(2x^6y\right)+6\left(4x^4y^2\right)+4\left(8x^2y^3\right)+16y^4=x^8+8x^6y+24x^4y^2+32x^2y^3+16y^4$$

19. Applying the Binomial Theorem to $(y-3)^4$, we have $a = y,\ b = -3,$ and $n = 4.$

$$(y-3)^4 = \binom{4}{0}y^4 + \binom{4}{1}y^3(-3) + \binom{4}{2}y^2(-3)^2 + \binom{4}{3}y(-3)^3 + \binom{4}{4}(-3)^4$$

$$= \frac{4!}{0!(4-0)!}y^4 - \frac{4!}{1!(4-1)!}3y^3 + \frac{4!}{2!(4-2)!}9y^2 - \frac{4!}{3!(4-3)!}27y + \frac{4!}{4!(4-4)!}81$$

$$= \frac{\cancel{4!}}{0!\cancel{4!}}y^4 - \frac{4!}{1!3!}3y^3 + \frac{4!}{2!2!}9y^2 - \frac{4!}{3!1!}27y + \frac{\cancel{4!}}{\cancel{4!}0!}81$$

$$= 1\left(y^4\right) - \frac{4\cdot\cancel{3!}}{1\cdot\cancel{3!}}3y^3 + \frac{4\cdot3\cdot\cancel{2!}}{2\cdot1\cdot\cancel{2!}}9y^2 - \frac{4\cdot\cancel{3!}}{\cancel{3!}\cdot1}27y + 81$$

$$= y^4 - 4\left(3y^3\right) + 6\left(9y^2\right) - 4(27y) + 81 = y^4 - 12y^3 + 54y^2 - 108y + 81$$

21. Applying the Binomial Theorem to $\left(2x^3-1\right)^4$, we have $a = 2x^3,\ b = -1,$ and $n = 4.$

$$\left(2x^3-1\right)^4 = \binom{4}{0}\left(2x^3\right)^4 + \binom{4}{1}\left(2x^3\right)^3(-1) + \binom{4}{2}\left(2x^3\right)^2(-1)^2 + \binom{4}{3}\left(2x^3\right)(-1)^3 + \binom{4}{4}(-1)^4$$

$$= \frac{4!}{0!(4-0)!}16x^{12} - \frac{4!}{1!(4-1)!}8x^9 + \frac{4!}{2!(4-2)!}4x^6 - \frac{4!}{3!(4-3)!}2x^3 + \frac{4!}{4!(4-4)!}$$

$$= \frac{\cancel{4!}}{0!\cancel{4!}}16x^{12} - \frac{4!}{1!3!}8x^9 + \frac{4!}{2!2!}4x^6 - \frac{4!}{3!1!}2x^3 + \frac{\cancel{4!}}{\cancel{4!}0!}$$

$$= 1\left(16x^{12}\right) - \frac{4\cdot\cancel{3!}}{1\cdot\cancel{3!}}8x^9 + \frac{4\cdot3\cdot\cancel{2!}}{2\cdot1\cdot\cancel{2!}}4x^6 - \frac{4\cdot\cancel{3!}}{\cancel{3!}\cdot1}2x^3 + 1$$

$$= 16x^{12} - 4\left(8x^9\right) + 6\left(4x^6\right) - 4\left(2x^3\right) + 1 = 16x^{12} - 32x^9 + 24x^6 - 8x^3 + 1$$

23. Applying the Binomial Theorem to $(c+2)^5$, we have $a = c,\ b = 2,$ and $n = 5.$

$$(c+2)^5 = \binom{5}{0}c^5 + \binom{5}{1}c^4(2) + \binom{5}{2}c^3(2)^2 + \binom{5}{3}c^2(2)^3 + \binom{5}{4}c(2)^4 + \binom{5}{5}2^5$$

$$= \frac{\cancel{5!}}{0!\cancel{5!}}c^5 + \frac{5!}{1!(5-1)!}2c^4 + \frac{5!}{2!(5-2)!}4c^3 + \frac{5!}{3!(5-3)!}8c^2 + \frac{5!}{4!(5-4)!}16c + \frac{5!}{5!(5-5)!}32$$

$$= 1c^5 + \frac{5\cdot\cancel{4!}}{1\cdot\cancel{4!}}2c^4 + \frac{5\cdot4\cdot\cancel{3!}}{2\cdot1\cdot\cancel{3!}}4c^3 + \frac{5\cdot4\cdot\cancel{3!}}{\cancel{3!}2\cdot1}8c^2 + \frac{5\cdot\cancel{4!}}{\cancel{4!}\cdot1}16c + \frac{\cancel{5!}}{\cancel{5!}0!}32$$

$$= c^5 + 5\left(2c^4\right) + 10\left(4c^3\right) + 10\left(8c^2\right) + 5(16c) + 1(32) = c^5 + 10c^4 + 40c^3 + 80c^2 + 80c + 32$$

25. Applying the Binomial Theorem to $(x-1)^5$, we have $a=x,\ b=-1,$ and $n=5$.

$$(x-1)^5=\binom{5}{0}x^5+\binom{5}{1}x^4(-1)+\binom{5}{2}x^3(-1)^2+\binom{5}{3}x^2(-1)^3+\binom{5}{4}x(-1)^4+\binom{5}{5}(-1)^5$$
$$=\frac{5!}{0!5!}x^5-\frac{5!}{1!(5-1)!}x^4+\frac{5!}{2!(5-2)!}x^3-\frac{5!}{3!(5-3)!}x^2+\frac{5!}{4!(5-4)!}x-\frac{5!}{5!(5-5)!}$$
$$=1x^5-\frac{5\cdot 4!}{1\cdot 4!}x^4+\frac{5\cdot 4\cdot 3!}{2\cdot 1\cdot 3!}x^3-\frac{5\cdot 4\cdot 3!}{3!2\cdot 1}x^2+\frac{5\cdot 4!}{4!\cdot 1}x-\frac{5!}{5!0!}$$
$$=x^5-5x^4+10x^3-10x^2+5x-1$$

27. Applying the Binomial Theorem to $(3x-y)^5$, we have $a=3x,\ b=-y,$ and $n=5$.

$$(3x-y)^5$$
$$=\binom{5}{0}(3x)^5+\binom{5}{1}(3x)^4(-y)+\binom{5}{2}(3x)^3(-y)^2+\binom{5}{3}(3x)^2(-y)^3+\binom{5}{4}3x(-y)^4+\binom{5}{5}(-y)^5$$
$$=\frac{5!}{0!5!}243x^5-\frac{5!}{1!(5-1)!}81x^4y+\frac{5!}{2!(5-2)!}27x^3y^2$$
$$-\frac{5!}{3!(5-3)!}9x^2y^3+\frac{5!}{4!(5-4)!}3xy^4-\frac{5!}{5!(5-5)!}y^5$$
$$=243x^5-\frac{5\cdot 4!}{1\cdot 4!}81x^4y+\frac{5\cdot 4\cdot 3!}{2\cdot 1\cdot 3!}27x^3y^2-\frac{5\cdot 4\cdot 3!}{3!2\cdot 1}9x^2y^3+\frac{5\cdot 4!}{4!\cdot 1}3xy^4-\frac{5!}{5!0!}y^5$$
$$=243x^5-5(81x^4y)+10(27x^3y^2)-10(9x^2y^3)+5(3xy^4)-y^5$$
$$=243x^5-405x^4y+270x^3y^2-90x^2y^3+15xy^4-y^5$$

29. Applying the Binomial Theorem to $(2a+b)^6$, we have $a=2a,\ b=b,$ and $n=6$.

$$(2a+b)^6$$
$$=\binom{6}{0}(2a)^6+\binom{6}{1}(2a)^5b+\binom{6}{2}(2a)^4b^2+\binom{6}{3}(2a)^3b^3+\binom{6}{4}(2a)^2b^4+\binom{6}{5}2ab^5+\binom{6}{6}b^6$$
$$=\frac{6!}{0!(6-0)!}64a^6+\frac{6!}{1!(6-1)!}32a^5b+\frac{6!}{2!(6-2)!}16a^4b^2+\frac{6!}{3!(6-3)!}8a^3b^3$$
$$+\frac{6!}{4!(6-4)!}4a^2b^4+\frac{6!}{5!(6-5)!}2ab^5+\frac{6!}{6!(6-6)!}b^6$$
$$=\frac{6!}{16!}64a^6+\frac{6\cdot 5!}{1\cdot 5!}32a^5b+\frac{6\cdot 5\cdot 4!}{2\cdot 1\cdot 4!}16a^4b^2+\frac{6\cdot 5\cdot 4\cdot 3!}{3\cdot 2\cdot 1\cdot 3!}8a^3b^3$$
$$+\frac{6\cdot 5\cdot 4!}{4!2\cdot 1}4a^2b^4+\frac{6\cdot 5!}{5!1}2ab^5+\frac{6!}{6!\cdot 1}b^6$$
$$=64a^6+6(32a^5b)+15(16a^4b^2)+20(8a^3b^3)+15(4a^2b^4)+6(2ab^5)+1b^6$$
$$=64a^6+192a^5b+240a^4b^2+160a^3b^3+60a^2b^4+12ab^5+b^6$$

31. $(x+2)^8$

First Term $(r=0)$: $\binom{n}{r}a^{n-r}b^r = \binom{8}{0}x^{8-0}2^0 = \frac{8!}{0!(8-0)!}x^8 \cdot 1 = \frac{\cancel{8!}}{0!\cancel{8!}}x^8 = x^8$

Second Term $(r=1)$: $\binom{n}{r}a^{n-r}b^r = \binom{8}{1}x^{8-1}2^1 = \frac{8!}{1!(8-1)!}2x^7 = \frac{8\cdot\cancel{7!}}{1\cdot\cancel{7!}}2x^7 = 8\cdot 2x^7 = 16x^7$

Third Term $(r=2)$: $\binom{n}{r}a^{n-r}b^r = \binom{8}{2}x^{8-2}2^2 = \frac{8!}{2!(8-2)!}4x^6 = \frac{8\cdot 7\cdot\cancel{6!}}{2\cdot 1\cdot\cancel{6!}}4x^6 = 28\cdot 4x^6 = 112x^6$

33. $(x-2y)^{10}$

First Term $(r=0)$: $\binom{n}{r}a^{n-r}b^r = \binom{10}{0}x^{10-0}(-2y)^0 = \frac{10!}{0!(10-0)!}x^{10}\cdot 1 = \frac{\cancel{10!}}{0!\cancel{10!}}x^{10} = x^{10}$

Second Term $(r=1)$: $\binom{n}{r}a^{n-r}b^r = \binom{10}{1}x^{10-1}(-2y)^1 = -\frac{10!}{1!(10-1)!}2x^9y = -\frac{10\cdot\cancel{9!}}{1\cdot\cancel{9!}}2x^9y = -10\cdot 2x^9y = -20x^9y$

Third Term $(r=2)$: $\binom{n}{r}a^{n-r}b^r = \binom{10}{2}x^{10-2}(-2y)^2 = \frac{10!}{2!(10-2)!}4x^8y^2 = \frac{10\cdot 9\cdot\cancel{8!}}{2\cdot 1\cdot\cancel{8!}}4x^8y^2 = 45\cdot 4x^8y^2 = 180x^8y^2$

35. $(x^2+1)^{16}$

First Term $(r=0)$: $\binom{n}{r}a^{n-r}b^r = \binom{16}{0}(x^2)^{16-0}1^0 = \frac{16!}{0!(16-0)!}x^{32}\cdot 1 = \frac{\cancel{16!}}{0!\cancel{16!}}x^{32} = x^{32}$

Second Term $(r=1)$: $\binom{n}{r}a^{n-r}b^r = \binom{16}{1}(x^2)^{16-1}1^1 = \frac{16!}{1!(16-1)!}x^{30}\cdot 1 = \frac{16\cdot\cancel{15!}}{1\cdot\cancel{15!}}x^{30} = 16x^{30}$

Third Term $(r=2)$: $\binom{n}{r}a^{n-r}b^r = \binom{16}{2}(x^2)^{16-2}1^2 = \frac{16!}{2!(16-2)!}x^{28}\cdot 1 = \frac{16\cdot 15\cdot\cancel{14!}}{2\cdot 1\cdot\cancel{14!}}x^{28} = 120x^{28}$

37. $(y^3-1)^{20}$

First Term $(r=0)$: $\binom{n}{r}a^{n-r}b^r = \binom{20}{0}(y^3)^{20-0}(-1)^0 = \frac{20!}{0!(20-0)!}y^{60}\cdot 1 = \frac{\cancel{20!}}{0!\cancel{20!}}y^{60} = y^{60}$

Second Term $(r=1)$: $\binom{n}{r}a^{n-r}b^r = \binom{20}{1}(y^3)^{20-1}(-1)^1 = \frac{20!}{1!(20-1)!}y^{57}\cdot(-1) = -\frac{20\cdot\cancel{19!}}{1\cdot\cancel{19!}}y^{57} = -20y^{57}$

Third Term $(r=2)$: $\binom{n}{r}a^{n-r}b^r = \binom{20}{2}(y^3)^{20-2}(-1)^2 = \frac{20!}{2!(20-2)!}y^{54}\cdot 1 = \frac{20\cdot 19\cdot\cancel{18!}}{2\cdot 1\cdot\cancel{18!}}y^{54} = 190y^{54}$

39. $(2x+y)^6$

Third Term $(r=2)$: $\binom{n}{r}a^{n-r}b^r = \binom{6}{2}(2x)^{6-2}\ y^2 = \frac{6!}{2!(6-2)!}16x^4y^2 = \frac{6\cdot 5\cdot \cancel{4!}}{2\cdot 1\cdot \cancel{4!}}16x^4y^2 = 15\left(16x^4y^2\right) = 240x^4y^2$

41. $(x-1)^9$

Fifth Term $(r=4)$: $\binom{n}{r}a^{n-r}b^r = \binom{9}{4}x^{9-4}(-1)^4 = \frac{9!}{4!(9-4)!}x^5\cdot 1 = \frac{9\cdot 8\cdot 7\cdot \cancel{6}\cdot \cancel{5!}}{4\cdot \cancel{3}\cdot \cancel{2}\cdot 1\cdot \cancel{5!}}x^5 = 126x^5$

43. $\left(x^2+y^3\right)^8$

Sixth Term $(r=5)$: $\binom{n}{r}a^{n-r}b^r = \binom{8}{5}\left(x^2\right)^{8-5}\left(y^3\right)^5 = \frac{8!}{5!(8-5)!}x^6y^{15} = \frac{8\cdot 7\cdot \cancel{6}\cdot \cancel{5!}}{\cancel{5!}\cdot \cancel{3}\cdot \cancel{2}\cdot 1}x^6y^{15} = 56x^6y^{15}$

45. $\left(x-\frac{1}{2}\right)^9$

Fourth Term $(r=3)$: $\binom{n}{r}a^{n-r}b^r = \binom{9}{3}x^{9-3}\left(-\frac{1}{2}\right)^3 = -\frac{9!}{3!(9-3)!}\cdot\frac{1}{8}x^6 = -\frac{9\cdot \cancel{8}\cdot 7\cdot \cancel{6!}}{3\cdot 2\cdot 1\cdot \cancel{6!}}\cdot\frac{1}{\cancel{8}}x^6 = -\frac{21}{2}x^6$

47. $\left(x^2+y\right)^{22}$

y^{14} will occur in the fifteenth term

Fifteenth Term $(r=14)$: $\binom{n}{r}a^{n-r}b^r = \binom{22}{14}\left(x^2\right)^{22-14}(y)^{14} = \frac{22!}{14!(22-14)!}\left(x^2\right)^8 y^{14} = 319{,}770x^{16}y^{14}$

49. $\left(x^3+x^{-2}\right)^4 = \binom{4}{0}\left(x^3\right)^4 + \binom{4}{1}\left(x^3\right)^3\left(x^{-2}\right) + \binom{4}{2}\left(x^3\right)^2\left(x^{-2}\right)^2 + \binom{4}{3}\left(x^3\right)^1\left(x^{-2}\right)^3 + \binom{4}{4}\left(x^{-2}\right)^4$

$$= \frac{4!}{0!(4-0)!}x^{12} + \frac{4!}{1!(4-1)!}x^9x^{-2} + \frac{4!}{2!(4-2)!}x^6x^{-4} + \frac{4!}{3!(4-3)!}x^3x^{-6} + \frac{4!}{4!(4-4)!}x^{-8}$$

$$= \frac{\cancel{4!}}{0!\cdot\cancel{4!}}x^{12} + \frac{4\cdot\cancel{3!}}{1!\cdot\cancel{3!}}x^7 + \frac{4\cdot 3\cdot\cancel{2!}}{2\cdot 1\cdot\cancel{2!}}x^2 + \frac{4\cdot\cancel{3!}}{\cancel{3!}\cdot 1!}x^{-3} + \frac{\cancel{4!}}{\cancel{4!}\cdot 0!}x^{-8}$$

$$= x^{12} + 4x^7 + 6x^2 + \frac{4}{x^3} + \frac{1}{x^8}$$

51. $\left(x^{\frac{1}{3}}-x^{-\frac{1}{3}}\right)^3=\left(x^{\frac{1}{3}}+\left(-x^{-\frac{1}{3}}\right)\right)^3$

$$=\binom{3}{0}\left(x^{\frac{1}{3}}\right)^3+\binom{3}{1}\left(x^{\frac{1}{3}}\right)^2\left(-x^{-\frac{1}{3}}\right)+\binom{3}{2}\left(x^{\frac{1}{3}}\right)^1\left(-x^{-\frac{1}{3}}\right)^2+\binom{3}{3}\left(-x^{-\frac{1}{3}}\right)^3$$

$$=\frac{3!}{0!(3-0)!}x+\frac{3!}{1!(3-1)!}x^{\frac{2}{3}}\left(-x^{-\frac{1}{3}}\right)+\frac{3!}{2!(3-2)!}x^{\frac{1}{3}}x^{-\frac{2}{3}}+\frac{3!}{3!(3-3)!}\left(-x^{-1}\right)$$

$$=x-3x^{\frac{1}{3}}+3x^{-\frac{1}{3}}-x^{-1}$$

$$=x-3x^{\frac{1}{3}}+\frac{3}{x^{\frac{1}{3}}}-\frac{1}{x}$$

53. $\left(-1+i\sqrt{3}\right)^3=\binom{3}{0}(-1)^3+\binom{3}{1}(-1)^2\left(i\sqrt{3}\right)+\binom{3}{2}(-1)^1\left(i\sqrt{3}\right)^2+\binom{3}{3}\left(i\sqrt{3}\right)^3$

$$=\frac{3!}{0!(3-0)!}\cdot-1+\frac{3!}{1!(3-1)!}1\cdot i\sqrt{3}+\frac{3!}{2!(3-2)!}\cdot-1\cdot-3+\frac{3!}{3!(3-3)!}\cdot-3i\sqrt{3}$$

$$=\frac{\cancel{3!}}{0!\cancel{3!}}(-1)+\frac{3\cdot\cancel{2!}}{1!\cancel{2!}}i\sqrt{3}+\frac{3\cdot\cancel{2!}}{\cancel{2!}1!}\cdot 3+\frac{\cancel{3!}}{\cancel{3!}0!}\cdot-3i\sqrt{3}$$

$$=-1+3i\sqrt{3}+9-3i\sqrt{3}$$

$$=-1+9+3i\sqrt{3}-3i\sqrt{3}$$

$$=8$$

55. $f(x)=x^4+7$

$$\frac{f(x+h)-f(x)}{h}=\frac{(x+h)^4+7-\left(x^4+7\right)}{h}=\frac{\binom{4}{0}x^4+\binom{4}{1}x^3h+\binom{4}{2}x^2h^2+\binom{4}{3}xh^3+\binom{4}{4}h^4+7-x^4-7}{h}$$

$$=\frac{\frac{4!}{0!(4-0)!}x^4+\frac{4!}{1!(4-1)!}x^3h+\frac{4!}{2!(4-2)!}x^2h^2+\frac{4!}{3!(4-3)!}xh^3+\frac{4!}{4!(4-4)!}h^4-x^4}{h}$$

$$=\frac{\frac{\cancel{4!}}{0!\cancel{4!}}x^4+\frac{4\cdot\cancel{3!}}{1!\cancel{3!}}x^3h+\frac{4\cdot3\cdot\cancel{2!}}{\cancel{2!}\cdot2\cdot1}x^2h^2+\frac{4\cdot\cancel{3!}}{\cancel{3!}1!}xh^3+\frac{\cancel{4!}}{\cancel{4!}0!}h^4-x^4}{h}$$

$$=\frac{\cancel{x^4}+4x^3h+6x^2h^2+4xh^3+h^4-\cancel{x^4}}{h}=\frac{h(4x^3+6x^2h+4xh^2+h^3)}{h}$$

$$=4x^3+6x^2h+4xh^2+h^3$$

57. We want to find the $(5+1)=6^{th}$ term.

$$\binom{n}{r}a^{n-r}b^r=\binom{10}{5}\left(\frac{3}{x}\right)^{10-5}\left(\frac{x}{3}\right)^5=\frac{10!}{5!(10-5)!}\left(\frac{3}{x}\right)^5\left(\frac{x}{3}\right)^5$$

$$=\frac{\cancel{10}\cdot9\cdot8\cdot7\cdot6\cdot\cancel{5!}}{\cancel{5!}\cdot\cancel{5}\cdot4\cdot3\cdot\cancel{2}\cdot1}\left(\frac{3}{x}\right)^5\left(\frac{x}{3}\right)^5=252\cdot\frac{3^5}{x^5}\cdot\frac{x^5}{3^5}=252$$

59. $(0.28+0.72)^5$

Third Term $(r=2)$: $\binom{n}{r}a^{n-r}b^r=\binom{5}{2}0.28^{5-2}0.72^2=\dfrac{5!}{2!(5-2)!}0.28^{5-2}0.72^2=\dfrac{5!}{2!3!}0.28^3 0.72^2\approx 0.1138$

61. – 67. Answers will vary.

69. $f_1(x)=(x+2)^3$
$f_2(x)=x^3$

$f_3(x)=x^3+6x^2$
$f_4(x)=x^3+6x^2+12x$
$f_5(x)=x^3+6x^2+12x+8$

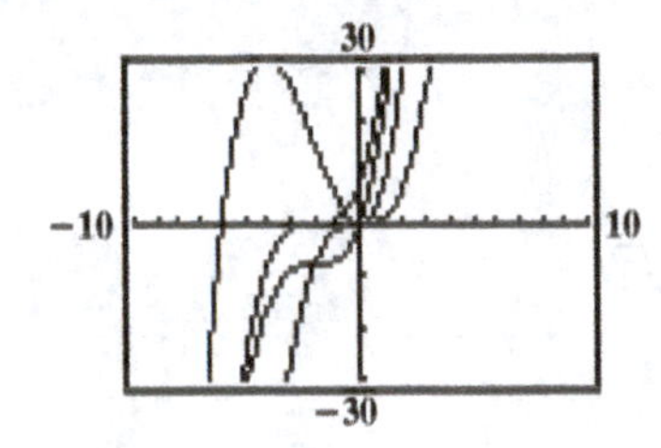

Graphs f_1 and f_5 are the same. This means that the functions are equivalent. Graphs f_2 through f_4 are increasingly similar to the graphs of f_1 and f_5.

71. Applying the Binomial Theorem to $(x-1)^3$, we have $a=x$, $b=-1$, and $n=3$.

$$(x-1)^3=\binom{3}{0}x^3+\binom{3}{1}x^2(-1)+\binom{3}{2}x(-1)^2+\binom{3}{3}(-1)^3$$
$$=\frac{3!}{0!(3-0)!}x^3-\frac{3!}{1!(3-1)!}x^2+\frac{3!}{2!(3-2)!}x(1)-\frac{3!}{3!(3-3)!}$$
$$=\frac{\cancel{3!}}{1\cdot\cancel{3!}}x^3-\frac{3\cdot\cancel{2!}}{1\cdot\cancel{2!}}x^2+\frac{3\cdot\cancel{2!}}{\cancel{2!}1!}x-\frac{\cancel{3!}}{\cancel{3!}0!}=x^3-3x^2+3x-1$$

Graph using the method from Exercises 69 and 70.

$f_1(x)=(x-1)^3$ $\quad$ $f_2(x)=x^3$
$f_3(x)=x^3+3x^2$ $\quad$ $f_4(x)=x^3+3x^2+3x$
$f_5(x)=x^3-3x^2+3x-1$

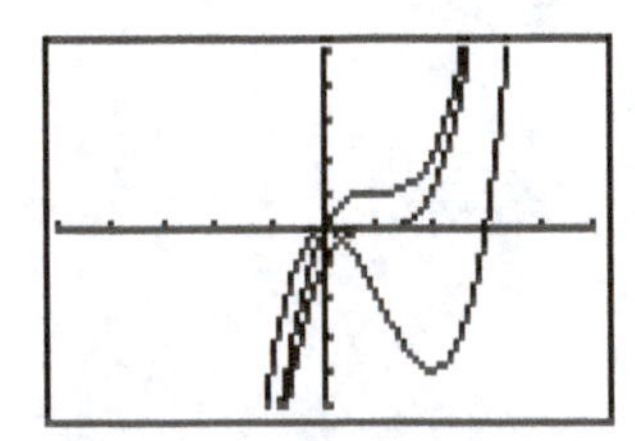

Graphs f_1 and f_5 are the same. This means that the functions are equivalent. Graphs f_2 through f_4 are increasingly similar to the graphs of f_1 and f_5.

73. Applying the Binomial Theorem to $(x+2)^6$, we have $a = x$, $b = 2$, and $n = 6$.

$$(x+2)^6 = \binom{6}{0}x^6 + \binom{6}{1}x^5 2 + \binom{6}{2}x^4 2^2 + \binom{6}{3}x^3 2^3 + \binom{6}{4}x^2 2^4 + \binom{6}{5}x2^5 + \binom{6}{6}2^6$$

$$= \frac{6!}{0!(6-0)!}x^6 + \frac{6!}{1!(6-1)!}2x^5 + \frac{6!}{2!(6-2)!}4x^4 + \frac{6!}{3!(6-3)!}8x^3 + \frac{6!}{4!(6-4)!}16x^2 + \frac{6!}{5!(6-5)!}32x + \frac{6!}{6!(6-6)!}64$$

$$= \frac{\not{6}!}{1\not{6}!}x^6 + \frac{6\cdot\not{5}!}{1\cdot\not{5}!}2x^5 + \frac{6\cdot5\cdot\not{4}!}{2\cdot1\cdot\not{4}!}4x^4 + \frac{\not{6}\cdot5\cdot4\cdot\not{3}!}{\not{3}\not{2}\cdot1\cdot\not{3}!}8x^3 + \frac{6\cdot5\cdot\not{4}!}{\not{4}!2\cdot1}16x^2 + \frac{6\cdot\not{5}!}{\not{5}!1}32x + \frac{\not{6}!}{\not{6}!\cdot1}64$$

$$= x^6 + 6(2x^5) + 15(4x^4) + 20(8x^3) + 15(16x^2) + 6(32x) + 1(64)$$

$$= x^6 + 12x^5 + 60x^4 + 160x^3 + 240x^2 + 192x + 64$$

Graph using the method from Exercises 69 and 70.

$f_1(x) = (x+2)^6$

$f_2(x) = x^6$

$f_3(x) = x^6 + 12x^5$

$f_4(x) = x^6 + 12x^5 + 60x^4$

$f_5(x) = x^6 + 12x^5 + 60x^4 + 160x^3$

$f_6(x) = x^6 + 12x^5 + 60x^4 + 160x^3 + 240x^2$

$f_7(x) = x^6 + 12x^5 + 60x^4 + 160x^3 + 240x^2 + 192x$

$f_8(x) = x^6 + 12x^5 + 60x^4 + 160x^3 + 240x^2 + 192x + 64$

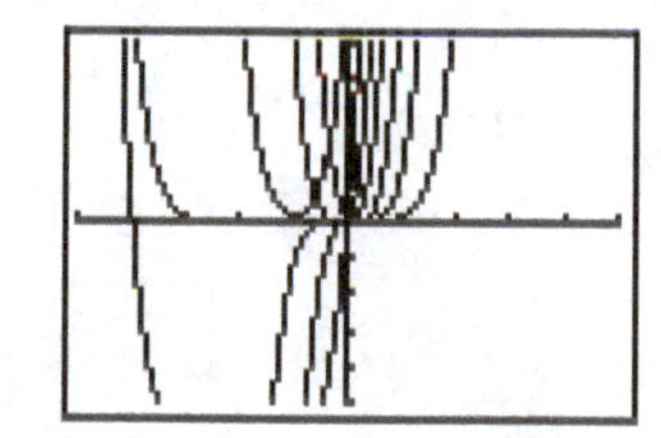

Graphs f_1 and f_8 are the same. This means that the functions are equivalent. Graphs f_2 through f_7 are increasingly similar to the graphs of f_1 and f_8.

75. makes sense

77. does not make sense; Explanations will vary. Sample explanation: $7 \neq 2+4$

79. true

81. false; Changes to make the statement true will vary. A sample change is: There are values of a and b for which $(a+b)^4 = a^4 + b^4$. Consider $a = 0$ and $b = 1$.

$$(0+1)^4 = 0^4 + 1^4$$
$$(1)^4 = 0 + 1$$
$$1 = 1$$

83. In $\left(x^2+y^2\right)^5$, the term containing x^4 is the term in which $a=x^2$ is squared. Applying the Binomial Theorem, the following pattern results. In the first term, x^2 is taken to the fifth power. In the second term, x^2 is taken to the fourth power. In the third term x^2 is taken to the third power. In the fourth term, x^2 is taken to the second power. This is the term we are looking for. Applying the Binomial Theorem to $\left(x^2+y^2\right)^5$, we have $a=x^2$, $b=y^2$, and $n=5$. We are looking for the 4th term where $r+1=4$ so $r=3$.

$$\begin{aligned}\binom{n}{r}a^{n-r}b^r &= \binom{5}{3}\left(x^2\right)^2\left(y^2\right)^3\\ &= \frac{5!}{3!(5-3)!}x^4y^6\\ &= \frac{5!}{3!2!}x^4y^6\\ &= \frac{5\cdot 4\cdot \cancel{3!}}{\cancel{3!}2\cdot 1}x^4y^6\\ &= 10x^4y^6\end{aligned}$$

84.
$$\begin{aligned}f(a+1) &= (a+1)^2+2(a+1)+3\\ &= a^2+2a+1+2a+2+3\\ &= a^2+4a+6\end{aligned}$$

85. $f(x)=x^2+5x \qquad g(x)=2x-3$

$$\begin{aligned}f(g(x)) &= f(2x-3)\\ &= (2x-3)^2+5(2x-3)\\ &= 4x^2-12x+9+10x-15\\ &= 4x^2-2x-6\end{aligned}$$

$$\begin{aligned}g(f(x)) &= g\left(x^2+5x\right)\\ &= 2\left(x^2+5x\right)-3\\ &= 2x^2+10x-3\end{aligned}$$

86. $\frac{x}{x+3}-\frac{x+1}{2x^2-2x-24}$

$=\frac{x}{x+3}-\frac{x+1}{2\left(x^2-x-12\right)}$

$=\frac{x}{x+3}-\frac{x+1}{2(x-4)(x+3)}$

$=\frac{x}{x+3}\cdot\frac{2(x-4)}{2(x-4)}-\frac{x+1}{2(x-4)(x+3)}$

$=\frac{x}{x+3}\cdot\frac{2x-8}{2(x-4)}-\frac{x+1}{2(x-4)(x+3)}$

$=\frac{2x^2-8x}{2(x-4)(x+3)}-\frac{x+1}{2(x-4)(x+3)}$

$=\frac{2x^2-8x-(x+1)}{2(x-4)(x+3)}$

$=\frac{2x^2-8x-x-1}{2(x-4)(x+3)}$

$=\frac{2x^2-9x-1}{2(x-4)(x+3)}$

Chapter 11 Review Exercises

1. $a_n=7n-4$

$a_1=7(1)-4=7-4=3$

$a_2=7(2)-4=14-4=10$

$a_3=7(3)-4=21-4=17$

$a_4=7(4)-4=28-4=24$

The first four terms are 3, 10, 17, 24.

2. $a_n=(-1)^n\frac{n+2}{n+1}$

$a_1=(-1)^1\frac{1+2}{1+1}=-\frac{3}{2}$

$a_2=(-1)^2\frac{2+2}{2+1}=\frac{4}{3}$

$a_3=(-1)^3\frac{3+2}{3+1}=-\frac{5}{4}$

$a_4=(-1)^4\frac{4+2}{4+1}=\frac{6}{5}$

The first four terms are $-\frac{3}{2},\frac{4}{3},-\frac{5}{4},\frac{6}{5}$.

3. $a_n = \dfrac{1}{(n-1)!}$

$a_1 = \dfrac{1}{(1-1)!} = \dfrac{1}{0!} = \dfrac{1}{1} = 1$

$a_2 = \dfrac{1}{(2-1)!} = \dfrac{1}{1!} = \dfrac{1}{1} = 1$

$a_3 = \dfrac{1}{(3-1)!} = \dfrac{1}{2!} = \dfrac{1}{2 \cdot 1} = \dfrac{1}{2}$

$a_4 = \dfrac{1}{(4-1)!} = \dfrac{1}{3!} = \dfrac{1}{3 \cdot 2 \cdot 1} = \dfrac{1}{6}$

The first four terms are $1, 1, \dfrac{1}{2}, \dfrac{1}{6}$.

4. $a_n = \dfrac{(-1)^{n+1}}{2^n}$

$a_1 = \dfrac{(-1)^{1+1}}{2^1} = \dfrac{(-1)^2}{2} = \dfrac{1}{2}$

$a_2 = \dfrac{(-1)^{2+1}}{2^2} = \dfrac{(-1)^3}{4} = -\dfrac{1}{4}$

$a_3 = \dfrac{(-1)^{3+1}}{2^3} = \dfrac{(-1)^4}{8} = \dfrac{1}{8}$

$a_4 = \dfrac{(-1)^{4+1}}{2^4} = \dfrac{(-1)^5}{16} = -\dfrac{1}{16}$

The first four terms are $\dfrac{1}{2}, -\dfrac{1}{4}, \dfrac{1}{8}, -\dfrac{1}{16}$.

5.
$$\begin{aligned}\sum_{i=1}^{5}(2i^2-3) &= \left(2(1)^2-3\right)+\left(2(2)^2-3\right)+\left(2(3)^2-3\right)+\left(2(4)^2-3\right)+\left(2(5)^2-3\right)+\left(2(6)^2-3\right)\\ &= \left(2(1)-3\right)+\left(2(4)-3\right)+\left(2(9)-3\right)+\left(2(16)-3\right)+\left(2(25)-3\right)\\ &= (2-3)+(8-3)+(18-3)+(32-3)+(50-3) = -1+5+15+29+47 = 95\end{aligned}$$

6.
$$\begin{aligned}\sum_{i=0}^{4}(-1)^{i+1}\, i! &= (-1)^{0+1}\, 0! + (-1)^{1+1}\, 1! + (-1)^{2+1}\, 2! + (-1)^{3+1}\, 3! + (-1)^{4+1}\, 4!\\ &= (-1)^1\, 1 + (-1)^2\, 1 + (-1)^3\, 2\cdot 1 + (-1)^4\, 3\cdot 2\cdot 1 + (-1)^5\, 4\cdot 3\cdot 2\cdot 1\\ &= -1+1-2+6-24 = -20\end{aligned}$$

7. $\dfrac{1}{3}+\dfrac{2}{4}+\dfrac{3}{5}+\ldots+\dfrac{15}{17} = \displaystyle\sum_{i=1}^{15}\dfrac{i}{i+2}$

8. $4^3+5^3+6^3+...+13^3=\sum_{i=4}^{13} i^3$ or $\sum_{i=1}^{10}(i+3)^3$

9. $a_1=7$

$a_2=7+4=11$

$a_3=11+4=15$

$a_4=15+4=19$

$a_5=19+4=23$

$a_6=23+4=27$

The first six terms are $7, 11, 15, 19, 23, 27$.

10. $a_1=-4$

$a_2=-4-5=-9$

$a_3=-9-5=-14$

$a_4=-14-5=-19$

$a_5=-19-5=-24$

$a_6=-24-5=-29$

The first six terms are $-4, -9, -14, -19, -24, -29$.

11. $a_1=\frac{3}{2}$

$a_2=\frac{3}{2}-\frac{1}{2}=\frac{2}{2}=1$

$a_3=1-\frac{1}{2}=\frac{1}{2}$

$a_4=\frac{1}{2}-\frac{1}{2}=0$

$a_5=0-\frac{1}{2}=-\frac{1}{2}$

$a_6=-\frac{1}{2}-\frac{1}{2}=-\frac{2}{2}=-1$

The first six terms are $\frac{3}{2}, 1, \frac{1}{2}, 0, -\frac{1}{2}, -1$.

12. $a_6=5+(6-1)3=5+(5)3=5+15=20$

13. $a_{12}=-8+(12-1)(-2)=-8+11(-2)=-8+(-22)=-30$

14. $a_{14}=14+(14-1)(-4)=14+13(-4)=14+(-52)=-38$

15.
$$d = -3-(-7) = 4$$
$$a_n = -7+(n-1)4 = -7+4n-4$$
$$= 4n-11$$
$$a_{20} = 4(20)-11 = 80-11 = 69$$

16.
$$a_n = 200+(n-1)(-20)$$
$$= 200-20n+20$$
$$= 220-20n$$
$$a_{20} = 220-20(20)$$
$$= 220-400 = -180$$

17.
$$a_n = -12+(n-1)\left(-\frac{1}{2}\right)$$
$$= -12-\frac{1}{2}n+\frac{1}{2}$$
$$= -\frac{24}{2}-\frac{1}{2}n+\frac{1}{2}$$
$$= -\frac{1}{2}n-\frac{23}{2}$$

$$a_{20} = -\frac{1}{2}(20)-\frac{23}{2}$$
$$= -\frac{20}{2}-\frac{23}{2} = -\frac{43}{2}$$

18.
$$d = 8-15 = -7$$
$$a_n = 15+(n-1)(-7) = 15-7n+7$$
$$= 22-7n$$
$$a_{20} = 22-7(20) = 22-140 = -118$$

19. First, find d.
$$d = 12-5 = 7$$

Next, find a_{22}.
$$a_{22} = 5+(22-1)7 = 5+(21)7$$
$$= 5+147 = 152$$

Now, find the sum.
$$S_{22} = \frac{22}{2}(5+152) = 11(157) = 1727$$

20. First, find d.

$d = -3-(-6) = 3$

Next, find a_{15}.

$$a_{15} = -6+(15-1)3 = -6+(14)3$$
$$= -6+42 = 36$$

Now, find the sum.

$$S_{15} = \frac{15}{2}(-6+36) = \frac{15}{2}(30) = 225$$

21. We are given that $a_{100} = 300$, $a_1 = 3$, and $n = 100$.

$$S_{100} = \frac{100}{2}(3+300)$$
$$= 50(303)$$
$$= 15{,}150$$

22. $$\sum_{i=1}^{16}(3i+2) = (3(1)+2)+(3(2)+2)+(3(3)+2)+\ldots+(3(16)+2)$$
$$= (3+2)+(6+2)+(9+2)+\ldots+(48+2)$$
$$= 5+8+11+\ldots+50$$

$$S_{16} = \frac{16}{2}(5+50) = 8(55) = 440$$

23. $$\sum_{i=1}^{25}(-2i+6) = (-2(1)+6)+(-2(2)+6)+(-2(3)+6)+\ldots+(-2(25)+6)$$
$$= (-2+6)+(-4+6)+(-6+6)+\ldots+(-50+6)$$
$$= 4+2+0+\ldots+(-44)$$

$$S_{25} = \frac{25}{2}(4+(-44)) = \frac{25}{2}(-40) = -500$$

24. $$\sum_{i=1}^{30}(-5i) = (-5(1))+(-5(2))+(-5(3))+\ldots+(-5(30))$$
$$= -5+(-10)+(-15)+\ldots+(-150)$$

$$S_{30} = \frac{30}{2}(-5+(-150)) = 15(-155) = -2325$$

25. **a.** $$a_n = 39+(n-1)(4.75)$$
$$= 39+4.75n-4.75$$
$$= 4.75n+34.25$$

b. $$a_n = 4.75n+34.25$$
$$a_{12} = 4.75(13)+34.25$$
$$= 96$$

The percentage of students ages 12 – 18 who will report seeing security cameras at school in the year 2013 will be approximately 96%.

26. $a_{10} = 31500 + (10-1)2300$
$= 31500 + (9)2300$
$= 31500 + 20700 = 52200$

$S_{10} = \frac{10}{2}(31500 + 52200)$
$= 5(83700) = 418500$

The total salary over a ten-year period is \$418,500.

27. $a_{35} = 25 + (35-1)1 = 25 + (34)1$
$= 25 + 34 = 59$

$S_{35} = \frac{35}{2}(25 + 59) = \frac{35}{2}(84) = 1470$

There are 1470 seats in the theater.

28. The first term is 3.
The second term is $3 \cdot 2 = 6$.
The third term is $6 \cdot 2 = 12$.
The fourth term is $12 \cdot 2 = 24$.
The fifth term is $24 \cdot 2 = 48$.

29. The first term is $\frac{1}{2}$.

The second term is $\frac{1}{2} \cdot \frac{1}{2} = \frac{1}{4}$.

The third term is $\frac{1}{4} \cdot \frac{1}{2} = \frac{1}{8}$.

The fourth term is $\frac{1}{8} \cdot \frac{1}{2} = \frac{1}{16}$.

The fifth term is $\frac{1}{16} \cdot \frac{1}{2} = \frac{1}{32}$.

30. The first term is 16.

The second term is $16 \cdot -\frac{1}{4} = -4$.

The third term is $-4 \cdot -\frac{1}{4} = 1$.

The fourth term is $1 \cdot -\frac{1}{4} = -\frac{1}{4}$.

The fifth term is $-\frac{1}{4} \cdot -\frac{1}{4} = \frac{1}{16}$.

31. The first term is -5.
The second term is $-5 \cdot -1 = 5$.
The third term is $5 \cdot -1 = -5$.
The fourth term is $-5 \cdot -1 = 5$.
The fifth term is $5 \cdot -1 = -5$.

32. $a_7 = 2(3)^{7-1} = 2(3)^6 = 2(729) = 1458$

33. $a_6 = 16\left(\frac{1}{2}\right)^{6-1} = 16\left(\frac{1}{2}\right)^5 = 16\left(\frac{1}{32}\right) = \frac{1}{2}$

34. $a_5 = -3(2)^{5-1} = -3(2)^4 = -3(16) = -48$

35. $a_n = a_1 r^{n-1} = 1(2)^{n-1}$
$a_8 = 1(2)^{8-1} = 1(2)^7 = 1(128) = 128$

36. $a_n = a_1 r^{n-1} = 100\left(\frac{1}{10}\right)^{n-1}$

$a_8 = 100\left(\frac{1}{10}\right)^{8-1} = 100\left(\frac{1}{10}\right)^7$

$= 100\left(\frac{1}{10,000,000}\right) = \frac{1}{100,000}$

37. $d = \frac{-4}{12} = -\frac{1}{3}$

$a_n = a_1 r^{n-1} = 12\left(-\frac{1}{3}\right)^{n-1}$

$a_8 = 12\left(-\frac{1}{3}\right)^{8-1} = 12\left(-\frac{1}{3}\right)^7$

$= 12\left(-\frac{1}{2187}\right) = -\frac{12}{2187} = -\frac{4}{729}$

38. $r = \frac{a_2}{a_1} = \frac{-15}{5} = -3$

$S_{15} = \frac{5\left(1-(-3)^{15}\right)}{1-(-3)} = \frac{5(1-(-14348907))}{4}$

$= \frac{5(14348908)}{4} = \frac{71744540}{4}$

$= 17,936,135$

39. $r = \frac{a_2}{a_1} = \frac{4}{8} = \frac{1}{2}$

$$S_7 = \frac{8\left(1-\left(\frac{1}{2}\right)^7\right)}{1-\frac{1}{2}} = \frac{8\left(1-\frac{1}{128}\right)}{\frac{1}{2}}$$

$$= \frac{8\left(\frac{128}{128}-\frac{1}{128}\right)}{\frac{1}{2}} = \frac{8\left(\frac{127}{128}\right)}{\frac{1}{2}}$$

$$= \frac{8}{1}\left(\frac{127}{128}\right) \div \frac{1}{2} = \frac{8}{1}\left(\frac{127}{128}\right) \cdot \frac{2}{1}$$

$$= \frac{2032}{128} = \frac{127}{8} = 15.875$$

40. $\sum_{i=1}^{6} 5^i = \frac{5\left(1-5^6\right)}{1-5} = \frac{5(1-15625)}{-4}$

$$= \frac{5(-15624)}{-4} = 5(3906)$$
$$= 19,530$$

41. $\sum_{i=1}^{7} 3(-2)^i = \frac{-6\left(1-(-2)^7\right)}{1-(-2)} = \frac{-6(1-(-128))}{3}$

$$= \frac{-6(129)}{3} = -2(129) = -258$$

42. $\sum_{i=1}^{5} 2\left(\frac{1}{4}\right)^{i-1} = \frac{2\left(1-\left(\frac{1}{4}\right)^5\right)}{1-\left(\frac{1}{4}\right)} = \frac{2\left(1-\frac{1}{1024}\right)}{\frac{3}{4}}$

$$= \frac{2\cdot\frac{1023}{1024}}{\frac{3}{4}} = \frac{1023}{512}\cdot\frac{4}{3} = \frac{341}{128}$$

43. $r = \frac{a_2}{a_1} = \frac{3}{9} = \frac{1}{3}$

$$S = \frac{9}{1-\frac{1}{3}} = \frac{9}{\frac{2}{3}} = 9 \div \frac{2}{3} = 9\cdot\frac{3}{2} = \frac{27}{2}$$

44. $r = \frac{a_2}{a_1} = \frac{-1}{2} = -\frac{1}{2}$

$$S = \frac{2}{1-\left(-\frac{1}{2}\right)} = \frac{2}{\frac{3}{2}} = 2 \div \frac{3}{2} = 2\cdot\frac{2}{3} = \frac{4}{3}$$

45. $r = \frac{a_2}{a_1} = \frac{4}{-6} = -\frac{2}{3}$

$$S = \frac{-6}{1-\left(-\frac{2}{3}\right)} = \frac{-6}{\frac{5}{3}} = -6 \div \frac{5}{3}$$

$$= -6\cdot\frac{3}{5} = -\frac{18}{5}$$

46. $\sum_{i=1}^{\infty} 5(0.8)^i = \frac{4}{1-0.8} = \frac{4}{0.2} = 20$

47. $0.\overline{6} = \frac{a_1}{1-r} = \frac{\frac{6}{10}}{1-\frac{1}{10}} = \frac{\frac{6}{10}}{\frac{9}{10}} = \frac{6}{10} \div \frac{9}{10}$

$$= \frac{6}{10}\cdot\frac{10}{9} = \frac{2}{3}$$

48. $0.\overline{47} = \frac{a_1}{1-r} = \frac{\frac{47}{100}}{1-\frac{1}{100}} = \frac{\frac{47}{100}}{\frac{99}{100}}$

$$= \frac{47}{100} \div \frac{99}{100} = \frac{47}{100}\cdot\frac{100}{99} = \frac{47}{99}$$

49. **a.** Divide each value by the previous value:

$$r_{2000-2010} \approx \frac{a_2}{a_1} = \frac{5.9}{4.2} \approx 1.4$$

$$r_{2010-2020} \approx \frac{a_3}{a_2} = \frac{8.3}{5.9} \approx 1.4$$

$$r_{2020-2030} \approx \frac{a_4}{a_3} = \frac{11.6}{8.3} \approx 1.4$$

$$r_{2030-2040} \approx \frac{a_5}{a_4} = \frac{16.2}{11.6} \approx 1.4$$

$$r_{2040-2050} \approx \frac{a_6}{a_5} = \frac{22.7}{16.2} \approx 1.4$$

r is approximately 1.4 for each division.

b. $a_n = 4.2(1.4)^{n-1}$

c. 2080 is 9 decades after 1990 so $n = 9$.

$a_n = 4.2(1.4)^{n-1}$

$a_9 = 4.2(1.4)^{9-1} \approx 62.0$

In 2080, the model predicts the U.S. population, ages 85 and older, will be 62.0 million.

50. $r = 1.06$

$a_n = a_1 r^{n-1} = 32000(1.06)^{n-1}$

$a_6 = \$32{,}000(1.06)^{6-1} \approx \$42{,}823$

The salary in the sixth year is approximately $42,823.

$$S_n = \frac{a_1(1-r^n)}{1-r}$$

$$S_6 = \frac{32000(1-(1.06)^6)}{1-1.06} \approx 223{,}210$$

The total salary over the six years is approximately $223,210.

51. a. $A = \dfrac{P\left[\left(1+\frac{r}{n}\right)^{nt}-1\right]}{\frac{r}{n}}$

$P = \$520,\ r = 0.06,\ n = 1,\ t = 20$

$$A = \frac{\$520\left[\left(1+\frac{0.06}{1}\right)^{1\cdot 20}-1\right]}{\frac{0.06}{1}}$$

$$= \frac{\$520\left[(1.06)^{20}-1\right]}{0.06} \approx \$19{,}129$$

The value of the annuity will be $19,129.

b. Interest = Value of annuity − Total deposits

$\approx \$19{,}129 - \$520 \cdot 20$

$\approx \$8729$

52. a. $A = \dfrac{P\left[\left(1+\frac{r}{n}\right)^{nt}-1\right]}{\frac{r}{n}}$

$P = 100,\ r = 0.055,\ n = 12,\ t = 30$

$$A = \frac{\$100\left[\left(1+\frac{0.055}{12}\right)^{12\cdot 30}-1\right]}{\frac{0.055}{12}} \approx \$91{,}361$$

The value of the IRA will be $91,361.

b. Interest = Value of IRA − Total deposits

$\approx \$91{,}361 - \$100 \cdot 12 \cdot 30$

$\approx \$55{,}361$

53. $r = 70\% = 0.7$

$a_1 = 4(.7) = 2.8$

$$S = \frac{2.8}{1-0.7} = \frac{2.8}{0.3} = 9\frac{1}{3}$$

The total spending in the town will be approximately $\$9\frac{1}{3}$ million each year.

54. $\dbinom{11}{8} = \dfrac{11!}{8!(11-8)!} = \dfrac{11!}{8!3!} = \dfrac{11\cdot 10\cdot 9\cdot \cancel{8!}}{\cancel{8!}3!} = 165$

55. $\dbinom{90}{2} = \dfrac{90!}{2!(90-2)!} = \dfrac{90\cdot 89\cdot \cancel{88!}}{2\cdot 1\cdot \cancel{88!}} = 4005$

56. Applying the Binomial Theorem to $(2x+1)^3$, we have $a=2x,\ b=1,$ and $n=3$.

$$\begin{aligned}(2x+1)^3 &= \binom{3}{0}(2x)^3+\binom{3}{1}(2x)^2\cdot 1+\binom{3}{2}(2x)\cdot 1^2+\binom{3}{3}1^3\\ &= \frac{3!}{0!(3-0)!}8x^3+\frac{3!}{1!(3-1)!}4x^2\cdot 1+\frac{3!}{2!(3-2)!}2x\cdot 1+\frac{3!}{3!(3-3)!}1\\ &= \frac{\cancel{3!}}{1\cdot\cancel{3!}}8x^3+\frac{3\cdot\cancel{2!}}{1\cdot\cancel{2!}}4x^2+\frac{3\cdot\cancel{2!}}{\cancel{2!}1!}2x+\frac{\cancel{3!}}{\cancel{3!}0!}\\ &= 8x^3+3\left(4x^2\right)+3(2x)+1\\ &= 8x^3+12x^2+6x+1\end{aligned}$$

57. Applying the Binomial Theorem to $\left(x^2-1\right)^4$, we have $a=x^2,\ b=-1,$ and $n=4$.

$$\begin{aligned}\left(x^2-1\right)^4 &= \binom{4}{0}\left(x^2\right)^4+\binom{4}{1}\left(x^2\right)^3(-1)+\binom{4}{2}\left(x^2\right)^2(-1)^2+\binom{4}{3}x^2(-1)^3+\binom{4}{4}(-1)^4\\ &= \frac{4!}{0!(4-0)!}x^8-\frac{4!}{1!(4-1)!}x^6+\frac{4!}{2!(4-2)!}x^4-\frac{4!}{3!(4-3)!}x^2+\frac{4!}{4!(4-4)!}1\\ &= \frac{\cancel{4!}}{0!\cancel{4!}}x^8-\frac{4!}{1!3!}x^6+\frac{4!}{2!2!}x^4-\frac{4!}{3!1!}x^2+\frac{\cancel{4!}}{\cancel{4!}0!}\\ &= 1\left(x^8\right)-\frac{4\cdot\cancel{3!}}{1\cdot\cancel{3!}}x^6+\frac{4\cdot 3\cdot\cancel{2!}}{2\cdot 1\cdot\cancel{2!}}x^4-\frac{4\cdot\cancel{3!}}{\cancel{3!}\cdot 1}x^2+1\\ &= x^8-4x^6+6x^4-4x^2+1\end{aligned}$$

58. Applying the Binomial Theorem to $(x+2y)^5$, we have $a=x,\ b=2y,$ and $n=5$.

$$\begin{aligned}(x+2y)^5 &= \binom{5}{0}x^5+\binom{5}{1}x^4(2y)+\binom{5}{2}x^3(2y)^2+\binom{5}{3}x^2(2y)^3+\binom{5}{4}x(2y)^4+\binom{5}{5}(2y)^5\\ &= \frac{\cancel{5!}}{0!\cancel{5!}}x^5+\frac{5!}{1!(5-1)!}2x^4y+\frac{5!}{2!(5-2)!}4x^3y^2+\frac{5!}{3!(5-3)!}8x^2y^3+\frac{5!}{4!(5-4)!}16xy^4+\frac{5!}{5!(5-5)!}32y^5\\ &= 1x^5+\frac{5\cdot\cancel{4!}}{1\cdot\cancel{4!}}2x^4y+\frac{5\cdot 4\cdot\cancel{3!}}{2\cdot 1\cdot\cancel{3!}}4x^3y^2+\frac{5\cdot 4\cdot\cancel{3!}}{\cancel{3!}2\cdot 1}8x^2y^3+\frac{5\cdot\cancel{4!}}{\cancel{4!}\cdot 1}16xy^4+\frac{\cancel{5!}}{\cancel{5!}0!}32y^5\\ &= x^5+5\left(2x^4y\right)+10\left(4x^3y^2\right)+10\left(8x^2y^3\right)+5\left(16xy^4\right)+1\left(32y^5\right)\\ &= x^5+10x^4y+40x^3y^2+80x^2y^3+80xy^4+32y^5\end{aligned}$$

59. Applying the Binomial Theorem to $(x-2)^6$, we have $a = x,\ b = -2,$ and $n = 6.$

$$
\begin{aligned}
(x-2)^6 &= \binom{6}{0}x^6 + \binom{6}{1}x^5(-2) + \binom{6}{2}x^4(-2)^2 + \binom{6}{3}x^3(-2)^3 + \binom{6}{4}x^2(-2)^4 + \binom{6}{5}x(-2)^5 + \binom{6}{6}(-2)^6 \\
&= \frac{6!}{0!(6-0)!}x^6 + \frac{6!}{1!(6-1)!}x^5(-2) + \frac{6!}{2!(6-2)!}x^4(-2)^2 + \frac{6!}{3!(6-3)!}x^3(-2)^3 \\
&\qquad + \frac{6!}{4!(6-4)!}x^2(-2)^4 + \frac{6!}{5!(6-5)!}x(-2)^5 + \frac{6!}{6!(6-6)!}(-2)^6 \\
&= \frac{6!}{1 6!}x^6 - \frac{6\cdot 5!}{1\cdot 5!}2x^5 + \frac{6\cdot 5\cdot 4!}{2\cdot 1\cdot 4!}4x^4 - \frac{6\cdot 5\cdot 4\cdot 3!}{3\cdot 2\cdot 1\cdot 3!}8x^3 \\
&\qquad + \frac{6\cdot 5\cdot 4!}{4!2\cdot 1}16x^2 - \frac{6\cdot 5!}{5!1}32x + \frac{6!}{6!\cdot 1}64 \\
&= x^6 - 6\left(2x^5\right) + 15\left(4x^4\right) - 20\left(8x^3\right) + 15\left(16x^2\right) - 6(32x) + 1\cdot 64 \\
&= x^6 - 12x^5 + 60x^4 - 160x^3 + 240x^2 - 192x + 64
\end{aligned}
$$

60. $\left(x^2+3\right)^8$

First Term $(r=0)$: $\binom{n}{r}a^{n-r}b^r = \binom{8}{0}\left(x^2\right)^{8-0}3^0 = \frac{8!}{0!(8-0)!}x^{16}\cdot 1 = \frac{8!}{0!8!}x^{16} = x^{16}$

Second Term $(r=1)$: $\binom{n}{r}a^{n-r}b^r = \binom{8}{1}\left(x^2\right)^{8-1}3^1 = \frac{8!}{1!(8-1)!}3x^{14} = \frac{8\cdot 7!}{1\cdot 7!}3x^{14} = 8\cdot 3x^{14} = 24x^{14}$

Third Term $(r=2)$: $\binom{n}{r}a^{n-r}b^r = \binom{8}{2}\left(x^2\right)^{8-2}3^2 = \frac{8!}{2!(8-2)!}9x^{12} = \frac{8\cdot 7\cdot 6!}{2\cdot 1\cdot 6!}9x^{12} = 28\cdot 9x^{12} = 252x^{12}$

61. $(x-3)^9$

First Term $(r=0)$: $\binom{n}{r}a^{n-r}b^r = \binom{9}{0}x^{9-0}(-3)^0 = \frac{9!}{0!(9-0)!}x^9\cdot 1 = \frac{9!}{0!9!}x^9 = x^9$

Second Term $(r=1)$: $\binom{n}{r}a^{n-r}b^r = \binom{9}{1}x^{9-1}(-3)^1 = -\frac{9!}{1!(9-1)!}3x^8 = -\frac{9\cdot 8!}{1\cdot 8!}3x^8 = -9\cdot 3x^8 = -27x^8$

Third Term $(r=2)$: $\binom{n}{r}a^{n-r}b^r = \binom{9}{2}x^{9-2}(-3)^2 = \frac{9!}{2!(9-2)!}9x^7 = \frac{9\cdot 8\cdot 7!}{2\cdot 1\cdot 7!}9x^7 = 36\cdot 9x^7 = 324x^7$

62. $(x+2)^5$

Fourth Term $(r=3)$: $\binom{n}{r}a^{n-r}b^r = \binom{5}{3}x^{5-3}(2)^3 = \frac{5!}{3!(5-3)!}8x^2 = \frac{5!}{3!2!}8x^2 = \frac{5\cdot 4\cdot 3!}{3!\cdot 2\cdot 1}8x^2 = (10)8x^2 = 80x^2$

63. $(2x-3)^6$

Fifth Term $(r=4)$: $\binom{n}{r}a^{n-r}b^r = \binom{6}{4}(2x)^{6-4}(-3)^4 = \frac{6!}{4!(6-4)!}4x^2(81)$

$$= \frac{6!}{4!2!}324x^2 = \frac{6\cdot 5\cdot 4!}{4!\cdot 2\cdot 1}324x^2 = (15)324x^2 = 4860x^2$$

Chapter 11 Test

1. $a_n = \dfrac{(-1)^{n+1}}{n^2}$

$a_1 = \dfrac{(-1)^{1+1}}{1^2} = \dfrac{(-1)^2}{1} = \dfrac{1}{1} = 1$

$a_2 = \dfrac{(-1)^{2+1}}{2^2} = \dfrac{(-1)^3}{4} = \dfrac{-1}{4} = -\dfrac{1}{4}$

$a_3 = \dfrac{(-1)^{3+1}}{3^2} = \dfrac{(-1)^4}{9} = \dfrac{1}{9}$

$a_4 = \dfrac{(-1)^{4+1}}{4^2} = \dfrac{(-1)^5}{16} = \dfrac{-1}{16} = -\dfrac{1}{16}$

$a_5 = \dfrac{(-1)^{5+1}}{5^2} = \dfrac{(-1)^6}{25} = \dfrac{1}{25}$

2. $\displaystyle\sum_{i=1}^{5}\left(i^2+10\right) = \left(1^2+10\right)+\left(2^2+10\right)+\left(3^2+10\right)+\left(4^2+10\right)+\left(5^2+10\right)$

$= (1+10)+(4+10)+(9+10)+(16+10)+(25+10)$

$= 11+14+19+26+35 = 105$

3. $\dfrac{2}{3}+\dfrac{3}{4}+\dfrac{4}{5}+\ldots+\dfrac{21}{22} = \displaystyle\sum_{i=2}^{21}\frac{i}{i+1}$ or $\displaystyle\sum_{i=1}^{20}\frac{i+1}{i+2}$

4. $d = 9-4 = 5$

$a_n = 4+(n-1)5 = 4+5n-5 = 5n-1$

$a_{12} = 5(12)-1 = 60-1 = 59$

5. $d = \dfrac{a_2}{a_1} = \dfrac{4}{16} = \dfrac{1}{4}$

$a_n = a_1 r^{n-1} = 16\left(\dfrac{1}{4}\right)^{n-1}$

$a_{12} = 16\left(\dfrac{1}{4}\right)^{12-1} = 16\left(\dfrac{1}{4}\right)^{11} = 16\left(\dfrac{1}{4,194,304}\right) = \dfrac{16}{4,194,304} = \dfrac{1}{262,144}$

6. First, find d.

$d = -14-(-7) = -7$

Next, find a_{10}.

$a_{10} = -7+(10-1)(-7) = -7-63 = -70$

Now, find the sum.

$S_{10} = \dfrac{10}{2}(-7+(-70)) = 5(-77) = -385$

7. $\sum_{i=1}^{20}(3i-4)=(3(1)-4)+(3(2)-4)+(3(3)-4)+...+(3(20)-4)$

$=(3-4)+(6-4)+(9-4)+...+(60-4)$

$=-1+2+5+...+56$

$S_{20}=\frac{20}{2}(-1+56)=10(55)=550$

8. $r=\frac{a_2}{a_1}=\frac{-14}{7}=-2$

$S_{10}=\frac{7\left(1-(-2)^{10}\right)}{1-(-2)}=\frac{7(1-1024)}{3}=\frac{7(-1023)}{3}=-2387$

9. $\sum_{i=1}^{15}(-2)^i=\frac{-2\left(1-(-2)^{15}\right)}{1-(-2)}=\frac{-2(1-(-32,768))}{3}=\frac{-2(32,769)}{3}=-21,846$

10. $r=\frac{1}{2}$

$S=\frac{4}{1-\frac{1}{2}}=\frac{4}{\frac{1}{2}}=4\div\frac{1}{2}=4\cdot\frac{2}{1}=8$

11. $0.\overline{73}=\frac{a_1}{1-r}=\frac{\frac{73}{100}}{1-\frac{1}{100}}=\frac{\frac{73}{100}}{\frac{99}{100}}=\frac{73}{100}\div\frac{99}{100}=\frac{73}{100}\cdot\frac{100}{99}=\frac{73}{99}$

12. $r=1.04$

$S_8=\frac{a_1\left(1-r^n\right)}{1-r}=\frac{30,000\left(1-(1.04)^8\right)}{1-1.04}\approx 276,427$

The total salary over the eight years is approximately $276,427.

13. $\binom{9}{2}=\frac{9!}{2!(9-2)!}=\frac{9!}{2!7!}=\frac{9\cdot 8\cdot \cancel{7!}}{2\cdot 1\cdot \cancel{7!}}=36$

14. Applying the Binomial Theorem to $\left(x^2-1\right)^5$, we have $a=x^2$, $b=-1$, and $n=5$.

$$\left(x^2-1\right)^5=\binom{5}{0}\left(x^2\right)^5+\binom{5}{1}\left(x^2\right)^4(-1)+\binom{5}{2}\left(x^2\right)^3(-1)^2+\binom{5}{3}\left(x^2\right)^2(-1)^3+\binom{5}{4}\left(x^2\right)(-1)^4+\binom{5}{5}(-1)^5$$
$$=\frac{\cancel{5!}}{0!\cancel{5!}}x^{10}-\frac{5!}{1!(5-1)!}x^8+\frac{5!}{2!(5-2)!}x^6-\frac{5!}{3!(5-3)!}x^4+\frac{5!}{4!(5-4)!}x^2-\frac{5!}{5!(5-5)!}$$
$$=1x^{10}-\frac{5\cdot\cancel{4!}}{1\cdot\cancel{4!}}x^8+\frac{5\cdot4\cdot\cancel{3!}}{2\cdot1\cdot\cancel{3!}}x^6-\frac{5\cdot4\cdot\cancel{3!}}{\cancel{3!}2\cdot1}x^4+\frac{5\cdot\cancel{4!}}{\cancel{4!}\cdot1}x^2-\frac{\cancel{5!}}{\cancel{5!}0!}$$
$$=x^{10}-5x^8+10x^6-10x^4+5x^2-1$$

15. $\left(x+y^2\right)^8$

First Term $(r=0)$: $\binom{n}{r}a^{n-r}b^r=\binom{8}{0}x^{8-0}\left(y^2\right)^0=\frac{8!}{0!(8-0)!}x^8\cdot1=\frac{\cancel{8!}}{0!\cancel{8!}}x^8=x^8$

Second Term $(r=1)$: $\binom{n}{r}a^{n-r}b^r=\binom{8}{1}x^{8-1}\left(y^2\right)^1=\frac{8!}{1!(8-1)!}x^7y^2=\frac{8\cdot\cancel{7!}}{1\cdot\cancel{7!}}x^7y^2=8x^7y^2$

Third Term $(r=2)$: $\binom{n}{r}a^{n-r}b^r=\binom{8}{2}x^{8-2}\left(y^2\right)^2=\frac{8!}{2!(8-2)!}x^6y^4=\frac{8\cdot7\cdot\cancel{6!}}{2\cdot1\cdot\cancel{6!}}x^6y^4=28x^6y^4$

Cumulative Review Exercises (Chapters 1 – 11)

1.
$$\begin{aligned}\sqrt{2x+5}-\sqrt{x+3}&=2\\ -\sqrt{x+3}&=2-\sqrt{2x+5}\\ x+3&=\left(2-\sqrt{2x+5}\right)^2\\ x+3&=4-4\sqrt{2x+5}+2x+5\\ x+3&=-4\sqrt{2x+5}+2x+9\\ -x-6&=-4\sqrt{2x+5}\\ (-x-6)^2&=16(2x+5)\\ x^2+12x+36&=32x+80\\ x^2-20x-44&=0\\ (x-22)(x+2)&=0\\ x=22 \text{ or } x&=-2\end{aligned}$$

The solution set is $\{22\}$.

2.
$$\begin{aligned}(x-5)^2&=-49\\ x-5&=\pm\sqrt{-49}\\ x&=5\pm\sqrt{-49}\\ x&=5\pm7i\end{aligned}$$

3. $x^2 + x > 6$

Solve the related quadratic equation.

$$\begin{aligned} x^2 + x &= 6 \\ x^2 + x - 6 &= 0 \\ (x+3)(x-2) &= 0 \end{aligned}$$

$$x+3=0 \quad \text{or} \quad x-2=0$$
$$x=-3 \qquad\qquad x=2$$

The boundary points are -3 and 2.

Interval	Test Value	Test	Conclusion
$(-\infty, -3)$	-4	$(-4)^2 + (-4) > 6$ $12 > 6$, true	$(-\infty, -3)$ does belong to the solution set.
$(-3, 2)$	0	$(0)^2 + 0 > 6$ $0 > 6$, false	$(-3, 2)$ does not belong to the solution set.
$(2, \infty)$	3	$(3)^2 + 3 > 6$ $12 > 6$, true	$(2, \infty)$ does belong to the solution set.

The solution set is $(-\infty, -3) \cup (2, \infty)$.

4.
$$\begin{aligned} 6x - 3(5x+2) &= 4(1-x) \\ 6x - 15x - 6 &= 4 - 4x \\ -9x - 6 &= 4 - 4x \\ -5x &= 10 \\ x &= -2 \end{aligned}$$

The solution set is $\{-2\}$.

5.
$$\begin{aligned} \frac{2}{x-3} - \frac{3}{x+3} &= \frac{12}{x^2-9} \\ \frac{2}{x-3} - \frac{3}{x+3} &= \frac{12}{(x-3)(x+3)} \\ (x-3)(x+3)\left(\frac{2}{x-3} - \frac{3}{x+3}\right) &= (x-3)(x+3)\left(\frac{12}{(x-3)(x+3)}\right) \\ 2(x+3) - 3(x-3) &= 12 \\ 2x + 6 - 3x + 9 &= 12 \\ -x + 15 &= 12 \\ -x &= -3 \\ x &= 3 \end{aligned}$$

Since 3 would make one or more of the denominators in the original equation zero, we disregard it and conclude that there is no solution. The solution set is $\varnothing$ or $\{\ \}$.

6. $3x+2<4$ and $4-x>1$

$$3x<2 \qquad -x>-3$$

$$x<\frac{2}{3} \qquad x<3$$

The solution set is $\left(-\infty, \frac{2}{3}\right)$.

7.
$$\begin{aligned} 3x-2y+z &= 7 \\ 2x+3y-z &= 13 \\ x-y+2z &= -6 \end{aligned}$$

Multiply the second equation by 2 and add to the third equation.

$$\begin{aligned} 4x+6y-2z &= 26 \\ x-y+2z &= -6 \\ \hline 5x+5y &= 20 \end{aligned}$$

Add the first equation to the second equation.

$$\begin{aligned} 3x-2y+z &= 7 \\ 2x+3y-z &= 13 \\ \hline 5x+y &= 20 \end{aligned}$$

We now have a system of two equations in two variables.

$$\begin{aligned} 5x+5y &= 20 \\ 5x+y &= 20 \end{aligned}$$

Multiply the second equation by -1 and add to the first equation.

$$\begin{aligned} 5x+5y &= 20 \\ -5x-y &= -20 \\ \hline 4y &= 0 \\ y &= 0 \end{aligned}$$

Back-substitute 0 for y to find x.

$$\begin{aligned} 5x+y &= 20 \\ 5x+0 &= 20 \\ 5x &= 20 \\ x &= 4 \end{aligned}$$

Back-substitute 4 for x and 0 for y to find z.

$$\begin{aligned} 3x-2y+z &= 7 \\ 3(4)-2(0)+z &= 7 \\ 12-0+z &= 7 \\ 12+z &= 7 \\ z &= -5 \end{aligned}$$

The solution set is $\{(4, 0, -5)\}$.

8.
$$\begin{aligned} \log_9 x+\log_9(x-8) &= 1 \\ \log_9\left(x(x-8)\right) &= 1 \\ x(x-8) &= 9^1 \\ x^2-8x &= 9 \\ x^2-8x-9 &= 0 \\ (x-9)(x+1) &= 0 \end{aligned}$$

$$x-9=0 \quad \text{or} \quad x+1=0$$

$$x=9 \qquad x=-1$$

Since we cannot take a log of a negative number, we disregard -1 and conclude that the solution set is $\{9\}$.

9.
$$\begin{aligned} 2x^2-3y^2 &= 5 \\ 3x^2+4y^2 &= 16 \end{aligned}$$

Multiply the first equation by -3 and the second equation by 2 and solve by addition.

$$\begin{aligned} -6x^2+9y^2 &= -15 \\ 6x^2+8y^2 &= 32 \\ \hline 17y^2 &= 17 \\ y^2 &= 1 \\ y &= \pm 1 \end{aligned}$$

Back-substitute ± 1 for y to find x.

$$\begin{aligned} y &= \pm 1 \\ 2x^2-3(\pm 1)^2 &= 5 \\ 2x^2-3(1) &= 5 \\ 2x^2-3 &= 5 \\ 2x^2 &= 8 \\ x^2 &= 4 \\ x &= \pm 2 \end{aligned}$$

The solutions are $(-2,-1), (-2,1)$, $(2,-1)$ and $(2,1)$ and the solution set is $\{(-2,-1),(-2,1),(2,-1),(2,1)\}$.

10. $2x^2 - y^2 = -8$
$x - y = 6$

Solve the second equation for x.

$$x - y = 6$$
$$x = y + 6$$

Substitute $y + 6$ for x.

$$\begin{aligned} 2(y+6)^2 - y^2 &= -8 \\ 2\left(y^2 + 12x + 36\right) - y^2 &= -8 \\ 2y^2 + 24x + 72 - y^2 &= -8 \\ y^2 + 24x + 72 &= -8 \\ y^2 + 24x + 80 &= 0 \\ (y+20)(y+4) &= 0 \end{aligned}$$

$$y + 20 = 0 \quad \text{or} \quad y + 4 = 0$$
$$y = -20 \qquad\qquad y = -4$$

Back-substitute –4 and –20 for y to find x.

$$y = -20 \quad \text{or} \quad y = -4$$
$$f(x) = (x+2)^2 - 4 \quad x = -20 + 6 \qquad x = -4 + 6$$
$$x = -14 \qquad\qquad x = 2$$

The solution set is $\{(-14, -20), (2, -4)\}$.

11.

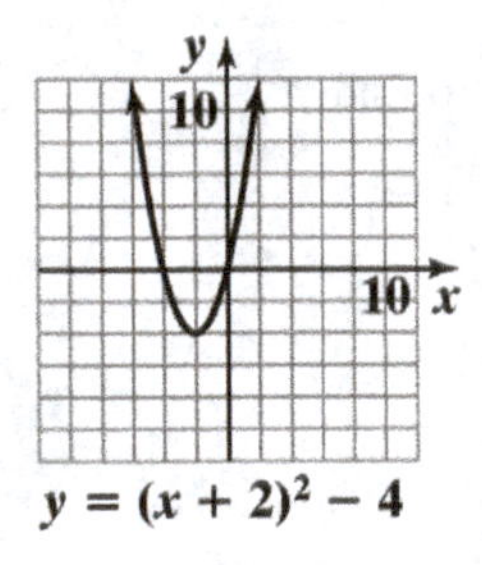

12. $y < -3x + 5$

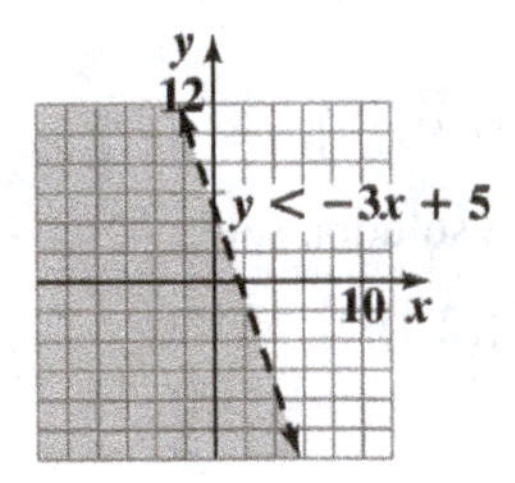

13. $f(x) = 3^{x-2}$

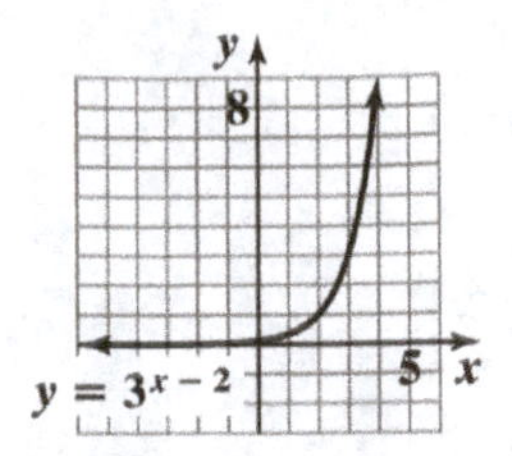

14. $\dfrac{x^2}{16} + \dfrac{y^2}{4} = 1$

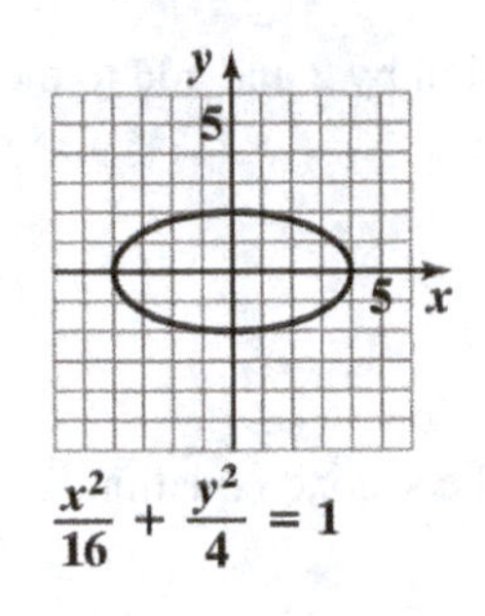

15. $x^2 - y^2 = 9$

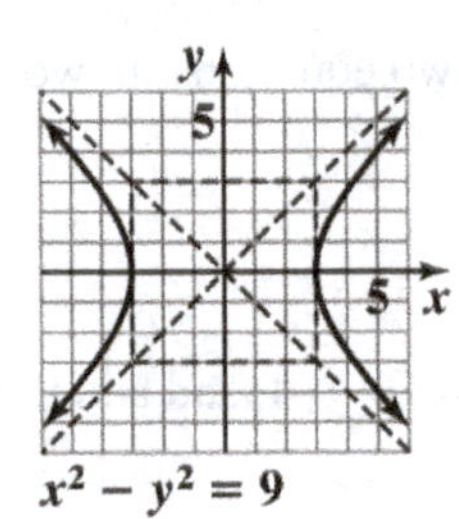

16.

$$\begin{aligned} &\frac{2x+1}{x-5} - \frac{4}{x^2 - 3x - 10} \\ &= \frac{2x+1}{x-5} - \frac{4}{(x-5)(x+2)} \\ &= \frac{2x+1}{x-5} \cdot \frac{x+2}{x+2} - \frac{4}{(x-5)(x+2)} \\ &= \frac{(2x+1)(x+2) - 4}{(x-5)(x+2)} \\ &= \frac{2x^2 + 5x + 2 - 4}{(x-5)(x+2)} \\ &= \frac{2x^2 + 5x - 2}{(x-5)(x+2)} \end{aligned}$$

17. $$\frac{\frac{1}{x-1}+1}{\frac{1}{x+1}-1}=\frac{\frac{1}{x-1}+\frac{x-1}{x-1}}{\frac{1}{x+1}-\frac{x+1}{x+1}}$$
$$=\frac{\frac{1+(x-1)}{x-1}}{\frac{1-(x+1)}{x+1}}$$
$$=\frac{\frac{x}{x-1}}{\frac{-x}{x+1}}=$$
$$\frac{x}{x-1}\cdot\frac{x+1}{-x}$$
$$=-\frac{x+1}{x-1}$$

18. $$\frac{6}{\sqrt{5}-\sqrt{2}}\cdot\frac{\sqrt{5}+\sqrt{2}}{\sqrt{5}+\sqrt{2}}=\frac{6\left(\sqrt{5}+\sqrt{2}\right)}{5-2}$$
$$=\frac{6\left(\sqrt{5}+\sqrt{2}\right)}{3}$$
$$=2\left(\sqrt{5}+\sqrt{2}\right)$$
$$=2\sqrt{5}+2\sqrt{2}$$

19. $$8\sqrt{45}+2\sqrt{5}-7\sqrt{20}$$
$$=8\sqrt{9\cdot5}+2\sqrt{5}-7\sqrt{4\cdot5}$$
$$=8\cdot3\sqrt{5}+2\sqrt{5}-7\cdot2\sqrt{5}$$
$$=24\sqrt{5}+2\sqrt{5}-14\sqrt{5}$$
$$=12\sqrt{5}$$

20. $$\frac{5}{\sqrt[3]{2x^2y}}=\frac{5}{\sqrt[3]{2x^2y}}\cdot\frac{\sqrt[3]{2^2xy^2}}{\sqrt[3]{2^2xy^2}}$$
$$=\frac{5\sqrt[3]{4xy^2}}{\sqrt[3]{2^3x^3y^3}}=\frac{5\sqrt[3]{4xy^2}}{2xy}$$

21. $$5ax+5ay-4bx-4by$$
$$=5a(x+y)-4b(x+y)$$
$$=(x+y)(5a-4b)$$

22. $$5\log x-\frac{1}{2}\log y=\log x^5-\log y^{\frac{1}{2}}$$
$$=\log\left(\frac{x^5}{y^{\frac{1}{2}}}\right)=\log\left(\frac{x^5}{\sqrt{y}}\right)$$

23. $$\frac{1}{p}+\frac{1}{q}=\frac{1}{f};$$
$$\frac{1}{p}=\frac{1}{f}-\frac{1}{q}$$
$$\frac{1}{p}=\frac{1}{f}\cdot\frac{q}{q}-\frac{1}{q}\cdot\frac{f}{f}$$
$$\frac{1}{p}=\frac{q}{qf}-\frac{f}{qf}$$
$$\frac{1}{p}=\frac{q-f}{qf}$$
$$p(q-f)=qf$$
$$p=\frac{qf}{q-f}$$

24. $$d=\sqrt{\left(6-(-3)\right)^2+\left(-1-(-4)\right)^2}$$
$$=\sqrt{9^2+3^2}=\sqrt{81+9}=\sqrt{90}$$
$$=\sqrt{9\cdot10}=3\sqrt{10}\approx9.49$$

The distance is $3\sqrt{10}$, or about 9.49, units.

25. $$\sum_{i=2}^{5}\left(i^3-4\right)$$
$$=\left(2^3-4\right)+\left(3^3-4\right)+\left(4^3-4\right)+\left(5^3-4\right)$$
$$=(8-4)+(27-4)+(64-4)+(125-4)$$
$$=4+23+60+121=208$$

26. First, find d.
$$d=6-2=4$$

Next, find a_{30}.
$$a_{30}=2+(30-1)4=2+(29)4$$
$$=2+116=118$$

Now, find the sum.
$$S_{30}=\frac{30}{2}(2+118)=15(120)=1800$$

27. $0.\overline{3} = \dfrac{a_1}{1-r} = \dfrac{\frac{3}{10}}{1-\frac{1}{10}} = \dfrac{\frac{3}{10}}{\frac{9}{10}} = \dfrac{3}{10} \div \dfrac{9}{10}$

$= \dfrac{3}{10} \cdot \dfrac{10}{9} = \dfrac{1}{3}$

28. Applying the Binomial Theorem to $\left(2x - y^3\right)^4$, we have $a = 2x,\ b = -y^3$, and $n = 4$.

$$\begin{aligned}
&\left(2x - y^3\right)^4 \\
&= \binom{4}{0}(2x)^4 + \binom{4}{1}(2x)^3\left(-y^3\right) + \binom{4}{2}(2x)^2\left(-y^3\right)^2 + \binom{4}{3}2x\left(-y^3\right)^3 + \binom{4}{4}\left(-y^3\right)^4 \\
&= \frac{4!}{0!(4-0)!}16x^4 - \frac{4!}{1!(4-1)!}8x^3y^3 + \frac{4!}{2!(4-2)!}4x^2y^6 - \frac{4!}{3!(4-3)!}2xy^9 + \frac{4!}{4!(4-4)!}y^{12} \\
&= \frac{\cancel{4!}}{0!\cancel{4!}}16x^4 - \frac{4!}{1!3!}8x^3y^3 + \frac{4!}{2!2!}4x^2y^6 - \frac{4!}{3!1!}2xy^9 + \frac{\cancel{4!}}{\cancel{4!}0!}y^{12} \\
&= 1\left(16x^4\right) - \frac{4\cdot\cancel{3!}}{1\cdot\cancel{3!}}8x^3y^3 + \frac{4\cdot 3\cdot\cancel{2!}}{2\cdot 1\cdot\cancel{2!}}4x^2y^6 - \frac{4\cdot\cancel{3!}}{\cancel{3!}\cdot 1}2xy^9 + y^{12} \\
&= 16x^4 - 4\left(8x^3y^3\right) + 6\left(4x^2y^6\right) - 4\left(2xy^9\right) + y^{12} \\
&= 16x^4 - 32x^3y^3 + 24x^2y^6 - 8xy^9 + y^{12}
\end{aligned}$$

29. $f(x) = \dfrac{2}{x^2 + 2x - 15}$

Set the denominator equal to 0 to find the domain:

$x^2 + 2x - 15 = 0$
$(x+5)(x-3) = 0$

$x = -5,\ x = 3$

So, $\{x \mid x$ is a real number and $x \neq -5$ and $x \neq 3\}$ or $(-\infty, -5) \cup (-5, 3) \cup (3, \infty)$.

30. $f(x) = \sqrt{2x-6}$;

We can not take the square root of a negative number.

$2x - 6 \geq 0$
$2x \geq 6$
$x \geq 3$

So, $\{x \mid x \geq 3\}$ or $[3, \infty)$.

31. $f(x) = \ln(1-x)$

We can only take the natural logarithm of positive numbers.

$$1-x > 0$$
$$-x > -1$$
$$x < 1$$

So, $\{x | x < 1\}$ or $(-\infty, 1)$.

32. Let w = width of the rectangle. Then $l = 2w+2$.

The perimeter of a rectangle is given by $P = 2w+2l$.

$$P = 2w+2l$$
$$22 = 2w+2(2w+2)$$
$$22 = 2w+4w+4$$
$$22 = 6w+4$$
$$18 = 6w$$
$$w = 3$$

Thus, $l = 2w+2 = 2(3)+2 = 8$

The dimension of the rectangle is 8 feet by 3 feet.

33.
$$A = P(1+r)^t$$
$$19610 = P(1+0.06)^1$$
$$19610 = 1.06P$$
$$P = 18500$$

Your salary before the raise is $18,500.

34. $F(t) = 1 - k\ln(t+1)$

$$\frac{1}{2} = 1 - k\ln(3+1)$$
$$\frac{1}{2} = 1 - k\ln 4$$
$$k\ln 4 = \frac{1}{2}$$
$$k = \frac{1}{2\ln 4} \approx 0.3607$$

$$F(t) \approx 1 - 0.3607\ln(t+1)$$
$$F(6) \approx 1 - 0.3607\ln(6+1)$$
$$\approx 1 - 0.3607\ln(7)$$
$$\approx 1 - 0.7019$$
$$\approx 0.298 \text{ or } \frac{298}{1000}$$